AF361025

Advances and Novel Treatment Options in Metastatic Melanoma

Advances and Novel Treatment Options in Metastatic Melanoma

Editors

Antonio Facchiano
Donatella Del Bufalo
Alessandra Carè

MDPI • Basel • Beijing • Wuhan • Barcelona • Belgrade • Manchester • Tokyo • Cluj • Tianjin

Editors

Antonio Facchiano
Laboratory of Molecular
Oncology
Istituto Dermopatico
dell'Immacolata, IDI-IRCCS
Rome
Italy

Donatella Del Bufalo
Research and Advanced
Technologies Department
Regina Elena National Cancer
Institute
Rome
Italy

Alessandra Carè
Center for Gender Medicine
Italian National Health Institute
Rome
Italy

Editorial Office
MDPI
St. Alban-Anlage 66
4052 Basel, Switzerland

This is a reprint of articles from the Special Issue published online in the open access journal *Cancers* (ISSN 2072-6694) (available at: www.mdpi.com/journal/cancers/special_issues/metastatic_melanoma).

For citation purposes, cite each article independently as indicated on the article page online and as indicated below:

LastName, A.A.; LastName, B.B.; LastName, C.C. Article Title. *Journal Name* **Year**, *Volume Number*, Page Range.

ISBN 978-3-0365-3304-9 (Hbk)
ISBN 978-3-0365-3303-2 (PDF)

Contents

About the Editors

Antonio Facchiano

Antonio Facchiano is an oncologist with long lasting experience in the field of new therapeutics and new diagnostic approaches in melanoma. He published several studies regarding the identifications of novel anti-angiogenic and anti-melanoma molecules, as well as studies clarifying pathogenic mechanisms controlling melanoma cells proliferation. He also identified several molecules acting as relevant diagnostic markers and therapeutic targets in melanoma, with particular focus on peptides, cytokines, ion-channels, and immune-system related molecules.

Donatella Del Bufalo

Dr Donatella Del Bufalo is a senior scientist working at the Regina Elena National Cancer Institute (Rome, Italy). During her career she focused her interest in the experimental chemotherapy of tumors and published about 110 papers demonstrating the in vitro and in vivo efficacy of several anti-neoplastic, anti-angiogenic, and epigenetic drugs, and oligonucleotide antisense as therapeutic agents able to induce apoptosis and/or autophagy in cancer cells. Dr Del Bufalo's group identified an important function for bcl-2, bcl-xL, and bcl2L10 antiapoptotic proteins in tumor progression and angiogenesis, highlighting a mechanism independent from the pro-survival functions of these proteins. Dr Del Bufalo is currently funded by the Italian Association for Cancer Research for a project "Shaping melanoma microenvironment by bcl-2: from novel mediators of tumor stroma crosstalk to new therapeutic approaches". Very recently, in collaboration with Sapienza University (Rome), Dr Del Bufalo group's identified IS21, as a new compound able to interact with bcl-2, bcl-xL, and mcl-1 proteins, and to induce in vitro and in vivo antitumoral activity.

Alessandra Carè

Dr. Alessandra Carè is a Research Director at the Italian National Health Institute where she is now leading the Reference Center for Gender-medicine.

She was was deeply involved since more than 30 years in basic and translational cancer studies, focusing on melanoma biology, and the functional roles played by microRNAs, exosomes, and, more recently, lipids. Since 2017, she has been working in the Oncology Unit of the Gender-specific Center looking at the functional roles of male and female specificities in all the steps going from experimental laboratory studies to patient care. As important differences have been described in cancers, including melanoma, she and her group are now including sex and gender determinants in their studies.

Editorial

Editorial on Special Issue "Advances and Novel Treatment Options in Metastatic Melanoma"

Alessandra Carè [1], **Donatella Del Bufalo** [2] and **Antonio Facchiano** [3,*]

1   Center for Gender-Specific Medicine, Istituto Superiore di Sanità, Viale Regina Elena 299, 00161 Rome, Italy; alessandra.care@iss.it
2   Preclinical Models and New Therapeutic Agents Unit, IRCCS Regina Elena National Cancer Institute, Via Elio Chianesi 53, 00144 Rome, Italy; donatella.delbufalo@ifo.it
3   Laboratory of Molecular Oncology, Istituto Dermopatico dell'Immacolata, IDI-IRCCS, Via Monti di Creta 104, 00167 Rome, Italy
*   Correspondence: a.facchiano@idi.it

Citation: Carè, A.; Del Bufalo, D.; Facchiano, A. Editorial on Special Issue "Advances and Novel Treatment Options in Metastatic Melanoma". *Cancers* **2022**, *14*, 707. https://doi.org/10.3390/cancers14030707

Received: 19 January 2022
Accepted: 27 January 2022
Published: 29 January 2022

Publisher's Note: MDPI stays neutral with regard to jurisdictional claims in published maps and institutional affiliations.

Investigating mechanisms controlling melanoma setup, development and progression is currently an extremely hot and rapidly evolving topic. Immunotherapy- and target therapy-based approaches are expected to substantially improve the survival in melanoma, as well as several other cancer types, changing the prognosis in many cancer patients. This is fostering a large interest in the detailed analysis of melanoma underlying mechanisms, to support the identification of more effective markers and therapeutic approaches. The Special Issue published in 2020–2021 by Cancers, entitled "Advances and Novel Treatment Options in Metastatic Melanoma", collects 24 manuscripts from Authors working in several different Institutions acting in different nations. According to the official page (https://www.mdpi.com/journal/cancers/special_issues/metastatic_melanoma, accessed on 18 January 2022), the Special Issue raised a very high interest in readers, and the number of views and citations of each manuscript is constantly increasing. The Special Issue addresses, under different aspects, the biology of melanoma and the way it affects its onset, development, and response to therapy. Manuscripts can be categorized under three main fields, namely: (1) mechanisms underlying melanoma setup and development, (2) melanoma therapeutic approaches, and (3) melanoma biomarkers.

An overview of each manuscript is reported below.

1. Mechanism Underlying Melanoma Setup and Development

Recently, there have been discoveries of new factors involved in melanoma progression and response to therapy, and their potential as pharmacologic targets. Among these, the myristoylated alanine-rich c-kinase substrate, MARCKS, which in addition to its role as an indicator of the protein kinase C activity, also plays a pivotal role as regulator of the metastatic behaviour of melanoma through its activation by WNT5A [1]. Also, BCL2L10, an antiapoptotic protein belonging to the bcl-2 family, has been identified both in melanoma cell lines and tumor samples, and its role as a pro-survival factor in melanoma has been reported [2]. Blockade of carbonic anhydrases XII (CAXII) under low oxygen condition induced a negative effect on melanoma cell migration and invasion, paralleled by a reduction of FAK phosphorylation and metalloprotease activities, thus suggesting CAXII as a possible therapeutic target for melanoma [3].

Melanoma pathobiology and response to target therapy or immunotherapy strictly depends on cells present in the tumor microenvironment (TME), such as fibroblasts, endothelial and immune cells [4]. A dynamic crosstalk between tumor cells and stroma plays a fundamental role in melanoma initiation, progression and metastasis [5]. TME affects melanoma progression also through the presence of basic or acidic extracellular matrix components released by melanoma or stromal cells. To this regard, in metastatic melanoma patients, plasmacytoid dendritic cells (pDCs), which play a relevant function in the anti-tumor immune response, are strongly impaired in terms of viability and function by lactic acidosis. BRAF (BRAFi) and MEK (MEKi) inhibitors partially rescued pDCs function in melanoma patients carrying BRAFV600 mutations [6].

Even if several mechanisms of intrinsic and acquired resistance to melanoma treatment have been identified, resistance still remains a threat for melanoma patient's survival [7]. Cancer stem cells (CSCs) represent one of the causes for development of resistance, thus contributing to the disease relapse following an initial response. A bioinformatic approach led to the identification of four major clusters of CSCs in BRAF mutated melanoma [8].

2. Melanoma Therapeutic Approaches

Treatment of metastatic melanoma still represents a challenge due to rapid dissemination and drug resistance. In the last decade, many novel therapeutic options have been developed and others are ongoing with important improved outcomes for patients. Major results are based on target- and immuno-therapies, including the option of novel combination treatments.

In the era of precision medicine, the first important novelty was represented by BRAF inhibitors acting on BRAF-mutated melanoma cells to selectively inhibit the MAPK pathway. More recent data evidenced the relevance of the immunomodulatory effects of BRAFi, including the capability to increase levels of immunostimulatory cytokines and decrease the immunosuppressive ones, enhance melanoma differentiation and presentation of tumor antigens, paralleled by increased intra-tumoral T-cell infiltration and activity [9]. Looking at adverse events, evaluated in the BRAFi plus MEKi combination, beside the most common pyrexia, fatigue, nausea, rash and hypertension, patients showed arthralgia. More than 30% of the treated patients reported arthralgia mostly affecting small joints and resembling symptoms associated with rheumatoid arthritis. It is interesting to note that occurrence of arthralgia was often associated with a better prognosis [10]. Going further in the development of additional strategies, promising approaches are based on the inhibition of protein tyrosine phosphatases, also in view of the effects on tumor infiltrates [11], and on the molecular targeting of the human antigen R (HuR), which stabilizes several oncoproteins. Interestingly, HuR inhibition in melanoma exhibited antitumor activity independently from the BRAF mutational status [12].

Based on recent progress in nanotechnology, nanoparticles represent another important therapeutic approach now moving to therapy. Indeed, nanovesicles can be loaded with a therapeutic cargo and modified in order to improve their entry and accumulation into the tumor, in turn preventing their content from degradation [13]. In addition, novel approaches based on so called "smart nanoparticles" are now under clinical trials in melanoma patients. These nanoparticles display higher affinity for melanoma components reducing toxic effects and aspecific targeting [14]. A growing interest is now based on natural products as antineoplastic drugs. Among them, promising results are showing the effectiveness of essential oil as complementary or alternative drugs for several diseases and as factors improving quality of life of cancer patients [15].

More complex is treatment of advanced mucosal and uveal melanomas still lacking effective therapeutic options. In advanced mucosal melanoma, immunotherapy and radiotherapy results, based on retrospective studies, showed the efficacy of anti-PD1 antibodies and the beneficial effects of combination with radiotherapy. Even so, median overall survival was 16.3 months [16]. Seemingly, the situation for uveal melanoma shows survival rates of approximately one year for nearly 50% of patients [17].

3. Melanoma Biomarkers

In the Special Issue several studies address the identification of markers able to effectively characterize melanoma patients. In the study published by Cesati et al. [18], the protein and gene expression of 27 cytokines, has been evaluated in the serum and in the tissue biopsies, respectively, of hundreds of patients and healthy controls. Different statistical approaches were applied leading to the identification of molecules in the serum able to significantly characterize patients vs. controls, as well as patients as function of gender and Breslow thickness. A striking feature is reported at the gene expression level. In fact, this study identifies, for the first time, a 4-gene signature able to discriminate patients form controls, with an extremely high accuracy (AUC = 0.98). A study by Aladowicz et al. [19] investigated metastatic melanoma, addressing metastasis-specific traits and metastasis-specific signaling pathways. From a patient derived xenografts collection, authors study the role of the adaptor protein ShcD, reporting its role in controlling melanoma spreading and its adhesion to the extracellular matrix. Overexpression of this factor has been found to increase melanoma cells movement via RAC1 signaling and DOCK4 confinement in the cytoplasm. Therefore, ShcD expression values are proposed to be investigated to predict the metastatic potential of melanoma cells. Metastatic features of melanoma cells were also related to the expression of the two-pore channel 2 (TPC2) [20]. The study shows that TPC2 knockout enhances metastatic potential of the melanoma cells consistently to increased expression of ZB-1, Vimentin, N-Cadherin and MMP9, and indicates the involvement of ORAI1/Ca^{2+}/PKC-βII pathway. The study shows that TPC2 reduction observed in metastatic patients opens new diagnostic hypotheses in metastatic patients. The manuscript by Pilla et al. [21] reviews the current status of investigation on biomarkers useful for melanoma patients stratification and to predict patients' responsiveness to therapy and other clinical outcomes. Biomarkers discussed belong to the classical melanoma-related markers, as well as to immune factors, microbiome-related molecules, mutations status, microenvironment-derived molecules, circulating DNA and others. The manuscript also discusses how biomarkers investigation may help the understanding of molecular basis of melanoma and the identification of new therapeutic approaches. The review by Lucianò and Tata [22] summarizes and points out recent findings regarding the role of cholinergic receptors outside the nervous system, namely in controlling the proliferation, apoptosis, angiogenesis, and epithelial mesenchymal transition in different cancer types including melanoma. The role of such receptors is also discussed in controlling metastatic features of melanoma. The review by Sacco et al. [23] highlights the role of circulating DNA in both early melanoma diagnosis and in the clinical management of more advanced phases. In fact, evaluation of circulating DNA can monitor aggressiveness of the resected melanoma, as well as minimal residual disease, recurrence, clinical progression, response and resistance to therapy. The importance of biomarkers such circulating DNA is also highlighted for patients stratification purposes. Finally, the review by Bellenghi et al. [24] underlines sex- and gender-related diversities in melanoma patients. The manuscript also highlights different incidence, anatomic localization, expression of sex-hormones as well as different expression of immune related genes located on the X chromosome. Additional evidences are discussed since different response to immunotherapy are reported, showing better improvements in men than in women. Such data highlight the possible consequences of unbalanced recruitment in clinical studies addressing the efficacy of new immunotherapy approaches.

Funding: The research leading to these results has received funding from AIRC under IG 2020-ID. 24315–P.I. Del Bufalo Donatella, and from Ministry of Health Italy, RC.3.4 2021 to Antonio Facchiano.

Conflicts of Interest: The authors declare no conflict of interest.

References

1. Mohapatra, P.; Yadav, V.; Toftdahl, M.; Andersson, T. WNT5A-Induced Activation of the Protein Kinase C Substrate MARCKS Is Required for Melanoma Cell Invasion. *Cancers* **2020**, *12*, 346. [CrossRef] [PubMed]
2. Quezada, M.J.; Picco, M.E.; Villanueva, M.B.; Castro, M.V.; Barbero, G.; Fernández, N.B.; Illescas, E.; Lopez-Bergami, P. BCL2L10 Is Overexpressed in Melanoma Downstream of STAT3 and Promotes Cisplatin and ABT-737 Resistance. *Cancers* **2020**, *13*, 78. [CrossRef] [PubMed]
3. Giuntini, G.; Monaci, S.; Cau, Y.; Mori, M.; Naldini, A.; Carraro, F. Inhibition of Melanoma Cell Migration and Invasion Targeting the Hypoxic Tumor Associated CAXII. *Cancers* **2020**, *12*, 3018. [CrossRef] [PubMed]
4. Falcone, I.; Conciatori, F.; Bazzichetto, C.; Ferretti, G.; Cognetti, F.; Ciuffreda, L.; Milella, M. Tumor Microenvironment: Implications in Melanoma Resistance to Targeted Therapy and Immunotherapy. *Cancers* **2020**, *12*, 2870. [CrossRef] [PubMed]

5. Bellei, B.; Migliano, E.; Picardo, M. A Framework of Major Tumor-Promoting Signal Transduction Pathways Implicated in Melanoma-Fibroblast Dialogue. *Cancers* **2020**, *12*, 3400. [CrossRef] [PubMed]

6. Monti, M.; Vescovi, R.; Consoli, F.; Farina, D.; Moratto, D.; Berruti, A.; Specchia, C.; Vermi, W. Plasmacytoid Dendritic Cell Impairment in Metastatic Melanoma by Lactic Acidosis. *Cancers* **2020**, *12*, 2085. [CrossRef]

7. Proietti, I.; Skroza, N.; Bernardini, N.; Tolino, E.; Balduzzi, V.; Marchesiello, A.; Michelini, S.; Volpe, S.; Mambrin, A.; Mangino, G.; et al. Mechanisms of Acquired BRAF Inhibitor Resistance in Melanoma: A Systematic Review. *Cancers* **2020**, *12*, 2801. [CrossRef]

8. Fattore, L.; Mancini, R.; Ciliberto, G. Cancer Stem Cells and the Slow Cycling Phenotype: How to Cut the Gordian Knot Driving Resistance to Therapy in Melanoma. *Cancers* **2020**, *12*, 3368. [CrossRef]

9. Proietti, I.; Skroza, N.; Michelini, S.; Mambrin, A.; Balduzzi, V.; Bernardini, N.; Marchesiello, A.; Tolino, E.; Volpe, S.; Maddalena, P.; et al. BRAF Inhibitors: Molecular Targeting and Immunomodulatory Actions. *Cancers* **2020**, *12*, 1823. [CrossRef]

10. Salzmann, M.; Benesova, K.; Buder-Bakhaya, K.; Papamichail, D.; Dimitrakopoulou-Strauss, A.; Lorenz, H.M.; Enk, A.H.; Hassel, J.C. Arthralgia Induced by BRAF Inhibitor Therapy in Melanoma Patients. *Cancers* **2020**, *12*, 3004. [CrossRef]

11. Pardella, E.; Pranzini, E.; Leo, A.; Taddei, M.L.; Paoli, P.; Raugei, G. Oncogenic Tyrosine Phosphatases: Novel Therapeutic Targets for Melanoma Treatment. *Cancers* **2020**, *12*, 2799. [CrossRef]

12. Ahmed, R.; Muralidharan, R.; Srivastava, A.; Johnston, S.E.; Zhao, Y.D.; Ekmekcioglu, S.; Munshi, A.; Ramesh, R. Molecular Targeting of HuR Oncoprotein Suppresses MITF and Induces Apoptosis in Melanoma Cells. *Cancers* **2021**, *13*, 166. [CrossRef] [PubMed]

13. Arasi, M.B.; Pedini, F.; Valentini, S.; Felli, N.; Felicetti, F. Advances in Natural or Synthetic Nanoparticles for Metastatic Melanoma Therapy and Diagnosis. *Cancers* **2020**, *12*, 2893. [CrossRef] [PubMed]

14. Chillà, A.; Margheri, F.; Biagioni, A.; Del Rosso, T.; Fibbi, G.; Del Rosso, M.; Laurenzana, A. Cell-Mediated Release of Nanoparticles as a Preferential Option for Future Treatment of Melanoma. *Cancers* **2020**, *12*, 1771. [CrossRef]

15. Di Martile, M.; Garzoli, S.; Ragno, R.; Del Bufalo, D. Essential Oils and Their Main Chemical Components: The Past 20 Years of Preclinical Studies in Melanoma. *Cancers* **2020**, *12*, 2650. [CrossRef] [PubMed]

16. Teterycz, P.; Czarnecka, A.M.; Indini, A.; Spalek, M.J.; Labianca, A.; Rogala, P.; Cybulska-Stopa, B.; Quaglino, P.; Ricardi, U.; Badellino, S.; et al. Multimodal Treatment of Advanced Mucosal Melanoma in the Era of Modern Immunotherapy. *Cancers* **2020**, *12*, 3131. [CrossRef]

17. Mallone, F.; Sacchetti, M.; Lambiase, A.; Moramarco, A. Molecular Insights and Emerging Strategies for Treatment of Metastatic Uveal Melanoma. *Cancers* **2020**, *12*, 2761. [CrossRef] [PubMed]

18. Cesati, M.; Scatozza, F.; D'Arcangelo, D.; Antonini-Cappellini, G.C.; Rossi, S.; Tabolacci, C.; Nudo, M.; Palese, E.; Lembo, L.; Di Lella, G.; et al. Investigating Serum and Tissue Expression Identified a Cytokine/Chemokine Signature as a Highly Effective Melanoma Marker. *Cancers* **2020**, *12*, 3680. [CrossRef]

19. Aladowicz, E.; Granieri, L.; Marocchi, F.; Punzi, S.; Giardina, G.; Ferrucci, P.F.; Mazzarol, G.; Capra, M.; Viale, G.; Confalonieri, S.; et al. ShcD Binds DOCK4, Promotes Ameboid Motility and Metastasis Dissemination, Predicting Poor Prognosis in Melanoma. *Cancers* **2020**, *12*, 3366. [CrossRef]

20. D'Amore, A.; Hanbashi, A.A.; Di Agostino, S.; Palombi, F.; Sacconi, A.; Voruganti, A.; Taggi, M.; Canipari, R.; Giovanni Blandino, G.; Parrington, J.; et al. Loss of Two-Pore Channel 2 (TPC2) Expression Increases the Metastatic Traits of Melanoma Cells by a Mechanism Involving the Hippo Signalling Pathway and Store-Operated Calcium Entry. *Cancers* **2020**, *12*, 2391. [CrossRef]

21. Pilla, L.; Alberti, A.; Di Mauro, P.; Gemelli, M.; Cogliati, V.; Cazzaniga, M.E.; Bidoli, P.; Maccalli, C. Molecular and Immune Biomarkers for Cutaneous Melanoma: Current Status and Future Prospects. *Cancers* **2020**, *12*, 3456. [CrossRef] [PubMed]

22. Lucianò, A.M.; Tata, A.M. Functional Characterization of Cholinergic Receptors in Melanoma Cells. *Cancers* **2020**, *12*, 3141. [CrossRef] [PubMed]

23. Sacco, A.; Forgione, L.; Carotenuto, M.; De Luca, A.; Ascierto, P.A.; Botti, G.; Normanno, N. Circulating Tumor DNA Testing Opens New Perspectives in Melanoma Management. *Cancers* **2020**, *12*, 2914. [CrossRef] [PubMed]

24. Bellenghi, M.; Puglisi, R.; Pontecorvi, G.; De Feo, A.; Carè, A.; Mattia, G. Sex and Gender Disparities in Melanoma. *Cancers* **2020**, *12*, 1819. [CrossRef] [PubMed]

cancers

Article

Molecular Targeting of HuR Oncoprotein Suppresses MITF and Induces Apoptosis in Melanoma Cells

Rebaz Ahmed [1,2], Ranganayaki Muralidharan [1,3], Akhil Srivastava [1,3], Sarah E. Johnston [4], Yan D. Zhao [3,4], Suhendan Ekmekcioglu [5], Anupama Munshi [3,6] and Rajagopal Ramesh [1,2,3,*]

[1] Department of Pathology, The University of Oklahoma Health Sciences Center, Oklahoma City, OK 73104, USA; rebaz-ahmed@ouhsc.edu (R.A.); mranga550@gmail.com (R.M.); akhil-srivastava@ouhsc.edu (A.S.)
[2] Graduate Program in Biomedical Sciences, The University of Oklahoma Health Sciences Center, Oklahoma City, OK 73104, USA
[3] Stephenson Cancer Center, The University of Oklahoma Health Sciences Center, Oklahoma City, OK 73104, USA; daniel-zhao@ouhsc.edu (Y.D.Z.); anupama-munshi@ouhsc.edu (A.M.)
[4] Department of Biostatistics and Epidemiology, The University of Oklahoma Health Sciences Center, Oklahoma City, OK 73104, USA; sarah-johnston@ouhsc.edu
[5] Department of Melanoma Medical Oncology, The University of Texas MD Anderson Cancer Center, Houston, TX 77030, USA; sekmekcioglu@mdanderson.org
[6] Department of Radiation Oncology, The University of Oklahoma Health Sciences Center, Oklahoma City, OK 73104, USA
* Correspondence: rajagopal-ramesh@ouhsc.edu; Tel.: +1-405-271-6101

check for
updates

Citation: Ahmed, R.; Muralidharan, R.; Srivastava, A.; Johnston, S.E.; Zhao, Y.D.; Ekmekcioglu, S.; Munshi, A.; Ramesh, R. Molecular Targeting of HuR Oncoprotein Suppresses MITF and Induces Apoptosis in Melanoma Cells. *Cancers* **2021**, *13*, 166. https://doi.org/10.3390/cancers13020166

Received: 18 December 2020
Accepted: 29 December 2020
Published: 6 January 2021

Publisher's Note: MDPI stays neutral with regard to jurisdictional claims in published maps and institutional affiliations.

Simple Summary: The human antigen R (HuR) protein regulates the expression of hundreds of proteins in a cell that support tumor growth, drug resistance, and metastases. HuR is overexpressed in several human cancers, including melanoma, and is a molecular target for cancer therapy. Our study objective, therefore, was to develop HuR-targeted therapy for melanoma. We identified that HuR regulates the microphthalmia-associated transcription factor (MITF) that has been implicated in both intrinsic and acquired drug resistance in melanoma and is a putative therapeutic target in melanoma. Using a gene therapeutic approach, we demonstrated silencing of HuR reduced MITF protein expression and inhibited the growth of melanoma cells but not normal melanocytes. However, combining HuR-targeted therapy with a small molecule MEK inhibitor suppressed MITF and produced a synergistic antitumor activity against melanoma cells. Our study results demonstrate that HuR is a promising target for melanoma treatment and offers new combinatorial treatment strategies for overriding MITF-mediated drug resistance.

Abstract: Background: Treatment of metastatic melanoma possesses challenges due to drug resistance and metastases. Recent advances in targeted therapy and immunotherapy have shown clinical benefits in melanoma patients with increased survival. However, a subset of patients who initially respond to targeted therapy relapse and succumb to the disease. Therefore, efforts to identify new therapeutic targets are underway. Due to its role in stabilizing several oncoproteins' mRNA, the human antigen R (HuR) has been shown as a promising molecular target for cancer therapy. However, little is known about its potential role in melanoma treatment. Methods: In this study, we tested the impact of siRNA-mediated gene silencing of HuR in human melanoma (MeWo, A375) and normal melanocyte cells in vitro. Cells were treated with HuR siRNA encapsulated in a lipid nanoparticle (NP) either alone or in combination with MEK inhibitor (U0126) and subjected to cell viability, cell-cycle, apoptosis, Western blotting, and cell migration and invasion assays. Cells that were untreated or treated with control siRNA-NP (C-NP) were included as controls. Results: HuR-NP treatment significantly reduced the expression of HuR and HuR-regulated oncoproteins, induced G1 cell cycle arrest, activated apoptosis signaling cascade, and mitigated melanoma cells' aggressiveness while sparing normal melanocytes. Furthermore, we demonstrated that HuR-NP treatment significantly reduced the expression of the microphthalmia-associated transcription factor (MITF) in both MeWo and MITF-overexpressing MeWo cells ($p < 0.05$). Finally, combining HuR-NP with U0126 resulted in synergistic antitumor activity against MeWo cells ($p < 0.01$). Conclusion:

HuR-NP exhibited antitumor activity in melanoma cells independent of their oncogenic B-RAF mutational status. Additionally, combinatorial therapy incorporating MEK inhibitor holds promise in overriding MITF-mediated drug resistance in melanoma.

Keywords: melanoma; HuR; MITF; metastases; siRNA; targeted therapy; nanoparticles

1. Introduction

Melanoma is the most aggressive form of skin cancer and is associated with the highest mortality. An estimated 100,350 new melanoma cases (60,190 in men and 40,160 in women) are expected to be diagnosed in 2020 in the United States, and roughly 8% of these patients are expected to die of the disease [1]. The treatment option depends on the stage of melanoma. Primary tumors and patients with limited loco-regional metastases are resected. Advanced metastasized patients received systemic therapy with immunotherapy or targeted therapy. Targeted therapies for melanoma include small molecule inhibitors towards B-RAF^{V600} and MEK kinases (e.g., Vemurafenib, Dabrafenib, Trametinib, Cobimetinib and Binimetinib) that have been tested as monotherapy or combination treatments [2–5]. B-RAF^{V600} mutation, which is very common among melanoma patients, leads to constitutive activation of the mitogen-activated protein kinase (MAPK), leading to the proliferation and growth of melanoma cells [6,7]. Despite the clinical efficacy of B-RAF^{V600}/MEK targeted inhibitors, a large subset of melanoma patients develop resistance due to the reactivation of other elements of the MAPK or the PI3K pathway and receptor tyrosine kinases (e.g., c-KIT) [8–11]. Therefore, a combination of B-RAF^{V600} and PI3K inhibitors were tested, but the clinical trial (NCT01616199, NCT01512251) showed no significant improvement of efficacy [12–14]. Although B-RAF/MEK combination therapy strategies have shown improvement in therapeutic outcomes, the development of acquired resistance is still a concern [15]. In addition to targeted therapy, immune therapy has gained attention as a promising therapeutic strategy for metastatic melanoma [16]. Recent advancements in immunotherapy for melanoma include the blockade of immune checkpoint proteins (ICPs), Programmed Death-1 (PD-1), Programmed Death-Ligand 1 (PD-L1), and cytotoxic T-lymphocyte-associated protein 4 (CTLA-4). Disrupting the interaction between the immune checkpoint proteins results in reactivating the immune system and eliminating tumor growth [17–19]. Despite the success of immunotherapy, obstacles still exist, including the inability to predict the treatment efficacy and development of resistance as well as the occurrence of immune-related adverse events (irAEs). Thus, challenges in effectively treating metastatic melanoma continue warranting testing of inhibitors against new therapeutic targets.

HuR is an RNA binding protein encoded by the embryonic lethal, abnormal vision like 1 (*ELAVL1*) gene located on chromosome 19p13.2 [20]. HuR protein binds the adenylate/uridylate (AU)- and U-rich elements (AREs) in the untranslated region (UTR) of mRNAs [21]. HuR stabilizes and shuttles mRNAs from the nucleus to the cytoplasm where they are translated. HuR has been shown to regulate the expression of many transcripts whose products are oncoproteins [22]. Therefore, overexpression of HuR in many cancer types, including oral, colorectal, gastric, lung, breast, ovarian, renal, and melanoma, has been identified and correlated with poor prognosis [23–29]. Consequently, HuR has gained attention as a target for cancer therapy. In recent years, our lab and others have demonstrated that the downregulation of HuR using small interfering RNA (siRNA) and small molecule inhibitors results in a global knockdown of proteins involved in cancer growth and metastases, consequently leading to suppression of tumor growth both in vitro and in vivo [30–38]. However, the role of targeting HuR and the effect of its inhibition on melanoma cell growth has not been reported.

In this study, we examined the inhibitory effect of human HuR-specific small interfering RNA (HuR-siRNA) encapsulated in a lipid nanoparticle (NP) on HuR in human melanoma cell lines in vitro. Our study showed that inhibiting HuR reduced HuR-regulated oncoproteins, including MITF, inhibited cell proliferation and cell cycle, diminished melanoma cell's migration and invasion, and culminated in apoptotic cell death. Additionally, combinatorial therapy of HuR-NP and MEK1/2 inhibitor, U0126, produced synergistic anticancer activity compared to individual treatments. In conclusion, HuR-targeted monotherapy and combinatorial therapy with MEK1/2 inhibitor is a new approach for melanoma treatment.

2. Materials and Methods

2.1. Cell Lines

Human melanoma cell lines (MeWo, A375, SK-MEL3, and WM39) and primary human melanocytes were obtained from the American Type Culture Collection (ATCC, Rockville, MD, USA). The cell lines were authenticated to be of human origin by single tandem repeat assay (STR; IDEXX Bioresearch, Columbia, MO, USA). Melanoma cells were maintained in Dulbecco's modified Eagle's medium (DMEM), and Minimum Essential Medium (MEM) supplemented with 10% fetal bovine serum (FBS; Sigma Aldrich, St. Louis, MO, USA) and 1% penicillin/streptomycin. Melanocytes (ATCC) were maintained in Dermal Cell Basal Medium (DCBM) supplemented with Melanocyte Growth Kit (ATCC). The passage number of the melanoma cell lines used in this study ranged from 4 to 35. The passage number of melanocytes used in the study ranged from 3 to 8.

For generating MeWo-EGFP and MeWo-MITF cell line, MeWo cells seeded in six-well plates were stably transfected with 2 µg of pEGFP × 2-N1 (Empty vector), and pEGFP-N1-MITF-M plasmid (Addgene, Watertown, MA, USA) encapsulated in a cationic lipid nanoparticle (NP). At 72 h after transfection, neomycin (G418; 400 µg/mL) (Sigma Aldrich) was added to the cells and selected for fourteen days. The surviving cell colonies were trypsinized, expanded, and maintained in neomycin (100 µg/mL) and used for experiments presented in this study.

2.2. Synthesis and Preparation of siRNA Containing Nanoparticles

The lipid-based cationic DOTAP:cholesterol nanoparticles (NP) formulation that comprises cationic DOTAP:cholesterol was synthesized and characterized as previously described [30,32,39–41]. The NPs were used to encapsulate human HuR-specific siRNA (Dharmacon, Lafayette, CO, USA) and scrambled control siRNA as previously described [30,32] and labeled as HuR-NP and C-NP, respectively, and used in the present study.

2.3. Cell Viability Assay

Trypan blue exclusion assay method was used to test the cytotoxic effect of HuR-siRNA containing nanoparticles (HuR-NP) as previously described [30,32,41]. Briefly, cells (MeWo, 7×10^4 cells/well; A375, 1×10^5 cells/well; melanocytes, 2×10^5 cells/well) were seeded in six-well plates and incubated overnight in a CO_2 incubator at 37 °C. The following day, the tissue culture medium from the plates was replaced with an appropriate fresh serum-free culture medium and treated with nanoparticles (NP) containing HuR-specific siRNA (100 nM; HuR-NP) or scrambled siRNA (100 nM; C-NP). At six hours after HuR-NP and C-NP treatment, the culture medium was replaced with fresh 2% serum containing culture medium, and incubation continued. At 24 h and 48 h after treatment, the cells were harvested, and the number of viable cells in each treatment group was determined. Cells that did not receive any treatment served as control. The results were expressed as the percentage of viable cells over untreated control cells. The experiments were repeated at least three separate times for reproducibility and were analyzed using appropriate statistical methods.

For testing the inhibitory activity of HuR-NP treatment on MeWo-MITF cells, cells were seeded in six-well plates and treated with HuR-NP and C-NP (100 nM siRNA). All other experimental conditions, including end-point analysis, were similar to those described above.

For determining the combinatorial treatment effect of HuR-NP and MEK inhibitor (U0126), MeWo or MeWo-MITF cells (7×10^4) seeded in six-well plates were treated with DMSO, C-NP, HuR-NP (100 nM siRNA), U0126 (20 μM; Cell Signaling Technology Inc., Beverly, MA, USA), C-NP plus U0126, and HuR-NP plus U0126. The treatment protocol was as follows: cells were first treated with C-NP and HuR-NP for six-hours in a serum-free culture medium. After six hours, the culture medium was replaced with a 2% FBS-containing culture medium, and U0126 was added. Cells were harvested at 24 h and 48 h post-treatment and analyzed for cell viability by trypan blue assay and molecular markers by Western blotting. The combination treatments' synergistic inhibitory effect was analyzed using SynergyFinder 1.0 tool [42].

2.4. Quantitative (q) RT-PCR Assay

Total RNA from HuR-NP- and C-NP-treated cells and untreated control cells were isolated using Trizol reagent (Life Technologies, Grand Island, NY, USA), and the RNA quality was determined using Denovix DS11 spectrophotometer. Complementary DNA (cDNA) was synthesized from 1 μg RNA/sample using a QuantScript cDNA synthesis kit (Bio-Rad, Richmond, CA, USA). An amount of 3 μg of synthesized cDNA, quantified using Denovix DS11 spectrophotometer, was subjected to real-time quantitative reverse transcriptase (qRT)-PCR (Bio-Rad CFX96™TouchReal-Time PCR Detection System; Richmond, CA, USA) using the premix iQ SYBR green qRT-PCR kit (Bio-Rad) as previously described [30,41]. The oligonucleotide primers (Integrated DNA Technology, Coralville, IA, USA) and their sequences for the amplification of HuR, BCL-2, and 18S RNA are shown below.

Human HuR
Forward—5′ATGAAGACCACATGGCCGAAGACT 3′
Reverse—5′ TGTGGTCATGAGTCCTTCCACGAT 3′
Human BCL-2
Forward—5′ ATG TGT GTG GAG AGC GTC AA 3′
Reverse—5′ ACA GTT CCA CAA AGG CAT CC 3′
Human 18S
Forward—5′ tagtagggacgggcggtgtg 3′
Reverse—5′ cagccacccgagattgagca 3′

The PCR cycling parameters and all other experimental conditions followed have previously been described [41]. The cycle threshold (Ct) value assessed by qRT-PCR was calculated for the transcripts and was normalized to a housekeeping gene. The changes in mRNA expression levels were expressed as fold change relative to control. Each sample was run in triplicate. The experiments were repeated at least three times for reproducibility and subjected to statistical analysis.

2.5. Western Blotting

Total cell lysates prepared from treated and untreated cells were harvested at defined time-points and subjected to Western blotting analysis as previously described [30,32,41]. Primary antibodies were purchased from commercial vendors and used for detecting human HuR, p27, BCL-2, alpha-tubulin (Santa Cruz Biotechnology, Dallas, TX, USA), cyclin D1, cyclin E1, HIF1-α, MITF, VEGF-A, Caspase-9, and PARP (Cell Signaling Technology Inc., Beverly, MA, USA) and beta-actin (Sigma Aldrich, St. Louis, MO, USA). Protein bands were detected using appropriate horseradish peroxidase (HRP)-tagged secondary antibodies (Santa Cruz Biotechnology) and an enhanced chemiluminescence kit (Thermo Scientific, MA, USA). Protein expression levels were detected on a chemiluminescence imaging system (Syngene, Frederick, MD, USA), and the relative protein expression compared to beta-actin or alpha-tubulin was quantified using Gene Tools software (Syngene) as previously described [30,32,41]. Experiments were repeated at least three separate times for reproducibility, and the data were analyzed for statistical significance.

2.6. Cell Cycle Analysis

Melanoma cells (MeWo, 4×10^4 cells/well, and A375, 6×10^4 cells/well) and melanocytes (2×10^5 cells/well) seeded in six-well plates were treated with HuR-NP and C-NP (100 nM siRNA). At 24 h and 48 h after treatment, the cells were harvested and washed with PBS, then fixed in absolute ice-cold ethanol for 30 min, followed by washing 2 times with PBS. Fixed cells were then resuspended in cell staining buffer (Invitrogen, Carlsbad, CA, USA) at a concentration of 2×10^5 cells/mL, and 100 μL of the cell suspension was incubated with 1 μL of 100 μg/mL propidium iodide (PI) (Invitrogen, Carlsbad, CA, USA) for 15 min at room temperature. The final washing was performed with PBS to wash out the unbound PI. Cells were subsequently subjected to flow cytometric analysis, as previously described [30,31]. Cells not receiving any treatment served as untreated controls. Experiments were conducted at least three separate times for melanoma cells and two times for melanocytes and subjected to statistical analysis. The data represented are the averages of two experiments.

2.7. Annexin V Assay

Cells seeded in six-well plates (MeWo, 4×10^4 cells/well, and A375, 6×10^4 cells/well) were treated with HuR-NP and C-NP (100 nM siRNA). At 24 h and 48 h after treatment, cells were harvested and stained with annexin V conjugated to fluorescein isothiocyanate (FITC) and propidium iodide (PI) using a dead cell apoptosis kit (Invitrogen) according to the manufacturer's protocol. Briefly, harvested cells were suspended in annexin V binding buffer at a concentration of 2×10^5 cells/mL, and 100 μL of the cell suspension was incubated with 5 μL of annexin V FITC and 1 μL of 100 μg/mL PI for 15 min of at room temperature. At the end of the incubation, the cells were processed and subjected to flow cytometric analysis (FACSCalibur™; BD Biosciences, Bedford, MA, USA). The number of viable (annexin V- and PI-negative), early apoptotic (annexin V-positive and PI-negative), and dead (annexin V- and PI-positive) cells was determined at excitation 488 nm and emission 530 nm using the Cell Quest software (FACSCalibur™; BD Biosciences). Cisplatin (CDDP; 1 μg)-treated cells served as a positive control, and the cells receiving no treatment served as a negative control. The results were plotted as the percentage of cells undergoing apoptosis. Experiments were repeated three times and subjected to statistical analysis.

2.8. Cell Migration and Invasion Assay

Cell migration was carried out as previously described [32,43]. Briefly, MeWo cells (3×10^4) were seeded in the upper chamber of 8 μm transwell (BD Biosciences) and placed in individual wells of a six-well plate filled with 1 mL of serum-free RPMI-1640 medium. At 24 h after seeding, the cells in the upper chamber were transfected with HuR-NP and C-NP (100 nM siRNA) in a serum-free medium. After 6 h of transfection, the upper and lower chambers' medium was replaced with 2% and 20% serum-containing media, respectively. Incubation was continued, and the experiment was terminated at 24 h and 48 h after transfection, at which time the inserts were removed and stained with crystal violet (Sigma Aldrich). The number of cells migrated to the lower chamber was counted using an inverted bright-field microscope, and the results were expressed as an average number of migrated cells per microscopic field and subjected to statistical analysis. The experiment was performed at least two separate times to ensure reproducibility.

Cell invasion assay was performed using Matrigel-coated 8 μm transwell chambers (BD Biosciences) as previously described. All of the experimental conditions and parameters followed were the same as described above for the cell migration assay. The number of invading cells to the lower chamber was counted using an inverted bright-field microscope, and the results were expressed as an average number of invaded cells per microscopic field and subjected to statistical analysis. The experiment was performed at least two separate times to ensure reproducibility.

2.9. Statistics

The study results were subjected to statistical analysis using SAS 9.4 statistical analysis software. Continuous outcome variables are summarized with means and standard deviations. One-way ANOVA model with Tukey's adjusted *p*-values was used to assess all pairwise comparisons. Adjusted *p*-values less than 0.05 were considered statistically significant.

3. Results

3.1. Genetic Knockdown of HuR Inhibited Cell Growth in Melanoma

To validate HuR as a potential target for melanoma therapy, we first examined the expression pattern of HuR in a panel of human melanoma cell lines (A375.S2, WM39, SK-MEL-3, OCM-1, A375, WM1316A, OMM2.3, and MeWo) and primary normal human melanocytes. Our data revealed high HuR expression in all the melanoma cell lines examined and were independent of the B-RAF mutation status compared to HuR expression in melanocytes (Figure S1). For all of the remaining studies, we selected MeWo (B-RAF^{wt}) and A375 (B-RAF^{V600E}) as representative cell lines for melanoma and used melanocytes as control. Genetic knockdown of HuR using HuR-NP resulted in the attenuation of HuR mRNA in both MeWo and A375 cell lines at 24 h and 48 h (Figure 1A; $p < 0.05$).

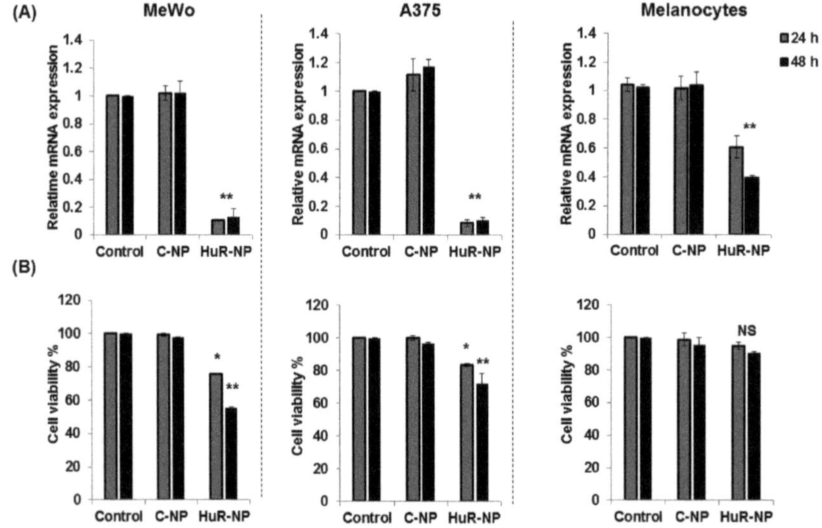

Figure 1. Effect of human antigen R (HuR)-nanoparticle (NP) treatment on HuR mRNA and cell viability. (**A**) HuR mRNA level and (**B**) cell viability was determined in human melanoma (MeWo, A375) cells and melanocytes treated with control siRNA-NP (C-NP) or HuR-NP (100 nM) for 24 h and 48 h. Untreated cells served as control. Error bar denotes SD; NS not significant; * $p < 0.05$; ** $p < 0.01$.

In melanocytes, HuR mRNA was also reduced upon HuR-NP treatment compared to C-NP and untreated control cells. However, the mRNA reduction in HuR-NP-treated melanocytes was markedly less compared to HuR-NP-treated melanoma cells. Next, we tested the inhibitory effect of HuR-NP treatment on the cell viability of melanoma cells and melanocytes. HuR-NP-treated MeWo cells showed approximately 25% and 45% inhibition at 24 h and 48 h, respectively, compared to C-NP-treated and untreated control cells (Figure 1B; $p < 0.05$). Similarly, HuR-NP treatment of A375 reduced cell viability up to 17% and 28% at 24 h and 48 h, respectively, compared to C-NP and untreated control ($p < 0.05$). To ensure the efficacy of HuR-NP is not limited to two melanoma (MeWo and A375) cell lines, cell viability was also evaluated in HuR-NP-treated SK-MEL-3 and WM39 melanoma cell lines. As shown in Figure S2, HuR-NP treatment significantly reduced SK-MEL-3 and WM39 cell viability at both 24 h and 48 h compared to C-NP-treated and untreated control cells. On the other hand, the treatment of melanocytes with HuR-NP exhibited a 6% and 10% reduction in cell viability at 24 h and 48 h, respectively, compared to C-NP and untreated controls (Figure 1B; $p > 0.05$). These results indicated that HuR-NP treatment, albeit reducing HuR mRNA in both melanoma cells and melanocytes, exerted selective and greater inhibitory activity on melanoma cell growth than in melanocytes.

3.2. Genetic Knockdown of HuR Using HuR-NP Reduced the Expression of HuR and HuR-Regulated Oncoproteins in Melanoma Cell Lines but Not in Melanocytes

HuR is known to stabilize many transcripts whose products are oncoproteins. To determine the role of HuR-NP treatment on the HuR target-oncoproteins, total cell lysates prepared from melanoma (MeWo, A375) cells and melanocytes receiving various treatments were examined by Western blot analysis. HuR-NP treatment significantly diminished protein expression of HuR, Cyclin D1, Cyclin E1, BCL-2, HIF1-α, VEGEF-A, and an increased expression of p27 at both 24 h and 48 h in melanoma cell lines (Figure 2 and Figure S3; $p < 0.05$). A similar observation of HuR silencing, resulting in BCL-2 reduction, was observed in SK-MEL-3 and WM39 melanoma cell lines (Figure S2).

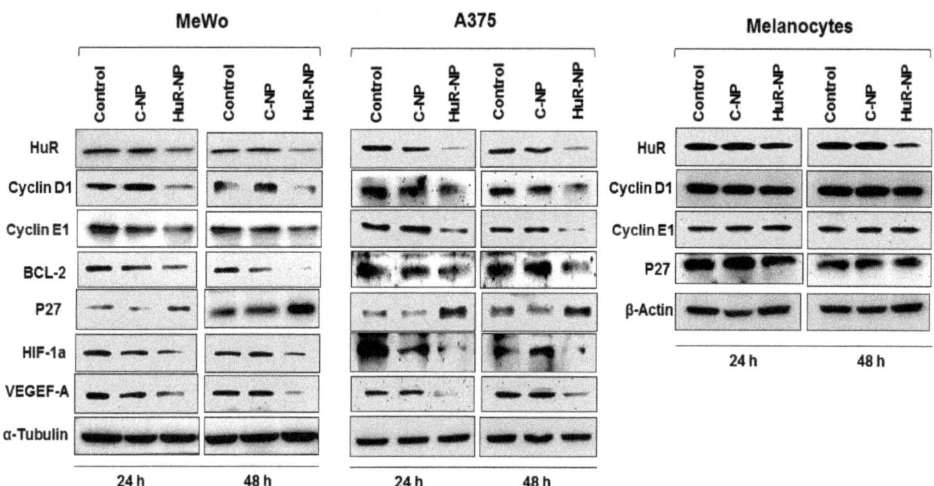

Figure 2. HuR-NP downregulates the expression of HuR and HuR-regulated proteins. Expression of HuR and HuR-regulated proteins in melanoma cell lines (MeWo, A375) and melanocytes at 24 h and 48 h after treatment with either C-NP or HuR-NP. Untreated cells served as control. β-actin and α-tubulin were used as loading controls.

We did not analyze for additional protein markers in SK-MEL-3 and WM39 melanoma cells that were examined in MeWo and A375 cells. Intriguingly, although HuR-NP treatment reduced HuR protein in melanocytes, no significant reduction in the expression of Cyclin E1, Cyclin D1, and p27 was observed (Figure 2 and Figure S3). The molecular mechanism for the observed differences in HuR-mediated downregulation of its targets between melanoma cell lines and melanocytes is unclear and warrants further investigation. Together, these results indicated that melanoma cell lines independent of the B-RAF mutational status were highly sensitive and responded to HuR-NP treatment compared to melanocytes.

3.3. Genetic Knockdown of HuR Induced G1 Cell Cycle Arrest in Melanoma

Since HuR knockdown reduced cyclin D1 and cyclin E1 and concomitantly increased the expression of p27, we evaluated the cell cycle profile of melanoma (MeWo and A375) and melanocytes with and without the HuR-NP exposure. Our results demonstrated that

HuR-NP treatment significantly enriched the MeWo and A375 melanoma cells in the G1 phase of the cell cycle compared to C-NP-treated and untreated control cells (Figure 3). In HuR-NP-treated MeWo cells, an increase of 9% and 11% of cells in the G1 phase was observed at 24 and 48 h, respectively, compared to untreated and C-NP-treated cells ($p < 0.05$). In A375, about 5% and 8% increase in the number of cells in the G1 phase at 24 h and 48 h, respectively, was observed upon HuR-NP treatment compared to untreated and C-NP-treated cells ($p < 0.05$). However, HuR-NP did not alter the cell cycle phases in melanocytes at 24 h and 48 h after treatment. Our results revealed that the HuR-NP treatment selectively induced a G1 phase cell-cycle arrest in melanoma cells but not in melanocytes and concurred with previous results reported for other solid tumors [30,31,44].

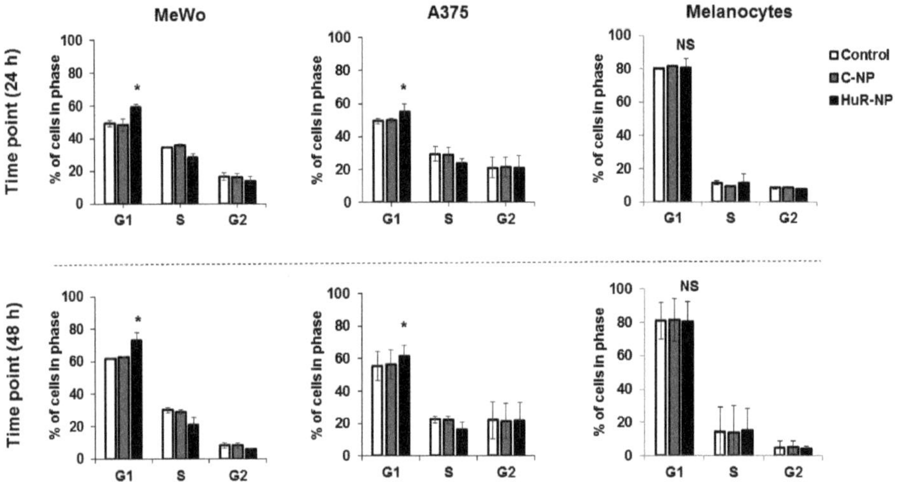

Figure 3. HuR-NP treatment induced G1 phase cell-cycle arrest in melanoma cell lines but not in melanocytes. Cell cycle analysis of melanoma cell lines (MeWo, A375) and melanocytes, at 24 h and 48 h after treatment with either C-NP or HuR-NP. Untreated cells served as control. Error bar denotes SD; NS not significant; * $p < 0.05$.

3.4. Genetic Knockdown of HuR in Melanoma Cells Activated the Apoptosis Cascade

To establish whether the reduction in anti-apoptotic protein BCL-2 in the HuR-NP-treated melanoma cells resulted in apoptosis, Western blot analysis for apoptotic proteins was performed. Cleavage of caspase-9, an indicator of activation of the caspase cascade, was greatly increased in both MeWo and A375 melanoma cells treated with HuR-NP compared to C-NP-treated and untreated control cells (Figure 4A and Figure S4). Accompanied by caspase 9 activation was the cleavage of its substrate, PARP. No significant activation of caspase-9 and PARP cleavage was observed in melanocytes following treatment with HuR-NP compared to HuR-NP-treated MeWo cells (Figure S5A,B).

Correlating with caspase activation in the melanoma cells was a significant increase in annexin-V-positive staining in HuR-NP-treated MeWo cells at 24 h (16% increase over controls) and 48 h (12% increase over controls), respectively, compared to C-NP-treated and untreated cells (Figure 4B; $p < 0.05$). In A375 cells, HuR-NP-treated cells showed a 25% and 33% increase in annexin-V-positive staining over controls at the two time-points tested (Figure 4B; $p < 0.05$). Cisplatin (CDDP; 10 μg)-treated cells were used as a positive control in the annexin V assay. The ability of HuR-NP treatment inducing apoptosis in melanoma cells but not in melanocytes is in agreement with prior reports showing HuR inhibition in tumor cells but not in normal cells results in apoptotic cell death [30,31,44].

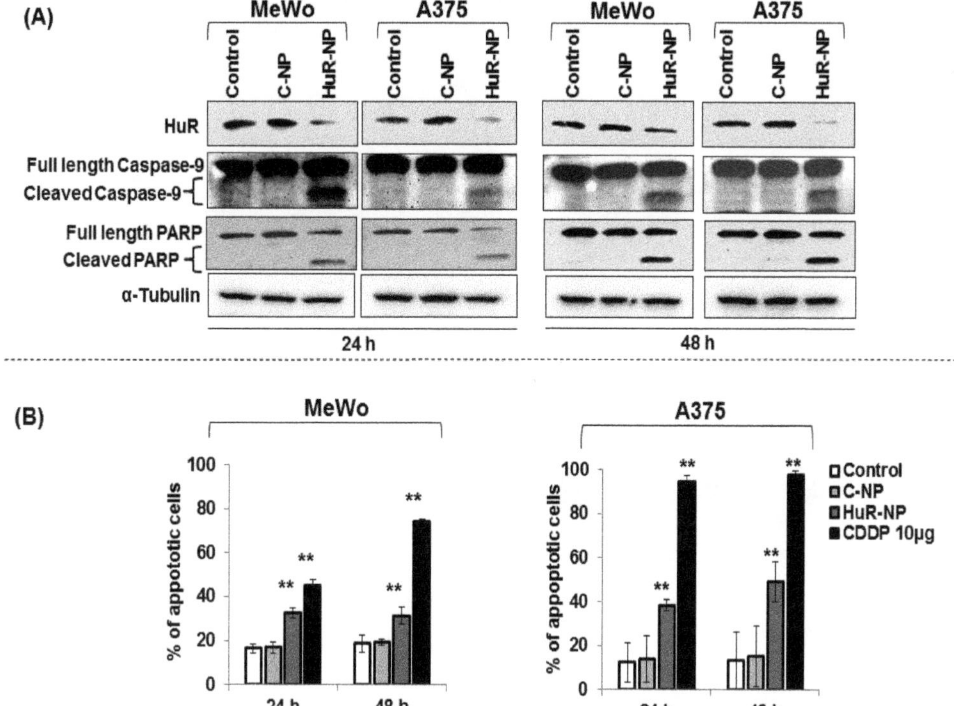

Figure 4. HuR-NP induced apoptosis in melanoma cell lines. Assessment of apoptosis in melanoma cell lines (MeWo, A375) at 24 h and 48 h after treatment with either C-NP or HuR-NP. Untreated cells served as control. (**A**), Western blotting showed caspase 9, and PARP cleavage was greater in HuR-NP-treated cells than in controls. (**B**), Annexin V staining was determined by flow cytometry. Cells treated with cisplatin (CDDP) served as a positive control for each cell line. Error bar denotes SD; ** $p < 0.01$.

3.5. Genetic Knockdown of HuR Inhibited Melanoma Cell Migration and Invasion

Migration and invasion are critical events in tumor metastases [45,46]. Previous studies have shown that HuR inhibition reduced tumor cell migration and invasion [30–32]. Therefore, we investigated the HuR-NP inhibitory effect on melanoma migration and invasion. HuR-NP treatment significantly reduced the number of migrated MeWo cells by 55% and 65% over C-NP-treated and untreated control cells at 24 h and 48 h, respectively (Figure 5A; $p < 0.05$). Similarly, testing the HuR-NP inhibitory effect on tumor cell invasion showed a significant reduction in the invasive ability of HuR-NP-treated MeWo cells by 46% and 60% over untreated and C-NP-treated cells at 24 h and 48 h, respectively (Figure 5B; $p < 0.05$).

3.6. HuR Inhibition Reduced MITF in Melanoma Cells

Studies have shown that MITF plays an important role in melanoma metastases and contributes to drug resistance [47–50]. Since HuR-NP treatment decreased cell invasion and migration, we investigated if HuR-NP treatment impacted MITF. Using gene sequence analysis, we analyzed for HuR-binding sites on MITF. As shown in Figure 6A, several HuR binding motifs were identified in the $3'$ untranslated region (UTR) and coding sequence (CDS) of the MITF gene using the RBP map (https://rbmap.technion.ac.il) [51] and WebLogo (https://weblogo.berkeley.edu) [52]. Next, alterations in the expression levels of MITF mRNA and its downstream target, BCL-2, were assessed in HuR-NP-treated cells compared to untreated and C-NP-treated cells. A significant reduction in the mRNA expression of

HuR, MITF, and BCL-2 was observed in HuR-NP-treated MeWo and A375 cells compared to their respective controls at the two time-points tested (Figure 6B; $p < 0.05$).

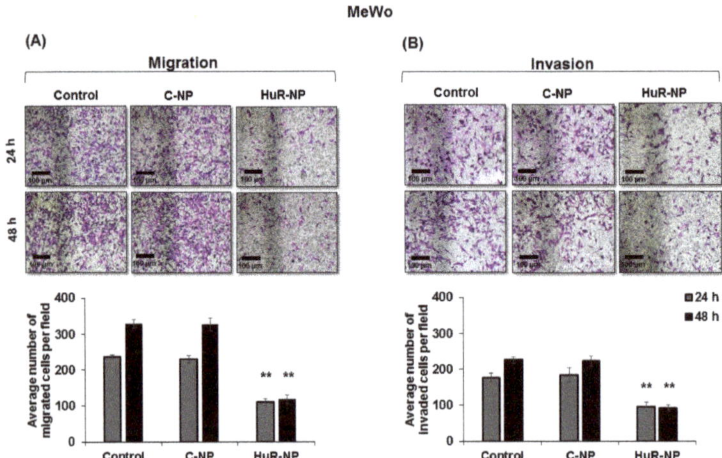

Figure 5. HuR-NP reduced melanoma cell migration and invasion. Representative images of crystal violet stained MeWo cells in (**A**), migration, and (**B**), invasion assay. Bar graphs represent the quantification of migrating and invading cells. Scale bar 100 μm. Error bar denotes SD; ** $p < 0.01$.

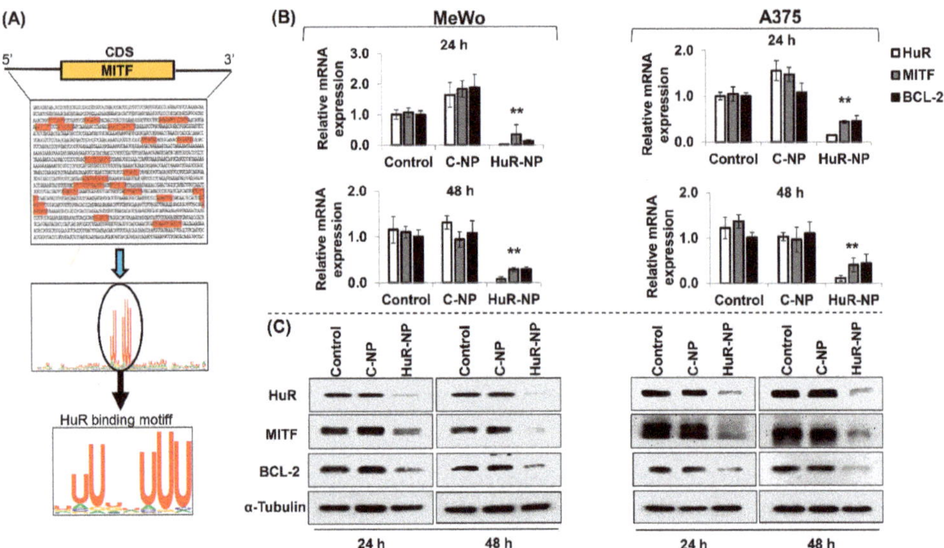

Figure 6. HuR-NP attenuates MITF mRNA and protein expression. (**A**), The MITF gene sequence, including promoter, untranslated regions (UTRs), and coding sequence (CDS), were scouted for HuR binding site. Multiple HuR binding sites were observed along the MITF gene. The conserved sequence search indicates HuR binding motif (5′NNUUNNUUU3′) conserved across the MITF gene sequence. RBPmap (http://rbpmap.technion.ac.il/) [51] and WebLogo (https://weblogo. berkeley.edu/) [52] available in the public domain as open resource tools were used for the analysis. The HuR-MITF binding was validated by (**B**), qRT-PCR, and (**C**), Western blot analysis in C-NP, and HuR-NP-treated MeWo and A375 melanoma cells. Untreated cells served as controls for each cell line. BCL-2 was assessed as positive control and downstream transcriptional target of MITF. 18S forward and reverse primers, and alpha α-Tubulin were used as internal controls in qRT-PCR and Western blotting, respectively. Error bar denotes SD; ** $p < 0.01$.

In accordance with the mRNA data, we observed a significant reduction in HuR, MITF, and BCL-2 protein expression levels in HuR-NP-treated MeWo and A375 cells at 24 h and 48 h compared to untreated and C-NP-treated cells (Figure 6C and Figure S6; $p < 0.05$). HuR-NP treatment of melanocytes showed no reduction in MITF compared to the untreated and C-NP-treated cells (Figure S7). To ensure consistency in our study results, MeWo cells receiving treatments identical to melanocytes were evaluated for MITF. HuR-NP treatment reduced MITF compared to untreated and C-NP-treated cells (Figure S7) and concurred with our data shown in Figure 6. Finally, immunostaining of a panel of human melanoma cell lines showed robust expression of MITF and HuR, albeit at varying levels for the two proteins in the cell lines (Figure S8). These results show that MITF is a molecular target of HuR and that inhibiting HuR concomitantly reduces MITF and its downstream BCL-2 expression. Furthermore, the MITF reduction in HuR-NP-treated cells likely contributes to the diminished cell migration and invasion observed in Figure 5.

3.7. HuR-NP Reduces U0126 Induced MITF in Melanoma Cells

Studies have shown that inhibiting the mitogen-activated protein kinase (MAPK) signaling pathway is beneficial in melanoma treatment [53–55]. Clinical studies testing inhibitors targeting MEK1/2, such as Trametinib alone and in combination with B-RAF inhibitor, Dabrafenib, have demonstrated clinical benefit in melanoma patients [56,57]. However, in the majority of the patients, the disease recurs and exhibits treatment resistance. Furthermore, high MITF expression in melanoma contributes to resistance towards MAPK inhibitors [58–60]. Based on these reports, we investigated whether incorporating HuR-NP with MEK1/2 inhibitor (U0126) will demonstrate improved efficacy.

Prior to conducting combinatorial efficacy studies, optimization studies testing different concentrations of U0126 (10 µM, 20 µM, and 30 µM) were performed and evaluated for cytotoxicity as well as MEK1/2 inhibition. Treatment of MeWo cells with U0126 resulted in a dose-dependent reduction in cell viability at 24 h and 48 h compared to DMSO-treated control cells (Figure S9; $p < 0.05$). Molecular analysis showed U0126 treatment while reducing phosphorylated (p)-MEK1/2$^{Ser217/221}$ significantly increased MITF expression in a dose-dependent manner at 24 h and 48 h (Figure S9; $p < 0.05$).

Next, we investigated whether HuR-NP treatment could override U0126-treatment-induced MITF and exhibit increased efficacy. MeWo cells treated with HuR-NP showed significant inhibition of cell viability and marked reduction in MITF and p-MEK1/2$^{Ser217/221}$ expression compared to DMSO- and C-NP-treated cells (Figure 7 and Figure S10; $p < 0.05$).

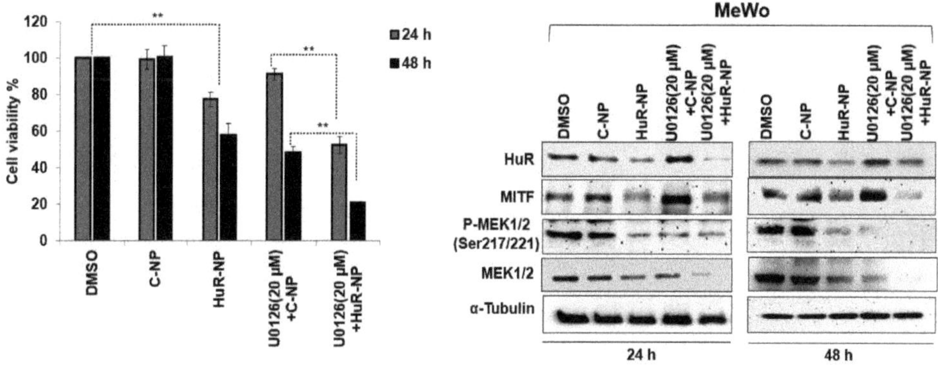

Figure 7. Cytotoxic effect of HuR-NP and U0126 combinatorial therapy in melanoma cell lines. Cells were treated with C-NP, HuR-NP, U0126, and a combination of C-NP or HuR-NP with U0126, and cell viability was assessed at 24 h and 48 h after treatment. DMSO-treated cells served as a control. HuR-NP plus U0126 treatment produced the greatest antitumor activity compared to all other treatment groups. Western blotting showed that HuR-NP plus U0126 combination treatment significantly reduced MITF, p-MEK1/2$^{Ser217/221}$, and total MEK1/2 compared to all other treatment groups. Error bar denotes SD; ** $p < 0.01$.

However, a synergistic effect in reducing cell viability and reduction in MITF and p-MEK1/2$^{Ser217/221}$ expression was observed when HuR-NP was combined with U0126 compared to all other treatment groups (Figure 7 and Figure S10; $p < 0.05$). C-NP, in combination with U0126, produced an inhibitory effect akin to that observed with HuR-NP treatment alone but less than that with the combination treatment of HuR-NP and U0126. Another important observation was that the increased MITF expression observed in C-NP and U0126 combination treatment and attributed to U0126 was almost completely eliminated in HuR-NP and U0126 combination treatment, especially at 48 h after treatment (Figure 7 and Figure S10; $p < 0.05$). Our study results showed the HuR-NP suppressed U0126 induced MITF and produced enhanced antitumor activity in the melanoma cell line.

3.8. HuR-NP Suppresses the Cell Viability of MITF Overexpressing Melanoma Cell Line

Since our results showed HuR-NP treatment alone reduced MITF and U0126 induced MITF, we investigated whether HuR-NP can suppress MITF when overexpressed in melanoma cells analogous to that seen in melanoma patients. For this purpose, we first generated MITF- and GFP-overexpressing MeWo cell lines and labeled them as MeWo-MITF-M and MeWo-GFP, respectively. The two cell lines were characterized for cell viability, MITF, and MITF-regulated downstream markers (Figure S11). MITF overexpression increased MeWo-MITF-M cell number indicative of MITF's ability to support cell proliferation compared to MeWo-GFP cells and parental MeWo cells (Figure S11). Furthermore, MITF overexpression greatly increased BCL-2 and HIF1-α expression in MeWo-MITF-M cells (Figure S11), both of which are downstream transcriptional targets of MITF [61,62].

Next, we tested HuR-NP's ability to reduce the viability of MeWo-MITF-M cells and suppress MITF and its downstream targets. As shown in Figure 8, HuR-NP treatment significantly reduced MeWo-MITF-M cell viability compared to controls and was comparable to the HuR-NP inhibitory effect on MeWo-GFP cells at the two time-points tested ($p < 0.05$). Molecular analysis revealed HuR-NP unequivocally and effectively reduced MITF and BCL-2 in both MeWo-GFP and MeWo-MITF-M cells compared to their untreated and C-NP-treated control cells at 24 h after treatment (Figure 8 and Figure S12).

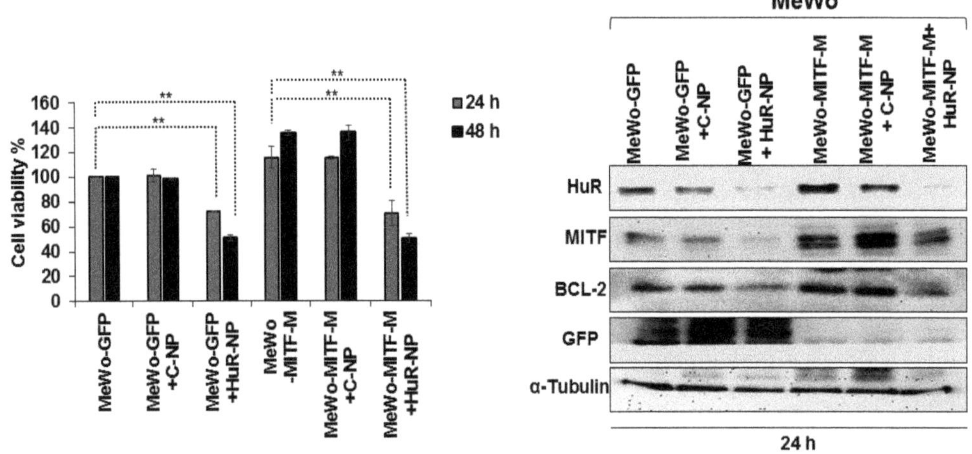

Figure 8. Cytotoxic effect of C-NP and HuR-NP treatment on MeWo-GFP and MeWo-MITF-M treated cells. HuR-NP significantly reduced the viability of both MeWo-GFP and MeWo-MITF-M cells at 24 h and 48 h compared to all other treatment groups. Western blotting showed HuR-NP reduced MITF, p-MEK1/2$^{Ser\,217/221}$ and total MEK1/2, and BCL-2 protein expression in both MeWo-GFP and MeWo-MITF-M cells. α-Tubulin was used as an internal loading control. Error bar denotes SD; ** $p < 0.01$.

Finally, we examined the combinatorial therapeutic efficacy of HuR-NP and U0126 on MeWo-MITF-M cells compared to individual treatments. HuR-NP and U0126 combination treatment produced a significant and greater inhibitory effect on MeWo-MITF-M cell viability with approximately 64% reduction at 24 h and 86% reduction at 48 h after treatment compared to all other treatment groups (Figure 9). The inhibitory effect on cell viability produced by U0126 treatment (46% inhibition) and C-NP and U0126 combination treatment (44% inhibition) was equivalent to the inhibitory effect produced by HuR-NP treatment alone (46% inhibition) at 48 h compared to DMSO-treated control cells (Figure 9). Molecular studies showed HuR-NP and U0126 combination treatment produced the greatest reduction in HuR, p-MEK1/2$^{Ser217/221}$, BCL-2 protein expression, and most importantly, almost completely eliminated MITF expression in MeWo-MITF-M cells compared to all other treatment groups (Figure 9 and Figure S13). These results demonstrate the effectiveness of combining HuR-NP with MEK inhibitors to overriding the oncogenic effects of MITF and potentially mitigating MITF-mediated resistance in melanoma (Figure S13).

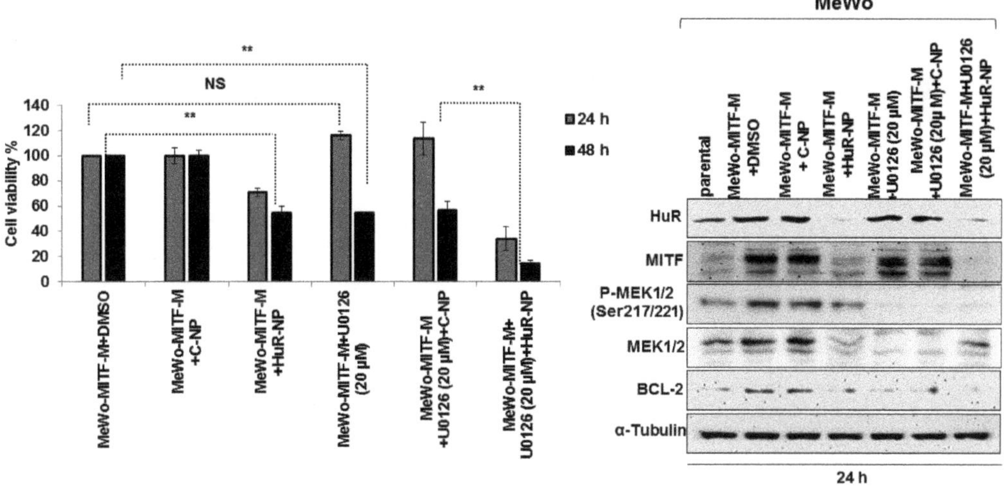

Figure 9. HuR-NP and U0126 combinatorial treatment produce enhanced antitumor activity in MeWo-MITF-M cells. Cells were treated with C-NP and HuR-NP in the presence or absence of U0126 and assessed for cell viability by trypan blue exclusion assay. DMSO-treated cells served as controls. Changes in molecular markers after treatment were examined by Western blotting assay. Combination treatment of HuR-NP and U0126 showed the greatest antitumor activity and maximum reduction in MITF. Reduction in p-MEK1/2$^{Ser\,217/221,}$ total MEK1/2, and BCL-2 observed in HuR-NP, and U0126 combination treatment was comparable to the reduction in these proteins in cells treated with C-NP plus U0126 and U0126 alone. Error bar denotes SD; ** $p < 0.01$.

4. Discussion

Tremendous efforts to improve treatment outcomes for melanoma patients have met with limited success until the recent development of immunotherapy. However, limitations continue to exist with immunotherapy, and reports of resistance to immunotherapy and disease recurrence are emerging [63–65]. Enthusiasm for targeted therapies, especially towards B-RAF inhibitors (Vemurafenib, Dabrafenib) for melanoma treatment, continue to persist, and several combinatorial treatments incorporating MAPK inhibitors (Trametinib, Cobimetinib) are being pursued [66,67]. While co-targeting B-RAF and MEK1/2 have shown to improve treatment response and provide clinical benefit, the manifestation of treatment-related acquired resistance continues to evolve [49]. Therefore, efforts to develop and test improved therapies for melanoma are in pursuit.

This study established the benefit of targeting human antigen R, HuR, in melanoma. Studies have shown that HuR is a molecular target for therapy [22,27,68,69], and inhibiting HuR resulted in antitumor and antimetastatic activity [30,31]. No studies, however, have evaluated the effect of targeting HuR on melanoma cell growth and metastases. Activating B-RAF mutations are common and occur in approximately 50% of cutaneous melanomas [5,14]. However, our results clearly demonstrated that HuR overexpression occurred independent of B-RAF mutation status in melanoma cell lines compared to melanocytes. This observation encouraged us to investigate targeting HuR in melanoma. Therefore, we tested the genetic inhibition of HuR using HuR-specific siRNA containing nanoparticles (HuR-NP) in two melanoma cell lines differing in their B-RAF status (MeWo, B-RAF^{wt}, and A375, B-RAF^{V600E}). HuR-NP inhibited cell proliferation of both the cell lines independent of their B-RAF status. Molecular studies demonstrated that HuR-NP significantly reduced HuR mRNA and protein. Furthermore, HuR-NP attenuated HuR-regulated oncoproteins in both melanoma cell lines. HuR-NP-mediated inhibition led to a G1 phase cell cycle arrest that subsequently led to apoptotic cell death, as evidenced by the activation of the caspase cascade. While our results concurred with other studies that used different tumor models, it is also suggested that HuR-NP could be an attractive target for melanoma therapy independent of oncogenic B-RAF mutation status. Considering the role of HuR in metastases, we tested the impact of HuR-NP on an invasive MeWo melanoma cell line. Interestingly, HuR diminished the migratory and invasive ability of MeWo cells. Most importantly, our results showed that HuR-NP exerted selective cytotoxicity towards melanoma cells but not towards normal melanocytes, a feature that is preferred in having effective cancer treatment.

Next, we investigated whether the inhibitory effect on melanoma migration and invasion following HuR-NP treatment could partly be due to MITF. MITF has been reported to contribute to melanoma cell survival, cell migration and invasion, drug resistance, and metastases [49,70–72]. To our surprise, a marked reduction in MITF and MITF-regulated BCL-2 and HIF-1α proteins was observed upon silencing of HuR. The ability of HuR to suppress MITF at both the mRNA and protein levels is an interesting observation that is hitherto not reported. Based on this initial observation of HuR-NP reducing MITF, we conducted combinatorial studies using U0126, a MEK1/2 inhibitor, to emulate clinical studies conducted for treating melanoma patients. Use of MEK1/2 inhibitors such as Trametinib, either alone or in combination with B-RAF inhibitor, Dabrafenib, while showing initial clinical benefit, has failed to demonstrate long-term efficacy due to disease recurrence and drug resistance. Furthermore, the expression of MITF has been reported to play a role in the failure of MEK1/2-targeted therapy [50,73,74]. In fact, our in vitro results showed HuR-NP treatment markedly reduced MITF, MEK1/2, and p-MEK1/2$^{Ser217/221}$ proteins, which concurred with previous study results that showed MEK1 mRNA as a HuR mRNA-target in intestinal epithelial cells [75]. Therefore, we speculated that incorporating U0126 into HuR-NP treatment will result in enhanced antitumor activity. We clearly and convincingly demonstrated that HuR-NP plus U0126 treatment in MeWo cells abolished MITF-induced MEK1/2 expression; suppressed MEK1/2 inhibitor-induced MITF; and overrides the anti-apoptotic effects of MITF in MITF overexpressing MeWo cells resulting in synergistic antitumor activity. Based on our study results, it is interesting to speculate that the combinatorial treatment of HuR-NP and U0126 will be very effective against melanoma cells that have developed acquired resistance to B-RAF/MEK inhibitors. However, the authors have not conducted this experiment as it was beyond the scope of the study but plan to test it in the future. Additionally, the study is limited to in vitro observations that need to be validated in in vivo melanoma models prior to advancing to clinical translation.

It is to be noted that while our study demonstrated the therapeutic benefits of HuR-targeted therapy for melanoma, several questions remain unanswered. For example, Slominski et al. reported melanogenesis regulated HIF-1α and HIF-1α-regulated target genes are involved in angiogenesis and cellular metabolism [76]. Similarly, the involvement of cyclic adenosine monophosphate (cAMP) and response binding protein element (CREB) in MITF activation and subsequent upregulation of melanogenesis-regulated gene coding for tyrosinase has been reported [77,78]. Since our study showed a link between HuR and MITF and that HuR is known to regulate HIF-1α and HIF-1α-regulated target genes such as VEGF, it will be of interest to investigate the role of HuR in melanogenesis and melanogenesis regulated tyrosinase in melanoma. A cross talk between MAPK and melanogenetic pathways are also strongly interconnected [79]. However, the exact role of HuR in regulating the MAPK pathway and its implications in melanogenesis is unknown. Finally, little to none is known about HuR expression and its role in uveal melanoma (UM). Interestingly, unlike cutaneous melanoma, UM is characterized by a very low mutational burden. In addition, B-RAF and N-RAS are rarely mutated, instead GNAQ or GNA11 mutations are frequently detected and known to activate the MAPK pathway also [80,81]. Hence, targeting HuR might be a treatment option in UM. However, laboratory studies on this are pending to date. Together, the results from the present study highlight the importance of targeting HuR in melanoma and concomitantly opening new avenues for investigating HuR in melanogenesis and UM.

5. Conclusions

We have established proof-of-concept and shown that targeting HuR represents a promising therapeutic option for melanoma treatment with or without oncogenic B-RAF mutation. Furthermore, inhibiting HuR offered the additional advantage of reducing MITF expression in melanoma cells, and combinatorial therapy targeting HuR and MEK1/2 produced synergistic antitumor activity. These results support additional combinatorial testing of HuR-targeted therapeutics in combination with B-RAF and MEK1/2 inhibitors for melanoma both in vitro and in vivo.

Supplementary Materials: The following are available online at https://www.mdpi.com/2072-669 4/13/2/166/s1. Figure S1: The baseline expression level of HuR, MITF, p-MEK1/2$^{Ser217/221}$, and total MEK1/2 along with B-RAF and N-RAS status in melanoma cell lines and melanocytes, Figure S2: Cytotoxicity activity of HuR-NP on melanoma cell lines, Figure S3: Semi-quantitative analysis of the protein expression detected by Western blot analysis in C-NP and HuR-NP-treated melanoma (MeWo, A375) cell lines and melanocytes at 24 h and 48 h after treatment, Figure S4: Semi-quantitative analysis of the cleaved Caspase-9 and PARP proteins detected by Western blot analysis in C-NP and HuR-NP-treated melanoma (MeWo, A375) cell lines at 24 h and 48 h after treatment, Figure S5: (A), HuR-NP induced apoptosis in MeWo cells but not in melanocytes. Gels of lighter and darker exposure for PARP and caspase 9 included. (B), Bar graphs represent the semi-quantitative analysis of the cleaved Caspase-9, PARP, and HuR proteins detected by Western blot analysis, Figure S6: Bar graphs represent the semi-quantitative analysis of the protein (HuR, MITF, BCL-2) expression in C-NP and HuR-NP-treated MeWo and A375 melanoma cells detected by Western blot analysis at 24 h and 48 h after treatment, Figure S7: HuR-NP treatment reduced MITF in melanoma (MeWo) cell line but not in melanocytes at 24 h and 48 h after treatment, Figure S8: Immunocytochemistry of HuR and MITF expression in a panel of human melanoma cell lines, Figure S9: Cytotoxic effect of U0126 on MeWo melanoma cell line, Figure S10: Bar graphs represent the semi-quantitative analysis of protein expression detected by Western blot analysis in MeWo melanoma cells treated with C-NP, HuR-NP, U0126, and a combination of C-NP or HuR-NP with U0126 at 24 h and 48 h after treatment, Figure S11: Cell viability of MeWo-MITF-M compared to MeWo-GFP and parental MeWo cells were assessed by trypan blue exclusion assay at 24 h and 48 h, Figure S12: Semi-quantitative analysis of the protein expression in C-NP- and HuR-NP-treated MeWo-GFP and MeWo-MITF-M cells detected by Western blot 24 h after treatment, Figure S13: Bar graphs represent the semi-quantitative analysis of the protein expression detected in MeWo-MITF-M cells treated with C-NP, HuR-NP, U0126 (20 µM), a combination of U0126 with C-NP or HuR-NP by Western blot 24 h post-treatment.

Author Contributions: R.A., R.M., A.S., S.E.,Y.D.Z., A.M. and R.R. conceptualized and designed the experiments; R.A., R.M. and A.S. conducted the experiments and acquired and assembled study results; R.A., A.S., A.M. and R.R. wrote the original manuscript draft; R.A., A.S., S.E., S.E.J., Y.D.Z., A.M., and R.R. reviewed and edited the manuscript; R.A., R.R., S.E.J. and Y.D.Z. analyzed the data and applied statistical methods; R.A., A.S., A.M. and R.R. funding acquisition; R.R. supervised the project. All authors have read and agreed to the published version of the manuscript.

Funding: This research was funded by the National Institutes of Health: R01 CA167516; National Institute of General Medical Sciences: P20 GM103639; National Cancer Institute: P30CA225520.

Institutional Review Board Statement: Not applicable.

Informed Consent Statement: Not applicable.

Data Availability Statement: Data is contained within the article or supplementary material.

Acknowledgments: This study was supported by grants received from the National Institutes of Health (NIH), (R01 CA167516); from the National Institute of General Medical Sciences (P20 GM103639) of the National Institutes of Health; a Pilot Grant, Seed Grant, and Student Trainee Grant funded by the National Cancer Institute Cancer Center Support Grant P30CA225520 awarded to the University of Oklahoma Stephenson Cancer Center; funds received from the Presbyterian Health Foundation (PHF) Seed Grant, Presbyterian Health Foundation Bridge Grant, Oklahoma Tobacco Settlement Endowment Trust (TSET) awarded to the University of Oklahoma Stephenson Cancer Center, and the Jim and Christy Everest Endowed Chair in Cancer Developmental Therapeutics. The content is solely the responsibility of the authors. The opinions, interpretations, conclusions, and recommendations are those of the author and not necessarily endorsed by or representative of the official views by the NIH, TSET, or PHF. Rajagopal Ramesh is an Oklahoma TSET Research Scholar and holds the Jim and Christy Everest Endowed Chair in Cancer Developmental Therapeutics.

Conflicts of Interest: The authors report no conflicts of interest in this work.

References

1. Wyant, T.; Alteri, R.; Kalidas, M.; Ogoro, C.; Lubejko, B.; Eidsmoe, K.; McDowell, S.; Greene, B.; Delfin-Davis, R.; Cance, G.W.; et al. Melanoma Skin Cancer Causes, RiskFactors, and Prevention. American Cancer Society. 2020. Available online: https://www.cancer.org/content/dam/CRC/PDF/Public/8824.00.pdf (accessed on 20 October 2020).
2. Kakavand, H.; Wilmott, J.S.; Long, G.V.; Scolyer, R.A. Targeted therapies and immune checkpoint inhibitors in the treatment of metastatic melanoma patients: A guide and update for pathologists. *Pathology* **2016**, *48*, 194–202. [CrossRef]
3. Wong, D.J.; Ribas, A. Targeted Therapy for Melanoma. *Cancer Treat. Res.* **2016**, *167*, 251–262. [CrossRef]
4. Kee, D.; McArthur, G. Targeted therapies for cutaneous melanoma. *Hematol. Oncol. Clin. N. Am.* **2014**, *28*, 491–505. [CrossRef] [PubMed]
5. Manzano, J.L.; Layos, L.; Buges, C.; de Los Llanos Gil, M.; Vila, L.; Martinez-Balibrea, E.; Martinez-Cardus, A. Resistant mechanisms to BRAF inhibitors in melanoma. *Ann. Transl. Med.* **2016**, *4*, 237. [CrossRef] [PubMed]
6. Davies, H.; Bignell, G.R.; Cox, C.; Stephens, P.; Edkins, S.; Clegg, S.; Teague, J.; Woffendin, H.; Garnett, M.J.; Bottomley, W.; et al. Mutations of the BRAF gene in human cancer. *Nature* **2002**, *417*, 949–954. [CrossRef] [PubMed]
7. Wellbrock, C.; Ogilvie, L.; Hedley, D.; Karasarides, M.; Martin, J.; Niculescu-Duvaz, D.; Springer, C.J.; Marais, R. V599EB-RAF is an oncogene in melanocytes. *Cancer Res.* **2004**, *64*, 2338–2342. [CrossRef] [PubMed]
8. Carvajal, R.D.; Antonescu, C.R.; Wolchok, J.D.; Chapman, P.B.; Roman, R.A.; Teitcher, J.; Panageas, K.S.; Busam, K.J.; Chmielowski, B.; Lutzky, J.; et al. KIT as a therapeutic target in metastatic melanoma. *JAMA* **2011**, *305*, 2327–2334. [CrossRef]
9. Hodi, F.S.; Friedlander, P.; Corless, C.L.; Heinrich, M.C.; Mac Rae, S.; Kruse, A.; Jagannathan, J.; Van den Abbeele, A.D.; Velazquez, E.F.; Demetri, G.D.; et al. Major response to imatinib mesylate in KIT-mutated melanoma. *J. Clin. Oncol.* **2008**, *26*, 2046–2051. [CrossRef]
10. Carvajal, R.D.; Hamid, O.; Antonescu, C.R. Selecting patients for KIT inhibition in melanoma. *Methods Mol. Biol.* **2014**, *1102*, 137–162. [CrossRef]
11. Woodman, S.E.; Trent, J.C.; Stemke-Hale, K.; Lazar, A.J.; Pricl, S.; Pavan, G.M.; Fermeglia, M.; Gopal, Y.N.; Yang, D.; Podoloff, D.A.; et al. Activity of dasatinib against L576P KIT mutant melanoma: Molecular, cellular, and clinical correlates. *Mol. Cancer Ther.* **2009**, *8*, 2079–2085. [CrossRef]
12. Busca, R.; Bertolotto, C.; Ortonne, J.P.; Ballotti, R. Inhibition of the phosphatidylinositol 3-kinase/p70(S6)-kinase pathway induces B16 melanoma cell differentiation. *J. Biol. Chem.* **1996**, *271*, 31824–31830. [CrossRef] [PubMed]
13. Trojaniello, C.; Festino, L.; Vanella, V.; Ascierto, P.A. Encorafenib in combination with binimetinib for unresectable or metastatic melanoma with BRAF mutations. *Expert. Rev. Clin. Pharmacol.* **2019**, *12*, 259–266. [CrossRef] [PubMed]
14. Spagnolo, F.; Ghiorzo, P.; Orgiano, L.; Pastorino, L.; Picasso, V.; Tornari, E.; Ottaviano, V.; Queirolo, P. BRAF-mutant melanoma: Treatment approaches, resistance mechanisms, and diagnostic strategies. *Onco Targets Ther.* **2015**, *8*, 157–168. [CrossRef]
15. Simeone, E.; Grimaldi, A.M.; Festino, L.; Vanella, V.; Palla, M.; Ascierto, P.A. Combination Treatment of Patients with BRAF-Mutant Melanoma: A New Standard of Care. *BioDrugs* **2017**, *31*, 51–61. [CrossRef] [PubMed]
16. Achkar, T.; Tarhini, A.A. The use of immunotherapy in the treatment of melanoma. *J. Hematol. Oncol.* **2017**, *10*, 88. [CrossRef]

17. Karlsson, A.K.; Saleh, S.N. Checkpoint inhibitors for malignant melanoma: A systematic review and meta-analysis. *Clin. Cosmet. Investig. Dermatol.* **2017**, *10*, 325–339. [CrossRef]

18. Ventola, C.L. Cancer Immunotherapy, Part 3: Challenges and Future Trends. *Pharm. Ther.* **2017**, *42*, 514–521.

19. Christiansen, S.A.; Khan, S.; Gibney, G.T. Targeted Therapies in Combination With Immune Therapies for the Treatment of Metastatic Melanoma. *Cancer J.* **2017**, *23*, 59–62. [CrossRef]

20. Peng, S.S.; Chen, C.Y.; Xu, N.; Shyu, A.B. RNA stabilization by the AU-rich element binding protein, HuR, an ELAV protein. *EMBO J.* **1998**, *17*, 3461–3470. [CrossRef]

21. Ripin, N.; Boudet, J.; Duszczyk, M.M.; Hinniger, A.; Faller, M.; Krepl, M.; Gadi, A.; Schneider, R.J.; Sponer, J.; Meisner-Kober, N.C.; et al. Molecular basis for AU-rich element recognition and dimerization by the HuR C-terminal RRM. *Proc. Natl. Acad. Sci. USA* **2019**, *116*, 2935–2944. [CrossRef]

22. Wang, J.; Guo, Y.; Chu, H.; Guan, Y.; Bi, J.; Wang, B. Multiple functions of the RNA-binding protein HuR in cancer progression, treatment responses and prognosis. *Int. J. Mol. Sci.* **2013**, *14*, 10015–10041. [CrossRef] [PubMed]

23. Heinonen, M.; Bono, P.; Narko, K.; Chang, S.H.; Lundin, J.; Joensuu, H.; Furneaux, H.; Hla, T.; Haglund, C.; Ristimaki, A. Cytoplasmic HuR expression is a prognostic factor in invasive ductal breast carcinoma. *Cancer Res.* **2005**, *65*, 2157–2161. [CrossRef] [PubMed]

24. Denkert, C.; Weichert, W.; Pest, S.; Koch, I.; Licht, D.; Kobel, M.; Reles, A.; Sehouli, J.; Dietel, M.; Hauptmann, S. Overexpression of the embryonic-lethal abnormal vision-like protein HuR in ovarian carcinoma is a prognostic factor and is associated with increased cyclooxygenase 2 expression. *Cancer Res.* **2004**, *64*, 189–195. [CrossRef] [PubMed]

25. Mrena, J.; Wiksten, J.P.; Thiel, A.; Kokkola, A.; Pohjola, L.; Lundin, J.; Nordling, S.; Ristimaki, A.; Haglund, C. Cyclooxygenase-2 is an independent prognostic factor in gastric cancer and its expression is regulated by the messenger RNA stability factor HuR. *Clin. Cancer Res.* **2005**, *11*, 7362–7368. [CrossRef]

26. Niesporek, S.; Kristiansen, G.; Thoma, A.; Weichert, W.; Noske, A.; Buckendahl, A.C.; Jung, K.; Stephan, C.; Dietel, M.; Denkert, C. Expression of the ELAV-like protein HuR in human prostate carcinoma is an indicator of disease relapse and linked to COX-2 expression. *Int. J. Oncol.* **2008**, *32*, 341–347. [CrossRef]

27. Stoppoloni, D.; Cardillo, I.; Verdina, A.; Vincenzi, B.; Menegozzo, S.; Santini, M.; Sacchi, A.; Baldi, A.; Galati, R. Expression of the embryonic lethal abnormal vision-like protein HuR in human mesothelioma: Association with cyclooxygenase-2 and prognosis. *Cancer* **2008**, *113*, 2761–2769. [CrossRef]

28. Liaudet, N.; Fernandez, M.; Fontao, L.; Kaya, G.; Merat, R. Hu antigen R (HuR) heterogeneous expression quantification as a prognostic marker of melanoma. *J. Cutan. Pathol.* **2018**, *45*, 333–336. [CrossRef]

29. Giaginis, C.; Alexandrou, P.; Tsoukalas, N.; Sfiniadakis, I.; Kavantzas, N.; Agapitos, E.; Patsouris, E.; Theocharis, S. Hu-antigen receptor (HuR) and cyclooxygenase-2 (COX-2) expression in human non-small-cell lung carcinoma: Associations with clinico-pathological parameters, tumor proliferative capacity and patients' survival. *Tumour Biol.* **2015**, *36*, 315–327. [CrossRef]

30. Muralidharan, R.; Babu, A.; Amreddy, N.; Srivastava, A.; Chen, A.; Zhao, Y.D.; Kompella, U.B.; Munshi, A.; Ramesh, R. Tumor-targeted Nanoparticle Delivery of HuR siRNA Inhibits Lung Tumor Growth In Vitro and In Vivo By Disrupting the Oncogenic Activity of the RNA-binding Protein HuR. *Mol. Cancer Ther.* **2017**, *16*, 1470–1486. [CrossRef]

31. Muralidharan, R.; Mehta, M.; Ahmed, R.; Roy, S.; Xu, L.; Aube, J.; Chen, A.; Zhao, Y.D.; Herman, T.; Ramesh, R.; et al. HuR-targeted small molecule inhibitor exhibits cytotoxicity towards human lung cancer cells. *Sci. Rep.* **2017**, *7*, 9694. [CrossRef]

32. Muralidharan, R.; Babu, A.; Amreddy, N.; Basalingappa, K.; Mehta, M.; Chen, A.; Zhao, Y.D.; Kompella, U.B.; Munshi, A.; Ramesh, R. Folate receptor-targeted nanoparticle delivery of HuR-RNAi suppresses lung cancer cell proliferation and migration. *J. Nanobiotechnol.* **2016**, *14*, 47. [CrossRef] [PubMed]

33. Jimbo, M.; Blanco, F.F.; Huang, Y.H.; Telonis, A.G.; Screnci, B.A.; Cosma, G.L.; Alexeev, V.; Gonye, G.E.; Yeo, C.J.; Sawicki, J.A.; et al. Targeting the mRNA-binding protein HuR impairs malignant characteristics of pancreatic ductal adenocarcinoma cells. *Oncotarget* **2015**, *6*, 27312–27331. [CrossRef]

34. Lang, M.; Berry, D.; Passecker, K.; Mesteri, I.; Bhuju, S.; Ebner, F.; Sedlyarov, V.; Evstatiev, R.; Dammann, K.; Loy, A.; et al. HuR Small-Molecule Inhibitor Elicits Differential Effects in Adenomatosis Polyposis and Colorectal Carcinogenesis. *Cancer Res.* **2017**, *77*, 2424–2438. [CrossRef] [PubMed]

35. Wu, X.; Lan, L.; Wilson, D.M.; Marquez, R.T.; Tsao, W.C.; Gao, P.; Roy, A.; Turner, B.A.; McDonald, P.; Tunge, J.A.; et al. Identification and validation of novel small molecule disruptors of HuR-mRNA interaction. *ACS Chem. Biol.* **2015**, *10*, 1476–1484. [CrossRef] [PubMed]

36. Blanco, F.F.; Preet, R.; Aguado, A.; Vishwakarma, V.; Stevens, L.E.; Vyas, A.; Padhye, S.; Xu, L.; Weir, S.J.; Anant, S.; et al. Impact of HuR inhibition by the small molecule MS-444 on colorectal cancer cell tumorigenesis. *Oncotarget* **2016**, *7*, 74043–74058. [CrossRef] [PubMed]

37. Romeo, C.; Weber, M.C.; Zarei, M.; DeCicco, D.; Chand, S.N.; Lobo, A.D.; Winter, J.M.; Sawicki, J.A.; Sachs, J.N.; Meisner-Kober, N.; et al. HuR Contributes to TRAIL Resistance by Restricting Death Receptor 4 Expression in Pancreatic Cancer Cells. *Mol. Cancer Res.* **2016**, *14*, 599–611. [CrossRef] [PubMed]

38. Mehta, M.; Basalingappa, K.; Griffith, J.N.; Andrade, D.; Babu, A.; Amreddy, N.; Muralidharan, R.; Gorospe, M.; Herman, T.; Ding, W.Q.; et al. HuR silencing elicits oxidative stress and DNA damage and sensitizes human triple-negative breast cancer cells to radiotherapy. *Oncotarget* **2016**, *7*, 64820–64835. [CrossRef] [PubMed]

39. Amreddy, N.; Ahmed, R.A.; Munshi, A.; Ramesh, R. Tumor-Targeted Dendrimer Nanoparticles for Combinatorial Delivery of siRNA and Chemotherapy for Cancer Treatment. *Methods Mol. Biol.* **2020**, *2059*, 167–189. [CrossRef]
40. Ramesh, R.; Saeki, T.; Templeton, N.S.; Ji, L.; Stephens, L.C.; Ito, I.; Wilson, D.R.; Wu, Z.; Branch, C.D.; Minna, J.D.; et al. Successful treatment of primary and disseminated human lung cancers by systemic delivery of tumor suppressor genes using an improved liposome vector. *Mol. Ther.* **2001**, *3*, 337–350. [CrossRef]
41. Ahmed, R.; Amreddy, N.; Babu, A.; Munshi, A.; Ramesh, R. Combinatorial Nanoparticle Delivery of siRNA and Antineoplastics for Lung Cancer Treatment. *Methods Mol. Biol.* **2019**, *1974*, 265–290. [CrossRef]
42. Ianevski, A.; Giri, A.K.; Aittokallio, T. SynergyFinder 2.0: Visual analytics of multi-drug combination synergies. *Nucleic Acids Res.* **2020**, *48*, W488–W493. [CrossRef] [PubMed]
43. Panneerselvam, J.; Srivastava, A.; Muralidharan, R.; Wang, Q.; Zheng, W.; Zhao, L.; Chen, A.; Zhao, Y.D.; Munshi, A.; Ramesh, R. IL-24 modulates the high mobility group (HMG) A1/miR222/AKT signaling in lung cancer cells. *Oncotarget* **2016**, *7*, 70247–70263. [CrossRef] [PubMed]
44. Andrade, D.; Mehta, M.; Griffith, J.; Oh, S.; Corbin, J.; Babu, A.; De, S.; Chen, A.; Zhao, Y.D.; Husain, S.; et al. HuR Reduces Radiation-Induced DNA Damage by Enhancing Expression of ARID1A. *Cancers* **2019**, *11*, 14. [CrossRef] [PubMed]
45. Komina, A.V.; Palkina, N.V.; Aksenenko, M.B.; Lavrentev, S.N.; Moshev, A.V.; Savchenko, A.A.; Averchuk, A.S.; Rybnikov, Y.A.; Ruksha, T.G. Semaphorin-5A downregulation is associated with enhanced migration and invasion of BRAF-positive melanoma cells under vemurafenib treatment in melanomas with heterogeneous BRAF status. *Melanoma Res.* **2019**, *29*, 544–548. [CrossRef] [PubMed]
46. Glitza Oliva, I.; Tawbi, H.; Davies, M.A. Melanoma Brain Metastases: Current Areas of Investigation and Future Directions. *Cancer J.* **2017**, *23*, 68–74. [CrossRef] [PubMed]
47. Yajima, I.; Kumasaka, M.Y.; Thang, N.D.; Goto, Y.; Takeda, K.; Iida, M.; Ohgami, N.; Tamura, H.; Yamanoshita, O.; Kawamoto, Y.; et al. Molecular Network Associated with MITF in Skin Melanoma Development and Progression. *J. Skin Cancer* **2011**, *2011*, 730170. [CrossRef] [PubMed]
48. Hsiao, J.J.; Fisher, D.E. The roles of microphthalmia-associated transcription factor and pigmentation in melanoma. *Arch. Biochem. Biophys.* **2014**, *563*, 28–34. [CrossRef] [PubMed]
49. Hartman, M.L.; Czyz, M. Pro-survival role of MITF in melanoma. *J. Investig. Dermatol.* **2015**, *135*, 352–358. [CrossRef]
50. Cheli, Y.; Giuliano, S.; Fenouille, N.; Allegra, M.; Hofman, V.; Hofman, P.; Bahadoran, P.; Lacour, J.P.; Tartare-Deckert, S.; Bertolotto, C.; et al. Hypoxia and MITF control metastatic behaviour in mouse and human melanoma cells. *Oncogene* **2012**, *31*, 2461–2470. [CrossRef]
51. Paz, I.; Kosti, I.; Ares, M., Jr.; Cline, M.; Mandel-Gutfreund, Y. RBPmap: A web server for mapping binding sites of RNA-binding proteins. *Nucleic Acids Res.* **2014**, *42*, W361–W367. Available online: https://rbmap.technion.ac.il (accessed on 8 September 2018). [CrossRef]
52. Crooks, G.E.; Hon, G.; Chandonia, J.M.; Brenner, S.E. WebLogo: A sequence logo generator. *Genome Res.* **2004**, *14*, 1188–1190. Available online: https://weblogo.berkeley.edu (accessed on 14 September 2018). [CrossRef] [PubMed]
53. Sullivan, R.J.; Atkins, M.B. Molecular targeted therapy for patients with melanoma: The promise of MAPK pathway inhibition and beyond. *Expert Opin. Investig. Drugs* **2010**, *19*, 1205–1216. [CrossRef] [PubMed]
54. Goldinger, S.M.; Zimmer, L.; Schulz, C.; Ugurel, S.; Hoeller, C.; Kaehler, K.C.; Schadendorf, D.; Hassel, J.C.; Becker, J.; Hauschild, A.; et al. Upstream mitogen-activated protein kinase (MAPK) pathway inhibition: MEK inhibitor followed by a BRAF inhibitor in advanced melanoma patients. *Eur. J. Cancer* **2014**, *50*, 406–410. [CrossRef] [PubMed]
55. Czyz, M.; Sztiller-Sikorska, M.; Gajos-Michniewicz, A.; Osrodek, M.; Hartman, M.L. Plasticity of Drug-Naive and Vemurafenib- or Trametinib-Resistant Melanoma Cells in Execution of Differentiation/Pigmentation Program. *J. Oncol.* **2019**, *2019*, 1697913. [CrossRef]
56. Robert, C.; Karaszewska, B.; Schachter, J.; Rutkowski, P.; Mackiewicz, A.; Stroiakovski, D.; Lichinitser, M.; Dummer, R.; Grange, F.; Mortier, L.; et al. Improved overall survival in melanoma with combined dabrafenib and trametinib. *N. Engl. J. Med.* **2015**, *372*, 30–39. [CrossRef]
57. Long, G.V.; Weber, J.S.; Infante, J.R.; Kim, K.B.; Daud, A.; Gonzalez, R.; Sosman, J.A.; Hamid, O.; Schuchter, L.; Cebon, J.; et al. Overall Survival and Durable Responses in Patients With BRAF V600-Mutant Metastatic Melanoma Receiving Dabrafenib Combined With Trametinib. *J. Clin. Oncol.* **2016**, *34*, 871–878. [CrossRef]
58. Pathria, G.; Garg, B.; Borgdorff, V.; Garg, K.; Wagner, C.; Superti-Furga, G.; Wagner, S.N. Overcoming MITF-conferred drug resistance through dual AURKA/MAPK targeting in human melanoma cells. *Cell Death Dis.* **2016**, *7*, e2135. [CrossRef]
59. Muller, J.; Krijgsman, O.; Tsoi, J.; Robert, L.; Hugo, W.; Song, C.; Kong, X.; Possik, P.A.; Cornelissen-Steijger, P.D.; Geukes Foppen, M.H.; et al. Low MITF/AXL ratio predicts early resistance to multiple targeted drugs in melanoma. *Nat. Commun.* **2014**, *5*, 5712. [CrossRef]
60. Kozar, I.; Margue, C.; Rothengatter, S.; Haan, C.; Kreis, S. Many ways to resistance: How melanoma cells evade targeted therapies. *Biochim. Biophys. Acta. Rev. Cancer* **2019**, *1871*, 313–322. [CrossRef]
61. Busca, R.; Berra, E.; Gaggioli, C.; Khaled, M.; Bille, K.; Marchetti, B.; Thyss, R.; Fitsialos, G.; Larribere, L.; Bertolotto, C.; et al. Hypoxia-inducible factor 1{alpha} is a new target of microphthalmia-associated transcription factor (MITF) in melanoma cells. *J. Cell Biol.* **2005**, *170*, 49–59. [CrossRef]

62. McGill, G.G.; Horstmann, M.; Widlund, H.R.; Du, J.; Motyckova, G.; Nishimura, E.K.; Lin, Y.L.; Ramaswamy, S.; Avery, W.; Ding, H.F.; et al. Bcl2 regulation by the melanocyte master regulator Mitf modulates lineage survival and melanoma cell viability. *Cell* **2002**, *109*, 707–718. [CrossRef]
63. Iwai, Y.; Ishida, M.; Tanaka, Y.; Okazaki, T.; Honjo, T.; Minato, N. Involvement of PD-L1 on tumor cells in the escape from host immune system and tumor immunotherapy by PD-L1 blockade. *Proc. Natl. Acad. Sci. USA* **2002**, *99*, 12293–12297. [CrossRef]
64. Zhao, F.; Evans, K.; Xiao, C.; DeVito, N.; Theivanthiran, B.; Holtzhausen, A.; Siska, P.J.; Blobe, G.C.; Hanks, B.A. Stromal Fibroblasts Mediate Anti-PD-1 Resistance via MMP-9 and Dictate TGFbeta Inhibitor Sequencing in Melanoma. *Cancer Immunol. Res.* **2018**, *6*, 1459–1471. [CrossRef] [PubMed]
65. Peng, W.; Chen, J.Q.; Liu, C.; Malu, S.; Creasy, C.; Tetzlaff, M.T.; Xu, C.; McKenzie, J.A.; Zhang, C.; Liang, X.; et al. Loss of PTEN Promotes Resistance to T Cell-Mediated Immunotherapy. *Cancer Discov.* **2016**, *6*, 202–216. [CrossRef] [PubMed]
66. Daud, A.; Gill, J.; Kamra, S.; Chen, L.; Ahuja, A. Indirect treatment comparison of dabrafenib plus trametinib versus vemurafenib plus cobimetinib in previously untreated metastatic melanoma patients. *J. Hematol. Oncol.* **2017**, *10*, 3. [CrossRef] [PubMed]
67. Ascierto, P.A.; Ferrucci, P.F.; Fisher, R.; Del Vecchio, M.; Atkinson, V.; Schmidt, H.; Schachter, J.; Queirolo, P.; Long, G.V.; Di Giacomo, A.M.; et al. Dabrafenib, trametinib and pembrolizumab or placebo in BRAF-mutant melanoma. *Nat. Med.* **2019**, *25*, 941–946. [CrossRef] [PubMed]
68. Ishimaru, D.; Ramalingam, S.; Sengupta, T.K.; Bandyopadhyay, S.; Dellis, S.; Tholanikunnel, B.G.; Fernandes, D.J.; Spicer, E.K. Regulation of Bcl-2 expression by HuR in HL60 leukemia cells and A431 carcinoma cells. *Mol. Cancer Res.* **2009**, *7*, 1354–1366. [CrossRef]
69. Blanco, F.F.; Jimbo, M.; Wulfkuhle, J.; Gallagher, I.; Deng, J.; Enyenihi, L.; Meisner-Kober, N.; Londin, E.; Rigoutsos, I.; Sawicki, J.A.; et al. The mRNA-binding protein HuR promotes hypoxia-induced chemoresistance through posttranscriptional regulation of the proto-oncogene PIM1 in pancreatic cancer cells. *Oncogene* **2016**, *35*, 2529–2541. [CrossRef]
70. Urban, P.; Rabajdova, M.; Velika, B.; Spakova, I.; Bolerazska, B.; Marekova, M. The Importance of MITF Signaling Pathway in the Regulation of Proliferation and Invasiveness of Malignant Melanoma. *Klin. Onkol.* **2016**, *29*, 347–350. [CrossRef]
71. Najem, A.; Krayem, M.; Sales, F.; Hussein, N.; Badran, B.; Robert, C.; Awada, A.; Journe, F.; Ghanem, G.E. P53 and MITF/Bcl-2 identified as key pathways in the acquired resistance of NRAS-mutant melanoma to MEK inhibition. *Eur. J. Cancer* **2017**, *83*, 154–165. [CrossRef]
72. Ennen, M.; Keime, C.; Gambi, G.; Kieny, A.; Coassolo, S.; Thibault-Carpentier, C.; Margerin-Schaller, F.; Davidson, G.; Vagne, C.; Lipsker, D.; et al. MITF-High and MITF-Low Cells and a Novel Subpopulation Expressing Genes of Both Cell States Contribute to Intra- and Intertumoral Heterogeneity of Primary Melanoma. *Clin. Cancer Res.* **2017**, *23*, 7097–7107. [CrossRef] [PubMed]
73. Wellbrock, C.; Arozarena, I. Microphthalmia-associated transcription factor in melanoma development and MAP-kinase pathway targeted therapy. *Pigment. Cell Melanoma Res.* **2015**, *28*, 390–406. [CrossRef] [PubMed]
74. Johannessen, C.M.; Johnson, L.A.; Piccioni, F.; Townes, A.; Frederick, D.T.; Donahue, M.K.; Narayan, R.; Flaherty, K.T.; Wargo, J.A.; Root, D.E.; et al. A melanocyte lineage program confers resistance to MAP kinase pathway inhibition. *Nature* **2013**, *504*, 138–142. [CrossRef] [PubMed]
75. Wang, P.Y.; Rao, J.N.; Zou, T.; Liu, L.; Xiao, L.; Yu, T.X.; Turner, D.J.; Gorospe, M.; Wang, J.Y. Post-transcriptional regulation of MEK-1 by polyamines through the RNA-binding protein HuR modulating intestinal epithelial apoptosis. *Biochem. J.* **2010**, *426*, 293–306. [CrossRef] [PubMed]
76. Slominski, A.; Kim, T.K.; Brozyna, A.A.; Janjetovic, Z.; Brooks, D.L.; Schwab, L.P.; Skobowiat, C.; Jozwicki, W.; Seagroves, T.N. The role of melanogenesis in regulation of melanoma behavior: Melanogenesis leads to stimulation of HIF-1alpha expression and HIF-dependent attendant pathways. *Arch. Biochem. Biophys.* **2014**, *563*, 79–93. [CrossRef]
77. Alam, M.B.; Ahmed, A.; Motin, M.A.; Kim, S.; Lee, S.H. Attenuation of melanogenesis by Nymphaea nouchali (Burm. f) flower extract through the regulation of cAMP/CREB/MAPKs/MITF and proteasomal degradation of tyrosinase. *Sci. Rep.* **2018**, *8*, 13928. [CrossRef]
78. Jean, D.; Bar-Eli, M. Regulation of tumor growth and metastasis of human melanoma by the CREB transcription factor family. *Mol. Cell. Biochem.* **2000**, *212*, 19–28. [CrossRef]
79. Padua, R.A.; Barrass, N.; Currie, G.A. A novel transforming gene in a human malignant melanoma cell line. *Nature* **1984**, *311*, 671–673. [CrossRef]
80. Piperno-Neumann, S.; Piulats, J.M.; Goebeler, M.; Galloway, I.; Lugowska, I.; Becker, J.C.; Vihinen, P.; Van Calster, J.; Hadjistilianou, T.; Proenca, R.; et al. Uveal Melanoma: A European Network to Face the Many Challenges of a Rare Cancer. *Cancers* **2019**, *11*, 817. [CrossRef]
81. Eagle, R.C., Jr. Ocular tumors: Triumphs, challenges and controversies. *Saudi. J. Ophthalmol.* **2013**, *27*, 129–132. [CrossRef]

Article

BCL2L10 Is Overexpressed in Melanoma Downstream of STAT3 and Promotes Cisplatin and ABT-737 Resistance

María Josefina Quezada [1,2,†], María Elisa Picco [3,†], María Belén Villanueva [1,2], María Victoria Castro [1,2], Gastón Barbero [1,2], Natalia Brenda Fernández [3], Edith Illescas [1,2] and Pablo Lopez-Bergami [1,2,*]

[1] Centro de Estudios Biomédicos, Básicos, Aplicados y Desarrollo (CEBBAD), Universidad Maimónides, C1405BCK Buenos Aires, Argentina; quezada.josefina@maimonides.edu (M.J.Q.); villanueva.belen@maimonides.edu (M.B.V.); castro.victoria@maimonides.edu (M.V.C.); barbero.gaston@maimonides.edu (G.B.); illescas.edith@maimonides.edu (E.I.)
[2] Consejo Nacional de Investigaciones Científicas y Técnicas (CONICET), C1405BCK Buenos Aires, Argentina
[3] Instituto de Medicina y Biología Experimental (IBYME), CONICET, C1428ADN Buenos Aires, Argentina; mariaelisapicco@gmail.com (M.E.P.); nfernandez@cerebro.fbmc.fcen.uba.ar (N.B.F.)
* Correspondence: lopezbergami.pablo@maimonides.edu; Tel.: +54-11-4905-1133
† Contributed equally to this work.

Simple Summary: BCL2L10 is the sixth and less studied protein from the group of Bcl-2 anti-apoptotic proteins. These proteins are important therapeutic targets since they convey resistance to anticancer regimens. We describe here for the first time the role of BCL2L10 in melanoma. We found that BCL2L10 is abundantly and frequently expressed both in melanoma cell lines and tumor samples. This increased expression is due to the activity of the transcription factor STAT3 that positively regulate BCL2L10 transcription. We describe that Bcl2l10 is a pro-survival factor in melanoma, being able to protect cells from the cytotoxic effect of different drugs, including cisplatin, dacarbazine, and ABT-737. BCL2L10 also inhibited the cell death upon combination treatments of PLX-4032, a BRAF inhibitor, with ABT-737 or cisplatin. In summary, we determined that BCL2L10 is expressed in melanoma and contributes to cell survival. Hence, targeting BCL2L10 may enhance the clinical efficacy of other therapies for malignant melanoma.

Abstract: The anti-apoptotic proteins from the Bcl-2 family are important therapeutic targets since they convey resistance to anticancer regimens. Despite the suspected functional redundancy among the six proteins of this subfamily, both basic studies and therapeutic approaches have focused mainly on BCL2, Bcl-xL, and MCL1. The role of BCL2L10, another member of this group, has been poorly studied in cancer and never has been in melanoma. We describe here that BCL2L10 is abundantly and frequently expressed both in melanoma cell lines and tumor samples. We established that BCL2L10 expression is driven by STAT3-mediated transcription, and by using reporter assays, site-directed mutagenesis, and ChIP analysis, we identified the functional STAT3 responsive elements in the BCL2L10 promoter. BCL2L10 is a pro-survival factor in melanoma since its expression reduced the cytotoxic effects of cisplatin, dacarbazine, and ABT-737 (a BCL2, Bcl-xL, and Bcl-w inhibitor). Meanwhile, both genetic and pharmacological inhibition of BCL2L10 sensitized melanoma cells to cisplatin and ABT-737. Finally, BCL2L10 inhibited the cell death upon combination treatments of PLX-4032, a BRAF inhibitor, with ABT-737 or cisplatin. In summary, we determined that BCL2L10 is expressed in melanoma and contributes to cell survival. Hence, targeting BCL2L10 may enhance the clinical efficacy of other therapies for malignant melanoma.

Keywords: BCL2L10; STAT3; melanoma; cytotoxicity; survival; ABT-737; Bcl-2 family; ML258

Citation: Quezada, M.J.; Picco, M.E.; Villanueva, M.B.; Castro, M.V.; Barbero, G.; Fernández, N.B.; Illescas, E.; Lopez-Bergami, P. BCL2L10 Is Overexpressed in Melanoma Downstream of STAT3 and Promotes Cisplatin and ABT-737 Resistance. *Cancers* **2021**, *13*, 78. https://doi.org/10.3390/cancers13010078

Received: 10 December 2020
Accepted: 18 December 2020
Published: 30 December 2020

Publisher's Note: MDPI stays neutral with regard to jurisdictional claims in published maps and institutional affiliations.

1. Introduction

Melanoma is the fifth and sixth most common cancer for men and women, respectively, and thus represents a major public health problem. In addition to its high incidence, melanoma is one of the most aggressive tumor types with a 5-year survival rate of around 20% [1]. At the molecular level, melanoma is characterized by the highly prevalent BRAF^{V600E} mutation that renders the MAPK/ERK pathway constitutively active and is critical for melanoma progression. Moreover, many other signaling pathways, such as PI3K/Akt, PKC, STAT3, Wnt, and Eph/ephrin, are also constitutively activated [2–5]. Targeted therapies toward BRAF and MEK and immunotherapy have shown promise in the management of this cancer, and currently, there are several single or combination therapies approved for first-line treatment of metastatic or unresectable disease [6]. However, melanoma remains difficult to treat due to innate and acquired resistance to these therapies. Sadly, this has been a recurrent problem in melanoma since the massive failure of standard chemotherapy nearly 50 years ago [7].

Melanoma is intrinsically resistant to diverse cytotoxic insults, such as DNA damage (e.g., by irradiation, alkylation, methylation or crosslinking), microtubule destabilization or topoisomerase inhibition. One of the underlying mechanisms is a profound dysregulation of cell death pathways in part due to the aberrant expression of proteins of the Bcl-2 family that regulates the intrinsic or mitochondrial apoptotic pathway [8]. There are at least 20 Bcl-2-related proteins that are categorized into one of the three subfamilies; anti-apoptotic proteins (BCL2, Bcl-xL/BCL2L1, Bcl-w/BCL2L2, MCL1, Bfl-1/BCL2A1, and BCL2L10), pro-apoptotic BH3-only proteins (BAD, BID, BIK, Bim/BCL2L11, BMF, HRK, NOXA, PUMA/BBC3, etc.) and pro-apoptotic pore-formers or executioners (BAX, BAK1, BOK). Proteins from the first two groups compete to influence the executioners that, if activated, form pores in the outer mitochondrial membrane and thus trigger mitochondrial outer membrane permeability (MOMP) and apoptosis [9,10]. Given its important role in cancer progression, anti-apoptotic members of the Bcl-2 family have been studied as therapeutic targets in cancer, first by using antisense oligonucleotides against BCL2 and Bcl-xL, and more recently by using the small molecule compounds called BH3-mimetics since they emulate the function of the BH3-only proteins. Currently, several BH3-mimetics are being evaluated as single agents or combined with other compounds in clinical trials for a wide number of tumor types, including melanoma [11].

BCL2L10 (also known as Bcl-b, NrH, Diva, or Boo) is the most recently identified and least studied protein of the Bcl-2 anti-apoptotic subfamily, and its function is only partially understood [12,13]. BCL2L10 has been classified into the anti-apoptotic group of Bcl-2 proteins since it contains all four BH domains, a distinctive feature that is missing in the other groups. Nevertheless, both pro-apoptotic [14–18] and anti-apoptotic [19–22] activities of BCL2L10 have been described. BCL2L10 was also found to have functions not related to apoptosis. In ovarian cancer and hepatocellular carcinoma, BCL2L10 acts as a tumor suppressor by negatively regulating cell proliferation [15,23]. Along this line, the BCL2L10 promoter was found to be aberrantly methylated in gastric cancer [14,18,24], hepatocellular carcinoma [17], and acute myeloid leukemia [25]. The ensuing inhibition of BCL2L10 expression correlated with a decreased overall survival and disease-free survival of hepatocellular carcinoma [17] and gastric cancer [26] patients. On the other side, BCL2L10 was found to be overexpressed in breast, prostate, colorectal, and lung cancers as well as in multiple myeloma [27–30]. In many of these tumor types, the elevated BCL2L10 expression was correlated with poor prognosis [27–30]. The reasons for these discrepancies in BCL2L10′s role in cancer have not been elucidated.

This study aimed to investigate the expression and function of BCL2L10 in melanoma since they have not been studied to the present. We found that BCL2L10 is expressed in both melanoma cell lines and patients and that its transcription is regulated by STAT3. In addition, we determined that BCL2L10 plays a pro-survival role in melanoma, protecting the melanoma cells from different cytotoxic insults.

2. Results

2.1. BCL2L10 Expression in Melanoma

We begin our study by determining the expression of BCL2L10 in a panel of eight melanoma cell lines by Western blot. Antibody #3869 detected the expression of BCL2L10 in all melanoma cell lines analyzed (Figure 1A).

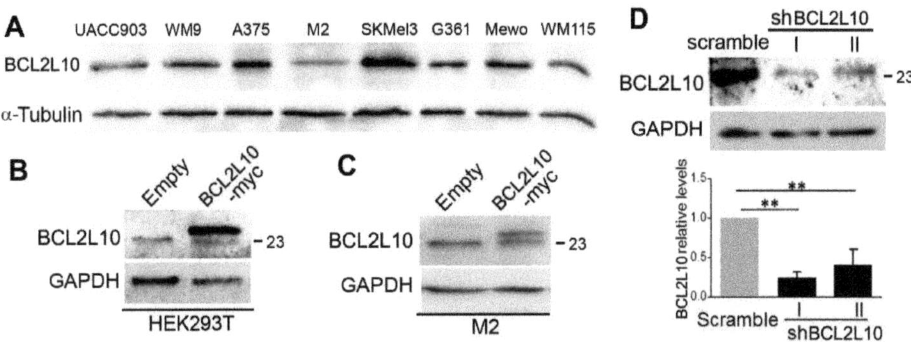

Figure 1. BCL2L10 expression in human melanoma cell lines and validation of BCL2L10 overexpression and silencing. (**A**) Protein extracts from the indicated melanoma cell lines were probed by Western blot with a 1:1000 dilution of the BCL2L10 antibody #3869. α-tubulin was used as the loading control. (**B**) Validation of the BCL2L10 antibody in HEK293 transfected with BCL2L10-myc. Protein extracts from HEK293T cells transfected with BCL2L10-myc or empty plasmid were assayed by Western blot using the anti-BCL2L10 PA5-22190 antibody (1:1000 dilution). (**C**) Validation of the BCL2L10 antibody in M2 cells stably transfected with BCL2L10-myc. Protein extracts from M2-empty and M2-BCL2L10-myc cells were assayed by Western blot using the PA5-22190 antibody (1:1000 dilution). (**D**) Silencing of BCL2L10 in A375 cells. Protein extracts from A375 cells transduced with two shRNA for BCL2L10 (I and II) or a scramble shRNA were analyzed for BCL2L10 expression by Western blot using the PA5-22190 antibody (1:1000 dilution). Bar graphs show the mean ± SD (from three independent experiments) of BCL2L10 levels normalized to the loading control and expressed as the fold change relative to scramble cells. The statistical analysis is described in Methods. **: $p < 0.01$, $n = 3$. α-tubulin was used as a loading control in panel A and GAPDH in panels B-D. The blots displayed in all panels are representative of three independent experiments. The 23 kDa marker is indicated in panels (**B–D**). Whole Western blot figures are provided in Supplemental Materials Figure S1.

Since the high frequency of BCL2L10 expression in our cell line collection may be due to a selection process inherent in creating tumor-derived cell lines or the conditions of "in vitro" cell culture, we sought to analyze BCL2L10 expression in melanoma specimens. However, this antibody did not result suitable this technique. Since BCL2L10 has been little studied, the detection of BCL2L10 in human tissues by immunohistochemistry (IHC) has been hindered by the lack of validated tools. The anti-BCL2L10 antibody PA5-22190 is one of the few commercial antibodies recommended for IHC, but to date, it has not been utilized in any scientific publication. Hence, to study BCL2L10 expression in melanoma patients, we first set to validate this antibody by Western blot and IHC analyses. An expression vector encoding human myc-tagged BCL2L10 was introduced into HEK293T cells, and BCL2L10 expression was evaluated in lysates from transfected cells. The PA5-22190 antibody detected a strong band corresponding to exogenous BCL2L10 at 24 kDa in the BCL2L10-myc transfected lysates, but not in lysates from cells transfected with empty plasmid (Figure 1B). Likewise, the antibody revealed a weaker band of about 23 kDa in both protein extracts, which corresponds with the expected molecular weight of the endogenous protein. The antibody detected the same two bands (of similar intensity in this case) in M2 melanoma cells stably transfected with BCL2L10-myc (Figure 1C). Importantly, no additional bands were detected, and the recognition pattern was similar to that of the BCL2L10 antibody #3869 (Supplementary Materials Figure S2). Expression of other Bcl-2 protein family members and that of other proteins implicated in apoptosis was not affected by BCL2L10 overexpression (Supplementary Materials Figure S3). To confirm the identity of the presumed endogenous BCL2L10 protein band, we transduced A375 melanoma cells with retroviral particles encoding two short-hairpin RNAs (shRNAs) specific for BCL2L10 or a scramble sequence (control shRNA), and the stable cell lines A375-shBCL2L10 I and II and A375-scramble were established. Western blot experiments revealed that the 23 kDa band is efficiently silenced in both A375-shBCL2L10 cell lines (Figure 1D). These results indicate that the PA5-22190 antibody is specific for BCL2L10.

To establish the conditions for IHC analysis of melanoma samples using the PA5-22190 antibody, we determined the antigen recovery method, antibody dilution, and other testing conditions (see Methods) using both positive (liver) and negative (placenta) control tissues that were selected based on BCL2L10 mRNA expression data from the Human Protein Atlas (www.proteinatlas.org). Then, we assayed samples from 20 melanoma patients from both primary and metastatic sites by IHC. Tissues were graded as strongly positive (+++), moderately positive (++), weakly positive (+), or negative (−). Robust expression of BCL2L10 was found in 90% of samples (scores ++ and +++), whereas the remaining 10% were negative (Figure 2 and Supplementary Materials Table S1).

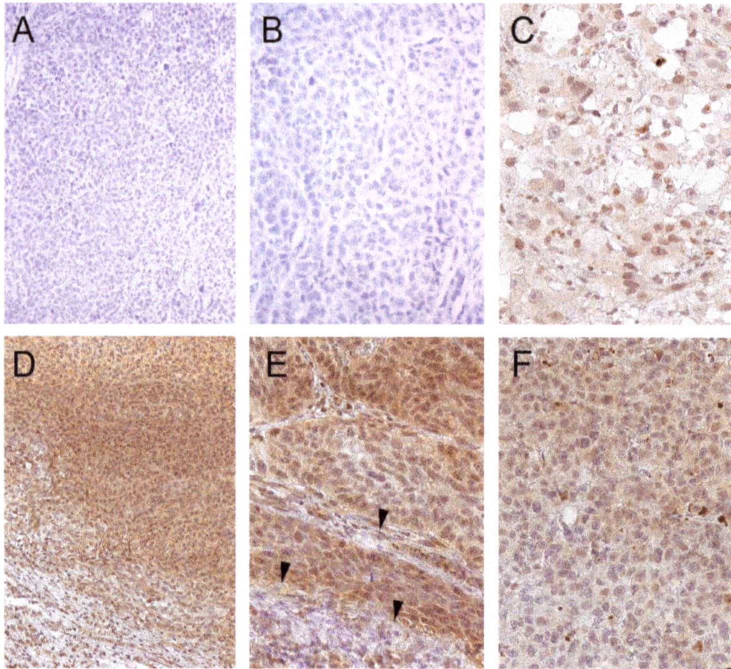

Figure 2. BCL2L10 is highly expressed in melanoma tumor samples. Representative images of immunohistochemistry staining for BCL2L10 using the PA5-22190 antibody (1:300 dilution) with hematoxylin counterstaining. (**A,B**) Representative images from a sample that scored negative for BCL2L10 staining (**A**, 100× magnification, **B**, 200× magnification). (**C**) Sample showing predominantly nuclear BCL2L10 staining (200× magnification). (**D,E**) Representative images from a sample showing nucleocytoplasmic BCL2L10 staining (**D**, 100× magnification, **E**, 200× magnification). BCL2L10 staining is not observed in the adjacent stroma (at the bottom on panel D) and in the septal stroma (arrows in panel E). (**F**) Sample showing predominantly cytoplasmic BCL2L10 staining (200× magnification).

BCL2L10 presented a heterogeneous pattern of immunoreactivity since both cytoplasmic and nuclear staining was observed both between- and within-samples. BCL2L10 staining was observed exclusively in the tumor tissue, whereas the immunoreactivity was not apparent in the adjacent noncancerous stroma (Figure 2D,E). These results demonstrate that BCL2L10 protein is frequently and strongly expressed in melanoma.

2.2. STAT3 Regulates BCL2L10 Expression

BCL2L10 was found to be overexpressed in several tumor types, but the mechanisms implicated in BCL2L10 expression have not been identified to date. Interestingly, analysis of microarray data generated from A375 melanoma cells treated with siRNAs against 45 transcription factors and signaling molecules (GSE31534) [31] revealed that the silencing of STAT3 markedly reduced BCL2L10 mRNA levels (Figure 3A). Furthermore, BCL2L10 mRNA levels were also reduced in three additional microarray studies that inhibited STAT3 expression or activity (GSE64536, GSE63092, and GSE48124, Supplementary Materials Figure S4). In line with these findings, we observed that in four out of the six patient samples tested, P-STAT3 staining overlapped with BCL2L10 staining. Unlike BCL2L10, which presented a heterogeneous pattern including both nuclear and cytoplasmic localization, P-STAT3 staining was almost exclusively nuclear in the melanoma samples (Supplementary Materials Figure S5).

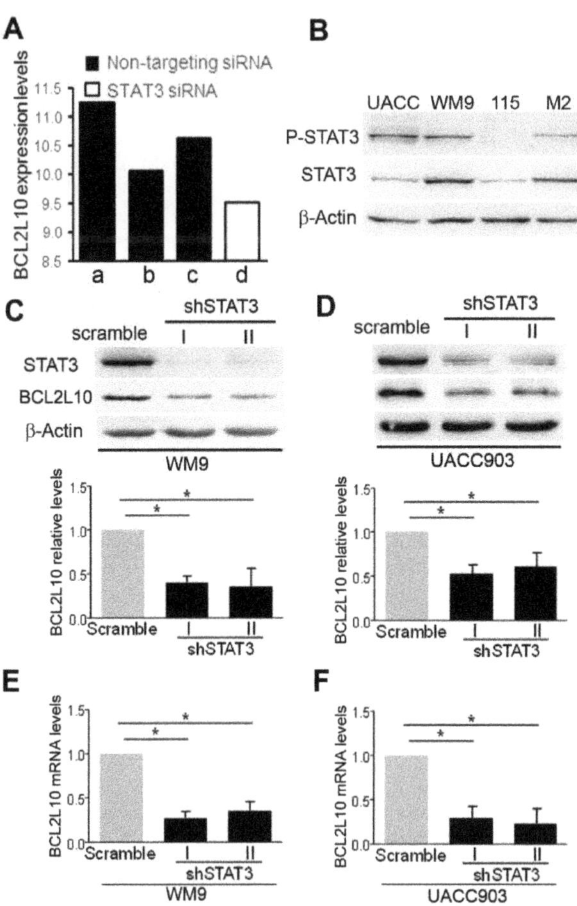

Figure 3. Silencing of STAT3-reduced BCL2L10 expression. (**A**) Regulation of BCL2L10 by STAT3 in dataset GSE31534. Analysis of dataset GSE31534 showed that a STAT3 siRNA (column d, GSM782740)

reduced BCL2L10 mRNA levels in A375 melanoma cells compared to non-targeting siRNA (columns a to c: GSM782733, GSM782734, and GSM782735, respectively). (**B**) Phosphorylation of STAT3 at Tyr^{705} in melanoma cell lines. Protein extracts from UACC903 (UACC), WM9, WM115 (115) and M2 were probed with the indicated antibodies. β-actin was used as a loading control. The blots displayed are representative of three independent experiments. (**C,D**) Silencing of STAT3 in WM9 (**C**) and UACC903 (**D**) cells reduced BCL2L10 protein expression. UACC903 and WM9 cells were transduced with retrovirus encoding two STAT3 shRNA (I and II) and a scramble sequence as a control. Protein extracts were probed with antibodies to STAT3, BCL2L10 and β-actin as a loading control. The blots displayed are representative of three independent experiments. Bar graphs show the mean ± SD (from three independent experiments) of BCL2L10 levels normalized to the loading control and expressed as the fold change relative to scramble cells. *: $p < 0.05$, $n = 3$. (**E,F**) STAT3 silencing decreased BCL2L10 mRNA levels in WM9 (**E**) and UACC903 (**F**). Relative levels of BCL2L10 mRNA were determined by real-time PCR. mRNA levels, normalized to internal RNPII levels and expressed as relative to control cells. The mean ± SD from three independent experiments is shown. The statistical analysis is described in Methods. *: $p < 0.05$, $n = 3$. Whole Western blot figures are provided in Supplemental Materials Figure S6.

The results above motivated us to determine whether STAT3 regulates BCL2L10 transcription. To perform this study, we used the cell lines UACC903 and WM9 since they have high levels of STAT3 phosphorylation at Tyr^{705}, a marker of STAT3 activation (Figure 3B), together with an abundant BCL2L10 expression as showed in Figure 1A. Both cell lines were transduced with a scramble sequence, or two STAT3 specific pRetroSuper-based shRNAs (short-hairpin RNA) labeled shSTAT3-I and -II that efficiently knockdown STAT3 protein levels (Figure 3C). Importantly, the two STAT3 shRNA significantly reduced BCL2L10 protein levels in both cell lines, as demonstrated by the quantification of Western blots from three independent biological replicates (Figure 3C). Real-time PCR experiments revealed that STAT3 silencing in both WM9 and UACC903 cells led to a significant decrease in BCL2L10 mRNA levels (Figure 3D,E). These results indicate that STAT3 regulates BCL2L10 expression, most likely through the transactivation of its promoter.

2.3. STAT3 Binds the Human BCL2L10 Promoter and Activates BCL2L10 Transcription

To study the regulation of BCL2L10 transcription by STAT3, bioinformatic analysis of the human BCL2L10 gene was performed at Dr. Joo-Yeon Yoo's laboratory. Analysis of the BCL2L10 promoter region (from −2.0 kb to +0.5 kb) using the STAT3-Finder software [32] revealed ten putative STAT3 responsive elements (SRE) with scores exceeding the cutoff value (>0.8, Figure 4A). The site at −559 was the only one conserved among mammals (data not shown). The sites at positions −559 (TTCTCAGAA) and −1306 (CCCCCAGAA) of the BCL2L10 promoter showed a score of 0.97 and 0.98, respectively and hence were considered to be the most likely functional SRE in the human BCL2L10 gene. Thus, the region between −1674 to +115 of the human BCL2L10 promoter that contains both candidate SREs and a cluster of SREs close to the transcription start was cloned into a luciferase reporter plasmid. Thereafter, we generated two deletion constructs that contained one (plasmid −678/+115) and none (plasmid −548/+115) of the two high ranked sites (Figure 4A). When transfected into UACC903 control cells, the three promoter fragments (−1674/+115, −678/+115 and −548/+115) displayed a high degree of activity (more than 200 times) compared with a promotorless pGL2 vector used as a control (data not shown). Then, we compared the reporter activity of these fragments in both UACC903-scramble and UACC903-shSTAT3 cells. As expected, the activity of Ly6E, a STAT reporter, is inhibited in UACC903-shSTAT3 compared with UACC903-scramble cells (Figure 4B). The activity elicited by constructs −1674/+115 and −678/+115 significantly decreased when transfected into UACC903-shSTAT3 cells compared to that observed when transfecting UACC903-scramble cells (Figure 4B).

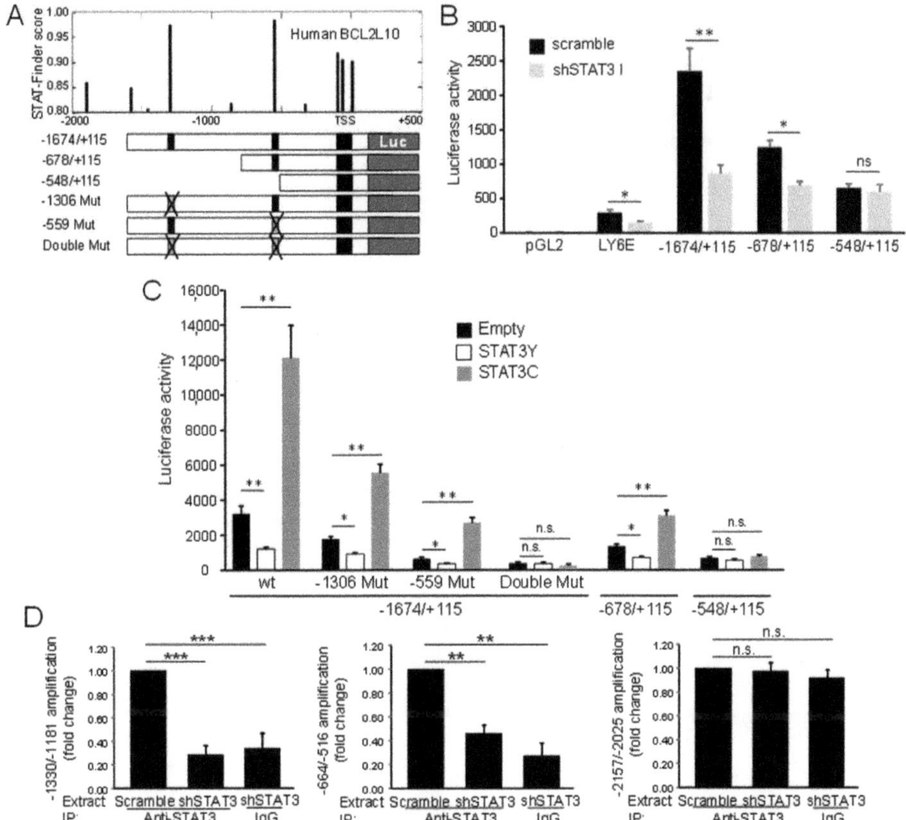

Figure 4. STAT3 binds to the BCL2L10 promoter and regulates its transcription. (**A**) Structure of the proximal region of the human BCL2L10 promoter. Putative STAT3 responsive elements (above) and fragments of the promoter that were cloned into pGL2 (below) are depicted. The sites at -1306 and -559 were deleted by site-directed mutagenesis, generating two plasmids with a single mutation and one plasmid with a double mutation. (**B**) Silencing of STAT3 diminished the reporter activity of the BCL2L10 promoter. The indicated reporter plasmids were transfected into UACC903-shSTAT3 and UACC903-scramble cells. Results are shown as the mean $\pm$ SD *: $p < 0.05$. **: $p < 0.01$, ns: not significant, $n = 3$. (**C**) STAT3 enhances BCL2L10 transactivation through the SREs at -1306 and -559. The indicated reporter plasmids were transfected into HEK293T cells together with STAT3Y and STAT3C plasmids. Results are shown as the mean $\pm$ SD *: $p < 0.05$, **: $p < 0.01$, ns: not significant, $n = 3$. (**D**) STAT3 silencing decreased binding of STAT3 to the BCL2L10 promoter. The plots show the relative level of BCL2L10 amplification (normalized to GAPDH levels) corresponding to region $-1330/-1181$ (left) and region $-664/-516$ (center) following a chromatin immunoprecipitation assay on UACC903-scramble and UACC903-shSTAT3 cells. Amplification of region $-2157/-2025$, used as a negative control, is shown on the right. The fold change compared to the level of amplification observed in UACC903-scramble cells is shown. The mean $\pm$ SD from three independent experiments is shown. The statistical analysis is described in Methods. **: $p < 0.01$ ***: $p < 0.001$, $n = 3$. n.s. = not significant.

The reduction seen upon STAT3 silencing was greater for plasmid $-1674/+115$ (70% of reduction) than for plasmid $-678/+115$ (43% of reduction). This result is consistent with the presence of two and one SREs in plasmids $-1674/+115$ and $-678/+115$, respectively. On the contrary, the shortest promoter fragment ($-548/+115$) showed similar luciferase activity both in UACC903-shSTAT3 and UACC903-scramble cells, indicating that the putative SREs in the cluster close to the transcription start are not functional. These results suggest that STAT3 regulates BCL2L10 transcription through responsive elements present in the region encompassed between nucleotides -1674 and -548. To confirm this result, we evaluated the luciferase activity in HEK293T cells cotransfected with plasmids encoding STAT3Y>F, a STAT3 dominant-negative mutant, STAT3C, a constitutively active STAT3 mutant, or empty plasmid. Expression of STAT3Y>F closely reproduced the reduction on BCL2L10 transcriptional activity seen in −1674/+115 and −678/+115 plasmids upon STAT3 silencing (Figure 4C). Meanwhile, transfection with STAT3C significantly increased the transcriptional activity driven by these two promoter constructs (−1674/+115 and −678/+115) over the level observed with empty plasmid (Figure 4C). To determine whether STAT3 mediates transcription of the BCL2L10 promoter through the putative SREs located at −1306 and −559, we generated two reporter plasmids with mutations that destroyed each of these sites (−1306 Mut and −559 Mut, displayed in Figure 3A). The mutation of either of these sites reduced the reporter activity compared to that observed with the wt BCL2L10 promoter (Figure 4C). Moreover, transfection with STAT3Y > F further reduced the activity of both promoter fragments indicating that the mutant reporters are still transactivated by STAT3. In line with this data, STAT3C increased the activity of both mutant promoters. In contrast, the plasmid with simultaneous mutation of both sites (double mut plasmid) presented not only a low luciferase activity but also a similar reporter activity in cells transfected with empty plasmid and the STAT3 mutants, indicating that this promoter fragment had completely lost STAT3 responsiveness (Figure 4C). Further support for the role of STAT3 in the regulation of BCL2L10 transcription comes from ChIP analysis. To this end, sheared chromatin was immunoprecipitated with antibodies to STAT3 (or control IgG) followed by real-time PCR amplification of BCL2L10 promoter sequences bearing putative SRE. Immunoprecipitation of STAT3 enabled specific amplification of DNA fragments corresponding to regions −1330 to −1181 (Figure 4D, above) and −664 to −516 (Figure 4D, below), but not of the fragment −2157 to −2025 that served as a negative control. The silencing of STAT3 reduced the amount of DNA amplified from the chromatin after STAT3 immunoprecipitation to values similar to those observed after immunoprecipitation with control IgG (Figure 4D). Altogether, these results indicate that STAT3 transactivates BCL2L10 through two SREs located at −1306 and −559 that display the classical STAT3 sequences CCCCCAGAA and TTCTCAGAA, respectively.

2.4. BCL2L10 Contributes to Melanoma Cell Survival

2.4.1. BCL2L10 Enhances the Survival of Melanoma Cells Treated with DNA-Damaging Agents

To investigate the role of BCL2L10 in cell proliferation and survival, both gain- and loss-of-function approaches were employed. For the gain of function strategy, we used the M2 cells stably transfected with BCL2L10-myc already described in Figure 1C. The M2 cell line was selected since it presented the lowest endogenous BCL2L10 levels among our cell lines (Figure 1A). It is important to note that the level of exogenous BCL2L10 expressed in this cell line was similar to that of the endogenous protein (Figure 1C), and therefore, the augmentation on BCL2L10 level was within a physiological range. For the loss of function approach, we silenced BCL2L10 expression by shRNA in A375 (Figure 1D), one of the cell lines that expressed the highest levels of endogenous BCL2L10 among the melanoma cell lines tested (Figure 1A). Since BCL2L10 was shown to regulate cell proliferation in both ovarian and hepatocellular carcinoma [15,23], we first evaluated the impact of altering BCL2L10 level on melanoma cell growth. We determined that neither BCL2L10 overexpression nor silencing affected cell growth (Figure 5A).

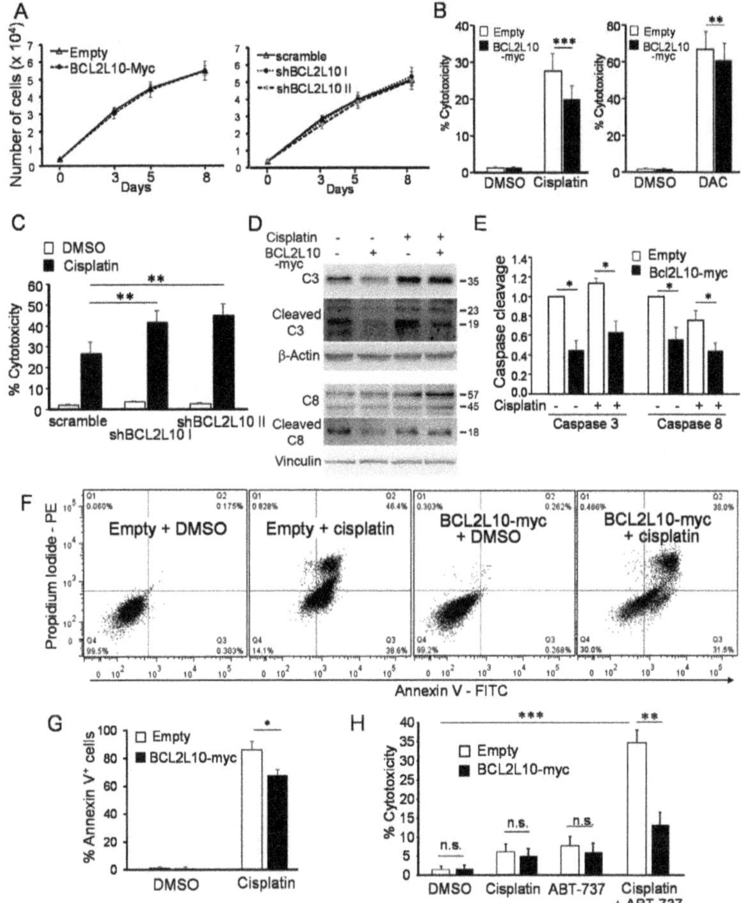

Figure 5. BCL2L10 enhances cell survival in melanoma upon treatment with DNA-damaging agents. (**A**) BCL2L10 does not affect cell growth. M2-empty and M2-BCL2L10-myc (left) and A375-scramble, A375-shBCL2L10 I, and A375-shBCL2L10 II cell lines (right) were grown for 8 days and the cell density was determined using a crystal violet assay. Results are shown as the mean ± SD, $n = 3$. (**B**) M2-empty and M2-BCL2L10-myc were treated with 40 µM cisplatin (left) or 100 µg/mL dacarbazine (DAC, right) for 48 h. Bar graph shows the mean ± SD ($n = 7$ for cisplatin and $n = 5$ for dacarbazine) of the percent of cytotoxicity. The percentage of cytotoxicity was calculated as the quotient between the number of cells in treated wells and the number of cells in non-treated wells times 100. (**C**) A375-scramble, A375-shBCL2L10 I, and A375-shBCL2L10 II cells were treated with cisplatin (40 µM) for 48 h. Bar graph shows the mean ± SD (from three independent experiments) of the percent of cytotoxicity. (**D**) M2-empty and M2-BCL2L10-myc were treated with 40 µM cisplatin for 24 h. Protein extracts were probed with antibodies against full-length caspase 3 (C3), cleaved caspase 3 (cleaved C3) or caspase 8 (C8). β-actin and vinculin were used as loading control. The blots displayed are representative of three independent experiments. (**E**) Bar graphs show the ratio (mean ± SD) between cleaved (19 kDa fragment for caspase 3 and 18 kDa fragment for caspase 8) and uncleaved full-length caspase 3 or caspase 8 from the experiment in (**D**). This ratio is expressed as the fold change relative to the ratio measured in DMSO-treated M2-empty cells. *: $p < 0.05$, $n = 3$. (**F**) M2-empty and M2-BCL2L10-myc were treated with 40 µM cisplatin for 24 h, stained with PI/annexin V and analyzed by flow cytometry. Representative histograms are shown. (**G**) Bar graph shows the mean ± SD ($n = 3$) of the percentage of annexin V positive cells from the experiment in (**F**). * $p < 0.05$, $n = 3$. (**H**) M2-empty and M2-BCL2L10-myc cells were treated with ABT-737 (2.5 µM), cisplatin (10 µM) or ABT-737 (2.5 µM) plus cisplatin (10 µM). Bar graph shows the mean ± SD (from three independent experiments) of the percent of cytotoxicity. The statistical analysis is described in Methods. **: $p < 0.01$, ***: $p < 0.001$, ns: not significant, $n = 3$. Whole Western blot figures are provided in Supplemental Materials Figure S7.

Next, we sought to evaluate changes in the cellular response to standard chemotherapeutic agents. For these experiments, we used a crystal violet cytotoxicity assay that, in agreement with previous publications [33–36], proved in our hands to be more reliable and sensitive than MTT and other methods to examine the impact of cytotoxic drugs on cell survival. Overexpression of BCL2L10-myc significantly reduced the cytotoxicity induced by cisplatin and dacarbazine (Figure 5B) in comparison to M2-empty cells. Similarly, the silencing of BCL2L10 sensitized A375 cells to cisplatin in comparison with A375-scramble cells (Figure 5C). To confirm these results, we analyzed cell death by both annexin V/PI staining and caspase cleavage in M2-empty and M2-BCL2L10-myc cells treated with cisplatin. Overexpression of BCL2L10-myc reduced both caspase 3 and caspase 8 cleavage, measured as the ratio of cleaved over uncleaved forms (Figure 5D,E). Further, BCL2L10-myc reduced the amount of annexin V positive cells upon cisplatin treatment (Figure 5F,G). These results indicate that BCL2L10 promotes cell survival in cells subjected to treatment with DNA-damaging agents.

Since Bcl-2 proteins are in many cases functionally redundant [37–39], we reasoned that the pro-survival activity of BCL2L10 in these experiments might have been curtailed by the expression of other anti-apoptotic Bcl-2 family members with a similar role. To address this question, we evaluated the effect of BCL2L10 expression in cisplatin-treated cells in which the contribution of other anti-apoptotic Bcl-2 proteins was neutralized by the addition of ABT-737 [40]. This compound is a BH3 mimetic that binds with high affinity to BCL2, Bcl-xL, and Bcl-w but does not inhibit MCL1, Bfl-1, and BCL2L10 [40]. Since the combination of ABT-737 and cisplatin has a synergistic killing effect in melanoma cells [41], we reduced the concentration of cisplatin from 40 to 10 μM to avoid excessive cell death. The treatment with 10 μM cisplatin or 2.5 μM ABT-737 induced less than 10% of cell death in M2-empty cells but adding both compounds together increased the cytotoxicity up to 34.7% in these cells (Figure 5H). However, the expression of BCL2L10 significantly reduced the cytotoxicity induced by the combination treatment to 13.7% (Figure 5H). This result confirms that BCL2L10 contributes to cisplatin resistance.

2.4.2. BCL2L10 Promotes Resistance to ABT-737

Since BCL2L10 (our results above), BCL2 [42,43], Bcl-xL [44], and MCL1 [45] have all been implicated in cisplatin resistance in melanoma and the partial protection to cisplatin provided by BCL2L10 was much greater in the presence of ABT-737 (34.7% vs. 13.7%, Figure 5H) than in its absence (27.6% vs. 19.7%, Figure 5B), it can be concluded that, at least in regard to cisplatin resistance, BCL2L10 can take over functions of the ABT-737 targets when they are inhibited by this drug. This conclusion evidences there are functional similarities between BCL2L10 and ABT-737 targets. To explore this in further depth without the confounding effect of cisplatin-induced cytotoxicity, we sought to assess the cell death induced exclusively by ABT-737 or TW-37 [46], a BH3-mimetic that targets MCL1 primarily. Although ABT-737 is usually poorly cytotoxic in melanoma (Figure 5H), it has been shown that higher doses can induce a modest, "on-target", cell death [47]. Increasing ABT-737 to 10 μM induced 41.5% of cytotoxicity in M2-Empty cells, but significantly less (25.2%) in M2-BCL2L10-myc cells (Figure 6A). In contrast, the cytotoxicity induced by TW-37 was not affected by BCL2L10 expression (Figure 6B). Similarly, BCL2L10-myc reduced the amount of annexin V positive cells (Figure 6C,D) following ABT-737 treatment. These results suggest that BCL2L10 mediates resistance to ABT-737. This antagonistic effect of BCL2L10 and ABT-737 reinforces the notion that BCL2L10 can play a similar role than the ABT-737 targets BCL2, Bcl-xL or Bcl-w.

Next, we studied this matter in BCL2L10-knockdown cells. We observed that ABT-737 induced greater cytotoxicity in A375-shBCL2L10 cells compared to A375-scramble cells (Figure 6E), confirming that BCL2L10 is implicated in the resistance of melanoma cells to ABT-737.

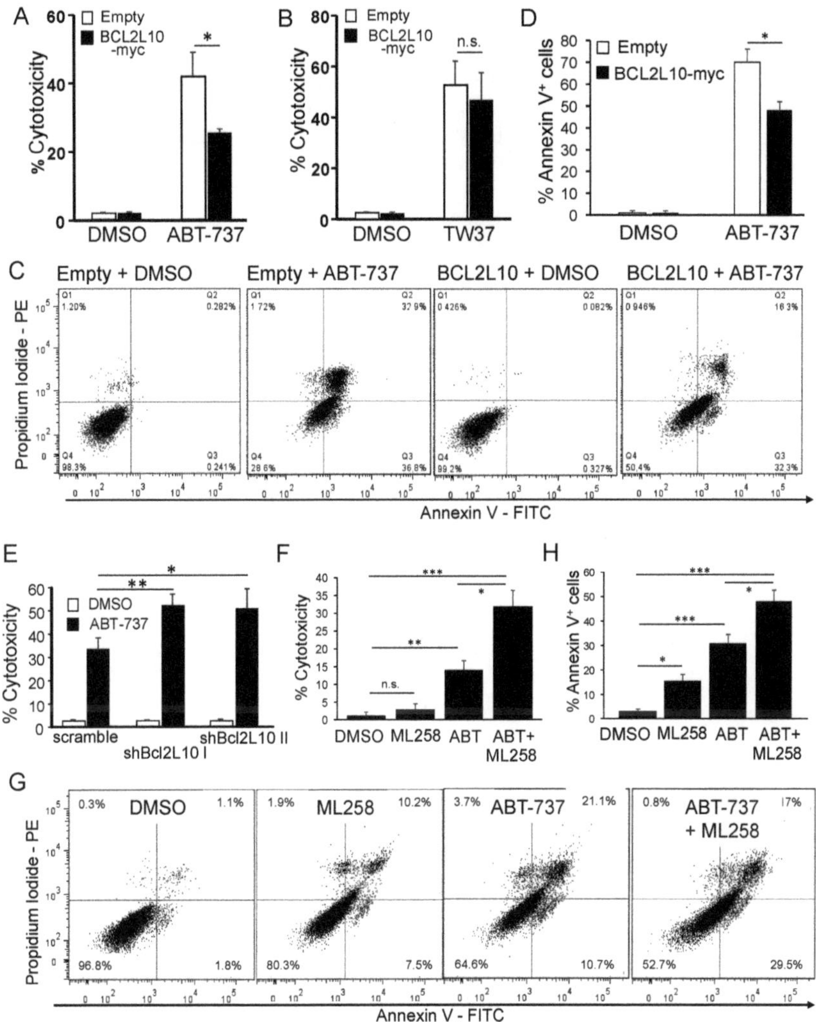

Figure 6. BCL2L10 promotes resistance to ABT-737. (**A,B**) M2-empty and M2-BCL2L10-myc cells were treated with 10 μM ABT-737 (**A**) or 5 μM TW-37 (**B**) for 48 h. Bar graph shows the mean ± SD (*n* = 4 for cisplatin and *n* = 5 for TW-37) of the percent of cytotoxicity. (**C**) M2-empty and M2-BCL2L10-myc cells were treated with 10 μM ABT-737 for 24 h, stained with PI/annexin V and analyzed by flow cytometry. Representative histograms are shown. (**D**) Bar graph shows the mean ± SD (*n* = 3) of the percentage of annexin V positive cells from the experiment in (**C**). * $p < 0.01$. ** $p < 0.001$, ns: not significant, *n* = 3. (**E**) A375-scramble and A375-shBCL2L10 cells were treated with ABT-737 (10 μM). Bar graph shows the mean ± SD (from three independent experiments) of the percent of cytotoxicity. (**F**) A375 cells were treated with ABT-737 (2.5 μM), ML258 (10 μM) or ABT-737 plus ML258 (same concentrations) and the percent of cytotoxicity was determined by using a crystal violet cytotoxicity assay. Bar graph shows the mean ± SD (from three independent experiments) of the percent of cytotoxicity. *: $p < 0.05$, *n* = 3. (**F**) A375 cells were treated with ABT-737 (2.5 μM), ML258 (10 μM) (alone or in combination) Bar graph shows the mean ± SD (*n* = 3) of the percent of cytotoxicity. (**G**) A375 cells were treated with ABT-737 (10 μM), ML258 (40 μM) (alone or in combination) for 24 h, stained with PI/annexin V and analyzed by flow cytometry. Representative histograms are shown. (**H**) Bar graph shows the mean ± SD (*n* = 3) of the percentage of annexin V positive cells from the experiment in (**G**). The statistical analysis is described in the Methods. *: $p < 0.05$, **: $p < 0.01$, ***: $p < 0.001$, ns: not significant, *n* = 3.

Since the concentration used of ABT-737 is considered a high one, it is improbable that this effect is due to a sub-optimal inhibition of Bcl-2 proteins by the drug. Instead, the additional cell death observed in A375-shBCL2L10 cells can be attributed to functions of BCL2L10 not fulfilled by the ABT-737 targets. To confirm this result by using an alternative approach to suppress the BCL2L10 function, we used the compound ML258. ML258 is a BH3 mimetic t as a highly specific inhibitor of the Bim:BCL2L10 interaction [41]. ML258 had a poor cytotoxic effect by itself but induced a significant increase in cell death in the presence of ABT-737 as determined by both crystal violet cytotoxicity assay (Figure 6F) and annexin V staining (Figure 6G,H), confirming the results obtained in cells with knockdown of BCL2L10. Altogether, these results indicate that BCL2L10 is implicated in ABT-737 resistance, and it has both shared and distinctive functions when compared with ABT-737 targets, BCL2, Bcl-xL, and Bcl-w.

2.4.3. BCL2L10 Enhances the Survival of Melanoma Cells Treated with a Combination of PLX-4032 and cisplatin or ABT-737

BRAF inhibition is one of the current approaches to treat melanoma patients. Since targeting BRAF in melanoma patients proved unsuccessful in the long term, the combined use of BRAF inhibitors with other therapeutic strategies (i.e., immunotherapy, standard chemotherapy or BH3-mimetics) is currently being studied. Following our finding that BCL2L10 is implicated in melanoma resistance to cisplatin and ABT-737, we wanted to evaluate whether the effect of BCL2L10 expression in response to these drugs was also observed in the context of simultaneous inhibition of BRAF by the BRAF^{V600E} inhibitor PLX-4032. As expected, both combination treatments had greater cytotoxicity than either drug alone. We found that M2-BCL2L10-myc cells presented a significantly reduced cytotoxicity compared with M2-empty cells upon treatment with both PLX-4032 plus cisplatin and PLX-4032 plus ABT-737 (Figure 7). This result indicates that BCL2L10 expression protects melanoma cells from cisplatin and ABT-737, even when these drugs are combined with BRAF inhibitors.

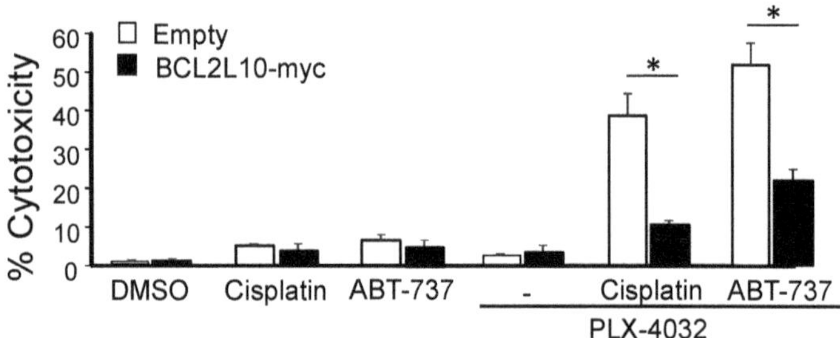

Figure 7. BCL2L10 protects melanoma cells from cell death upon treatment with PLX-4032 plus cisplatin or ABT-737. M2-empty and M2-BCL2L10-myc were treated for 48 h with the indicated drugs at the following concentrations: PLX-4032, 1 μM; cisplatin, 10 μM; ABT-737, 2.5 μM. The percent of cytotoxicity was determined by using a crystal violet cytotoxicity assay. Bar graph shows the mean ± SD (from three independent experiments) of the percent of cytotoxicity. The statistical analysis is described in Methods. *: $p < 0.05$.

3. Discussion

Mitochondrial-mediated apoptosis is regulated by a delicate balance of the opposing actions of pro-apoptotic and anti-apoptotic Bcl-2 family members. Therefore, the life/death decision is determined by the relative abundance of the Bcl-2 proteins on either "side" and their binding profiles [10]. Thus, the anti-apoptotic Bcl-2 proteins represent an interesting target for cancer therapy since its inhibition will unleash the pro-apoptotic function. This goal has been achieved using small molecule inhibitors generically named BH3-mimetics that compete with BH3-only proteins for the same hydrophobic groove on the anti-apoptotic proteins. As a result, the pro-apoptotic proteins are no longer inhibited and can promote the release and activation of Bax and/or Bak, the executors of MOMP [9,10]. Since different BH3-mimetic drugs target different anti-apoptotic Bcl-2 proteins and the Bcl-2 proteins are known to be functionally redundant [37–39], any rational attempt to inhibit these proteins must ideally begin by understanding which of the anti-apoptotic Bcl-2 proteins are expressed in a given tumor type, in a given patient, and under which circumstances. Otherwise, these therapeutic attempts are bound to fail, as happened with the therapies targeting BCL2 or BCL2/Bcl-xL in tumors rich in MCL1 [48–51]. The surveying of anti-apoptotic Bcl-2 proteins in different tumor types led to a focus on the members more frequently expressed: BCL2, Bcl-xL, and MCL1 [11,52]. In contrast, our knowledge of the function of the other three members of the family (Bcl-w, Bfl-1, and BCL2L10) has lagged behind. The study of Bcl-2 proteins in melanoma followed the same trend and also focused on BCL2, Bcl-xL, and MCL1 [53,54] although, to our knowledge, a comprehensive study of the relative expression of all six anti-apoptotic Bcl-2 proteins in melanoma cells or tumors has not been performed yet. In the present study, we have determined that BCL2L10 is expressed both frequently and at elevated levels in both melanoma cell lines and tissues from melanoma patients. One limitation of our study is that the reduced size of our patient cohort prevents us from analyzing changes in BCL2L10 expression in different stages of melanoma. It is important to mention that Placzek et al. in 2010 used quantitative PCR techniques to evaluate the mRNA expression levels of all six anti-apoptotic Bcl-2 subfamily members in 69 cancer cell lines, including 12 melanoma cell lines [55]. The expression of BCL2L10 in this study was disparaged since all except one cell line (68 out of the 69, including all 12 from melanoma) presented BCL2L10 levels below the baseline. The absence of BCL2L10 expression in cell lines from leukemia and from breast, prostate, colorectal, and lung cancers in this study is difficult to reconcile with the later observation that BCL2L10 is overexpressed in all these tumor types [27–30]. Moreover, four of the 12 melanoma cell lines were included in our study, and they all showed strong expression of BCL2L10 as assessed by Western blot. This important discrepancy most likely reflects the inaccuracy of profiling BCL2L10 in cancer at the mRNA levels, as was remarked in recent publications [28,56].

Despite the fact that BCL2L10 was found to be overexpressed in several tumor types, the mechanisms underlying BCL2L10 overexpression have not been investigated. We have found that BCL2L10 expression is enhanced by STAT3, a transcription factor that has a critical role in the development and progression of human tumors by promoting uncontrolled cell proliferation and growth, cell survival, induction of angiogenesis, and the suppression of host immune surveillance [57]. STAT3 is a critical point of convergence downstream of several hyperactive tyrosine kinase receptors that are either mutated or amplified in melanoma, such as KIT, ERBB4, EPH, FGFR, EGFR and PDGRFA, among others [58]. Accordingly, persistent phosphorylation of STAT3 at Tyr^{705} and elevated STAT3-dependent transactivation of target genes has been documented and implicated in melanoma progression [59,60]. Here, we have identified two STAT3 responsive elements in the BCL2L10 promoter that are responsible for strong STAT3-dependent BCL2L10 transcription. Interestingly, STAT3 has been implicated in cancer cell survival by regulating the transcription of other anti-apoptotic members of the Bcl-2 family, such as BCL2, Bcl-xL, and MCL1 [61–63]. Since we observed BCL2L10 transactivation by STAT3 in the model cell line HEK293, it is very likely that STAT3 drives BCL2L10 upregulation in other tumor types presenting constitutive activity of STAT3.

Considering that anti-apoptotic Bcl-2 proteins were and are evaluated as prospective therapeutic targets in melanoma [64], it is surprising that the role of BCL2L10 in melanoma has never been studied. Since BCL2L10 was shown to both promote and inhibit cell death in different systems, to elucidate the role of BCL2L10 in melanoma is not trivial. Our data established that BCL2L10 is a pro-survival protein that contributes to protecting melanoma cells from the DNA-damaging agents cisplatin and dacarbazine. In this regard, this function of BCL2L10 is similar to that of BCL2, Bcl-xL, and MCL1 since they are both abundantly expressed in melanoma and implicated in cisplatin resistance [42–45].

In addition, the three approaches we used through this work (BCL2L10 overexpression and BCL2L10 genetic and chemical inhibition) indicate that BCL2L10 is an important factor contributing to ABT-737 resistance. This observation is in agreement with the previous finding that BCL2L10 expression caused ABT-737 resistance in the acute lymphoblastic leukemia cell lines J16 and MOLT-4 [65]. Altogether, these findings position BCL2L10 as an important new pro-survival factor in melanoma.

A critical step to further understand BCL2L10's role is to determine whether BCL2L10 has either distinctive or overlapping functions with the other five Bcl-2 anti-apoptotic proteins. Our data show that BCL2L10 expression inhibits cytotoxicity induced by ABT-737 but not by TW-37. This observation suggests that there is no overlap between MCL1 and BCL2L10 function. The most likely reason is that they partner with different BH3-only sensitizers proteins; Noxa and Hrk interact with MCL1 and BIK with BCL2L10. Another implication of the finding above is that BCL2L10 shares functions with ABT-737 targets. Similar to BCL2L10, Bcl-xL and Bcl-w also interact with Bik, apart from other sensitizers. Hence, it is tempting to speculate that these two proteins, rather than BCL2 (that interacts with Bmf and Bad, but not with Bik), are the ones that present some functional similitudes with BCL2L10. Therefore, the very restricted (Bik and Bim) and unique BH3-only protein binding profile of BCL2L10 may determine its biological function. Moreover, unlike the other five Bcl-2 anti-apoptotic proteins, BCL2L10 exclusively inhibits Bax-dependent cell death and is not involved in Bid or Bak-dependent apoptosis. These observations are consistent with sequence and phylogenetic analysis that suggest BCL2L10 is the most divergent Bcl-2 anti-apoptotic member [66]. On the other hand, when we inhibited BCL2L10 by either shRNA or the ML258 inhibitor, we observed increased cell death, suggesting that BCL2L10 has distinct functions from that of ABT-737 targets. Although this observation does not contradict the previous conclusion that BCL2L10 functions overlap with that of ABT-737 targets, it adds another layer of complexity to the BCL2L10 function. Notwithstanding, these experiments just represent an initial exploration of the similitudes and differences of BCL2L10 with the other five Bcl-2 anti-apoptotic proteins.

The mechanisms underlying the pro-survival activity of BCL2L10 in melanoma have not been explored in this work. The conventional interpretation is that BCL2L10 regulates apoptosis by interacting with BH3-only proteins and inhibiting the intrinsic pathway of apoptosis [12,20,22,67]. However, it was also demonstrated that BCL2L10 prevents apoptosis by BH3-independent mechanisms and through the binding of Apaf-1 [19,21,68], a critical component of the apoptosome implicated in the cleavage of Procaspase-9. The anti-apoptotic effect of BCL2L10 was also shown to be the consequence of the inhibition of mitophagy [69] or the release of Ca2+ from the endoplasmic reticulum [29]. Future studies will determine specifically which of these mechanisms are operating in melanoma cells.

The constitutive activation of the BRAF/MEK/ERK pathway is a hallmark of melanoma. Pharmacological inhibition of this pathway strongly enhances the expression or activity of almost all Bcl-2 pro-apoptotic proteins [70]. However, the initial effect of BRAF inhibitors is usually not cytotoxic but cytostatic (i.e., G1 cell arrest) since the accumulation of pro-apoptotic BH3-only proteins is neutralized by the abundant amounts of anti-apoptotic proteins usually expressed by melanoma cells. However, the induction of BH3-only proteins by BRAF inhibition efficiently primes tumor cells for apoptosis, allowing them to straightforwardly tilt the balance toward cell death by co-targeting the pro-survival Bcl-2 family members. In agreement with this model, it has been demonstrated that ABT-737 and other BH3 mimetics combine synergistically with MAPK inhibitors in the killing of BRAF mutant melanoma cells [71–73]. These observations have provided the rationale to use this combination in melanoma patients (ABT-263 [74] plus dabrafenib and trametinib, clinical trial NCT01989585) with the intent to delay the onset of resistance to BRAF inhibitors. This phenomenon of apoptotic priming by BRAF inhibitors can also enhance the cytotoxicity of standard chemotherapy, suggesting that this combination still needs to be further investigated [75–78]. We evaluated the role of BCL2L10 in these two possible therapeutic scenarios. Even though we have demonstrated that BCL2L10 can prevent cell death by both cisplatin and ABT-737, in the presence of PLX-4032, the protective role of BCL2L10 was much more pronounced (compare Figure 7 vs. Figure 5B for cisplatin and Figure 7 vs. Figure 6A for ABT-737). This difference may be due to the upregulation of Bik, one of the few BCL2L10 partners, following ERK pathway inhibition [79,80]. If this hypothesis holds true, it may indicate that inhibition of BCL2L10 can be envisioned as an additional way to enhance the clinical efficacy of BRAF inhibition.

Altogether, our results indicate that the elevated expression of BCL2L10 in melanoma contributes to cell survival upon treatment with different cytotoxic compounds, making BCL2L10 a promising target in the context of inhibiting anti-apoptotic Bcl-2 proteins in malignant melanoma treatment.

4. Materials and Methods

4.1. Cell Culture

Melanoma cell lines were kindly provided by Dr. Zeev Ronai (Sanford Burnham Prebys Medical Discovery Institute, San Diego, CA, USA) except for the M2 cell line that was provided by Dr. S. Alvarez (IMIBIO-SL) [81]. All cell lines were maintained in DMEM supplemented with 10% fetal bovine serum (FBS, Invitrogen, Carlsbad, CA, USA) 100 U/mL penicillin and 100 mg/mL streptomycin (Invitrogen, Carlsbad, CA, USA) at 37 °C and 5% CO_2. Cells were transfected with calcium phosphate or by lipofectamine PLUS reagent (Invitrogen) following the manufacturer's protocol. The cell lines are free of mycoplasma contamination and were authenticated as described [82].

4.2. Plasmids and Viral Constructs

The BCL2L10 expression plasmid incorporating the N-terminal Myc epitope tag has been previously described [83]. This vector was introduced into M2 cells, and transfected cells were selected with 100 ug/mL Hygromycin B (Gibco). The oligonucleotides targeting STAT3 (5′-GCAGCAGCTGAACAACATG-3′ and 5′-GATTGACCTAGAGACCCAC-3′) [82] and scramble (5′-GAAACTGCTGACCGTTAAT-3′) were cloned into the pRetroSuper vector. Silencing of BCL2L10 was performed using pLKO.1 shRNA clones TRCN0000033595 and TRCN0000033596 designed by The RNAi Consortium (TRC) and previously validated [84]. Viral particles were generated as described [85].

4.3. Real-Time PCR

Real-time PCR was performed as described [86]. Specific primers used for PCR were as follows: BCL2L10 forward 5′ GCCTTCATTTATCTCTGGACAC3′; BCL2L10 reverse, 5′AAGGTGCTTTCCCTCAGTTC3′, RNPII forward 5′GCTGTGTCTGCTTCTTCTG3′, RNPII reverse 5′CGAACTTGTTGTCCATCTCC3′ RNPII (RNA Polymerase II) served as an endogenous control. Reactions were run in triplicate. The target mRNA concentration of control cells, normalized to the level of RNPII mRNA, was set to 1.

4.4. Chromatin Immunoprecipitation (ChIP)

For ChIP analysis, UACC903 cells were fixed with 11% formaldehyde and sheared chromatin was immunoprecipitated with a STAT3 antibody (sc-482, Santa Cruz Biotechnology, Dallas, TX) or control IgG and subjected to real-time PCR. The following primers corresponding to the proximal region of the BCL2L10 promoter were used: BCL2L10 (−1330/−1181) forward 5′GGCCTAGTAGCAAGGCAGAA3′ and BCL2L10 (−1330/−1181) reverse 5′GGCCTAGTAGCAAGGCAGAA3′ and BCL2L10 (−664/−516) forward 5′CTAAG ACAGCTGCCAAGTGC3′ and BCL2L10 (−664/−516) reverse 5′TCCATTCTGCATCAGTC TGG3′. The primers BCL2L10 (−2157/−2025) forward 5′CTTTGGAGGGAGAATTCCA3′ and BCL2L10 (−2157/−2025) reverse 5′CAGATGGACAGAATTACATGC3′ were used as a control.

4.5. Luciferase Assays

The BCL2L10 promoter was amplified from genomic DNA from HUVEC cells using the primers: −1674 (forward) 5′CTCTTTCATGTGGTACCAGCAC3′ and +234 (reverse) 5′TTACGGCAGATTCACCGGTC3′. The purified product was amplified in a semi-nested PCR using the same forward primer and the reverse primer +115 5′GGGAGCGCACTCGAG CTGTTG3′. The smaller fragments were generated using the +115 reverse primer and the following forward primers: −703 5′ACCCAGTCTATGGCATTCTGC3′, −678 5′ GCCGC-CTGGTACCAGACTAAGAC 3′ and −548 5′ CCACTGCTGGTA CCATTCTGC 3′. The three promoter fragments were cloned into the Xho I y Kpn I sites of the pGL2-Basic plasmid. Site-directed mutagenesis of STAT3 sites was performed using the QuikChange II kit (Stratagene, San Diego, CA, USA) following the manufacturer's protocol. Cell lysates were prepared from lipofectamine-transfected cells after 24 or 48 h. Luciferase activity was measured with the luciferase assay system (Promega, Madison, WI, USA) in a Berthold luminometer (Berthold Technologies, Bad Wildbad, Germany, Germany) and was normalized with β-galactosidase activity measured in the same sample. Results are shown as the mean (bar) ± SD

4.6. Proliferation Assays

Cells (5×10^3/well) were plated in a 96-well plate (8 wells per time-point) and incubated for 72 h with DMEM 10% FBS. Cells were fixed (4% PFA), washed and stained with 0.1% crystal violet in 10% ethanol for 30 min. Then, the crystal violet solution was recovered, and plates were washed by immersion and dried at 37 °C. Crystal violet was dissolved in 10% Acetic Acid. The absorbance (optical density (OD)) was detected at 590 nm with a μQuant microplate reader (Biotek Instruments, Winooski, VT, USA). A plate fixed at 6 h was used as a control of seeding. A standard calibration curve was used to convert OD to the number of cells.

4.7. Immunohistochemistry

All the melanoma samples were from the Hospital Israelita archives, currently at the Universidad Maimonides. Samples were deparaffinized, rehydrated, and subjected to heat-induced epitope retrieval using citrate buffer (10 mM, pH 6). The endogenous peroxidase activity was quenched by placing them in methanol, 1.5% H_2O_2 for 30 min. After washing, the slides were blocked with blocking solution (PBS, 3% BFS) for 30 min at room temperature. The slides were incubated with a 1:300 dilution of the anti-BCL2L10 antibody (PA5-22190, Invitrogen) for 1 h at 30 °C in a humidified chamber. The signal was detected using VECTASTAIN® Elite ABC Kit according to manufacturer protocol, followed by the staining with the DAB Substrate kit, Peroxidase (HRP), with nickel (Vector, Burlingame, CA, USA), as indicated by the kit. Sections were counterstained with hematoxylin and analyzed by a pathologist.

4.8. Crystal Violet Cytotoxicity Assay

Cells (5×10^3/well) were plated in 96-well plates and incubated for 24 h with DMEM 10% FBS. Thereafter, the testing compounds were added to the plate (in quadruplicates) and left for 48 h. The concentration of the drugs is indicated in the corresponding figure legend. Cells were incubated with DMSO as a control. After removal of the medium, the plates were rinsed with 100 μL PBS/well, fixed and stained with 200 μL of 0.1% crystal violet in 10% ethanol for 30 min. Plates were rinsed 4 times in tap water and dried at 37 °C. Crystal violet was dissolved in 10% acetic acid. The absorbance (optical density (OD)) was detected at 590 nm with a μQuant microplate reader (Biotek Instruments). A plate fixed at 6 h was used as a control of seeding. A standard calibration curve was used to convert OD to the number of cells. The percentage of cytotoxicity was calculated as the quotient between the number of cells in treated wells and the number of cells in untreated wells times 100.

4.9. Quantification of Apoptotic Cell Death

The M2 cells were seeded on 6-well plates at a density of 1.25×10^5 cells per well. The following day they were exposed to 40 µM cisplatin or 10 µM ABT-737 for 24 h. A375 cells were treated with 10 µM ABT-737 and 40 µM ML258, either alone or in combination for 24 h. Cells were washed twice with PBS and resuspended in 100 µL of annexin V binding buffer (pH 7.4) (BD Biosciences, Franklin Lakes, NJ, USA). Then, annexin V-Alexa Fluor 488 (BD Biosciences) was added and incubated for 15 min under dark conditions. Propidium iodide (0.1 µg/mL; Sigma-Aldrich; Merck KGaA, Darmstadt, Germany) was added just prior to signal acquisition. Cells were analyzed using a FACSAria flow cytometer (BD Biosciences, San Jose, CA) and analyzed with FACSDiva 7.6.1 software (BD Biosciences).

4.10. Western Blotting

For the Western blotting analysis, cell lysates were collected by the addition of lysis buffer supplemented with protease and phosphatase inhibitors for 10 min on ice [87]. The cell lysates were centrifuged at 13,000 rpm for 15 min at 4 °C, and the supernatants were collected and quantified using the Bradford method. Between 20 and 50 µg of proteins were diluted in $6\times$ Lemmli buffer, boiled at 95 °C for 5 min, separated on 8–12% SDS–PAGE gels and then transferred to nitrocellulose membrane. The membranes were blocked with 5% milk in 0.05% Tween-PBS at room temperature for 1 h and then incubated with the primary antibodies at 4 °C overnight. The following antibodies were used: GAPDH (sc-25,778), STAT3 (sc-482), pSTAT3 (sc-8059), and caspase-3 (sc-7148) from Santa Cruz Biotechnologies, cleaved caspase-3 (9664) from Cell Signaling (Danvers, MA, USA) and caspase-8 (66231A) from BD Pharmingen. The primary antibodies anti-BCL2L10 were from Invitrogen (PA5-22190) and from Cell Signaling (CS #3869) and were both used at a 1:1000 dilution. Antibodies to β-actin (A5441) and α-tubulin (T9026) were from Sigma. The corresponding HRP-conjugated secondary antibodies: anti-mouse (GE NA931V), anti-rabbit (GE NA934) or anti-goat (sc-2020) were incubated for 1 h at room temperature. Immunoreactive bands were detected by an ECL system (Amersham Biosciences, Buckinghamshire, UK) using an image reader (ImageQuant 350, GE Healthcare, Chicago, IL, USA). Quantification of band intensities was performed using ImageJ (NIH). The intensity of each band was normalized to GAPDH or another housekeeping gene (i.e., Tubulin or actin) and the fold change (FC) relative to control cells was calculated. To draw a conclusion on a particular experiment, at least three biological (independent) replicates of paired samples were examined to calculate the mean and standard deviation. The log transformation of FC values was calculated to obtain a more symmetric distribution that better suits the normality assumptions of the subsequent statistical tests.

4.11. Statistics

Except when indicated, experiments were performed at least 3 times. Mean differences between groups were determined using either Student's *t*-tests (Figures 4B and 5A (M2 cells), Figure 5B,E,G,H, Figure 6A,B,D and Figure 7) or one-way ANOVAs followed by *post hoc* tests (Figure 1D, Figure 3C–F, Figure 4C,D, Figure 5A (A375 cells), Figures 5C and 6E,F,H). Values of $p < 0.05$ were considered statistically significant. Statistical analyses were conducted using software from GraphPad Prism.

5. Conclusions

The data presented here allow us to conclude that BCL2L10 is frequently and abundantly expressed in melanoma. BCL2L10 plays a pro-survival role by contributing to the resistance of melanoma cells to DNA-damaging agents and ABT-737. These functions were also observed in the context of BRAF inhibition, indicating that targeting BCL2L10 may enhance the clinical efficacy of other therapies against melanoma.

Supplementary Materials: The following are available online at https://www.mdpi.com/2072-6694/13/1/78/s1, Figure S1: Whole Western blot figures for Figure 1, Figure S2: Antibodies PA5-

22190 and #3869 have the same pattern of reaction against BCL2L10. Figure S3: Expression of other Bcl-2 protein family members was not affected by BCL2L10 overexpression. Figure S4: Analysis of GSE64536, GSE63092, and GSE48124 demonstrated that STAT3 inhibition reduced BCL2L10 mRNA levels. Figure S5: Overlap of P-STAT3 and BCL2L10 staining in melanoma samples, Figure S6: Whole Western blot figures for Figure 3, Figure S7: Whole Western blot figures for Figure 5 Table S1: Immunohistochemical analysis of BCL2L10 staining in samples from melanoma patients.

Author Contributions: Conceptualization, P.L.-B.; formal analysis, M.J.Q., M.E.P., E.I. and P.L.-B.; funding acquisition, P.L.-B.; investigation, M.J.Q. and M.E.P.; methodology, M.J.Q., M.E.P., M.V.C., G.B., M.B.V., N.B.F. and E.I.; resources, E.I. and P.L.-B.; visualization, M.V.C.; writing—original draft, P.L.-B.; writing—review and editing, P.L.-B. All authors have read and agreed to the published version of the manuscript.

Funding: This work was supported by grants BID-PICT-2007-1010 and BID-PICT2011-1605 from the Agencia Nacional de Promoción Científica y Tecnológica and grants from Fundación Alberto Roemmers and the Instituto Nacional de Cancer. Consejo Nacional de Investigaciones Científicas y Técnicas (CONICET) provided fellowships to M.J.Q., M.E.P., G.B., M.B.V., N.B.F. and M.V.C.

Institutional Review Board Statement: Not applicable.

Informed Consent Statement: The requirement for obtaining informed patient consent was waived because of the retrospective nature of the investigation and the deidentified nature of the data collected.

Data Availability Statement: All the data is contained within the article or supplementary material.

Acknowledgments: We thank Shu-ichi Matsuzawa (Kyoto University Graduate School of Medicine, Kyoto, Japan) for providing ML258. We acknowledge Young Min Oh and Joo-Yeon Yoo (Pohang University of Science and Technology, Republic of Korea) for the analysis of the BCL2L10 promoter. We thank Alejandra Fisz for her administrative assistance. We thank Yanina Moller, Sara Orrea and Engr. Alberto Varela and his crew for technical assistance.

Conflicts of Interest: The authors declare no conflict of interest.

Abbreviations

STAT3	Signal transducer and activator of transcription 3
BCL2	B-cell lymphoma 2
BCL2L10	BCL2 Like 10
SRE	STAT3 responsive elements
GAPDH	Glyceraldehyde-3-phosphate dehydrogenase
PI3K	Phosphoinositide-3-kinase
shRNA	Short hairpin RNA
RNPII	RNA polymerase II
MOMP	Mitochondrial outer membrane permeability
IHC	Immunohistochemistry

References

1. Siegel, R.L.; Miller, K.D.; Jemal, A. Cancer statistics, 2020. *CA A Cancer J. Clin.* **2020**, *70*, 7–30. [CrossRef] [PubMed]
2. Lopez-Bergami, P.; Ronai, Z. Requirements for PKC-augmented JNK activation by MKK4/7. *Int. J. Biochem. Cell Biol.* **2008**, *40*, 1055–1064. [CrossRef] [PubMed]
3. Paluncic, J.; Kovacevic, Z.; Jansson, P.J.; Kalinowski, D.; Merlot, A.M.; Huang, M.L.-H.; Lok, H.C.; Sahni, S.; Lane, D.J.R.; Richardson, D.R. Roads to melanoma: Key pathways and emerging players in melanoma progression and oncogenic signaling. *Biochim. Biophys. Acta BBA Mol. Cell Res.* **2016**, *1863*, 770–784. [CrossRef] [PubMed]
4. Yang, N.-Y.; Lopez-Bergami, P.; Goydos, J.S.; Yip, D.; Walker, A.M.; Pasquale, E.B.; Ethell, I.M. The EphB4 receptor promotes the growth of melanoma cells expressing the ephrin-B2 ligand. *Pigment. Cell Melanoma Res.* **2010**, *23*, 684–687. [CrossRef] [PubMed]
5. Lopez-Bergami, P.; Fitchman, B.; Ronai, Z. Understanding signaling cascades in melanoma. *Photochem. Photobiol.* **2008**, *84*, 289–306. [CrossRef]
6. Bhatia, S.; Tykodi, S.S.; Thompson, J.A. Treatment of metastatic melanoma: An overview. *Oncology* **2009**, *23*, 488–496.
7. Lee, C.; Collichio, F.; Ollila, D.; Moschos, S. Historical review of melanoma treatment and outcomes. *Clin. Dermatol.* **2013**, *31*, 141–147. [CrossRef]
8. Soengas, M.S.; Lowe, S.W. Apoptosis and melanoma chemoresistance. *Oncogene* **2003**, *22*, 3138–3151. [CrossRef]
9. Kale, J.; Osterlund, E.J.; Andrews, D.W. BCL-2 family proteins: Changing partners in the dance towards death. *Cell Death Differ.* **2018**, *25*, 65–80. [CrossRef]

10. Shamas-Din, A.; Kale, J.; Leber, B.; Andrews, D.W. Mechanisms of action of Bcl-2 family proteins. *Cold Spring Harb. Perspect. Biol.* **2013**, *5*, a008714. [CrossRef]
11. D'Aguanno, S.; Del Bufalo, D. Inhibition of Anti-Apoptotic Bcl-2 Proteins in Preclinical and Clinical Studies: Current Overview in Cancer. *Cells* **2020**, *9*, 1287. [CrossRef] [PubMed]
12. Ke, N.; Godzik, A.; Reed, J.C. Bcl-B, a novel Bcl-2 family member that differentially binds and regulates Bax and Bak. *J. Biol. Chem.* **2001**, *276*, 12481–12484. [CrossRef] [PubMed]
13. Zhai, D.; Ke, N.; Zhang, H.; Ladror, U.; Joseph, M.; Eichinger, A.; Godzik, A.; Ng, S.-C.; Reed, J.C. Characterization of the anti-apoptotic mechanism of Bcl-B. *Biochem. J.* **2003**, *376*, 229–236. [CrossRef] [PubMed]
14. Mikata, R.; Fukai, K.; Imazeki, F.; Arai, M.; Fujiwara, K.; Yonemitsu, Y.; Zhang, K.; Nabeya, Y.; Ochiai, T.; Yokosuka, O. BCL2L10 is frequently silenced by promoter hypermethylation in gastric cancer. *Oncol. Rep.* **2010**, *23*, 1701–1708. [CrossRef] [PubMed]
15. Bai, Y.; Wang, J.; Han, J.; Xie, X.-L.; Ji, C.-G.; Yin, J.; Chen, L.; Wang, C.-K.; Jiang, X.-Y.; Qi, W.; et al. BCL2L10 inhibits growth and metastasis of hepatocellular carcinoma both in vitro and in vivo. *Mol. Carcinog.* **2017**, *56*, 1137–1149. [CrossRef]
16. Lee, R.; Chen, J.; Matthews, C.P.; McDougall, J.K.; Neiman, P.E. Characterization of NR13-related human cell death regulator, Boo/Diva, in normal and cancer tissues. *Biochim. Biophys. Acta* **2001**, *1520*, 187–194. [CrossRef]
17. Liu, X.; Hu, X.; Kuang, Y.; Yan, P.; Li, L.; Li, C.; Tao, Q.; Cai, X. BCLB, methylated in hepatocellular carcinoma, is a starvation stress sensor that induces apoptosis and autophagy through the AMPK-mTOR signaling cascade. *Cancer Lett.* **2017**, *395*, 63–71. [CrossRef]
18. Xu, J.D.; Cao, X.X.; Long, Z.W.; Liu, X.P.; Furuya, T.; Xu, J.W.; Liu, X.L.; De Xu, Z.; Sasaki, K.; Li, Q.Q. BCL2L10 protein regulates apoptosis/proliferation through differential pathways in gastric cancer cells. *J. Pathol.* **2011**, *223*, 400–409. [CrossRef]
19. Inohara, N.; Gourley, T.S.; Carrio, R.; Muñiz, M.; Merino, J.; Garcia, I.; Koseki, T.; Hu, Y.; Chen, S.; Núñez, G. Diva, a Bcl-2 Homologue that Binds Directly to Apaf-1 and Induces BH3-independent Cell Death. *J. Biol. Chem.* **1998**, *273*, 32479–32486. [CrossRef]
20. Aouacheria, A.; Arnaud, E.; Venet, S.; Lalle, P.; Gouy, M.; Rigal, D.; Gillet, G. Nrh, a human homologue of Nr-13 associates with Bcl-Xs and is an inhibitor of apoptosis. *Oncogene* **2001**, *20*, 5846–5855. [CrossRef]
21. Song, Q. Boo, a novel negative regulator of cell death, interacts with Apaf-1. *EMBO J.* **1999**, *18*, 167–178. [CrossRef] [PubMed]
22. Zhang, H.; Holzgreve, W.; De Geyter, C. BCL2-L-10, a novel anti-apoptotic member of the Bcl-2 family, blocks apoptosis in the mitochondria death pathway but not in the death receptor pathway. *Hum. Mol. Genet.* **2001**, *10*, 2329–2339. [CrossRef] [PubMed]
23. Lee, S.-Y.; Kwon, J.; Woo, J.H.; Kim, K.-H.; Lee, K.-A. BCL2L10 mediates the proliferation, invasion and migration of ovarian cancer cells. *Int. J. Oncol.* **2020**, *56*, 618–629. [CrossRef] [PubMed]
24. Mikata, R.; Yokosuka, O.; Fukai, K.; Imazeki, F.; Arai, M.; Tada, M.; Kurihara, T.; Zhang, K.; Kanda, T.; Saisho, H. Analysis of genes upregulated by the demethylating agent 5-aza-2′-deoxycytidine in gastric cancer cell lines. *Int. J. Cancer* **2006**, *119*, 1616–1622. [CrossRef]
25. Fabiani, E.; Leone, G.; Giachelia, M.; D'alo', F.; Greco, M.; Criscuolo, M.; Guidi, F.; Rutella, S.; Hohaus, S.; Teresa voso, M. Analysis of genome-wide methylation and gene expression induced by 5-aza-2′-deoxycytidine identifies BCL2L10 as a frequent methylation target in acute myeloid leukemia. *Leuk. Lymphoma* **2010**, *51*, 2275–2284. [CrossRef]
26. Xu, J.D.; Furuya, T.; Cao, X.X.; Liu, X.L.; Li, Q.Q.; Wang, W.J.; Xu, J.W.; Xu, Z.D.; Sasaki, K.; Liu, X.P. Loss of BCL2L10 protein expression as prognostic predictor for poor clinical outcome in gastric carcinoma. *Histopathology* **2010**, *57*, 814–824. [CrossRef]
27. Krajewska, M.; Kitada, S.; Winter, J.N.; Variakojis, D.; Lichtenstein, A.; Zhai, D.; Cuddy, M.; Huang, X.; Luciano, F.; Baker, C.H.; et al. Bcl-B expression in human epithelial and nonepithelial malignancies. *Clin. Cancer Res.* **2008**, *14*, 3011–3021. [CrossRef]
28. Hamouda, M.-A.; Jacquel, A.; Robert, G.; Puissant, A.; Richez, V.; Cassel, R.; Fenouille, N.; Roulland, S.; Gilleron, J.; Griessinger, E.; et al. BCL-B (BCL2L10) is overexpressed in patients suffering from multiple myeloma (MM) and drives an MM-like disease in transgenic mice. *J. Exp. Med.* **2016**, *213*, 1705–1722. [CrossRef]
29. Nougarede, A.; Popgeorgiev, N.; Kassem, L.; Omarjee, S.; Borel, S.; Mikaelian, I.; Lopez, J.; Gadet, R.; Marcillat, O.; Treilleux, I.; et al. Breast Cancer Targeting through Inhibition of the Endoplasmic Reticulum-Based Apoptosis Regulator Nrh/BCL2L10. *Cancer Res.* **2018**, *78*, 1404–1417. [CrossRef]
30. Cluzeau, T.; Robert, G.; Mounier, N.; Karsenti, J.M.; Dufies, M.; Puissant, A.; Jacquel, A.; Renneville, A.; Preudhomme, C.; Cassuto, J.-P.; et al. BCL2L10 is a predictive factor for resistance to azacitidine in MDS and AML patients. *Oncotarget* **2012**, *3*, 490–501. [CrossRef]
31. Wang, L.; Hurley, D.G.; Watkins, W.; Araki, H.; Tamada, Y.; Muthukaruppan, A.; Ranjard, L.; Derkac, E.; Imoto, S.; Miyano, S.; et al. Cell cycle gene networks are associated with melanoma prognosis. *PLoS ONE* **2012**, *7*, e34247. [CrossRef] [PubMed]
32. Oh, Y.M.; Kim, J.K.; Choi, Y.; Choi, S.; Yoo, J.-Y. Prediction and Experimental Validation of Novel STAT3 Target Genes in Human Cancer Cells. *PLoS ONE* **2009**, *4*, e6911. [CrossRef] [PubMed]
33. Aslantürk, Ö.S. In Vitro Cytotoxicity and Cell Viability Assays: Principles, Advantages, and Disadvantages. In *Genotoxicity—A Predictable Risk to Our Actual World*; Larramendy, M.L., Soloneski, S., Eds.; InTech: London, UK, 2018; ISBN 978-1-78923-418-3.
34. Martin, A.; Clynes, M. Comparison of 5 microplate colorimetric assays forin vitro cytotoxicity testing and cell proliferation assays. *Cytotechnology* **1993**, *11*, 49–58. [CrossRef]
35. Papadimitriou, M.; Hatzidaki, E.; Papasotiriou, I. Linearity Comparison of Three Colorimetric Cytotoxicity Assays. *JCT* **2019**, *10*, 580–590. [CrossRef]

36. Śliwka, L.; Wiktorska, K.; Suchocki, P.; Milczarek, M.; Mielczarek, S.; Lubelska, K.; Cierpiał, T.; Łyżwa, P.; Kiełbasiński, P.; Jaromin, A.; et al. The Comparison of MTT and CVS Assays for the Assessment of Anticancer Agent Interactions. *PLoS ONE* **2016**, *11*, e0155772. [CrossRef]

37. Eichhorn, J.M.; Alford, S.E.; Sakurikar, N.; Chambers, T.C. Molecular analysis of functional redundancy among anti-apoptotic Bcl-2 proteins and its role in cancer cell survival. *Exp. Cell Res.* **2014**, *322*, 415–424. [CrossRef]

38. Campbell, K.J.; Tait, S.W.G. Targeting BCL-2 regulated apoptosis in cancer. *Open Biol.* **2018**, *8*, 180002. [CrossRef]

39. Carrington, E.M.; Zhan, Y.; Brady, J.L.; Zhang, J.-G.; Sutherland, R.M.; Anstee, N.S.; Schenk, R.L.; Vikstrom, I.B.; Delconte, R.B.; Segal, D.; et al. Anti-apoptotic proteins BCL-2, MCL-1 and A1 summate collectively to maintain survival of immune cell populations both in vitro and in vivo. *Cell Death Differ.* **2017**, *24*, 878–888. [CrossRef]

40. Oltersdorf, T.; Elmore, S.W.; Shoemaker, A.R.; Armstrong, R.C.; Augeri, D.J.; Belli, B.A.; Bruncko, M.; Deckwerth, T.L.; Dinges, J.; Hajduk, P.J.; et al. An inhibitor of Bcl-2 family proteins induces regression of solid tumours. *Nature* **2005**, *435*, 677–681. [CrossRef]

41. Zou, J.; Ardecky, R.; Pinkerton, A.B.; Sergienko, E.; Su, Y.; Stonich, D.; Curpan, R.F.; Simons, P.C.; Zhai, D.; Diaz, P.; et al. Selective Bcl-2 Inhibitor Probes. In *Probe Reports from the NIH Molecular Libraries Program*; National Center for Biotechnology Information: Bethesda, MD, USA, 2010.

42. Wacheck, V.; Losert, D.; Günsberg, P.; Vornlocher, H.-P.; Hadwiger, P.; Geick, A.; Pehamberger, H.; Müller, M.; Jansen, B. Small Interfering RNA Targeting Bcl-2 Sensitizes Malignant Melanoma. *Oligonucleotides* **2003**, *13*, 393–400. [CrossRef]

43. Zupi, G. Antitumor Efficacy of bcl-2 and c-myc Antisense Oligonucleotides in Combination with Cisplatin in Human Melanoma Xenografts: Relevance of the Administration Sequence. *Clin. Cancer Res.* **2005**, *11*, 1990–1998. [CrossRef] [PubMed]

44. Heere-Ress, E.; Thallinger, C.; Lucas, T.; Schlagbauer-Wadl, H.; Wacheck, V.; Monia, B.P.; Wolff, K.; Pehamberger, H.; Jansen, B. Bcl-XL is a chemoresistance factor in human melanoma cells that can be inhibited by antisense therapy. *Int. J. Cancer* **2002**, *99*, 29–34. [CrossRef] [PubMed]

45. Sinnberg, T.; Lasithiotakis, K.; Niessner, H.; Schittek, B.; Flaherty, K.T.; Kulms, D.; Maczey, E.; Campos, M.; Gogel, J.; Garbe, C.; et al. Inhibition of PI3K-AKT-mTOR Signaling Sensitizes Melanoma Cells to Cisplatin and Temozolomide. *J. Investig. Dermatol.* **2009**, *129*, 1500–1515. [CrossRef] [PubMed]

46. Wang, G.; Nikolovska-Coleska, Z.; Yang, C.-Y.; Wang, R.; Tang, G.; Guo, J.; Shangary, S.; Qiu, S.; Gao, W.; Yang, D.; et al. Structure-Based Design of Potent Small-Molecule Inhibitors of Anti-Apoptotic Bcl-2 Proteins. *J. Med. Chem.* **2006**, *49*, 6139–6142. [CrossRef] [PubMed]

47. Reuland, S.N.; Goldstein, N.B.; Partyka, K.A.; Smith, S.; Luo, Y.; Fujita, M.; Gonzalez, R.; Lewis, K.; Norris, D.A.; Shellman, Y.G. ABT-737 synergizes with Bortezomib to kill melanoma cells. *Biol. Open* **2012**, *1*, 92–100. [CrossRef] [PubMed]

48. Rudin, C.M.; Hann, C.L.; Garon, E.B.; Ribeiro de Oliveira, M.; Bonomi, P.D.; Camidge, D.R.; Chu, Q.; Giaccone, G.; Khaira, D.; Ramalingam, S.S.; et al. Phase II study of single-agent navitoclax (ABT-263) and biomarker correlates in patients with relapsed small cell lung cancer. *Clin. Cancer Res.* **2012**, *18*, 3163–3169. [CrossRef]

49. Mazumder, S.; Choudhary, G.S.; Al-Harbi, S.; Almasan, A. Mcl-1 Phosphorylation defines ABT-737 resistance that can be overcome by increased NOXA expression in leukemic B cells. *Cancer Res.* **2012**, *72*, 3069–3079. [CrossRef]

50. Yecies, D.; Carlson, N.E.; Deng, J.; Letai, A. Acquired resistance to ABT-737 in lymphoma cells that up-regulate MCL-1 and BFL-1. *Blood* **2010**, *115*, 3304–3313. [CrossRef]

51. Geserick, P.; Wang, J.; Feoktistova, M.; Leverkus, M. The ratio of Mcl-1 and Noxa determines ABT737 resistance in squamous cell carcinoma of the skin. *Cell Death Dis.* **2014**, *5*, e1412. [CrossRef]

52. Warren, C.F.A.; Wong-Brown, M.W.; Bowden, N.A. BCL-2 family isoforms in apoptosis and cancer. *Cell Death Dis.* **2019**, *10*, 177. [CrossRef]

53. Eberle, J.; Hossini, A. Expression and Function of Bcl-2 Proteins in Melanoma. *CG* **2008**, *9*, 409–419. [CrossRef] [PubMed]

54. Hartman, M.L.; Czyz, M. Anti-apoptotic proteins on guard of melanoma cell survival. *Cancer Lett.* **2013**, *331*, 24–34. [CrossRef] [PubMed]

55. Placzek, W.J.; Wei, J.; Kitada, S.; Zhai, D.; Reed, J.C.; Pellecchia, M. A survey of the anti-apoptotic Bcl-2 subfamily expression in cancer types provides a platform to predict the efficacy of Bcl-2 antagonists in cancer therapy. *Cell Death Dis.* **2010**, *1*, e40. [CrossRef] [PubMed]

56. Beverly, L.J.; Lockwood, W.W.; Shah, P.P.; Erdjument-Bromage, H.; Varmus, H. Ubiquitination, localization, and stability of an anti-apoptotic BCL2-like protein, BCL2L10/BCLb, are regulated by Ubiquilin1. *Proc. Natl. Acad. Sci. USA* **2012**, *109*, E119–E126. [CrossRef] [PubMed]

57. Yu, H.; Lee, H.; Herrmann, A.; Buettner, R.; Jove, R. Revisiting STAT3 signalling in cancer: New and unexpected biological functions. *Nat. Rev. Cancer* **2014**, *14*, 736–746. [CrossRef] [PubMed]

58. Chin, L.; Garraway, L.A.; Fisher, D.E. Malignant melanoma: Genetics and therapeutics in the genomic era. *Genes Dev.* **2006**, *20*, 2149–2182. [CrossRef]

59. Messina, J.L.; Yu, H.; Riker, A.I.; Munster, P.N.; Jove, R.L.; Daud, A.I. Activated stat-3 in melanoma. *Cancer Control.* **2008**, *15*, 196–201. [CrossRef]

60. Lee, I.; Fox, P.S.; Ferguson, S.D.; Bassett, R.; Kong, L.-Y.; Schacherer, C.W.; Gershenwald, J.E.; Grimm, E.A.; Fuller, G.N.; Heimberger, A.B. The expression of p-STAT3 in stage IV melanoma: Risk of CNS metastasis and survival. *Oncotarget* **2012**, *3*, 336–344. [CrossRef]

61. Zushi, S.; Shinomura, Y.; Kiyohara, T.; Miyazaki, Y.; Kondo, S.; Sugimachi, M.; Higashimoto, Y.; Kanayama, S.; Matsuzawa, Y. STAT3 mediates the survival signal in oncogenic ras-transfected intestinal epithelial cells. *Int. J. Cancer* **1998**, *78*, 326–330. [CrossRef]

62. Karni, R.; Jove, R.; Levitzki, A. Inhibition of pp60c-Src reduces Bcl-XL expression and reverses the transformed phenotype of cells overexpressing EGF and HER-2 receptors. *Oncogene* **1999**, *18*, 4654–4662. [CrossRef]

63. Liu, H.; Ma, Y.; Cole, S.M.; Zander, C.; Chen, K.-H.; Karras, J.; Pope, R.M. Serine phosphorylation of STAT3 is essential for Mcl-1 expression and macrophage survival. *Blood* **2003**, *102*, 344–352. [CrossRef] [PubMed]

64. Mukherjee, N.; Schwan, J.V.; Fujita, M.; Norris, D.A.; Shellman, Y.G. Alternative Treatments For Melanoma: Targeting BCL-2 Family Members to De-Bulk and Kill Cancer Stem Cells. *J. Investig. Dermatol.* **2015**, *135*, 2155–2161. [CrossRef] [PubMed]

65. Rooswinkel, R.W.; van de Kooij, B.; Verheij, M.; Borst, J. Bcl-2 is a better ABT-737 target than Bcl-xL or Bcl-w and only Noxa overcomes resistance mediated by Mcl-1, Bfl-1, or Bcl-B. *Cell Death Dis.* **2012**, *3*, e366. [CrossRef] [PubMed]

66. Aouacheria, A.; Brunet, F.; Gouy, M. Phylogenomics of life-or-death switches in multicellular animals: Bcl-2, BH3-Only, and BNip families of apoptotic regulators. *Mol. Biol. Evol.* **2005**, *22*, 2395–2416. [CrossRef] [PubMed]

67. Guillemin, Y.; Cornut-Thibaut, A.; Gillet, G.; Penin, F.; Aouacheria, A. Characterization of unique signature sequences in the divergent maternal protein BCL2L10. *Mol. Biol. Evol.* **2011**, *28*, 3271–3283. [CrossRef]

68. Naumann, U.; Weit, S.; Wischhusen, J.; Weller, M. Diva/Boo is a negative regulator of cell death in human glioma cells. *FEBS Lett.* **2001**, *505*, 23–26. [CrossRef]

69. Ding, Q.; Xie, X.-L.; Wang, M.-M.; Yin, J.; Tian, J.-M.; Jiang, X.-Y.; Zhang, D.; Han, J.; Bai, Y.; Cui, Z.-J.; et al. The role of the apoptosis-related protein BCL-B in the regulation of mitophagy in hepatic stellate cells during the regression of liver fibrosis. *Exp. Mol. Med.* **2019**, *51*, 1–13. [CrossRef]

70. Sale, M.J.; Cook, S.J. That which does not kill me makes me stronger; combining ERK1/2 pathway inhibitors and BH3 mimetics to kill tumour cells and prevent acquired resistance. *Br. J. Pharmacol.* **2013**, *169*, 1708–1722. [CrossRef]

71. Cragg, M.S.; Jansen, E.S.; Cook, M.; Harris, C.; Strasser, A.; Scott, C.L. Treatment of B-RAF mutant human tumor cells with a MEK inhibitor requires Bim and is enhanced by a BH3 mimetic. *J. Clin. Investig.* **2008**, *118*, 3651–3659. [CrossRef]

72. Wroblewski, D.; Mijatov, B.; Mohana-Kumaran, N.; Lai, F.; Gallagher, S.J.; Haass, N.K.; Zhang, X.D.; Hersey, P. The BH3-mimetic ABT-737 sensitizes human melanoma cells to apoptosis induced by selective BRAF inhibitors but does not reverse acquired resistance. *Carcinogenesis* **2013**, *34*, 237–247. [CrossRef]

73. Serasinghe, M.N.; Missert, D.J.; Asciolla, J.J.; Podgrabinska, S.; Wieder, S.Y.; Izadmehr, S.; Belbin, G.; Skobe, M.; Chipuk, J.E. Anti-apoptotic BCL-2 proteins govern cellular outcome following B-RAF(V600E) inhibition and can be targeted to reduce resistance. *Oncogene* **2015**, *34*, 857–867. [CrossRef] [PubMed]

74. Tse, C.; Shoemaker, A.R.; Adickes, J.; Anderson, M.G.; Chen, J.; Jin, S.; Johnson, E.F.; Marsh, K.C.; Mitten, M.J.; Nimmer, P.; et al. ABT-263: A potent and orally bioavailable Bcl-2 family inhibitor. *Cancer Res.* **2008**, *68*, 3421–3428. [CrossRef] [PubMed]

75. Mattila, K.E.; Vihinen, P.; Ramadan, S.; Skyttä, T.; Tiainen, L.; Vuoristo, M.-S.; Tyynelä-Korhonen, K.; Koivunen, J.; Kohtamäki, L.; Mäkelä, S.; et al. Combination chemotherapy with temozolomide, lomustine, vincristine and interferon-alpha (TOL-IFN) plus vemurafenib or TOL-IFN as first-line treatment for patients with advanced melanoma. *Acta Oncol.* **2020**, *59*, 310–314. [CrossRef] [PubMed]

76. Makino, E.; Gutmann, V.; Kosnopfel, C.; Niessner, H.; Forschner, A.; Garbe, C.; Sinnberg, T.; Schittek, B. Melanoma cells resistant towards MAPK inhibitors exhibit reduced TAp73 expression mediating enhanced sensitivity to platinum-based drugs. *Cell Death Dis.* **2018**, *9*, 930. [CrossRef]

77. Bhatty, M.; Kato, S.; Piha-Paul, S.A.; Naing, A.; Subbiah, V.; Huang, H.J.; Karp, D.D.; Tsimberidou, A.M.; Zinner, R.G.; Hwu, W.; et al. Phase 1 study of the combination of vemurafenib, carboplatin, and paclitaxel in patients with *BRAF*-mutated melanoma and other advanced malignancies. *Cancer* **2019**, *125*, 463–472. [CrossRef]

78. Simon, A.; Kourie, H.R.; Kerger, J. Is there still a role for cytotoxic chemotherapy after targeted therapy and immunotherapy in metastatic melanoma? A case report and literature review. *Chin. J. Cancer* **2017**, *36*, 10. [CrossRef]

79. Borst, A.; Haferkamp, S.; Grimm, J.; Rösch, M.; Zhu, G.; Guo, S.; Li, C.; Gao, T.; Meierjohann, S.; Schrama, D.; et al. BIK is involved in BRAF/MEK inhibitor induced apoptosis in melanoma cell lines. *Cancer Lett.* **2017**, *404*, 70–78. [CrossRef]

80. Sale, M.J.; Cook, S.J. The increase in BIK expression following ERK1/2 pathway inhibition is a consequence of G1 cell-cycle arrest and not a direct effect on BIK protein stability. *Biochem. J.* **2014**, *459*, 513–524. [CrossRef]

81. Campos, L.S.; Rodriguez, Y.I.; Leopoldino, A.M.; Hait, N.C.; Lopez Bergami, P.; Castro, M.G.; Sanchez, E.S.; Maceyka, M.; Spiegel, S.; Alvarez, S.E. Filamin A Expression Negatively Regulates Sphingosine-1-Phosphate-Induced NF-κB Activation in Melanoma Cells by Inhibition of Akt Signaling. *Mol. Cell. Biol.* **2016**, *36*, 320–329. [CrossRef]

82. Picco, M.E.; Castro, M.V.; Quezada, M.J.; Barbero, G.; Villanueva, M.B.; Fernández, N.B.; Kim, H.; Lopez-Bergami, P. STAT3 enhances the constitutive activity of AGC kinases in melanoma by transactivating PDK1. *Cell Biosci.* **2019**, *9*, 3. [CrossRef]

83. Guo, B.; Zhai, D.; Cabezas, E.; Welsh, K.; Nouraini, S.; Satterthwait, A.C.; Reed, J.C. Humanin peptide suppresses apoptosis by interfering with Bax activation. *Nature* **2003**, *423*, 456–461. [CrossRef] [PubMed]

84. Bhatnagar, S.; Gazin, C.; Chamberlain, L.; Ou, J.; Zhu, X.; Tushir, J.S.; Virbasius, C.-M.; Lin, L.; Zhu, L.J.; Wajapeyee, N.; et al. TRIM37 is a new histone H2A ubiquitin ligase and breast cancer oncoprotein. *Nature* **2014**, *516*, 116–120. [CrossRef] [PubMed]

85. Barbero, G.; Castro, M.V.; Villanueva, M.B.; Quezada, M.J.; Fernández, N.B.; DeMorrow, S.; Lopez-Bergami, P. An Autocrine Wnt5a Loop Promotes NF-κB Pathway Activation and Cytokine/Chemokine Secretion in Melanoma. *Cells* **2019**, *8*, 1060. [CrossRef]

86. Fernández, N.B.; Lorenzo, D.; Picco, M.E.; Barbero, G.; Dergan-Dylon, L.S.; Marks, M.P.; García-Rivello, H.; Gimenez, L.; Labovsky, V.; Grumolato, L.; et al. ROR1 contributes to melanoma cell growth and migration by regulating N-cadherin expression via the PI3K/Akt pathway: ROR1 INCREASES MELANOMA CELL GROWTH AND MIGRATION. *Mol. Carcinog.* **2016**, *55*, 1772–1785. [CrossRef] [PubMed]
87. Lopez-Bergami, P.; Huang, C.; Goydos, J.S.; Yip, D.; Bar-Eli, M.; Herlyn, M.; Smalley, K.S.M.; Mahale, A.; Eroshkin, A.; Aaronson, S.; et al. Rewired ERK-JNK Signaling Pathways in Melanoma. *Cancer Cell* **2007**, *11*, 447–460. [CrossRef]

Article

Investigating Serum and Tissue Expression Identified a Cytokine/Chemokine Signature as a Highly Effective Melanoma Marker

Marco Cesati [1], Francesca Scatozza [2], Daniela D'Arcangelo [2], Gian Carlo Antonini-Cappellini [2], Stefania Rossi [3], Claudio Tabolacci [3], Maurizio Nudo [2], Enzo Palese [2], Luigi Lembo [2], Giovanni Di Lella [2], Francesco Facchiano [3,*] and Antonio Facchiano [2,*]

1   Department of Civil Engineering and Computer Science Engineering, University of Rome Tor Vergata, 00133 Rome, Italy; cesati@uniroma2.it
2   Istituto Dermopatico dell'Immacolata, IDI-IRCCS, via Monti di Creta 104, 00167 Rome, Italy; f.scatozza@idi.it (F.S.); d.darcangelo@idi.it (D.D.); giancarlo.antoninic@aslroma2.it (G.C.A.-C.); nudomaurizio@gmail.com (M.N.); e.palese@idi.it (E.P.); l.lembo@idi.it (L.L.); g.dilella@idi.it (G.D.L.)
3   Department of Oncology and Molecular Medicine, Istituto Superiore di Sanità, Viale Regina Elena 299, 00161 Rome, Italy; stefania.rossi@iss.it (S.R.); claudiotabolacci@tiscali.it (C.T.)
*   Correspondence: francesco.facchiano@iss.it (F.F.); a.facchiano@idi.it (A.F.)

Received: 2 November 2020; Accepted: 4 December 2020; Published: 8 December 2020

Simple Summary: In this study, we investigated the expression of 27 cytokines/chemokines in the serum of 232 individuals (136 melanoma patients vs. 96 controls). It identified several cytokines/chemokines differently expressed in melanoma patients as compared to the healthy controls, as a function of the presence of the melanoma, age, tumor thickness, and gender, indicating different systemic responses to the melanoma presence. We also analyzed the gene expression of the same 27 molecules at the tissue level in 511 individuals (melanoma patients vs. controls). From the gene expression analysis, we identified several cytokines/chemokines showing strongly different expression in melanoma as compared to the controls, and the 4-gene signature "*IL-1Ra, IL-7, MIP-1a*, and *MIP-1b*" as the best combination to discriminate melanoma samples from the controls, with an extremely high accuracy (AUC = 0.98). These data indicate the molecular mechanisms underlying melanoma setup and the relevant markers potentially useful to help the diagnosis of biopsy samples.

Abstract: The identification of reliable and quantitative melanoma biomarkers may help an early diagnosis and may directly affect melanoma mortality and morbidity. The aim of the present study was to identify effective biomarkers by investigating the expression of 27 cytokines/chemokines in melanoma compared to healthy controls, both in serum and in tissue samples. Serum samples were from 232 patients recruited at the IDI-IRCCS hospital. Expression was quantified by xMAP technology, on 27 cytokines/chemokines, compared to the control sera. RNA expression data of the same 27 molecules were obtained from 511 melanoma- and healthy-tissue samples, from the GENT2 database. Statistical analysis involved a 3-step approach: analysis of the single-molecules by Mann–Whitney analysis; analysis of paired-molecules by Pearson correlation; and profile analysis by the machine learning algorithm Support Vector Machine (SVM). Single-molecule analysis of serum expression identified IL-1b, IL-6, IP-10, PDGF-BB, and RANTES differently expressed in melanoma ($p < 0.05$). Expression of IL-8, GM-CSF, MCP-1, and TNF-α was found to be significantly correlated with Breslow thickness. Eotaxin and MCP-1 were found differentially expressed in male vs. female patients. Tissue expression analysis identified very effective marker/predictor genes, namely, IL-1Ra, IL-7, MIP-1a, and MIP-1b, with individual AUC values of 0.88, 0.86, 0.93, 0.87, respectively. SVM analysis of the tissue expression data identified the combination of these four molecules as the most effective signature to discriminate melanoma patients (AUC = 0.98). Validation, using the GEPIA2 database on an additional 1019 independent samples, fully confirmed these observations. The present study demonstrates, for the first time, that the *IL-1Ra, IL-7, MIP-1a,* and *MIP-1b* gene signature discriminates melanoma from control tissues with extremely high efficacy. We therefore propose this 4-molecule combination as an effective melanoma marker.

Keywords: melanoma markers; cytokines; machine learning; Support Vector Machine; principal component analysis

---

1. Introduction

Melanoma is the most aggressive skin cancer with a good prognosis when early diagnosis is achieved. While relevant advances come from newly available therapies, novel approaches are necessary to improve early diagnosis and therapeutic efficacy. Several studies addressed the complex role that specific cytokines and growth factors may play in melanoma biology, acting either as pro- or anti-proliferation and either positively or negatively regulating the immune response [1]. For instance, CXCL10 (IP-10) exerts both pro-and anti-melanoma effects, mostly due to splice variants of its CXCR3 receptors [2]. Several cytokines/chemokines and corresponding receptors are known to be expressed in melanoma tissue, to regulate the multifaceted machinery coordinating the proliferation rate, the angiogenic response, the inflammatory response, the immune response, and the metastatic diffusion [1,3]. Simultaneous quantification of several cytokine/chemokine analytes has recently become available in serum as well as in tissue samples. Previous studies report gene expression profiles identifying low- vs. high-risk patients. For instance, the 31 GEP prognostic classifier identifies BAP1b, MGP, SPP1, CXCL14, CLCA2, S100A8, BTG1, SAP130, ARG1, KRT6B, GJA1, ID2, EIF1B, S100A9, CRABP2, KRT14, ROBO1, RBM23, TACSTD2, DSC1, SPRR1B, TRIM29, AQP3, TYRP1, PPL, LTA4H, and CST6 [4]. Other studies investigated the expression of orphan receptors as well as known chemokine receptors and chemokine ligands in melanoma metastases, leading to the identification of several molecules differentially expressed in metastatic melanoma, such as GPR18, GPR34, GPR119, GPR160, GPR183, P2RY10, CCR5, CXCR4, CXCR6, CCL4, CCL5, CCL14/15, CXCL8, CXCL9, CXCL14, and XCL1/2 [5]. An additional study reports a significant expression change of six chemokines (namely, CCL2, CCL3, CCL4, CCL5, CXCL9, and CXCL10) related to the lymphocyte infiltration in the melanoma tissue [6]. Statistically significant differential plasma expression in melanoma patients vs. controls has been reported for IL-2, IL-6, and IL-10 [7]. Despite the high statistical significance of the differences, none of these molecules show relevant AUC values according to ROC analyses; therefore, to date, they cannot be proposed as markers with clinical relevance.

Additional studies carried out in melanoma patients identified the serum expression level of proinflammatory cytokines, such as IL-2Ra, IL-12-p40, and IFN-α, as good predictors of relapse-free survival [8]. In another study carried out in 40 patients, the serum expression levels of 115 analytes were investigated, including most of the known cytokines and chemokines, such as IL-6, IL-7, IL-10, IL-16, TNF- α trimer, IL-1b, IFN-γ, IL-4R, IL-18, RANK-L, IL-1b, IL-2R, IL-6R, MPIF-1, Leptin, MIG, GDNF, MIP-1 alpha, MIP-1b, MIP-1 delta, ITAC, GM-CSF, MCP-4, MIP-3a, MIP-3b, MMP-1, SP-C, amphiregulin, RANK, MCP-2, IP-10, OPG, FGF-2, and many others. The serum expression profile of TNF-α receptor II, TGF-a, TIMP-1, and C-reactive protein was identified as a profile with prognostic value to predict overall survival in melanoma patients, with an Area Under the Curve (AUC) of 0.89 reduced to 0.72 when the leave-one-out cross-validation technique was applied [9]. An additional study indicates the expression of IL-1Ra, IL-2, and IFNa2 as pro-inflammatory cytokines related to the cytotoxicity associated with anti-CTL4 and anti-PD1 combined therapy [10]. Tissue expression of CCR6 and its ligand CCL20 (MIP-3a) were identified as progression predictors in primary melanoma patients [11]. Prostate-specific membrane antigen (PSMA) was identified by immunohistochemistry analysis as a good marker of metastatic melanoma, in 41 Stage III/IV melanoma human specimens [12]. The ROC analysis measured an AUC = 0.82, i.e., a good performance but not good enough to allow its clinical application as a melanoma marker. Consensus among different studies is often difficult given experimental discrepancies on serum/plasma handling or the antibodies' sensibility/specificity. Using multiplex immune-based technology may overcome these issues, at least in part, by measuring many different analytes within the same sample.

We have previously investigated melanoma markers by in vitro screening [13], as well as by investigating the ion channels [14,15], autophagy-related molecules [16,17], or molecules related to lipid metabolism [18] in populations composed of hundreds/thousands of controls and patients. In the present study, we investigated the cytokine/chemokine protein expression in the serum of 232 controls/melanoma patients recruited in our hospital, and the gene expression on 511 melanoma tissues selected from the GENT2 database. We report here, for the first time, significant differences related to gender, age, and Breslow thickness in the serum-expression dataset. In the tissue-expression dataset, we report, for the first time, a highly relevant gene marker combination, discriminating healthy controls from melanoma patients with an extremely high accuracy, and reaching an AUC = 0.982, according to the ROC analysis.

2. Results

The cytokine/chemokine expression in melanoma patients was analyzed to identify molecules with strong and significant differential expression in patients vs. controls. The cytokine/chemokine protein expression in the serum of 232 patients recruited at the IDI hospital and their RNA expression in tissue biopsies of 511 samples from the GENT2 public database were evaluated. The serum expression and tissue expression of the same 27 human chemokines/cytokines were analyzed as a single-molecule analysis, as a paired-molecule analysis, or as a profile analysis, as reported in the cartoon depicted in Figure 1.

2.1. Serum Expression: Single-Molecule Analysis of the Cytokines/Chemokines in Melanoma Patients vs. Controls

The serum dataset included the following information: histopathological diagnosis (96 pathological subjects versus 136 controls), sex (112 male and 120 female), age (median 46.5 years, and mean 48.54 years), Breslow's depth (minimum value 0 mm, maximum 12 mm, median 0.7 mm, and mean 1.34 mm), and the expression values of the 27 cytokines/chemokines expressed as pg/mL. Table 1 summarizes the information on the serum dataset.

Table 1. Descriptive statistics of the population for the serum-expression analysis.

Patient Type	Number	Mean Age	Mean Thickness (mm)	Thickness Distribution	
				<1 mm * Number	≥1 mm * Number
Female controls	72	41.3	0.00	0	0
Male controls	64	45.3	0.00	0	0
Female melanoma	48	54.5	1.60	23	22
Male melanoma	48	58.0	1.08	31	14
Total	232				

* The 1 mm limit is consistent with the current threshold used for staging of T1 melanoma patients and allowed the best case distribution. Not all pathological samples report the thickness value.

Tables S1 and S2 report more general data (number of samples for each molecule, minimum value, 25% percentile, median, 75% percentile, maximum, mean, standard deviation, and having passed the normality test (or not)) for all controls and all melanoma patients, respectively.

Table 2 reports the mean values of serum expression of the 27 cytokines/chemokines, the statistical significance of the differences, and the AUC according to the ROC analyses. Five molecules show a significantly ($p < 0.05$) different expression in melanoma vs. the controls, namely, IL-1b, IL-6, IP-10, PDGF-BB, and RANTES. The ROC analyses indicated that none shows a good ability to act as a serum marker of melanoma; in fact, an AUC < 0.70 was found in all cases. Nevertheless, the following Breslow-, age-, and gender-specific characterization indicated many statistically significant differences.

STUDY WORKFLOW

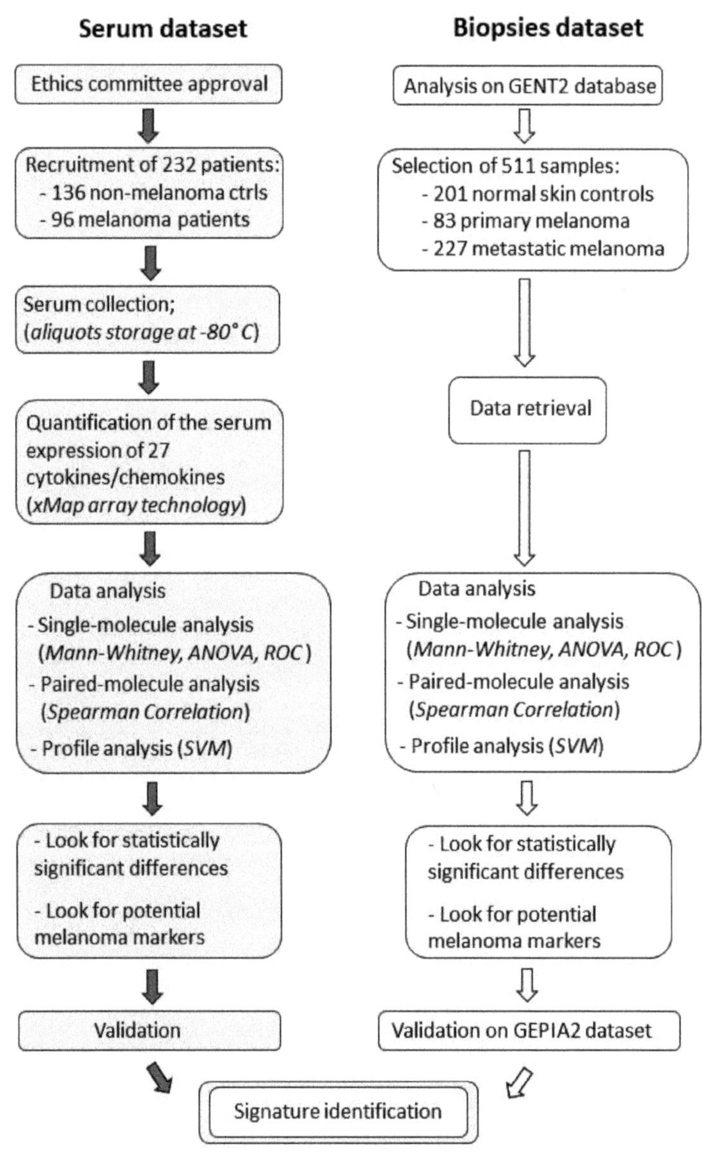

Figure 1. The cartoon reports the study workflow for the serum and tissue samples.

Table 2. Serum expression. Medians of the expression values of the 27 cytokines/chemokines for the control and melanoma patients. The table also reports the significance of the differences according to Mann–Whitney analyses. For p-values < 0.05, the null hypothesis that the control and melanoma patients have the same median should be rejected (significant values reported in bold and are underlined); i.e., a p < 0.05 indicates a significant difference. Moreover, the table reports the classification performances as the AUC values from the ROC analyses.

Cytokines	Controls Median ($n = 136$)	Melanoma Median ($n = 96$)	p-Value Mann–Whitney Controls vs. Melanoma	AUC $\pm$ S.E. by ROC Analysis
IL-1b	0.53	0.65	**0.04**	0.61 ± 0.05
IL-1Ra	26.77	17.83	0.14	0.56 ± 0.04
IL-2	3.45	2.14	0.14	0.66 ± 0.10
IL-4	2.95	2.88	0.47	0.53 ± 0.04
IL-5	2.77	2.34	0.22	0.60 ± 0.08
IL-6	5.37	3.17	**0.04**	0.70 ± 0.09
IL-7	2.24	2.24	0.72	0.52 ± 0.05
IL-8	6.63	6.4	0.55	0.52 ± 0.04
IL-9	45.58	42.05	0.29	0.54 ± 0.04
IL-10	7.08	4.56	0.10	0.63 ± 0.08
IL-12(p70)	15.74	16.07	1.00	0.50 ± 0.05
IL-13	2.43	2.99	0.83	0.52 ± 0.09
IL-15	52.22	30.11	1.00	0.50 ± 0.27
IL-17	16.72	13.1	0.37	0.54 ± 0.04
Eotaxin	95.28	106.28	0.38	0.53 ± 0.04
FGF-2	32.68	30.01	0.26	0.55 ± 0.04
G-CSF	4.73	5.41	0.33	0.55 ± 0.05
GM-CSF	10.21	10.63	0.67	0.53 ± 0.06
IFN-γ	19.1	23.2	0.48	0.53 ± 0.04
IP-10 (CXCL10)	438.69	501.41	**0.04**	0.58 ± 0.04
MCP-1(MCAF)	18.59	12.49	0.24	0.58 ± 0.07
MIP-1a (CCL3)	1.78	1.74	0.59	0.52 ± 0.04
MIP-1b (CCL4)	54.36	56.26	0.43	0.53 ± 0.04
PDGF-BB	1603.74	1033.41	**0.01**	0.61 ± 0.05
RANTES (CCL5)	11,353.34	8735.27	**0.01**	0.57 ± 0.06
TNF-α	16.4	18.06	0.26	0.52 ± 0.06
VEGF	59.75	57.98	0.88	0.54 ± 0.06

Bold underlined: highlight the result.

The Breslow thickness-related differences are reported in Tables 3 and 4. Table 3 reports the mean expression of the 27 cytokines in all melanoma patients as a function of Breslow thickness <1 mm vs. >1mm. Three molecules, namely IL-8, MCP-1, and RANTES, show a statistically significant differential expression. As a further characterization, the correlation of Breslow thickness with serum expression was then investigated in all melanoma patients. Expression of IL-8, GM-CSF, and MCP-1 on one site, and TNF-α on the other, shows a significant negative and positive correlation, respectively (Table 4).

Table S3 reports the correlations with Breslow thickness in male melanoma and in female melanoma patients and shows that the cytokines with significant correlations are different in males vs. females.

Table 3. Serum expression as a function of Breslow thickness. Count indicates the number of patients analyzed. The median expression values of the 27 molecules with a Breslow thickness <1 mm or >1 mm are reported. The significance of the difference according to the Mann–Whitney analyses is also reported: for p-values > 0.05, the null hypothesis that the two distributions of cytokine expressions have the same median should be rejected (significant values are reported in bold and are underlined). In simpler words, when the p-value is < 0.05, the cytokine expressions in patients with a Breslow thickness <1 mm and expression in patients with a thickness >1 mm have significantly different medians.

Cytokines	Melanoma Breslow Thickness <1 mm		Melanoma Breslow Thickness ≥1 mm		Mann-Whitney <1 mm vs. ≥1 mm
	Count	Median	Count	Median	p Value
IL-1b	26	0.65	16	0.72	0.66
IL-1Ra	44	18.24	27	17.83	0.59
IL-2	5	1.9	7	2.38	0.25
IL-4	54	2.85	35	3.06	0.81
IL-5	15	2.34	9	1.7	0.86
IL-6	7	0.83	8	3.34	0.18
IL-7	31	2.24	15	2.9	0.34
IL-8	47	7.8	30	5.78	**0.01**
IL-9	52	42.55	35	41.91	0.64
IL-10	11	3.84	10	5.12	0.92
IL-12(p70)	40	20.48	20	14.64	0.51
IL-13	11	2.43	8	3.2	0.60
IL-15	2	13.02	1	42.31	0.67
IL-17	38	13.15	30	11.79	0.94
Eotaxin	54	108.94	35	93.24	0.71
FGF-2	49	29.73	34	31.46	0.77
G-CSF	27	5.45	10	4.12	0.30
GM-CSF	17	14.06	25	9.78	0.22
IFN-γ	51	23.49	31	22.13	0.51
IP-10 (CXCL10)	53	491.81	35	574.3	0.88
MCP-1(MCAF)	18	10.56	11	24.83	**0.02**
MIP-1a (CCL3)	54	1.82	35	1.71	0.56
MIP-1b (CCL4)	53	55.51	35	53.52	0.73
PDGF-BB	53	1048.16	35	1080.98	0.47
RANTES (CCL5)	53	10,341.43	35	7534.18	**0.03**
TNF-α	43	16.65	19	22.78	0.06
VEGF	51	63.58	35	52.26	0.45

Bold underlined: highlight the result.

Table 4. Serum expression in all melanoma patients (male + female): correlation with Breslow thickness. For *p*-values < 0.05 (reported in bold and underlined), the null hypothesis (i.e., the Spearman's correlation coefficient R is 0) should be rejected; i.e., cytokine distributions with *p*-values < 0.05 are significantly correlated with Breslow thickness.

Cytokines	No. of Pairs	Spearman R Correlation of Serum Expression with Breslow Thickness	*p*-Value (2-Tails)
IL-1b	42	0.04	0.80
IL-1Ra	71	0.04	0.75
IL-2	12	0.01	0.97
IL-4	89	−0.02	0.88
IL-5	24	0.01	0.96
IL-6	15	0.24	0.40
IL-7	46	0.28	0.06
IL-8	77	−0.23	**0.05**
IL-9	87	−0.09	0.42
IL-10	21	−0.14	0.55
IL-12(p70)	60	−0.02	0.86
IL-13	19	−0.09	0.72
IL-15	3	1.00	0.33
IL-17	68	0.05	0.67
Eotaxin	89	−0.01	0.96
FGF-2	83	0.03	0.81
G-CSF	37	0.04	0.83
GM-CSF	42	−0.40	**0.01**
IFN-γ	82	0.07	0.53
IP-10 (CXCL10)	88	0.04	0.72
MCP-1(MCAF)	29	0.39	**0.04**
MIP-1a (CCL3)	89	−0.09	0.40
MIP-1b (CCL4)	88	−0.02	0.86
PDGF-BB	88	−0.10	0.35
RANTES (CCL5)	88	−0.20	0.06
TNF-α	62	0.31	**0.01**
VEGF	86	−0.02	0.87

Bold underlined: highlight the result.

A further analysis was carried out as a function of age, in all melanoma patients and all controls. The Spearman's correlation index was computed between the age and the expression value of each cytokine. Seven significant correlations were found in the controls involving IL-7, IL-12(p70), IL-13, IP-10, MIP-1a, MIP-1b, and VEGF. Such correlation were mostly lost in the melanoma patients; in fact, the patients showed only two significant correlations with age, namely, IP-10 and G-CSF (Table 5). A similar finding in male controls vs. male melanoma and in female controls vs. female melanoma is reported in Tables S4 and S5, showing a strong reduction in the correlation with age in melanoma samples compared to the controls.

Table 5. Serum expression in all melanoma patients (male + female): correlation with age. For *p*-values < 0.05 (reported in bold and underlined), the null hypothesis (i.e., the Spearman correlation coefficient R is 0) should be rejected; i.e., cytokine distributions with *p*-values < 0.05 are significantly correlated with age.

Cytokines	All Controls (Male + Female)			All Melanoma (Male + Female)		
	No. of Pairs	Spearman R Correlation of Serum Expression with Age	*p* Value (2 Tails)	No. of Pairs	Spearman R Correlation of Serum Expression with Age	*p*-Value (2 Tails)
IL-1b	85	0.14	0.21	44	−0.13	0.40
IL-1Ra	121	0.15	0.10	75	−0.08	0.50
IL-2	19	−0.21	0.38	13	0.23	0.46
IL-4	135	0.08	0.36	95	−0.05	0.66
IL-5	30	0.34	0.07	25	−0.20	0.33
IL-6	21	0.00	0.99	16	−0.01	0.96
IL-7	81	0.35	**0.001**	48	−0.09	0.53
IL-8	128	0.12	0.17	83	−0.20	0.07
IL-9	135	0.12	0.17	93	0.05	0.61
IL-10	39	0.05	0.77	22	−0.09	0.69
IL-12(p70)	110	0.30	**0.002**	62	−0.08	0.55
IL-13	27	0.39	**0.04**	19	0.21	0.38
IL-15	2	-	-	4	-	-
IL-17	108	0.12	0.23	73	−0.01	0.93
Eotaxin	132	0.13	0.13	95	−0.01	0.97
FGF-2	129	0.02	0.85	88	0.01	0.94
G-CSF	92	−0.03	0.76	40	−0.38	**0.02**
GM-CSF	48	−0.05	0.72	46	−0.10	0.53
IFN-γ	136	0.09	0.31	86	−0.12	0.26
IP-10 (CXCL10)	136	0.22	**0.01**	94	0.20	**0.05**
MCP-1(MCAF)	47	0.11	0.45	30	−0.17	0.38
MIP-1a (CCL3)	133	0.23	**0.01**	95	−0.16	0.13
MIP-1b (CCL4)	136	0.30	**0.0005**	94	−0.15	0.15
PDGF-BB	135	0.02	0.86	94	−0.11	0.29
RANTES (CCL5)	136	−0.02	0.84	94	−0.08	0.42
TNF-α	120	0.17	0.06	65	−0.14	0.28
VEGF	135	0.17	**0.05**	92	0.09	0.41

Bold underlined: highlight the result.

Then, a gender-specific analysis was carried out. Namely, expression levels of the 27 cytokines in male melanoma were compared to female melanoma. Table 6 shows that Eotaxin is significantly increased in male vs. female melanoma, and MCP-1 expression is significantly reduced in male vs. female melanoma, highlighting gender-related differences in cytokine/chemokine serum expression.

Table 6. Serum expression in male melanoma compared to female melanoma. For *p*-values < 0.05 (reported in bold and underlined), the null hypothesis (i.e., the two distributions have the same median) should be rejected. Briefly, the cytokine distributions for the control and melanoma patients with *p*-values < 0.05 have significantly different medians.

Cytokines	Male Melanoma	Female Melanoma	Mann–Whitney
	Median Value	**Median Value**	***p*-Value**
IL-1b	0.63	0.785	0.07
IL-1Ra	15.17	24.125	0.25
IL-2	1.9	3.09	0.14
IL-4	3.05	2.75	0.06
IL-5	2.34	2.02	0.36
IL-6	3.34	3.0	0.73
IL-7	2.24	2.31	0.49
IL-8	6.57	6.32	0.64
IL-9	44.35	38.84	0.14
IL-10	3.84	6.31	0.13
IL-12(p70)	16.94	14.57	0.98
IL-13	2.19	3.23	0.60
IL-15	-	30.11	-
IL-17	13.1	12.96	0.67
Eotaxin	135.15	91.8	**<u>0.002</u>**
FGF-2	32.55	29.73	0.25
G-CSF	5.09	5.45	0.60
GM-CSF	11.71	9.97	0.47
IFN-γ	20.83	27.33	0.26
IP-10 (CXCL10)	567.19	468.93	0.13
MCP-1(MCAF)	10.995	25.6	**<u>0.05</u>**
MIP-1a (CCL3)	1.86	1.7	0.49
MIP-1b (CCL4)	56.89	51.09	0.30
PDGF-BB	1145.64	900.76	0.06
RANTES (CCL5)	9536.67	7879.5	0.19
TNF-α	17.355	20.11	0.34
VEGF	65.345	50.515	0.49

Bold underlined: highlight the result.

Altogether, the results reported in Tables 2–6 indicate strong and significant differential expression of several cytokines/chemokines, as a function of Breslow thickness, age, and gender. Such differences support the known role of these molecules in controlling proliferation, immune response, chemotaxis, and inflammation in melanoma samples, and provide molecular insights into the systemic response to melanoma (see the Discussion section).

2.2. Serum Expression: Analysis of Paired Molecules by a Correlation Matrix

A correlation analysis was then carried out. Namely, Spearman's correlations between the expression values of all pairs of molecules were investigated in control and in melanoma patients. Figures 2 and 3 show the molecule pairs exhibiting the highest correlation coefficients in the control and in melanoma patients, respectively. Specifically, Figure 2A,B shows the heatmap of the intersections having a $p < 0.05$ and Spearman's rank coefficient >0.60, in the control and melanoma samples, respectively. Figure 3A,B show the molecules pair exhibiting a more severe selection, i.e., a correlation with $p < 0.05$ and a coefficient >0.7. In Figure 2, the 27 molecules were roughly clustered according to their biological functions. A higher number of strong correlations appear in the melanoma samples as compared to the controls, involving either immune/inflammatory molecules, chemokines, and angiogenic factors, indicating that the cytokines/chemokines expression network appears to be more reciprocally correlated in the melanoma samples.

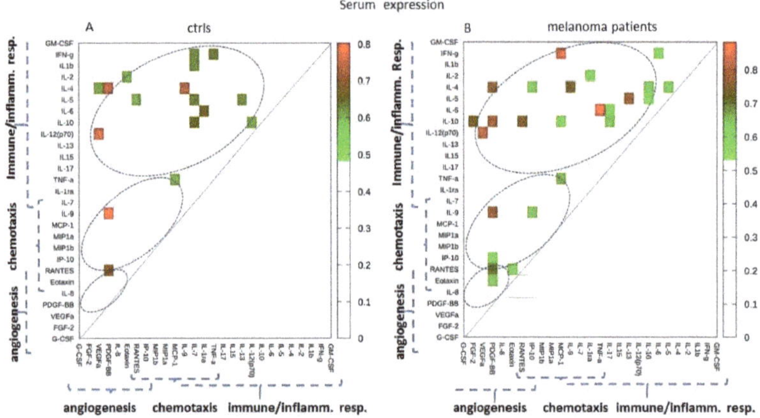

Figure 2. Correlation analysis of the serum expression values. The heatmaps in (**A**,**B**) show the correlation of pairs of expression values having a p-value <0.05 and Spearman's rank coefficient >0.60, in the control and in melanoma samples, respectively.

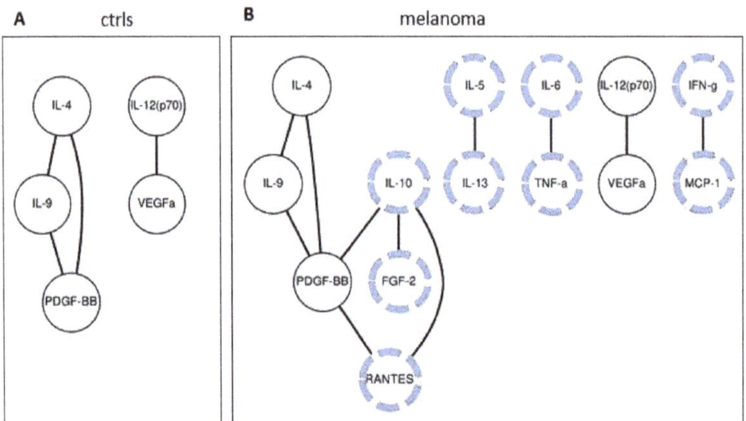

Figure 3. Correlation analysis of the serum expression values of all the control (**A**) and melanoma (**B**) samples. The connected circles show the paired molecules having $p < 0.05$ and Spearman's R coefficient >0.70, in the control and melanoma samples, respectively. The correlations specifically present in the melanoma samples are highlighted with blue dashed lines.

2.3. Serum Expression: Profile Analysis by SVM

The serum expression of the 27 chemokines/cytokines was then analyzed as a global profile. The SVM supervised learning algorithm was used, performing a simultaneous analysis of all molecules as predictors of melanoma state. A 10-fold cross-validated, linear-kernel SVM search method was carried out, and the missing values were handled in two alternative ways, as specified in the Methods section. The SVM analysis of the sera data improved the classification efficacy as compared to the single-molecule analysis reported in Table 2, leading to an AUC value = 0.761 and an average accuracy of 0.724 with a $p = 0.108$ (reported in bold and underlined in Table 7). This is slightly higher than 0.7, i.e., the best AUC value obtained by analysis in the single-molecule data (Table 2). Such a result was obtained by removing the missing values and considering as predictors the age and the expression values of the molecules IL-4, IL-8, IL-9, Eotaxin, FGF-2, IFN-γ, IP10, MIP-1a, MIP-1b, PDGF-BB, RANTES, and VEGF.

Table 7. Results of the SVM method applied to the serum expression dataset. Missing values are either removed or assigned as zero. In the first case, some molecules are removed from the predictor values (remaining molecules are listed in the "Molecules" column). Sex and/or age are or are not regarded as predictors. The p-value is the probability that the "Accuracy" value is not significantly above the "No Info Rate" value.

Missing Values	Num. Melanoma	Num. Controls	Training Set Size	Testing Set Size	Predictors: Sex or Age	AUC (ROC)	Accuracy	No Info Rate	p-Value
Removed *	72	124	138	58	Sex, Age	0.674	0.621	0.64	0.66
					Sex	0.658	0.638	0.64	0.56
					Age	0.761	0.724	0.64	0.11
					None	0.615	0.586	0.64	0.83
Set to 0 **	96	136	164	68	Sex, Age	0.621	0.588	0.59	0.55
					Sex	0.510	0.588	0.59	0.55
					Age	0.704	0.662	0.59	0.13
					None	0.619	0.588	0.59	0.55

* When the missing values were removed, IL-4, IL-8, IL-9, Eotaxin, FGF-2, IFN-γ, IP-10, MIP-1a, MIP-1b, PDGF-BB, RANTES, and VEGF were simultaneously analyzed by the SVM. ** When the missing values were set to 0, all 27 cytokines/chemokines were simultaneously analyzed by SVM.

The SVM analysis on the serum expression data show that the profile analysis may improve the classification efficacy as compared to the single-molecule analysis, but unfortunately not enough to reach clinically relevant values.

2.4. Tissues Expression: Single-Molecule Analysis of 27 Cytokines/Chemokines in Melanoma Patients and Controls

Gene expression of the 27 chemokines/cytokines was then evaluated in tissue biopsies of melanoma samples and in control samples. Expression data were derived from the skin cancers section of the GENT2 database. The interface available at the link http://gent2.appex.kr/gent2/ presents data from all skin cancers combined, reporting analyses of Basal Cell Carcinoma (BCC) pooled with Squamous Cell Carcinoma (SCC), Merkel carcinoma primary, Merkel carcinoma metastatic, primary and metastatic melanoma data, for a total of 810 samples. We therefore extracted data referring to normal skin, primary melanoma, and all other melanoma data, excluding all other skin cancers from the analysis. After such a selection, 511 samples were considered, namely, 201 normal skins, 83 primary/primary in-transit melanoma patients, and 227 metastatic melanoma patients. The median gene expression values of the 27 molecules are reported in Table 8 for the three categories. No other stratifications (such as sex, age, or Breslow thickness) were carried out, given the database limitations reported in the Material and Methods section. Differences of the medians in the categories assessed by Mann–Whitney analysis revealed that most molecules show significantly different median values ($p < 0.05$). ANOVA analysis was also carried out on the three groups, indicating similarly strong significant differences for most molecules investigated (see Table S6).

Table 8. Tissue expression: medians of the expression values of the 27 chemokines/cytokines for the control and for melanoma patients (all, primary, and metastatic). The table also reports the significance of the difference according to Mann–Whitney analyses (Wilcoxon two-sample rank-sum test): for *p*-values < 0.05 the null hypothesis (i.e., the corresponding sets of samples have the same median) should be rejected (values in bold and underlined).

	Median					**p-Value (Mann–Whitney)**			
						(with Bonferroni Correction)			
Cytokines	**Ctrls (201)**	**Melanoma**			**Ctrls vs. all**	**Ctrls vs. Prim.**	**Ctrls vs. Metast.**	**Prim. vs. Metast.**
		All (310)	**Prim. (83)**	**Metast. (227)**				
IL-1b	**6.66**	7.04	6.73	7.15	**<0.0001**	0.15	**<0.0001**	0.71
IL-1Ra	9.38	7.02	7.03	7.01	**<0.0001**	**<0.0001**	**<0.0001**	1.31
IL-2	2.58	2.81	2.58	3.00	0.59	0.96	1.00	1.00
IL-4	3.46	3.46	3.17	3.58	0.96	0.45	1.00	0.15
IL-5	2.81	3.00	3.00	3.00	0.48	1.00	1.00	1.00
IL-6	5.61	6.39	5.98	6.64	**<0.0001**	0.06	**<0.0001**	0.06
IL-7	7.22	5.49	5.17	5.73	**<0.0001**	**<0.0001**	**<0.0001**	**0.01**
IL-8	3.32	3.32	3.17	3.32	0.99	1.00	1.00	1.00
IL-9	2.58	2.81	2.58	2.81	0.76	1.00	1.00	0.48
IL-10	3.70	4.88	4.75	5.04	**<0.0001**	**<0.0001**	**<0.0001**	0.66
IL-12(p70)	4.25	2.32	2.81	2.00	**<0.0001**	**<0.0001**	**<0.0001**	0.12
IL-13	5.64	5.52	5.55	5.49	0.96	1.00	1.00	1.00
IL-15	6.95	6.83	6.30	6.97	0.44	**0.001**	1.00	**0.01**
IL-17	5.67	6.83	6.07	7.37	**<0.0001**	0.08	**<0.0001**	**0.05**
Eotaxin	4.75	5.29	4.95	5.39	**<0.0001**	0.21	**<0.0001**	0.06
FGF-2	7.11	7.03	6.25	7.24	0.51	**<0.0001**	0.53	**<0.0001**
G-CSF	11.83	11.55	12.25	11.37	0.95	**0.001**	0.30	**0.006**
GM-CSF	4.81	4.64	4.52	4.64	0.83	1.00	1.00	0.99
IFN-γ	4.70	5.58	4.91	5.83	**<0.0001**	**0.03**	**<0.0001**	**0.01**
IP-10 (CXCL10)	2.81	3.86	3.17	4.25	**<0.0001**	0.36	**<0.0001**	**0.01**
MCP-1(MCAF)	3.32	3.46	3.00	3.58	**0.03**	1.00	**0.03**	**0.01**
MIP-1a (CCL3)	5.13	8.11	7.76	8.24	**<0.0001**	**<0.0001**	**<0.0001**	0.07
MIP-1b (CCL4)	5.29	7.90	7.35	8.13	**<0.0001**	**<0.0001**	**<0.0001**	0.12
PDGF-BB	13.21	13.01	13.21	12.83	**0.02**	0.60	**0.001**	**0.003**
RANTES (CCL5)	6.86	8.16	7.81	8.22	**<0.0001**	**<0.0001**	**<0.0001**	1.00
TNF-α	6.13	7.54	7.27	7.57	**<0.0001**	**<0.0001**	**<0.0001**	**0.003**
VEGF	8.95	8.97	8.67	9.03	0.37	**0.001**	1.00	**0.002**

Bold underlined: highlight the result; Italics: Genes.

A ROC analysis was then carried out for every molecule by comparing the control vs. all melanoma, control vs. primary melanoma, control vs. metastatic, and primary melanoma vs. metastatic melanoma samples. The results are reported in Table 9. Four molecules were found to be very good classifiers of the control vs. melanoma samples, namely, IL-1Ra, IL-7, MIP-1a, and MIP-1b, with AUC values of 0.88, 0.86, 0.93, and 0.87, respectively. The corresponding ROC curves for the control vs. all melanoma samples are shown in Figure 4.

Table 9. ROC analysis of the tissue expression data. For each molecule, the AUC values for the ROC analysis was calculated for the following sets of expression values: controls vs. all melanoma, controls vs. primary melanoma, controls vs. metastatic melanoma, and primary melanoma vs. metastatic melanoma samples. For every AUC value, the standard error of the measure obtained by the cross-validation procedure is also shown.

Cytokines	AUC ± S.E. of ROC Analysis			
	Ctrls vs. all Melanoma	Ctrls vs. Primary	Ctrls vs. Metastatic	Primary vs. Metastatic
IL-1b	0.62 ± 0.03	0.57 ± 0.04	0.63 ± 0.03	0.54 ± 0.04
IL-1Ra	**0.88 ± 0.02**	**0.88 ± 0.03**	**0.88 ± 0.02**	0.53 ± 0.04
IL-2	0.51 ± 0.03	0.54 ± 0.04	0.51 ± 0.03	0.53 ± 0.04
IL-4	0.50 ± 0.03	0.55 ± 0.04	0.52 ± 0.03	0.57 ± 0.04
IL-5	0.52 ± 0.03	0.51 ± 0.04	0.52 ± 0.03	0.51 ± 0.04
IL-6	0.64 ± 0.03	0.59 ± 0.04	0.66 ± 0.03	0.59 ± 0.04
IL-7	**0.86 ± 0.02**	**0.91 ± 0.03**	**0.85 ± 0.02**	0.61 ± 0.04
IL-8	0.50 ± 0.03	0.52 ± 0.04	0.51 ± 0.03	0.52 ± 0.04
IL-9	0.51 ± 0.03	0.53 ± 0.04	0.52 ± 0.03	0.55 ± 0.04
IL-10	0.68 ± 0.03	0.65 ± 0.03	0.69 ± 0.03	0.55 ± 0.04
IL-12(p70)	0.78 ± 0.02	0.77 ± 0.03	0.79 ± 0.02	0.58 ± 0.04
IL-13	0.50 ± 0.03	0.50 ± 0.04	0.50 ± 0.03	0.50 ± 0.04
IL-15	0.52 ± 0.03	0.64 ± 0.01	0.52 ± 0.03	0.61 ± 0.04
IL-17	0.61 ± 0.03	0.54 ± 0.04	0.63 ± 0.03	0.59 ± 0.04
Eotaxin	0.63 ± 0.03	0.57 ± 0.04	0.65 ± 0.03	0.59 ± 0.04
FGF-2	0.52 ± 0.03	0.67 ± 0.04	0.54 ± 0.03	0.66 ± 0.04
G-CSF	0.50 ± 0.03	0.63 ± 0.04	0.55 ± 0.03	0.61 ± 0.04
GM-CSF	0.51 ± 0.03	0.52 ± 0.04	0.52 ± 0.03	0.54 ± 0.04
IFN-γ	0.69 ± 0.03	0.60 ± 0.04	0.72 ± 0.03	0.61 ± 0.04
IP-10 (CXCL10)	0.64 ± 0.03	0.56 ± 0.04	0.67 ± 0.03	0.61 ± 0.04
MCP-1(MCAF)	0.56 ± 0.03	0.54 ± 0.04	0.59 ± 0.03	0.62 ± 0.04
MIP-1a (CCL3)	**0.93 ± 0.01**	**0.91 ± 0.02**	**0.93 ± 0.02**	0.58 ± 0.04
MIP-1b (CCL4)	**0.87 ± 0.02**	**0.87 ± 0.02**	**0.86 ± 0.02**	0.58 ± 0.04
PDGF-BB	0.56 ± 0.03	0.55 ± 0.04	0.60 ± 0.03	0.62 ± 0.04
RANTES (CCL5)	0.73 ± 0.03	0.73 ± 0.03	0.72 ± 0.02	0.53 ± 0.04
TNF-α	0.77 ± 0.03	0.68 ± 0.03	0.80 ± 0.02	0.62 ± 0.04
VEGF	0.52 ± 0.03	0.63 ± 0.04	0.52 ± 0.03	0.63 ± 0.04

Bold: highlight the result; Italics: Genes.

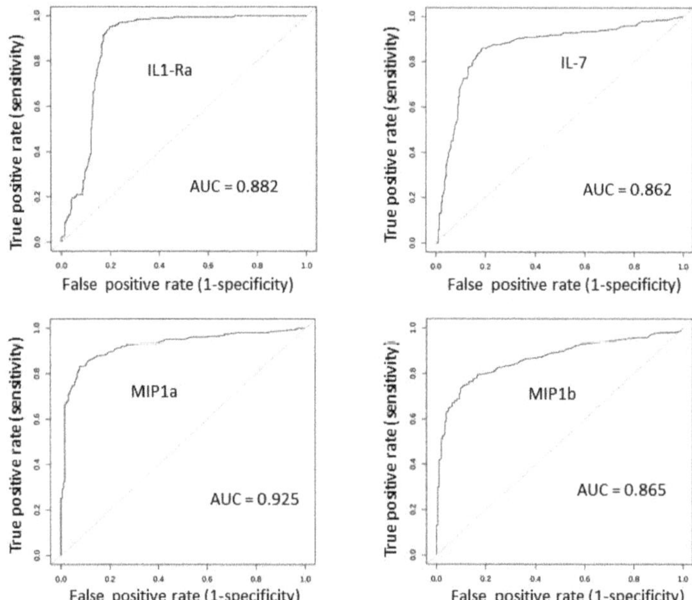

Figure 4. AUC plots of the IL-1Ra, IL-7, MIP-1a, and MIP-1b gene expression in the controls vs. all melanoma samples, for the tissue-expression values.

These results indicate that analyzing the gene expression of single molecules identifies relevant and significant differences; in this case, the ability to discriminate melanoma from the controls is much higher than the serum-expression data. Nevertheless, such values are below the threshold commonly indicated for potential clinical application. We then carried out the paired-molecule and profile analysis.

2.5. Tissue Expression: Analyzing Paired Molecules by a Matrix Correlation

The analysis of the paired-molecule correlations was then carried out, similarly to what we have done for the sera dataset. The correlations between the expression values of all pairs of molecules in the control and melanoma patients were analyzed by computing Spearman's rank correlation coefficient.

Figures 5 and 6 show the molecule pairs exhibiting high correlation coefficients in the control and melanoma patients. The heatmap in Figure 5 highlights the pairs with significant correlation coefficients of R > 0.6 with $p < 0.05$. Figure 6 shows a more severe selection, i.e., the pairs with a correlation p-value < 0.05 and coefficient >0.70. As observed in the serum dataset, this analysis indicates that many strong correlations appear in the melanoma samples as compared to the controls, involving immune/inflammatory molecules, chemokines, and angiogenic factors.

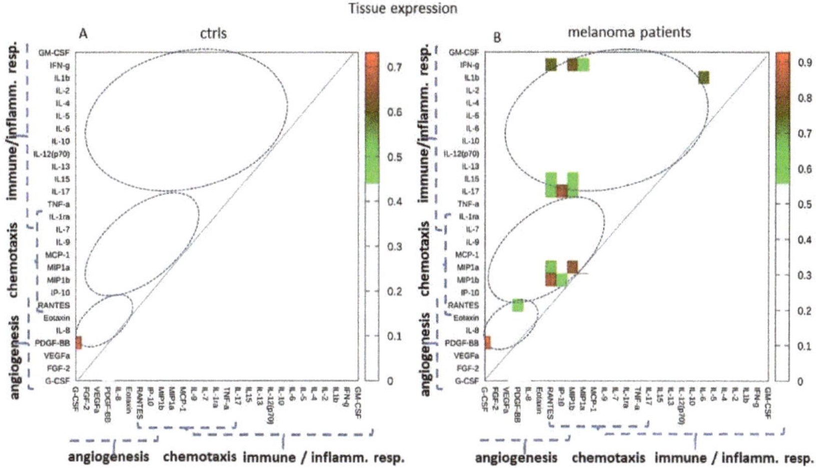

Figure 5. Correlations of the tissue expression values. The heatmap shows the correlation among pairs of expression values with a $p < 0.05$ and Spearman's rank coefficient >0.60.

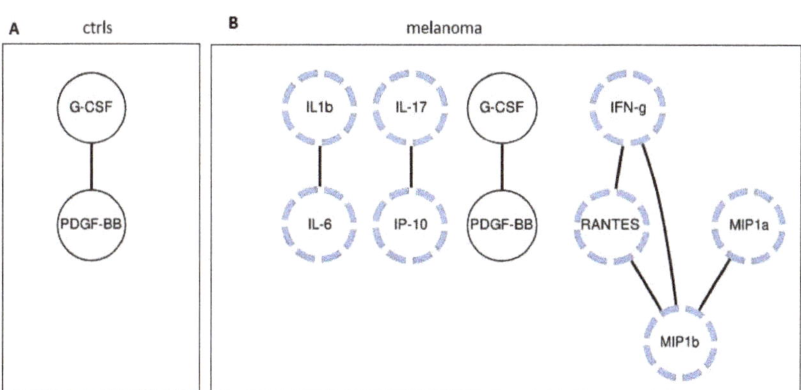

Figure 6. Correlation analysis of the tissue expression values of all the control (**A**) and melanoma (**B**) samples. The connected circles show the paired molecules with a $p < 0.05$ and Spearman's R coefficient >0.70, in the control and melanoma samples, respectively. The correlations specifically present in the melanoma samples are highlighted with blue dashed lines.

2.6. Tissue Expression: Profile Analysis by SVM

As in the case of the serum expression data, the SVM supervised learning algorithm was used to investigate all molecules simultaneously as melanoma predictors. Impressive results were achieved by analyzing the tissue-expression data. The results are shown in Table 10. The average of the AUC values obtained in the 10 iterations of the cross-validation procedure is 0.99, much higher than the highest AUC value obtained in the ROC analysis of the single molecules (namely, 0.93; see Figure 4 and Table 9). The p-value is <0.00001; hence, we are highly confident about the statistical significance of this observation.

Table 10. SVM method applied to the tissue-expression dataset. Sex and age were not considered as predictors for the lack of data in the dataset. The *p*, i.e., the probability that the "Accuracy" value is not significantly higher than the "No Info Rate" value, is lower than 0.00001. Hence, we can safely reject the null hypothesis and we can assume that the accuracy of the predictive model is higher than the value of the No Info Rate (0.61), corresponding to the performance of a dummy, fixed-answer predictor.

Num. Melanoma	Num. Controls	Training Set Size	Testing Set Size	AUC (ROC)	Accuracy	No Info Rate	*p*-Value
310	201	358	153	0.99	0.95	0.61	<0.00001

We then conclude that by using all molecules as melanoma classifiers, the accuracy of the prediction is extremely strong. The ROC curve of the predictive model based on the simultaneous analysis of all molecules is shown in Figure 7.

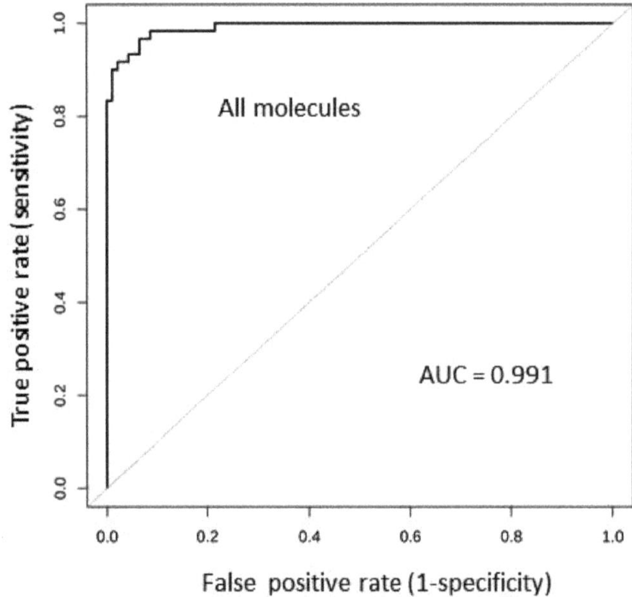

Figure 7. ROC analysis of the SVM-based model, classifying the melanoma samples against the controls.

A model based on many predictors (in this case 27) may present some practical issues. We thus performed a Recursive Feature Elimination (RFE) procedure (summarized in the Methods section), to select the most relevant molecules of the predictive SVM-based model. According to this analysis, the most sensible predictors are, in order, *MIP-1a, IL-1RA, IL-7, MIP-1b, IL-12(p70)*, and *TNF-α*. The best four molecules correspond to the ones shown in Table 9, obtained with ROC analyses of the single molecules. As shown in Figure 8, a model based on two molecules, namely, *MIP-1a* and *IL-1RA*, achieves an AUC value = 0.965, while a 4-predictor model *(MIP-1a, IL-1RA, IL-7,* and *MIP-1b)* reaches an AUC = 0.982. These molecules stably represent the best 4-marker combination with the highest AUC value. Further increasing the number of predictors does not add any relevant improvement (see Figure 8).

We therefore conclude that combined analysis of the expression of the *MIP-1a, IL-1RA, IL-7,* and *MIP-1b* genes represents the best combination within the 27 investigated, able to very effectively discriminate the control from the melanoma samples.

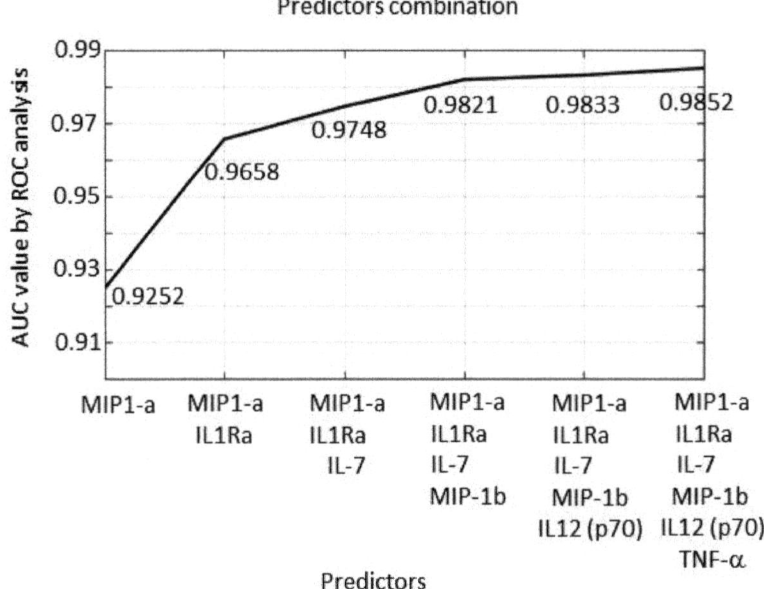

Figure 8. AUC values obtained by the SVM algorithm using one predictor (*MIP-1a*), two predictors (*MIP-1a + IL-1RA*), three predictors (*MIP-1a + IL-1RA + IL-7*), four predictors (*MIP-1a + IL-1RA + IL-7 + MIP-1b*), etc., up to six predictors.

We finally investigated the role of the expression of these four genes as a prognostic factor. According to the survival analysis tool available in the GEPIA2 database, 3 out of 4 show significant Hazard Ratios. Namely, *IL7*, *MIP-1a* (*CCL3*), and MIP-1b (*CCL4*) show a HR of 0.71, 0.65, and 0.5, respectively, with $p = 0.01$, 0.002, and 1×10^{-7}, respectively. These data indicate significantly improved survival in patients with high expression values for these three genes.

2.7. Results Validation

The expression of the four molecules reported in Figure 6 was then investigated in an independent database, namely, GEPIA2 (found at http://gepia2.cancer-pku.cn/). Expression was confirmed to be significantly different in melanoma compared to the healthy controls, for *IL-1Ra* (recognized as *ILRn* by GEPIA2), *IL-7*, *MIP-1a* (recognized as *CCL3* by GEPIA2), and *MIP-1b* (recognized as *CCL4* by GEPIA2) (see Figure 9).

The combined expression of these four molecules was then subject to a PCA analysis carried out by the "Dimensionality reduction" tool in GEPIA2. The three most relevant components very effectively differentiated melanoma from controls (Figure 10), indicating that the combined analysis of these four molecules may represent an effective melanoma marker.

This observation fully validated the SVM analysis reported in Figure 6.

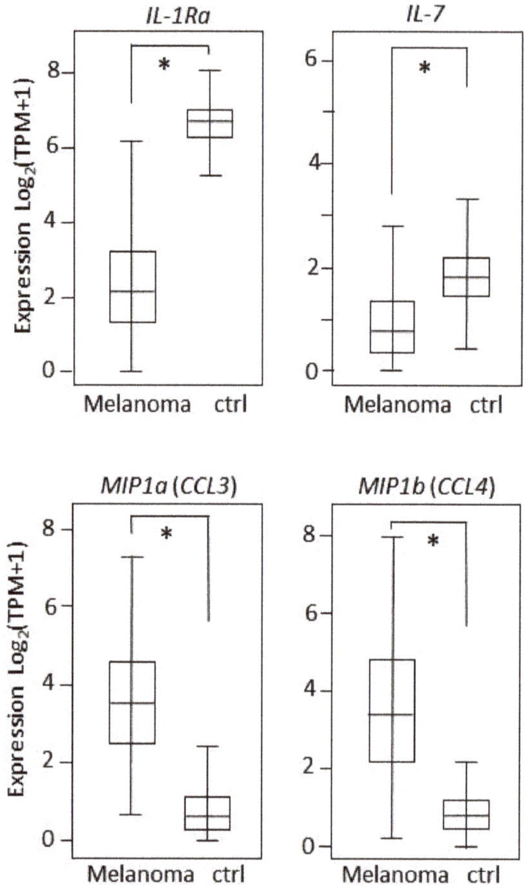

Figure 9. Tissue expression according to the GEPIA2 database. An asterisk (*) indicates $p < 0.0001$.

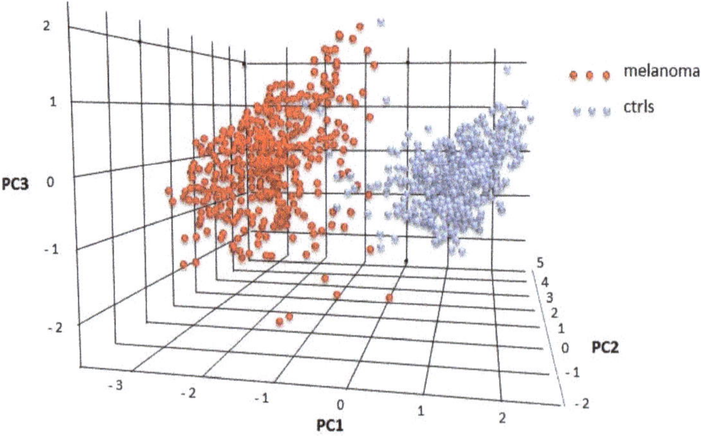

Figure 10. PCA analysis on the tissue expression values of *IL-1Ra*, *IL-7*, *MIP-1a*, and *MIP-1b*.

3. Discussion

While several serum biomarkers are investigated in melanoma patients [19], diagnostic markers currently applied in clinics are restricted to S-100, HMB-45, Melan-A, and SM5-1 [20,21], and the prognostic markers to monitor melanoma progression are S100B, MART1, PMEL, and S100A13 [22]. As recently reported [23], potential markers in melanoma are mutations (on BRAF, NRAS, KIT, GNA11/GNAQ, NF1, CDKN2AI, immunohistochemical biomarkers (such as PD-11 and PD-L1, as well as mutated BRAF and NY-ESO-1), miRNAs, and other serum molecules. The key role cytokines/chemokines play in the immune/inflammatory response and in proliferation and chemotaxis control has been largely investigated; nevertheless, their role as diagnostic or prognostic markers remains to be elucidated. We and others demonstrated the key role of growth factors such as FGF-2, PDGF, and TNF-α in controlling melanoma growth [24–26] and melanoma aggressiveness [27]. The present study is the first, to our knowledge, presenting a signature of four cytokines/chemokines as an extremely effective melanoma marker, in a large patient collection. The present study measures cytokine/chemokine expression in serum and in tumor biopsies. We did not expect that the same cytokines/chemokines would be modified in serum and in tissues. In fact, the molecules measured in the serum are likely produced as a systemic response, while the molecules measured within the biopsies are directly produced in the tumor or in the regions immediately close to it. Therefore, the cytokines/chemokines measured within the biopsies reflect more directly the tumor biology and its aggressive behavior. On the contrary, the cytokines/chemokines measured in the serum reflect more how the organism responds to the tumor from an inflammatory/immunological point of view. We cannot exclude that molecules produced within the primary tumor may reach the blood. However, such signals may be not measured due to the large dilution in the blood stream and their expression values may fall below the detection limit. We used the xMAP technology for quantification in serum samples, to minimize as much as possible the sensitivity limitations.

For the sake of clarity, we will discuss below the results of serum expression separately from the results on tissue expression.

Serum expression: Analyzing the serum expression of 27 cytokines/chemokines did not identify any relevant marker when individually analyzed. Nevertheless, significant and strong differential expressions were found in melanoma vs. controls (see Table 2), as well as in melanoma samples as a function of Breslow thickness (see Tables 3 and 4) or age (Table 5). Furthermore, significant gender-specific differences were identified in Eotaxin and MCP-1 expression (Table 6), as well as in GM-CSF, TNF-α, IL-9, MIP-1a, IL-8, PDGF-BB, and MCP-1 (Tables S3–S5). This finding indicated the molecular bases possibly underlying the different incidence and different mortality rates in male vs. female melanoma [28–32], as well as the unexpected better response of immunotherapies in men than in women [33].

Serum expression of IL-1b, IL-6, IP-10, PDGF-BB, and RANTES was found to be significantly different in melanoma vs. controls (Table 2), reinforced by a much larger patient cohort, since previous observations were carried out on much smaller patient cohorts [34,35]. These molecules make a proinflammatory milieu previously found in uveal melanoma [36] and such findings agree with recent data showing that serum inflammation markers are strongly associated with melanoma progression [37]. Cytokines and chemokines are closely engaged within a large network where several ligands share few receptors [38], therefore reciprocally modulating their pro-, anti-inflammatory, chemotactic, and angiogenic functions [39–41]. The chemokines network is known to mediate melanoma interaction with the surrounding tissues [42]. Investigating cytokine/chemokine serum expression may therefore reveal the coordinated tissue reaction to the presence of melanoma. In the present study, Spearman's correlation matrix revealed for the first time that strong and significant correlations of the expression values are more numerous in melanoma samples than in healthy controls (Figures 2 and 3). In the melanoma samples, molecules involved in inflammation, chemotaxis, and angiogenesis had strong and significant correlations, namely, according to Figure 3, strong correlations of IL-10 with FGF-2, IL-10 with RANTES, IL-5 with IL-13, IL-6 with TNF-α, and of IFN-γ with MCP-1 appear in melanoma samples.

Particularly interesting is the strong correlation of IL-6 to TNF-α in the melanoma samples; these two molecules are known to control the ability to evade the immune system control in a PDL-1-dependent manner [43]. Such correlation data indicated that the cross-talk of cytokines and chemokines is altered in melanoma and this may help in defining a melanoma-specific, correlation-matrix fingerprint. We therefore analyzed the entire panel of 27 cytokines/chemokines with the SVM machine learning algorithm, to investigate simultaneously all molecules as predictors of melanoma state. In other studies, SVM effectively discriminated melanoma on the basis of dermoscopic images [44], ultrasonic and spectrophotometric images [45], BRAF status [46], or dermo-fluorescence spectra [47], with a reported accuracy up to 90%. SVM was previously used for prognostic purposes in melanoma patients [48] but, to our knowledge, the present study is the first applying the SVM analysis to cytokine/chemokine-expression values to discriminate melanoma from controls, both in serum and in tissue, in a large group of controls and patients. The SVM procedure was indeed able to improve the ability to classify the serum samples, from AUC = 0.70 for IL-6 expression (see Table 2) up to AUC = 0.761 for the combined indicators (see Table 7). However, this is still not good enough to propose a clinical diagnostic application. We then concluded that the serum expression data of these molecules, while showing strong and significant differences, may not be good classifiers. Several reasons may underlie this result, such as biological reasons (namely, the large serum dilution) as well as technical reasons (namely, samples storage or antibody cross-specificity). We cannot exclude that the cytokine serum expression will give improved information with an improved technology. Protein quantification is a rapidly evolving technology with the continuous upgrading of antibody combinations, sensitivity improvement, and protocol optimization. As an example, a 2007 report [49] identified several cytokines differently expressed in melanoma serum using multiplex xMAP technology, in 179 melanoma patients and 378 healthy controls. However, those data merit to be re-evaluated in the light of the currently available multiplex xMAP technology and new antibodies. As an alternative, quantitative proteomics approaches, based on mass spectrometry, indicated proteins differently expressed in melanoma compared to the controls [50]. However, the sensitivity of the latter techniques limits their application for cytokine/chemokine quantification, as compared to immunometric methods. Analyzing serum from melanoma patients aimed at identifying markers suitable for the early diagnosis, using a minimally invasive technique, expressions significantly different were indeed identified. However, we could not identify good markers within the 27 molecules investigated. This may depend on the molecules chosen (i.e., we should probably change targets and focus on other molecules), or it may depend on the high dilution factor in the serum samples.

We should address briefly the age-matching issue. As reported in Table 1, the mean age in the healthy groups and melanoma groups is different. Such difference reflects what the reality is, i.e., cancer patients are generally older than healthy controls, since increased age is a specific cancer risk factor. In the present study, individuals were sequentially enrolled, and controls were individuals with a suspect lesion removed and diagnosed by the pathologist as a not-cancer lesion. To have a similar age distribution in patients and in the control groups, one would be forced to remove several young healthy controls from the dataset (to match the rarely present young melanoma patients) and to remove several old melanoma patients (to match the rarely present old healthy controls). Age-matching would, therefore, strongly decrease the number of individuals analyzed, and alter the actual patient and control age distribution. The SVM analysis reported in Table 7 was also carried out on the age-matched groups, and the results are similar to the ones obtained from unmatched groups (See Table S7). Furthermore, the matrix analysis reported in Figure 3 was carried out also on the age-matched groups, and the results are identical to the ones obtained from the unmatched groups. We then conclude that age-matching, required for correct statistical analysis, would strongly reduce the group numerosity; it also would abolish a specific risk factor. In addition, the results achieved on the age-matched groups appear very similar or identical to those obtained on the age-unmatched groups, under our experimental conditions.

Tissue expression: Analyzing the tissue expression was much more effective to discriminate melanoma from controls. This is likely related to biological reasons (i.e., the direct analysis of the melanoma bearing tissue), as well as technical reasons (i.e., using a more stable quantification technique). Very high AUC values were calculated, up to 0.92 (see Figure 4), with the single-molecule analysis. Most of the investigated molecules showed a significant differential expression. This finding indicated that, at the tissue level, most of the cytokines/chemokines investigated are strongly altered in melanoma, prompting us to look for a further improvement of their classifier ability. Analysis of paired molecules reinforced what was observed in the sera dataset, showing that high-correlation pairs appear in melanoma while they are almost absent in the controls (Figures 5 and 6). Particularly interesting are the strong correlations involving the chemokines RANTES, MIP-1a, and MIP-1b (Figures 5 and 6).

Then, an SVM analysis on all molecules simultaneously analyzed was carried out. This analysis strongly improved the classification ability as compared to the single molecules. AUC reached an extremely high value of 0.991 when all 27 cytokines/chemokines were simultaneously considered as melanoma predictors. Use of the Recursive Feature Elimination (RFE) [51] procedure allowed us to identify the four best-performing molecules. The combination of IL-1Ra/IL-7/MIP-1a/MIP-1b shows the relevant AUC = 0.982. Interestingly to notice, the expression of IL-1Ra and IL-7 (known anti-inflammatory cytokines) is significantly reduced in the melanoma samples, while expression of MIP-1a and MIP-1b (known inflammatory chemokines) is significantly increased. Previous data demonstrate that CCL3 (MIP-1α) and CCL4 (MIP-1b) control the infiltration of the immune cells by recruiting antigen-presenting cells, including dendritic cells (DCs), to the tumor site via IFN-γ [52]. However, the specific signature made of the two anti-inflammatory and two pro-inflammatory molecules is a novel finding to our knowledge.

Full validation of these results was achieved on an independent dataset, the GEPIA2 database, reporting expression data from 1019 control and melanoma samples. The four molecules IL-1Ra, IL-7, MIP-1a, and MIP-1b were confirmed to have a significant ($p < 0.0001$) differential expression in melanoma (Figure 9), and their combined analysis with the PCA methodology (a different methodology compared to SVM) was found to effectively discriminate the controls from the melanoma samples (Figure 10).

The identification of the relevant gene markers from the tissue-expression data by using a quantitative technique may help improve histological diagnoses. Identification of a 4-gene signature may be a relevant help for pathologists. Measuring expression of these genes represents a quantitative approach that is operator-independent and may be part of an automatic process useful to identify suspect samples.

4. Materials and Methods

4.1. Patients Selection and Recruitment

Melanoma patients were consecutively recruited at the hospital sections of IDI-IRCCS, according to the procedure approved by the IDI Ethics Committee (CE 287/1 approved 7/04/2009) based on a suspect skin lesion. All patients gave written informed consent. Patients under any pharmacological melanoma therapy were excluded. Serum was collected before the biopsy procedure, aliquoted, and stored at −80 °C. According to the histological analysis, patients were then assigned to the control arm or to the melanoma arm. A total of 232 patients were recruited in the present study.

4.2. Serum Handling

A total of 7 mL of peripheral blood were collected from patients. Blood was collected in tubes with no additives of any type. Tubes were taken at room temperature for 2 to 3 h; they were then centrifuged at 15,000 rpm for 15 min; clear yellow color serum was stored in 100 μL aliquots and stored at −80 °C. The red color was considered a hemolysis sign and such sera were then not analyzed in the current study.

4.3. Cytokines Quantification in Sera Samples

Sera were obtained from melanoma patients ($n = 96$) and from patients with non-melanoma suspect lesions ($n = 136$) and were analyzed using xMAP technology on the Luminex platform (X200 Instrumentation equipped with a magnetic washer workstation and software Manager 6.1), which allows the simultaneous quantification of many molecules. The commercial kit used was Bio-Plex Pro human cytokine 27-plex panel (Bio-Rad Laboratories, Hercules, CA, USA), able to measure the following analytes: IL-1Ra, IL-1b, IL-2, IL-4, IL-5, IL-6, IL-7, IL-8, IL-9, IL-10, IL-12(p70), IL-13, IL-15, IL-17, TNF-α, IFN-γ, Eotaxin, Macrophage Inflammatory Protein *(MIP)-1a (MIP-1a; CCL3)*, Macrophage Inflammatory Protein *(MIP)-1b (MIP-1b; CCL4)*, Monocyte Chemoattractant Protein *(MCP)-1 (CCL2)*, Granulocyte Colony stimulating factor (G-CSF), GM-CSF, Basic Fibroblast growth factor (FGF-2), Interferon γ-induced protein 10 (IP-10; CXCL10), Regulated on Activation, Normal T cell Expressed and Secreted (RANTES), Platelet-Derived Growth Factor (PDGF-BB), and Vascular Endothelial Growth Factor (VEGF). Samples were handled according to the manufacturer's instructions and as previously reported [27].

4.4. Serum Expression Data

The serum dataset was composed of 232 records corresponding to 232 different individuals. The recorded data included in each case the histopathological diagnosis, sex, age, Breslow's thickness, and the expression values of the 27 chemokines/cytokines, reported in pg/mL. The expression of a few molecules was undetectable in some patients. Specifically, expressions of IL-2, IL-5, IL-6, IL-10, IL-13, and IL-15 were undetected in most cases and were measured in less than 30 controls and/or less than 30 melanoma samples. We handled missing values in two alternative ways, i.e., we removed the missing data point from the analysis (either the whole cytokine from all samples or the whole sample containing the missing value, depending on the performed analysis while trying to maximize the size of the resulting dataset), or, alternatively, we assumed the missing values are equal to zero, thus assuming that all missing values were caused by expression values lower than the measurement thresholds of the diagnostic kits.

4.5. Tissue Expression Data from GENT2 Database

The 27 molecules measured in the sera were investigated in control and melanoma samples taken from the GENT2 database, according to transcriptomic data; data of 201 normal skin biopsies were from 11 independent studies (referred to as GSE39612, GSE30355, GSE14905, GSE13355, GSE7553, GSE42109, GSE16161, GSE15605, GSE7307, GSE46239, and GSE7307); data of 83 primary melanoma biopsies were from 4 independent studies (referred to as GSE10282, GSE15605, GSE7553, and GSE62837); data of 227 metastatic melanoma biopsies were from 11 independent studies (referred to as GSE62837, GSE7307, GSE31879, GSE38312, GSE15605, GSE35640, GSE7553, GSE4587, GSE19293, GSE19234, and GSE22968). The tissue dataset was composed of 511 records, each describing a single subject: the recorded data included the histopathological diagnosis, sex, age, melanoma stage, and the expression values of the 27 cytokines. Each record always states whether the subject has been clinically diagnosed as affected by melanoma (310 patients) or not (201 controls). In this dataset, there were no missing values within the expression data. However, sex and age data were not recorded in most cases: only 217 records reported sex (138 males and 79 females) and only 125 records indicated the age (minimum 20 years, maximum 92 years, median 56 years, mean 59.56 years), and the majority of the control subjects had no sex and age data (about 70% of total). Therefore, sex and age stratifications were not carried out in the tissue-expression data.

4.6. Statistical Analyses

All statistical analyses were carried out on the "R" package version 4.0.0 [53].

4.6.1. Single-Molecule Analysis

Expression data of the single molecules were analyzed as an evaluation of the statistical significance of the median differences, by the Mann–Whitney test with Bonferroni correction. ANOVA analysis of the differences between the controls, primary melanoma, and metastatic melanoma samples in the tissue data was also performed and reported in Supplementary Material. In this case, normal distribution was evaluated by the Shapiro–Wilk test, and homoscedasticity (homogeneity of variance) was evaluated by the nonparametric Levene test. When the ANOVA analysis showed a significant difference between the medians, Mann–Whitney with Bonferroni correction and Tukey's honest significance tests were applied as post-hoc tests.

4.6.2. Paired-Molecule Analysis

A data-matrix analysis investigated whether the expression correlations of all molecule pairs show any relevant difference between the control and melanoma samples. Spearman's rank correlation coefficient was calculated.

4.6.3. Profile Analysis

Profile analysis was based on the Support Vector Machine (SVM) supervised learning algorithm, using a linear kernel [54,55]. Briefly, the method finds the best separation hyperplane between the set of control samples and the set of melanoma samples. Each sample is assumed as a single point in the hyperspace of dimensions n, where n is the number of features that can be used as predictors (specifically, the expression values of the molecules, and optionally age and sex). The result of the SVM algorithm is a separation hyperplane that maximizes the cumulative quadratic distance between the boundary points and the hyperplane itself. A parameter C plays a crucial role when the points are not linearly separable: C represents the tradeoff between decreasing the quadratic distance and ensuring that the boundary points are properly classified. We tuned the parameter C by testing 40 values between approximately 10^{-14} and 10^5 and selecting the value yielding the largest Area Under Curve (AUC) of the Receiver Operating Characteristic (ROC).

The missing expression values were removed from the dataset, according to two alternative approaches. In the first approach, we removed either the entire sample or, if more convenient in terms of resulting dataset size, all expression values of a specific molecule from the dataset. In the second approach, we assigned all missing values to zero [56]. The expression values of each molecule were then transformed to have average = 0 and standard deviation = 1, according to standard methods for this kind of analysis [57,58]. To validate the results, a 10-fold cross-validation procedure was applied. The SVM algorithm considers all predictors as coordinates in a multidimensional space, hence the prediction model is based on the whole set of 27 expression values of the molecules. For the analysis of the tissue-expression data, a Recursive Feature Elimination procedure [51] was applied to identify the molecules having the greatest impact. Therefore, the most relevant molecules of the predictive SVM-based model were identified. This method essentially repeats several times the cross-validated SVM analysis by excluding one of the predictors at a time, then discarding the weakest one, and restarts the whole process on the set of remaining predictors. By this procedure, the molecules having the weakest impact on the performance of the SVM model were identified and removed from the feature set.

4.7. Results Validation

4.7.1. Cross-Validation Procedure

All statistical results involving random selections of samples (namely SVM analyses) were validated using "cross-validation" methods to reduce errors due to overfitting. Overfitting errors are caused by an over-optimization of the parameters of a statistical method that achieves an optimal result on the available dataset but poor results on a dataset built from a different set of observations. In a typical k-fold cross-validation procedure, the dataset is randomly partitioned in k subsets of approximately equal size. The statistical method is then repeated k times: in every execution, k-1 subsets are used as "training sets" to optimize the parameters of the method, while the remaining "testing'" subset is used to evaluate the performances of the method. Every execution of the method uses a different subset as the "testing" set. Eventually, the performances of the method are taken as the average performances of all k executions. In this work we selected k = 10; thus, every measure is the net results of 10 experimental runs.

4.7.2. Validation

The tissue-expression results obtained from the data collected from the GENT2 database were validated on an independent database, GEPIA2 (available at http://gepia2.cancer-pku.cn/#index), reporting the RNA expression data from 461 controls and from 558 melanoma patients. Expression and dimensionality reduction by PCA analysis were carried out by the specific tools available at the GEPIA2 database.

5. Conclusions

We report here, for the first time, significant differences in cytokine expression as a function of the pathological state and gender, or age, or Breslow thickness, in the serum expression of a large patient dataset. Such differences are likely related to the systemic response to the tumor and may help, at least in part, investigating the known heterogeneity of this tumor. Furthermore, by analyzing gene expression in a large tissue expression dataset, we report, for the first time, a highly relevant 4-gene signature that discriminates the controls from the melanoma patients. We also show here that the machine learning algorithm SVM appears to be very effective in improving the classification ability for potentially diagnostic purposes and clinical applications.

Supplementary Materials: The following are available online at http://www.mdpi.com/2072-6694/12/12/3680/s1, Table S1: Serum expression general data in all controls (male + female). Table S2: General data on serum expression in all melanoma patients (male + female). Table S3: Correlation of the serum expression with Breslow thickness in male melanoma and in female melanoma. Table S4: Correlation of the serum expression with age, in female controls compared to female melanoma. Table S5: Correlation of the serum expression with age, in male controls compared to male melanoma. Table S6: Anova analysis of tissue expression data. Table S7: Results of the SVM method applied to the serum expression dataset after age-matching. Results are similar to the analysis performed on the unmatched data (see Table 7).

Author Contributions: M.C.: conceptualization; data analysis; draft writing; final writing: F.S.: sample handling; database updating; draft writing; D.D.: sample handling; database updating; draft writing; G.C.A.-C.: patients recruitment; informed-consent; S.R.: expression analysis; C.T.: expression analysis; M.N.: patients recruitment; informed-consent; E.P.: patients recruitment; informed-consent; L.L.: patients recruitment; informed-consent; G.D.L.: patients recruitment; informed-consent; F.F.: conceptualization; study design; draft writing; final writing; A.F.: conceptualization; study design; draft writing; data analysis; final writing; study supervision. All authors have read and agreed to the published version of the manuscript.

Funding: MOH: RC2018: RC2019 to A.F. MOH: Programma Italia-USA, and Oncotecnologico to FF.

Acknowledgments: We gratefully thank the support from the Facility of Complex Protein Mixture (CPM) analysis at ISS, Rome, and the entire sections of Oncology- Dermatology and Surgery at IDI-IRCSS. T.C. was supported by Fondazione Umberto Veronesi that is gratefully acknowledged.

Conflicts of Interest: The authors declare no conflict of interest.

References

1. Richmond, A.; Yang, J.; Su, Y. The good and the bad of chemokines/chemokine receptors in melanoma. *Pigment. Cell Melanoma Res.* **2009**, *22*, 175–186. [CrossRef] [PubMed]

2. Bagheri, H.; Pourhanifeh, M.H.; Derakhshan, M.; Mahjoubin-Tehran, M.; Ghasemi, F.; Mousavi, S.; Rafiei, R.; Abbaszadeh-Goudarzi, K.; Mirzaei, H.R.; Mirzaei, H. CXCL-10: A new candidate for melanoma therapy? *Cell Oncol. (Dordr)* **2020**, *43*, 353–365. [CrossRef] [PubMed]

3. Payne, A.S.; Cornelius, L.A. The Role of Chemokines in Melanoma Tumor Growth and Metastasis. *J. Investig. Dermatol.* **2002**, *118*, 915–922. [CrossRef] [PubMed]

4. Gerami, P.; Cook, R.W.; Wilkinson, J.; Russell, M.C.; Dhillon, N.; Amaria, R.N.; Gonzalez, R.; Lyle, S.; Johnson, C.E.; Oelschlager, K.M.; et al. Development of a Prognostic Genetic Signature to Predict the Metastatic Risk Associated with Cutaneous Melanoma. *Clin. Cancer Res.* **2015**, *21*, 175–183. [CrossRef]

5. Qin, Y.; Verdegaal, E.M.; Siderius, M.; Bebelman, J.P.; Smit, M.J.; Leurs, R.; Willemze, R.; Tensen, C.P.; Osanto, S. Quantitative expression profiling of G-protein-coupled receptors (GPCRs) in metastatic melanoma: The constitutively active orphan GPCR GPR18 as novel drug target. *Pigment. Cell Melanoma Res.* **2010**, *24*, 207–218. [CrossRef]

6. Harlin, H.; Meng, Y.; Peterson, A.C.; Zha, Y.; Tretiakova, M.; Slingluff, C.; McKee, M.; Gajewski, T.F. Chemokine Expression in Melanoma Metastases Associated with CD8+ T-Cell Recruitment. *Cancer Res.* **2009**, *69*, 3077–3085. [CrossRef]

7. Kucera, R.; Topolcan, O.; Treskova, I.; Kinkorová, J.; Windrichova, J.; Fuchsova, R.; Svobodova, S.; Třeška, V.; Babuska, V.; Novák, J.; et al. Evaluation of IL-2, IL-6, IL-8 and IL-10 in Malignant Melanoma Diagnostics. *Anticancer. Res.* **2015**, *35*, 3537–3542.

8. Tarhini, A.A.; Lin, Y.; Zahoor, H.; Shuai, Y.; Butterfield, L.H.; Ringquist, S.; Gogas, H.; Sander, C.; Lee, S.; Agarwala, S.S.; et al. Pro-Inflammatory Cytokines Predict Relapse-Free Survival after One Month of Interferon-α but Not Observation in Intermediate Risk Melanoma Patients. *PLoS ONE* **2015**, *10*, e0132745. [CrossRef]

9. Tarhini, A.A.; Lin, Y.; Yeku, O.; LaFramboise, W.A.; Ashraf, M.; Sander, C.; Kirkwood, J.M.; Lee, S.J. A four-marker signature of TNF-RII, TGF-α, TIMP-1 and CRP is prognostic of worse survival in high-risk surgically resected melanoma. *J. Transl. Med.* **2014**, *12*, 19. [CrossRef]

10. Lim, S.Y.; Lee, J.H.; Gide, T.N.; Menzies, A.M.; Guminski, A.; Carlino, M.S.; Breen, E.J.; Yang, J.Y.; Ghazanfar, S.; Kefford, R.F.; et al. Circulating Cytokines Predict Immune-Related Toxicity in Melanoma Patients Receiving Anti-PD-1–Based Immunotherapy. *Clin. Cancer Res.* **2019**, *25*, 1557–1563. [CrossRef]

11. Samaniego, R.; Gutiérrez-González, A.; Gutiérrez-Seijo, A.; Sánchez-Gregorio, S.; García-Giménez, J.; Mercader, E.; Márquez-Rodas, I.; Avilés, J.A.; Relloso, M.; Sánchez-Mateos, P. CCL20 Expression by Tumor-Associated Macrophages Predicts Progression of Human Primary Cutaneous Melanoma. *Cancer Immunol. Res.* **2018**, *6*, 267–275. [CrossRef]

12. Snow, H.; Hazell, S.; Francis, N.; Mohammed, K.; O'Neill, S.; Davies, E.; Mansfield, D.; Messiou, C.; Hujairi, N.; Nicol, D.; et al. Prostate-specific membrane antigen expression in melanoma metastases. *J. Cutan. Pathol.* **2020**, *47*, 1115–1122. [CrossRef] [PubMed]

13. Verdoliva, V.; Senatore, C.; Polci, M.L.; Rossi, S.; Cordella, M.; Carlucci, G.; Marchetti, P.; Antonini-Cappellini, G.; Facchiano, A.; D'Arcangelo, D.; et al. Differential Denaturation of Serum Proteome Reveals a Significant Amount of Hidden Information in Complex Mixtures of Proteins. *PLoS ONE* **2013**, *8*, e57104. [CrossRef] [PubMed]

14. D'Arcangelo, D.; Scatozza, F.; Giampietri, C.; Marchetti, P.; Facchiano, F.; Facchiano, A. Ion Channel Expression in Human Melanoma Samples: In Silico Identification and Experimental Validation of Molecular Targets. *Cancers* **2019**, *11*, 446. [CrossRef] [PubMed]

15. Biasiotta, A.; D'Arcangelo, D.; Passarelli, F.; Nicodemi, E.M.; Facchiano, A. Ion channels expression and function are strongly modified in solid tumors and vascular malformations. *J. Transl. Med.* **2016**, *14*, 285. [CrossRef] [PubMed]

16. D'Arcangelo, D.; Giampietri, C.; Muscio, M.; Scatozza, F.; Facchiano, F.; Facchiano, A. WIPI1, BAG1, and PEX3 Autophagy-Related Genes Are Relevant Melanoma Markers. *Oxidative Med. Cell. Longev.* **2018**, *2018*, 1–12. [CrossRef] [PubMed]

17. Scatozza, F.; D'Arcangelo, D.; Giampietri, C.; Facchiano, F.; Facchiano, A. Melanogenesis and autophagy in melanoma. *Melanoma Res.* **2020**, *30*, 530–531. [CrossRef]

18. Giampietri, C.; Tomaipitinca, L.; Scatozza, F.; Facchiano, A. Expression of genes related to lipid-handling may underlie the "obesity paradox" in melanoma: A public database-based approach. *JMIR Cancer* **2020**, *6*, e16974. [CrossRef]

19. Vereecken, P.; Cornélis, F.; Van Baren, N.; Vandersleyen, V.; Baurain, J.-F. A Synopsis of Serum Biomarkers in Cutaneous Melanoma Patients. *Dermatol. Res. Pr.* **2012**, *2012*, 1–7. [CrossRef]

20. Steppert, C.; Krugmann, J.; Sterlacci, W. Simultaneous endocrine expression and loss of melanoma markers in malignant melanoma metastases, a retrospective analysis. *Pathol. Oncol. Res.* **2019**, *26*, 1777–1779. [CrossRef]

21. Weinstein, D.; Leininger, J.; Hamby, C.; Safai, B. Diagnostic and Prognostic Biomarkers in Melanoma. *J. Clin. Aesthetic Dermatol.* **2014**, *7*, 13–24.

22. Welinder, C.; Pawłowski, K.; Sugihara, Y.; Yakovleva, M.; Jönsson, G.; Ingvar, C.; Lundgren, L.; Baldetorp, B.; Olsson, H.; Rezeli, M.; et al. A Protein Deep Sequencing Evaluation of Metastatic Melanoma Tissues. *PLoS ONE* **2015**, *10*, e0123661. [CrossRef] [PubMed]

23. Donnelly, D.; Aung, P.P.; Jour, G. The "OMICS" facet of melanoma: Heterogeneity of genomic, proteomic and metabolomic biomarkes. *Semin. Cancer Biol.* **2019**, *59*, 165–174. [CrossRef]

24. Faraone, D.; Aguzzi, M.S.; Toietta, G.; Facchiano, A.M.; Facchiano, F.; Magenta, A.; Martelli, F.; Truffa, S.; Cesareo, E.; Ribatti, D.; et al. Platelet Derived Growth Factor- Receptor alpha strongly inhibits melanoma growth in vitro and in vivo. *Neoplasia* **2009**, *11*, 732–742. [CrossRef] [PubMed]

25. Aguzzi, M.S.; Faraone, D.; De Marchis, F.; Toietta, G.; Ribatti, D.; Parazzoli, A.; Colombo, P.; Capogrossi, M.C.; Facchiano, A. The FGF-2 Derived Peptide FREG Inhibits Melanoma Growth In Vitro And In Vivo". *Mol. Ther.* **2011**, *19*, 266–273.

26. D'Arcangelo, D.; Facchiano, F.; Nassa, G.; Stancato, A.; Antonini, A.; Rossi, S.; Senatore, C.; Cordella, M.; Tabolacci, C.; Salvati, A.; et al. PDGFR-alpha inhibits melanoma growth via CXCL10/IP-10: A multi-omics approach. *Oncotarget* **2016**, *7*, 77257–77275. [CrossRef]

27. Rossi, S.; Cordella, M.; Tabolacci, C.; Nassa, G.; D'Arcangelo, D.; Senatore, C.; Pagnotto, P.; Magliozzi, R.; Salvati, A.; Weisz, A.; et al. TNF-alpha and metalloproteases as key players in melanoma cells aggressiveness. *J. Exp. Clin. Cancer Res.* **2018**, *37*, 1–17. [CrossRef]

28. Behbahani, S.; Bs, S.M.; Ms, J.B.C.; Lambert, W.C.; Schwartz, R.A. Gender differences in cutaneous melanoma: Demographics, prognostic factors, and survival outcomes. *Dermatol. Ther.* **2020**, *5*, e14131. [CrossRef]

29. U.S. Cancer Statistics Data Brief. *Melanoma Incidence and Mortality, United States-2012–2016*; U.S. Cancer Statistics Data Brief: Atlanta, GA, USA, 2019; Volume 9.

30. Joosse, A.; De Vries, E.; Eckel, R.; Nijsten, T.; Eggermont, A.M.; Hölzel, D.; Coebergh, J.W.W.; Engel, J. Gender Differences in Melanoma Survival: Female Patients Have a Decreased Risk of Metastasis. *J. Investig. Dermatol.* **2011**, *131*, 719–726. [CrossRef]

31. El Sharouni, M.-A.; Witkamp, A.J.; Sigurdsson, V.; Van Diest, P.J.; Louwman, M.W.J.; Kukutsch, N. Sex matters: Men with melanoma have a worse prognosis than women. *J. Eur. Acad. Dermatol. Venereol.* **2019**, *33*, 2062–2067. [CrossRef]

32. Enninga, E.A.L.; Moser, J.C.; Weaver, A.L.; Markovic, S.N.; Brewer, J.D.; Leontovich, A.A.; Hieken, T.J.; Shuster, L.; Kottschade, L.A.; Olariu, A.; et al. Survival of cutaneous melanoma based on sex, age, and stage in the United States, 1992–2011. *Cancer Med.* **2017**, *6*, 2203–2212. [CrossRef] [PubMed]

33. Bellenghi, M.; Puglisi, R.; Pontecorvi, G.; De Feo, A.; Carè, A.; Mattia, G. Sex and Gender Disparities in Melanoma. *Cancers* **2020**, *12*, 1819. [CrossRef] [PubMed]

34. Kučera, J.; Strnadová, K.; Dvořánková, B.; Lacina, L.; Krajsová, I.; Štork, J.; Kovářová, H.; Skalníková, H.K.; Vodička, P.; Motlík, J.; et al. Serum proteomic analysis of melanoma patients with immunohistochemical profiling of primary melanomas and cultured cells: Pilot study. *Oncol. Rep.* **2019**, *42*, 1793–1804. [CrossRef] [PubMed]

35. Paganelli, A.; Garbarino, F.; Toto, P.; Di Martino, G.; D'Urbano, M.; Auriemma, M.; Di Giovanni, P.; Panarese, F.; Staniscia, T.; Amerio, P.; et al. Serological landscape of cytokines in cutaneous melanoma. *Cancer Biomark.* **2019**, *26*, 333–342. [CrossRef]

36. Nagarkatti-Gude, N.; Bronkhorst, I.H.G.; Van Duinen, S.G.; Luyten, G.P.M.; Jager, M.J. Cytokines and Chemokines in the Vitreous Fluid of Eyes with Uveal Melanoma. *Investig. Opthalmology Vis. Sci.* **2012**, *53*, 6748–6755. [CrossRef]

37. Reinert, C.P.; Gatidis, S.; Sekler, J.; Dittmann, H.; Pfannenberg, C.; la Fougere, C.; Nikolaou, K.; Forschner, A. Clinical and prognostic value of tumor volumetric parameters in melanoma patients undergoing 18 F-FDG-PET/CT: A comparison with serologic markers of tumor burden and inflammation. *Cancer Imaging* **2020**, *20*, 44. [CrossRef]

38. Nomiyama, H.; Osada, N.; Yoshie, O. Systematic classification of vertebrate chemokines based on conserved synteny and evolutionary history. *Genes Cells* **2013**, *18*, 1–16. [CrossRef]

39. De Marchis, F.; Ribatti, D.; Giampietri, C.; Lentini, A.; Faraone, D.; Scoccianti, M.; Capogrossi, M.C.; Facchiano, A. Platelet-Derived Growth Factor Inhibits Basic Fibroblast Growth Factor Angiogenic Properties in vitro and in vivo, via its alpha receptor. *Blood* **2002**, *99*, 2045–2053. [CrossRef]

40. Faraone, D.; Aguzzi, M.S.; Ragone, G.; Russo, K.; Capogrossi, M.C.; Facchiano, A. Heterodimerization of FGF-receptor 1 and PDGF-receptor-α: A novel mechanism underlying the inhibitory effect of PDGF-BB on FGF-2 in human cells. *Blood* **2006**, *107*, 1896–1902. [CrossRef]

41. Zhang, J.M.; An, J. Cytokines, Inflammation and Pain. *Int. Anesth. Clin.* **2007**, *45*, 27–37. [CrossRef]

42. Spranger, S.; Bao, R.; Gajewski, T.F. Melanoma-intrinsic β-catenin signalling prevents anti-tumour immunity. *Nat. Cell Biol.* **2015**, *523*, 231–235. [CrossRef] [PubMed]

43. Li, Z.; Zhang, C.; Du, J.-X.; Zhao, J.; Shi, M.-T.; Jin, M.-W.; Liu, H. Adipocytes promote tumor progression and induce PD-L1 expression via TNF-α/IL-6 signaling. *Cancer Cell Int.* **2020**, *20*, 1–9. [CrossRef] [PubMed]

44. Seeja, R.D.; Suresh, A. Deep Learning Based Skin Lesion Segmentation and Classification of Melanoma Using Support Vector Machine (SVM). *Asian Pac. J. Cancer Prev.* **2019**, *20*, 1555–1561. [CrossRef]

45. Tiwari, K.A.; Raišutis, R.; Liutkus, J.; Valiukevičienė, S. Diagnostics of Melanocytic Skin Tumours by a Combination of Ultrasonic, Dermatoscopic and Spectrophotometric Image Parameters. *Diagnostics* **2020**, *10*, 632. [CrossRef]

46. Shofty, B.; Artzi, M.; Shtrozberg, S.; Fanizzi, C.; DiMeco, F.; Haim, O.; Hason, S.P.; Ram, Z.; Ben Bashat, D.; Grossman, R. Virtual biopsy using MRI radiomics for prediction of BRAF status in melanoma brain metastasis. *Sci. Rep.* **2020**, *10*, 1–7. [CrossRef]

47. Szyc, Ł.; Hillen, U.; Scharlach, C.; Kauer, F.; Garbe, C. Diagnostic Performance of a Support Vector Machine for Dermatofluoroscopic Melanoma Recognition: The Results of the Retrospective Clinical Study on 214 Pigmented Skin Lesions. *Diagnostics* **2019**, *9*, 103. [CrossRef]

48. Mancuso, F.; Lage, S.; Rasero, J.; Díaz-Ramón, J.L.; Apraiz, A.; Pérez-Yarza, G.; Ezkurra, P.A.; Penas, C.; Sánchez-Diez, A.; García-Vazquez, M.D.; et al. Serum markers improve current prediction of metastasis development in early-stage melanoma patients: A machine learning-based study. *Mol. Oncol.* **2020**, *14*, 1705–1718. [CrossRef]

49. Yurkovetsky, Z.R.; Kirkwood, J.M.; Edington, H.D.; Marrangoni, A.M.; Velikokhatnaya, L.; Winans, M.T.; Gorelik, E.; Lokshin, A.E. Multiplex Analysis of Serum Cytokines in Melanoma Patients Treated with Interferon-2b. *Clin. Cancer Res.* **2007**, *13*, 2422–2428. [CrossRef]

50. Byrum, S.D. Quantitative Proteomics Identifies Activation of Hallmark Pathways of Cancer in Patient Melanoma. *J. Proteom. Bioinform.* **2013**, *6*, 043–050. [CrossRef]

51. Guyon, I.; Weston, J.; Barnhill, S.; Vapnik, V. Gene Selection for Cancer Classification using Support Vector Machines. *Mach. Learn.* **2002**, *46*, 389–422. [CrossRef]

52. Allen, F.; Bobanga, I.D.; Rauhe, P.; Barkauskas, D.S.; Teich, N.; Tong, C.; Myers, J.; Huang, A.Y. CCL3 augments tumor rejection and enhances CD8+ T cell infiltration through NK and CD103+ dendritic cell recruitment via IFNγ. *OncoImmunology* **2018**, *7*, e1393598. [CrossRef] [PubMed]

53. R Foundation for Statistical Computing. R: A Language and Environment for Statistical Computing. R Core Team. 2020. Available online: http://www.r-project.org (accessed on 4 December 2020).

54. Boser, B.E.; Guyon, I.M.; Vapnik, V.N. *A Training Algorithm for Optimal Margin Classifiers*; ACM: New York, NY, USA, 1992; pp. 144–152.

55. Cortes, C.; Vapnik, V.N. Support-Vector Networks. *Mach. Learn.* **1995**, *20*, 273–297. [CrossRef]

56. Little, R.; Rubin, D. *Statistical Analysis with Missing Data*, 2nd ed.; Wiley-Interscience: Hoboken, NJ, USA, 2002.

57. Hsu, C.W.; Chang, C.C.; Lin, C.J. *A Practical Guide to Support Vector Classification*; Bioinformatics 1; Oxford University Press: Oxford, UK, 2010.

58. Luor, D.-C. A comparative assessment of data standardization on support vector machine for classification problems. *Intell. Data Anal.* **2015**, *19*, 529–546. [CrossRef]

Publisher's Note: MDPI stays neutral with regard to jurisdictional claims in published maps and institutional affiliations.

Review

Molecular and Immune Biomarkers for Cutaneous Melanoma: Current Status and Future Prospects

Lorenzo Pilla [1,*], **Andrea Alberti** [2], **Pierluigi Di Mauro** [1], **Maria Gemelli** [1], **Viola Cogliati** [1], **Marina Elena Cazzaniga** [1], **Paolo Bidoli** [1] and **Cristina Maccalli** [3]

[1] Division of Medical Oncology, San Gerardo Hospital, University of Milano-Bicocca School of Medicine, 20900 Monza, Italy; p.dimauro001@unibs.it (P.D.M.); maria.gemelli@asst-monza.it (M.G.); viola.cogliati@unimi.it (V.C.); marina.cazzaniga@asst-monza.it (M.E.C.); p.bidoli@asst-monza.it (P.B.)

[2] Medical Oncology Unit, Department of Medical and Surgical Specialties, Radiological Health Science and Public Health, University of Brescia, ASST Ospedali Civili, 25123 Brescia, Italy; a.alberti015@unibs.it

[3] Laboratory of Immune and Biological Therapy, Research Department, Sidra Medicine, Doha 26999, Qatar; cmaccalli@sidra.org

* Correspondence: l.pilla@asst-monza.it; Tel.: +39-392-9167530

Received: 28 September 2020; Accepted: 11 November 2020; Published: 20 November 2020

Simple Summary: The prognosis and treatment of metastatic melanoma have changed substantially since the advent of target therapy and immune checkpoint inhibitors. Thus, strategies must be developed to identify responder patients, reduce toxicities, and investigate target and immune based therapy ideal sequencing. To this aim, the determinants driving response, resistance, and adverse events, should be defined. In addition, novel oncogenic drivers should be discovered to provide new therapeutic targets. Current methods of detection, prognosis and monitoring of melanoma are based on clinical, morphological and histopathologic characteristics of the tumor. This review provides an update on prognostic and predictive biomarkers with a potential application in melanoma patients' clinical management.

Abstract: Advances in the genomic, molecular and immunological make-up of melanoma allowed the development of novel targeted therapy and of immunotherapy, leading to changes in the paradigm of therapeutic interventions and improvement of patients' overall survival. Nevertheless, the mechanisms regulating either the responsiveness or the resistance of melanoma patients to therapies are still mostly unknown. The development of either the combinations or of the sequential treatment of different agents has been investigated but without a strongly molecularly motivated rationale. The need for robust biomarkers to predict patients' responsiveness to defined therapies and for their stratification is still unmet. Progress in immunological assays and genomic techniques as long as improvement in designing and performing studies monitoring the expression of these markers along with the evolution of the disease allowed to identify candidate biomarkers. However, none of them achieved a definitive role in predicting patients' clinical outcomes. Along this line, the cross-talk of melanoma cells with tumor microenvironment plays an important role in the evolution of the disease and needs to be considered in light of the role of predictive biomarkers. The overview of the relationship between the molecular basis of melanoma and targeted therapies is provided in this review, highlighting the benefit for clinical responses and the limitations. Moreover, the role of different candidate biomarkers is described together with the technical approaches for their identification. The provided evidence shows that progress has been achieved in understanding the molecular basis of melanoma and in designing advanced therapeutic strategies. Nevertheless, the molecular determinants of melanoma and their role as biomarkers predicting patients' responsiveness to therapies warrant further investigation with the vision of developing more effective precision medicine.

Keywords: biomarkers; melanoma; checkpoint inhibitor; PD-1; target therapy

---

1. Introduction

Malignant melanoma is the 5th most common cancer and represents 1.5% of all cancer deaths. However, during the last 20 years, the incidence was increased, with more than 9000 melanoma-related deaths registered in 2018 [1].

The prognosis of melanoma, particularly in the metastatic setting, over the last ten years was significantly improved [2]. This progress is mainly due to the clinical approval of target therapy with BRAF and MEK inhibitors and of immunotherapy with immune checkpoint inhibitors (ICIs), specifically those targeting the Cytotoxic T-Lymphocyte Associated Protein 4 (CTLA-4) and Programmed Death 1 (PD-1)/Ligand-1 (PD-L1) pathways. Different studies demonstrated an enhanced clinical benefit from these treatments in specific patient populations, such as early-stage patients and stage IV patients with limited tumor burden [3,4].

This suggests that in the upcoming years, it will be crucial to identify patients at higher recurrence risk and monitor for early relapse, in order to assign the best preventive approach, sparing patients at very low risk from over-treatment side effects.

A biomarker can be anything from a serum protein, detectable genetic alteration, pathology finding, or imaging finding that helps to predict the presence of disease or guide its therapeutic options. The use of biomarkers has tremendously improved predictivity of response to treatment of some specific melanoma subtypes. Indeed, genetic testing looking for BRAF mutations, has a proven role in the choice of targeted therapy. However, the identification of definitive clinically useful biomarkers for the response to ICIs of melanoma patients is still under investigation. There is no blood test that can be used for diagnosis to detect melanoma recurrence, although lactate dehydrogenase (LDH) and S-100B can be useful for monitoring.

This review examines the past, current, and future role of biomarkers in melanoma detection, treatment selection, and treatment monitoring.

2. Biomarkers in Target Therapy

Melanoma is a highly heterogeneous disease from a genetic point of view. However, the identification of specific subtypes of melanoma based on distinct molecular alterations enables critical information on patient prognosis and potential treatment options [5] (Table 1).

Table 1. Selected Biomarkers in Melanoma-Targeted Therapy.

Gene	Incidence	Comments	References
BRAF	40–60%	Correlated with response to BRAF-targeted therapies. has led to FDA approval of amplification and sequencing technologies, and multiple laboratory tests to assess BRAF mutation status	[3]
NRAS	20%	Correlated with response to MEK inhibitors	[6]
C-KIT	3% of melanoma; 20–30% melanomas arising from (CSD) skin, acral and mucosal sites	KIT-inhibitors have shown activity in patients with specific mutations	[7–11]
GNAQ	80% of uveal melanoma	MEK inhibitors failed to show efficacy in Phase III trials	[12]
NF1	46% of cases with BRAF and NRAS wild type	Early clinical trials	NCT02645149
PTEN	25–30%	Implicated in mechanism of resistance to MAPK inhibition	[13]
CDK2	11%	Early clinical trials	NCT02645149

The identification of the mutations in the V600 codon of BRAF (35–50% of melanomas) and Q61 codons (less frequently, the G12 or G13 codon) of NRAS (10–25% of melanomas) prompted the era of target therapy [13].

Several next-generation sequencing (NGS) studies showed that the mutational rate of melanoma is by far higher than that reported for other solid tumors, probably as a consequence of ultraviolet radiation exposure. However, a large proportion of these mutations are bystanders, not involved in the neoplastic process [13].

Interestingly, those subtypes of melanoma in which UV does not have any involvement in the pathogenesis, display a different pattern of specific mutations. In Uveal melanoma, the most frequent driver mutation is associated with the G protein subunits. In mucosal melanoma, BRAF and NRAS mutations are far less frequent compared to cutaneous melanoma, while mutations in c-KIT are observed in 7–25% of cases [14].

The Cancer Genome Atlas Network has recently provided a potential step ahead in the comprehension of melanoma genesis, proposing a classification of melanoma based on protein, RNA and DNA analysis from over 300 melanoma patients, in addition to BRAF, NRAS, and NF1 mutational profile [15]. This genomic classification provides a step further to identify new subtypes of melanoma which can help to understand different clinical behaviors and resistance to therapy.

2.1. BRAF

BRAF is the most common mutated gene in cutaneous melanoma, ranging from 40–60% of cases, and leading to uncontrolled activation of the mitogen-activated protein kinases (MAPK) pathway. The two most frequent mutations of this gene are V600E and V600K [16,17], and are associated with different factors: BRAF V600E is associated with younger onset age, superficial spreading subtype and skin without chronic sun damage (CSD) (e.g., extremities), while BRAF V600K with older age, and skin with CSD (e.g., head and neck) [18,19]. The development of BRAF inhibitors such as vemurafenib, dabrafenib, and encorafenib dramatically improved the overall response rate (ORR) and the overall survival (OS) in patients with this melanoma subtype. However, despite encouraging clinical results with monotherapy [13], early development of acquired resistance through several mechanisms was noted, such as the upregulation of MEK, ERK or NRAS [20]. Combination therapy with BRAF- and MEK-inhibitors, including trametinib, cobimetinib and binimetinib, reported a prolongation in both the progression-free survival (PFS) and OS compared with single-agent BRAF inhibitors [21–23]. In particular, the newest doublet therapy, encorafenib/binimetinib showed even greater efficacy, improving the PFS from 7 to 15 months and the OS from 16.9 to 33.6 months [23].

Rare BRAF mutations (V600 non-E/K and non-V600) account for 5% of BRAF-mutated melanoma, and their role in tumorigenesis and response to target therapy is still to be elucidated. Despite most of them, such as L597V, K601E, G469A, showed BRAF and MAPK pathway activation, their prognostic and predictive role is uncertain [24]. Among these, the V600R, L597P/Q/R/S, and K601E are the most common rare BRAF mutations.

To date, melanoma with non-V600 BRAF mutations has been mostly excluded from enrollment in clinical studies investigating the efficacy of BRAF and MEK inhibitors. A phase 2 trial, evaluating the activity and the efficacy of trametinib in patients with advanced melanoma not mutated in V600 BRAF (NCT02296112), could help to shed some light on the role of MEK inhibitors in this rare subset of melanoma.

2.2. NRAS

The gene N-RAS encodes for a GTPase, which plays a crucial role in the signal transduction of both the MAPK and phosphatidyl inositol 3 kinases (PI3K) pathways [25]. NRAS mutations account for 20% of cases [26] and are generally mutually exclusive with BRAF mutations, even if there is evidence of cases in which both mutations can co-exist in the same lesions [27]. However, melanoma with N-RAS mutation is frequently detected in CSD skin and associated with the nodular subtype [28,29].

The association between N-RAS alteration and worse survival for metastatic melanoma is not ultimately demonstrated, although some studies reported it to be a negative prognostic factor [26]. Several studies did not show any difference in OS; an association between NRAS mutation and poor survival or a positive correlation among the two factors were observed [26,30–32].

MEK-inhibition has been identified as a potential therapy in this melanoma subtype, since improved response rate and PFS were found upon treatment with binimetinib as compared to chemotherapy [6]. Nonetheless, the clinical risk/benefit ratio was not sufficient to support the approval of this compound for this indication. The combination of MEK inhibitors with PI3K, RAF, and other cell cycle inhibitors is the object of ongoing investigations [33].

2.3. KIT

Among the multiple genes affected by alterations that could potentially lead to melanoma development, mutations in c-KIT are becoming an appealing target for personalized therapy. This proto-oncogene encodes for a receptor tyrosine kinase (RTK), and its mutations were found in other cancer types [34]. Although alterations in KIT were identified in only 3% of all melanomas, they were frequently (28–39%) detected in melanoma arising from CSD skin, acral and mucosal sites [35,36]. Several trials based on targeting with kinase inhibitors c-KIT showed consistent results, obtaining an ORR of 16–29% and a median OS of 12–13 months [7–10]. Other studies investigating nilotinib, a small molecule more potent than imatinib in inhibiting KIT, showed promising clinical activity [11,37,38]. The global, single-arm, phase II TEAM trial showed 26.2% ORR (all partial responses) and OS of 18 months [11]. The kinase inhibitors dasatinib and sunitinib were also tested in this setting, with mixed results [39,40]. Superior response rates in patients harboring KIT mutations in exon 11 or 13 were noted, likely representing driver mutations responsible for melanoma growth.

2.4. Other Single Gene Candidate Predictive Biomarkers

Advances in genomic sequencing technologies, coupled with the development of effective melanoma therapies, have led to the identification of genes bearing specific driver mutations and displaying the role of predictors of response.

NF1 gene encodes for neurofibromin, a protein that negatively regulates the MAPK pathway: it is the third most commonly mutated gene in melanoma, found in 46% of cases with BRAF and NRAS wild type, and often co-mutated with RAS-associated genes [41].

NF1 mutant melanoma failed to show specific clinicopathological characteristics, other than the association with advanced patients' age [15].

Currently, there's no clinical trial investigating NF-1 mutant melanoma, while a combination of MEK and PI3K or mTOR inhibitors showed promising activity in mouse models [42].

A large proportion (85–90%) of uveal melanoma harbors a driver mutation in GNAQ or GNA11: these mutations are mutually exclusive and cause an overamplification of the downstream signaling through the MAPK pathway [43,44]. Therapies specifically targeting this altered gene are not available. Treatment with selumetinib, a competitive MEK-inhibitor, failed to show any improvement in overall survival when either compared to or added to dacarbazine mono-chemotherapy [12,45]. Multiple clinical trials are ongoing to evaluate the clinical efficacy of the inhibition of mitogenic signaling mediated by GNAQ/GNA11. These include PKC inhibitors (NCT01430416, NCT01801358) and MTOR inhibitors (NCT01430416; NCT01801358.)

CDKN2A loss is found in 50% of malignant melanoma [46]. CDKN2A and CDKN2B genes block cell cycle progression through inhibition of CDK4 and retinoblastoma protein. CDK4 mutations are associated with 20% of familial melanoma [47], and it is generally associated with a worse prognosis [48,49].

Alterations of the CDK4/6 signaling pathway is observed in different types of melanoma, concomitantly with NRAS, KRAS or BRAF; thus the genetic profiling of CDK4/6 may provide insights for the targeted treatment of various types of melanoma. Preclinical studies demonstrated that

CDK4/6 inhibitors could overcome the resistance to the treatment with RAS/RAF/MEK/ERK inhibitors and anti-PD-1 immunotherapy [50].

2.5. Resistance to Target Therapy

The combination of BRAF inhibitors and MEK inhibitors showed clinical efficacy and long-lasting disease control in metastatic melanoma whit an OS rate of 30% at five years [22]. However, resistance mechanisms, leading to disease progression, may occur during the treatment, and about 15% of patients are refractory to the treatment [22]. Different mechanisms may be involved: genetic causes with sustained activation of the MAPK pathway, the emergence of alternative pro-oncogenic pathways, epigenetic alterations, and microenvironment modulation [51].

Genetic alterations of genes involved in the MAPK pathway are reported in about 50% of patients progressing to the treatment with BRAFi and MEKi; among these NRAS mutations accounted for 17% and MAP2KI-2 mutations of 15% and 8%, respectively. NRAS point mutations are detected at an early stage of treatment, usually within 12 weeks, while the presence of MAP2KI P124S and P124L in pre-treatment specimens correlated with rapid progression [52,53]. Nevertheless, genetic alterations could not be identified in 40% of patients with disease progression [54]. BRAF V600E/K amplifications have been reported in 8–13% of resistance to BRAFi and could coexist with other genetic alterations as NRAS mutations [52]. Another mechanism of escape is the onset of BRAF alternate splicing variants, which determine aberrant BRAF proteins' dimerization and unable the BRAFi to bind it [54]. Amplifications of cMET and MITF have been associated with resistance to MAPK pathway inhibition: overexpression of these genes enhances transcription factor stimulating melanoma cell growth [55,56].

The occurrence of an alternative pathway(s) is another possible resistance mechanism. Activation of the PI3K-AKT pathway, through the loss of PTEN and BIM down-modulation, prevents melanoma cell apoptosis and stimulates cell growth, thus leading to BRAF inhibitor resistance [57]. Loss of PTEN correlates with worse PFS in patients treated with dabrafenib monotherapy [58]. Clinical trials evaluating the association of PI3K inhibitors with MAPK inhibitors are ongoing [59].

Furthermore, several pieces of evidence showed the reciprocal influence between BRAFi therapy and the immune system. On the one hand, BRAFi therapy modulates the tumor microenvironment through up-regulating the expression of melanoma antigens (e.g., MART, gp100) and peritumoral CD8+ T lymphocytes and reduction of immunosuppression cytokines and Treg [60]. On the other hand, increased expression of PD-L1 and PD-1 in tumor cells, of exhaustion markers (like TIM3 and FOXP3) on T cells, and the reduction of the frequency of CD8+ T cell infiltrate have been reported in histologic samples of patients progressing to BRAF inhibitors [61,62]. These observations were also matched with the detection of increased levels of pro-tumorigenic type 2 tumor-associated macrophages [61,62]. The combination of immune checkpoint inhibitors and target therapy might be a promising strategy to enhance immune responses and prevent immune-mediated resistance; this hypothesis is currently under evaluation in phase III clinical trials.

Finally, given the paucity of therapeutic alternatives for metastatic melanoma patients, understanding the mechanisms of resistance to target therapies remains crucial to implement the efficacy of the current treatments and develop new strategies for a prolonged clinical benefit.

3. Biomarkers in Immunotherapy

Together with target therapy, immunotherapy development, contributed to the rapid and dramatic change in melanoma's therapeutic landscape [63]. Initially, two classes of ICIs were approved: 1. anti-CTLA-4 antibody, and 2. anti-PD-1 [64,65] antibodies, showing, when used in monotherapy, durable benefit in survival, both in BRAF mutated and wild-type patients [64].

In addition, the combination of anti-PD-1 and anti-CTLA-4 antibodies demonstrated a sharp increase in response rate and progression-free survival compared to either drug alone [66]. However, a long term clinical benefit was limited to a fraction of patients, and the significant toxicity reported in the case of combination therapy highlights the importance of the accurate selection of patients [66].

Biomarkers are still an unmet need to predict patients who will most likely benefit from different immunotherapies, those who will experience toxicities, and to identify the mechanisms to overcome the resistance to therapies (Table 2).

Table 2. Selected Biomarkers in Immunotherapy.

Biomarker	Clinical Validation	Tissue for Assessment	Assay	Comments	References
PD-L1	Yes; Phase III Trial	Tumor; TME	IHC	Clinical responses in PD-L1 negative tumors. Variability of the assays	[64,67]
TMB	Yes; Phase III Trial	Blood; TME	NGS; WES	Lack of standardized TMB thresholds. Variability in quantification methods.	[68]
GEP	No; early clinical development	Tumor	IMPRES (RNA-seq)	Costs	[69]
TIL	No; early clinical development	Tumor	IF, IHC	Tumor tissue availability	[70]
Peripheral lymphocytes	No; early clinical development	Blood	IF	Role of T-cell subpopulations in predicticting clinical benefit	[71–73]
Gut Microbiota	No; early clinical development	Oral, gut	PCR; NGS	Inter-patients variability. Role in predicting toxicity	[74,75]

Abbreviations: TME, Tumor; Microenvironment; IHC, Immunohistochemistry; ICIs, immune checkpoint inhibitors; NGS, next-generation sequencing; WES, whole exome sequencing; GEP, gene expression profile; TMB, Tumor Mutational Burden; IMPRES, immune-predictive score; TIL, Tumor-infiltrating lymphocytes; IF, Immunofluorescence; PCR, Polymerase Chain Reaction.

3.1. PD-1/PD-L1 Signaling

PDL-1 expression on tumor cells, assessed by immunohistochemical staining, has been extensively evaluated as a predictor of clinical response to anti-PD-1/PD-L1 therapy.

Several studies, including different tumor histology, revealed a positive correlation between PD-L1 expression and response to ICIs [64,76,77], while others did not report any significant association [78]. These contradictory results might be, in part, attributed to the lack of a clear definition of threshold and significance of PDL-1 positivity. In addition, different PD-1 and-PD-L1 specific antibodies have different complementary diagnostic tests, which also have different thresholds of positivity. In the studies with pembrolizumab, the positivity of tissues for PD-L1 is defined, using the 22C3 antibody (Merck, Co., Inc., Kenilworth, NJ, USA), as at least ≥1% positive tumor cells. For nivolumab, the 28-8 antibodies (Dako, Santa Clara, CA, USA) were used, with a threshold of positivity of at least 5% of tumor cells. This variability was reflected in the results of the clinical trials. The KEYNOTE-006 clinical trial, for instance, showed that the expression of PD-L1 was found in 80.5% of cancer cases [64], while in the Checkmate-067 study, only 23.6% of cases displayed positivity [66].

Moreover, while PD-L1 positivity is associated with the clinical benefit to ICIs, clinical responses could also be observed in patients with PD-L1 negative tumors. Interestingly, in clinical trials combining nivolumab and ipilimumab [66], for PD-L1 positive patients, the PFS was comparable between the two arms of single agents and the combination. In PD-L1 negative patients, the RR, PFS, and OS were significantly higher in the group treated with the combination. Altogether, these evidences highlight that the solely expression of PD-L1 might be inappropriate as predictive biomarker. The potential variability of PD-1 expression between primary and metastatic lesions [67] and its modulation over time, indicating that a single small biopsy not representative of the actual extension of the disease, represents the major limitations for the assessment of PD-L1 as a standalone biomarker [67].

On the other hand, the prognostic value of the PD-L1 expression in patients with metastatic melanoma remains even more controversial, having been associated with better or worst prognosis in different studies [79–83]. Indeed, the results of 2 different studies in the adjuvant setting, one comparing nivolumab to ipilimumab and pembrolizumab to placebo, respectively, showed that recurrence-free survival was longer in patients with PD-L1 positive tumors, irrespectively from the treatment [4,84]. These findings remain hypothesis generating and, further studies are required to validate the role of PD-L1 in defining patients' prognosis and in driving clinical decisions.

3.2. Baseline Immune Factors

The considerable variability in intra-patient and inter-patient immunogenicity of melanoma reflects the evolutionary process between the immune system and melanoma cells, which cannot be limited to PD-1/PD-L1 interaction but must be viewed as an evolutionary process defined by a different phenotypic and functional subset of cells.

Since 2006, when Galon and co-workers demonstrated that the qualitative, the quantitative, and the spatial localization of CD3+ CD45RO+ memory T cells were independent prognostic factors for the survival in colorectal cancer (CRC) [85,86], many efforts have been placed to determine the role of immune infiltrating cells as biomarkers of immunotherapy.

Herbst et al. showed that the density of CD8+ T cells represented a reliable predictive indicator of response to PD-1 inhibition [87]. In particular, in the setting of anti-PD1/PDL1 therapy, where a possible mechanism of action is the revitalization of pre-existing T cell-mediated responses [88], the phenotype of tumor-infiltrating lymphocytes (TILs) could play a central role in predicting patients' clinical outcome. Two different studies documented the association between the presence of effector T cells, assessed by perforin and granzyme levels [89,90], and the anti-PD-1 response was observed, highlighting the importance of the characterization of intratumoral immune cells.

In melanoma patients, PD-1 is expressed by heterogeneous populations of T cells. The use of additional T cell-associated markers, such as CTLA-1, LAG-3, TIGIT, can lead to the discrimination between an effector and exhausted cell populations. Daud et al. [70] found that TILs expressing CTLA-4 and PD-1 represent a subset of exhausted T cells associated with ICI's clinical response.

A similar observation was performed by Huang et al., showing that the phenotype of circulating KI67+ T cells normalized to tumor burden, could be rescued toward effector cells by ICI therapy in association with a favorable clinical outcome of melanoma patients [91]. Maccalli and colleagues characterized more deeply the immunological responses in advanced melanoma patients treated with the combination of chemo-immunotherapy (fotemustine and ipilimumab). The levels of central memory T cells in the peripheral blood, expressing either co-stimulatory and activatory molecules (CD45RA−CD62L+ CCR7+ CD27+ CD28+ BTLA+/PD-1+), were associated with clinical responses [71].

Two different reports from Weide et al. assessed the role of 2 different baseline signatures in the peripheral blood of patients treated with ICIs. In the first one, the improved OS and PFS of metastatic melanoma patients treated with ipilimumab were associated with baseline determinations of high absolute eosinophil counts (AEC), relative lymphocyte counts (RLC), absolute monocyte counts, CD4(+)CD25(+)FoxP3(+) cells, myeloid-derived suppressor cells (MDSC) level, and low baseline LDH [72]. In the second study, the clinical benefit in patients treated with pembrolizumab was associated with low LDH and high relative lymphocyte and eosinophil counts [73].

Little attention has been given to the role of innate immunity in the context of the anti-PD1 blockade. Indeed PD-1 is expressed on natural killer (NK) cells and dendritic cells (DCs). This is particularly relevant in tumors with loss of HLA expression [92], in which tumor rejection is dependent on NK cells, and is enhanced after PD-1 blockade. Indeed, PD-1 is expressed on natural killer (NK) cells and DCs.

PD-1 blockade can also trigger or enhance the secretion of activating cytokine by tumor-infiltrating DCs, such as IL-12, IFN-γ, and CXCL9/CXCL10 [92,93]. The complexity of this information explains, at least in part, that the activity of ICIs is exploited through a series of events, with many variables influencing the impact in a heterogeneous tumor microenvironment (TME), which is rarely identical to itself.

3.3. Tumor Mutational Burden

Tumors with a high frequency of non-synonymous somatic mutations, such as melanoma and lung cancer, can generate an increased number of neoantigens, resulting in 'novel' and highly immunogenic targets for the immune system. Importantly, this altered immunogenic molecular profile can induce potent immune responses and clinically activity of ICIs [94–97].

The tumor mutational burden (TMB) is identified as the number of non-synonymous mutations in the coding area of the tumor genome.

In general, high TMB is associated with better survival [96], almost in most histologies [98]. Snyder et al. evaluated the presence of somatic mutations and neo-antigen load in 64 tumor samples derived from patients with melanoma treated with an anti-CTLA-4 antibody [68,99]. The mutational load was associated with the expression of neo-antigens and a subsequent correlation with the clinical response to ICIs.

Similar observations were obtained in NSCLC [100]; more recently, other studies extended these evidences to tumors with diverse histologies treated with immunotherapy. However, many challenges need to be resolved in order to use the TMB in the clinical practice as a predictive biomarker for immunotherapy. First of all, the lack of agreement on TMB assessment, the multiple platforms use for its assessment (DNA amounts, the extension of genome analyzed), and the different cut-off values render this evaluation contradictory [101,102].

3.4. Tumor Microenvironment

As described in the previous paragraphs, single factors, such as the presence of either TILs or PD-L1 in the tumor, are not sufficient to predict patients' responsivness to immunotherapy. The lack of direct correlation mirrors the complexity of tumor-host interactions in the microenvironment, with the interplay of cancer cells with different subpopulations of immune cells, such as MDSCs, fibroblastsa, and a variety of paracrine signals. The tumor microenvironment evolves depending on both tumor-related factors, such as tumor histology and host factors, shaping the characteristics of immune infiltrate. For instance, a correlation between PD-L1 expression on immune cells (e.g., DCs, macrophages, and T lymphocytes) and response to ICIs was reported in different tumor types, including melanoma [87,103]. Additionally, there is a growing body of evidences that tumor-associated macrophages (TAM) may abrogate anti–PD-1 response in patient cohorts with advanced melanoma through different mechanisms. TAMs might weaken antitumor immune responses, probably by sequestering therapeutic anti-PD-1 antibodies from T cells [104]. Nuebert et al., demonstrated that the secretion of colony-stimulating factor-1(CSF-1) by melanoma cells upon exposure to T cell-derived cytokines recruits TAMs to the tumor site and consequently hampers antitumor immune responses [105]. Similarly, the lack of response to atezolizumab was associated with transforming growth factor β (TGFβ) signaling in tumor-associated fibroblasts, which prevent T-cell differentiation and infiltration into the tumor [106]. In this context, a possible solution to reset antitumor immune responses and to improve clinical outcomes in melanoma patients unresponsive to immunotherapy, could be the combination of immune checkpoint blockade with macrophage elimination. In conclusion, a deeper understanding of non-tumoral cells and their paracrine signaling within the TME could facilitate the discovery of biomarkers associated with the efficacy of for ICIs.

The immune infiltrate represents a multifaceted component of TME. Indeed, different preclinical and pathological studies demonstrated that the development and progression of melanoma are associated with changes in cell metabolism involving a shift from oxidative phosphorylation to aerobic glycolysis [107,108]. In BRAF-mutated melanoma, the constitutive activation of BRAF kinase triggers a metabolic rewiring which involves a cascade of consequences, including HIF-1 activation [109] which is normally induced in response to low oxygen levels and increased expression of glucose transporters and glycolytic enzymes [108] (e.g., SLC7A11). In addition, the oxygen concentration in the micro-environment depends on melanin concentration in the lesion: a retrospective analysis showed that its synthesis is related to worst disease advancement and radiotherapy outcomes [107].

In other preclinical works, it was shown that the inhibition of RAS/RAF pathway results in metabolic reprogramming, which can ultimately determine a decreased level of Reactive Oxygen Species (ROS) and hence a diminished antitumor effect of inhibitors of RAS/RAF pathway causing resistance to BRAF inhibition [110].

While this is still an emerging area of investigation, a growing number of preclinical works support the role of oxidative metabolism in melanoma progression and resistance to target [111] and immunotherapy [112], and a deeper understanding of metabolic alterations of cancer cells would be critical to achieving greater therapeutic success.

3.5. Microbiome

In the past decade, many studies have highlighted the central role of intestinal microbiota in the metabolic process and immunity [113]. Accumulated evidence indicates a crucial contribution of the microbiome also in the disease process, including carcinogenesis [114], in relationship with the establishment of a pro- or antitumor inflammatory milieu. These effects can be exploited both locally and at long distances [115].

For example, in CRC, the levels of *Fusobacterium nucleatum* (Fn) are increased in tumor tissues vs. normal tissues and are also increased in metastatic lesions compared to the primary tumor. Fn might mediate resistance to chemotherapy through a toll like receptor 4 (TLR4) mechanism [116]. In animal models, specific alterations in gut microbiota are associated with spontaneous antitumor immunity and different responses to CTLA-4 and PD-L1 blockade. The effect of immunotherapy with CpG oligodeoxynucleotide and anti-IL-10 mAb was dependent on a functional microbiota.

Moreover, myeloablative regimens increase their effectiveness when they are associated with total body irradiation through augmented levels of endotoxins and, thus, of pro-inflammatory cytokines influencing the microbiome [117].

Two studies showed an association between intestinal microbiota composition and response to ICI therapy in melanoma. Gopalakrishnan et al. found that the clinical responses in 112 melanoma patients treated with anti-PD-1 therapy were associated with microbiota diversity and enriched specific subspecies of the Ruminococcaceae family [74]. This is probably because microbiota diversity is associated with an increased immune infiltrate of CD8+ T cells. Finally, the composition of the gut microbiota may also be associated with ICI-induced side effects. Dubin et al. found that in 34 patients treated with Ipilimumab, the presence of species from the Bacteroidetes phylum was associated with decreased risk of ICI-induced colitis [75].

The study of the human microbiota and its genetic composition (microbiome) is still in its early stages. Microbic taxa that can influence specific metabolic processes, therapeutic response to anti-cancer treatment, and related toxicity remain to be defined; nonetheless, microbiology in precision medicine will play a central role in the upcoming years.

4. Circulating Total DNA

Circulating total DNA (ctDNA) consists of soluble short nucleic fragments (~166 bp) released in the plasma due to cell apoptosis and necrosis [118,119]. Several studies have shown that in cancer patients, the ctDNA carries genetic information specifically present in tumor cells, providing information on cancer cells' clonal heterogeneity and their evolution over time [118,120] (Table 3). Multiple assays can be used with different sensitivity levels to identify alterations in ctDNA. NGS-based methods may detect novel genetic aberrations or multiple co-existing mutations; in contrast, single or multiplexed locus assays identify only one or a few genetic mutations.

Table 3. Circulating nucleic acids or tumor cells as biomarkers for melanoma.

Biomarker	Predictive/Prognostic	Clinical Validation	Assay	Comments	References
ctDNA	Prognostic	Advanced clinical investigation	PCR based BEAMing	ctDNA has the potential to anticipate clinical progression. Need of a specific gene target	[121–123]
ctDNA	Predictive	Advanced clinical investigation	PCR based	Prognostic and predictive to dabrafenib and Trametinib. Limited sensitivity for brain metastasis	[124]
ctDNA	Prognostic	Advanced clinical investigation	PCR based	Prognostic and predictive to target therapy and Immunotherapy	[125,126]
MicroRNAs	Prognostic/Predictive	Pre-clinical	Luciferase assay	miRNAs are more stable compared to ctDNA. Low Tumor specificity	[127,128]
CTC	Prognostic/Predictive	Pre-clinical	PCR based	Lack of standardized technology	[129,130]

Abbreviations: ctDNA, cellular tumor DNA; PCR, Polymerase Chain Reaction; BEAMing beads, emulsion, amplification and magnetics; miRNA, microRNA; CTC, circulating tumor cells.

Beads, emulsion, amplification and magnetics (BEAMing) and droplet digital PCR (ddPCR), two PCR-based techniques, have very high sensitivity [121,131–133], but they are limited by the need for a specific gene target and hence used mostly in wild type melanoma for BRAF, NRAS or c-KIT. In this subpopulation of patients, which accounts for 20% of melanoma patients, TERT promoter or TP53 might identify an additional 15% of the cases [15].

The existing knowledge, derived from ctDNA studies in melanoma, favors the demonstration that ctDNA levels function as a prognostic biomarker. Levels of ctDNA were found to significantly correlate with serological markers of disease burden, like lactate dehydrogenase (LDH), S100 calcium-binding protein B (S100B), and melanoma inhibitory activity (MIA) in melanoma specimen [134]. In addition, baseline ctDNA levels were significantly associated with tumor burden and progression-free survival (PFS) [133–135].

Lee et al. demonstrated that pre-operative ctDNA predicts metastasis-free survival in high-risk stage III melanoma patients undergoing complete lymph nodes dissection, independent of stage III substage136. Detectable ctDNA before complete surgical resection in patients with AJCC stage IIIB/C/D (high-risk stage III) with a BRAF, NRAS, or KIT mutant melanoma is an independent predictor of worse melanoma-specific survival (MSS) in patients receiving no systemic adjuvant therapy [122,136].

Thus, these biomarkers might be a critical tool in determining patients' prognosis and stratifying patients for adjuvant treatment clinical trials. The size of the largest tumor-involved lymph nodes did not correlate with MSS, indicating that the prognostic significance of ctDNA detectability was not solely due to tumor volume [136]. ctDNA collected after surgery (median of two weeks after surgery) was only detectable in 12–36% of high-risk melanoma patients [122,137], but it was predictive of recurrence and survival. Median DFI was four months (95% CI 0.1–1.0) in patients with detectable ctDNA compared with 4.2 years (95% CI 2.5–limit not reached) in those where ctDNA was not detected. Sensitivity for predicting relapse was 18–55% and specificity 95%, with a positive predictive value of 79% and a negative predictive value of 51% [122,123]. The majority of patients with detectable ctDNA relapsed within one year from surgery, suggesting that ctDNA in the plasma can reveal occult metastatic disease that is not evident in radiological imaging [123].

The presence of ctDNA bearing BRAF mutation provided information regarding the responsiveness of malignant melanoma to targeted therapy that complements the usual tissue biopsy results [124,125,138–140]. Not only the presence but also the level of ctDNA has predictive value. In the BREAK-2 study, a phase II trial, aimed at evaluating the safety and clinical activity of the dabrafenib, showed that high basal ctDNA levels correlated with lower overall response rate and lower progression-free survival to targeted therapy [124]. These results were further confirmed in a large study that included the BREAK-3, BREAK-MB, and METRIC clinical trials [141]. Overall, these studies support the predictive value of ctDNA for the response to targeted therapy in melanoma patients.

Similar results were also observed with immunotherapy. Gray et al. showed that baseline ctDNA levels were predictive of clinical response and long term clinical benefit to anti-CTLA-4 antibodies [125]. More recently, Lee et al. reported that also in the setting of anti-PD-1 therapy, ctDNA levels at baseline provide an accurate prediction of tumor response, progression-free survival, and overall survival [139]. High levels of ctDNA can precede radiological evidence of disease progression and acquired resistance to targeted therapy. Finally, two recent studies highlighted ctDNA quantification as a suitable complementary modality to functional imaging for real-time monitoring of tumor burden [126]. Imaging integration could be useful, in particular, to differentiate pseudo-progression from real progression [136].

However, ctDNA is not able to detect or monitor intracranial disease activity [126]. The blood-brain barrier may restrict the release of ctDNA into the circulation. This has been shown in patients with brain metastases who had undetectable peripheral blood circulating ctDNA although the ctDNA was detectable in cerebral spinal fluid [142,143]. Finally, ctDNA might be a potential tool to monitor disease-clonal evolution. Anti-BRAF and anti-MEK therapies lead melanoma cells to a positive clonal selection, driving acquired mutation resistance. Studies have demonstrated that the identification of specific mutations of resistance, such as NRAS at codon 61 (p.Q61K/R), is associated with resistance to several drugs [125,140]. In this setting, an early therapy switches to immunotherapy, before an uncontrolled increase in tumor burden, might increase the subsequent treatment response rate. Previous studies have shown that a low tumor burden correlates with response to immunotherapy.

4.1. Micro RNAs

MicroRNAs (miRNAs) are short (20–200 nucleotides) non-coding RNA molecules that regulate and modulate gene transcription processes and epigenetic processes involved in cell proliferation, differentiation, and apoptosis. miRNAs are secreted by cells into the circulation, but compared to ctDNA, they are more stable [144,145].

miRNAs have been implicated in the regulation of tumor development, progression, and metastasis, and as such, have been proposed as potential cancer biomarkers [146,147]. The expression of miRNAs has shown diagnostic, prognostic, and predictive value in melanoma [127]. Elevated levels of miRNA-221 have been observed in melanoma compared to healthy controls, and their levels correlate with the stage of the disease [148]. However, the diagnostic accuracy of miRNA can be superior when a panel of miRNAs is evaluated [128]. The major limitation of the role of miRNAs as prognostic/predictive tool is the lack of tumor-specificity [127]. Behind these favorable results, further studies are warranted to confirm the predictive and prognostic value of miRNAs in melanoma.

4.2. Circulating Tumor Cells

The detection of circulating tumor cells (CTC) in patients with melanoma is challenging because of the low concentration of these cells in peripheral blood and the lack of common CTC markers, as in the case of epithelial cancers (e.g., EpCAM) [149]. In addition, melanoma CTCs represent highly heterogeneous cell subpopulations [150,151], and the simultaneous usage of multiple markers is required for the isolation of CTCs.

The variety of techniques used to identify CTCs [152–154] resulted in high variability of information that might limit this methodology's current clinical application.

In some studies, monitoring the levels of CTCs before and during melanoma treatment has been shown to be informative with respect to prognosis and therapy response in melanoma [129,155].

Using real time-PCR to detect transcripts in blood, Reid and colleagues showed that in 230 patients, the presence of two transcripts (MLANA and ABCB5) was associated with disease recurrence and the expression of one other (MCAM) was significantly more common in patients with poor treatment outcomes [130]. Also, the detection of multiple melanoma markers in patients' blood was associated with disease stage [156], disease-free survival (DFS), and overall survival (OS) [129,157].

Despite some encouraging results, CTCs clinical use as a reliable biomarker will remain limited because of the uncertain biology of these cells and the lack of standardized technology.

5. Conclusions

In the last decade, an extraordinary leap forward in the treatment of melanoma occurred, taking advantage of the advent of targeted and immunotherapies. Hence, from now on, preclinical and clinical investigations will have to face new emerging needs.

While these approaches have provided a radical improvement in RR, DFS, and OS, although a relevant proportion of the patients experienced limited clinical benefits. The "classical" clinical-histological classification of T-, N- and M-stage might no longer be appropriate in the light of new therapeutic approaches. The availability of predictive biomarkers has the potential to develop more personalized therapeutic approaches.

For oncogene-addicted cancers, such as BRAF+ melanoma, the presence of BRAF V600E or V600K mutation represents an effective predictive tool for the choice of specific target therapies.

However, instead, for others, such as NRAS or c-KIT mutated melanoma, the mutation's predictive power is not sufficient to identify true responders' subpopulations.

Immunotherapy represents the only therapeutic option for wild-type melanomas and an alternative option for BRAF mutated melanoma. However, only a fraction of these patients can show clinical benefit. In this setting, despite numerous candidate biomarkers have demonstrated a correlation with clinical response, the composite nature of the factors involved in treatment efficacy requires alternative approaches for proper patient selection.

PD-L1 expression by IHC is not sufficient to stratify or to withhold patients from potentially beneficial therapy. Other characteristics may then help to select patients for ICIs treatment. These include genomic features, such as TMB, immune infiltrate, lymphocytes subpopulations, and also the composition of gut microbiota.

The combinatorial use of these parameters might improve to define the dynamic nature of tumor/host interactions and might ultimately improve our capacity to predict response to immunotherapy.

Another unmet need in the management of melanoma patients is to develop biomarkers that predict toxicity, mostly in the case of treatments of similar efficacy or when there is the need to identify a subgroup of patients able to tolerate significant toxicities as for Nivolumab and Ipilimumab combination [158].

Although many patients experience an initial clinical benefit to anti-melanoma therapies, acquired resistance ultimately develops. In this setting, biomarkers may provide valuable information to develop a more personalized approach. On-treatment biomarkers could provide information on early signs of resistance and allow a correct and efficient sequencing of target and immune therapy or identify patients who most likely will experience a long-term benefit (Figure 1).

To overcome the aforementioned hurdles, the clinical validation and application of biomarkers need to integrate a vast amount of information into a clinically applicable setting. This will imply systematic specimen collection and handling, the use of standardized assays available in a large number of laboratories, the development of appropriate clinical endpoints for determining early efficacy, and clinical development, integrating existing preclinical work with different platforms in development and appropriate clinical tools able to give specific answers.

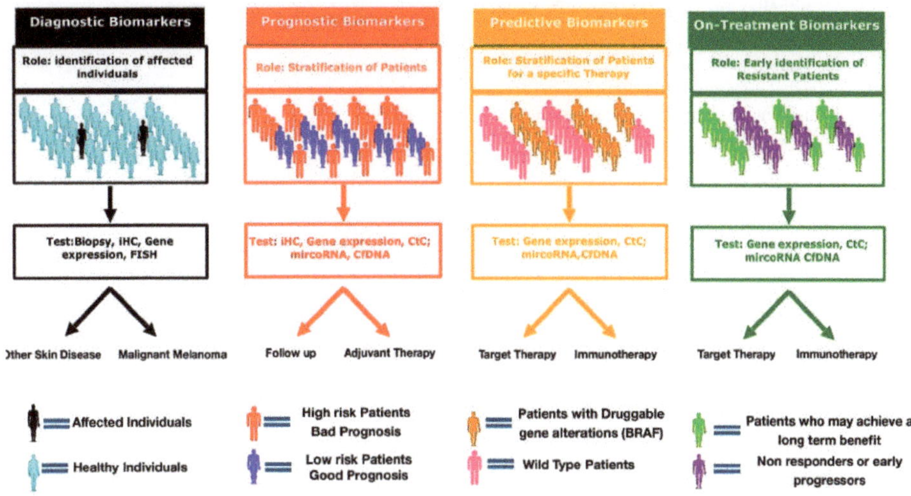

Figure 1. Clinical applications of cancer biomarkers. Genetic, protein and cellular components can serve as diagnostic, prognostic, predictive and/or on-treatment biomarkers. diagnostic biomarker are used identify and detect the presence of cancer in individuals. Prognostic biomarkers provide information on the risk of recurrence and expected outcomes. Predictive biomarkers forecast the potential benefit of a specific treatment. On-treatment biomarker help to identify early progressors from long responders.

Funding: The APC was funded by A&Q Polo per la qualificazione del Sistema Agro-Industriale.

Conflicts of Interest: The authors declare no conflict of interest

Abbreviations

Ab	Antibody
AEC	Absolute eosinophil count
AMC	Absolute monocyte count
AJCC	American Joint Committee for Cancer
AKT	Protein Kinase B
BEAMing	Beads, emulsion, amplification and magnetics
BIM	Bcl-2-like protein 11
BTLA	B-T lymphocyte attenuator
BRAF	v-Raf murine sarcoma viral oncogene homolog B1
CCR	Cytokeratin receptor
CDKN	Cyclin-Dependent Kinase Inhibitor
CDK4	Cyclin-Dependent Kinase 4
CpG	Cytosine linked to a guanine by a phosphate bond
CSD	Chronic sun damage
CSF-1	Colony-Stimulating Factor-1
CTC	Circulating tumor cells
ctDNA	Cellular tumor DNA
CTLA-4	Cyotoxic T-Lymphocyte associated protein 4
CXCL	Chemokine (C-X-C motif) ligand
DC	Dendritic cells
DFI	Disease-free survival
DNA	Desosiiribonucleic acid
ddPCR	Digital droplets polymerase chain reaction

EpCAM	Epithelial cell adhesion
Fn	*Fusobacterium nucleatum*
FOXP3	Forkhead box P3
GNAQ	G protein subunit alpha q
GNA11	Guanine nucleotide-binding protein subunit alpha-11
HLA	Human leukocyte antigen
ICIs	Immune checkpoint inhibitors
IFN	Interferon
IHC	Immunohistochemistry
IL-12	Interleukin
KIT	Proto-oncogene receptor tyrosine kinase
LAG-3	Lymphocyte-activation gene 3
LDH	Lactate dehydrogenase
MAPK	Mitogen-activated protein kinases
MDSC	Myeloid-derived suppressor cell
MET	MET proto-oncogene
MIA	Melanoma inhibitory activity
miRNA	Microrna
MITF	Melanocyte inducing transcription factor
MSS	Melanoma-specific survival
mTOR	Mammalian target of rapamycin
NF1	Neurofibromin 1
NGS	Next-generation sequencing
NK	Natural killer
NRAS	Neuroblastoma RAS Viral Oncogene Homolog
NSCLC	Non-small cell lung cancer
ORR	Overall response rate
OS	Overall survival
PD-L1	Programmed Death—Ligand 1
PCR	Polymerase chain reaction
PFS	Progression Free Survival
PI3K	Phosphatidyl inositol 3 kinases
PKC	Protein kinase c
PTEN	Phosphatase and TENsin homolog
RLC	Relative lymphocyte count
RNA	Ribonucleic acid
RTK	Receptor tyrosine kinase
TAM	Tumor-Associated Macrophages
TGFβ	Transforming Growth factor β
TIGIT	T-cell immunoreceptor with immunoglobulin and ITIM domains
TIL	Tumor-infiltrating lymphocytes
TLR	Toll-like receptor
TMB	Tumor mutational burden
TME	Tumor Microenvironment
Treg	Regulatory T-cell
WES	Whole exome sequencing
Bi	Bliography

References

1. Siegel, R.L.; Mph, K.D.M.; Jemal, A. Cancer statistics, 2017. *CA Cancer J. Clin.* **2017**, *67*, 7–30. [CrossRef] [PubMed]
2. Ascierto, P.A.; Flaherty, K.; Goff, S. Emerging strategies in systemic therapy for the treatment of melanoma. *Am. Soc. Clin. Oncol. Educ. B* **2018**, *38*, 751–758. [CrossRef] [PubMed]
3. Long, G.V.; Hauschild, A.; Santinami, M.; Atkinson, V.; Mandalà, M.; Chiarion-Sileni, V.; Larkin, J.; Nyakas, M.; Dutriaux, C.; Haydon, A.; et al. Adjuvant Dabrafenib plus Trametinib in stage IIIBRAF-mutated melanoma. *N. Engl. J. Med.* **2017**, *377*, 1813–1823. [CrossRef] [PubMed]
4. Weber, J.; Mandalà, M.; Del Vecchio, M.; Gogas, H.; Arance, A.M.; Cowey, C.L.; Dalle, S.; Schenker, M.; Chiarion-Sileni, V.; Marquez-Rodas, I.; et al. Adjuvant Nivolumab vs. Ipilimumab in resected stage III or IV melanoma. *N. Engl. J. Med.* **2017**, *377*, 1824–1835. [CrossRef] [PubMed]
5. Curtin, J.A.; Fridlyand, J.; Kageshita, T.; Patel, H.N.; Busam, K.J.; Kutzner, H.; Cho, K.-H.; Aiba, S.; Bröcker, E.-B.; LeBoit, P.E.; et al. Distinct sets of genetic alterations in melanoma. *N. Engl. J. Med.* **2005**, *353*, 2135–2147. [CrossRef] [PubMed]
6. Dummer, R.; Schadendorf, D.; Ascierto, P.A.; Arance, A.; Dutriaux, C.; Di Giacomo, A.M.; Rutkowski, P.; Del Vecchio, M.; Gutzmer, R.; Mandalà, M.; et al. Binimetinib versus dacarbazine in patients with advanced NRAS-mutant melanoma (NEMO): A multicentre, open-label, randomised, phase 3 trial. *Lancet Oncol.* **2017**, *18*, 435–445. [CrossRef]
7. Carvajal, R.D.; Antonescu, C.R.; Wolchok, J.D.; Chapman, P.B.; Roman, R.-A.; Teitcher, J.; Panageas, K.S.; Busam, K.J.; Chmielowski, B.; Lutzky, J.; et al. KIT as a therapeutic target in metastatic melanoma. *JAMA* **2011**, *305*, 2327–2334. [CrossRef]
8. Guo, J.; Si, L.; Kong, Y.; Flaherty, K.T.; Xu, X.; Zhu, Y.; Corless, C.L.; Li, L.; Li, H.; Sheng, X.; et al. Phase II, open-label, single-arm trial of imatinib mesylate in patients with metastatic melanoma harboring C-kit mutation or amplification. *J. Clin. Oncol.* **2011**, *29*, 2904–2909. [CrossRef]
9. Hodi, F.S.; Corless, C.L.; Giobbie-Hurder, A.; Fletcher, J.A.; Zhu, M.; Marino-Enriquez, A.; Friedlander, P.; Gonzalez, R.; Weber, J.S.; Gajewski, T.F.; et al. Imatinib for melanomas harboring mutationally activated or amplified KIT arising on mucosal, acral, and chronically sun-damaged skin. *J. Clin. Oncol.* **2013**, *31*, 3182–3190. [CrossRef]
10. Wei, X.; Mao, L.; Chi, Z.; Sheng, X.; Cui, C.; Kong, Y.; Dai, J.; Wang, X.; Li, S.; Tang, B.; et al. Efficacy evaluation of Imatinib for the treatment of melanoma: Evidence from a retrospective study. *Oncol. Res.* **2019**, *27*, 495–501. [CrossRef]
11. Guo, J.; Carvajal, R.D.; Dummer, R.; Hauschild, A.; Daud, A.; Bastian, B.C.; Markovic, S.N.; Queirolo, P.; Arance, A.; Berking, C.; et al. Efficacy and safety of nilotinib in patients with KIT-mutated metastatic or inoperable melanoma: Final results from the global, single-arm, phase II TEAM trial. *Ann. Oncol.* **2017**, *28*, 1380–1387. [CrossRef] [PubMed]
12. Carvajal, R.D.; Sosman, J.A.; Quevedo, J.F.; Milhem, M.M.; Joshua, A.M.; Kudchadkar, R.R.; Linette, G.P.; Gajewski, T.F.; Lutzky, J.; Lawson, D.H.; et al. Effect of selumetinib vs chemotherapy on progression-free survival in uveal melanoma: A randomized clinical trial. *JAMA* **2014**, *311*, 2397–2405. [CrossRef] [PubMed]
13. Shtivelman, E.; Davies, M.A.; Hwu, P.; Yang, J.; Lotem, M.; Oren, M.; Flaherty, K.T.; Fisher, D.E. Pathways and therapeutic targets in melanoma. *Oncotarget* **2014**, *5*, 1701–1752. [CrossRef] [PubMed]
14. Furney, S.J.; Turajlic, S.; Stamp, G.; Nohadani, M.; Carlisle, A.; Thomas, J.M.; Hayes, A.; Strauss, D.; Gore, M.; Oord, J.V.D.; et al. Genome sequencing of mucosal melanomas reveals that they are driven by distinct mechanisms from cutaneous melanoma. *J. Pathol.* **2013**, *230*, 261–269. [CrossRef] [PubMed]
15. Akbani, R.; Akdemir, K.C.; Aksoy, B.A.; Albert, M.; Ally, A.; Amin, S.B.; Arachchi, H.M.; Arora, A.; Auman, J.T.; Ayala, B.; et al. Genomic classification of cutaneous melanoma. *Cell* **2015**, *161*, 1681–1696. [CrossRef]
16. Long, G.V.; Menzies, A.M.; Nagrial, A.M.; Haydu, L.E.; Hamilton, A.L.; Mann, G.J.; Hughes, T.M.; Thompson, J.F.; Scolyer, R.A.; Kefford, R.F. Prognostic and clinicopathologic associations of oncogenic BRAF in metastatic melanoma. *J. Clin. Oncol.* **2011**, *29*, 1239–1246. [CrossRef]
17. Davies, H.N.; Bignell, G.R.; Cox, C.E.; Stephens, P.J.; Edkins, S.; Clegg, S.; Teague, J.W.; Woffendin, H.; Garnett, M.J.; Bottomley, W.; et al. Mutations of the BRAF gene in human cancer. *Nat. Cell Biol.* **2002**, *417*, 949–954. [CrossRef]

18. Kim, S.; Hahn, H.J.; Lee, Y.W.; Choe, Y.B.; Ahn, K.J.; Kim, S.-N. Metaanalysis of BRAF mutations and clinicopathologic characteristics in primary melanoma. *J. Am. Acad. Derm.* **2015**, *72*, 1036–1046. [CrossRef]

19. Menzies, A.M.; Haydu, L.E.; Visintin, L.; Carlino, M.S.; Howle, J.R.; Thompson, J.F.; Kefford, R.F.; Scolyer, R.A.; Long, G.V. Distinguishing clinicopathologic features of patients with V600E and V600K BRAF-mutant metastatic melanoma. *Clin. Cancer Res.* **2012**, *18*, 3242–3249. [CrossRef]

20. Nazarian, R.; Shi, H.; Wang, Q.; Kong, X.; Koya, R.C.; Lee, H.; Chen, Z.; Lee, M.-K.; Attar, N.; Sazegar, H.; et al. Melanomas acquire resistance to B-RAF(V600E) inhibition by RTK or N-RAS upregulation. *Nat. Cell Biol.* **2010**, *468*, 973–977. [CrossRef]

21. Ascierto, P.A.; McArthur, G.A.; Dréno, B.; Atkinson, V.; Liszkay, G.; Di Giacomo, A.M.; Mandalà, M.; Demidov, L.; Stroyakovskiy, D.; Thomas, L.; et al. Cobimetinib combined with vemurafenib in advanced BRAFV600-mutant melanoma (coBRIM): Updated efficacy results from a randomised, double-blind, phase 3 trial. *Lancet Oncol.* **2016**, *17*, 1248–1260. [CrossRef]

22. Robert, C.; Grob, J.J.; Stroyakovskiy, D.; Karaszewska, B.; Hauschild, A.; Levchenko, E.; Sileni, V.C.; Schachter, J.; Garbe, C.; Bondarenko, I.; et al. Five-Year Outcomes Dabrafenib Plus Trametinib Metastatic Melanoma. *N. Engl. J. Med.* **2019**, *381*, 626–636. [CrossRef] [PubMed]

23. Dummer, R.; Ascierto, P.A.; Gogas, H.J.; Arance, A.; Mandala, M.; Liszkay, G.; Garbe, C.; Schadendorf, D.; Krajsova, I.; Gutzmer, R.; et al. Overall survival in patients with BRAF-mutant melanoma receiving encorafenib plus binimetinib versus vemurafenib or encorafenib (COLUMBUS): A multicentre, open-label, randomised, phase 3 trial. *Lancet Oncol.* **2018**, *19*, 1315–1327. [CrossRef]

24. Wan, P.T.; Garnett, M.J.; Roe, S.M.; Lee, S.; Niculescu-Duvaz, D.; Good, V.M.; Project, C.G.; Jones, C.M.; Marshall, C.J.; Springer, C.J.; et al. Mechanical acts RAF-Erk signal pathway by oncological mutations B-Raf. *Cell* **2004**, *116*, 855–867. [CrossRef]

25. Hall, A.; Marshall, C.J.; Spurr, N.K.; Weiss, R.A. Identification of transforming gene in two human sarcoma cell lines as a new member of the ras gene family located on chromosome 1. *Nat. Cell Biol.* **1983**, *303*, 396–400. [CrossRef]

26. Jakob, J.A.; Bassett, R.L.; Ng, C.S.; Curry, J.L.; Joseph, R.W.; Alvarado, G.C.; Apn, M.L.R.; Richard, J.; Gershenwald, J.E.; Kim, K.B.; et al. NRAS mutation status is an independent prognostic factor in metastatic melanoma. *Cancer* **2012**, *118*, 4014–4023. [CrossRef]

27. Chiappetta, C.; Proietti, I.; Soccodato, V.; Puggioni, C.; Zaralli, R.; Pacini, L.; Porta, N.; Skroza, N.; Petrozza, V.; Potenza, C.; et al. BRAF and NRAS mutations are heterogeneous and not mutually exclusive in nodular melanoma. *Appl. Immunohistochem. Mol. Morphol.* **2015**, *23*, 172–177. [CrossRef]

28. Lee, J.-H.; Choi, J.-W.; Kim, Y.-S. Frequencies of BRAF and NRAS mutations are different in histological types and sites of origin of cutaneous melanoma: A meta-analysis. *Br. J. Dermatol.* **2011**, *164*, 776–784. [CrossRef]

29. Lattanzi, M.; Lee, Y.; Simpson, D.; Moran, U.; Darvishian, F.; Kim, R.H.; Hernando, E.; Polsky, D.; Hanniford, D.; Shapiro, R.; et al. Primary melanoma histologic subtype: Impact on survival and response to therapy. *J. Natl. Cancer Inst.* **2018**, *111*, 180–188. [CrossRef]

30. Ugurel, S.; Thirumaran, R.K.; Bloethner, S.; Gast, A.; Sucker, A.; Mueller-Berghaus, J.; Rittgen, W.; Hemminki, K.; Becker, J.C.; Kumar, R.; et al. B-RAF and N-RAS mutations are preserved during short time in vitro propagation and differentially impact prognosis. *PLoS ONE* **2007**, *2*, e236. [CrossRef]

31. Heppt, M.V.; Siepmann, T.; Engel, J.; Schubert-Fritschle, G.; Eckel, R.; Mirlach, L.; Kirchner, T.; Jung, A.; Gesierich, A.; Ruzicka, T.; et al. Prognostic significance of BRAF and NRAS mutations in melanoma: A German study from routine care. *BMC Cancer* **2017**, *17*, 1–12. [CrossRef] [PubMed]

32. Thomas, N.E.; Edmiston, S.N.; Alexander, A.; Groben, P.A.; Parrish, E.; Kricker, A.; Armstrong, B.K.; Anton-Culver, H.; Gruber, S.B.; From, L.; et al. Association between NRAS and BRAF mutational status and melanoma-specific survival among patients with higher-risk primary melanoma. *JAMA Oncol.* **2015**, *1*, 359–368. [CrossRef] [PubMed]

33. Vu, H.L.; Aplin, A.E. Targeting mutant NRAS signaling pathways in melanoma. *Pharm. Res.* **2016**, *107*, 111–116. [CrossRef] [PubMed]

34. Pham, D.; Daniel, M.; Guhan, S.M.; Tsao, H. KIT and melanoma: Biological insights and clinical implications. *Yonsei Med. J.* **2020**, *61*, 562–571. [CrossRef]

35. Curtin, J.A.; Busam, K.; Pinkel, D.; Bastian, B.C. Somatic activation of KIT in distinct subtypes of melanoma. *J. Clin. Oncol.* **2006**, *24*, 4340–4346. [CrossRef]

36. Beadling, C.; Jacobson-Dunlop, E.; Hodi, F.S.; Le, C.; Warrick, A.; Patterson, J.; Town, A.; Harlow, A.; Cruz, F.; Azar, S.; et al. KIT gene mutations and copy number in melanoma subtypes. *Clin. Cancer Res.* **2008**, *14*, 6821–6828. [CrossRef]

37. Lee, S.J.; Kim, T.M.; Kim, Y.J.; Jang, K.; Lee, H.J.; Lee, S.N.; Ahn, M.S.; Hwang, I.G.; Lee, S.; Lee, M.; et al. Phase II trial of Nilotinib in patients with metastatic malignant melanoma harboring KIT gene aberration: A multicenter trial of Korean cancer study group (UN10-06). *Oncology* **2015**, *20*, 1312–1319. [CrossRef]

38. Delyon, J.; Chevret, S.; Jouary, T.; Dalac, S.; Dalle, S.; Guillot, B.; Arnault, J.-P.; Avril, M.-F.; Bedane, C.; Bens, G.; et al. STAT3 mediates Nilotinib response in KIT-altered melanoma: A phase II multicenter trial of the French skin cancer network. *J. Investig. Dermatol.* **2018**, *138*, 58–67. [CrossRef]

39. Kluger, H.M.; Dudek, A.Z.; McCann, C.; Ritacco, J.; Southard, N.; Jilaveanu, L.B.; Molinaro, A.; Sznol, M.; Dudek, A.Z.; Jilaveanu, L.B. A phase 2 trial of dasatinib in advanced melanoma. *Cancer* **2010**, *117*, 2202–2208. [CrossRef]

40. DeCoster, L.; Broek, I.V.; Neyns, B.; Majois, F.; Baurain, J.F.; Rottey, S.; Rorive, A.; Anckaert, E.; De Mey, J.; De Brakeleer, S.; et al. Biomarker analysis in a phase II Study of Sunitinib in patients with advanced melanoma. *Anticancer Res.* **2015**, *35*, 6893–6899.

41. Krauthammer, M.; Kong, Y.; Bacchiocchi, A.; Evans, P.; Pornputtapong, N.; Wu, C.; McCusker, J.P.; Ma, S.; Cheng, E.; Straub, R.; et al. Exome sequencing identifies recurrent mutations in NF1 and RASopathy genes in sun-exposed melanomas. *Nat. Genet.* **2015**, *47*, 996–1002. [CrossRef]

42. Whittaker, S.R.; Theurillat, J.-P.; Van Allen, E.; Wagle, N.; Hsiao, J.; Cowley, G.S.; Schadendorf, D.; Root, D.E.; Garraway, L.A. A genome-scale RNA interference screen implicates NF1 loss in resistance to RAF inhibition. *Cancer Discov.* **2013**, *3*, 350–362. [CrossRef] [PubMed]

43. Kivela, T.; Simpson, E.R.; Grossniklaus, H.E.; Jager, M.J.; Singh, A.D.; Caminal, J.M.; Pavlick, A.C.; Kujala, E.; Coupland, S.E.; Finger, P. Uveal melanoma. In *AJCC Cancer Staging Manual*; Springer: New York, NY, USA, 2017; pp. 805–817.

44. Bol, K.F.; Donia, M.; Heegaard, S.; Kiilgaard, J.F.; Svane, I.M. Genetic biomarkers in melanoma of the ocular region: What the medical oncologist should know. *Int. J. Mol. Sci.* **2020**, *21*, 5231. [CrossRef] [PubMed]

45. Carvajal, R.D.; Piperno-Neumann, S.; Kapiteijn, E.; Chapman, P.B.; Frank, S.; Joshua, A.M.; Piulats, J.M.; Wolter, P.; Cocquyt, V.; Chmielowski, B.; et al. Selumetinib in combination with dacarbazine in patients with metastatic uveal melanoma: A phase III, multicenter, randomized trial (SUMIT). *J. Clin. Oncol.* **2018**, *36*, 1232–1239. [CrossRef] [PubMed]

46. Hussussian, C.J.; Struewing, J.P.; Goldstein, A.M.; Higgins, P.A.T.; Ally, D.S.; Sheahan, M.D.; Clark, W.H.; Tucker, M.A.; Dracopoli, N.C. Germline p16 mutations in familial melanoma. *Nat. Genet.* **1994**, *8*, 15–21. [CrossRef] [PubMed]

47. Bishop, D.T. Geographical Variation in the Penetrance of CDKN2A Mutations for Melanoma. *J. Natl. Cancer Inst.* **2002**, *94*, 894–903. [CrossRef] [PubMed]

48. Conway, C.; Beswick, S.; Elliott, F.; Chang, Y.-M.; Randerson-Moor, J.; Harland, M.; Affleck, P.; Marsden, J.; Sanders, D.S.; Boon, A.; et al. Deletion at chromosome arm 9p in relation to BRAF/NRAS mutations and prognostic significance for primary melanoma. *Genes Chromosom. Cancer* **2010**, *49*, 425–438. [CrossRef]

49. Cachia, A.R.; Indsto, J.O.; McLaren, K.M.; Mann, G.J.; Arends, M.J. CDKN2A mutation and deletion status in thin and thick primary melanoma. *Clin. Cancer Res.* **2000**, *6*, 3511–3515.

50. Guo, L.; Qi, J.; Wang, H.; Chen, Y.; Liu, Y. Getting under the skin: The role of CDK4/6 in melanomas. *Eur. J. Med. Chem.* **2020**, *204*, 112531. [CrossRef]

51. Buchbinder, E.I.; Flaherty, K.T. Biomarkers in melanoma: Lessons from translational medicine. *Trends Cancer* **2016**, *2*, 305–312. [CrossRef]

52. Van Allen, E.M.; Wagle, N.; Sucker, A.; Treacy, D.J.; Johannessen, C.M.; Goetz, E.M. The genetic landscape of clinical resistance to RAF inhibition in metastatic melanoma. *Cancer Discov.* **2014**, *4*, 94–109. [CrossRef] [PubMed]

53. Johnson, D.; Menzies, A.M.; Zimmer, L.; Eroglu, Z.; Ye, F.; Zhao, S.; Rizos, H.; Sucker, A.; Scolyer, R.A.; Gutzmer, R.; et al. Acquired BRAF inhibitor resistance: A multicenter meta-analysis of the spectrum and frequencies, clinical behaviour, and phenotypic associations of resistance mechanisms. *Eur. J. Cancer* **2015**, *51*, 2792–2799. [CrossRef] [PubMed]

54. Poulikakos, P.I.; Persaud, Y.; Janakiraman, M.; Kong, X.; Ng, C.; Moriceau, G.; Shi, H.; Atefi, M.; Titz, B.; Gabay, M.T.; et al. RAF inhibitor resistance is mediated by dimerization of aberrantly spliced BRAF(V600E). *Nat. Cell Biol.* **2011**, *480*, 387–390. [CrossRef] [PubMed]

55. Hugo, W.; Shi, H.; Sun, L.; Piva, M.; Song, C.; Kong, X.; Moriceau, G.; Hong, A.; Dahlman, K.B.; Johnson, D.B.; et al. Non-genomic and immune evolution of melanoma acquiring MAPKi resistance. *Cell* **2015**, *162*, 1271–1285. [CrossRef]

56. Garraway, L.A.; Widlund, H.R.; Rubin, M.A.; Getz, G.; Berger, A.J.; Ramaswamy, S.; Beroukhim, R.; Milner, D.A.; Granter, S.R.; Du, J.; et al. Integrative genomic analyses identify MITF as a lineage survival oncogene amplified in malignant melanoma. *Nat. Cell Biol.* **2005**, *436*, 117–122. [CrossRef]

57. Paraiso, K.H.T.; Xiang, Y.; Rebecca, V.W.; Abel, E.V.; Chen, Y.A.; Munko, A.C.; Wood, E.; Fedorenko, I.V.; Sondak, V.K.; Anderson, A.R.; et al. PTEN loss confers BRAF inhibitor resistance to melanoma cells through the suppression of BIM expression. *Cancer Res.* **2011**, *71*, 2750–2760. [CrossRef]

58. Nathanson, K.L.; Martin, A.-M.; Wubbenhorst, B.; Greshock, J.; Letrero, R.; D'Andrea, K.; O'Day, S.; Infante, J.R.; Falchook, G.S.; Arkenau, H.-T.; et al. Tumor genetic analyses of patients with metastatic melanoma treated with the BRAF inhibitor Dabrafenib (GSK2118436). *Clin. Cancer Res.* **2013**, *19*, 4868–4878. [CrossRef]

59. Shimizu, T.; Tolcher, A.W.; Papadopoulos, K.P.; Beeram, M.; Rasco, D.W.; Smith, L.S.; Gunn, S.; Smetzer, L.; Mays, T.A.; Kaiser, B.; et al. The clinical effect of the dual-targeting strategy involving PI3K/AKT/mTOR and RAS/MEK/ERK pathways in patients with advanced cancer. *Clin. Cancer Res.* **2012**, *18*, 2316–2325. [CrossRef]

60. Frederick, D.T.; Piris, A.; Cogdill, A.P.; Cooper, Z.A.; Lezcano, C.; Ferrone, C.R.; Mitra, D.; Boni, A.; Newton, L.P.; Liu, C.; et al. BRAF inhibition is associated with enhanced melanoma antigen expression and a more favorable tumor microenvironment in patients with metastatic melanoma. *Clin. Cancer Res.* **2013**, *19*, 1225–1231. [CrossRef]

61. Cooper, Z.A.; Reuben, A.; Spencer, C.N.; Prieto, P.A.; Austin-Breneman, J.L.; Jiang, H.; Haymaker, C.; Gopalakrishnan, V.; Tetzlaff, M.T.; Frederick, D.T.; et al. Distinct clinical patterns and immune infiltrates are observed at time of progression on targeted therapy versus immune checkpoint blockade for melanoma. *OncoImmunology* **2016**, *5*, e1136044. [CrossRef]

62. Gazzaniga, S.; Bravo, A.I.; Guglielmotti, A.; Van Rooijen, N.; Maschi, F.; Vecchi, A.; Mantovani, A. Targeting tumor-associated macrophages and inhibition of MCP-1 reduce angiogenesis and tumor growth in a human melanoma xenograft. *J. Investig. Dermatol.* **2007**, *127*, 2031–2041. [CrossRef] [PubMed]

63. Hodi, F.S.; O'Day, S.J.; McDermott, D.F.; Weber, R.W.; Sosman, J.A.; Haanen, J.B.; Gonzalez, R.; Robert, C.; Schadendorf, D.; Hassel, J.C.; et al. Improved survival with ipilimumab in patients with metastatic melanoma. *N. Engl. J. Med.* **2010**, *363*, 711–723. [CrossRef] [PubMed]

64. Robert, C.; Schachter, J.; Long, G.V.; Arance, A.; Grob, J.J.; Mortier, L.; Daud, A.; Carlino, M.S.; McNeil, C.; Lotem, M.; et al. Pembrolizumab versus ipilimumab in advanced melanoma. *N. Engl. J. Med.* **2015**, *372*, 2521–2532. [CrossRef] [PubMed]

65. Robert, C.; Long, G.V.; Brady, B.; Dutriaux, C.; Maio, M.; Mortier, L.; Hassel, J.C.; Rutkowski, P.; McNeil, C.; Kalinka-Warzocha, E.; et al. Nivolumab previously untreated melanoma without BRAF mutation. *N. Engl. J. Med.* **2015**, *372*, 320–330. [CrossRef]

66. Larkin, J.; Chiarion-Sileni, V.; Gonzalez, R.; Grob, J.J.; Cowey, C.L.; Lao, C.D.; Schadendorf, D.; Dummer, R.; Smylie, M.; Rutkowski, P.; et al. Combined nivolumab and ipilimumab or monotherapy in untreated melanoma. *N. Engl. J. Med.* **2015**, *373*, 23–34. [CrossRef]

67. Madore, J.; Vilain, R.E.; Menzies, A.M.; Kakavand, H.; Wilmott, J.S.; Hyman, J.; Yearley, J.H.; Kefford, R.F.; Thompson, J.F.; Long, G.V.; et al. PD-L1 expression in melanoma shows marked heterogeneity within and between patients: Implications for anti-PD-1/PD-L1 clinical trials. *Pigment. Cell Melanoma Res.* **2014**, *28*, 245–253. [CrossRef]

68. Snyder, A.; Makarov, V.; Merghoub, T.; Yuan, J.; Zaretsky, J.M.; Desrichard, A.; Walsh, L.A.; Postow, M.A.; Wong, P.; Ho, T.S.; et al. Genetic basis for clinical response to CTLA-4 blockade in melanoma. *N. Engl. J. Med.* **2014**, *371*, 2189–2199. [CrossRef]

69. Luke, J.J.; Flaherty, K.T.; Ribas, A.; Long, G.V. Targeted agents and immunotherapies: Optimizing outcomes in melanoma. *Nat. Rev. Clin. Oncol.* **2017**, *14*, 463–482. [CrossRef]

70. Daud, A.I.; Loo, K.; Pauli, M.L.; Sanchez-Rodriguez, R.; Sandoval, P.M.; Taravati, K.; Tsai, K.; Nosrati, A.; Nardo, L.; Alvarado, M.D.; et al. Tumor immune profiling predicts response to anti–PD-1 therapy in human melanoma. *J. Clin. Investig.* **2016**, *126*, 3447–3452. [CrossRef]

71. Maccalli, C.; Giannarelli, D.; Capocefalo, F.; Pilla, L.; Fonsatti, E.; Di Giacomo, A.M.; Parmiani, G.; Maio, M. Immunological markers and clinical outcome of advanced melanoma patients receiving ipilimumab plus fotemustine in the NIBIT-M1 study. *OncoImmunology* **2015**, *5*, e1071007. [CrossRef]

72. Martens, A.; Wistuba-Hamprecht, K.; Foppen, M.G.; Yuan, J.; Postow, M.A.; Wong, P.; Romano, E.; Khammari, A.; Dreno, B.; Capone, M.; et al. Peripheral CD8 effector-memory type 1 T-cells correlate with outcome in ipilimumab-treated stage IV melanoma patients. *Clin. Cancer Res.* **2016**, *22*, 2908–2918. [CrossRef] [PubMed]

73. Weide, B.; Martens, A.; Hassel, J.C.; Berking, C.; Postow, M.A.; Bisschop, K.; Simeone, E.; Mangana, J.; Schilling, B.; Di Giacomo, A.M.; et al. Baseline biomarkers for outcome of melanoma patients treated with pembrolizumab. *Clin. Cancer Res.* **2016**, *22*, 5487–5496. [CrossRef] [PubMed]

74. Gopalakrishnan, V.; Spencer, C.N.; Nezi, L.; Reuben, A.; Andrews, M.C.; Karpinets, T.V.; Prieto, P.A.; Vicente, D.A.; Hoffman, K.; Wei, S.C.; et al. Gut microbiome modulates response to anti–PD-1 immunotherapy in melanoma patients. *Science* **2018**, *359*, 97–103. [CrossRef] [PubMed]

75. Dubin, K.; Callahan, M.K.; Ren, B.; Khanin, R.; Viale, A.; Ling, L.; No, D.; Gobourne, A.; Littmann, E.; Huttenhower, B.R.C.; et al. Intestinal microbiome analyses identify melanoma patients at risk for checkpoint-blockade-induced colitis. *Nat. Commun.* **2016**, *7*, 10391. [CrossRef]

76. Topalian, S.L.; Hodi, F.S.; Brahmer, J.R.; Gettinger, S.N.; Smith, D.C.; McDermott, D.F.; Powderly, J.D.; Carvajal, R.D.; Sosman, J.A.; Atkins, M.B.; et al. Safety, activity, and immune correlates of anti–PD-1 antibody in cancer. *N. Engl. J. Med.* **2012**, *366*, 2443–2454. [CrossRef]

77. Reck, M.; Rodríguez-Abreu, D.; Robinson, A.G.; Hui, R.; Csőszi, T.; Fülöp, A.; Gottfried, M.; Peled, N.; Tafreshi, A.; Cuffe, S.; et al. Pembrolizumab versus chemotherapy for PD-L1–positive non–small-cell lung cancer. *N. Engl. J. Med.* **2016**, *375*, 1823–1833. [CrossRef]

78. Nishino, M.; Ramaiya, N.H.; Hatabu, M.N.N.H.R.H.; Hodi, F.S. Monitoring immune-checkpoint blockade: Response evaluation and biomarker development. *Nat. Rev. Clin. Oncol.* **2017**, *14*, 655–668. [CrossRef]

79. Bence, C.; Hofman, V.; Chamorey, E.; Long-Mira, E.; Lassalle, S.; Albertini, A.; Liolios, I.; Zahaf, K.; Picard, A.; Montaudié, H.; et al. Association of combined PD -L1 expression and tumour-infiltrating lymphocyte features with survival and treatment outcomes in patients with metastatic melanoma. *J. Eur. Acad. Dermatol. Venereol.* **2019**, *34*, 984–994. [CrossRef]

80. Wang, Q.; Liu, F.; Liu, L. Prognostic significance of PD-L1 in solid tumor. *Medicine* **2017**, *96*, e6369. [CrossRef]

81. Hino, R.; Kabashima, K.; Kato, Y.; Yagi, H.; Nakamura, M.; Honjo, T.; Okazaki, T.; Tokura, Y. Tumor cell expression of programmed cell death-1 ligand 1 is a prognostic factor for malignant melanoma. *Cancer* **2010**, *116*, 1757–1766. [CrossRef]

82. Kluger, H.M.; Zito, C.R.; Barr, M.L.; Baine, M.K.; Chiang, V.L.; Sznol, M.; Rimm, D.L.; Chen, L.; Jilaveanu, L.B. Characterization of PD-L1 expression and associated T-cell infiltrates in metastatic melanoma samples from variable anatomic sites. *Clin. Cancer Res.* **2015**, *21*, 3052–3060. [CrossRef] [PubMed]

83. Massi, D.; Brusa, D.; Merelli, B.; Ciano, M.; Audrito, V.; Serra, S.; Buonincontri, R.; Baroni, G.; Nassini, R.; Minocci, D.; et al. PD-L1 marks a subset of melanomas with a shorter overall survival and distinct genetic and morphological characteristics. *Ann. Oncol.* **2014**, *25*, 2433–2442. [CrossRef] [PubMed]

84. Eggermont, A.M.; Blank, C.U.; Mandalà, M.; Long, G.V.; Atkinson, V.; Dalle, S.; Haydon, A.; Lichinitser, M.; Khattak, A.; Carlino, M.S.; et al. Adjuvant pembrolizumab versus placebo in resected stage III melanoma. *N. Engl. J. Med.* **2018**, *378*, 1789–1801. [CrossRef] [PubMed]

85. Galon, J.; Costes, A.; Sanchez-Cabo, F.; Kirilovsky, A.; Mlecnik, B.; Lagorce-Pagès, C.; Tosolini, M.; Camus, M.; Berger, A.; Wind, P.; et al. Type, density, and location of immune cells within human colorectal tumors predict clinical outcome. *Science* **2006**, *313*, 1960–1964. [CrossRef] [PubMed]

86. Mlecnik, B.; Bindea, G.; Angell, H.K.; Maby, P.; Angelova, M.; Tougeron, D.; Church, S.E.; Lafontaine, L.; Fischer, M.; Fredriksen, T.; et al. Integrative analyses of colorectal cancer show immunoscore is a stronger predictor of patient survival than microsatellite instability. *Immunity* **2016**, *44*, 698–711. [CrossRef]

87. Herbst, R.S.; Soria, J.-C.; Kowanetz, M.; Fine, G.D.; Hamid, O.; Gordon, M.S.; Sosman, J.A.; McDermott, D.F.; Powderly, J.D.; Gettinger, S.N.; et al. Predictive correlates of response to the anti-PD-L1 antibody MPDL3280A in cancer patients. *Nat. Cell Biol.* **2014**, *515*, 563–567. [CrossRef]

88. Tumeh, P.C.; Harview, C.L.; Yearley, J.H.; Shintaku, I.P.; Taylor, E.J.M.; Robert, L.; Chmielowski, B.; Spasić, M.; Henry, G.; Ciobanu, V.; et al. PD-1 blockade induces responses by inhibiting adaptive immune resistance. *Nature* **2014**, *515*, 568–571. [CrossRef]

89. Riaz, N.; Havel, J.J.; Makarov, V.; Desrichard, A.; Urba, W.J.; Sims, J.S.; Hodi, F.S.; Martín-Algarra, S.; Mandal, R.; Sharfman, W.H.; et al. Tumor and microenvironment evolution during immunotherapy with nivolumab. *Cell* **2017**, *171*, 934–949. [CrossRef]

90. Rooney, M.S.; Shukla, S.A.; Wu, C.J.; Getz, G.; Hacohen, N. Molecular and genetic properties of tumors associated with local immune cytolytic activity. *Cell* **2015**, *160*, 48–61. [CrossRef]

91. Huang, A.C.; Postow, M.A.; Orlowski, R.J.; Mick, R.; Bengsch, B.; Manne, S.; Xu, W.; Harmon, S.; Giles, J.R.; Wenz, B.; et al. T-cell invigoration to tumour burden ratio associated with anti-PD-1 response. *Nat. Cell Biol.* **2017**, *545*, 60–65. [CrossRef]

92. Barry, K.C.; Hsu, J.; Broz, M.L.; Cueto, F.J.; Binnewies, M.; Combes, A.J.; Nelson, A.E.; Loo, K.; Kumar, R.; Rosenblum, M.D.; et al. A natural killer—Dendritic cell axis defines checkpoint therapy-responsive tumor microenvironments. *Nat. Med.* **2018**, *24*, 1178–1191. [CrossRef]

93. Garris, C.S.; Arlauckas, S.P.; Kohler, R.H.; Trefny, M.P.; Garren, S.; Piot, C.; Engblom, C.; Pfirschke, C.; Siwicki, M.; Gungabeesoon, J.; et al. Successful Anti-PD-1 cancer immunotherapy requires T cell-dendritic cell crosstalk involving the cytokines IFN-γ and IL-12. *Immunity* **2018**, *49*, 1148–1161. [CrossRef] [PubMed]

94. Carbone, D.P.; Reck, M.; Paz-Ares, L.; Creelan, B.; Horn, L.; Steins, M.; Felip, E.; van den Heuvel, M.M.; Ciuleanu, T.E.; Badin, F.; et al. First-line nivolumab in stage IV or recurrent non–small-cell lung cancer. *N. Engl. J. Med.* **2017**, *376*, 2415–2426. [CrossRef] [PubMed]

95. Le, D.T.; Durham, J.N.; Smith, K.N.; Wang, H.; Bartlett, B.R.; Aulakh, L.K.; Lu, S.; Kemberling, H.; Wilt, C.; Luber, B.S.; et al. Mismatch repair deficiency predicts response of solid tumors to PD-1 blockade. *Science* **2017**, *357*, 409–413. [CrossRef]

96. Samstein, R.M.; Lee, C.-H.; Shoushtari, A.N.; Hellmann, M.D.; Shen, R.; Janjigian, Y.Y.; Barron, D.A.; Zehir, A.; Jordan, E.J.; Omuro, A.; et al. Tumor mutational load predicts survival after immunotherapy across multiple cancer types. *Nat. Genet.* **2019**, *51*, 202–206. [CrossRef] [PubMed]

97. Gubin, M.M.; Zhang, X.; Schuster, H.; Caron, E.; Ward, J.P.; Noguchi, T.; Ivanova, Y.; Hundal, J.; Arthur, C.D.; Krebber, W.-J.; et al. Checkpoint blockade cancer immunotherapy targets tumour-specific mutant antigens. *Nat. Cell Biol.* **2014**, *515*, 577–581. [CrossRef] [PubMed]

98. Cao, D.; Xu, H.; Xu, X.; Guo, T.; Ge, W. High tumor mutation burden predicts better efficacy of immunotherapy: A pooled analysis of 103078 cancer patients. *OncoImmunology* **2019**, *8*, e1629258. [CrossRef]

99. Van Allen, E.M.; Miao, D.; Schilling, B.; Shukla, S.A.; Blank, C.; Zimmer, L.; Sucker, A.; Hillen, U.; Foppen, M.H.G.; Goldinger, S.M.; et al. Genomic correlates of response to CTLA-4 blockade in metastatic melanoma. *Science* **2015**, *350*, 207–211. [CrossRef]

100. Rizvi, N.A.; Hellmann, M.D.; Snyder, A.; Kvistborg, P.; Makarov, V.; Havel, J.J.; Lee, W.; Yuan, J.; Wong, P.; Ho, T.S.; et al. Mutational landscape determines sensitivity to PD-1 blockade in non–small cell lung cancer. *Science* **2015**, *348*, 124–128. [CrossRef]

101. Goodman, A.M.; Kato, S.; Bazhenova, L.; Patel, S.P.; Frampton, G.M.; Miller, V.; Stephens, P.J.; Daniels, G.A.; Kurzrock, R. Tumor mutational burden as an independent predictor of response to immunotherapy in diverse cancers. *Mol. Cancer Ther.* **2017**, *16*, 2598–2608. [CrossRef]

102. Cristescu, R.; Mogg, R.; Ayers, M.; Albright, A.; Murphy, E.; Yearley, J.; Sher, X.; Liu, X.Q.; Lu, H.; Nebozhyn, M.; et al. Pan-tumor genomic biomarkers for PD-1 checkpoint blockade–based immunotherapy. *Science* **2018**, *362*, eaar3593. [CrossRef] [PubMed]

103. Rosenberg, J.E.; Hoffman-Censits, J.; Powles, T.; Van Der Heijden, M.S.; Balar, A.V.; Necchi, A.; Dawson, N.; O'Donnell, P.H.; Balmanoukian, A.; Loriot, Y.; et al. Atezolizumab in patients with locally advanced and metastatic urothelial carcinoma who have progressed following treatment with platinum-based chemotherapy: A single-arm, multicentre, phase 2 trial. *Lancet* **2016**, *387*, 1909–1920. [CrossRef]

104. Arlauckas, S.P.; Garris, C.S.; Kohler, R.H.; Kitaoka, M.; Cuccarese, M.F.; Yang, K.S.; Miller, M.A.; Carlson, J.C.; Freeman, G.J.; Anthony, R.M.; et al. In vivo imaging reveals a tumor-associated macrophage–mediated resistance pathway in anti–PD-1 therapy. *Sci. Transl. Med.* **2017**, *9*, eaal3604. [CrossRef] [PubMed]

105. Neubert, N.J.; Schmittnaegel, M.; Bordry, N.; Nassiri, S.; Wald, N.; Martignier, C.; Tillé, L.; Homicsko, K.; Damsky, W.; Hajjami, H.M.-E.; et al. T cell–induced CSF1 promotes melanoma resistance to PD1 blockade. *Sci. Transl. Med.* **2018**, *10*, eaan3311. [CrossRef] [PubMed]

106. Mariathasan, S.; Turley, S.J.; Nickles, D.; Castiglioni, A.; Yuen, K.; Wang, Y.; Iii, E.E.K.; Koeppen, H.; Astarita, J.L.; Cubas, R.; et al. TGFβ attenuates tumour response to PD-L1 blockade by contributing to exclusion of T cells. *Nature* **2018**, *554*, 544–548. [CrossRef]

107. Brożyna, A.A.; Jóźwicki, W.; Roszkowski, K.; Filipiak, J.; Slominski, A.T. Melanin content in melanoma metastases affects the outcome of radiotherapy. *Oncotarget* **2016**, *7*, 17844–17853. [CrossRef]

108. Galván, I.; Inácio, Â.; Dañino, M.; Corbí-Llopis, R.; Monserrat, M.T.; Bernabeu-Wittel, J. High SLC7A11 expression in normal skin of melanoma patients. *Cancer Epidemiol.* **2019**, *62*, 101582. [CrossRef]

109. Osrodek, M.; Hartman, M.L.; Czyz, M. Physiologically relevant oxygen concentration (6% O2) as an important component of the microenvironment impacting melanoma phenotype and melanoma response to targeted therapeutics in vitro. *Int. J. Mol. Sci.* **2019**, *20*, 4203. [CrossRef]

110. Cesi, G.; Walbrecq, G.; Zimmer, A.; Kreis, S.; Haan, C. ROS production induced by BRAF inhibitor treatment rewires metabolic processes affecting cell growth of melanoma cells. *Mol. Cancer* **2017**, *16*, 1–16. [CrossRef]

111. Ratnikov, B.I.; Scott, D.A.; Osterman, A.L.; Smith, J.W.; Ronai, Z.A. Metabolic rewiring in melanoma. *Oncogene* **2017**, *36*, 147–157. [CrossRef]

112. DeNardo, D.G.; Ruffell, B. Macrophages as regulators of tumour immunity and immunotherapy. *Nat. Rev. Immunol.* **2019**, *19*, 369–382. [CrossRef] [PubMed]

113. Rooks, M.G.; Garrett, W.S. Gut microbiota, metabolites and host immunity. *Nat. Rev. Immunol.* **2016**, *16*, 341–352. [CrossRef] [PubMed]

114. Chen, J.; Domingue, J.C.; Sears, C.L. Microbiota dysbiosis in select human cancers: Evidence of association and causality. *Semin. Immunol.* **2017**, *32*, 25–34. [CrossRef] [PubMed]

115. Schwabe, R.F.; Jobin, C. The microbiome and cancer. *Nat. Rev. Cancer* **2013**, *13*, 800–812. [CrossRef] [PubMed]

116. Ramos, A.; Hemann, M.T. Drugs, bugs, and cancer: Fusobacterium nucleatum promotes chemoresistance in colorectal cancer. *Cell* **2017**, *170*, 411–413. [CrossRef] [PubMed]

117. Paulos, C.M.; Kaiser, A.; Wrzesinski, C.; Hinrichs, C.S.; Cassard, L.; Boni, A.; Muranski, P.; Sanchez-Perez, L.; Palmer, D.C.; Yu, Z.; et al. Toll-like receptors in tumor immunotherapy. *Clin. Cancer Res.* **2007**, *13*, 5280–5289. [CrossRef] [PubMed]

118. Bettegowda, C.; Sausen, M.; Leary, R.J.; Kinde, I.; Wang, Y.; Agrawal, N.; Bartlett, B.R.; Wang, H.; Luber, B.; Alani, R.M.; et al. Detection of circulating tumor DNA in early- and late-stage human malignancies. *Sci. Transl. Med.* **2014**, *6*, 224ra24. [CrossRef]

119. Wan, J.C.M.; Massie, C.; Garcia-Corbacho, J.; Mouliere, F.; Brenton, J.D.; Caldas, C.M.; Pacey, S.; Baird, C.C.S.P.R.; Rosenfeld, N. Liquid biopsies come of age: Towards implementation of circulating tumour DNA. *Nat. Rev. Cancer* **2017**, *17*, 223–238. [CrossRef]

120. Murtaza, M.; Dawson, S.; Pogrebniak, K.; Rueda, O.M.; Provenzano, E.; Grant, J.; Chin, S.-F.; Tsui, D.W.Y.; Marass, F.; Gale, D.; et al. Multifocal clonal evolution characterized using circulating tumour DNA in a case of metastatic breast cancer. *Nat. Commun.* **2015**, *6*, 8760. [CrossRef]

121. Chang, G.A.; Tadepalli, J.S.; Shao, Y.; Zhang, Y.; Weiss, S.A.; Robinson, E.; Spittle, C.; Furtado, M.; Shelton, D.N.; Karlin-Neumann, G.; et al. Sensitivity of plasma BRAFmutant and NRASmutant cell-free DNA assays to detect metastatic melanoma in patients with low RECIST scores and non-RECIST disease progression. *Mol. Oncol.* **2016**, *10*, 157–165. [CrossRef]

122. Tan, L.; Sandhu, S.; Lee, R.; Li, J.; Callahan, J.; Ftouni, S.; Dhomen, N.; Middlehurst, P.; Wallace, A.; Raleigh, J.; et al. Prediction and monitoring of relapse in stage III melanoma using circulating tumor DNA. *Ann. Oncol.* **2019**, *30*, 804–814. [CrossRef] [PubMed]

123. Lee, R.; Gremel, G.; Marshall, A.; Myers, K.; Fisher, N.; Dunn, J.; Dhomen, N.; Corrie, P.; Middleton, M.; Lorigan, P.; et al. Circulating tumor DNA predicts survival in patients with resected high-risk stage II/III melanoma. *Ann. Oncol.* **2018**, *29*, 490–496. [CrossRef] [PubMed]

124. Ascierto, P.A.; Minor, D.; Ribas, A.; Lebbé, C.; O'Hagan, A.; Arya, N.; Guckert, M.; Schadendorf, D.; Kefford, R.F.; Grob, J.-J.; et al. Phase II trial (BREAK-2) of the BRAF inhibitor Dabrafenib (GSK2118436) in patients with metastatic melanoma. *J. Clin. Oncol.* **2013**, *31*, 3205–3211. [CrossRef] [PubMed]

125. Gray, E.S.; Rizos, H.; Reid, A.L.; Boyd, S.C.; Pereira, M.R.; Lo, J.; Tembe, V.; Freeman, J.; Lee, J.H.; Scolyer, R.A.; et al. Circulating tumor DNA to monitor treatment response and detect acquired resistance in patients with metastatic melanoma. *Oncotarget* **2015**, *6*, 42008–42018. [CrossRef] [PubMed]

126. Wong, S.Q.; Raleigh, J.M.; Callahan, J.; Vergara, I.A.; Ftouni, S.; Hatzimihalis, A.; Colebatch, A.J.; Li, J.; Semple, T.; Doig, K.; et al. Circulating tumor DNA analysis and functional imaging provide complementary approaches for comprehensive disease monitoring in metastatic melanoma. *JCO Precis. Oncol.* **2017**, 1–14. [CrossRef]

127. Cortez, M.A.; Buesoramos, C.E.; Ferdin, J.; Lopez-Berestein, G.; Sood, A.K.; Calin, G.A. MicroRNAs in body fluids—The mix of hormones and biomarkers. *Nat. Rev. Clin. Oncol.* **2011**, *8*, 467–477. [CrossRef]

128. Friedman, E.B.; Shang, S.; De Miera, E.V.-S.; Fog, J.U.; Teilum, M.W.; Ma, M.W.; Berman, R.; Shapiro, R.; Pavlick, A.C.; Hernando, E.; et al. Serum microRNAs as biomarkers for recurrence in melanoma. *J. Transl. Med.* **2012**, *10*, 155. [CrossRef]

129. Gray, E.; Reid, A.L.; Bowyer, S.; Calapre, L.; Siew, K.; Pearce, R.; Cowell, L.; Frank, M.H.; Millward, M.; Ziman, M. Circulating melanoma cell subpopulations: Their heterogeneity and differential responses to treatment. *J. Investig. Dermatol.* **2015**, *135*, 2040–2048. [CrossRef]

130. Reid, A.; Millward, M.; Pearce, R.; Lee, M.; Frank, M.; Ireland, A.; Monshizadeh, L.; Rai, T.; Heenan, P.; Medic, S.; et al. Markers of circulating tumour cells in the peripheral blood of patients with melanoma correlate with disease recurrence and progression. *Br. J. Dermatol.* **2013**, *168*, 85–92. [CrossRef]

131. Thress, K.S.; Brant, R.; Carr, T.H.; Dearden, S.; Jenkins, S.; Brown, H.; Hammett, T.; Cantarini, M.; Barrett, J.C. EGFR mutation detection in ctDNA from NSCLC patient plasma: A cross-platform comparison of leading technologies to support the clinical development of AZD9291. *Lung Cancer* **2015**, *90*, 509–515. [CrossRef]

132. Hayashi, M.; Chu, D.; Meyer, C.F.; Llosa, N.J.; Bs, G.M.; Morris, C.D.; Levin, A.S.; Wolinsky, J.-P.; Albert, C.M.; Steppan, D.A.; et al. Highly personalized detection of minimal Ewing sarcoma disease burden from plasma tumor DNA. *Cancer* **2016**, *122*, 3015–3023. [CrossRef] [PubMed]

133. Reid, A.L.; Freeman, J.B.; Millward, M.; Ziman, M.; Gray, E.S. Detection of BRAF-V600E and V600K in melanoma circulating tumour cells by droplet digital PCR. *Clin. Biochem.* **2015**, *48*, 999–1002. [CrossRef] [PubMed]

134. Sanmamed, M.F.; Fernández-Landázuri, S.; Rodríguez, C.; Zárate, R.; Lozano, M.D.; Zubiri, L.; Perez-Gracia, J.L.; Martín-Algarra, S.; González, Á. Quantitative cell-free circulating BRAFV600E mutation analysis by use of droplet digital PCR in the follow-up of patients with melanoma being treated with BRAF inhibitors. *Clin. Chem.* **2015**, *61*, 297–304. [CrossRef] [PubMed]

135. Bidard, F.-C.; Madic, J.; Mariani, P.; Piperno-Neumann, S.; Rampanou, A.; Servois, V.; Cassoux, N.; Desjardins, L.; Milder, M.; Vaucher, I.; et al. Detection rate and prognostic value of circulating tumor cells and circulating tumor DNA in metastatic uveal melanoma. *Int. J. Cancer* **2014**, *134*, 1207–1213. [CrossRef]

136. Lee, J.; Saw, R.; Thompson, J.; Lo, S.; Spillane, A.; Shannon, K.; Stretch, J.; Howle, J.; Menzies, A.; Carlino, M.; et al. Pre-operative ctDNA predicts survival in high-risk stage III cutaneous melanoma patients. *Ann. Oncol.* **2019**, *30*, 815–822. [CrossRef]

137. Newman, A.M.; Bratman, S.V.; To, J.; Wynne, J.F.; Eclov, N.C.W.; Modlin, L.A.; Liu, C.L.; Neal, J.W.; Wakelee, H.A.; Merritt, R.E.; et al. An ultrasensitive method for quantitating circulating tumor DNA with broad patient coverage. *Nat. Med.* **2014**, *20*, 548–554. [CrossRef]

138. Schreuer, M.; Meersseman, G.; Herrewegen, S.V.D.; Jansen, Y.; Chevolet, I.; Bott, A.; Wilgenhof, S.; Seremet, T.; Jacobs, B.; Buyl, R.; et al. Quantitative assessment of *BRAF V600* mutant circulating cell-free tumor DNA as a tool for therapeutic monitoring in metastatic melanoma patients treated with BRAF/MEK inhibitors. *J. Transl. Med.* **2016**, *14*, 95. [CrossRef]

139. Lee, J.H.; Long, G.V.; Boyd, S.; Lo, S.; Menzies, A.M.; Tembe, V.; Guminski, A.; Jakrot, V.; Scolyer, R.A.; Mann, G.J.; et al. Circulating tumour DNA predicts response to anti-PD1 antibodies in metastatic melanoma. *Ann. Oncol.* **2017**, *28*, 1130–1136. [CrossRef]

140. Girotti, M.R.; Gremel, G.; Lee, R.; Galvani, E.; Rothwell, D.; Viros, A.; Mandal, A.K.; Lim, K.H.J.; Saturno, G.; Furney, S.J.; et al. Application of sequencing, liquid biopsies, and patient-derived xenografts for personalized medicine in melanoma. *Cancer Discov.* **2015**, *6*, 286–299. [CrossRef]

141. Santiago-Walker, A.; Gagnon, R.; Mazumdar, J.; Casey, M.; Long, G.V.; Schadendorf, D.; Flaherty, K.T.; Kefford, R.F.; Hauschild, A.; Hwu, P.; et al. Correlation of BRAF mutation status in circulating-free DNA and tumor and association with clinical outcome across four BRAFi and MEKi clinical trials. *Cancer Res.* **2015**, *22*, 567–574. [CrossRef]

142. Redzic, Z.B. Molecular biology of the blood-brain and the blood-cerebrospinal fluid barriers: Similarities and differences. *Fluids Barriers CNS* **2011**, *8*, 3. [CrossRef] [PubMed]

143. De Mattos-Arruda, L.; Mayor, R.; Ng, C.K.Y.; Weigelt, B.; Martínez-Ricarte, F.; Torrejon, D.; Oliveira, M.; Arias, A.; Raventos, C.; Tang, J.; et al. Cerebrospinal fluid-derived circulating tumour DNA better represents the genomic alterations of brain tumours than plasma. *Nat. Commun.* **2015**, *6*, 8839. [CrossRef] [PubMed]

144. Vickers, K.C.; Palmisano, B.T.; Shoucri, B.M.; Shamburek, R.D.; Remaley, A.T. MicroRNAs are transported in plasma and delivered to recipient cells by high-density lipoproteins. *Nat. Cell Biol.* **2011**, *13*, 423–433. [CrossRef] [PubMed]

145. Mitchell, P.S.; Parkin, R.K.; Kroh, E.M.; Fritz, B.R.; Wyman, S.K.; Pogosova-Agadjanyan, E.L.; Peterson, A.; Noteboom, J.; O'Briant, K.C.; Allen, A.; et al. Circulating microRNAs as stable blood-based markers for cancer detection. *Proc. Natl. Acad. Sci. USA* **2008**, *105*, 10513–10518. [CrossRef] [PubMed]

146. Shi, T.; Gao, G.; Cao, Y. Long noncoding RNAs as novel biomarkers have a promising future in cancer diagnostics. *Dis. Markers* **2016**, *2016*, 1–10. [CrossRef]

147. Deng, H.; Wang, J.M.; Li, M.; Tang, R.; Tang, K.; Su, Y.; Hou, Y.; Zhang, J. Long non-coding RNAs: New biomarkers for prognosis and diagnosis of colon cancer. *Tumor Biol.* **2017**, *39*, 1010428317706332. [CrossRef]

148. Kanemaru, H.; Fukushima, S.; Yamashita, J.; Honda, N.; Oyama, R.; Kakimoto, A.; Masuguchi, S.; Ishihara, T.; Inoue, Y.; Jinnin, M.; et al. The circulating microRNA-221 level in patients with malignant melanoma as a new tumor marker. *J. Dermatol. Sci.* **2011**, *61*, 187–193. [CrossRef]

149. Dupin, E.; Le Douarin, N.M. Development of melanocyte precursors from the vertebrate neural crest. *Oncogene* **2003**, *22*, 3016–3023. [CrossRef]

150. Rosenberg, R.; Gertler, R.; Friederichs, J.; Fuehrer, K.; Dahm, M.; Phelps, R.; Thorban, S.; Nekarda, H.; Siewert, J.R. Comparison of two density gradient centrifugation systems for the enrichment of disseminated tumor cells in blood. *Cytol. J.* **2002**, *49*, 150–158. [CrossRef]

151. Freeman, J.B.; Gray, E.; Millward, M.; Pearce, R.; Ziman, M. Evaluation of a multi-marker immunomagnetic enrichment assay for the quantification of circulating melanoma cells. *J. Transl. Med.* **2012**, *10*, 192. [CrossRef]

152. Khoja, L.; Lorigan, P.C.; Zhou, C.; Lancashire, M.; Booth, J.; Cummings, J.; Califano, R.; Clack, G.; Hughes, A.; Dive, C. Biomarker utility of circulating tumor cells in metastatic cutaneous melanoma. *J. Investig. Dermatol.* **2013**, *133*, 1582–1590. [CrossRef] [PubMed]

153. De Souza, L.M.; Robertson, B.M.; Robertson, G.P. Future of circulating tumor cells in the melanoma clinical and research laboratory settings. *Cancer Lett.* **2017**, *392*, 60–70. [CrossRef] [PubMed]

154. Lim, S.Y.; Lee, J.H.; Diefenbach, R.J.; Kefford, R.F.; Rizos, H. Liquid biomarkers in melanoma: Detection and discovery. *Mol. Cancer* **2018**, *17*, 1–14. [CrossRef] [PubMed]

155. Klinac, D.; Gray, E.; Freeman, J.B.; Reid, A.L.; Bowyer, S.; Millward, M.; Ziman, M. Monitoring changes in circulating tumour cells as a prognostic indicator of overall survival and treatment response in patients with metastatic melanoma. *BMC Cancer* **2014**, *14*, 423. [CrossRef] [PubMed]

156. Koyanagi, K.; Kuo, C.; Nakagawa, T.; Mori, T.; Ueno, H.; Lorico, A.R.; Wang, H.-J.; Hseuh, E.; O'Day, S.J.; Hoon, D.S. Multimarker quantitative real-time PCR detection of circulating melanoma cells in peripheral blood: Relation to disease stage in melanoma patients. *Clin. Chem.* **2005**, *51*, 981–988. [CrossRef]

157. Terstappen, L.W.M.M.; Rao, C.; Bui, T.; Connelly, M.; Doyle, G.; Karydis, I.; Middleton, M.R.; Clack, G.; Malone, M.; Coumans, F.A.W. Circulating melanoma cells and survival in metastatic melanoma. *Int. J. Oncol.* **2011**, *38*, 755–760. [CrossRef]

158. Anagnostou, V.; Yarchoan, M.; Hansen, A.R.; Wang, H.; Verde, F.; Sharon, E.; Collyar, D.; Chow, L.Q.; Forde, P.M. Immuno-oncology trial endpoints: Capturing clinically meaningful activity. *Clin. Cancer Res.* **2017**, *23*, 4959–4969. [CrossRef]

Publisher's Note: MDPI stays neutral with regard to jurisdictional claims in published maps and institutional affiliations.

Review

A Framework of Major Tumor-Promoting Signal Transduction Pathways Implicated in Melanoma-Fibroblast Dialogue

Barbara Bellei [1,*], Emilia Migliano [2] and Mauro Picardo [1]

[1] Laboratory of Cutaneous Physiopathology and Integrated Center of Metabolomics Research, San Gallicano Dermatological Institute, IRCCS, 00144 Rome, Italy; mauro.picardo@ifo.gov.it

[2] Department of Plastic and Regenerative Surgery, San Gallicano Dermatological Institute, IRCCS, 00144 Rome, Italy; emilia.migliano@ifo.gov.it

* Correspondence: barbara.bellei@ifo.gov.it; Tel.: +39-0652666246

Received: 18 September 2020; Accepted: 13 November 2020; Published: 17 November 2020

Simple Summary: Melanoma cells reside in a complex stromal microenvironment, which is a critical component of disease onset and progression. Mesenchymal or fibroblastic cell type are the most abundant cellular element of tumor stroma. Factors secreted by melanoma cells can activate non-malignant associated fibroblasts to become melanoma associate fibroblasts (MAFs). MAFs promote tumorigenic features by remodeling the extracellular matrix, supporting tumor cells proliferation, neo-angiogenesis and drug resistance. Additionally, environmental factors may contribute to the acquisition of pro-tumorigenic phenotype of fibroblasts. Overall, in melanoma, perturbed tissue homeostasis contributes to modulation of major oncogenic intracellular signaling pathways not only in tumor cells but also in neighboring cells. Thus, targeted molecular therapies need to be considered from the reciprocal point of view of melanoma and stromal cells.

Abstract: The development of a modified stromal microenvironment in response to neoplastic onset is a common feature of many tumors including cutaneous melanoma. At all stages, melanoma cells are embedded in a complex tissue composed by extracellular matrix components and several different cell populations. Thus, melanomagenesis is not only driven by malignant melanocytes, but also by the altered communication between melanocytes and non-malignant cell populations, including fibroblasts, endothelial and immune cells. In particular, cancer-associated fibroblasts (CAFs), also referred as melanoma-associated fibroblasts (MAFs) in the case of melanoma, are the most abundant stromal cells and play a significant contextual role in melanoma initiation, progression and metastasis. As a result of dynamic intercellular molecular dialogue between tumor and the stroma, non-neoplastic cells gain specific phenotypes and functions that are pro-tumorigenic. Targeting MAFs is thus considered a promising avenue to improve melanoma therapy. Growing evidence demonstrates that aberrant regulation of oncogenic signaling is not restricted to transformed cells but also occurs in MAFs. However, in some cases, signaling pathways present opposite regulation in melanoma and surrounding area, suggesting that therapeutic strategies need to carefully consider the tumor–stroma equilibrium. In this novel review, we analyze four major signaling pathways implicated in melanomagenesis, TGF-β, MAPK, Wnt/β-catenin and Hyppo signaling, from the complementary point of view of tumor cells and the microenvironment.

Keywords: melanoma; cancer associated fibroblast; tumor microenvironment; melanomagenesis

1. Introduction

Melanoma represents approximatively 4% of skin cancer cases but is the deadliest one, corresponding to 80% of skin cancer deaths and about 1–2% of all cancer deaths [1,2]. Its incidence is rising in most countries of the western world [3]. The transformation of melanocytes into melanoma requires a burden of mutations that can be caused by both endogenous and exogenous cues [4–6]. However, genetic studies demonstrated that sporadic melanoma are associated to allele variants with high prevalence and low penetrance indicating that environmental factors play a key role in melanoma development [7–9]. Among them, the exposure to ultraviolet (UV) rays has a significant impact on skin biology and homeostasis [10]. The most deleterious effect of UV radiations is the direct damage to DNA. In addition, UVA not only contributes to the direct formation of DNA lesions but also impairs the removal of UV photoproducts from genomic DNA through oxidation and damage to DNA repair proteins [11,12]. In the epidermis, melanocytes, which are classified as intermittent mitotic cells [13] and normally divide only on demand, are more prone to accumulate damage than rapidly dividing cells such as keratinocytes. Likewise, melanocytes have a reduced repair capacity for oxidative DNA damage than skin fibroblasts [14]. Moreover, frequent excessive exposure to UV light impacts on melanocyte microenvironment within the epidermis, contributing to melanoma onset [15,16]. For example, alterations in the composition of basement membrane and dermal extracellular matrix might anticipate melanomagenesis, facilitating disease occurrence. In chronically sun-exposed skin, qualitative and quantitative alterations of dermal extracellular matrix proteins causing loss of tensile strength, increase fragility and impair wound healing [17]. The expression of type VII collagen that anchors fibrils at the dermal–epidermal junction by keratinocytes is decreased in UV-irradiated skin areas. UV-irradiated skin produces several enzymes such as matrix metalloproteinases (MMPs), which degrade dermal collagen fibers (especially type I collagen) and elastic fibers. This process causes an overall modification of mechanical properties of tissues that is part of the photoaging process [18,19]. Instead, in the contest of melanoma, MMPs altering the basement membrane and dermal ECM architecture can facilitate invasion of tumor cells. Accordingly, excessive production of MMP1, MMP2 and MMP9 has been frequently observed in melanoma patients [20–22]. Tissue surrounding benign nevi and melanomas display greater stiffness than normal skin indicating that mechanical properties of the matrix impact on melanoma initiation or progression [23]. Increasing evidence suggests that tissue rigidity or matrix stiffness controls phenotypic states and contributes to the invasive process in advanced melanoma [24]. Thus, in the skin, due to continuous extrinsic stimulation, persistent alteration in the extracellular matrix can act independently to tumor onset moving as a driver of the tumorigenic process. In addition, biological behavior of normal dermal and epidermal cells is constantly influenced by external agents. Since fibroblasts are long lived cells constantly undergoing damage accumulation, they are considered a relevant player of skin carcinogenesis. Following UV light irradiation, keratinocytes secrete melanocyte growth factor (including α-melanocytes stimulating hormone, α-MSH and endothelin-1, EDN1), which increase cytokine and the melanin production and transfer preventing further damage caused by UV [25–27]. In vitro, UVA and UVB activate bFGF production by both skin fibroblasts and keratinocytes [28]. Keratinocytes additionally increase the secretion of hepatocyte growth factor (HGF) and transforming growth factor-β1 (TGF-β1) inducing proliferation in melanocytes that in turn become more susceptible to transformation [29].

Transitory activation of melanocytes is part of the tanning response, the main physiological process protecting the skin from UV light. However, aberrant increased production of growth factors is considered part of the acquisition of stress-induced premature senescent phenotype and the corresponding senescence-associated secretory phenotype (SASP) as those promoted by chronic sun exposure. Correspondingly, growth factors hyperproducing senescent fibroblasts are frequently described in age spot and melasma, two pathological condition characterized by hyperpigmentation [30–33]. It is extensively documented that senescent fibroblasts accumulate in habitually sun-exposed skin and orchestrate stroma modification into a tumor-promoting one [34–37]. In line with other cancer types, the secretory profile of melanoma-associated fibroblasts largely overlaps with that of senescent fibroblasts [38,39]. Properly, multivariate analyses demonstrate increasing age is the strongest independent adverse prognostic factor together with Breslow thickness [40]. Tumor cells utilize fibroblast-secreted growth factors to facilitate their own survival and proliferation. The secretory profile of fibroblasts is also critical in metabolic and immune reprogramming of the tumor microenvironment with an impact on angiogenesis regulation and adaptive resistance to therapy [41–44]. Due to the intense melanoma-stoma crosstalk, fibroblasts progressively modify their biological feature, presenting a molecular signature that partially overlaps with the cancer one.

This review focuses on major intracellular signaling pathways deregulated in melanoma analyzed from the reciprocal point of view of melanoma cells and MAFs.

2. Transformation of Normal Fibroblasts to Melanoma-Associated Fibroblasts

Physiologically, melanocytes reside within the basal layer of the epidermis and interplay tight contacts with epidermal keratinocytes through E- and P-cadherin adhesion proteins, whereas an intense communication networks mediated by soluble factors explains the deep influence of dermal compartment in melanocyte biology [45–47]. However, during melanoma progression, there is a progressive loss of E-cadherin [48–50] and gain of N-cadherin [49,51–53], which not only frees melanoma cells from control by keratinocytes, but also provides new adhesion characteristics [54–56]. Consequently, melanoma cells acquire invasive properties, violate the basement membrane and invade the underlying dermis establishing unusual homotypic interaction between melanoma cells and heterotypic cell–cell contact with fibroblasts, endothelial and immunocompetent cells. All these elements synergistically play a specific role in disease progression. Cadherin switch as well as integrins expression profile modification also implies the activation of inappropriate survival signals [57,58] and thus enhances the malignant phenotype [51,55,59].

Melanoma cells actively interact with stromal cells, not only through direct cell–cell but also through cell–matrix interactions and secreted growth factors and cytokines. A complex network of soluble bioactive molecules contributes to the alteration of the host tissue and to the definition of malignant behavior of melanoma. An exclusive feature distinguishing melanoma from other tumors is the communication through melanosome, tissue-specific organelles deputies to the extracellular melanin distribution [60]. Fibroblasts around melanoma contain in their cytoplasm a significantly higher density of melanosome [61]. Melanosomes released from melanoma cells carrying pro-inflammatory molecule and several microRNA are able to transform dermal fibroblasts into pro-tumorigenic [61]. Evidence that fibroblasts begin to aggregate in the dermis at early stages of melanoma initiation, before melanoma cells invade, underlies the importance of paracrine communication between melanoma cells and surrounding. Thus, starting from early tumor stage, due to continuous paracrine stimulation by transformed cells, surrounding stromal fibroblasts are induced to initiate phenotypic, molecular and biochemical transitions and to transdifferentiate into cancer-associated fibroblasts (CAFs). Cancer-associated fibroblasts represent one of the major players in tumor–stroma network. CAFs acquire myofibroblast features and produce several growth factors that contribute to tumor cells proliferation, survival and metastasis [42,62]. CAFs are similar to myofibroblasts present during wound healing or the fibrotic conditions. In fact, desmoplastic wound healing-like tumor stroma is frequently referred as a consequence of mutual interaction of tumor cells and CAFs [63]. CAFs are distinguished from their normal counterparts by the expression of several markers such as alpha-smoot muscle actin (α-SMA), fibroblast specific protein-1 (FSP-1 also referred as S100A4), fibroblast-activating protein (FAP), platelet derived growth factor receptor-alpha/beta (PDGR α/β), tenascin-C, collagen 11-α1 (COL11A1), vimentin and fibronectin. However, a univocal molecular definition of CAFs profile is yet lacking. Clinically, the presence of many myofibroblasts in the tumor microenvironment has been associated with elevated risk of invasion, metastasis and a poor prognosis [34,64]. In addition to resident fibroblasts, there are several sources of CAFs, including bone marrow mesenchymal cells and endothelial cells [62,65]. In the case of melanoma, due to the prevalent localization at the junction of the epidermis and the dermis, dermal fibroblasts are considered the major source of CAFs, also referred to as melanoma-associated fibroblasts (MAFs) [42,66–68] (Figure 1). MAFs are less frequent compared to CAFs in other solid tumors [66]. Fibroblasts are associated to melanoma cells at all stages of disease and their functional contribution to disease progression has been largely documented but now increasing data also highlight their antitumor actions [69,70]. As an extreme consequence of the intimate relationship between melanoma and fibroblast, an original study described cell fusion events capable of generating tumor–stroma cell hybrid clones [44]. It is not fully clear if the dual nature of cancer microenvironment reflects the contemporary presence of heterogenic populations or if differences reside in disease evolution. Since MAFs co-evolve with tumorigenic cells it is possible that an early anti-tumor phenotype is replaced by a pro-tumorigenic one during disease progression. In line with the idea that MAFs co-evolve with melanoma cells, several research papers demonstrated an inhibitory function of dermal fibroblasts during tumor onset. This is largely because normal dermal fibroblasts, which are mostly quiescent cells in healthy condition, function as controller of tissue homeostasis. In vitro, co-culture experiments using normal fibroblasts and cells isolated from primary melanoma evidenced repressive influence on melanoma cells [71]. Multiple factors have been implicated in the transition of normal tumor-suppressive fibroblasts into reactive and tumor-promoting CAFs [72].

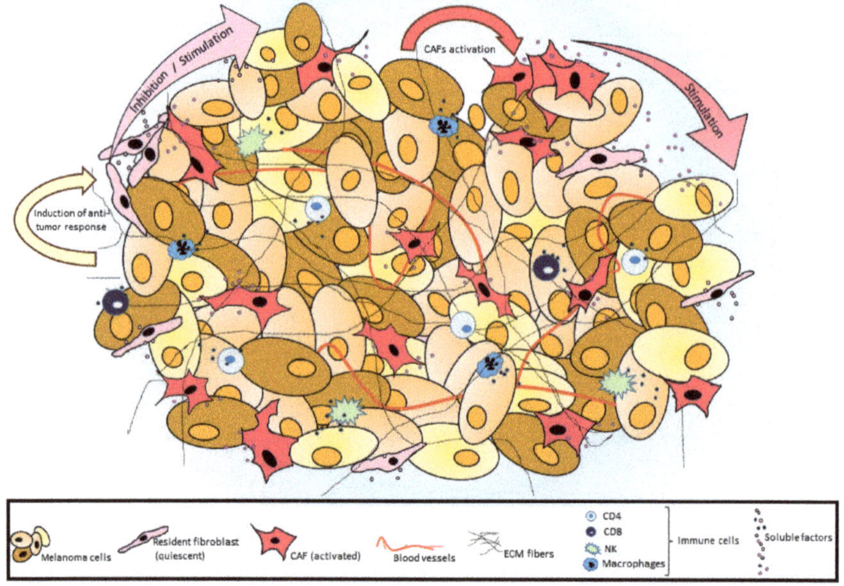

Figure 1. Schematic representation of tumor–stroma cross-talk. In melanoma, tumor cells share their microenvironment with MAFs, immune cells and blood vessels. Resident fibroblasts might oppose an early anti-tumor activity or facilitate tumor development, whereas, during disease progression, activated MAFs gradually acquire a marked pro-tumorigenic phenotype. An intense bi-directional exchange of soluble factors between melanoma and surrounding cells significantly modifies in MAFs several intracellular signaling pathways including oncogenic pathways. Extracellular matrix supports tumor architecture and influences various signal transduction pathways in both tumor and associated cells.

One of the biological behaviors of activated fibroblasts is increased proliferation rate [73]. A possible explanation of augmented proliferation of stromal cells might reside in the pro-mitogeic tumor milieu. At least during early stage of disease, the simple nearness to transformed cells could explain the involvement of non-cancerous cells that endure the extraordinary tumor secretory activity of melanoma cells. Proteomic analysis revealed that factors released in the culture medium by melanoma cells stimulate in dermal fibroblasts a general increase in protein synthesis. In particular, the biological process involved in fibroblast reprograming are those related to mRNA catabolic process, translation initiation, protein targeting to membrane and synthesis of metabolic-related small molecules [74,75]. Paired analysis of melanoma cells and associated MAFs revealed a trend of functionally coordinated reorganization of metabolic pathways, cytoskeleton reorganization and ECM remodeling [74]. Fibroblast reprograming is a crucial step for melanoma progression as demonstrated by the correlation between capacity of melanoma cells to alter fibroblast gene expression and the invasive potential in vitro [76]. Specifically, metastatic melanoma increases the production of cytokines and chemokines by MAFs more efficiently than nonmetastatic melanoma [76]. Among the immunomodulators involved in MAFs activation, IL1β seems to be a driver of melanoma invasion both in vitro and in vivo [76]. Cytokines produced by dermal fibroblasts, such as interleukin-6 and -8 (IL-6 and IL-8), interferon gamma (INFγ), tumor necrosis factor-alpha (TNF-α) [77,78] and a variety of CXCLs [37], have the capacity to mobilize immune cells. The balance between pro-inflammatory and anti-inflammatory cytokines in tumor area strongly impact on patient's prognosis. Cytokines profile supporting M1 macrophage differentiation sustain CD4+ and CD8+ T-cells infiltrating the tumor microenvironment and favorable clinical outcome, whereas alternative M2 macrophage polarization present immune-suppressive function. Since MAFs are the main producer of suppressive cytokines [46], their role in immune escape and tumor progression is relevant. MAFs suppress NK-cell activity and CD8+ cytotoxic activity [66,79]. Release of IL-8, CCL2/MCP1 (monocyte chemoattractant protein-1) and tissue inhibitor of metalloproteinase 2 (TIMP-2) by fibroblasts when co-cultured with melanoma cells has also been implicated in the angiogenic process indicating that recruitment of microvascular endothelial cells depends on the synergic melanoma-fibroblast network [80].

Considering all these pleiotropic roles, MAFs are considered a promising target for melanoma therapy. Essentially, targeting CAFs has been investigated by using agents aiming to eliminate or reprogram CAFs. However, in mouse model, reduced stromal content accelerates tumor growth and angiogenesis and full depletion of CAFs induces immunosuppression [81,82]. Thus, based on fibroblasts exceptional phenotypic plasticity, therapeutic re-orientation of CAFs to an anti-tumor phenotype seems to be more promising. Therefore, to efficiently reprogram CAFs' activity against tumor, we need to exactly know the contribution of stromal fibroblasts in a specific tumor type and to promote exclusively anti-neoplastic properties.

3. Significance of Altered Intracellular Signal Transduction Pathways in Melanoma and Associated Fibroblasts

3.1. Wnt Signaling

Signal activation due to the Wnt family of secreted glycoproteins is involved in embryogenic development, cell polarity, tissue homeostasis and cell proliferation in adult stage [83,84]. β-catenin is the key protein regulating Wnt signaling-mediated gene expression [85]. Moreover, β-catenin is deeply implicated in cadherin-based cell adhesion [86,87]. In the absence of Wnt ligands, β-catenin is recruited into a destruction complex that contains adenomatous polyposis coli (APC) and AXIN, which facilitates the phosphorylation of β-catenin by casein kinase 1 (CK1) and then glycogen synthase kinase-3 beta (GSK3β) leading to its ubiquitylation and proteasomal degradation. Following the binding of Wnt factors to their receptors (frizzled, FZD and low-density lipoprotein receptor protein 5/6 (LPR5/6)), cytosolic GSK-3β is sequestered, and the phosphorylation of β-catenin is prevented. The blocking of β-catenin degradation leads to its stabilization in the cytosol and consequent translocation into the nucleus, where it binds to members of lymphoid enhancer-binding factor (LEF)/T-cell specific factor (TCF) family and some other co-regulators to promote the transcription of ubiquitous genes such as *Jun*, *c-Myc* and *CyclinD-1*, most of which encode oncoproteins [88,89]. In addition, β-catenin is a co-activator for the expression of melanocyte-lineage restricted genes including Microphthalmia-associated Transcription Factor-M (*MITF-M*) [90–92], Dopachrome Tautomerase (*DCT*) [91,93–95] and *Brn-2* [96]. The ability of Wnt/β-catenin signaling to drive the expression of differentiation-related genes reflects its critical role in melanocyte development [91,97,98] and adult melanocyte stem cells mobilization [99,100]. Given peculiar involvement of β-catenin in both proliferation and differentiation of melanocytes, it is not surprising that in this type of cells its expression level is subjected to tight regulation. For example, in normal melanocytes in vitro, reduced β-catenin gene transcript by RNA interfering leads to a rapid stabilization of the corresponding protein capable of restoring the physiological level of expression [101].

As demonstrated in numerous studies, the aberrant activation of Wnt signaling contributes to malignant cell transformation and neoplastic proliferation with further metastatic dissemination and resistance to treatment [102–105]. Deregulations in the canonical Wnt signaling in cancer may result from gain-of-function gene alterations and epigenetic mechanisms. Nonetheless, β-catenin, *APC* and *AXIN2* mutations are rare in primary melanoma specimens [101,106–111], in comparison to well-characterized melanoma cell lines cultured in vitro for a long period [112]. This suggests that elevated β-catenin protein level confers proliferative advantage under a selective pressure such as in vitro cell growth. In melanoma, epigenetic regulation of Wnt antagonists such as Dickkopf proteins (DKKs), Wnt inhibitor factor-1 (WIF1) and soluble frizzled-related protein-2 sFRP2 contributes substantially to cell-autonomous activation of Wnt/β-catenin signaling [113,114]. *DKK2* and *DKK3* have been found upregulated in the good-prognosis melanoma patients presenting basal high immune signature [115]. Unlike most cancers, where Wnt signaling is considered a driver of both tumor formation and progression, in human melanoma, there are contradictory results [116–118]. The major discrepancies emerged from the comparison between studies performed with melanoma cell culture model and investigations based on immunohistochemical analyses in skin biopsy and clinical outcome. In vitro studies proposed that an increased nuclear translocation and activity of β-catenin promote melanoma proliferation [119] and invasion [105]. By contrast, diverse studies linked the activation of Wnt/β-catenin signaling to decreased proliferation [120] and repressed invasion [121] and migration [122]. In vivo observations from melanoma patients indicated that nuclear β-catenin correlates with improved survival [120–123]. Furthermore, almost all benign nevi are positive for nuclear β-catenin [124,125]. Adding complexity to this scenario, changes in Wnt signaling pathway have been linked to phenotype switching of melanoma cells between a highly proliferative/non-invasive (high β-catenin expressing cells) and a slow proliferative/metastatic (low β-catenin expressing cells) condition [126–129]. These data collectively suggest that the ambiguous role of Wnt pathway activation in melanoma strongly depends on the combination of both intracellular and microenvironmental contexts. The expression and transcriptional activity of β-catenin inversely correlate to immune activation and it has been proposed as a predictive marker of immunotherapy response [115,116,130–134]. The mechanism by which β-catenin promotes resistance to immunotherapy involves the reduced secretion of attractant chemokines that allows impaired infiltration and activation of dendritic and T cells [130]. Conversely, MAPK inhibitors demonstrated an enhanced efficacy in cultured melanoma cell lines with activated Wnt/β-catenin signaling [135,136]. However, in line with the idea that patient's immune activity contributes to MAPK inhibitors outcome, in vivo studies did not confirmed improvements in patient's survival presenting intrinsic β-catenin activation [137].

In the tumor background, in addition to genetic and epigenetic mechanisms, increased paracrine factors from the surrounding tissue might contribute for the activation of Wnt signaling in tumor cells [138]. Among these, Wnt ligands and several growth factors are frequently hyperproduced by CAFs [139,140] such as hepatocyte growth factor (HGF) [141] and platelet-derived growth factor (PDGF) [142]. It has been proposed that elevated level of sFRP2, a Wnt antagonist, secreted by aged fibroblasts could facilitate the acquisition of a metastatic, therapy-resistant state of melanoma [143]. Modification of Wnt pathway modulators in melanoma cells, including Wnt5a [128,144,145], Wnt7b, Wnt10b [146], Frizzled-3 (FZD3) [130] and DKKs [113] has been largely investigated as a cell autonomous mechanism responsible of signaling activation in tumor cells. However, this intense secretory activity might also deeply influence neighboring cell populations. Interestingly, Wnt ligands secreted by tumor cells could stimulate the polarization of tumor-associated macrophages (TAMs) to M2 tumor promoting subtype via Wnt signaling [147]. Thus, tumor paracrine activity might also play a relevant role in Wnt signaling regulation of other skin cell types, including mesenchymal cells. In dermal fibroblasts, transient stimulation of Wnt/β-catenin signaling is associated to an activated state of this type of cells during tissue repair [148–151]. Sustained Wnt/β-catenin activation in dermal fibroblasts is involved in pathogenesis of fibrotic diseases including hyperplastic wounds [152] and keloids [153–155]. Stroma recruited by melanoma resembles fibrotic microenvironment of persistent wound healing, since the entire tissue experiences a chronic injury due to the damage caused by tumor growth. As with other cancers, nuclear and cytoplasmic β-catenin has been demonstrated highly expressed in MAFs located around and in the melanoma tissue [72]. In 3D multicellular tumor spheroid model and in in vivo mouse melanoma model, skin fibroblasts ablated for *β-catenin* gene demonstrated reduced ability to support the growth of B16F10 melanoma cells [70,72]. In this case, the observed reduced number of stromal fibroblasts partially explain the diminished inhibitory effect of dermal fibroblast on melanoma formation. However, deactivation of β-catenin also weakens the expression of HGF and ECM proteins in residual fibroblasts [72,156]. Interestingly, β-catenin loss in fibroblast affects the RAF-MEK-ERK signaling cascade suppressing melanoma cell proliferation and cells death at the same time [72,157]. Shao and co-workers reported that the expression of Wnt-induced secreted protein-1 (WISP-1), a β-catenin target gene, is almost undetectable in areas with melanoma and in the surrounding tissue, whereas strong expression has been observed in non-activated fibroblasts of uninvolved skin [158] indicating a negative correlation between WISP-1 expression and a permissive tumor microenvironment. However, in this case, deregulated WISP-1 expression has been linked to the suppression of Notch signaling in MAFs rather than to the activity of Wnt pathway. As observed for other mesenchymal cells, independently of the presence of Wnts ligand at the extracellular level, Wnt signaling could be coordinately modulated in melanoma and stromal cells also by tissue stiffness [159]. Activation Wnt/β-catenin signaling in response to change in substrate stiffness might be stabilized or reinforced by a positive feedback loop based on the transcription of the β-catenin target gene *wnt-1* [159].

3.2. Hippo Signaling

Hippo signaling is an important regulator of cell proliferation and survival in animals playing a critical role in organ size control, stem cells homeostasis, cell polarity and shape [160,161]. The Hippo pathway is regulated upstream by extracellular mechanosensory signals arising from perturbation of actin cytoskeleton and adhesion change, as well as by a variety of extracellular signaling molecules. Hippo activity is deregulated in many cancers, despite mutations of pathway components are uncommon especially in melanoma [162,163]. Alteration of gene copy number among Hippo pathway elements have been frequently observed [164]. The Hippo signaling pathway includes a kinase cascade that modulates different proteins in order to phosphorylate and inactivate its main downstream cytosolic effectors, yes-associated protein (YAP) and tafazzin (TAZ), which direct gene expression via control of the Transcriptional enhancer factor (TEAD) family of transcription factors [165,166]. Nuclear YAP/TAZ interact with several important transcription factor including TCF/LEF, small mother against decapentaplegic factors (SMADs), cAMP response element-binding protein (CREB), myoblast determination protein 1 (MyoD) and tumor protein p73 (TP73) controlling cell proliferation and apoptosis [167–169]. This pathway is thought to be central to uveal melanomagenesis as YAP is hyperactive in uveal melanoma cells and mediates the oncogenic effect of guanine nucleotide binding protein (G protein) q polypeptide (GNAQ), or G protein α 11 (GNA11) mutations, which occur in approximately 80% of these type of melanoma [170,171]. YAP protein expression is elevated in most benign nevi and primary cutaneous melanomas but present at only very low levels in normal melanocytes [163]. Furthermore, since Hippo pathway is modulated by adhesion change and mechanical signaling, there is a strong possibility that extracellular stimuli from the melanoma microenvironment such as ECM modification might therefore impact on its activation. In melanoma patients, increased collagen and fibronectin abundance correlates with YAP nuclear localization [24]. Further analysis of the same patient cohort evidenced that melanoma cells positive for YAP nuclear staining present elevated MITF expression and a proliferation/differentiation signature. Proliferation and differentiation are not mutually exclusive events in the melanocyte lineage and are both promoted by MITF. By contrast, low level of MITF are associated to a dedifferentiated/invasive phenotype [126,172]. Several other studies indicated that hyperactive YAP is sufficient to drive cells switch from proliferative to invasive phenotype [173,174]. In murine xenograft model silencing the expression of YAP and TAZ results in reduced proliferation of human melanoma cell lines and decreases lung metastasis [133]. The E-cadherin/catenin complex functions as an upstream regulator of the Hippo signaling combining loss of E-cadherin at the cell surface with β-catenin and YAP nuclear accumulation [175]. Added evidence demonstrated cross-regulation between Wnt/β-catenin and Hippo signaling: TAZ is targeted for degradation by the β-catenin destruction complex [176,177]. On the other hand, YAP and TAZ can retain β-catenin in cytoplasm limiting Wnt/β-catenin signaling [178]. Thus, YAP and TAZ can be viewed as integral components of the Wnt/β-catenin signaling pathway in addition to their role in Hippo signaling [178,179]. Interestingly, TGF-β produced by stromal fibroblasts might exert a fine regulation of YAP transcriptional activity promoting YAP/SMAD interaction instead of YAP/PAX3 transcription complex redirecting cells to a less differentiated more aggressive state similar to melanocyte stem cells [24].

In cutaneous melanoma, evidence from cell lines supports not only a role of YAP/TAZ in cell invasion but also in resistance to target therapy and immunotherapy [171,180,181]. YAP promotes PD-1 expression driving immune evasion in BRAF inhibitors-resistant melanoma [181,182]. However, opposite to tumor cells, activation of YAP and TAZ in CAF exerts a tumor-suppressive function. In fact, deletion of *YAP* and *TAZ* in these peritumoral cells accelerated primary liver tumor growth. Experimental hyperactivation of YAP in peritumoral hepatocytes triggered regression of melanoma-derived liver metastasis. Since tumor cell survival depend on the relative activity of YAP and TAZ tumor and surrounding tissue, it has been hypothesized that the major function of YAP/TAZ in tumor cells is to elevate their competitive fitness and to "protect" them from the tumor-suppressive action of the surrounding parenchyma [182]. According to the concept of cell competition, an interesting point resides in the concept that it is the relative level and not the absolute level of a molecular pathway (not only Hippo pathway) that determines which cells (peritumoral or tumoral) lose competition. On the other hand, acquired mutations render cancer cells more competitive than non-neoplastic cells compromising tumor elimination [183].

3.3. TGF-β Signaling

The transforming growth factors (TGF)-β family of growth factors are secreted multifunctional cytokines that signals via plasma membrane TGF-β type I and type II receptors and intercellular small mother against decapentaplegic (SMADs) transcriptional effectors [184]. TGF-β controls tissue remodeling during embryonic development, angiogenesis, tissue repair and several cellular functions, such as cell growth, adhesion, recognition, cell fate determination and apoptosis [185,186]. In the skin, TGF-β is important for the wound healing process, especially in burn wounds [187]. TGF-β has a dual action in cancer as a tumor suppressor and a tumor promoter. As a tumor suppressor, it inhibits tumorigenesis by inducing growth arrest and apoptosis. As a tumor promoter, it induces tumor cell migration and stimulates epithelial to mesenchymal transition, a process during which cancer cells lose epithelial features and activate genes that increases cell motility and dissemination [35,36,188].

Alterations of TGF-β signaling, including loss-of-function mutations in genes encoding TGF-β receptors or SMAD proteins, confer escape from the antiproliferative activity of TGF-β [189]. Melanoma produces increasing amounts of TGF-β with disease progression [190,191], providing an optimal microenvironment for undisturbed tumor growth. Contrary to other tumor types, no genetic alteration of TGF-β signaling molecules has been identified in melanoma [192]. Among different forms, TGF-β1 is secreted by normal melanocytes and melanomas at various stages, while TGF-β2 and TGF-β3 levels rise early in melanoma and increase with tumor progression [193]. In addition, a correlation between TGF-β2 expression and tumor thickness has been reported [194]. Although melanoma cells efficiently respond to TGF-β at the receptor level, in contrast to normal melanocytes, melanoma cells display various degree of desensitization to the growth inhibitory activity of TGF-β. Resistance to the growth inhibitory activity of TGF-β has been explained by the frequent aberrant activation of MAPK pathway in melanoma that is capable to reprogram intracellular TGF-β cascade [192]. Similar to wound-healing process, tumor-derived TGF-β is likely to recruit other stromal cells. It is well demonstrated that transforming growth factor-β produced by tumor cells may promote tumor growth and advancement by modifying the microenvironment [56]. Forced overexpression of TGF-β1 by melanoma cells activates stromal fibroblasts, leading to augmented collagen, fibronectin, tenascin and α2 integrin expression. In experimental mouse model, tumors generated by subcutaneous co-injection of fibroblasts with melanoma cells demonstrated that TGF-β-overexpressing melanoma cells exhibit fewer necrotic and apoptotic cells and form more lung metastases than control melanoma cells [22]. Thus, activation of stromal fibroblasts by tumor-derived TGF-β provides an optimal microenvironment for tumor progression and metastasis [56,64]. Autocrine loop based on TGF-β and stromal derived factor 1 (SDF-1) is necessary for the consolidation of activated CAF phenotype [43,195]. Mechanical stress is the second major factor for myofibroblasts activation [64,196]. Initial small changes in tissue stiffness occur during the inflammatory response in tumor development and seem to be induced by increased collagen production and crosslinking [197,198]. Secondarily, stiff ECM promotes myofibroblast phenotypic conversion improving the efficiency of latent TGFβ1 activation [64]. Metastatic and primary melanoma cell lines overexpress collagen VI and in vivo the level of this type of collagen positively correlates to advanced stages [197,199]. Hyperactive TGF-β signaling associated to loss of caveolin-1 promotes tumorigenesis by shifting fibroblasts toward catabolic metabolism, a mechanism that generates energy-rich metabolites [43,200]. Another member of the TGF superfamily, Nodal, an embryonic morphogen not expressed in healthy adult tissues, has been demonstrated highly present in MAFs in vivo and in vitro [201,202]. Fibroblasts activated by Nodal, promote melanoma proliferation in vitro and in xenograft tumor models. A recent study demonstrated that the extremely high level of TGF-β1 produced by melanoma cells and detected in patients' sera is also capable to activate fibroblasts of distant uninvolved skin [203]. This is of particular interest since recruitment of MAFs also supports the generation of metastatic niche necessary for invasion of distant organs [204]. Based on this observation, it is possible to extend, at least in advanced disease stage, the concept of melanoma microenvironment on the entire body.

3.4. MAPK Signaling

The mitogen-activated protein kinase (MAPK) pathway activation is a critical player in the biology of different types of cancer and is the most frequent pathway aberrantly activated in melanoma [205]. Up to 70% of melanomas exhibit activating mutations within the kinases *BRAF* gene [206–209] and approximately 15% of melanomas within *NRAS* gene [210,211], resulting in constitutive and sustained activation of downstream targets, RAS–MEK–ERK1/2 axis, in addition to unresponsive negative feedback mechanisms [212]. In BRAF and NRAS mutated cells, MAPK cascade is turned on without the need of ECM signaling or growth factors thereby allowing to proliferation, survival and cell transformation [213]. However, by itself oncogenic BRAF is not sufficient for melanoma and must cooperate with other processes to induce the fully cancerous state. In fact, *BRAF* is mutated in up to 80% of the benign nevi [214]. Indeed, nevi remain growth-arrested for decades and rarely progress into melanomas [215,216] presumably because aberrant BRAF signaling induces a robust senescence response mediated by upregulation of the cell cycle inhibitor p16 [217–219]. Escape from BRAF-induced senescence requires cooperation with other oncogenic process including additional DNA damage, epigenetic mechanisms, loss of PTEN, activation of PI3K/AKT and mTOR signaling, as well as metabolic reprogramming. In addition, microenvironmental mediators might directly and indirectly influence MAPK pathway activity in melanocyte lineage. Nevus melanocytes secrete several molecules belonging the SASP [220], a powerful autocrine/paracrine mechanism for the maintenance of the senescent state. Although *BRAF* mutation and activation of the MAPK pathway is important in nevogenesis, MAPK pathway activation do not appear to persist at high levels in nevi after growth arrest [214–221]. However, due to the potent re-activation of MAPK pathway in melanoma [222] selective BRAF inhibitors are used in the treatment of patients with BRAF-mutant advanced melanoma [223,224]. Unfortunately, patients frequently develop mutation-independent resistance to this therapy. Acquired resistance is mostly driven by the secretory activity of TAMs and MAFs [225–227].

Although TGF-β released locally from BRAF-inhibitor treated melanoma cells appeared to constitute an important mechanism of fibroblast activation, there is also evidence that the introduction of mutant *BRAF* into melanoma cells increases their secretion of interleukin (IL)-1α that causes tumor-associated fibroblasts to induce immune suppression [228]. In fibroblasts, the expression of mitogen-activated protein kinase kinase 1 (MAPKK1) is strongly induced by melanoma cells secretome [75]. Several studies underlined the key role of fibroblast-derived cytokines in MAPK inhibitor tolerance [225–229]. This effect is mostly due to the non-negligible direct effect of this class of compound on *BRAF* wild type surrounding non-melanoma cells. BRAF inhibition might lead to a paradoxical activation of MAPK in fibroblasts increasing the production of survival factor as neuregulin (NRG) and HGF [229]. Further, as an indirect effect, MAPKi-treated melanoma cells stimulate macrophages to produce IL-1β that in turn lead to the conversion of fibroblast into a melanoma-protective phenotype [225]. In a very interesting study, Hirata and collaborators demonstrated that BRAF inhibitors promote the formation of dense collagen fibrils and an overall increased matrix deposition by MAFs that render BRAF-mutant melanoma cells insensitive to treatment. Remodeled ECM leads to adhesion-dependent (integrin and focal adhesion-dependent) signaling to ERK that negates the effect of BRAF inhibition in the melanoma cells. Histological examination of melanoma sample from vemurafenib-resistant patients confirmed increased fibroblastic stroma and stiffer matrix once resistance to BRAF inhibitors had developed [230].

4. Conclusions

A huge number of studies documented that the imbalance of cellular homeostasis during melanomagenesis combines oncogenic transformation of melanocytes within altered tumor stroma. Thus, from the therapeutic point of view, tumor and stroma might be considered a unique functional unit. In melanoma, most of the targetable signal transduction pathways are correspondingly altered in MAFs, suggesting that target therapies presumably deeply impact on microenvironment biological behavior. Contextual modulation of oncogenic signal cascades in MAFs might arise from extrinsic stimuli (e.g., preexisting chronic tissue damage), melanoma-induced activation and stimulation by therapeutic pressure. On the other hand, in line with symmetric bi-directional cancer-fibroblast crosstalk, the competition imposed by antitumor function of tumor microenvironment might elicit activation of intracellular signaling in tumor cells exacerbating neoplastic phenotype. ECM remodeling seems to play a central role in melanoma biology since the activation of important signal transduction pathways involved in melanomagenesis are extremely sensible to mechanotransduction. This point of view implies a shift of the therapeutic approach from the neoplastic cell-centric to a stroma-centric consideration (Figure 2). Full characterization of CAFs might be considered part of the customization of healthcare. As discussed in this review, molecules already used in clinical practice, such as MAPKi, or which are in the preclinical study phase should be reevaluated considering the intercellular molecular dialogue of neoplastic and non-neoplastic cells. In our opinion, the development of strategies able to simultaneously target melanoma cells and MAFs represents an extraordinary opportunity in the current setting of precision cancer medicine.

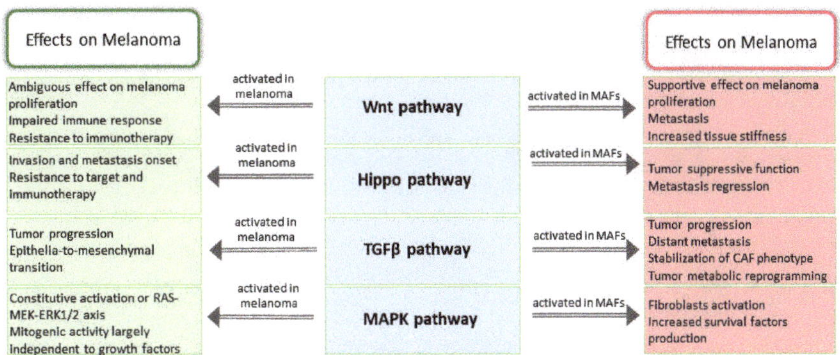

Figure 2. Tumor promoting function of oncogenic pathways deregulated in CAFs. Scheme recapitulates data presented concerning Wnt, Hippo, TGFβ and MAPK pathways deregulation in MAFs and melanoma cells. Mostly, activation of these pathways in MAFs exerts a pro-tumorigenic effect with the exception of Hippo signaling that trigger a competition between tumor and stroma characterized by a tumor-suppressive function.

Author Contributions: B.B. planned the review and prepared the manuscript. E.M. contributed to analyzing clinical data. M.P. coordinated all the activities. All authors have read and agreed to the published version of the manuscript.

Funding: This research received no external funding.

Conflicts of Interest: The authors declare that they have no known competing financial interests or personal relationships that could have appeared to influence the work reported in this paper.

References

1. Miller, A.J.; Mihm, M.C., Jr. Melanoma. *N. Engl. J. Med.* **2006**, *355*, 51–65. [CrossRef]
2. Houghton, A.N.; Polsky, D. Focus on Melanoma. *Cancer Cell.* **2002**, *2*, 275–278. [CrossRef]
3. Arnold, M.; Holterhues, C.; Hollestein, L.M.; Coebergh, J.W.; Nijsten, T.; Pukkala, E.; Holleczek, B.; Tryggvadóttir, L.; Comber, H.; Bento, M.J.; et al. Trends in incidence and predictions of cutaneous melanoma across Europe up to 2015. *J. Eur. Acad. Dermatol. Venereol.* **2014**, *28*, 1170–1178. [CrossRef] [PubMed]
4. Sun, X.; Zhang, N.; Yin, C.; Zhu, B.; Li, X. Ultraviolet radiation and melanomagenesis: From mechanism to immunotherapy. *Front. Oncol.* **2020**, *10*, 951. [CrossRef] [PubMed]
5. Palmieri, G.; Colombino, M.; Casula, M.; Manca, A.; Mandalà, M.; Cossu, A. Italian Melanoma Intergroup (IMI). Molecular pathways in melanomagenesis: What we learned from next-generation sequencing approaches. *Curr. Oncol. Rep.* **2018**, *20*, 86. [CrossRef] [PubMed]
6. Shain, A.H.; Bastian, B.C. From melanocytes to melanomas. *Nat. Rev. Cancer* **2016**, *16*, 345–358. [CrossRef] [PubMed]
7. Eggermont, A.M.; Spatz, A.; Robert, C. Cutaneous melanoma. *Lancet* **2014**, *383*, 816–827. [CrossRef]
8. Hawryluk, E.B.; Tsao, H. Melanoma: Clinical features and genomic insights. *Cold Spring Harb Perspect. Med.* **2014**, *4*, a015388. [CrossRef] [PubMed]
9. Landi, M.T.; Kanetsky, P.A.; Tsang, S.; Gold, B.; Munroe, D.; Rebbeck, T.; Swoyer, J.; Ter-Minassian, M.; Hedayati, M.; Grossman, L.; et al. MC1R, ASIP, and DNA repair in sporadic and familial melanoma in a mediterranean population. *J. Natl. Cancer Inst.* **2005**, *97*, 998–1007. [CrossRef]
10. Sample, A.; He, Y.Y. Mechanisms and prevention of UV-induced melanoma. *Photodermatol. Photoimmunol. Photomed.* **2018**, *34*, 13–24. [CrossRef]
11. Schulman, J.M.; Fisher, D.E. Indoor ultraviolet tanning and skin cancer: Health risks and opportunities. *Curr. Opin. Oncol.* **2009**, *21*, 144–149. [CrossRef] [PubMed]
12. Sarkar, S.; Gaddameedhi, S. Solar ultraviolet-induced DNA damage response: Melanocytes story in transformation to environmental melanomagenesis. *Environ. Mol. Mutagen.* **2020**, *61*, 736–751. [CrossRef] [PubMed]

13. Cichorek, M.; Wachulska, M.; Stasiewicz, A.; Tymińska, A. Skin melanocytes: Biology and development. *Postepy Dermatol. Alergol.* **2013**, *30*, 30–41. [CrossRef] [PubMed]

14. Wang, H.T.; Choi, B.; Tang, M.S. Melanocytes are deficient in repair of oxidative DNA damage and UV-induced photoproducts. *Proc. Natl. Acad. Sci. USA* **2010**, *107*, 12180–12185. [CrossRef]

15. Khan, A.Q.; Travers, J.B.; Kemp, M.G. Roles of UVA radiation and DNA damage responses in melanoma pathogenesis. *Env. Mol. Mutagen.* **2018**, *59*, 438–460. [CrossRef]

16. Mo, X.; Preston, S.; Zaidi, M.R. Macroenvironment-gene-microenvironment interactions in ultraviolet radiation-induced melanomagenesis. *Adv. Cancer Res.* **2019**, *144*, 1–54.

17. Zhang, S.; Duan, E. Fighting against skin aging: The way from bench to bedside. *Cell Transplant.* **2018**, *27*, 729–738. [CrossRef]

18. Naylor, E.C.; Watson, R.E.; Sherratt, M.J. Molecular aspects of skin ageing. *Maturitas* **2011**, *69*, 249–256. [CrossRef]

19. Yaar, M.; Gilchrest, B.A. Photoageing: Mechanism, prevention and therapy. *Br. J. Dermatol.* **2007**, *157*, 874–887. [CrossRef]

20. Chen, Y.; Chen, Y.; Huang, L.; Yu, J. Evaluation of heparanase and matrix metalloproteinase-9 in patients with cutaneous malignant melanoma. *J. Dermatol.* **2012**, *39*, 339–343. [CrossRef]

21. Rotte, A.; Martinka, M.; Li, G. MMP2 expression is a prognostic marker for primary melanoma patients. *Cell Oncol. (Dordr.)* **2012**, *35*, 207–216. [CrossRef]

22. Berking, C.; Takemoto, R.; Schaider, H.; Showe, L.; Satyamoorthy, K.; Robbins, P.; Herlyn, M. Transforming growth factor-beta1 increases survival of human melanoma through stroma remodeling. *Cancer Res.* **2001**, *61*, 8306–8316.

23. Kirkpatrick, S.J.; Wang, R.K.; Duncan, D.D.; Kulesz-Martin, M.; Lee, K. Imaging the mechanical stiffness of skin lesions by in vivo acousto-optical elastography. *Opt. Express* **2006**, *14*, 9770–9779. [CrossRef]

24. Miskolczi, Z.; Smith, M.P.; Rowling, E.J.; Ferguson, J.; Barriuso, J.; Wellbrock, C. Collagen abundance controls melanoma phenotypes through lineage-specific microenvironment sensing. *Oncogene* **2018**, *37*, 3166–3182. [CrossRef]

25. Natarajan, V.T.; Ganju, P.; Ramkumar, A.; Grover, R.; Gokhale, R.S. Multifaceted pathways protect human skin from UV radiation. *Nat. Chem. Biol.* **2014**, *10*, 542–551. [CrossRef]

26. Wang, J.X.; Fukunaga-Kalabis, M.; Herlyn, M. Crosstalk in skin: Melanocytes, keratinocytes, stem cells, and melanoma. *J. Cell. Commun. Signal.* **2016**, *10*, 191–196. [CrossRef]

27. Maresca, V.; Flori, E.; Bellei, B.; Aspite, N.; Kovacs, D.; Picardo, M. MC1R stimulation by alpha-MSH induces catalase and promotes its Re-distribution to the cell periphery and dendrites. *Pigment Cell. Melanoma Res.* **2010**, *23*, 263–275. [CrossRef]

28. Brenner, M.; Degitz, K.; Besch, R.; Berking, C. Differential expression of melanoma-associated growth factors in keratinocytes and fibroblasts by ultraviolet A and ultraviolet B radiation. *Br. J. Dermatol.* **2005**, *153*, 733–739. [CrossRef]

29. Berking, C.; Takemoto, R.; Satyamoorthy, K.; Elenitsas, R.; Herlyn, M. Basic fibroblast growth factor and ultraviolet B transform melanocytes in human skin. *Am. J. Pathol.* **2001**, *158*, 943–953. [CrossRef]

30. Kovacs, D.; Cardinali, G.; Aspite, N.; Cota, C.; Luzi, F.; Bellei, B.; Briganti, S.; Amantea, A.; Torrisi, M.R.; Picardo, M. Role of fibroblast-derived growth factors in regulating hyperpigmentation of solar lentigo. *Br. J. Dermatol.* **2010**, *163*, 1020–1027. [CrossRef]

31. Lin, C.B.; Hu, Y.; Rossetti, D.; Chen, N.; David, C.; Slominski, A.; Seiberg, M. Immuno-histochemical evaluation of solar lentigines: The association of KGF/KGFR and other factors with lesion development. *J. Dermatol. Sci.* **2010**, *59*, 91–97. [CrossRef]

32. Passeron, T.; Picardo, M. Melasma, a photoaging disorder. *Pigment Cell. Melanoma Res.* **2018**, *31*, 461–465. [CrossRef]

33. Imokawa, G. Autocrine and paracrine regulation of melanocytes in human skin and in pigmentary disorders. *Pigment Cell Res.* **2004**, *17*, 96–110. [CrossRef]

34. Mellone, M.; Hanley, C.J.; Thirdborough, S.; Mellows, T.; Garcia, E.; Woo, J.; Tod, J.; Frampton, S.; Jenei, V.; Moutasim, K.A.; et al. Induction of fibroblast senescence generates a non-fibrogenic myofibroblast phenotype that differentially impacts on cancer prognosis. *Aging (Albany N. Y.)* **2016**, *9*, 114–132. [CrossRef]

35. Ahmadi, A.; Najafi, M.; Farhood, B.; Mortezaee, K. Transforming growth factor-β signaling: Tumorigenesis and targeting for cancer therapy. *J. Cell. Physiol.* **2019**, *234*, 12173–12187. [CrossRef]

36. Miyazono, K.; Katsuno, Y.; Koinuma, D.; Ehata, S.; Morikawa, M. Intracellular and extracellular TGF-β signaling in cancer: Some recent topics. *Front. Med.* **2018**, *12*, 387–411. [CrossRef]
37. Alspach, E.; Fu, Y.; Stewart, S.A. Senescence and the pro-tumorigenic stroma. *Crit. Rev. Oncog.* **2013**, *18*, 549–558. [CrossRef]
38. Kim, E.; Rebecca, V.; Fedorenko, I.V.; Messina, J.L.; Mathew, R.; Maria-Engler, S.S.; Basanta, D.; Smalley, K.S.; Anderson, A.R. Senescent fibroblasts in melanoma initiation and progression: An integrated theoretical, experimental, and clinical approach. *Cancer Res.* **2013**, *73*, 6874–6885. [CrossRef]
39. Moinfar, F.; Beham, A.; Friedrich, G.; Deutsch, A.; Hrzenjak, A.; Luschin, G.; Tavassoli, F.A. Macro-environment of breast carcinoma: Frequent genetic alterations in the normal appearing skins of patients with breast cancer. *Mod. Pathol.* **2008**, *21*, 639–646. [CrossRef]
40. Balch, C.M.; Soong, S.J.; Gershenwald, J.E.; Thompson, J.F.; Reintgen, D.S.; Cascinelli, N.; Urist, M.; McMasters, K.M.; Ross, M.I.; Kirkwood, J.M.; et al. Prognostic factors analysis of 17,600 melanoma patients: Validation of the american joint committee on cancer melanoma staging system. *J. Clin. Oncol.* **2001**, *19*, 3622–3634. [CrossRef]
41. Coppé, J.P.; Desprez, P.Y.; Krtolica, A.; Campisi, J. The senescence-associated secretory phenotype: The dark side of tumor suppression. *Annu. Rev. Pathol.* **2010**, *5*, 99–118. [CrossRef]
42. Cirri, P.; Chiarugi, P. Cancer associated fibroblasts: The dark side of the coin. *Am. J. Cancer Res.* **2011**, *1*, 482–497.
43. Guido, C.; Whitaker-Menezes, D.; Capparelli, C.; Balliet, R.; Lin, Z.; Pestell, R.G.; Howell, A.; Aquila, S.; Andò, S.; Martinez-Outschoorn, U.; et al. Metabolic reprogramming of cancer-associated fibroblasts by TGF-β drives tumor growth: Connecting TGF-β signaling with "warburg-like" cancer metabolism and L-lactate production. *Cell. Cycle* **2012**, *11*, 3019–3035. [CrossRef]
44. Micke, P.; Ostman, A. Exploring the tumour environment: Cancer-associated fibroblasts as targets in cancer therapy. *Expert Opin. Ther. Targets* **2005**, *9*, 1217–1233. [CrossRef]
45. Bellei, B.; Picardo, M. Premature cell senescence in human skin: Dual face in chronic acquired pigmentary disorders. *Ageing Res. Rev.* **2020**, *57*, 100981. [CrossRef]
46. Almeida, F.V.; Douglass, S.M.; Fane, M.E.; Weeraratna, A.T. Bad Company: Microenvironmentally mediated resistance to targeted therapy in melanoma. *Pigment Cell. Melanoma Res.* **2019**, *32*, 237–247. [CrossRef]
47. Tsang, M.; Quesnel, K.; Vincent, K.; Hutchenreuther, J.; Postovit, L.M.; Leask, A. Insights into fibroblast plasticity: Cellular communication network 2 is required for activation of cancer-associated fibroblasts in a murine model of melanoma. *Am. J. Pathol.* **2020**, *190*, 206–221. [CrossRef]
48. Danen, E.H.; de Vries, T.J.; Morandini, R.; Ghanem, G.G.; Ruiter, D.J.; van Muijen, G.N. E-cadherin expression in human melanoma. *Melanoma Res.* **1996**, *6*, 127–131. [CrossRef]
49. Hsu, M.Y.; Meier, F.E.; Nesbit, M.; Hsu, J.Y.; Van Belle, P.; Elder, D.E.; Herlyn, M. E-cadherin expression in melanoma cells restores keratinocyte-mediated growth control and down-regulates expression of invasion-related adhesion receptors. *Am. J. Pathol.* **2000**, *156*, 1515–1525. [CrossRef]
50. Silye, R.; Karayiannakis, A.J.; Syrigos, K.N.; Poole, S.; van Noorden, S.; Batchelor, W.; Regele, H.; Sega, W.; Boesmueller, H.; Krausz, T.; et al. E-cadherin/catenin complex in benign and malignant melanocytic lesions. *J. Pathol.* **1998**, *186*, 350–355. [CrossRef]
51. Hsu, M.Y.; Meier, F.; Herlyn, M. Melanoma development and progression: A conspiracy between tumor and host. *Differentiation* **2002**, *70*, 522–536. [CrossRef] [PubMed]
52. Sanders, D.S.; Blessing, K.; Hassan, G.A.; Bruton, R.; Marsden, J.R.; Jankowski, J. Alterations in cadherin and catenin expression during the biological progression of melanocytic tumours. *Mol. Pathol.* **1999**, *52*, 151–157. [CrossRef] [PubMed]
53. Haass, N.K.; Smalley, K.S.; Li, L.; Herlyn, M. Adhesion, migration and communication in melanocytes and melanoma. *Pigment Cell Res.* **2005**, *18*, 150–159. [CrossRef] [PubMed]
54. Tang, A.; Eller, M.S.; Hara, M.; Yaar, M.; Hirohashi, S.; Gilchrest, B.A. E-cadherin is the major mediator of human melanocyte adhesion to keratinocytes in vitro. *J. Cell. Sci.* **1994**, *107*, 983–992.
55. Li, G.; Satyamoorthy, K.; Herlyn, M. N-cadherin-mediated intercellular interactions promote survival and migration of melanoma cells. *Cancer Res.* **2001**, *61*, 3819–3825.
56. Li, G.; Satyamoorthy, K.; Meier, F.; Berking, C.; Bogenrieder, T.; Herlyn, M. Function and regulation of melanoma-stromal fibroblast interactions: When seeds meet soil. *Oncogene* **2003**, *22*, 3162–3171. [CrossRef]

57. Kuphal, S.; Poser, I.; Jobin, C.; Hellerbrand, C.; Bosserhoff, A.K. Loss of E-cadherin leads to upregulation of NFkappaB activity in malignant melanoma. *Oncogene* **2004**, *23*, 8509–8519. [CrossRef]

58. McGary, E.C.; Lev, D.C.; Bar-Eli, M. Cellular adhesion pathways and metastatic potential of human melanoma. *Cancer Biol. Ther.* **2002**, *1*, 459–465. [CrossRef]

59. Hazan, R.B.; Phillips, G.R.; Qiao, R.F.; Norton, L.; Aaronson, S.A. Exogenous expression of N-cadherin in breast cancer cells induces cell migration, invasion, and metastasis. *J. Cell Biol.* **2000**, *148*, 779–790. [CrossRef]

60. Raposo, G.; Marks, M.S. Melanosomes—dark organelles enlighten endosomal membrane transport. *Nat. Rev. Mol. Cell Biol.* **2007**, *8*, 786–797. [CrossRef]

61. Dror, S.; Sander, L.; Schwartz, H.; Sheinboim, D.; Barzilai, A.; Dishon, Y.; Apcher, S.; Golan, T.; Greenberger, S.; Barshack, I.; et al. Melanoma miRNA trafficking controls tumour primary niche formation. *Nat. Cell Biol.* **2016**, *18*, 1006–1017. [CrossRef] [PubMed]

62. Anderberg, C.; Pietras, K. On the origin of cancer-associated fibroblasts. *Cell. Cycle* **2009**, *8*, 1461–1462. [CrossRef] [PubMed]

63. Hauge, A.; Rofstad, E.K. Antifibrotic therapy to normalize the tumor microenvironment. *J. Transl. Med.* **2020**, *18*, 207–218. [CrossRef]

64. Otranto, M.; Sarrazy, V.; Bonté, F.; Hinz, B.; Gabbiani, G.; Desmoulière, A. The role of the myofibroblast in tumor stroma remodeling. *Cell. Adh. Migr.* **2012**, *6*, 203–219. [CrossRef] [PubMed]

65. Shimoda, M.; Mellody, K.T.; Orimo, A. Carcinoma-associated fibroblasts are a rate-limiting determinant for tumour progression. *Semin. Cell Dev. Biol.* **2010**, *21*, 19–25. [CrossRef] [PubMed]

66. Érsek, B.; Silló, P.; Cakir, U.; Molnár, V.; Bencsik, A.; Mayer, B.; Mezey, E.; Kárpáti, S.; Pós, Z.; Németh, K. Melanoma-associated fibroblasts impair CD8+ T cell function and modify expression of immune checkpoint regulators via increased arginase activity. *Cell Mol. Life Sci.* **2020**, *77*, 1–13.

67. Zhou, L.; Yang, K.; Andl, T.; Wickett, R.R.; Zhang, Y. Perspective of targeting cancer-associated fibroblasts in melanoma. *J. Cancer* **2015**, *6*, 717–726. [CrossRef]

68. García-Silva, S.; Peinado, H. Melanosomes foster a tumour niche by activating CAFs. *Nat. Cell Biol.* **2016**, *18*, 911–913. [CrossRef]

69. Gieniec, K.A.; Butler, L.M.; Worthley, D.L.; Woods, S.L. Cancer-associated fibroblasts-heroes or villains? *Br. J. Cancer* **2019**, *121*, 293–302. [CrossRef]

70. Zhou, L.; Yang, K.; Randall Wickett, R.; Zhang, Y. Dermal fibroblasts induce cell cycle arrest and block epithelial-mesenchymal transition to inhibit the early stage melanoma development. *Cancer Med.* **2016**, *5*, 1566–1579. [CrossRef]

71. Cornil, I.; Theodorescu, D.; Man, S.; Herlyn, M.; Jambrosic, J.; Kerbel, R.S. Fibroblast cell interactions with human melanoma cells affect tumor cell growth as a function of tumor progression. *Proc. Natl. Acad. Sci. USA* **1991**, *88*, 6028–6032. [CrossRef] [PubMed]

72. Zhou, L.; Yang, K.; Wickett, R.R.; Kadekaro, A.L.; Zhang, Y. Targeted deactivation of cancer-associated fibroblasts by β-catenin ablation suppresses melanoma growth. *Tumour Biol.* **2016**, *37*, 14235–14248. [CrossRef] [PubMed]

73. Madar, S.; Brosh, R.; Buganim, Y.; Ezra, O.; Goldstein, I.; Solomon, H.; Kogan, I.; Goldfinger, N.; Klocker, H.; Rotter, V. Modulated expression of WFDC1 during carcinogenesis and cellular senescence. *Carcinogenesis* **2009**, *30*, 20–27. [CrossRef] [PubMed]

74. Liberato, T.; Pessotti, D.S.; Fukushima, I.; Kitano, E.S.; Serrano, S.M.T.; Zelanis, A. Signatures of protein expression revealed by secretome analyses of cancer associated fibroblasts and melanoma cell lines. *J. Proteomics* **2018**, *174*, 1–8. [CrossRef] [PubMed]

75. Pessotti, D.S.; Andrade-Silva, D.; Serrano, S.M.T.; Zelanis, A. Heterotypic signaling between dermal fibroblasts and melanoma cells induces phenotypic plasticity and proteome rearrangement in malignant cells. *Biochim. Biophys. Acta Proteins Proteom.* **2020**, *1868*, 140525. [CrossRef] [PubMed]

76. Li, L.; Dragulev, B.; Zigrino, P.; Mauch, C.; Fox, J.W. The invasive potential of human melanoma cell lines correlates with their ability to alter fibroblast gene expression in vitro and the stromal microenvironment in vivo. *Int. J. Cancer* **2009**, *125*, 1796–1804. [CrossRef]

77. Lu, C.; Vickers, M.F.; Kerbel, R.S. Interleukin 6: A fibroblast-derived growth inhibitor of human melanoma cells from early but not advanced stages of tumor progression. *Proc. Natl. Acad. Sci. USA* **1992**, *89*, 9215–9219. [CrossRef]

78. Jobe, N.P.; Rösel, D.; Dvořánková, B.; Kodet, O.; Lacina, L.; Mateu, R.; Smetana, K.; Brábek, J. Simultaneous blocking of IL-6 and IL-8 is sufficient to fully inhibit CAF-induced human melanoma cell invasiveness. *Histochem. Cell Biol.* **2016**, *146*, 205–217. [CrossRef]

79. Balsamo, M.; Scordamaglia, F.; Pietra, G.; Manzini, C.; Cantoni, C.; Boitano, M.; Queirolo, P.; Vermi, W.; Facchetti, F.; Moretta, A.; et al. Melanoma-associated fibroblasts modulate NK cell phenotype and antitumor cytotoxicity. *Proc. Natl. Acad. Sci. USA* **2009**, *106*, 20847–20852. [CrossRef]

80. Goldstein, L.J.; Chen, H.; Bauer, R.J.; Bauer, S.M.; Velazquez, O.C. Normal human fibroblasts enable melanoma cells to induce angiogenesis in type I collagen. *Surgery* **2005**, *138*, 439–449. [CrossRef]

81. Rhim, A.D.; Oberstein, P.E.; Thomas, D.H.; Mirek, E.T.; Palermo, C.F.; Sastra, S.A.; Dekleva, E.N.; Saunders, T.; Becerra, C.P.; Tattersall, I.W.; et al. Stromal elements act to restrain, rather than support, pancreatic ductal adenocarcinoma. *Cancer Cell* **2014**, *25*, 735–747. [CrossRef] [PubMed]

82. Özdemir, B.C.; Pentcheva-Hoang, T.; Carstens, J.L.; Zheng, X.; Wu, C.C.; Simpson, T.R.; Laklai, H.; Sugimoto, H.; Kahlert, C.; Novitskiy, S.V.; et al. Depletion of carcinoma-associated fibroblasts and fibrosis induces immunosuppression and accelerates pancreas cancer with reduced survival. *Cancer Cell* **2015**, *28*, 831–833. [CrossRef] [PubMed]

83. Kikuchi, A.; Yamamoto, H.; Sato, A.; Matsumoto, S. New insights into the mechanism of Wnt signaling pathway activation. *Int. Rev. Cell. Mol. Biol.* **2011**, *291*, 21–71. [PubMed]

84. van Amerongen, R.; Nusse, R. Towards an integrated view of Wnt signaling in development. *Development* **2009**, *136*, 3205–3214. [CrossRef]

85. MacDonald, B.T.; Tamai, K.; He, X. Wnt/beta-catenin signaling: Components, mechanisms, and diseases. *Dev. Cell.* **2009**, *17*, 9–26. [CrossRef]

86. Gloushankova, N.A.; Rubtsova, S.N.; Zhitnyak, I.Y. Cadherin-mediated cell-cell interactions in normal and cancer cells. *Tissue Barriers* **2017**, *5*, e1356900. [CrossRef]

87. Tafrihi, M.; Nakhaei Sistani, R. E-cadherin/β-catenin complex: A target for anticancer and antimetastasis plants/plant-derived compounds. *Nutr. Cancer* **2017**, *69*, 702–722. [CrossRef]

88. Behrens, J.; Jerchow, B.A.; Würtele, M.; Grimm, J.; Asbrand, C.; Wirtz, R.; Kühl, M.; Wedlich, D.; Birchmeier, W. Functional interaction of an axin homolog, conductin, with beta-catenin, APC, and GSK3beta. *Science* **1998**, *280*, 596–599. [CrossRef]

89. Hrckulak, D.; Kolar, M.; Strnad, H.; Korinek, V. TCF/LEF transcription factors: An update from the internet resources. *Cancers* **2016**, *8*, 70. [CrossRef]

90. Dorsky, R.I.; Raible, D.W.; Moon, R.T. Direct regulation of nacre, a zebrafish MITF homolog required for pigment cell formation, by the Wnt pathway. *Genes Dev.* **2000**, *14*, 158–162.

91. Schepsky, A.; Bruser, K.; Gunnarsson, G.J.; Goodall, J.; Hallsson, J.H.; Goding, C.R.; Steingrimsson, E.; Hecht, A. The microphthalmia-associated transcription factor Mitf interacts with beta-catenin to determine target gene expression. *Mol. Cell. Biol.* **2006**, *26*, 8914–8927. [CrossRef] [PubMed]

92. Saito, H.; Yasumoto, K.; Takeda, K.; Takahashi, K.; Yamamoto, H.; Shibahara, S. Microphthalmia-associated transcription factor in the Wnt signaling pathway. *Pigment Cell Res.* **2003**, *16*, 261–265. [CrossRef] [PubMed]

93. Larue, L.; Delmas, V. The WNT/beta-catenin pathway in melanoma. *Front. Biosci.* **2006**, *11*, 733–742. [CrossRef] [PubMed]

94. Bellei, B.; Flori, E.; Izzo, E.; Maresca, V.; Picardo, M. GSK3beta inhibition promotes melanogenesis in mouse B16 melanoma cells and normal human melanocytes. *Cell. Signal.* **2008**, *20*, 1750–1761. [CrossRef] [PubMed]

95. Bellei, B.; Pitisci, A.; Catricalà, C.; Larue, L.; Picardo, M. Wnt/β-catenin signaling is stimulated by α-melanocyte-stimulating hormone in melanoma and melanocyte cells: Implication in cell differentiation. *Pigment Cell. Melanoma Res.* **2011**, *24*, 309–325. [CrossRef]

96. Larue, L.; Kumasaka, M.; Goding, C.R. Beta-catenin in the melanocyte lineage. *Pigment Cell Res.* **2003**, *16*, 312–317. [CrossRef]

97. Hari, L.; Brault, V.; Kléber, M.; Lee, H.Y.; Ille, F.; Leimeroth, R.; Paratore, C.; Suter, U.; Kemler, R.; Sommer, L. Lineage-specific requirements of beta-catenin in neural crest development. *J. Cell Biol.* **2002**, *159*, 867–880. [CrossRef]

98. De Melker, A.A.; Desban, N.; Duband, J.L. Cellular localization and signaling activity of beta-catenin in migrating neural crest cells. *Dev. Dyn.* **2004**, *230*, 708–726. [CrossRef]

99. Yamada, T.; Hasegawa, S.; Inoue, Y.; Date, Y.; Yamamoto, N.; Mizutani, H.; Nakata, S.; Matsunaga, K.; Akamatsu, H. Wnt/β-catenin and kit signaling sequentially regulate melanocyte stem cell differentiation in UVB-induced epidermal pigmentation. *J. Investig. Dermatol.* **2013**, *133*, 2753–2762. [CrossRef]

100. Goldstein, N.B.; Koster, M.I.; Jones, K.L.; Gao, B.; Hoaglin, L.G.; Robinson, S.E.; Wright, M.J.; Birlea, S.I.; Luman, A.; Lambert, K.A.; et al. Repigmentation of human vitiligo skin by NBUVB is controlled by transcription of GLI1 and activation of the β-catenin pathway in the hair follicle bulge stem cells. *J. Investig. Dermatol.* **2018**, *138*, 657–668. [CrossRef]

101. Bellei, B.; Pacchiarotti, A.; Perez, M.; Faraggiana, T. Frequent beta-catenin overexpression without exon 3 mutation in cutaneous lymphomas. *Mod. Pathol.* **2004**, *17*, 1275–1281. [CrossRef] [PubMed]

102. Clevers, H. Wnt/beta-catenin signaling in development and disease. *Cell* **2006**, *127*, 469–480. [CrossRef] [PubMed]

103. Ring, A.; Kim, Y.M.; Kahn, M. Wnt/catenin signaling in adult stem cell physiology and disease. *Stem Cell. Rev. Rep.* **2014**, *10*, 512–525. [CrossRef] [PubMed]

104. Galluzzi, L.; Spranger, S.; Fuchs, E.; López-Soto, A. WNT signaling in cancer immunosurveillance. *Trends Cell Biol.* **2019**, *29*, 44–65. [CrossRef] [PubMed]

105. Sinnberg, T.; Menzel, M.; Ewerth, D.; Sauer, B.; Schwarz, M.; Schaller, M.; Garbe, C.; Schittek, B. B-catenin signaling increases during melanoma progression and promotes tumor cell survival and chemoresistance. *PLoS ONE* **2011**, *6*, e23429. [CrossRef] [PubMed]

106. Rimm, D.L.; Caca, K.; Hu, G.; Harrison, F.B.; Fearon, E.R. Frequent nuclear/cytoplasmic localization of beta-catenin without exon 3 mutations in malignant melanoma. *Am. J. Pathol.* **1999**, *154*, 325–329. [CrossRef]

107. Omholt, K.; Platz, A.; Ringborg, U.; Hansson, J. Cytoplasmic and nuclear accumulation of beta-catenin is rarely caused by CTNNB1 exon 3 mutations in cutaneous malignant melanoma. *Int. J. Cancer* **2001**, *92*, 839–842. [CrossRef]

108. Demunter, A.; Libbrecht, L.; Degreef, H.; De Wolf-Peeters, C.; van den Oord, J.J. Loss of membranous expression of beta-catenin is associated with tumor progression in cutaneous melanoma and rarely caused by exon 3 mutations. *Mod. Pathol.* **2002**, *15*, 454–461. [CrossRef]

109. Worm, J.; Christensen, C.; Grønbaek, K.; Tulchinsky, E.; Guldberg, P. Genetic and epigenetic alterations of the APC gene in malignant melanoma. *Oncogene* **2004**, *23*, 5215–5226. [CrossRef]

110. Reifenberger, J.; Knobbe, C.B.; Wolter, M.; Blaschke, B.; Schulte, K.W.; Pietsch, T.; Ruzicka, T.; Reifenberger, G. Molecular genetic analysis of malignant melanomas for aberrations of the WNT signaling pathway genes CTNNB1, APC, ICAT and BTRC. *Int. J. Cancer* **2002**, *100*, 549–556. [CrossRef]

111. Castiglia, D.; Bernardini, S.; Alvino, E.; Pagani, E.; De Luca, N.; Falcinelli, S.; Pacchiarotti, A.; Bonmassar, E.; Zambruno, G.; D'Atri, S. Concomitant activation of Wnt pathway and loss of mismatch repair function in human melanoma. *Genes Chromosomes Cancer* **2008**, *47*, 614–624. [CrossRef] [PubMed]

112. Rubinfeld, B.; Robbins, P.; El-Gamil, M.; Albert, I.; Porfiri, E.; Polakis, P. Stabilization of beta-catenin by genetic defects in melanoma cell lines. *Science* **1997**, *275*, 1790–1792. [CrossRef] [PubMed]

113. Kuphal, S.; Lodermeyer, S.; Bataille, F.; Schuierer, M.; Hoang, B.H.; Bosserhoff, A.K. Expression of dickkopf genes is strongly reduced in malignant melanoma. *Oncogene* **2006**, *25*, 5027–5036. [CrossRef] [PubMed]

114. Huynh, K.T.; Takei, Y.; Kuo, C.; Scolyer, R.A.; Murali, R.; Chong, K.; Takeshima, L.; Sim, M.S.; Morton, D.L.; Turner, R.R.; et al. Aberrant hypermethylation in primary tumours and sentinel lymph node metastases in paediatric patients with cutaneous melanoma. *Br. J. Dermatol.* **2012**, *166*, 1319–1326. [CrossRef] [PubMed]

115. Massi, D.; Romano, E.; Rulli, E.; Merelli, B.; Nassini, R.; De Logu, F.; Bieche, I.; Baroni, G.; Cattaneo, L.; Xue, G.; et al. Baseline β-catenin, programmed death-ligand 1 expression and tumour-infiltrating lymphocytes predict response and poor prognosis in BRAF inhibitor-treated melanoma patients. *Eur. J. Cancer* **2017**, *78*, 70–81. [CrossRef]

116. Gajos-Michniewicz, A.; Czyz, M. WNT signaling in melanoma. *Int. J. Mol. Sci.* **2020**, *21*, 4852. [CrossRef]

117. Lucero, O.M.; Dawson, D.W.; Moon, R.T.; Chien, A.J. A re-evaluation of the "oncogenic" nature of Wnt/beta-catenin signaling in melanoma and other cancers. *Curr. Oncol. Rep.* **2010**, *12*, 314–318. [CrossRef]

118. Larue, L.; Delmas, V. Secrets to developing Wnt-age melanoma revealed. *Pigment Cell. Melanoma Res.* **2009**, *22*, 520–521. [CrossRef]

119. Widlund, H.R.; Horstmann, M.A.; Price, E.R.; Cui, J.; Lessnick, S.L.; Wu, M.; He, X.; Fisher, D.E. Beta-Catenin-induced melanoma growth requires the downstream target microphthalmia-associated transcription factor. *J. Cell Biol.* **2002**, *158*, 1079–1087. [CrossRef]

120. Chien, A.J.; Moore, E.C.; Lonsdorf, A.S.; Kulikauskas, R.M.; Rothberg, B.G.; Berger, A.J.; Major, M.B.; Hwang, S.T.; Rimm, D.L.; Moon, R.T. Activated Wnt/beta-catenin signaling in melanoma is associated with decreased proliferation in patient tumors and a murine melanoma model. *Proc. Natl. Acad. Sci. USA* **2009**, *106*, 1193–1198. [CrossRef]

121. Arozarena, I.; Bischof, H.; Gilby, D.; Belloni, B.; Dummer, R.; Wellbrock, C. In melanoma, beta-catenin is a suppressor of invasion. *Oncogene* **2011**, *30*, 4531–4543. [CrossRef] [PubMed]

122. Gallagher, S.J.; Rambow, F.; Kumasaka, M.; Champeval, D.; Bellacosa, A.; Delmas, V.; Larue, L. Beta-catenin inhibits melanocyte migration but induces melanoma metastasis. *Oncogene* **2013**, *32*, 2230–2238. [CrossRef] [PubMed]

123. Bachmann, I.M.; Straume, O.; Puntervoll, H.E.; Kalvenes, M.B.; Akslen, L.A. Importance of p-cadherin, beta-catenin, and Wnt5a/frizzled for progression of melanocytic tumors and prognosis in cutaneous melanoma. *Clin. Cancer Res.* **2005**, *11*, 8606–8614. [CrossRef] [PubMed]

124. Kageshita, T.; Hamby, C.V.; Ishihara, T.; Matsumoto, K.; Saida, T.; Ono, T. Loss of beta-catenin expression associated with disease progression in malignant melanoma. *Br. J. Dermatol.* **2001**, *145*, 210–216. [CrossRef]

125. De la Fouchardière, A.; Caillot, C.; Jacquemus, J.; Durieux, E.; Houlier, A.; Haddad, V.; Pissaloux, D. B-catenin nuclear expression discriminates deep penetrating nevi from other cutaneous melanocytic tumors. *Virchows Arch.* **2019**, *474*, 539–550. [CrossRef]

126. Hoek, K.S.; Eichhoff, O.M.; Schlegel, N.C.; Döbbeling, U.; Kobert, N.; Schaerer, L.; Hemmi, S.; Dummer, R. In vivo switching of human melanoma cells between proliferative and invasive states. *Cancer Res.* **2008**, *68*, 650–656. [CrossRef]

127. Eichhoff, O.M.; Zipser, M.C.; Xu, M.; Weeraratna, A.T.; Mihic, D.; Dummer, R.; Hoek, K.S. The immunohistochemistry of invasive and proliferative phenotype switching in melanoma: A case report. *Melanoma Res.* **2010**, *20*, 349–355. [CrossRef]

128. Kovacs, D.; Migliano, E.; Muscardin, L.; Silipo, V.; Catricalà, C.; Picardo, M.; Bellei, B. The role of Wnt/β-catenin signaling pathway in melanoma epithelial-to-mesenchymal-like switching: Evidences from patients-derived cell lines. *Oncotarget* **2016**, *7*, 43295–43314. [CrossRef]

129. Webster, M.R.; Kugel, C.H., III; Weeraratna, A.T. The Wnts of change: How Wnts regulate phenotype switching in melanoma. *Biochim. Biophys. Acta* **2015**, *1856*, 244–251. [CrossRef]

130. Spranger, S.; Bao, R.; Gajewski, T.F. Melanoma-intrinsic β-catenin signalling prevents anti-tumour immunity. *Nature* **2015**, *523*, 231–235. [CrossRef]

131. Nsengimana, J.; Laye, J.; Filia, A.; O'Shea, S.; Muralidhar, S.; Poźniak, J.; Droop, A.; Chan, M.; Walker, C.; Parkinson, L.; et al. B-catenin-mediated immune evasion pathway frequently operates in primary cutaneous melanomas. *J. Clin. Investig.* **2018**, *128*, 2048–2063. [CrossRef] [PubMed]

132. Li, X.; Xiang, Y.; Li, F.; Yin, C.; Li, B.; Ke, X. WNT/β-catenin signaling pathway regulating T cell-inflammation in the tumor microenvironment. *Front. Immunol.* **2019**, *10*, 2293. [CrossRef] [PubMed]

133. Luke, J.J.; Bao, R.; Sweis, R.F.; Spranger, S.; Gajewski, T.F. WNT/β-catenin pathway activation correlates with immune exclusion across human cancers. *Clin. Cancer Res.* **2019**, *25*, 3074–3083. [CrossRef] [PubMed]

134. Ganesh, S.; Shui, X.; Craig, K.P.; Park, J.; Wang, W.; Brown, B.D.; Abrams, M.T. RNAi-mediated β-catenin inhibition promotes T cell infiltration and antitumor activity in combination with immune checkpoint blockade. *Mol. Ther.* **2018**, *26*, 2567–2579. [CrossRef]

135. Biechele, T.L.; Kulikauskas, R.M.; Toroni, R.A.; Lucero, O.M.; Swift, R.D.; James, R.G.; Robin, N.C.; Dawson, D.W.; Moon, R.T.; Chien, A.J. Wnt/β-catenin signaling and AXIN1 regulate apoptosis triggered by inhibition of the mutant kinase BRAFV600E in human melanoma. *Sci. Signal.* **2012**, *5*, ra3. [CrossRef]

136. Conrad, W.H.; Swift, R.D.; Biechele, T.L.; Kulikauskas, R.M.; Moon, R.T.; Chien, A.J. Regulating the response to targeted MEK inhibition in melanoma: Enhancing apoptosis in NRAS- and BRAF-mutant melanoma cells with Wnt/β-catenin activation. *Cell. Cycle* **2012**, *11*, 3724–3730. [CrossRef]

137. Chien, A.J.; Haydu, L.E.; Biechele, T.L.; Kulikauskas, R.M.; Rizos, H.; Kefford, R.F.; Scolyer, R.A.; Moon, R.T.; Long, G.V. Targeted BRAF inhibition impacts survival in melanoma patients with high levels of Wnt/β-catenin signaling. *PLoS ONE* **2014**, *9*, e94748. [CrossRef]

138. Fodde, R.; Brabletz, T. Wnt/beta-catenin signaling in cancer stemness and malignant behavior. *Curr. Opin. Cell Biol.* **2007**, *19*, 150–158. [CrossRef]

139. Unterleuthner, D.; Neuhold, P.; Schwarz, K.; Janker, L.; Neuditschko, B.; Nivarthi, H.; Crncec, I.; Kramer, N.; Unger, C.; Hengstschläger, M.; et al. Cancer-associated fibroblast-derived WNT2 increases tumor angiogenesis in colon cancer. *Angiogenesis* **2020**, *23*, 159–177. [CrossRef]

140. Kramer, N.; Schmöllerl, J.; Unger, C.; Nivarthi, H.; Rudisch, A.; Unterleuthner, D.; Scherzer, M.; Riedl, A.; Artaker, M.; Crncec, I.; et al. Autocrine WNT2 signaling in fibroblasts promotes colorectal cancer progression. *Oncogene* **2017**, *36*, 5460–5472. [CrossRef]

141. Rasola, A.; Fassetta, M.; De Bacco, F.; D'Alessandro, L.; Gramaglia, D.; Di Renzo, M.F.; Comoglio, P.M. A positive feedback loop between hepatocyte growth factor receptor and beta-catenin sustains colorectal cancer cell invasive growth. *Oncogene* **2007**, *26*, 1078–1087. [CrossRef] [PubMed]

142. Yang, L.; Lin, C.; Liu, Z.R. P68 RNA helicase mediates PDGF-induced epithelial mesenchymal transition by displacing axin from beta-catenin. *Cell* **2006**, *127*, 139–155. [CrossRef]

143. Kaur, A.; Webster, M.R.; Marchbank, K.; Behera, R.; Ndoye, A.; Kugel, C.H., III; Dang, V.M.; Appleton, J.; O'Connell, M.P.; Cheng, P.; et al. SFRP2 in the aged microenvironment drives melanoma metastasis and therapy resistance. *Nature* **2016**, *532*, 250–254. [CrossRef] [PubMed]

144. Bittner, M.; Meltzer, P.; Chen, Y.; Jiang, Y.; Seftor, E.; Hendrix, M.; Radmacher, M.; Simon, R.; Yakhini, Z.; Ben-Dor, A.; et al. Molecular classification of cutaneous malignant melanoma by gene expression profiling. *Nature* **2000**, *406*, 536–540. [CrossRef] [PubMed]

145. Weeraratna, A.T.; Jiang, Y.; Hostetter, G.; Rosenblatt, K.; Duray, P.; Bittner, M.; Trent, J.M. Wnt5a signaling directly affects cell motility and invasion of metastatic melanoma. *Cancer Cell* **2002**, *1*, 279–288. [CrossRef]

146. Pham, K.; Milovanovic, T.; Barr, R.J.; Truong, T.; Holcombe, R.F. Wnt ligand expression in malignant melanoma: Pilot study indicating correlation with histopathological features. *Mol. Pathol.* **2003**, *56*, 280–285. [CrossRef] [PubMed]

147. Yang, Y.; Ye, Y.C.; Chen, Y.; Zhao, J.L.; Gao, C.C.; Han, H.; Liu, W.C.; Qin, H.Y. Crosstalk between hepatic tumor cells and macrophages via Wnt/β-catenin signaling promotes M2-like macrophage polarization and reinforces tumor malignant behaviors. *Cell. Death Dis.* **2018**, *9*, 793. [CrossRef]

148. Katoh, M. Multi-layered prevention and treatment of chronic inflammation, organ fibrosis and cancer associated with canonical WNT/β-catenin signaling activation (review). *Int. J. Mol. Med.* **2018**, *42*, 713–725. [CrossRef]

149. Baarsma, H.A.; Spanjer, A.I.; Haitsma, G.; Engelbertink, L.H.; Meurs, H.; Jonker, M.R.; Timens, W.; Postma, D.S.; Kerstjens, H.A.; Gosens, R. Activation of WNT/β-catenin signaling in pulmonary fibroblasts by TGF-β_1 is increased in chronic obstructive pulmonary disease. *PLoS ONE* **2011**, *6*, e25450. [CrossRef]

150. Chilosi, M.; Poletti, V.; Zamò, A.; Lestani, M.; Montagna, L.; Piccoli, P.; Pedron, S.; Bertaso, M.; Scarpa, A.; Murer, B.; et al. Aberrant Wnt/beta-catenin pathway activation in idiopathic pulmonary fibrosis. *Am. J. Pathol.* **2003**, *162*, 1495–1502. [CrossRef]

151. Cheon, H.; Boyle, D.L.; Firestein, G.S. Wnt1 inducible signaling pathway protein-3 regulation and microsatellite structure in arthritis. *J. Rheumatol.* **2004**, *31*, 2106–2114. [PubMed]

152. Cheon, S.; Poon, R.; Yu, C.; Khoury, M.; Shenker, R.; Fish, J.; Alman, B.A. Prolonged beta-catenin stabilization and Tcf-dependent transcriptional activation in hyperplastic cutaneous wounds. *Lab. Investig.* **2005**, *85*, 416–425. [CrossRef] [PubMed]

153. Akhmetshina, A.; Palumbo, K.; Dees, C.; Bergmann, C.; Venalis, P.; Zerr, P.; Horn, A.; Kireva, T.; Beyer, C.; Zwerina, J.; et al. Activation of canonical Wnt signalling is required for TGF-β-mediated fibrosis. *Nat. Commun.* **2012**, *3*, 735. [CrossRef] [PubMed]

154. Jahoda, C.A.B.; Gilmore, A.C. What lies beneath: Wnt/β-catenin signaling and cell fate in the lower dermis. *J. Investig. Dermatol.* **2016**, *136*, 1084–1087. [CrossRef] [PubMed]

155. Mastrogiannaki, M.; Lichtenberger, B.M.; Reimer, A.; Collins, C.A.; Driskell, R.R.; Watt, F.M. B-catenin stabilization in skin fibroblasts causes fibrotic lesions by preventing adipocyte differentiation of the reticular dermis. *J. Investig. Dermatol.* **2016**, *136*, 1130–1142. [CrossRef] [PubMed]

156. Liu, T.; Zhou, L.; Yang, K.; Iwasawa, K.; Kadekaro, A.L.; Takebe, T.; Andl, T.; Zhang, Y. The β-catenin/YAP signaling axis is a key regulator of melanoma-associated fibroblasts. *Signal. Transduct. Target. Ther.* **2019**, *4*, 63–78. [CrossRef]

157. Zhou, L.; Yang, K.; Dunaway, S.; Abdel-Malek, Z.; Andl, T.; Kadekaro, A.L.; Zhang, Y. Suppression of MAPK signaling in BRAF-activated PTEN-deficient melanoma by blocking β-catenin signaling in cancer-associated fibroblasts. *Pigment Cell. Melanoma Res.* **2018**, *31*, 297–307. [CrossRef]

158. Shao, H.; Kong, R.; Ferrari, M.L.; Radtke, F.; Capobianco, A.J.; Liu, Z.J. Notch1 pathway activity determines the regulatory role of cancer-associated fibroblasts in melanoma growth and invasion. *PLoS ONE* **2015**, *10*, e0142815. [CrossRef]

159. Du, J.; Zu, Y.; Li, J.; Du, S.; Xu, Y.; Zhang, L.; Jiang, L.; Wang, Z.; Chien, S.; Yang, C. Extracellular matrix stiffness dictates Wnt expression through integrin pathway. *Sci. Rep.* **2016**, *6*, 20395. [CrossRef]

160. Yeung, Y.T.; Guerrero-Castilla, A.; Cano, M.; Muñoz, M.F.; Ayala, A.; Argüelles, S. Dysregulation of the hippo pathway signaling in aging and cancer. *Pharmacol. Res.* **2019**, *143*, 151–165. [CrossRef]

161. Yu, F.X.; Zhao, B.; Guan, K.L. Hippo pathway in organ size control, tissue homeostasis, and cancer. *Cell* **2015**, *163*, 811–828. [CrossRef] [PubMed]

162. Harvey, K.F.; Zhang, X.; Thomas, D.M. The hippo pathway and human cancer. *Nat. Rev. Cancer* **2013**, *13*, 246–257. [CrossRef] [PubMed]

163. Zhang, X.; Tang, J.Z.; Vergara, I.A.; Zhang, Y.; Szeto, P.; Yang, L.; Mintoff, C.; Colebatch, A.; McIntosh, L.; Mitchell, K.A.; et al. Somatic hypermutation of the YAP oncogene in a human cutaneous melanoma. *Mol. Cancer Res.* **2019**, *17*, 1435–1449. [CrossRef] [PubMed]

164. Menzel, M.; Meckbach, D.; Weide, B.; Toussaint, N.C.; Schilbach, K.; Noor, S.; Eigentler, T.; Ikenberg, K.; Busch, C.; Quintanilla-Martinez, L.; et al. In melanoma, hippo signaling is affected by copy number alterations and YAP1 overexpression impairs patient survival. *Pigment Cell. Melanoma Res.* **2014**, *27*, 671–673. [CrossRef] [PubMed]

165. Sarmasti Emami, S.; Zhang, D.; Yang, X. Interaction of the hippo pathway and phosphatases in tumorigenesis. *Cancers* **2020**, *12*, 2438. [CrossRef] [PubMed]

166. Dey, A.; Varelas, X.; Guan, K.L. Targeting the hippo pathway in cancer, fibrosis, wound healing and regenerative medicine. *Nat. Rev. Drug Discov.* **2020**, *19*, 480–494. [CrossRef]

167. Huh, H.D.; Kim, D.H.; Jeong, H.S.; Park, H.W. Regulation of TEAD transcription factors in cancer biology. *Cells* **2019**, *8*, 600. [CrossRef]

168. Kim, M.K.; Jang, J.W.; Bae, S.C. DNA binding partners of YAP/TAZ. *BMB Rep.* **2018**, *51*, 126–133. [CrossRef]

169. Corvaisier, M.; Bauzone, M.; Corfiotti, F.; Renaud, F.; El Amrani, M.; Monté, D.; Truant, S.; Leteurtre, E.; Formstecher, P.; Van Seuningen, I.; et al. Regulation of cellular quiescence by YAP/TAZ and cyclin E1 in colon cancer cells: Implication in chemoresistance and cancer relapse. *Oncotarget* **2016**, *7*, 56699–56712. [CrossRef]

170. Perez, D.E.; Henle, A.M.; Amsterdam, A.; Hagen, H.R.; Lees, J.A. Uveal melanoma driver mutations in GNAQ/11 yield numerous changes in melanocyte biology. *Pigment Cell. Melanoma Res.* **2018**, *31*, 604–613. [CrossRef]

171. Lyubasyuk, V.; Ouyang, H.; Yu, F.X.; Guan, K.L.; Zhang, K. YAP inhibition blocks uveal melanogenesis driven by GNAQ Or GNA11 mutations. *Mol. Cell. Oncol.* **2014**, *2*, e970957. [CrossRef] [PubMed]

172. Javelaud, D.; Alexaki, V.I.; Pierrat, M.J.; Hoek, K.S.; Dennler, S.; Van Kempen, L.; Bertolotto, C.; Ballotti, R.; Saule, S.; Delmas, V.; et al. GLI2 and M-MITF transcription factors control exclusive gene expression programs and inversely regulate invasion in human melanoma cells. *Pigment Cell. Melanoma Res.* **2011**, *24*, 932–943. [CrossRef] [PubMed]

173. Zhang, X.; Yang, L.; Szeto, P.; Abali, G.K.; Zhang, Y.; Kulkarni, A.; Amarasinghe, K.; Li, J.; Vergara, I.A.; Molania, R.; et al. The hippo pathway oncoprotein YAP promotes melanoma cell invasion and spontaneous metastasis. *Oncogene* **2020**, *39*, 5267–5281. [CrossRef]

174. Nallet-Staub, F.; Marsaud, V.; Li, L.; Gilbert, C.; Dodier, S.; Bataille, V.; Sudol, M.; Herlyn, M.; Mauviel, A. Pro-invasive activity of the hippo pathway effectors YAP and TAZ in cutaneous melanoma. *J. Investig. Dermatol.* **2014**, *134*, 123–132. [CrossRef] [PubMed]

175. Kim, N.G.; Koh, E.; Chen, X.; Gumbiner, B.M. E-cadherin mediates contact inhibition of proliferation through hippo signaling-pathway components. *Proc. Natl. Acad. Sci. USA* **2011**, *108*, 11930–11935. [CrossRef] [PubMed]

176. Azzolin, L.; Zanconato, F.; Bresolin, S.; Forcato, M.; Basso, G.; Bicciato, S.; Cordenonsi, M.; Piccolo, S. Role of TAZ as mediator of Wnt signaling. *Cell* **2012**, *151*, 1443–1456. [CrossRef]

177. Azzolin, L.; Panciera, T.; Soligo, S.; Enzo, E.; Bicciato, S.; Dupont, S.; Bresolin, S.; Frasson, C.; Basso, G.; Guzzardo, V.; et al. YAP/TAZ incorporation in the β-catenin destruction complex orchestrates the Wnt response. *Cell* **2014**, *158*, 157–170. [CrossRef]

178. Imajo, M.; Miyatake, K.; Iimura, A.; Miyamoto, A.; Nishida, E. A molecular mechanism that links hippo signalling to the inhibition of Wnt/β-catenin signalling. *EMBO J.* **2012**, *31*, 1109–1122. [CrossRef]

179. Kim, M.; Jho, E.H. Cross-talk between Wnt/β-catenin and hippo signaling pathways: A brief review. *BMB Rep.* **2014**, *47*, 540–545. [CrossRef]

180. Johannessen, C.M.; Johnson, L.A.; Piccioni, F.; Townes, A.; Frederick, D.T.; Donahue, M.K.; Narayan, R.; Flaherty, K.T.; Wargo, J.A.; Root, D.E.; et al. A melanocyte lineage program confers resistance to MAP kinase pathway inhibition. *Nature* **2013**, *504*, 138–142. [CrossRef]

181. Kim, M.H.; Kim, C.G.; Kim, S.K.; Shin, S.J.; Choe, E.A.; Park, S.H.; Shin, E.C.; Kim, J. YAP-induced PD-L1 expression drives immune evasion in BRAFi-resistant melanoma. *Cancer Immunol. Res.* **2018**, *6*, 255–266. [CrossRef] [PubMed]

182. Moya, I.M.; Castaldo, S.A.; Van den Mooter, L.; Soheily, S.; Sansores-Garcia, L.; Jacobs, J.; Mannaerts, I.; Xie, J.; Verboven, E.; Hillen, H.; et al. Peritumoral activation of the hippo pathway effectors YAP and TAZ suppresses liver cancer in mice. *Science* **2019**, *366*, 1029–1034. [CrossRef] [PubMed]

183. Lawlor, K.; Pérez-Montero, S.; Lima, A.; Rodríguez, T.A. Transcriptional versus metabolic control of cell fitness during cell competition. *Semin. Cancer Biol.* **2020**, *63*, 36–43. [CrossRef] [PubMed]

184. Wu, M.Y.; Hill, C.S. Tgf-beta superfamily signaling in embryonic development and homeostasis. *Dev. Cell.* **2009**, *16*, 329–343. [CrossRef]

185. Shi, Y.; Massagué, J. Mechanisms of TGF-beta signaling from cell membrane to the nucleus. *Cell* **2003**, *113*, 685–700. [CrossRef]

186. Schmierer, B.; Hill, C.S. TGFbeta-SMAD signal transduction: Molecular specificity and functional flexibility. *Nat. Rev. Mol. Cell Biol.* **2007**, *8*, 970–982. [CrossRef]

187. Weber, C.E.; Li, N.Y.; Wai, P.Y.; Kuo, P.C. Epithelial-mesenchymal transition, TGF-β, and osteopontin in wound healing and tissue remodeling after injury. *J. Burn Care. Res.* **2012**, *33*, 311–318. [CrossRef]

188. Colak, S.; Ten Dijke, P. Targeting TGF-β signaling in cancer. *Trends Cancer* **2017**, *3*, 56–71. [CrossRef]

189. De Caestecker, M.P.; Piek, E.; Roberts, A.B. Role of transforming growth factor-beta signaling in cancer. *J. Natl. Cancer Inst.* **2000**, *92*, 1388–1402. [CrossRef]

190. Krasagakis, K.; Garbe, C.; Schrier, P.I.; Orfanos, C.E. Paracrine and autocrine regulation of human melanocyte and melanoma cell growth by transforming growth factor beta in vitro. *Anticancer Res.* **1994**, *14*, 2565–2571.

191. Rodeck, U.; Bossler, A.; Graeven, U.; Fox, F.E.; Nowell, P.C.; Knabbe, C.; Kari, C. Transforming growth factor beta production and responsiveness in normal human melanocytes and melanoma cells. *Cancer Res.* **1994**, *54*, 575–581. [PubMed]

192. Javelaud, D.; Alexaki, V.I.; Mauviel, A. Transforming growth factor-beta in cutaneous melanoma. *Pigment Cell Melanoma Res.* **2008**, *21*, 123–132. [CrossRef] [PubMed]

193. Van Belle, P.; Rodeck, U.; Nuamah, I.; Halpern, A.C.; Elder, D.E. Melanoma-associated expression of transforming growth factor-beta isoforms. *Am. J. Pathol.* **1996**, *148*, 1887–1894. [PubMed]

194. Reed, J.A.; McNutt, N.S.; Prieto, V.G.; Albino, A.P. Expression of transforming growth factor-beta 2 in malignant melanoma correlates with the depth of tumor invasion. Implications for tumor progression. *Am. J. Pathol.* **1994**, *145*, 97–104. [PubMed]

195. Polanska, U.M.; Orimo, A. Carcinoma-associated fibroblasts: Non-neoplastic tumour-promoting mesenchymal cells. *J. Cell. Physiol.* **2013**, *228*, 1651–1657. [CrossRef]

196. Wipff, P.J.; Rifkin, D.B.; Meister, J.J.; Hinz, B. Myofibroblast contraction activates latent TGF-beta1 from the extracellular matrix. *J. Cell Biol.* **2007**, *179*, 1311–1323. [CrossRef]

197. Pasco, S.; Brassart, B.; Ramont, L.; Maquart, F.X.; Monboisse, J.C. Control of melanoma cell invasion by type IV collagen. *Cancer Detect. Prev.* **2005**, *29*, 260–266. [CrossRef]

198. Najafi, M.; Farhood, B.; Mortezaee, K. Extracellular matrix (ECM) stiffness and degradation as cancer drivers. *J. Cell. Biochem.* **2019**, *120*, 2782–2790. [CrossRef]

199. Burchardt, E.R.; Hein, R.; Bosserhoff, A.K. Laminin, hyaluronan, tenascin-C and type VI collagen levels in sera from patients with malignant melanoma. *Clin. Exp. Dermatol.* **2003**, *28*, 515–520. [CrossRef]

200. Khan, H.Y.; Orimo, A. Transforming growth factor-β: Guardian of catabolic metabolism in carcinoma-associated fibroblasts. *Cell. Cycle* **2012**, *11*, 4302–4303. [CrossRef]

201. Li, Z.; Zhang, J.; Zhou, J.; Lu, L.; Wang, H.; Zhang, G.; Wan, G.; Cai, S.; Du, J. Nodal facilitates differentiation of fibroblasts to cancer-associated fibroblasts that support tumor growth in melanoma and colorectal cancer. *Cells* **2019**, *8*, 538. [CrossRef] [PubMed]

202. Topczewska, J.M.; Postovit, L.M.; Margaryan, N.V.; Sam, A.; Hess, A.R.; Wheaton, W.W.; Nickoloff, B.J.; Topczewski, J.; Hendrix, M.J. Embryonic and tumorigenic pathways converge via nodal signaling: Role in melanoma aggressiveness. *Nat. Med.* **2006**, *12*, 925–932. [CrossRef] [PubMed]
203. Kodet, O.; Dvořánková, B.; Bendlová, B.; Sýkorová, V.; Krajsová, I.; Štork, J.; Kučera, J.; Szabo, P.; Strnad, H.; Kolář, M.; et al. Microenvironment-driven resistance to B-raf inhibition in a melanoma patient is accompanied by broad changes of gene methylation and expression in distal fibroblasts. *Int. J. Mol. Med.* **2018**, *41*, 2687–2703. [CrossRef] [PubMed]
204. Pramong, N.; Gojaseni, P.; Suttipongkeat, S.; Kiattisunthorn, K.; Chittinandana, A. Diagnostic accuracy of fibroblast growth factor 23 for predicting acute kidney injury in patients with acute decompensated heart failure. *Nephrology (Carlton)* **2020**, e13780. [CrossRef]
205. Carlino, M.S.; Long, G.V.; Kefford, R.F.; Rizos, H. Targeting oncogenic BRAF and aberrant MAPK activation in the treatment of cutaneous melanoma. *Crit. Rev. Oncol. Hematol.* **2015**, *96*, 385–398. [CrossRef]
206. Daniotti, M.; Oggionni, M.; Ranzani, T.; Vallacchi, V.; Campi, V.; Di Stasi, D.; Torre, G.D.; Perrone, F.; Luoni, C.; Suardi, S.; et al. BRAF alterations are associated with complex mutational profiles in malignant melanoma. *Oncogene* **2004**, *23*, 5968–5977. [CrossRef]
207. Uribe, P.; Wistuba, I.I.; González, S. BRAF mutation: A frequent event in benign, atypical, and malignant melanocytic lesions of the skin. *Am. J. Dermatopathol.* **2003**, *25*, 365–370. [CrossRef]
208. Vanni, I.; Tanda, E.T.; Spagnolo, F.; Andreotti, V.; Bruno, W.; Ghiorzo, P. The current state of molecular testing in the BRAF-mutated melanoma landscape. *Front. Mol. Biosci.* **2020**, *7*, 113. [CrossRef]
209. Halaban, R.; Krauthammer, M. RASopathy gene mutations in melanoma. *J. Investig. Dermatol.* **2016**, *136*, 1755–1759. [CrossRef]
210. Vanni, I.; Tanda, E.T.; Dalmasso, B.; Pastorino, L.; Andreotti, V.; Bruno, W.; Boutros, A.; Spagnolo, F.; Ghiorzo, P. Non-BRAF mutant melanoma: Molecular features and therapeutical implications. *Front. Mol. Biosci.* **2020**, *7*, 172. [CrossRef]
211. Platz, A.; Egyhazi, S.; Ringborg, U.; Hansson, J. Human cutaneous melanoma; A review of NRAS and BRAF mutation frequencies in relation to histogenetic subclass and body site. *Mol. Oncol.* **2008**, *1*, 395–405. [CrossRef] [PubMed]
212. Dantonio, P.M.; Klein, M.O.; Freire, M.R.V.B.; Araujo, C.N.; Chiacetti, A.C.; Correa, R.G. Exploring major signaling cascades in melanomagenesis: A rationale route for targetted skin cancer therapy. *Biosci. Rep.* **2018**, *38*. [CrossRef] [PubMed]
213. Dhomen, N.; Marais, R. New insight into BRAF mutations in cancer. *Curr. Opin. Genet. Dev.* **2007**, *17*, 31–39. [CrossRef] [PubMed]
214. Damsky, W.E.; Bosenberg, M. Melanocytic nevi and melanoma: Unraveling a complex relationship. *Oncogene* **2017**, *36*, 5771–5792. [CrossRef]
215. Kuwata, T.; Kitagawa, M.; Kasuga, T. Proliferative activity of primary cutaneous melanocytic tumours. *Virchows Arch. A Pathol. Anat. Histopathol.* **1993**, *423*, 359–364. [CrossRef]
216. Maldonado, J.L.; Timmerman, L.; Fridlyand, J.; Bastian, B.C. Mechanisms of cell-cycle arrest in spitz nevi with constitutive activation of the MAP-kinase pathway. *Am. J. Pathol.* **2004**, *164*, 1783–1787. [CrossRef]
217. Michaloglou, C.; Vredeveld, L.C.; Soengas, M.S.; Denoyelle, C.; Kuilman, T.; van der Horst, C.M.; Majoor, D.M.; Shay, J.W.; Mooi, W.J.; Peeper, D.S. BRAFE600-associated senescence-like cell cycle arrest of human naevi. *Nature* **2005**, *436*, 720–724. [CrossRef]
218. Gray-Schopfer, V.C.; Cheong, S.C.; Chong, H.; Chow, J.; Moss, T.; Abdel-Malek, Z.A.; Marais, R.; Wynford-Thomas, D.; Bennett, D.C. Cellular senescence in naevi and immortalisation in melanoma: A role for p16? *Br. J. Cancer* **2006**, *95*, 496–505. [CrossRef]
219. Dhomen, N.; Reis-Filho, J.S.; da Rocha Dias, S.; Hayward, R.; Savage, K.; Delmas, V.; Larue, L.; Pritchard, C.; Marais, R. Oncogenic BRAF induces melanocyte senescence and melanoma in mice. *Cancer Cell* **2009**, *15*, 294–303. [CrossRef]
220. Huang, J.M.; Chikeka, I.; Hornyak, T.J. Melanocytic nevi and the genetic and epigenetic control of oncogene-induced senescence. *Dermatol. Clin.* **2017**, *35*, 85–93. [CrossRef]
221. Uribe, P.; Andrade, L.; Gonzalez, S. Lack of association between BRAF mutation and MAPK ERK activation in melanocytic nevi. *J. Investig. Dermatol.* **2006**, *126*, 161–166. [CrossRef] [PubMed]

222. Damsky, W.; Micevic, G.; Meeth, K.; Muthusamy, V.; Curley, D.P.; Santhanakrishnan, M.; Erdelyi, I.; Platt, J.T.; Huang, L.; Theodosakis, N.; et al. MTORC1 activation blocks BRAFV600E-induced growth arrest but is insufficient for melanoma formation. *Cancer Cell* **2015**, *27*, 41–56. [CrossRef] [PubMed]

223. Proietti, I.; Skroza, N.; Michelini, S.; Mambrin, A.; Balduzzi, V.; Bernardini, N.; Marchesiello, A.; Tolino, E.; Volpe, S.; Maddalena, P.; et al. BRAF Inhibitors: Molecular targeting and immunomodulatory actions. *Cancers* **2020**, *12*, 1823. [CrossRef] [PubMed]

224. Croce, L.; Coperchini, F.; Magri, F.; Chiovato, L.; Rotondi, M. The multifaceted anti-cancer effects of BRAF-inhibitors. *Oncotarget* **2019**, *10*, 6623–6640. [CrossRef] [PubMed]

225. Young, H.L.; Rowling, E.J.; Bugatti, M.; Giurisato, E.; Luheshi, N.; Arozarena, I.; Acosta, J.C.; Kamarashev, J.; Frederick, D.T.; Cooper, Z.A.; et al. An adaptive signaling network in melanoma inflammatory niches confers tolerance to MAPK signaling inhibition. *J. Exp. Med.* **2017**, *214*, 1691–1710. [CrossRef]

226. Diazzi, S.; Tartare-Deckert, S.; Deckert, M. Bad neighborhood: Fibrotic stroma as a new player in melanoma resistance to targeted therapies. *Cancers* **2020**, *12*, 1364. [CrossRef]

227. Atzori, M.G.; Ceci, C.; Ruffini, F.; Trapani, M.; Barbaccia, M.L.; Tentori, L.; D'Atri, S.; Lacal, P.M.; Graziani, G. Role of VEGFR-1 in melanoma acquired resistance to the BRAF inhibitor vemurafenib. *J. Cell. Mol. Med.* **2020**, *24*, 465–475. [CrossRef]

228. Khalili, J.S.; Liu, S.; Rodríguez-Cruz, T.G.; Whittington, M.; Wardell, S.; Liu, C.; Zhang, M.; Cooper, Z.A.; Frederick, D.T.; Li, Y.; et al. Oncogenic BRAF(V600E) promotes stromal cell-mediated immunosuppression via induction of interleukin-1 in melanoma. *Clin. Cancer Res.* **2012**, *18*, 5329–5340. [CrossRef]

229. Fedorenko, I.V.; Wargo, J.A.; Flaherty, K.T.; Messina, J.L.; Smalley, K.S.M. BRAF inhibition generates a host-tumor niche that mediates therapeutic escape. *J. Investig. Dermatol.* **2015**, *135*, 3115–3124. [CrossRef]

230. Hirata, E.; Girotti, M.R.; Viros, A.; Hooper, S.; Spencer-Dene, B.; Matsuda, M.; Larkin, J.; Marais, R.; Sahai, E. Intravital imaging reveals how BRAF inhibition generates drug-tolerant microenvironments with high integrin β1/FAK signaling. *Cancer Cell* **2015**, *27*, 574–588. [CrossRef]

Publisher's Note: MDPI stays neutral with regard to jurisdictional claims in published maps and institutional affiliations.

Article

ShcD Binds DOCK4, Promotes Ameboid Motility and Metastasis Dissemination, Predicting Poor Prognosis in Melanoma

Ewa Aladowicz [1,†], Letizia Granieri [2,†], Federica Marocchi [2,†], Simona Punzi [2,5], Giuseppina Giardina [2], Pier Francesco Ferrucci [2], Giovanni Mazzarol [3], Maria Capra [3], Giuseppe Viale [3,4], Stefano Confalonieri [2], Sara Gandini [2], Fiorenza Lotti [2,*] and Luisa Lanfrancone [2,*]

[1] Sarcoma Molecular Pathology Team, Divisions of Molecular Pathology and Cancer Therapeutics, The Institute of Cancer Research (ICR), 15 Cotswold Road, Sutton SM2 5NG, UK; ewa.aladowicz@icr.ac.uk

[2] Department of Experimental Oncology, European Institute of Oncology IRCCS (IEO), Via Adamello, 16, 20139 Milan, Italy; letizia.granieri@ieo.it (L.G.); federica.marocchi@ieo.it (F.M.); simona.punzi@ieo.it (S.P.); giuseppina.giardina@ieo.it (G.G.); pier.ferrucci@ieo.it (P.F.F.); stefano.confalonieri@ieo.it (S.C.); sara.gandini@ieo.it (S.G.)

[3] Division of Pathology, European Institute of Oncology IRCCS (IEO), Via Ripamonti 435, 20141 Milan, Italy; giovanni.mazzarol@ieo.it (G.M.); maria.capra@ieo.it (M.C.); giuseppe.viale@ieo.it (G.V.)

[4] Department of Oncology and Hemato-oncology, University of Milan, 20141 Milan, Italy

[5] Candiolo Cancer Institute—FPO IRCCS, 10060 Candiolo Torino, Italy

* Correspondence: fiorenza.lotti@ieo.it (F.L.); luisa.lanfrancone@ieo.it (L.L.)

† These authors contributed equally to the work.

Received: 5 November 2020; Accepted: 10 November 2020; Published: 13 November 2020

Simple Summary: Metastasis formation and dissemination is a complex process that relies on several steps. Even though highly inefficient, metastasis spreading is the primary cause of cancer morbidity and *mortality* in patients. The aim of our study was to investigate the molecular pathways leading to metastases making use of human-in-mouse melanoma models of patient-derived xenografts. We demonstrate that the modulation of the expression of an adaptor protein of the Shc family, ShcD, can change the phenotype and the invasive properties of melanoma cells when highly expressed. We also show that ShcD binds DOCK4 and confines it into the cytoplasm, blocking the Rac1 signaling pathways, thus leading to metastasis development. Moreover, our results indicate that melanoma cells are more sensitive to therapeutic treatments when the ShcD molecular pathway is inactivated, suggesting that new therapeutic strategies can be designed in melanomas.

Abstract: Metastases are the primary cause of cancer-related deaths. The underlying molecular and biological mechanisms remain, however, elusive, thus preventing the design of specific therapies. In melanomas, the metastatic process is influenced by the acquisition of metastasis-associated mutational and epigenetic traits and the activation of metastatic-specific signaling pathways in the primary melanoma. In the current study, we investigated the role of an adaptor protein of the Shc family (ShcD) in the acquisition of metastatic properties by melanoma cells, exploiting our cohort of patient-derived xenografts (PDXs). We provide evidence that the depletion of ShcD expression increases a spread cell shape and the capability of melanoma cells to attach to the extracellular matrix while its overexpression switches their morphology from elongated to rounded on 3D matrices, enhances cells' invasive phenotype, as observed on collagen gel, and favors metastasis formation in vivo. ShcD overexpression sustains amoeboid movement in melanoma cells, by suppressing the Rac1 signaling pathway through the confinement of DOCK4 in the cytoplasm. Inactivation of the ShcD signaling pathway makes melanoma cells more sensitive to therapeutic treatments. Consistently, ShcD expression predicts poor outcome in a cohort of 183 primary melanoma patients.

Keywords: melanoma metastasis; ShcD adaptor protein; amoeboid motility; Rac1; DOCK4; melanoma PDX

---

1. Introduction

Over the last decade, the worldwide incidence of cutaneous melanoma raised more rapidly than that of any other cancer type [1]. Though early-stage melanoma is curable by surgical excision in most cases, when tumor thickness increases, melanomas tend to spread and colonize other organs, leading to dismal prognosis.

Melanoma metastatic dissemination is a complex process primarily involving the phenotypic plasticity of melanoma cells, a process that includes adaptive modifications of gene expression, changes of cell morphology, detachment from the extracellular matrix (ECM), loss of cell–cell contacts and spreading [2,3]. Though the metastatic process is per se highly inefficient, in melanomas it is favored by the high plasticity and strong motility of melanoma cells [3,4]

Depending on the matrix composition of the tumor microenvironment, migrating melanoma cells can adopt two mutually exclusive phenotypes, corresponding to different modalities of motility, either the so-called rounded (amoeboid) or the elongated (mesenchymal) phenotype, and can switch between them [4,5]. Both phenotypes are driven by regulators of actin contractility belonging to the Rho family of small GTPases [6,7]. The rapid, amoeboid motility is regulated by RhoA, and is based on high actomyosin contractility enabling cells to squeeze through the matrix. It is characterized by weak or no adhesion to ECM, with little or no proteolysis of the matrix, and is driven by membrane blebs, actin-rich pseudopodia and/or highly contractile uropods [8–11]. On the contrary, elongated movement, characterized by actin assembly, cell polarization and the digestion of an extracellular matrix by proteases, is driven by Rac1 [6]. These two pathways show inhibitory effects on each other. Notably, amoeboid movement is prominent at the invasive front of melanoma in animal models [6,12,13], as well as in human melanoma lesions [7,13],suggesting that it may favor tumor invasion.

The switch between the amoeboid and elongated movements is based on the activation of the small GTPases cycle [14]. In melanoma, it was previously shown that DOCK3, a Rac1-specific guanine nucleotide exchange factor (GEF), is required for mesenchymal movement, whereas the Rac1 suppressor ARHGAP22 has an inhibitory effect. [12,14]. Another member of the DOCK GEF family, DOCK10, drives the amoeboid motility of melanoma cells through the activation of Cdc42, another Rho-family GEF [15]. DOCK4 expression has been previously linked to cell migration in vitro and to the risk of developing bone metastasis in breast cancer patients [16–18] as well as correct axon guidance in vivo during embryo development [17], thus suggesting a prominent role of DOCK4 in regulating cell motility.

ShcD/Shc4/RaLP is a cytosolic protein belonging to the SHC adaptors family. During mouse development, this plays a critical role in the transition from embryonic to epiblast stem cells; in the adult it is expressed in the developing nervous system, heart, muscle and skin [19]. In humans, ShcD expression is restricted to invasive and metastatic melanomas, and its silencing reduces cancer cell migration [20]. ShcD phosphorylation on specific tyrosine residues by activated receptor tyrosine kinases leads to the transient stimulation of MAPK signaling. MAPK activation by phosphorylated ShcD, however, is not sufficient to support the migration of metastatic melanomas, suggesting that ShcD activates other critical, MAPK-independent migratory pathways [20].

Here, we show that the modulation of ShcD protein levels affects the cell morphology and invasiveness in melanoma patient-derived xenografts (PDXs) [21]. ShcD expressing cells detach from ECM, acquire rounded morphology and invasive traits, as observed in in vitro assays on collagen substrates invading collagen. Moreover, in three-dimensional (3D) assays, cells switch from a mesenchymal to amoeboid phenotype. In vivo amoeboid ShcD overexpressing cells disseminate to lymph nodes and seed distant metastases. Consistently, analyses of a large cohort of melanoma patients demonstrated that ShcD is a prognostic factor in melanoma. Notably, ShcD silencing sensitizes BRAF-mutant melanoma cells to targeted therapy [22,23], suggesting that the regulation of ShcD expression influences both the cell motility and drug sensitivity of melanoma cells.

2. Results

2.1. ShcD Impairs the Adhesive Properties of Melanoma Cells

Using metastatic melanoma cell lines (WM266-4 and IGR-37), we previously showed that decreased ShcD expression reduces migration without interfering with cell proliferation in vitro [20]. We confirmed these data in PDX cells (MM27) by the lentiviral transduction of ShcD-targeting shRNAs (shShcD#1 and shShcD#2 pooled together; ShShcD) or control (ShLuc) vectors (Figure S1A–C). Since tumor cell migration involves modifications of cell- and stroma-interactions, we analyzed whether ShcD regulated the adhesive properties of MM27 PDX cells. ShShcD and ShLuc cells were plated on different extracellular matrices, including fibronectin, collagen and matrigel, and counted by Crystal Violet at different time points after the extensive wash-out of floating cells. As shown in Figure 1A, ShcD depletion significantly increased numbers of adherent cells to all tested matrices. To visualize the morphology of ShShcD MM27 adherent cells, cell spreading was analyzed upon adhesion to fibronectin. A higher percentage of spread cells was observed in ShShcD cells, as compared to control shLuc cells (Figure 1B), indicating that ShcD silencing increases cell attachment and spreading capacities.

Cell spreading depends on the formation of focal adhesions (FA), multi-protein complexes that serve to connect the cellular cytoskeleton with components of the extracellular matrix. We analyzed FA formation by staining MM27 cells with antibodies against known components of the complex, e.g., vinculin, paxillin and focal adhesion kinase (FAK) (Figure S1D) and their phosphorylated counterparts (Figure 1C) [24]. After adhesion to fibronectin, we observed a significant increase in the number and intensity of phospho-FA staining in ShcD knockdown cells (Figure 1C). Together, these results demonstrate that ShcD impairs the ability of melanoma cells to adhere to extracellular matrix components, through the modulation of FA formation, thus favoring cell migration.

2.2. ShcD Regulates Melanoma Cell Morphology and Sustains Amoeboid Movement of Melanoma Cells in 3D Matrix

The capacity of melanoma cells to switch to different morphologies can be visualized in vitro by culturing cells in 3D matrix conditions. We first analyzed the morphology of MM27 PDX cells overexpressing ShcD plated on thick collagen layers (Figure 2A). While control cells (PincoPuro (PP)-vector) showed mixed morphologies when plated on thick collagen layers (65% rounded and 35% elongated) (Figure 2B), rounded cells raised to 87% in ShcD overexpressing cells (PP-ShcD), suggesting that ShcD drives morphological changes in melanoma cells. Similar results were obtained in WM115 and WM266.4 cells (Figure S2), two independent cell lines isolated, respectively, from the primary and metastatic tumors of the same patient. Both cell lines were transduced with a control vector (ShLuc), shShcD#1 and shShcD#2 vectors. The WM115 cell line consists mainly of rounded cells (79%), while WM266.4 is composed of a mixed population of rounded and elongated cells, as in the MM27 PDX. In WM115, ShcD silencing decreased the population of rounded cells to 27% for shShcD#1 ($p < 0.0001$) and 48% for shShcD#2 ($p < 0.0001$) (Figure S2A). Similarly, in WM266.4, ShcD silencing reduced rounded cells to 38% and 42% ($p < 0.001$) with shShcD#1 and shShcD#2, respectively. Morphology of elongated cells within the two ShcD-interfered melanoma cell populations is shown in Figure S2B.

We then investigated whether ShcD influences melanoma cell invasion using melanoma spheroids embedded in collagen gel. As shown in Figure 2C, ShcD overexpression increased significantly the invasive properties of MM27 cells, as compared to the control. Both control (PP) and ShcD overexpressing cells (PP-ShcD) were capable of invading the matrix over time, but the PP-ShcD cells resulted as significantly more invasive than the control cells at all time points (Figure 2C). Notably, time-lapse analyses showed that invasive PP-ShcD cells move faster than the controls (Figure 2D, Movie S1).

Melanoma cells invade the surrounding tissue either collectively or as single-cells [25]. When tumor cells detach from ECM, they start invading the surrounding tissue as single cells and switch to rounded/amoeboid movement [25]. The amoeboid phenotype allows cells to rapidly move out of the tumor: indeed, amoeboid cells are found at the invasive fronts of primary tumors close to tumor-associated macrophages and vessels [26]. We therefore analyzed the invasive front of melanoma spheroids in 3D assays. The invasion of control (PP) and ShcD overexpressing (PP-ShcD) collagen-embedded PDX cells were visualized by time-lapse for 24 h. We observed significantly increased numbers of amoeboid cells among the ShcD overexpressing cells (Figure 2E, Movie S1).

These data indicate that ShcD expression favors the rounded morphology of melanoma cells on thick collagen layers, adhesion, migration and invasion in vitro, suggesting that ShcD is a critical determinant of the plasticity and migratory movement of melanoma cells.

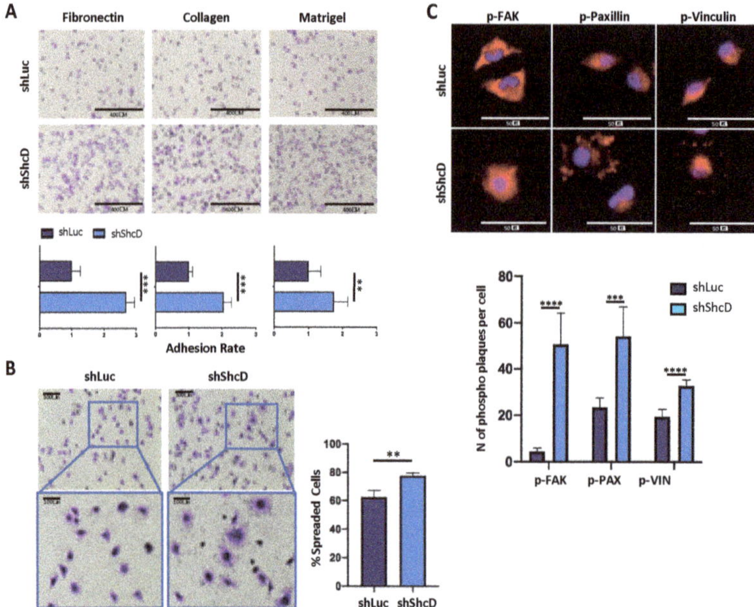

Figure 1. ShcD depletion modifies the adhesive phenotype of melanoma cells. (**A**) ShLuc and ShShcD (pool of ShShcD#1 and #2) MM27 cells' adhesion to different extracellular matrices (fibronectin, collagen, matrigel). Cells were detected by Crystal Violet staining. Adhesion rate is calculated as the ratio of relative numbers of adherent cells in shShcD vs. shLuc. Five images per well were acquired in duplicate experiments. Student *t-test* (***, $p < 0.001$; **, $p < 0.01$, ****, $p < 0.0001$) was applied to assess the significance. Representative images are shown (20×). (**B**) ShLuc and ShShcD MM27 cells spreading evaluation on fibronectin. Cells were stained with Crystal Violet. Images were quantified with ImageJ software. Data are shown as the mean ±SD of 3 fields of 3 different cover slips. Student *t-test* (**, $p < 0.01$). (**C**) ShLuc and ShShcD MM27 cells' focal adhesion analysis by immunofluorescence. Cells were treated as in (**B**) and the protein expression of p-vinculin, p-paxillin and p-FAK (red) was detected. Nuclei were counterstained with DAPI (blue). Representative images are shown (63× magnification).

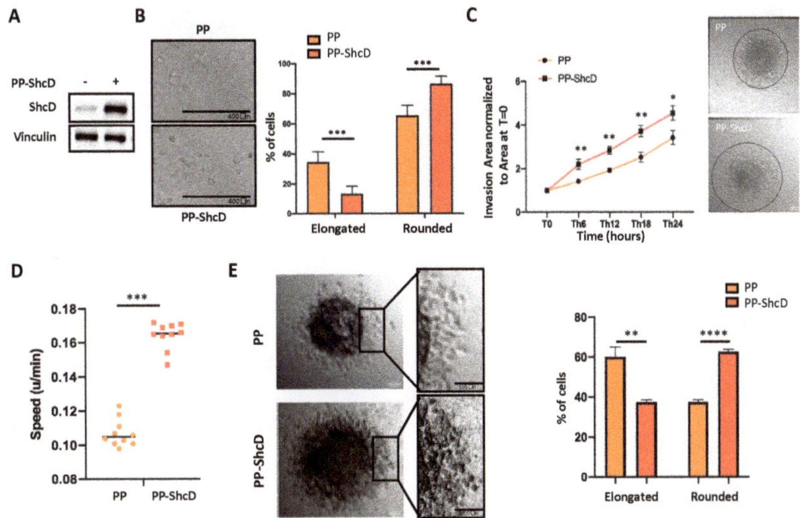

Figure 2. ShcD overexpression increases the invasion and amoeboid movement of melanoma cells. (**A**) ShcD overexpression was estimated by immunoblot analysis in MM27 cells transduced to overexpress ShcD (PincoPuro (PP)-ShcD) or a neutral control (PP, empty vector). Vinculin was used as a loading control. (**B**) PP/PP-ShcD MM27 cell morphology assessment. Cells were plated on a thick collagen layer and the percentage of elongated/rounded cells was calculated after 24 h. Three images per well were acquired in triplicate experiments. Student *t-test* (***, $p < 0.001$). (**C**) Spheroid collagen invasion assay in PP and PP-ShcD MM27 cells. The area of invasion was monitored by time-lapse microscopy for 24 h. Data are shown as the mean ±SD of 10 different spheroids per group. Student *t-test* (*, $p < 0.05$; **, $p < 0.01$). Representative images are shown. (**D**) Spheroid invasion speed in PP and PP-ShcD MM27 cells at the front of invasion. Cell speed was analyzed with the manual tracking plugin of the ImageJ software. Data are shown as the mean ±SD of 2 different cells of 5 spheroids per group. Student *t-test* (***, $p < 0.0001$). (**E**) Assessment of cell shape at the front of invasion in PP and PP-ShcD MM27 cells. Cell shape was analyzed and quantified by scanning 2 high-power fields of 5 spheroids per sample using ImageJ software. Student *t-test* (**, $p < 0.01$; ****, $p < 0.0001$). Representative images are shown.

2.3. ShcD Is Crucial for Metastasis Dissemination

To investigate the potential of ShcD in promoting metastasis formation in vivo, MM13 and MM27 PDXs were chosen as representatives of melanoma subtypes with invasive or proliferative phenotypes, respectively [21]. MM13 PDX carries the NRAS mutation and displays a more invasive phenotype, with high levels of AXL, PDGF-Rβ, EGFR and EPHA2 proteins and very low levels of MITF, Brn2 and SOX10. MM27, instead, harbors the mutated BRAF gene and expresses MITF, BRN2 and SOX10, thus suggesting a proliferative phenotype [27] (Figure 3A). The retroviral transduction of MM13 and MM27 PDX cells with the PP-ShcD vector allowed for high levels of ShcD expression in both PDXs, as compared to the PP-vector transduced cells (Figure 3B). Analyses of the migratory potential of SchD overexpressing MM13 and MM27 cells showed increased migratory capability (Figure 3C). To investigate the metastatic potential of ShcD overexpressing cells, transduced cells were injected into the back dermis of five NSG mice and the tumors were surgically removed to prolong animal survival and allow for metastasis detection. Under these experimental conditions, we showed that MM13 and MM27 can metastasize to different organs (including lymph nodes, lung, liver and spleen) (Figure 3C–H). Mean latency to a volume of the primary tumor of 0.35–0.5 cm^3 was similar in the two PDXs, ranging from 21 to 25 days (MM27 and MM13, respectively), with no statistically significant difference with the control PP-vector tumors (Figure 3D). At early time points, after the surgical resection of the primary tumor (approx. 1 month), both PP and PP-ShcD mice developed metastases at lymph nodes, with comparable frequencies (4/5 vs. 5/5 in the MM13 group; 2/2 vs. 2/2 in the MM27 group). However, the size of the metastatic lymph nodes was significantly enlarged in the MM13 PP-ShcD mice (~4–5 folds; Figure 3E, upper panel; $p < 0.0001$). A similar trend was observed in MM27 (~3–4 folds; Figure 3E, lower panel), though the small number of mice developing metastases did not enable a statistical representation.

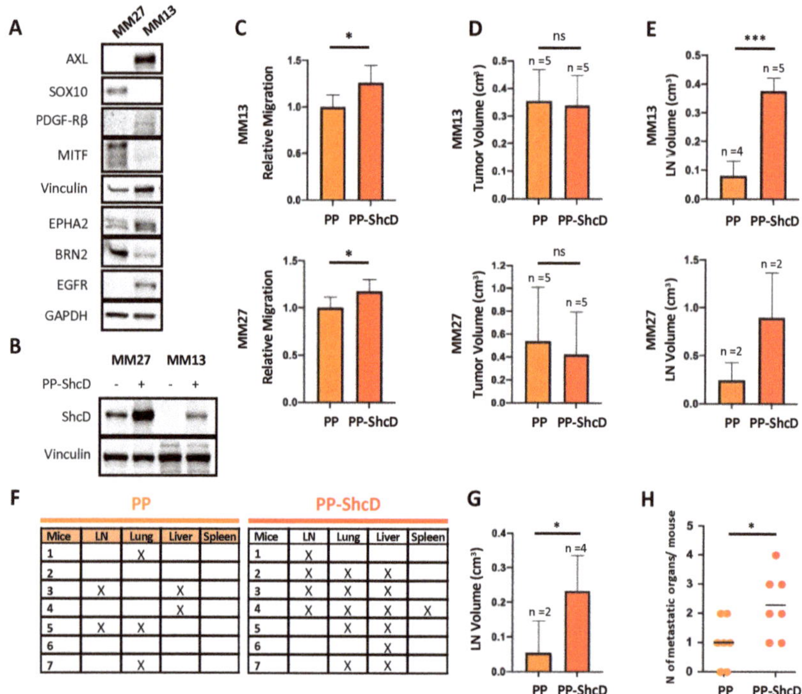

Figure 3. ShcD overexpression promotes metastasis formation at lymph nodes and distant sites. (**A**) Phenotype characterization of MM27 and MM13 patient-derived xenografts (PDXs) cells by Western blotting. Immunoblots show the protein expression of invasive markers (AXL, PDGFR-β, EGFR, EPHA2) and proliferative markers (SOX10, MITF, BRN2). Vinculin and GAPDH were used as loading controls. (**B**) The assessment of ShcD overexpression in MM27 and MM13 PDX cells by Western blot analysis, before mice transplantation. MM27 and MM13 PDX cells were transduced to overexpress ShcD (PP-ShcD) or a neutral control (PP, empty vector). Vinculin was used as a loading control. (**C**) Transwell migration assay in PP/PP-ShcD MM13 and MM27 cells, at the 48 h time point. Relative migration (mean ± SD) is expressed as a ratio of PP-ShcD vs. PP cell migration values (calculated with ImageJ). Student *t*-test (*, $p < 0.05$). (**D–H**) In vivo metastatic potential of PP and PP-ShcD MM13 and MM27 cells. Transduced cells were intradermally injected in NSG mice (D–E n = 5 per group; F–H MM13 n = 8 per group) and the tumors were surgically resected. The early time point evaluation of metastatic dissemination (MM13, MM27): (**D**) tumor volumes at resection. Student *t*-test (not significant, n.s.). (**E**) Lymph nodes (LNs) volumes at mouse sacrifice. MM13 PP, n = 4; PP-ShcD, n = 5; MM13 PP, n = 2; PP-ShcD, n = 2. Student *t*-test (***, $p < 0.0001$). Late time point evaluation of metastatic dissemination (MM13): **F**) metastatic organs evaluation of MM13 PP and PP-ShcD injected mice. "X" indicates presence of metastatic foci in the reported organs (n = 7). (**G**) Lymph nodes' (LNs) volumes at mouse sacrifice. MM13 PP, n = 2; PP-ShcD, n = 4. Student *t*-test (*, $p < 0.05$). (**H**) Number of metastatic organs (LNs and distant metastasis) per mouse in PP and PP-ShcD MM13 injected mice. PP, n = 7; PP-ShcD, n = 16. Student *t*-test (*, $p < 0.05$).

We then investigated whether ShcD overexpression led to metastasis formation at distant organs, by observing the animals up to 2 months from the resection of the primary tumor and using MM13 as model. As shown in Figure 3F, lymph nodes, lung, liver and spleen were colonized by PDX cells. At 2 months, the size of the metastatic lymph nodes was significantly enlarged in the MM13 PP-ShcD mice, similar to what we had previously observed at earlier time points (Figure 3G). Most notably, the numbers of metastatic events, regardless of the specific sites, were markedly higher in PP-ShcD mice, as compared to the controls (seven independent metastatic events in the seven control mice; 16 in the seven PP-ShcD mice; $p = 0.0298$) (Figure 3H). Thus, ShcD may activate molecular pathways that are crucial for the metastasization of melanoma cells.

2.4. ShcD Expression Correlates with Melanoma Progression and Patient Outcome in Melanoma Patients

We sought to interrogate how the ShcD RNA transcript is regulated in a cohort of primary and metastatic melanomas. A total of 256 archival paraffin-embedded melanoma patients were included in the study, 183 patients carrying primary melanomas and 94 melanoma metastatic tissues (Table 1). We could retrieve matched primary and metastatic samples from 21 patients. The ShcD positive rate was significantly higher in the metastatic tissue than in primary melanoma (54% and 34% for primary and metastatic tissues, respectively; $p = 0.0003$), as previously shown in a small cohort of patients [20].

Demographics, as well as the clinical and histo-pathologic features of the melanoma patients and follow-up events are listed in Table 1. The ShcD RNA transcript showed statistically significant associations with all the main prognostic factors. ShcD-positive melanomas were significantly thicker (Breslow's thickness >4 mm was 35% among ShcD-positive vs. 8% among ShcD-negative; $p < 0.0001$ overall), with more frequent ulceration (51% among ShcD-positive vs. 12% among ShcD-negative; $p < 0.0001$), with a greater number of mitotic events (94% among ShcD-positive vs. 46% among ShcD-negative patients; $p < 0.0001$) and they were more often nodular melanomas (46% in ShcD-positive vs. 20% in ShcD-negative; $p = 0.001$). ShcD expression was also associated with the stage ($p = 0.0004$) and lymph nodes status ($p = 0.02$): the majority of advanced stage patients were ShcD-positive (30% in stage III and IV vs. 15%, $p = 0.0004$), and the frequency of positive lymph nodes was significantly greater in the positive (28%) rather than negative (14%) ShcD expressing patients ($p = 0.02$). Multivariate logistic models, adjusted for confounders, confirmed these associations. These findings strengthen the association of ShcD expression with the increased aggressiveness of the tumor.

Overall, with a median follow-up of 100 months, 35% of the patients within our cohort of primary melanoma patients died, and we found a significant worse overall survival (log-rank $p = 0.05$) among the ShcD-positive as compared to the ShcD-negative patients: at two years, we had 52% vs. 72% probability of survival, respectively (Figure 4A). Disease-free survival at two years was also significantly worse (66%) in ShcD positive than in ShcD-negative patients (78%; log-rank test $p = 0.04$, Figure 4B; stage IV patients were not included in this analysis). The Cox regression model, adjusted for age and gender, indicated a trend toward a significant association of ShcD expression with disease-free survival ($p = 0.08$).

ShcD expression was then evaluated in metastatic melanoma tissue in 94 patients (Table 1). The analysis of the prognostic factors in the antecedent melanoma showed no significant association with ShcD expression. Of note, ShcD-positive patients were younger than the ShcD-negative (median age at diagnosis 45 vs. 50; $p = 0.02$), more frequently carrying ulcerated melanomas ($p = 0.04$). In these patients, ShcD expression is associated with significantly more distant metastases ($p = 0.01$), suggesting that ShcD might predict for metastasis dissemination.

Comparison between the evaluation of ShcD in primary and metastatic tissues confirms the hypothesis that ShcD is a prognostic marker, even if probably not independent from other prognostic markers, and that it is associated with an unfavorable outcome in melanoma patients.

Table 1. Patient demographics and clinical characteristics of the primary cutaneous melanoma according to the ShcD status (AJCC VII).

		ShcD in Primary Tissue				ShcD in Metastatic Tissue		
	No. of Patients, n (%)	Negative	Positive	*p*-Values	No. of Patients, n (%)	Negative	Positive	*p*-Values
No. of patients, n (%)	183 (100%)	120 (66%)	63 (34%)		94 (100%)	40 (43%)	54 (57%)	<001
At diagnosis					Antecedent CM			
Sex				0.46				
Female	97 (54%)	67 (56%)	30 (50%)		29 (31%)	13 (32%)	16 (30%)	0.76
Men	83 (46%)	53 (44%)	30 (50%)		65 (69%)	27 (68%)	38 (70%)	
Age at diagnosis				0.21				
Median	56	54	59		48	50	45	0.02
Interquartile range	(44, 67)	(43, 65)	(46, 69)		(32, 59)	(41, 62)	(24, 55)	
Breslow thickness mm				<0001				
Median	1.0	0.5	2.4		2.8	1.9	3.1	0.14
Interquartile range	(0.5, 3.1)	(0.3, 1.6)	(1.2, 4.7)		(1.5, 5.0)	(0.9,4.8)	(1.8, 5.4)	
Breslow thickness mm				<0001				0.12
0.01–1.0	91 (51)	80 (67)	11 (18)	(<0001 †)	9 (25)	7 (67)	2 (4)	(0.02)
1.01–2.0	26 (14)	14 (12)	12 (20)		20 (21)	8 (17)	12 (22)	
2.01–4.0	31 (17)	15 (13)	16 (27)		16 (17)	3 (20)	13 (24)	
>4.0	31 (17)	10 (8)	21 (35)		25 (27)	10 (8)	15 (28)	
Unknown	1 (1)	1 (1)	0 (0)		24 (1)	12 (30)	12 (22)	
Site of primary				0.75				0.23
Extremity	83 (45%)	51 (43%)	30 (50%)	(0.28 ‡)	30 (32%)	12 (30%)	18 (33%)	(0.73 ‡)
Trunk	78 (43%)	54 (45%)	23 (38%)		38 (41%)	16 (40%)	22 (41%)	
Head and neck	13 (7%)	9 (8%)	4 (7%)		6 (6%)	1 (3%)	5 (9%)	
Other	7 (4%)	4 (3%)	3 (5%)		5 (5%)	4 (10%)	1 (2%)	
Occults	2 (1%)	2 (1%)	0 (0%)		15 (16%)	7 (18%)	8 (15%)	
Ulceration				<0001				0.04
Yes	46 (25%)	14 (12%)	32 (51%)	(<0001)	32 (34%)	9 (23%)	23 (43%)	(0.04)
No	131 (72%)	102 (85%)	29 (46%)		23 (24%)	13 (32%)	10 (18%)	
Unknown	6 (3%)	4 (3%)	2 (3%)		39 (42%)	18 (45%)	21 (39%)	
Mitotic rate				<0001				0.27
Mitoses <1 mm^2	64 (35%)	60 (50%)	4 (6%)	(<0001)	32 (34%)	16 (40%)	16 (30%)	(0.06)
Mitoses ≥1 mm^2	114 (62%)	55 (46%)	59 (94%)		48 (51%)	18 (45%)	30 (55%)	
Unknown	5 (3%)	5 (4%)	0 (0%)		14 (15%)	6 (15%)	8 (15%)	

Table 1. *Cont.*

	No. of Patients, n (%)	ShcD in Primary Tissue			No. of Patients, n (%)	ShcD in Metastatic Tissue		
		Negative	Positive	*p*-Values		Negative	Positive	*p*-Values
Histo-pathologic subtype				0.001				0.83
NM	53 (29%)	24 (20%)	29 (46%)	(0.0004¥)	12 (13%)	4 (10%)	8 (15%)	(0.63¥)
SSM	125 (68%)	93 (77%)	32 (51%)		37 (39%)	16 (40%)	21 (39%)	
Other	4 (2%)	2 (2%)	2 (3%)		5 (5%)	2 (5%)	3 (5%)	
Unknown	1 (1%)	1 (1%)	0 (0%)		40 (43%)	18 (45%)	22 (41%)	
AJCC stage				0.0004				0.33
I	105 (58%)	83 (69%)	22 (35%)	(0.01)	7 (7%)	4 (10%)	3 (5%)	(0.02§)
II	33 (18%)	16 (13%)	17 (27%)		19 (20%)	10 (25%)	9 (17%)	
III	31 (17%)	16 (13%)	15 (24%)		46 (49%)	18 (45%)	28 (52%)	
IV	6 (3%)	2 (2%)	4 (6%)		3 (3%)	0 (0%)	3 (5%)	
Unknown	8 (4%)	3 (3%)	5 (8%)		19 (20%)	8 (20%)	11 (21%)	
Lymph-nodes status				0.02				0.33
Positive	35 (19%)	17 (14%)	18 (28%)	(0.03)	46 (49%)	17 (43%)	18 (54%)	(0.02)
Negative	136 (74%)	94 (78%)	42 (67%)		26 (28%)	15 (37%)	11 (20%)	
Unknown	12 (7%)	9 (8%)	3 (5%)		22 (23%)	8 (20%)	14 (26%)	
ShcD evaluation								
First metastasis								0.01
Regional metastasis					76 (81%)	28 (70%)	48 (89%)	(0.94)
Distant metastasis					17 (18%)	12 (30%)	5 (9%)	
Unknown					1 (1%)	0 (0%)	1 (2%)	
Events during follow-up								
First recurrence *				0.04				
Yes	41 (23%)	22 (19%)	19 (32%)	(0.08)				
No	136 (77%)	96 (81%)	40 (68%)					
Deaths				0.05				0.32
Yes	64 (35%)	36 (30%)	28 (44%)	(0.39)	54 (57%)	11 (69%)	43 (55%)	(0.40)
No	119 (65%)	84 (70%)	35 (56%)		40 (43%)	5 (31%)	35 (45%)	

p-values from chi-square, Fisher exact, Mantel–Haenszel or Wilcoxon tests. In parentheses, *p*-values from the multivariate logistic model. For an event during follow-up, *p* = value from Log rank tests and among parentheses from multivariate Cox regression model. § stage III or IV vs. I or II; ¥ SSM (superficial spreading melanoma) vs. NM (nodular melanoma); ‡ trunk vs. other; † Breslow >1 mm vs. ≤1 mm; * any type of first recurrence among non-metastatic patients (stage IV excluded) and adjusted for age, gender, and ulceration.

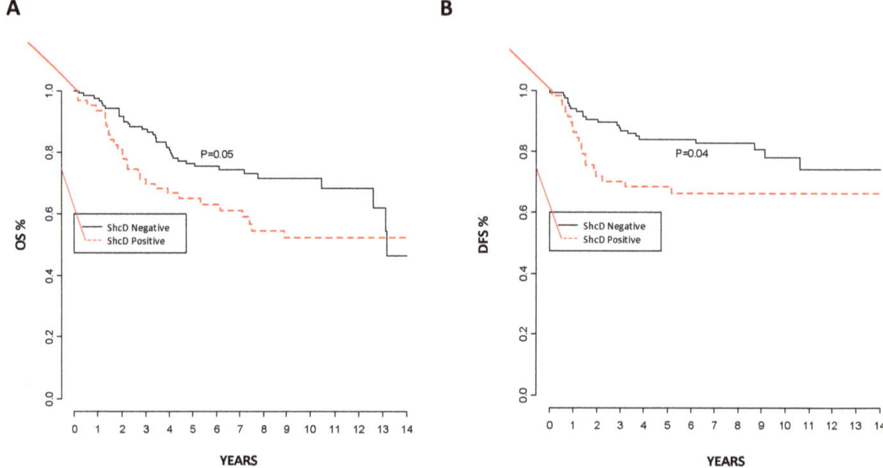

Figure 4. ShcD expression correlates with melanoma progression and poor prognosis in melanoma patients. (**A**) ShcD-positive patients show a significant worse overall survival than the ShcD-negative patients (log-rank p = 0.05). (**B**) Disease-free survival at two years is significantly worse (66%) in ShcD-positive patients than in ShcD-negative patients (78%; log-rank test p = 0.04, stage IV patients not included in the analysis).

2.5. Rac1 Molecular Pathway Is Altered in ShcD Expressing Melanoma Cells

The metastatic process is mainly driven by aberrant cell migration, which relies on specific actin cytoskeleton dynamics and cell phenotypes activated by microenvironmental signals. In melanoma, the elongated/mesenchymal phenotype of cells is driven by activated Rac1 [6]. As shown in Figure S2A, ShcD silencing switches the morphology of WM115 cells from rounded to elongated. To investigate whether the elongated phenotype of ShcD-depleted cells depends on Rac1, we either treated WM115 cells with the NSC23766 Rac1 specific-inhibitor [28] or we silenced Rac1 by means of Rac1 SmartPool siRNA (Figure S2B). As expected, NSC23766-treated control WM115 cells or Rac1-silenced cells showed significantly increased numbers of rounded cells (p < 0.01) (Figure 5A,B). Notably, NSC23766 treatment rescued the rounded morphology of ShcD-silenced cells (from 28% to 61% for shShcD#1 and from 46% to 75% for shShcD#2 (Figure 5A)), demonstrating that cell elongation after ShcD knockdown is Rac1-dependent and suggesting that ShcD acts as a suppressor of the Rac1 pathway.

We therefore investigated whether the downregulation of ShcD leads to Rac1 activation. Pull-down assays of Rac1-GTP in WM115 cells upon adhesion to fibronectin showed higher levels of Rac1-GTP in ShcD knockdown cells, as compared to controls (Figure 5C). Consistently, the phosphorylation levels of N-WASP and Cofilin, two Rac1 downstream effectors [29], were markedly increased in ShcD downregulated cells (Figure 5D), demonstrating that ShcD silencing leads to an increased activation of the Rac1 signaling pathway.

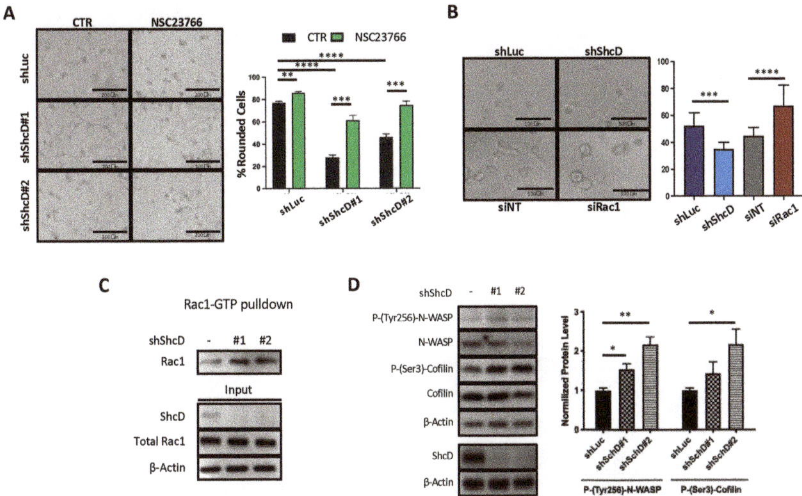

Figure 5. ShcD-silencing cells show a deregulated Rac1 molecular pathway. (**A**) shLuc and shShcD WM115 cells morphology assessment upon Rac1 inhibitor treatment (NSC23766). Cells were transduced with shLuc, shShcD#1 and shShcD#2, plated on a thick collagen layer and treated with either vehicle (CTR, control, DMSO) or 50 μM NSC23766. After 24 h treatment, the percentage of elongated/rounded cells was calculated. Three images per well were acquired in triplicate experiments. Student *t-test* (***, $p < 0.001$; ****, $p < 0.0001$). (**B**) MM27 cells morphology assessment upon Rac1 silencing. Cells were infected with shLuc and shShcD (pool of shShcD#1 and #2) or transfected with the non-targeting and Rac1 SmartPool siRNA and after 48 h plated on a thick collagen layer. After 24 h, the percentage of elongated/rounded cells was calculated. Three images per well were acquired in triplicate experiments. Student *t-test* (****, $p < 0.0001$). (**C**) Rac1 activation assay in shLuc and shShcD WM115 cells. Rac1 activity was assayed by GTP-bound Rac1 pulled-down and analyzed by Western blotting. WM115 cells were transduced with shLuc, shShcD#1 and shShcD#2 and plated on fibronectin. Rac1 indicates active Rac1 in the pull-down assay, and Total Rac1 represents the total amount of Rac1 in the cell lysates (input). Actin was used as a loading control. (**D**) Rac1 pathway regulation upon ShcD silencing in WM115 cells. WM115 cells were transduced with shLuc, shShcD#1 and shShcD#2 and plated on fibronectin. The Rac1 molecular pathway was investigated at the protein level by Western blotting in ShcD-depleted WM115 cells. Both the phosphorylation status of N-WASP (P-Tyr256) and Cofilin (P-Ser3) and their total expression levels were detected and ShcD silencing was also checked. Actin was used as a loading control. The quantification of two experiments was performed with ImageJ normalizing for the total protein level and house-keeping gene. Student *t-test* (*, $p < 0.05$; **, $p < 0.01$).

2.6. ShcD Negatively Regulates Rac1 Activation by Preventing Its Binding to DOCK4

Rac1 is activated by different upstream regulators, including DOCK4, a guanine nucleotide exchange factor (GEF) GTPase that promotes Rac-dependent cell migration [18]. To investigate its relationship with ShcD, we assessed the presence of ShcD and DOCK4 in the same protein complex, using ShcD overexpressing MM27 PDX cells. As shown in Figure 6A, ShcD protein was detected in the DOCK4 immuno-precipitate. These data were also confirmed in ShcD overexpressing WM115 cells (Figure S3A).

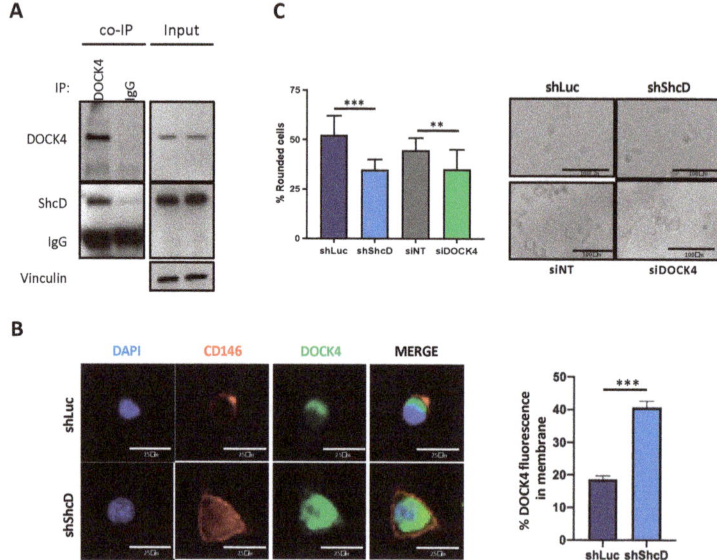

Figure 6. Rac1 molecular pathway is altered in ShcD expressing melanoma cells via DOCK4. (**A**) DOCK4 immunoprecipitation in ShcD overexpressing MM27 cells. Cell lysates of PP-ShcD MM27 cells were immunoprecipitated (IP) with anti-DOCK4 antibody or anti-IgG antibody and subsequently immunoblotted with anti-ShcD antibody. Expression of vinculin was used as a loading control in total cell lysates (input). (**B**) DOCK4 localization in shLuc and shShcD MM27 cells. Cells were plated on fibronectin and co-stained for DOCK4 and CD146 plasma membrane marker by immunofluorescence (red). Nuclei were counterstained with DAPI (blue). Representative images are shown (63× magnification). Quantification was done using ImageJ and the DOCK4 plasma membrane signal was normalized to the total fluorescence. Student *t-test* (***, $p < 0.001$). (**C**) The analysis of cell morphology in DOCK4 siRNA and ShShcD-silenced MM27 cells. DOCK4-silenced cells and their controls together with ShLuc and ShShcD cells were plated on a thick layer of collagen and rounded vs. the elongated morphology assessed after 24 h. The percentage of elongated/rounded cells was calculated. Three images per well were acquired in triplicate experiments. Student *t-test* (**, $p < 0.01$).

Based on the demonstrated ShcD–DOCK4 interaction and the role of DOCK4 in mediating Rac1 activation, we hypothesized that ShcD may negatively regulate Rac1 activation by preventing its binding to DOCK4. To test this hypothesis, we analyzed DOCK4 localization in the control and ShcD-silenced MM27 and WM115 cells after adhesion to fibronectin by immunofluorescence. In shLuc control cells, DOCK4 was found in the cytoplasm, while in ShShcD cells, DOCK4 was also partly localized at the plasma membrane (Figure S3B,C). To precisely quantify the percentage of DOCK4 confined at the plasma membrane, the control and ShcD-silenced MM27 were co-stained with DOCK4 and CD146, a known cell adhesion molecule present on the plasma membrane of melanoma cells (Figure 6B). All together, these results suggest that when ShcD is overexpressed, it sequesters DOCK4 in the cytoplasm, thus blocking its recruitment to the plasma-membrane where it would activate Rac1. To strengthen the molecular value of our model, we silenced DOCK4 in MM27 cells using SmartPool siRNA. Once the silencing was confirmed by Western blot (Figure S3D), we plated the DOCK4-silenced, control, and ShcD-silenced cells on collagen and we evaluated the effect of the silencing on cell morphology (Figure 6C). DOCK4-silenced cells recapitulated the elongated morphology of ShcD-silenced cells confirming the ShcD role in melanoma cell invasiveness through DOCK4 regulation.

2.7. ShcD Contributes to Resistance of Melanoma Cells to Treatment

Highly invasive cells are frequently resistant to standard care therapy. To investigate whether ShcD expression modulates the drug sensitivity of melanoma cells, MM27 PDX cells, carrying the BRAF^{V600E} mutation, were transduced with ShLuc and ShShcD vectors and treated with Dabrafenib (BRAF inhibitor), Trametinib (MEK inhibitor) or Everolimus (mTOR inhibitor). Analyses of cell viability at three different doses of the three drugs showed no difference between the control and ShcD-silenced cells (Figure 7A). We then analyzed the effects on the migration of sub-apoptotic doses of each of the three drugs. As expected, the reduction in ShcD protein levels lowered the migration rate in untreated cells (Figure S1). The rate of migration was significantly affected by all three drugs in the control cells and to a similar extent, also in ShcD-silenced cells (Figure 7B). Finally, we tested the effects on the migration of combinations of Dabrafenib, Trametinib and Everolimus. Treatment with the combination of Dabrafenib and Trametinib (Dab/Trame; a standard treatment in BRAF-mutant melanomas) had no effect on the migration of ShShcD cells compared to the control cells, showing that the drug response to the inhibition of the BRAF/MEK pathway is not influenced by the depletion of ShcD. Notably, treatment with the combination of Dabrafenib and Everolimus (Dab/Eve; a new combination therapy used to bypass the induction of resistance to BRAF inhibitors, [30]) significantly reduced the migration of ShShcD MM27 cells, as compared to the ShLuc control cells (Figure 7B). This indicates that the ShcD depletion increases the drug sensitivity of melanoma cells to this combination. All together, these data suggest that ShcD affects intracellular pathways that can, in turn, sensitize melanoma cells to new combination therapies.

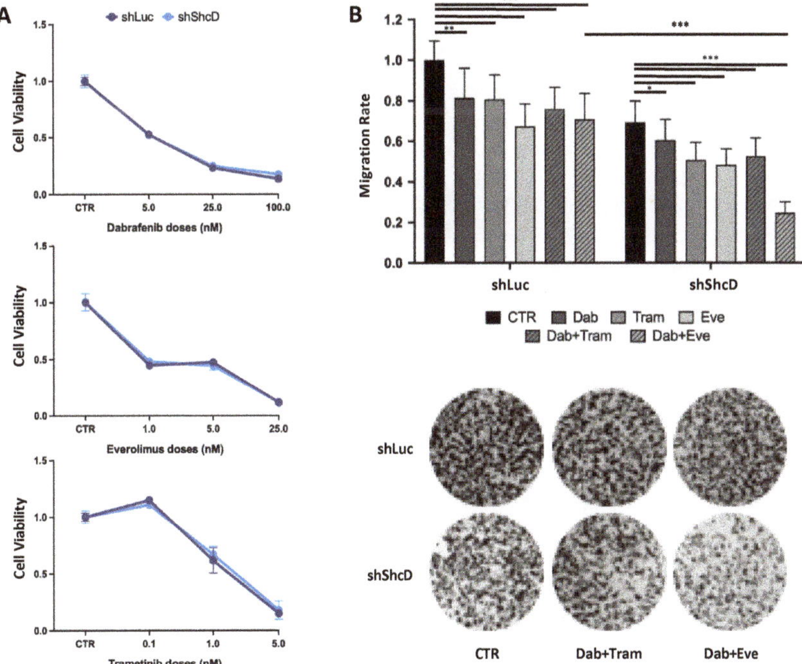

Figure 7. ShcD silencing sensitizes melanoma cells to targeted therapy combination. (**A**) ShLuc and ShShcD MM27 cell viability upon treatment with Dabrafenib (BRAF inhibitor), Trametinib (MEK inhibitor) and Everolimus (mTOR inhibitor). Cells were treated for 72 h and cell viability was assessed by CyQuant. The inhibition of viability is indicated as a percent over control (GraphPad Prism

software). Student *t*-test (not statistically significant). (**B**) Transwell migration assay upon specific targeted therapy in ShLuc and ShShcD MM27 cells. Cells were plated in the upper chamber in a serum-free medium with Dabrafenib (2.5 nM), Trametinib (0.5 nM), Everolimus (2.5 µM), Dabrafenib/Trametinib and Dabrafenib/Everolimus combinations (same doses used for the single drug testing) or vehicle (CTR, DMSO). Complete medium was added to the lower chamber. After 48 h, the cells were stained with Crystal Violet and the migration rate was quantified using ImageJ. Histograms represent the relative cell migration (mean ± SD) of a triplicate experiment, normalized to ShLuc as the control. Differences among the groups were calculated by applying the Student *t-test*, using migration rate values normalized to the CTR of the corresponding group (*, $p < 0.05$; ***, $p < 0.001$; **, $p < 0.01$).

3. Discussion

Melanoma cells are characterized by a high propensity to migrate and metastasize, which is probably linked to the embryologic derivation of its cell of origin. Cutaneous melanoma is derived from melanocytes, which originate from the neural crest. During development, cells exiting from the neural crest undergo morphological changes accompanied by altered adhesion, increased migratory capacity and spreading throughout the embryo before they reach target tissues and differentiate. The process of melanoma migration and spreading is associated with the intrinsic properties of neural crest-derived melanocytes and is tightly regulated by the surrounding environment. Notably, metastatic melanoma cells can be reprogrammed to neural crest-like cells by the embryonic microenvironment [31]. Moreover, during the invasion of the epidermal/dermal layer of the skin, melanoma cells go through an EMT-like program and at least partially exploit the same migratory transcriptional program used by neural crest cells during the development to colonize the skin [32]. Most notably, ShcD is expressed in neural crest cells [33] and melanoblasts [34], thus suggesting that it represents a lineage-specific factor critical for the migration of melanocyte precursors, whose expression and function can be reacquired by invasive melanoma cells.

We previously showed that ShcD positively regulates melanoma cell migration in the 2D cultures of melanoma cell lines [19]. Here, firstly, we showed that PDX models, which more faithfully phenocopy patients' biological features, recapitulate the ShcD migratory function formerly documented in cell lines (Figure S1). Furthermore, we extended our analyses and demonstrated that ShcD silencing increased the capability of melanoma cells to adhere to different extracellular matrices, suggesting that ShcD expression favors the detachment of melanoma cells from the surrounding microenvironment, thus favoring dissemination (Figure 1).

During the last decade, the invasive properties of melanoma cells have been extensively studied using 3D model systems, which better mimic the tumor environment, as compared to the traditional 2D cell cultures. Most notably, our 3D approach showed that invasive melanoma cells are featured by high plasticity, with cells capable of modulating either RhoA or Rac1 small GTPases and thus adopting two alternative modes of invasion: amoeboid (rounded) or mesenchymal (elongated), respectively [6,8]. Interestingly, the MITF expression in the WM266.4 melanoma cell line increased the numbers of elongated cells and decreased their invasiveness, while MITF silencing in 501Mel cells induced rounded morphology, consistently with the capacity of MITF to negatively regulate the invasive phenotype of melanoma cells [35]. Consistently, MITF has been reported to downregulate ShcD expression by directly binding to at least five sites of the genomic locus [36], suggesting that the MITF/ShcD axis is a critical regulator of invasiveness phenotypes of melanoma cells. Notably, ShcD overexpression increased the invasion properties of melanoma cells in 3D assays (Figure 2). Since adhesion, in vitro migration and invasion are all pivotal steps in metastasis formation in vivo, we analyzed the metastatic properties of ShcD overexpressing cells in two different PDX models and showed that ShcD endowed melanoma cells with high metastatic potential (Figure 3). Notably, the metastatic tumor burden was increased by ShcD, as lymph nodes display a larger volume and a higher number of distant metastases is detected throughout the body of the animal. The metastatic tumor burden is significantly associated with the clinical response and outcomes in melanoma patients [37]. Not surprisingly, we showed that high ShcD expression is associated with a poor outcome in a cohort of melanoma patients (Figure 4).

The mechanisms regulating the different modalities of cell motility involve several signaling pathways. RhoA signaling activates ARHGAP22 Rho-GAP, which keeps Rac1 in an inactive state leading to high actomyosin contractility [6]. On the other hand, RhoA signaling is inhibited by WAVE2, which is activated by Rac1 [6]. Consistently, cells that adopt the mesenchymal mode of motility are characterized by the high activity of Rac1 [6]. The treatment of ShcD-knockdown cells with the selective Rac1 inhibitor NSC23766, as well as the silencing of Rac1, rescued the rounded phenotype of melanoma cells (Figure 5A,B), indicating that ShcD actively inhibits Rac1, inducing the rounded morphology of melanoma cells. In fact, ShcD-silenced cells showed increased Rac1 activation and elongated morphology. Based on these results, we hypothesized that ShcD inhibited the Rac1 signaling pathway, enabling active RhoA to guide amoeboid motility. Indeed, Rac1 inhibition in ShcD-knockdown cells reverted them to an amoeboid phenotype, suggesting that RhoA was active and did not need ShcD to execute its functions. We investigated the mechanism through which ShcD blocks Rac1 activation and we showed that ShcD formed a stable complex with DOCK4 (Figure 6A), a protein belonging to the DOCK180-related family of guanine nucleotide exchange factors (GEFs), whose function was already linked to cell invasion [16]. All Shc family members are equipped with different domains (CH2, PTB, CH1 and SH2) that enable them to recruit different signaling molecules to properly convey and organize the extracellular signals and that, at the same time, allow them to translate signals into specific biological responses [38]. ShcD differs from the other members of the family being selectively expressed in the neuronal-derived tissues only [38]. The specificity of expression suggests that ShcD drives crucial functions in these cells. Here, we demonstrate for the first time a direct binding of ShcD with the guanine nucleotide exchange factor DOCK4 and we show that after ShcD silencing, DOCK4 is recruited to the plasma membrane during cell adhesion to fibronectin (Figure 6B). In the same experimental conditions, Rac1 was highly activated. Conversely, in the presence of ShcD, the translocation of DOCK4 to the membrane did not occur and Rac1 was less active than in the ShcD knockdown cells. The recruitment of DOCKs to the membrane was previously shown to be an important step in the activation of Rac1 after the adhesion of epithelial cells to fibronectin or stimulation with growth factors [39,40]. Our results suggest that in our model system, ShcD prevents Rac1 activation by blocking DOCK4 recruitment to the plasma membrane (by a not yet identified mechanism). This would be an additional mechanism to keep the RhoA signaling pathway activated and promote amoeboid motility.

Having shown that ShcD overexpressing cells migrate and invade the surrounding tissue and metastasize to lymph nodes and distant organs, we sought to investigate whether ShcD depletion correlates with an increased sensitivity to targeted therapy. To this end, we used a BRAF-mutated melanoma (MM27). In patients, these melanomas are mostly treated with combination therapy based on the use of BRAF/MEK inhibitors, but the duration of response is variable and patients frequently experience recurrence: new treatment strategies, especially different combinations, are therefore needed. It was previously shown that Dabrafenib/Everolimus combination was effective in melanoma cell lines to bypass the induction of resistance to BRAF inhibitors [30]. Here, we report that this combination can become even more efficacious by preventing ShcD activation. Indeed, ShcD depletion does not influence the MAPK intracellular pathway [20], in line with our finding that shShcD cells are as sensitive to the standard of care therapy (Dabrafenib + Tramentib) as control cells, but it does activate the Rac1 signaling pathway, which in turn activates the mTOR pathway [41]. The upregulation of mTOR signaling makes melanoma cells more sensitive to Everolimus, thus suggesting that ShcD depletion can potentiate the effect of combined targeted therapy in melanoma cells.

Our findings demonstrate a key role of the adaptor protein ShcD in the process of metastasis formation and drug resistance. We propose to take into consideration ShcD as a putative therapeutic target to improve the prognosis of melanoma patients.

4. Materials and Methods

4.1. Plasmids, Lentiviral Transduction and Cell Culture

4.1.1. Plasmids

Human ShcD cDNA was cloned in the Pinco plasmid [42]. The original backbone was modified to a new version, PincoPuro (PP) plasmid, containing the puromycin cassette under the 5′LTR and GFP (PP, empty vector) or ShcD cDNA (PP-ShcD, ShcD overexpressing vector) under the CMV promoter. LL3.7 Puro2GFP (LLPG) plasmid and its modified version with shRNA targeting Luciferase (shLuc) were a generous gift of Dr. Bruno Amati. Two different short hairpin RNAs (shRNAs) targeting ShcD (shShcD#1 and shShcD#2) were inserted in the XbaI and NotI restriction site of the LLPG plasmid. The following pairs of shRNA sequences were annealed and cloned into the vector backbone:

shShcD#1 forward 5′TCAATGAGATCACTGGATTTTTCAAGAGAAAATCCAGTGATCTCATTG TTTTTTC3′; shShcD# reverse 5′TCGAGAAAAAACAATGAGATCACTGGATTTTCTCTTGAAAAATC CAGTGATCTCATTGA3′;
shShcD#2 forward 5′TTGAGGAGGTGCATATTGATTTCAAGAGAATCAATATGCACCTCCTCAT TTTTTC3′; shShcD#2 reverse 5′TCGAGAAAAAATGAGGAGGTGCATATTGATTCTCTTGAAATCAAT ATGCACCTCCTCAA3′

4.1.2. Retroviral and Lentiviral Transfections and Infections

Phoenix-AMPHO helper cell line was transfected with PincoPuro constructs for retroviral production as previously described [42]. Three cycles of infection (3 h each) with supernatant retroviral production, together with 4 µg/mL of polybrene, were performed on MM13 and MM27 PDX cells. The 293T cell line was transfected with LLPG constructs for lentiviral production using calcium phosphate with pMD2.G-VSVG and pCD-NLBH packaging plasmids (generous gifts from Prof. Colin Goding). One cycle of infection (16 h) with lentiviral vectors, with 4 µg/mL of polybrene, was performed on MM13 and MM27 PDX cells and on WM115 and WM266.4 cell lines. Cells were puromycin (2 µg/mL) selected for 3 days and the surviving cells were used for subsequent experiments.

4.1.3. siRNA Transfection Protocol

MM27 PDX cells were transfected following the DharmaFECT transfection protocol (Dharmacon Horizon) with ON-TARGET plus Human DOCK4 (9732) siRNA-SMARTpool (L-017968-01-0005, Dharmacon), ON-TARGET plus Human Rac1 (5879) siRNA-SMARTpool (L-003560-00-0005, Dharmacon) and ON-TARGET plus Non-targeting Pool (D-001810-10-05, Dharmacon). Briefly, the cells were transfected with siRNA-SMARTpool at the final concentration of 25 nM and DharmaFECT for 48 h.

4.1.4. Cell Cultures

WM115 and WM266.4 melanoma cell lines were maintained in RPMI (Euroclone, Cat#ECM2001) supplemented with 10% FBS, 200 mmol/L glutamine, 100 U/mL penicillin and 100 µg/mL streptomycin. MM13 and MM27 PDX cells were maintained in Iscove's modified Dulbecco's medium (IMDM, Sigma Aldrich-Merck, Cat#I3390) supplemented with 10% North American origin FBS, 200 mmol/L glutamine, 100 U/mL penicillin and 100 g/mL streptomycin.

4.2. PDX Generation and In Vivo Studies

PDX were generated as previously described [21]. All the in vivo studies were performed after approval from our fully authorized animal facility and the notification of the experiments to the Ministry of Health (as required by the Italian Law; IACUCs N° 29/2013, N° 326/16 and N° 758/2015-PR), and in accordance with EU directive 2010/63.

Then, 7×10^5 MM13 and MM27 PDX cells transduced with PP and PP-ShcD were transplanted by intradermal injection into 34 NOD.Cg-Prkdcscid Il2rgtm1Wjl/SzJ (NSG) mice. Mice were monitored for tumor development and excised when they reached ~0.5 cm^3 in volume (calculated using the modified ellipsoid formula 1/2 (length $\times$ width^2)). Mice were then monitored weekly for metastasis formation by lymph node detection. After 1–2 months from resection, all mice were sacrificed and enlarged lymph nodes collected, eventually together with lung, liver, kidney and spleen. The lymph node volume was measured, with the same formula applied to primary tumors, as evidence of metastatic burden. All the organs were analyzed for the presence of macroscopic metastatic lesions.

4.3. Cell Adhesion and Spreading Assays

For the adhesion assay, ShLuc and ShShcD MM27 cells (2.5×10^5 cells/well, 6-well-plate format, in duplicate) were plated onto different extracellular matrices: Fibronectin 5 µg/mL (Roche, Cat#11080938001), Collagen Type I Rat Tail 20 µg/mL (Corning, Cat#354249), Matrigel 20 µg/mL (Corning, Cat#356231). After 120 (Fibronectin), 100 (Collagen) and 80 (Matrigel) minutes, cells were fixed and stained with 0.5% Crystal Violet (Sigma Aldrich-Merck, Cat# V5265). Five images per well were acquired and images were analyzed (cell number) using ImageJ Software.

For the spreading assays, 2.5×10^4 shLuc and shShcD cells were plated in triplicates on a 5 µg /mL fibronectin-coated (1 h at 37 °C) 24-well plate. After 75 min, the cells were fixed and stained with 0.5% Crystal Violet. Three images per well were acquired at EVOS XL CORE microscope, Life Technologies (20× and 40× magnification) and cell morphology and numbers were analyzed and counted using ImageJ Software.

4.4. Cell Morphology Assessment

In addition, 2.5×10^5 MM27 PP and PP-ShcD cells/well (6-well-plate format) were plated on a thick collagen layer (2.3 mg/mL, Collagen Type I Rat Tail 20 µg/mL). Three images per well were acquired and the cells were counted as rounded or elongated on the basis of their shape. Experiments were carried out in triplicate.

The 2.5×10^5 WM115 shLuc and shShcD#1 or #2 cells/well (6-well-plate format) were plated as previously described and treated for 24 h with 50 µM NSC23766 Rac1 inhibitor (Sigma Aldrich-Merck, Cat#SML0952). Five images per well were acquired and the cells were counted as described above.

4.5. Cell Migration Assay

The migration assay was performed using 8.0 µm pore size, fibronectin-coated inserts in 24-well plates (Corning-Falcon cat#353097). Triplicates of 3×10^4 MM27 shLuc and ShcD-silenced cells were plated in the upper chamber in serum-free medium. Vehicle (DMSO) or Dabrafenib 2.5 nM (GSK2118436A, Active Biochem, Cat#A-1220), Trametinib 0.5 nM (GSK-1120212, Active Biochem, Cat#A-1258), and Everolimus 2.5 µM (RAD001, MedChemExpress, Cat#HY-10218) were added alone or in combination (Dabrafenib/Trametinib and Dabrafenib/Everolimus, same doses as single testing) to the medium. The complete medium was added to the lower chamber. After 48 h, cells that migrated to the lower surface of the inserts were stained with 0.5% Crystal Violet. Four images of each insert were acquired and analyzed with the ImageJ Software.

4.6. Spheroid Invasion Assay and Time-Lapse Analysis

For the spheroid formation, 2×10^3 PP and PP-ShcD MM27 cells were grown as hanging drops in IMDM complete medium (Sigma Aldrich-Merck, Cat#I3390) with 0.4% methylcellulose for 48 h at 37 °C. Single spheroids were harvested, resuspended in 1.5 mg/mL Collagen Type I Rat Tail and plated on a 96 well-plate pre-coated with a thin collagen layer. After 1 h at 37 °C, the solidified collagen gel was covered with complete IMDM medium. The ability of cells to invade the area was monitored and imaged for 24 h on a TIRF Leica DMI6000B camera, temperature controller (37 °C), 5% CO_2 incubation chamber (Leica), Okolab incubator and LAS AF Software. Images of 10 spheroids per group were collected every 15 min for 24 h using a dry 4× objective lens, and phase-contrast optics with a Z-stack ± 45 μm and 15 μm step. The invasion area was analyzed with the ImageJ Software and normalized against the area at time 0. The speed of the cells (μ/min) at the invasive front was analyzed by the manual tracking plugin of the ImageJ Software and calculated as a mean of 10 cells' tracking. The rounded vs. elongated cell shape at the front of the invasion was carried out by the manual counting of 3 different fields of 5 spheroids per group on the ImageJ Software.

4.7. Immunofluorescence Staining

Furthermore, 2.5×10^4 MM27 and WM115 cells infected with shLuc and shShcD were plated on 5 μg/mL fibronectin pre-coated (O/N at 4 °C) slides and allowed to attach for 75 min in complete IMDM. Cells were fixed with 4% paraformaldehyde for 10 min, permeabilized with 0.1% Triton-X, blocked for 1 h with 5% bovine serum albumin and immune-stained with Vinculin (Sigma Aldrich-Merck, Cat#V9131), p-Vinculin (Tyr-1065) (Invitrogen, Cat#44-10786), Paxillin (Invitrogen, Cat#03-6100), p-Paxillin (Tyr118) (Invitrogen, Cat# 44-7226), FAK (c-20, Santa Cruz, Cat#sc-558), p-FAK (Tyr397) (Upstate, Cat#07-012), DOCK4 (Novus Biologicals, Cat#NBP1-266-48) and CD146 (Novocastra, Cat#NCL-L-U). Secondary detection was done using secondary fluorescently labeled antibodies (Life Technologies Italia) for 45 min at room temperature. Slides were counterstained with 4′,6-diamidino-2-phenylindole (DAPI, Life Technologies Italia) for nuclei labelling and mounted on glass with Mowiol (Merk Life Science, Cat#81838). Images were collected at 63× magnification by motorized Leica DM6B fluorescence microscope, equipped with a Zyla camera, LASX Software, and by a Leica TCS SP5II confocal microscope, equipped with a Leica camera, LASX Software.

4.8. Immunoblot and Immunoprecipitation

For the immunoblot analysis, cells were rinsed with cold PBS and lysed in RIPA lysis buffer (150 mM sodium chloride, 1% NP-40, 0.5% sodium deoxycholate, 0.1% SDS, 50 mM Tris pH 8.0) plus protease and phosphatase inhibitor cocktail (Roche, Cat#11697498001 and ThermoScientific Cat#A32957), and sonicated twice for 20 s. Cell debris was removed by centrifugation at 13,000 rpm for 20 min at 4 °C. Protein concentration was determined using the Bradford Assay (Bio-Rad, Cat#500-0006). Protein lysates were separated on SDS-PAGE gel and transferred onto nitrocellulose membrane. Membranes were blocked with 5% nonfat milk or BSA in Tris-buffered saline with 0.1% Tween 20 (TBST) for 1 h. Membranes were immunoblotted with the following appropriately diluted primary antibodies overnight at 4 °C or 1 h at room temperature: ShcD anti-CH2 Clone 47-10-21 (home-made, see below), AXL H3 (Santa-Cruz, Cat#sc-166269), Sox10 (D5V9L) (Cell Signaling, Cat#89356), EGFR (E114) (Abcam, Cat#ab32562), EphA2 (D4A2) (Cell Signaling, Cat#6997), Brn2/POU3F2 (D2C1L) (Cell Signaling, Cat#12137), MITF antibody (C5) (Abcam, Cat#ab12039), PDGF-Rβ (R&D Systems, Cat#AF385), p-(Tyr-256) N-WASP (ECM Biosciences, Cat#WP2601), N-WASP (Novus Biologicals, #CatSC66-05), P-(Ser3)-Cofilin (Cell Signaling, Cat#3311), Cofilin (D3F9) (Cell Signaling, Cat#5175), ß-Actin (Novus Biologicals, Cat#NB600-503), Vinculin (Sigma Aldrich-Merck, Cat#V9131), GAPDH (Cell Signaling, Cat#2118). Membranes were washed three times with TBST and incubated with the appropriate horseradish peroxidase-conjugated secondary antibody for 1 h at room temperature. Finally, the expression of protein was detected by enhanced-chemiluminescence solutions (ECL-BioRad, Cat#170-5061) and captured by Amersham Hyperfilm (Cat#GEH28906837) or ChemiDoc XRS.

For immunoprecipitation, cells were harvested by scraping in PLC lysis buffer (50 mM HEPES pH 7.5, 150 mM NaCl, 10% glycerol, 0.5% Triton X-100, 1.5 mM MgCl2, 1 mM EGTA) plus protease and phosphatase inhibitor cocktail. Cells were lysed for 30 min on ice, centrifuged at 13,000 rpm for 30 min and the protein concentration was detected as above. Cell lysates (3 mg) were incubated with anti-DOCK4 Antibody (Novus Biologicals, Cat#NBP1-26648) for 3 h at 4 °C followed by 1 h incubation with protein A Sepharose CL-4B beads (GE-Healthcare, Cat#GE17-0780-01). As the negative control, cell lysates were immunoprecipitated with anti-IgG antibody. Immunocomplexes were washed

5 times with lysis buffer and boiled with sample buffer. Immunoblot analysis was performed with the indicated antibodies.

4.9. ShcD Antibody Generation (Anti-CH2 Clone 47-10-21)

The ShcD antibody was generated in collaboration with the Biochemistry Unit at the IFOM-European Institute of Oncology (IEO) campus. Antibodies against the CH2 domain of murine ShcD was raised using the peptide sequence of 16 amino acids: QPYRKYDNTGLLPPKK. This synthesized peptide sequence corresponds to the carboxy-terminal part of the CH2 region, which is specific for ShcD and not present in the other three members of the family. The immunized sera were tested on RaLP overexpressing cell lines by Western blotting and the reactive sera were immunopurified.

4.10. RAC1 Activation Assay

Rac1 activity was measured with Rac1 Activation Assay Biochem kit (Cytoskeleton, Cat#BK035) according to the manufacturer's protocol. Cells were transduced with shLuc, shShcD#1 and shShcD#2 and plated on 5 µg/mL fibronectin-coated plates (as described above) for 15 min in complete RPMI medium. Cells were first washed with phosphate-buffered saline (PBS) and then lysed with a proper volume of lysis buffer provided by the manufacturer and supplemented with protease inhibitors. The lysates were precleared and after measuring protein concentration with the provided Precision Red Advanced Protein Assay, and 50 µg of lysates were saved for Western quantitation of total Rac1. About 800 µg of lysate was incubated with 15 µL of GST–PAK–PBD (Rac1 effector protein, p21 activated kinase 1) for specific pull-down of activated Rac1. After rocking at 4 °C for 1 h, beads were washed once and boiled at 95 °C for 2 min. Protein lysates were subsequently resolved in 14% SDS-PAGE gel and transferred to polyvinylidene fluoride membrane (PVDF).

The total and activated Rac1 was detected by Western blotting using an anti-Rac1 monoclonal antibody provided in the kit, followed by incubation with goat anti-mouse secondary antibody at room temperature for 1 h. Immunoblotting and detection were carried out as previously indicated.

4.11. Tissue Microarrays (TMA)

Tissue Microarrays were performed in collaboration with the Division of Pathology and the Molecular Pathology Unit at the European Institute of Oncology (IEO). Human specimens derived from formalin fixed and paraffin embedded melanocytic lesions were arrayed as previously described [43]. The choice of the samples was based on tumor availability but also on reliable and sufficient clinical information. Briefly, for each sample, two 0.6 mm cylinders from previously identified on hematoxylin-eosin stained sections were removed from the donor blocks and deposited on the recipient block using a custom-built precision instrument (Tissue Arrayer, Beecher Instruments, Sun Prairie, WI, USA). Two-micrometer sections of the resulting recipient block were cut, mounted on glass slides, and processed for in situ hybridization (ISH), as previously reported [44]. TMAs contained 183 primary melanomas, subdivided according to TNM staging and 94 metastatic melanomas. Patients were divided according to the diagnosis and ShcD status evaluated in primary and metastatic tissues. Patients carrying primary melanomas had a median age of 56 years at diagnosis, 97 were female (54%) and 83 were male (46%). The majority of primary melanomas were SSM (skin-sparing mastectomy) (68%, n = 125 patients, pts), thin (51%, n = 91 pts Breslow < 1 mm), non-ulcerated (72%, n = 131 pts) and with a mitotic rate ≥ 1 mm^2 (62%, n = 114). Lymph node status was positive in 35 patients (19%) and not available in 12 patients (7%). In the metastatic group, 68% of the patients were male. Fifteen patients (15%) had an occult primary melanoma, while for the remaining 79 patients with a known antecedent primary melanoma, some but not all primary tumor parameters were available. The majority of the melanomas were thick (75% with Breslow > 1 mm), with a high mitotic rate (51% with mitoses >1 mm^2) and a positive lymph node status (49% positive, 28% negative and 23% unknown). The sites of primary tumor were often trunk (43% and 41% for primary and metastatic tissue, respectively) and extremities (45% and 32% for primary and metastatic tissue, respectively).

4.12. In Situ Hybridization (ISH)

ShcD expression levels were assessed by ISH as previously reported [44]. TMA sections were deparaffinized, digested with Proteinase K (20 μg/mL), postfixed, acetylated and dried. After overnight hybridization at 50 °C, the sections were washed in 50% formamide, 2 × SSC, 20 mM 2-mercaptoethanol at 60°, coated with Kodak NTB-2 photographic emulsion and exposed for three weeks. The slides were lightly H&E counterstained and analyzed under microscope with a dark-field condenser for the silver grains.

All TMA were first analyzed for the expression of the housekeeping gene β-actin, against which the specific signals were normalized. Gene expression levels were evaluated by three independent operators, counting the number of grains per cell and were expressed in a semi-quantitative scale (ISH score): 0 (no staining), 1 (1–25 grains; weak staining), 2 (26–50 grains; moderate staining), and 3 (>50 grains, strong staining). ISH scores 2 and 3 were considered to represent an unequivocal positive signal.

4.13. Drug Treatment

In vitro drug sensitivity was assessed by CyQuant (Invitrogen, Cat#C35012) in MM27 shLuc and shShcD cells. Briefly, the cells were plated in triplicate in 96 wells (1500 cells/well) and treated for 72 h by a single exposure to either the vehicle or increasing concentrations of Dabrafenib (BRAF inhibitor, in the range of 5–100 nM), Trametinib (MEK inhibitor, in the range of 0.1–5 nM), Everolimus (mTOR inhibitor, in the range of 1–25 μM). The inhibition of viability is indicated as a percent over control (GraphPad Prism software).

4.14. Statistical Methods

Clinical and pathologic features were tested for association with ShcD status using simple cross tabulations, chi-square test, Mantel–Haenszel and Fisher's exact test; median values and range inter-quartiles were presented for continuous variables and compared with non-parametric Wilcoxon tests.

Time to death and time to recurrence were defined as the time from the first diagnosis of melanoma until the event of interest. For the analyses among metastatic patients with ShcD evaluated in the metastatic tissue, time to death was defined as the time from the diagnosis of first distant metastasis until death. All patients, alive or free of disease at last follow-up date, were considered right censored. Multivariable binary logistic regression was used to assess the statistically significant features associated with the ShcD status. Disease-free survival was estimated by the Kaplan–Meier method to evaluate the time to events. The log-rank test was used to compare the survival time between groups. Cox proportional hazards models were used to assess if the ShcD was associated with survival, after adjustments for confounders (age and gender), for disease-free survival and also chemotherapy for overall survival. Prognostic factors, such as Breslow thickness, ulceration and lymph node involvement, were included, if significant, in multivariate models for metastatic tissue. In primary tissue, they were not included because they were considered intermediate steps in the casual path between ShcD and recurrence or death. In fact, we hypothesized that ShcD had either a direct influence on cell invasion or it indirectly regulated target molecules that lead to reduced invasive potential.

All statistical tests were two-sided, and $p < 0.05$ was considered statistically significant. The statistical analyses were performed with the Statistical Analysis System Version 9.2 (SAS Institute, Cary, NC 27513, USA).

All in vitro and in vivo data were represented as the mean ± SD of biological triplicates (if not diversely indicated in the text). Comparisons between two or more groups were assessed by using two-tailed Student's *t* test. $p < 0.05$ and lower were considered significant.

5. Conclusions

Comprehension of how melanoma infiltrates the surrounding tissues and disseminate is key to designing new and more effective therapeutic strategies. The metastatic cascade is a multi-step process still not completely dissected. Our results indicate that ShcD, a signaling adaptor protein, drives changes in melanoma plasticity, enabling cells to switch between different types of motility and therefore invade and metastasize. The uncovering of this novel signal transduction pathway may contribute to the understanding of how metastases develop and evolve, and how the different phenotypes leading to metastasis can be either reversed or efficiently inhibited. Our results may open new therapeutic perspectives.

Supplementary Materials: The following are available online at http://www.mdpi.com/2072-6694/12/11/3366/s1, Figure S1: Evaluation of the impact of ShcD silencing on MM27 cells, Figure S2: Cell morphology assessment in ShcD-silenced WM266.4 and WM115 cells, Figure S3: ShcD form a complex with DOCK4 and confines it in the cytoplasm, and Movie S1: Time-lapse analysis of ShcD overexpressing MM27 cells.

Author Contributions: Conceptualization, E.A. and L.L.; methodology, E.A., L.G., F.L., F.M., S.P. and G.G.; clinical investigation, P.F.F., G.M., M.C., G.V., S.C. and S.G.; validation, E.A., L.G., F.L., F.M. and S.P.; statistical analysis, E.A., L.G., F.L., F.M., S.P., S.C. and S.G.; resources, E.A., L.G., F.L., F.M., S.P. and G.G.; data curation, L.L.; writing—original draft preparation, E.A., L.G., F.L., F.M., L.L.; writing—review and editing, L.L.; supervision, L.L.; project administration, L.L.; funding acquisition, L.L. All authors have read and agreed to the published version of the manuscript.

Funding: L.G. is supported by a Fondazione Umberto Veronesi fellowship. This work was partially supported by the AIRC grant IG 2017 Id 20508 to L.L., and the Italian Ministry of Health with Ricerca Corrente and 5 × 1000 funds.

Acknowledgments: Federica Marocchi is a student within the European School of Molecular Medicine (SEMM). We thank G. Ossolengo for the production and S. Pasqualato and S. Monzani for the purification of ShcD monoclonal antibody. We thank F. Pisati for technical support in immunohistochemistry (Cogentech) and Simona Rodighiero for imaging supervision (IEO). We also thank A. Gobbi and M. Capillo and the Mouse Facility (Cogentech) for excellent support in animal work. We wish to thank all members of the Department of Experimental Oncology for discussion and reagents. In particular, we thank Stefania Averaimo for critical reading and revision of the manuscript, and Elisabetta Venditti, Micaela Quarto, Giulio Cesare Vitali, Francesco Cataldo and Carolina D'Alesio (IEO) that were instrumental in the first set-up phase of the project.

References

1. Bray, F.; Me, J.F.; Soerjomataram, I.; Siegel, R.L.; Torre, L.A.; Jemal, A. Global cancer statistics 2018: GLOBOCAN estimates of incidence and mortality worldwide for 36 cancers in 185 countries. *CA A Cancer J. Clin.* **2018**, *68*, 394–424. [CrossRef] [PubMed]

2. Arozarena, I.; Wellbrock, C. Phenotype plasticity as enabler of melanoma progression and therapy resistance. *Nat. Rev. Cancer* **2019**, *19*, 377–391. [CrossRef] [PubMed]

3. Rambow, F.; Marine, J.-C.; Goding, C.R. Melanoma plasticity and phenotypic diversity: Therapeutic barriers and opportunities. *Genes Dev.* **2019**, *33*, 1295–1318. [CrossRef] [PubMed]

4. Friedl, P.; Wolf, K. Plasticity of cell migration: A multiscale tuning model. *J. Exp. Med.* **2010**, *207*, 11–19. [CrossRef]

5. Wolf, K.; Friedl, P. Extracellular matrix determinants of proteolytic and non-proteolytic cell migration. *Trends Cell Biol.* **2011**, *21*, 736–744. [CrossRef]

6. Sanz-Moreno, V.; Gadea, G.; Ahn, J.; Paterson, H.; Marra, P.; Pinner, S.; Sahai, E.; Marshall, C.J. Rac Activation and Inactivation Control Plasticity of Tumor Cell Movement. *Cell* **2008**, *135*, 510–523. [CrossRef]

7. Orgaz, J.L.; Sanz-Moreno, V. Emerging molecular targets in melanoma invasion and metastasis. *Pigment. Cell Melanoma Res.* **2013**, *26*, 39–57. [CrossRef]

8. Sahai, E. Mechanisms of cancer cell invasion. *Curr. Opin. Genet. Dev.* **2005**, *15*, 87–96. [CrossRef]

9. Paluch, E.K.; Raz, E. The role and regulation of blebs in cell migration. *Curr. Opin. Cell Biol.* **2013**, *25*, 582–590. [CrossRef]

10. Ruprecht, V.; Wieser, S.; Callan-Jones, A.; Smutny, M.; Morita, H.; Sako, K.; Barone, V.; Ritsch-Marte, M.; Sixt, M.; Voituriez, R.; et al. Cortical Contractility Triggers a Stochastic Switch to Fast Amoeboid Cell Motility. *Cell* **2015**, *160*, 673–685. [CrossRef]

11. Liu, Y.-J.; Le Berre, M.; Lautenschlaeger, F.; Maiuri, P.; Callan-Jones, A.; Heuzé, M.; Takaki, T.; Voituriez, R.; Piel, M. Confinement and Low Adhesion Induce Fast Amoeboid Migration of Slow Mesenchymal Cells. *Cell* **2015**, *160*, 659–672. [CrossRef] [PubMed]

12. Sanz-Moreno, V.; Gaggioli, C.; Yeo, M.; Albrengues, J.; Wallberg, F.; Virós, A.; Hooper, S.; Mitter, R.; Féral, C.C.; Cook, M.; et al. ROCK and JAK1 Signaling Cooperate to Control Actomyosin Contractility in Tumor Cells and Stroma. *Cancer Cell* **2011**, *20*, 229–245. [CrossRef] [PubMed]

13. Tozluoğlu, M.; Tournier, A.L.; Jenkins, R.P.; Hooper, S.; Bates, P.A.; Sahai, E. Matrix geometry determines optimal cancer cell migration strategy and modulates response to interventions. *Nat. Cell Biol.* **2013**, *15*, 751–762. [CrossRef] [PubMed]

14. Sanz-Moreno, V.; Marshall, C.J. Rho-GTPase signaling drives melanoma cell plasticity. *Cell Cycle* **2009**, *8*, 1484–1487. [CrossRef] [PubMed]

15. Gadea, G.; Sanz-Moreno, V.; Self, A.; Godi, A.; Marshall, C.J. DOCK10-Mediated Cdc42 Activation Is Necessary for Amoeboid Invasion of Melanoma Cells. *Curr. Biol.* **2008**, *18*, 1456–1465. [CrossRef]

16. Westbrook, J.A.; Wood, S.L.; Cairns, D.A.; McMahon, K.; Gahlaut, R.; Thygesen, H.; Shires, M.; Roberts, S.; Marshall, H.; Oliva, M.R.; et al. Identification and validation of DOCK4 as a potential biomarker for risk of bone metastasis development in patients with early breast cancer. *J. Pathol.* **2019**, *247*, 381–391. [CrossRef]

17. Kobayashi, M.; Harada, K.; Negishi, M.; Katoh, H. Dock4 forms a complex with SH3YL1 and regulates cancer cell migration. *Cell. Signal.* **2014**, *26*, 1082–1088. [CrossRef]

18. Hiramoto, K.; Negishi, M.; Katoh, H. Dock4 is regulated by RhoG and promotes Rac-dependent cell migration. *Exp. Cell Res.* **2006**, *312*, 4205–4216. [CrossRef]

19. Turco, M.Y.; Furia, L.; Dietze, A.; Diaz, L.F.; Ronzoni, S.; Sciullo, A.; Simeone, A.; Constam, D.B.; Faretta, M.; Lanfrancone, L. Cellular Heterogeneity During Embryonic Stem Cell Differentiation to Epiblast Stem Cells Is Revealed by the ShcD/RaLP Adaptor Protein. *STEM CELLS* **2012**, *30*, 2423–2436. [CrossRef]

20. Fagiani, E.; Giardina, G.; Luzi, L.; Cesaroni, M.; Quarto, M.; Capra, M.; Germano, G.; Bono, M.; Capillo, M.; Pelicci, P.; et al. RaLP, a New Member of the Src Homology and Collagen Family, Regulates Cell Migration and Tumor Growth of Metastatic Melanomas. *Cancer Res.* **2007**, *67*, 3064–3073. [CrossRef]

21. Bossi, D.; Cicalese, A.; Dellino, G.I.; Luzi, L.; Riva, L.; D'Alesio, C.; Diaferia, G.R.; Carugo, A.; Cavallaro, E.; Piccioni, R.; et al. In Vivo Genetic Screens of Patient-Derived Tumors Revealed Unexpected Frailty of the Transformed Phenotype. *Cancer Discov.* **2016**, *6*, 650–663. [CrossRef] [PubMed]

22. Tanda, E.T.; Vanni, I.; Boutros, A.; Andreotti, V.; Bruno, W.; Ghiorzo, P.; Spagnolo, F. Current State of Target Treatment in BRAF Mutated Melanoma. *Front. Mol. Biosci.* **2020**, *7*, 154. [CrossRef] [PubMed]

23. Jenkins, R.W.; Fisher, D.E. Treatment of Advanced Melanoma in 2020 and Beyond. *J. Investig. Dermatol.* **2020**, 1–9. [CrossRef]

24. Geiger, B.; Yamada, K.M. Molecular Architecture and Function of Matrix Adhesions. *Cold Spring Harb. Perspect. Biol.* **2011**, *3*, a005033. [CrossRef]

25. Pandya, P.; Orgaz, J.L.; Sanz-Moreno, V. Modes of invasion during tumour dissemination. *Mol. Oncol.* **2016**, *11*, 5–27. [CrossRef]

26. Georgouli, M.; Herraiz, C.; Crosas-Molist, E.; Fanshawe, B.; Maiques, O.; Perdrix, A.; Pandya, P.; Rodriguez-Hernandez, I.; Ilieva, K.M.; Cantelli, G.; et al. Regional Activation of Myosin II in Cancer Cells Drives Tumor Progression via a Secretory Cross-Talk with the Immune Microenvironment. *Cell* **2019**, *176*, 757–774.e23. [CrossRef]

27. Verfaillie, A.; Imrichova, H.; Atak, Z.K.Z.K.; Dewaele, M.; Rambow, F.; Hulselmans, G.; Christiaens, V.; Svetlichnyy, D.; Luciani, F.; Mooter, L.L.V.D.; et al. Decoding the regulatory landscape of melanoma reveals TEADS as regulators of the invasive cell state. *Nat. Commun.* **2015**, *6*, 6683. [CrossRef]

28. Gao, Y.; Dickerson, J.B.; Guo, F.; Zheng, J.; Zheng, Y. Rational design and characterization of a Rac GTPase-specific small molecule inhibitor. *Proc. Natl. Acad. Sci. USA* **2004**, *101*, 7618–7623. [CrossRef]

29. Tomasevic, N.; Jia, Z.; Russell, A.J.; Fujii, T.; Hartman, J.J.; Clancy, S.; Wang, M.; Béraud, C.; Wood, K.W.; Sakowicz, R. Differential Regulation of WASP and N-WASP by Cdc42, Rac1, Nck, and PI(4,5)P2. *Biochemistrty* **2007**, *46*, 3494–3502. [CrossRef]

30. Ruzzolini, J.; Peppicelli, S.; Andreucci, E.; Bianchini, F.; Margheri, F.; Laurenzana, A.; Fibbi, G.; Pimpinelli, N.; Calorini, L. Everolimus selectively targets vemurafenib resistant BRAFV600E melanoma cells adapted to low pH. *Cancer Lett.* **2017**, *408*, 43–54. [CrossRef]

31. Kulesa, P.M.; Kasemeier-Kulesa, J.C.; Teddy, J.M.; Margaryan, N.V.; Seftor, E.A.; Seftor, R.E.; Hendrix, M.J. Reprogramming metastatic melanoma cells to assume a neural crest cell-like pnenotype in an embryonic microenvironment. *Proc. Natl. Acad. Sci. USA* **2006**, *103*, 3752–3757. [CrossRef] [PubMed]

32. Bailey, C.M.; Morrison, J.A.; Kulesa, P.M. Melanoma revives an embryonic migration program to promote plasticity and invasion. *Pigment. Cell Melanoma Res.* **2012**, *25*, 573–583. [CrossRef]

33. Hawley, S.P.; Wills, M.K.B.; Rabalski, A.J.; Bendall, A.J.; Jones, N. Expression patterns of ShcD and Shc family adaptor proteins during mouse embryonic development. *Dev. Dyn.* **2010**, *240*, 221–231. [CrossRef] [PubMed]

34. Colombo, S.; Champeval, D.; Rambow, F.; LaRue, L. Transcriptomic Analysis of Mouse Embryonic Skin Cells Reveals Previously Unreported Genes Expressed in Melanoblasts. *J. Investig. Dermatol.* **2012**, *132*, 170–178. [CrossRef] [PubMed]

35. Arozarena, I.; Bischof, H.; Gilby, D.; Belloni, B.; Dummer, R.; Wellbrock, C. In melanoma, beta-catenin is a suppressor of invasion through a cell-type specific mechanism. *Oncogene* **2011**, *30*, 4531–4543. [CrossRef] [PubMed]

36. Strub, T.; Kobi, D.; Koludrovic, D.; Davidson, I. A POU3F2-MITF-SHC4 Axis in Phenotype Switching of Melanoma Cells. In *Research on Melanoma—A Glimpse into Current Directions and Future Trends*; IntechOpen: London, UK, 2011.

37. Weichenthal, M.; Senel, G.; Both, M.; Senkpiehl, I.; Lekic, D.; Kähler, K.C.; Ugurel, S.; Leiter, U.M.; Mohr, P.; Hauschild, A.; et al. Evaluation of the Melanoma Tumor Burden Score (MTBS) in a real-world setting. *J. Clin. Oncol.* **2017**, *35*, 9565. [CrossRef]

38. Ahmed, S.B.; Prigent, S.A. Insights into the Shc Family of Adaptor Proteins. *J. Mol. Signal.* **2017**, *12*, 2. [CrossRef]

39. Katoh, H.; Negishi, M. RhoG activates Rac1 by direct interaction with the Dock180-binding protein Elmo. *Nat. Cell Biol.* **2003**, *424*, 461–464. [CrossRef]

40. Hiramoto-Yamaki, N.; Takeuchi, S.; Ueda, S.; Harada, K.; Fujimoto, S.; Negishi, M.; Katoh, H. Ephexin4 and EphA2 mediate cell migration through a RhoG-dependent mechanism. *J. Cell Biol.* **2010**, *190*, 461–477. [CrossRef]

41. Saci, A.; Cantley, L.C.; Carpenter, C.L. Rac1 Regulates the Activity of mTORC1 and mTORC2 and Controls Cellular Size. *Mol. Cell* **2011**, *42*, 50–61. [CrossRef]

42. Kinsella, T.; Mencarelli, A.; Valtieri, M.; Riganelli, D.; Grignani, F.; Lanfrancone, L.; Peschle, C.; Nolan, G.P.; Pelicci, P.G. High-efficiency gene transfer and selection of human hematopoietic progenitor cells with a hybrid EBV/retroviral vector expressing the green fluorescence protein. *Cancer Res.* **1998**, *58*, 14–19.

43. Confalonieri, S.; Quarto, M.; Goisis, G.; Nuciforo, P.; Donzelli, M.; Jodice, G.; Pelosi, G.; Viale, G.; Pece, S.; Di Fiore, P.P. Alterations of ubiquitin ligases in human cancer and their association with the natural history of the tumor. *Oncogene* **2009**, *28*, 2959–2968. [CrossRef] [PubMed]

44. Capra, M.; Nuciforo, P.G.; Confalonieri, S.; Quarto, M.; Bianchi, M.; Nebuloni, M.; Boldorini, R.; Pallotti, F.; Viale, G.; Gishizky, M.L.; et al. Frequent Alterations in the Expression of Serine/Threonine Kinases in Human Cancers. *Cancer Res.* **2006**, *66*, 8147–8154. [CrossRef] [PubMed]

Publisher's Note: MDPI stays neutral with regard to jurisdictional claims in published maps and institutional affiliations.

 *cancers*

Article

Cancer Stem Cells and the Slow Cycling Phenotype: How to Cut the Gordian Knot Driving Resistance to Therapy in Melanoma

Luigi Fattore [1,2], Rita Mancini [3,†] and Gennaro Ciliberto [4,*,†]

1   Department of Research, Advanced Diagnostics and Technological Innovation, SAFU Laboratory, Translational Research Area, IRCCS Regina Elena National Cancer Institute, 00144 Rome, Italy; luigi.fattore@ifo.gov.it
2   Department of Melanoma, Cancer Immunotherapy and Development Therapeutics, Istituto Nazionale Tumori IRCCS, "Fondazione G. Pascale", 80131 Naples, Italy
3   Department of Clinical and Molecular Medicine, Sant' Andrea Hospital, Sapienza University of Rome, 00161 Rome, Italy; rita.mancini@uniroma1.it
4   Scientific Directorate, IRCSS Regina Elena National Cancer Institute, 00144 Rome, Italy
*   Correspondence: gennaro.ciliberto@ifo.gov.it
†   These authors contributed equally to this work.

Received: 17 September 2020; Accepted: 11 November 2020; Published: 13 November 2020

Simple Summary: Cancer stem cells play a central role in the development of cancer and are poorly sensitive to standard chemotherapy and radiotherapy. Furthermore, they are also responsible for the onset of drug resistance. This also occurs in malignant melanoma, the deadliest form of skin cancer. Hence, cancer stem cells eradication is one of the main challenges for medical oncology. Here, we conducted a bioinformatics approach aimed to identify the main circuits and proteins underpinning cancer stem cell fitness in melanoma. Several lessons emerged from our work and may help to conceptualize future therapeutic approaches to prolong the efficacy of current therapies.

Abstract: Cancer stem cells (CSCs) have historically been defined as slow cycling elements that are able to differentiate into mature cells but without dedifferentiation in the opposite direction. Thanks to advances in genomic and non-genomic technologies, the CSC theory has more recently been reconsidered in a dynamic manner according to a "phenotype switching" plastic model. Transcriptional reprogramming rewires this plasticity and enables heterogeneous tumors to influence cancer progression and to adapt themselves to drug exposure by selecting a subpopulation of slow cycling cells, similar in nature to the originally defined CSCs. This model has been conceptualized for malignant melanoma tailored to explain resistance to target therapies. Here, we conducted a bioinformatics analysis of available data directed to the identification of the molecular pathways sustaining slow cycling melanoma stem cells. Using this approach, we identified a signature of 25 genes that were assigned to four major clusters, namely (1) kinases and metabolic changes, (2) melanoma-associated proteins, (3) Hippo pathway and (4) slow cycling/CSCs factors. Furthermore, we show how a protein-protein interaction network may be the main driver of these melanoma cell subpopulations. Finally, mining The Cancer Genome Atlas (TCGA) data we evaluated the expression levels of this signature in the four melanoma mutational subtypes. The concomitant alteration of these genes correlates with the worst overall survival (OS) for melanoma patients harboring BRAF-mutations. All together these results underscore the potentiality to target this signature to selectively kill CSCs and to achieve disease control in melanoma.

Keywords: melanoma; target therapy; cancer stem cells; slow cycling phenotype; drug resistance; OXPHOS; lipid metabolism

1. Introduction

Melanoma remains one of the most widespread types of cancer in western countries, and its incidence is rapidly increasing [1,2]. In recent years, immunotherapy and targeted therapies have changed the treatment scenario for advanced melanoma [3–6].

Selective inhibitors of V600E BRAF-mutated melanoma, such as vemurafenib, dabrafenib and encorafenib, prolong survival of patients harboring the V600E mutation [3,4,7,8]. However, the onset of tumor resistance observed following this treatment, which was found to be related to the emergence of bypass mutations in resistant tumors that often cause reactivation of the RAS/BRAF/MEK pathway [9,10], led to the development of combo therapies with BRAF and MEK inhibitors as the current standard of care [11–13]. Unfortunately, dual therapy, although being able to provide more durable disease control compared to BRAF inhibitors alone, is also plagued by the development of drug resistance [14,15].

Beside genetic mechanisms, a plethora of non-genetic changes have been identified to be involved in the evolution of resistance to target therapy in melanoma [16–18], the main ones being: (1) the induction of changes in the inflammatory niche driving drug tolerance [19]; (2) the displacement of the bioenergetic equilibrium [20] and (3) the involvement of receptor tyrosine kinases (RTKs). As for the last mechanisms, for example, our group and others have demonstrated a key role of ErbB3 receptor up-regulation upon exposure to BRAF and MEK inhibitors [21,22]. In addition, also non-coding RNAs, such as microRNAs are emerging as key players able to orchestrate epigenetic and non-genomic mechanisms of resistance to target therapy in melanoma [23–28].

Similarly, multiple reports document the development of resistance to immune checkpoint inhibitors [29–31]. Therefore, strategies aimed to reduce the onset of resistance are of the utmost importance in the therapy of melanoma [31].

Cancer stem cells (CSCs) are known to be involved in the development of resistance to treatment, thereby contributing to disease relapse after an initial response [32–35]. CSCs, also known as tumor-initiating cells, are cells that can perpetuate themselves by self-renewal [36], and present peculiar characteristics, including the expression of specific surface markers [34,37]. The theory of CSCs has originally been described as a bone fide biological phenomenon in hematologic tumors, such as leukemia. From here, they have also been identified in solid malignancies, such as lung, breast and colon cancers. Differently, in the case of melanoma, some debates on the existence of CSCs are still ongoing mostly due to the difficulty to identify reliable markers for their identification [38].

CSCs are clearly linked to tumor heterogeneity, which is also a hallmark of cancer development and at the basis of resistance to anti-neoplastic treatments [39]. Heterogeneity is observed at different levels, for example, within a single tumor (intra-tumoral) or between tumor masses of the same histopathological subtype in the same patient (inter-tumoral) or between tumors of the same histotype deriving from different patients (inter-patient) [40,41]. Moreover, heterogeneity can occur at a spatial level (uneven distribution of genetically and/or epigenetically different subpopulations within the same or synchronous lesions in the same patient), or at a temporal level (i.e., dynamic variations of tumor cells over time) [39,41]. Notably, temporal heterogeneity may also encompass the existence of a tiny pre-existing subpopulation of cells, which emerge as the dominant population under the pressure imposed by a given therapy. Tumor heterogeneity can be influenced in response to the selective pressure of the immune system or of antineoplastic treatments [42].

According to the CSC model, tumors are per se heterogeneous and are organized in a hierarchical manner [43]. At the top of the hierarchy is a small fraction of cells called CSCs, which are endowed with the ability to undergo both symmetrical and asymmetrical divisions. These cells can differentiate into "non-CSCs", with the acquisition of stable genetic and/or epigenetic changes [44]. According to the model, non-CSCs represent the largest fraction of the tumor. Nowadays the alternative models postulated to explain tumor evolution are: linear, branching, neutral and punctuated [45]. According to the first model tumor cells acquire mutations linearly in a step-wise process leading to more malignant stages of cancer. The branching model predicts that single clones originate from a common ancestor, and evolve simultaneously in the tumor mass because of their increased fitness. Neutral evolution is a case of branching evolution for which no selection or fitness changes occur during the lifetime of the tumor and influence the evolution of the clones. Finally, and differently from the other models, the theory of "punctuate equilibrium" assumes that mutations are not acquired gradually and sequentially over time but in short bursts of tumor progression. According to this model, intratumor heterogeneity is higher in the early stages in which one or a few dominant clones stably expand to form the bulk of tumor mass. Interestingly, a recent mathematical modeling has revealed that bursts of mutations are the best models able to recapitulate the long-stemmed clonal trees of the evolution of different cancers. For further details we recommend several excellent reviews on this topic such as the work by Davis and colleagues [45].

More recently, those models have been challenged by several pieces of evidence showing a plastic interconversion of epigenetic changes underpinning stem phenotype. Thereby, the theory of CSCs has been reconciled with the evidence of a dynamic interplay between slow cycling (and drug-resistant) and fast cycling (drug-sensitive) states [46]. Although the correspondence of CSCs with slow cycling cells within a tumor is still debated, it is widely accepted that tumor heterogeneity is linked to a slow cycling and plastic sub-population of cells able to challenge therapeutic efforts and to emerge as drug-resistant survivors.

In this context, the "phenotype switching" model has been proposed [47]. It has been conceptualized for malignant melanoma, which is notorious for the high level of genetic and non-genetic heterogeneity [48,49]. Therefore, it is thought that melanoma cells have the possibility to shift between different transcriptional programs mostly depending on the oscillation of the microphthalmia-associated transcription factor MITF [46]. These states are the "proliferative/differentiative" or MITF^{high} and the "invasive" or MITF^{low} (also known as rheostat model) [50]. Furthermore, molecular changes leading to tumor heterogeneity are also regulated by local microenvironment cues (e.g., interactions with non-tumor cells, hypoxia, stroma-derived factors HGF, TGF-β) [47]. For example, it has been recently reported that an acidic tumor microenvironment influences a stem-like phenotype in melanoma [51]. According to "phenotype switching" model, non-hierarchical plasticity may lead to the transient existence of slow-cycling cells able to survive to therapeutic pressures activating compensatory signaling pathways. Thereby, it has been named "dynamic stemness". Those aspects and their therapeutic implications will be discussed later in our review. Noteworthily, those melanoma cell subpopulations possess neural crest stem cell features and are dedifferentiated further than the canonical invasive phenotype [52,53].

Coherently, one of the main markers of this state, namely the nerve growth factor receptor (NGFR) has been described as a putative melanoma CSC marker [54].

Along the same topic, H3K4 demethylase JARID1B has been identified as a marker of slow cycling cells in melanoma. According to the notion of CSCs, the subpopulation of cells express this protein cycle very slowly (times of >4 weeks) compared to the rest of the rapidly proliferating main population [55].

Recently, the development of single cell (sc) analysis approaches, which are able to better characterize tumor heterogeneity and CSCs have represented a major technological breakthrough [42]. For example, sc-RNAseq studies have confirmed the presence of a small population of non-cycling cells in both melanoma cell lines in vitro, as well as from those derived from freshly processed melanomas. Coherently, those cells are enriched for CSC markers like JARID1B and NGFR [56,57].

In this paper we have tried to interpret available data using a bioinformatic approach directed to identify novel targets to selectively hit CSCs in BRAF mutated cutaneous melanomas. This led us to identify a common interaction network encompassing the existence of four major clusters which may be at the basis of CSC fitness. Importantly, the signature of genes belonging to this network showed a prognostic potential for BRAF-mutant melanoma patients based on The Cancer Genome Atlas (TCGA) data. Finally, we discuss how to potentially tackle it at multiple levels to selectively block the spread of those cells and to prolong the efficacy of target therapy in melanoma.

2. Results

2.1. Identifying Common Denominators for Melanoma Stem Cell Survival through Bibliographic Search

CSCs are charged to be among the main determinants of the failure of anticancer treatments. This is also the case of target therapy in *BRAF*-mutant melanomas. An open question remains whether common molecular pathways sustain slow cycling cells under selective pressure by therapeutic agents and foster their growth as drug-resistant survivors. In this regard, there is a general agreement to consider metabolic rewiring as a central process responsible for promoting CSC fitness [20]. Notwithstanding the involvement of these common pathways, the specific players identified differ among different studies.

In order to identify common denominators, we carried out the following effort in three steps (Figure 1).

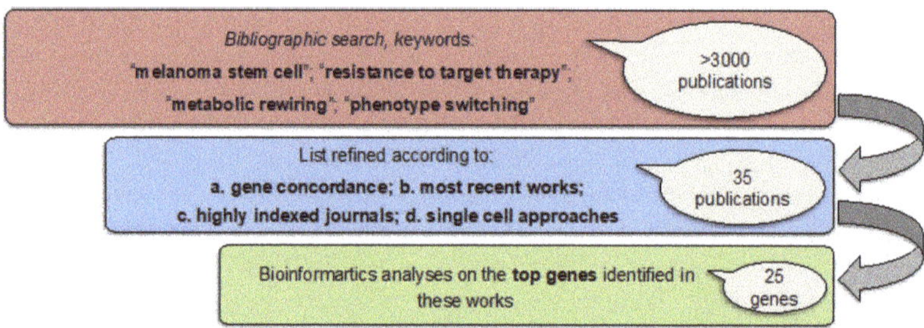

Figure 1. Schematic workflow of the study divided into three steps to obtain a list of 25 top genes for bioinformatics analyses.

(1) We first conducted an extensive bibliographic search using as keywords "melanoma stem cells", "resistance to target therapy", "metabolic rewiring" and "phenotype switching" and we found more than 3000 publications. (2) We refined this list according to the: (a) potential concordance in the genes identified among studies, i.e., for their presence in at least two of them (b) most recent works (c) highly indexed journals (impact factor >7) and (d) studies taking advantages of single-cell approaches. These parameters allowed us to restrict the list of publication up to 35. (3) These publications were screened to finally obtain a list of 25 top genes, which potentially sustain melanoma stem cell fitness. In detail, the complete list is available as Table 1. This signature will be named from now simply "MSCsign".

Table 1. List of 25 top genes relevant for melanoma stem cells and resistance to target therapy.

Gene	Description	Reference #	Cluster
MAP2K1	Dual specificity mitogen-activated protein kinase kinase 1	[9,11–15]	
JUN	Transcription factor AP-1	[15,42,46,56,58]	
RXRG	Retinoic acid receptor RXR-gamma	[42,53]	
CPT1A	Carnitine O-palmitoyltransferase 1	[59]	
AKT1	RAC-alpha serine/threonine-protein kinase	[9,16,20]	
BRAF	Serine/threonine-protein kinase B-raf	[4,7,9,11–14]	**Kinase and metabolic signature**
PPARA	Peroxisome proliferator-activated receptor alpha	[59]	
ATF4	Cyclic AMP-dependent transcription factor ATF-4	[20,60]	
PPARGC1A	Peroxisome proliferator-activated receptor gamma coactivator 1-alpha	[20,61–63]	

Table 1. *Cont.*

Gene	Description	Reference #	Cluster
SCD	Acyl-CoA desaturase	[60,64]	
TFAM	Transcription factor A, mitochondrial	[65]	
HIF1A	Hypoxia-inducible factor 1-alpha	[20,66]	
MITF	Microphthalmia-associated transcription factor	[20,42,46,47,50,52,53,56,57, 60–62,64,67]	
MTOR	Serine/threonine-protein kinase mTOR	[62,67]	
SERPINE2	Glia-derived nexin	[68]	
AXL	Tyrosine-protein kinase receptor UFO	[20,42,46,47,50,52,53,56–58, 60,66]	**Melanoma-associated signature**
MLANA	Melanoma antigen recognized by T-cells 1	[52,53,56,66]	
NFKB1	Nuclear factor NF-kappa-B p105 subunit	[52,66]	
SOX10	Transcription factor SOX-10	[53,58]	
TP53	Cellular tumor antigen p53	[69]	
YAP1	Transcriptional coactivator YAP1	[47,64]	
TEAD1	Transcriptional enhancer factor TEF-1	[47,58]	**Hippo pathway signature**
TAZ	Tafazzin	[47,64]	
NGFR	Tumor necrosis factor receptor superfamily member 16	[52–54,56–58,66,70,71]	**Slow cycling signature**
KDM5B	Lysine-specific demethylase 5B	[55,56,63,64,66,70,72]	

Bold: Genes divided into four major clusters.

Here below we recapitulate the main findings of the major studies emerging from our analysis. Rambow et al. have reported that BRAF/MEK inhibition enriches multiple therapy-resistant slow cycling populations, which retain the ability to proliferate in the absence of therapeutic stress [53]. Those cells exhibited neural crest stem cell transcriptional programs largely driven by the nuclear receptor RXRG. Another mechanism proposed to support the onset of drug resistance is the epigenetic reprogramming induced by therapy [58]. Reprogramming may be initiated with a loss of SOX10-mediated differentiation, and then followed by a multi-stage process involving the activation of new signaling pathways, such as Jun-AP-1 and TEAD [58]. These events lead to post-treatment transition to stable resistance phenotypes characterized by well-known markers, such as the receptor tyrosine kinase AXL. Furthermore, it has been reported in in vivo mouse models that slow cycling melanoma cells which adaptively resist to BRAF/MEK inhibitors (MAPKi) are also capable of reentering the cell cycle and give rise to highly metastatic subclones that invade different tissues [70]. Importantly, these cells show a dedifferentiated state characterized by high levels of cancer stem cell markers, such as NGFR and JARID1B. Notably, all these studies converge to support the notion that drug treatment initially induces a fast increase in de-differentiation toward a slow cycling CSC state characterized by high NFGR levels [53,66]. In essence, the neural crest stem cell signature could be defined as "point of entry" for the development of resistance to target therapy in melanoma [71].

While CSCs are poorly sensitive to chemotherapy and target therapies that mainly act by blocking cell cycle, several studies have shown that they are sensitive to the interference with signaling pathways [68,73]. Zakaria et al. have suggested that NF-κB inhibition reduces the ability of CSCs to maintain their population within the tumor mass [73]. Along the same line, Su and colleagues have recently demonstrated that the switch from rapidly dividing drug-responsive to drug-tolerant/slow cycling states early occurs upon exposure to MAPKi in melanoma. This event mostly encompasses the activation of NF-κB. Coherently, its inhibition together with MAPKi keeps melanoma cells in a drug-sensitive state [66].

In a similar study, the same result was achieved by modulating the p53, NF-κB and HIF-1α pathways [74]. Paradoxically, inhibition of TP53 was shown to sensitize melanoma cells to BRAF/MEK inhibition [69]. In addition, SerpinE2 appears to be involved in the maintenance and invasiveness of CSCs in melanoma, and the inhibition of this proteolytic enzyme has been speculated to be a potential therapeutic target [68].

As previously stated, the alteration of the metabolic status is a hallmark of CSC maintenance and mostly encompasses: (1) increased oxidative phosphorylation (OXPHOS) and (2) lipidome alterations.

As to the first, it is widely accepted that therapy resistant slow-cycling melanoma cells are addicted to mitochondrial OXPHOS: this vulnerability represents a source of therapeutic opportunities [75]. Roesch et al. firstly described through proteome profiling, which the JARID1B^{high} subpopulation of cells upregulate several enzymes involved in mitochondrial oxidative ATP synthesis [55,72]. Coherently, inhibition of mitochondrial respiration (using oligomycin, rotenone and phenformin) blocked the emergence of these slow cycling cells and potentiated the tumor-suppressive potential of BRAFi in vivo. In a subsequent study, the same group demonstrated that MAPKi induce the upregulation of mitochondrial biogenesis in intrinsically resistant melanoma cells [65]. The authors found an increase in mitochondrial DNA copy number, mitochondrial mass, maximal oxygen consumption rate, and reactive oxygen species production in these cells upon exposure to the drugs. In contrast, therapy sensitive cells show the opposite pattern, i.e., they downregulated transcriptional signatures associated with MitoBiogenesis. Therefore, the specific mitochondrial HSP90 inhibitor, namely Gamitrinib was demonstrated to be effective in eradicating intrinsically resistant cells and increased the efficacy of MAPKi in vitro and in vivo [65].

Along the same topic, Vazquez et al. have demonstrated the existence of an additional adaptive metabolic program in melanoma that is dependent upon the translocation to the nucleus of the master regulator of melanocyte MITF, which, in turn, activates PGC1α, a key regulator of mitochondrial respiration [61]. Mechanistically, MITF^{high}/PGC1α^{high} cells exhibit increased OXPHOS coupled with ROS detoxification capacities enabling them to survive under oxidative stress conditions [61]. Hence, it is not surprising that the mTORC1/2 inhibitor AZD8055, which triggers MITF cytoplasmic retention, was able to decrease PGC1α expression and OXPHOS in melanoma [62]. Thereby, this compound potentiated the efficacy of MAPKi in BRAF-mutated melanoma cells in vitro and in vivo [62]. In line with this finding the novel mitochondrial complex I inhibitor IACS-010759 demonstrated a significant anti-tumor activity as single-agent of in high OXPHOS MAPKi-resistant melanoma models in vivo [67].

Noteworthy, the existence of JARID1B^{high}/PGC1α^{high} cells has been reported in melanoma according to their relevance in sustaining high-OXPHOS metabolism [63].

CSCs maintenance also seems to depend upon the size of the pool of monounsaturated fatty acids (MUFAs) generated by the activity of the stearoyl-CoA desaturase 1 (SCD1) [37,76–78] because SCD1 inhibition was shown to selectively eliminate CSCs in lung cancer, both alone and in synergy with chemotherapy [79,80].

Pisanu et al. investigated the role of SCD1 and its inhibition by a specific compound, MF-438, in melanoma CSCs, by a comprehensive approach employing bioinformatics and 2D and 3D cultures [64]. In line with the initial hypothesis of the importance of SCD1 in maintaining the CSCs pool in melanoma, the expression of this gene increased during melanoma progression. Moreover, *BRAF*-mutated melanoma cell cultures enriched in CSCs showed an overexpression of SCD1 and were more resistant to BRAF and MEK inhibitors than non-enriched cultures. Exposure of *BRAF*-mutated melanoma cells to inhibitors of the MAPK pathway enhanced stemness features by increasing the expression of YAP/TAZ and downstream genes, but not SCD1. However, the pharmacological inhibition of SCD1 by MF-438 downregulated YAP/TAZ and was able to revert CSC enrichment and resistance to MAPK inhibitors [64]. These findings, albeit limited to in vitro studies, underscore the potential role of SCD1 in melanoma progression and suggest the opportunity to further SCD1 inhibitors in combination with MAPK inhibitors for the control of resistance to targeted therapy.

Very recently, Vivas-Garcia et al. investigated the impact of the tumor microenvironment on phenotype switching focusing on fatty acid metabolism [60]. The authors observed the downregulation of both MITF and SCD in melanoma cells following glutamine deprivation with an alteration of the balance of saturated fatty acids and MUFAs [60]. As expected, MITF was accompanied by a differentiation/proliferative-to-invasive switch characterized also by hallmarks of drug resistance (i.e., AXL). However, in apparent contrast with Pisanu et al., in this model, loss of SCD1 was accompanied by an increase in cells with an invasive phenotype. Further studies are needed to deepen understanding of the role of SCD1 in melanoma, for example, taking advantage of single cell approaches.

In addition, the metabolism of fatty acid oxidation (FAO) has been reported to govern stem cell balance between quiescent and proliferative states [81]. In this context, Aloia et al. have recently demonstrated that an FAO metabolic shift early occurs in BRAF-mutant melanoma cells upon exposure to MAPKi [59]. Mechanically, this is characterized by the upregulation of peroxisome proliferator-activated receptor α (PPARα) and carnitine palmitoyltransferase 1A (CPT1A) enzymes. Coherently, the upfront inhibition of FAO and MAPK synergistically inhibits tumor cell growth in vitro and in vivo [59].

2.2. Identification of a Common Interactome Sustaining Melanoma Stem Cell Fitness Divided into Four Major Clusters

First of all we subjected the 25 genes of "MSCsign" to Markov Cluster Algorithm for bioinformatics clustering based on protein-protein interaction (PPI) and similarity networks [82]. This led to the identification of four major clusters (represented in Table 1). On the basis of the genes present in these clusters, these signatures were named: (1) kinase and metabolic, (2) melanoma-associated, (3) Hippo pathway and (4) slow cycling/CSCs. The first cluster includes nearly 50% of the genes (i.e., 12 out of 25). Among them, there are BRAF, MEK, AKT kinases as well as metabolic enzymes, such as SCD, CPT1A and PPARGC1A. The melanoma-associated cluster includes lineage specific genes, such as MITF and MLANA, as well as markers associated with resistance to target therapy in melanoma, such as AXL and NFKB. The third cluster encompasses three genes all belonging to the Hippo oncogenic signaling (i.e., YAP, TAZ and TEAD). Finally, the last one includes NGFR and KDM5B (or JARID1B). For this reason, it was named the slow cycling/CSCs cluster.

In the next step, we used "MSCsign" to build a PPI network using the STRING online database [83]. This software allows us to complement available information of PPI with computational predictions to generate a global network including direct (physical), as well as indirect (functional) interactions. Using a minimum required interaction score (>0.4, medium confidence) we plotted the interactome of the 25 genes as connected by 89 hedges with a PPI enrichment $p < 1.0 \times 10^{-16}$. The displayed networks available are based on: (1) evidence, as multiple lines where the color indicates the type of interaction; (2) confidence, where line thickness denotes the strength of data support and (3) molecular action, in which different lines represent the predicted mode of action (Figure 2). Again, we obtained four clusters with a partial overlap between signatures 1 and 2, i.e., the kinase/metabolic and melanoma-associated, respectively. Differently, the hippo and slow cycling/CSCs clusters segregated separately from all the others. Furthermore, the same list was also subjected to gene-set enrichment analyses using well-known classification systems, such as Gene Ontology, KEGG (Kyoto Encyclopedia of Genes and Genomes) and Reactome (Table S1). Notwithstanding the different clustering, all PPI networks converge in demonstrating that these 25 genes are closely interconnected (Figure 2). This suggests that a synchronous interactomic profile may exist and can be potentially tackled as will be discussed later.

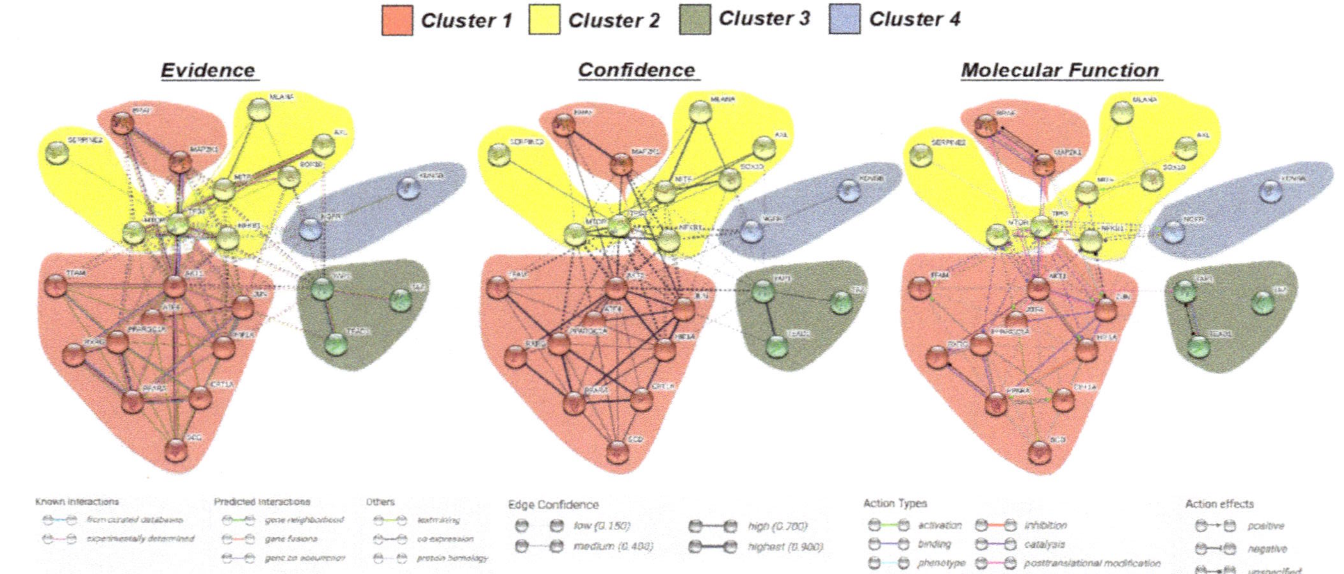

Figure 2. Evidence, confidence and molecular function-based protein-protein interaction (PPI) networks performed using the list of 25 top genes relevant for melanoma cancer stem cells (CSCs) and resistance to target therapy. The legends indicate the meaning of the lines. Interaction score applied >0.4 (medium confidence). A total of 89 were hedges obtained with a PPI enrichment $p < 1.0 \times 10^{-16}$. https://string-db.org.

2.3. TCGA Data Mining Uncovers the Prognostic Value of "MSCsign" in BRAF-Mutant Melanomas

Next, we addressed the potential prognostic value of the "MSCsign" for skin cutaneous melanoma patients. Towards this goal one of the most reliable methods is The Cancer Genome Atlas (TCGA) data mining. Hence, first of all we checked the expression levels of the aforementioned 25 genes in 480 samples of skin cutaneous melanoma using UALCAN software [84]. The results underscore the high levels of heterogeneity of gene expression among the different melanoma specimens, as already well-known for this tumor [41]. Coherently, the most expressed markers belong to cluster 2 according to its melanoma-associated signature (Figure 3). It is important to point out that certain genes especially those belonging to cluster 3 and 4 are expressed at low levels. These data probably reflect a pre-therapy scenario and the activation of specific clusters of genes might occur following treatment in order to escape from the therapeutic pressure. Notably, gene expression levels were also normalized considering the maximum median expression value across all the blocks (Table S2).

Thereafter, we decided to investigate the biological meaningfulness of "MSCsign" stratifying melanoma patients according to the mutational subtype again through TCGA data. To this aim we used the skin cutaneous melanoma dataset (TCGA, PanCancer Atlas) [85] available on cBioPortal website [86,87]. In detail, this contains data of 448 melanoma samples of which 440 were profiled for the mutational status and 443 further subjected to RNA-Seq analyses.

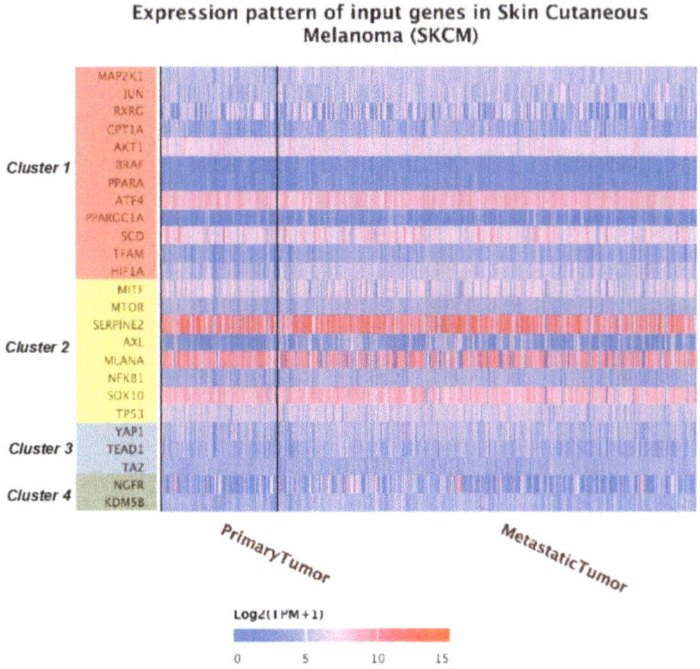

Figure 3. Expression levels of the 25 top genes relevant for melanoma CSCs and resistance to target therapy on 480 samples of skin cutaneous melanoma from The Cancer Genome Atlas. http://ualcan.path.uab.edu/index.html.

According to the genomic classification and mutational status [85] this tumor is categorized into four main molecular subtypes: BRAF^{mut}, RAS^{mut}, NF1^{mut}, or triple wild type. The incidence of these mutational states [10,85] is represented as a cake graph in Figure 4A. It is important to point out that the effects of the three most common driver mutations (BRAF^{mut}, RAS^{mut}, NF1^{mut}) are influenced by additional mutations in other genes, such as CDKN2A and PTEN [10]. This was confirmed by the OncoPrint evaluation of melanoma genomic alterations obtained through the cBioPortal online tool (Figure 4B). Following this approach, we extracted gene expression data of "MSCsign" of 237 samples from BRAF^{mut}, 140 from RAS^{mut}, 76 from NF1^{mut} and 205 from triple wild type subtypes. This latter subtype is the most overlapping to the others as evident by the Venn Diagrams shown in Figure 4C. These data were then subjected to hierarchical clustering to investigate whether a specific melanoma subtype might particularly express one of the four clusters of "MSCsign". Results shown as heat-maps in Figure 4D demonstrate that the expression levels of the 25 genes are similar among the four mutational subtypes. These findings were confirmed by principal component analyses which demonstrated that the "MSCsign" is not able to distinguish a specific mutational subtype from the others (Figure 4E). The raw data relative to gene expression levels are reported in Table S3.

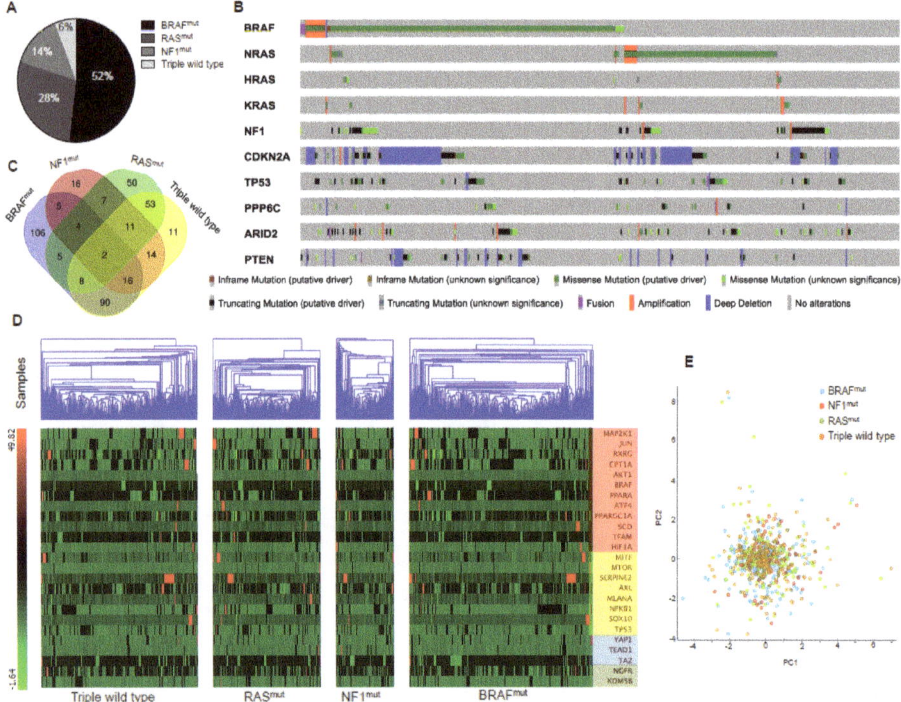

Figure 4. Melanoma stem cell signature ("MSCsign") expression levels evaluated according to skin cutaneous melanoma mutational subtypes. (**A**) Cake graph representing the most common mutational subsets of metastatic melanoma. (**B**) OncoPrint evaluation of distinct genomic alterations. (**C**) Venn diagrams showing the overlapping of the patients belonging to the four mutational subtypes. (**D**) Heat-maps of the expression levels of the 25 genes of "MSCsign" clustered according to the mutational subsets. (**E**) Principal component analysis performed on the expression levels of "MSCsign" in the four mutational subtypes. Data from https://www.cbioportal.org/.

Finally, we decided to investigate the prognostic potential of "MSCsign" in the aforementioned mutational subtypes. Our results shown as Kaplan–Meyer curves demonstrated that the alteration of the 25 genes is statistically correlated with the worst overall survival (OS) only in the specific subset of BRAF^{mut} melanoma patients (Figure 5). All data relative to OS are available as Table S4. Again, these findings suggest the possibility that BRAF^{mut} melanomas tend to activate specific clusters of genes in order to escape treatments.

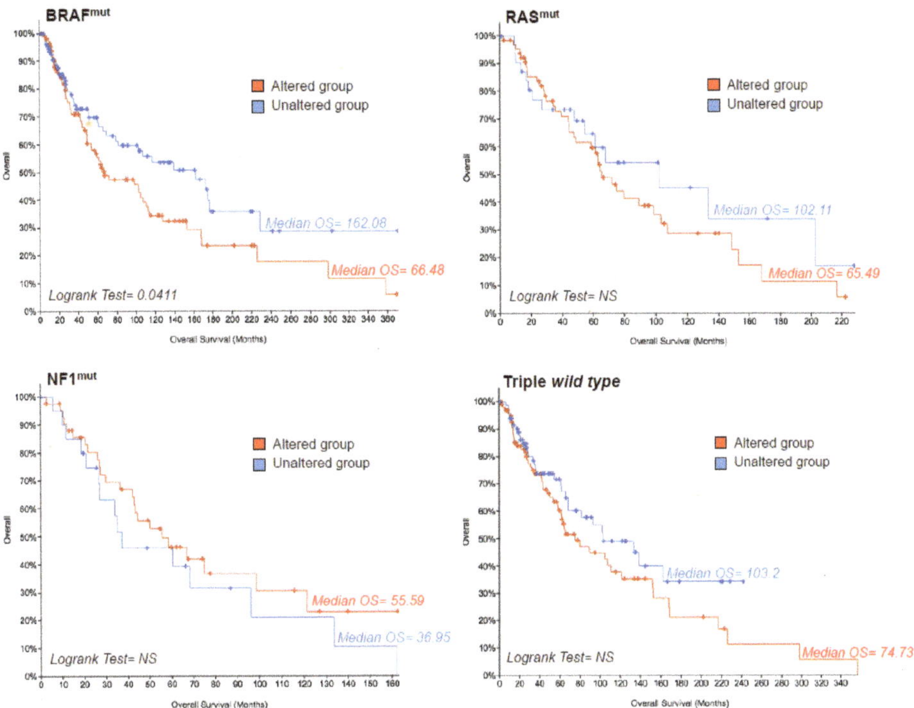

Figure 5. Kaplan–Meyer curves evaluating the prognostic value of "MSCsign" in the four mutational subtypes of skin cutaneous melanoma. Data plotted from https://www.cbioportal.org/.

Hereafter, we performed a principal component analysis using the expression levels of "MSCsign" from the skin cutaneous melanoma dataset in comparison with those of other six solid tumors (all data coming from The Cancer Genome Atlas). They are: breast invasive carcinoma, lung adenocarcinoma, ovarian serous cystadenocarcinoma, colon adenocarcinoma, glioblastoma multiforme and head and neck squamous cell carcinoma. Principal component analysis results show that our molecular signature distinguishes melanomas from all the other tumors (Figure 6A). Data were plotted through the online software GEPIA [88]. Finally, we further refined the principal component analyses to the single clusters belonging to "MSCsign". Results shown in Figure 4B clearly demonstrate that cluster 2 genes are the only ones able to fully distinguish cutaneous melanomas from all the other tumors. This is coherent with the "melanoma-associated" feature of this signature. Differently, the other three clusters of "MSCsign" commonly are segregated in the graphs for all the seven solid tumors tested (Figure 6B).

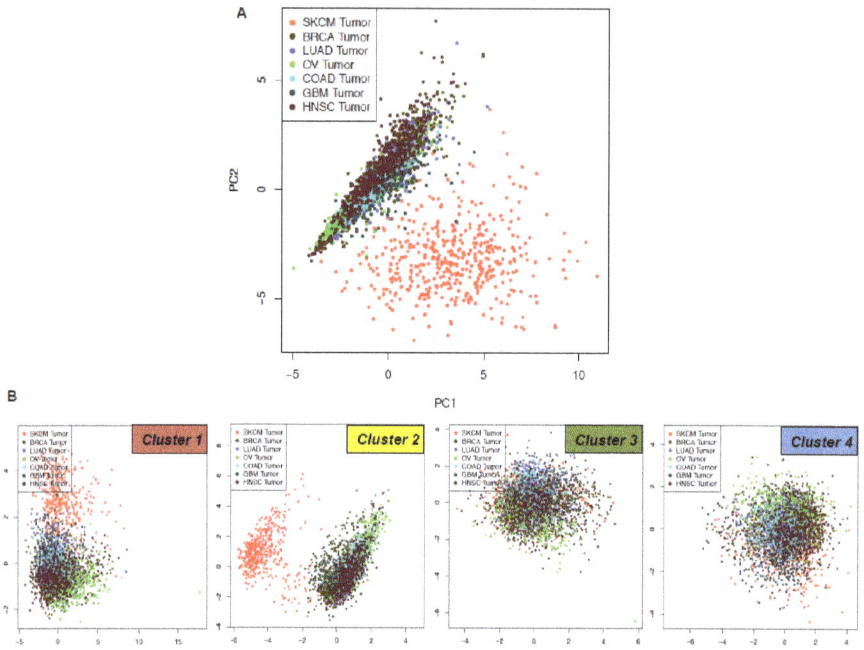

Figure 6. Principal component analyses performed using the expression levels of (**A**) all the 25 genes of "MSCsign" based on The Cancer Genome Atlas (TCGA) data and (**B**) divided according to the four different clusters identified. SKCM = Skin Cutaneous Melanoma, BRCA = Breast invasive carcinoma, LUAD = Lung adenocarcinoma, OV = Ovarian serous cystadenocarcinoma, COAD = colon adenocarcinoma, GBM = Glioblastoma multiforme, HNSC = Head and Neck squamous cell carcinoma. http://gepia.cancer-pku.cn/index.html.

3. Conclusions

Drug resistance virtually frustrates every kind of anti-neoplastic treatment. A paradigmatic example is the development of resistance to MAPKi in *BRAF*-mutant melanomas. This phenomenon is largely driven by the selection of cells with a slow cycling phenotype (which can alternatively be called CSCs), which have therefore emerged as the key therapeutic targets for intervention.

In the present work we carried out a novel approach which allowed to identify a signature of genes relevant for melanoma stem cell fitness, namely "MSCsign". Those genes are correlated to each other in a complex interactome and organized into four different clusters. We postulate that the simultaneous inhibition of multiple effectors belonging to the aforementioned four clusters may be a successful strategy to eradicate CSCs molecular roots. This hypothesis is supported by the finding that alterations of the expression of genes belonging to "MSCsign" is associated with the worst OS for BRAF mutated melanoma patients based on TCGA data.

Metabolic rewiring is a hallmark of resistance to MAPKi in melanoma. Consistent with this, many genes of cluster 1 belong to metabolic signatures. As a matter of fact, the use of OXPHOS and fatty acid inhibitors (see previous paragraph) potentiate MAPKi bypassing the dynamics between fast cycling and slow cycling phenotypes. However, two important limitations emerged: (1) low concordance in the molecular targets among studies and (2) results obtained only at single drug level. To overcome this issue, a combined approach of mitochondrial inhibitors together with inhibitors of MUFAs and FAO could be a successful strategy.

Very recently, it has been reported that the inhibition of the fatty acid transporter FATP2 using the specific inhibitor lipofermata reduces the accumulation of lipids and also challenges the mitochondrial metabolism in an aged melanoma microenvironment. This allows the overcoming of age-related resistance to MAPKi in melanoma mouse models [89]. This study paves the way to additional combinatorial strategies using FATP2 inhibitors together with inhibitors of enzymes involved in MUFAs synthesis, such as SCD itself.

Metabolic inhibitors may also be combined with additional druggable targets belonging to cluster 2, such as NFκB and mTOR. Interestingly, mitochondrial hyperfusion, a process that antagonizes apoptosis is an adaptive response to mTOR inhibition [90] and may be overcome by the combination with mitochondrial inhibitors. It is important to point out that while mTOR inhibitors, such as rapamycin analogs (i.e., rapalog), have been approved in the clinic to treat cancer, NFκB inhibitors efficacy as therapeutics is limited because of the host toxicity [91].

It is also worth considering the therapeutic opportunities provided by YAP/TAZ-TEAD inhibitors, many of which are currently under clinical development. These compounds are divided into three groups: (1) inhibitors of YAP/TAZ stimulators, (2) direct inhibitors of YAP/TAZ-TEAD and (3) drugs blocking the oncogenic downstream YAP/TAZ transcriptional target genes [92].

Finally, CSCs may also be directly tackled thanks to the use of specific inhibitors of JARID1B [93] despite in this case their clinical development being far from being successfully accomplished.

In summary, we believe that in order to cut the Gordian knot linking drug resistance with CSCs' fitness in metastatic melanomas new therapeutic strategies will have to be rationally developed and take into account the simultaneous targeting of multiple nodes in the limited key pathways identified.

4. Methods

4.1. Bibliographic Search

A bibliographic search was conducted through "PubMed" free resource supporting the biomedical and life science research. Keywords used were: "melanoma stem cells", "resistance to target therapy", "metabolic rewiring" and "phenotype switching". In this way we found more than 3000 publications. The list was refined according to the: (a) potential concordance in the genes identified among studies, (b) most recent works (c) highly indexed journals (impact factor >7) and (d) studies taking advantages of single-cell approaches. Thereby, a list of 25 top genes was obtained, namely "MSCsign".

4.2. Interactomic and Clustering Plots

"MSCsign" was subjected to the Markov Cluster Algorithm [82] and four clusters identified: (1) kinase and metabolic, (2) melanoma-associated, (3) Hippo pathway and (4) slow cycling/CSCs. The same gene list was used to build a protein-protein interaction (PPI) network using the STRING online database [83]. The minimum required interaction score used was >0, 4, 89 hedges were identified. p-value $< 1.0 \times 10^{-16}$.

4.3. Mining TCGA Data of SKCM Dataset

The expression levels of the genes belonging to "MSCsign" were evaluated in 480 samples coming from the skin cutaneous melanoma dataset using UALCAN software [84]. Those analyses were refined according to the mutational status of melanoma patients extracting RNA-Seq data from skin cutaneous melanoma dataset (TCGA, PanCancer Atlas) [85] available on the cBioPortal website [86,87]. mRNA Expression, RSEM (RNA-seq by expectation-maximization) (batch normalized from Illumina HiSeq_RNASeqV2). Heatmaps wereplotted through the online tool Orange. Kaplan–Meyer curves were used to estimate the prognostic values of "MSCsign" for each mutational subset.

4.4. Principal Component Analyses

The expression levels of "MSCsign" from skin cutaneous melanoma dataset were subjected to Principal Component Analysis (PCA) in comparison with other six solid cancers through the online software GEPIA [88]. SKCM = Skin Cutaneous Melanoma, BRCA = Breast invasive carcinoma, LUAD = Lung adenocarcinoma, OV = Ovarian serous cystadenocarcinoma, COAD = colon adenocarcinoma, GBM = Glioblastoma multiforme, HNSC = Head and Neck squamous cell carcinoma.

4.5. Statistical Analyses

p-values were estimated using the log-rank test (significance $p < 0.05$).

4.6. List of the Online Tools Used

https://pubmed.ncbi.nlm.nih.gov/
https://string-db.org/
http://ualcan.path.uab.edu/index.html
https://www.cbioportal.org/
http://gepia.cancer-pku.cn/index.html
https://orange.biolab.si/

Supplementary Materials: The following are available online at http://www.mdpi.com/2072-6694/12/11/3368/s1. Table S1. Enrichment analyses of Gene Ontology, KEGG and Reactome on the list of 25 top genes performed using STRING online database. Table S2. Gene expression levels of "MSCsign" normalized considering the maximum median expression value coming from SKCM dataset. Table S3. Raw data of gene expression levels obtained from https://www.cbioportal.org/. Table S4. OS data obtained from https://www.cbioportal.org/.

Author Contributions: L.F. wrote the manuscript and performed the bioinformatics analyses; R.M. revised the manuscript; G.C. conceived and revised the review. All authors have read and agreed to the published version of the manuscript.

Funding: This work was supported by: (1) Italian Association for Cancer Research (AIRC) grants IG15216 to G. Ciliberto and IG17009 to Rita Mancini; (2) the Lazioinova grant 2018 n.85-2017-13750 to Rita Mancini; (3) PRIN Bando 2017 (Prot. 2017HWTP2K) to G. Ciliberto and R. Mancini.

Acknowledgments: Editorial assistance was provided by Luca Giacomelli and Aashni Shah (Polistudium SRL, Milan, Italy). This assistance was supported by internal funds.

Conflicts of Interest: The authors declare no conflict of interest. The funders had no role in the design of the study; in the collection, analyses, or interpretation of data; in the writing of the manuscript, or in the decision to publish the results.

Abbreviations

MAPKi	BRAF/MEK inhibitors
CSCs	cancer stem cells
sc	Single Cell
OXPHOS	Oxidative Phosphorylation
MUFAs	Monounsaturated Fatty Acids
FAO	Fatty Acid Oxidation
PPI	Protein-Protein Interaction
MSCsign	melanoma stem cells signature

References

1. Siegel, R.L.; Miller, K.D.; Jemal, A. Cancer Statistics, 2017. *CA Cancer J. Clin.* **2017**, *67*, 7–30. [CrossRef] [PubMed]
2. Marzagalli, M.; Moretti, R.M.; Messi, E.; Marelli, M.M.; Fontana, F.; Anastasia, A.; Bani, M.R.; Beretta, G.; Limonta, P. Targeting melanoma stem cells with the Vitamin E derivative δ-tocotrienol. *Sci. Rep.* **2018**, *8*, 587. [CrossRef] [PubMed]

3. Ottaviano, M.; De Placido, S.; Ascierto, P.A. Recent success and limitations of immune checkpoint inhibitors for cancer: A lesson from melanoma. *Virchows Arch.* **2019**, *474*, 421–432. [CrossRef] [PubMed]

4. Ugurel, S.; Röhmel, J.; Ascierto, P.A.; Flaherty, K.T.; Grob, J.J.; Hauschild, A.; Larkin, J.; Long, G.V.; Lorigan, P.; McArthur, G.A.; et al. Survival of patients with advanced metastatic melanoma: The impact of novel therapies. *Eur. J. Cancer* **2016**, *53*, 125–134. [CrossRef] [PubMed]

5. Pelster, M.S.; Amaria, R.N. Combined targeted therapy and immunotherapy in melanoma: A review of the impact on the tumor microenvironment and outcomes of early clinical trials. *Ther. Adv. Med. Oncol.* **2019**, *11*. [CrossRef]

6. Madonna, G.; Ballesteros-Merino, C.; Feng, Z.; Bifulco, C.; Capone, M.; Giannarelli, D.; Mallardo, D.; Simeone, E.; Grimaldi, A.M.; Caracò, C.; et al. PD-L1 expression with immune-infiltrate evaluation and outcome prediction in melanoma patients treated with ipilimumab. *Oncoimmunology* **2018**, *7*, e1405206. [CrossRef]

7. Pavlick, A.C.; Fecher, L.; Ascierto, P.A.; Sullivan, R.J. Frontline Therapy for BRAF-Mutated Metastatic Melanoma: How Do You Choose, and Is There One Correct Answer? *Am. Soc. Clin. Oncol. Educ. Book* **2019**, *39*, 564–571. [CrossRef]

8. Fattore, L.; Mancini, R.; Ascierto, P.A.; Ciliberto, G. The potential of BRAF-associated non-coding RNA as a therapeutic target in melanoma. *Expert Opin. Ther. Targets* **2019**, *23*, 53–68. [CrossRef]

9. Shi, H.; Hugo, W.; Kong, X.; Hong, A.; Koya, R.C.; Moriceau, G.; Chodon, T.; Guo, R.; Johnson, D.B.; Dahlman, K.B.; et al. Acquired resistance and clonal evolution in melanoma during BRAF inhibitor therapy. *Cancer Discov.* **2014**, *4*, 80–93. [CrossRef]

10. Luke, J.J.; Flaherty, K.T.; Ribas, A.; Long, G. V Targeted agents and immunotherapies: Optimizing outcomes in melanoma. *Nat. Rev. Clin. Oncol.* **2017**, *14*, 463–482. [CrossRef]

11. Robert, C.; Karaszewska, B.; Schachter, J.; Rutkowski, P.; Mackiewicz, A.; Stroiakovski, D.; Lichinitser, M.; Dummer, R.; Grange, F.; Mortier, L.; et al. Improved Overall Survival in Melanoma with Combined Dabrafenib and Trametinib. *N. Engl. J. Med.* **2014**, *372*, 30–39. [CrossRef] [PubMed]

12. Dummer, R.; Ascierto, P.A.; Gogas, H.J.; Arance, A.; Mandala, M.; Liszkay, G.; Garbe, C.; Schadendorf, D.; Krajsova, I.; Gutzmer, R.; et al. Encorafenib plus binimetinib versus vemurafenib or encorafenib in patients with BRAF-mutant melanoma (COLUMBUS): A multicentre, open-label, randomised phase 3 trial. *Lancet Oncol.* **2018**, *19*, 603–615. [CrossRef]

13. Long, G.V.; Eroglu, Z.; Infante, J.; Patel, S.; Daud, A.; Johnson, D.B.; Gonzalez, R.; Kefford, R.; Hamid, O.; Schuchter, L.; et al. Long-Term Outcomes in Patients with BRAF V600-Mutant Metastatic Melanoma Who Received Dabrafenib Combined with Trametinib. *J. Clin. Oncol.* **2018**, *36*, 667–673. [CrossRef] [PubMed]

14. Moriceau, G.; Hugo, W.; Hong, A.; Shi, H.; Kong, X.; Yu, C.C.; Koya, R.C.; Samatar, A.A.; Khanlou, N.; Braun, J.; et al. Tunable-Combinatorial Mechanisms of Acquired Resistance Limit the Efficacy of BRAF/MEK Cotargeting but Result in Melanoma Drug Addiction. *Cancer Cell* **2015**, *27*, 240–256. [CrossRef] [PubMed]

15. Hong, A.; Moriceau, G.; Sun, L.; Lomeli, S.; Piva, M.; Damoiseaux, R.; Holmen, S.L.; Sharpless, N.E.; Hugo, W.; Lo, R.S. Exploiting Drug Addiction Mechanisms to Select against MAPKi-Resistant Melanoma. *Cancer Discov.* **2018**, *8*, 74–93. [CrossRef]

16. Hugo, W.; Shi, H.; Sun, L.; Piva, M.; Song, C.; Kong, X.; Moriceau, G.; Hong, A.; Dahlman, K.B.; Johnson, D.B.; et al. Non-genomic and Immune Evolution of Melanoma Acquiring MAPKi Resistance. *Cell* **2015**, *162*, 1271–1285. [CrossRef]

17. Fattore, L.; Sacconi, A.; Mancini, R.; Ciliberto, G. MicroRNA-driven deregulation of cytokine expression helps development of drug resistance in metastatic melanoma. *Cytokine Growth Factor Rev.* **2017**, *36*, 39–48. [CrossRef] [PubMed]

18. Kozar, I.; Margue, C.; Rothengatter, S.; Haan, C.; Kreis, S. Many ways to resistance: How melanoma cells evade targeted therapies. *Biochim. Biophys. Acta Rev. Cancer* **2019**, *1871*, 313–322. [CrossRef]

19. Young, H.L.; Rowling, E.J.; Bugatti, M.; Giurisato, E.; Luheshi, N.; Arozarena, I.; Acosta, J.-C.; Kamarashev, J.; Frederick, D.T.; Cooper, Z.A.; et al. An adaptive signaling network in melanoma inflammatory niches confers tolerance to MAPK signaling inhibition. *J. Exp. Med.* **2017**, *214*, 1691–1710. [CrossRef]

20. Bristot, I.J.; Kehl Dias, C.; Chapola, H.; Parsons, R.B.; Klamt, F. Metabolic rewiring in melanoma drug-resistant cells. *Crit. Rev. Oncol. Hematol.* **2020**, *153*, 102995. [CrossRef]

21. Ruggiero, C.F.; Malpicci, D.; Fattore, L.; Madonna, G.; Vanella, V.; Mallardo, D.; Liguoro, D.; Salvati, V.; Capone, M.; Bedogni, B.; et al. ErbB3 Phosphorylation as Central Event in Adaptive Resistance to Targeted

Therapy in Metastatic Melanoma: Early Detection in CTCs during Therapy and Insights into Regulation by Autocrine Neuregulin. *Cancers* **2019**, *11*, 1425. [CrossRef]

22. Abel, E.V.; Basile, K.J.; Kugel, C.H., 3rd; Witkiewicz, A.K.; Le, K.; Amaravadi, R.K.; Karakousis, G.C.; Xu, X.; Xu, W.; Schuchter, L.M.; et al. Melanoma adapts to RAF/MEK inhibitors through FOXD3-mediated upregulation of ERBB3. *J. Clin. Investig.* **2013**, *123*, 2155–2168. [CrossRef] [PubMed]

23. Fattore, L.; Campani, V.; Ruggiero, C.F.; Salvati, V.; Liguoro, D.; Scotti, L.; Botti, G.; Ascierto, P.A.; Mancini, R.; De Rosa, G.; et al. In Vitro Biophysical and Biological Characterization of Lipid Nanoparticles Co-Encapsulating Oncosuppressors miR-199b-5p and miR-204-5p as Potentiators of Target Therapy in Metastatic Melanoma. *Int. J. Mol. Sci.* **2020**, *21*, 1930. [CrossRef] [PubMed]

24. Fattore, L.; Costantini, S.; Malpicci, D.; Ruggiero, C.F.; Ascierto, P.A.; Croce, C.M.; Mancini, R.; Ciliberto, G. MicroRNAs in melanoma development and resistance to target therapy. *Oncotarget* **2017**, *8*, 22262–22278. [CrossRef] [PubMed]

25. Fattore, L.; Ruggiero, C.F.; Pisanu, M.E.; Liguoro, D.; Cerri, A.; Costantini, S.; Capone, F.; Acunzo, M.; Romano, G.; Nigita, G.; et al. Reprogramming miRNAs global expression orchestrates development of drug resistance in BRAF mutated melanoma. *Cell Death Differ.* **2019**, *26*, 1267–1282. [CrossRef] [PubMed]

26. Tupone, M.G.; D'Aguanno, S.; Di Martile, M.; Valentini, E.; Desideri, M.; Trisciuoglio, D.; Donzelli, S.; Sacconi, A.; Buglioni, S.; Ercolani, C.; et al. microRNA-378a-5p iS a novel positive regulator of melanoma progression. *Oncogenesis* **2020**, *9*, 22. [CrossRef]

27. Caporali, S.; Amaro, A.; Levati, L.; Alvino, E.; Lacal, P.M.; Mastroeni, S.; Ruffini, F.; Bonmassar, L.; Antonini Cappellini, G.C.; Felli, N.; et al. miR-126-3p down-regulation contributes to dabrafenib acquired resistance in melanoma by up-regulating ADAM9 and VEGF-A. *J. Exp. Clin. Cancer Res.* **2019**, *38*, 272. [CrossRef]

28. Fattore, L.; Mancini, R.; Acunzo, M.; Romano, G.; Laganà, A.; Pisanu, M.E.; Malpicci, D.; Madonna, G.; Mallardo, D.; Capone, M.; et al. miR-579-3p controls melanoma progression and resistance to target therapy. *Proc. Natl. Acad. Sci. USA* **2016**, *113*, E5005–E5013. [CrossRef]

29. Ascierto, P.A.; Bifulco, C.; Buonaguro, L.; Emens, L.A.; Ferris, R.L.; Fox, B.A.; Delgoffe, G.M.; Galon, J.; Gridelli, C.; Merlano, M.; et al. Perspectives in immunotherapy: Meeting report from the "Immunotherapy Bridge 2018" (28–29 November 2018, Naples, Italy). *J. Immunother. Cancer* **2019**, *7*, 332. [CrossRef]

30. Weiss, S.A.; Wolchok, J.D.; Sznol, M. Immunotherapy of Melanoma: Facts and Hopes. *Clin. Cancer Res.* **2019**, *25*, 5191–5201. [CrossRef]

31. Liu, A.; Curran, M.A. Tumor hypermetabolism confers resistance to immunotherapy. *Semin. Cancer Biol.* **2020**, *65*, 155–163. [CrossRef] [PubMed]

32. Abbaszadegan, M.R.; Bagheri, V.; Razavi, M.S.; Momtazi, A.A.; Sahebkar, A.; Gholamin, M. Isolation, identification, and characterization of cancer stem cells: A review. *J. Cell. Physiol.* **2017**, *232*, 2008–2018. [CrossRef] [PubMed]

33. Eun, K.; Ham, S.W.; Kim, H. Cancer stem cell heterogeneity: Origin and new perspectives on CSC targeting. *BMB Rep.* **2017**, *50*, 117–125. [CrossRef] [PubMed]

34. Atashzar, M.R.; Baharlou, R.; Karami, J.; Abdollahi, H.; Rezaei, R.; Pourramezan, F.; Zoljalali Moghaddam, S.H. Cancer stem cells: A review from origin to therapeutic implications. *J. Cell. Physiol.* **2020**, *235*, 790–803. [CrossRef] [PubMed]

35. Bruschini, S.; Ciliberto, G.; Mancini, R. The emerging role of cancer cell plasticity and cell-cycle quiescence in immune escape. *Cell Death Dis.* **2020**, *11*, 471. [CrossRef]

36. De Vitis, C.; Corleone, G.; Salvati, V.; Ascenzi, F.; Pallocca, M.; De Nicola, F.; Fanciulli, M.; di Martino, S.; Bruschini, S.; Napoli, C.; et al. B4GALT1 Is a New Candidate to Maintain the Stemness of Lung Cancer Stem Cells. *J. Clin. Med.* **2019**, *8*, 1928. [CrossRef]

37. Bruschini, S.; di Martino, S.; Pisanu, M.E.; Fattore, L.; De Vitis, C.; Laquintana, V.; Buglioni, S.; Tabbì, E.; Cerri, A.; Visca, P.; et al. CytoMatrix for a reliable and simple characterization of lung cancer stem cells from malignant pleural effusions. *J. Cell. Physiol.* **2020**, *235*, 1877–1887. [CrossRef] [PubMed]

38. Brinckerhoff, C.E. Cancer Stem Cells (CSCs) in melanoma: There's smoke, but is there fire? *J. Cell. Physiol.* **2017**, *232*, 2674–2678. [CrossRef]

39. Dagogo-Jack, I.; Shaw, A.T. Tumour heterogeneity and resistance to cancer therapies. *Nat. Rev. Clin. Oncol.* **2018**, *15*, 81–94. [CrossRef]

40. Ahmed, F.; Haass, N.K. Microenvironment-Driven Dynamic Heterogeneity and Phenotypic Plasticity as a Mechanism of Melanoma Therapy Resistance. *Front. Oncol.* **2018**, *8*, 173. [CrossRef]

41. Grzywa, T.M.; Paskal, W.; Włodarski, P.K. Intratumor and Intertumor Heterogeneity in Melanoma. *Transl. Oncol.* **2017**, *10*, 956–975. [CrossRef] [PubMed]

42. Fattore, L.; Ruggiero, C.F.; Liguoro, D.; Mancini, R.; Ciliberto, G. Single cell analysis to dissect molecular heterogeneity and disease evolution in metastatic melanoma. *Cell Death Dis.* **2019**, *10*, 827. [CrossRef] [PubMed]

43. Reya, T.; Morrison, S.J.; Clarke, M.F.; Weissman, I.L. Stem cells, cancer, and cancer stem cells. *Nature* **2001**, *414*, 105–111. [CrossRef] [PubMed]

44. Toh, T.B.; Lim, J.J.; Chow, E.K.-H. Epigenetics in cancer stem cells. *Mol. Cancer* **2017**, *16*, 29. [CrossRef] [PubMed]

45. Davis, A.; Gao, R.; Navin, N. Tumor evolution: Linear, branching, neutral or punctuated? *Biochim. Biophys. Acta Rev. Cancer* **2017**, *1867*, 151–161. [CrossRef]

46. Bai, X.; Fisher, D.E.; Flaherty, K.T. Cell-state dynamics and therapeutic resistance in melanoma from the perspective of MITF and IFNγ pathways. *Nat. Rev. Clin. Oncol.* **2019**, *16*, 549–562. [CrossRef]

47. Arozarena, I.; Wellbrock, C. Phenotype plasticity as enabler of melanoma progression and therapy resistance. *Nat. Rev. Cancer* **2019**, *19*, 377–391. [CrossRef]

48. Andor, N.; Graham, T.A.; Jansen, M.; Xia, L.C.; Aktipis, C.A.; Petritsch, C.; Ji, H.P.; Maley, C.C. Pan-cancer analysis of the extent and consequences of intratumor heterogeneity. *Nat. Med.* **2016**, *22*, 105–113. [CrossRef]

49. Sini, M.C.; Doneddu, V.; Paliogiannis, P.; Casula, M.; Colombino, M.; Manca, A.; Botti, G.; Ascierto, P.A.; Lissia, A.; Cossu, A.; et al. Genetic alterations in main candidate genes during melanoma progression. *Oncotarget* **2018**, *9*, 8531–8541. [CrossRef]

50. Rambow, F.; Marine, J.-C.; Goding, C.R. Melanoma plasticity and phenotypic diversity: Therapeutic barriers and opportunities. *Genes Dev.* **2019**, *33*, 1295–1318. [CrossRef]

51. Andreucci, E.; Peppicelli, S.; Ruzzolini, J.; Bianchini, F.; Biagioni, A.; Papucci, L.; Magnelli, L.; Mazzanti, B.; Stecca, B.; Calorini, L. The acidic tumor microenvironment drives a stem-like phenotype in melanoma cells. *J. Mol. Med.* **2020**, *98*, 1431–1446. [CrossRef] [PubMed]

52. Tsoi, J.; Robert, L.; Paraiso, K.; Galvan, C.; Sheu, K.M.; Lay, J.; Wong, D.J.L.; Atefi, M.; Shirazi, R.; Wang, X.; et al. Multi-stage Differentiation Defines Melanoma Subtypes with Differential Vulnerability to Drug-Induced Iron-Dependent Oxidative Stress. *Cancer Cell* **2018**, *33*, 890–904.e5. [CrossRef] [PubMed]

53. Rambow, F.; Rogiers, A.; Marin-Bejar, O.; Aibar, S.; Femel, J.; Dewaele, M.; Karras, P.; Brown, D.; Chang, Y.H.; Debiec-Rychter, M.; et al. Toward Minimal Residual Disease-Directed Therapy in Melanoma. *Cell* **2018**, *174*, 843–855.e19. [CrossRef] [PubMed]

54. Boiko, A.D.; Razorenova, O.V.; van de Rijn, M.; Swetter, S.M.; Johnson, D.L.; Ly, D.P.; Butler, P.D.; Yang, G.P.; Joshua, B.; Kaplan, M.J.; et al. Human melanoma-initiating cells express neural crest nerve growth factor receptor CD271. *Nature* **2010**, *466*, 133–137. [CrossRef]

55. Roesch, A.; Fukunaga-Kalabis, M.; Schmidt, E.C.; Zabierowski, S.E.; Brafford, P.A.; Vultur, A.; Basu, D.; Gimotty, P.; Vogt, T.; Herlyn, M. A temporarily distinct subpopulation of slow-cycling melanoma cells is required for continuous tumor growth. *Cell* **2010**, *141*, 583–594. [CrossRef]

56. Tirosh, I.; Izar, B.; Prakadan, S.M.; Wadsworth, M.H., 2nd; Treacy, D.; Trombetta, J.J.; Rotem, A.; Rodman, C.; Lian, C.; Murphy, G.; et al. Dissecting the multicellular ecosystem of metastatic melanoma by single-cell RNA-seq. *Science* **2016**, *352*, 189–196. [CrossRef]

57. Gerber, T.; Willscher, E.; Loeffler-Wirth, H.; Hopp, L.; Schadendorf, D.; Schartl, M.; Anderegg, U.; Camp, G.; Treutlein, B.; Binder, H.; et al. Mapping heterogeneity in patient-derived melanoma cultures by single-cell RNA-seq. *Oncotarget* **2017**, *8*, 846–862. [CrossRef]

58. Shaffer, S.M.; Dunagin, M.C.; Torborg, S.R.; Torre, E.A.; Emert, B.; Krepler, C.; Beqiri, M.; Sproesser, K.; Brafford, P.A.; Xiao, M.; et al. Rare cell variability and drug-induced reprogramming as a mode of cancer drug resistance. *Nature* **2017**, *546*, 431–435. [CrossRef]

59. Aloia, A.; Müllhaupt, D.; Chabbert, C.D.; Eberhart, T.; Flückiger-Mangual, S.; Vukolic, A.; Eichhoff, O.; Irmisch, A.; Alexander, L.T.; Scibona, E.; et al. A Fatty Acid Oxidation-dependent Metabolic Shift Regulates the Adaptation of BRAF-mutated Melanoma to MAPK Inhibitors. *Clin. Cancer Res.* **2019**, *25*, 6852–6867. [CrossRef]

60. Vivas-García, Y.; Falletta, P.; Liebing, J.; Louphrasitthiphol, P.; Feng, Y.; Chauhan, J.; Scott, D.A.; Glodde, N.; Chocarro-Calvo, A.; Bonham, S.; et al. Lineage-Restricted Regulation of SCD and Fatty Acid Saturation by MITF Controls Melanoma Phenotypic Plasticity. *Mol. Cell* **2020**, *77*, 120–137.e9. [CrossRef]

61. Vazquez, F.; Lim, J.-H.; Chim, H.; Bhalla, K.; Girnun, G.; Pierce, K.; Clish, C.B.; Granter, S.R.; Widlund, H.R.; Spiegelman, B.M.; et al. PGC1α expression defines a subset of human melanoma tumors with increased mitochondrial capacity and resistance to oxidative stress. *Cancer Cell* **2013**, *23*, 287–301. [CrossRef] [PubMed]

62. Gopal, Y.N.V.; Rizos, H.; Chen, G.; Deng, W.; Frederick, D.T.; Cooper, Z.A.; Scolyer, R.A.; Pupo, G.; Komurov, K.; Sehgal, V.; et al. Inhibition of mTORC1/2 overcomes resistance to MAPK pathway inhibitors mediated by PGC1α and oxidative phosphorylation in melanoma. *Cancer Res.* **2014**, *74*, 7037–7047. [CrossRef] [PubMed]

63. Vogel, F.C.E.; Bordag, N.; Zügner, E.; Trajkovic-Arsic, M.; Chauvistré, H.; Shannan, B.; Váraljai, R.; Horn, S.; Magnes, C.; Thomas Siveke, J.; et al. Targeting the H3K4 Demethylase KDM5B Reprograms the Metabolome and Phenotype of Melanoma Cells. *J. Investig. Dermatol.* **2019**, *139*, 2506–2516.e10. [CrossRef] [PubMed]

64. Pisanu, M.E.; Maugeri-Saccà, M.; Fattore, L.; Bruschini, S.; De Vitis, C.; Tabbì, E.; Bellei, B.; Migliano, E.; Kovacs, D.; Camera, E.; et al. Inhibition of Stearoyl-CoA desaturase 1 reverts BRAF and MEK inhibition-induced selection of cancer stem cells in BRAF-mutated melanoma. *J. Exp. Clin. Cancer Res.* **2018**, *37*, 318. [CrossRef]

65. Zhang, G.; Frederick, D.T.; Wu, L.; Wei, Z.; Krepler, C.; Srinivasan, S.; Chae, Y.C.; Xu, X.; Choi, H.; Dimwamwa, E.; et al. Targeting mitochondrial biogenesis to overcome drug resistance to MAPK inhibitors. *J. Clin. Investig.* **2016**, *126*, 1834–1856. [CrossRef] [PubMed]

66. Su, Y.; Ko, M.E.; Cheng, H.; Zhu, R.; Xue, M.; Wang, J.; Lee, J.W.; Frankiw, L.; Xu, A.; Wong, S.; et al. Multi-omic single-cell snapshots reveal multiple independent trajectories to drug tolerance in a melanoma cell line. *Nat. Commun.* **2020**, *11*, 2345. [CrossRef]

67. Vashisht Gopal, Y.N.; Gammon, S.; Prasad, R.; Knighton, B.; Pisaneschi, F.; Roszik, J.; Feng, N.; Johnson, S.; Pramanik, S.; Sudderth, J.; et al. A Novel Mitochondrial Inhibitor Blocks MAPK Pathway and Overcomes MAPK Inhibitor Resistance in Melanoma. *Clin. Cancer Res.* **2019**, *25*, 6429–6442. [CrossRef] [PubMed]

68. Perego, M.; Maurer, M.; Wang, J.X.; Shaffer, S.; Müller, A.C.; Parapatics, K.; Li, L.; Hristova, D.; Shin, S.; Keeney, F.; et al. A slow-cycling subpopulation of melanoma cells with highly invasive properties. *Oncogene* **2018**, *37*, 302–312. [CrossRef]

69. Webster, M.R.; Fane, M.E.; Alicea, G.M.; Basu, S.; Kossenkov, A.V.; Marino, G.E.; Douglass, S.M.; Kaur, A.; Ecker, B.L.; Gnanapradeepan, K.; et al. Paradoxical Role for Wild-Type p53 in Driving Therapy Resistance in Melanoma. *Mol. Cell* **2020**, *77*, 633–644.e5. [CrossRef]

70. Zubrilov, I.; Sagi-Assif, O.; Izraely, S.; Meshel, T.; Ben-Menahem, S.; Ginat, R.; Pasmanik-Chor, M.; Nahmias, C.; Couraud, P.-O.; Hoon, D.S.B.; et al. Vemurafenib resistance selects for highly malignant brain and lung-metastasizing melanoma cells. *Cancer Lett.* **2015**, *361*, 86–96. [CrossRef]

71. Liguoro, D.; Fattore, L.; Mancini, R.; Ciliberto, G. Drug tolerance to target therapy in melanoma revealed at single cell level: What next? *Biochim. Biophys. Acta Rev. Cancer* **2020**, *1874*, 188440. [CrossRef]

72. Roesch, A.; Vultur, A.; Bogeski, I.; Wang, H.; Zimmermann, K.M.; Speicher, D.; Körbel, C.; Laschke, M.W.; Gimotty, P.A.; Philipp, S.E.; et al. Overcoming intrinsic multidrug resistance in melanoma by blocking the mitochondrial respiratory chain of slow-cycling JARID1B(high) cells. *Cancer Cell* **2013**, *23*, 811–825. [CrossRef]

73. Zakaria, N.; Mohd Yusoff, N.; Zakaria, Z.; Widera, D.; Yahaya, B.H. Inhibition of NF-κB Signaling Reduces the Stemness Characteristics of Lung Cancer Stem Cells. *Front. Oncol.* **2018**, *8*, 166. [CrossRef] [PubMed]

74. Moriyama, H.; Moriyama, M.; Ozawa, T.; Tsuruta, D.; Iguchi, T.; Tamada, S.; Nakatani, T.; Nakagawa, K.; Hayakawa, T. Notch Signaling Enhances Stemness by Regulating Metabolic Pathways Through Modifying p53, NF-κB, and HIF-1α. *Stem Cells Dev.* **2018**, *27*, 935–947. [CrossRef]

75. Fattore, L.; Malpicci, D.; Milite, C.; Castellano, S.; Sbardella, G.; Botti, G.; Ascierto, P.A.; Mancini, R.; Ciliberto, G. Reverse transcriptase inhibition potentiates target therapy in BRAF-mutant melanomas: Effects on cell proliferation, apoptosis, DNA-damage, ROS induction and mitochondrial membrane depolarization. *Cell Commun. Signal.* **2020**, *18*, 150. [CrossRef] [PubMed]

76. Mancini, R.; Giarnieri, E.; De Vitis, C.; Malanga, D.; Roscilli, G.; Noto, A.; Marra, E.; Laudanna, C.; Zoppoli, P.; De Luca, P.; et al. Spheres derived from lung adenocarcinoma pleural effusions: Molecular characterization and tumor engraftment. *PLoS ONE* **2011**, *6*, e21320. [CrossRef] [PubMed]

77. Noto, A.; De Vitis, C.; Pisanu, M.E.; Roscilli, G.; Ricci, G.; Catizone, A.; Sorrentino, G.; Chianese, G.; Taglialatela-Scafati, O.; Trisciuoglio, D.; et al. Stearoyl-CoA-desaturase 1 regulates lung cancer stemness via stabilization and nuclear localization of YAP/TAZ. *Oncogene* **2017**, *36*, 4573–4584. [CrossRef] [PubMed]

78. Noto, A.; Raffa, S.; De Vitis, C.; Roscilli, G.; Malpicci, D.; Coluccia, P.; Di Napoli, A.; Ricci, A.; Giovagnoli, M.R.; Aurisicchio, L.; et al. Stearoyl-CoA desaturase-1 is a key factor for lung cancer-initiating cells. *Cell Death Dis.* **2013**, *4*, e947. [CrossRef] [PubMed]

79. Mancini, R.; Noto, A.; Pisanu, M.E.; De Vitis, C.; Maugeri-Saccà, M.; Ciliberto, G. Metabolic features of cancer stem cells: The emerging role of lipid metabolism. *Oncogene* **2018**, *37*, 2367–2378. [CrossRef]

80. Pisanu, M.E.; Noto, A.; De Vitis, C.; Morrone, S.; Scognamiglio, G.; Botti, G.; Venuta, F.; Diso, D.; Jakopin, Z.; Padula, F.; et al. Blockade of Stearoyl-CoA-desaturase 1 activity reverts resistance to cisplatin in lung cancer stem cells. *Cancer Lett.* **2017**, *406*, 93–104. [CrossRef]

81. Knobloch, M.; Pilz, G.-A.; Ghesquière, B.; Kovacs, W.J.; Wegleiter, T.; Moore, D.L.; Hruzova, M.; Zamboni, N.; Carmeliet, P.; Jessberger, S. A Fatty Acid Oxidation-Dependent Metabolic Shift Regulates Adult Neural Stem Cell Activity. *Cell Rep.* **2017**, *20*, 2144–2155. [CrossRef] [PubMed]

82. Enright, A.J.; Van Dongen, S.; Ouzounis, C.A. An efficient algorithm for large-scale detection of protein families. *Nucleic Acids Res.* **2002**, *30*, 1575–1584. [CrossRef] [PubMed]

83. Szklarczyk, D.; Gable, A.L.; Lyon, D.; Junge, A.; Wyder, S.; Huerta-Cepas, J.; Simonovic, M.; Doncheva, N.T.; Morris, J.H.; Bork, P.; et al. STRING v11: Protein-protein association networks with increased coverage, supporting functional discovery in genome-wide experimental datasets. *Nucleic Acids Res.* **2019**, *47*, D607–D613. [CrossRef] [PubMed]

84. Chandrashekar, D.S.; Bashel, B.; Balasubramanya, S.A.H.; Creighton, C.J.; Ponce-Rodriguez, I.; Chakravarthi, B.V.S.K.; Varambally, S. UALCAN: A Portal for Facilitating Tumor Subgroup Gene Expression and Survival Analyses. *Neoplasia* **2017**, *19*, 649–658. [CrossRef] [PubMed]

85. Network, C.G.A. Genomic Classification of Cutaneous Melanoma. *Cell* **2015**, *161*, 1681–1696. [CrossRef]

86. Gao, J.; Aksoy, B.A.; Dogrusoz, U.; Dresdner, G.; Gross, B.; Sumer, S.O.; Sun, Y.; Jacobsen, A.; Sinha, R.; Larsson, E.; et al. Integrative Analysis of Complex Cancer Genomics and Clinical Profiles Using the cBioPortal. *Sci. Signal.* **2013**, *6*, pl1. [CrossRef]

87. Cerami, E.; Gao, J.; Dogrusoz, U.; Gross, B.E.; Sumer, S.O.; Aksoy, B.A.; Jacobsen, A.; Byrne, C.J.; Heuer, M.L.; Larsson, E.; et al. The cBio Cancer Genomics Portal: An Open Platform for Exploring Multidimensional Cancer Genomics Data. *Cancer Discov.* **2012**, *2*, 401–404. [CrossRef]

88. Tang, Z.; Li, C.; Kang, B.; Gao, G.; Li, C.; Zhang, Z. GEPIA: A web server for cancer and normal gene expression profiling and interactive analyses. *Nucleic Acids Res.* **2017**, *45*, W98–W102. [CrossRef]

89. Alicea, G.M.; Rebecca, V.W.; Goldman, A.R.; Fane, M.E.; Douglass, S.M.; Behera, R.; Webster, M.R.; Kugel, C.H., 3rd; Ecker, B.L.; Caino, M.C.; et al. Changes in Aged Fibroblast Lipid Metabolism Induce Age-Dependent Melanoma Cell Resistance to Targeted Therapy via the Fatty Acid Transporter FATP2. *Cancer Discov.* **2020**. [CrossRef]

90. Hua, H.; Kong, Q.; Zhang, H.; Wang, J.; Luo, T.; Jiang, Y. Targeting mTOR for cancer therapy. *J. Hematol. Oncol.* **2019**, *12*, 71. [CrossRef]

91. Tornatore, L.; Sandomenico, A.; Raimondo, D.; Low, C.; Rocci, A.; Tralau-Stewart, C.; Capece, D.; D'Andrea, D.; Bua, M.; Boyle, E.; et al. Cancer-selective targeting of the NF-κB survival pathway with GADD45β/MKK7 inhibitors. *Cancer Cell* **2014**, *26*, 495–508. [CrossRef] [PubMed]

92. Pobbati, A.V.; Hong, W. A combat with the YAP/TAZ-TEAD oncoproteins for cancer therapy. *Theranostics* **2020**, *10*, 3622–3635. [CrossRef] [PubMed]

93. Pippa, S.; Mannironi, C.; Licursi, V.; Bombardi, L.; Colotti, G.; Cundari, E.; Mollica, A.; Coluccia, A.; Naccarato, V.; La Regina, G.; et al. Small Molecule Inhibitors of KDM5 Histone Demethylases Increase the Radiosensitivity of Breast Cancer Cells Overexpressing JARID1B. *Molecules* **2019**, *24*, 1739. [CrossRef] [PubMed]

Review

Functional Characterization of Cholinergic Receptors in Melanoma Cells

Anna Maria Lucianò [1] and Ada Maria Tata [1,2,*] 

1   Department of Biology and Biotechnologies Charles Darwin, Sapienza University of Rome, 00185 Rome, Italy; anna.luciano94@gmail.com
2   Research Centre of Neurobiology Daniel Bovet, Sapienza University of Rome, 00185 Rome, Italy
*   Correspondence: adamaria.tata@uniroma1.it

Received: 29 September 2020; Accepted: 23 October 2020; Published: 27 October 2020

Simple Summary: Interest in the involvement of the cholinergic system at the non-neuronal level emerged in recent years, thereby allowing to identify different physiological and pathological processes that acetylcholine and its receptors modulate in different tissues. The cholinergic system can also take part in the activation and maintenance of pathological processes such as neurodegenerative diseases and cancer. This review summarizes studies concerning the involvement of the cholinergic system in different tumor isotypes, with a special focus on melanoma, the most lethal form of skin cancer.

Abstract: In the last two decades, the scientific community has come to terms with the importance of non-neural acetylcholine in light of its multiple biological and pathological functions within and outside the nervous system. Apart from its well-known physiological role both in the central and peripheral nervous systems, in the autonomic nervous system, and in the neuromuscular junction, the expression of the acetylcholine receptors has been detected in different peripheral organs. This evidence has contributed to highlight new roles for acetylcholine in various biological processes, (e.g., cell viability, proliferation, differentiation, migration, secretion). In addition, growing evidence in recent years has also demonstrated new roles for acetylcholine and its receptors in cancer, where they are involved in the modulation of cell proliferation, apoptosis, angiogenesis, and epithelial mesenchymal transition. In this review, we describe the functional characterization of acetylcholine receptors in different tumor types, placing attention on melanoma. The latest set of data accessible through literature, albeit limited, highlights how cholinergic receptors both of muscarinic and nicotinic type can play a relevant role in the migratory processes of melanoma cells, suggesting their possible involvement in invasion and metastasis.

Keywords: cholinergic system; acetylcholine; muscarinic receptors; nicotinic receptors; melanoma; cancer

1. Introduction

Non-Neuronal Acetylcholine

Acetylcholine (ACh) was the first molecule identified as a neurotransmitter operating through two subgroups of receptors, namely muscarinic receptors (mAChRs) and nicotinic receptors (nAChRs) (Tables 1 and 2). The presence of this neurotransmitter in different primitive organisms (e.g., bacteria, protozoa, fungi, algae) highlights its relevance not only as mediator in the nervous system, but also as a non-neuronal signaling molecule involved in the regulation of basic cell functions like proliferation, differentiation, cell–cell contact, and secretion, thus making ACh phylogenetically, the oldest signaling molecule [1–5].

Table 1. Acetylcholine nicotinic receptors.

Types	Class	Subunits	Structure	Localization	Signaling Transduction
Nicotinic Receptors	Ionotropic	10 α subunits (α1-α10), 4 β subunits (β1 -β4), 1 δ, 1 ε, 1γ subunits	Pentameric transmembrane protein complex	Neurons, glia, ganglia, interneurons, motor endplate, immune cells	Voltage-gated ion channels, metabotropic signals (MAPK, PKC)

Table 2. Acetylcholine muscarinic receptors.

Subtype	Signal Transduction	Localization	Second Messengers	Function
M1	Gq, Gs, Gi	Brain, autonomic ganglia, secretory glands	IP3/DAG, NO	slow EPSP and decreased K+ conductance
M2	Gi, G0	Brain, heart, sympathetic ganglia, lung	cAMP ($\downarrow$)	reduce heart rate, control contractile response, thermoregulation, cognitive functions
M3	Gi, Gq	Brain, secretory glands, smooth muscles	IP3/DAG, NO	activation of phospholipase C, phosphoinositide breakdown, and inhibition of calcium-regulated potassium channels
M4	Gi, Go	Brain, lung	cAMP ($\downarrow$)	adenylate cyclase inhibition and potassium channel regulation, acts as inhibitory autoreceptor
M5	Gq	Midbrain, the ventral tegmental area (VTA)	IP3/DAG, NO	phospholipid turnover, decrease of cyclic AMP levels and downregulation of the activity of protein kinase A (PKA)

The importance of ACh as a modulator of different biological processes and the concept itself of non-neuronal acetylcholine function, has been growing since the early 1970s, capitalizing on the discoveries of functional cholinergic receptors in non-neural cells [6]. In particular, the muscarinic receptors have been identified in various organs such as retina [7], lymphocytes [8], spleen [9], adrenal gland [10] and stomach [11]. In the same years, the expression of nicotinic receptors was first documented outside of the nervous system [12]. Locating acetylcholine receptors in different districts has moved in synchrony with identifying the roles played by the latter at non-neuronal level. In this context, it was demonstrated during sea urchin embryogenesis how ACh acts as a morphogen, controlling cell migration and regulating basic cell functions including growth, survival, differentiation, and apoptosis [13,14]. It is also relevant to note that in several tissues, AChRs are not just a target of ACh-deriving from nervous system innervation. In fact, several non-neuronal cells appear able to synthesize and release ACh autonomously, highlighting possible autocrine/paracrine effects mediated by non-neuronal ACh [15,16]. The involvement of ACh in non-neuronal functions was also supported by the demonstration of the cholinergic modulation of the inflammatory states. Several authors reported lymphocytes, dendritic cells, and macrophages clearly express cholinergic components sufficient to constitute their own cholinergic system, and that immune cells and peritoneal macrophages express all five muscarinic receptors and several nAChR subunits. Furthermore, during immunological reaction, stimulated T cells and dendritic cells have the ability to synthesize and release ACh, which in turns acts in an autocrine–paracrine manner on muscarinic or nicotinic receptors expressed in the immune cells, modulating the immune response and inflammatory processes. The same evidence indicates that ACh is able to inhibit the release of tumor necrosis factor and IL1 from macrophages via the activation of nicotinic receptors, thus highlighting the existence of a cholinergic anti-inflammatory pathway [17–19]. Lastly, the microglial cells in the central nervous system are also able to respond to ACh stimulation, inhibiting the release of TNF-α. These data also suggest the existence of a "brain cholinergic anti-inflammatory pathway," mediated by α7 nicotinic receptors, which may become of interest in the treatment of the neuroinflammation, typically established by neurodegenerative processes [20,21]. Similarly, in the tumor cells, the ACh can be synthesized and released locally in the tumor bulk and actively participate in tumorigenesis processes [22,23]. Specifically for the epithelial cells, the first discovery of acetylcholine production in the epithelial cells came in 1983, thanks to the study conducted by Mark [24] on salivary glands of rats, which continued to produce large amounts of ACh despite denervation. Subsequently, it has been proved that human epidermal keratinocytes possess cholinergic enzymes, which synthesize and degrade acetylcholine, and express both nicotinic and muscarinic classes of cholinergic receptors on their cell surfaces. Consequent to ACh release, these receptors are able to initiate a cell response [25]. Furthermore, epidermal and gingival keratinocytes express the M1–M5 and M2–M5 mAChR subtypes, respectively, as well as the α3, α5, α7, α9, and β2 nAChR subunits [26–28]. ACh synthesized by and released from keratinocytes acts on both mAChRs and nAChRs in an autocrine and/or paracrine manner, modulating basic cell functions in the skin, including keratinocyte proliferation, differentiation, adhesion and migration, as well as epidermal barrier formation. Similarly, ACh is required to sustain viability also of the keratinocytes in vitro [29]. Chernyavsky and collaborators demonstrated that keratinocyte migration is modulated by distinct muscarinic acetylcholine receptor subtypes. Additional studies showed that M4 receptor increased the expression of "migratory" integrins, whereas M3 receptor upregulated "sedentary" integrins, demonstrating that the M3 muscarinic receptor is able to inhibit the migration via the guanylyl cyclase–cyclic GMP–protein kinase G signaling pathway [30]. Based on these data, it appears evident that the role of ACh should be largely reconsidered, both at physiological and pathological level. A deeper knowledge of the mechanisms through which ACh mediates its effects may open to new perspectives for the development of therapeutic protocols for different pathologies.

2. Cholinergic Receptor Subtypes: Structure and Function

2.1. Nicotinic Acetylcholine Receptors

The nAChRs are fast ionotropic cationic acetylcholine receptors that mediate the fast-synaptic transmission, whose activation is extremely quick (msec to sub-msec) [31,32]. They are pentameric transmembrane protein complexes, formed by five receptor subunits that enclose a central ion channel [33]. At present, several subunits have been identified and characterized, comprising 10 α subunits (from α1 to α10), 4 β subunits (from β1 to β4), 1 δ, and 1 ε or 1γ subunits [17]. Among all nAChR subunits, the α8 subunit is the only one not present in human cells [34]. In neurons, the nAChRs can be homopentamers formed by α7, α8, and α9 subunits or heteropentamers formed by combination of α and β subunits. In muscle, functional heteropentamer consists of two α subunits and one of each β, γ, and δ subunits. All the nAChRs show permeability to various cations, such as Na^+, Ca^{2+} and K^+, with α7 nAChR displaying the highest permeability for the Ca^{2+}. In fact, the binding of nAChRs to acetylcholine or nAChRs agonists leads to membrane depolarization and opening of voltage-gated ion channels, resulting in ion influx and efflux [31]. The channel activity is the first step in the activation of these receptors, which leads subsequently to a number of other intracellular events and various downstream signaling pathways, such as activation of mitogen-activated protein kinase (MAPK), protein kinase C, and vascular endothelial growth factor, depending on the nAChR subtype and the cell-type involved [35].

Involvement of Nicotinic Acetylcholine Receptors in Cancer

As previously described, the nicotinic acetylcholine receptors are expressed not only in the neuronal system, but also in numerous non-neuronal tissues such as skin, pancreas, and lung, thus suggesting a role in other biological processes in addition to synaptic transmission. Specifically, these receptors were found to play a role in the regulation of cellular processes such as cell proliferation and cell death [36].

It is known that a dysregulation of nAChRs is associated with neurological disorders such as Alzheimer, Parkinson's disease, and schizophrenia [37]. Moreover, studies performed in lung cancer highlighted that the activation of the nAChRs pathways may affect cell proliferation, apoptosis, angiogenesis, and epithelia mesenchymal transition (EMT), supporting a pro-invasive phenotype [38]. Crucial for this discussion is that cancer cells can synthesize ACh, promoting tumorigenesis through AChRs in the absence of nicotine or other agonists. The release of acetylcholine was first demonstrated in small-cell lung cancer (SCLC) and expression of choline acetyltrasferase was found in different SCLC cell lines. Acetylcholine released was inhibited by vesicular ACh transporter (VAChT) inhibitor, vesamicol, in a dose-dependent manner, suggesting a vesicular ACh release also in non-neuronal tissues [39]. Similar findings were also reported in colon and gastric cancer models. In particular, in gastric cancer, acetylcholine production promoted cell growth in a dose-dependent manner, and acetylcholine-stimulated cell proliferation was abolished by AChRs antagonists. Similarly, colon cancer cells can release the self-produced acetylcholine, which promoted cell viability in an autocrine manner [40]. The pathways activated in cancer processes by α7-nAChR activation-nicotine mediated are generally Ras/ERK/MAPK and JAK2/STAT/-PI3K pathways, leading to cancer cell proliferation and migration as demonstrated in lung cancer cells [41,42].

2.2. Muscarinic Acetylcholine Receptors

The other class of acetylcholine receptors is represented by muscarinic receptors, which belong to the class of heptaelical G protein-coupled receptors and consist of five subtypes (M1–M5). Like all the G protein-coupled receptors, they share a structure composed of seven transmembrane helices (TM1–TM7) connected by three intracellular (i1–i3) and three extracellular (e1–e3) loops. The receptor activation occurs always via coupling to heterotrimeric guanine nucleotide-binding proteins and subsequent kinases activation or ion channel activity regulation [43,44].

The odd muscarinic receptors, M1, M3, and M5, are coupled to the Gq11 protein, which stimulates the IP3 hydrolysis and the intracellular calcium mobilization. They also modulate phospholipase A2 and phospholipase D in different cell lines and primary cultures [45–47]. M2 and M4 receptors are coupled with the Gi protein, which inhibits adenylate cyclase activity, thus reducing cAMP intracellular levels. They are also able to modulate the activity of K^+ channels. Muscarinic receptor subtypes can also activate small G proteins, such as Rho GTPase and recruit new effectors including IP3K and MAPK/ERK kinases. These last signaling pathways appear to play important roles in the control of cell growth and proliferation, as suggested by studies on glial cell primary cultures [48–54]. Moreover these receptors play a strategic role in the control of relevant physiological functions [55,56]. Generally, M2Rs in central nervous system (CNS) are inhibitory auto- or heteroreceptors, regulating several processes such as thermoregulation, cognitive functions, behavioral flexibility, and memory [57–60]. In the last years, the involvement of these receptors has also been investigated in different neurological pathologies such as Alzheimer's and Parkinson's diseases, which are characterized by significant reduction of M1 mAChRs expression [59,60]. The M3 muscarinic receptor (M3R) is located in different districts in the body, e.g., smooth muscles, endocrine and exocrine glands, lungs, pancreas, and brain. Their localization in the brain is relevant in hypothalamus and brainstem [61]. These receptors are highly expressed on pancreatic beta cells and are critical regulators of glucose homeostasis by modulating insulin secretion [62]. M3R mediates the ACh functions in the central and peripheral nervous system, with various relevant physiological functions. However, peripheral M3Rs are critical in mediating ACh functions in smooth muscle and glandular tissues [63–65]. The M4R is primarily located in the CNS and belongs to the family of Gi/o protein-coupled receptors [66,67]. M4R activation through adenylate cyclase inhibition, and potassium channel regulation, acts as inhibitory autoreceptor [68,69]. Dysregulation of M4R may cause different types of mental disorders such as schizophrenia, and neurodegenerative disorders, such as Parkinson's and Alzheimer's diseases [60,68]. The last muscarinic subtype, M5 receptor (M5R), belongs to the Gq protein-coupled receptor family [70]. The M5R is expressed in terminals of dopaminergic neurons. It is the only muscarinic receptor located in the ventral tegmental area (VTA), where it regulates the dopamine and glutamate release from midbrain projections [71–75]. It has also been proved via M5R-deficient mouse that the activation of M5R causes reduction of cerebral vascular tone, demonstrating the ability of AChRs to mediate the cerebral vessel dilation [76–78].

Involvement of Muscarinic Acetylcholine Receptors in Cancer

The distribution of all mAChRs in different tissues, such as nervous system [79], gastrointestinal tract [80], urinary bladder [81,82], heart [83], lung [84], vessels [85], and smooth muscle [86], are largely documented. Considering their involvement in different physiological processes, the alterations of their distribution or function is frequently associated with various pathologies including cancer. In particular, evidence produced in recent years has indicated a dysregulation of mAChRs in various types of human malignancies including breast, colon, prostate, and brain cancers (Table 3) [21,87–90].

The M1 receptor is involved in the progression of the prostate cancer, where it regulated cell migration and invasion through hedgehog signaling regulation [91–94].

Table 3. Muscarinic receptors involved in cancer.

Subtype	Type of Cancer	Function	Signaling Pathways	References
M1	Prostate Cancer	Tumor cell migration and invasion	Hedgehog	[91–94]
M2	Glioblastoma, Neuroblastoma, Breast Cancer, Urothelial Bladder Cancer	Inhibition of cell proliferation, survival, migration, and chemoresistance	Notch1/EGFR	[95–101]
M3	Colon Cancer, Ovarian Cancer, Breast Cancer	Growth and progression of tumor, poor prognosis	ERK1/2 and AKT; EGFR	[102–105]
M4	Oral cancer	Induction of cell migration	SFKs and ERK1/2	[106]

An inhibitory effect on tumor progression was demonstrated for M2 receptor subtype. In particular in human glioblastoma, the M2 receptor stimulation by the preferential orthosteric agonist arecaidine propargyl ester, was reported to inhibit cell proliferation and survival in stable cell lines and in glioblastoma cancer stem cells isolated from human biopsies [92–98]. The same effect has been described in neuroblastoma cell lines [99], in urothelial bladder cancer cells [100], and in breast cancer [101].

Several data were reported on the role of the M3 muscarinic receptor in promoting tumor cell growth, invasion, migration and angiogenesis in different tumor types such as lung, breast, ovarian, and brain cancers [107]. In particular, data on lung cancer demonstrated a direct correlation between lung cancer metastasis/poor prognosis and the levels of M3 receptor [108]. Interestingly, research by Hua and collaborators proved that R2HBJJ, a novel mAChRs antagonist, can block the cholinergic autocrine loop in non-small-cell lung cancer, antagonizing the M3 receptors [109]. M3R is also involved in growth and progression of several tumors when co-expressed with epidermal growth factor receptor (EGFR) [92–94]. The activation of M3R results in transactivation of EGFR, thus promoting cell proliferation and inducing phosphorylation of ERK1/2 and AKT; conversely EGFR inhibition arrests the acetylcholine-induced gastric cancer proliferation [110–112]. In ovarian and breast cancers, the expression of M3R promotes cell viability via ERK1/2 activation; in fact, its expression is directly correlated with poor prognosis [102–105]. To date, there are limited data on M4 and M5 receptors in cancer. M4R has been described to be involved in the migration of oral cancer cells through the downstream signaling effectors, including SFKs and ERK1/2 [102]. Conversely for M5R, no data are at least available. Probably the lack of information concerning the role of M5R in cancer may in part depend on the absence of selective pharmacological ligands able to activate or block the activity of this receptor.

3. Melanoma

In 2012, 67,753 people were diagnosed with melanoma and 9251 individuals died from this type of tumor [103]. At the start of 21st century, melanoma remains a potentially fatal malignancy. Deaths from melanoma represent more than 75% of all skin cancer deaths [104–106,113]. Although the incidence of melanoma is low, as it represents only 1% of all malignant skin tumors, malignant cutaneous melanoma still represents the most aggressive and deadly form of skin cancers, affecting mainly the Caucasian population. Moreover, once metastatic, the prognosis is usually rather poor [105]. Fortunately, the increase of the molecular knowledge concerning the development of this type of cancer is allowing enormous progress towards the improvement of new targets and immunotherapies.

Melanoma arises from genetic mutations in melanocytes, the cells producing the cutaneous pigment, which can be found in the skin, eye, inner ear, and leptomeninges [114,115].

In relation to clinical and histological features, melanoma can be divided into three main subtypes: superficial spreading melanoma (SSM), nodular melanoma (NMM), and lentigo malignant melanoma (LMM) [114].

Superficial spreading melanoma is the most common type that accounts for 70% of the cases approximately. SSM may arise de novo or in association with a nevus [114]. Nodular melanoma accounts for 5% of melanomas and is more common in males than females. NMMs are often ulcerated. It does not have a radial growth phase, but only a vertical growth phase correlated with more rapid growth and higher rate of metastasis. Histologically, NMM is characterized by a predominance of dermal invasive tumor cells [114]. Lentigo malignant melanoma accounts for between 4% and 15% of cutaneous melanomas and, unlike NMM and SSM, correlates with long-term sun exposure and increasing age. LMM may evolve for decades before invading the papillary dermis. Histologically, it is characterized by a proliferation of cells that are in the basal layers of the epidermis [114]. Over the past years, several therapies have been approved by the US Food and Drug Administration (FDA) for melanoma treatment [116–118]. Depending on the features of the tumor (e.g., location, stage, genetic profile), the therapeutic protocol may require surgical resection, chemotherapy, radiotherapy, photodynamic therapy (PDT), and immunotherapy. For patients with stage I–IIIB melanoma, surgery is the primary treatment [118]. At present, there are two kinds of limitations in melanoma therapy. The first one is associated with adverse events, which are commonly linked to toxicity at the level of the skin or gastrointestinal tract, as well as immune reactions. The second one could be a reduced efficiency, which can occur due to resistance to chemo- or intra-lesion therapies [119]. Recently, new therapeutic targets have emerged from studies of the genetic profile of melanocytes and from the identification of molecular factors involved in the pathogenesis of the malignant transformation of the melanocytic cells [120–122]. In this context, it is critical to further investigate how those mechanisms may contribute to melanoma pathogenesis and progression. Hereafter, we summarize studies reporting the role of the ACh and cholinergic receptors in melanoma (Table 4), in order to better define the molecular mechanisms influencing melanoma pathobiology and to identify new possible targets for the treatment of this neoplasia.

Table 4. Receptor subtypes involved in melanoma progression.

Receptor Subtype	Cell Lines	Pathway Involved	Function	Reference
α5	M14 and A375	AKT/Notch-Hes ERK1/2	Cell proliferation Cell migration	[123]
α9	A375, A2058, and MDA-MB 435	STAT3 PD-L1	Cell proliferation Cell migration	[124]
M3	SK-Mel 28	Not investigated	Positive modulation of migration	[125]

4. Acetylcholine Receptors in Melanoma

4.1. The Nicotinic Receptors in Melanoma

The nicotinic receptors have been studied in melanoma cells, highlighting the presence of multiple subunits such as α5/α9 / α3/ β4 [123,124,126,127]. Alpha-5 nicotinic receptors (α5-nAChR), one of the homopentameric nicotinic receptors, have been identified in different forms of solid tumors. Performing an immunocytochemical analysis, it has been demonstrated that the expression of α5-nAChR is significantly increased in human melanoma tissue and melanoma cell lines, compared with normal human skin tissue [123]. The mechanism of action has been identified very recently; via the knockdown of the α5-nAChR expression in melanoma cells, it has been demonstrated that this receptor positively modulates cell proliferation, migration, and invasion. Moreover, the silencing of α5-nAChR has also demonstrated that PI3K-AKT and ERK1 are the signaling pathways activated downstream from the α5 receptors activation. Interestingly, a cross-interaction between α5 and Notch-1 pathways has also been reported. In particular, in absence of α5 receptors, the expression of Notch-1 and Hes1, the main gene downstream Notch activation, was significantly downregulated [124]. These data suggest that α5 nAChR may promote melanoma cell proliferation and metastasis through activation of AKT pathway—Notch1 mediated [124]. In addition, α9-nAChR has proven to be largely involved in promoting cancer progression both in vitro and in vivo [126,127]. In agreement with that observed in the breast cancer, α9-nAChR was also studied in melanoma. Similar to that observed in the case of α5 receptors, α9-nAChR expression was also expressed at higher levels in melanoma cells than in melanocytes. Interestingly, a first association of α9-nAChR expression with clinical-pathological features of patients affected by melanoma (e.g., clinical staging, lymph node status) was recently described [126]. In comparison with a proliferative phenotype, the mRNA levels for α9-subunit were higher in melanoma cells with the invasive phenotype than in those with the proliferative phenotype; this result was confirmed by comparing a metastatic patient group with a primary patient group. The α9-nAChR overexpression in melanoma cells has also demonstrated an increase in cell proliferation and migration; conversely, the suppression of α9-nAChR expression or activity significantly inhibited these events. Taken together, these findings suggest that α9-nAChR expression is not only relevant to melanoma growth and metastasis, but it represents a potentially unfavorable prognostic factor in melanoma patients [126].

Microarray studies and database analysis have suggested in melanoma a co-expressed upregulation of α9-nAChR and PD-L1, a protein overexpressed in melanoma cells and involved in in the regulation of EMT, cancer stemness, tumor development, metastasis formation, and resistance to therapy [127]. The silencing of α9-nAChR and overexpression studies have demonstrated a tight correlation between α9 and PD-L1 expression, suggesting that α9-nAChR may promote melanoma migration through the regulation of PD-L1 [126,127].

Albeit α9-nAChR receptor was not indicated as a prognostic factor for melanoma, different missense mutations causing an altered function of this receptor have been identified. At least 82 different mutations (in particular, 62 are nonsense mutations) have been described in various kinds of melanoma tumors (see www.cBioPortal.org).

4.2. Muscarinic Receptors in Melanoma

The muscarinic receptors are expressed during embryonic development by migrating neural crest cells, some of which are precursors of melanocytes [128]. Once the melanocytes differentiate, they lose the expression of the mAChRs [129] and, in the skin, only the keratinocytes maintain the expression of these receptors [29]. Interestingly, the expression of mAChRs has been demonstrated in human melanomas [129], suggesting that the tumorigenesis processes in melanoma could require the reactivated expression of these receptors.

Immunolocalization data obtained in melanoma cell lines demonstrated the presence of cholinergic markers such as choline acetyltransferase (ChAT) and acetylcholinesterase (AChE) (ACh biosynthetic and degradative enzymes, respectively), demonstrating that melanoma cells may synthesize and degrade ACh in an autonomous manner. Moreover, M3R expression at protein level was evidenced by the use of a specific antibody for M3 receptor. In addition, the presence of mRNA for M3 and M5 subtypes was reported, indicating that melanoma cells express mainly M3 subtype and probably also M5 receptor [129].

Calcium is the second messenger associated with the transduction pathway downstream from the cholinergic activation of the M1, M3, and M5 receptor subtypes. In melanoma cells, fluorometric measurements with Fura-2 have shown that in the presence of the ACh mimetic carbachol (CCh), there was an increase of intracellular Ca^{2+} concentration, indicating the functional activity of these receptors [125].

A chemotaxis measurement assay was also performed on these cells using the Boyden chamber. The melanoma cells were placed in the upper chamber in the absence or in the presence of carbachol and/or atropine (selective muscarinic antagonist), while the chemotactic agent fibronectin was placed in the lower chamber. The time taken by the cells to migrate from the upper to the lower chamber was measured as a chemotactic power of the agent used. It was observed that the addition of carbachol in the upper chamberm increased chemotaxis by 30%, while atropine inhibited it [130]. The migration of these cells was also observed in the presence of CCh or ACh using digital video microscopy; in this condition over a defined time interval, 30% of the cells showed cell body contraction and cell process retraction of more than 5 μm. Once again, the administration of atropine inhibited this contraction and retraction, indicating that the effects observed were mediated by muscarinic receptors [130].

All these evidences clearly suggest that the cholinergic activation of mAChRs on melanoma cells modulates cell migration and that, in general, their expression may be required to control invasion and metastasis formation. This may prompt the potential use of M3 antagonists to inhibit melanoma growth and metastasis, albeit this remains to be further determined. Although the M3 muscarinic receptor subtype function has been well characterized, the presence and role of other muscarinic receptor subtypes have not been thoroughly investigated in melanoma. Whilst a faint expression of M5 subtype has been described in the melanoma cells, its function remains at least unknown [130].

5. Conclusions

The data presented in this review clearly indicate the relevant roles played by cholinergic receptors in the cancer [21]. Comprehensive data have been reported in relation to different tumor types, such as colon and urothelial cancers [100], glioblastoma [92–98], neuroblastoma [99], and breast cancer [101]. Unfortunately, studies reporting the role of cholinergic receptors in melanoma are, at this stage, still limited. However, the effects mediated by M3/M5 muscarinic receptors and α5/α9 nicotinic receptors in different melanoma cell lines suggest that these receptors are involved in the modulation of cell proliferation and migration. In particular, the cell migration mediated by cholinergic receptor appears to be a function that melanoma cells recover from embryogenesis, considering that the precursors of melanocytes use the cholinergic stimuli to modulate their migration [125,128–130]. Although the knowledge on cholinergic system functions should be further investigated, in particular for melanoma cells obtained from human biopsies, it appears evident that acetylcholine and its receptors play an important role in tumor processes, whose better understanding could help to identify new and interesting therapeutic tools for different tumors, including melanoma.

Author Contributions: Conceptualization, A.M.T. and A.M.L.; writing—original draft preparation, A.M.L.; writing—review and editing, A.M.T. All authors have read and agreed to the published version of the manuscript.

Funding: This research received no external funding.

Acknowledgments: This work was supported by Ateneo Sapienza Funds.

Conflicts of Interest: The authors declare no conflict of interest.

References

1. Wessler, I.; Kirkpatrick, C.J. Acetylcholine beyond neurons: The non-neuronal cholinergic system in humans. *Br. J. Pharmacol.* **2008**, *154*, 1558–1571. [CrossRef] [PubMed]
2. Chen, J.; Cheuk, I.W.; Shin, V.Y.; Kwong, A. Acetylcholine receptors: Key players in cancer development. *Surg. Oncol.* **2019**, *31*, 46–53. [CrossRef] [PubMed]
3. Wessler, I.; Kilbinger, H.; Bittinger, F.; Kirkpatrick, C.J. The Biological Role of Non-neuronal Acetylcholine in Plants and Humans. *Jpn. J. Pharmacol.* **2001**, *85*, 2–10. [CrossRef] [PubMed]
4. Horiuchi, Y.; Kimura, R.; Kato, N.; Fujii, T.; Seki, M.; Endo, T.; Kato, T.; Kawashima, K. Evolutional study on acetylcholine expression. *Life Sci.* **2003**, *72*, 1745–1756. [CrossRef]
5. Wessler, I.; Kirkpatrick, C.; Racké, K. The cholinergic 'pitfall': Acetylcholine, a universal cell molecule in biological systems, including humans. *Clin. Exp. Pharmacol. Physiol.* **1999**, *26*, 198–205. [CrossRef]
6. Grando, S.A.; Kawashima, K.; Wessler, I. A historic perspective on the current progress in elucidation of the biologic significance of non-neuronal acetylcholine. *Int. Immunopharmacol.* **2020**, *81*, 106289. [CrossRef] [PubMed]
7. Salceda, R. Muscarinic receptors binding in retinal pigment epithelium during rat development. *Neurochem. Res.* **1994**, *19*, 1207–1210. [CrossRef] [PubMed]
8. Genaro, A.M.; Cremaschi, G.A.; Borda, E.S. Muscarinic cholinergic receptors on murine lympocyte subpopulations. *Immunopharmacology* **1993**, *26*, 21–29. [CrossRef]
9. Lane, M.-A. Muscarinic Cholinergic Activation of Mouse Spleen Cells Cytotoxic to Tumor Cells In Vitro2. *J. Natl. Cancer Inst.* **1978**, *61*. [CrossRef]
10. Yanagihara, N.; Isosaki, M.; Ohuchi, T.; Oka, M. Muscarinic receptor-mediated increase in cyclic GMP level in isolated bovine adrenal medullary cells. *FEBS Lett.* **1979**, *105*, 296–298. [CrossRef]
11. Hammer, R. Muscarinic receptors in the stomach. *Scand. J. Gastroenterol. Suppl.* **1980**, *66*, 5–11. [PubMed]
12. Engel, W.; Trotter, J.; McFarlin, D.; McIntosh, C. Thymic epithelial cell contains acetylcholine receptor. *Lancet* **1977**, *309*, 1310–1311. [CrossRef]
13. Oettling, G.; Götz, U.; Drews, U. Characterization of the Ca^{2+} influx into embryonic cells after stimulation of the embryonic muscarinic receptor. *J. Dev. Physiol.* **1992**, *17*, 147–155. [PubMed]
14. Buznikov, G.A.; Shmukler, Y.B.; Lauder, J.M. From oocyte to neuron: Do neurotransmitters function in the same way throughout development? *Cell. Mol. Neurobiol.* **1996**, *16*, 533–559. [CrossRef]
15. Grando, S.A.; Kawashima, K.; Kirkpatrick, C.J.; Wessler, I. Recent progress in understanding the non-neuronal cholinergic system in humans. *Life Sci.* **2007**, *80*, 2181–2185. [CrossRef]
16. Piovesana, R.; Melfi, S.; Fiore, M.; Emagnaghi, V.; Tata, A.M. M2 muscarinic receptor activation inhibits cell proliferation and migration of rat adipose-mesenchymal stem cells. *J. Cell. Physiol.* **2018**, *233*, 5348–5360. [CrossRef]
17. Fujii, T.; Mashimo, M.; Moriwaki, Y.; Misawa, H.; Ono, S.; Horiguchi, K.; Kawashima, K. Expression and Function of the Cholinergic System in Immune Cells. *Front. Immunol.* **2017**, *8*, 1085. [CrossRef]
18. Tracey, K.J. The inflammatory reflex. *Nat. Cell Biol.* **2002**, *420*, 853–859. [CrossRef]
19. Kawashima, K.; Fujii, T. Extraneuronal cholinergic system in lymphocytes. *Pharmacol. Ther.* **2000**, *86*, 29–48. [CrossRef]
20. De Simone, R.; Ajmone-Cat, M.A.; Carnevale, D.; Minghetti, L. Activation of α7 nicotinic acetylcholine receptor by nicotine selectively up-regulates cyclooxygenase-2 and prostaglandin E2 in rat microglial cultures. *J. Neuroinflamm.* **2005**, *2*, 4. [CrossRef]
21. Tata, A.M. Muscarinic Acetylcholine Receptors: New Potential Therapeutic Targets in Antinociception and in Cancer Therapy. *Recent Patents CNS Drug Discov.* **2008**, *3*, 94–103. [CrossRef]
22. Cheng, K.; Samimi, R.; Xie, G.; Shant, J.; Drachenberg, C.; Wade, M.; Davis, R.J.; Nomikos, G.; Raufman, J.-P. Acetylcholine release by human colon cancer cells mediates autocrine stimulation of cell proliferation. *Am. J. Physiol. Liver Physiol.* **2008**, *295*, G591–G597. [CrossRef]
23. Wang, L.; Zhi, X.; Zhang, Q.; Wei, S.; Li, Z.; Zhou, J.; Jianguo, J.; Zhu, Y.; Yang, L.; Xu, H.; et al. Muscarinic receptor M3 mediates cell proliferation induced by acetylcholine and contributes to apoptosis in gastric cancer. *Tumor Biol.* **2015**, *37*, 2105–2117. [CrossRef] [PubMed]

24. Mark, M.R.; Domino, E.F.; Han, S.S.; Ortiz, A.; Mathews, B.N.; Tait, S.K. Effect of parasympathetic denervation on acetylcholine levels in the rat parotid gland. Is there an extraneuronal pool of acetylcholine? *Life Sci.* **1983**, *33*, 1191–1197. [CrossRef]

25. Grando, S.A. Biological functions of keratinocyte cholinergic receptors. *J. Investig. Dermatol. Symp. Proc.* **1997**, *2*, 41–48. [CrossRef] [PubMed]

26. Nguyen, V.T.; Hall, L.; Gallacher, G.; Ndoye, A.; Jolkovsky, D.; Webber, R.; Buchli, R.; Grando, S. Choline Acetyltransferase, Acetylcholinesterase, and Nicotinic Acetylcholine Receptors of Human Gingival and Esophageal Epithelia. *J. Dent. Res.* **2000**, *79*, 939–949. [CrossRef]

27. Arredondo, J.; Hall, L.L.; Ndoye, A.; Chernyavsky, A.I.; Jolkovsky, D.L.; Grando, S.A. Muscarinic acetylcholine receptors regulating cell cycle progression are expressed in human gingival keratinocytes. *J. Periodontal Res.* **2003**, *38*, 79–89. [CrossRef]

28. Grando, S.A.; Kist, D.A.; Qi, M.; Dahl, M.V. Human Keratinocytes Synthesize, Secrete, and Degrade Acetylcholine. *J. Investig. Dermatol.* **1993**, *101*, 32–36. [CrossRef]

29. Chernyavsky, A.I.; Arredondo, J.; Wess, J.; Karlsson, E.; Grando, S.A. Novel signaling pathways mediating reciprocal control of keratinocyte migration and wound epithelialization through M3 and M4 muscarinic receptors. *J. Cell Biol.* **2004**, *166*, 261–272. [CrossRef]

30. Dang, N.; Meng, X.; Song, H. Nicotinic acetylcholine receptors and cancer. *Biomed. Rep.* **2016**, *4*, 515–518. [CrossRef]

31. Egleton, R.D.; Brown, K.C.; Dasgupta, P. Nicotinic acetylcholine receptors in cancer: Multiple roles in proliferation and inhibition of apoptosis. *Trends Pharmacol. Sci.* **2008**, *29*, 151–158. [CrossRef]

32. Albuquerque, E.X.; Pereira, E.F.R.; Alkondon, M.; Rogers, S.W. Mammalian Nicotinic Acetylcholine Receptors: From Structure to Function. *Physiol. Rev.* **2009**, *89*, 73–120. [CrossRef]

33. Karlin, A. Emerging structure of the Nicotinic Acetylcholine receptors. *Nat. Rev. Neurosci.* **2002**, *3*, 102–114. [CrossRef]

34. Heusch, W.L. Signalling pathways involved in nicotine regulation of apoptosis of human lung cancer cells. *Carcinogenesis* **1998**, *19*, 551–556. [CrossRef]

35. Nakayama, H.; Numakawa, T.; Ikeuchi, T. Nicotine-induced phosphorylation of Akt through epidermal growth factor receptor and Src in PC12h cells. *J. Neurochem.* **2002**, *83*, 1372–1379. [CrossRef]

36. Catassi, A.; Servent, D.; Paleari, L.; Cesario, A.; Russo, P. Multiple roles of nicotine on cell proliferation and inhibition of apoptosis: Implications on lung carcinogenesis. *Mutat. Res. Mutat. Res.* **2008**, *659*, 221–231. [CrossRef]

37. Sacco, K.A.; Bannon, K.L.; George, T.P. Nicotinic receptor mechanisms and cognition in normal states and neuropsychiatric disorders. *J. Psychopharmacol.* **2004**, *18*, 457–474. [CrossRef]

38. Medjber, K.; Freidja, M.L.; Grelet, S.; Lorenzato, M.; Maouche, K.; Nawrocki-Raby, B.; Birembaut, P.; Polette, M.; Tournier, J.-M. Role of nicotinic acetylcholine receptors in cell proliferation and tumour invasion in broncho-pulmonary carcinomas. *Lung Cancer* **2015**, *87*, 258–264. [CrossRef]

39. Song, P.; Sekhon, H.S.; Jia, Y.; Keller, J.A.; Blusztajn, J.K.; Mark, G.P.; Spindel, E. Acetylcholine is synthesized by and acts as an autocrine growth factor for small cell lung carcinoma. *Cancer Res.* **2003**, *63*, 214–221.

40. Yu, H.; Xia, H.; Tang, Q.; Xu, H.; Wei, G.; Chen, Y.; Dai, X.; Gong, Q.; Bi, F. Acetylcholine acts through M3 muscarinic receptor to activate the EGFR signaling and promotes gastric cancer cell proliferation. *Sci. Rep.* **2017**, *7*, 40802. [CrossRef]

41. Grando, S.A. Connections of nicotine to cancer. *Nat. Rev. Cancer* **2014**, *14*, 419–429. [CrossRef]

42. Schuller, H.M. Is cancer triggered by altered signalling of nicotinic acetylcholine receptors? *Nat. Rev. Cancer* **2009**, *9*, 195–205. [CrossRef]

43. Caulfield, M.P.; Birdsall, N.J. International Union of Pharmacology. XVII. Classification of muscarinic acetylcholine receptors. *Pharmacol. Rev.* **1998**, *50*, 279–290.

44. Hulme, E.C.; Birdsall, N.J.M.; Buckley, N.J. Muscarinic Receptor Subtypes. *Annu. Rev. Pharmacol. Toxicol.* **1990**, *30*, 633–673. [CrossRef]

45. Peralta, E.G.; Ashkenazi, A.; Winslow, J.W.; Smith, D.H.; Ramachandran, J.; Capon, D.J. Distinct primary structures, ligand-binding properties and tissue-specific expression of four human muscarinic acetylcholine receptors. *EMBO J.* **1987**, *6*, 3923–3929. [CrossRef]

46. Caulfield, M. Muscarinic Receptors—Characterization, coupling and function. *Pharmacol. Ther.* **1993**, *58*, 319–379. [CrossRef]

47. Felder, C.C. Muscarinic acetylcholine receptors: Signal transduction through multiple effectors. *FASEB J.* **1995**, *9*, 619–625. [CrossRef]

48. Horih, S.I. Basic Neurochemistry: Molecular, Cellular, and Medical Aspects, 4th Ed. *Neurology* **1989**, *39*, 460. [CrossRef]

49. Van Koppen, C.J.; Kaiser, B. Regulation of muscarinic acetylcholine receptor signaling. *Pharmacol. Ther.* **2003**, *98*, 197–220. [CrossRef]

50. Cui, Q.-L.; Fogle, E.; Almazan, G. Muscarinic acetylcholine receptors mediate oligodendrocyte progenitor survival through Src-like tyrosine kinases and PI3K/Akt pathways. *Neurochem. Int.* **2006**, *48*, 383–393. [CrossRef]

51. Loreti, S.; Vilaró, M.T.; Visentin, S.; Rees, H.; Levey, A.I.; Tata, A.M. Rat Schwann cells express M1–M4 muscarinic receptor subtypes. *J. Neurosci. Res.* **2006**, *84*, 97–105. [CrossRef]

52. Loreti, S.; Ricordy, R.; De Stefano, M.E.; Augusti-Tocco, G.; Tata, A.M. Acetylcholine inhibits cell cycle progression in rat Schwann cells by activation of the M2 receptor subtype. *Neuron Glia Biol.* **2007**, *3*, 269–279. [CrossRef]

53. De Angelis, F.; Bernardo, A.; Magnaghi, V.; Minghetti, L.; Tata, A.M. Muscarinic receptor subtypes as potential targets to modulate oligodendrocyte progenitor survival, proliferation, and differentiation. *Dev. Neurobiol.* **2012**, *72*, 713–728. [CrossRef] [PubMed]

54. Uggenti, C.; De Stefano, M.E.; Costantino, M.; Loreti, S.; Pisano, A.; Avallone, B.; Talora, C.; Emagnaghi, V.; Tata, A.M. M2 muscarinic receptor activation regulates schwann cell differentiation and myelin organization. *Dev. Neurobiol.* **2014**, *74*, 676–691. [CrossRef] [PubMed]

55. Wess, J.; Zhang, W.; Duttaroy, A.; Miyakawa, T.; Gomeza, J.; Cui, Y.; Basile, A.S.; Bymaster, F.P.; McKinzie, D.L.; Felder, C.C.; et al. Muscarinic Acetylcholine Receptor Knockout Mice. In *Transgenic Models in Pharmacology. Handbook of Experimental Pharmacology*; Offermanns, S., Hein, L., Eds.; Springer: Berlin, Germany, 2004; Volume 159, pp. 65–93. [CrossRef]

56. Brodde, O.E.; Michel, M.C. Adrenergic and muscarinic receptors in the human heart. *Pharmacol. Rev.* **1999**, *51*, 651–690.

57. Höglund, A.U.; Baghdoyan, H.A. M2, M3 and M4, but not M1, muscarinic receptor subtypes are present in rat spinal cord. *J. Pharmacol. Exp. Ther.* **1997**, *281*, 470–477. [PubMed]

58. Levey, A.I. Immunological localization of m1–m5 muscarinic acetylcholine receptors in peripheral tissues and brain. *Life Sci.* **1993**, *52*, 441–448. [CrossRef]

59. Manuel, I.; Lombardero, L.; LaFerla, F.M.; Giménez-Llort, L.; Rodríguez-Puertas, R. Activity of muscarinic, galanin and cannabinoid receptors in the prodromal and advanced stages in the triple transgenic mice model of Alzheimer's disease. *Neuroscience* **2016**, *329*, 284–293. [CrossRef]

60. Bertrand, D.; Wallace, T.L. A Review of the Cholinergic System and Therapeutic Approaches to Treat Brain Disorders. *Biol. Basis Sex Differ. Psychopharmacol.* **2020**, *45*, 1–28. [CrossRef]

61. Weston-Green, K.; Huang, X.-F.; Lian, J.; Deng, C. Effects of olanzapine on muscarinic M3 receptor binding density in the brain relates to weight gain, plasma insulin and metabolic hormone levels. *Eur. Neuropsychopharmacol.* **2012**, *22*, 364–373. [CrossRef]

62. Gautam, D.; Han, S.-J.; Hamdan, F.F.; Jeon, J.; Li, B.; Li, J.H.; Cui, Y.; Mears, D.; Lu, H.; Deng, C.; et al. A critical role for β cell M3 muscarinic acetylcholine receptors in regulating insulin release and blood glucose homeostasis in vivo. *Cell Metab.* **2006**, *3*, 449–461. [CrossRef]

63. Carroll, R.C.; Peralta, E.G. The m3 muscarinic acetylcholine receptor differentially regulates calcium influx and release through modulation of monovalent cation channels. *EMBO J.* **1998**, *17*, 3036–3044. [CrossRef]

64. Tran, J.A.; Matsui, M.; Ehlert, F.J. Differential Coupling of Muscarinic M1, M2, and M3 Receptors to Phosphoinositide Hydrolysis in Urinary Bladder and Longitudinal Muscle of the Ileum of the Mouse. *J. Pharmacol. Exp. Ther.* **2006**, *318*, 649–656. [CrossRef] [PubMed]

65. Kan, W.; Adjobo-Hermans, M.J.W.; Burroughs, M.; Faibis, G.; Malik, S.; Tall, G.G.; Smrcka, A.V. M3 Muscarinic Receptor Interaction with Phospholipase C β3 Determines Its Signaling Efficiency*. *J. Biol. Chem.* **2014**, *289*, 11206–11218. [CrossRef] [PubMed]

66. Weiner, D.M.; Levey, A.I.; Brann, M.R. Expression of muscarinic acetylcholine and dopamine receptor mRNAs in rat basal ganglia. *Proc. Natl. Acad. Sci. USA* **1990**, *87*, 7050–7054. [CrossRef] [PubMed]

67. Yasuda, R.P.; Ciesla, W.; Flores, L.R.; Wall, S.J.; Li, M.; A Satkus, S.; Weisstein, J.S.; Spagnola, B.V.; Wolfe, B.B. Development of antisera selective for m4 and m5 muscarinic cholinergic receptors: Distribution of m4 and m5 receptors in rat brain. *Mol. Pharmacol.* **1993**, *43*, 149–157.

68. Levey, A.I.; Edmunds, S.M.; Koliatsos, V.; Wiley, R.G.; Heilman, C.J. Expression of m1-m4 muscarinic acetylcholine receptor proteins in rat hippocampus and regulation by cholinergic innervation. *J. Neurosci.* **1995**, *15*, 4077–4092. [CrossRef]

69. D'Agostino, G.; Barbieri, A.; Chiossa, E.; Tonini, M. M4 muscarinic autoreceptor-mediated inhibition of -3H-acetylcholine release in the rat isolated urinary bladder. *J. Pharmacol. Exp. Ther.* **1997**, *283*, 750–756.

70. Langmead, C.J.; Watson, J.; Reavill, C. Muscarinic acetylcholine receptors as CNS drug targets. *Pharmacol. Ther.* **2008**, *117*, 232–243. [CrossRef]

71. Scarr, E.; Dean, B. Muscarinic receptors: Do they have a role in the pathology and treatment of schizophrenia? *J. Neurochem.* **2008**, *107*, 1188–1195. [CrossRef]

72. Qin, K.; Dong, C.; Wu, G.; Lambert, N.A. Inactive-state preassembly of Gq-coupled receptors and Gq heterotrimers. *Nat. Chem. Biol.* **2011**, *7*, 740–747. [CrossRef] [PubMed]

73. Zhu, X.; Jiang, M.; Peyton, M.; Boulay, G.; Hurst, R.; Stefani, E.; Birnbaumer, L. trp, a Novel Mammalian Gene Family Essential for Agonist-Activated Capacitative Ca^{2+} Entry. *Cell* **1996**, *85*, 661–671. [CrossRef]

74. Shin, J.H.; Adrover, M.F.; Wess, J.; Alvarez, V.A. Muscarinic regulation of dopamine and glutamate transmission in the nucleus accumbens. *Proc. Natl. Acad. Sci. USA* **2015**, *112*, 8124–8129. [CrossRef] [PubMed]

75. Garzón, M.; Pickel, V.M. Somatodendritic targeting of M5 muscarinic receptor in the rat ventral tegmental area: Implications for mesolimbic dopamine transmission. *J. Comp. Neurol.* **2013**, *521*, 2927–2946. [CrossRef]

76. Phillips, J.K.; Vidovic, M.; Hill, C.E. Variation in mRNA expression of alpha-adrenergic, neurokinin and muscarinic receptors amongst four arteries of the rat. *J. Auton. Nerv. Syst.* **1997**, *62*, 85–93. [CrossRef]

77. Sato, A.; Sato, Y. Cholinergic Neural Regulation of Regional Cerebral Blood Flow. *Alzheimer Dis. Assoc. Disord.* **1995**, *9*, 28–38. [CrossRef]

78. Scremin, O.U.; Jenden, D.J. Cholinergic control of cerebral blood flow in stroke, trauma and aging. *Life Sci.* **1996**, *58*, 2011–2018. [CrossRef]

79. Abrams, P.; Andersson, K.-E.; Buccafusco, J.J.; Chapple, C.; De Groat, W.C.; Fryer, A.D.; Kay, G.; Laties, A.; Nathanson, N.M.; Pasricha, P.J.; et al. Muscarinic receptors: Their distribution and function in body systems, and the implications for treating overactive bladder. *Br. J. Pharmacol.* **2006**, *148*, 565–578. [CrossRef]

80. Gomez, A.; Martos, F.; Bellido, I.; Marquez, E.; Garcia, A.J.; Pavia, J.; De La Cuesta, F.S. Muscarinic receptor subtypes in human and rat colon smooth muscle. *Biochem. Pharmacol.* **1992**, *43*, 2413–2419. [CrossRef]

81. Hegde, S.S.; Eglen, R.M. Muscarinic receptor subtypes modulating smooth muscle contractility in the urinary bladder. *Life Sci.* **1999**, *64*, 419–428. [CrossRef]

82. Wang, P.; Luthin, G.R.; Ruggieri, M.R. Muscarinic acetylcholine receptor subtypes mediating urinary bladder contractility and coupling to GTP binding proteins. *J. Pharmacol. Exp. Ther.* **1995**, *273*, 959–966. [PubMed]

83. Wang, Z.; Shi, H.; Wang, H. Functional M3 muscarinic acetylcholine receptors in mammalian hearts. *Br. J. Pharmacol.* **2004**, *142*, 395–408. [CrossRef] [PubMed]

84. Fryer, A.D.; Jacoby, D.B. Muscarinic Receptors and Control of Airway Smooth Muscle. *Am. J. Respir. Crit. Care Med.* **1998**, *158*, S154–S160. [CrossRef] [PubMed]

85. Yamada, M.; Lamping, K.G.; Duttaroy, A.; Zhang, W.; Cui, Y.; Bymaster, F.P.; McKinzie, D.L.; Felder, C.C.; Deng, C.-X.; Faraci, F.M.; et al. Cholinergic dilation of cerebral blood vessels is abolished in M5 muscarinic acetylcholine receptor knockout mice. *Proc. Natl. Acad. Sci. USA* **2001**, *98*, 14096–14101. [CrossRef] [PubMed]

86. Kitazawa, T.; Hirama, R.; Masunaga, K.; Nakamura, T.; Asakawa, K.; Cao, J.; Teraoka, H.; Unno, T.; Komori, S.-I.; Yamada, M.; et al. Muscarinic receptor subtypes involved in carbachol-induced contraction of mouse uterine smooth muscle. *Naunyn-Schmiedeberg's Arch. Pharmacol.* **2007**, *377*, 503–513. [CrossRef]

87. Shah, N.; Khurana, S.; Cheng, K.; Raufman, J.-P. Muscarinic receptors and ligands in cancer. *Am. J. Physiol. Physiol.* **2009**, *296*, C221–C232. [CrossRef]

88. Spindel, E.R. Muscarinic Receptor Agonists and Antagonists: Effects on Cancer. *Novel Antischizophrenia Treatments* **2011**, 451–468. [CrossRef]

89. Baig, A.M.; Khan, N.A.; Effendi, V.; Rana, Z.; Ahmad, H.; Abbas, F. Differential receptor dependencies. *Anti-Cancer Drugs* **2017**, *28*, 75–87. [CrossRef]

90. Magnon, C.; Hall, S.J.; Lin, J.; Xue, X.; Gerber, L.; Freedland, S.J.; Frenette, P.S. Autonomic Nerve Development Contributes to Prostate Cancer Progression. *Science* **2013**, *341*, 1236361. [CrossRef]

91. Xu, C.; Yin, Q.-Q.; Xu, L.-H.; Zhang, M. Muscarinic acetylcholine receptor M1 mediates prostate cancer cell migration and invasion through hedgehog signaling. *Asian J. Androl.* **2018**, *20*, 608–614. [CrossRef]

92. Cristofaro, I.; Alessandrini, F.; Spinello, Z.; Guerriero, C.; Fiore, M.; Caffarelli, E.; Laneve, P.; Dini, L.; Conti, L.; Tata, A.M. Cross Interaction between M2 Muscarinic Receptor and Notch1/EGFR Pathway in Human Glioblastoma Cancer Stem Cells: Effects on Cell Cycle Progression and Survival. *Cells* **2020**, *9*, 657. [CrossRef] [PubMed]

93. Cristofaro, I.; Limongi, C.; Piscopo, P.; Crestini, A.; Guerriero, C.; Fiore, M.; Conti, L.; Confaloni, A.; Tata, A.M. M2 Receptor Activation Counteracts the Glioblastoma Cancer Stem Cell Response to Hypoxia Condition. *Int. J. Mol. Sci.* **2020**, *21*, 1700. [CrossRef] [PubMed]

94. Di Bari, M.; Bevilacqua, V.; De Jaco, A.; Laneve, P.; Piovesana, R.; Trobiani, L.; Talora, C.; Caffarelli, E.; Tata, A.M. Mir-34a-5p Mediates Cross-Talk between M2 Muscarinic Receptors and Notch-1/EGFR Pathways in U87MG Glioblastoma Cells: Implication in Cell Proliferation. *Int. J. Mol. Sci.* **2018**, *19*, 1631. [CrossRef] [PubMed]

95. Cristofaro, I.; Spinello, Z.; Matera, C.; Fiore, M.; Conti, L.; De Amici, M.; Dallanoce, C.; Tata, A.M. Activation of M2 muscarinic acetylcholine receptors by a hybrid agonist enhances cytotoxic effects in GB7 glioblastoma cancer stem cells. *Neurochem. Int.* **2018**, *118*, 52–60. [CrossRef]

96. Di Bari, M.; Tombolillo, V.; Conte, C.; Castigli, E.; Sciaccaluga, M.; Iorio, E.; Carpinelli, G.; Ricordy, R.; Fiore, M.; Degrassi, F.; et al. Cytotoxic and genotoxic effects mediated by M2 muscarinic receptor activation in human glioblastoma cells. *Neurochem. Int.* **2015**, *90*, 261–270. [CrossRef]

97. Alessandrini, F.; Cristofaro, I.; Di Bari, M.; Zasso, J.; Conti, L.; Tata, A.M. The activation of M2 muscarinic receptor inhibits cell growth and survival in human glioblastoma cancer stem cells. *Int. Immunopharmacol.* **2015**, *29*, 105–109. [CrossRef]

98. Ferretti, M.; Fabbiano, C.; Di Bari, M.; Conte, C.; Castigli, E.; Sciaccaluga, M.; Ponti, D.; Ruggieri, P.; Raco, A.; Ricordy, R.; et al. M2 receptor activation inhibits cell cycle progression and survival in human glioblastoma cells. *J. Cell. Mol. Med.* **2013**, *17*, 552–566. [CrossRef]

99. Lucianò, A.M.; Mattei, F.; Damo, E.; Panzarini, E.; Dini, L.; Tata, A.M. Effects mediated by M2 muscarinic orthosteric agonist on cell growth in human neuroblastoma cell lines. *Pure Appl. Chem.* **2019**, *91*, 1641–1650. [CrossRef]

100. Pacini, L.; De Falco, E.; Di Bari, M.; Coccia, A.; Siciliano, C.; Ponti, D.; Pastore, A.L.; Petrozza, V.; Carbone, A.; Tata, A.M.; et al. M2muscarinic receptors inhibit cell proliferation and migration in urothelial bladder cancer cells. *Cancer Biol. Ther.* **2014**, *15*, 1489–1498. [CrossRef]

101. Español, A.J.; Salem, A.; Di Bari, M.; Cristofaro, I.; Sanchez, Y.; Tata, A.M.; Sales, M.E. The metronomic combination of paclitaxel with cholinergic agonists inhibits triple negative breast tumor progression. Participation of M2 receptor subtype. *PLoS ONE* **2020**, *15*, e0226450. [CrossRef]

102. Chiu, C.-C.; Chen, B.-H.; Hour, T.-C.; Chiang, W.-F.; Wu, Y.-J.; Chen, C.-Y.; Chen, H.-R.; Chan, P.-T.; Liu, S.-Y.; Chen, J.Y.-F. Betel quid extract promotes oral cancer cell migration by activating a muscarinic M4 receptor-mediated signaling cascade involving SFKs and ERK1/2. *Biochem. Biophys. Res. Commun.* **2010**, *399*, 60–65. [CrossRef]

103. U.S. Census Bureau. United States Cancer Statistics: 1999–2012 Incidence and Mortality Web-Based Report. In Invasive Cancer Counts. 2015. Available online: https://wonder.cdc.gov/wonder/help/cancer-v2013.html (accessed on 19 November 2019).

104. Ward, W.H.; Farma, J.M. *Cutaneous Melanoma: Etiology and Therapy*; Ward, W.H., Farma, J.M., Eds.; Codon Publications: Brisbane, Australia, 2017. [CrossRef]

105. Ferlay, J.; Soerjomataram, I.; Dikshit, R.; Eser, S.; Mathers, C.; Rebelo, M.; Parkin, D.M.; Forman, D.; Bray, F. Cancer incidence and mortality worldwide: Sources, methods and major patterns in GLOBOCAN 2012. *Int. J. Cancer* **2015**, *136*, E359–E386. [CrossRef] [PubMed]

106. Grayschopfer, V.C.; Wellbrock, C.; Marais, R. Melanoma biology and new targeted therapy. *Nat. Cell Biol.* **2007**, *445*, 851–857. [CrossRef]

107. Sehdev, V.; Betzu, N.J.J.M.; Desai, V.S.B. Selective Targeting of M3 Muscarinic Receptors: An Opportunity for Improved Treatment of Upper Gastrointestinal Carcinomas. *Pharm. Pharmacol. Int. J.* **2014**, *1*. [CrossRef]

108. Lin, G.; Sun, L.; Wang, R.; Guo, Y.; Xie, C. Overexpression of Muscarinic Receptor 3 Promotes Metastasis and Predicts Poor Prognosis in Non–Small-Cell Lung Cancer. *J. Thorac. Oncol.* **2014**, *9*, 170–178. [CrossRef]

109. Hua, N.; Wei, X.; Liu, X.; Ma, X.; He, X.; Zhuo, R.; Zhao, Z.; Wang, L.; Yan, H.; Zhong, B.; et al. A Novel Muscarinic Antagonist R2HBJJ Inhibits Non-Small Cell Lung Cancer Cell Growth and Arrests the Cell Cycle in G0/G1. *PLoS ONE* **2012**, *7*, e53170. [CrossRef]

110. Oppitz, M.; Möbus, V.; Brock, S.; Drews, U. Muscarinic Receptors in Cell Lines from Ovarian Carcinoma: Negative Correlation with Survival of Patients. *Gynecol. Oncol.* **2002**, *85*, 159–164. [CrossRef]

111. Jiménez, E.; Montiel, M. Activation of MAP kinase by muscarinic cholinergic receptors induces cell proliferation and protein synthesis in human breast cancer cells. *J. Cell. Physiol.* **2005**, *204*, 678–686. [CrossRef]

112. Schmitt, J.M.; Abell, E.; Wagner, A.; Davare, M.A. ERK activation and cell growth require CaM kinases in MCF-7 breast cancer cells. *Mol. Cell. Biochem.* **2009**, *335*, 155–171. [CrossRef] [PubMed]

113. Tolleson, W.H. Human Melanocyte Biology, Toxicology, and Pathology. *J. Environ. Sci. Health Part C* **2005**, *23*, 105–161. [CrossRef] [PubMed]

114. Rastrelli, M.; Tropea, S.; Rossi, C.R.; Alaibac, M. Melanoma: Epidemiology, risk factors, pathogenesis, diagnosis and classification. *In Vivo* **2014**, *28*, 1005–1011. [PubMed]

115. Markovic, S.N.; Erickson, L.A.; Rao, R.D.; McWilliams, R.R.; Kottschade, L.A.; Creagan, E.T.; Weenig, R.H.; Hand, J.L.; Pittelkow, M.R.; Pockaj, B.A.; et al. Malignant Melanoma in the 21st Century, Part 1: Epidemiology, Risk Factors, Screening, Prevention, and Diagnosis. In *Proceedings of the Mayo Clinic Proceedings*; Elsevier B.V.: Amsterdam, The Netherlands, 2007; Volume 82, pp. 364–380.

116. Batus, M.; Waheed, S.; Ruby, C.; Petersen, L.; Bines, S.D.; Kaufman, H.L. Optimal Management of Metastatic Melanoma: Current Strategies and Future Directions. *Am. J. Clin. Dermatol.* **2013**, *14*, 179–194. [CrossRef]

117. Domingues, B.; Lopes, J.M.; Soares, P.; Populo, H. Melanoma treatment in review. *ImmunoTargets Ther.* **2018**, *7*, 35–49. [CrossRef]

118. Huang, Y.-Y.; Vecchio, D.; Avci, P.; Yin, R.; Garcia-Diaz, M.; Hamblin, M.R. Melanoma resistance to photodynamic therapy: New insights. *Biol. Chem.* **2013**, *394*, 239–250. [CrossRef] [PubMed]

119. Widakowich, C.; de Castro, G., Jr.; De Azambuja, E.; Dinh, P.; Awada, A. Review: Side Effects of Approved Molecular Targeted Therapies in Solid Cancers. *Oncologist* **2007**, *12*, 1443–1455. [CrossRef] [PubMed]

120. Ko, J.M.; Fisher, D.E. A new era: Melanoma genetics and therapeutics. *J. Pathol.* **2010**, *223*, 242–251. [CrossRef]

121. Song, F.; Qureshi, A.A.; Gao, X.; Li, T.; Han, J. Smoking and risk of skin cancer: A prospective analysis and a meta-analysis. *Int. J. Epidemiol.* **2012**, *41*, 1694–1705. [CrossRef] [PubMed]

122. Wu, W.; Liu, H.; Song, F.; Chen, L.-S.; Kraft, P.; Wei, Q.; Han, J. Associations between smoking behavior-related alleles and the risk of melanoma. *Oncotarget* **2016**, *7*, 47366–47375. [CrossRef]

123. Dang, N.; Meng, X.; Qin, G.; An, Y.; Zhang, Q.; Cheng, X.; Huang, S. α5-nAChR modulates melanoma growth through the Notch1 signaling pathway. *J. Cell. Physiol.* **2020**, *235*, 7816–7826. [CrossRef]

124. Lee, C.-H.; Huang, C.-S.; Chen, C.-S.; Tu, S.-H.; Wang, Y.-J.; Chang, Y.-J.; Tam, K.-W.; Wei, P.-L.; Cheng, T.-C.; Chu, J.-S.; et al. Overexpression and Activation of the α9-Nicotinic Receptor During Tumorigenesis in Human Breast Epithelial Cells. *J. Natl. Cancer Inst.* **2010**, *102*, 1322–1335. [CrossRef]

125. Boss, A.; Oppitz, M.; Drews, U. Muscarinic cholinergic receptors in the human melanoma cell line SK-Mel 28: Modulation of chemotaxis. *Clin. Exp. Dermatol.* **2005**, *30*, 557–564. [CrossRef]

126. Huang, L.-C.; Lin, C.-L.; Qiu, J.-Z.; Lin, C.-Y.; Hsu, K.-W.; Tam, K.-W.; Lee, J.-Y.; Yang, J.-M.; Lee, C.-H. Nicotinic Acetylcholine Receptor Subtype Alpha-9 Mediates Triple-Negative Breast Cancers Based on a Spontaneous Pulmonary Metastasis Mouse Model. *Front. Cell. Neurosci.* **2017**, *11*. [CrossRef]

127. Nguyen, H.D.; Liao, Y.-C.; Ho, Y.-S.; Chen, L.-C.; Chang, H.-W.; Cheng, T.-C.; Liu, D.; Lee, W.-R.; Shen, S.-C.; Wu, C.-H.; et al. The α9 Nicotinic Acetylcholine Receptor Mediates Nicotine-Induced PD-L1 Expression and Regulates Melanoma Cell Proliferation and Migration. *Cancers* **2019**, *11*, 1991. [CrossRef] [PubMed]

128. Buchli, R.; Ndoye, A.; Arredondo, J.; Webber, R.J.; Grando, S.A. Identification and characterization of muscarinic acetylcholine receptor subtypes expressed in human skin melanocytes. *Mol. Cell. Biochem.* **2001**, *228*, 57–72. [CrossRef] [PubMed]

129. Lammerding-Köppel, M.; Noda, S.; Blum, A.; Schaumburg-Lever, G.; Rassner, G.; Drews, U. Immunohistochemical localization of muscarinic acetylcholine receptors in primary and metastatic malignant melanomas. *J. Cutan. Pathol.* **1997**, *24*, 137–144. [CrossRef] [PubMed]

130. Sailer, M.; Oppitz, M.; Drews, U. Induction of cellular contractions in the human melanoma cell line SK-mel 28 after muscarinic cholinergic stimulation. *Anat. Embryol.* **2000**, *201*, 27–37. [CrossRef] [PubMed]

Publisher's Note: MDPI stays neutral with regard to jurisdictional claims in published maps and institutional affiliations.

Article

Multimodal Treatment of Advanced Mucosal Melanoma in the Era of Modern Immunotherapy

Pawel Teterycz [1,†], Anna M. Czarnecka [1,*,†], Alice Indini [2], Mateusz J. Spałek [1], Alice Labianca [2], Pawel Rogala [1], Bożena Cybulska-Stopa [3], Pietro Quaglino [4], Umberto Ricardi [5], Serena Badellino [6], Anna Szumera-Ciećkiewicz [7,8], Slawomir Falkowski [1], Mario Mandala [2,‡] and Piotr Rutkowski [1,‡]

[1] Department of Soft Tissue/Bone Sarcoma and Melanoma, Maria Sklodowska-Curie National Research Institute of Oncology, 02-781 Warsaw, Poland; pawel.tetrycz@pib-nio.pl (P.T.); mateusz@spalek.co (M.J.S.); pan.rogal@gmail.com (P.R.); slaw.falkowski@gmail.com (S.F.); piotr.rutkowski@coi.pl (P.R.)

[2] Melanoma Unit, Department of Oncology and Hematology, Papa Giovanni XXIII Hospital, 24127 Bergamo, Italy; alice.indini@gmail.com (A.I.); alabianca@asst-pg23.it (A.L.); mmandala@asst-pg23.it (M.M.)

[3] Maria Skłodowska-Curie National Research Institute—Oncology Center, Krakow Branch, 31-115 Krakow, Poland; bcybulskastopa@vp.pl

[4] Department of Medical Sciences, Dermatologic Clinic, University of Turin, 10126 Turin, Italy; pietro.quaglino@unito.it

[5] Department of Oncology, Radiation Oncology, University of Turin, 10126 Turin, Italy; umberto.ricardi@unito.it

[6] Department of Oncology, Radiotherapy Unit, AOU Città della Salute e della Scienza di Torino, 10126 Turin, Italy; serena.badellino@tiscali.it

[7] Department of Pathology and Laboratory Medicine Maria Sklodowska-Curie Memorial Cancer Center and Institute of Oncology, 02-781 Warsaw, Poland; szumann@gmail.com

[8] Department of Diagnostic Hematology, Institute of Hematology and Transfusion Medicine, 02-776 Warsaw, Poland

* Correspondence: anna.czarnecka@gmail.com or am.czarnecka@pib-nio.pl

† These authors contributed equally to this work.

‡ These authors contributed equally to this work, senior authors.

Received: 30 August 2020; Accepted: 19 October 2020; Published: 26 October 2020

Simple Summary: Immunotherapy revolutionized the treatment of cutaneous melanoma and greatly improved treatment outcomes in this group of patients. Mucosal melanoma is a rare disease, biologically distinct from the cutaneous subtype. There is little real-world data on immunotherapy efficacy in mucosal melanoma. Therefore, we aimed to analyze and describe experiences in mucosal melanoma treatment in five high volume oncology centers in Europe. Furthermore, we evaluated if concomitant radiotherapy may improve the outcomes of these patients. We conclude that immunotherapy with anti-PD1 antibodies is a safe and effective treatment of mucosal melanoma. Concomitant radiotherapy may be beneficial in a selected subgroup of patients with advanced mucosal melanoma.

Abstract: Mucosal melanoma is a rare disease epidemiologically and molecularly distinct from cutaneous melanoma developing from melanocytes located in mucosal membranes. Little is known about its therapy. In this paper, we aimed to evaluate the results of immunotherapy and radiotherapy in a group of patients with advanced mucosal melanoma, based on the experience of five high-volume centers in Poland and Italy. There were 82 patients (53 female, 29 male) included in this retrospective study. The median age in this group was 67.5 (IQR: 57.25–75.75). All patients received anti-PD1 or anti-CTLA4 antibodies in the first or second line of treatment. Twenty-three patients received radiotherapy during anti-PD1 treatment. In the first-line treatment, the median progression-free survival (PFS) reached six months in the anti-PD1 group, which was statistically better than 3.1 months in the other modalities group ($p = 0.004$). The median overall survival (OS) was 16.3 months

(CI: 12.1–22.3) in the whole cohort. Patients who received radiotherapy (RT) during the anti-PD1 treatment had a median PFS of 8.9 months (CI: 7.4–NA), whereas patients treated with single-modality anti-PD1 therapy had a median PFS of 4.2 months (CI: 3.0–7.8); this difference was statistically significant ($p = 0.047$). Anti-PD1 antibodies are an effective treatment option in advanced mucosal melanoma (MM). The addition of RT may have been beneficial in the selected subgroup of mucosal melanoma patients.

Keywords: mucosal melanoma; nivolumab; pembrolizumab; ipilimumab; radiotherapy

1. Introduction

Mucosal melanoma (MM) is a rare disease epidemiologically and molecularly distinct from cutaneous melanoma (CM) developing from melanocytes located in mucosal membranes. Overall, MM represents about 1 to 1.5% of all melanoma cases and 0.03% of all diagnosed cancers [1–3]. MM incidence has been reported globally stable over the last 20 years [4,5]. MM is diagnosed twice more often in Caucasians than in populations with darker skin colors, including African Americans [6,7], but much rarer than in the Asian population. The risk of developing MM increases with age. The majority of patients diagnosed are 60 years of age and older. The median age at diagnosis is 70 years, except for MMs arising in the mouth that affect younger patients more frequently [6,8]. Anatomically, it is mostly diagnosed on the mucous membranes of the head and neck (31% to 55%), anus and rectum (17% to 24%) or the vulva and vagina (18% to 40%) and less frequently in the colon, throat, larynx, lungs, urinary tract, cervix, esophagus or gallbladder [3,8,9]. The incidence of MM is over 80% higher in women than in men due to the relatively high number of genital melanomas diagnosed in women [3]. Risk factors for developing MM are currently unknown since ultraviolet (UV) radiation and viral etiology—cytomegalovirus, Eppstein-Barr virus, human papillomavirus, human herpesvirus—have also been excluded. In MM, no environmental exposures nor carcinogenic viruses have been found pathogenic [10–14]. Unlike cutaneous melanoma, which is characterized by a UV signature, MM harbors distinct molecular features, including a lower incidence of v-Raf murine sarcoma viral oncogene homolog B (*BRAF*) oncogene mutations but a higher incidence of tyrosine-protein kinase KIT (CD117) oncogene mutations, suggesting different genetic etiologies. In general, MM harbors fewer nucleotide substitutions per cell than CM, but more gene amplifications and structural variants than CM; the cause of this chromosomal instability has not yet been clarified [3].

Mucosal melanomas are generally diagnosed in an advanced stage, though they are more aggressive and carry a worse prognosis regardless of the stage at diagnosis. The five-year overall survival (OS) rate for MM is only 25% regardless of stage [15]. Poor treatment results and shorter survival may be associated with a lack of early symptoms or signs, resulting in advanced disease at the time of diagnosis. Insidious anatomical localization, and often amelanotic presentation, result in difficult visual detection and challenging resections with wide, or even negative, margins impossible to achieve. Moreover, rich lymphatic drainage from the mucosal surfaces promote metastases [3,8,15]. MM metastases most often develop in the lungs (54%), liver (35%) and bone (25%) [16].

Due to its rare occurrence, undefined etiopathogenesis and unpredictable clinical course, there are no specific recommendations of MM treatment. Both the European Society for Medical Oncology (ESMO) and National Comprehensive Cancer Network (NCCN) guidelines focus on the important role of surgery and radiotherapy (RT) in this group of patients [2,17]. The preferred therapeutic strategy in MM is still surgical excision. Perioperative RT improves local control but does not improve overall survival (OS), probably because of the high rate of distant relapse [18–20]. RT with definitive intent can also provide satisfactory local control. Definitive RT should be considered in patients who are not candidates for extensive surgery or in cases where adequate resection margin cannot be achieved [21]. Proton therapy and heavy ions deserve special attention [22]. In a retrospective analysis of a cohort of patients with sinonasal MM, the authors observed a 62% five-year local control rate for proton therapy [23]. The understanding of multimodal treatment in MM comes mostly from CM data extrapolation. Phase III clinical trials focused only on MM patients are lacking. Phase III trials with all melanoma patients, regardless of subtype, do not provide satisfactory data on MM due to the low number of cases enrolled. The real-world data on the efficacy of antiprogram med cell death 1 (PD-1) therapies in patients with MM are scarce. Although a growing number of studies suggest the significant benefit of RT as a boost for immunotherapy (ITH) in CM, such observations in MM have not been reported. Therefore, this study aims to assess the efficacy of systemic therapy with the emphasis on ITH in the MM patient population treated outside of clinical trials, and to define the efficacy of the immunotherapy-radiotherapy combination in MM.

2. Results

2.1. Cohort

Eighty-two patients from the participating centers who met inclusion criteria were enrolled. The clinicopathological characteristics of patients and tumors are summarized in Table 1. The median follow-up, as estimated by the reverse Kaplan-Meier method, was 29.0 months (CI: 19.1–41.7). At the time of data analysis—August 2020, 30 (37%) patients remained alive on the treatment or in follow up. It is worth mentioning that in fourteen (17.1%) cases, the systemic treatment was initiated due to unresectable tumors without distant metastases (with the majority of cases in the head and neck region—ten). MM, located in the genitourinary system, was predominantly present in females (female to male ratio = 21:1); no such differences were observed in any other location. The BRAF V600 mutation was detected in 4/33 patients from the head and neck region and in 1/49 patients in other locations (rectum).

Table 1. The clinicopathological characteristics of patients.

Variable	Value	Number (Percentage)
n		82 (100)
Sex (%)	female	53 (64.6)
	male	29 (35.4)
Age (median [IQR])		67.50 [57.25, 75.75]
BRAF V600 mutation(%)	negative	77 (93.9)
	positive	5 (6.1)
Disease stage at the start of the treatment (%)	Localized, nonresectable disease	14 (17.1)
	M1a	6 (7.3)
	M1b	14 (17.1)
	M1c	43 (52.4)
	M1d	5 (6.1)
LDH at the start of the first-line (%)	Normal	43 (52.4)
	Over ULN	31 (37.8)
	Unknown	8 (9.8)
ECOG score at the start of first-line(%)	0	40 (48.8)
	1+	42 (51.2)
Localization (%)	Gastrointestinal system *	27 (33.0)
	Genitourinary tract	22 (26.8)
	Head and neck region	33 (40.2)
First-line treatment (%)	BRAFi+/-MEKi inhibitor	1 (1.2)
	Chemotherapy	5 (6.1)
	Anti-CTLA-4 antibody	13 (15.9)
	Anti-PD1 antibody	63 (76.8)
	Nivolumab	39 (47.6)
	Pembrolizumab	24 (29.2)

Table 1. *Cont.*

Variable	Value	Number (Percentage)
The best response to the first-line (%)	Complete response	4 (4.9)
	Partial response	15 (18.3)
	Stable disease	25 (30.5)
	Progressive disease	37 (45.1)
	Not evaluable	1 (1.2)
LDH at the start of the second-line (%)	Normal	14 (32.6)
	Over ULN	17 (39.5)
	Unknown	12 (27.9)
ECOG score at the start of second-line (%)	0	14 (32.6)
	1+	23 (53.4)
	Unknown	6 (14.0)
Second-line treatment (%)	Chemotherapy	6 (7.3)
	Anti-CTLA-4 antibody	21 (25.6)
	Imatinib	1 (1.2)
	Anti-PD1 antibody	15 (18.3)
	None	39 (47.6)
The best response to the second-line (%)	Complete response	2 (4.7)
	Partial response	1 (9.3)
	Stable disease	11 (25.6)
	Progressive disease	25 (58.1)
	Not evaluable	1 (1.2)
Brain metastases at the start of treatment (%)	Absent	77 (93.9)
	Present	5 (6.1)
Liver metastases at the start of treatment (%)	Absent	61 (74.3)
	Present	21 (25.6)
Number of organs involved at the start of treatment (%)	1	38 (46.3)
	2	23 (28.0)
	≥3	21 (25.6)
Radiotherapy during anti-PD1 treatment (%)	Not performed	65 (79.3)
	Performed	17 (20.7)

* comprised of 23 cases of anorectal mucosal melanoma (MM), two cases of esophageal MM, one case of stomach MM, and once case of MM in the gall bladder. IQR—interquartile range, LDH—lactate dehydrogenase, MEKi—MEK inhibitor, ULN—upper limit of normal, ECOG—Eastern Cooperative Oncology Group.

2.2. Treatment

The summary of administered systemic treatment is shown in Table 1. Most of the patients received anti-PD1 ITH in the first line ($n = 63$, 76.8%). Progression of the disease on the first-line treatment was observed in 63 patients, including 44 treated with anti-PD1 ITH. After progression of the disease (PD), forty-three patients (68%) were eligible for second-line therapy. Among those patients, 21 received ipilimumab, while 15 patients received anti-PD1 ITH antibodies. Overall, only five patients did not undergo anti-PD1 treatment.

2.3. Response

In the first-line treatment, OR and disease control (DC) were observed in 16 (25%) and 37 (59%) patients in the anti-PD1 ITH group ($n = 63$); and 3 (15%) and 7 (37%) patients in the other treatments group ($n = 19$), respectively. In the second and subsequent treatment lines, OR lasting longer than one year was observed only in two patients in the ipilimumab group ($n = 21$), and in four patients treated with anti-PD1 ITH ($n = 15$).

In the second and subsequent treatment lines, DC was observed in six patients in the ipilimumab group ($n = 21$) in eight patients in the anti-PD1 ITH group ($n = 15$), two patients treated with dacarbazine-based chemotherapy and one treated with imatinib. The duration of response to anti-PD1 ITH has been visualized as a swimmer plot in Figure 1.

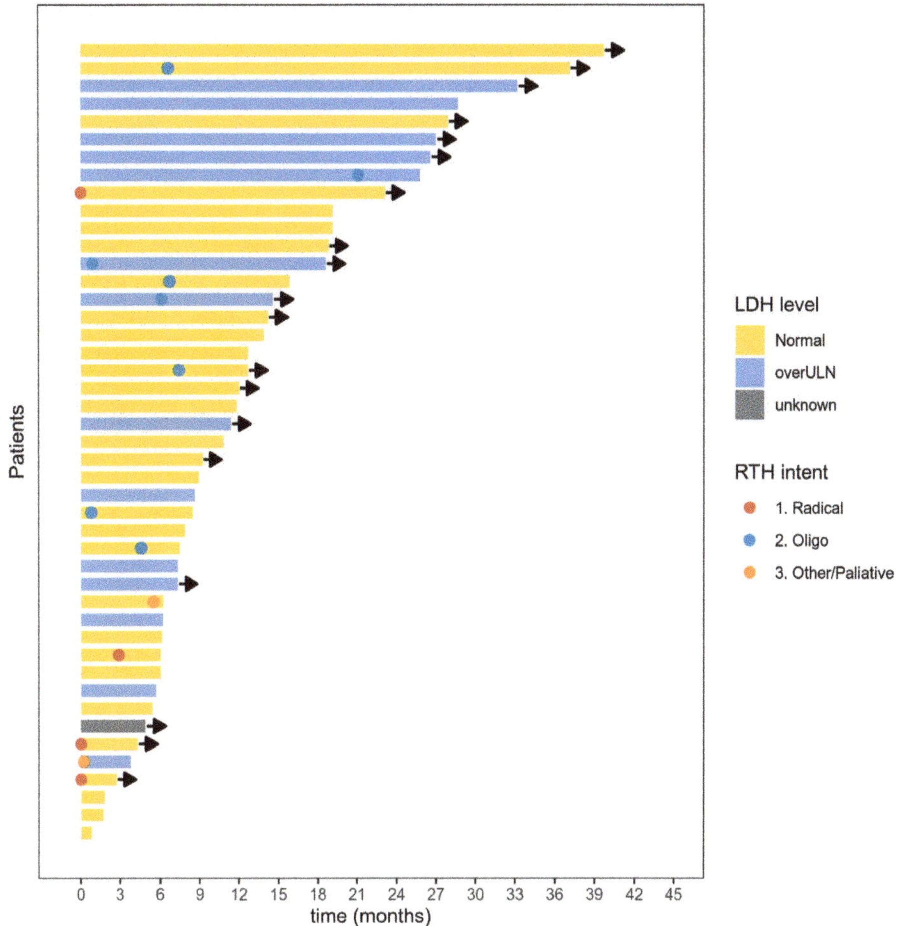

Figure 1. Swimmer plot for the duration of response to the anti-PD1 immunotherapy. Each bar represents a case; arrows represent an ongoing response. LDH level = lactic dehydrogenase activity at the start of treatment. overULN = over upper limit normal. RTH intent = intention of radiotherapy. Oligo = radiotherapy due to oligometastatic/oligoprogressive disease. Radiotherapy after progressive disease according to Response Evaluation Criteria in Solid Tumours (RECIST) 1.1 is not shown.

2.4. Progression-Free Survival

In the first-line treatment, the median progression-free survival (PFS), 12-month and 18-month PFS rate reached six months (CI: 3.8–10.8), 33% (CI: 23–48%) and 25% (CI: 16–41%), respectively, in the anti-PD1 ITH group. The same parameters were equal to 3.1 months (CI: 2.3–6.7), 5% (CI: 1–36%) and 5% (CI: 1–36%), respectively, in the other treatments group. These differences were statistically significant by the log-rank test with $p = 0.004$ (Figure 2A). Long-term disease stabilization was not achieved in patients treated with other systemic treatments. None of the preselected prognostic factors had a significant influence on PFS. Patients who received treatment other than anti-PD1 ITH had HR 2.18 (CI: 1.26–3.76, $p = 0.005$) (Table 2) of progression.

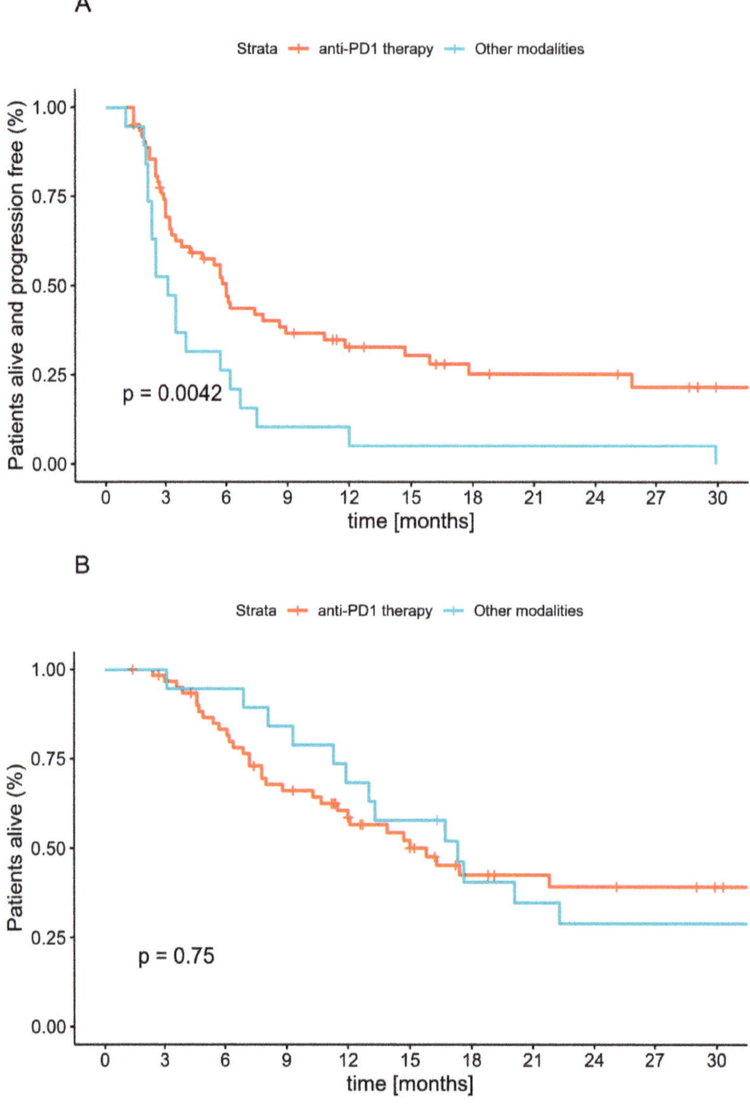

Figure 2. Progression-free survival (**A**) and overall survival (**B**) in mucosal melanoma patients in subgroups treated with anti-PD1 therapy (red curve) or other modality (blue curve) in first-line.

Table 2. Univariable Cox models for overall and progression-free survival. The 95% confidence intervals and *p* values are listed in brackets.

Variable	Value	HR for OS (Univariable)	HR for PFS (Univariable)
Age	Per one-year change	1.01 (0.98–1.03, p = 0.579)	1.01 (0.99–1.03, p = 0.436)
Sex	female	-	-
	male	1.83 (1.05–3.21, p = 0.034)	1.35 (0.81–2.26, p = 0.255)
BRAF V600 mutation	negative	-	-
	positive	0.62 (0.15–2.58, p = 0.516)	0.50 (0.15–1.58, p = 0.236)
Localization	Gastrointestinal system	-	-
	Genitourinary system	0.90 (0.43–1.88, p = 0.781)	1.28 (0.66–2.46, p = 0.462)
	Head and Neck Region	1.07 (0.57–2.02, p = 0.834)	1.39 (0.76–2.53, p = 0.290)
Disease stage at the start of the treatment	Localized, nonoperable disease	-	-
	M1a	3.11 (0.77–12.60, p = 0.112)	1.60 (0.54–4.68, p = 0.394)
	M1b	2.54 (0.82–7.85, p = 0.105)	1.61 (0.71–3.62, p = 0.252)
	M1c	1.89 (0.66–5.44, p = 0.238)	1.05 (0.51–2.15, p = 0.897)
	M1d	1.77 (0.39–8.00, p = 0.459)	1.57 (0.48–5.08, p = 0.455)
Number of organs involved at the start of treatment	1	-	-
	2	0.95 (0.48–1.88, p = 0.893)	1.26 (0.69–2.29, p = 0.453)
	3+	1.30 (0.65–2.58, p = 0.457)	1.54 (0.84–2.83, p = 0.162)
LDH at the start of the first-line	Normal	-	-
	overULN	2.11 (1.16–3.84, p = 0.014)	1.40 (0.82–2.37, p = 0.215)
	(Missing)	1.61 (0.67–3.89, p = 0.290)	1.45 (0.64–3.31, p = 0.375)
ECOG score at the start of first-line	0	-	-
	1+	1.19 (0.68–2.08, p = 0.553)	1.01 (0.61–1.65, p = 0.978)
The first line of treatment	PD1	-	-
	Other	0.90 (0.49–1.68, p = 0.747)	2.18 (1.26–3.76, p = 0.005)

In the second line of treatment, the median PFS reached 6.2 months (CI: 2.8–NA) for anti-PD1 ITH, 3.0 months (CI: 2.6–10) for ipilimumab and 4.6 months (CI: 1.8–NA) for other treatments. These differences were not statistically significant (p = 0.51).

2.5. Overall Survival

The median OS was 16.3 months (CI: 12.1–22.3) in the whole cohort. The 12-month and 18-month OS rates reached 61% (CI: 51–73%) and 41% (CI: 31–55%), respectively. There was no OS difference between patients who received anti-PD1 ITH or other treatments as first-line therapy (p = 0.7). Kaplan-Meier curves for OS are presented in Figure 2B. Men and patients with elevated LDH levels had a worse prognosis. None of the other preselected clinical factors had a significant influence on OS in this group of patients (Table 2).

2.6. Radiotherapy

Twenty-three patients received RT during anti-PD1 ITH (Supplement A). The one-year local control rate (LCR) in the subgroup who received RT with anti-PD1 ITH for unresectable locally advanced disease reached 75.0% (CI: 42.6–100%). This value was stable throughout the two years of follow-up. The one-year LCR after RT for the oligometastatic disease during anti-PD1 ITH was 85.7% (CI: 63.3–100%). Patients who received RT during the anti-PD1 treatment had a median PFS of 8.9 months (CI: 7.4–NA), whereas in patients who received anti-PD1 ITH as the only treatment the PFS was 4.2 months (CI: 3.0–7.8); the difference was statistically significant (HR 1.8. CI: 1.0–3.3, p = 0.05). The median OS since the start of anti-PD1 ITH was 32.2 months (12.0–NA) in patients who received RT, and 12.1 months (CI: 7.8–21.8) in nonirradiated patients. This was not statistically significant (p = 0.11) (Figure 3, Supplement A).

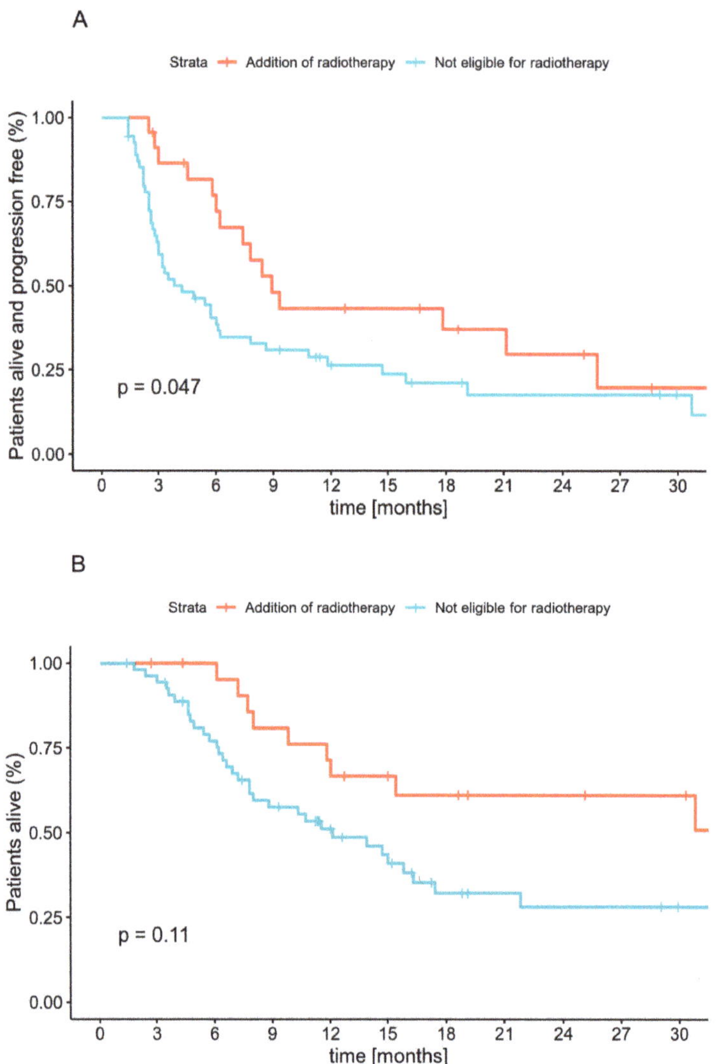

Figure 3. Progression-free survival (**A**) and overall survival (**B**) since the start of anti-PD1 therapy according to radiotherapy use.

3. Discussion

Until now, limited evidence has supported the efficacy of anti-PD-1 ITH in MM. In this population of patients, in individuals with the KIT mutation, imatinib was reported to result in significant clinical benefits, while for MM patients whose tumors harbor *BRAF* mutations, treatment with a combination of BRAF and MEK inhibitors was shown to be effective [8,24]. The efficacy of anti-PD1 ITH remains unclear in MM patients since a low number of patients have been enrolled in clinical trials. Conflicting data on MM patients' responses to nivolumab, pembrolizumab and ipilimumab have been published. Some reports suggest that anti-PD ITH is significantly less effective in MM than in CM patients, while other report durable anti-tumor effects [25–27].

We analyzed anti-PD-1 ITH efficacy in routine clinical practice outside of clinical trials in a nonpreselected population of subsequent patients with MM, and report ORR and PFS similar to patients with CM, also treated outside of clinical trials. In a retrospective analysis covering 25 dermatology departments in France, 75 MM patients were treated with first-line nivolumab or pembrolizumab and, out of these, fifteen achieved OR, which corresponds to an ORR of 20% (95% CI: 11.6–30.8) [28], which is lower than the 25% reported by us. In a Japanese study of 24 MM cases, the ORR was 20.8%, and for 17 cases with visceral metastases—17.6%. For the 17 cases, OS and PFS periods were 422 days and 226 days [29]. In another small Japanese trial, 17 MM patients were treated with nivolumab, and ORR was reported at 23.5%. One patient achieved a CR, three—PR, and five SD as their best response. The median PFS was 1.4 months (95% CI: 1.2–2.8) [30], which is again inferior to the results reported by us. High OR was reported for the first-line pembrolizumab—35% accompanied with five-months of median PFS (n = 20) [31].

We previously described the results of anti-PD1 therapy as well as BRAF/MEK inhibitors treatment in the general melanoma population in Poland [32]. While comparing data between the general population of patients receiving anti-PD1 ITH in the first line and an analogous subpopulation of mucosal melanoma patients, no differences were seen in ORR (objective response rate), PFS and OS. The ORR was equal to 28% in the mucosal melanoma subgroup, and 30% in the general population (p = 0.85 by Fisher's exact test). The median PFS and OS for mucosal melanoma was 6.0 (CI: 3.8–10.8) and 15.8 (CI: 11.5–NA) months, while for the general population—6.9 (CI: 5.3–9.0) and 20.5 (CI: 15.3–NA), respectively. The HR for mucosal melanoma was 0.97 (CI: 0.68–1.38, p = 0.86) for PFS and 0.91 (CI: 0.60–1.38, p = 0.65) for OS (see Supplement B). Therefore, our results confirm the efficacy of immunotherapy, especially anti-PD1 antibodies, in MM treatment [31].

Due to strict inclusion criteria in clinical trials, covering mostly patients without brain metastases and patients with good performance status, corresponding to low tumor burden, we expected longer PFS and OS for patients treated in the trials than for patients treated in our study. In a meta-analysis of major ITH clinical trials by Shoushtari, A.N. et al., 35 MM patients were identified. This analysis covered patients treated within NCT02083484 (MK-3475), NCT01295827 (KEYNOTE-001), NCT01927419 (CheckMate 069), NCT01024231 (CA209-004), and NCT01721746 (CheckMate 037) trials. Therefore, the majority of patients were treated with pembrolizumab or nivolumab, not in the first line but after previous therapy. In these trials the majority of patients received ipilimumab before anti-PD-1 ITH. These authors reported that for MM treated with anti-PD-1 ORR was 23% (95% CI: 10–40%) with PD as the best response for 57% of patients (95% CI: 39–74%), which represents ORR similar as that reported by us (25% for anti-PD1 in the first line). In the same meta-analysis, MM patients treated with nivolumab/pembrolizumab monotherapy achieved PFS of 3.9 months, which is shorter than the PFS of our patients treated in the first line. ORR reported by Shoushtari, A.N. et al. is also generally numerically similar to the 23% ORR reported by D'Angelo, S.P et al. for 86 MM patients included in another meta-analysis of multiple prospective trials of nivolumab; but seven patients—that is almost 10% of analyzed cases—were included in both described analyses, and these data may not be interpreted independently [25,33]. In the analysis by D'Angelo, S.P et al. covering, again, mostly patients treated with anti-PD1 ITH in further lines of therapy in NCT00730639 (CA209-003), NCT01621490 (CA209-038), NCT01721772 (CheckMate 066), NCT01721746 (CheckMate 037), and NCT01844505 (CheckMate 067) trials, ORR was—37.1% (95% CI, 21.5% to 55.1%) and the median PFS was 3.0 months (95% CI, 2.2 to 5.4 months), which represents higher ORR, but shorter PFS than reported by us [25]. For pembrolizumab only trials—KEYNOTE-001 (NCT01295827), -002 (NCT01704287), and -006 (NCT01866319), the ORR was 22% (95% CI: 11–35) and 15% (95% CI: 5–32) in ipilimumab-naive and ipilimumab-treated MM patients, which is concordant with our treatment results. At the same time, the median PFS was 2.8 months (95% CI: 2.7–2.8) for KEYNOTE-001/-002/-006 patients, which is less than half that of the six months reported by us. Similar ITH efficacy was reported in eight patients pretreated with dacarbazine before nivolumab administration [34]. The administration of nivolumab was listed as long-term for three patients with 13–17 cycles given, and over 30 cycles for one case. In this group, ORR was 37.5% (CR—25.0%, PR—12.5%), median PFS—10.2 months [35].

It should be noted that in the analysis by D'Angelo, S.P et al., median PFS in patients treated with nivolumab combined with ipilimumab (n = 35) was 5.9 months (95% CI: 2.8 months—not reached), while our patients (n = 63) achieved median PFS of six months during anti-PD-1 monotherapy therapy. In concordance with our report, in the pooled analysis CM patients achieved PFS of 6.2 months with nivolumab monotherapy and 11.7 months with nivolumab-ipilimumab combination therapy [25]. Nevertheless, in the pooled analysis of major clinical trials, MM patients achieved reduced clinical benefits in comparison to CM patients during anti-PD-1 ITH treatment [25]. This was not the case in routine clinical practice reported by us with PFS in a general population of 6.9 months and OS of 20.5 and resultant HR for MM of 0.95 for PFS and 0.89 for OS. On the contrary, in clinical trial data, MM patients who were treated with nivolumab monotherapy received a median of 7.0 doses (1–34) and CM patients 11.0 doses (1–61) [25]. Moreover, in MM patients, the median reduction in tumor burden in the target lesions was reported only at 1.4% for nivolumab monotherapy and as high as 34.2% for combined therapy [25], which may indicate the specific biology of the disease requiring dual pathway activation for MM cell elimination.

In terms of OS of MM, patients treated with ITH were reported to achieve a median OS of 15.97 months and a one-year OS rate of 57.8% (95% CI: 49.5–67.5) while no specific data for nivolumab, pembrolizumab and ipilimumab differences were reported [28]. In a Japanese trial, median OS was 12.0 months (95% CI: 3.5—not reached) for nivolumab treatment, while for pembrolizumab pooled analysis the median OS was 11.3 months (7.7–16.6) [30,34]. In general, despite the limitations of the studies discussed, data suggest that in MM patients anti-PD1 therapy may be effective both in terms of OR, PFS and OS. Nevertheless, it should be considered that in all of the multi-trial analyses, the number of MM cases was about 10% of CM patients, which may influence statistical calculations [25]. With our report, we confirm the efficacy of ITH in MM patients with real-world data.

The efficacy of ITH may be enhanced by RT, which was reported before (ORR = 57.1%) in the case of seven patients [36]. Another analysis of 10 patients covered mucosal melanoma of the nasal cavity or maxillary sinus, in which patients were treated with nivolumab or pembrolizumab and concomitant radiotherapy. In this group of patients, after a median follow-up period of 46 weeks, the local control rate of the primary lesion and regional lymph nodes was 100% with a median PFS of 7.4 months (range 2–82 weeks). In this group of patients, the six-month PFS rate was 60% [37], which is lower than in our study—8.9 months. The largest report published until now was a retrospective study of 23 patients out of whom 12 patients were treated with pembrolizumab and RT, 11 patients were treated with RT alone and the others were treated with pembrolizumab monotherapy. It was shown that pembrolizumab with concomitant radiotherapy enabled achievement of a one-year target lesion control rate in 94.1% cases, while radiotherapy without immunotherapy enabled the control in 57.1% and pembrolizumab in 25%. Treatment-related AEs were not significantly different between radiotherapy with or without pembrolizumab [38]. In our study all 23 patients received concomitant RT and immunotherapy (Supplement C). Generally, concomitant RT is not allowed in the majority of clinical trials. Thus, a separate study with RT-ITH is required to confirm the expected benefit of irradiation during ITH. While we report the data, a separate trial is also ongoing (NCT04017897).

At the same time, no significant association is known between PFS duration and other clinical factors including primary tumor localization, *BRAF* gene mutation status, stage at treatment start, presence of CNS or liver metastases, type of prior therapy, or response to first-line (ipilimumab) therapy if used in the second-line, as in the case of our analysis (Table 2) [31,33]. The duration of OS and PFS are correlated with irAE development and high levels of PD-L1 expression (>5% of cells in the tumors) [37,38]. In fact in patients with mucosal melanoma and tumor PD-L1 expression ≥ 5% (n = 15), ORR was 53.3% (95% CI: 26.6–78.7%), while in patients with PD-L1 expression < 5% (n = 49)—ORR is 12.2% (95% CI: 4.6–24.8%), and in these patients median PFS with PD-L1 expression ≥ 5% was 12.2 months (95% CI: 3.0 months—not reached), while for PD-L1 expression < 5%—median

PFS was < 3 months while the role of tumor PD-L1 expression as a response biomarker was not fully defined [25,39]. In fact, the significantly lower mutational burden in MM in comparison to CM may explain the decreased efficacy of the immune checkpoint blockade in many MM patients [40,41]. In general, not only PD-1/PD-L1 expression level, but also immune cells infiltration and transcriptional immunoscore, may be correlated with immunotherapy response. Multiple genetic and genomic factors, including mutation burden, mismatch repair deficiency, somatic copy-number variation burden and neoantigen load are being suggested to predict immunotherapy response [42]. While somatic mutation clonality was reported to positively correlate with immunotherapy response, copy number variation (CNV) was shown to negatively correlate with efficacy of response to the PD-1 checkpoint blockade and to be associated with downregulation of immune-related pathways expression [43,44]. Moreover, the expression of melanin pigment and active melanogenesis, which are often found at high levels in MM, may decrease the sensitivity of MM cells to immunotherapy [45,46]. Melanin also has radioprotective and scavenging properties and may decrease the efficacy of radiotherapy [47]. The relationship had not been finally defined as in survival meta-analysis pigmentation; that is, melanin level was not significantly correlated with survival (HR = 0.87; 95% CI, 0.66–1.15; p = 0.34) in mucosal melanoma [48].

On the other hand, it has been suggested that RT is an immune adjuvant boosting the antitumor immune response. At the same time, it is also possible that ITH has a radiosensitizing effect and increases the efficacy of radiotherapy [49,50]. In general, our MM study population is as large as the group that was identified in a pulled analysis of KEYNOTE-001, KEYNOTE-002 and KEYNOTE-006 trials [50]. We confirm the effect of RTH on PFS duration in MM patients, although larger prospective trials would be needed to validate our findings. In the case of MM, only international multicenter studies would be able to recruit a large number of patients due to the epidemiology of the disease.

4. Materials and Methods

4.1. Cohort

In this international retrospective study, we included consecutive patients with confirmed histopathological diagnosis of unresectable or metastatic MM. Patients must have been treated with anti-PD1 (nivolumab or pembrolizumab) or anti-CLTA-4 (ipilimumab) antibodies as a first or second-line treatment. Cases were recruited in five high-volume centers in Poland (two centers) and Italy (three centers) between July 2013 and August 2020.

4.2. Treatment

We analyzed the treatment sequence and response to systemic therapy, as well as factors that may influence on patients' survival, including: patients' age, sex, primary tumor site, the mutation in BRAF V600, tumor stage at the start of first-line treatment, number of metastatic sites (organs) at the start of first-line treatment, lactate dehydrogenase (LDH) levels at the start of first-line treatment, performance status at the start of first-line treatment and treatment sequence (ITH as a first or second-line). To define possible benefits from the addition of RT, we analyzed concurrent RT during anti-PD1 ITH and its parameters, regardless of the treatment line.

4.3. Response and Survival Analysis

The response was evaluated every 12 weeks in the contrast-enhanced CT scan of the primary tumor site, thorax, abdomen, and pelvis, and contrast-enhanced magnetic resonance imaging (MRI) for lesions in the central nervous system. Response to the treatment was assessed using response evaluation criteria in Solid Tumors 1.1 (RECIST). Objective response (OR) was defined as the sum of the number of patients who archived partial response (PR) or complete response (CR) as their best response. Disease control (DC) was defined as a sum of patients who archived stable disease (SD), PR or CR as their best response. Progression-free survival (PFS) was calculated from the start of treatment first dose to the disease progression, as assessed according to RECIST 1.1, or death. Patients who had at least stable disease (SD) at the last follow-up were censored. The overall survival (OS) was calculated from the start date of first-line treatment to the date of death. The OS from the start of anti-PD1 was also calculated. In both cases, patients alive at the last follow-up were censored.

4.4. Statistical Analysis

The continuous variables were summarized by median and interquantile range (IQR), while categorical variables were summarized by count and percentage of total cases. All point estimates were reported with a 95% confidence interval (CI) unless stated otherwise.

The Kaplan-Meier estimator with the log-rank test, as well as the Cox proportional hazard model, were used for the survival analysis. Fisher's exact test was used to assess for independence between categorical variables.

All analyses were performed in the R language environment version 3.6.3 (The R Foundation for Statistical Computing) with abundant use of tidyverse and survminer packages [51–54]. A $p \leq 0.05$ was deemed statistically significant.

4.5. Ethical Statement

This study was approved by Bioethical Committee at Maria Sklodowska-Curie National Research Institute of Oncology in Warsaw under the registration number 73/2018.

5. Conclusions

To our knowledge, this is the largest analysis of data to date for anti–PD-1 therapy combined with analysis of the radiation therapy impact on immunotherapy efficacy in mucosal melanoma. Anti-PD1 antibodies are an effective treatment in advanced MM. The addition of RT to ITH may be beneficial in the selected subgroup of MM patients.

Supplementary Materials: The following are available online at http://www.mdpi.com/2072-6694/12/11/3131/s1. Supplement A table: Details regarding Radiotherapy during anti-PD1 treatment. Supplement B: Comparison of (A) progression-free survival and (B) overall survival between mucosal melanoma and Polish general melanoma cohort. Supplement C: PFS and OS in patients treated with radiotherapy with concomitant immunotherapy, chemotherapy of BRAFi/MEKi.

Author Contributions: Conceptualization—A.M.C., M.M., and P.R. (Piotr Rutkowski); methodology—A.M.C., P.R. (Piotr Rutkowski), M.J.S. and P.T.; software—P.T. and M.J.S.; validation—A.M.C., P.T., M.J.S.; formal analysis—P.T.; project administration—A.M.C.; investigation—A.M.C., P.T., A.I., M.J.S., A.L, P.R. (Pawel Rogala), B.C.-S., P.Q., U.R., S.B., A.S.-C., S.F., M.M., and P.R. (Piotr Rutkowski); resources—A.M.C., P.T., A.I., M.J.S., A.L., B.C.-S., P.Q., U.R., S.B., P.R. (Pawel Rogala), A.S.-C., S.F., M.M., and P.R. (Piotr Rutkowski); data curation—A.M.C., P.T., M.J.S., P.R. (Pawel Rogala), A.L., A.I., R.Q., B.C.-S.; writing—original draft preparation, A.M.C. and P.T.; writing—review and editing, A.M.C., P.T., M.J.S., M.M., and P.R. (Piotr Rutkowski); visualization—P.T.; supervision—M.M. and P.R. (Piotr Rutkowski) All authors have read and agreed to the published version of the manuscript.

Funding: This work has been supported by Maria-Sklodowska Curie National Research Institute of Oncology statutory funding.

Conflicts of Interest: Pawel Teterycz travel, accommodations, expenses—Bristol-Myers Squibb, MSD, Novartis, Roche. Anna Malgorzata Czarnecka Speakers' Bureau—Bristol-Myers Squibb, Pierre Fabre. Research Funding—Bristol-Myers Squibb; Travel, Accommodations, Expenses—Bristol-Myers Squibb, MSD, Novartis, Pierre Fabre, Roche. Pawel Rogala Travel, Accommodations, Expenses—Bristol-Myers Squibb, MSD, Novartis, Roche. Mario Mandala: Consulting or Advisory Role—Bristol-Myers Squibb, MSD Brazil, Pierre Fabre, Roche; Research Funding—Bristol-Myers Squibb, Novartis pharma SAS; Roche. Pietro Quaglino is on the advisory board of Novartis, BMS, MSD, and Pierre Fabre and received fees for lectures from Novartis, BMS, MSD, and Pierre Fabre. Piotr Rutkowski has received honoraria for lectures and Advisory Boards from BMS, MSD, Novartis, Roche, Pierre Fabre, Amgen, Merck, Sanofi, Blueprint Medicines outside of the scope of the study. The funders had no role in the design of the study; in the collection, analyses, or interpretation of data; in the writing of the manuscript, or in the decision to publish the results. Mateusz Spalek, Sławomir Falkowski, Alice Indini, Umberto Ricardi, Serena Badellino, Anna Szumera-Ciećkiewicz, Alice Labiance declare no conflict of interest.

References

1. Topić, B.; Mašić, T.; Radović, S.; Lincender, I.; Muhić, E. Primary Oral Mucosal Melanomas-Two Case Reports and Comprehensive Literature Review. *Acta Clin. Croat* **2017**, *56*, 323–330. [CrossRef] [PubMed]
2. Mihajlovic, M.; Vlajkovic, S.; Jovanovic, P.; Stefanovic, V. Primary mucosal melanomas: A comprehensive review. *Int. J. Clin. Exp. Pathol.* **2012**, *5*, 739–753. [PubMed]
3. Lerner, B.A.; Stewart, L.A.; Horowitz, D.P.; Carvajal, R.D. Mucosal Melanoma: New Insights and Therapeutic Options for a Unique and Aggressive Disease. *Oncology* **2017**, *31*, e23–e32. [PubMed]
4. Ss, T. Re: Cancers with increasing incidence trends in the United States: 1999 through 2008. *J. Urol.* **2012**, *188*, 1120–1121. [CrossRef]
5. Simard, E.P.; Ward, E.M.; Siegel, R.; Jemal, A. Cancers with increasing incidence trends in the United States: 1999 through 2008. *CA Cancer J. Clin.* **2012**, *62*, 118–128. [CrossRef]
6. Yde, S.S.; Sjoegren, P.; Heje, M.; Stolle, L.B. Mucosal Melanoma: A Literature Review. *Curr. Oncol. Rep.* **2018**, *20*, 28. [CrossRef]
7. Zhang, Y.; Fu, X.; Qi, Y.; Gao, Q. A study of the clinical characteristics and prognosis of advanced mucosal and cutaneous melanoma in a Chinese population. *Immunotherapy* **2018**, *11*, 91–99. [CrossRef]
8. Alicea, G.M.; Rebecca, V.W. Emerging strategies to treat rare and intractable subtypes of melanoma. *Pigment Cell Melanoma Res.* **2020**. [CrossRef]
9. Werdin, C.; Limas, C.; Knodell, R.G. Primary malignant melanoma of the rectum. Evidence for origination from rectal mucosal melanocytes. *Cancer* **1988**, *61*, 1364–1370. [CrossRef]
10. Lundberg, R.; Brytting, M.; Dahlgren, L.; Kanter-Lewensohn, L.; Schloss, L.; Dalianis, T.; Ragnarsson-Olding, B. Human herpes virus DNA is rarely detected in non-UV light-associated primary malignant melanomas of mucous membranes. *Anticancer Res.* **2006**, *26*, 3627–3631.
11. Giraud, G.; Ramqvist, T.; Ragnarsson-Olding, B.; Dalianis, T. DNA from BK virus and JC virus and from KI, WU, and MC polyomaviruses as well as from simian virus 40 is not detected in non-UV-light-associated primary malignant melanomas of mucous membranes. *J. Clin. Microbiol.* **2008**, *46*, 3595–3598. [CrossRef]
12. Lourenço, S.V.; Fernandes, J.D.; Hsieh, R.; Coutinho-Camillo, C.M.; Bologna, S.; Sangueza, M.; Nico, M.M. Head and Neck Mucosal Melanoma: A Review. *Am. J. Dermatopathol.* **2014**, *36*, 578–587. [CrossRef] [PubMed]
13. Axéll, T.; Hedin, C.A. Epidemiologic study of excessive oral melanin pigmentation with special reference to the influence of tobacco habits. *Scand. J. Dent. Res.* **1982**, *90*, 434–442. [CrossRef] [PubMed]
14. Purdue, M.P. Re: Determinants of BRAF mutations in primary melanomas. *J. Natl. Cancer Inst.* **2005**, *97*, 401–402. [CrossRef] [PubMed]
15. Kirchoff, D.D.; Deutsch, G.B.; Foshag, L.J.; Lee, J.H.; Sim, M.-S.; Faries, M.B. Evolving Therapeutic Strategies in Mucosal Melanoma Have Not Improved Survival Over Five Decades. *Am. Surg.* **2016**, *82*, 1–5. [CrossRef] [PubMed]
16. DeMatos, P.; Tyler, D.S.; Seigler, H.F. Malignant melanoma of the mucous membranes: A review of 119 cases. *Ann. Surg. Oncol.* **1998**, *5*, 733–742. [CrossRef]
17. Pittaka, M.; Kardamakis, D.; Spyropoulou, D. Comparison of International Guidelines on Mucosal Melanoma of the Head and Neck: A Comprehensive Review of the Role of Radiation Therapy. *In Vivo* **2016**, *30*, 165–170.
18. Yii, N.W.; Eisen, T.; Nicolson, M.; A'Hern, R.; Rhys-Evans, P.; Archer, D.; Henk, J.M.; Gore, M.E. Mucosal malignant melanoma of the head and neck: The Marsden experience over half a century. *Clin. Oncol. (R. Coll. Radiol.)* **2003**, *15*, 199–204. [CrossRef]
19. Temam, S.; Mamelle, G.; Marandas, P.; Wibault, P.; Avril, M.F.; Janot, F.; Julieron, M.; Schwaab, G.; Luboinski, B. Postoperative radiotherapy for primary mucosal melanoma of the head and neck. *Cancer* **2005**, *103*, 313–319. [CrossRef]
20. Moreno, M.A.; Roberts, D.B.; Kupferman, M.E.; DeMonte, F.; El-Naggar, A.K.; Williams, M.; Rosenthal, D.S.; Hanna, E.Y. Mucosal melanoma of the nose and paranasal sinuses, a contemporary experience from the M. D. Anderson Cancer Center. *Cancer* **2010**, *116*, 2215–2223. [CrossRef]
21. Krengli, M.; Masini, L.; Kaanders, J.H.; Maingon, P.; Oei, S.B.; Zouhair, A.; Ozyar, E.; Roelandts, M.; Amichetti, M.; Bosset, M.; et al. Radiotherapy in the treatment of mucosal melanoma of the upper aerodigestive tract: Analysis of 74 cases. A Rare Cancer Network study. *Int. J. Radiat. Oncol. Biol. Phys.* **2006**, *65*, 751–759. [CrossRef]

22. Koto, M.; Demizu, Y.; Saitoh, J.I.; Suefuji, H.; Tsuji, H.; Okimoto, T.; Ohno, T.; Shioyama, Y.; Ikawa, H.; Nemoto, K.; et al. Definitive Carbon-Ion Radiation Therapy for Locally Advanced Sinonasal Malignant Tumors: Subgroup Analysis of a Multi-center Study by the Japan Carbon-Ion Radiation Oncology Study Group (J-CROS). *Int. J. Radiat. Oncol. Biol. Phys.* **2018**, *102*, 353–361. [CrossRef] [PubMed]

23. Fuji, H.; Yoshikawa, S.; Kasami, M.; Murayama, S.; Onitsuka, T.; Kashiwagi, H.; Kiyohara, Y. High-dose proton beam therapy for sinonasal mucosal malignant melanoma. *Radiat. Oncol.* **2014**, *9*, 162. [CrossRef]

24. Kim, K.B.; Alrwas, A. Treatment of KIT-mutated metastatic mucosal melanoma. *Chin. Clin. Oncol.* **2014**, *3*, 12. [CrossRef]

25. D'Angelo, S.P.; Larkin, J.; Sosman, J.A.; Lebbé, C.; Brady, B.; Neyns, B.; Schmidt, H.; Hassel, J.C.; Hodi, F.S.; Savage, K.J.; et al. Efficacy and Safety of Nivolumab Alone or in Combination with Ipilimumab in Patients With Mucosal Melanoma: A Pooled Analysis. *JCO* **2016**, *35*, 226–235. [CrossRef] [PubMed]

26. Studentova, H.; Kalabova, H.; Koranda, P.; Chytilova, K.; Kucerova, L.; Melichar, B.; Vrana, D. Immunotherapy in mucosal melanoma: A case report and review of the literature. *Oncotarget* **2018**, *9*, 17971–17977. [CrossRef] [PubMed]

27. Napierała, M.J.; Czarnecka, A.M. Mucosal melanoma—clinical presentation and treatment based on a case series. *Oncol. Clin. Pract.* **2019**, *15*, 223–230. [CrossRef]

28. Mignard, C.; Deschamps-Huvier, A.; Duval-Modeste, A.B.; Dutriaux, C.; Khammari, A.; Avril, M.F.; Kramkimel, N.; Machet, L.; Marcant, P.; Lesimple, T.; et al. Efficacy of Immunotherapy in Patients with Metastatic Mucosal or Uveal Melanoma. *J. Oncol.* **2018**, *2018*, 1908065. [CrossRef]

29. Maeda, T.; Yoshino, K.; Nagai, K.; Oaku, S.; Kato, M.; Hiura, A.; Hata, H. Efficacy of nivolumab monotherapy against acral lentiginous melanoma and mucosal melanoma in Asian patients. *Br. J. Dermatol.* **2019**, *180*, 1230–1231. [CrossRef]

30. Nomura, M.; Oze, I.; Masuishi, T.; Yokota, T.; Satake, H.; Iwasawa, S.; Kato, K.; Andoh, M. Multi-center prospective phase II trial of nivolumab in patients with unresectable or metastatic mucosal melanoma. *Int. J. Clin. Oncol.* **2020**, *25*, 972–977. [CrossRef]

31. Moya-Plana, A.; Gómez, R.G.H.; Rossoni, C.; Dercle, L.; Ammari, S.; Girault, I.; Roy, S.; Scoazec, J.-Y.; Vagner, S.; Janot, F.; et al. Evaluation of the efficacy of immunotherapy for non-resectable mucosal melanoma. *Cancer Immunol. Immunother.* **2019**, *68*, 1171–1178. [CrossRef] [PubMed]

32. Czarnecka, A.M.; Teterycz, P.; Mariuk-Jarema, A.; Lugowska, I.; Rogala, P.; Dudzisz-Sledz, M.; Switaj, T.; Rutkowski, P. Treatment Sequencing and Clinical Outcomes in BRAF-Positive and BRAF-Negative Unresectable and Metastatic Melanoma Patients Treated with New Systemic Therapies in Routine Practice. *Target Oncol.* **2019**, *14*, 729–742. [CrossRef]

33. Shoushtari, A.N.; Munhoz, R.R.; Kuk, D.; Ott, P.A.; Johnson, D.B.; Tsai, K.K.; Rapisuwon, S.; Eroglu, Z.; Sullivan, R.J.; Luke, J.J.; et al. Efficacy of Anti-PD-1 Agents in Acral and Mucosal Melanoma. *Cancer* **2016**, *122*, 3354–3362. [CrossRef] [PubMed]

34. Hamid, O.; Robert, C.; Ribas, A.; Hodi, F.S.; Walpole, E.; Daud, A.; Arance, A.S.; Brown, E.; Hoeller, C.; Mortier, L.; et al. Antitumour activity of pembrolizumab in advanced mucosal melanoma: A post-hoc analysis of KEYNOTE-001, 002, 006. *British J. Cancer* **2018**, *119*, 670–674. [CrossRef]

35. Urasaki, T.; Ono-Fuchiwaki, M.; Tomomatsu, J.; Nakano, K.; Taira, S.; Takahashi, S. Eight patients with mucosal malignant melanoma treated by nivolumab: A retrospective analysis in a single institution. *Ann. Oncol.* **2017**, *28*, ix102–ix103. [CrossRef]

36. Kato, J.; Hida, T.; Someya, M.; Sato, S.; Sawada, M.; Horimoto, K.; Fujioka, M.; Minowa, T.; Matsui, Y.; Tsuchiya, T.; et al. Efficacy of combined radiotherapy and anti-programmed death 1 therapy in acral and mucosal melanoma. *J. Dermatol.* **2019**, *46*, 328–333. [CrossRef] [PubMed]

37. Hanaoka, Y.; Tanemura, A.; Takafuji, M.; Kiyohara, E.; Arase, N.; Suzuki, O.; Isohashi, F.; Ogawa, K.; Fujimoto, M. Local and disease control for nasal melanoma treated with radiation and concomitant anti-programmed death 1 antibody. *J. Dermatol.* **2020**, *47*, 423–425. [CrossRef]

38. Kim, H.J.; Chang, J.S.; Roh, M.R.; Oh, B.H.; Chung, K.Y.; Shin, S.J.; Koom, W.S. Effect of Radiotherapy Combined With Pembrolizumab on Local Tumor Control in Mucosal Melanoma Patients. *Front. Oncol.* **2019**, *9*, 835. [CrossRef]

39. Otsuka, M.; Sugihara, S.; Mori, S.; Hamada, K.; Sasaki, Y.; Yoshikawa, S.; Kiyohara, Y. Immune-related adverse events correlate with improved survival in patients with advanced mucosal melanoma treated with nivolumab: A single-center retrospective study in Japan. *J. Dermatol.* **2020**, *47*, 356–362. [CrossRef]

40. Tyrrell, H.; Payne, M. Combatting mucosal melanoma: Recent advances and future perspectives. *Melanoma Manag.* **2018**, *5*, MMT11. [CrossRef]

41. Johnson, D.B.; Carlson, J.A.; Elvin, J.A.; Vergilio, J.A.; Suh, J.; Ramkissoon, S.; Daniel, S.; Fabrizio, D.; Frampton, G.; Ali, S.M.; et al. Landscape of genomic alterations (GA) and tumor mutational burden (TMB) in different metastatic melanoma (MM) subtypes. *JCO* **2017**, *35*, 9536. [CrossRef]

42. Furney, S.J.; Turajlic, S.; Stamp, G.; Nohadani, M.; Carlisle, A.; Thomas, J.M.; Hayes, A.; Strauss, D.; Gore, M.; van den Oord, J.; et al. Genome sequencing of mucosal melanomas reveals that they are driven by distinct mechanisms from cutaneous melanoma. *J. Pathol.* **2013**, *230*, 261–269. [CrossRef] [PubMed]

43. Wu, C.-C.; Wang, Y.A.; Livingston, J.A.; Zhang, J.; Futreal, P.A. A computational network approach to identify predictive biomarkers and therapeutic combinations for anti-PD-1 immunotherapy in cancer. *bioRxiv* **2020**. [CrossRef]

44. Cristescu, R.; Mogg, R.; Ayers, M.; Albright, A.; Murphy, E.; Yearley, J.; Sher, X.; Liu, X.Q.; Lu, H.; Nebozhyn, M.; et al. Pan-tumor genomic biomarkers for PD-1 checkpoint blockade–based immunotherapy. *Science* **2018**, *362*. [CrossRef]

45. Roh, W.; Chen, P.-L.; Reuben, A.; Spencer, C.N.; Prieto, P.A.; Miller, J.P.; Gopalakrishnan, V.; Wang, F.; Cooper, Z.A.; Reddy, S.M.; et al. Integrated molecular analysis of tumor biopsies on sequential CTLA-4 and PD-1 blockade reveals markers of response and resistance. *Sci. Transl. Med.* **2017**, *9*. [CrossRef] [PubMed]

46. Slominski, A.; Zbytek, B.; Slominski, R. Inhibitors of melanogenesis increase toxicity of cyclophosphamide and lymphocytes against melanoma cells. *Int. J. Cancer* **2009**, *124*, 1470–1477. [CrossRef]

47. Feller, L.; Khammissa, R.A.G.; Lemmer, J. A Review of the Aetiopathogenesis and Clinical and Histopathological Features of Oral Mucosal Melanoma. *Sci. World J.* **2017**, *2017*, 9189812. [CrossRef] [PubMed]

48. Brozyna, A.A.; Jozwicki, W.; Roszkowski, K.; Filipiak, J.; Slominski, A.T. Melanin content in melanoma metastases affects the outcome of radiotherapy. *Oncotarget* **2016**, *7*, 17844–17853. [CrossRef] [PubMed]

49. Hahn, H.M.; Lee, K.G.; Choi, W.; Cheong, S.H.; Myung, K.B.; Hahn, H.J. An updated review of mucosal melanoma: Survival meta-analysis. *Mol. Clin. Oncol.* **2019**, *11*, 116–126. [CrossRef]

50. Demaria, S.; Formenti, S.C. Radiation as an immunological adjuvant: Current evidence on dose and fractionation. *Front. Oncol.* **2012**, *2*, 153. [CrossRef]

51. Yentz, S.; Lao, C.D. Immunotherapy for mucosal melanoma. *Ann. Transl. Med.* **2019**, *7*, S118. [CrossRef] [PubMed]

52. R Core Team. *R: A Language and Environment for Statistical Computing*; R Foundation for Statistical Computing: Vienna, Austria, 2017.

53. Wickham, H. tidyverse: Easily Install and Load "Tidyverse" Packages. *R Package Version* **2017**, *1*, 51.

54. Kassambara, A.; Kosinski, M. *survminer: Drawing Survival Curves Using "ggplot2"*; R Package Version 0.4; R Foundation for Statistical Computing: Vienna, Austria, 2018; Volume 3.

Publisher's Note: MDPI stays neutral with regard to jurisdictional claims in published maps and institutional affiliations.

Article

Inhibition of Melanoma Cell Migration and Invasion Targeting the Hypoxic Tumor Associated CAXII

Gaia Giuntini [1], Sara Monaci [1], Ylenia Cau [2], Mattia Mori [2], Antonella Naldini [1] and Fabio Carraro [3,*]

[1] Department of Molecular and Developmental Medicine, Cellular and Molecular Physiology Unit, University of Siena, 53100 Siena, Italy; gaia.giuntini@student.unisi.it (G.G.); sara.monaci@student.unisi.it (S.M.); Antonella.Naldini@unisi.it (A.N.)

[2] Department of Biotechnology, Chemistry and Pharmacy, University of Siena, 53100 Siena, Italy; Cau.Ylenia@gmail.com (Y.C.); mattia.mori@unisi.it (M.M.)

[3] Department of Medical Biotechnologies, Cellular and Molecular Physiology Unit, University of Siena, 53100 Siena, Italy

* Correspondence: Fabio.Carraro@unisi.it; Tel.: +39-0577-235774; Fax: +39-0577-234217

Received: 16 September 2020; Accepted: 14 October 2020; Published: 17 October 2020

Simple Summary: Melanoma is a potential metastatic cancer with a poor prognosis and a low free-survival rate. Thus, discovering new therapeutic strategies will be helpful to fight against it. Carbonic anhydrases IX (CAIX) and XII (CAXII) along with Hedgehog pathway aberrations are already demonstrated to be involved in melanoma progression. Here we investigated the effects of either an indirect or direct CAXII inhibition on cell migration and invasion in three melanoma cell lines. First we indirectly inhibited CAXII through the smoothened antagonist cyclopamine, which resulted in a decreased CAXII protein expression and cell migration. Thereafter, we directly blocked CAXII using new small molecules. This resulted in reducing not only cell migration, but also the invasiveness ability of highly aggressive melanoma cell lines. This evidence may contribute to further exploiting the therapeutic role of CAXII in melanoma progression and invasiveness.

Abstract: Background: Intratumoral hypoxia contributes to cancer progression and poor prognosis. Carbonic anhydrases IX (CAIX) and XII (CAXII) play pivotal roles in tumor cell adaptation and survival, as aberrant Hedgehog (Hh) pathway does. In malignant melanoma both features have been investigated for years, but they have not been correlated before and/or identified as a potential pharmacological target. Here, for the first time, we demonstrated that malignant melanoma cell motility was impaired by targeting CAXII via either CAs inhibitors or through the inhibition of the Hh pathway. Methods: We tested cell motility in three melanoma cell lines (WM-35, SK-MEL28, and A375), with different invasiveness capabilities. To this end we performed a scratch assay in the presence of the smoothened (SMO) antagonist cyclopamine (cyclo) or CAs inhibitors under normoxia or hypoxia. Then, we analyzed the invasiveness potential in the cell lines which were more affected by cyclo and CAs inhibitors (SK-MEL28 and A375). Western blot was employed to assess the expression of the hypoxia inducible factor 1α, CAXII, and FAK phosphorylation. Immunofluorescence staining was performed to verify the blockade of CAXII expression. Results: Hh inhibition reduced melanoma cell migration and CAXII expression under both normoxic and hypoxic conditions. Interestingly, basal CAXII expression was higher in the two more aggressive melanoma cell lines. Finally, a direct CAXII blockade impaired melanoma cell migration and invasion under hypoxia. This was associated with a decrease of FAK phosphorylation and metalloprotease activities. Conclusions: CAXII may be used as a target for melanoma treatment not only through its direct inhibition, but also through Hh blockade.

Keywords: carbonic anhydrase; hedgehog; cyclopamine; small molecules; acetazolamide; motility; metalloproteinases; FAK; cancer

1. Introduction

Melanoma is one of the most aggressive skin cancers and represents the fifth most common cancer type in men and the sixth in women [1]. Its incidence has enhanced in recent decades [2–4] but unfortunately, the prognosis is quite good only for localized melanoma after surgery [5]. Indeed melanoma often metastasizes and the prognosis remains poor, along with a low free-disease survival rate [6,7]. The key role in the metastatic transition is due to increased cell motility caused by cytoskeletal changes and altered interactions with extra-cellular matrix (ECM) components [2,8].

Within this context, many new strategies have been identified in an attempt to manage malignant melanoma. This includes the targeting of the strategical FAK/paxillin pathway, whose inhibition resulted in the reduction of melanoma cell migration [9,10]. Still, with regard to ECM, the role of matrix metalloproteinase (MMPs), including MMP-9, in the rapid progression of metastatic melanoma has been also reported [11,12]. MMPs are typical components of the tumor microenvironment (TME) and the TME itself could be one of the potential targets for blocking the metastatic capability of melanoma. Nevertheless, the TME is characterized by hypoxia [13]. Indeed, the local imbalance of O_2 consumption and supply results in the development of adaptive strategies/mechanisms that promote cell survival and progression to metastatic phenotype [14]. Among the above adaptive strategies, hypoxia triggers the expression of carbonic anhydrases (CAs) [15], a family of metalloenzymes which catalyze the reversible hydration of CO_2 to HCO_3^- and H^+. [16]. Eight different classes have been described but only α-CAs isoforms have been characterized in humans [17]. Many roles have been attributed to CAs, from intra- and extracellular pH regulation, to homeostasis maintenance and cell survival and migration [18]. However, only human isoforms IX (CAIX) and XII (CAXII) play crucial roles in tumorigenicity and metastatic progression [19,20]. While CAIX has a limited expression in normal tissues and is upregulated in cancer, CAXII has several properties both in physiological and tumorigenic conditions, thus gaining wide interest among several researchers [21–24]. Unlike CAIX, whose role in cancer has been largely demonstrated and which has been already identified as a target in antitumor therapy [25–29], there are only a few reports regarding CAXII [30–32]. The fact that CAXII inhibition may be effective in anticancer therapy led to the synthesis of newly CAs inhibitors. In this regard, we took advantage of new available small chemical molecules which were designed starting from the pan-CAs inhibitor GV2-20 [33,34]. Novel compounds, with a different inhibitory potency against CAIX and CAXII, were identified based on an in silico drug design approach. Among them, we selected two compounds with a high specificity for the tumor-associated isoforms IX and XII: C-7 (CAIX K_i = 27.6 nM; CAXII K_i >50,000 nM) and C-10 (CAIX K_i = 16 nM; CAXII K_i = 82.1 nM). While C-7 is a selective inhibitor only for CAIX, C-10 has the most potent effect not only on CAIX but also on CAXII [32]. We have recently shown that CAXII is regulated by the Hedgehog (Hh) pathway in breast cancer [35,36]. Physiologically, the Hh pathway controls organ development during embryogenesis, whereas in adults remains quiescent, except for tissue repairing [37,38]. Of interest, aberrations of this pathway occur in tumors, being responsible for tumorigenesis and cancer maintenance [39,40]. Indeed, its role in controlling the proliferation of stem cells and tissue progenitors means the Hh pathway is increasingly the subject of cancer management studies [41–45].

Recent reports have shown that activation of the Hh pathway promotes metastasis in melanoma [46]. In addition, melanoma cells adapt to the acidic TME resulting in a higher tumor initiation potential [47]. Indeed, the role of CAs has already been reported in melanoma progression. However, while the involvement of CAIX has been associated with worse overall survival in patients with melanoma [48], information regarding CAXII is still scant.

For the above reasons, we decided to investigate the role of CAXII in melanoma cell migration and invasiveness. Our results, obtained by inhibiting CAXII either directly or indirectly through the Hh pathway, may contribute to the identification of novel therapeutic strategies for future management of melanoma.

2. Results

2.1. Inhibition of the Hh Pathway Affects Melanoma Cell Migration and CAXII Protein Expression

Recently, our research group discovered a correlation between the Hh pathway and CAXII on breast cancer cell migration [36]. The Hh pathway was already explored in melanoma [49,50], however its involvement in CAs modulation has not been investigated yet. Thus, we decided to analyze this possible interaction in two human melanoma cell lines, WM35 and SK-MEL-28. It should be underlined that WM35 are characterized by reduced migratory capability and aggressivity [51], while SK-MEL-28 cells are associated with intermediate aggressive phenotype, along with high cell migration and invasiveness [52]. Cells were treated with the natural smoothened (SMO) antagonist cyclopamine (cyclo) and we evaluated cell migration and CAXII protein expression. We performed experiments in both normoxic (20% $O_2 \simeq$ pO_2 of 140 mmHg) and hypoxic conditions (2% O_2 = pO_2 of 14 mmHg) in order to mimic the TME pO_2. As shown in Figure 1A,B, both cell lines expressed significantly higher protein levels of HIF-1α when they were exposed to hypoxia for 24 h. These results confirm that the pO_2 employed in our experiments was adequate to induce a hypoxic state in both cell lines. We therefore investigated whether the Hh pathway was involved in melanoma cell migration, either under normoxia or hypoxia. To this end, we performed a wound healing assay, where confluent monolayers of WM35 and SK-MEL-28 were scratched after an overnight treatment with cyclo and cell motility was monitored for 24 h incubation either under normoxic or hypoxic conditions. Of interest, inhibition of the Hh pathway did not significantly affect the WM35 cell migration rate (Figure 1B,C), either in normoxia or hypoxia. However, Hh inhibition resulted in a significant reduction of SK-MEL-28 cell migration in a normoxic environment. Indeed, we observed a five-fold inhibition of cell migration after 24 h treatment when compared with the untreated cells (Figure 1B). Interestingly, the impairment of cell migration was significant also under hypoxia (Figure 1C), indicating that Hh inhibition was also effective at a pO_2 which resembles the TME. To investigate whether the Hh pathway may be involved in modulating the expression of CAXII in melanoma cells, we then analyzed the protein levels of CAXII in both cell lines exposed either under normoxia or hypoxia, after 24 h treatment in the presence or not of cyclo. Figure 1D clearly shows that WM35 cells expressed low levels of CAXII. Thereafter, we did not observe any significant differences upon cyclo treatment in both experimental conditions. In contrast, CAXII expression was significantly increased in hypoxic SK-MEL-28, and according with the cell migration results, cyclo treatment significantly reduced the level of CAXII (Figure 1D). The latter results indicate that in SK-MEL-28 cell line the expression of CAXII is regulated by the Hh pathway. The same regulatory effect may be present also in WM35, but the low protein levels do not allow the detection of a significant reduction in protein expression.

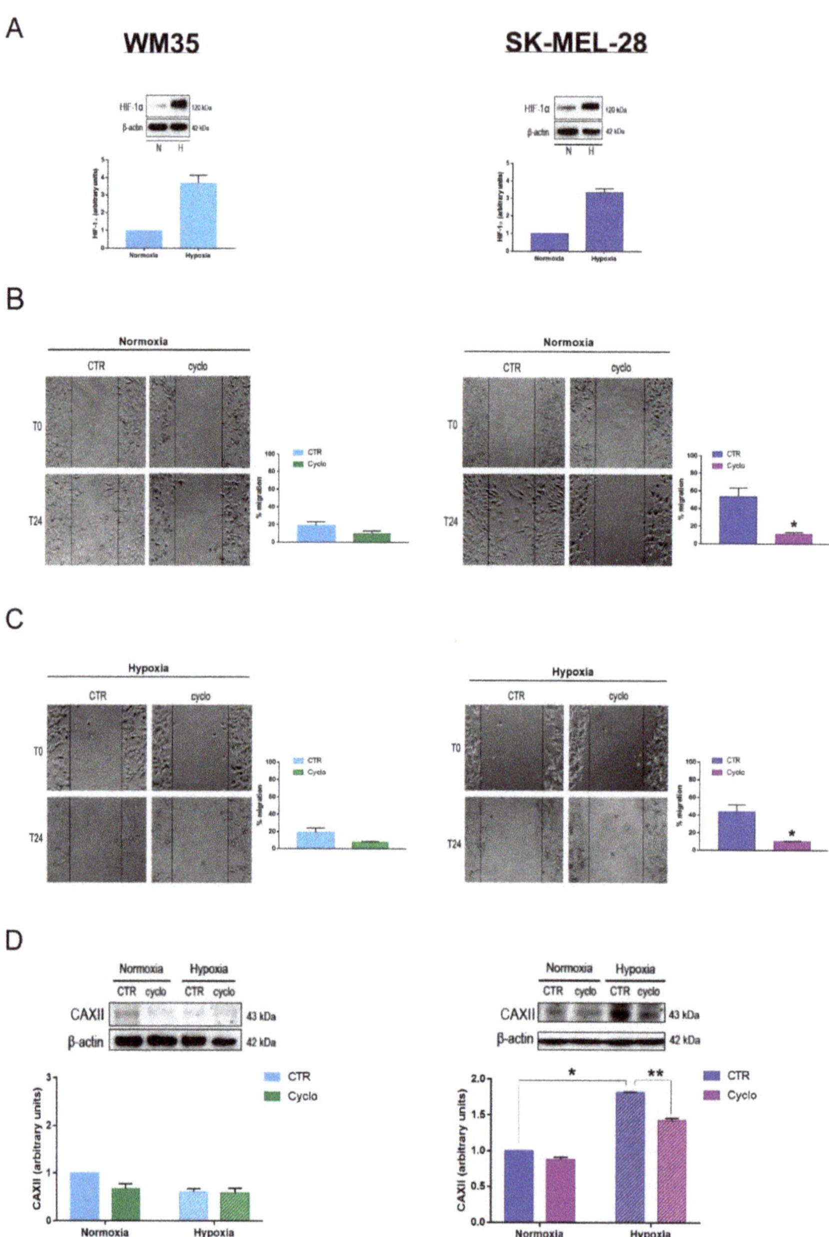

Figure 1. Inhibition of the Hedgehog pathway affects melanoma cell migration and CAXII protein expression. (**A**) HIF-1α protein expression in WM35 and SK-MEL-28 after 24 h exposure to normoxia or hypoxia. Blots are representations of the means of three independent experiments and β-actin was used as loading control. (**B**) Representative image of WM35 and SK-MEL-28 scratch assay after 24 h treatment with 20 µM cyclopamine (cyclo) under normoxia. Phase contrast microscopy images were taken with Olympus IX81 at a 10× magnification. Quantifications represent the means and standard errors of the

mean of three independent experiments. (**C**) Representative image of WM35 and SK-MEL-28 scratch assay after 24 h treatment with 20 µM cyclo under hypoxia. Phase contrast microscopy images were taken with Olympus IX81 at a 10× magnification. Quantifications represent the means and standard errors of the mean of three independent experiments. (**D**) CAXII protein expression in WM35 and SK-MEL-28 after 24 h treatment with 20 µM cyclo exposed to normoxia or hypoxia. Blots are representation of three independent experiments and β-actin was used as loading control. * $p \leq 0.05$, ** $p \leq 0.01$ indicate statistically significant differences. The uncropped blots of Figure 1A,D are shown in Figure S1.

2.2. Effects of CAs Inhibitors on Melanoma Cell Migration

Since inhibition of the Hh pathway by cyclo impaired both cell migration and CAXII expression in SK-MEL-28, we next evaluated whether a direct inhibition of CAXII could affect melanoma cell migration as well. To this end, we used not only WM35 and SK-MEL28 but also a third cell line, A375, which is characterized by a highly aggressive phenotype.

First of all, we measured CAXII protein expression in all the cell lines, under normoxia or hypoxia, to ascertain a different expression within the two microenvironments. Figure 2A shows that only SK-MEL-28 expressed high level of CAXII under normoxic condition. As expected, under hypoxia, CAXII expression in WM35 remained low; however, its expression was still high in SK-MEL-28 and, of interest, it was significantly upregulated in A375. Thus, the different CAXII expression, especially in hypoxic conditions, seemed to be related to the different aggressiveness of the cell lines. Thereafter, we used two new small molecules, C-7 and C-10, along with the pan-CAs inhibitors GV2-20 and acetazolamide (AAZ), which were employed as reference compounds. As described above, C-7 is a selective CAIX inhibitor, while C-10 inhibits both CAIX and CAXII at nanomolar concentrations [33]. First of all, we tested the effects of these new compounds on WM35, SK-MEL-28, and A375 viability up to 72 h incubation. As shown in Figure 2B, CAs inhibitors did not significantly affect cell viability in all cell lines at 10 nM concentration, which was the same concentration employed in the following assays. This was evident when the cell lines were exposed either to normoxia or hypoxia. These results were important to rule out the possibility that the migration and invasion assays might be related to cell viability variations.

Thus, we evaluated the effect of CAs inhibitors on cell migration by performing a scratch assay. Similarly to the Hh inhibition results, pretreatment with CAs inhibitors did not significantly affect WM35 migration either under normoxia or hypoxia (Figure 2C,D). However, we observed a significant reduction in SK-MEL-28 and A375 cell migration, after pretreatment with compound C-10. Again these effects were similar under either normoxic or hypoxic conditions (Figure 2C,D). These results suggest that the inhibition of CAIX alone is not sufficient to regulate cell migration. In contrast, CAXII appears to be a key regulator in the migration of the intermediate and high aggressive melanoma cell lines SK-MEL-28 and A375. Indeed, C-10 is characterized as one of the more specific inhibitors of CAXII, when compared not only to C-7 but also to the pan-inhibitors GV2-20 and AAZ.

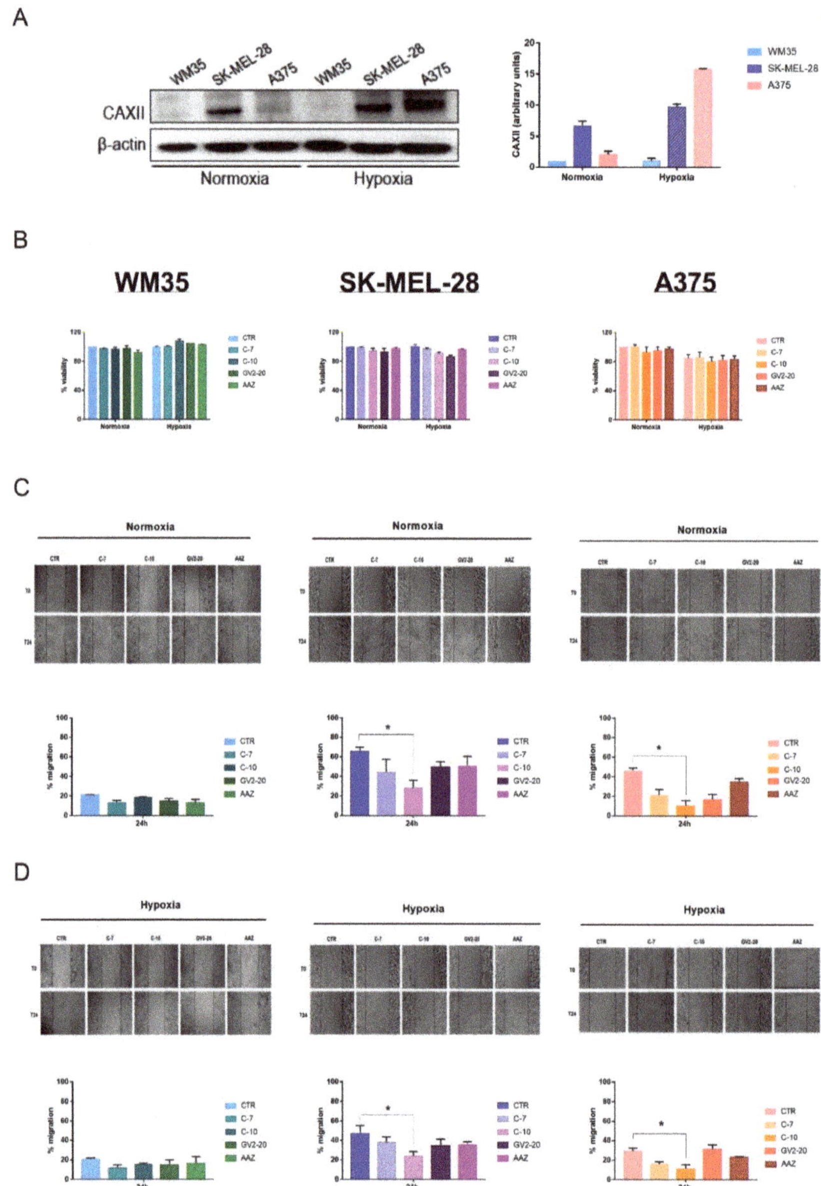

Figure 2. Effects of CAs inhibitors on cell migration. (**A**) CAXII protein expression in WM35, SK-MEL-28, and A375 under normoxia and hypoxia. Blot is a representation of the means and standard errors of the mean of three independent experiments. β-actin was used as loading control. (**B**) WM35, SK-MEL-28, and A375 viability assays after 72 h treatment with 10 nM of each compound and exposed to normoxia or hypoxia. Data are reported as the percentage of the ratio between the treated groups and controls. Graphs represent the means and standard errors of the mean of three independent experiments performed in triplicates. (**C**) WM35, SK-MEL-28, and A375 scratch assays after an overnight pretreatment and 24 h treatment with 10 nM of each compound and exposed to normoxia. Phase contrast

microscopy images were taken with Olympus IX81 at a 10× magnification. Data are represented as the percentage of migration rate between treated groups and controls. Graphs represent the means and standard errors of the mean of three independent experiments performed in duplicates. (**D**) WM35, SK-MEL-28 and A375 scratch assays after an overnight pretreatment and 24 h treatment with 10 nM of each compound and exposed to hypoxia. Phase contrast microscopy images were taken with Olympus IX81 at a 10× magnification. Data are represented as the percentage of migration rate between treated groups and controls. Graphs represent the means and standard errors of the mean of three independent experiments performed in duplicates. * indicates statistically significant differences ($p \leq 0.05$). The uncropped blots of Figure 2A is shown in Figure S2.

2.3. SK-MEL-28 and A375 Invasion Is Reduced by Direct CAXII Inhibition under Hypoxia

Based on the above results, showing that CAs inhibitors significantly affected only SK-MEL-28 and A375 migration, we next focused our experiments on these two, more aggressive cell lines, which were exposed to a hypoxic condition that resembled the TME. It should be underlined that previous results have shown that hypoxia significantly affects the invasiveness of melanoma cells [53–57].

At first, we evaluated the effect of CAs inhibition on hypoxic SK-MEL-28 and A375 invasiveness. Figure 3A clearly shows that C-10, unlike the pan-inhibitor AAZ, was able to reduce SK-MEL-28 invasion in a significant manner. A375 invasion was also significantly impaired by both C-7 and C-10 inhibitors but the last resulted in the most potent effect. These results indicate that the inhibition of CAXII plays a crucial role in regulating the invasiveness of both intermediate and highly aggressive melanoma cell lines. Furthermore, we analyzed the effects of C-7 and C-10 on the phosphorylated focal adhesion kinase (phFAK), as its role in the promotion of the aggressive melanoma phenotype has been previously established in several reports [58]. In line with the migration and invasion results, inhibition by C-10 resulted in a significant decrease of phFAK levels both in SK-MEL-28 and A375 (Figure 3B). In addition, C-10 treatment resulted also in a significant decrease of MMP-9 activity, as shown by the zymographic analyses in Figure 3C. The relevance of MMP-9 in melanoma malignancy and progression has been already documented [11]. However, such an inhibition was exerted also by C-7, indicating that CAIX and CAXII were similarly involved in the regulation of MMP-9. Finally, to further confirm the inhibitory effect of C-7 and C-10 on CAIX and CAXII, we analyzed their protein expression in hypoxic SK-MEL28 by fluorescence microscopy. As shown in Figure 3D, C-7 and C-10 were able to reduce CAIX expression. However, the expression of CAXII was significantly reduced only by C-10. The fluorescence data supported our results which demonstrated that C10 was able to reduce the expression of CAXII, which could be crucial in the regulation of melanoma cell migration and invasion.

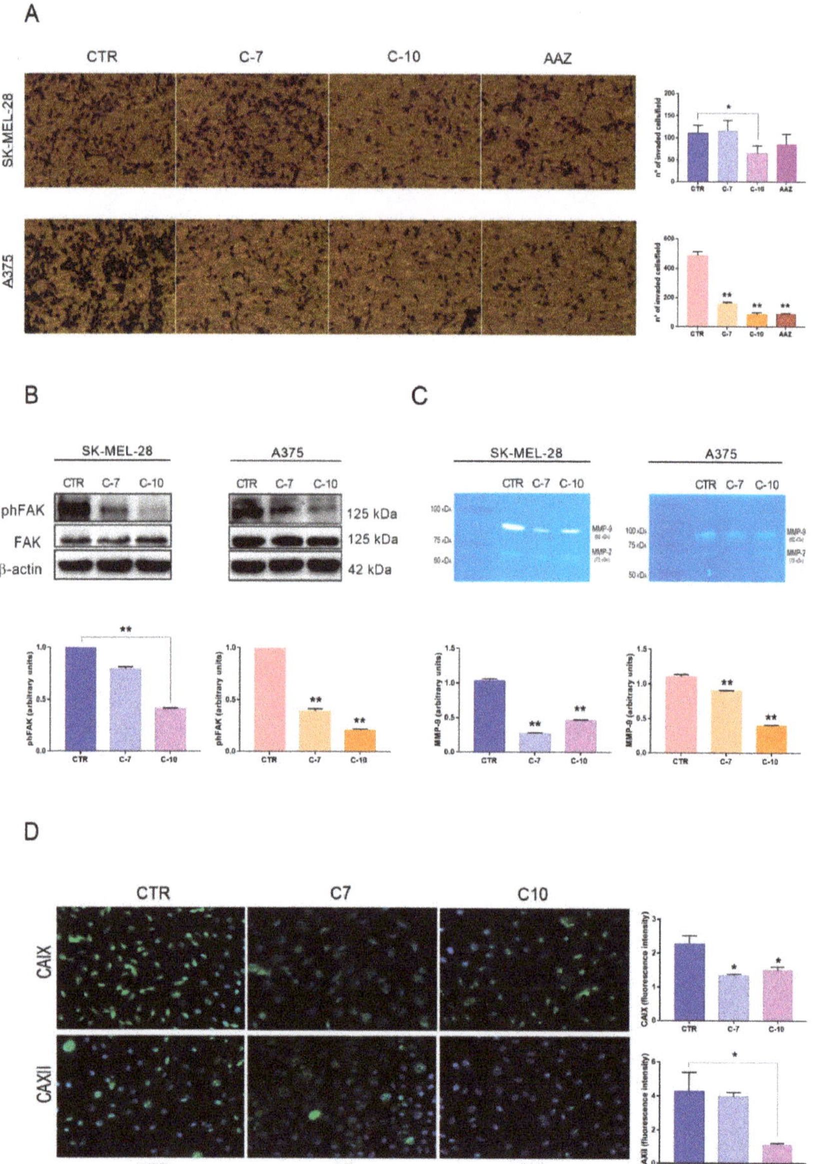

Figure 3. SK-MEL-28 and A375 invasion is reduced by direct CAXII inhibition under hypoxia. (**A**) Modified Boyden chamber invasion assays after pretreatment and 24 h treatment under hypoxia (10× magnification). Data are presented as the means and standard errors of the mean of three independent experiments performed in triplicates. Acetazolamide (AAZ) was used as control. (**B**) Phosphorylated focal adhesion kinase (phFAK) protein expressions after 24 h treatment with C-7 and C-10 under hypoxia as determined by western blot analyses. Data are presented as the means and standard errors of the mean of three independent experiments. B-actin was used as loading control. (**C**) Matrix metalloproteinase 9 (MMP-9) enzymatic activity after 24 h treatment upon hypoxia as determined by

zymography. Data are presented as the means and standard errors of the mean of three independent experiments. (**D**) CAXII protein expression in SK-MEL-28 after 24 h treatment with C-7 and C-10 under hypoxia as determined by immunofluorescence microscopy analyses. (25× objective 50× magnification). Data are presented as the means and standard errors of the mean of three independent experiments. * $p \leq 0.05$, ** $p \leq 0.01$ indicate statistically significant differences. The uncropped blots of Figure 3B is shown in Figure S3.

3. Discussion

CAXII is a CAs isoform strictly related to cancer and hypoxic responses. In fact, when oxygen is lacking, CAXII expression is modified, as tumor cells might still proliferate and survive [26]. CAXII has been recently proposed as a possible target for anticancer therapy [31] and novel and specific inhibitors of this enzyme, along with CAIX, have been recently identified [33]. Furthermore, we have recently reported that CAXII expression is regulated by the Hh pathway in breast cancer, indicating that this enzyme plays a potential role in breast cancer cell migration and invasiveness [36]. We here report, for the first time, that the Hh pathway is involved in the modulation of CAXII also in melanoma cells, in particular, in a moderate aggressive melanoma cell line and, more interestingly, in hypoxic conditions.

Previous reports clearly demonstrated that hypoxia was critical for tumor progression in melanoma [57]. In the present manuscript, we show that the more aggressive melanoma cell lines, SK-MEL-28 and A375, expressed higher protein levels of CAXII under hypoxia. This is in agreement with previous reports showing that CAXII expression is increased in several aggressive tumors along with a poor prognosis [27]. Furthermore, hypoxia has been reported to trigger the Hh pathway in colorectal cancer [59], suggesting that its inhibition may be relevant for tumor progression. In the present manuscript, we show that Hh inhibition, under hypoxia, resulted in a downregulation of CAXII. Moreover, we here report that CAXII is overexpressed in SK-MEL-28 and A375 in comparison to WM35 and that its expression correlates with higher migration capability. Thus, CAXII appears to be crucial for cell migration. It should be underlined that cell migration is associated with epithelial–mesenchymal transition (EMT) in several neoplasias, including the metastatic progression in melanoma [60–63]. The involvement of the Hh pathway was already explored in melanoma and its blockade reduced cell viability and migration [49,50]. The fact that, in the present manuscript, Hh inhibition resulted in CAXII downregulation in melanoma cells, and in particular in the more aggressive cell line, opens a new scenario for a possible therapeutic targeting. In fact, C-10, which is active at nanomolar concentrations on both CAXII and CAIX, was able to reduce cell migration while the inhibition of only CAIX by C-7 was not sufficient. It should be pointed out that in our experimental protocol C-10 was even more effective than the pan-inhibitors GV2-20 and AAZ. We and others have previously reported that CAXII promotes not only the migration but also invasive capability of breast cancer cells [36,64]. Indeed, the decreased invasiveness and migration ability of CAXII-knockdown cells were restored by an overexpression of CAXII [64]. Here, we demonstrated that C-10 was particularly effective in reducing the invasion ability of SK-MEL-28 and A375 under a hypoxic condition which mimics the TME. Furthermore, we here report that C-10 reduced the phosphorylation of FAK as well as the activity of MMP-9. Since both FAK and MMP-9 have been previously related to the invasive properties of melanoma cells, our results further support the anti-invasive properties of CAXII inhibition.

4. Materials and Methods

4.1. Cell Cultures

WM35 and SK-MEL-28 were kindly donated by Dr. Raffaella Giavazzi (Department of Oncology, Istituto di Ricerche Farmacologiche Mario Negri IRCCS, Milan, Italy) and Dr. Francesca Chiarini (CNR Institute of Molecular Genetics "Luigi Luca Cavalli-Sforza", Bologna, Italy.) respectively. A375 were purchased from American Type Culture Collection (ATCC, Manassas, VA, USA). WM35 and SK-MEL-28 were maintained in RPMI 1640 (Euroclone, Devon, UK) and A375 were cultured in

Dulbecco's modified Eagle's medium (DMEM, Euroclone, Devon, UK). All media were supplemented with antibiotics, L-glutamine 2 mM and 10% fetal bovine serum (FBS) (Euroclone, Devon, UK), at 37 °C in humidified 5% CO_2.

The hypoxic treatment was performed using a workstation In Vivo 400 (Ruskinn, Pencoed, UK) providing a customized and stable humidified environment through electronic control of 2% O_2 and 5% CO_2.

4.2. Chemical Compounds

C-7 and C-10 are new small molecules targeting CAIX or both CAIX and CAXII respectively, which were kindly provided by Department of Biotechnology, Chemistry and Pharmacy (University of Siena, Siena, Italy) [33].

We used pan-CAs inhibitor GV2-20 (a benzoic acid derivative previously identified as CAs inhibitor by a computational-driven target fishing approach [34] (Molport, Riga, Latvia)) and AAZ (Sigma-Aldrich, St.Louis, MO, USA) as reference compounds.

All the inhibitors were resuspended in dimethyl sulfoxide (DMSO) and diluted in water and used at a final concentration of 10 nM.

Cyclo (AlphaAesar, Haverhill, MA, USA) was resuspended in DMSO and used at the final concentration of 20 µM.

4.3. Cell Viability

Cells were seeded in 96-well plates at a density of 2×10^4 cells per well, treated with CAs inhibitors and incubated under normoxia or hypoxia. After 72 h, they were counted using fluorescein diacetate (1 mg/mL, diluted 1:200). The viability index was calculated as the percentage of the ratio between treated groups and control. All the experiments were run in triplicate and repeated at least three times.

4.4. Western Blot

Cells were seeded in 60 mm petri dishes at a density of 5×10^5 cells per dish and, following specific treatments, were incubated for 24 h in normoxic or hypoxic condition. Thereafter, cells were washed with cold PBS, were lysed with Laemmli Buffer containing protease inhibitors (1 mM PMSF, 1 mg/mL aprotinin, 1 mg/mL leupeptin) and frozen at −20 °C. Samples were sonicated for three times and the quantification of total protein concentration was determined using Micro BCA Protein Assay Reagent kit (Thermo Fisher Scientific, Cleveland, OH, USA). Sample measures of 40 µg per lane were resolved on 10% acrylamide gel, transferred to nitrocellulose membrane, and incubated for 1 h with 5% nonfat powdered milk. We used HIF-1α (BD Biosciences, San Jose, CA, USA), phFAK, FAK (Cell Signaling, Denver, CO, USA), CAXII (Santa Cruz, CA, USA), and β-Actin (Sigma-Aldrich, St. Louis, MO, USA). The specific secondary antibodies, anti-rabbit IgG-HRP and anti-mouse IgG-HRP (Cell Signaling, Denver, CO, USA), were incubated for 1 h. Chemiluminescence of immunoreactive bands was detected with ChemiDoc™ MP System (Bio-Rad, Hercules, CA, USA) and quantified with Image Lab software. The original Western blot figures can be found in the supplementary file.

4.5. Scratch Wound Healing Assay

Cells were seeded in 24-well plates at a density of 1×10^5 cells per well, pretreated with CAs inhibitors overnight and the following day the confluent monolayer was scratched with a 200 µL tip. Washing twice with PBS and refreshing medium with 10% FBS was followed. Each well was treated again with the respective inhibitor and incubated either under normoxic or hypoxic microenvironments. After 0 h and 24 h snapshots of the scratch were taken with the microscope (10×) (Olympus IX81, Tokyo, Japan) with cell F© software. Migration was calculated as $\left(1 - \frac{Ax}{A0}\right)\%$, where $A0$ and Ax represented the empty scratch area at 0 h and 24 h, respectively.

4.6. Boyden Chamber

Chemoinvasion was analyzed employing a modified Boyden 48-well micro chemotaxis chamber (Neuro Probe, Gaithersburg, MD, USA) with 8 µm pore size polycarbonate polyvinylpyrrolidone-free nucleopore filters, as previously described [65,66]. Cells were pretreated overnight, the following day were resuspended in RPMI 0,1% bovine serum albumin (BSA) and 50 µL of an amount of 2.7×10^5 cell/mL were loaded in triplicate on each upper well, which was previously coated with 0.5 mg/mL of Matrigel (Corning, Life Science, Corning, Tewksbury, MA, USA). NIH3T3 supernatant was added in the lower chamber and used as chemoattractant. After 24 h incubation under hypoxic condition, cells on the lower surface were stained using Diff Quick (Merz-Dade, Düdingen, Switzerland) and photographed by an OLYMPUS IX81 Research Microscope with a 10× magnification. Invaded cells in four high-power fields were counted and data were expressed as number of invaded cells per field.

4.7. Zymography

Cells were incubated with C-7 and C-10 inhibitors for 24 h under hypoxia and then media were collected and centrifuged at 300× g for 10 min to remove cells debris. Total protein amount was analyzed using Micro BCA Protein Assay Reagent kit (Thermo Fisher Scientific, Cleveland, OH, USA) and 15 µg of each appropriately diluted media was loaded into 0.1% gelatin-8% acrylamide gels. After electrophoresis, gelatin gels were washed with 2.5% Triton x-100 for 30 min in agitation and an overnight incubation at 37 °C in agitation with developing buffer was followed. Then, gels were stained with Blue Comassie R-250 (Sigma Aldrich, St.Louis, MO, USA) and, after destaining, images were taken with ChemiDoc™ MP System (Bio-Rad, Hercules, CA, USA) and quantified with ImageJ software.

4.8. Immunofluorescence Staining

SK-MEL-28 were plated on 11 mm diameter glass coverslips in a 24-well plate at 10^5 cells per well and treated with C-7 and C-10 inhibitors for 24 h before fixation. Cells were then fixed in cold methanol for 10 min at −20 °C, were then washed three times for 10 min with PBS and incubated in blocking buffer (1% BSA in PBS) for 30 min at room temperature. An overnight incubation in CAIX (Cell Signaling, Denver, CO, USA, diluted 1:10 in PBS/BSA) or CAXII primary antibody (Santa Cruz, CA, USA; diluted 1:50 in PBS/BSA) at 4 °C in humid chambers was followed. Coverslips were then washed for 10 min in PBS/BSA and incubated with secondary antibody (Alexa Fluor-555-conjugated anti-rabbit-IgG and Alexa Fluor-488-conjugated anti-mouse-IgG, 1:800; Invitrogen, Carlsbad, CA, USA) for 1 h at room temperature. After a final washing in PBS, the coverslips were mounted on slides with 90% glycerol in PBS. DNA was visualized with incubation of 3–4 min in Hoechst (Sigma Aldrich, St.Louis, MO, USA).

Images were taken by using an Axio Imager Z1 (Carl Zeiss, Jena, Germany) microscope equipped with an AxioCam HR cooled charge-coupled camera (Carl Zeiss, Jena, Germany). Grayscale digital images were collected separately and pseudocolored and merged using Adobe Photoshop 7.0 software (Adobe Incorporated, San Jose, CA, USA). Fluorescence intensity was measured with ImageJ software.

4.9. Statistical Analyses

Data are presented as means ± standard error of the means of at least three independent experiments performed in duplicates or triplicates. Statistical analyses were conducted using unpaired t-test and ANOVA test with GraphPad Prism 7 software. Values of $p \leq 0.05$ and $p \leq 0.01$ were conventionally considered statistically significant.

5. Conclusions

The identification of new small molecules with specific inhibitory activities on different CAs, especially CAXII, opens a new scenario in the therapeutic approach in melanoma patients. Previous reports indicate the relevance of hypoxia and CAs in the TME and metastatic progression. Overall, our results suggest that CAXII inhibition under hypoxia leads to a reduction of both cell migration and invasion in aggressive melanoma cell lines. This, along with our data regarding Hh inhibition, may contribute to possible alternative treatment by acting either directly or indirectly on CAXII to reduce the metastatic capability of melanoma.

Supplementary Materials: The following are available online at http://www.mdpi.com/2072-6694/12/10/3018/s1, Figure S1: The uncropped blots of Figure 1A,D, Figure S2: The uncropped blots of Figure 2A, Figure S3: The uncropped blots of Figure 3B.

Author Contributions: Experimental procedures, writing–original draft preparation and editing, G.G.; Experimental support, S.M.; Chemical support, Y.C. and M.M.; Writing—review and editing, F.C. and A.N.; Supervision, F.C. All authors have read and agreed to the published version of the manuscript.

Funding: This study has received financial support from the Istituto Toscano Tumori (ITT) and MIUR (PRIN 2017NTK4HY_002 to A.N.).

Acknowledgments: We thank Carlo Aldinucci for technical support. In memory of Maurizio Botta for his precious support in the development of the present study.

Conflicts of Interest: The authors declare no conflict of interest.

References

1. Rastrelli, M.; Tropea, S.; Rossi, C.R.; Alaibac, M. Melanoma: Epidemiology, risk factors, pathogenesis, diagnosis and classification. *In Vivo* **2014**, *28*, 1005–1011. [PubMed]

2. Rossi, S.; Cordella, M.; Tabolacci, C.; Nassa, G.; D'Arcangelo, D.; Senatore, C.; Pagnotto, P.; Magliozzi, R.; Salvati, A.; Weisz, A.; et al. TNF-alpha and metalloproteases as key players in melanoma cells aggressiveness. *J. Exp. Clin. Cancer Res.* **2018**, *37*, 326. [CrossRef] [PubMed]

3. Eggermont, A.M.; Spatz, A.; Robert, C. Cutaneous melanoma. *Lancet* **2014**, *383*, 816827. [CrossRef]

4. Matthews, N.H.; Li, W.-Q.; Qureshi, A.A.; Weinstock, M.A.; Cho, E. Epidemiology of melanoma. In *Cutaneous Melanoma: Etiology and Therapy*; Ward, W.H., Farma, J.M., Eds.; Codon Publications: Brisbane, Austrilia, 2017.

5. Jansen, B.; Wacheck, V.; Heere-Ress, E.; Schlagbauer-Wadl, H.; Hoeller, C.; Lucas, T.; Hoermann, M.; Hollenstein, U.; Wolff, K.; Pehamberger, H. Chemosensitisation of malignant melanoma by BCL2 antisense therapy. *Lancet* **2000**, *356*, 1728–1733. [CrossRef]

6. Balch, C.M.; Gershenwald, J.E.; Soong, S.-J.; Thompson, J.F.; Atkins, M.B.; Byrd, D.R.; Buzaid, A.C.; Cochran, A.J.; Coit, D.G.; Ding, S.; et al. Final version of 2009 AJCC melanoma staging and classification. *J. Clin. Oncol.* **2009**, *27*, 6199–6206. [CrossRef]

7. Santini, R.; Vinci, M.C.; Pandolfi, S.; Penachioni, J.Y.; Montagnani, V.; Olivito, B.; Gattai, R.; Pimpinelli, N.; Gerlini, G.; Borgognoni, L.; et al. Hedgehog-gli signaling drives self-renewal and tumorigenicity of human melanoma-initiating cells. *Stem Cells* **2012**, *30*, 1808–1818. [CrossRef]

8. Gaggioli, C.; Sahai, E.M. Melanoma invasion? Current knowledge and future directions. *Pigment. Cell Res.* **2007**, *20*, 161–172. [CrossRef]

9. Sieg, D.J.; Hauck, C.R.; Ilic, D.; Klingbeil, C.K.; Schaefer, E.; Damsky, C.H.; Schlaepfer, D.D. FAK integrates growth-factor and integrin signals to promote cell migration. *Nat. Cell Biol.* **2000**, *2*, 249–256. [CrossRef]

10. Liu, J.-F.; Lai, K.; Peng, S.-F.; Maraming, P.; Huang, Y.-P.; Huang, A.-C.; Chueh, F.-S.; Huang, W.-W.; Chung, J.-G. Berberine inhibits human melanoma A375.S2 cell migration and invasion via affecting the FAK, uPA, and NF-κB signaling pathways and inhibits PLX4032 resistant A375.S2 cell migration in vitro. *Molecules* **2018**, *23*, 2019. [CrossRef]

11. Nikkola, J.; Vihinen, P.; Vuoristo, M.-S.; Kellokumpu-Lehtinen, P.; Kähäri, V.-M.; Pyrhönen, S. High serum levels of matrix metalloproteinase-9 and matrix metalloproteinase-1 are associated with rapid progression in patients with metastatic melanoma. *Clin. Cancer Res.* **2005**, *11*, 5158–5166. [CrossRef]

12. Malaponte, G.; Zacchia, A.; Bevelacqua, Y.; Marconi, A.; Perrotta, R.; Mazzarino, M.C.; Cardile, V.; Stivala, F. Co-regulated expression of matrix metalloproteinase-2 and transforming growth factor-β in melanoma development and progression. *Oncol. Rep.* **2010**, *24*. [CrossRef]

13. Semenza, G.L. Hypoxia and cancer. *Cancer Metastasis Rev.* **2007**, *26*, 223. [CrossRef] [PubMed]

14. Matolay, O.; Méhes, G. Sustain, adapt and overcome-Hypoxia associated changes in the progression of lymphatic neoplasia. *Front. Oncol.* **2019**, *9*, 1277. [CrossRef] [PubMed]

15. Morris, J.C.; Chiche, J.; Grellier, C.; Lopez, M.; Bornaghi, L.F.; Maresca, A.; Supuran, C.T.; Pouysségur, J.; Poulsen, S.-A. Targeting hypoxic tumor cell viability with carbohydrate-based carbonic anhydrase IX and XII inhibitors. *J. Med. Chem.* **2011**, *54*, 6905–6918. [CrossRef]

16. Mboge, M.Y.; Mahon, B.P.; Agbandje-McKenna, M.; Frost, S.C. Carbonic anhydrases: Role in pH control and cancer. *Metabolites* **2018**, *8*, 19. [CrossRef]

17. Haapasalo, J.; Nordfors, K.; Haapasalo, H.; Parkkila, S. The expression of carbonic anhydrases II, IX and XII in brain tumors. *Cancers* **2020**, *12*, 1723. [CrossRef] [PubMed]

18. Lomelino, C.L.; Supuran, C.T.; McKenna, R. Non-Classical inhibition of carbonic anhydrase. *Int. J. Mol. Sci.* **2016**, *17*, 1150. [CrossRef] [PubMed]

19. Supuran, C.T.; Winum, J.-Y. Designing carbonic anhydrase inhibitors for the treatment of breast cancer. *Expert Opin. Drug Discov.* **2015**, *10*, 591–597. [CrossRef]

20. McDonald, P.C.; Winum, J.-Y.; Supuran, C.T.; Dedhar, S. Recent developments in targeting carbonic anhydrase IX for cancer therapeutics. *Oncotarget* **2012**, *3*, 84–97. [CrossRef]

21. Kivelä, A.J.; Parkkila, S.; Saarnio, J.; Karttunen, T.J.; Kivelä, J.; Parkkila, A.-K.; Pastoreková, S.; Pastorek, J.; Waheed, A.; Sly, W.S.; et al. Expression of transmembrane carbonic anhydrase isoenzymes IX and XII in normal human pancreas and pancreatic tumours. *Histochem. Cell Biol.* **2000**, *114*, 197–204. [CrossRef]

22. Ilie, M.I.; Hofman, V.; Ortholan, C.; El Ammadi, R.; Bonnetaud, C.; Havet, K.; Vénissac, N.; Mouroux, J.; Mazure, N.M.; Pouysségur, J.; et al. Overexpression of carbonic anhydrase XII in tissues from resectable non-small cell lung cancers is a biomarker of good prognosis. *Int. J. Cancer* **2010**, *128*, 1614–1623. [CrossRef]

23. Kopecka, J.; Campia, I.; Jacobs, A.; Frei, A.P.; Ghigo, D.; Wollscheid, B.; Riganti, C. Carbonic anhydrase XII is a new therapeutic target to overcome chemoresistance in cancer cells. *Oncotarget* **2015**, *6*, 6776–6793. [CrossRef] [PubMed]

24. Supuran, C.T. Carbonic anhydrases: Novel therapeutic applications for inhibitors and activators. *Nat. Rev. Drug Discov.* **2008**, *7*, 168–181. [CrossRef] [PubMed]

25. Kim, J.-Y.; Shin, H.-J.; Kim, T.H.; Cho, K.-H.; Shin, K.-H.; Kim, B.-K.; Roh, J.-W.; Lee, S.; Park, S.-Y.; Hwang, Y.-J.; et al. Tumor-associated carbonic anhydrases are linked to metastases in primary cervical cancer. *J. Cancer Res. Clin. Oncol.* **2006**, *132*, 302–308. [CrossRef]

26. Chiche, J.; Ilc, K.; Laferrière, J.; Trottier, E.; Dayan, F.; Mazure, N.M.; Brahimi-Horn, M.C.; Pouysségur, J. Hypoxia-Inducible carbonic anhydrase IX and XII promote tumor cell growth by counteracting acidosis through the regulation of the intracellular pH. *Cancer Res.* **2009**, *69*, 358–368. [CrossRef] [PubMed]

27. Chen, Z.; Ai, L.; Mboge, M.Y.; Tu, C.; McKenna, R.; Brown, K.D.; Heldermon, C.D.; Frost, S.C. Differential expression and function of CAIX and CAXII in breast cancer: A comparison between tumorgraft models and cells. *PLoS ONE* **2018**, *13*, e0199476. [CrossRef] [PubMed]

28. Robertson, N.; Potter, C.; Harris, A.L. Role of carbonic anhydrase IX in human tumor cell growth, survival, and invasion. *Cancer Res.* **2004**, *64*, 6160–6165. [CrossRef]

29. Lou, Y.; McDonald, P.C.; Oloumi, A.; Chia, S.; Ostlund, C.; Ahmadi, A.; Kyle, A.; Keller, U.A.D.; Leung, S.; Huntsman, D.; et al. Targeting tumor hypoxia: Suppression of breast tumor growth and metastasis by novel carbonic anhydrase IX inhibitors. *Cancer Res.* **2011**, *71*, 3364–3376. [CrossRef]

30. Doyen, J.; Parks, S.K.; Marcié, S.; Pouysségur, J.; Chiche, J. Knock-down of hypoxia-induced carbonic anhydrases IX and XII radiosensitizes tumor cells by increasing intracellular acidosis. *Front. Oncol.* **2013**, *2*. [CrossRef]

31. Von Neubeck, B.; Gondi, G.; Riganti, C.; Pan, C.; Damas, A.P.; Scherb, H.; Ertürk, A.; Zeidler, R. An inhibitory antibody targeting carbonic anhydrase XII abrogates chemoresistance and significantly reduces lung metastases in an orthotopic breast cancer model in vivo. *Int. J. Cancer* **2018**, *143*, 2065–2075. [CrossRef]

32. Li, Y.; Lei, B.; Zou, J.; Wang, W.; Chen, A.; Zhang, J.; Fu, Y.; Li, Z. High expression of carbonic anhydrase 12 (CA12) is associated with good prognosis in breast cancer. *Neoplasma* **2019**, *66*, 420–426. [CrossRef] [PubMed]

33. Cau, Y.; Vullo, D.; Mori, M.; Dreassi, E.; Supuran, C.T.; Botta, M. Potent and selective carboxylic acid inhibitors of tumor-associated carbonic anhydrases IX and XII. *Molecules* **2017**, *23*, 17. [CrossRef]

34. Mori, M.; Cau, Y.; Vignaroli, G.; Laurenzana, I.; Caivano, A.; Vullo, D.; Supuran, C.T.; Botta, M. Hit recycling: Discovery of a potent carbonic anhydrase inhibitor by in silico target fishing. *ACS Chem. Biol.* **2015**, *10*, 1964–1969. [CrossRef] [PubMed]

35. Guerrini, G.; Durivault, J.; Filippi, I.; Criscuoli, M.; Monaci, S.; Pouyssegur, J.; Naldini, A.; Carraro, F.; Parks, S. Carbonic anhydrase XII expression is linked to suppression of Sonic hedgehog ligand expression in triple negative breast cancer cells. *Biochem. Biophys. Res. Commun.* **2019**, *516*, 408–413. [CrossRef] [PubMed]

36. Guerrini, G.; Criscuoli, M.; Filippi, I.; Naldini, A.; Carraro, F. Inhibition of smoothened in breast cancer cells reduces CAXII expression and cell migration. *J. Cell. Physiol.* **2018**, *233*, 9799–9811. [CrossRef]

37. Scales, S.J.; De Sauvage, F.J. Mechanisms of Hedgehog pathway activation in cancer and implications for therapy. *Trends Pharmacol. Sci.* **2009**, *30*, 303–312. [CrossRef] [PubMed]

38. Varjosalo, M.; Taipale, J. Hedgehog: Functions and mechanisms. *Genes Dev.* **2008**, *22*, 2454–2472. [CrossRef]

39. Zeng, X.; Ju, D. Hedgehog signaling pathway and autophagy in cancer. *Int. J. Mol. Sci.* **2018**, *19*, 2279. [CrossRef]

40. Briscoe, J.; Thérond, P.P. The mechanisms of hedgehog signalling and its roles in development and disease. *Nat. Rev. Mol. Cell Biol.* **2013**, *14*, 416–429. [CrossRef]

41. Feldmann, G.; Dhara, S.; Fendrich, V.; Bedja, D.; Beaty, R.; Mullendore, M.; Karikari, C.; Alvarez, H.; Iacobuzio-Donahue, C.; Jimeno, A.; et al. Blockade of hedgehog signaling inhibits pancreatic cancer invasion and metastases: A new paradigm for combination therapy in solid cancers. *Cancer Res.* **2007**, *67*, 2187–2196. [CrossRef]

42. Rubin, L.L.; De Sauvage, F.J. Targeting the hedgehog pathway in cancer. *Nat. Rev. Drug Discov.* **2006**, *5*, 1026–1033. [CrossRef]

43. Qualtrough, D.; Buda, A.; Gaffield, W.; Williams, A.C.; Paraskeva, C. Hedgehog signalling in colorectal tumour cells: Induction of apoptosis with cyclopamine treatment. *Int. J. Cancer* **2004**, *110*, 831–837. [CrossRef] [PubMed]

44. Sanchez, P.; Hernández, A.M.; Stecca, B.; Kahler, A.J.; DeGueme, A.M.; Barrett, A.; Beyna, M.; Datta, M.W.; Datta, S.; i Altaba, A.R. Inhibition of prostate cancer proliferation by interference with SONIC HEDGEHOG-GLI1 signaling. *Proc. Natl. Acad. Sci. USA* **2004**, *101*, 12561–12566. [CrossRef] [PubMed]

45. Pavlick, A.; Zhang, X.; Chamness, G.; Wong, H.; Rosen, J.; Chang, J.C.; Varmus, H.E. PMCID: PMC1635021. *Breast Cancer Res Treat.* **2009**, *115*, 505–521.

46. Chen, J.; Zhou, X.; Yang, J.; Sun, Q.; Liu, Y.; Li, N.; Zhang, Z.; Xu, H. Circ-GLI1 promotes metastasis in melanoma through interacting with p70S6K2 to activate Hedgehog/GLI1 and Wnt/β-catenin pathways and upregulate Cyr61. *Cell Death Dis.* **2020**, *11*, 1–16. [CrossRef]

47. Andreucci, E.; Peppicelli, S.; Ruzzolini, J.; Bianchini, F.; Biagioni, A.; Papucci, L.; Magnelli, L.; Mazzanti, B.; Stecca, B.; Calorini, L. The acidic tumor microenvironment drives a stem-like phenotype in melanoma cells. *J. Mol. Med.* **2020**, *2020*, 1–16. [CrossRef]

48. Chafe, S.C.; McDonald, P.C.; Saberi, S.; Nemirovsky, O.; Venkateswaran, G.; Burugu, S.; Gao, D.; Delaidelli, A.; Kyle, A.H.; Baker, J.H.; et al. Targeting hypoxia-induced carbonic anhydrase IX enhances immune-checkpoint blockade locally and systemically. *Cancer Immunol. Res.* **2019**, *7*, 1064–1078. [CrossRef]

49. O'Reilly, K.E.; De Miera, E.V.-S.; Segura, M.F.; Friedman, E.; Poliseno, L.; Han, S.W.; Zhong, J.; Zavadil, J.; Pavlick, A.C.; Hernando, E.; et al. Hedgehog pathway blockade inhibits melanoma cell growth in vitro and in vivo. *Pharmaceuticals* **2013**, *6*, 1429–1450. [CrossRef]

50. Duan, F.; Lin, M.; Li, C.; Ding, X.; Qian, G.; Zhang, H.; Ge, S.; Fan, X.; Li, J. Effects of inhibition of hedgehog signaling on cell growth and migration of uveal melanoma cells. *Cancer Biol. Ther.* **2014**, *15*, 544–559. [CrossRef]

51. Oliveira, A.D.S.D.; Yang, L.; Echevarria-Lima, J.; Monteiro, R.Q.; Rezaie, A.R. Thrombomodulin modulates cell migration in human melanoma cell lines. *Melanoma Res.* **2014**, *24*, 11–19. [CrossRef]

52. Xiao, P.; Zheng, B.; Sun, J.; Yang, J. Biochanin A induces anticancer effects in SK-Mel-28 human malignant melanoma cells via induction of apoptosis, inhibition of cell invasion and modulation of NF-κB and MAPK signaling pathways. *Oncol. Lett.* **2017**, *14*, 5989–5993. [CrossRef] [PubMed]

53. Sun, B.; Zhang, D.; Zhang, S.; Zhang, W.; Guo, H.; Zhao, X. Hypoxia influences vasculogenic mimicry channel formation and tumor invasion-related protein expression in melanoma. *Cancer Lett.* **2007**, *249*, 188–197. [CrossRef]

54. Hanna, S.C.; Krishnan, B.; Bailey, S.T.; Moschos, S.J.; Kuan, P.-F.; Shimamura, T.; Osborne, L.D.; Siegel, M.B.; Duncan, L.M.; O'Brien, E.T.; et al. HIF1α and HIF2α independently activate SRC to promote melanoma metastases. *J. Clin. Investig.* **2013**, *123*, 2078–2093. [CrossRef] [PubMed]

55. Widmer, D.S.; Hoek, K.S.; Cheng, P.F.; Eichhoff, O.M.; Biedermann, T.; Raaijmakers, M.I.; Hemmi, S.; Dummer, R.; Levesque, M.P. Hypoxia contributes to melanoma heterogeneity by triggering HIF1α-dependent phenotype switching. *J. Investig. Dermatol.* **2013**, *133*, 2436–2443. [CrossRef] [PubMed]

56. Asnaghi, L.; Lin, M.H.; Lim, K.S.; Lim, K.J.; Tripathy, A.; Wendeborn, M.; Merbs, S.L.; Handa, J.T.; Sodhi, A.; Bar, E.E.; et al. Hypoxia promotes uveal melanoma invasion through enhanced notch and mapk activation. *PLoS ONE* **2014**, *9*, e105372. [CrossRef]

57. Bedogni, B.; Powell, M.B. Hypoxia, melanocytes and melanoma-survival and tumor development in the permissive microenvironment of the skin. *Pigment. Cell Melanoma Res.* **2009**, *22*, 166–174. [CrossRef]

58. Hess, A.R.; Postovit, L.-M.; Margaryan, N.V.; Seftor, E.A.; Schneider, G.; Nickoloff, B.J.; Hendrix, M.J. Focal adhesion kinase promotes the aggressive melanoma phenotype. *Cancer Res.* **2005**, *65*, 9851–9860. [CrossRef]

59. Park, S.H.; Jeong, S.; Kim, B.R.; Jeong, Y.A.; Kim, J.L.; Na, Y.J.; Jo, M.J.; Yun, H.K.; Kim, D.Y.; Kim, B.G.; et al. Activating CCT2 triggers Gli-1 activation during hypoxic condition in colorectal cancer. *Oncogene* **2019**, *39*, 136–150. [CrossRef]

60. Li, H.; Yue, D.-S.; Jin, J.Q.; Woodard, G.A.; Tolani, B.; Luh, T.M.; Giroux-Leprieur, E.; Mo, M.; Chen, Z.; Che, J.; et al. Gli promotes epithelial-mesenchymal transition in human lung adenocarcinomas. *Oncotarget* **2016**, *7*, 80415–80425. [CrossRef]

61. Aiello, N.M.; Maddipati, R.; Norgard, R.J.; Balli, D.; Li, J.; Yuan, S.; Yamazoe, T.; Black, T.; Sahmoud, A.; Furth, E.E.; et al. EMT subtype influences epithelial plasticity and mode of cell migration. *Dev. Cell* **2018**, *45*, 681–695. [CrossRef]

62. Byles, V.; Zhu, L.; Lovaas, J.D.; Chmilewski, L.K.; Wang, J.; Faller, D.V.; Dai, Y. SIRT1 induces EMT by cooperating with EMT transcription factors and enhances prostate cancer cell migration and metastasis. *Oncogene* **2012**, *31*, 4619–4629. [CrossRef]

63. Yee, V.S.; Thompson, J.F.; McKinnon, J.G.; Scolyer, R.A.; Li, L.-X.L.; McCarthy, W.H.; O'Brien, C.J.; Quinn, M.J.; Saw, R.P.; Shannon, K.F.; et al. Outcome in 846 cutaneous melanoma patients from a single center after a negative sentinel node biopsy. *Ann. Surg. Oncol.* **2005**, *12*, 429–439. [CrossRef] [PubMed]

64. Hsieh, M.-J.; Chen, K.-S.; Chiou, H.-L.; Hsieh, Y.-S. Carbonic anhydrase XII promotes invasion and migration ability of MDA-MB-231 breast cancer cells through the p38 MAPK signaling pathway. *Eur. J. Cell Biol.* **2010**, *89*, 598–606. [CrossRef] [PubMed]

65. Naldini, A.; Filippi, I.; Ardinghi, C.; Silini, A.R.; Giavazzi, R.; Carraro, F. Identification of a functional role for the protease-activated receptor-1 in hypoxic breast cancer cells. *Eur. J. Cancer* **2009**, *45*, 454–460. [CrossRef]

66. Albini, A.; Iwamoto, Y.; Kleinman, H.K.; Martin, G.R.; Aaronson, S.A.; Kozlowski, J.M.; McEwan, R.N. A rapid in vitro assay for quantitating the invasive potential of tumor cells. *Cancer Res.* **1987**, *47*, 3239–3245.

Publisher's Note: MDPI stays neutral with regard to jurisdictional claims in published maps and institutional affiliations.

Article

Arthralgia Induced by BRAF Inhibitor Therapy in Melanoma Patients

Martin Salzmann [1], Karolina Benesova [2], Kristina Buder-Bakhaya [1], Dimitrios Papamichail [3], Antonia Dimitrakopoulou-Strauss [3], Hanns-Martin Lorenz [2], Alexander H. Enk [1] and Jessica C. Hassel [1,*]

[1]   Department of Dermatology and National Center for Tumor Diseases, University Hospital Heidelberg, Im Neuenheimer Feld 460, 69120 Heidelberg, Germany; Martin.Salzmann@med.uni-heidelberg.de (M.S.); kristina_buder@web.de (K.B.-B.); alexander.enk@med.uni-heidelberg.de (A.H.E.)

[2]   Division of Rheumatology, Department of Medicine V, University Hospital Heidelberg, Im Neuenheimer Feld 410, 69120 Heidelberg, Germany; karolina.benesova@med.uni-heidelberg.de (K.B.); Hannes.Lorenz@med.uni-heidelberg.de (H.-M.L.)

[3]   Clinical Cooperation Unit Nuclear Medicine, German Cancer Research Center, Im Neuenheimer Feld 280, 69120 Heidelberg, Germany; dimitris.papamihail@gmail.com (D.P.); ads@ads-net.de (A.D.-S.)

*   Correspondence: jessica.hassel@med.uni-heidelberg.de; Tel.: +49-62215638503

Received: 19 August 2020; Accepted: 15 October 2020; Published: 16 October 2020

Simple Summary: BRAF inhibitors (BRAFi) are standard of care for BRAF-mutated metastatic melanoma (MM). One of the most common side effects is arthralgia, for which a high incidence has been described, but whose clinical presentation and management have not yet been characterized. The aim of this retrospective study was to assess the patterns and clinical course of this drug-induced joint pain and to discuss a potential pathogenesis based on our clinical findings. In our cohort of patients treated with BRAFi between 2010 and 2018, 48 of 154 (31%) patients suffered from new-onset joint pain, which primarily affected small joints with a symmetrical pattern, as can be observed in patients affected by rheumatoid arthritis, the most frequent rheumatic and musculoskeletal disease. Most cases were sufficiently treated by non-steroidal anti-inflammatory drugs; however, some patients required dose reduction or permanent discontinuation of the BRAFi. Interestingly, we found that the occurrence of arthralgia was associated with better tumor control.

Abstract: Introduction: BRAF inhibitors (BRAFi), commonly used in BRAF-mutated metastatic melanoma (MM) treatment, frequently cause arthralgia. Although this is one of the most common side effects, it has not been characterized yet. Methods: We retrospectively included all patients treated with BRAFi +/− MEK inhibitors (MEKi) for MM at the National Center for Tumor Diseases (Heidelberg) between 2010 and 2018 and reviewed patient charts for the occurrence and management of arthralgia. The evaluation was supplemented by an analysis of frozen sera. Results: We included 154 patients (63% males); 31% (48/154) of them reported arthralgia with a median onset of 21 days after the start of the therapy. Arthralgia mostly affected small joints (27/36, 75%) and less frequently large joints (19/36, 53%). The most commonly affected joints were in fingers (19/36, 53%), wrists (16/36, 44%), and knees (12/36, 33%). In 67% (24/36) of the patients, arthralgia occurred with a symmetrical polyarthritis, mainly of small joints, resembling the pattern typically observed in patients affected by rheumatoid arthritis (RA), for which a role of the MAPK signaling pathway was previously described. Patients were negative for antinuclear antibodies, anti-citrullinated protein antibodies, and rheumatoid factor; arthritis was visible in 10 of 13 available PET–CT scans. The development of arthralgia was linked to better progression-free survival and overall survival. Conclusion: Arthralgia is a common side effect in patients receiving BRAFi +/− MEKi therapy and often presents a clinical pattern similar to that observed in RA patients. Its occurrence was associated with longer-lasting tumor control.

Keywords: melanoma; BRAF; BRAF inhibitor; arthralgia; rheumatoid arthritis

1. Introduction

BRAF inhibitors (BRAFi), which are usually used in combination with MEK inhibitors (MEKi), are important in the therapy of BRAF-mutated metastatic melanoma (MM). Arthralgia is acknowledged as one of the most common adverse events (AE) of BRAFi therapy. However, dedicated research on its clinical presentation, relevance, pathogenesis, and management is currently not available.

As shown in Table 1, up to about two-thirds of patients treated with vemurafenib monotherapy experience arthralgia of any grade; however, severe arthralgia (common terminology criteria for adverse events (CTC-AE) grade 3–4 [1]) was reported in less than 3–7% of patients [2–10]. In contrast, lower incidences of arthralgia were reported for dabrafenib (25–35% of patients) [11–14] and for encorafenib monotherapy (44% of patients) [9]. Only few publications report the incidence of arthralgia in patients treated with a combination therapy of BRAF and MEK inhibitors [9,12,14,15]. An analysis of real-world patients revealed arthralgia or myalgia in 32.9% of patients, with higher rates in those receiving BRAFi monotherapy [16]. Since rheumatologic AEs are frequently seen in patients treated with various anticancer medications, the CTC-AE v5.0 have now been adapted for their proper evaluation, after previous versions likely underestimated the impact of this side effect on patients' activities of daily living [17,18].

Table 1. Reported incidence of arthralgia in major publications on BRAF inhibitors (BRAFi) therapy.

Drug	Study	n	Arthralgia Any Grade (%)	Arthralgia Grade 3/4 (%)	Comments
Vemurafenib	Chapman et al. 2011 [2]	336	n/a	11 (3)	grade 1 not documented; 60 patients (18%) experienced grade 2 arthralgia
	Sosman et al. 2012 [3]	132	78 (59)	8 (6)	
	Larkin et al. 2014 [4]	3222	1259 (39)	106 (3)	
	Kim et al. 2014 [5]	336	180 (54)	15 (5)	vs. dacarbacine (3% overall)
	Arance et al. 2016 [6]	301	134 (44)	16 (5)	
	Blank et al. 2017 [7]	3219	1201 (37)	102 (3)	
	Maio et al. 2018 [8]	247	151 (61)	17 (7)	adjuvant; vs. placebo (22% overall)
	Dummer et al. 2018 [9]	186	85 (46)	11 (6)	
	Si et al. 2018 [10]	46	30 (65)	0 (0)	
Vemurafenib + Cobimetinib	Ribas et al. 2014 [15]	63	30 (48)	7 (11)	
	(Ribas et al. 2014 [15])	66	8 (12)	1 (2)	patients had progressed under previous vemurafenib therapy
Dabrafenib	Hauschild et al. 2012 [11]	100	n/a	1 (1)	Grade 1 not documented; 9 (5%) patients experienced grade 2 arthralgia
	Flaherty et al. 2012 [12]	53	18 (34)	0 (0)	
	Ascierto et al. 2013 [13]	92	30 (33)	1 (1)	
	Long et al. 2014 [14]	211	58 (27)	0 (0)	
Dabrafenib + Trametinib	Flaherty et al. 2012 [12]	55	15 (27)	0 (0)	44% any grade in 1 mg trametinib group
	Long et al. 2014 [14]	209	52 (24)	1 (0)	
Encorafenib	Dummer et al. 2018 [9]	192	85 (44)	18 (9)	
Encorafenib + Binimetinib	Dummer et al. 2018 [9]	192	54 (28)	2 (1)	

Collectively, arthralgia under BRAFi is commonly seen for all mono- and combination treatments as one of the most prevalent, class-specific side effects. However, no study has ever described its clinical presentation and relevance in patient care, with the exception of some case reports [19–21]. Furthermore, no study has been investigating its pathogenesis.

The aim of our analysis was to better describe the clinical patterns of arthralgia and its management during treatment with BRAFi and propose mechanisms of pathogenesis based on the clinical findings.

2. Methods

2.1. Patients

We systematically included all patients treated with BRAFi with or without MEKi in the approved doses for MM at the National Center for Tumor diseases (NCT) at the University Hospital of Heidelberg, Germany, between 01/2010 and 10/2018 and retrospectively screened clinical data for the incidence and localization of arthralgia and its management. Data were collected during clinical routine, no data were collected for the purpose of this study. Since the study was designed to describe a class-specific side effect occurring with both monotherapy and combination therapy [22], patients receiving monotherapy were included, even though this is no longer the treatment standard. All conclusions on progression-free survival (PFS) and overall survival (OS) were backed by multivariate analyses due to potential bias of patients receiving monotherapy.

The patient selection process is displayed in Figure 1. Final follow-up was completed on 31 January 2019. Patients administered concurrent immune checkpoint blockers and patients not evaluable due to external treatment monitoring or duration of therapy of less than 15 days were excluded. Patients with known joint disease before the start of the therapy or arthralgia due to metastatic disease of the joint region were removed from further analysis of the arthralgia group. The BRAFi encorafenib in combination with the MEKi binimetinib has been approved for clinical use in Germany in 2018; due to the different power of this drug combination and its peculiar affinity during binding to melanoma cells, the only patient treated with it was excluded from the analysis.

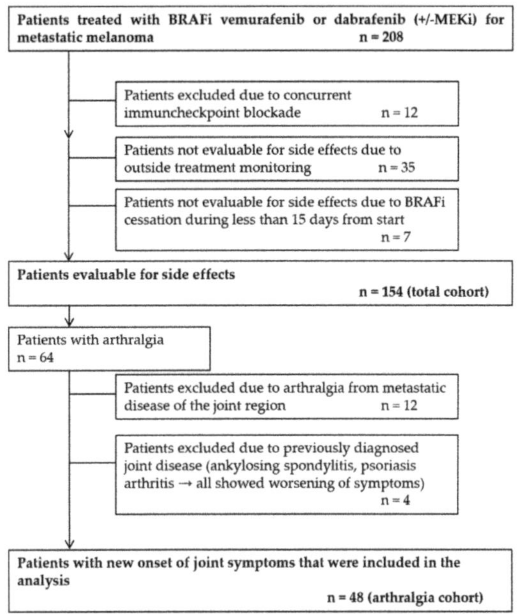

Figure 1. Flow diagram illustrating the process of patient selection. Cessation of BRAFi therapy in non-evaluable patients was due to unacceptable toxicity or performed at patient's wish.

2.2. Data Collection

Clinical and laboratory checks were routinely performed before treatment start, 2 weeks and 4 weeks after start, and then every 4 weeks. Imaging was performed every 12 weeks using CT of the neck/chest/abdomen or whole-body FDG-PET/CT scan, each combined with an MRI of the brain. Response to treatment was defined according to the Response Evaluation Criteria in Solid Tumors (RECIST) version 1.1 [23] and the PET Response Evaluation Criteria for Immunotherapy (PERCIMT) [24]. FDG-PET/CT was not done regularly, but if available, PET-CT scans were retrospectively evaluated for radiologic signs of arthritis, indicated by a newly emerging or enhanced diffuse periarticular radiotracer uptake in the joints.

We retrospectively analyzed frozen blood samples of patients for laboratory parameters of inflammation, including anti-citrullinated protein antibodies (ACPA), antinuclear antibodies (ANA), and rheumatoid factor (RF). C-reactive protein (CRP) levels were assessed regularly, and the value at the date of onset of arthralgia retrospectively documented. All patients with available samples had given their consent to retrospective research projects at the time of their blood draw.

2.3. Statistical Analysis

Statistical analysis was performed using IBM SPSS Statistics, version 27. PFS and OS were calculated from initiation of BRAFi treatment until progression or death from any cause, respectively. In patients with no events of progression or death at the time of the final data analysis, the date of last contact was used for censored calculation. Survival was estimated by the Kaplan–Meier method. Univariate comparisons of Kaplan–Meier estimators were performed using the log-rank test. Two-sided Fisher's exact and Chi-square tests were used for comparisons among groups with categorical variables; p values were considered significant if $p < 0.05$.

2.4. Ethical Approval

Retrospective analyses of clinical data were approved by the institutional review board of the Medical Faculty of the University Hospital Heidelberg (no. S-069/2010). The ethical committee had agreed to the retrospective analysis of routinely collected clinical data without prior informed consent of patients.

3. Results

3.1. Patient Characteristics

We included 154 patients, 63% (97/154) male and 37% female, with a median age of 56 years (range 21–86 years). The identified BRAF mutation was V600E in 128 patients (83%), V600K in 15 (10%), V600R in 2 (1%), and V600G in 1 patient (1%); the type of V600-mutation was unknown in 8 patients (5%). Eighty-five patients (55%) were treated with vemurafenib monotherapy, 4 patients (3%) with vemurafenib + cobimetinib, 13 patients (8%) with dabrafenib monotherapy, and 52 patients (34%) with dabrafenib + trametinib. Overall, 98 patients (64%) were treated with BRAFi monotherapy, and 56 (36%) with combination therapy of BRAFi and MEKi. The median duration of therapy was 4.2 months (0.5–93.5). In addition, 88 patients (57%) received no previous systemic treatment for MM, 44 (29%) patients were administered one previous line of therapy, and 22 (14%) two or more previous lines, with a maximum of four. Forty-five patients (29%) were previously treated with immune checkpoint inhibitor (ICI) therapy, 22 of which received PD1-directed therapy as monotherapy or in combination with ipilimumab. The last line of ICI treatment was discontinued due to progression in 36 patients (80%) and due to unacceptable toxicity in 9 patients (20%); 44 patients (29%) were administered previous adjuvant interferon therapy. The response rate (RR) of our cohort was 55% (79/154), and the disease control rate (DCR) 75% (109/154). Median PFS was 5.3 months, median OS was 10.1 months; further results on treatment efficacy are described in the respective sections. All 154 patients had adverse events due to BRAFi treatment. In 8 cases (5%), therapy was ongoing at the time of data collection. Patient characteristics are shown in Table 2.

Table 2. Patient characteristics. MEKi, MEK inhibitors.

Parameter	Number of Patients in Total Cohort		Number of Patients in Arthralgia Cohort	
		(%)		(%)
Total number of patients	154	100	48	31.2
Age in years, median (range)	56 (21–86)		53 (28–79)	
Gender				
Male	97	63.0	28	58.3
Female	57	37.0	20	41.7
BRAF mutation				
V600E	128	83.1	42	89.4
V600K	15	9.7	4	8.5
V600R	2	1.3	0	0
V600G	1	0.6	0	0
unknown	8	5.2	2	4.2
Type of BRAFi used				
Vemurafenib	85	55.2	31	64.6
Vemurafenib + Cobimetinib	4	2.6	0	0
Dabrafenib	13	8.4	3	6.3
Dabrafenib + Trametinib	52	33.8	14	29.2
BRAFi monotherapy	98	63.6	34	70.8
BRAFi + MEKi	56	36.4	14	29.2
Duration of BRAFi treatment, median in months, (range in months)	4.2	(0.5–93.5)	7.5	(1.4–93.5)
Number of prior treatments (range)	(0–4)		(0–4)	
0	88	57.1	32	66.7
1	44	28.6	9	18.8
>1	22	14.3	7	14.6
Previous checkpoint inhibitor therapy	45	29.2	7	14.6
Previous PD1-inhibitor	22	14.2	3	6.3
Previous ipilimumab	38	24.7	5	10.4
Previous ipilimumab + nivolumab	10	6.5	1	2.1
Discontinued due to progression	36 (of 45)	80.0	6 (of 7)	85.7
Discontinued due to toxicity	9 (of 45)	20.0	1 (of 7)	14.3
Previous adjuvant interferon	44	28.6	13	27.1
Response	79	54.5	31	64.6
Disease control	109	75.2	41	85.4
Progression-free survival (median in months) [95% CI]	5.3	[4.7–5.9]	7.9	[5.7–10.1]
Overall survival (median in months) [95% CI]	10.1	[8.8–11.4]	14.9	[12.4–17.4]
Any adverse event	154	100	48	100
Treatment discontinued	146	94.8	45	93.8

3.2. Arthralgia

Arthralgia occurred in 48 of the 154 patients, with an incidence of 31.2% in our cohort. The median onset of arthralgia was at 21 days (range 1–338), and the median duration was 65 days (range 2–2623). We found that 26 patients (54.2%) reported spontaneous cessation of arthralgia despite ongoing therapy, and 21 patients (43.8%) had symptoms until discontinuation; for one patient, the exact date of arthralgia cessation is unknown. No patient reported symptoms beyond the discontinuation of BRAFi therapy.

The localization of arthralgia was documented in detail in 36 of the 48 patients; 7 patients reported generalized arthralgia, and in 5 cases, the localization was unknown. Overall, small joints were predominantly affected (27/36, 75%); less patients reported involvement of large joints (19/36, 53%). Arthralgia affected most frequently the finger joints (19/36, 53%) and wrists (16/36, 44%), followed by knees (12/36, 33%), ankles (10/36, 28%), shoulders (8/36, 25%), and feet/toes (6/36, 17%). Hips or back were involved in one case each. In the majority of patients, arthralgia occurred symmetrically (32/41, 78%, 7 not evaluable). All patients with the involvement of small joints presented arthralgia in several small joints. A summary of the presentation of arthralgia is shown in Table 3.

Table 3. Clinical presentation of arthralgia.

Parameter	Number of Patients	(%)
Arthralgia with detailed information	36	100
Involvement of small joints	27	75
Finger joints	19	52.8
Wrists	16	44.4
Feet/toes	6	16.7
Involvement of large joints	19	52.8
Knees	12	33.3
Ankles	10	27.8
Shoulders	8	25
Hips	1	2.8
Back	1	2.8
Involvement of small joints only	17	47.2
Involvement of large joints only	9	25.0
Involvement of small and large joints	10	27.8
Symmetrical involvement	32 (of 41 evaluable)	78.0
Symmetrical polyarthritis of primarily small joints	24	66.7

Cases and data were critically discussed and compared to those of known diseases. With symmetrical polyarticular involvement of primarily small joints, the pattern of involvement resembled the typical presentation of rheumatoid arthritis (RA) in 24 of 36 cases (67%). As the regular rheumatic conditions' symptom of morning stiffness was not documented thoroughly in clinical routine and not included in the analysis, the 1987 American College of Rheumatology (ACR) classification criteria for rheumatoid arthritis were not applicable [25]. Only 19% of patients (7/36) fulfilled the 2010 European League Against Rheumatism (EULAR)/ACR classification criteria [26]. However, the exact number of involved joints may be underestimated in clinical routine documentation without rheumatologic examination, and serology did not provide points in the scoring system.

3.3. Influencing Factors in the Development of Arthralgia

We found that 35% (34/98) of patients treated with BRAFi monotherapy developed arthralgia, compared to 25% (14/56) of those undergoing treatment with BRAFi + MEKi. This difference in the occurrence of arthralgia between patients treated with BRAFi monotherapy and those receiving combination treatment was not statistically significant in our cohort (p = 0.254). Considering all studied agents, there was a trend towards the highest incidence of arthralgia in patients treated with vemurafenib (36% (30/84 patients) vs. 26% (18/70 patients), p = 0.182). Patients treated with ICI before initiation of the BRAFi treatment had significantly lower rates of de novo arthralgia (38% (41/109) vs. 16% (7/45), p = 0.009). However, five of these patients required prednisolone treatment at a dose of 5–80 mg per day for immune-related adverse events (irAE), and none of these patients developed arthralgia during BRAFi treatment. Of four patients with previous PD1 inhibitor therapy-induced arthralgia, only one developed arthralgia also during BRAFi treatment, which additionally affected small joints. The development of arthralgia was also suppressed by dexamethasone treatment for brain metastases, with arthralgia documented in only 4 of 28 patients (14%, vs. 35% (44/126), $p = 0.033$). One patient developed arthralgia after discontinuing dexamethasone 7 weeks into BRAFi treatment.

3.4. Imaging

For 13 of 48 symptomatic patients, FDG-PET/CT scans during BRAFi treatment were available. Imaging confirmed arthritis in 77% (10/13) of patients, which corresponded to the clinically described joint region of arthralgia in 7 of them (58%). Eight patients (62%) had also increased FDG uptake in asymptomatic joints. Figure 2 shows an example of enhanced FDG uptake in the joints of a symptomatic 70-year-old patient treated with dabrafenib and trametinib. Data on visible arthritis of asymptomatic patients were not collected.

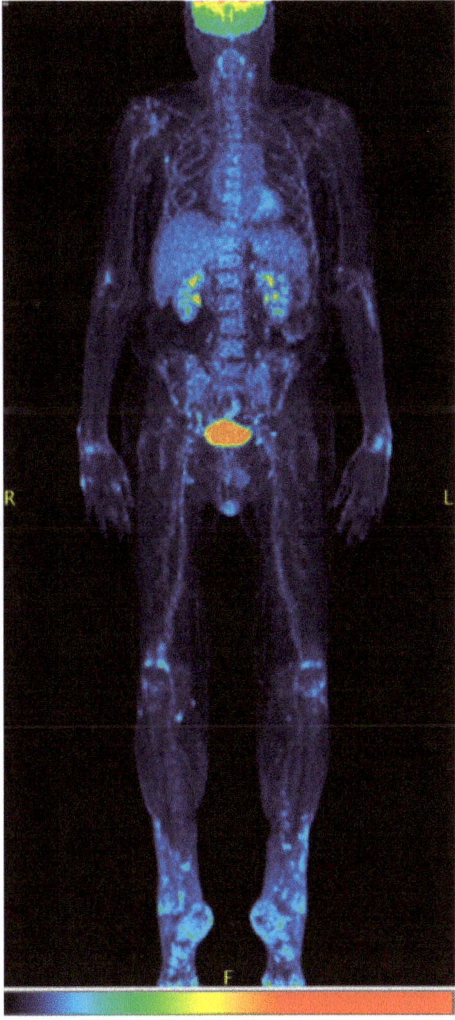

Figure 2. A 70-year old patient after 3 months of therapy with dabrafenib + trametinib, demonstrating radiologic signs of arthritis in PET/CT, reflected by a mostly symmetrical FDG uptake in elbows, wrists, fingers, knees, ankles, and feet (of note, the patient was in remission extracranially). The patient reported pain and swelling of fingers, wrists, and ankles.

3.5. Serology

The levels of CRP, as an unspecific inflammatory parameter, were elevated in 79% of the cases (34/43). Confronted with symptoms mimicking rheumatic diseases, and RA in particular, we analyzed the sera of symptomatic patients for typical serological markers of RA. This included ACPA, RF, and ANA titers. Frozen sera were available for 36 of the 48 patients (79%) in the arthralgia cohort. ACPA were not detected in any patient, RF was elevated (30.1 IU/mL) in one patient, ANA were significantly elevated in one patient (1:10,000) with an unspecific speckled pattern. Slightly elevated ANA titers of 1:320 to 1:1280 were found in seven patients, with unspecific fluorescence patterns and without positive extractable nuclear antigen antibody (ENA) titers or further clinical signs of connective tissue disease.

3.6. Management of Arthralgia

Fourteen of the 48 patients (38%) did not require or refused any treatment for arthralgia. The majority of patients were sufficiently managed with only metamizole sodium or non-steroidal anti-inflammatory drugs (NSAIDs) like etoricoxib, indomethacin, diclofenac, or ibuprofen (22 of 48 patients, 46%).

In 6 of 48 patients (13%) low dose steroids ($\leq$5 mg prednisone equivalent) were added to metamizole sodium or NSAIDs. In 6 of 48 patients (13%) with refractory arthralgia, dose reduction of the BRAFi was necessary in the course of treatment, in 2 cases BRAFi therapy had to be discontinued.

3.7. Treatment Efficacy

The median PFS was 5.3 months (95% confidence interval (CI) 4.7–5.9). This regarded patients treated by BRAFi monotherapy as well as combination therapy. For patients receiving combination treatment, the median PFS was 5.4 months (95% CI 4.6–6.2). An overview of factors influencing PFS is shown in Table 4; all of them were subjected to multivariate Cox regression due to the potential bias of different treatment regimens and disease status. PFS was significantly higher in the arthralgia cohort (median 7.9 months (95% CI 5.7–10.1) vs. 4.2 months (95% CI 3.3–5.1 months), $p = 0.001$, Figure 3), as confirmed by multivariate analysis. Patients requiring dose reduction or steroid treatment for arthralgia did not have a shortened PFS compared to patients with no treatment or NSAIDs only ($p = 0.950$). Also, glucocorticoid intake (5.7 months vs. 3.2 months, $p = 0.000$) and elevated lactate dehydrogenase (LDH) levels (6.3 months vs. 3.6 months, $p = 0.000$) had an impact on PFS. Median OS was 10.1 months (95% CI 8.8–11.4) and significantly higher in the arthralgia cohort (14.9 months (95% CI 12.4–17.4) vs. 8.7 months (95% CI 7.8–9.6), $p = 0.006$). For OS, an influence of arthralgia (odds ratio 0.62 (95% CI 0.40–0.96), $p = 0.033$), elevated LDH (odds ratio 2.59 (95% CI 1.76–3.68), $p < 0.001$), and concurrent steroid treatment (odds ratio 1.65 (95% CI 1.04–2.62), $p = 0.033$) was found in multivariate analysis.

Table 4. Influence of age, sex, tumor stage, previous immune checkpoint inhibitor (ICI) treatment, glucocorticoid intake, use of MEKi, BRAF mutation, elevated lactate dehydrogenase (LDH), and arthralgia on progression-free survival (PFS). PFS was available for 147 patients. Multivariate analysis was performed by Cox regression. Bold print indicates significant *p* values. * Two patients had advanced locoregional disease without distant metastases.

Parameter	Category	n	Univariate Analysis			Multivariate Analysis		
			Median PFS (Months)	CI (95%) (Months)	*p*	Odds Ratio	CI (95%)	*p*
Age	<60	89	5.3	4.0–6.6	0.603	1.03	0.71–1.48	0.881
	>60	58	4.8	4.1–5.5				
Sex	Male	91	4.8	3.7–5.9	0.114	0.82	0.56–1.20	0.302
	Female	56	6.2	4.9–7.5				
Stage (brain metastases)	M0 *–M1c	86	6.2	5.4–7.0	**0.000**	1.63	1.10–2.45	**0.017**
	M1d	61	4.2	3.0–5.4				
Previous ICI treatment	no	104	5.3	4.0–6.6	0.948	0.74	0.49–1.13	0.487
	yes	43	5.3	4.6–6.0				
Glucocorticoid intake	no	111	5.7	4.9–6.5	**0.000**	1.92	1.23–3.00	**0.004**
	yes (any dose)	36	3.2	2.6–3.8				
MEK inhibitor used	no	93	4.8	3.2–6.4	0.367	0.76	0.52–1.12	0.165
	yes	54	5.4	4.6–6.2				
BRAF mutation	V600E	122	5.4	4.6–6.2	**0.002**	1.93	1.20–3.11	**0.007**
	other	25	3.1	2.5–3.7				
Elevated LDH	no	74	6.3	4.8–7.8	**0.000**	2.09	1.44–3.02	**0.000**
	yes	69	3.6	2.8–4.4				
Arthralgia	no	101	4.2	3.3–5.1	**0.001**	0.52	0.34–0.78	**0.002**
	yes	46	7.9	5.7–10.1				

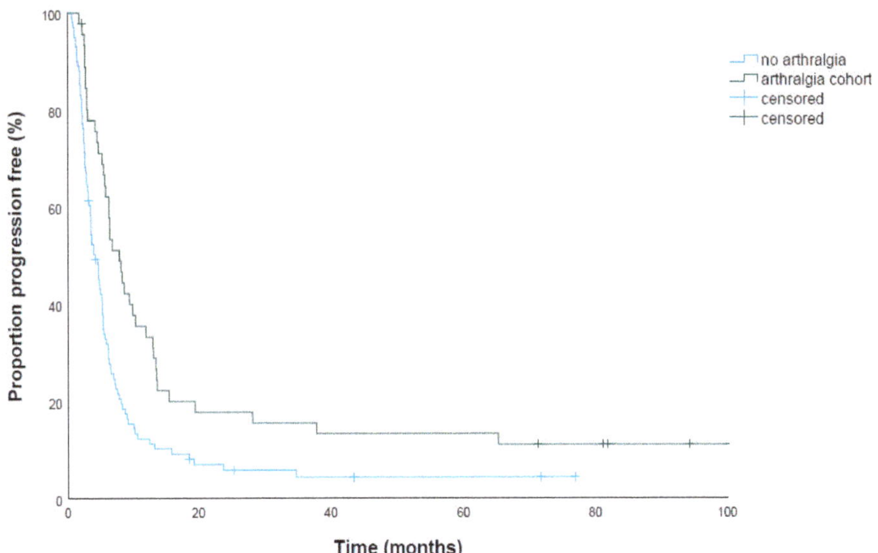

Figure 3. Kaplan–Meier estimated PFS shows improved PFS in patients with arthralgia (median 7.9 months (95% CI 5.7–10.1) vs. 4.2 months (95% CI 3.3–5.1 months), *p* = 0.001 (log rank)).

4. Discussion

Arthralgia induced by BRAFi therapy has not been characterized in detail previously, despite its clinical relevance. Our results suggest a pattern with primarily symmetrical involvement of small joints as observed in patients affected by RA, the most common rheumatic disease in adults. However, only a minority of patients fulfilled the 2010 EULAR/ACR classification criteria of RA, with routine clinical documentation likely underestimating the number of joints involved when no rheumatologic assessment was performed. Similarly, for rheumatic immune-related adverse events induced by checkpoint inhibitors, only 20% of patients fulfilled the respective criteria [27]. In contrast to irAEs, no patient showed other symptoms of rheumatic disease beyond arthralgia, such as sicca or vasculitis. Unspecific inflammatory parameters were found in 79% of patients with BRAFi-induced arthralgia, as found for almost three-quarters of ICI-treated patients with irAEs showing elevated CRP levels. Specific markers of rheumatic and connective tissue diseases were not detected in symptomatic patients, as is usually the case for irAEs [17,18,27]. This is of particular interest, as seropositivity for RF and ACPA is closely linked to a more aggressive course of disease and radiographic signs of bone damage [28]. The negativity of serological markers suggests that preformed autoantibodies for a rheumatic condition are not a prerequisite to develop the respective symptoms. However, a significant proportion of RA patients also remains seronegative throughout the course of the disease. Although new autoantibodies continue to be discovered in RA and are not yet used in clinical routine [29], the self-limiting course of BRAFi arthralgia supports a pro-inflammatory potential of BRAFi different from the pathogenic mechanisms of autoimmune diseases.

Previous research into the role of the MAPK signaling pathway has proposed BRAF abnormalities as a potential pathomechanism in RA: Weisbart et al. reported the BRAF V600R mutation in synovial fibroblasts in two of nine RA patients [30] and aberrant BRAF splice variants in six of nine RA patients [31]. Interestingly, mutated BRAF was detected more frequently in the peripheral blood lymphocytes of RA patients than in controls [32] and was independent of treatment modalities. BRAF was also found to be a target of autoantibodies in RA patients [33,34], but in contrast to RF and ACPA, these are not screened for in clinical routine. An established testing kit was not available for analysis in sera obtained from our patients for further investigation of this interesting link.

Thus, a possible hypothesis for arthralgia during BRAFi treatment is a paradoxical activation of the MAPK pathway in synovial tissue and lymphocytes by BRAF V600E inhibition, with other MAPK activating mutations possibly taking place. This effect might be comparable to the well-described mechanism of development of cutaneous epithelial tumors during BRAFi treatment caused by cutaneous RAS mutations [35,36]. Relevant MAPK pathway mutations will likely include additional mutations to BRAF V600R, which is supported by the fact that in a reported case, Babacan et al. did not find this mutation in their patient [21]. The hypothesis of paradoxical MAPK activation would be in accordance with the early onset of the joint symptoms and their quick resolution after discontinuation of the therapy. The idea of targeting the MAPK signaling pathway as a predominant cause of inflammation in RA was already successfully tested by Thiel et al. [37] and later by Yamaguchi et al. [38] but has not yet brought to a novel treatment option. Recently, a case of ICI-induced polymyalgia rheumatica controlled by the MEK inhibitor cobimetinib showed the in vivo anti-inflammatory potential of targeting the MAPK signaling pathway [39]. Our data show that symptoms similar to those of RA can be triggered by interfering with or potentially activating the MAPK signaling pathway, once again supporting the idea of inhibiting MEK as a potential treatment. We would like to clearly point out that this is a hypothesis generated from clinical observations described in this study. A direct association with RA cannot be proven based on our clinical data; however, it may be of value for future research by oncologists and rheumatologists alike.

Another hypothesis is the development of an immune-related adverse event leading to arthritis due to autoreactivity similar to the side effects of ICI, as described by Ben-Betzalel et al. [40]. An indicator could be the longer-lasting tumor response in patients suffering from arthralgia, as the authors linked the development of irAEs in patients receiving BRAFi therapy with tumor control. The same was found for arthritis induced by ICI [17]. However, the clinical presentation of arthralgia is different, as ICI-induced arthritis had a median onset of 100 days and predominantly affected large joints. Furthermore, in contrast to ICI-induced irAEs, no patient developed a persisting rheumatic condition, as arthralgia in BRAFi patients was self-limiting after drug discontinuation [41]. Collectively, the clinical difference between ICI- and BRAFi-induced arthritis suggests a difference in pathogenesis. In this context, the joint symptoms can be interpreted as a surrogate parameter for enhanced proinflammatory activity that also leads to improved tumor control.

Although our study was not explicitly designed to test the diagnostic performance of FDG PET/CT in BRAFi-induced arthritis and only few cases were available, we note the relatively high incidence (10/13) of positive molecular imaging findings in symptomatic patients. Interestingly, 8 of 13 of the patients studied with PET/CT demonstrated radiologic signs of arthritis without corresponding clinical symptoms in the respective joints. These findings are in line with recent results of our group in 16 BRAF mutation-positive, metastatic melanoma patients receiving a combination therapy of vemurafenib and ipilimumab [42]. Seven patients of that cohort developed radiologic signs of irAEs, four of which presenting arthritis without clinical symptoms; importantly, these patients had a significantly longer PFS than those without radiologic irAEs, highlighting a potential relation between the appearance of irAEs signs in PET/CT and the clinical benefit of the treatment [24].

There is currently no consensus on the management of arthralgia. Welsh et al. proposed a therapy algorithm for different grades of arthralgia [43], and a recommendation is available for the management of rheumatic irAEs, including arthralgia/arthritis [18]. A standard approach to treatment is the addition of NSAID to BRAFi therapy without dose reduction for low-grade arthralgia, which was sufficient for many of our studied patients. If NSAID are not sufficient, either low-dose steroids can be added or dose reduction of BRAFi treatment can be performed. Based on our experiences and data, we propose to avoid dose reduction or discontinuation in grade 2 and, if possible, in grade 3 patients, as most cases are sufficiently treated with NSAID or, in severe cases, with concurrent low-dose glucocorticoids, which can be reduced over the course of treatment. There are no comparison data available on how treatment efficacy of BRAFi is affected by either dose reduction or addition of low-dose glucocorticoids to define the preferred way of AE management in refractory cases. Of interest, no patient received disease-modifying antirheumatic drugs (DMARDs) such as methotrexate or infliximab, which are regularly needed for adequate management of ICI-induced irAEs [18]. However, in this retrospective cohort, none of the patients was examined by a rheumatologist to discuss an indication for DMARD treatment. Possibly, the initiation of DMARDs, that may enable BRAFi continuation despite this side effect, may be of advantage for tumor control. Especially in adjuvant treatment, risks and benefits should be considered and discussed critically with the patient and a rheumatologist, as arthralgia is usually not life-threatening but can result in considerable impairment in activities of daily living and reduced quality of life.

We are aware of the multiple limitations of our study. These mainly include its retrospective nature, the limited number of patients studied, and the lack of documentation during clinical routine, leading to a reduced number of cases studied in detail (of 48 patients with arthralgia, detailed documentation on its localization was available for only 36 patients, 25% were lost). Patients were treated by clinical standards, which led to a heterogeneous distribution of the used drugs, as well as to the majority of patients receiving BRAFi monotherapy. Also, rheumatological examination was not performed collectively. Therefore, a prospective study has been launched to systematically examine and document all patients with rheumatological side effects of novel systemic treatments at our institution [44].

5. Conclusions

Arthralgia is a common side effect of BRAFi treatment and mainly affects small joints. Patients are typically seronegative for immunological markers of rheumatic disease; however, they present with a pattern resembling RA. A longer-lasting tumor control was observed in patients experiencing arthralgia under BRAFi treatment. Symptoms can usually be controlled by symptomatic treatment with NSAIDs or low-dose steroids. To further address the mechanisms underlying arthralgia in BRAFi-treated patients, a prospective study to characterize arthralgia clinically and by imaging (ultrasound, MRI), as well as by rheumatologic assessment, is under way.

Author Contributions: Conceptualization, M.S., K.B., K.B.-B., and J.C.H.; Data curation, M.S., K.B.-B., and J.C.H.; Formal analysis, M.S. and J.C.H.; Investigation, M.S., K.B., K.B.-B., D.P., A.D.-S., H.-M.L., A.H.E., and J.C.H.; Methodology, M.S., K.B., K.B.-B., and J.C.H.; Project administration, M.S., K.B.-B., and J.C.H.; Supervision, A.D.-S., H.-M.L., A.H.E., and J.C.H.; Validation, M.S., K.B., K.B.-B. and J.C.H.; Visualization, M.S., D.P., A.D.-S., and J.C.H.; Writing—original draft, M.S.; Writing—review & editing, M.S., K.B., K.B.-B., D.P., A.D.-S., H.-M.L., A.H.E., and J.C.H. All authors have read and agreed to the published version of the manuscript.

Funding: This research received no external funding.

Conflicts of Interest: M.S.: honoraria and travel grants from Abbvie, Bristol-Myers Squibb (BMS), Merck, Merck Sharp & Dohme (MSD), Novartis and Pfizer. K.B.: Honoraria, research funding or travel grants from Abbvie, BMS, Gilead, Janssen, Lilly, Medac, MSD, Mundipharma, Novartis, Pfizer, Rheuma-Liga Baden-Württemberg e.V., Roche and UCB. K.B.-B.: honoraria and travel grants from Merck Sharp & Dohme (MSD), and Novartis. D.P.: No conflicts of interest to declare. A.D.-S.: No conflicts of interest to declare. H.-M.L.: consultancy fees and/or honoraria for lectures and/or travel reimbursements and/or support for scientific projects and/or educational seminars and/or clinical studies from Abbvie, Astra-Zeneca, Actelion, Alexion, Amgen, Bayer Vital, Baxter, Biogen, Boehringer Ingelheim, BMS, Celgene, Fresenius, Genzyme, GSK, Gilead, Hexal, Janssen-Cilag, Lilly, Medac, MSD, Mundipharm, Mylan, Novartis, Octapharm, Pfizer, Roche/Chugai, Sandoz, Sanofi, Shire, SOBI, Thermo Fisher, UCB. A.H.E.: advisory honoraria from Biotest AG, MSD Oncology, Galderma, Janssen Cilag, AbbVie as well as speaker's honoraria from Roche Pharma. J.C.H.: honoraria for talks from BMS, MSD, Roche, Novartis; advisory board member for MSD, Pierre Fabre; scientific grant support from BMS; travel grants from BMS, Pierre Fabre.

Abbreviations

ACPA	Anti-citrullinated protein antibodies
ANA	Antinuclear antibodies
BRAFi	BRAF inhibitor
CI	Confidence interval
CRP	C-reactive protein
CT	Computed tomography
DMARD	Disease-modifying antirheumatic drug
FDG-PET	Fludeoxyglucose positron emission tomography
ICI	Immune-checkpoint inhibitor
irAE	Immune-related adverse event
MEKi	MEK inhibitor
MM	Metastatic melanoma
MRI	Magnetic resonance imaging
NCT	National Center for Tumor Diseases, Heidelberg
NSAID	Non-steroidal antirheumatic drug
OS	Overall survival
PFS	Progression-free survival

References

1. US Department of Health and Human Services. *Common Terminology Criteria for Adverse Events (CTCAE), Version 4.03*; US Department of Health and Human Services: Washington, DC, USA; Bethesda: Rockville, MD, USA, 2010.
2. Chapman, P.B.; Hauschild, A.; Robert, C.; Haanen, J.B.; Ascierto, P.; Larkin, J.; Dummer, R.; Garbe, C.; Testori, A.; Maio, M.; et al. Improved survival with vemurafenib in melanoma with BRAF V600E mutation. *N. Engl. J. Med.* **2011**, *364*, 2507–2516. [CrossRef] [PubMed]
3. Sosman, J.A.; Kim, K.B.; Schuchter, L.; Gonzalez, R.; Pavlick, A.C.; Weber, J.S.; McArthur, G.A.; Hutson, T.E.; Moschos, S.J.; Flaherty, K.T.; et al. Survival in BRAF V600-mutant advanced melanoma treated with vemurafenib. *N. Engl. J. Med.* **2012**, *366*, 707–714. [CrossRef] [PubMed]
4. Larkin, J.; Del Vecchio, M.; Ascierto, P.A.; Krajsova, I.; Schachter, J.; Neyns, B.; Espinosa, E.; Garbe, C.; Sileni, V.C.; Gogas, H.; et al. Vemurafenib in patients with BRAF(V600) mutated metastatic melanoma: An open-label, multicentre, safety study. *Lancet Oncol.* **2014**, *15*, 436–444. [CrossRef]

5. Kim, G.; McKee, A.E.; Ning, Y.M.; Hazarika, M.; Theoret, M.; Johnson, J.R.; Xu, Q.C.; Tang, S.; Sridhara, R.; Jiang, X.; et al. FDA approval summary: Vemurafenib for treatment of unresectable or metastatic melanoma with the BRAFV600E mutation. *Clin. Cancer Res.* **2014**, *20*, 4994–5000. [CrossRef] [PubMed]

6. Arance, A.M.; Berrocal, A.; Lopez-Martin, J.A.; de la Cruz-Merino, L.; Soriano, V.; Martin Algarra, S.; Alonso, L.; Cerezuela, P.; La Orden, B.; Espinosa, E. Safety of vemurafenib in patients with BRAF (V600) mutated metastatic melanoma: The Spanish experience. *Clin. Transl. Oncol.* **2016**, *18*, 1147–1157. [CrossRef]

7. Blank, C.U.; Larkin, J.; Arance, A.M.; Hauschild, A.; Queirolo, P.; Del Vecchio, M.; Ascierto, P.A.; Krajsova, I.; Schachter, J.; Neyns, B.; et al. Open-label, multicentre safety study of vemurafenib in 3219 patients with BRAF(V600) mutation-positive metastatic melanoma: 2-year follow-up data and long-term responders' analysis. *Eur. J. Cancer* **2017**, *79*, 176–184. [CrossRef]

8. Maio, M.; Lewis, K.; Demidov, L.; Mandala, M.; Bondarenko, I.; Ascierto, P.A.; Herbert, C.; Mackiewicz, A.; Rutkowski, P.; Guminski, A.; et al. Adjuvant vemurafenib in resected, BRAF(V600) mutation-positive melanoma (BRIM8): A randomised, double-blind, placebo-controlled, multicentre, phase 3 trial. *Lancet Oncol.* **2018**, *19*, 510–520. [CrossRef]

9. Dummer, R.; Ascierto, P.A.; Gogas, H.J.; Arance, A.; Mandala, M.; Liszkay, G.; Garbe, C.; Schadendorf, D.; Krajsova, I.; Gutzmer, R.; et al. Overall survival in patients with BRAF-mutant melanoma receiving encorafenib plus binimetinib versus vemurafenib or encorafenib (COLUMBUS): A multicentre, open-label, randomised, phase 3 trial. *Lancet Oncol.* **2018**, *19*, 1315–1327. [CrossRef]

10. Si, L.; Zhang, X.; Xu, Z.; Jiang, Q.; Bu, L.; Wang, X.; Mao, L.; Zhang, W.; Richie, N.; Guo, J. Vemurafenib in Chinese patients with BRAF(V600) mutation-positive unresectable or metastatic melanoma: An open-label, multicenter phase I study. *BMC Cancer* **2018**, *18*, 520. [CrossRef]

11. Hauschild, A.; Grob, J.J.; Demidov, L.V.; Jouary, T.; Gutzmer, R.; Millward, M.; Rutkowski, P.; Blank, C.U.; Miller, W.H., Jr.; Kaempgen, E.; et al. Dabrafenib in BRAF-mutated metastatic melanoma: A multicentre, open-label, phase 3 randomised controlled trial. *Lancet* **2012**, *380*, 358–365. [CrossRef]

12. Flaherty, K.T.; Infante, J.R.; Daud, A.; Gonzalez, R.; Kefford, R.F.; Sosman, J.; Hamid, O.; Schuchter, L.; Cebon, J.; Ibrahim, N.; et al. Combined BRAF and MEK inhibition in melanoma with BRAF V600 mutations. *N. Engl. J. Med.* **2012**, *367*, 1694–1703. [CrossRef] [PubMed]

13. Ascierto, P.A.; Minor, D.; Ribas, A.; Lebbe, C.; O'Hagan, A.; Arya, N.; Guckert, M.; Schadendorf, D.; Kefford, R.F.; Grob, J.J.; et al. Phase II trial (BREAK-2) of the BRAF inhibitor dabrafenib (GSK2118436) in patients with metastatic melanoma. *J. Clin. Oncol.* **2013**, *31*, 3205–3211. [CrossRef] [PubMed]

14. Long, G.V.; Stroyakovskiy, D.; Gogas, H.; Levchenko, E.; de Braud, F.; Larkin, J.; Garbe, C.; Jouary, T.; Hauschild, A.; Grob, J.J.; et al. Combined BRAF and MEK inhibition versus BRAF inhibition alone in melanoma. *N. Engl. J. Med.* **2014**, *371*, 1877–1888. [CrossRef] [PubMed]

15. Ribas, A.; Gonzalez, R.; Pavlick, A.; Hamid, O.; Gajewski, T.F.; Daud, A.; Flaherty, L.; Logan, T.; Chmielowski, B.; Lewis, K.; et al. Combination of vemurafenib and cobimetinib in patients with advanced BRAF(V600)-mutated melanoma: A phase 1b study. *Lancet Oncol.* **2014**, *15*, 954–965. [CrossRef]

16. Mackin, A.G.; Pecen, P.E.; Dinsmore, A.L.; Patnaik, J.L.; Gonzalez, R.; Robinson, W.A.; Palestine, A.G. Inflammatory side effects of BRAF and MEK inhibitors. *Melanoma Res.* **2019**, *29*, 522–526. [CrossRef] [PubMed]

17. Buder-Bakhaya, K.; Benesova, K.; Schulz, C.; Anwar, H.; Dimitrakopoulou-Strauss, A.; Weber, T.F.; Enk, A.; Lorenz, H.M.; Hassel, J.C. Characterization of arthralgia induced by PD-1 antibody treatment in patients with metastasized cutaneous malignancies. *Cancer Immunol. Immunother.* **2018**, *67*, 175–182. [CrossRef]

18. Benesova, K.L.; Lorenz, H.-M.; Leipe, J.; Jordan, K. How I treat cancer: Treatment of rheumatological side effects of immunotherapy. *ESMO Open* **2019**, *4*, e000529. [CrossRef]

19. Zimmer, L.; Livingstone, E.; Hillen, U.; Domkes, S.; Becker, A.; Schadendorf, D. Panniculitis with arthralgia in patients with melanoma treated with selective BRAF inhibitors and its management. *Arch. Dermatol.* **2012**, *148*, 357–361. [CrossRef]

20. Mossner, R.; Zimmer, L.; Berking, C.; Hoeller, C.; Loquai, C.; Richtig, E.; Kahler, K.C.; Hassel, J.C.; Gutzmer, R.; Ugurel, S. Erythema nodosum-like lesions during BRAF inhibitor therapy: Report on 16 new cases and review of the literature. *J. Eur. Acad. Dermatol. Venereol.* **2015**, *29*, 1797–1806. [CrossRef]

21. Babacan, T.; Turkbeyler, I.H.; Balakan, O.; Pehlivan, Y.; Suner, A.; Kisacik, B. A case of vemurafenib-induced polyarhritis in a patient with melanoma: How to manage it? *Int. J. Rheum. Dis.* **2017**, *20*, 398–401. [CrossRef]

22. Heinzerling, L.; Eigentler, T.K.; Fluck, M.; Hassel, J.C.; Heller-Schenck, D.; Leipe, J.; Pauschinger, M.; Vogel, A.; Zimmer, L.; Gutzmer, R. Tolerability of BRAF/MEK inhibitor combinations: Adverse event evaluation and management. *ESMO Open* **2019**, *4*, e000491. [CrossRef]

23. Eisenhauer, E.A.; Therasse, P.; Bogaerts, J.; Schwartz, L.H.; Sargent, D.; Ford, R.; Dancey, J.; Arbuck, S.; Gwyther, S.; Mooney, M.; et al. New response evaluation criteria in solid tumours: Revised RECIST guideline (version 1.1). *Eur. J. Cancer* **2009**, *45*, 228–247. [CrossRef] [PubMed]

24. Dimitrakopoulou-Strauss, A. Monitoring of patients with metastatic melanoma treated with immune checkpoint inhibitors using PET-CT. *Cancer Immunol. Immunother.* **2019**, *68*, 813–822. [CrossRef] [PubMed]

25. Arnett, F.C.; Edworthy, S.M.; Bloch, D.A.; McShane, D.J.; Fries, J.F.; Cooper, N.S.; Healey, L.A.; Kaplan, S.R.; Liang, M.H.; Luthra, H.S.; et al. The American Rheumatism Association 1987 revised criteria for the classification of rheumatoid arthritis. *Arthritis Rheum.* **1988**, *31*, 315–324. [CrossRef] [PubMed]

26. Aletaha, D.; Neogi, T.; Silman, A.J.; Funovits, J.; Felson, D.T.; Bingham, C.O., 3rd; Birnbaum, N.S.; Burmester, G.R.; Bykerk, V.P.; Cohen, M.D.; et al. 2010 Rheumatoid arthritis classification criteria: An American College of Rheumatology/European League Against Rheumatism collaborative initiative. *Arthritis Rheum.* **2010**, *62*, 2569–2581. [CrossRef]

27. Kostine, M.; Finckh, A.; Bingham, C.O., 3rd; Visser, K.; Leipe, J.; Schulze-Koops, H.; Choy, E.H.; Benesova, K.; Radstake, T.; Cope, A.P.; et al. EULAR points to consider for the diagnosis and management of rheumatic immune-related adverse events due to cancer immunotherapy with checkpoint inhibitors. *Ann. Rheum. Dis.* **2020**. [CrossRef]

28. Mouterde, G.; Rincheval, N.; Lukas, C.; Daien, C.; Saraux, A.; Dieudé, P.; Morel, J.; Combe, B. Outcome of patients with early arthritis without rheumatoid factor and ACPA and predictors of rheumatoid arthritis in the ESPOIR cohort. *Arthritis Res. Ther.* **2019**, *21*, 140. [CrossRef]

29. Charpin, C.; Arnoux, F.; Martin, M.; Toussirot, E.; Lambert, N.; Balandraud, N.; Wendling, D.; Diot, E.; Roudier, J.; Auger, I. New autoantibodies in early rheumatoid arthritis. *Arthritis Res. Ther.* **2013**, *15*, R78. [CrossRef]

30. Weisbart, R.H.; Chan, G.; Heinze, E.; Mory, R.; Nishimura, R.N.; Colburn, K. BRAF drives synovial fibroblast transformation in rheumatoid arthritis. *J. Biol. Chem.* **2010**, *285*, 34299–34303. [CrossRef]

31. Weisbart, R.H.; Chan, G.; Li, E.; Farmani, N.; Heinze, E.; Rubell, A.; Nishimura, R.N.; Colburn, K. BRAF splice variants in rheumatoid arthritis synovial fibroblasts activate MAPK through CRAF. *Mol. Immunol.* **2013**, *55*, 247–252. [CrossRef]

32. Arnoux, F.; Fina, F.; Lambert, N.; Balandraud, N.; Martin, M.; Ouafik, L.; Kanaan, S.B.; Roudier, J.; Auger, I. Newly Identified BRAF Mutation in Rheumatoid Arthritis. *Arthritis Rheumatol.* **2016**, *68*, 1377–1383. [CrossRef] [PubMed]

33. Auger, I.; Balandraud, N.; Rak, J.; Lambert, N.; Martin, M.; Roudier, J. New autoantigens in rheumatoid arthritis (RA): Screening 8268 protein arrays with sera from patients with RA. *Ann. Rheum. Dis.* **2009**, *68*, 591–594. [CrossRef] [PubMed]

34. Charpin, C.; Martin, M.; Balandraud, N.; Roudier, J.; Auger, I. Autoantibodies to BRAF, a new family of autoantibodies associated with rheumatoid arthritis. *Arthritis Res. Ther.* **2010**, *12*, R194. [CrossRef] [PubMed]

35. Peng, L.; Wang, Y.; Hong, Y.; Ye, X.; Shi, P.; Zhang, J.; Zhao, Q. Incidence and relative risk of cutaneous squamous cell carcinoma with single-agent BRAF inhibitor and dual BRAF/MEK inhibitors in cancer patients: A meta-analysis. *Oncotarget* **2017**, *8*, 83280–83291. [CrossRef]

36. Hassel, J.C.; Groesser, L.; Herschberger, E.; Weichert, W.; Hafner, C. RAS mutations in benign epithelial tumors associated with BRAF inhibitor treatment of melanoma. *J. Investig. Dermatol.* **2015**, *135*, 636–639. [CrossRef] [PubMed]

37. Thiel, M.J.; Schaefer, C.J.; Lesch, M.E.; Mobley, J.L.; Dudley, D.T.; Tecle, H.; Barrett, S.D.; Schrier, D.J.; Flory, C.M. Central role of the MEK/ERK MAP kinase pathway in a mouse model of rheumatoid arthritis: Potential proinflammatory mechanisms. *Arthritis Rheum.* **2007**, *56*, 3347–3357. [CrossRef]

38. Yamaguchi, T.; Kakefuda, R.; Tanimoto, A.; Watanabe, Y.; Tajima, N. Suppressive effect of an orally active MEK1/2 inhibitor in two different animal models for rheumatoid arthritis: A comparison with leflunomide. *Inflamm. Res.* **2012**, *61*, 445–454. [CrossRef]

39. Chan, K.K.; Bass, A.R. Checkpoint inhibitor-induced polymyalgia rheumatica controlled by cobimetinib, a MEK 1/2 inhibitor. *Ann. Rheum. Dis.* **2019**, *78*, e70. [CrossRef]

40. Ben-Betzalel, G.; Baruch, E.N.; Boursi, B.; Steinberg-Silman, Y.; Asher, N.; Shapira-Frommer, R.; Schachter, J.; Markel, G. Possible immune adverse events as predictors of durable response to BRAF inhibitors in patients with BRAF V600-mutant metastatic melanoma. *Eur. J. Cancer* **2018**, *101*, 229–235. [CrossRef]

41. Braaten, T.J.; Brahmer, J.R.; Forde, P.M.; Le, D.; Lipson, E.J.; Naidoo, J.; Schollenberger, M.; Zheng, L.; Bingham, C.O.; Shah, A.A.; et al. Immune checkpoint inhibitor-induced inflammatory arthritis persists after immunotherapy cessation. *Ann. Rheum. Dis.* **2020**, *79*, 332–338. [CrossRef]

42. Sachpekidis, C.; Kopp-Schneider, A.; Hakim-Meibodi, L.; Dimitrakopoulou-Strauss, A.; Hassel, J.C. 18F-FDG PET/CT longitudinal studies in patients with advanced metastatic melanoma for response evaluation of combination treatment with vemurafenib and ipilimumab. *Melanoma Res.* **2019**, *29*, 178–186. [CrossRef] [PubMed]

43. Welsh, S.J.; Corrie, P.G. Management of BRAF and MEK inhibitor toxicities in patients with metastatic melanoma. *Ther. Adv. Med. Oncol.* **2015**, *7*, 122–136. [CrossRef] [PubMed]

44. Benesova, K.; Diekmann, L.; Lorenz, H.M.; Jordan, K.; Leipe, J. OP0270 trheuma registry explores characteristics and suitable diagnostic and therapeutic management of rheumatic immune-related adverse events (IRAES). *Ann. Rheum. Dis.* **2020**, *79*, 168–169. [CrossRef]

Publisher's Note: MDPI stays neutral with regard to jurisdictional claims in published maps and institutional affiliations.

 *cancers*

Review

Circulating Tumor DNA Testing Opens New Perspectives in Melanoma Management

Alessandra Sacco [1,†] , Laura Forgione [1,†], Marianeve Carotenuto [1], Antonella De Luca [1], Paolo A. Ascierto [2] , Gerardo Botti [3] and Nicola Normanno [1,*]

1. Cell Biology and Biotherapy Unit, Istituto Nazionale Tumori-IRCCS-Fondazione G. Pascale, 80131 Naples, Italy; a.sacco@istitutotumori.na.it (A.S.); l.forgione@istitutotumori.na.it (L.F.); m.carotenuto@istitutotumori.na.it (M.C.); a.deluca@istitutotumori.na.it (A.D.L.)
2. Department of Melanoma, Cancer Immunotherapy and Development Therapeutics, Istituto Nazionale Tumori IRCCS Fondazione Pascale, 80131 Napoli, Italy; p.ascierto@istitutotumori.na.it
3. Scientific Direction, Istituto Nazionale Tumori IRCCS Fondazione Pascale, 80131 Napoli, Italy; g.botti@istitutotumori.na.it
* Correspondence: n.normanno@istitutotumori.na.it; Tel.: +39-081-5903-826
† These authors contributed equally to this work.

Received: 15 September 2020; Accepted: 8 October 2020; Published: 10 October 2020

Simple Summary: Melanoma, like other solid tumors, releases DNA molecules that are referred to as circulating tumor DNA (ctDNA), into the blood and other biological fluids. ctDNA analysis performed with molecular biology techniques can provide important information on the aggressiveness of the disease and its genetic characteristics. This review aims to highlight all the possible clinical applications of ctDNA analysis that can contribute to an improvement in the diagnosis and therapy of melanoma.

Abstract: Malignant melanoma accounts for about 1% of all skin cancers, but it causes most of the skin cancer-related deaths. Circulating tumor DNA (ctDNA) testing is emerging as a relevant tool for the diagnosis and monitoring of cancer. The availability of highly sensitive techniques, including next generation sequencing (NGS)-based panels, has increased the fields of application of ctDNA testing. While ctDNA-based tests for the early detection of melanoma are not available yet, perioperative ctDNA analysis in patients with surgically resectable melanoma offers relevant prognostic information: (i) the detection of ctDNA before surgery correlates with the extent and the aggressiveness of the disease; (ii) ctDNA testing after surgery/adjuvant therapy identifies minimal residual disease; (iii) testing ctDNA during the follow-up can detect a tumor recurrence, anticipating clinical/radiological progression. In patients with advanced melanoma, several studies have demonstrated that the analysis of ctDNA can better depict tumor heterogeneity and provides relevant prognostic information. In addition, ctDNA testing during treatment allows assessing the response to systemic therapy and identifying resistance mechanisms. Although validation in prospective clinical trials is needed for most of these approaches, ctDNA testing opens up new scenarios in the management of melanoma patients that could lead to improvements in the diagnosis and therapy of this disease.

Keywords: ctDNA; melanoma; liquid biopsy; prognosis; prediction; patient stratification

---

1. Introduction

Melanoma is an aggressive and deadly disease that is responsible for the largest number of skin cancer-related deaths, although it comprises less than 10% of skin cancers. The high mortality rate of melanoma is due to the late diagnosis and to the highly metastatic potential of melanoma cells, which typically spread in different organs [1].

The therapeutic strategies for advanced melanoma have significantly improved in the past few years thanks to the introduction of targeted agents and immune checkpoint inhibitors. More recently, both targeted therapy and immunotherapy have shown to reduce the rate of recurrence in patients with resectable, locally advanced disease. However, mechanisms of intrinsic and acquired resistance greatly limit the activity of therapies in melanoma, especially in the most advanced phases of disease. For this reason, the identification of non-invasive biomarkers could be key for facilitating early detection, patient stratification, and monitoring the response and resistance to therapy.

The testing of circulating tumor DNA (ctDNA) is emerging as a relevant tool for the diagnosis and monitoring of cancer [2]. Almost every tumor type releases DNA that can be isolated from the peripheral blood or other body fluids. Increasing evidence suggests that ctDNA recapitulates the genomic complexity of the tumor, and, therefore, it might represent a non-invasive tool for assessing its genomic profile. In addition, the non-invasive or minimally invasive nature of ctDNA testing allows repeated measurements over time, thus ensuring the possibility to evaluate the response to therapy and to monitor the genomic evolution of the disease under the pressure of the therapies [3]. Importantly, in patients with advanced disease and multiple localizations, the analysis of ctDNA might allow a better profiling of the heterogeneity of the disease as compared to the testing of tumor tissue derived from single metastases.

The availability of new methods for the genotyping of ctDNA and the development of next generation sequencing (NGS)-based panels with increased sensitivity have significantly amplified the applications of ctDNA testing in the management of cancer patients [4,5]. In this review, we will describe and discuss the current knowledge as well as the potential and future applications of ctDNA analysis in the different stages of melanoma, from early diagnosis to the genomic profiling of the tumor and the monitoring of the response to therapy (Figure 1).

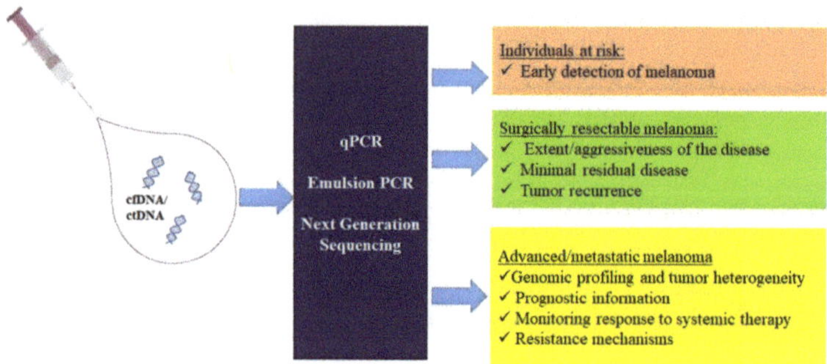

Figure 1. Applications of circulating tumor DNA (ctDNA) analysis in melanoma.

2. Circulating Tumor DNA and Circulating Cell-Free DNA

Circulating cell-free DNA (cfDNA) is highly fragmented (~166 bp) double-stranded DNA that freely circulates in body fluids including the plasma, serum, urine and cerebrospinal fluid. The release of cfDNA from damaged or dead cells occurs in normal physiological conditions, with most of the cfDNA being shed by white blood cells [6]. cfDNA has a relatively short half-life, ranging from 16 to 139 min.

An increase in cfDNA levels is observed in physiological conditions, as well as in pathological conditions, including cardiovascular diseases, heart failure, infections and cancer [7]. In cancer patients, a proportion of cfDNA, defined as ctDNA, comes from primary tumors, metastatic sites and/or circulating tumor cells. The ctDNA is released into the systemic circulation as a result of tumor cell apoptosis and/or necrosis, but the exact mechanism by which tumor cells release DNA is not yet fully clarified [2]. In the blood of cancer patients, ctDNA is only a fraction of the cfDNA, which includes DNA released from non-tumor cells, cells of the tumor microenvironment, and other cell types including stromal cells, endothelial cells and lymphocytes [7].

In this regard, the ctDNA fraction of the total cfDNA isolated from the peripheral blood can vary from <1% to 90% and is generally correlated with the tumor burden, although the localization of the tumor and the histological type might affect ctDNA release in the bloodstream (Figure 2) [8]. For this reason, the analysis of ctDNA in a background of cfDNA might be difficult, particularly in early stage cancer.

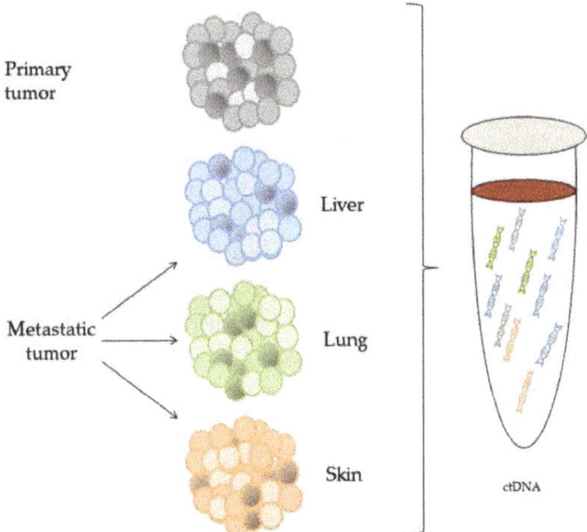

Figure 2. Circulating tumor DNA (ctDNA) reflects the genomic profile of different tumor localizations in metastatic melanoma and better recapitulates the heterogeneity of the disease.

cfDNA can be isolated from different body fluids. When cfDNA is obtained from the peripheral blood, plasma is preferred to serum, especially because the large amounts of wild-type DNA released by white cell lysis during clotting can cause a further dilution of ctDNA. To prevent clotting and DNase I activity, it is recommended to use ethylenediamine-tetraacetic acid (EDTA) tubes or tubes containing formaldehyde-free preservative reagents for blood collection [9]. When EDTA-containing tubes are used to collect blood, plasma separation must be performed within 4 h of drawing to prevent the lysis of leukocytes [10,11].

3. Methods for cfDNA Testing

Because of the possible dilution of tumor DNA in normal DNA, highly sensitive methods are needed to detect variants at low mutant allelic frequency (MAF) in the cfDNA. In this regard, sensitive methodologies have been developed to detect and quantify rare variants in the blood of cancer patients, with analytical sensitivity ranging between 0.005 and 5% [12]. These technologies can be classified into three groups: quantitative PCR (qPCR), emulsion-PCR and massively parallel sequencing, more commonly defined as NGS.

Melanoma most frequently harbors genomic alterations in *BRAF, NRAS, KIT* or *NF1* [13,14]. In particular, several efforts have been made to develop technologies capable of detecting *BRAF V600* variants in cfDNA, because of the availability in clinical research and clinical practice of drugs targeting tumors carrying these mutations.

A number of different approaches based on qPCR have been developed in order to improve its sensitivity for the analysis of rare variants. Allele-specific PCR (AS-PCR), allele-specific amplification refractory mutation system PCR (ARMS) and peptide nucleic acid (PNA)-based PCR have demonstrated, in different studies, analytical sensitivity up to 0.001% for the detection of the *BRAF-V600E* mutation [15].

Comparing the performance of some qPCR tests in detecting mutant *BRAF* in the plasma from stage IIIc–IV melanoma patients, Denis et al. observed that the Therascreen *BRAF* RGQ kit (Qiagen) was able to detect *BRAF* mutations in 73.7% of plasma samples from patients with BRAF-positive tumors. Similar results were obtained with the ct*BRAF* Mutation Detection Kit (Entrogen), which showed a positive percent agreement (PPA) of 68.4%. In all patients with wild-type *BRAF* tumors, the test was negative for cfDNA with both techniques [16].

In a recent study, the Cobas *BRAF/NRAS* (Roche) mutation test was used to analyze the ctDNA of 68 patients with stage III or IV melanoma. The test is an allele-specific, real-time PCR assay for the qualitative detection and identification of exon 11 and 15 mutations in the *BRAF* gene and exon 2, 3 and 4 variants of the *NRAS* gene from tumor tissue or plasma samples. The expected mutations were detected in the plasma of 34/68 patients (50% sensitivity), and the results obtained by the Cobas analysis were similar to those obtained by digital PCR (dPCR) [17]. Importantly, the sensitivity of the test significantly correlated with the stage. In fact, 64% of the tumor-positive stage IV patients were also positive at ctDNA analysis, while the sensitivity dropped to 19% for stage III melanoma patients. The specificity was 100%.

Recently, Biocartis (Belgium) developed a new fully automated platform, designated Idylla, for detecting the major *NRAS*, *BRAF*, *KRAS* or *EGFR* mutations in either ctDNA or genomic DNA isolated from tissue. The test allows the detection of *BRAF V600* mutations in plasma with up to three mutant copies per PCR reaction with an analytical sensitivity of 0.01% [18]. A 75% clinical sensitivity has been reported in stage IV melanoma patients [19].

Emulsion-PCR-based methodologies, such as dPCR, droplet digital PCR (ddPCR) and BEAming (which stands for beads, emulsion, amplification and magnetics) have a greater sensitivity as compared with qPCR, overcoming the problems due to the low levels of cfDNA and the low MAF of the variants [2,5]. In fact, BEAming- and ddPCR-based assays can detect and enumerate mutant and wild-type DNA at ratios greater than 0.01% [15]. An additional advantage of the ddPCR technique is the lower sensitivity to clinically relevant inhibitors (SDS, EDTA and heparin) than qPCR [20,21]. A ddPCR *BRAF-V600E* test detected the variant in 84.3% of plasma-derived cfDNA obtained from advanced melanoma patients [22], while a sensitivity >75% for the detection *of BRAF-V600E/V600K* mutations in cfDNA was observed using a BEAming-based assay [23].

Although qPCR and emulsion-PCR have a high sensitivity and specificity, both these methodologies have the limitation of being able to interrogate only a few loci for analysis. Therefore, the use of these techniques is limited to the screening of the most frequent known mutations. In addition, they cannot provide information on the genetic evolution of the tumor that might be relevant for therapeutic intervention, such as the identification of resistance mutations.

Many of these limitations are overcome by NGS, which is a high-throughput screening method that allows the simultaneous analysis of multiple genes and the detection of novel alterations, including low-frequency mutations. However, sensitivity represented, for a long time, a limiting factor for using NGS to test cfDNA. In this respect, targeted-sequencing panels are more suitable compared with whole-genome sequencing (WGS) or whole-exome sequencing (WES) due to their higher sensitivity, and reduced turnaround time and cost [5]. Novel NGS technologies have been recently developed for the specific analysis of cfDNA. These methods, by using molecular barcoding and improved bioinformatics pipelines [24–27], can detect variants at MAF < 1%. Tagged-Amplicon deep sequencing (TAm-seq), Safe-Sequencing System (Safe-SeqS), CAncer Personalized Profiling by deep sequencing (CAPP-Seq), and Ampliseq are examples of NGS technologies used for the targeted sequencing of cfDNA [4]. The limit of detection of these target panels depends on the quantity of cfDNA used for the preparation of libraries. To obtain high-quality libraries, about 20–30 ng of cfDNA is recommended, which can be commonly obtained from 4–5 mL of plasma.

Targeted-sequencing panels of different sizes, from a few mutations to hundreds of genes, are available both as a service and sold by different providers for independent laboratories [12]. Different labs developed custom NGS panels for cfDNA testing in melanoma patient. Although the specificity of NGS panels has significantly improved over time, the concordance between cfDNA and tissue testing ranges between 60% and 80% in most of the studies conducted in different tumor types. Several factors, including sequencing artifacts, tumor heterogeneity and clonal hematopoiesis, are involved in this relatively low concordance [28–30].

4. ctDNA for Early Detection of Melanoma

The chances of survival for melanoma patients significantly increase when the disease is diagnosed at an early clinical stage. Therefore, the improvement of melanoma early detection methods is fundamental for a proper management of this disease. However, several issues including the tumor heterogeneity, the tumor growth dynamics and the timing of metastasis, as well as the feasibility and cost of routinely applying liquid-biopsy techniques in clinical practice, make early detection challenging [31].

In a study aiming to define whether *BRAF V600E* could represent a suitable marker for melanoma detection, this mutation was identified only in ctDNA from stage III–IV melanoma patients, whereas, in early stages, it was undetectable in the majority of the cases [32]. While technical improvements might increase the sensitivity of mutation detection, the biology underlying ctDNA release and the relative low specificity of genetic alterations found in melanoma might still limit its potential use for early detection in an unselected population. For example, *BRAF-V600* mutations are present in only 50% of melanomas, while they can be found in a number of different tumor types [33]. Prospective studies might elucidate whether ctDNA analysis might be useful for monitoring melanocyte cell transformation in individuals at high risk.

NGS-based technologies might improve the use of cfDNA testing for the early detection of cancer. By testing a number of genetic alterations in a single analysis, these methods can significantly increase the sensitivity of the test. However, the detection of genetic alterations in the cfDNA might not have sufficient specificity for the early diagnosis of cancer. In this respect, the integration of information deriving from cfDNA analysis with other circulating biomarkers might significantly increase the sensitivity and specificity for the detection of melanoma in the early stages of development, as demonstrated for other cancer types [34]. In addition, incorporating such data into multiscale computational modeling platforms that can elaborate and integrate different types of information might allow individualized prediction for different cancer types.

5. Prognostic Value of ctDNA Testing in Patients with Surgically Resectable Melanoma

A considerable fraction of patients with localized melanoma who undergo surgery with curative intent experience a relapse of the disease. Local recurrence and distant metastasis in melanoma patients are heavily dependent on tumor size. They occur in 50% of patients with a tumor thickness larger than 4 mm [35], in regional lymph nodes (50%), as local recurrence (20%), or at distant sites (30%) [36]. The recurrence of apparently localized disease following radical surgery and adjuvant therapy (if administered) is likely due to the persistence of minimal residual disease (MRD), a potential source of subsequent metastatic dissemination.

In this scenario, perioperative ctDNA testing in patients with surgically resectable melanoma can offer different, relevant prognostic information: (i) the detection of ctDNA before surgery might correlate with the extent and the aggressiveness of the disease; (ii) ctDNA testing after surgery and, eventually, adjuvant therapy might identify MRD; (iii) testing ctDNA during the follow-up after curative resection might detect a tumor recurrence at the molecular level, thus anticipating clinical or radiological recurrence. Indeed, given the possibility of minimally invasive repeated sampling, ctDNA testing can allow real-time monitoring during the course of the disease.

The detection and monitoring of MRD are widely established in patients with hematological malignancies [37]. More recently, several studies have shown that the perioperative detection of ctDNA in colorectal, breast and lung cancer patients has a strong negative prognostic significance [38–40].

Similar findings have been reported in patients with high-risk stage III melanoma [41]. In particular, pre-operative ctDNA levels were assessed by dPCR in 174 patients (119 in the discovery cohort and 55 in the validation cohort) with stage III melanoma who underwent complete lymph node dissection. Patients with a mutation in *BRAF*, *NRAS* or *cKIT* in tumor tissue were included in this study. ctDNA was detected in 34% of the patients in the discovery cohort and 33% in the validation cohort. The presence of ctDNA was significantly associated with tumor burden. More importantly, patients with detectable ctDNA had worse melanoma-specific survival in both the discovery cohort (Hazard Ratio, HR, 2.11) and the validation cohort (HR, 2.29), and this difference was confirmed via multivariate analysis (HR, 1.85).

The prognostic value of pre-operative ctDNA detection was confirmed in an additional study that enrolled stage III melanoma patients who were tested for ctDNA by dPCR [42]. However, this latter study demonstrated that in patients who did not receive adjuvant therapy, the post-operative detection of ctDNA was an even stronger predictor of shorter relapse-free survival (HR, 10). Interestingly, serial ctDNA testing in a cohort of patients with post-operative negative ctDNA could detect somatic mutations in plasma samples prior to clinical recurrence in 48% of the cases, with a 2 month median lead time. Similar findings have been recently reported in a small cohort of melanoma patients whose cfDNA was retrospectively analyzed [43].

The prognostic value of post-operative ctDNA testing was also explored in a retrospective analysis of plasma samples from 161 stage II/III patients carrying either a *BRAF* or *NRAS* mutation in their baseline-resected tumors. Patients with detectable ctDNA after surgery had a significantly increased risk of death compared to those with undetectable ctDNA, with HR = 2.50 for overall survival (OS) after adjustment for performance status [44]. The aforementioned studies are summarized in Table 1.

Table 1. Main studies that addressed the prognostic value of ctDNA testing in patients with early (surgically resectable) melanoma.

Target Genes	AJCC Stage	Pre-/Post-Operative ctDNA Testing	N. of Patients	Clinical Outcomes	Reference
BRAF, *NRAS*, *cKIT*	III	Pre-operative	174	Presence of ctDNA associated with tumor burden and worse melanoma-specific survival.	Lee et al. [41]
BRAF, *NRAS*, *TERT*	III	Pre-operative	99	- Pre- and post-operative ctDNA detection correlates with shorter RFS and DMFS; - Post-operative ctDNA is an independent predictor of RFS and DMFS.	Tan et al. [42]
		Post-operative	68		
BRAF, *NRAS*, *TERT*	0–III	Post-operative	30	ctDNA detected at or before disease recurrence.	McEvoy et al. [43]
BRAF, *NRAS*	II/III	Post-operative	161	Presence of ctDNA predicts shorter DFI, DMFI and OS.	Lee et al. [44]

AJCC: American Joint Committee on Cancer; RFS: relapse-free survival; DMFS: distant metastasis-free survival; DFI: disease-free interval; DMFI: distant metastasis-free interval; OS: overall survival; ctDNA: circulating tumor DNA.

Although the above-summarized findings suggest a possible prognostic role of cfDNA testing in patients with resectable melanoma, these studies have several limitations. First, they included only patients with known mutations in a limited number of genes, thus excluding those patients that carry rare variants. In addition, while the specificity of the test is high, its sensitivity is relatively low because a significant fraction of ctDNA-negative patients have recurrences of the disease. In this respect, the use of NGS-based assays might increase the sensitivity of the test, being able to detect multiple mutations at the same time. Alternatively, the NGS testing of tumor tissue could be used to identify an adequate number of variants to test for in the cfDNA with highly sensitive dPCR-based assays.

6. cfDNA Testing as Tool to Support Treatment Decisions in Metastatic Melanoma Patients

The availability of different therapies makes highly relevant the use of biomarkers to stratify melanoma patients and identify the best therapeutic strategy for each individual patient. In this respect, cfDNA testing can provide relevant information at different levels: (i) the genomic profile of the disease and tumor heterogeneity, (ii) the prognosis, (iii) the response to therapy, and (iv) the development of resistance mechanisms.

6.1. cfDNA Testing for Genomic Profiling and Assessment of Tumor Heterogeneity

The analysis of cfDNA might represent an alternative to tumor tissue testing for the detection of predictive biomarkers. In this respect, a number of studies have demonstrated that qPCR and emulsion-PCR-based techniques can detect *BRAF* and *NRAS* mutations in plasma-derived cfDNA from patients with advanced melanoma (Table 2).

Table 2. Diagnostic performance of detecting *BRAF* and *NRAS* mutations in plasma using quantitative real-time PCR and emulsion-PCR-based techniques.

Reference	N° of patients	AJCC stage	Tissue mutation	Method	Specificity (%)	Sensitivity (%)
Ascierto et al. [45]	91	IV	*BRAF V600E* ($n = 72$) *BRAF V600K* ($n = 19$)	Digital PCR (BEAMing Inostics)	- -	79.2 89.5
Sanmamed et al. [22]	20	IIIc–IV	*BRAF V600* ($n = 20$)	Droplet digital PCR (Biorad)	-	84.3
Gray et al. [46]	48	IV	*BRAF V600E* ($n = 34$) Healthy patients ($n = 22$) *BRAF V600K* ($n = 8$) Healthy patients ($n = 23$) *BRAF V600R* ($n = 2$) Healthy patients ($n = 10$) *NRAS Q61K* ($n = 1$) Healthy patients ($n = 19$) *NRAS Q61R* ($n = 2$) Healthy patients ($n = 13$) *NRAS Q61L* ($n = 1$) Healthy patients ($n = 12$)	Droplet digital PCR (Biorad)	100 100 100 100 84.6 75	64.7 87.5 100 100 100 100
Gonzalo-Cao et al. [47]	22	IV	*BRAF V600E* ($n = 22$)	qPCR LNA PNA clamp (in-house)	-	57.7
Santiago-Walker et al. [23]	661	IV	*BRAF V600E* ($n = 661$)	Digital PCR (BEAMing Inostics)	97.6	76.2
Chang et al. [48]	43	IIIc–IV	*BRAF V600E* ($n = 20$) *BRAF V600K* ($n = 2$) *NRAS Q61K* ($n = 4$) *NRAS Q61R* ($n = 3$) *NRAS Q61L* ($n = 2$)	Droplet digital PCR (Biorad)	-	80
Schreuer et al. [19]	16	IV	*BRAF V600* ($n = 16$)	Idylla (Biocartis)	100	75
Knol et al. [49]	29	IIIc–IV	*BRAF V600E* ($n = 29$)	Therascreen *BRAF* RGQ kit (Qiagen)	-	75.9
Denis et al. [16]	54	IIIc–IV	*BRAF V600* ($n = 38$) *BRAF WT* ($n = 16$)	Therascreen *BRAF* RGQ kit (Qiagen) ct*BRAF* Mutation Detection Kit (Entrogen) QuantStudio 3D system (Life Technologies)	100	73.7 68.4 58.8
Tang et al. [50]	58	I–II–III–IV	*BRAF V600E* ($n = 58$)	QuantStudio 3D system (Life Technologies)	-	74.1
Haselmann et al. [51]	187	III–IV	*BRAF V600* ($n = 62$) *BRAF WT* ($n = 125$)	Digital PCR (BEAMing Inostics)	91.2	90.3
Long-Mira et al. [52]	19	IV	*BRAF V600* ($n = 10$) *NRAS Q61/G12/G13* ($n = 5$) Double WT ($n = 4$)	Idylla (Biocartis)	89 100	80 79
Seremet et al. [18]	85	III–IV1a, IV1b, IV–M1c	*BRAF V600* ($n = 68$) *NRAS Q61/G12/G13* ($n = 22$)	Idylla (Biocartis)	- -	47 37.5
Herbreteau et al. [17]	48	III–IV	*BRAF V600* ($n = 32$) *NRAS* ($n = 36$)	Cobas *BRAF/NRAS* Mutation Test LSR kit (Roche)	- -	50

AJCC: American Joint Committee on Cancer.

The sensitivity of the test ranged from 37.5% to 100%, although some of these studies enrolled a limited number of patients. By contrast, the specificity of the *BRAF/NRAS* testing of cfDNA was reported in a few studies and ranged between 75% and 100%. Taken together, these studies suggest that *BRAF* and *NRAS* mutations can be detected in the cfDNA from metastatic melanoma patients with a good specificity and an acceptable sensitivity.

The sensitivity of the cfDNA test might be affected by the localization of the tumor. In fact, patients with visceral, bone or lymph node involvement often display higher levels of ctDNA as compared with patients with extensive subcutaneous disease or brain metastases [53]. In agreement with these findings, the sensitivity of the *BRAF/NRAS* testing of cfDNA was found to correlate with the stage of the disease (higher in stage IV as compared with stage III) and the number and type of metastatic sites [17].

It has also been reported that the detectability of ctDNA is related to the nature of the mutated gene. Herbretau described a sensitivity of the cfDNA test of 36% for NRAS mutations and 66% for *BRAF* mutations [17]. In addition, mutations in the promoter region of the TERT gene were found at lower concentrations in the cfDNA as compared with other driver mutations, suggesting that they might be underrepresented in the cfDNA [54].

The sensitivity of cfDNA testing is also limited by the frequency of the mutant alleles in the context of wild-type DNA derived from normal cells. On the other hand, in selected cases, *BRAF* mutations have been identified through cfDNA analysis but not in the corresponding tumor tissue [49]. Such 'discordance is likely due to the heterogeneous expression of *BRAF* mutations in melanoma cells.

It has been demonstrated that driver mutations, including *BRAF* mutations in melanoma, are often clonal but can occasionally be subclonal [55]. This might lead to discordant results when different sites of the disease or different areas of the same tumor lesion are analyzed. In agreement with this hypothesis, discordant *BRAF* mutational statuses have been found between different sites of a primary tumor (intratumor heterogeneity), between a primary tumor and metastases, and between different metastases of the same patient [56,57]. In this scenario, the analysis of cfDNA is an approach that allows identifying mutations present in all the tumor sites of a given patient, thus better representing tumor heterogeneity [58]. Indeed, in cases with cfDNA positive for *BRAF* and BRAF mutation not detected in tumor tissue, the testing of additional samples from a different tumor area confirmed the presence of the *BRAF* variant [49].

Analysis of longitudinal samples from melanoma patients revealed that recurrent lesions might show a different *BRAF* mutational status over time as a consequence of tumor heterogeneity [59]. In the case of tumor relapse, the assessment of *BRAF* mutation status with a liquid biopsy might represent a non-invasive approach to confirming the mutational status of the disease.

Because the *BRAF-V600* mutations are the only approved biomarker in melanoma, the majority of data on genome profiling with liquid biopsies are based on the use of assays specific for these variants. However, NGS-based technologies might provide a better portrait of the genomic landscape and heterogeneity of melanoma, which might be relevant for therapeutic purposes. For example, it has been described that MITF and TP53 alterations are more frequent in patients with rapid progression of the disease following targeted therapy, while NF1 alterations are more common in cases with complete responses [60].

6.2. Prognostic Value of cfDNA Testing in Metastatic Melanoma

The identification of genetic alterations in cfDNA and the assessment of ctDNA levels before the administration of any systemic treatment may provide important prognostic information for patients with metastatic melanoma. In fact, a number of studies have demonstrated that high baseline levels of ctDNA correlate with a worse prognosis in metastatic melanoma patients treated with targeted therapy [22,23,45]. While the correlation between baseline ctDNA levels and survival is in line with previous findings in other tumor types including lung and colon carcinoma, in melanoma patients, a correlation of ctDNA levels with the response to treatment has been also reported. For example, in the phase II trial BREAK-2 enrolling *BRAF-V600E/K* metastatic melanoma patients to evaluate the clinical activity and the safety of the BRAF inhibitor dabrafenib, a correlation was found between the basal levels of *BRAF-V600E* in the cfDNA and tumor burden but not for the *V600K* variant, as assessed by using BEAMing technology [45]. Interestingly, patients with higher basal levels of cfDNA *BRAF-V600E* mutation showed a lower response rate and a shorter progression free survival (PFS) when treated with dabrafenib. In line with these findings, similar results were obtained in a larger cohort of 732 patients from four clinical studies of targeted therapy in metastatic melanoma (BREAK-2, BREAK-3, BREAK-MB and METRIC). Patients negative for *BRAF* mutations in the cfDNA had longer PFS and OS and higher rates of response to dabrafenib and trametinib, as compared with patients with detectable cfDNA *BRAF* mutations [23]. In multivariate analysis, the presence of *BRAF* mutations in the cfDNA was an independent predictive factor for shorter PFS in three out of four studies, and for shorter OS in one study.

Studies in smaller cohorts of patients further confirmed these findings. The quantification with ddPCR of mutant *BRAF-V600E* copies in cfDNA from 20 patients with metastatic melanoma showed a direct relationship between the *BRAF-V600E* copy numbers and clinical outcomes, where basal concentrations <216 copies/mL were significantly associated with better outcomes (OS = 27.7 months; PFS = 9 months) as compared with higher concentrations (OS = 8.6 months; PFS = 3 months) [22]. However, Schreuer did not find any correlation between the baseline levels of *BRAF* mutations in cfDNA and either survival or tumor responses in 25 patients receiving a combination of dabrafenib plus trametinib [61]. Importantly, this latter study enrolled patients who progressed on a previous line of therapy with a *BRAF* inhibitor, and the response rate was only 32%. The presence of a mechanism of resistance to a targeted therapy might represent a confounding factor for the prognostic role of ctDNA.

The basal levels of ctDNA are also a relevant prognostic marker in patients treated with immunotherapy. Seremet assessed ctDNA levels in 85 advanced melanoma patients using either the Idylla assay or a ddPCR test [18]. They found that patients in which the ctDNA was not detectable at baseline had longer PFS (HR, 0.47) and OS (HR, 0.37) as compared with patients with detectable ctDNA. Furthermore, high ctDNA levels (>500 copies/mL) at baseline in the group of patients with progressive disease were indicative of a very poor clinical outcome [18]. However, an additional study exploring the prognostic role of ctDNA in melanoma patients receiving immune-checkpoint inhibitors suggested that the dynamics of the ctDNA were better associated with outcome than was the baseline ctDNA status [62].

Low baseline ctDNA levels were associated with higher response rates and longer PFS and OS in studies that included cohorts of patients treated with either targeted therapy or immunotherapy [17,46]. Importantly, the baseline levels of ctDNA before first-line therapy were confirmed via multivariate analysis to be an independent prognostic factor for OS, irrespective of treatment, in patients with stage IV or unresectable stage III metastatic melanoma [17]. Interestingly, ctDNA analysis in recurrent patients or during treatment (non-first line) was not associated with either PFS or OS. However, this latter analysis was limited to a very small cohort of patients.

The strong prognostic value of ctDNA levels in metastatic melanoma patients suggests that this parameter might better recapitulate the disease burden as compared with other prognostic factors. In this respect, several studies have demonstrated a correlation between ctDNA levels and the American Joint Committee on Cancer (AJCC) stage, number of metastatic sites, and serum levels of Lactate dehydrogenase (LDH) and S100 [49,63]. A significant correlation has also been reported between baseline ctDNA levels and metabolic tumor activity [43,53,63]. Interestingly, baseline ctDNA levels presented a better correlation with disease burden as measured by 2-[fluorine-18]-fluoro-2-deoxy-d-glucose positron emission tomography (FDG-PET) when compared with LDH [53]. However, subcutaneous and cerebral disease localization were associated with lower ctDNA levels, suggesting that in these cases, FDG-PET might better recapitulate tumor burden [53].

6.3. Monitoring Response to Therapy

The availability of predictive and prognostic biomarkers significantly improved the selection of appropriate therapeutic approaches in melanoma patients. Nevertheless, the response to both targeted therapy and immunotherapy is often heterogeneous, even in molecularly selected cohorts of patients. Therefore, the availability of tools to assess the response to therapy could further improve the therapeutic approaches for patients with metastatic melanoma. In this respect, the analysis of cfDNA could represent a non-invasive and repeatable technique for the early detection and monitoring of melanoma response and/or progression.

A number of studies have addressed the possibility of using the *BRAF* mutation testing of cfDNA to monitor the response to *BRAF* inhibitors [22] or combinations of *BRAF* and *MEK* inhibitors [51] in patients with *BRAF*-mutant melanoma. Although these studies enrolled a limited number of patients and different qPCR [19] or emulsion-PCR [22,51]-based techniques, they consistently found that response to therapy was associated with a significant decrease in the levels of *BRAF* mutations in cfDNA, while an increase in *BRAF* mutant DNA was observed at progression. Interestingly, the increase in ctDNA levels preceded the clinical progression of the disease in a significant fraction of patients, with a lead time up to 110 days [51,61].

The predictive value of monitoring the *BRAF* mutations in cfDNA using the Idylla test was also explored in a phase II trial of a combination of dabrafenib plus trametinib in patients with advanced *BRAF-V600*-mutant melanoma pre-treated with targeted therapy [61]. Upon analyzing plasma samples from 25 patients enrolled in this study, patients responding to therapy showed a significantly lower level of *BRAF-V600*-mutant ctDNA after 2 weeks of treatment as compared with non-responding patients. In addition, the persistent detection of *BRAF-V600*-mutant ctDNA after 2 weeks of therapy was correlated with a shorter PFS as compared with that for patients with undetectable ctDNA (1.8 months vs. 5.9 months; $p = 0.001$).

Some studies have reported a very early spike in ctDNA concentration occurring 24/48 h after starting treatment with *BRAF* inhibitors [19,64]. This early spike is probably related to a massive release of tumor DNA due to treatment-induced tumor cell lysis.

Taken together, these findings strongly suggest that monitoring *BRAF* mutations in the cfDNA could represent a valuable biomarker for assessing the early response to targeted therapy in melanoma patients.

Testing cfDNA could also provide important information for the evaluation of the response to immune-checkpoint inhibitors in melanoma and other cancers. In fact, the assessment of the response to immune therapies is sometimes difficult, due to the possible increase in the tumor lesions because of the immune reaction, which mimics a progression of the disease. Such pseudo-progression might lead to the suspension of an active treatment.

Few studies have addressed the potential of cfDNA testing for monitoring the response to immune-checkpoint inhibitors in patients with advanced melanoma. Two initial studies in small cohorts of patients receiving different immune-therapeutics (either PD-1 inhibitors or CTLA inhibitors or combinations) demonstrated a correlation between ctDNA dynamics and response to therapy [64,65]. In particular, levels of mutant DNA were found to decrease in the cfDNA of patients responding to treatment and to increase at or before tumor progression, using either NGS [65] or ddPCR [64].

The predictive value of ctDNA monitoring was next explored in a cohort of 76 patients with metastatic melanoma who received treatment with pembrolizumab or nivolumab monotherapy or in combination with ipilimumab [62]. Although ddPCR was employed in this study for the detection of the *BRAF*, *NRAS* and *cKIT* most frequent variants, the criterion used by the authors was only whether the ctDNA was detectable or not. Interestingly, the response rate in patients with detectable ctDNA at baseline but undetectable after 12 weeks of therapy was similar to that in patients with undetectable ctDNA at baseline and after 12 weeks (77% and 72%, respectively), and much higher as compared with that in patients with detectable ctDNA at baseline and after 12 weeks (6%). In addition, the first two groups had significantly longer PFS and OS as compared with the third group. The predictive value of ctDNA monitoring at 12 weeks was confirmed via multivariate analysis, thus suggesting that the clearance of ctDNA is a relevant predictive marker of response to immune-checkpoint inhibitors in metastatic melanoma. The same research group assessed the ability of cfDNA testing to identify pseudo-progression in a study including 125 metastatic melanoma patients who received anti-PD-1 antibodies alone or in combination with ipilimumab [66]. In this study, all nine patients with confirmed pseudo-progression had a favorable ctDNA profile, defined as ctDNA undetectable at baseline that remained undetectable, or detectable at baseline that became undetectable or decreased by at least 10-fold during treatment, thus introducing a quantitative criterion. By contrast, 18/20 patients with true progressive disease had an unfavorable ctDNA profile, with ctDNA detectable at baseline and during treatment. The dynamics of the ctDNA levels in the first month after treatment were also found to predict progressive disease or response to therapy in 21/24 stage III/IV melanoma patients receiving either immunotherapy or targeted therapy [63].

Finally, some studies in small cohorts of patients tried to establish quantitative thresholds to define patients with response or progression at ctDNA analysis [67,68]. Both these studies found a good correlation between an increase in ctDNA levels and progression of the disease, while ctDNA reduction was also observed in patients who experienced progressive disease. One of the limitations of these studies was a focus only on the few most frequent mutations in *BRAF*, *NRAS* and *cKIT*. By assessing a single mutation, the test might follow the dynamics of a single cell clone, which might not represent the behavior of the entire tumor in the case of tumor heterogeneity.

The above-summarized data suggest that cfDNA testing might represent an adequate tool for monitoring responses to immune-checkpoint inhibitors in melanoma. However, prospective clinical trials are required to validate this approach in the clinic and to demonstrate that the early detection of tumor progression might allow a better therapeutic strategy.

The use of NGS-based techniques might significantly increase the fraction of patients who can be monitored, improve the specificity and sensitivity of the test and provide information on tumor heterogeneity and the clonal evolution of the disease. For example, the NGS analysis of cfDNA from a patient with vaginal mucosa melanoma revealed the presence of two subclones with different responses to imatinib [69]. One subclone carried a KIT mutation and responded to imatinib, while the other had a KIT wild-type gene and did not respond to targeted therapy.

6.4. Identification of Mechanisms of Resistance

Combined targeted therapy with *BRAF* and *MEK* inhibitors is associated with a high response rate in melanoma patients who carry a *BRAF-V600* mutation [70]. However, most patients who initially respond will relapse during therapy due to mechanisms of acquired resistance. In many cases, acquired resistance to anti-*BRAF* therapy in melanoma patients is due to the reactivation of the *MAPK* pathway by genetic or epigenetic mechanisms [71]. However, other mechanisms of resistance have been described, including genetic alterations leading to the activation of the *PIK3CA* signaling pathway.

The feasibility of cfDNA testing in the assessment of acquired resistance to melanoma has been explored in a few studies. In a study that employed ddPCR to test *BRAF* and *NRAS* variants, *NRAS* mutations were detected in 3/7 melanoma patients progressing on targeted therapy with vemurafenib, dabrafenib or a dabrafenib/trametinib combination [46]. These data are in agreement with previous reports that described the frequent involvement of *NRAS* mutations, in particular, at codon 61 (*p.Q61K/R*), in acquired resistance to dabrafenib/trametinib combination therapy in *BRAF*-mutant metastatic melanoma patients [72]. Interestingly, *NRAS* mutations were detected in the cfDNA before the clinical and radiological progression of the disease.

By using the WES analysis of cfDNA, *NRAS* and *PIK3CA* mutations not present prior to therapy were identified in melanoma patients progressing on targeted therapy with *BRAF* and/or *MEK* inhibitors [73]. Multiple *NRAS* mutations were also found in the same cfDNA sample from a patient with progression, thus confirming the likely multi-clonal origin of acquired resistance to targeted therapies.

Acquired mutations in *NRAS* were also identified in metastatic melanoma patients at progression following dabrafenib-and-trametinib treatment using the targeted sequencing of cfDNA [53]. Interestingly, targeted-sequencing analysis identified two variants in *MAP2K1* and *PTEN* in a patient at progression, suggesting that these two genetic alterations might both contribute to resistance.

In addition to point mutations, other genomic alterations including *BRAF* gene amplification have been described as driving acquired resistance to *BRAF/MEK* inhibitors [74]. Indeed, the whole-exome sequencing and low-coverage whole-genome sequencing of cfDNA identified *BRAF* gene amplification in 2/3 patients at progression [53].

Evidence suggests that acquired resistance to targeted therapy is often polyclonal. In agreement with this hypothesis, distinct molecular alterations have been detected concurrently in the same tumor sample or among multiple tumor sites from the same melanoma patient progressing on targeted therapy [75]. Therefore, resistance to targeted therapy is associated with an increase in tumor heterogeneity and branched evolution that might be better depicted by using cfDNA testing as compared with tissue analysis. However, the ability of plasma-derived cfDNA testing to represent spatial heterogeneity was found to be limited in the presence of subcutaneous or brain metastases that shed limited amounts of ctDNA into the blood flow [53].

7. Conclusions and Future Perspectives

The data summarized in this article clearly show that cfDNA analysis can offer important information on the prognosis of patients with melanoma in different stages of the disease. Furthermore, cfDNA analysis could allow a more accurate identification of patient candidates for targeted therapy or immunotherapy, through the representation of tumor heterogeneity and the early identification of patients who do not respond to therapies. However, the possibility of transferring this information into daily clinical practice depends on the resolution of a series of technological, biological and clinical issues.

The introduction of NGS technologies for cfDNA analysis has increased the possibility of extending the test to a larger percentage of patients and has increased its clinical and analytical sensitivity. However, several NGS panels for cfDNA analysis are commercially available. For many of these panels, validation studies on adequate cohorts of biological samples are lacking, and external quality control programs are not yet available for these specific tests. Therefore, their introduction into clinical practice must be carried out with extreme caution, in order to avoid problems of false positives or negatives.

The use of NGS panels increases the sensitivity of the cfDNA test but also increases the possibility of identifying mutations associated with clonal hematopoiesis [29]. This possibility considerably limits the use of NGS techniques for applications such as early diagnosis or even monitoring of the disease. Although *BRAF* mutations have not been associated with clonal hematopoiesis to date, the identification of other genetic alterations could still pose problems in the clinical interpretation of the data. At present, the contemporary analysis of DNA derived from leukocytes appears as the only approach to overcoming this limit.

The analysis of cfDNA can provide information only on genetic alterations, which, alone, are not sufficient to represent the biological variability of neoplasms. In the era of precision medicine, the integration of the genomic profile with other biological omics, including transcriptomics, proteomics, metabolomics and epigenomics, that can be determined in biological fluids, represents a fundamental element for the advancement of knowledge on the pathogenesis and progression of cancer [76]. A biological multiomic pattern combined with clinical and radiological information, including radiomics, will allow the development of novel strategies for diagnostic, prognostic and therapeutic purposes [77]. In particular, the possibility of integrating different information could then be important for the early diagnosis of cancer, given the low specificity of genetic alterations for this specific application. In addition, the addition of multiple layers of clinical and biological information will allow a better stratification of patients, thus improving the clinical implementation of precision medicine.

It must be emphasized that the integration of multiple omics information for patients' stratification will require the development of appropriate disease modeling systems based on the use of machine learning [78].

The various potential applications of liquid biopsies in the early diagnosis of recurrence and in monitoring the response to therapy must be validated in prospective clinical trials. In fact, it will be necessary to demonstrate that the early detection of disease recurrence allows the development of therapeutic strategies that result in a decrease in the mortality of patients with early stage melanoma. At the same time, it must be demonstrated that the modification of therapy in patients with metastatic melanoma who do not respond in terms of reduced ctDNA results in better survival.

However, although there are many issues to be solved for the clinical implementation of cfDNA analysis, it can certainly be said that this technology opens up new scenarios in the management of patients with melanoma that could lead to important improvements in the diagnosis and therapy of this disease. In particular, the possibility of determining the overall profile of the genetic alterations of the neoplasm and of being able to evaluate its progression over time represents an important opportunity to improve the stratification of patients with melanoma for the purposes of therapeutic planning.

Author Contributions: A.S.: conceptualization and writing original draft; L.F.: conceptualization and writing original draft; M.C.: conceptualization and writing original draft A.D.L.: supervision, visualization and writing—review; P.A.A.: resources and visualization; G.B.: resources and visualization; N.N.: project coordination, conceptualization, visualization, supervision and writing—review and editing. All authors have read and agreed to the published version of the manuscript.

Funding: This paper was not funded.

Conflicts of Interest: N.N.: Personal financial interests (speaker's fees and/or advisory boards): MSD, QIAGEN, Bayer, Biocartis, Incyte, Roche, BMS, MERCK, Thermo Fisher, Boehringer Ingelheim, AstraZeneca, Sanofi, Eli Lilly, Illumina, and Amgen Institutional; financial interests (financial support to research projects): MERCK, Sysmex, Thermo Fisher, QIAGEN, Roche, AstraZeneca, Biocartis, and Illumina.

Non-financial interests: N.N.: President, International Quality Network for Pathology (IQN Path); President Elect, Italian Cancer Society (SIC).

References

1. Matthews, N.H.; Li, W.Q.; Qureshi, A.A.; Weinstock, M.A.; Cho, E. Epidemiology of Melanoma. In *Cutaneous Melanoma: Etiology and Therapy*; Ward, W.H., Farma, J.M., Eds.; Codon Publications: Brisbane, Australia, 2017. [CrossRef]
2. Wan, J.C.M.; Massie, C.; Garcia-Corbacho, J.; Mouliere, F.; Brenton, J.D.; Caldas, C.; Pacey, S.; Baird, R.; Rosenfeld, N. Liquid biopsies come of age: Towards implementation of circulating tumour DNA. *Nat. Rev. Cancer* **2017**, *17*, 223–238. [CrossRef] [PubMed]
3. Corcoran, R.B.; Chabner, B.A. Application of Cell-free DNA Analysis to Cancer Treatment. *N. Engl. J. Med.* **2018**, *379*, 1754–1765. [CrossRef] [PubMed]
4. Esposito Abate, R.; Pasquale, R.; Fenizia, F.; Rachiglio, A.M.; Roma, C.; Bergantino, F.; Forgione, L.; Lambiase, M.; Sacco, A.; Piccirillo, M.C.; et al. The role of circulating free DNA in the management of NSCLC. *Expert Rev. Anticancer Ther.* **2019**, *19*, 19–28. [CrossRef] [PubMed]
5. Normanno, N.; Cervantes, A.; Ciardiello, F.; De Luca, A.; Pinto, C. The liquid biopsy in the management of colorectal cancer patients: Current applications and future scenarios. *Cancer Treat. Rev.* **2018**, *70*, 1–8. [CrossRef] [PubMed]
6. Sun, K.; Jiang, P.; Chan, K.C.; Wong, J.; Cheng, Y.K.; Liang, R.H.; Chan, W.K.; Ma, E.S.; Chan, S.L.; Cheng, S.H.; et al. Plasma DNA tissue mapping by genome-wide methylation sequencing for noninvasive prenatal, cancer, and transplantation assessments. *Proc. Natl. Acad. Sci. USA* **2015**, *112*, E5503–E5512. [CrossRef] [PubMed]
7. Thierry, A.R.; El Messaoudi, S.; Gahan, P.B.; Anker, P.; Stroun, M. Origins, structures, and functions of circulating DNA in oncology. *Cancer Metastasis Rev.* **2016**, *35*, 347–376. [CrossRef] [PubMed]
8. Abbosh, C.; Birkbak, N.J.; Wilson, G.A.; Jamal-Hanjani, M.; Constantin, T.; Salari, R.; Le Quesne, J.; Moore, D.A.; Veeriah, S.; Rosenthal, R.; et al. Phylogenetic ctDNA analysis depicts early-stage lung cancer evolution. *Nature* **2017**, *545*, 446–451. [CrossRef]

9. Barra, G.B.; Santa Rita, T.H.; de Almeida Vasques, J.; Chianca, C.F.; Nery, L.F.; Santana Soares Costa, S. EDTA-mediated inhibition of DNases protects circulating cell-free DNA from ex vivo degradation in blood samples. *Clin. Biochem.* **2015**, *48*, 976–981. [CrossRef]

10. Norton, S.E.; Lechner, J.M.; Williams, T.; Fernando, M.R. A stabilizing reagent prevents cell-free DNA contamination by cellular DNA in plasma during blood sample storage and shipping as determined by digital PCR. *Clin. Biochem.* **2013**, *46*, 1561–1565. [CrossRef]

11. Normanno, N.; Denis, M.G.; Thress, K.S.; Ratcliffe, M.; Reck, M. Guide to detecting epidermal growth factor receptor (EGFR) mutations in ctDNA of patients with advanced non-small-cell lung cancer. *Oncotarget* **2017**, *8*, 12501–12516. [CrossRef]

12. Diefenbach, R.J.; Lee, J.H.; Rizos, H. Monitoring Melanoma Using Circulating Free DNA. *Am. J. Clin. Dermatol.* **2019**, *20*, 1–12. [CrossRef] [PubMed]

13. Curtin, J.A.; Fridlyand, J.; Kageshita, T.; Patel, H.N.; Busam, K.J.; Kutzner, H.; Cho, K.H.; Aiba, S.; Brocker, E.B.; LeBoit, P.E.; et al. Distinct sets of genetic alterations in melanoma. *N. Engl. J. Med.* **2005**, *353*, 2135–2147. [CrossRef] [PubMed]

14. Bastian, B.C. The molecular pathology of melanoma: An integrated taxonomy of melanocytic neoplasia. *Annu. Rev. Pathol.* **2014**, *9*, 239–271. [CrossRef] [PubMed]

15. Busser, B.; Lupo, J.; Sancey, L.; Mouret, S.; Faure, P.; Plumas, J.; Chaperot, L.; Leccia, M.T.; Coll, J.L.; Hurbin, A.; et al. Plasma Circulating Tumor DNA Levels for the Monitoring of Melanoma Patients: Landscape of Available Technologies and Clinical Applications. *Biomed. Res. Int.* **2017**, *2017*, 5986129. [CrossRef]

16. Denis, M.G.; Knol, A.-C.; Vallee, A.; Theoleyre, S.; Herbreteau, G.; Khammar, i.; Dréno, B. Cross-platform comparison of techniques to detect BRAF mutations in circulating tumor DNA of melanoma patients. *J. Clin. Oncol.* **2016**, *34*, e21026. [CrossRef]

17. Herbreteau, G.; Vallee, A.; Knol, A.C.; Theoleyre, S.; Quereux, G.; Frenard, C.; Varey, E.; Hofman, P.; Khammari, A.; Dreno, B.; et al. Circulating Tumour DNA Is an Independent Prognostic Biomarker for Survival in Metastatic BRAF or NRAS-Mutated Melanoma Patients. *Cancers* **2020**, *12*, 1871. [CrossRef]

18. Seremet, T.; Jansen, Y.; Planken, S.; Njimi, H.; Delaunoy, M.; El Housni, H.; Awada, G.; Schwarze, J.K.; Keyaerts, M.; Everaert, H.; et al. Undetectable circulating tumor DNA (ctDNA) levels correlate with favorable outcome in metastatic melanoma patients treated with anti-PD1 therapy. *J. Transl. Med.* **2019**, *17*, 303. [CrossRef]

19. Schreuer, M.; Meersseman, G.; Van Den Herrewegen, S.; Jansen, Y.; Chevolet, I.; Bott, A.; Wilgenhof, S.; Seremet, T.; Jacobs, B.; Buyl, R.; et al. Quantitative assessment of BRAF V600 mutant circulating cell-free tumor DNA as a tool for therapeutic monitoring in metastatic melanoma patients treated with BRAF/MEK inhibitors. *J. Transl. Med.* **2016**, *14*, 95. [CrossRef]

20. Dingle, T.C.; Sedlak, R.H.; Cook, L.; Jerome, K.R. Tolerance of droplet-digital PCR vs real-time quantitative PCR to inhibitory substances. *Clin. Chem.* **2013**, *59*, 1670–1672. [CrossRef]

21. Huggett, J.F.; Whale, A. Digital PCR as a novel technology and its potential implications for molecular diagnostics. *Clin. Chem.* **2013**, *59*, 1691–1693. [CrossRef]

22. Sanmamed, M.F.; Fernandez-Landazuri, S.; Rodriguez, C.; Zarate, R.; Lozano, M.D.; Zubiri, L.; Perez-Gracia, J.L.; Martin-Algarra, S.; Gonzalez, A. Quantitative cell-free circulating BRAFV600E mutation analysis by use of droplet digital PCR in the follow-up of patients with melanoma being treated with BRAF inhibitors. *Clin. Chem.* **2015**, *61*, 297–304. [CrossRef] [PubMed]

23. Santiago-Walker, A.; Gagnon, R.; Mazumdar, J.; Casey, M.; Long, G.V.; Schadendorf, D.; Flaherty, K.; Kefford, R.; Hauschild, A.; Hwu, P.; et al. Correlation of BRAF Mutation Status in Circulating-Free DNA and Tumor and Association with Clinical Outcome across Four BRAFi and MEKi Clinical Trials. *Clin. Cancer. Res.* **2016**, *22*, 567–574. [CrossRef] [PubMed]

24. Phallen, J.; Sausen, M.; Adleff, V.; Leal, A.; Hruban, C.; White, J.; Anagnostou, V.; Fiksel, J.; Cristiano, S.; Papp, E.; et al. Direct detection of early-stage cancers using circulating tumor DNA. *Sci. Transl. Med.* **2017**, *9*, 403. [CrossRef]

25. Newman, A.M.; Liu, C.L.; Green, M.R.; Gentles, A.J.; Feng, W.; Xu, Y.; Hoang, C.D.; Diehn, M.; Alizadeh, A.A. Robust enumeration of cell subsets from tissue expression profiles. *Nat. Methods* **2015**, *12*, 453–457. [CrossRef] [PubMed]

26. Goldberg, S.B.; Narayan, A.; Kole, A.J.; Decker, R.H.; Teysir, J.; Carriero, N.J.; Lee, A.; Nemati, R.; Nath, S.K.; Mane, S.M.; et al. Early Assessment of Lung Cancer Immunotherapy Response via Circulating Tumor DNA. *Clin. Cancer. Res.* **2018**, *24*, 1872–1880. [CrossRef] [PubMed]

27. Newman, A.M.; Bratman, S.V.; To, J.; Wynne, J.F.; Eclov, N.C.; Modlin, L.A.; Liu, C.L.; Neal, J.W.; Wakelee, H.A.; Merritt, R.E.; et al. An ultrasensitive method for quantitating circulating tumor DNA with broad patient coverage. *Nat. Med.* **2014**, *20*, 548–554. [CrossRef] [PubMed]

28. Stetson, D.; Ahamed, A.; Xu, X.; Nuttall, B.R.B.; Lubinski, T.J.; Johnson, J.H.; Barrett, J.C.; Dougherty, B.A. Orthogonal Comparison of Four Plasma NGS Tests With Tumor Suggests Technical Factors are a Major Source of Assay Discordance. *JCO Precis. Oncol.* **2019**, *3*, 1–9. [CrossRef]

29. Watson, C.J.; Papula, A.L.; Poon, G.Y.P.; Wong, W.H.; Young, A.L.; Druley, T.E.; Fisher, D.S.; Blundell, J.R. The evolutionary dynamics and fitness landscape of clonal hematopoiesis. *Science* **2020**, *367*, 1449–1454. [CrossRef]

30. Pasquale, R.; Forgione, L.; Roma, C.; Fenizia, F.; Bergantino, F.; Rachiglio, A.M.; De Luca, A.; Gallo, M.; Maiello, M.R.; Palumbo, G.; et al. Targeted sequencing analysis of cell-free DNA from metastatic non-small-cell lung cancer patients: Clinical and biological implications. *Transl. Lung Cancer Res.* **2020**, *9*, 61–70. [CrossRef]

31. Pashayan, N.; Pharoah, P.D.P. The challenge of early detection in cancer. *Science* **2020**, *368*, 589–590. [CrossRef]

32. Daniotti, M.; Vallacchi, V.; Rivoltini, L.; Patuzzo, R.; Santinami, M.; Arienti, F.; Cutolo, G.; Pierotti, M.A.; Parmiani, G.; Rodolfo, M. Detection of mutated BRAFV600E variant in circulating DNA of stage III-IV melanoma patients. *Int. J. Cancer* **2007**, *120*, 2439–2444. [CrossRef] [PubMed]

33. Consortium, I.T.P.-C.A.o.W.G. Pan-cancer analysis of whole genomes. *Nature* **2020**, *578*, 82–93.

34. Cohen, J.D.; Li, L.; Wang, Y.; Thoburn, C.; Afsari, B.; Danilova, L.; Douville, C.; Javed, A.A.; Wong, F.; Mattox, A.; et al. Detection and localization of surgically resectable cancers with a multi-analyte blood test. *Science* **2018**, *359*, 926–930. [CrossRef] [PubMed]

35. Leiter, U.; Meier, F.; Schittek, B.; Garbe, C. The natural course of cutaneous melanoma. *J. Surg. Oncol.* **2004**, *86*, 172–178. [CrossRef]

36. Benvenuto-Andrade, C.; Oseitutu, A.; Agero, A.L.; Marghoob, A.A. Cutaneous melanoma: Surveillance of patients for recurrence and new primary melanomas. *Dermatol. Ther.* **2005**, *18*, 423–435. [CrossRef]

37. van der Velden, V.H.; Hochhaus, A.; Cazzaniga, G.; Szczepanski, T.; Gabert, J.; van Dongen, J.J. Detection of minimal residual disease in hematologic malignancies by real-time quantitative PCR: Principles, approaches, and laboratory aspects. *Leukemia* **2003**, *17*, 1013–1034. [CrossRef]

38. Garcia-Murillas, I.; Schiavon, G.; Weigelt, B.; Ng, C.; Hrebien, S.; Cutts, R.J.; Cheang, M.; Osin, P.; Nerurkar, A.; Kozarewa, I.; et al. Mutation tracking in circulating tumor DNA predicts relapse in early breast cancer. *Sci. Transl. Med.* **2015**, *7*, 302ra133. [CrossRef]

39. Chaudhuri, A.A.; Chabon, J.J.; Lovejoy, A.F.; Newman, A.M.; Stehr, H.; Azad, T.D.; Khodadoust, M.S.; Esfahani, M.S.; Liu, C.L.; Zhou, L.; et al. Early Detection of Molecular Residual Disease in Localized Lung Cancer by Circulating Tumor DNA Profiling. *Cancer Discov.* **2017**, *7*, 1394–1403. [CrossRef]

40. Tie, J.; Wang, Y.; Tomasetti, C.; Li, L.; Springer, S.; Kinde, I.; Silliman, N.; Tacey, M.; Wong, H.L.; Christie, M.; et al. Circulating tumor DNA analysis detects minimal residual disease and predicts recurrence in patients with stage II colon cancer. *Sci. Transl. Med.* **2016**, *8*, 346ra392. [CrossRef]

41. Lee, J.H.; Saw, R.P.; Thompson, J.F.; Lo, S.; Spillane, A.J.; Shannon, K.F.; Stretch, J.R.; Howle, J.; Menzies, A.M.; Carlino, M.S.; et al. Pre-operative ctDNA predicts survival in high-risk stage III cutaneous melanoma patients. *Ann. Oncol.* **2019**, *30*, 815–822. [CrossRef]

42. Tan, L.; Sandhu, S.; Lee, R.J.; Li, J.; Callahan, J.; Ftouni, S.; Dhomen, N.; Middlehurst, P.; Wallace, A.; Raleigh, J.; et al. Prediction and monitoring of relapse in stage III melanoma using circulating tumor DNA. *Ann. Oncol.* **2019**, *30*, 804–814. [CrossRef] [PubMed]

43. McEvoy, A.C.; Pereira, M.R.; Reid, A.; Pearce, R.; Cowell, L.; Al-Ogaili, Z.; Khattak, M.A.; Millward, M.; Meniawy, T.M.; Gray, E.S.; et al. Monitoring melanoma recurrence with circulating tumor DNA: A proof of concept from three case studies. *Oncotarget* **2019**, *10*, 113–122. [CrossRef] [PubMed]

44. Lee, R.J.; Gremel, G.; Marshall, A.; Myers, K.A.; Fisher, N.; Dunn, J.A.; Dhomen, N.; Corrie, P.G.; Middleton, M.R.; Lorigan, P.; et al. Circulating tumor DNA predicts survival in patients with resected high-risk stage II/III melanoma. *Ann. Oncol.* **2018**, *29*, 490–496. [CrossRef] [PubMed]

45. Ascierto, P.A.; Minor, D.; Ribas, A.; Lebbe, C.; O'Hagan, A.; Arya, N.; Guckert, M.; Schadendorf, D.; Kefford, R.F.; Grob, J.J.; et al. Phase II trial (BREAK-2) of the BRAF inhibitor dabrafenib (GSK2118436) in patients with metastatic melanoma. *J. Clin. Oncol.* **2013**, *31*, 3205–3211. [CrossRef] [PubMed]

46. Gray, E.S.; Rizos, H.; Reid, A.L.; Boyd, S.C.; Pereira, M.R.; Lo, J.; Tembe, V.; Freeman, J.; Lee, J.H.; Scolyer, R.A.; et al. Circulating tumor DNA to monitor treatment response and detect acquired resistance in patients with metastatic melanoma. *Oncotarget* **2015**, *6*, 42008–42018. [CrossRef]

47. Gonzalez-Cao, M.; Mayo-de-Las-Casas, C.; Molina-Vila, M.A.; De Mattos-Arruda, L.; Munoz-Couselo, E.; Manzano, J.L.; Cortes, J.; Berros, J.P.; Drozdowskyj, A.; Sanmamed, M.; et al. BRAF mutation analysis in circulating free tumor DNA of melanoma patients treated with BRAF inhibitors. *Melanoma Res.* **2015**, *25*, 486–495. [CrossRef]

48. Chang, G.A.; Tadepalli, J.S.; Shao, Y.; Zhang, Y.; Weiss, S.; Robinson, E.; Spittle, C.; Furtado, M.; Shelton, D.N.; Karlin-Neumann, G.; et al. Sensitivity of plasma BRAFmutant and NRASmutant cell-free DNA assays to detect metastatic melanoma in patients with low RECIST scores and non-RECIST disease progression. *Mol. Oncol.* **2016**, *10*, 157–165. [CrossRef]

49. Knol, A.C.; Vallee, A.; Herbreteau, G.; Nguyen, J.M.; Varey, E.; Gaultier, A.; Theoleyre, S.; Saint-Jean, M.; Peuvrel, L.; Brocard, A.; et al. Clinical significance of BRAF mutation status in circulating tumor DNA of metastatic melanoma patients at baseline. *Exp. Dermatol.* **2016**, *25*, 783–788. [CrossRef]

50. Tang, H.; Kong, Y.; Si, L.; Cui, C.; Sheng, X.; Chi, Z.; Dai, J.; Yu, S.; Ma, M.; Wu, X.; et al. Clinical significance of BRAF(V600E) mutation in circulating tumor DNA in Chinese patients with melanoma. *Oncol. Lett.* **2018**, *15*, 1839–1844.

51. Haselmann, V.; Gebhardt, C.; Brechtel, I.; Duda, A.; Czerwinski, C.; Sucker, A.; Holland-Letz, T.; Utikal, J.; Schadendorf, D.; Neumaier, M. Liquid Profiling of Circulating Tumor DNA in Plasma of Melanoma Patients for Companion Diagnostics and Monitoring of BRAF Inhibitor Therapy. *Clin. Chem.* **2018**, *64*, 830–842. [CrossRef]

52. Long-Mira, E.; Ilie, M.; Chamorey, E.; Leduff-Blanc, F.; Montaudie, H.; Tanga, V.; Allegra, M.; Lespinet-Fabre, V.; Bordone, O.; Bonnetaud, C.; et al. Monitoring BRAF and NRAS mutations with cell-free circulating tumor DNA from metastatic melanoma patients. *Oncotarget* **2018**, *9*, 36238–36249. [CrossRef]

53. Wong, S.Q.; Raleigh, J.M.; Callahan, J.; Vergara, I.A.; Ftouni, S.; Hatzimihalis, A.; Colebatch, A.J.; Li, J.; Semple, T.; Doig, K.; et al. Circulating Tumor DNA Analysis and Functional Imaging Provide Complementary Approaches for Comprehensive Disease Monitoring in Metastatic Melanoma. *JCO Precis. Oncol.* **2017**. [CrossRef]

54. Calapre, L.; Giardina, T.; Robinson, C.; Reid, A.L.; Al-Ogaili, Z.; Pereira, M.R.; McEvoy, A.C.; Warburton, L.; Hayward, N.K.; Khattak, M.A.; et al. Locus-specific concordance of genomic alterations between tissue and plasma circulating tumor DNA in metastatic melanoma. *Mol. Oncol.* **2019**, *13*, 171–184. [CrossRef] [PubMed]

55. McGranahan, N.; Furness, A.J.; Rosenthal, R.; Ramskov, S.; Lyngaa, R.; Saini, S.K.; Jamal-Hanjani, M.; Wilson, G.A.; Birkbak, N.J.; Hiley, C.T.; et al. Clonal neoantigens elicit T cell immunoreactivity and sensitivity to immune checkpoint blockade. *Science* **2016**, *351*, 1463–1469. [CrossRef] [PubMed]

56. Yancovitz, M.; Litterman, A.; Yoon, J.; Ng, E.; Shapiro, R.L.; Berman, R.S.; Pavlick, A.C.; Darvishian, F.; Christos, P.; Mazumdar, M.; et al. Intra- and inter-tumor heterogeneity of BRAF(V600E))mutations in primary and metastatic melanoma. *PLoS ONE* **2012**, *7*, e29336. [CrossRef] [PubMed]

57. Lin, J.; Goto, Y.; Murata, H.; Sakaizawa, K.; Uchiyama, A.; Saida, T.; Takata, M. Polyclonality of BRAF mutations in primary melanoma and the selection of mutant alleles during progression. *Br. J. Cancer* **2011**, *104*, 464–468. [CrossRef]

58. Diaz, L.A., Jr.; Bardelli, A. Liquid biopsies: Genotyping circulating tumor DNA. *J. Clin. Oncol.* **2014**, *32*, 579–586. [CrossRef]

59. Heinzerling, L.; Baiter, M.; Kuhnapfel, S.; Schuler, G.; Keikavoussi, P.; Agaimy, A.; Kiesewetter, F.; Hartmann, A.; Schneider-Stock, R. Mutation landscape in melanoma patients clinical implications of heterogeneity of BRAF mutations. *Br. J. Cancer* **2013**, *109*, 2833–2841. [CrossRef]

60. Yan, Y.; Wongchenko, M.J.; Robert, C.; Larkin, J.; Ascierto, P.A.; Dreno, B.; Maio, M.; Garbe, C.; Chapman, P.B.; Sosman, J.A.; et al. Genomic Features of Exceptional Response in Vemurafenib +/- Cobimetinib-treated Patients with BRAF (V600)-mutated Metastatic Melanoma. *Clin. Cancer Res.* **2019**, *25*, 3239–3246. [CrossRef]

61. Schreuer, M.; Jansen, Y.; Planken, S.; Chevolet, I.; Seremet, T.; Kruse, V.; Neyns, B. Combination of dabrafenib plus trametinib for BRAF and MEK inhibitor pretreated patients with advanced BRAF(V600)-mutant melanoma: An open-label, single arm, dual-centre, phase 2 clinical trial. *Lancet Oncol.* **2017**, *18*, 464–472. [CrossRef]

62. Lee, J.H.; Long, G.V.; Boyd, S.; Lo, S.; Menzies, A.M.; Tembe, V.; Guminski, A.; Jakrot, V.; Scolyer, R.A.; Mann, G.J.; et al. Circulating tumour DNA predicts response to anti-PD1 antibodies in metastatic melanoma. *Ann. Oncol.* **2017**, *28*, 1130–1136. [CrossRef] [PubMed]

63. Braune, J.; Keller, L.; Schiller, F.; Graf, E.; Rafei-Shamsabadi, D.; Wehrle, J.; Follo, M.; Philipp, U.; Hussung, S.; Pfeifer, D.; et al. Circulating Tumor DNA Allows Early Treatment Monitoring in BRAF- and NRAS-Mutant Malignant Melanoma. *JCO Precis. Oncol.* **2020**, *4*, 20–31. [CrossRef]

64. Tsao, S.C.; Weiss, J.; Hudson, C.; Christophi, C.; Cebon, J.; Behren, A.; Dobrovic, A. Monitoring response to therapy in melanoma by quantifying circulating tumour DNA with droplet digital PCR for BRAF and NRAS mutations. *Sci. Rep.* **2015**, *5*, 11198. [CrossRef] [PubMed]

65. Lipson, E.J.; Velculescu, V.E.; Pritchard, T.S.; Sausen, M.; Pardoll, D.M.; Topalian, S.L.; Diaz, L.A., Jr. Circulating tumor DNA analysis as a real-time method for monitoring tumor burden in melanoma patients undergoing treatment with immune checkpoint blockade. *J. Immunother. Cancer* **2014**, *2*, 42. [CrossRef] [PubMed]

66. Lee, J.H.; Long, G.V.; Menzies, A.M.; Lo, S.; Guminski, A.; Whitbourne, K.; Peranec, M.; Scolyer, R.; Kefford, R.F.; Rizos, H.; et al. Association Between Circulating Tumor DNA and Pseudoprogression in Patients With Metastatic Melanoma Treated With Anti-Programmed Cell Death 1 Antibodies. *JAMA Oncol.* **2018**, *4*, 717–721. [CrossRef]

67. Herbreteau, G.; Vallee, A.; Knol, A.C.; Theoleyre, S.; Quereux, G.; Varey, E.; Khammari, A.; Dreno, B.; Denis, M.G. Quantitative monitoring of circulating tumor DNA predicts response of cutaneous metastatic melanoma to anti-PD1 immunotherapy. *Oncotarget* **2018**, *9*, 25265–25276. [CrossRef]

68. Keller, L.; Guibert, N.; Casanova, A.; Brayer, S.; Farella, M.; Delaunay, M.; Gilhodes, J.; Martin, E.; Balague, G.; Favre, G.; et al. Early Circulating Tumour DNA Variations Predict Tumour Response in Melanoma Patients Treated with Immunotherapy. *Acta Derm. Venereol.* **2019**, *99*, 206–210. [CrossRef]

69. Gremel, G.; Lee, R.J.; Girotti, M.R.; Mandal, A.K.; Valpione, S.; Garner, G.; Ayub, M.; Wood, S.; Rothwell, D.G.; Fusi, A.; et al. Distinct subclonal tumour responses to therapy revealed by circulating cell-free DNA. *Ann. Oncol.* **2016**, *27*, 1959–1965. [CrossRef]

70. Robert, C.; Grob, J.J.; Stroyakovskiy, D.; Karaszewska, B.; Hauschild, A.; Levchenko, E.; Chiarion Sileni, V.; Schachter, J.; Garbe, C.; Bondarenko, I.; et al. Five-Year Outcomes with Dabrafenib plus Trametinib in Metastatic Melanoma. *N. Engl. J. Med.* **2019**, *381*, 626–636. [CrossRef]

71. Amaral, T.; Sinnberg, T.; Meier, F.; Krepler, C.; Levesque, M.; Niessner, H.; Garbe, C. MAPK pathway in melanoma part II-secondary and adaptive resistance mechanisms to BRAF inhibition. *Eur. J. Cancer* **2017**, *73*, 93–101. [CrossRef]

72. Long, G.V.; Fung, C.; Menzies, A.M.; Pupo, G.M.; Carlino, M.S.; Hyman, J.; Shahheydari, H.; Tembe, V.; Thompson, J.F.; Saw, R.P.; et al. Increased MAPK reactivation in early resistance to dabrafenib/trametinib combination therapy of BRAF-mutant metastatic melanoma. *Nat. Commun.* **2014**, *5*, 5694. [CrossRef] [PubMed]

73. Girotti, M.R.; Gremel, G.; Lee, R.; Galvani, E.; Rothwell, D.; Viros, A.; Mandal, A.K.; Lim, K.H.; Saturno, G.; Furney, S.J.; et al. Application of Sequencing, Liquid Biopsies, and Patient-Derived Xenografts for Personalized Medicine in Melanoma. *Cancer Discov.* **2016**, *6*, 286–299. [CrossRef] [PubMed]

74. Van Allen, E.M.; Wagle, N.; Sucker, A.; Treacy, D.J.; Johannessen, C.M.; Goetz, E.M.; Place, C.S.; Taylor-Weiner, A.; Whittaker, S.; Kryukov, G.V.; et al. The genetic landscape of clinical resistance to RAF inhibition in metastatic melanoma. *Cancer Discov.* **2014**, *4*, 94–109. [CrossRef]

75. Shi, H.; Hugo, W.; Kong, X.; Hong, A.; Koya, R.C.; Moriceau, G.; Chodon, T.; Guo, R.; Johnson, D.B.; Dahlman, K.B.; et al. Acquired resistance and clonal evolution in melanoma during BRAF inhibitor therapy. *Cancer Discov.* **2014**, *4*, 80–93. [CrossRef]

76. Lu, M.; Zhan, X. The crucial role of multiomic approach in cancer research and clinically relevant outcomes. *EPMA J.* **2018**, *9*, 77–102. [CrossRef] [PubMed]

77. Gerner, C.; Costigliola, V.; Golubnitschaja, O. Multiomic Patterns in Body Fluids: Technological Challenge with a Great Potential to Implement the Advanced Paradigm of 3p Medicine. *Mass. Spectrom. Rev.* **2020**, *39*, 442–451. [CrossRef]

78. Goldstein, E.; Yeghiazaryan, K.; Ahmad, A.; Giordano, F.A.; Frohlich, H.; Golubnitschaja, O. Optimal multiparametric set-up modelled for best survival outcomes in palliative treatment of liver malignancies: Unsupervised machine learning and 3 PM recommendations. *EPMA J.* **2020**, *11*, 505–515. [CrossRef]

Review

Advances in Natural or Synthetic Nanoparticles for Metastatic Melanoma Therapy and Diagnosis

Maria Beatrice Arasi [†], Francesca Pedini [†], Sonia Valentini, Nadia Felli [*,‡] and Federica Felicetti [‡]

Department of Oncology and Molecular Medicine, Istituto Superiore di Sanità, 00161 Rome, Italy; mariabeatrice.arasi@guest.iss.it (M.B.A.); francesca.pedini@iss.it (F.P.); sonia.valentini@guest.iss.it (S.V.); federica.felicetti@iss.it (F.F.)
* Correspondence: nadia.felli@iss.it; Tel.: +39-06-4990-2145
† Co-first authors.
‡ Co-last authors.

Received: 10 September 2020; Accepted: 7 October 2020; Published: 9 October 2020

Simple Summary: Malignant melanoma is the most aggressive skin cancer; its incidence is constantly growing in the white population. In the advanced stage of the disease, after metastatic dissemination, patients have a poor prognosis. Nanomedicine represents a new frontier in cancer treatment; in this field, synthetic and natural nanoparticles (NPs) may represent an important therapeutic and diagnostic opportunity. This review provides an overview of current knowledge in this area: several kinds of NPs, like PLGA, chitosan, liposome and gold-NPs, are used to increase the specificity of drug delivery, allowing a dose reduction and, consequently, a lower toxic effect. Particular attention is given to exosomes (EXOs), an example of natural NPs, important both in conveying molecules with a therapeutic function and in the diagnostic field.

Abstract: Advanced melanoma is still a major challenge in oncology. In the early stages, melanoma can be treated successfully with surgery and the survival rate is high, nevertheless the survival rate drops drastically after metastasis dissemination. The identification of parameters predictive of the prognosis to support clinical decisions and of new efficacious therapies are important to ensure patients the best possible prognosis. Recent progress in nanotechnology allowed the development of nanoparticles able to protect drugs from degradation and to deliver the drug to the tumor. Modification of the nanoparticle surface by specific molecules improves retention and accumulation in the target tissue. In this review, we describe the potential role of nanoparticles in advanced melanoma treatment and discuss the current efforts of designing polymeric nanoparticles for controlled drug release at the site upon injection. In addition, we highlight the advances as well as the challenges of exosome-based nanocarriers as drug vehicles. We place special focus on the advantages of these natural nanocarriers in delivering various cargoes in advanced melanoma treatment. We also describe the current advances in knowledge of melanoma-related exosomes, including their biogenesis, molecular contents and biological functions, focusing our attention on their utilization for early diagnosis and prognosis in melanoma disease.

Keywords: nanoparticles; exosomes; extracellular vesicles; melanoma; therapy; diagnosis; prognosis

---

1. Introduction

Advanced cutaneous melanoma is a highly aggressive and drug-resistant cancer [1,2]. The main problem associated with treating melanoma is its low response rate to conventional therapy. Nevertheless, the rapidly developed fields of melanoma research are introducing new potential therapeutic approaches, focusing on developing new efficient drugs and delivery systems. Before 2011, metastatic melanoma was considered incurable, leading to death within 18 months of diagnosis [3,4]. The therapies used were based on the administration of chemotherapeutic agents such as dacarbazine and temozolomide [5]. In order to circumvent the intrinsic resistance against chemotherapy, researchers explored the possibility of potentiating the immune response by using immune stimulators, such as interleukin 2 (IL-2), interferon alpha (IFN-α), ipilimumab and thymosin alpha 1 [6,7]. Afterwards, as result of genome sequencing and identification of the main mutation, a series of therapeutic agents have been discovered; about half of all melanomas have mutations in the BRAF (v-raf murine sarcoma viral oncogene homolog B1) gene [8,9]. These mutations constitutively activate the MAPK (mitogen-activated protein kinase) pathway, resulting in uncontrolled cell proliferation, tumor development and growth [1,10]. Several tyrosine kinase inhibitors (TKIs) have been developed that block the MAPK pathway and inhibit cell growth and survival. The use of TKIs has shown a significant improvement of patient overall survival and the combinations therapy with BRAF/MEK inhibitors is now accepted as the standard care for BRAF-muted advanced melanoma. Unfortunately, in most treated patients, responses to these inhibitory molecules are transient, and, in a few months, a therapy-resistant phenotype occurs. Recently, the inhibition of the programmed cell death 1 (PD-1)/programmed cell death 1 ligand 1 (PD-L1) axis, a suppressor of T-cell response, turned out to be an effective therapeutic strategy with a low toxicity profile and contributed to better clinical outcomes compared to chemotherapy and also to other immune mediated approaches with cytotoxic T lymphocyte-associated 4 (CTLA-4) inhibition [11–13]. However, substantial heterogeneity exists in metastatic melanoma response to treatments; therefore, the development of multimodality strategies is the major challenge to fight this deadly disease. Several types of nanoparticles (NPs) and nanovesicles have been explored for their employment as drug carriers in melanoma therapy and their functionalization with specific molecules, such as antibodies, can generate different smart nanodrugs for application in melanoma therapy (Figure 1). Herein, we will discuss the new frontiers in the treatment of patients with unresectable or metastatic melanoma, paying special attention to new anticancer molecules and their delivery through natural or synthetic nanoparticles.

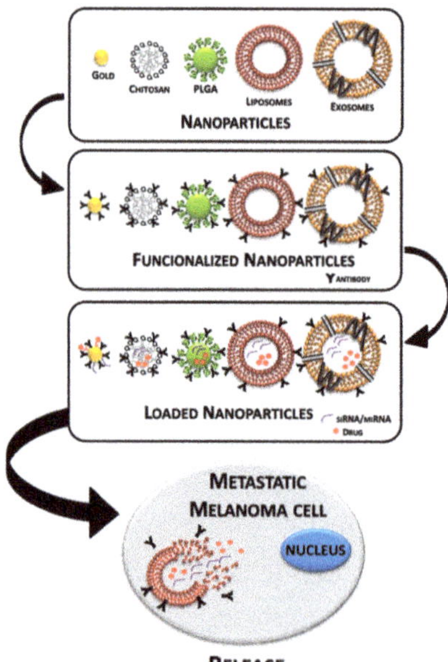

Figure 1. Step-by-step representation of a nanoparticles-based delivery system for melanoma therapy.

2. Nanoparticles and Drug Delivery

Nanomaterials are delivery vehicles used to protect the intended drug against degradation and enhance its stability. Using a nanotechnology treatment for several types of cancer enhances targeted delivery to cancer cells and uptake, as well as reducing cytotoxic side effects on normal tissues [14].

Chemotherapeutic agents generally exhibit a short plasma half-life due to rapid clearance and fast biodegradation; these elements, together with the low specificity, make it necessary to use high doses, resulting in severe toxic effects. In addition, since cancer diagnosis is often late, cancer cells have already invaded other parts of the body [15].

Nanotechnologies are based on the use of different polymeric systems whose modifiable physical and chemical systems provide tools to improve different aspects of the therapy. Some examples are the release of the drug in a tissue-specific way, the decrease in side effects, the protection from drug degradation, the possibility to reduce the frequency of administrations and the toxicity of the substance itself [16]. Depending on their specific structural characteristics, nanoparticles can easily penetrate tissues and cells, circumscribe the biological effect of the drug on a specific cell type and modify the pharmacokinetic properties, thus allowing a prolonged drug release over time. The Food and Drug Administration (FDA) approved only a few nanodrugs for use and clinical trials, as they have been shown to either target and directly kill tumor cells or improve, overall, targeted chemotherapy drug delivery. The formulation of these nanoparticles increases the drug concentration into the tumor by 100–400% [17].

2.1. PLGA

The FDA and the European Medicine Agency (EMA) approved PLGA nanoparticles (PLGA-NPs) in 2011 as a system for conveying nucleic acids, drugs and other molecules for cancer and other diseases. PLGA is a biocompatible and biodegradable polymer (poly-lactic acid and poly-glycolic acid) whose degradation produces non-toxic substances, such as water and carbon dioxide [18]. The PLGA-NPs' cellular uptake depends on both the size of the NP and the composition of the plasma membrane of the cells themselves [19]. The different physical characteristics of PLGA-NPs, such as their size and morphology, can be determined by varying the parameters of the synthesis method used. Nanoparticles have a large surface area that can be functionalized with different agents to make the vehicle target-specific [20].

In the last period, several publications highlighted the efficiency of PLGA-NPs in delivering anticancer agents; in a recent work, Arruda and his co-workers showed that the encapsulation of a peptide with antitumoral activity, P20 (CSSRTMHH), within the PLGA-NPs conjugated to the C peptide with a homing function (CVNHPAFACGYGHTMYYHHYQHHL), strongly increases their functionality. The therapeutic effect of these NPs was evaluated in a syngeneic model of metastasis and was shown to produce the same inhibitory effect of free P20 at fivefold higher concentration, reducing by 28% the number of lung nodules [21].

In another recent study, trastuzumab has been used as a recognizing unit attached to NPs for Her-2 positive breast cancer therapy. Targeted nanoparticles were prepared by combining PLGA/ polyethylenimine (PEI)/lipid nanoparticles (PPLNs) with trastuzumab through the electrostatic adsorption method to deliver the antitumor drug docetaxel. The results showed that trastuzumab-PPLNs had a cell-targeting effect and could effectively inhibit the proliferation of cancer cells. This work demonstrates that trastuzumab-PPLNs are a promising treatment for breast cancer and can be considered a proof of concept for anticancer therapies, including melanoma [22].

Following a similar approach, Kou et al. selected paclitaxel as anti-metastatic and anti-angiogenic model drug to produce paclitaxel-loaded PLGA-NPs (Ptx-NPs), coated with a polylysine cationic SM5-1 single-chain antibody (Ptx-NP-S) which binds to a membrane protein specifically expressed in melanoma, hepatocellular carcinoma (HCC) and breast cancer cells. The authors demonstrated that Ptx-NP-S not only showed very high affinity to SM5-1 but also was internalized by the target cells. They proved a significantly enhanced cytotoxicity of Ptx-NP-S when compared to non-targeted paclitaxel-loaded PLGA-NPs, suggesting SM5-1 single-chain antibody-polylys as a good candidate to synthesize cancer-targeted PLGA nanoparticles [23].

Another important field of application of PLGA-NPs is as peptide transporters in vaccines. Vaccines could be promising tools for treating cancers, including melanoma [24]; however, several studies obtained unsatisfying results, possibly because of the low stability of peptides and delivery approaches. Thus, the appropriate and efficient peptide carrier systems still continue to be one of the major obstacles. In order to overcome this difficulty, Li and coworkers recently used PLGA-NPs as a new method to deliver tumor-associated antigen (TAA) peptides to dendritic cells (DC) [25].

2.2. Chitosan

Chitosan is a bio-poly-amino-saccharide cationic polymer obtained from chitin deacetylation moieties formed by the repetition of two units: N-acetyl-D-glucosamine and D- glucosamine [26]. The molecular weight and the extent of deacetylation of the chitosan influence its properties; this is employed specially to prepare micro and nanoparticles capable of conveying nucleic acids or medicines. The chitosan backbone consists of multiple free amino and hydroxyl groups that can be functionalized. Versatile approaches have been reported for the construction of chitosan-based nanomaterials, such as nanogels, NPs, micelles, liposomes, nanofibers and nanospheres [27]. The polymer's success is due to its properties such as: biocompatibility and biodegradability, FDA approval in clinical use, the possibility of modifying its surface and to complex it in nanoparticles conveyed against cells and/or target tissues. Chitosan exhibits no or minimal toxicity, which made it widely regarded as a safe material in drug delivery [28]. Various cargos, hydrophilic and also hydrophobic, can be incorporated into chitosan-NPs. Their release is influenced by solubility, diffusion, and size of the nanoparticles [29]. Many publications show innovative ways for antitumor drug delivery based on environmental response and targeting principles [30,31].

A recent study evaluated the effect of doxorubicin encapsulated into chitosan/alginate nanoparticles to inhibit viability and tumor growth of melanoma cells in a mouse melanoma model. The experiments revealed that the encapsulation of doxorubicin could be considered advantageous because it increases intracellular accumulation and produces a long-lasting cytotoxic effect [32]. The cationic nature of chitosan makes it suitable to complex anionic nucleic acids, protecting them from degradation by nucleases in serum [33].

One of the possible applications of this property is the transport of two specific categories of non-coding RNAs called microRNA (miRs) and short interfering RNA (siRNAs) [34,35]. MiRs regulate the expression of messenger RNAs (mRNAs) at a post-transcriptional level. They are implicated in various aspects of tumor progression, including proliferation, apoptosis, and epithelial–mesenchymal transition (EMT), and they can promote or suppress tumorigenesis and metastatization, thus acting either as onco-suppressors or oncogenes [14,36–38].

The possibility of protecting these molecules from enzymatic degradation using chitosan-NPs represents a very important therapeutic option. SiRNAs have broad potential as therapeutic agents because they are able to silence any gene in a reversible way. To achieve the clinical potential of siRNAs, delivery materials are required to transport them in the cells of target tissues.

Rostami and coworkers, in a recent study, developed the alginate-conjugated trimethyl chitosan (ATMC) nanoparticles loaded with specific siRNAs. In particular, in the cancer cells the activation of S1P/sphingosine-1-phosphate receptor 1 (S1PR1) and IL-6/glycoprotein 130 (GP130) genes induces STAT3 phosphorylation, which in turn stimulates their expression.

On this basis, the authors investigated the potential anticancer effect by silencing the S1PR1 and GP130 genes, through ATMC-NPs containing specific siRNAs; the results obtained in different cancer cell lines, including melanoma B16F10, although in need of further study, have been encouraging [35].

2.3. Liposomes

Liposomes are small lipid vesicles (50 to 100 nm) generated from cholesterol, which reduces their permeability and increases their in vivo and in vitro stability [39]. Liposomes, categorized by the composition and mechanism of intracellular delivery, fall under five types: conventional liposomes (CL), pH-sensitive liposomes, cationic liposomes, immunoliposomes, and long-circulating liposomes (LCL). Liposomes with modified surfaces have also been developed using several molecules, such as glycolipids or sialic acid [40]. To prolong their circulation time, liposome surfaces have been coated with a hydrophilic polymer, such as polyethylene glycol (PEG), thus increasing repulsive forces between liposomes and plasma components [41]. Liposomes seem to be an almost ideal drug-carrier system due to their morphology being similar to that of cellular membranes and for their ability to incorporate various substances.

Recently, Gao et al. demonstrated that liposome-PEI encapsulation improved the uptake of n-Butylidenephthalide (BP) through enhancement of cell endocytosis, indicating that the BP/lipo PEG PEI complex (LPPC) has great potential in melanoma treatment. In particular, because BP is metabolized very rapidly, its antitumor role, already demonstrated in various types of cancer, including melanoma, would be limited. To avoid this problem, the researchers developed the complex BP/LPPC, improving both the uptake of this drug by the cells and its stability over time [42].

Yin et al. prepared R8-dGR peptide modified paclitaxel (PTX) and hydroxychloroquine (HCQ) co-loaded liposomes (PTX/HCQ-R8-dGR-Lip) to enhance delivery through recognition of integrin $\alpha v \beta 3$ receptors and neuropilin-1 receptors on B16F10 melanoma cells. The results showed that R8-dGR modified liposomes (R8-dGR-Lip) enhanced tumor-targeting delivery in vitro and in vivo [43].

Health authorities have approved different liposomal formulations of small molecule therapeutics; their ability to deliver macromolecules, such as nucleic acid-based therapeutics plasmid DNA (pDNA) and siRNA, to disease sites is under investigation. In a recent study, lipid nanocapsules were used to combine two anti-melanoma strategies, gene therapy by using Bcl-2 siRNA and ferrocifens as novel anticancer agents. The results showed a 50% reduction in tumor volume in tumor-bearing animals compared to the control group [44].

Sadhu et al. utilized liposomes to deliver glutathione disulfide to melanoma cells. These liposomes slowed down tumor proliferation by 85–90% in melanoma-bearing mice and significantly improved the survival rate [45]. Interestingly, lipid constituents of liposomal vesicles can be tailored to achieve particular immunogenic responses. Manipulation of surface charge density in cationic liposomes can regulate immune response [46].

Fu and coworkers developed paclitaxel (PTX)-loaded liposomes functionalized with trans activator of transcription (TAT), the most frequently used cell-penetrating peptide, and cleavable PEG. Under physiological conditions, the TAT peptide is shielded by a PEG layer and so liposomes exhibit a good stability in the bloodstream. When liposomes arrive at the tumor site, exogenous reducing agents could remove PEG, causing TAT exposure and facilitating cell internalization. In this study, the authors showed a tumor inhibition of around 70% in B16F1-bearing mice [47].

Lastly, the liposomes may also be used as an adjuvant to boost the protective or therapeutic immune response [48].

2.4. Gold Nanoparticles

Gold nanoparticles (AuNPs) are a type of nanoparticles produced in various sizes and shapes (i.e., gold nanospheres, nanorods, nanocages, nanoshells, and nanostars). They are dependent on the synthetic methods adopted for their preparation [49]. These nanoparticles can be prepared in a broad range of core sizes (1 to 150 nm), which makes it easier to control their dispersion. The presence of a negative charge on the surface of gold nanoparticles makes them easily modifiable.

Literature reports that various types of gold nanoparticles are effective tools in interfering with human cancer cell biology, including melanoma [50]. Indeed, their application in therapeutic development and in cancer diagnosis could provide a potential step forward in cancer theranostics [51].

Thanks to their physical, chemical and biological features, AuNPs can be used in imaging technology, such as computed tomography, ultrasound and magnetic resonance and a few of them have been approved by the FDA in medicine [52].

Kim et al. demonstrated the use of AuNPs as a contrast agent for the quantitative molecular combining of high-resolution photoacoustic tomography (PAT) of melanomas in vivo. To avoid the limitations of conventional diagnostic techniques, the authors exploited the enormous optical absorbing gold nanocages (AuNCs), obtaining a 300% increase in the contrast when AuNCs were bioconjugated with [Nle4,d-Phe7]-α-melanocyte-stimulating hormone compared to the control nanocages [53].

Through the addition of various biomolecules such as drugs, target ligands and genes, the nanoparticles could function in an enhanced way. To overcome in vivo delivery barriers, Au nanocomplexes are normally modified with functional moieties such as stabilizing materials, target ligands, and bio responsive linkers [54,55].

Conjugation with the anti-nucleosome monoclonal antibody (mAb 2C5) of PEGylated liposomes improves cancer cell-specific delivery targeting the matrix metalloprotease 2 in the tumor microenvironment [56]. Many studies showed that the use of magnetic fields or altered temperatures and light improve the delivery of gold particles including anticancer agents [57].

2.5. Other Nano-Agents

Other nanoparticles have a potential use as anti-melanoma agents (Table 1) [58–65].

Table 1. Several types of nanoparticles used in melanoma treatment.

Nanoparticle	Results	Cell Line Used	Reference
Albumin hybrid nanoparticles	Better tumor-targeting capacity and significantly increased drug accumulation in tumor	B16F10 melanoma-bearing mice	[58]
Micelles	High loading efficiency of drug	B16F10	[59]
Dendrimers (PAMAM DAB, PEA) grafted with PEG, acetyl groups, carbohydrates	Increased the bioavailability and efficiency of transported compounds	B16F10	[60]
Copper nanoparticles	Induced cell death by inducing oxidative stress	B16F10 melanoma-bearing mice	[61]
Iron oxide nanoparticles Carbon nanotube	All nanoparticles induced selective toxicity and caspase 3 activation through mitochondria pathway; Caused generation of ROS, mitochondrial membrane potential decline, mitochondria swelling and cytochrome c release	F10	[62]
Nanoemulsions (Coffee oil- algae oil-based)	Effective inhibition of melanoma cell growth; Cell cycle arrested at G2/M phase	B16F10	[63]
Nanoemulsion of 5-FU	Much more efficacious than free 5-FU when used for topical delivery	SK-MEL 5	[64]
Multi-peptide and toll-like receptor 4 agonist co-delivery system based on lipid coated Zinc-phosphate hybrid nanoparticles	Exhibited anti-tumor immunity evident by secretion of cytokines *in vitro* and increased CD8+ T-cell response from IFN-γ ELISPOT analysis ex vivo; Improved anti-tumor effects evidenced from prophylactic, therapeutic and metastatic melanoma tumor models compared with free antigens and single peptide-loaded nano-vaccines	B16F10	[65]

Abbreviations: PAMAM, poly-amidoamine; DAB, diaminobenzidine; PEA, palmitoylethanolamide; PEG, polyethylene glycol; ROS, reactive oxygen species; 5-FU, 5-Fluorouracil; IFN-γ, interferon-gamma.

2.6. Specific Drug Delivery

The ideal drug carrier would have the ability to deliver the drug to the cell/tissue/organ in a specific way to make the cancer treatment more efficient. A potentially powerful technology is represented by targeted nanoparticles. The most used approach is to conjugate to nanoparticles antibodies or antibody-derivatives which must be able to overcome the physiological barriers and conditions that limit access to the target, and to bind target cells through specific antibody–receptor interactions [66].

Primary tumors as well as metastatic tumors overexpress some antigens on their surfaces. Particularly, for melanoma treatment, the literature shows the specificity of many antigens.

Secreted protein acidic and rich in cysteine (SPARC) was deepened in the tumorigenicity and progression of melanoma, resulting as a nonstructural matricellular protein [67]. The conjugation of nanoparticles with a peptide able to specifically recognizing the SPARC protein, has made it possible to obtain an important tool that exhibited high tumor affinity with a very low binding in negative cell lines. In in vivo targeting experiments, an increase of SPARC-targeted nanoparticles was observed in tumor tissue as compared with the control group. Moreover, a very important result was the detection of both primary tumors and metastatic growth obtained by using the fluorescent version of these nanoparticles [68].

Melanoma antigen genes (MAGE)-A proteins are a melanoma-specific-antigen family [69]. Saeed and coworkers in a recent work described that the single-chain variable fragment (scFv) antibodies directed against MAGE-A1 peptide, presented by human leukocyte antigen A1 (HLA-A1), in short M1/A1, have been coupled to liposomes to achieve specific melanoma targeting. Through experiments by flow cytometry and confocal microscopy, the authors demonstrated the bind between M1/A1-positive and B-cells as well as with melanoma cells and the internalization by these cells [70].

The chondroitin sulfate proteoglycan 4 (CSPG4) is a highly specific marker of the nevomelanocyte lineage and has been utilized to target melanoma. In a recent work, Falvo and coworkers produced nanoparticles conjugated to a monoclonal antibody specific for CSPG4 encapsulating cisplatin molecules. Flow cytometry showed a specific binding to CSPG4 (+) melanoma cell lines. Selectivity towards melanoma was also confirmed in xenograft models: targeted nanoparticles significantly inhibited growth of melanoma xenografts whereas modestly affected breast carcinoma xenografts [71].

These studies highlight the importance of the potential active targeting to improve the outcome of a nanoparticle-based therapy.

2.7. Natural Nanoparticles

All cell types of the body release extracellular Vesicles (EVs), which are present in all body fluids. They are defined by their mechanism of cellular production and include exosomes (EXOs), microvesicles (MVs) and apoptotic bodies [72]. The diameter of the EXOs ranges from 30 to 150 nm; they are secreted by most cell types into the extracellular microenvironment under normal and pathological conditions and can be found in blood, sweat, saliva, urine, amniotic fluids, breast milk, cerebrospinal fluids and malignant ascites [73–76]. EXOs have a membrane consisting of a lipid bilayer and contain various biomolecules, including proteins, lipids, RNA and DNA, suggesting their involvement in the regulation of various biological functions through different molecular mechanisms [77–79].

When EXOs are taken up by other cells, their cargo (mRNAs, miRs and protein) is transferred and influences the phenotype of recipient cells [80,81]. EXOs mediate cell-to-cell communication and are involved in different physiological and pathological processes, also playing an important role in cancer [82]. Like other tumors, melanoma cells modify their microenvironment to enable growth and metastasis via direct cellular interactions or the indirect release of soluble factors and/or EXOs.

Similarly to synthetic nanoparticles, EXOs can be considered drug delivery systems and possess important advantages such as long-term circulatory capability, good biocompatibility, minimal toxicity and intrinsic ability to target specific cells [83]. Moreover, one of the most useful properties of EXOs is their ability to cross the cytoplasmic membrane [84] and the blood/brain barrier [85].

Altogether, these characteristics make EXOs a very interesting drug delivery system [86]. Recently, several studies have focused on specific modification of EXOs surface to make them a proper candidate for cellular uptake and target specificity [87]. In particular, it has been proved that modification of EXO–tetraspanin complexes strongly influences target cell selection both in vitro and in vivo [88], improving the targeting of tissues and cell types of interest. Other modified transmembrane proteins are the platelet-derived growth factor receptors [89]. In 2018, Kim et al. engineered macrophage-derived EXOs loaded with the anticancer drug paclitaxel to improve their circulation time in the blood and allow them to target pulmonary metastases. By using this modified EXOs, the drug can selectively deliver to target cancer cells and also can increase the survival rate in an in vivo lung cancer model [90].

In the last decade, several studies have focused on immune system stimulation for the treatment of different cancers. Dendritic cell-derived EXOs (dexosome) have been successfully engineered to target helper T cells to stimulate cytotoxic T cell proliferation, influence T cell differentiation and create an anti-tumor environment [91]. Dexosomes have entered clinical trials for colorectal cancer [92], non-small cell lung cancer [93] and metastatic melanoma [94]. Particularly in melanoma cancer, EXOs are studied as therapeutic cell-free vaccines, e.g., dendritic cell-derived EXOs loaded with MAGE3 antibodies were administered subcutaneously or intradermally to stimulate the immune response of melanoma stage IIIb/IV patients obtaining positive clinical effects [95].

Moreover, Zhu et al. analyzed the fundamental role of EXOs derived from natural killer cells to exert cytotoxic effects on melanoma cells, suggesting this could be a potential immunotherapeutic strategy for cancer [96]. More recently, Luigini L. et al. performed a homemade test for the analysis of natural killer exosomes derived from plasma of melanoma patients and healthy donors. Their results showed that these vesicles could be used to improve cancer treatment in combination with NK cells or immune checkpoint-based therapies [97].

Among the new therapeutic approaches, very interesting results were reviewed by Chillà A. et al. In their paper, they refer to the therapeutic use of some particular cells used as natural nanoparticles [98]. For example, the possibility to target an experimental melanoma with red blood cells (RBCs) and platelet membrane-coated gold nanoparticles containing curcumin was described. In this study, the natural cell membranes have been used as biointerfaces that interact with the host environment. Platelet membrane coating has been used to target cancer cells while RBC membrane coating has been used to provide clearance by macrophages [99]. This cell-mediated delivery would provide controlled dosing, no systemic toxicity, and high tolerance by the patient and could be possibly evaluated as a therapeutic strategy in melanoma [98].

2.8. Natural Nanoparticles for Metastatic Melanoma Diagnosis

Early diagnosis and effective therapy are undoubtedly the primary issues for prolonging the survival of cancer patients. At present, the solid biopsy used for diagnosis is mostly the basis for treatment of cancer. Unfortunately, this method has considerable limitations as it is invasive and sometimes not feasible. Accumulating evidence over the past few decades has highlighted the potential use of circulating molecular biomarkers as a promising platform for the noninvasive diagnosis and definition of prognosis. As previously reported, the cargos incorporated in EXOs reflect the contents and the pathological status or signaling alterations of the original cells. Moreover, the cargos are protected from degradation and are isolable from patients' body fluids. Thus, analyses of circulating EXOs and their contents in blood could represent a promising strategy in cancer diagnosis advancement and monitor patients' therapeutic response.

For instance, PD-L1 expression in tumor tissues is widely used in selecting patients who will benefit from treatment with immunotherapy. Recently, evaluating the role of exosomal PD-L1 expressed in melanoma patients treated with anti-PD-1/PD-L1 antibodies, Danesi R. and his group showed that exosomal PD-L1 expression increased in subjects with disease progression and decreased in patients responding to treatment, while no significant changes were observed in patients with disease stabilization [100]. In the same year, Chen G. et al. [101] showed that Interferon-γ (IFN-γ) up-regulates PD-L1-containing EXOs secretion and changes during the course of anti-PD-1 therapy. In accordance, a very recent prospective study to evaluate exosomal-PD-L1 expression from 100 melanoma patients treated with immune checkpoint inhibitors and BRAF/MEK inhibitors, showed that tumor EXOs carrying PD-L1 had immunosuppressive properties since they were as efficient as the melanoma cell at inhibiting T-cell activation. This study provided a rationale for monitoring the EXO-PD-L1 level as a potential predictor of treatment response in melanoma patients [102].

Functional nucleic acids contained in EXO (miRs, long non-coding RNAs, circulating RNAs, mRNAs, and DNA) are chemically stable and resistant to degradation by RNAases. Numerous previous studies have shown that exosomal miRs could be potential diagnostic and prognostic biomarkers for melanoma [103]. In particular, González and his group, in one of the first publications that identified circulating exosomal miRs in patients with melanoma, reported that the EXOs from patients with melanoma contain a level of miR-125b that is significantly lower compared to that contained in EXOs from disease-free melanoma patients or healthy controls [104]. Consistently, miR-125b appears to function as a tumor suppressor in melanoma, by direct regulation of c-Jun and MLK3 protein expressions [105,106] and inducing cell senescence [107]. Recently, Huber et al. have described seven miRs, including miR-125b-5p, that were released in extracellular vesicles and associated with myeloid-derived suppressor cells (MDSCs) and resistance to treatment with immune checkpoint inhibitors in melanoma patients [108]. Moreover, it was reported that the regulation of the miR-125b-5p/EIF5A2 axis in melanoma, through the long intergenic non-protein coding RNA 520 (LICN00520), promotes the proliferation, invasion and migration of melanoma cells. The authors suggest that long non-coding RNA (lncRNA) expression level may be possibly evaluated in blood circulation as a therapeutic biomarker [109].

Another report by Tengda et al. reported that the analysis of a panel of five exosomal miRs, in particular miR-532-5p and miR-106b, was able to distinguish patients with or without metastases, patients with stage I–II from patients with III–IV of disease and patients who received pembrolizumab treatment from those not treated [110].

Most studies have shown the potential role of tumor-derived EVs, mainly EXOs, as biomarkers to predict cutaneous malignant melanoma (CMM) outcome and resistance to MAPKis [110–112]. A recent work published by Svedman has led to the identification of the EV-derived miRs, let-7g-5p and miR-497-5p, that can be used as putative good predictive biomarkers after MAPKi treatment in metastatic CMM patients [113]. As reported by Lunavat et al., the expression of miR-211–5p is up-regulated upon vemurafenib and dabrafenib treatment in EXOs and melanoma cells. Interestingly, the authors suggested that stable expression of miR-211–5p reduces sensitivity to BRAF inhibition by regulating cellular proliferation. The authors indicate this miR as a promising prognostic biomarker in melanoma [112].

MiR-21, one of the most frequently up-regulated miRs in solid tumors, controls important tumor suppressor genes as well as genes involved in carcinogenesis, such as phosphatase and tensin homolog (PTEN), programmed cell death protein 4 (PDCD4), phosphoinositide 3-kinase (PI3K), Sprouty and reversion inducing cysteine rich protein with Kazal motifs (RECK) [114–118]. As a result, miR-21 has been proposed as a plausible diagnostic and prognostic biomarker, as well as a therapeutic target for several types of cancer [119]. Moreover, the use of exosomal miR-21 as a potential biomarker has been proposed. Specifically, a systematic review provided evidence that is consistent with this hypothesis although the limited number of each cancer type analyzed should be considered. In addition, the author suggested that a combination of miR panels or cancer antigens may be a good strategy to have better diagnostic results or prognostic predictions in most circumstances [120].

MiR-21 plays a pivotal role in melanomagenesis and is released in EXOs produced from melanoma cells [121]. In an interesting work, Pfeffer et al. performed a miR profiling on RNA prepared from plasma-derived EXOs from specific patient cohorts [122]. As stated in the cited review [120], they found that a panel of five miRs (miR-17, miR-19a, miR-21, miR-126, and miR-149) was expressed at higher levels in plasma-derived EXOs from patients with metastatic melanoma. In addition, these miRs were found to be over-expressed in patients with metastatic sporadic melanoma compared to familial melanoma patients or healthy controls. This panel is proposed as a predictive biomarker to monitor remission as well as relapse following therapeutic intervention [122]. Of interest, miR-21 and miR-146a up-regulations were found in vitreal EXOs of patients with uveal melanoma (UM) as a consequence of dysregulation arising from tumor cells. Data shown in that paper also suggest the possibility of detecting these miRs in VH (vitreous humor) and that serum of UM patients could be considered a potential circulating marker [123]. In addition, the plasma level of miR-21 was positively correlated with tumor burden in metastatic melanoma patients [124] and its relative expression was positively correlated with the tumor nodes and metastasis (TNM) stage of melanoma [125]. All these studies confirmed that the expression of miR-21 in melanoma patients can be used as a potential serum biomarker for melanoma metastasis diagnosis and disease assessment.

Several proteins involved in melanoma progression and metastasis have been identified in EXOs and could be considered as possible prognostic biomarkers [126].

A large increase in total exosomal protein derived from malignant melanoma occurs during disease progression. Specifically, each melanoma cell line appears to exhibit an individual type of heat shock protein (HSP)70 expression, likely reflecting a selection during tumor progression and therapy [127]. Very interesting results were obtained from Peinado and coworkers who identified a melanoma-specific exosomal signature in blood from patients with stage IV melanoma that included expression of tyrosinase-related protein-2 (TYRP-2), very late antigen-4 (VLA-4), HSP70, HSP90 and mesenchymal epithelial transition (MET) oncoprotein [128]. More recently, Alegre et al. evaluated the presence of the melanoma biomarkers melanoma inhibitory activity (MIA) protein, S100B and tyrosinase-related protein 2 (TYRP2) in EXOs derived from serum obtained from stage IV melanoma patients, melanoma-free patients and healthy controls. They observed that MIA and S100B can be detected in EXOs and their quantification presents diagnostic and prognostic utility in melanoma patients [129].

Recently, lncRNAs have emerged as an important regulatory factor in tumor growth and transformation [130]. In particular, Cantile et al. have analyzed the fundamental role played by HOTAIR in the malignant transformation and progression of melanoma cells. Moreover, they have identified this lncRNA in the blood and have suggested its potential role as circulating marker [131]. Numerous reports have shown that exosomal lncHOTAIR could be a sensitive liquid biomarker of different types of cancers, such as glioblastoma multiform, breast cancer and non-small-cell lung cancer (NSCLC) [132–134]. However, the potential use of EXOs for HOTAIR detection in melanoma cells has not yet been explored.

In a very recent and interesting study, the authors describe a comprehensive proteomic analysis of 426 human samples from tissue explants (TEs), plasma and other body fluids. They reported that the "tiny packages of materials released by tumors", called EVPs (extracellular vesicles and non-vesicular particles <50 nm), may serve as biomarkers for detecting a number of different types of cancers (including melanoma) in its early stages. In particular, the analysis of some specific proteins (e.g., versican, tenascin C, and thrombospondin 2 can distinguish tumors from normal tissues with 90% sensitivity and 94% specificity. Moreover, they have defined a panel of tumor-type-specific EVP proteins in TEs and plasma, which can classify tumors of unknown primary origin [135].

3. Conclusions

Advanced melanoma is a rapid metastasizing and drug-resistant cancer [1,2]. Despite the availability of new therapies discovered in recent years, metastatic melanoma remains a highly lethal disease. Recent advances in nanotechnologies opened up new perspectives for treating melanoma, potentially effective in overcoming the limitations of conventional approaches: toxicity to normal tissues, development of drug resistance and rapid blood clearance. Nanotechnologies are based on the use of different polymers whose specific characteristics are adapted to different aspects of therapies. Both the Food and Drug Administration (FDA) and European Medicines Agency (EMA) recently approved several types of nanoparticles as a system for conveying nucleic acids, drugs and other molecules for cancer and other disease treatment, including melanoma [17,18,28]. More optimized nanocarriers for the delivery of drugs in a specific way that is able to make the melanoma treatment more efficient is represented by targeted nanoparticles. Melanoma-specific antibodies are conjugated to nanoparticles, thus overcoming the physiological barriers and conditions to reach the target and to recognize and actively bind to target cells through specific antibody–receptor interactions [66].

The discovery of EXOs as natural carriers of RNA, DNA, proteins and lipids have highlighted the potential use of these vesicles as circulating molecular biomarkers for both diagnostic and therapeutic purposes. Due to their easy handling and their ability to transfer molecules between cells, EXOs may be a very performant delivery system [86]. Moreover, naturally secreted EXOs loaded with therapeutic molecules may optimize efficacy of treatment while also reducing off-target delivery. Specifically, molecules may be included in EXOs that are difficult to deliver intracellularly without the use of a carrier (ncRNA, recombinant proteins, small-molecule drugs). Particularly in melanoma, EXOs are also studied as a therapeutic cell-free vaccine [95]. The purification of pure populations of EXOs from EXO-secreting cell lines, unlike those released from autologous primary cells, have immunogenic and oncogenic potential. However, the modification of the complex structure of naturally secreted EXOs may limit their functional activity and their pharmaceutical acceptability. As a potential alternative, some authors have proposed the development of EXO-mimetics manufactured using selected components of EXOs that can be incorporated into synthetic nanoparticles to enhance their stability, immunogenicity, targeting, and uptake [136]. However, the main and useful components of exosomes to therapeutic delivery have not been identified yet.

Although the high number of publications in this field suggests the high diagnostic and prognostic potential of EVs in melanoma, our knowledge remains limited. Several factors hinder the real use of EVs or EXOs as biomarkers. For example, the differences in the methods used by researchers to isolate EXOs from patients' blood introduce heterogeneity in the composition of the different preparations, thus precluding the identification of the active components. Furthermore, different populations of vesicles are found in the plasma of patients and the isolation of those released by melanoma tumor cells would be of great scientific importance. In this regard, a recent publication by Ferrone et al. suggests the use of an antibody directed against CSPG4 antigen to separate exosomes released by melanoma from those secreted by normal cells [137].

In conclusion, natural or synthetic nanoparticles are promising tools to design new therapeutic, diagnostic and prognostic approaches to cancer and, in particular, to melanoma. Studies are needed to fill the gap between the large amount of evidence available in the preclinical setting and the limited evidence obtained in the real clinical situation.

Funding: This work was supported by the following grants: MAECI, Progetti di Grande Rilevanza, Domanda di contributo; L. 401/90 (PGR01049) to N.F.; Regione Lazio "Progetto di Gruppo di Ricerca finanziato ai sensi della LR Lazio 13/08" (Protocollo 15263) to Mauro Biffoni (M.B.) for an M.B.A. fellowship.

Acknowledgments: We thank Mauro Biffoni (M.B.) for critical reading of the manuscript and Deborah Paradiso for English-language editing and proofreading.

Conflicts of Interest: The authors declare no conflict of interest.

References

1. Liu, Y.; Sheikh, M.S. Melanoma: Molecular pathogenesis and therapeutic management. *Mol. Cell. Pharmacol.* **2014**, *6*, 228. [CrossRef]
2. Gray-Schopfer, V.; Wellbrock, C.; Marais, R. Melanoma biology and new targeted therapy. *Nature* **2007**, *445*, 851–857. [CrossRef]
3. Luke, J.J.; Schwartz, G.K. Chemotherapy in the management of advanced cutaneous malignant melanoma. *Clin. Dermatol.* **2013**, *31*, 290–297. [CrossRef] [PubMed]
4. Atkins, M.B.; Lotze, M.T.; Dutcher, J.P.; Fisher, R.I.; Weiss, G.; Margolin, K.; Abrams, J.; Sznol, M.; Parkinson, D.; Hawkins, M.; et al. High-Dose Recombinant Interleukin 2 Therapy for Patients With Metastatic Melanoma: Analysis of 270 Patients Treated Between 1985 and 1993. *J. Clin. Oncol.* **1999**, *17*, 2105. [CrossRef]
5. Bhatia, S.; Tykodi, S.S.; Thompson, J.A. Treatment of metastatic melanoma: An overview. *Oncology* **2009**, *23*, 488–496. [PubMed]

6. Di Franco, S.; Turdo, A.; Todaro, M.; Stassi, G. Role of Type I and II Interferons in Colorectal Cancer and Melanoma. *Front. Immunol.* **2017**, *8*. [CrossRef] [PubMed]

7. Chen, J.; Shao, R.; Zhang, X.D.; Chen, C. Applications of nanotechnology for melanoma treatment, diagnosis, and theranostics. *Int. J. Nanomed.* **2013**, *8*, 2677–2688. [CrossRef] [PubMed]

8. Flaherty, K.T.; Puzanov, I.; Kim, K.B.; Ribas, A.; McArthur, G.A.; Sosman, J.A.; O'Dwyer, P.J.; Lee, R.J.; Grippo, J.F.; Nolop, K.; et al. Inhibition of Mutated, Activated BRAF in Metastatic Melanoma. *N. Engl. J. Med.* **2010**, *363*, 809–819. [CrossRef] [PubMed]

9. Viros, A.; Sanchez-Laorden, B.; Pedersen, M.; Furney, S.J.; Rae, J.; Hogan, K.; Ejiama, S.; Girotti, M.R.; Cook, M.; Dhomen, N.; et al. Ultraviolet radiation accelerates BRAF-driven melanomagenesis by targeting TP53. *Nat. Cell Biol.* **2014**, *511*, 478–482. [CrossRef]

10. Dankort, D.; Curley, D.P.; Cartlidge, R.A.; Nelson, B.; Karnezis, A.N.; Damsky, W.E., Jr.; You, M.J.; Depinho, R.A.; McMahon, M.; Bosenberg, M. BrafV600E cooperates with Pten loss to induce metastatic melanoma. *Nat. Genet.* **2009**, *41*, 544–552. [CrossRef] [PubMed]

11. Pardoll, D. The blockade of immune checkpoints in cancer immunotherapy. *Nat. Rev. Cancer* **2012**, *12*, 252–264. [CrossRef]

12. Hugo, W.; Zaretsky, J.M.; Sun, L.; Song, C.; Moreno, B.H.; Hu-Lieskovan, S.; Berent-Maoz, B.; Pang, J.; Chmielowski, B.; Cherry, G.; et al. Genomic and Transcriptomic Features of Response to Anti-PD-1 Therapy in Metastatic Melanoma. *Cell* **2016**, *165*, 35–44. [CrossRef] [PubMed]

13. Domingues, B.; Lopes, J.M.; Soares, P.; Populo, H. Melanoma treatment in review. *Immuno Targets Ther.* **2018**, *7*, 35–49. [CrossRef]

14. Gmeiner, W.H.; Ghosh, S. Nanotechnology for cancer treatment. *Nanotechnol. Rev.* **2014**, *3*, 111–122. [CrossRef]

15. Martin, T.A.; Ye, L.; Sanders, A.J.; Lane, J.; Jiang, W.G. Cancer Invasion and Metastasis: Molecular and Cellular Perspective. In *Madame Curie Bioscience Database*; Landes Bioscience: Austin, TX, USA, 2013.

16. Chowdhury, A.; Kunjiappan, S.; Panneerselvam, T.; Somasundaram, B.; Bhattacharjee, C. Nanotechnology and nanocarrier-based approaches on treatment of degenerative diseases. *Int. Nano Lett.* **2017**, *7*, 91–122. [CrossRef]

17. Park, K. Facing the Truth about Nanotechnology in Drug Delivery. *ACS Nano* **2013**, *7*, 7442–7447. [CrossRef] [PubMed]

18. Rezvantalab, S.; Drude, N.I.; Moraveji, M.K.; Güvener, N.; Koons, E.K.; Shi, Y.; Lammers, T.; Kiessling, F. PLGA-Based Nanoparticles in Cancer Treatment. *Front. Pharmacol.* **2018**, *9*, 1260. [CrossRef]

19. Palocci, C.; Valletta, A.; Chronopoulou, L.; Donati, L.; Bramosanti, M.; Brasili, E.; Baldan, B.; Pasqua, G. Endocytic pathways involved in PLGA nanoparticle uptake by grapevine cells and role of cell wall and membrane in size selection. *Plant Cell Rep.* **2017**, *36*, 1917–1928. [CrossRef] [PubMed]

20. Yameen, B.; Choi, W.I.; Vilos, C.; Swami, A.; Shi, J.; Farokhzad, O.C. Insight into nanoparticle cellular uptake and intracellular targeting. *J. Control. Release* **2014**, *190*, 485–499. [CrossRef]

21. Arruda, D.C.; De Oliveira, T.D.; Cursino, P.H.F.; Maia, V.S.C.; Berzaghi, R.; Travassos, L.R.; Tada, D.B. Inhibition of melanoma metastasis by dual-peptide PLGA NPS. *Pept. Sci.* **2017**, *108*, e23029. [CrossRef] [PubMed]

22. Zhang, X.; Liu, J.; Li, X.; Li, F.; Lee, R.J.; Sun, F.; Li, Y.; Liu, Z.; Teng, L. Trastuzumab-Coated Nanoparticles Loaded With Docetaxel for Breast Cancer Therapy. *Dose-Response* **2019**, *17*. [CrossRef] [PubMed]

23. Kou, G.; Gao, J.; Wang, H.; Chen, H.; Li, B.; Zhang, D.; Wang, S.; Hou, S.; Qian, W.; Dai, J.; et al. Preparation and Characterization of Paclitaxel-loaded PLGA nanoparticles coated with cationic SM5-1 single-chain antibody. *J. Biochem. Mol. Biol.* **2007**, *40*, 731–739. [CrossRef]

24. Rodríguez-Cerdeira, C.; Gregorio, M.C.; López-Barcenas, A.; Sánchez-Blanco, E.; Sánchez-Blanco, B.; Fabbrocini, G.; Bardhi, B.; Sinani, A.; Arenas, R. Advances in Immunotherapy for Melanoma: A Comprehensive Review. *Mediat. Inflamm.* **2017**, *2017*, 1–14. [CrossRef] [PubMed]

25. Li, H.; Ma, W. Intracellular Delivery of Tumor Antigenic Peptides in Biodegradablepolymer Adjuvant for Enhancing Cancer Immunotherapy. *Curr. Med. Chem.* **2014**, *21*, 2357–2366. [CrossRef]

26. Ahmed, T.A.; Aljaeid, B.M. Preparation, characterization, and potential application of chitosan, chitosan derivatives, and chitosan metal nanoparticles in pharmaceutical drug delivery. *Drug Des. Dev. Ther.* **2016**, *10*, 483–507. [CrossRef] [PubMed]

27. Li, J.; Cai, C.; Li, J.; Li, J.; Li, J.; Sun, T.; Wang, L.; Wu, H.; Yu, G. Chitosan-Based Nanomaterials for Drug Delivery. *Molecules* **2018**, *23*, 2661. [CrossRef] [PubMed]

28. Smith, A.; Perelman, M.; Hinchcliffe, M. Chitosan a promising safe and immune-enhancing adjuvant for intranasal vaccines. *Hum. Vaccines Immunother.* **2014**, *10*, 797–807. [CrossRef]

29. Naskar, A.; Kim, K.-S. Nanomaterials as Delivery Vehicles and Components of New Strategies to Combat Bacterial Infections: Advantages and Limitations. *Microorganisms* **2019**, *7*, 356. [CrossRef] [PubMed]

30. Vahed, S.Z.; Fathi, N.; Samiei, M.; Dizaj, S.M.; Sharifi, S. Targeted cancer drug delivery with aptamer-functionalized polymeric nanoparticles. *J. Drug Target.* **2018**, *27*, 292–299. [CrossRef] [PubMed]

31. Cosco, D.; Cilurzo, F.; Maiuolo, J.; Federico, C.; Di Martino, M.T.; Cristiano, M.C.; Tassone, P.; Paolino, D.; Paolino, D. Delivery of miR-34a by chitosan/PLGA nanoplexes for the anticancer treatment of multiple myeloma. *Sci. Rep.* **2015**, *5*, 17579. [CrossRef]

32. Yoncheva, K.; Merino, M.; Shenol, A.; Daskalov, N.T.; Petkov, P.S.; Vayssilov, G.N.; Garrido, M.J. Optimization and in-vitro/in-vivo evaluation of doxorubicin-loaded chitosan-alginate nanoparticles using a melanoma mouse model. *Int. J. Pharm.* **2019**, *556*, 1–8. [CrossRef] [PubMed]

33. Rudzinski, W.E.; Aminabhavi, T. Chitosan as a carrier for targeted delivery of small interfering RNA. *Int. J. Pharm.* **2010**, *399*, 1–11. [CrossRef] [PubMed]

34. Cryan, S.-A.; McKiernan, P.J.; Cunningham, O.; Greene, C.M. Targeting miRNA-based medicines to cystic fibrosis airway epithelial cells using nanotechnology. *Int. J. Nanomed.* **2013**, *8*, 3907–3915. [CrossRef] [PubMed]

35. Rostami, N.; Nikkhoo, A.; Khazaei-Poul, Y.; Farhadi, S.; Haeri, M.S.; Ardebili, S.M.; Vanda, N.A.; Atyabi, F.; Namdar, A.; Baghaei, M.; et al. Coinhibition of S1PR1 and GP130 by siRNA-loaded alginate-conjugated trimethyl chitosan nanoparticles robustly blocks development of cancer cells. *J. Cell. Physiol.* **2020**, *235*, 9702–9717. [CrossRef] [PubMed]

36. Felli, N.; Errico, M.C.; Pedini, F.; Petrini, M.; Puglisi, R.; Bellenghi, M.; Boe, A.; Felicetti, F.; Mattia, G.; De Feo, A.; et al. AP2α controls the dynamic balance between miR-126&126* and miR-221&222 during melanoma progression. *Oncogene* **2015**, *35*, 3016–3026. [CrossRef] [PubMed]

37. Lin, S.; Gregory, R.I. MicroRNA biogenesis pathways in cancer. *Nat. Rev. Cancer* **2015**, *15*, 321–333. [CrossRef] [PubMed]

38. Michael, I.P.; Saghafinia, S.; Hanahan, D. A set of microRNAs coordinately controls tumorigenesis, invasion, and metastasis. *Proc. Natl. Acad. Sci. USA* **2019**, *116*, 24184–24195. [CrossRef]

39. Yingchoncharoen, P.; Kalinowski, D.S.; Richardson, D.R. Lipid-Based Drug Delivery Systems in Cancer Therapy: What Is Available and What Is Yet to Come. *Pharmacol. Rev.* **2016**, *68*, 701–787. [CrossRef]

40. Akbarzadeh, A.; Rezaei-Sadabady, R.; Davaran, S.; Joo, S.; Zarghami, N.; Hanifehpour, Y.; Samiei, M.; Kouhi, M.; Nejati-Koshki, K. Liposome: Classification, preparation, and applications. *Nanoscale Res. Lett.* **2013**, *8*, 102. [CrossRef]

41. Hatakeyama, H.; Akita, H.; Harashima, H. The Polyethyleneglycol Dilemma: Advantage and Disadvantage of PEGylation of Liposomes for Systemic Genes and Nucleic Acids Delivery to Tumors. *Biol. Pharm. Bull.* **2013**, *36*, 892–899. [CrossRef]

42. Gao, H.-W.; Chang, K.-F.; Huang, X.-F.; Lin, Y.-L.; Weng, J.-C.; Liao, K.-W.; Tsai, N.-M. Antitumor Effect of n-Butylidenephthalide Encapsulated on B16/F10 Melanoma Cells In Vitro with a Polycationic Liposome Containing PEI and Polyethylene Glycol Complex. *Molecules* **2018**, *23*, 3224. [CrossRef]

43. Yin, S.; Xia, C.; Wang, Y.; Wan, D.; Rao, J.; Tang, X.; Wei, J.; Wang, X.; Li, M.; Zhang, Z.; et al. Dual receptor recognizing liposomes containing paclitaxel and hydroxychloroquine for primary and metastatic melanoma treatment via autophagy-dependent and independent pathways. *J. Control. Release* **2018**, *288*, 148–160. [CrossRef] [PubMed]

44. Resnier, P.; Galopin, N.; Sibiril, Y.; Clavreul, A.; Cayon, J.; Briganti, A.; Legras, P.; Vessières, A.; Montier, T.; Jaouen, G.; et al. Efficient ferrocifen anticancer drug and Bcl-2 gene therapy using lipid nanocapsules on human melanoma xenograft in mouse. *Pharmacol. Res.* **2017**, *126*, 54–65. [CrossRef] [PubMed]

45. Sadhu, S.S.; Wang, S.; Dachineni, R.; Averineni, R.K.; Yang, Y.; Yin, H.; Bhat, G.J.; Guan, X. In Vitro and In Vivo Tumor Growth Inhibition by Glutathione Disulfide Liposomes. *Cancer Growth Metastasis* **2017**, *10*. [CrossRef] [PubMed]

46. Abu Lila, A.S.; Ishida, T. Liposomal Delivery Systems: Design Optimization and Current Applications. *Biol. Pharm. Bull.* **2017**, *40*, 1–10. [CrossRef] [PubMed]

47. Fu, H.; Shi, K.; Hu, G.; Yang, Y.; Kuang, Q.; Lu, L.; Zhang, L.; Chen, W.; Dong, M.; Chen, Y.; et al. Tumor-Targeted Paclitaxel Delivery and Enhanced Penetration Using TAT-Decorated Liposomes Comprising Redox-Responsive Poly(Ethylene Glycol). *J. Pharm. Sci.* **2015**, *104*, 1160–1173. [CrossRef]

48. De Serrano, L.O.; Burkhart, D.J. Liposomal vaccine formulations as prophylactic agents: Design considerations for modern vaccines. *J. Nanobiotechnol.* **2017**, *15*, 83. [CrossRef]

49. Hwang, S.; Nam, J.; Jung, S.; Song, J.; Doh, H.; Kim, S. Gold nanoparticle-mediated photothermal therapy: Current status and future perspective. *Nanomedicine* **2014**, *9*, 2003–2022. [CrossRef]

50. Bagheri, S.; Yasemi, M.; Safaie-Qamsari, E.; Rashidiani, J.; Abkar, M.; Hassani, M.; Mirhosseini, S.A.; Kooshki, H. Using gold nanoparticles in diagnosis and treatment of melanoma cancer. *Artif. Cells Nanomed. Biotechnol.* **2018**, *46*, 462–471. [CrossRef]

51. Singh, P.; Pandit, S.; Mokkapati, V.; Garg, A.; Ravikumar, V.; Mijakovic, I. Gold Nanoparticles in Diagnostics and Therapeutics for Human Cancer. *Int. J. Mol. Sci.* **2018**, *19*, 1979. [CrossRef]

52. Jing, L.; Liang, X.; Deng, Z.; Feng, S.; Li, X.; Huang, M.; Li, C.; Dai, Z. Prussian blue coated gold nanoparticles for simultaneous photoacoustic/CT bimodal imaging and photothermal ablation of cancer. *Biomaterials* **2014**, *35*, 5814–5821. [CrossRef] [PubMed]

53. Kim, C.; Cho, E.C.; Chen, J.; Song, K.H.; Au, L.; Favazza, C.; Zhang, Q.; Cobley, C.M.; Gao, F.; Xia, Y.; et al. In Vivo Molecular Photoacoustic Tomography of Melanomas Targeted by Bioconjugated Gold Nanocages. *ACS Nano* **2010**, *4*, 4559–4564. [CrossRef] [PubMed]

54. Ajnai, G.; Chiu, A.; Kan, T.; Cheng, C.-C.; Tsai, T.-H.; Chang, J. Trends of Gold Nanoparticle-based Drug Delivery System in Cancer Therapy. *J. Exp. Clin. Med.* **2014**, *6*, 172–178. [CrossRef]

55. Guo, J.; Rahme, K.; He, Y.; Li, L.-L.; Holmes, J.D.; O'Driscoll, C.M. Gold nanoparticles enlighten the future of cancer theranostics. *Int. J. Nanomed.* **2017**, *12*, 6131–6152. [CrossRef]

56. Zhu, L.; Kate, P.; Torchilin, V.P. Matrix Metalloprotease 2-Responsive Multifunctional Liposomal Nanocarrier for Enhanced Tumor Targeting. *ACS Nano* **2012**, *6*, 3491–3498. [CrossRef]

57. Zhu, L.; Torchilin, V.P. Stimulus-responsive nanopreparations for tumor targeting. *Integr. Biol.* **2013**. [CrossRef]

58. Zhang, Q.; Zhang, L.; Li, Z.; Xie, X.; Gao, X.; Xu, X. Inducing Controlled Release and Increased Tumor-Targeted Delivery of Chlorambucil via Albumin/Liposome Hybrid Nanoparticles. *AAPS PharmSciTech* **2017**, *18*, 2977–2986. [CrossRef]

59. Varshosaz, J.; Taymouri, S.; Hassanzadeh, F.; Javanmard, S.H.; Rostami, M. Self-assembly micelles with lipid core of cholesterol for docetaxel delivery to B16F10 melanoma and HepG2 cells. *J. Liposome Res.* **2014**, *25*, 157–165. [CrossRef]

60. Janaszewska, A.; Lazniewska, J.; Trzepiński, P.; Marcinkowska, M.; Klajnert-Maculewicz, B.; Klajnert-Maculewicz, B. Cytotoxicity of Dendrimers. *Biomology* **2019**, *9*, 330. [CrossRef]

61. Mukhopadhyay, R.; Kazi, J.; Debnath, M.C. Synthesis and characterization of copper nanoparticles stabilized with Quisqualis indica extract: Evaluation of its cytotoxicity and apoptosis in B16F10 melanoma cells. *Biomed. Pharmacother.* **2018**, *97*, 1373–1385. [CrossRef]

62. Naserzadeh, P.; Esfeh, F.A.; Kaviani, M.; Ashtari, K.; Kheirbakhsh, R.; Salimi, A.; Pourahmad, J. Single-walled carbon nanotube, multi-walled carbon nanotube and Fe_2O_3 nanoparticles induced mitochondria mediated apoptosis in melanoma cells. *Cutan. Ocul. Toxicol.* **2017**, *37*, 157–166. [CrossRef] [PubMed]

63. Yang, C.-C.; Hung, C.-F.; Chen, B.-H. Preparation of coffee oil-algae oil-based nanoemulsions and the study of their inhibition effect on UVA-induced skin damage in mice and melanoma cell growth. *Int. J. Nanomed.* **2017**, *12*, 6559–6580. [CrossRef] [PubMed]

64. Shakeel, F.; Haq, N.; Al-Dhfyan, A.; Alanazi, F.K.; Alsarra, I.A. Chemoprevention of skin cancer using low HLB surfactant nanoemulsion of 5-fluorouracil: A preliminary study. *Drug Deliv.* **2013**, *22*, 573–580. [CrossRef] [PubMed]

65. Zhuang, X.; Wu, T.; Zhao, Y.; Hu, X.; Bao, Y.; Guo, Y.; Song, Q.; Li, G.; Tan, S.; Zhang, Z. Lipid-enveloped zinc phosphate hybrid nanoparticles for codelivery of H-2Kb and H-2Db-restricted antigenic peptides and monophosphoryl lipid A to induce antitumor immunity against melanoma. *J. Control. Release* **2016**, *228*, 26–37. [CrossRef]

66. Huang, H.C.; Barua, S.; Sharma, G.; Dey, S.K.; Rege, K. Inorganic nanoparticles for cancer imaging and therapy. *J. Control. Release* **2011**, *155*, 344–357. [CrossRef]

67. Wong, G.S.; Rustgi, A.K. Matricellular proteins: Priming the tumour microenvironment for cancer development and metastasis. *Br. J. Cancer* **2013**, *108*, 755–761. [CrossRef]

68. Thomas, S.; Waterman, P.; Chen, S.; Marinelli, B.; Seaman, M.; Rodig, S.; Ross, R.W.; Josephson, L.; Weissleder, R.; Kelly, K.A. Development of Secreted Protein and Acidic and Rich in Cysteine (SPARC) Targeted Nanoparticles for the Prognostic Molecular Imaging of Metastatic Prostate Cancer. *J. Nanomed. Nanotechnol.* **2011**, *2*, 1–7. [CrossRef]

69. Weon, J.L.; Potts, P.R. The MAGE protein family and cancer. *Curr. Opin. Cell Biol.* **2015**, *37*, 1–8. [CrossRef]

70. Saeed, M.; Van Brakel, M.; Zalba, S.; Schooten, E.; Rens, J.A.P.; Debets, R.; Koning, G.A. Targeting melanoma with immunoliposomes coupled to anti-MAGE A1 TCR-like single-chain antibody. *Int. J. Nanomed.* **2016**, 955–975. [CrossRef]

71. Falvo, E.; Tremante, E.; Fraioli, R.; Leonetti, C.; Zamparelli, C.; Boffi, A.; Morea, V.; Ceci, P.; Giacomini, P. Antibody–drug conjugates: Targeting melanoma with cisplatin encapsulated in protein-cage nanoparticles based on human ferritin. *Nanoscale* **2013**, *5*, 12278–12285. [CrossRef]

72. Shao, H.; Im, H.; Castro, C.M.; Breakefield, X.; Weissleder, R.; Lee, H. New Technologies for Analysis of Extracellular Vesicles. *Chem. Rev.* **2018**, *118*, 1917–1950. [CrossRef] [PubMed]

73. Zonneveld, M.I.; Brisson, A.R.; Van Herwijnen, M.J.C.; Tan, S.; Van De Lest, C.H.A.; Redegeld, F.A.; Garssen, J.; Wauben, M.H.M.; Nolte-'t Hoen, E.N. Recovery of extracellular vesicles from human breast milk is influenced by sample collection and vesicle isolation procedures. *J. Extracell. Vesicles* **2014**, *3*. [CrossRef] [PubMed]

74. Chen, W.W.; Balaj, L.; Liau, L.M.; Samuels, M.L.; Kotsopoulos, S.K.; Maguire, C.A.; Loguidice, L.; Soto, H.; Garrett, M.; Zhu, L.D.; et al. BEAMing and Droplet Digital PCR Analysis of Mutant IDH1 mRNA in Glioma Patient Serum and Cerebrospinal Fluid Extracellular Vesicles. *Mol. Ther. Nucleic Acids* **2013**, *2*, e109. [CrossRef] [PubMed]

75. Wu, C.-X.; Liu, Z.-F. Proteomic Profiling of Sweat Exosome Suggests its Involvement in Skin Immunity. *J. Investig. Dermatol.* **2018**, *138*, 89–97. [CrossRef]

76. Yang, J.; Wei, F.; Schafer, C.; Wong, D.T.W. Detection of Tumor Cell-Specific mRNA and Protein in Exosome-Like Microvesicles from Blood and Saliva. *PLoS ONE* **2014**, *9*, e110641. [CrossRef]

77. Mathieu, M.; Martin-Jaular, L.; Lavieu, G.; Théry, C. Specificities of secretion and uptake of exosomes and other extracellular vesicles for cell-to-cell communication. *Nat. Cell Biol.* **2019**, *21*, 9–17. [CrossRef]

78. Keeerthikumar, S.; Chisanga, D.; Ariyaratne, D.; Al Saffar, H.; Anand, S.; Zhao, K.; Samuel, M.; Pathan, M.; Jois, M.; Chilamkurti, N.; et al. ExoCarta: A Web-Based Compendium of Exosomal Cargo. *J. Mol. Biol.* **2016**, *428*, 688–692. [CrossRef]

79. Pathan, M.; Fonseka, P.; Chitti, S.V.; Kang, T.; Sanwlani, R.; Van Deun, J.; Hendrix, A.; Mathivanan, S. Vesiclepedia 2019: A compendium of RNA, proteins, lipids and metabolites in extracellular vesicles. *Nucleic Acids Res.* **2018**, *47*, D516–D519. [CrossRef]

80. Friedman, R.C.; Farh, K.K.-H.; Burge, C.B.; Bartel, D.P. Most mammalian mRNAs are conserved targets of microRNAs. *Genome Res.* **2008**, *19*, 92–105. [CrossRef]

81. Ogawa, Y.; Taketomi, Y.; Murakami, M.; Tsujimoto, M.; Yanoshita, R. Small RNA transcriptomes of two types of exosomes in human whole saliva determined by next generation sequencing. *Biol. Pharm. Bull.* **2013**, *36*, 66–75. [CrossRef]

82. Skog, J.; Würdinger, T.; Van Rijn, S.; Meijer, D.H.; Gainche, L.; Curry, W.T., Jr.; Carter, R.S.; Krichevsky, A.M.; Breakefield, X.O. Glioblastoma microvesicles transport RNA and proteins that promote tumour growth and provide diagnostic biomarkers. *Nat. Cell Biol.* **2008**, *10*, 1470–1476. [CrossRef]

83. Ha, D.; Yang, N.; Nadithe, V. Exosomes as therapeutic drug carriers and delivery vehicles across biological membranes: Current perspectives and future challenges. *Acta Pharm. Sin. B* **2016**, *6*, 287–296. [CrossRef]

84. Van Niel, G.; Raposo, G.; Candalh, C.; Boussac, M.; Hershberg, R.; Cerf–Bensussan, N.; Heyman, M. Intestinal epithelial cells secrete exosome–like vesicles. *Gastroenterology* **2001**, *121*, 337–349. [CrossRef] [PubMed]

85. Zhuang, X.; Xiang, X.; Grizzle, W.; Sun, D.; Zhang, S.; Axtell, R.C.; Ju, S.; Mu, J.; Zhang, L.; Steinman, L.; et al. Treatment of Brain Inflammatory Diseases by Delivering Exosome Encapsulated Anti-inflammatory Drugs From the Nasal Region to the Brain. *Mol. Ther.* **2011**, *19*, 1769–1779. [CrossRef]

86. Zöller, M. Tetraspanins: Push and pull in suppressing and promoting metastasis. *Nat. Rev. Cancer* **2008**, *9*, 40–55. [CrossRef] [PubMed]

87. Stickney, Z.; Losacco, J.; McDevitt, S.; Zhang, Z.; Lu, B. Development of exosome surface display technology in living human cells. *Biochem. Biophys. Res. Commun.* **2016**, *472*, 53–59. [CrossRef] [PubMed]

88. Rana, S.; Yue, S.; Stadel, D.; Zöller, M. Toward tailored exosomes: The exosomal tetraspanin web contributes to target cell selection. *Int. J. Biochem. Cell Biol.* **2012**, *44*, 1574–1584. [CrossRef]

89. Ohno, S.-I.; Takanashi, M.; Sudo, K.; Ueda, S.; Ishikawa, A.; Matsuyama, N.; Fujita, K.; Mizutani, T.; Ohgi, T.; Ochiya, T.; et al. Systemically Injected Exosomes Targeted to EGFR Deliver Antitumor MicroRNA to Breast Cancer Cells. *Mol. Ther.* **2013**, *21*, 185–191. [CrossRef]

90. Kim, M.S.; Haney, M.J.; Zhao, Y.; Yuan, D.; Deygen, I.; Klyachko, N.L.; Kabanov, A.V.; Batrakova, E.V. Engineering macrophage-derived exosomes for targeted paclitaxel delivery to pulmonary metastases: In Vitro and in vivo evaluations. *Nanomed. Nanotechnol. Biol. Med.* **2018**, *14*, 195–204. [CrossRef]

91. Hao, S.; Liu, Y.; Yuan, J.; Zhang, X.; He, T.; Wu, X.; Wei, Y.; Sun, D.; Xiang, J. Novel Exosome-Targeted CD4+ T Cell Vaccine Counteracting CD4+25+ Regulatory T Cell-Mediated Immune Suppression and Stimulating Efficient Central Memory CD8+ CTL Responses. *J. Immunol.* **2007**, *179*, 2731–2740. [CrossRef]

92. Dai, S.; Wei, D.; Wu, Z.; Zhou, X.; Wei, X.; Huang, H.; Li, G. Phase I Clinical Trial of Autologous Ascites-derived Exosomes Combined With GM-CSF for Colorectal Cancer. *Mol. Ther.* **2008**, *16*, 782–790. [CrossRef] [PubMed]

93. Morse, M.A.; Garst, J.; Osada, T.; Khan, S.; Hobeika, A.C.; Clay, T.M.; Valente, N.; Shreeniwas, R.; Sutton, M.A.; Delcayre, A.; et al. A phase I study of dexosome immunotherapy in patients with advanced non-small cell lung cancer. *J. Transl. Med.* **2005**, *3*, 9. [CrossRef] [PubMed]

94. Boorn, J.G.V.D.; Dassler, J.; Coch, C.; Schlee, M.; Hartmann, G. Exosomes as nucleic acid nanocarriers. *Adv. Drug Deliv. Rev.* **2013**, *65*, 331–335. [CrossRef] [PubMed]

95. Escudier, B.; Dorval, T.; Chaput, N.; Andre, F.; Caby, M.-P.; Novault, S.; Flament, C.; Leboulaire, C.; Borg, C.; Amigorena, S.; et al. Vaccination of metastatic melanoma patients with autologous dendritic cell (DC) derived-exosomes: Results of thefirst phase I clinical trial. *J. Transl. Med.* **2005**, *3*, 10. [CrossRef]

96. Zhu, L.; Kalimuthu, S.; Gangadaran, P.; Oh, J.M.; Lee, H.W.; Baek, S.H.; Jeong, S.Y.; Lee, S.-W.; Lee, J.; Ahn, B.-C. Exosomes Derived From Natural Killer Cells Exert Therapeutic Effect in Melanoma. *Theranostics* **2017**, *7*, 2732–2745. [CrossRef]

97. Federici, C.; Shahaj, E.; Cecchetti, S.; Camerini, S.; Casella, M.; Iessi, E.; Camisaschi, C.; Paolino, G.; Calvieri, S.; Ferro, S.; et al. Natural-Killer-Derived Extracellular Vesicles: Immune Sensors and Interactors. *Front. Immunol.* **2020**, *11*. [CrossRef]

98. Chillà, A.; Margheri, F.; Biagioni, A.; Del Rosso, T.; Fibbi, G.; Del Rosso, M.; Laurenzana, A. Cell-Mediated Release of Nanoparticles as a Preferential Option for Future Treatment of Melanoma. *Cancers* **2020**, *12*, 1771. [CrossRef]

99. Kim, M.W.; Lee, G.; Niidome, T.; Komohara, Y.; Lee, R.; Park, Y.I. Platelet-Like Gold Nanostars for Cancer Therapy: The Ability to Treat Cancer and Evade Immune Reactions. *Front. Bioeng. Biotechnol.* **2020**, *8*. [CrossRef]

100. Del Re, M.; Marconcini, R.; Pasquini, G.; Rofi, E.; Vivaldi, C.; Bloise, F.; Restante, G.; Arrigoni, E.; Caparello, C.; Bianco, M.G.; et al. PD-L1 mRNA expression in plasma-derived exosomes is associated with response to anti-PD-1 antibodies in melanoma and NSCLC. *Br. J. Cancer* **2018**, *118*, 820–824. [CrossRef]

101. Chen, G.; Huang, A.C.; Zhang, W.; Zhang, G.; Wu, M.; Xu, W.; Yu, Z.; Yang, J.; Wang, B.; Sun, H.; et al. Exosomal PD-L1 contributes to immunosuppression and is associated with anti-PD-1 response. *Nat. Cell Biol.* **2018**, *560*, 382–386. [CrossRef]

102. Cordonnier, M.; Nardin, C.; Chanteloup, G.; Derangere, V.; Algros, M.-P.; Arnould, L.; Garrido, C.; Aubin, F.; Gobbo, J. Tracking the evolution of circulating exosomal-PD-L1 to monitor melanoma patients. *J. Extracell. Vesicles* **2020**, *9*, 1710899. [CrossRef] [PubMed]

103. Xiao, D.; Ohlendorf, J.; Chen, Y.; Taylor, D.D.; Rai, S.N.; Waigel, S.; Zacharias, W.; Hao, H.; McMasters, K.M. Identifying mRNA, MicroRNA and Protein Profiles of Melanoma Exosomes. *PLoS ONE* **2012**, *7*, e46874. [CrossRef] [PubMed]

104. Alegre, E.; Sanmamed, M.F.; Rodriguez, C.; Carranza, O.; Martin-Algarra, S.; González, A. Study of Circulating MicroRNA-125b Levels in Serum Exosomes in Advanced Melanoma. *Arch. Pathol. Lab. Med.* **2014**, *138*, 828–832. [CrossRef] [PubMed]

105. Kappelmann, M.; Kuphal, S.; Meister, G.; Vardimon, L.; Bosserhoff, A.-K. MicroRNA miR-125b controls melanoma progression by direct regulation of c-Jun protein expression. *Oncogene* **2012**, *32*, 2984–2991. [CrossRef]

106. Zhang, J.; Lu, L.; Xiong, Y.; Qin, W.; Zhang, Y.; Qian, Y.; Jiang, H.; Liu, W. MLK3 promotes melanoma proliferation and invasion and is a target of microRNA-125b. *Clin. Exp. Dermatol.* **2014**, *39*, 376–384. [CrossRef]

107. Nyholm, A.M.; Lerche, C.M.; Manfé, V.; Biskup, E.; Johansen, P.; Morling, N.; Thomsen, B.M.; Glud, M.; Gniadecki, R. miR-125b induces cellular senescence in malignant melanoma. *BMC Dermatol.* **2014**, *14*, 8. [CrossRef]

108. Huber, V.; Vallacchi, V.; Fleming, V.; Hu, X.; Cova, A.; Dugo, M.; Shahaj, E.; Sulsenti, R.; Vergani, E.; Filipazzi, P.; et al. Tumor-derived microRNAs induce myeloid suppressor cells and predict immunotherapy resistance in melanoma. *J. Clin. Investig.* **2018**, *128*, 5505–5516. [CrossRef]

109. Luan, W.; Ding, Y.; Yuan, H.; Ma, S.; Ruan, H.; Wang, J.; Lu, F.; Bu, X. Long non-coding RNA LINC00520 promotes the proliferation and metastasis of malignant melanoma by inducing the miR-125b-5p/EIF5A2 axis. *J. Exp. Clin. Cancer Res.* **2020**, *39*, 96. [CrossRef]

110. Tengda, L.; Shuping, L.; Mingli, G.; Jie, G.; Yun, L.; Weiwei, Z.; Anmei, D. Serum exosomal microRNAs as potent circulating biomarkers for melanoma. *Melanoma Res.* **2018**, *28*, 295–303. [CrossRef]

111. Valadi, H.; Ekström, K.; Bossios, A.; Sjöstrand, M.; Lee, J.J.; Lötvall, J.O. Exosome-mediated transfer of mRNAs and microRNAs is a novel mechanism of genetic exchange between cells. *Nat. Cell Biol.* **2007**, *9*, 654–659. [CrossRef]

112. Lunavat, T.R.; Echeng, L.; Einarsdottir, B.O.; Bagge, R.O.; Muralidharan, S.V.; Sharples, R.A.; Lässer, C.; Gho, Y.S.; Hill, A.F.; Nilsson, J.A.; et al. BRAFV600 inhibition alters the microRNA cargo in the vesicular secretome of malignant melanoma cells. *Proc. Natl. Acad. Sci. USA* **2017**, *114*, E5930–E5939. [CrossRef] [PubMed]

113. Svedman, F.C.; Lohcharoenkal, W.; Bottai, M.; Brage, S.E.; Sonkoly, E.; Hansson, J.; Pivarcsi, A.; Eriksson, H. Extracellular microvesicle microRNAs as predictive biomarkers for targeted therapy in metastastic cutaneous malignant melanoma. *PLoS ONE* **2018**, *13*, e0206942. [CrossRef] [PubMed]

114. Meng, F.; Henson, R.; Wehbe–Janek, H.; Ghoshal, K.; Jacob, S.T.; Patel, T. MicroRNA-21 Regulates Expression of the PTEN Tumor Suppressor Gene in Human Hepatocellular Cancer. *Gastroenterology* **2007**, *133*, 647–658. [CrossRef] [PubMed]

115. Asangani, I.A.; Rasheed, S.A.K.; Nikolova, D.A.; Leupold, J.H.; Colburn, N.H.; Post, S.; Allgayer, H. MicroRNA-21 (miR-21) post-transcriptionally downregulates tumor suppressor Pdcd4 and stimulates invasion, intravasation and metastasis in colorectal cancer. *Oncogene* **2007**, *27*, 2128–2136. [CrossRef]

116. Xiong, B.; Cheng, Y.; Ma, L.; Zhang, C. MiR-21 regulates biological behavior through the PTEN/PI-3 K/Akt signaling pathway in human colorectal cancer cells. *Int. J. Oncol.* **2012**, *42*, 219–228. [CrossRef]

117. Sayed, D.; Rane, S.; Lypowy, J.; He, M.; Chen, I.-Y.; Vashistha, H.; Yan, L.; Malhotra, A.; Vatner, D.; Abdellatif, M. MicroRNA-21 Targets Sprouty2 and Promotes Cellular Outgrowths. *Mol. Biol. Cell* **2008**, *19*, 3272–3282. [CrossRef]

118. Leite, K.R.; Reis, S.T.; Viana, N.I.; Morais, D.R.; Moura, C.M.; Silva, I.A.; Pontes, J.; Katz, B.; Srougi, M. Controlling RECK miR21 Promotes Tumor Cell Invasion and Is Related to Biochemical Recurrence in Prostate Cancer. *J. Cancer* **2015**, *6*, 292–301. [CrossRef]

119. Bautista-Sánchez, D.; Arriaga-Canon, C.; Pedroza-Torres, A.; De La Rosa-Velázquez, I.A.; González-Barrios, R.; Contreras-Espinosa, L.; Montiel-Manríquez, R.; Castro-Hernández, C.; Fragoso-Ontiveros, V.; Álvarez-Gómez, R.M.; et al. The Promising Role of miR-21 as a Cancer Biomarker and Its Importance in RNA-Based Therapeutics. *Mol. Ther. Nucleic Acids* **2020**, *20*, 409–420. [CrossRef]

120. Shi, J. Considering Exosomal miR-21 as a Biomarker for Cancer. *J. Clin. Med.* **2016**, *5*, 42. [CrossRef]

121. Melnik, B.C. MiR-21: An environmental driver of malignant melanoma? *J. Transl. Med.* **2015**, *13*, 202. [CrossRef]

122. Pfeffer, S.R.; Grossmann, K.F.; Cassidy, P.B.; Yang, C.H.; Fan, M.; Kopelovich, L.; Leachman, S.A.; Pfeffer, L.M. Detection of Exosomal miRNAs in the Plasma of Melanoma Patients. *J. Clin. Med.* **2015**, *4*, 2012–2027. [CrossRef] [PubMed]

123. Ragusa, M.; Barbagallo, C.; Statello, L.; Caltabiano, R.; Russo, A.; Puzzo, L.; Avitabile, T.; Longo, A.; Toro, M.D.; Barbagallo, D.; et al. miRNA profiling in vitreous humor, vitreal exosomes and serum from uveal melanoma patients: Pathological and diagnostic implications. *Cancer Biol. Ther.* **2015**, *16*, 1387–1396. [CrossRef] [PubMed]

124. Saldanha, G.; Potter, L.; Shendge, P.; Osborne, J.; Nicholson, S.; Yii, N.; Varma, S.; Aslam, M.I.; Elshaw, S.; Papadogeorgakis, E.; et al. Plasma MicroRNA-21 Is Associated with Tumor Burden in Cutaneous Melanoma. *J. Investig. Dermatol.* **2013**, *133*, 1381–1384. [CrossRef] [PubMed]

125. Mo, H.; Guan, J.; Yuan, Z.-C.; Lin, X.; Wu, Z.-J.; Liu, B.; He, J.-L. Expression and predictive value of miR-489 and miR-21 in melanoma metastasis. *World J. Clin. Cases* **2019**, *7*, 2930–2941. [CrossRef] [PubMed]

126. Lazar, I.; Clement, E.; Ducoux-Petit, M.; Denat, L.; Soldan, V.; Dauvillier, S.; Balor, S.; Burlet-Schiltz, O.; LaRue, L.; Muller, C.; et al. Proteome characterization of melanoma exosomes reveals a specific signature for metastatic cell lines. *Pigment. Cell Melanoma Res.* **2015**, *28*, 464–475. [CrossRef]

127. Dressel, R.; Johnson, J.P.; Günther, E. Heterogeneous patterns of constitutive and heat shock induced expression of HLA-linked HSP70-1 and HSP70-2 heat shock genes in human melanoma cell lines. *Melanoma Res.* **1998**, *8*, 482–492. [CrossRef]

128. Peinado, H.; Alečković, M.; Lavotshkin, S.; Matei, I.; Costa-Silva, B.; Moreno-Bueno, G.; Hergueta-Redondo, M.; Williams, C.; García-Santos, G.; Ghajar, C.M.; et al. Melanoma exosomes educate bone marrow progenitor cells toward a pro-metastatic phenotype through MET. *Nat. Med.* **2012**, *18*, 883–891. [CrossRef]

129. Alegre, E.; Zubiri, L.; Pérez-Gracia, J.L.; María, G.-C.; Soria, L.; Martín-Algarra, S.; González, A. Circulating melanoma exosomes as diagnostic and prognosis biomarkers. *Clin. Chim. Acta* **2016**, *454*, 28–32. [CrossRef]

130. Pardini, B.; Sabo, A.A.; Birolo, G.; Calin, G.A. Noncoding RNAs in Extracellular Fluids as Cancer Biomarkers: The New Frontier of Liquid Biopsies. *Cancers* **2019**, *11*, 1170. [CrossRef]

131. Cantile, M.; Scognamiglio, G.; Marra, L.; Aquino, G.; Botti, C.; Falcone, M.R.; Malzone, M.G.; Liguori, G.; Di Bonito, M.; Franco, R.; et al. HOTAIR role in melanoma progression and its identification in the blood of patients with advanced disease. *J. Cell. Physiol.* **2017**, *232*, 3422–3432. [CrossRef]

132. Tan, S.K.; Pastori, C.; Penas, C.; Komotar, R.J.; Ivan, M.E.; Wahlestedt, C.; Ayad, N.G. Serum long noncoding RNA HOTAIR as a novel diagnostic and prognostic biomarker in glioblastoma multiforme. *Mol. Cancer* **2018**, *17*, 74. [CrossRef] [PubMed]

133. Wang, Y.-L.; Liu, L.-C.; Hung, Y.; Chen, C.-J.; Lin, Y.-Z.; Wu, W.-R.; Wang, S.-C. Long non-coding RNA HOTAIR in circulatory exosomes is correlated with ErbB2/HER2 positivity in breast cancer. *Breast* **2019**, *46*, 64–69. [CrossRef] [PubMed]

134. Wu, F.; Yin, Z.; Yang, L.; Fan, J.; Xu, J.; Jin, Y.; Yu, J.; Zhang, D.; Yang, G. Smoking Induced Extracellular Vesicles Release and Their Distinct Properties in Non-Small Cell Lung Cancer. *J. Cancer* **2019**, *10*, 3435–3443. [CrossRef] [PubMed]

135. Hoshino, A.; Kim, H.S.; Bojmar, L.; Gyan, K.E.; Cioffi, M.; Hernandez, J.; Zambirinis, C.P.; Rodrigues, G.; Molina, H.; Heissel, S.; et al. Extracellular Vesicle and Particle Biomarkers Define Multiple Human Cancers. *Cell* **2020**, *182*, 1044–1061.e18. [CrossRef]

136. Barile, L.; Vassalli, G. Exosomes: Therapy delivery tools and biomarkers of diseases. *Pharmacol. Ther.* **2017**, *174*, 63–78. [CrossRef] [PubMed]

137. Ferrone, S.; Whiteside, T.L. Targeting CSPG4 for isolation of melanoma cell-derived exosomes from body fluids. *HNO* **2020**, *68*, 100–105. [CrossRef]

Review

Tumor Microenvironment: Implications in Melanoma Resistance to Targeted Therapy and Immunotherapy

Italia Falcone [1,*], Fabiana Conciatori [1], Chiara Bazzichetto [1], Gianluigi Ferretti [1], Francesco Cognetti [1], Ludovica Ciuffreda [2,†] and Michele Milella [3,†]

[1] Medical Oncology, IRCCS—Regina Elena National Cancer Institute, 00144 Rome, Italy; fabiana.conciatori@ifo.gov.it (F.C.); chiara.bazzichetto@ifo.gov.it (C.B.); gianluigi.ferretti@ifo.gov.it (G.F.); francesco.cognetti@ifo.gov.it (F.C.)

[2] SAFU, Department of Research, Advanced Diagnostics, and Technological Innovation, IRCCS—Regina Elena National Cancer Institute, 00144 Rome, Italy; ludovica.ciuffreda@ifo.gov.it

[3] Section of Oncology, Department of Medicine, University of Verona School of Medicine and Verona University Hospital Trust, 37126 Verona, Italy; michele.milella@univr.it

* Correspondence: italia.falcone@ifo.gov.it; Tel.: +39-06-5266-5185

† These authors contributed equally to this work.

Received: 29 August 2020; Accepted: 3 October 2020; Published: 6 October 2020

Simple Summary: The response to pharmacological treatments is deeply influenced by the tight interactions between the tumor cells and the microenvironment. In this review we describe, for melanoma, the most important mechanisms of resistance to targeted therapy and immunotherapy mediated by the components of the microenvironment. In addition, we briefly describe the most recent therapeutic advances for this pathology. The knowledge of molecular mechanisms, which are underlying of drug resistance, is fundamental for the development of new therapeutic approaches for the treatment of melanoma patients.

Abstract: Antitumor therapies have made great strides in recent decades. Chemotherapy, aggressive and unable to discriminate cancer from healthy cells, has given way to personalized treatments that, recognizing and blocking specific molecular targets, have paved the way for targeted and effective therapies. Melanoma was one of the first tumor types to benefit from this new care frontier by introducing specific inhibitors for v-Raf murine sarcoma viral oncogene homolog B (BRAF), mitogen-activated protein kinase (MEK), v-kit Hardy–Zuckerman 4 feline sarcoma viral oncogene homolog (KIT), and, recently, immunotherapy. However, despite the progress made in the melanoma treatment, primary and/or acquired drug resistance remains an unresolved problem. The molecular dynamics that promote this phenomenon are very complex but several studies have shown that the tumor microenvironment (TME) plays, certainly, a key role. In this review, we will describe the new melanoma treatment approaches and we will analyze the mechanisms by which TME promotes resistance to targeted therapy and immunotherapy.

Keywords: melanoma; targeted therapy; immunotherapy; tumor microenvironment; therapeutic resistance

---

1. Introduction

Melanoma is one of most aggressive human tumors, arising from the uncontrolled proliferation of melanocytes, the skin cells responsible for the production of melanin. In terms of incidence, malignant melanoma accounts for approximately 5% of all malignant tumors and its incidence is highly variable, depending on race and geographical variations: It is predominantly diagnosed in Caucasians and 85% of cases occur in North America, Europe, and Oceania. Although highly curable when diagnosed in an early phase, melanoma is an aggressive disease with five years' relative survival of only 25%, when diagnosed at an advanced metastatic stage [1,2]. Like many other solid tumors, malignant melanoma is highly heterogeneous and substantially resistant to unselective treatments, such as chemotherapy. In the past few years, mutational analysis and next-generation sequencing (NGS) approaches have shown that somatic mutations in *BRAF* or neuroblastoma RAS viral oncogene homolog (*NRAS*) genes promote deregulated survival and migration when combined with genetic alterations and/or epigenetic events that support senescence bypass [3,4].

Moreover, metastatic melanoma is considered a perfect example of immunogenic tumor because it is characterized by the consistent presence of lymphocid infiltrate, as compared to other cancers [5]. Based on these observations, molecularly targeted therapy and immunotherapy have revolutionized the approach to melanoma treatment and overall management. Clinical evidence has shown extremely encouraging results in terms of overall survival (OS) in patients treated with targeted therapy and immunotherapy [6–9]. Despite many important advances, however, development of the resistance remains a significant obstacle to melanoma curability and can be modulated by several factors, both intrinsic and extrinsic to the cancer cell. One such important factor is certainly the tumor microenvironment (TME), an intricate and complex network of cells, molecules, and paracrine factors that are tightly interconnected with melanoma cells, thereby influencing their initiation, progression, and sensitivity/resistance to therapeutic interventions.

In this review, we focused on melanoma microenvironment, analyzing its implications in therapy resistance.

2. Melanoma Targeted Therapy and Immunotherapy: An Overview

The identification of new molecular targets and the availability of modern immunotherapeutic approaches have revolutionized the treatment of advanced melanoma. By enabling the detection of genomic, transcriptional and epigenetic changes, NGS has allowed the identification of specific targets, thereby allowing for the development and optimization of treatments interfering with specific molecular targets [10,11]. Although such novel therapeutic approaches have demonstrated clinical efficacy and rapid responses in most patients, acquired resistance represents a significant challenge and has yet to be overcome. The most important therapeutic approaches to contemporary melanoma treatments are summarized in Figure 1.

2.1. Targeted Therapy

2.1.1. BRAF Inhibitors

The Mitogen-activated protein kinase (MAPK) pathway represents the signaling pathway most frequently dysregulated in melanoma and many inhibitors against this cascade have been developed at preclinical and clinical levels [12,13].

The serine/threonine protein kinase BRAF, physiologically involved in the control of cellular growth, is mutated in about 50% of all melanomas [14]. Although 20 individual BRAF mutations have been described, approximately 90% of BRAF-mutant melanomas present a mutation leading to the substitution of the valine 600 residue by glutamic acid (V600E) [15]. BRAF mutation per se is not sufficient to promote melanoma formation, but several studies have highlighted a crucial role of the mutant protein in disease progression [16]. In experiments in vitro, BRAF inhibition blocks melanoma growth and stimulates apoptosis, while, in vivo, reduced tumor formation is observed in mouse models [15–20]. Several BRAF inhibitors have been developed and approved for the treatment of melanoma and other BRAF-mutant tumors. Vemurafenib (or PLX4032) and dabrafenib (or GSK2118436) are potent and selective BRAF kinase inhibitors approved by the Food and Drug Administration (FDA) in 2011 and 2013, respectively, for metastatic melanoma. They shut down signaling through extracellular signal-regulated kinase (ERK) and, consequently, inhibit cellular growth in BRAF-mutant melanoma cells and induce tumor regression in xenograft models [21–23].

To date, these two drugs are employed, alone or in combination with chemotherapy, other targeted agents and immunotherapy in many clinical trials because they have shown to induce a substantial clinical responses and to prolong OS and progression-free survival (PFS) in melanoma patients [24,25]. Encorafenib (or LGX818) is a relatively novel BRAF inhibitor, approved by the FDA in June 2018. At preclinical level, it inhibits BRAF^{V600E} kinase activity determining cell growth inhibition in vitro and tumor regression in vivo in mouse models of BRAF-mutant melanoma [26]. A recent clinical trial, conducted on 577 BRAF-mutant melanoma patients, showed that encorafenib alone or in combination with binimetinib (MEK inhibitor) brings benefits, in terms of survival and tolerability, as compared to treatment with vemurafenib alone [7].

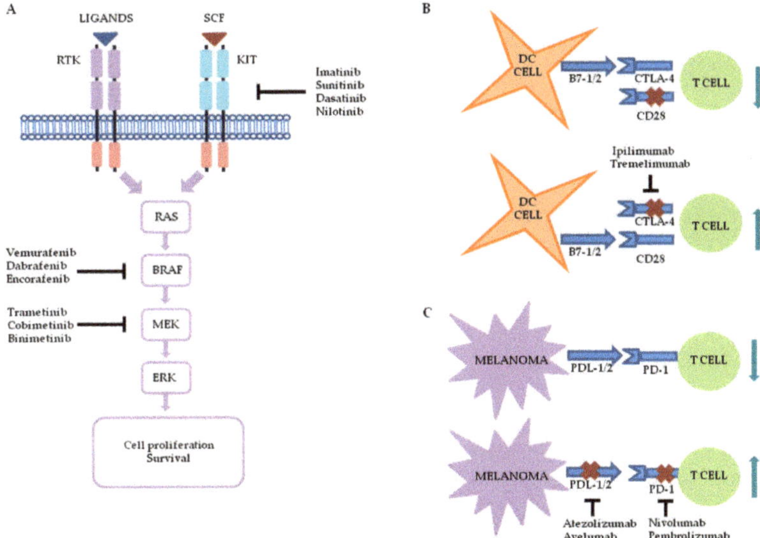

Figure 1. Schematic representation of the most important approaches to melanoma treatment. (**A**) Mitogen-activated protein kinase (MAPK) signaling is often deregulated in melanoma and, for this reason, several drugs against the components of the pathway have been developed. Although less frequently mutated, the v-kit Hardy–Zuckerman 4 feline sarcoma viral oncogene homolog (KIT) receptor tyrosine kinase, represents another molecular target for certain melanomas, and KIT inhibitors can be used in combination with chemotherapy or immunotherapy. (**B**) Cytotoxic T-lymphocyte-associated protein 4 (CTLA-4), interacting with B7 ligands present on the surface of dendritic cells (DCs), prevents the activation of T lymphocytes. The functional block of CTLA-4 mediated by monoclonal antibodies supports the interaction between B7 ligands and CD28, positive regulator receptor of the T lymphocyte activity. (**C**) Melanoma cells usually express elevated levels of programmed death ligand (PDL) 1 and 2, which, through interaction with programmed cell death protein 1 (PD-1) receptor on T cells, block their activation. Pharmacological inhibition of the PD-1 axis restores T cells' ability to recognize and kill tumor cells.

It is now well known that BRAF inhibition alone in melanoma can be overcome by many different mechanisms of resistance, some of which encompass downstream reactivation of MAPK pathway [27]. This phenomenon, called "paradoxical effect", is determined by dimerization between wild-type (wt) or kinase-inhibited BRAF and v-raf1 murine leukemia viral oncogene homolog 1 (CRAF), causing RAF signaling reactivation [28,29]. Even in tumor models different from melanoma, such as BRAF- wt)/RAS-mutant (mut) lung and pancreatic cancers, we confirmed paradoxical MAPK reactivation upon pharmacological BRAF kinase inhibition; however, simultaneous treatment with MEK inhibitors switched MAPK off again and induced a synergistic reduction of cell growth both in vitro and in vivo [30]. Indeed, in the past few years, clinical treatment of BRAF-mutant melanoma patients shifted toward a vertical combination of BRAF and MEK inhibitors, able to determine major improvements in terms of OS and PFS [8,30–33].

2.1.2. Mitogen-Activated Protein Kinase (MEK) Inhibitors

Trametinib (GSK1120212), cobimetinib (GDC-0973), and binimetinib (MEK162) are all potent selective inhibitors of MEK1 and MEK2 and are associated with significant cell growth inhibition in in vitro experiments and antitumor activity in mouse models of BRAF-mutant melanoma [34]. As described above, they are used as either monotherapy or, more often, in combination with dabrafenib, vemurafenib, and encorafenib, respectively, in patients affected by BRAF-mutant advanced melanoma [15,35,36]. MEK inhibitors also determine modest benefits in terms of PFS for melanoma patients whose tumors carry missense mutations in *NRAS* (occurring in about 20% of melanoma cases) [37–41].

2.1.3. V-kit Hardy–Zuckerman 4 Feline Sarcoma Viral Oncogene Homolog (KIT) Inhibitors

Activating somatic mutations in the *KIT* proto-oncogene are found in approximately 2–8% of melanomas, especially in those arising in mucosal and acral localizations (10–20% of the cases, respectively) [42,43]. When *KIT* is mutated, in exons 11 and 13, the regular growth and differentiation of melanocytes becomes uncontrolled; moreover, these mutations are generally mutually exclusive with the more frequent ones, such as those in *BRAF* and *NRAS* [13,44]. Many inhibitors, developed to block KIT and other tyrosine kinase receptors (RTKs), were analyzed in different clinical trials for melanoma such as imatinib, sunitinib, dasatinib, and nilotinib in combination with chemotherapy and immunotherapy [45,46].

2.2. Immunotherapy

Given its immunogenic characteristics, melanoma has been one of the solid tumors in which immunotherapy, using many different strategies aimed at stimulating the patient's immune system to recognize and eliminate cancer cells, has been most intensively studied [5]. Current immunotherapy approaches to human malignant melanoma include: monoclonal antibodies against immune checkpoint (ICIs), T-cell therapy, and cancer vaccines. Monoclonal antibodies inhibiting specific ICIs, including anti-programmed cell death protein 1 (PD-1), anti-programmed death ligand-1 (PDL-1), and cytotoxic T-lymphocyte-associated protein 4 (CTLA-4), alone or in combination, have been tested with great success in clinical trials and approved by the FDA for the treatment of advanced melanoma [47,48].

2.2.1. Anti-CTLA-4

CTLA-4, present on the surface of cluster of differentiation 4 (CD4$^+$) and CD8$^+$ lymphocytes, is another important pharmacological target for the treatment of several neoplastic forms, including metastatic melanoma [49]. Upon binding to the B7-1 (CD-80) and B7-2 (CD86) ligands on dendritic cells (DCs), CTLA-4 prevents their binding to the CD28 co-stimulatory receptor, which positively regulates lymphocyte activity, thereby triggering inhibitory signals that negatively regulate T-lymphocyte activation. Unlike the PD-1 axis (see below), which operates during the effector phase of the immune response, CTLA-4 and its inhibitors are implicated during the early stages of antigen presentation, leading to the first activation of T cells and immune recognition of the tumor. This prerogative is one of the reasons why combined checkpoint inhibition (with anti-CTLA-4 and anti-PD-1 agents) results in synergistic antitumor efficacy in the clinical setting [50]. Ipilimumab (MDX-010) is a humanized antibody against CTLA-4, currently approved by the FDA for the treatment of metastatic melanoma, either alone or in combination with PD-1 inhibitors. Ipilimumab significantly improved OS, as compared to cytotoxic chemotherapy, in metastatic melanoma, resulting in a proportion of patients experiencing prolonged disease control and causing a plateau in the survival curve at three years [51–53]. Tremelimumab (CP-675,206) is another monoclonal antibody against CTLA-4, which promotes important and durable tumor regressions in approximately 10% of metastatic melanoma patients; however, unlike ipilimumab, no significant changes in terms of survival were observed between patients treated with tremelimumab and those treated with chemotherapy [54]. Both of the two CTLA-4 antibodies are currently being studied in over 300 clinical trials involving patients with malignant melanoma [45].

2.2.2. Anti-PD-1

The PD-1 receptor, expressed on the surface of several immune cells, physiologically inhibits T cell activity upon binding to its ligands PDL-1 and -2. Activation of the PD-1/PDL-1/2 axis is frequently used by cancer cells to escape immune-mediated killing, often through suppression of downstream effectors the phosphatidylinositol 3-kinase (PI3K) pathway and cell cycle arrest in cytotoxic lymphocytes (CTL) [55]. Melanoma is generally characterized by high levels of PDL-1 expression, which correlates with poor prognosis; based on this finding, several monoclonal antibodies directed against the PD-1 axis have been developed and are used for melanoma treatment [56–60]. Nivolumab (BMS-936558, MDX-1106) and pembrolizumab (MK-3475) represent the two most important monoclonal antibodies against PD-1. They positively regulate the reactivation of T cells by blocking the interaction between the PD-1 receptor and its ligands, and have been studied in clinical trials, either alone or in combination with other ICIs, such as ipilimumab (CTLA-4 inhibitor, see above), chemotherapy, and targeted therapy. Preclinical studies have shown impressive results in terms of tumor growth inhibition; most importantly, clinical studies conducted in metastatic melanoma patients confirmed a clinically and statistically significant impact of these agents in terms of PFS and OS prolongation [21,61–65]. The phase III clinical trial CheckMate 067, completed in 2015, has shown a significant survival benefit (in terms of both PFS and OS) for metastatic melanoma patients treated with nivolumab, either alone or combined with ipilimumab. Compared with ipilimumab monotherapy the risk of death was reduced by 48% ($p < 0.001$) by the combination of nivolumab plus ipilimumab and by 36% ($p < 0.001$) by nivolumab alone [66,67]. Pembrolizumab revolutionized the treatment of patients with advanced melanoma versus ipilimumab, significantly improving PFS and OS [68]. Recently, the phase III clinical study KEYNOTE-006 further confirmed this finding, showing the superiority of pembrolizumab even after five years of follow-up. The median OS was 32.7 months for patients treated with pembrolizumab and 15.9 months for groups of patients treated with ipilimumab. The PFS was 8.4 months and 3.4 months for patients treated with pembrolizumab and ipilimumab, respectively [65,69]. Several PDL-1 inhibitors are now involved in clinical trials for melanoma. Atezolizumab has shown promising results in monotherapy for patients with metastatic melanoma [70]. Recently, the results of a triple combination of atezolizumab, vemurafenib, and cobimetinib, in patients with BRAF-mutant melanoma, were reported: the phase III clinical study IMspire 150 showed a significant PFS benefit (15.1 vs. 10.6 months) for patients treated with the triple combination, as compared to those who received vemurafenib and cobimetinib only [71]. Avelumab is another human anti-PDL-1 antibody involved in a phase I clinical trial (JAVELIN) for previously treated metastatic melanoma patients. The trial showed long-lasting and clinically meaningful disease control, with promising PFS and OS duration [72].

2.2.3. Alternative Melanoma Immunotherapies

Immunological therapies for melanoma are not limited to the use of checkpoint inhibitors. Isolation and ex vivo expansion of tumor-infiltrating lymphocytes (TILs) and their reintroduction in patients subjected to surgical removal of melanoma lesions has shown potential benefits; however, such complex approach remains limited to a research setting and is available only in a few specialized centers [73]. Chimeric antigen receptor T-cells (CAR-T) are another therapeutic approach that involves the use of engineered T cells to promote their ability to recognize cancer cells. Although this therapeutic field has made great strides and achieved unique results in the treatment of hematological malignancies, it has not produced the same positive effects in melanoma and other solid tumors, due to the presence of a highly immunosuppressive microenvironment in these contexts [74]. New treatment scenarios involve the development of vaccines that can stimulate the patient's immune system against tumor-associated antigens. Although preclinical research has obtained promising results in several cancer models, including melanoma, clinical results to date have been largely unsatisfactory [75,76]. Glycoprotein (Gp)-100, as an example, is a synthetic peptide encompassing a few amino acid residues of the trans-membrane Gp-100 protein expressed by melanoma cells. It has been formulated into a vaccine developed for advanced melanoma and utilized in clinical trials [77]. Vaccination with this peptide stimulates the host immune system inducing a CTL response that recognizes and kills melanoma cells in vitro. In a clinical study conducted in 185 patients with advanced melanoma, the immune-stimulating cytokine interleukin-2 (IL-2) was given alone or in combination with the Gp-110 peptide; patients treated with this combination presented a significant increase in terms of OS [78]. Vitespen is another vaccine developed for melanoma treatment, encompassing a heat shock protein/peptide complex (Gp-96) obtained and purified from surgically excised tumors. In advanced melanoma it has failed to show survival benefits but continues to stimulate the attention of researchers due to its extremely favorable toxicity profile [79,80].

3. TME Implications in Drug Resistance for Melanoma

Development of therapeutic resistance arguably represents the most important challenge in cancer therapy. Such phenomenon is associated with disease progression and low survival rates and is promoted by the ability of cancer cells to activate both intrinsic (i.e., dependent on genetic changes occurring in the cancer cell itself) and extrinsic (i.e., mediated by cross-talk mechanisms occurring between cancerous and noncancerous cells) escape mechanisms. Tumors are characterized by high genomic instability and heterogeneity and these prerogatives may lead to both primary (or de novo) or acquired resistance (i.e., occurring in cells previously responsive to the same treatment). Most importantly, the selective pressure applied by treatment itself may select out specific mechanisms of resistance. In some instances, therapeutic resistance occurs independent of genetic changes modifying cancer cells' acquired capabilities: In these cases, the insurgence of drug resistance can be attributed to changes occurring in different compartments of the TME. Indeed, tumor masses, including those that form in metastatic melanoma, should not be considered as isolated contexts without interactions; indeed, it has long been known that tumor cells "cross talk" continuously with many cellular and acellular components of the tumor stroma, which surrounds and penetrates the tumor mass. Such intricate structures constitute the TME and are characterized by mutual and continuous interactions between tumor and nontumor cells. TME is composed of cells and extracellular components of different origins, which contribute in several ways to the various stages of tumor progression (Figure 2) [81,82].

3.1. Cellular Components

3.1.1. Cancer-Associated Fibroblasts (CAFs)

Fibroblasts are important components of the stroma and their physiological functions encompass synthesis of extracellular matrix (ECM) and regulation of the inflammatory process. Upon tight (direct or mediated by soluble factors) interaction with cancer cells, fibroblasts differentiate into CAFs, which are characterized by specific markers: α smooth muscle actin (α-SMA), fibroblast activation protein (FAP), vimentin, fibroblast specific protein 1 (FSP1), and platelet-derived growth factor receptor (PDGFR)-α and β [83–85]. CAFs are involved in many cellular processes, including ECM remodeling, angiogenesis, and cell-to-cell interactions; in vivo, their activation is fundamental for tumor neo-vascularization [86,87]. In vitro and in vivo studies have shown that the continuous and persistent interactions between tumor cells and CAFs promote many aspects of the tumorigenic process, such as tumor progression, metastasis, and drug resistance. Melanoma cells, co-cultured with CAFs or grown in their conditioned media, display greater invasion and migration capabilities, as compared to the same cells cultured in isolation [88,89]. Recent studies also confirm that CAFs' activation is probably a crucial step for melanoma metastasis formation. Indeed, mice, in which the CAFs are inhibited by β-catenin suppression, displayed markedly decreased tumor-mediated vascularization [87,90]. CAFs and tumor cells reciprocally influence each other's biological behavior, and such cross talk is finely regulated by specific molecular mechanisms. As described in Figure 3A, the co-regulation system can involve tumor necrosis factor receptor-associated factor 6 (TRAF6), expressed in CAFs' activated and melanoma cells [91].

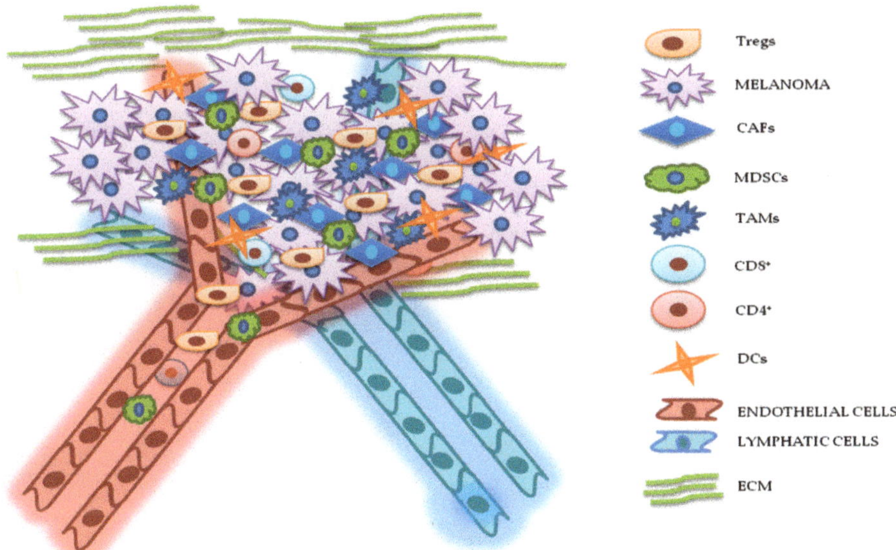

Figure 2. Relationship between melanoma and tumor microenvironment (TME). In this figure, is illustrated schematically the reciprocal interactions between melanoma cells and the other components of TME. Melanoma's TME, involved in tumor growth, progression, and drug resistance, is essentially represented by regulatory T cells (Tregs), cancer-associated fibroblasts (CAFs), myeloid-derived suppressor cells (MDSCs), tumor-associated macrophages (TAMs), cluster differentiation 4 (CD4⁺)/CD8⁺ lymphocytes, dendritic cells (DCs), endothelial and lymphatic cells, and extracellular matrix (ECM).

In melanoma cells, TRAF6 promotes nuclear factor kappa-light-chain-enhancer of activated B cells (NFkB)-dependent release of fibroblast growth factor 19 (FGF19), implicated in the transformation and activation of fibroblasts. FGF19-mediated CAFs' activation supports, in turn, the malignant and invasive phenotype of melanoma cells and their drugs resistance. On the other hand, TRAF6 upregulation in fibroblasts results in ECM remodeling through the release of matrix metalloproteinases (MMPs) 2 and 9 [91].

The mutual interaction between melanoma and CAFs can promote drug resistance in different ways. Straussman and collaborators highlighted the role of hepatocyte growth factor (HGF) in the development of acquired resistance to BRAF inhibitors (Figure 3B). Co-culture systems and proteomics analysis showed that HGF secreted by fibroblasts, by interacting with its mesenchymal ephitelial transition (MET) receptor on melanoma, induced MAPK and PI3K pathways' activation, thereby promoting resistance to RAF inhibition. Simultaneous downregulation of both RAF and MET reverted resistance in vitro, and it has been proposed as a possible therapeutic approach for the treatment of BRAF-mutant melanomas. Most importantly, the authors confirmed increased HGF expression in stromal cells of BRAF-mutant melanoma patients undergoing BRAF-targeted treatment in vivo, which resulted in poor prognosis and decreased response to treatments [92]. Neuregulin 1 (NRG1) is another paracrine factor through which CAFs may influence melanoma response to MAPK inhibitors (Figure 3C). NRG1 is the ligand of v-erb-b2 avian erythroblastic leukemia viral oncogene homolog3 (ErbB3), which is upregulated in melanoma cells after treatment with BRAF inhibitors. The use of ErbB3/ErbB2 antibodies restores the cytotoxic activity of these drugs in BRAF-mutant melanoma cell lines [93]. Furthermore, vemurafenib treatment increases the production of transforming growth factor β (TGF-β) by melanoma cells; TGF-β, in turn, causes CAFs' activation and increased fibronectin production, involved in BRAF inhibitors' resistance (Figure 3D) [94].

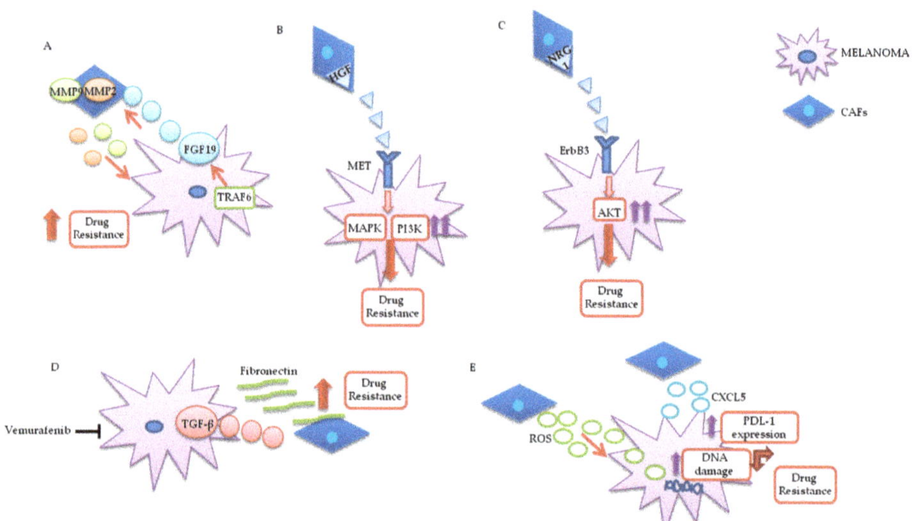

Figure 3. Melanoma/CAFs' paracrine interconnections. (**A–E**) In this figure, is illustrated schematically the mutual interactions between melanoma cells and cancer-associated fibroblasts (CAFs). Several factors are implicated in these intricate interconnections at the basis of drugs resistance: Tumor necrosis factor receptor-associated factor 6 (TRAF6), fibroblast growth factor 19 (FGF19), metalloproteinases 2 and 9 (MMP2 and MMP9), hepatocyte growth factor (HGF), neuregulin 1 (NRG1), V-erb-b2 avian erythroblastic leukemia viral oncogene homolog3 (ErbB3), transforming growth factor β (TGF-β), reactive oxygen species (ROS), CXC motif chemokine 5 (CXCL5), programmed death-ligand 1 (PDL-1).

Moreover, paradoxical MAPK activation, induced by BRAF inhibitors in genetically "normal" stromal cells, promotes a "therapy-resistant" microenvironment: Intravital imaging analyses conducted in melanoma have shown major paradox MAPK reactivation, especially in areas with high stromal density. Such activated CAFs, in turn, promoted matrix remodeling and ERK reactivation in melanoma, through integrin β1/focal adhesion kinase (FAK)/v-src sarcoma (Schmidt–Ruppin A-2) viral oncogene homolog avian (Src) signaling [95].

Uncontrolled production of reactive oxygen species (ROS) by TME fibroblasts is also associated with resistance to BRAF-targeted agents in melanoma (Figure 3E). Aging fibroblasts tend to release high levels of ROS in the TME, thereby modulating MAPK and PI3K pathways' activation in tumor cells and promoting cells' growth and drug resistance [96]. Such phenomenon could potentially be reversed by treating melanoma cells with antioxidants, thereby restoring drug responsiveness [97]. Along these lines, a recent preclinical study analyzed the role of secreted frizzled-related protein 2 (sFRP2), a wingless type MMTV integration site family member (Wnt) antagonist, in vemurafenib resistance: The sFRP2, produced and released in the TME by aged fibroblasts, actives a cascade of events in melanoma cells, ultimately leading to the loss of the redox effector apurinic/apyrimidinic endonuclease 1 (APE1). This condition significantly reduces the ability of melanoma cells to overcome ROS-induced DNA damage. Moreover, sFRP2-mediated inhibition of β-catenin leads to reduced melanoma response to vemurafenib [97].

It has recently been shown that CAFs are involved in the induction of a protumor immune microenvironment in many cancer models, favoring tumor growth and pharmacological resistance [98–100]. CAFs-mediated CXC chemokine ligand CXCL-2 production promotes regulatory T cells' (Tregs) growth and recruitment in the tumor stroma [100,101]. Melanoma-associated fibroblasts also directly influence tumor cells' ability to adapt and modify the response to immunotherapy. Experiments conducted on melanoma cell cultures have shown that CAFs release CXC motif chemokine 5 (CXCL5), which, in turn, induces PI3K/ protein kinase B (AKT)-dependent PDL-1 expression and resistance to immunotherapy in melanoma cells (Figure 3E) [102]. In addition, TGF-β secreted by CAFs also promotes resistance to PD-1 inhibitors: Transcriptomic and flow cytometric analysis, conducted on biopsies from 94 melanoma patients at various treatment stages, revealed a subset of patients characterized by loss of major histocompatibility complex 1 (MCH-I) and disease progression: Such phenomenon, induced by TGF-β released by CAFs, promotes a microphthalmia-associated transcription factor $(MITF)_{low}/AXL_{high}$ phenotype in melanoma cells, associated with resistance to MAPK pathway and PD-1 inhibitors [103].

Finally, CAFs' involvement in drug resistance is not limited to the production of paracrine factors but is also associated with cell-to-cell contact with cancer cells. Several studies have shown that fibroblasts create a physical barrier around the tumor mass and directly activate survival pathways in cancer cells [104–106]. In particular, these interactions are promoted by N-cadherin, expressed by both melanoma cells and fibroblasts, and are involved in tumor activation of the survival pathway PI3K/AKT/ BCL2 associated agonist of cell death (BAD) [104].

3.1.2. Lymphocytes

The immuno-microenvironment is characterized by T lymphocytes that recognize antigenic peptides presented by other components of the immune system [107]. $CD4^+$ T cells act as immune response "adjuvants" through the secretion of specific cytokines. $CD8^+$ T lymphocytes, on the other hand, are responsible for direct antigen/tumor cell individuation/elimination and are considered the most important mediators of tumor immune surveillance [82,108].

Depending on their genetic background, melanoma cells can influence the development of an immunosuppressive microenvironment. Phosphatase and tensin homolog deleted on chromosome 10 (*PTEN*), for example, is an important tumor suppressor gene often mutated/deleted in several cancer types, including melanoma; indeed, PTEN loss is present and concomitant with BRAF mutations in about 44% of melanomas and is associated with reduced OS [109]. PTEN loss promotes the formation of TME with low levels of cytotoxic T and natural killer (NK) cells and high concentrations of immunosuppressive elements, such as myeloid-derived suppressor (MDSCs) Tregs cells [110]. As described in in vitro and in vivo studies, PTEN-null melanoma cells inhibit antitumor activity of T cells and, consequently, response to immunotherapy. Through its negative regulation of PI3K and signal transducer and activator of transcription (STAT) 3 pathways, PTEN inhibits the production of immunosuppressive cytokines, such as interleukins (IL) 6 and 10 and vascular endothelial growth factor (VEGF). In melanoma, PTEN loss promotes STAT3 activation and, consequently, overproduction of these cytokines [111]. Moreover, PTEN loss is associated with reduced T cells' recruitment to the tumor site and cytotoxic activity [112].

An immunosuppressive TME influences the differentiation of dysfunctional $CD8^+$ T lymphocytes, i.e., T cells with reduced growth and effectors' cell recognition capacity and high concentrations of PD-1 and CTLA-4 receptors. If physiological conditions such as this status are necessary for immune homeostasis and to avoid self-reactive phenomena, in tumor contexts it may be an escape route that cancer cells use to evade immune response and promote resistance to immunotherapy [113]. A study conducted in patients with advanced melanoma demonstrated the presence of a subpopulation of T cells with high levels of PD-1 and immunoglobulin and mucin domain-containing molecule 3 (Tim3), another inhibitory receptor. Tim3 inhibition partially reverted the dysfunctional condition of T cells and increased their antitumor abilities. These results form the rationale for simultaneous blockade of PD-1 and Tim3 as a possible therapeutic approach to restore CD8$^+$ T lymphocytes' functionality in context of melanoma [114]. The same research group identified an additional inhibitory receptor, called T cell immunoglobulin (Ig) and immunoreceptor tyrosine-based inhibition motif (ITIM) domain (TIGIT). Inhibition of this receptor together with PD-1 may counteract dysregulated T cells' activity in a manner similar to Tim3 inhibition [115]. Extensive transcriptional profiling of the tumor infiltrate in 25 melanoma patients recently showed clonal expansion of dysfunctional CD8$^+$ T cell subset. The authors highlighted the reactivity and differentiation of these cells, which are likely involved in the regulation of antitumor activity and resistance to immunotherapeutic agents, making them an attractive target for more targeted and effective immunotherapeutic treatments in melanoma [116].

B lymphocytes are the cells responsible for humoral and acquired immunity. Their main function is to produce specific antibodies against foreign antigens, but they are also involved in maintenance of immune memory [117]. In melanoma, tumor-associated B cells (TAB) account for up to 33% of TME immune cells and are involved in resistance to targeted therapy by promoting angiogenesis and chronic inflammation. In addition, the presence of B cells in the tumor infiltrate is associated with increased metastatic capacity of melanoma cells and reduced patients' OS [118]. Recently, an interesting study analyzed the cross talk between melanoma cells and TAB and identified specific stimulating factors involved in the modulation of tumor response to different drugs. Melanoma secretes fibroblast growth factor 2 (FGF2), which actives B cells through its binding to fibroblast growth factor receptor 3 (FGFR-3) and promotes the release of insulin-like growth factor 1 (IGF-1). This factor, on tumor cells, induces proliferation and drug resistance. IGF-1, in turn, induces tumor cell proliferation and drug resistance. High levels of IGF-1 and FGFR-3 have been found in biopsies of melanoma patients treated with BRAF inhibitors in monotherapy or in combination with MEK inhibitors and IGF-1, and its receptor (IGF-1R) are associated with resistance to MAPK inhibitors [118]. However, TABs may have an opposite function in response to immunotherapy in melanoma. Indeed, a particular subtype of TABs can instead promote melanoma response to ICIs, by promoting the recruitment of CD8$^+$ T cells in the tumor compartment. The authors observed that the presence of higher concentrations of these B cells, in pretreated melanoma patients, is associated with a better response to future immunotherapy treatments [119]. More recently, analysis of metastatic melanoma samples showed that the co-occurrence of tumor- associated CD8$^+$ T cells and CD20$^+$ B cells is associated with improved survival [120]. The formation of tertiary lymphoid structures in these CD8$^+$/CD20$^+$ tumors is associated with a gene signature, which predicts clinical outcomes in melanoma patients treated with ICIs. Moreover, B cell-rich melanomas displayed increased levels of transcription factor 7 (TCF7)+ naive and/or memory T cells, whereas T cells in tumors without tertiary lymphoid structures had a dysfunctional molecular phenotype. In another study, it was shown that B cell signatures are enriched in human melanoma samples from patients who responded to neoadjuvant ICI treatment [121]. B cell markers were, indeed, the most differentially expressed genes in the tumors of responders versus non responders [122]. Histological evaluation again highlighted the localization of B cells within tertiary lymphoid structures, while RNA sequencing demonstrated clonal expansion and unique functional states of B cells (switched memory B cells) in responder.

NKs are an important subclass of granular lymphocytes, involved in the recognition and elimination of virus-infected and transformed cells [123,124]. In general, cancer promotes several mechanisms that destabilize the functionality of NKs, determining immune evasion: (1) Hyperproduction of activating ligands that paradoxically block NKs' receptors and (2) release of immunosuppressive factors, such as TGF-β and prostaglandin E [125]. Moreover, vemurafenib treatment of melanoma cells induces suppression of NKs activity in vitro, through downregulation of natural killer group 2D (NKG2D) and DNAX accessory molecule-1 (DNAM-1) activating receptors and simultaneous upregulation of major histocompatibility complex (MHC) I, which plays an inhibitory effect on NK cells [126].

Tregs represent a CD4$^+$ T cell's subpopulation with immunosuppressive properties [127,128]. In different cancer types, including melanoma, Tregs are able to promote immune evasion and cancer progression and are associated with poor prognosis [129–131]. In an analysis conducted on peripheral blood mononuclear cells (PBMCs) collected from healthy volunteers, Baumgartner and collaborators observed that melanoma evades the immune system by activation of Treg cells. Indeed, PBMCs exposed to melanoma-conditioned medium for a week presented an increase in Tregs' induction and a major presence of IL-10 and TGF-β in the supernatant, as compared to the same PBMCs grown in control medium [132]. In BRAF-mutant melanomas, uncontrolled MAPK activation leads to an increased production of different ILsand VEGF that influence the activity of the immune system toward a protumor condition. Sumimoto and collaborators showed that in BRAF-mutant melanomas Tregs are activated and suppress the antitumor function of T lymphocytes. Moreover, pharmacological blockades or genetic manipulation of key components of MAPK pathway drastically decrease tumor production of immunosuppressive cytokines, allowing for the development of an immune microenvironment favorable to tumor suppression [133].

Regulation of Tregs' differentiation and function could, therefore, be considered a valid therapeutic target for many cancers, including melanoma. Tregs are characterized by constitutive upregulation of PD-1 and CTLA-4 receptors and this condition leads to the hypothesis that Tregs could be the actual targets of ICI-based immunotherapy [134]. Unfortunately, results obtained in different studies are conflicting. Indeed, some studies have confirmed the inhibitory action of ICI on Tregs' functionality, while others have reported opposite results that could support the hypothesis of an involvement of ICI-mediated activation of these cells in immune-resistance [135–137]. Analysis conducted on murine models of autoimmune pancreatitis have partly elucidated the suppressive role of PD-1 on Treg cells activity. Indeed, mice characterized by PD1-deficient Tregs showed greater immunosuppressive capacities and rapid development of autoimmune disease [135]. On the basis of these results, it can be speculated that, physiologically, the PD-1 axis plays an important role in the regulation of Tregs' functionality and its inhibition may result in their increased activity. In vitro and in vivo experiments showed that, after treatments with nivolumab, Tregs proliferate and are functionally activated, resulting in the inhibition of antitumor activity [137]. Although with somewhat conflicting results, anti-CTLA-4 therapy would seem to bring more favorable effects on Tregs inhibition. Melanoma patients treated with ipilimumab showed a reduction in Tregs' levels and major benefits in terms of decreased tumor growth and survival [138]. In mice models of melanoma, CTLA-4 blockade increases the intratumor effector T cells/Tregs ratio, through fragment crystallizable (Fc)-gamma receptor (FcγR)-dependent mechanism. FcγR is expressed by several immune cells, such as macrophages, neutrophils and NK cells and, therefore, TME composition may influence the response to CTLA-4 inhibitors. Melanomas presenting low concentrations of macrophages or immune cells deficient for FcγR tend to respond less to therapy [139].

3.1.3. MDSCs

MDSCs are immature myeloid cells, with immunosuppressive functions, associated to tumor progression, metastasis, angiogenesis, and drug resistance. Absent or present in small concentrations in physiological conditions, they are recruited by tumor cells or by inflammatory stimuli and are responsible for the production of several factors involved in tumor growth and immune evasion, such as IL-10, TGF-β, and VEGF [140,141]. In melanoma, chronic inflammation promotes MSDCs' accumulation and activation in TME; thus, these cells are considered a possible therapeutic target in melanoma treatment [142]. A recent study identified a set of microRNAs (miRNAs) that regulate the differentiation and polarization of MDSCs in melanoma [143]. The authors found a significant association between the levels of these circulating miRNAs and reduced PFS and OS for melanoma patients treated with PD-1 and CTLA-4 inhibitors. This relationship was not reproduced in liquid biopsies of patients treated with MAPK pathway inhibitors, indicating a possible specific correlation with resistance to immunotherapy, as opposed to targeted therapy. These data suggest the possibility that combined treatments to inhibit myeloid dysfunctions could be able to overcome resistance to ICI in melanoma [143].

As reported by Gebhardt C. and collaborators, high levels of MDSCs in the TME are associated with ipilimumab resistance. Analysis of peripheral blood of 59 metastatic melanoma patients showed increased levels of MDSCs and their chemoattractant factors in patients poorly responsive or resistant to treatment with the CTLA-4 inhibitor. In addition, MDSCs exhibited a higher production of nitric oxide and were characterized by a higher expression of PDL-1, as compared to those isolated from responsive patients [144].

In melanoma, high levels of MDSCs are also associated with resistance to BRAF inhibitors. A recent preclinical study conducted in BRAF-inhibitor resistant mouse models showed that, after an initial response to treatment, these mice develop acquired resistance associated to an increase of MDSCs in TME. MAPK signaling reactivation in BRAF-resistant mice promotes the release of a complex system of stimulating cytokines, including C-C Motif Chemokine Ligand 2 (CCL2), that attract MDSCs and suppress the immune response [145].

3.1.4. Tumor-Associated Macrophages (TAMs)

Macrophages are immune cells involved in phagocytosis, pro-inflammatory cytokines' production, and specific immunity. The tumor-associated macrophages (TAMs), under the influence of cancer cells and the other microenvironment components, may be promoters or repressors of the tumorigenic process [146]. They are divided into two categories. M1-like macrophages (M1-TAMs), with antitumor activity, are important for the early stages of the inflammatory response. M2-like macrophages (M2-TAMs), predominant in TME, are correlated with tumor progression [82,147,148]. Different studies confirmed the key role of TAMs in tumor progression and have highlighted the significant correlation between the high levels of TAMs in the TME and poor prognosis for the patients [149,150]. In particular, melanoma cells, releasing miRNA-125b-5p in the microenvironment, inhibit the lysosomal acid lipase A (LIPA) and promote M2-macrophages' phenotype and their survival [151].

TAMs induce resistance to MAPK inhibitors by favoring the expression of tumor resistance factors or through their direct paradoxical activation of the pathway, driven by BRAF inhibitors. In in vivo melanoma models, MAPK-targeted agents induce tumor necrosis factor α (TNFα) production by macrophages, which promotes NFkB pathway activation and higher MITF expression. To overcome the TNFα- and MITF-mediated resistance to MAPK inhibitors, the authors proposed a selective inhibition of NFkB pathway that, in in vitro and in vivo analyses, synergized with MEK blockade and decreased the TNFα production [152]. Moreover, TAMs suffer the paradoxical reactivation of the MAPK pathway under the influence of BRAF inhibitors, with increased production of pro-angiogenic factors, such as VEGF and IL-8 and resistance to treatments [153].

TAMs express high levels of V-domain Ig suppressor of T cell activation (VISTA), another negative immune checkpoint, which is correlated with resistance to immunotherapy [154,155]. VISTA, associated with a significant decrease of survival in primary melanomas, in vivo, promotes a protumoral microenvironment mediated by upregulation of Tregs' levels and PDL-1 expression on macrophages' surface [155–158].

The switch of TAMs to an antitumor phenotype is considered an alternative approach to evade TAM-mediated resistance and to reconstitute the response to PD-1 axis inhibitors. TAMs' transformation process, mediated by STAT6 inactivation and NFkB phosphorylation, results in an increase of interleukin 12 (IL-12) production by TAMs and a reduction in inhibitory cytokine levels, such as IL-10 and C-C motif chemokine 22 (CCL22) [159]. TAMs induce immunotherapy resistance also by inhibiting the recruitment of CD8[+] T lymphocytes in the tumor site. Indeed, TAMs, by stable and durable interactions with T cells, lead to the maintenance of an immunosuppressive microenvironment not responsive to the inhibitory activity of anti-PD1 molecules [160].

3.1.5. DCs

DCs are immune cells derived from myeloid precursors and are implicated in recognition and capture of antigens considered "foreign", such as pathogens or cancer cells. They are antigen presenting cells (APCs) and interact with T lymphocytes through MHC present on their surface. DCs are involved in the production of cytokines and chemokines with anti- or pro-inflammatory function according to the stimuli received from the surrounding environment [161–163].

Unlike other immune cells, the involvement of DC in resistance to targeted and immunotherapy is mainly associated with the absence in tumor infiltrate of this type of cell. Melanoma is able to elude the complex mechanism of T cell activation by influencing DCs' maturation. Tumor cells produce inhibitory cytokines such as IL-8, IL-10, and VEGF and create an unfavorable environment for DCs' maturation. This condition negatively affects the DCs' ability to present antigen to T cells and, therefore, determines a reduced immune response [164]. Moreover, López González and collaborators demonstrated that, if inhibition of glycogen synthase kinase 3 beta (GSK3β) obstructs DC differentiation, a constitutively active GSK3β overcomes the IL-10 inhibition, leading to DC maturation [165]. In addition, melanoma promotes the switch of myeloid cells through immuno-suppressive macrophage-like cells rather than DCs [166]. The use of oncolytic virus (i.e., ORCA-010) could stimulate a specific differentiation of DCs and T cell priming by producing tumor-associated neo-antigen in order to increase the response to ICIs [167]. The therapeutic potential of DC vaccines was, recently, supported in an interesting preclinical work by Zhou and collaborators. The authors produced in vitro CD103+ murine and evaluated its activity in murine models of melanoma and osteosarcoma. CD103+ stimulated a favorable environment to the action of T lymphocytes, resulting in a reduced primary and metastatic tumor growth [168]. Further and recent results have also been obtained by direct DC targeting with molecular inhibitors. A recent study demonstrated that dasatinib (TK inhibitor) induces the activation of allogenic T cells by impairing the phosphorylation and metabolism of tryptophan induced by Indoleamine-2,3-dioxygenase (IDO), one of the most important intermediary cancer tolerants [169]. In addition, the same RAF kinase inhibitors could induce acquired resistance by influencing the differentiation and activation of DCs. Preclinical experiments, carried out on human and mouse DC cells, have detected a reduced or lack of DCs' ability to recruit T cells after treatment with RAF kinase inhibitors. These experiments, therefore, open possible new therapeutic scenarios, not only in melanoma, considering the negative effects of pan-RAF inhibitors on the immune response modulation [170].

3.2. ECM

ECM is an intricate network of proteins, proteoglycans, and glycoconjugates produced by TME cells and involved in the adhesion and support of the cellular compartment [171]. In different cancer contexts, the tumor matrix, creating a physical barrier, blocks drugs and inhibits their action. The mutual interactions between tumor and stroma cells induce rearrangements of the matrix architecture; this remodeling promotes tumor progression and modulates response to treatment [81].

Fibronectin, produced by CAFs, represents the major component of ECM and seems to play a key role in a decrease of sensitivity to therapies in different solid tumors [172,173]. BRAF-mutant and PTEN-loss melanomas show, after an initial response to treatments, the development of resistance to BRAF inhibitors mediated by reactivation of MAPK and PI3K pathways. In this molecular background, drug resistance seems to be mediated also by the protective effect of fibronectin, upregulated by BRAF inhibition [174]. Phosphoproteomic analysis conducted on BRAF-mutant melanoma cell lines showed, after treatment with vemurafenib or after BRAF gene silencing, an increased expression of fibronectin, only in PTEN-loss contexts. This evidence and further experiments conducted in cells genetically manipulated for PTEN have promoted the idea that regulation of fibronectin expression is associated with PTEN. Moreover, clinical data confirmed the higher expression of fibronectin in tissue of melanoma patients with PTEN loss [174]. The interaction between fibronectin and its receptor integrin α5β1 leads to AKT phosphorylation and decreases the apoptotic capacity of melanoma cells as a result of higher activity of induced myeloid leukemia cell differentiation protein 1 (MCL-1). Therefore, BRAF inhibition promotes a remodeling of melanoma microenvironment, by which the tumor cells escape to pharmacological blockade. In the molecular contexts analyzed, the study proposed BRAF/PI3K inhibitor combinations as an alternative to overcome the development of secondary resistance to BRAF-targeted agents [174].

Integrins are transmembrane receptors that physically regulate the interaction between cells and the ECM components and promote the development of intracellular signals. To date, 24 receptors have been identified, consisting of 18 subunits of α and eight of β, and several integrins are associated with melanoma progression and metastasis [175,176]. Integrin $\alpha5\beta1$ is the most important fibronectin receptor and their interaction modulates several cellular processes, such as adhesion, migration, and cellular differentiation [173,177]. An interesting work of intravital imaging of BRAF-mutant melanoma cells showed that, in co-culture systems, the treatment with BRAF inhibitors determines reactivation of MAPK pathway in areas with high stromal density; this condition is influenced by matrix remodeling associated to integrin $\beta1$/FAK/Src signaling [95]. Under the influence of treatment, the tumor fibroblasts suffer a paradoxical activation of ERK, resulting in higher fibronectin production and interaction with its receptor on tumor cells. Integrin $\alpha5\beta1$ promotes in melanoma cells the FAK-mediated ERK reactivation and resistance to BRAF inhibitors [95]. Integrin $\alpha v\beta3$ is another receptor involved, in melanoma context, in immunotherapy resistance through its regulation of the PDL-1 expression [178]. In in vitro and in vivo melanoma models, several evidences showed that integrin $\alpha v\beta3$ promotes the expression of PDL-1 through activation of STAT1 [179].

The matrix remodeling is influenced by a family of metalloproteinases, enzymes involved in hydrolysis of the other proteins and in mobility of tumor cells. In melanoma, several MMPs are involved in the different aspects of tumorigenic process, such as drug resistance [180]. Vemurafenib treatment in melanoma-resistant cells supports the paradoxical reactivation of ERK, a significant increase of IL-8 levels, and activation of MMPs, especially MMP2. This condition promotes the matrix disorganization, tumor motility, and immune evasion [181].

4. Conclusions

Currently, targeted therapy and immunotherapy represent a consolidated reality for the treatment of many aggressive tumors, such as metastatic melanoma. However, like previous therapeutic approaches, they are not exempt from the problem of innate and acquired resistance, which determines the failure of treatment in an important cohort of patients. Until a few years ago, scientific research was focused on the genetic and somatic changes by which cancer cells evaded drug inhibition. However, the new evidences on TME have revealed the existence of an intricate network of interconnections between the tumor and the surrounding microenvironment that massively influences all phases of the tumorigenic process. Through direct contact or release of soluble factors, the components of TME continuously influence tumor cells' activity modulating the drug response and therapeutic outcome. For these reasons, the research activities have focused on identifying the mechanisms and phenomena that TME implements to induce resistance to treatments. Based on the molecular mechanisms described in this review, it is evident that CAFs play a key role in the development of resistance to targeted therapy, especially through the production of many paracrine factors. Instead, the recruitment of immunosuppressive components, such as Tregs and MDSCs, in the TME is the primary mechanism of resistance to immunotherapy.

Funding: This research was funded by Fondo Sperimentazioni Medical Oncology 1(OM1). Chiara Bazzichetto was supported by an Italian Association for Cancer Research (AIRC) fellowship for Italy.

Acknowledgments: The authors wish to thank IRCCS Scientific Director office for supporting the manuscript.

Conflicts of Interest: The authors declare no conflict of interest.

Abbreviations

AKT	Protein kinase B
APC	Antigen presenting cell
APE1	Apurinic/apyrimidinic endonuclease 1
BAD	BCL2 associated agonist of cell death
BRAF	v-Raf murine sarcoma viral oncogene homolog B
CAFs	Cancer-associated fibroblasts
CAR-T	Chimeric antigen receptor T-cells
CCL2	C-C Motif Chemokine Ligand 2
CCL-22	C-C motif chemokine 22
CD	Cluster differentiation
CRAF	v-raf1 murine leukemia viral oncogene homolog 1
CTL	Cytotoxic lymphocytes
CTLA-4	cytotoxic T-lymphocyte-associated protein 4
CXCL-2	CXC chemokine ligand 2
CXCL5	CXC motif chemokine 5
DCs	Dendritic cells
DNAM-1	DNAX accessory molecule-1
ECM	Extracellular matrix
EMT	Epithelial mesenchymal transition
ErbB3	V-erb-b2 avian erythroblastic leukemia viral oncogene homolog3
ERK	Extracellular signal-regulated kinase
FAK	Focal adhesion kinase
FAP	Fibroblast activation protein
FcγR	Fragment crystallizable-gamma receptor
FDA	Food and Drug Administration
FGF 2-19	Fibroblast growth factor 2-19
FGFR-3	Fibroblast growth factor receptor 3
FSP1	Fibroblast specific protein 1
GSK3β	Glycogen synthase kinase 3 beta
Gp	Glycoprotein
HER3	V-erb-b2 avian erythroblastic leukemia viral oncogene homolog 3
HGF	Hepatocyte growth factor
ICIs	Immune checkpoint inhibitors
IDO	Indoleamine-2,3-dioxygenase
Ig	Immunoglobulin
IGF-1	Insulin-like growth factor 1
IGF-1R	Insulin-like growth factor receptor 1
IL	Interleukin
ITIM	Immunoreceptor tyrosine-based inhibition motif
KIT	v-kit Hardy–Zuckerman 4 feline sarcoma viral oncogene homolog
LIPA	Lysosomal acid lipase A
MAPK	Mitogen-activated protein kinase
MCL-1	Induced myeloid leukemia cell differentiation protein 1
MCH-I	Major histocompatibility complex 1
MDSCs	Myeloid-derived suppressor cells
MEK	Mitogen-activated protein kinase kinase
MET	Mesenchymal epithelial transition receptor
MHC	Major histocompatibility complex
miRNA	microRNA
MITF	Microphthalmia-associated transcription factor
MMPs	Matrix metalloproteinases
NFkB	Nuclear factor kappa-light-chain-enhancer of activated B cells

NGS	Next-generation sequencing
NK	Natural killer
NKG2D	Natural killer group 2D
NRAS	Neuroblastoma RAS viral oncogene homolog
NRG-1	Neuregulin 1
ORR	Objective response rate
OS	Overall survival
PBMC	Peripheral blood mononuclear cells
PD-1	Programmed cell death protein 1
PDGFR	Platelet-derived growth factor receptor
PDL-1/2	Programmed death ligands 1/2
PI3K	Phosphatidyl Inositol 3-kinase
PFS	Progression-free survival
PTEN	Phosphatase and tensin homolog on chromosome 10
ROS	Reactive oxygen species
RTK	Tyrosine kinase receptor
sFRP2	Secreted frizzled-related protein 2
STAT	Signal transducer and activator of transcription
SRC	V-src sarcoma (Schmidt–Ruppin A-2) viral oncogene homolog avian
TAB	Tumor-associated B cells
TAMs	Tumor-associated macrophages
TCF7	Transcription factor 7
TGF-β	Transforming growth factor β
TIGIT	T cell Ig and ITIM domain
TIL	Tumor-infiltrating lymphocytes
Tim3	Immunoglobulin and mucin domain-containing molecule 3
TME	Tumor microenvironment
TNFα	Tumor necrosis factor α
TRAF6	Tumor necrosis factor receptor-associated factor 6
Tregs	Regulatory T cells
VEGF	Vascular endothelial growth factor
VISTA	V-domain Ig suppressor of T cell activation
Wnt	wingless type MMTV integration site family member
α-SMA	α smooth muscle actin

References

1. Ferlay, J.; Shin, H.-R.; Bray, F.; Forman, D.; Mathers, C.; Parkin, D.M. Estimates of worldwide burden of cancer in 2008: GLOBOCAN 2008. *Int. J. Cancer* **2010**, *127*, 2893–2917. [CrossRef] [PubMed]
2. Siegel, R.L.; Mph, K.D.M.; Jemal, A. Cancer statistics, 2020. *CA A Cancer J. Clin.* **2020**, *70*, 7–30. [CrossRef] [PubMed]
3. Bennett, D.C. REVIEW ARTICLE: How to make a melanoma: What do we know of the primary clonal events? *Pigment. Cell Melanoma Res.* **2007**, *21*, 27–38. [CrossRef] [PubMed]
4. Shain, A.H.; Yeh, I.; Kovalyshyn, I.; Sriharan, A.; Talevich, E.; Gagnon, A.; Dummer, R.; North, J.; Pincus, L.; Ruben, B.; et al. The Genetic Evolution of Melanoma from Precursor Lesions. *N. Engl. J. Med.* **2015**, *373*, 1926–1936. [CrossRef]
5. Sanlorenzo, M.; Vujic, I.; Posch, C.; Dajee, A.; Yen, A.; Kim, S.; Ashworth, M.; Rosenblum, M.D.; Algazi, A.; Osella-Abate, S.; et al. Melanoma immunotherapy. *Cancer Biol. Ther.* **2014**, *15*, 665–674. [CrossRef]
6. Wolchok, J.D.; Chiarion-Sileni, V.; Gonzalez, R.; Rutkowski, P.; Grob, J.-J.; Cowey, C.L.; Lao, C.D.; Wagstaff, J.; Schadendorf, D.; Ferrucci, P.F.; et al. Overall Survival with Combined Nivolumab and Ipilimumab in Advanced Melanoma. *N. Engl. J. Med.* **2017**, *377*, 1345–1356. [CrossRef]

7.   Dummer, R.; Ascierto, P.A.; Gogas, H.; Arance, A.; Mandalà, M.; Liszkay, G.; Garbe, C.; Schadendorf, D.; Krajsová, I.; Gutzmer, R.; et al. Encorafenib plus binimetinib versus vemurafenib or encorafenib in patients with BRAF -mutant melanoma (COLUMBUS): A multicentre, open-label, randomised phase 3 trial. *Lancet Oncol.* **2018**, *19*, 603–615. [CrossRef]

8.   Long, G.V.; Flaherty, K.T.; Stroyakovskiy, D.; Gogas, H.; Levchenko, E.; De Braud, F.; Larkin, J.; Garbe, C.; Jouary, T.; Hauschild, A.; et al. Dabrafenib plus trametinib versus dabrafenib monotherapy in patients with metastatic BRAF V600E/K-mutant melanoma: Long-term survival and safety analysis of a phase 3 study. *Ann. Oncol.* **2017**, *28*, 1631–1639. [CrossRef]

9.   Ascierto, P.A.; Ferrucci, P.F.; Fisher, R.; Del Vecchio, M.; Atkinson, V.; Schmidt, H.; Schachter, J.; Queirolo, P.; Long, G.V.; Di Giacomo, A.M.; et al. Dabrafenib, trametinib and pembrolizumab or placebo in BRAF-mutant melanoma. *Nat. Med.* **2019**, *25*, 941–946. [CrossRef]

10.   Saito, M.; Momma, T.; Kono, K. Targeted therapy according to next generation sequencing-based panel sequencing. *FUKUSHIMA J. Med Sci.* **2018**, *64*, 9–14. [CrossRef]

11.   Tsimberidou, A.M. Targeted therapy in cancer. *Cancer Chemother. Pharmacol.* **2015**, *76*, 1113–1132. [CrossRef] [PubMed]

12.   Chappell, W.H.; Steelman, L.S.; Long, J.M.; Kempf, R.C.; Abrams, S.L.; Franklin, R.A.; Basecke, J.; Stivala, F.; Donia, M.; Fagone, P.; et al. Ras/Raf/MEK/ERK and PI3K/PTEN/Akt/mTOR Inhibitors: Rationale and Importance to Inhibiting These Pathways in Human Health. *Oncotarget* **2011**, *2*, 135–164. [CrossRef] [PubMed]

13.   Broussard, L.; Howland, A.; Ryu, S.; Song, K.; Norris, D.; Armstrong, C.A.; Song, P.I. Melanoma Cell Death Mechanisms. *Chonnam Med J.* **2018**, *54*, 135–142. [CrossRef] [PubMed]

14.   Amann, V.; Ramelyte, E.; Thurneysen, S.; Pitocco, R.; Bentele-Jaberg, N.; Goldinger, S.; Dummer, R.; Mangana, J. Developments in targeted therapy in melanoma. *Eur. J. Surg. Oncol. (EJSO)* **2017**, *43*, 581–593. [CrossRef]

15.   Savoia, P.; Fava, P.; Casoni, F.; Cremona, O. Targeting the ERK Signaling Pathway in Melanoma. *Int. J. Mol. Sci.* **2019**, *20*, 1483. [CrossRef]

16.   Hoeflich, K.P.; Gray, D.C.; Eby, M.T.; Tien, J.Y.; Wong, L.; Bower, J.; Gogineni, A.; Zha, J.; Cole, M.J.; Stern, H.M.; et al. Oncogenic BRAF Is Required for Tumor Growth and Maintenance in Melanoma Models. *Cancer Res.* **2006**, *66*, 999–1006. [CrossRef]

17.   Hingorani, S.R.; Jacobetz, M.A.; Robertson, G.P.; Herlyn, M.; Tuveson, D.A. Suppression of BRAF(V599E) in human melanoma abrogates transformation. *Cancer Res.* **2003**, *63*, 5198–5202.

18.   Hoeflich, K.P.; Jaiswal, B.; Davis, D.P.; Seshagiri, S. Inducible BRAF Suppression Models for Melanoma Tumorigenesis. *Methods Enzymol.* **2008**, *439*, 25–38. [CrossRef]

19.   Tsao, H.; Chin, L.; Garraway, L.A.; Fisher, D.E. Melanoma: From mutations to medicine. *Genes Dev.* **2012**, *26*, 1131–1155. [CrossRef]

20.   Leonardi, G.C.; Falzone, L.; Salemi, R.; Zanghì, A.; Spandidos, D.A.; McCubrey, J.A.; Candido, S.; Libra, M. Cutaneous melanoma: From pathogenesis to therapy (Review). *Int. J. Oncol.* **2018**, *52*, 1071–1080. [CrossRef]

21.   Tsai, K.K.; Zarzoso, I.; Daud, A.I. PD-1 and PD-L1 antibodies for melanoma. *Hum. Vaccines Immunother.* **2014**, *10*, 3111–3116. [CrossRef] [PubMed]

22.   Abraham, J.; Stenger, M. Dabrafenib in advanced melanoma with BRAF V600E mutation. *J. Community Support. Oncol.* **2014**, *12*, 48–49. [CrossRef] [PubMed]

23.   Roskoski, R. Targeting oncogenic Raf protein-serine/threonine kinases in human cancers. *Pharmacol. Res.* **2018**, *135*, 239–258. [CrossRef] [PubMed]

24.   Chapman, P.B.; Hauschild, A.; Robert, C.; Haanen, J.B.; Ascierto, P.; Larkin, J.; Dummer, R.; Garbe, C.; Testori, A.; Maio, M.; et al. Improved Survival with Vemurafenib in Melanoma with BRAF V600E Mutation. *N. Engl. J. Med.* **2011**, *364*, 2507–2516. [CrossRef]

25.   Morales, D.; Lombart, F.; Truchot, A.; Maire, P.; Hussein, M.; Hamitou, W.; Vigneron, P.; Galmiche, A.; Lok, C.; Vayssade, M. 3D Coculture Models Underline Metastatic Melanoma Cell Sensitivity to Vemurafenib. *Tissue Eng. Part A* **2019**, *25*, 1116–1126. [CrossRef]

26.   Delord, J.-P.; Robert, C.; Nyakas, M.; McArthur, G.A.; Kudchakar, R.; Mahipal, A.; Yamada, Y.; Sullivan, R.J.; Arance, A.; Kefford, R.F.; et al. Phase I Dose-Escalation and -Expansion Study of the BRAF Inhibitor Encorafenib (LGX818) in Metastatic BRAF -Mutant Melanoma. *Clin. Cancer Res.* **2017**, *23*, 5339–5348. [CrossRef]

27. Saei, A.; Eichhorn, P.J.A. Saei Adaptive Responses as Mechanisms of Resistance to BRAF Inhibitors in Melanoma. *Cancers* **2019**, *11*, 1176. [CrossRef]

28. Heidorn, S.J.; Milagre, C.; Whittaker, S.R.; Nourry, A.; Niculescu-Duvas, I.; Dhomen, N.; Hussain, J.; Reis-Filho, J.S.; Springer, C.; Pritchard, C.A.; et al. Kinase-Dead BRAF and Oncogenic RAS Cooperate to Drive Tumor Progression through CRAF. *Cell* **2010**, *140*, 209–221. [CrossRef]

29. Poulikakos, P.I.; Zhang, C.; Bollag, G.; Shokat, K.M.; Rosen, N. RAF inhibitors transactivate RAF dimers and ERK signalling in cells with wild-type BRAF. *Nat. Cell Biol.* **2010**, *464*, 427–430. [CrossRef]

30. Del Curatolo, A.; Conciatori, F.; Incani, U.C.; Bazzichetto, C.; Falcone, I.; Corbo, V.; D'Agosto, S.L.; Eramo, A.; Sette, G.; Sperduti, I.; et al. Therapeutic potential of combined BRAF/MEK blockade in BRAF-wild type preclinical tumor models. *J. Exp. Clin. Cancer Res.* **2018**, *37*, 140. [CrossRef]

31. Hatzivassiliou, G.; Song, K.; Yen, I.; Brandhuber, B.J.; Anderson, D.J.; Alvarado, R.; Ludlam, M.J.C.; Stokoe, D.; Gloor, S.L.; Vigers, G.; et al. RAF inhibitors prime wild-type RAF to activate the MAPK pathway and enhance growth. *Nat. Cell Biol.* **2010**, *464*, 431–435. [CrossRef] [PubMed]

32. Long, G.V.; Stroyakovskiy, D.; Gogas, H.; Levchenko, E.; De Braud, F.; Larkin, J.; Garbe, C.; Jouary, T.; Hauschild, A.; Grob, J.J.; et al. Combined BRAF and MEK Inhibition versus BRAF Inhibition Alone in Melanoma. *N. Engl. J. Med.* **2014**, *371*, 1877–1888. [CrossRef] [PubMed]

33. Trojaniello, C.; Festino, L.; Vanella, V.; Ascierto, P.A. Encorafenib in combination with binimetinib for unresectable or metastatic melanoma with BRAF mutations. *Expert Rev. Clin. Pharmacol.* **2019**, *12*, 259–266. [CrossRef] [PubMed]

34. Gilmartin, A.G.; Bleam, M.R.; Groy, A.; Moss, K.G.; Minthorn, E.A.; Kulkarni, S.G.; Rominger, C.M.; Erskine, S.; Fisher, K.E.; Yang, J.; et al. GSK1120212 (JTP-74057) Is an Inhibitor of MEK Activity and Activation with Favorable Pharmacokinetic Properties for Sustained In Vivo Pathway Inhibition. *Clin. Cancer Res.* **2011**, *17*, 989–1000. [CrossRef] [PubMed]

35. Ascierto, P.A.; McArthur, G.A.; Dréno, B.; Atkinson, V.; Liszkay, G.; Di Giacomo, A.M.; Mandalà, M.; Demidov, L.; Stroyakovskiy, D.; Thomas, L.; et al. Cobimetinib combined with vemurafenib in advanced BRAFV600-mutant melanoma (coBRIM): Updated efficacy results from a randomised, double-blind, phase 3 trial. *Lancet Oncol.* **2016**, *17*, 1248–1260. [CrossRef]

36. Liu, F.; Yang, X.; Geng, M.; Huang, M. Targeting ERK, an Achilles' Heel of the MAPK pathway, in cancer therapy. *Acta Pharm. Sin. B* **2018**, *8*, 552–562. [CrossRef]

37. Ascierto, P.A.; Schadendorf, D.; Berking, C.; Agarwala, S.S.; Van Herpen, C.M.; Queirolo, P.; Blank, C.U.; Hauschild, A.; Beck, J.T.; St-Pierre, A.; et al. MEK162 for patients with advanced melanoma harbouring NRAS or Val600 BRAF mutations: A non-randomised, open-label phase 2 study. *Lancet Oncol.* **2013**, *14*, 249–256. [CrossRef]

38. Stephen, A.G.; Esposito, D.; Bagni, R.K.; McCormick, F. Dragging Ras Back in the Ring. *Cancer Cell* **2014**, *25*, 272–281. [CrossRef]

39. Dummer, R.; Schadendorf, D.; A Ascierto, P.; Arance, A.; Dutriaux, C.; Di Giacomo, A.M.; Rutkowski, P.; Del Vecchio, M.; Gutzmer, R.; Mandalà, M.; et al. Binimetinib versus dacarbazine in patients with advanced NRAS-mutant melanoma (NEMO): A multicentre, open-label, randomised, phase 3 trial. *Lancet Oncol.* **2017**, *18*, 435–445. [CrossRef]

40. Munoz-Couselo, E.; Adelantado, E.Z.; Vélez, C.O.; García, J.S.; Perez-Garcia, J.M.; Ortiz, C. NRAS-mutant melanoma: Current challenges and future prospect. *OncoTargets Ther.* **2017**, *10*, 3941–3947. [CrossRef]

41. Sarkisian, S.; Davar, D. MEK inhibitors for the treatment of NRAS mutant melanoma. *Drug Des. Dev. Ther.* **2018**, *12*, 2553–2565. [CrossRef] [PubMed]

42. Beadling, C.; Jacobson-Dunlop, E.; Hodi, F.S.; Le, C.; Warrick, A.; Patterson, J.; Town, A.; Harlow, A.; Cruz, F.; Azar, S.; et al. KIT Gene Mutations and Copy Number in Melanoma Subtypes. *Clin. Cancer Res.* **2008**, *14*, 6821–6828. [CrossRef] [PubMed]

43. Handolias, D.; Salemi, R.; Murray, W.; Tan, A.; Liu, W.; Viros, A.; Dobrovic, A.; Kelly, J.; McArthur, G.A. Mutations in KIT occur at low frequency in melanomas arising from anatomical sites associated with chronic and intermittent sun exposure. *Pigment. Cell Melanoma Res.* **2010**, *23*, 210–215. [CrossRef] [PubMed]

44. Goldinger, S.M.; Murer, C.; Stieger, P.; Dummer, R. Targeted therapy in melanoma – the role of BRAF, RAS and KIT mutations. *Eur. J. Cancer Suppl.* **2013**, *11*, 92–96. [CrossRef]

45. ClinicalTrials.gov. Available online: https://clinicaltrials.gov (accessed on 28 September 2020).

46. Meng, D.; Carvajal, R.D. KIT as an Oncogenic Driver in Melanoma: An Update on Clinical Development. *Am. J. Clin. Dermatol.* **2019**, *20*, 315–323. [CrossRef]

47. Lugowska, I.; Teterycz, P.; Rutkowski, P. Immunotherapy of melanoma. *Współczesna Onkol.* **2018**, *22*, 61–67. [CrossRef]

48. Şimşek, M.; Tekin, S.B.; Bilici, M. Immunological Agents Used in Cancer Treatment. *Eurasian J. Med.* **2019**, *51*, 90–94. [CrossRef]

49. Franklin, C.; Livingstone, E.; Roesch, A.; Schilling, B.; Schadendorf, D. Immunotherapy in melanoma: Recent advances and future directions. *Eur. J. Surg. Oncol. (EJSO)* **2017**, *43*, 604–611. [CrossRef]

50. Buchbinder, E.I.; Desai, A. CTLA-4 and PD-1 Pathways. *Am. J. Clin. Oncol.* **2016**, *39*, 98–106. [CrossRef]

51. Schadendorf, D.; Hodi, F.S.; Robert, C.; Weber, J.S.; Margolin, K.; Hamid, O.; Patt, D.; Chen, T.-T.; Berman, D.M.; Wolchok, J.D. Pooled Analysis of Long-Term Survival Data From Phase II and Phase III Trials of Ipilimumab in Unresectable or Metastatic Melanoma. *J. Clin. Oncol.* **2015**, *33*, 1889–1894. [CrossRef]

52. Ramagopal, U.A.; Liu, W.; Garrett-Thomson, S.C.; Bonanno, J.B.; Yan, Q.; Srinivasan, M.; Wong, S.C.; Bell, A.; Mankikar, S.; Rangan, V.S.; et al. Structural basis for cancer immunotherapy by the first-in-class checkpoint inhibitor ipilimumab. *Proc. Natl. Acad. Sci. USA* **2017**, *114*, E4223–E4232. [CrossRef] [PubMed]

53. Eroglu, Z.; Kim, D.W.; Wang, X.; Camacho, L.H.; Chmielowski, B.; Seja, E.; Villanueva, A.; Ruchalski, K.; Glaspy, J.A.; Kim, K.B.; et al. Long term survival with cytotoxic T lymphocyte-associated antigen 4 blockade using tremelimumab. *Eur. J. Cancer* **2015**, *51*, 2689–2697. [CrossRef] [PubMed]

54. Ribas, A. Clinical Development of the Anti–CTLA-4 Antibody Tremelimumab. *Semin. Oncol.* **2010**, *37*, 450–454. [CrossRef] [PubMed]

55. Parry, R.V.; Chemnitz, J.M.; Frauwirth, K.A.; Lanfranco, A.R.; Braunstein, I.; Kobayashi, S.V.; Linsley, P.S.; Thompson, C.B.; Riley, J.L. CTLA-4 and PD-1 Receptors Inhibit T-Cell Activation by Distinct Mechanisms. *Mol. Cell. Biol.* **2005**, *25*, 9543–9553. [CrossRef] [PubMed]

56. Sunshine, J.; Taube, J.M. PD-1/PD-L1 inhibitors. *Curr. Opin. Pharmacol.* **2015**, *23*, 32–38. [CrossRef]

57. Li, Y.; Li, F.; Jiang, F.; Lv, X.; Zhang, R.; Lu, A.; Zhang, G. A Mini-Review for Cancer Immunotherapy: Molecular Understanding of PD-1/PD-L1 Pathway & Translational Blockade of Immune Checkpoints. *Int. J. Mol. Sci.* **2016**, *17*, 1151. [CrossRef]

58. Francisco, L.M.; Sage, P.T.; Sharpe, A.H. The PD-1 pathway in tolerance and autoimmunity. *Immunol. Rev.* **2010**, *236*, 219–242. [CrossRef]

59. Pardoll, D.M. The blockade of immune checkpoints in cancer immunotherapy. *Nat. Rev. Cancer* **2012**, *12*, 252–264. [CrossRef]

60. Hino, R.; Kabashima, K.; Kato, Y.; Yagi, H.; Nakamura, M.; Honjo, T.; Okazaki, T.; Tokura, Y. Tumor cell expression of programmed cell death-1 ligand 1 is a prognostic factor for malignant melanoma. *Cancer* **2010**, *116*, 1757–1766. [CrossRef]

61. Selby, M.J.; Engelhardt, J.J.; Johnston, R.J.; Lu, L.-S.; Han, M.; Thudium, K.; Yao, D.; Quigley, M.; Valle, J.; Wang, C.; et al. Preclinical Development of Ipilimumab and Nivolumab Combination Immunotherapy: Mouse Tumor Models, In Vitro Functional Studies, and Cynomolgus Macaque Toxicology. *PLoS ONE* **2016**, *11*, e0161779. [CrossRef]

62. Curran, M.A.; Montalvo, W.; Yagita, H.; Allison, J.P. PD-1 and CTLA-4 combination blockade expands infiltrating T cells and reduces regulatory T and myeloid cells within B16 melanoma tumors. *Proc. Natl. Acad. Sci. USA* **2010**, *107*, 4275–4280. [CrossRef] [PubMed]

63. Liang, Z.; Li, Y.; Tian, Y.; Zhang, H.; Cai, W.; Chen, A.; Chen, L.; Bao, Y.; Xiang, B.; Kan, H.; et al. High-affinity human programmed death-1 ligand-1 variant promotes redirected T cells to kill tumor cells. *Cancer Lett.* **2019**, *447*, 164–173. [CrossRef] [PubMed]

64. Simeone, E.; Ascierto, P.A. Anti-PD-1 and PD-L1 antibodies in metastatic melanoma. *Melanoma Manag.* **2017**, *4*, 175–178. [CrossRef] [PubMed]

65. Robert, C.; Schachter, J.; Long, G.V.; Arance, A.; Grob, J.J.; Mortier, L.; Daud, A.; Carlino, M.S.; McNeil, C.; Lotem, M.; et al. Pembrolizumab versus Ipilimumab in Advanced Melanoma. *N. Engl. J. Med.* **2015**, *372*, 2521–2532. [CrossRef] [PubMed]

66. Larkin, J.; Chiarion-Sileni, V.; Gonzalez, R.; Grob, J.-J.; Rutkowski, P.; Lao, C.D.; Cowey, C.L.; Schadendorf, D.; Wagstaff, J.; Dummer, R.; et al. Five-Year Survival with Combined Nivolumab and Ipilimumab in Advanced Melanoma. *N. Engl. J. Med.* **2019**, *381*, 1535–1546. [CrossRef]

67. Larkin, J.; Chiarion-Sileni, V.; Gonzalez, R.; Grob, J.J.; Cowey, C.L.; Lao, C.D.; Schadendorf, D.; Dummer, R.; Smylie, M.; Rutkowski, P.; et al. Combined Nivolumab and Ipilimumab or Monotherapy in Untreated Melanoma. *N. Engl. J. Med.* **2015**, *373*, 23–34. [CrossRef]

68. Ribas, A.; Hamid, O.; Daud, A.I.; Hodi, F.S.; Wolchok, J.D.; Kefford, R.F.; Joshua, A.M.; Patnaik, A.; Hwu, W.-J.; Weber, J.S.; et al. Association of Pembrolizumab With Tumor Response and Survival Among Patients With Advanced Melanoma. *JAMA* **2016**, *315*, 1600. [CrossRef]

69. Robert, C.; Ribas, A.; Schachter, J.; Arance, A.; Grob, J.-J.; Mortier, L.; Daud, A.; Carlino, M.S.; McNeil, C.M.; Lotem, M.; et al. Pembrolizumab versus ipilimumab in advanced melanoma (KEYNOTE-006): Post-hoc 5-year results from an open-label, multicentre, randomised, controlled, phase 3 study. *Lancet Oncol.* **2019**, *20*, 1239–1251. [CrossRef]

70. Hamid, O.; Molinero, L.; Bolen, C.R.; Sosman, J.A.; Muñoz-Couselo, E.; Kluger, H.M.; McDermott, D.F.; Powderly, J.D.; Sarkar, I.; Ballinger, M.; et al. Safety, Clinical Activity, and Biological Correlates of Response in Patients with Metastatic Melanoma: Results from a Phase I Trial of Atezolizumab. *Clin. Cancer Res.* **2019**, *25*, 6061–6072. [CrossRef]

71. Gutzmer, R.; Stroyakovskiy, D.; Gogas, H.; Robert, C.; Lewis, K.; Protsenko, S.; Pereira, R.P.; Eigentler, T.; Rutkowski, P.; Demidov, L.; et al. Atezolizumab, vemurafenib, and cobimetinib as first-line treatment for unresectable advanced BRAFV600 mutation-positive melanoma (IMspire150): Primary analysis of the randomised, double-blind, placebo-controlled, phase 3 trial. *Lancet* **2020**, *395*, 1835–1844. [CrossRef]

72. Keilholz, U.; Mehnert, J.M.; Bauer, S.; Bourgeois, H.; Patel, M.R.; Gravenor, D.; Nemunaitis, J.; Taylor, M.; Wyrwicz, L.; Lee, K.-W.; et al. Avelumab in patients with previously treated metastatic melanoma: Phase 1b results from the JAVELIN Solid Tumor trial. *J. Immunother. Cancer* **2019**, *7*, 12. [CrossRef] [PubMed]

73. Saint-Jean, M.; Knol, A.-C.; Volteau, C.; Quéreux, G.; Peuvrel, L.; Brocard, A.; Pandolfino, M.-C.; Saiagh, S.; Nguyen, J.-M.; Bedane, C.; et al. Adoptive Cell Therapy with Tumor-Infiltrating Lymphocytes in Advanced Melanoma Patients. *J. Immunol. Res.* **2018**, *2018*, 1–10. [CrossRef] [PubMed]

74. Simon, B.; Uslu, U. CAR -T cell therapy in melanoma: A future success story? *Exp. Dermatol.* **2018**, *27*, 1315–1321. [CrossRef] [PubMed]

75. Eggermont, A.M.; Blank, C.U.; Mandalà, M.; Long, G.V.; Atkinson, V.; Dalle, S.; Haydon, A.; Lichinitser, M.; Khattak, A.; Carlino, M.S.; et al. Adjuvant Pembrolizumab versus Placebo in Resected Stage III Melanoma. *N. Engl. J. Med.* **2018**, *378*, 1789–1801. [CrossRef]

76. Rosenberg, S.A.; Yang, J.C.; Restifo, N.P. Cancer immunotherapy: Moving beyond current vaccines. *Nat. Med.* **2004**, *10*, 909–915. [CrossRef]

77. Vigneron, N.; Ooms, A.; Morel, S.; Ma, W.; DeGiovanni, G.; Eynde, B.J.V.D. A peptide derived from melanocytic protein gp100 and presented by HLA-B35 is recognized by autologous cytolytic T lymphocytes on melanoma cells. *Tissue Antigens* **2005**, *65*, 156–162. [CrossRef]

78. Schwartzentruber, D.J.; Lawson, D.H.; Richards, J.M.; Conry, R.M.; Miller, D.M.; Treisman, J.; Gailani, F.; Riley, L.; Conlon, K.; Pockaj, B.; et al. gp100 Peptide Vaccine and Interleukin-2 in Patients with Advanced Melanoma. *N. Engl. J. Med.* **2011**, *364*, 2119–2127. [CrossRef]

79. Testori, A.; Richards, J.; Whitman, E.; Mann, G.B.; Lutzky, J.; Camacho, L.H.; Parmiani, G.; Tosti, G.; Kirkwood, J.M.; Hoos, A.; et al. Phase III Comparison of Vitespen, an Autologous Tumor-Derived Heat Shock Protein gp96 Peptide Complex Vaccine, With Physician's Choice of Treatment for Stage IV Melanoma: The C-100-21 Study Group. *J. Clin. Oncol.* **2008**, *26*, 955–962. [CrossRef]

80. Tosti, G.; Di Pietro, A.; Ferrucci, P.F.; Testori, A. HSPPC-96 vaccine in metastatic melanoma patients: From the state of the art to a possible future. *Expert Rev. Vaccines* **2009**, *8*, 1513–1526. [CrossRef]

81. Wu, T.; Dai, Y. Tumor microenvironment and therapeutic response. *Cancer Lett.* **2017**, *387*, 61–68. [CrossRef]

82. Conciatori, F.; Bazzichetto, C.; Falcone, I.; Pilotto, S.; Bria, E.; Cognetti, F.; Milella, M.; Ciuffreda, L. Role of mTOR Signaling in Tumor Microenvironment: An Overview. *Int. J. Mol. Sci.* **2018**, *19*, 2453. [CrossRef] [PubMed]

83. Kubo, N.; Araki, K.; Kuwano, H.; Shirabe, K. Cancer-associated fibroblasts in hepatocellular carcinoma. *World J. Gastroenterol.* **2016**, *22*, 6841–6850. [CrossRef] [PubMed]

84. Yuan, Y.; Jiang, Y.-C.; Sun, C.; Chen, Q. Role of the tumor microenvironment in tumor progression and the clinical applications (Review). *Oncol. Rep.* **2016**, *35*, 2499–2515. [CrossRef] [PubMed]

85. Hu, B.; Wu, Z.; Jin, H.; Hashimoto, N.; Liu, T.; Phan, S.H. CCAAT/Enhancer-Binding Protein β Isoforms and the Regulation of α-Smooth Muscle Actin Gene Expression by IL-1β. *J. Immunol.* **2004**, *173*, 4661–4668. [CrossRef] [PubMed]

86. Shiga, K.; Hara, M.; Nagasaki, T.; Sato, T.; Takahashi, H.; Takeyama, H. Cancer-Associated Fibroblasts: Their Characteristics and Their Roles in Tumor Growth. *Cancers* **2015**, *7*, 2443–2458. [CrossRef]

87. Hutchenreuther, J.; Vincent, K.; Norley, C.; Racanelli, M.; Gruber, S.B.; Johnson, T.M.; Fullen, D.R.; Raskin, L.; Perbal, B.; Holdsworth, D.W.; et al. Activation of cancer-associated fibroblasts is required for tumor neovascularization in a murine model of melanoma. *Matrix Biol.* **2018**, *74*, 52–61. [CrossRef]

88. Cornil, I.; Theodorescu, D.; Man, S.; Herlyn, M.; Jambrosic, J.; Kerbel, R.S. Fibroblast cell interactions with human melanoma cells affect tumor cell growth as a function of tumor progression. *Proc. Natl. Acad. Sci. USA* **1991**, *88*, 6028–6032. [CrossRef]

89. Jobe, N.P.; Rösel, D.; Dvořánková, B.; Kodet, O.; Lacina, L.; Mateu, R.; Smetana, K.; Brábek, J.; Smetana, K. Simultaneous blocking of IL-6 and IL-8 is sufficient to fully inhibit CAF-induced human melanoma cell invasiveness. *Histochem. Cell Biol.* **2016**, *146*, 205–217. [CrossRef]

90. Zhou, L.; Yang, K.; Wickett, R.R.; Kadekaro, A.L.; Zhang, Y. Targeted deactivation of cancer-associated fibroblasts by β-catenin ablation suppresses melanoma growth. *Tumor Biol.* **2016**, *37*, 14235–14248. [CrossRef]

91. Guo, Y.; Zhang, X.; Zeng, W.; Zhang, J.; Cai, L.; Wu, Z.; Su, J.; Xiao, Y.; Liu, N.; Tang, L.; et al. TRAF6 Activates Fibroblasts to Cancer-Associated Fibroblasts through FGF19 in Tumor Microenvironment to Benefit the Malignant Phenotype of Melanoma Cells. *J. Investig. Dermatol.* **2020**. [CrossRef]

92. Straussman, R.; Morikawa, T.; Shee, K.; Barzily-Rokni, M.; Qian, Z.R.; Du, J.; Davis, A.; Mongare, M.M.; Gould, J.; Frederick, D.T.; et al. Tumour micro-environment elicits innate resistance to RAF inhibitors through HGF secretion. *Nat. Cell Biol.* **2012**, *487*, 500–504. [CrossRef] [PubMed]

93. Capparelli, C.; Rosenbaum, S.; Berger, A.C.; Aplin, A.E. Fibroblast-derived Neuregulin 1 Promotes Compensatory ErbB3 Receptor Signaling in Mutant BRAF Melanoma*. *J. Biol. Chem.* **2015**, *290*, 24267–24277. [CrossRef] [PubMed]

94. Fedorenko, I.V.; Wargo, J.A.; Flaherty, K.T.; Messina, J.L.; Smalley, K.S. BRAF Inhibition Generates a Host-Tumor Niche that Mediates Therapeutic Escape. *J. Investig. Dermatol.* **2015**, *135*, 3115–3124. [CrossRef] [PubMed]

95. Hirata, E.; Girotti, M.R.; Viros, A.; Hooper, S.; Spencer-Dene, B.; Matsuda, M.; Larkin, J.; Marais, R.; Sahai, E. Intravital imaging reveals how BRAF inhibition generates drug-tolerant microenvironments with high integrin β1/FAK signaling. *Cancer Cell* **2015**, *27*, 574–588. [CrossRef]

96. Grivennikov, S.I.; Greten, F.R.; Karin, M. Immunity, Inflammation, and Cancer. *Cell* **2010**, *140*, 883–899. [CrossRef]

97. Kaur, A.; Webster, M.R.; Marchbank, K.; Behera, R.; Ndoye, A.; Kugel, C.H.; Dang, V.M.; Appleton, J.; O'Connell, M.P.; Cheng, P.; et al. sFRP2 in the aged microenvironment drives melanoma metastasis and therapy resistance. *Nat. Cell Biol.* **2016**, *532*, 250–254. [CrossRef]

98. Takahashi, H.; Sakakura, K.; Kawabata-Iwakawa, R.; Rokudai, S.; Toyoda, M.; Nishiyama, M.; Chikamatsu, K. Immunosuppressive activity of cancer-associated fibroblasts in head and neck squamous cell carcinoma. *Cancer Immunol. Immunother.* **2015**, *64*, 1407–1417. [CrossRef]

99. Zhang, A.; Qian, Y.; Ye, Z.; Chen, H.; Xie, H.; Zhou, L.; Shen, Y.; Zheng, S. Cancer-associated fibroblasts promote M2 polarization of macrophages in pancreatic ductal adenocarcinoma. *Cancer Med.* **2017**, *6*, 463–470. [CrossRef]

100. Costa, A.; Kieffer, Y.; Scholer-Dahirel, A.; Pelon, F.; Bourachot, B.; Cardon, M.; Sirven, P.; Magagna, I.; Fuhrmann, L.; Bernard, C.; et al. Fibroblast Heterogeneity and Immunosuppressive Environment in Human Breast Cancer. *Cancer Cell* **2018**, *33*, 463–479.e10. [CrossRef]

101. Ziani, L.; Ben Safta-Saadoun, T.; Gourbeix, J.; Cavalcanti, A.; Robert, C.; Favre, G.; Chouaib, S.; Thiery, J. Melanoma-associated fibroblasts decrease tumor cell susceptibility to NK cell-mediated killing through matrix-metalloproteinases secretion. *Oncotarget* **2017**, *8*, 19780–19794. [CrossRef]

102. Li, Z.; Zhou, J.; Zhang, J.; Li, S.; Wang, H.; Du, J. Cancer-associated fibroblasts promote PD-L1 expression in mice cancer cells via secreting CXCL5. *Int. J. Cancer* **2019**, *145*, 1946–1957. [CrossRef] [PubMed]

103. Lee, J.H.; Shklovskaya, E.; Lim, S.Y.; Carlino, M.S.; Menzies, A.M.; Stewart, A.; Pedersen, B.; Irvine, M.; Alavi, S.; Yang, J.; et al. Transcriptional downregulation of MHC class I and melanoma de- differentiation in resistance to PD-1 inhibition. *Nat. Commun.* **2020**, *11*, 1–12. [CrossRef] [PubMed]

104. Li, G.; Satyamoorthy, K.; Herlyn, M. N-cadherin-mediated intercellular interactions promote survival and migration of melanoma cells. *Cancer Res.* **2001**, *61*, 3819–3825.

105. Flach, E.H.; Rebecca, V.W.; Herlyn, M.; Smalley, K.S.; Anderson, A.R. Fibroblasts Contribute to Melanoma Tumor Growth and Drug Resistance. *Mol. Pharm.* **2011**, *8*, 2039–2049. [CrossRef] [PubMed]

106. Tiago, M.; De Oliveira, E.M.; Brohem, C.A.; Pennacchi, P.C.; Paes, R.D.; Haga, R.B.; Campa, A.; Barros, S.B.D.M.; Smalley, K.S.; Maria-Engler, S.S. Fibroblasts Protect Melanoma Cells from the Cytotoxic Effects of Doxorubicin. *Tissue Eng. Part A* **2014**, *20*, 2412–2421. [CrossRef] [PubMed]

107. Singer, A.; Bosselut, R. CD4/CD8 Coreceptors in Thymocyte Development, Selection, and Lineage Commitment: Analysis of the CD4/CD8 Lineage Decision. *Adv. Immunol.* **2004**, *83*, 91–131. [CrossRef]

108. Luckheeram, R.V.; Zhou, R.; Verma, A.D.; Xia, B. CD4+T Cells: Differentiation and Functions. *Clin. Dev. Immunol.* **2012**, *2012*, 1–12. [CrossRef]

109. Bazzichetto, C.; Conciatori, F.; Pallocca, M.; Falcone, I.; Fanciulli, M.; Cognetti, F.; Milella, M.; Ciuffreda, L. PTEN as a Prognostic/Predictive Biomarker in Cancer: An Unfulfilled Promise? *Cancers* **2019**, *11*, 435. [CrossRef]

110. Cetintas, V.B.; Batada, N.N. Is there a causal link between PTEN deficient tumors and immunosuppressive tumor microenvironment? *J. Transl. Med.* **2020**, *18*, 45. [CrossRef]

111. Dong, Y.; Richards, J.-A.; Gupta, R.; Aung, P.P.; Emley, A.; Kluger, Y.; Dogra, S.K.; Mahalingam, M.; Wajapeyee, N. PTEN functions as a melanoma tumor suppressor by promoting host immune response. *Oncogene* **2013**, *33*, 4632–4642. [CrossRef]

112. Peng, W.; Chen, J.Q.; Liu, C.; Malu, S.; Creasy, C.; Tetzlaff, M.T.; Xu, C.; McKenzie, J.A.; Zhang, C.; Liang, X.; et al. Loss of PTEN Promotes Resistance to T Cell-Mediated Immunotherapy. *Cancer Discov.* **2015**, *6*, 202–216. [CrossRef] [PubMed]

113. Xia, A.; Zhang, Y.; Xu, J.; Yin, T.; Lu, X.-J. T Cell Dysfunction in Cancer Immunity and Immunotherapy. *Front. Immunol.* **2019**, *10*, 1719. [CrossRef] [PubMed]

114. Fourcade, J.; Sun, Z.; Benallaoua, M.; Guillaume, P.; Luescher, I.F.; Sander, C.; Kirkwood, J.M.; Kuchroo, V.; Zarour, H.M. Upregulation of Tim-3 and PD-1 expression is associated with tumor antigen–specific CD8+ T cell dysfunction in melanoma patients. *J. Exp. Med.* **2010**, *207*, 2175–2186. [CrossRef]

115. Chauvin, J.-M.; Pagliano, O.; Fourcade, J.; Sun, Z.; Wang, H.; Sander, C.; Kirkwood, J.M.; Chen, T.-H.T.; Maurer, M.; Korman, A.J.; et al. TIGIT and PD-1 impair tumor antigen-specific CD8$^+$ T cells in melanoma patients. *J. Clin. Investig.* **2015**, *125*, 2046–2058. [CrossRef] [PubMed]

116. Li, H.; Van Der Leun, A.M.; Yofe, I.; Lubling, Y.; Gelbard-Solodkin, D.; Van Akkooi, A.C.; Braber, M.V.D.; Rozeman, E.A.; Haanen, J.B.; Blank, C.U.; et al. Dysfunctional CD8 T Cells Form a Proliferative, Dynamically Regulated Compartment within Human Melanoma. *Cell* **2019**, *176*, 775–789.e18. [CrossRef]

117. Marshall, J.S.; Warrington, R.; Watson, W.; Kim, H.L. An introduction to immunology and immunopathology. *Allergy, Asthma Clin. Immunol.* **2018**, *14*, 49. [CrossRef]

118. Somasundaram, R.; Zhang, G.; Fukunaga-Kalabis, M.; Perego, M.; Krepler, C.; Xu, X.; Wagner, C.; Hristova, D.; Zhang, J.; Tian, T.; et al. Tumor-associated B-cells induce tumor heterogeneity and therapy resistance. *Nat. Commun.* **2017**, *8*, 607. [CrossRef]

119. Griss, J.; Bauer, W.; Wagner, C.; Simon, M.; Chen, M.; Grabmeier-Pfistershammer, K.; Maurer-Granofszky, M.; Roka, F.; Penz, T.; Bock, C.; et al. B cells sustain inflammation and predict response to immune checkpoint blockade in human melanoma. *Nat. Commun.* **2019**, *10*, 1–14. [CrossRef]

120. Cabrita, R.; Lauss, M.; Sanna, A.; Donia, M.; Larsen, M.S.; Mitra, S.; Johansson, I.; Phung, B.; Harbst, K.; Vallon-Christersson, J.; et al. Tertiary lymphoid structures improve immunotherapy and survival in melanoma. *Nature* **2020**, *577*, 561–565. [CrossRef]

121. Amaria, R.N.; Reddy, S.M.; Tawbi, H.A.; Davies, M.A.; Ross, M.I.; Glitza, I.C.; Cormier, J.N.; Lewis, C.; Hwu, W.-J.; Hanna, E.; et al. Neoadjuvant immune checkpoint blockade in high-risk resectable melanoma. *Nat. Med.* **2018**, *24*, 1649–1654. [CrossRef]

122. Helmink, B.A.; Reddy, S.M.; Gao, J.; Zhang, S.; Basar, R.; Thakur, R.; Yizhak, K.; Sade-Feldman, M.; Blando, J.; Han, G.; et al. B cells and tertiary lymphoid structures promote immunotherapy response. *Nat. Cell Biol.* **2020**, *577*, 549–555. [CrossRef] [PubMed]

123. Pahl, J.H.; Cerwenka, A. Tricking the balance: NK cells in anti-cancer immunity. *Immunobiol.* **2017**, *222*, 11–20. [CrossRef] [PubMed]

124. Freud, A.G.; Mundy-Bosse, B.L.; Yu, J.; Caligiuri, M.A. The Broad Spectrum of Human Natural Killer Cell Diversity. *Immunity* **2017**, *47*, 820–833. [CrossRef]

125. Cristiani, C.M.; Garofalo, C.; Passacatini, L.C.; Carbone, E. New avenues for melanoma immunotherapy: Natural Killer cells? *Scand. J. Immunol.* **2020**, *91*, e12861. [CrossRef] [PubMed]

126. López-Cobo, S.; Pieper, N.; Campos-Silva, C.; García-Cuesta, E.M.; Reyburn, H.T.; Paschen, A.; Valés-Gómez, M. Impaired NK cell recognition of vemurafenib-treated melanoma cells is overcome by simultaneous application of histone deacetylase inhibitors. *OncoImmunology* **2017**, *7*, e1392426. [CrossRef]

127. Kondělková, K.; Vokurková, D.; Krejsek, J.; Borska, L.; Fiala, Z.; Andrys, C. Regulatory T cells (Treg) and Their Roles in Immune System with Respect to Immunopathological Disorders. *Acta Medica (Hradec Kralove, Czech Republic)* **2010**, *53*, 73–77. [CrossRef]

128. Chaudhary, B.; Elkord, E. Regulatory T Cells in the Tumor Microenvironment and Cancer Progression: Role and Therapeutic Targeting. *Vaccines* **2016**, *4*, 28. [CrossRef]

129. Ascierto, P.A.; Napolitano, M.; Celentano, E.; Simeone, E.; Gentilcore, G.; Daponte, A.; Capone, M.; Caracò, C.; Calemma, R.; Beneduce, G.; et al. Regulatory T cell frequency in patients with melanoma with different disease stage and course, and modulating effects of high-dose interferon-α 2b treatment. *J. Transl. Med.* **2010**, *8*, 76. [CrossRef]

130. Shang, B.; Liu, Y.; Jiang, S.-J.; Liu, Y. Prognostic value of tumor-infiltrating FoxP3+ regulatory T cells in cancers: A systematic review and meta-analysis. *Sci. Rep.* **2015**, *5*, srep15179. [CrossRef]

131. Leslie, C.; Bowyer, S.E.; White, A.; Grieu-Iacopetta, F.; Trevenen, M.; Iacopetta, B.; Amanuel, B.; Millward, M. FOXP3+ T regulatory lymphocytes in primary melanoma are associated with BRAF mutation but not with response to BRAF inhibitor. *Pathology* **2015**, *47*, 557–563. [CrossRef]

132. Baumgartner, J.; Wilson, C.; Palmer, B.; Richter, D.; Banerjee, A.; McCarter, M. Melanoma Induces Immunosuppression by Up-Regulating FOXP3+ Regulatory T Cells. *J. Surg. Res.* **2007**, *141*, 72–77. [CrossRef] [PubMed]

133. Sumimoto, H.; Imabayashi, F.; Iwata, T.; Kawakami, Y. The BRAF–MAPK signaling pathway is essential for cancer-immune evasion in human melanoma cells. *J. Exp. Med.* **2006**, *203*, 1651–1656. [CrossRef] [PubMed]

134. Zappasodi, R.; Budhu, S.; Hellmann, M.D.; Postow, M.A.; Senbabaoglu, Y.; Manne, S.; Gasmi, B.; Liu, C.; Zhong, H.; Li, Y.; et al. Non-conventional Inhibitory CD4+Foxp3−PD-1hi T Cells as a Biomarker of Immune Checkpoint Blockade Activity. *Cancer Cell* **2018**, *33*, 1017–1032.e7. [CrossRef] [PubMed]

135. Zhang, B.; Chikuma, S.; Hori, S.; Fagarasan, S.; Honjo, T. Nonoverlapping roles of PD-1 and FoxP3 in maintaining immune tolerance in a novel autoimmune pancreatitis mouse model. *Proc. Natl. Acad. Sci. USA* **2016**, *113*, 8490–8495. [CrossRef]

136. Gianchecchi, E.; Fierabracci, A. Inhibitory Receptors and Pathways of Lymphocytes: The Role of PD-1 in Treg Development and Their Involvement in Autoimmunity Onset and Cancer Progression. *Front. Immunol.* **2018**, *9*, 2374. [CrossRef]

137. Kamada, T.; Togashi, Y.; Tay, C.; Ha, D.; Sasaki, A.; Nakamura, Y.; Sato, E.; Fukuoka, S.; Tada, Y.; Tanaka, A.; et al. PD-1+ regulatory T cells amplified by PD-1 blockade promote hyperprogression of cancer. *Proc. Natl. Acad. Sci. USA* **2019**, *116*, 9999–10008. [CrossRef]

138. Simeone, E.; Gentilcore, G.; Giannarelli, D.; Grimaldi, A.M.; Caracò, C.; Curvietto, M.; Esposito, A.; Paone, M.; Palla, M.; Cavalcanti, E.; et al. Immunological and biological changes during ipilimumab treatment and their potential correlation with clinical response and survival in patients with advanced melanoma. *Cancer Immunol. Immunother.* **2014**, *63*, 675–683. [CrossRef]

139. Simpson, T.R.; Li, F.; Montalvo-Ortiz, W.; Sepulveda, M.A.; Bergerhoff, K.; Arce, F.; Roddie, C.; Henry, J.Y.; Yagita, H.; Wolchok, J.D.; et al. Fc-dependent depletion of tumor-infiltrating regulatory T cells co-defines the efficacy of anti–CTLA-4 therapy against melanoma. *J. Exp. Med.* **2013**, *210*, 1695–1710. [CrossRef]

140. Solito, S.; Marigo, I.; Pinton, L.; Damuzzo, V.; Mandruzzato, S.; Bronte, V. Myeloid-derived suppressor cell heterogeneity in human cancers. *Ann. N. Y. Acad. Sci.* **2014**, *1319*, 47–65. [CrossRef]

141. Gabrilovich, D.I. Myeloid-Derived Suppressor Cells. *Cancer Immunol. Res.* **2017**, *5*, 3–8. [CrossRef]

142. Umansky, V.; Sevko, A.; Gebhardt, C.; Utikal, J. Myeloid-derived suppressor cells in malignant melanoma. *J. Dtsch. Dermatol. Ges.* **2014**, *12*, 1021–1027. [CrossRef] [PubMed]

143. Huber, V.; Vallacchi, V.; Fleming, V.; Hu, X.; Cova, A.; Dugo, M.; Shahaj, E.; Sulsenti, R.; Vergani, E.; Filipazzi, P.; et al. Tumor-derived microRNAs induce myeloid suppressor cells and predict immunotherapy resistance in melanoma. *J. Clin. Investig.* **2018**, *128*, 5505–5516. [CrossRef] [PubMed]

144. Gebhardt, C.; Sevko, A.; Jiang, H.; Lichtenberger, R.; Reith, M.; Tarnanidis, K.; Holland-Letz, T.; Umansky, L.; Beckhove, P.; Sucker, A.; et al. Myeloid Cells and Related Chronic Inflammatory Factors as Novel Predictive Markers in Melanoma Treatment with Ipilimumab. *Clin. Cancer Res.* **2015**, *21*, 5453–5459. [CrossRef] [PubMed]

145. Steinberg, S.M.; Shabaneh, T.B.; Zhang, P.; Martyanov, V.; Li, Z.; Malik, B.T.; Wood, T.A.; Boni, A.; Molodtsov, A.; Angeles, C.V.; et al. Myeloid Cells That Impair Immunotherapy Are Restored in Melanomas with Acquired Resistance to BRAF Inhibitors. *Cancer Res.* **2017**, *77*, 1599–1610. [CrossRef] [PubMed]

146. Yuan, A.; Chen, J.J.; Yang, P. Pathophysiology of Tumor-Associated Macrophages. *Adv. Clin. Chem.* **2008**, *45*, 199–223. [CrossRef]

147. Yahaya, M.A.F.; Lila, M.A.M.; Ismail, S.; Zainol, M.; Afizan, N.A.R.N.M. Tumour-Associated Macrophages (TAMs) in Colon Cancer and How to Reeducate Them. *J. Immunol. Res.* **2019**, *2019*, 1–9. [CrossRef]

148. Donzelli, S.; Milano, E.; Pruszko, M.; Sacconi, A.; Masciarelli, S.; Iosue, I.; Melucci, E.; Gallo, E.; Terrenato, I.; Mottolese, M.; et al. Expression of ID4 protein in breast cancer cells induces reprogramming of tumour-associated macrophages. *Breast Cancer Res.* **2018**, *20*, 59. [CrossRef]

149. Liu, H.; Yang, L.; Qi, M.; Zhang, J. NFAT1 enhances the effects of tumor-associated macrophages on promoting malignant melanoma growth and metastasis. *Biosci. Rep.* **2018**, *38*, 38. [CrossRef]

150. Wanderley, C.W.; Colón, D.; Luiz, J.P.M.; Oliveira, F.F.; Viacava, P.R.; A Leite, C.; A Pereira, J.; Silva, C.M.; Silva, C.R.; Silva, R.L.; et al. Paclitaxel reduces tumor growth by reprogramming tumor-associated macrophages to an M1- profile in a TLR4-dependent manner. *Cancer Res.* **2018**, *78*, 5891–5900. [CrossRef]

151. Gerloff, D.; Lützkendorf, J.; Moritz, R.K.; Wersig, T.; Mäder, K.; Müller, L.P.; Sunderkötter, C. Melanoma-Derived Exosomal miR-125b-5p Educates Tumor Associated Macrophages (TAMs) by Targeting Lysosomal Acid Lipase A (LIPA). *Cancers* **2020**, *12*, 464. [CrossRef]

152. Smith, M.P.; Sanchez-Laorden, B.; O'Brien, K.; Brunton, H.; Ferguson, J.; Young, H.L.; Dhomen, N.; Flaherty, K.T.; Frederick, D.T.; Cooper, Z.A.; et al. The immune microenvironment confers resistance to MAPK pathway inhibitors through macrophage-derived TNFα. *Cancer Discov.* **2014**, *4*, 1214–1229. [CrossRef] [PubMed]

153. Wang, T.; Xiao, M.; Ge, Y.; Krepler, C.; Belser, E.; Coral, A.L.; Xu, X.; Zhang, G.; Azuma, R.; Liu, Q.; et al. BRAF Inhibition Stimulates Melanoma-Associated Macrophages to Drive Tumor Growth. *Clin. Cancer Res.* **2015**, *21*, 1652–1664. [CrossRef] [PubMed]

154. Gordon, S.R.; Maute, R.L.; Dulken, B.W.; Hutter, G.; George, B.M.; McCracken, M.N.; Gupta, R.; Tsai, J.M.; Sinha, R.; Corey, D.; et al. PD-1 expression by tumour-associated macrophages inhibits phagocytosis and tumour immunity. *Nat. Cell Biol.* **2017**, *545*, 495–499. [CrossRef] [PubMed]

155. Kuklinski, L.F.; Yan, S.; Li, Z.; Fisher, J.L.; Cheng, C.; Noelle, R.J.; Angeles, C.V.; Turk, M.J.; Ernstoff, M.S. VISTA expression on tumor-infiltrating inflammatory cells in primary cutaneous melanoma correlates with poor disease-specific survival. *Cancer Immunol. Immunother.* **2018**, *67*, 1113–1121. [CrossRef]

156. Lines, J.L.; Sempere, L.F.; Wang, L.; Pantazi, E.; Mak, J.; O'Connell, S.; Ceeraz, S.; Suriawinata, A.A.; Yan, S.; Ernstoff, M.S.; et al. VISTA is an immune checkpoint molecule for human T cells. *Cancer Res.* **2014**, *74*, 1924–1932. [CrossRef]

157. Kakavand, H.; A Jackett, L.; Menzies, A.M.; Gide, T.N.; Carlino, M.S.; Saw, R.P.M.; Thompson, J.F.; Wilmott, J.S.; Long, G.V.; Scolyer, R.A. Negative immune checkpoint regulation by VISTA: A mechanism of acquired resistance to anti-PD-1 therapy in metastatic melanoma patients. *Mod. Pathol.* **2017**, *30*, 1666–1676. [CrossRef]

158. Rosenbaum, S.R.; Knecht, M.; Mollaee, M.; Zhong, Z.; Erkes, D.A.; McCue, P.A.; Chervoneva, I.; Berger, A.C.; Lo, J.A.; Fisher, D.E.; et al. FOXD3 Regulates VISTA Expression in Melanoma. *Cell Rep.* **2020**, *30*, 510–524.e6. [CrossRef]

159. Zhang, Y.; Wu, L.; Li, Z.; Zhang, W.; Luo, F.; Chu, Y.; Chen, G. Glycocalyx-Mimicking Nanoparticles Improve Anti-PD-L1 Cancer Immunotherapy through Reversion of Tumor-Associated Macrophages. *Biomacromolecules* **2018**, *19*, 2098–2108. [CrossRef]

160. Peranzoni, E.; Lemoine, J.; Vimeux, L.; Feuillet, V.; Barrin, S.; Kantari-Mimoun, C.; Bercovici, N.; Guérin, M.; Biton, J.; Ouakrim, H.; et al. Macrophages impede CD8 T cells from reaching tumor cells and limit the efficacy of anti–PD-1 treatment. *Proc. Natl. Acad. Sci. USA* **2018**, *115*, E4041–E4050. [CrossRef]

161. Klarquist, J.S.; Janssen, E.M. Melanoma-infiltrating dendritic cells. *OncoImmunology* **2012**, *1*, 1584–1593. [CrossRef]

162. Álvarez-Domínguez, C.; Calderón-González, R.; Terán-Navarro, H.; Salcines-Cuevas, D.; García-Castaño, A.; Freire, J.; Gómez-Román, J.; Rivera, F. Dendritic cell therapy in melanoma. *Ann. Transl. Med.* **2017**, *5*, 386. [CrossRef] [PubMed]

163. Wu, L.; Dakic, A. Development of dendritic cell system. *Cell. Mol. Immunol.* **2004**, *1*, 112–118. [PubMed]

164. Passarelli, A.; Mannavola, F.; Stucci, L.S.; Tucci, M.; Silvestris, F. Immune system and melanoma biology: A balance between immunosurveillance and immune escape. *Oncotarget* **2017**, *8*, 106132–106142. [CrossRef] [PubMed]

165. González, M.L.; Oosterhoff, D.; Lindenberg, J.J.; Milenova, I.; Lougheed, S.M.; Martiáñez, T.; Dekker, H.; Quixabeira, D.C.A.; Hangalapura, B.; Joore, J.; et al. Constitutively active GSK3β as a means to bolster dendritic cell functionality in the face of tumour-mediated immune suppression. *OncoImmunology* **2019**, *8*, e1631119-18. [CrossRef]

166. Van De Ven, R.; Lindenberg, J.J.; Oosterhoff, D.; De Gruijl, T.D. Dendritic Cell Plasticity in Tumor-Conditioned Skin: CD14+ Cells at the Cross-Roads of Immune Activation and Suppression. *Front. Immunol.* **2013**, *4*, 403. [CrossRef]

167. González, M.L.; Van De Ven, R.; De Haan, H.; Sluijs, J.V.E.V.D.; Dong, W.; Van Beusechem, V.W.; De Gruijl, T.D. Oncolytic adenovirus ORCA-010 increases the type 1 T cell stimulatory capacity of melanoma-conditioned dendritic cells. *Clin. Exp. Immunol.* **2020**, *201*, 145–160. [CrossRef]

168. Zhou, Y.; Slone, N.; Chrisikos, T.T.; Kyrysyuk, O.; Babcock, R.L.; Medik, Y.B.; Li, H.S.; Kleinerman, E.S.; Watowich, S.S. Vaccine efficacy against primary and metastatic cancer with in vitro-generated CD103+conventional dendritic cells. *J. Immunother. Cancer* **2020**, *8*, e000474. [CrossRef]

169. Chu, C.-L.; Lee, Y.-P.; Pang, C.-Y.; Lin, H.-R.; Chen, C.-S.; Wen-Sheng, W. Tyrosine kinase inhibitors modulate dendritic cell activity via confining c-Kit signaling and tryptophan metabolism. *Int. Immunopharmacol.* **2020**, *82*, 106357. [CrossRef]

170. Riegel, K.; Schlöder, J.; Sobczak, M.; Jonuleit, H.; Thiede, B.; Schild, H.; Rajalingam, K. RAF kinases are stabilized and required for dendritic cell differentiation and function. *Cell Death Differ.* **2019**, *27*, 1300–1315. [CrossRef]

171. Botti, G.; Cerrone, M.; Scognamiglio, G.; Anniciello, A.; Ascierto, P.A.; Cantile, M. Microenvironment and tumor progression of melanoma: New therapeutic prospectives. *J. Immunotoxicol.* **2012**, *10*, 235–252. [CrossRef]

172. You, D.; Jung, S.P.; Jeong, Y.; Bae, S.Y.; Lee, J.E.; Kim, S. Fibronectin expression is upregulated by PI-3K/Akt activation in tamoxifen-resistant breast cancer cells. *BMB Rep.* **2017**, *50*, 615–620. [CrossRef] [PubMed]

173. Erdogan, B.; Ao, M.; White, L.M.; Means, A.L.; Brewer, B.M.; Yang, L.; Washington, M.K.; Shi, C.; Franco, O.E.; Weaver, A.M.; et al. Cancer-associated fibroblasts promote directional cancer cell migration by aligning fibronectin. *J. Cell Biol.* **2017**, *216*, 3799–3816. [CrossRef] [PubMed]

174. Fedorenko, I.V.; Abel, E.V.; Koomen, J.M.; Fang, B.; Wood, E.R.; Chen, Y.A.; Fisher, K.J.; Iyengar, S.; Dahlman, K.B.; Wargo, J.A.; et al. Fibronectin induction abrogates the BRAF inhibitor response of BRAF V600E/PTEN-null melanoma cells. *Oncogene* **2015**, *35*, 1225–1235. [CrossRef] [PubMed]

175. Takada, Y.; Ye, X.; Simon, S. The integrins. *Genome Biol.* **2007**, *8*, 1–9. [CrossRef]

176. Huang, R.; Rofstad, E.K. Integrins as therapeutic targets in the organ-specific metastasis of human malignant melanoma. *J. Exp. Clin. Cancer Res.* **2018**, *37*, 92. [CrossRef]

177. Jang, I.; Beningo, K.A. Integrins, CAFs and Mechanical Forces in the Progression of Cancer. *Cancers* **2019**, *11*, 721. [CrossRef]

178. Felding-Habermann, B.; Ruggeri, Z.M.; A Cheresh, D. Distinct biological consequences of integrin alpha v beta 3-mediated melanoma cell adhesion to fibrinogen and its plasmic fragments. *J. Biol. Chem.* **1992**, *267*, 5070–5077.

179. Vannini, A.; Leoni, V.; Barboni, C.; Sanapo, M.; Zaghini, A.; Malatesta, P.; Campadelli-Fiume, G.; Gianni, T. αvβ3-integrin regulates PD-L1 expression and is involved in cancer immune evasion. *Proc. Natl. Acad. Sci. USA* **2019**, *116*, 20141–20150. [CrossRef]

180. Hofmann, U.B.; Westphal, J.R.; Van Muijen, G.N.; Ruiter, D.J. Matrix Metalloproteinases in Human Melanoma. *J. Investig. Dermatol.* **2000**, *115*, 337–344. [CrossRef]

181. Sandri, S.; Faião-Flores, F.; Tiago, M.; Pennacchi, P.C.; Massaro, R.R.; Alves-Fernandes, D.K.; Berardinelli, G.N.; Evangelista, A.F.; Vazquez, V.D.L.; Reis, R.M.; et al. Vemurafenib resistance increases melanoma invasiveness and modulates the tumor microenvironment by MMP-2 upregulation. *Pharmacol. Res.* **2016**, *111*, 523–533. [CrossRef]

Review

Oncogenic Tyrosine Phosphatases: Novel Therapeutic Targets for Melanoma Treatment

Elisa Pardella [1,†], Erica Pranzini [1,†], Angela Leo [1], Maria Letizia Taddei [2], Paolo Paoli [1,*] and Giovanni Raugei [1]

1. Department of Experimental and Clinical Biomedical Sciences "Mario Serio" University of Florence, Viale Morgagni 50, 50134 Florence, Italy; elisa.pardella@student.unisi.it (E.P.); erica.pranzini@unifi.it (E.P.); angela.leo@student.unisi.it (A.L.); giovanni.raugei@unifi.it (G.R.)
2. Department of Experimental and Clinical Medicine, University of Florence, Viale Morgagni 50, 50134 Florence, Italy; marialetizia.taddei@unifi.it
* Correspondence: paolo.paoli@unifi.it; Tel.: +39-055-275-1248
† These authors equally contributed to the work.

Received: 28 August 2020; Accepted: 28 September 2020; Published: 29 September 2020

Simple Summary: Targeting oncogenic protein tyrosine phosphatases (PTPs) with specific pharmacological approaches has been considered for a long time a hard challenge, earning these PTPs the reputation of "undruggable" enzymes. Nevertheless, PTPs have been recognized as main targets for several diseases, including cancer, and great efforts have been made to identify novel PTPs inhibitors to fight cancer progression and metastasis formation. Here, we summarize recent evidence underlining the efficacy of this strategy for melanoma treatment. In particular, we illustrate how this approach could be applied to target both cancer cells and the immune infiltrate of tumors, providing a new promising adjuvant therapy for the treatment of melanoma.

Abstract: Despite a large number of therapeutic options available, malignant melanoma remains a highly fatal disease, especially in its metastatic forms. The oncogenic role of protein tyrosine phosphatases (PTPs) is becoming increasingly clear, paving the way for novel antitumor treatments based on their inhibition. In this review, we present the oncogenic PTPs contributing to melanoma progression and we provide, where available, a description of new inhibitory strategies designed against these enzymes and possibly useful in melanoma treatment. Considering the relevance of the immune infiltrate in supporting melanoma progression, we also focus on the role of PTPs in modulating immune cell activity, identifying interesting therapeutic options that may support the currently applied immunomodulating approaches. Collectively, this information highlights the value of going further in the development of new strategies targeting oncogenic PTPs to improve the efficacy of melanoma treatment.

Keywords: melanoma; protein tyrosine phosphatase; PTPs inhibitors; melanoma immune infiltrate

---

1. Introduction

Reversible tyrosine phosphorylation is one of the most important post-translational modifications, which regulates key aspects of cellular biology, such as protein stability, protein–protein interactions, and enzyme activity [1], thereby modulating the functionality of fundamental elements involved in signaling transduction of mammalian cells [2]. The intracellular tyrosine phosphorylation level is maintained by a strict balance between the activities of protein tyrosine kinases (PTKs) and protein tyrosine phosphatases (PTPs), which catalyze respectively the addition or the removal of phosphate from tyrosyl residues of their substrates [3]. An imbalance in this regulation is a well-recognized feature of several diseases, including cancer, since it induces alterations in cell growth and survival, cell migration, and tissue differentiation [4,5]. While the role of PTKs as oncogenic proteins has been largely described, allowing the development of a wide range of inhibitors already accepted in clinical use [6,7], an effective strategy to modulate PTPs for cancer therapy has yet to be identified [8].

PTPs are a large family of proteins consisting of 107 members that can be classified into four families (class I, II, III, and IV) according to the amino acid sequence at the catalytic domains [9]. Originally, PTPs were described to exclusively have a role as tumor suppressors, counteracting the activity of PTKs. However, further studies brought to light that some PTPs can also function as oncogenes depending on the availability of their functional partners and tumor type [5]. Such flexibility in functions can be explained considering that, differently from PTKs, PTPs can act both as negative and positive regulators of signal transduction pathways and can either activate or inhibit the oncogenic role of PTKs. Indeed, PTPs can exert their function by directly dephosphorylating PTKs, or, indirectly, by interfering with their downstream targets [10].

Starting from these bases, a deeper understanding of the dual role of PTPs in affecting tumor progression could lead to the development of new therapeutic strategies aimed at targeting different classes of tumors [11].

Despite the increasing evidence demonstrating the contribution of oncogenic PTPs in supporting tumor progression, the number of inhibitors available to date is still extremely limited. Primarily, this is due to the fact that PTPs have been considered for a long time as "undruggable targets", delaying the design of pharmacological inhibitors [12–14]. In particular, the nature of PTP active sites represents a big challenge for the development of specific inhibitors active inside the cells [14,15]. First of all, in order to target the highly positively charged PTP active site, it would be necessary to develop negatively charged molecules, a characteristic that unfortunately strongly limits their cell permeability and bioavailability. However, promising prospects have come from the recent discovery of nonhydrolyzable, polar, and cell-permeable pTyr mimetics that gave a new chance to overcome these problems [12]. Moreover, designing specific PTP inhibitors is further complicated by the excellent conservation of the amino acid sequence inside the active site among PTPs [13]. Remarkably, although crystal structures of different members of the class-I PTPs revealed a common Cα-backbone signature, the surfaces around the PTPs' catalytic site are characterized by distinct properties, including different topology, electrostatic potential, and lipophilic features, that can be addressed when designing novel selective inhibitors [16–19]. Furthermore, the identification of allosteric inhibitors is crucial to avoid targeting of the charged and conserved PTP active site, allowing the development of cell-permeable and selective compounds [20–22]. Another central obstacle is the presence of a shallow pocket at the PTPs' catalytic site. This issue could be solved by proposing bidentate inhibitors, targeting both the active site and proximal non-conserved binding sites that are present in most of the PTPs [23,24].

Finally, another major challenge that has emerged during the screening of PTP inhibitors is the susceptibility of these enzymes to radical oxygen species (ROS), a phenomenon that in the past has often led to erroneous conclusions during investigations about possible inhibitors. Indeed, it is known that several anticancer drugs increase ROS production through redox cycling or the inhibition of ROS scavenger enzymes [25]. In analogy to other cysteine-based enzymes, the activity of PTPs is highly susceptible to oxidation of catalytic cysteine residues, which leads to the complete inactivation of these enzymes [26]. This evidence highlighted that, in order to avoid an erroneous interpretation of the mechanism of action of PTP inhibitors, their ability to generate ROS should always be evaluated both in vitro and in vivo. As far as this topic is concerned, the studies conducted in the past to analyze the mechanism of action of different types of cysteine protease inhibitors could be kept in mind as pivotal examples [27–29].

Noteworthy, some of the PTPs recognized to have oncogenic properties have been found to be overexpressed in highly metastatic melanoma [30–35], providing the opportunity to develop new strategies to fight this disease.

Malignant melanoma is an aggressive form of cutaneous neoplasia that derives from a series of alterations occurring in melanocytes, the melanin-producing cells resident in the basal layer of the epidermidis [36]. Melanoma retains the highest mortality rate among skin cancers and the highest potential of dissemination [37]. The majority of melanoma patients develop the cutaneous form of the disease, while non-cutaneous melanomas (which include tumors of the ocular and mucosal sites, such as anorectal, vaginal, nasal, and gastrointestinal tract) are relatively rare [38]. Both classes of melanoma are considered as a multi-factorial disease, whose pathogenesis is affected by environmental and genetic factors. However, the differential incidence of genetic alterations among melanoma subtypes and the unequal exposure to UV radiation, according to the anatomic site, strongly influence the molecular pathways involved in tumorigenesis, ultimately leading to the need for specific therapies against the different subtypes [39].

Data published in the Cancer Genome Atlas (TCGA) Network classify cutaneous melanomas into four genetic subgroups on the basis of the most frequently mutated genes involved in the mitogen-activated protein kinase (MAPK) pathway: BRAF, RAS (N-H-K), NF1, and triple wild-type (WT) melanomas [40,41]. Mutations in BRAF and NRAS are most commonly detected in primary cutaneous melanomas [42]. In particular, BRAF is mutated in about 50% of cutaneous melanomas, and among these, in 80–90% of the cases, the missense activating mutation V600E is present. Besides, NRAS mutations occur in about 20–25% of melanomas [43,44]. Following BRAF and NRAS, NF1 is the third gene most commonly mutated in cutaneous melanoma (in about 17% of the cases) and frequently co-occurs with mutations in the RASA2 gene [41,45,46]. NF1 mostly displays point mutations, which determine a loss of function with consequent constitutive activation of the MAPK and phosphoinositide 3-kinase (PI3K) pathways [47]. Interestingly, all these mutations finally result in a constitutive activation of MAPK/ERK signaling, which indeed is present in 98% of melanomas, promoting cellular proliferation, survival, and angiogenesis [48,49]. Finally, the loss of function of PTEN is observed in about 10–35% of cutaneous melanomas, where it confers resistance to BRAF inhibitors. This mutation results in a constitutive activation of the PI3K/AKT pathways, which, in turn, leads to cell growth and proliferation and to the inhibition of apoptosis [50].

Conversely, non-cutaneous melanomas have significantly lower numbers of mutations: Acral melanomas have, in about 15–20% of the cases, mutations in BRAF, NRAS, and KIT [51]; mucosal melanomas display KIT mutations in about 15% of the cases (primarily in genitourinary or anal forms) but rarely present mutations in BRAF and NRAS [52]; and uveal melanomas have distinct genomic patterns, presenting mutations either in GNAQ or GNA11 in > 90% of the cases while BAP1, SF3B1, and EIFAX are distinct subsets [53–55].

The involvement of protumoral PTPs in the oncogenic signaling pathways that characterize malignant melanoma may pave the way for new possible combination therapies based on pharmacological inhibition of oncogenic PTPs. Indeed, this approach could provide longer lasting therapeutic benefits through the inhibition of multiple nodes in the main oncogenic signaling pathways. In agreement with this hypothesis, Prahallad and co-authors found that the suppression of Src homology region 2 domain-containing phosphatase-2 (SHP-2) in BRAF mutant and in Vemurafenib-sensitive melanoma cells inhibits growth factor-induced drug resistance and delays the onset of spontaneous resistance [56]. Moreover, they identified the activating phosphorylation site on Tyr542 of SHP-2 as a valid biomarker to recognize patients with melanoma who have acquired Vemurafenib resistance due to receptor tyrosine kinases (RTKs) activation [56]. Moreover, due to the central role of SHP-2 in mutant KRAS-driven carcinogenesis, it has been demonstrated that the synergic inhibition of SHP-2 and MAPK/ERK kinase (MEK) results in decreased tumor growth in xenograft models of pancreatic ductal adenocarcinoma and non-small cell lung cancer, sustaining the utility of the dual SHP-2/MEK inhibition in KRAS mutant cancers [57].

The overall survival of patients diagnosed with advanced melanoma has strongly increased over recent years, thanks to the latest development of therapeutic strategies [58]. However, recurrence frequently occurs due to therapy failure leading to metastasis formation, representing the main cause of patient death [59]. For this reason, many efforts have been made to design new therapeutic approaches aimed at targeting the most aggressive stages of melanoma [60].

The data presented above underline that even if the PTP inhibitors available to date show only a mild effect on cell proliferation, future efforts could be made to use these compounds in combination with other pathway-targeted drugs to fight melanoma progression.

In this review, we will present a detailed overview of PTPs reported, up to date, to function as oncogenes in melanoma, either facilitating tumor progression or dampening the immune response. This information lays the foundation for the design of new therapeutic strategies specifically directed against oncogenic PTPs in melanoma [61].

2. Oncogenic Protein Tyrosine Phosphatases in Melanoma

Among the 107 known PTPs, several of them have been identified to have an oncogenic role in different types of cancers [10]. Interestingly, recent evidence highlights their importance in supporting melanoma progression, as discussed in the following sections.

2.1. Cell Division Cycle 25 Proteins (CDC25s)

CDC25s are a family of dual-specificity phosphatases (DSPs), known to act as key regulators of cell cycle progression and DNA damage handling, through the activation of cyclin-dependent kinase (CDK) complexes. CDC25s remove inhibitory phosphates from both threonine and tyrosine residues present on the phosphate-binding loop of CDKs [62]. Their inactivation or degradation leads to cell cycle arrest [63,64]. Hence, it is not surprising that CDC25 deregulation may lead to genomic instability and cancer transformation [64]. In humans, there are three distinct CDC25 genes (CDC25A, B, and C), which specifically dephosphorylate and activate various targets [63]. All the three isoforms are involved in tumorigenesis, even if with varying extent. Specifically, the overexpression of either CDC25A and/or CDC25B has been observed in multiple tumor types, including melanoma, where it is frequently associated with a more aggressive and metastatic phenotype [32,62,65–67]. By contrast, the role of CDC25C seems to be less important for tumorigenesis [68–70]. Interestingly, Kaplan–Meier curves of melanoma patients confirm the correlation between the higher expression of all the three isoforms and a worse clinical outcome [71–74].

Considering the importance of CDC25s in facilitating cell cycle progression and cell proliferation, these proteins are potentially very interesting targets for melanoma treatment. Determination of the crystal structures of the catalytic domain of CDC25A (PDB: 1C25) [75] and CDC25B (PDB: 1QB0) [76], alongside recent progression in bioinformatic approaches, facilitated the identification of numerous compounds with inhibitory effects on these enzymes [61].

In particular, as far as melanoma is concerned, triptolide (TPL), a diterpene triepoxide natural compound, has been demonstrated to be active in different cancers and, specifically, on the A375.S2 melanoma cell line. Treatment with this compound induces cyclin A and CDC25A inhibition, thereby causing arrest of the cell cycle in S phase. In addition, the exposure of A375.S2 melanoma cells to TPL leads to apoptosis through caspase-8, -9, and -3 activation [77]. Moreover, cantharidin (CTD), another natural compound that shares many characteristics with TPL, shows a similar ability to inhibit cyclin A and CDC25A in the A375.S2 melanoma cell line [78].

Noteworthy, Capasso and co-workers, using both experimental and bioinformatic methods, developed several quinonoid derivatives, acting as CDC25 irreversible inhibitors [63]. The mechanism of action of these compounds involves electrophilic modification or ROS-induced oxidation [79,80] of the catalytic cysteine residue in the PTP active site [81,82]. Among the molecules selected with this strategy, nine were identified with K_i values in the range of micro- and nano-molar and one of them (referred to as "compound 7") showed efficacy on melanoma cells (A2058 and SAN), arresting their proliferation in G2/M phase and inducing a strong antiproliferative effect [63]. In the same paper, the authors also assessed that treatment with compound 7 stimulates the intrinsic apoptosis pathway in a caspase-dependent manner and leads to a reduction of the CDC25C protein level (and, at a lower extent, of CDC25A) [63]. However, the mechanism of action of these quinonoid-based agents may cause many different unrelated events, due to ROS reaction with other phosphatases and with unrelated enzymes. These possibilities represent a serious limit to their therapeutic applications, due to the potential toxicity.

In order to circumvent this issue, more recently, Cerchia and co-authors performed a screening of different classes of molecules, starting from the lead inhibitor NSC28620. This approach allowed them to identify naphthylphenylketone and naphthylphenylamine derivatives, acting as CDC25 inhibitors in two aggressive human melanoma cell lines, namely A2058 and A375. In contrast with quinonoid derivatives, these compounds reversibly inhibit the enzymes (in particular, the CDC25B isoform) without generating ROS. Altogether, these characteristics make these inhibitors more interesting for possible anticancer therapy. In agreement, the reported results indicate that the treatment with these compounds affects cell cycle progression, with increased G2/M phase and reduced G0/G1 phase accumulation, by causing an increase in the phosphorylated form of cyclin-dependent kinase 1 (CDK1) [83] (Figure 1).

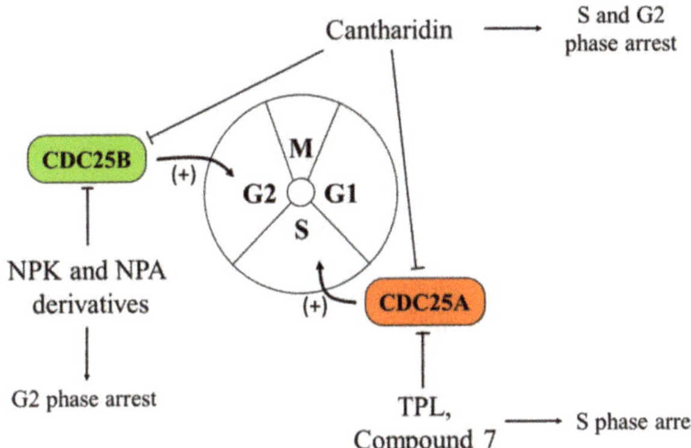

Figure 1. Effects of Cell Division Cycle 25 Proteins (CDC25s) targeting on melanoma cells. CDC25 phosphatases act as key regulators of the cell cycle, dephosphorylating cyclin-dependent kinases (CDK1, CDK2, CDK4, and CDK6) and cyclins (cyclin D, B, A, and E complexes). Several quinonoid derivatives, naphthylphenylketones and naphthylphenylamine derivatives, act as CDC25 inhibitors, arrest cells in the G0/G1 and G2/M phases of the cell cycle, and significantly inhibit the proliferation and colony formation ability of melanoma cells.

It should be underlined that the inhibitors so far cited are not closely specific for CDC25s and the question on their toxicity still remains open. Indeed, all the presented experiments, although showing clear cut results, are limited to in vitro melanoma models. More efforts are needed to confirm the efficacy of these new compounds as tools for melanoma treatment.

2.2. Low-Molecular-Weight Protein Tyrosine Phosphatase (LMW-PTP)

LMW-PTP belongs to the non-transmembrane PTPs sub-family and consists of 157 amino acids [84]. Two isoforms, generated by alternative splicing of a single gene and characterized by different activities and substrate specificities, have been found in mammalian cells [85].

Previous studies revealed that this enzyme displays a wide number of substrates, including the platelet derived growth factor (PDGF) receptor [86], insulin receptor [87], Ephrin A2 (EphA2) [88,89], and several non-receptor proteins, such as proto-oncogene tyrosine-protein kinase Src (Src) [90], focal adhesion kinase (FAK) [91], caveolin [92], signal transducer and activator of transcription 5 (STAT5) [93], β-catenin [94], and p190RhoGAP [95], thereby modulating key signaling pathways involved in tumor growth, differentiation, migration, and invasion [85,89]. In this context, it is not surprising that LMW-PTP has been found to be overexpressed in several types of human [96] and rat tumors [97], where it promotes an aggressive and malignant phenotype [98–100].

Recently, it has been demonstrated that LMW-PTP is overexpressed in melanoma cells, contributing to the regulation of cancer cell sensitivity toward chemo- and radiotherapy. Interestingly, it has been highlighted that the treatment of melanoma cells with morin, a non-toxic natural LMW-PTP inhibitor, is able to increase the sensitivity of tumor cells toward both dacarbazine and radiotherapy [31]. Coherently, data reported in The Human Protein Atlas database show that in melanoma patients, unfavorable prognosis is associated with high LMW-PTP expression levels, thereby confirming the role of this enzyme in regulating the in vivo survival and proliferation rate of melanoma cells [71,101]. Collectively, these findings suggest that LMW-PTP could be an interesting target to improve the effectiveness of anticancer treatment for melanoma patients that are naturally refractory to the therapies.

2.3. FAS-Associated Phosphatase 1 (FAP-1)

FAP-1 (or PTPN13/PTP-BAS) is a protein tyrosine phosphatase that interacts with the cytosolic portion of the Fas cell surface death receptor (FAS), whose activation leads to cell apoptosis.

FAP-1 interaction negatively regulates FAS-initiated apoptosis, preventing FAS export from the cytoplasm to the cell surface [102]. Other reported FAP-1 binding partners include the nuclear factor of kappa light polypeptide gene enhancer in B-cells inhibitor, α (IκBα), the Rho GTPase activated protein 1 (RhoGAP1), Ephrin B1, and the transient receptor potential cation channel M2 (TRPM2). In particular, IκBα is a putative FAP-1 substrate, being the only FAP-1-binding protein that is also dephosphorylated by this phosphatase [103].

Since FAP-1 negatively regulates FAS-initiated cell apoptosis, it has been suggested to positively affect tumor progression. Accordingly, FAP-1 has been reported to inhibit FAS-mediated apoptosis in pancreatic adenocarcinoma [104,105] and melanoma [106]. Interestingly, human melanoma cells silenced for FAP-1 show increased surface FAS expression and respond to recombinant FAS ligand (FasL) treatment by the induction of apoptosis [106]. Contrary to the possibility of blocking FAP-1 for the treatment of melanoma, there is also evidence that FAP-1 can act as a tumor suppressor in some cancer types. For example, reduced PTPN13 mRNA expression due to promoter hypermethylation or allelic loss has been observed in gastric and hepatocellular carcinomas [107,108]. Such a diversity of functions described for FAP-1, with positive and negative roles in a context-dependent manner, could be explained considering that it is among the largest intracellular PTPs, containing eight domains [109]. Despite the demonstrated role of this PTP in melanoma, to date, at least to our knowledge, no therapeutic approaches have been developed to inhibit FAP-1 in this tumor type, leaving open the possibility of designing new inhibitors.

2.4. Mitogen-Activated Protein Kinase Phosphatase-1 (MKP1)

MKP1 is a member of the threonine-tyrosine dual-specificity phosphatase family. MKP1 targets different members of the MAPK family that regulate cell proliferation and apoptosis, including extracellular signal-regulated kinase (ERK), c-Jun N-terminal kinase (JNK), and p38 MAPK [110]. Different evidence demonstrates that MKP1-mediated JNK dephosphorylation/inactivation is essential to protect tumor cells from anticancer drug-induced apoptosis [111,112]. Interestingly, melanoma patient survival analyzed by the Kaplan–Meier curve describes a correlation between high levels of MKP1 and poor prognosis [71,113]. Therefore, the inhibition of MKP1 may be an effective strategy to enhance the activity of antitumor therapy [114,115]. Promising results in this context come from a study by Kundu and co-workers, who demonstrated that by combining the interferon-α2b (IFN-α2b) with a new selective MKP1 inhibitor, tyrosine phosphatase inhibitor-3 (TPI-3), it is possible to obtain better results than those achieved with IFN-α2b or TPI-3 alone in inhibiting melanoma growth both in vitro and in a xenograft nude mice model. Interestingly, the authors reported that TPI-3 is a well-tolerated compound and that mice treated with TPI-3 alone did not experience loss of weight, abnormalities in behaviors, or anatomic alterations [116] (Figure 2). All together, these findings suggest that therapeutic strategies based on the treatment with MKP1 inhibitors could contribute to improve the prognosis of patients affected by tumors expressing high MKP1 levels, such as melanoma.

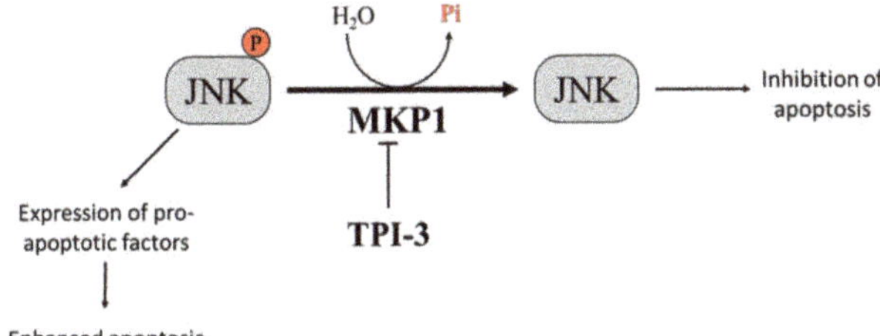

Figure 2. Effect of Mitogen-Activated Protein Kinase Phosphatase-1 (MKP1) targeting on melanoma cells. Overexpression of MKP1 phosphatase in melanoma cells contributes to enhance resistance toward anticancer drugs. For this reason, MKP1 inhibition is sufficient to enhance cancer cell death in culture and to sensitize cancer cells towards cytotoxic drugs. Among the substrates of MKP1, there is JNK. MKP1 dephosphorylates JNK, inhibiting its activation, and thereby avoiding apoptosis.

2.5. Phosphatase of Regenerating Liver (PRL)

The family of PRL consists of three members, namely PRL1, PRL2, and PRL3, and is a unique class of oncogenic dual-specificity phosphatases (alternatively known as protein tyrosine phosphatase 4A, PTP4A). Kaplan–Meier curve analysis of patients affected with melanoma shows a clear correlation between PRL phosphatase expression and poor survival [71,117,118].

Despite their role in cancer being well documented, the molecular functions of these proteins are still not completely understood [119]. PRL3 modulates different signaling pathways involving p53, MAPK, protein kinase B (PKB, also known as AKT), mammalian target of rapamycin (mTOR), signal transducer and activator of transcription 3 (STAT3), FAK, and vascular endothelial growth factor (VEGF), hence positively acting on tumor cell proliferation and aggressiveness [120]. PRL3 is also able to dephosphorylate the PI (4,5) P2 phosphoinositide, thereby contributing to modulation of the tumoral phenotype [121].

In addition, PRL3 promotes cell motility, invasion, and metastasis formation through different mechanisms, including its mutual activation involving the kinase Src [122], the accumulation of MMO14 matrix metalloprotease [123], and the downregulation of the tumor suppressor phosphatase and tensin homolog (PTEN), with consequent epithelial–mesenchymal transition [124]. Interestingly, in a very recent paper, PRL2 overexpression was also correlated with PTEN downregulation and poor patient survival. In particular, the authors proved that PRL2 directly downregulates PTEN by dephosphorylating its Tyr336 residue, thus increasing PTEN ubiquitination and degradation [125].

Overexpression of PRL3 has been demonstrated in many different solid tumors, including metastatic melanoma [30]. This finding was further confirmed by Wu and colleagues, who demonstrated a higher expression of PRL3 in the metastatic melanoma cell line B16-BL6 with respect to its less metastatic counterpart B16 cells, highlighting a clear role of PRL3 in promoting metastasis formation [126]. Daouti and co-authors established that while ectopic PRL3 overexpression induces cell transformation and increases motility and invasiveness, PRL3 silencing prevents anchorage-independent cell growth in soft agar [30]. The authors suggested that the adaptor protein p130Crk-associated substrate (p130Cas) is involved in this mechanism. Specifically, they demonstrated that treatment with thienopyridone, a selective inhibitor of all the three PRL isoforms, induces p130Cas and FAK cleavage, leading to the induction of caspase-mediated cell apoptosis and cancer cell *anoikis* [30].

Coherently, it was also shown that siRNA-mediated PRL3 depletion is able to inhibit the metastatic potential of B16-BL6 mouse melanoma cells both in vitro and in vivo [127].

Even if it has been known for a long time that a correlation exists between high PRL3 expression and metastatic risk in patients with uveal melanoma [128], only recently has a specific role for PRLs been recognized in this aggressive and metastatic tumor. In particular, collapsin response mediator protein 2 (CRMP2), a protein affecting microtubule dynamics, protein endocytosis, and vesicle recycling, has been described as a new target for PRL3 phosphatase activity. Specifically, PRL3 dephosphorylates CRMP2 on Thr514, thus enhancing cell invasiveness [129].

Considering the key role of PRL3 in mediating melanoma cell motility and metastasis formation, several attempts have been performed in order to select specific PRL3 inhibitors [119]. Pathak and colleagues identified pentamidine [1,5-di(4-amidinophenoxy)pentane] as a relatively specific inhibitor of PRLs and tested its activity on several cancer cell lines, including the WM9 melanoma-derived cell line. Interestingly, pentamidine was also tested in nude mice, where it was able to induce marked tumor cell necrosis in engrafted WM9 human melanoma cells, without any obvious side effects [130]. In addition, the previously mentioned thienopyridone is another promising inhibitor of PRLs that has been demonstrated to be effective in reducing the aggressiveness of melanoma cells by affecting their metastatic potential [30] (Figure 3).

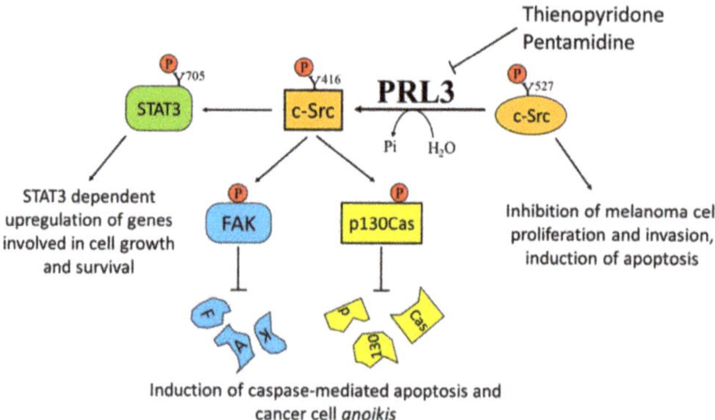

Figure 3. Effects of Phosphatase of Regenerating Liver-3 (PRL-3) targeting on melanoma progression. Elevated PRL-3 leads to Src activation through the downregulation of the synthesis of C-terminal Src kinase protein, which in turn leads to tyrosine phosphorylation of several proteins, including STAT3, FAK, and p130Cas. Thienopyridone and pentamidine derivatives, which act as PRL3 inhibitors, are effective in inhibiting melanoma cell proliferation, survival, and migration.

A possible alternative approach is based on the targeting of PRL1 trimer formation, a mechanism necessary for PRL1-mediated cell proliferation and migration [131]. Using a computer-based virtual screening, different specific compounds were selected as PRL1 trimerization inhibitors. Interestingly, one of these compounds, referred to as "Cmpd-43", displayed a strong anticancer activity both in vitro and in vivo in a murine xenograft model of melanoma [132].

Even if further efforts are needed to improve both the effectiveness of the inhibitors described and to reduce their side effects, the reported results suggest that PRLs could be an optimal target to reduce melanoma aggressiveness.

An event that should be considered when developing new treatments targeting PRLs is its interaction with the CNNM complexes (cyclin-M family, also termed cyclin and cystathionine β-synthase (CBS) domain magnesium transport mediators). This interaction is a key node in the regulation of magnesium homeostasis [133], and cells that overexpress PRLs in complex with CNNMs accumulate intracellular magnesium [134], which favors tumor proliferation and migration [135].

A possible more promising approach has very recently been developed using a humanized antibody (PRL3-zumab) that is able to target externalized PRL3 protein on different human liver and gastric tumor cell lines, used in an orthotopic tumor model in nude mice [136]. This antibody is currently under investigation in a phase 1 clinical trial on a wide range of solid tumors and hematological malignancies (Trial Number: NCT03191682; Table 1).

Table 1. Protein tyrosine phosphatase (PTP) inhibitors involved in clinical trials for melanoma treatment.

Trial Number	Compound	Target	Disease	Status
NCT03191682	PRL3-ZUMAB	PRL3	Solid Tumors and Hematologic Malignancies	Phase I
NCT03114319	TNO155	SHP-2	Non-Small Cell Lung Cancer; Esophageal Squamous Cell Cancer (SCC); Head/Neck SCC; Melanoma	Phase I
NCT00629200	Sodium stibogluconate	SHP-1	Malignant melanoma	Phase I completed
NCT00498979	Sodium stibogluconate	SHP-1	Malignant melanoma	Phase I completed

PRL3: Phosphatase of Regenerating Liver-3; SHP-2: Src Homology Region 2 Domain-Containing Phosphatase-2; SHP-1: Src Homology Region 2 Domain-Containing Phosphatase-1.

2.6. Src Homology Region 2 Domain-Containing Phosphatase-2 (SHP-2)

SHP-2, also termed tyrosine-protein phosphatase non-receptor type 11, is encoded by the PTPN11 gene [137]. SHP-2 contains two tandem SH2 domains, which act as phospho-tyrosine-binding domains and mediate the interaction of the tyrosine phosphatase with its substrates [137]. SHP-2 is auto-inhibited in the resting state, since the N-terminal SH2 domain binds to the catalytic cleft of the PTP domain, thereby blocking the access of SHP-2 substrates to the active site. Upon binding to target phospho-tyrosine residues, the N-terminal SH2 domain is released from the PTP domain and thus the enzyme is catalytically activated by reverting its auto-inhibited conformation [138].

SHP-2 is ubiquitously expressed and plays a key role in different cell signaling events, such as mitogenic activation, metabolic control, transcriptional regulation, and cell migration [139].

SHP-2 is the first tyrosine phosphatase identified as an oncogene in juvenile myelomonocytic leukemia, myelodysplastic syndromes, and acute myeloid leukemia [140]. It is remarkable that the gain-of-function mutations in SHP-2, leading to hyper-activated/deregulated mutants of the enzyme, occur in about 50% of Noonan syndrome patients [141]. Importantly, increased SHP-2 expression is a prognostic and a predictive marker of several malignancies and plays a key role in melanoma [142–149]. Indeed, this PTP has been found to be overexpressed and mutated in samples derived from melanoma patients, correlating with a strong metastatic phenotype and a poorer prognosis [33–35]. These findings were further confirmed by the Kaplan–Meier curve analysis, which revealed a strong correlation between higher expression levels of SHP-2 and poor overall survival in melanoma patients [71,150].

Due to the involvement of SHP-2 in multiple growth factor-mediated oncogenic pathways, such as the Ras/ERK1/2 pathway, and to its fundamental role in several tumors, inhibition of SHP-2 is considered to have broad therapeutic applications in the treatment of various cancers, including melanoma [151].

The PTP inhibitor sodium stibogluconate (SSG), a drug used in the treatment of leishmaniasis [152] and identified as an inhibitor of both SHP-1 and SHP-2 [153], increases interferon-α (IFN-α)-induced signal transducer and activator of transcription 1 (STAT1) tyrosine phosphorylation and has been shown to synergize with IFN-α to inhibit WM9 human melanoma tumor growth in nude mice [154]. Accordingly, Win-Piazza and co-workers demonstrated that suppression of SHP-2 increases the antitumor activity of IFN-α2b in A375 melanoma tumor xenografts. Indeed, IFN-α2b exerts antiproliferative effects on A375 cells through STAT1/STAT2 tyrosine phosphorylation, which is negatively regulated by SHP-2. In keeping with these data, treatment with the SHP-2 inhibitor, SPI-112, increases the IFN-α2b-stimulated STAT1 phosphorylation and inhibits A375 cell growth [155].

Furthermore, Soong and colleagues revealed a peculiar role of SHP-2 in melanocytes. Specifically, Plexin B1 and tyrosine protein kinase Met (MET) assemble in an oligomeric receptor-receptor complex in melanocytes and Semaphorin-4D (Sema4D) increases this association. The consequent MET activation correlates with the transformation of melanocytes to melanoma [156]. SHP-2 mediates, at least in part, the effects downstream to the MET receptor, and this phosphatase is required for the activation of the MAPK and AKT signaling pathways in response to hepatocyte growth factor (HGF) [157,158]. The blockade of SHP-2 phosphatase activity with the inhibitor NSC-87877 reduces HGF-induced MET activation and subsequently ERK1/ERK2 and AKT phosphorylation, suggesting an important role for SHP-2 in transducing proliferative and prosurvival signals in melanocytes [156]. Consequently, inhibition of SHP-2 can be proposed as a novel target to halt the transformation of melanocytes in melanoma.

Furthermore, SHP-2 acts as an oncogene in BRAF wild-type (either NRAS mutant or wild-type) melanoma cells. Indeed, both silencing of the activated SHP-2 E76K mutant or the administration of the allosteric SHP-2 inhibitor, SHP099, causes regression of the established melanoma, thereby suggesting that SHP-2 could be considered as a therapeutic target for BRAF wild-type melanoma [34].

It is widely described that HGF confers resistance to the BRAF inhibitor Vemurafenib in BRAF-mutant melanoma cells [159]. Interestingly, recent evidence underlines that SHP-2 is necessary to mediate this mechanism of resistance. Indeed, Prahallad and co-workers revealed that SHP-2 knockout clones of SK-Mel888 BRAF(V600E) mutant melanoma cells were unable to confer Vemurafenib resistance, following HGF, fibroblast growth factor 9 (FGF9), and stem cell factor (SCF) exposure [56]. Furthermore, it has been demonstrated that SHP-2 also drives adaptive resistance to RAS viral (v-raf) oncogene homolog (RAF) and MEK inhibitors in other tumor types [160]. Accordingly, Ahmed and co-authors proved that co-targeting of MEK and SHP-2 could serve as a powerful therapeutic approach in triple-negative breast cancer and showed that SHP-2 inhibition impairs adaptive resistance to Vemurafenib in a subset of BRAF(V600E) colorectal and thyroid cancers. These results suggest that SHP-2 blockade successfully overcomes adaptive resistance to BRAF and MEK inhibitors in a defined subgroup of ERK-dependent tumors, keeping the possibility open for exploiting this strategy for melanoma treatment.

Moreover, SHP-2 acts as a scaffold protein recruiting growth factor receptor-bound protein 2/Son of Sevenless (GRB2/SOS) complex to the membrane and promoting RTK-mediated RAS activation [161,162]. It is noteworthy that the allosteric SHP-2 inhibitor SHP099 stabilizes the phosphatase in its inactive conformation [22], thus preventing the assembly of SHP-2 with other adaptor proteins to achieve the complete activation of RTK signaling. In keeping with this, Zhang and colleagues demonstrated that SHP-2 overexpression enhances melanoma MeWo cell viability, motility, and anchorage-independent growth, through positive regulation of the ERK1/2 and PI3K/AKT pathway [35]. Accordingly, SHP-2 knockdown is able to revert these effects. Indeed, the specific SHP-2 inhibitor 11a-1 [163], an indole salicylic acid inhibitor, reduces the aforementioned phenomena in melanoma cells by downregulating the SHP-2-mediated ERK1/2 and AKT signaling pathways. Moreover, in vivo experiments demonstrated that 11a-1 significantly reduces xenografted melanoma tumor growth (Figure 4).

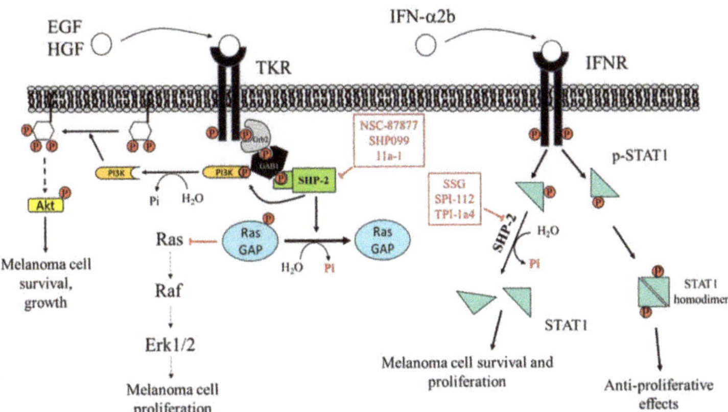

Figure 4. Effects of Src Homology Region 2 Domain-Containing Phosphatase-2 (SHP-2) targeting on melanoma cells. Left: SHP-2 has a key role in promoting the proliferation and survival of melanoma cells. SHP-2 dephosphorylates RasGAP, an inhibitor of Ras. Therefore, SHP-2, by activating Ras, promotes the RAS/RAF/ERK1/2 pathway, which sustains melanoma cell proliferation. Right: SHP-2 dephosphorylates GRB2-associated-binding protein 1 (GAB1), which releases PI3K, promoting activation of the PI3K/AKT pathway, melanoma cell growth, and survival. NSC-87877 and SHP099 inhibit SHP-2, impairing melanoma cell proliferation. SHP-2 also has an important role in regulating signaling activated by IFN-α2b. SHP-2 dephosphorylates STAT1, hindering its dimerization and its migration into the nucleus, where it stimulates the transcription of several genes, resulting in melanoma cell growth arrest. Overexpression of SHP-2 in melanoma cells blunts the response to IFN-α, favoring melanoma cell survival and dissemination. SSG, SPI-112, and TPI-1a4 are potent inhibitors of SHP-2, enhancing the anti-proliferative activity of IFN-α.

Overall, following the clear correlation between high expression levels of SHP2 and poorer survival of melanoma patients, several findings strongly suggest that SHP-2 may act as a targetable substrate against melanoma. In this perspective, SHP-2 inhibitors can be proposed as novel therapeutic approaches for melanoma treatment [35].

In keeping, TNO155, a recently discovered orally bioavailable SHP-2 inhibitor with antitumor activity in xenograft models [164], is currently in clinical trials for the treatment of solid tumors, including melanoma (Trial Number: NCT03114319; Table 1).

3. Role of PTPs in Immune Melanoma Cell Infiltrate

Melanoma is reported as one of the most immunogenic tumors characterized by a crosstalk between cancer cells and immune cells, which strongly affects cancer progression and metastasis [165]. Indeed, during melanomagenesis, activated T cells are recruited to the melanoma microenvironment through a "homing" mechanism, in order to recognize melanoma antigens and attack tumor cells, finally inducing cell death by the apoptosis pathway or the granule exocytosis pathway [166,167]. Interestingly, patients affected by melanoma with high CD8+ T cell infiltrate both in primary tumor and metastasis have better outcomes [168]. Targeting of the host immune response in melanoma with immunotherapies is one of the most interesting approaches used in the last decades to fight this malignancy [169]. Metastatic melanoma is mainly treated with Food and Drug Administration (FDA)-approved immune checkpoint inhibitors, such as Ipilimumab, Pembrolizumab, and Nivolumab [170–173]. However, many patients fail to respond to immunotherapy and frequently develop primary or secondary resistance, highlighting the need for supportive therapies to fight melanoma [174,175]. Several studies underlined that PTPs can exert an important role in modulating the immune infiltrate, thereby affecting melanoma growth.

Hence, inhibition of these PTPs might represent a novel strategy to be combined with the already consolidated immunotherapies.

3.1. Src Homology Region 2 Domain-Containing Phosphatase-1 (SHP-1)

SHP-1, also known as protein tyrosine phosphatase non-receptor type 6 (PTPN6), encoded by the PTPN6 gene, belongs to the family of non-receptor PTPs. This enzyme localizes to the cytosol and it is primarily expressed in hematopoietic cells, whereas it is present only at low levels in epithelial cells [176,177]. Similarly to SHP-2, SHP-1 is characterized by two tandem N-terminal SH2 domains that regulate the enzyme activity. The three-dimensional crystal structure of ligand-free SHP-1 revealed that this tyrosine phosphatase displays an auto-inhibited conformation [178].

SHP-1 was initially classified as a tumor suppressor phosphatase [179] since its silencing, due to hypermethylation of CpG islands in the PTPN6 promoter region, is frequently associated with a poor prognosis and worse outcome in different types of cancers [180–189]. Indeed, several studies underlined that this enzyme controls cell cycle progression by impairing PDGF and insulin receptor signaling [190,191]. Moreover, Nagakami and co-authors demonstrated that in endothelial cells, SHP-1 activation by tumor necrosis factor-alpha (TNF-α) inhibits VEGF- and epidermal growth factor (EGF)-mediated proliferation [192].

Despite the widely described tumor suppressor activity of SHP-1, recent studies demonstrated that this phosphatase could also act as an oncogene [193,194]. In particular, growing evidence highlights that SHP-1 is involved in modulation of the tumor microenvironment and impacts on immune infiltrate activation.

Specifically, SHP-1 is a key negative regulator of cytokine signaling and immune cell activity [195,196]. This tyrosine phosphatase limits T cell responsiveness by directly targeting T cell receptor (TCR) chain-ζ or downstream effectors, including the lymphocyte-specific protein tyrosine kinase (Lck), the zeta-chain-associated protein kinase 70 (ZAP70), the proto-oncogene vav (Vav), and PI3K [176,197]. Recently, Watson and co-authors highlighted that SHP-1 deficiency increases the ability of adoptively transferred CD8[+]T cells to impair tumor growth [198]. In addition, an enhanced in vivo cytotoxicity of naive SHP-1-deficient T cells was observed, highlighting the validity of targeting SHP-1 expression to boost human CD8[+]T cell functionality [199,200] (Figure 5).

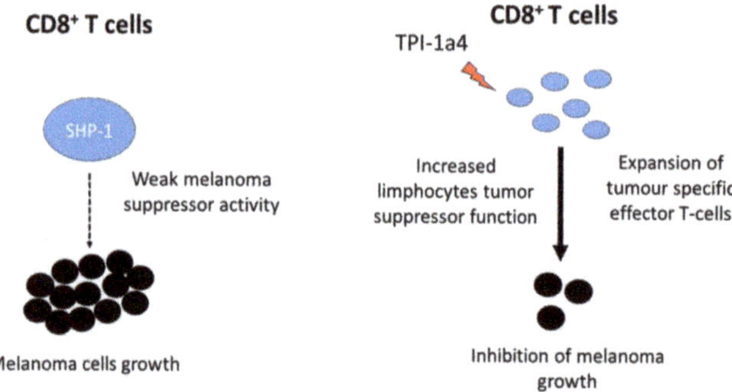

Figure 5. Effects of Src Homology Region 2 Domain-Containing Phosphatase-1 (SHP-1) targeting in T cells. Tumor-infiltrating lymphocytes express SHP-1, which regulates the TCR-driven T cell activation threshold and inhibits early events after TCR triggering. The SHP-1 inhibition increases interaction of CD8[+] T-cells with antigen presenting cells (APCs), leading to reduced activation thresholds and increased proliferation of antitumor T cells. Therefore, the inhibition of SHP-1 improves the ability of immune cells to suppress melanoma cell growth.

Considering the relevance of the immune infiltrate in supporting melanoma progression, targeting SHP-1 with specific inhibitors could be a promising novel strategy to improve the efficacy of cytokine therapy and immunotherapy, which are among the most successful approaches in clinical use for melanoma treatment [201,202]. For this purpose, different SHP-1 inhibitors have been developed, displaying antitumor potential in advanced cancer patients, including melanoma patients [153,203–205], as demonstrated by two different clinical trials (Trial Number: NCT00629200 and NCT00498979; Table 1). In this context, Yi and colleagues demonstrated that SSG inhibitor synergizes with IFN-α to overcome IFN-α resistance in different human cancer cell lines and extinguishes IFN-α-refractory WM9 human melanoma tumors in nude mice without displaying adverse effects [154].

More recently, a novel SHP-1 tyrosine phosphatase inhibitor-1 (TPI-1) was identified through the screening of a library of 34,000 drug-like compounds. TPI-1 was more effective than SSG in SHP-1 inhibition, immune cell activation, and antitumor potential [205]. Furthermore, Kundu and co-workers revealed that TPI-1 increases the expression of interferon-γ (IFN-γ), a cytokine produced during the activation of antitumor cells, in vitro, in mouse splenocytes, and in human peripheral blood. Interestingly, these authors also found that TPI-1 has anticancer potential on B16 melanoma thanks to the described induction of IFN-γ and demonstrated that B16 tumor growth is inhibited following TPI-1 treatment. Moreover, the antitumor activity of TPI-1a4, a TPI active analogue, was evaluated on the UV-induced K1735 murine melanoma with promising results [205].

Furthermore, Ramachandran and co-workers described the central role of SHP-1 in dendritic cell (DC) function. Indeed, this phosphatase inhibits numerous downstream effectors of multiple receptors, such as the nuclear factor kappa-light-chain-enhancer of activated B cells (NFkB), the activator protein 1 (AP-1), ERK, and JNK, as well as cytokine production in DCs. Accordingly, SSG-mediated inhibition of SHP-1 promotes proinflammatory cytokine production and increases the survival and migration of dendritic cells. Since DC signaling is required for the initiation of T cell immunity, SHP-1 is also able to reduce the ability of DCs to induce antigen-specific T cell proliferation. Finally, these authors assessed that SHP-1 inhibition in DCs could enhance their potency as antitumor vaccines. It is notable that mice bearing B16F10 melanoma, vaccinated with SHP-1-silenced DCs or with SHP-1 dominant negative DCs, showed a significantly slower tumor growth if compared to controls. Interestingly, this result suggests that inhibition of SHP-1 in DCs can improve their efficacy as in vivo antitumor vaccines against melanoma [206].

Moreover, a recent study demonstrated that SHP-1 inhibition in combination with immune checkpoint blockade treatment (anti-PD1 and anti-CTLA4) increases the recruitment and the effectors' function of low-affinity T cells, finally leading to melanoma tumor regression by increasing the frequency of IFN-γ-producing endogenous antitumor CD8$^+$T cells [207]. Collectively, this evidence highlights the role of SHP-1 as an oncogene in the tumor microenvironment, specifically in the immune cell population, suggesting that it may be a promising target to enlarge the repertoire of T cells sensitive to checkpoint blockade, finally leading to enhanced control of melanoma growth.

Notably, new efforts are needed to modulate SHP-1 activity in the selected immune cell population, which would be a significantly better strategy to impair tumor growth, rather than globally inhibit SHP-1 activity, independently of the cell subset. This approach could be particularly suitable in melanoma treatment, where immunotherapy is becoming increasingly important [208].

3.2. Src Homology Region 2 Domain-Containing Phosphatase-2 (SHP-2)

Recent evidence suggests a role of SHP-2 in tumor immunity. Specifically, following engagement with its ligands, mainly PD-L1, programmed cell death protein 1 (PD-1) is activated and recruits SHP-2 in the proximity of TCR and CD28. Consequently, SHP-2 dephosphorylates and decreases TCR and CD28 pathways, leading to the inhibition of T cell proliferation and activation, ultimately causing activated T cell death [209]. In keeping with this, Wu and co-workers recently showed that B16 melanoma cell-derived exosomes deliver SHP-2 to tumor-infiltrating lymphocytes, thus suppressing their function and inhibiting their proliferation [210]. Moreover, it has been demonstrated that

SHP-2 suppression in macrophages may promote a Th1-dominant tumor immune microenvironment, which is advantageous to suppress melanoma growth. Indeed, the deletion of SHP-2 increases macrophage production of chemokine (C-X-C motif) ligand 9 (CXCL9) in response to IFN-γ and tumor cell-derived cytokines, promoting T cell infiltration into the tumor [211]. Other evidence highlighting the importance of SHP-2 targeting in tumor-based immunotherapy comes from the studies of Ramesh and co-workers [212]. These authors demonstrated that the treatment with dual-inhibitor-loaded nanoparticles (DNTs), which simultaneously inhibit macrophage colony-stimulating factor 1 receptor (CSF1R) and SHP-2 pathways, results in an efficient re-polarization of M2 macrophages to their active M1 phenotype. Moreover, DNTs display antitumor activity without any toxicity in melanoma in both in vitro and in vivo settings [212]. In conflict with previous evidence, Zhang and co-authors highlighted that the expression of SHP-2 in CD4$^+$ T cells exerts tumor-suppressing effects on melanoma. Indeed, they demonstrated that SHP-2 deletion in CD4$^+$ T cells potentiates melanoma progression and promotes metastasis in mice [213] (Figure 6).

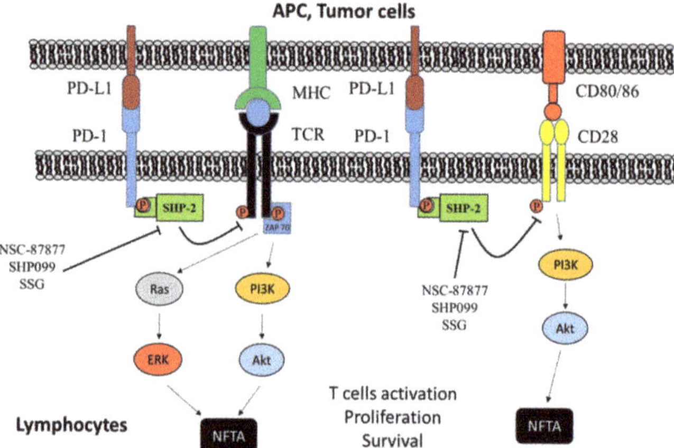

Figure 6. Targeting of SHP-2 in immune cells. After binding with PD-L1 ligand exposed by cancer cells or APC cells, PD-1 undergoes activation and recruits SHP-2, which can dephosphorylate ZAP70 or activate CD28 immune receptor. By this mechanism, SHP-2 inhibits signaling originated by TCR and CD28, thereby leading to a reduction in TCR-mediated interleukin-2 (IL-2) production and the impairment of T cell proliferation. Conversely, the treatment with SHP-2 inhibitors, such as NSC-87877, SHP099, and SSG, enhances lymphocyte activation and cytokine release, and stimulates lymphocyte proliferation, inducing melanoma cell death.

Even if the role of SHP-2 in modulating antitumor activity of immune infiltrate needs to be better investigated, reported results suggest that targeting SHP-2 in immune cells may be a promising approach for melanoma treatment.

3.3. Tyrosine-Protein Phosphatase Non-Receptor Type 2 (PTPN2)

PTPN2, also known as T-cell protein tyrosine phosphatase (TC-PTP), is ubiquitously expressed and has important functions [214] in modulating immune cell signaling [215], inflammatory responses [216], hematopoietic stem cell renewal [217], insulin signaling [218], and leptin regulation [219]. Importantly, several recent studies outlined a key role of PTPN2 in modulating oncogenic signaling. While few papers attribute a tumor-suppressive function to PTPN2 [220–223], many studies support its oncogenic role. For example, PTPN2 has been found to be overexpressed in B cell lymphomas, where its upregulation is under the control of MYC [224]. In addition, PTPN2 promotes IGF-2 induced migration in MCF-7 breast cancer cells [225].

Using Clustered Regularly Interspaced Short Palindromic Repeats/CRISPR associated protein 9 (CRISPR/Cas9) genome editing, Manguso and co-workers recently performed a pooled loss-of-function genetic screening in mice tumors derived from transplantable B16 melanoma cells. Mice were treated with immunotherapy in order to discover new potential targets involved in the resistance to this approach. Interestingly, among the genes depleted in immunotherapy-treated mice, PTPN2 was one of the most representative. Specifically, loss-of-function of PTPN2 sensitized B16 cells to immunotherapy in vivo and increased the sensitivity to T cell immunity in a *Braf/Pten* melanoma model, without affecting cell viability and growth in the absence of T cells. It is noteworthy that the re-expression of PTPN2 in PTPN2-null tumors abrogated the response to immunotherapy in vivo and the overexpression of PTPN2 in B16 cells promoted immunotherapy resistance. Moreover, the composition of the immune cell population in the tumor microenvironment was strongly altered in PTPN2-null tumors displaying a higher number of activated and cytotoxic CD8$^+$T cells and $\gamma\delta^+$ T cells. In addition, loss of PTPN2 increased antigen presentation and sensitivity to CD8$^+$T cells. In order to dissect the mechanism by which PTPN2 acts as an oncogene in melanoma and how its loss correlates with increased sensitivity of tumor cells to immunotherapy, it is important to underline that the tyrosine phosphatase inhibits IFNγ signaling by dephosphorylating STAT1 and JAK1. Indeed, PTPN2 deletion in B16 tumor cells enhanced IFNγ signaling, caused a strong change in the expression profile of IFNγ response gene, and increased phosphorylation of STAT1, thereby reducing tumor growth of PTPN2-null mice [226].

Moreover, a very recent study disclosed a new role of PTPN2 as a key regulator of the differentiation of the terminally exhausted CD8$^+$T cell subpopulation by attenuating type I interferon signaling. Indeed, PTPN2 loss in the immune system subpopulation, CD8$^+$T cells, induces PD-1 checkpoint blockade response to B16 tumors [227].

Despite no PTPN2 inhibitors having been extensively studied to date, the previously reported promising results revealing the oncogenic role of PTPN2 strongly suggest the necessity to design selective inhibitors of this phosphatase in order to potentiate the sensitivity of melanoma to the effects of immunotherapy.

4. Discussion

The discovery that mutated or constitutively activated tyrosine kinases contribute to melanocyte transformation pointed the attention to the importance of altered tyrosine phosphorylation in supporting different aspects of melanoma progression. This evidence paved the way for the development of tyrosine kinase inhibitors for the treatment of the unresectable form of melanoma [228]. Hence, many tyrosine kinase inhibitors have been synthesized in the past decades, and quickly approved for the treatment of patients affected by advanced metastatic melanoma. Although this therapeutic approach has produced clear benefits [229], clinical data revealed that a significant percentage of patients expressing mutated forms of oncogenic tyrosine kinases develop resistance to the treatment over the time, highlighting the need to find alternative therapeutic approaches [230].

A possible solution to this issue has recently emerged from the modulation of the activity of PTPs. Indeed, although many members of the PTP family behave as a tumor suppressor, several studies underscored a key role for PTPs as tumor promoters in different types of cancer [8]. This scenario suggested that the targeting of PTPs could be a promising alternative strategy to fight cancer progression. Despite the difficulties arising from the fact that all PTPs share the same charged active site, many advances have been made in the development of specific, cell-permeable, and bioavailable PTP inhibitors, mainly allosteric ones [231]. The first study about the possibility of inhibiting PTPs for melanoma treatment came from the late 1990s when Steinman and collaborators reported that the administration of different PTPs inhibitors, such as sodium orthovanadate and phenylarsine oxide, strongly inhibited melanoma metastasis formation, thus reducing tumor aggressiveness [232]. Preclinical studies subsequently revealed that some molecules targeting PTPs are able to slow down proliferation and reduce the metastatic dissemination of many aggressive forms of cancers without producing side effects, suggesting the possibility of introducing these compounds in the therapy against melanoma [21,231].

The evidence reported in this review confirms this hypothesis and shows that some of the PTPs previously described to have oncogenic functions also play an important role in promoting melanoma cell proliferation and survival [30–35]. In addition, some of these PTPs are implicated in the inactivation of the immune response, contributing to immune surveillance evasion of melanoma cells [195,196,209,226,227]. The relevance of PTPs in sustaining melanoma aggressiveness is further confirmed by tests carried out on mice xenografted with melanoma cells, revealing that treatment with PTPs inhibitors impairs proliferation, migration, and invasiveness of cancer cells [116,130], and enhances the antitumoral immune response [154,205,206,226]. Interestingly, many of the PTP inhibitors proposed above have been designed to overcome the highly conserved and positively charged nature of PTP active sites, thus ensuring a good specificity for their targets. For example, the SHP-2 inhibitor 11a-1, demonstrated to be effective against melanoma both in vitro and in vivo [163], has an IC50 over 5-fold more selective for SHP2 than any other PTPs tested. This specificity is correlated to its bidentate structure, which facilitates the access and the anchorage to both the SHP2 active site and a unique peripheral binding site, increasing its selectivity and potency [163]. The step-by-step modification strategy is also successful for the identification of new inhibitors. This approach, by optimizing the previously proposed CDC25B inhibitor NSC28620, led Cerchia and co-workers to the identification of the main structural requirements necessary to obtain the optimal inhibitory activity and specificity to design selective inhibitors against CDC25B [83].

Moreover, it is important to notice that some PTPs are simultaneously overexpressed in both cancer and immune cells, often displaying opposite functions. In these cases, it would be necessary to have some caution before proposing them as a possible therapy, evaluating the possibility of targeting the enzymes exclusively in one of the two populations.

For instance, it has been reported that LMW-PTP has a positive role in regulating the activation of lymphocytes [233,234], while it also promotes resistance of melanoma cells toward cytotoxic drugs [31]. As a consequence, the treatment of patients affected by melanoma with LMW-PTP inhibitors could induce a paradoxical effect, inhibiting on the one side, the proliferation of cancer cells, but on the other, impairing the immune response.

Different is the case of SHP-2, which plays a key role in sustaining melanoma cell survival, and behaves, at the same time, as a negative regulator of T lymphocytes [235]. Therefore, in this case, treatment with SHP-2 inhibitors could generate a synergistic effect, dampening melanoma cell growth [22] and at the same time potentiating the activation of immune cells [236].

In conclusion, in our opinion, future studies should be focused, on the one hand, on clarifying the physiological roles of PTPs in order to predict the efficacy of new targeted therapies based on their inhibition and, on the other, to synthesize more specific PTP inhibitors with a view to generating novel potential antitumoral drugs.

5. Conclusions

In synthesis, information collected in this review demonstrates that PTPs represent interesting molecules that can be targeted to fight the most aggressive forms of melanoma. PTP inhibitors could be successfully used, alone or in combination, to arrest melanoma progression while also potentiating antitumor immune surveillance. Based on the promising results obtained so far, PTP inhibitors could be proposed as a potential adjuvant therapy for the treatment of melanoma.

Funding: This work was funded by Associazione Italiana Ricerca sul Cancro (AIRC) (project code: 19515) for the project "Assaying tumor metabolic deregulation in live cells", by University of Florence (Fondo ex-60%), and by AIRC fellowship to Erica Pranzini (project code: 24132) for the project "Metabolic adaptations driving epigenetics of 5-Fluorouracil-resistant colon cancer: the role of one-carbon metabolism".

Conflicts of Interest: The authors declare no conflict of interest.

References

1. Hunter, T. Tyrosine phosphorylation: Thirty years and counting. *Curr. Opin. Cell Biol.* **2009**, *21*, 140–146. [CrossRef] [PubMed]
2. Ardito, F.; Giuliani, M.; Perrone, D.; Troiano, G.; Muzio, L. Lo The crucial role of protein phosphorylation in cell signaling and its use as targeted therapy (Review). *Int. J. Mol. Med.* **2017**, *40*, 271–280. [CrossRef] [PubMed]
3. Östman, A. Regulation of receptor tyrosine kinase signaling by protein tyrosine phosphatases. *Trends Cell Biol.* **2001**, *11*, 258–266. [CrossRef]
4. Blume-Jensen, P.; Hunter, T. Oncogenic kinase signalling. *Nature* **2001**, *411*, 355–365. [CrossRef] [PubMed]
5. Motiwala, T.; Jacob, S.T. Role of Protein Tyrosine Phosphatases in Cancer. *Prog. Nucleic Acid Res. Mol. Biol.* **2006**, *81*, 297–329. [PubMed]
6. Pottier, C.; Fresnais, M.; Gilon, M.; Jérusalem, G.; Longuespée, R.; Sounni, N.E. Tyrosine Kinase Inhibitors in Cancer: Breakthrough and Challenges of Targeted Therapy. *Cancers* **2020**, *12*, 731. [CrossRef]
7. Angelucci, A. Targeting Tyrosine Kinases in Cancer: Lessons for an Effective Targeted Therapy in the Clinic. *Cancers* **2019**, *11*, 490. [CrossRef]
8. Kim, M.; Baek, M.; Kim, D.J. Protein Tyrosine Signaling and its Potential Therapeutic Implications in Carcinogenesis. *Curr. Pharm. Des.* **2017**, *23*, 4226–4246. [CrossRef]
9. Alonso, A.; Nunes-Xavier, C.E.; Bayón, Y.; Pulido, R. The extended family of protein tyrosine phosphatases. In *Protein Tyrosine Phosphatases*; Humana Press: New York, NY, USA, 2016; pp. 1–23.
10. Julien, S.G.; Dubé, N.; Hardy, S.; Tremblay, M.L. Inside the human cancer tyrosine phosphatome. *Nat. Rev. Cancer* **2011**, *11*, 35–49. [CrossRef]
11. Ventura, J.-J.; Nebreda, Á.R. Protein kinases and phosphatases as therapeutic targets in cancer. *Clin. Transl. Oncol.* **2006**, *8*, 153–160. [CrossRef]
12. Zhang, Z.-Y. Drugging the Undruggable: Therapeutic Potential of Targeting Protein Tyrosine Phosphatases. *Acc. Chem. Res.* **2017**, *50*, 122–129. [CrossRef] [PubMed]
13. Scott, L.M.; Lawrence, H.R.; Sebti, S.M.; Lawrence, N.J.; Wu, J. Targeting protein tyrosine phosphatases for anticancer drug discovery. *Curr. Pharm. Des.* **2010**, *16*, 1843–1862. [CrossRef] [PubMed]
14. Stanford, S.M.; Bottini, N. Targeting Tyrosine Phosphatases: Time to End the Stigma. *Trends Pharmacol. Sci.* **2017**, *38*, 524–540. [CrossRef] [PubMed]
15. Barr, A.J. Protein tyrosine phosphatases as drug targets: Strategies and challenges of inhibitor development. *Future Med. Chem.* **2010**, *2*, 1563–1576. [CrossRef]
16. Alonso, A.; Sasin, J.; Bottini, N.; Friedberg, I.; Friedberg, I.; Osterman, A.; Godzik, A.; Hunter, T.; Dixon, J.; Mustelin, T. Protein tyrosine phosphatases in the human genome. *Cell* **2004**, *117*, 699–711. [CrossRef]
17. Andersen, J.N.; Mortensen, O.H.; Peters, G.H.; Drake, P.G.; Iversen, L.F.; Olsen, O.H.; Jansen, P.G.; Andersen, H.S.; Tonks, N.K.; Møller, N.P. Structural and evolutionary relationships among protein tyrosine phosphatase domains. *Mol. Cell. Biol.* **2001**, *21*, 7117–7136. [CrossRef]
18. Barr, A.J.; Ugochukwu, E.; Lee, W.H.; King, O.N.F.; Filippakopoulos, P.; Alfano, I.; Savitsky, P.; Burgess-Brown, N.A.; Müller, S.; Knapp, S. Large-scale structural analysis of the classical human protein tyrosine phosphatome. *Cell* **2009**, *136*, 352–363. [CrossRef]
19. Lawrence, H.R.; Pireddu, R.; Chen, L.; Luo, Y.; Sung, S.-S.; Szymanski, A.M.; Yip, M.L.R.; Guida, W.C.; Sebti, S.M.; Wu, J.; et al. Inhibitors of Src homology-2 domain containing protein tyrosine phosphatase-2 (Shp2) based on oxindole scaffolds. *J. Med. Chem.* **2008**, *51*, 4948–4956. [CrossRef]
20. Wiesmann, C.; Barr, K.J.; Kung, J.; Zhu, J.; Erlanson, D.A.; Shen, W.; Fahr, B.J.; Zhong, M.; Taylor, L.; Randal, M.; et al. Allosteric inhibition of protein tyrosine phosphatase 1B. *Nat. Struct. Mol. Biol.* **2004**, *11*, 730–737. [CrossRef]
21. Krishnan, N.; Koveal, D.; Miller, D.H.; Xue, B.; Akshinthala, S.D.; Kragelj, J.; Jensen, M.R.; Gauss, C.-M.; Page, R.; Blackledge, M.; et al. Targeting the disordered C terminus of PTP1B with an allosteric inhibitor. *Nat. Chem. Biol.* **2014**, *10*, 558–566. [CrossRef]
22. Chen, Y.-N.P.; LaMarche, M.J.; Chan, H.M.; Fekkes, P.; Garcia-Fortanet, J.; Acker, M.G.; Antonakos, B.; Chen, C.H.-T.; Chen, Z.; Cooke, V.G.; et al. Allosteric inhibition of SHP2 phosphatase inhibits cancers driven by receptor tyrosine kinases. *Nature* **2016**, *535*, 148–152. [CrossRef] [PubMed]

23. Puius, Y.A.; Zhao, Y.; Sullivan, M.; Lawrence, D.S.; Almo, S.C.; Zhang, Z.Y. Identification of a second aryl phosphate-binding site in protein-tyrosine phosphatase 1B: A paradigm for inhibitor design. *Proc. Natl. Acad. Sci. USA* **1997**, *94*, 13420–13425. [CrossRef] [PubMed]

24. Sun, J.-P.; Fedorov, A.A.; Lee, S.-Y.; Guo, X.-L.; Shen, K.; Lawrence, D.S.; Almo, S.C.; Zhang, Z.-Y. Crystal structure of PTP1B complexed with a potent and selective bidentate inhibitor. *J. Biol. Chem.* **2003**, *278*, 12406–12414. [CrossRef] [PubMed]

25. Trachootham, D.; Alexandre, J.; Huang, P. Targeting cancer cells by ROS-mediated mechanisms: A radical therapeutic approach? *Nat. Rev. Drug Discov.* **2009**, *8*, 579–591. [CrossRef] [PubMed]

26. Ostman, A.; Frijhoff, J.; Sandin, A.; Böhmer, F.-D. Regulation of protein tyrosine phosphatases by reversible oxidation. *J. Biochem.* **2011**, *150*, 345–356. [CrossRef] [PubMed]

27. Ohayon, S.; Refua, M.; Hendler, A.; Aharoni, A.; Brik, A. Harnessing the oxidation susceptibility of deubiquitinases for inhibition with small molecules. *Angew. Chem. Int. Ed. Engl.* **2015**, *54*, 599–603. [CrossRef]

28. Gopinath, P.; Mahammed, A.; Ohayon, S.; Gross, Z.; Brik, A. Understanding and predicting the potency of ROS-based enzyme inhibitors, exemplified by naphthoquinones and ubiquitin specific protease-2. *Chem. Sci.* **2016**, *7*, 7079–7086. [CrossRef]

29. Pereyra, C.E.; Dantas, R.F.; Ferreira, S.B.; Gomes, L.P.; Silva, F.P., Jr. The diverse mechanisms and anticancer potential of naphthoquinones. *Cancer Cell Int.* **2019**, *19*, 207. [CrossRef]

30. Daouti, S.; Li, W.; Qian, H.; Huang, K.-S.; Holmgren, J.; Levin, W.; Reik, L.; McGady, D.L.; Gillespie, P.; Perrotta, A.; et al. A Selective Phosphatase of Regenerating Liver Phosphatase Inhibitor Suppresses Tumor Cell Anchorage-Independent Growth by a Novel Mechanism Involving p130Cas Cleavage. *Cancer Res.* **2008**, *68*, 1162–1169. [CrossRef]

31. Lori, G.; Paoli, P.; Caselli, A.; Cirri, P.; Marzocchini, R.; Mangoni, M.; Talamonti, C.; Livi, L.; Raugei, G. Targeting LMW-PTP to sensitize melanoma cancer cells toward chemo- and radiotherapy. *Cancer Med.* **2018**, *7*, 1933–1943. [CrossRef]

32. Tang, L.; Li, G.; Tron, V.A.; Trotter, M.J.; Ho, V.C. Expression of cell cycle regulators in human cutaneous malignant melanoma. *Melanoma Res.* **1999**, *9*, 148. [CrossRef] [PubMed]

33. Cheng, Y.-P.; Chiu, H.-Y.; Hsiao, T.-L.; Hsiao, C.-H.; Lin, C.-C.; Liao, Y.-H. Scalp melanoma in a woman with LEOPARD syndrome: Possible implication of PTPN11 signaling in melanoma pathogenesis. *J. Am. Acad. Dermatol.* **2013**, *69*, e186–e187. [CrossRef] [PubMed]

34. Hill, K.S.; Roberts, E.R.; Wang, X.; Marin, E.; Park, T.D.; Son, S.; Ren, Y.; Fang, B.; Yoder, S.; Kim, S.; et al. PTPN11 Plays Oncogenic Roles and Is a Therapeutic Target for BRAF Wild-Type Melanomas. *Mol. Cancer Res.* **2019**, *17*, 583–593. [CrossRef] [PubMed]

35. Zhang, R.-Y.; Yu, Z.-H.; Zeng, L.; Zhang, S.; Bai, Y.; Miao, J.; Chen, L.; Xie, J.; Zhang, Z.-Y. SHP2 phosphatase as a novel therapeutic target for melanoma treatment. *Oncotarget* **2016**, *7*, 73817–73829. [CrossRef] [PubMed]

36. Jackett, L.A.; Scolyer, R.A. A Review of Key Biological and Molecular Events Underpinning Transformation of Melanocytes to Primary and Metastatic Melanoma. *Cancers* **2019**, *11*, 2041. [CrossRef] [PubMed]

37. Hartman, R.I.; Lin, J.Y. Cutaneous Melanoma-A Review in Detection, Staging, and Management. *Hematol. Oncol. Clin. North. Am.* **2019**, *33*, 25–38. [CrossRef] [PubMed]

38. McLaughlin, C.C.; Wu, X.-C.; Jemal, A.; Martin, H.J.; Roche, L.M.; Chen, V.W. Incidence of noncutaneous melanomas in the U.S. *Cancer* **2005**, *103*, 1000–1007. [CrossRef]

39. Wilkins, D.K.; Nathan, P.D. Therapeutic opportunities in noncutaneous melanoma. *Ther. Adv. Med. Oncol.* **2009**, *1*, 29–36. [CrossRef]

40. Akbani, R.; Akdemir, K.C.; Aksoy, B.A.; Albert, M.; Ally, A.; Amin, S.B.; Arachchi, H.; Arora, A.; Auman, J.T.; Ayala, B.; et al. Cancer Genome Atlas Network Genomic Classification of Cutaneous Melanoma. *Cell* **2015**, *161*, 1681–1696. [CrossRef]

41. Hayward, N.K.; Wilmott, J.S.; Waddell, N.; Johansson, P.A.; Field, M.A.; Nones, K.; Patch, A.-M.; Kakavand, H.; Alexandrov, L.B.; Burke, H.; et al. Whole-genome landscapes of major melanoma subtypes. *Nature* **2017**, *545*, 175–180. [CrossRef]

42. Rajkumar, S.; Watson, I.R. Molecular characterisation of cutaneous melanoma: Creating a framework for targeted and immune therapies. *Br. J. Cancer* **2016**, *115*, 145–155. [CrossRef] [PubMed]

43. Ribas, A.; Flaherty, K.T. BRAF targeted therapy changes the treatment paradigm in melanoma. *Nat. Rev. Clin. Oncol.* **2011**, *8*, 426–433. [CrossRef] [PubMed]

44. Wan, P.T.C.; Garnett, M.J.; Roe, S.M.; Lee, S.; Niculescu-Duvaz, D.; Good, V.M.; Jones, C.M.; Marshall, C.J.; Springer, C.J.; Barford, D.; et al. Mechanism of activation of the RAF-ERK signaling pathway by oncogenic mutations of B-RAF. *Cell* **2004**, *116*, 855–867. [CrossRef]

45. Peltonen, S.; Kallionpää, R.A.; Peltonen, J. Neurofibromatosis type 1 (NF1) gene: Beyond café au lait spots and dermal neurofibromas. *Exp. Dermatol.* **2017**, *26*, 645–648. [CrossRef] [PubMed]

46. Krauthammer, M.; Kong, Y.; Bacchiocchi, A.; Evans, P.; Pornputtapong, N.; Wu, C.; McCusker, J.P.; Ma, S.; Cheng, E.; Straub, R.; et al. Exome sequencing identifies recurrent mutations in NF1 and RASopathy genes in sun-exposed melanomas. *Nat. Genet.* **2015**, *47*, 996–1002. [CrossRef] [PubMed]

47. Kiuru, M.; Busam, K.J. The NF1 gene in tumor syndromes and melanoma. *Lab. Investig.* **2017**, *97*, 146–157. [CrossRef] [PubMed]

48. Chiappetta, C.; Proietti, I.; Soccodato, V.; Puggioni, C.; Zaralli, R.; Pacini, L.; Porta, N.; Skroza, N.; Petrozza, V.; Potenza, C.; et al. BRAF and NRAS mutations are heterogeneous and not mutually exclusive in nodular melanoma. *Appl. Immunohistochem. Mol. Morphol. AIMM* **2015**, *23*, 172–177. [CrossRef]

49. Griffin, M.; Scotto, D.; Josephs, D.H.; Mele, S.; Crescioli, S.; Bax, H.J.; Pellizzari, G.; Wynne, M.D.; Nakamura, M.; Hoffmann, R.M.; et al. BRAF inhibitors: Resistance and the promise of combination treatments for melanoma. *Oncotarget* **2017**, *8*, 78174–78192. [CrossRef]

50. Paraiso, K.H.T.; Xiang, Y.; Rebecca, V.W.; Abel, E.V.; Chen, Y.A.; Munko, A.C.; Wood, E.; Fedorenko, I.V.; Sondak, V.K.; Anderson, A.R.A.; et al. PTEN loss confers BRAF inhibitor resistance to melanoma cells through the suppression of BIM expression. *Cancer Res.* **2011**, *71*, 2750–2760. [CrossRef]

51. Yeh, I.; Jorgenson, E.; Shen, L.; Xu, M.; North, J.P.; Shain, A.H.; Reuss, D.; Wu, H.; Robinson, W.A.; Olshen, A.; et al. Targeted Genomic Profiling of Acral Melanoma. *J. Natl. Cancer Inst.* **2019**, *111*, 1068–1077. [CrossRef]

52. Curtin, J.A.; Busam, K.; Pinkel, D.; Bastian, B.C. Somatic activation of KIT in distinct subtypes of melanoma. *J. Clin. Oncol.* **2006**, *24*, 4340–4346. [CrossRef] [PubMed]

53. Van Raamsdonk, C.D.; Griewank, K.G.; Crosby, M.B.; Garrido, M.C.; Vemula, S.; Wiesner, T.; Obenauf, A.C.; Wackernagel, W.; Green, G.; Bouvier, N.; et al. Mutations in GNA11 in uveal melanoma. *N. Engl. J. Med.* **2010**, *363*, 2191–2199. [CrossRef] [PubMed]

54. Onken, M.D.; Worley, L.A.; Long, M.D.; Duan, S.; Council, M.L.; Bowcock, A.M.; Harbour, J.W. Oncogenic mutations in GNAQ occur early in uveal melanoma. *Investig. Ophthalmol. Vis. Sci.* **2008**, *49*, 5230–5234. [CrossRef] [PubMed]

55. Field, M.G.; Durante, M.A.; Anbunathan, H.; Cai, L.Z.; Decatur, C.L.; Bowcock, A.M.; Kurtenbach, S.; Harbour, J.W. Punctuated evolution of canonical genomic aberrations in uveal melanoma. *Nat. Commun.* **2018**, *9*, 116. [CrossRef]

56. Prahallad, A.; Heynen, G.J.; Germano, G.; Willems, S.M.; Evers, B.; Vecchione, L.; Gambino, V.; Lieftink, C.; Beijersbergen, R.L.; Di Nicolantonio, F.; et al. PTPN11 is a Central Node in Intrinsic and Acquired Resistance to Targeted Cancer Drugs. *Cell Rep.* **2015**, *12*, 1978–1985. [CrossRef]

57. Ruess, D.A.; Heynen, G.J.; Ciecielski, K.J.; Ai, J.; Berninger, A.; Kabacaoglu, D.; Görgülü, K.; Dantes, Z.; Wörmann, S.M.; Diakopoulos, K.N.; et al. Mutant KRAS-driven cancers depend on PTPN11/SHP2 phosphatase. *Nat. Med.* **2018**, *24*, 954–960. [CrossRef]

58. Luke, J.J.; Flaherty, K.T.; Ribas, A.; Long, G.V. Targeted agents and immunotherapies: Optimizing outcomes in melanoma. *Nat. Rev. Clin. Oncol.* **2017**, *14*, 463–482. [CrossRef]

59. Damsky, W.E.; Theodosakis, N.; Bosenberg, M. Melanoma metastasis: New concepts and evolving paradigms. *Oncogene* **2014**, *33*, 2413–2422. [CrossRef]

60. Mouawad, R.; Sebert, M.; Michels, J.; Bloch, J.; Spano, J.-P.; Khayat, D. Treatment for metastatic malignant melanoma: Old drugs and new strategies. *Crit. Rev. Oncol. Hematol.* **2010**, *74*, 27–39. [CrossRef]

61. Jiang, Z.-X.; Zhang, Z.-Y. Targeting PTPs with small molecule inhibitors in cancer treatment. *Cancer Metastasis Rev.* **2008**, *27*, 263–272. [CrossRef]

62. Boutros, R.; Lobjois, V.; Ducommun, B. CDC25 phosphatases in cancer cells: Key players? Good targets? *Nat. Rev. Cancer* **2007**, *7*, 495–507. [CrossRef] [PubMed]

63. Capasso, A.; Cerchia, C.; Di Giovanni, C.; Granato, G.; Albano, F.; Romano, S.; De Vendittis, E.; Ruocco, M.R.; Lavecchia, A. Ligand-based chemoinformatic discovery of a novel small molecule inhibitor targeting CDC25 dual specificity phosphatases and displaying in vitro efficacy against melanoma cells. *Oncotarget* **2015**, *6*, 40202–40222. [CrossRef] [PubMed]

64. Karlsson-Rosenthal, C.; Millar, J.B.A. Cdc25: Mechanisms of checkpoint inhibition and recovery. *Trends Cell Biol.* **2006**, *16*, 285–292. [CrossRef] [PubMed]

65. Liu, J.C.; Granieri, L.; Shrestha, M.; Wang, D.-Y.; Vorobieva, I.; Rubie, E.A.; Jones, R.; Ju, Y.; Pellecchia, G.; Jiang, Z.; et al. Identification of CDC25 as a Common Therapeutic Target for Triple-Negative Breast Cancer. *Cell Rep.* **2018**, *23*, 112–126. [CrossRef]

66. Cangi, M.G.; Cukor, B.; Soung, P.; Signoretti, S.; Moreira, G.; Ranashinge, M.; Cady, B.; Pagano, M.; Loda, M. Role of the Cdc25A phosphatase in human breast cancer. *J. Clin. Investig.* **2000**, *106*, 753–761. [CrossRef]

67. Ma, Z.-Q.; Chua, S.S.; DeMayo, F.J.; Tsai, S.Y. Induction of mammary gland hyperplasia in transgenic mice over-expressing human Cdc25B. *Oncogene* **1999**, *18*, 4564–4576. [CrossRef]

68. Kristjánsdóttir, K.; Rudolph, J. Cdc25 Phosphatases and Cancer. *Chem. Biol.* **2004**, *11*, 1043–1051. [CrossRef]

69. Albert, H.; Santos, S.; Battaglia, E.; Brito, M.; Monteiro, C.; Bagrel, D. Differential expression of CDC25 phosphatases splice variants in human breast cancer cells. *Clin. Chem. Lab. Med.* **2011**, *49*. [CrossRef]

70. Bahassi, E.M.; Hennigan, R.F.; Myer, D.L.; Stambrook, P.J. Cdc25C phosphorylation on serine 191 by Plk3 promotes its nuclear translocation. *Oncogene* **2004**, *23*, 2658–2663. [CrossRef]

71. Uhlen, M.; Zhang, C.; Lee, S.; Sjöstedt, E.; Fagerberg, L.; Bidkhori, G.; Benfeitas, R.; Arif, M.; Liu, Z.; Edfors, F.; et al. A pathology atlas of the human cancer transcriptome. *Science* **2017**, *357*. [CrossRef]

72. The Human Protein Atlas. Available online: https://www.proteinatlas.org/ENSG00000164045-CDC25A/pathology/melanoma (accessed on 27 August 2020).

73. The Human Protein Atlas. Available online: https://www.proteinatlas.org/ENSG00000101224-CDC25B/pathology/melanoma (accessed on 27 August 2020).

74. The Human Protein Atlas. Available online: https://www.proteinatlas.org/ENSG00000158402-CDC25C/pathology/melanoma (accessed on 27 August 2020).

75. Fauman, E.B.; Cogswell, J.P.; Lovejoy, B.; Rocque, W.J.; Holmes, W.; Montana, V.G.; Piwnica-Worms, H.; Rink, M.J.; Saper, M.A. Crystal Structure of the Catalytic Domain of the Human Cell Cycle Control Phosphatase, Cdc25A. *Cell* **1998**, *93*, 617–625. [CrossRef]

76. Reynolds, R.A.; Yem, A.W.; Wolfe, C.L.; Deibel, M.R.; Chidester, C.G.; Watenpaugh, K.D. Crystal structure of the catalytic subunit of Cdc25B required for G 2 /M phase transition of the cell cycle 1 1Edited by I. A. Wilson. *J. Mol. Biol.* **1999**, *293*, 559–568. [CrossRef] [PubMed]

77. Hung, F.-M.; Chen, Y.-L.; Huang, A.-C.; Hsiao, Y.-P.; Yang, J.-S.; Chung, M.-T.; Chueh, F.-S.; Lu, H.-F.; Chung, J.-G. Triptolide induces S phase arrest via the inhibition of cyclin E and CDC25A and triggers apoptosis via caspase- and mitochondrial-dependent signaling pathways in A375.S2 human melanoma cells. *Oncol. Rep.* **2013**, *29*, 1053–1060. [CrossRef] [PubMed]

78. Hsiao, Y.-P.; Tsai, C.-H.; Wu, P.-P.; Hsu, S.-C.; Liu, H.-C.; Huang, Y.-P.; Yang, J.-H.; Chung, J.-G. Cantharidin induces G2/M phase arrest by inhibition of Cdc25c and Cyclin A and triggers apoptosis through reactive oxygen species and the mitochondria-dependent pathways of A375.S2 human melanoma cells. *Int. J. Oncol.* **2014**, *45*, 2393–2402. [CrossRef] [PubMed]

79. Kar, S.; Lefterov, I.M.; Wang, M.; Lazo, J.S.; Scott, C.N.; Wilcox, C.S.; Carr, B.I. Binding and Inhibition of Cdc25 Phosphatases by Vitamin K Analogues †. *Biochemistry* **2003**, *42*, 10490–10497. [CrossRef] [PubMed]

80. Pu, L.; Amoscato, A.A.; Bier, M.E.; Lazo, J.S. Dual G 1 and G 2 Phase Inhibition by a Novel, Selective Cdc25 Inhibitor 7-Chloro-6-(2-morpholin-4-ylethylamino)- quinoline-5,8-dione. *J. Biol. Chem.* **2002**, *277*, 46877–46885. [CrossRef]

81. Brisson, M.; Nguyen, T.; Wipf, P.; Joo, B.; Day, B.W.; Skoko, J.S.; Schreiber, E.M.; Foster, C.; Bansal, P.; Lazo, J.S. Redox Regulation of Cdc25B by Cell-Active Quinolinediones. *Mol. Pharmacol.* **2005**, *68*, 1810–1820. [CrossRef]

82. Zhou, Y.; Feng, X.; Wang, L.; Du, J.; Zhou, Y.; Yu, H.; Zang, Y.; Li, J.; Li, J. LGH00031, a novel ortho-quinonoid inhibitor of cell division cycle 25B, inhibits human cancer cells via ROS generation. *Acta Pharmacol. Sin.* **2009**, *30*, 1359–1368. [CrossRef]

83. Cerchia, C.; Nasso, R.; Mori, M.; Villa, S.; Gelain, A.; Capasso, A.; Aliotta, F.; Simonetti, M.; Rullo, R.; Masullo, M.; et al. Discovery of Novel Naphthylphenylketone and Naphthylphenylamine Derivatives as Cell Division Cycle 25B (CDC25B) Phosphatase Inhibitors: Design, Synthesis, Inhibition Mechanism, and in Vitro Efficacy against Melanoma Cell Lines. *J. Med. Chem.* **2019**, *62*, 7089–7110. [CrossRef]

84. Tonks, N.K. Protein tyrosine phosphatases: From genes, to function, to disease. *Nat. Rev. Mol. Cell Biol.* **2006**, *7*, 833–846. [CrossRef]

85. Caselli, A.; Paoli, P.; Santi, A.; Mugnaioni, C.; Toti, A.; Camici, G.; Cirri, P. Low molecular weight protein tyrosine phosphatase: Multifaceted functions of an evolutionarily conserved enzyme. *Biochim. Biophys. Acta* **2016**, *1864*, 1339–1355. [CrossRef] [PubMed]

86. Chiarugi, P.; Cirri, P.; Raugei, G.; Manao, G.; Taddei, L.; Ramponi, G. Low M(r) phosphotyrosine protein phosphatase interacts with the PDGF receptor directly via its catalytic site. *Biochem. Biophys. Res. Commun.* **1996**, *219*, 21–25. [CrossRef] [PubMed]

87. Chiarugi, P.; Cirri, P.; Marra, F.; Raugei, G.; Camici, G.; Manao, G.; Ramponi, G. LMW-PTP is a negative regulator of insulin-mediated mitotic and metabolic signalling. *Biochem. Biophys. Res. Commun.* **1997**, *238*, 676–682. [CrossRef] [PubMed]

88. Kikawa, K.D.; Vidale, D.R.; Van Etten, R.L.; Kinch, M.S. Regulation of the EphA2 kinase by the low molecular weight tyrosine phosphatase induces transformation. *J. Biol. Chem.* **2002**, *277*, 39274–39279. [CrossRef]

89. Chiarugi, P.; Taddei, M.L.; Schiavone, N.; Papucci, L.; Giannoni, E.; Fiaschi, T.; Capaccioli, S.; Raugei, G.; Ramponi, G. LMW-PTP is a positive regulator of tumor onset and growth. *Oncogene* **2004**, *23*, 3905–3914. [CrossRef]

90. Zambuzzi, W.F.; Granjeiro, J.M.; Parikh, K.; Yuvaraj, S.; Peppelenbosch, M.P.; Ferreira, C.V. Modulation of Src activity by low molecular weight protein tyrosine phosphatase during osteoblast differentiation. *Cell. Physiol. Biochem.* **2008**, *22*, 497–506. [CrossRef]

91. Rigacci, S.; Rovida, E.; Dello Sbarba, P.; Berti, A. Low Mr phosphotyrosine protein phosphatase associates and dephosphorylates p125 focal adhesion kinase, interfering with cell motility and spreading. *J. Biol. Chem.* **2002**, *277*, 41631–41636. [CrossRef]

92. Caselli, A.; Taddei, M.L.; Bini, C.; Paoli, P.; Camici, G.; Manao, G.; Cirri, P.; Ramponi, G. Low molecular weight protein tyrosine phosphatase and caveolin-1: Interaction and isoenzyme-dependent regulation. *Biochemistry* **2007**, *46*, 6383–6392. [CrossRef]

93. Rigacci, S.; Talini, D.; Berti, A. LMW-PTP associates and dephosphorylates STAT5 interacting with its C-terminal domain. *Biochem. Biophys. Res. Commun.* **2003**, *312*, 360–366. [CrossRef]

94. Taddei, M.L.; Chiarugi, P.; Cirri, P.; Buricchi, F.; Fiaschi, T.; Giannoni, E.; Talini, D.; Cozzi, G.; Formigli, L.; Raugei, G.; et al. B-Catenin Interacts With Low-Molecular-Weight Protein Tyrosine Phosphatase Leading to Cadherin-Mediated Cell-Cell Adhesion Increase. *Cancer Res.* **2002**, *62*, 6489–6499.

95. Chiarugi, P.; Cirri, P.; Taddei, L.; Giannoni, E.; Camici, G.; Manao, G.; Raugei, G.; Ramponi, G. The low M(r) protein-tyrosine phosphatase is involved in Rho-mediated cytoskeleton rearrangement after integrin and platelet-derived growth factor stimulation. *J. Biol. Chem.* **2000**, *275*, 4640–4646. [CrossRef] [PubMed]

96. Malentacchi, F.; Marzocchini, R.; Gelmini, S.; Orlando, C.; Serio, M.; Ramponi, G.; Raugei, G. Up-regulated expression of low molecular weight protein tyrosine phosphatases in different human cancers. *Biochem. Biophys. Res. Commun.* **2005**, *334*, 875–883. [CrossRef] [PubMed]

97. Marzocchini, R.; Malentacchi, F.; Biagini, M.; Cirelli, D.; Luceri, C.; Caderni, G.; Raugei, G. The expression of low molecular weight protein tyrosine phosphatase is up-regulated in 1,2-dimethylhydrazine-induced colon tumours in rats. *Int. J. Cancer* **2008**, *122*, 1675–1678. [CrossRef] [PubMed]

98. Ferreira, P.A.; Ruela-de-Sousa, R.R.; Queiroz, K.C.S.; Souza, A.C.S.; Milani, R.; Pilli, R.A.; Peppelenbosch, M.P.; den Hertog, J.; Ferreira, C.V. Knocking down low molecular weight protein tyrosine phosphatase (LMW-PTP) reverts chemoresistance through inactivation of Src and Bcr-Abl proteins. *PLoS ONE* **2012**, *7*, e44312. [CrossRef]

99. Capitani, N.; Lori, G.; Paoli, P.; Patrussi, L.; Troilo, A.; Baldari, C.T.; Raugei, G.; D'Elios, M.M. LMW-PTP targeting potentiates the effects of drugs used in chronic lymphocytic leukemia therapy. *Cancer Cell Int.* **2019**, *19*, 67. [CrossRef]

100. Alho, I.; Costa, L.; Bicho, M.; Coelho, C. Low molecular weight protein tyrosine phosphatase isoforms regulate breast cancer cells migration through a RhoA dependent mechanism. *PLoS ONE* **2013**, *8*, e76307. [CrossRef]

101. The Human Protein Atlas. Available online: https://www.proteinatlas.org/ENSG00000143727-ACP1/pathology/melanoma (accessed on 27 August 2020).

102. Sato, T.; Irie, S.; Kitada, S.; Reed, J. FAP-1: A protein tyrosine phosphatase that associates with Fas. *Science* **1995**, *268*, 411–415. [CrossRef]

103. Nakai, Y.; Irie, S.; Sato, T.-A. Identification of IκBα as a substrate of Fas-associated phosphatase-1. *Eur. J. Biochem.* **2000**, *267*, 7170–7175. [CrossRef]

104. Ungefroren, H.; Voss, M.; Jansen, M.; Roeder, C.; Henne-Bruns, D.; Kremer, B.; Kalthoff, H. Human pancreatic adenocarcinomas express Fas and Fas ligand yet are resistant to Fas-mediated apoptosis. *Cancer Res.* **1998**, *58*, 1741–1749.

105. Ungefroren, H.; Kruse, M.L.; Trauzold, A.; Roeschmann, S.; Roeder, C.; Arlt, A.; Henne-Bruns, D.; Kalthoff, H. FAP-1 in pancreatic cancer cells: Functional and mechanistic studies on its inhibitory role in CD95-mediated apoptosis. *J. Cell Sci.* **2001**, *114*, 2735–2746.

106. Ivanov, V.N.; Lopez Bergami, P.; Maulit, G.; Sato, T.-A.; Sassoon, D.; Ronai, Z. FAP-1 association with Fas (Apo-1) inhibits Fas expression on the cell surface. *Mol. Cell. Biol.* **2003**, *23*, 3623–3635. [CrossRef] [PubMed]

107. Ying, J.; Li, H.; Cui, Y.; Wong, A.H.Y.; Langford, C.; Tao, Q. Epigenetic disruption of two proapoptotic genes MAPK10/JNK3 and PTPN13/FAP-1 in multiple lymphomas and carcinomas through hypermethylation of a common bidirectional promoter. *Leukemia* **2006**, *20*, 1173–1175. [CrossRef] [PubMed]

108. Yeh, S.-H.; Wu, D.-C.; Tsai, C.-Y.; Kuo, T.-J.; Yu, W.-C.; Chang, Y.-S.; Chen, C.-L.; Chang, C.-F.; Chen, D.-S.; Chen, P.-J. Genetic characterization of fas-associated phosphatase-1 as a putative tumor suppressor gene on chromosome 4q21.3 in hepatocellular carcinoma. *Clin. Cancer Res.* **2006**, *12*, 1097–1108. [CrossRef] [PubMed]

109. Abaan, O.D.; Toretsky, J.A. PTPL1: A large phosphatase with a split personality. *Cancer Metastasis Rev.* **2008**, *27*, 205–214. [CrossRef] [PubMed]

110. Shen, J.; Zhang, Y.; Yu, H.; Shen, B.; Liang, Y.; Jin, R.; Liu, X.; Shi, L.; Cai, X. Role of DUSP1/MKP1 in tumorigenesis, tumor progression and therapy. *Cancer Med.* **2016**, *5*, 2061–2068. [CrossRef] [PubMed]

111. Chattopadhyay, S.; Machado-Pinilla, R.; Manguan-García, C.; Belda-Iniesta, C.; Moratilla, C.; Cejas, P.; Fresno-Vara, J.A.; de Castro-Carpeño, J.; Casado, E.; Nistal, M.; et al. MKP1/CL100 controls tumor growth and sensitivity to cisplatin in non-small-cell lung cancer. *Oncogene* **2006**, *25*, 3335–3345. [CrossRef]

112. Wang, Z.; Xu, J.; Zhou, J.-Y.; Liu, Y.; Wu, G.S. Mitogen-Activated Protein Kinase Phosphatase-1 Is Required for Cisplatin Resistance. *Cancer Res.* **2006**, *66*, 8870–8877. [CrossRef]

113. The Human Protein Atlas. Available online: https://www.proteinatlas.org/ENSG00000120129-DUSP1/pathology/melanoma (accessed on 27 August 2020).

114. Liao, Q.; Guo, J.; Kleeff, J.; Zimmermann, A.; Büchler, M.W.; Korc, M.; Friess, H. Down-regulation of the dual-specificity phosphatase MKP-1 suppresses tumorigenicity of pancreatic cancer cells. *Gastroenterology* **2003**, *124*, 1830–1845. [CrossRef]

115. Mizuno, R.; Oya, M.; Shiomi, T.; Marumo, K.; Okada, Y.; Murai, M. Inhibition of MKP-1 expression potentiates JNK related apoptosis in renal cancer cells. *J. Urol.* **2004**, *172*, 723–727. [CrossRef]

116. Kundu, S.; Fan, K.; Cao, M.; Lindner, D.J.; Tuthill, R.; Liu, L.; Gerson, S.; Borden, E.; Yi, T. Tyrosine phosphatase inhibitor-3 sensitizes melanoma and colon cancer to biotherapeutics and chemotherapeutics. *Mol. Cancer Ther.* **2010**, *9*, 2287–2296. [CrossRef] [PubMed]

117. The Human Protein Atlas. Available online: https://www.proteinatlas.org/ENSG00000184007-PTP4A2/pathology/melanoma (accessed on 27 August 2020).

118. The Human Protein Atlas. Available online: https://www.proteinatlas.org/ENSG00000184489-PTP4A3/pathology/melanoma (accessed on 27 August 2020).

119. Wei, M.; Korotkov, K.V.; Blackburn, J.S. Targeting phosphatases of regenerating liver (PRLs) in cancer. *Pharmacol. Ther.* **2018**, *190*, 128–138. [CrossRef] [PubMed]

120. Duciel, L.; Monraz Gomez, L.C.; Kondratova, M.; Kuperstein, I.; Saule, S. The Phosphatase PRL-3 Is Involved in Key Steps of Cancer Metastasis. *J. Mol. Biol.* **2019**, *431*, 3056–3067. [CrossRef] [PubMed]

121. McParland, V.; Varsano, G.; Li, X.; Thornton, J.; Baby, J.; Aravind, A.; Meyer, C.; Pavic, K.; Rios, P.; Köhn, M. The metastasis-promoting phosphatase PRL-3 shows activity toward phosphoinositides. *Biochemistry* **2011**, *50*, 7579–7590. [CrossRef] [PubMed]

122. Fiordalisi, J.J.; Dewar, B.J.; Graves, L.M.; Madigan, J.P.; Cox, A.D. Src-Mediated Phosphorylation of the Tyrosine Phosphatase PRL-3 Is Required for PRL-3 Promotion of Rho Activation, Motility and Invasion. *PLoS ONE* **2013**, *8*, e64309. [CrossRef] [PubMed]

123. Maacha, S.; Anezo, O.; Foy, M.; Liot, G.; Mery, L.; Laurent, C.; Sastre-Garau, X.; Piperno-Neumann, S.; Cassoux, N.; Planque, N.; et al. Protein Tyrosine Phosphatase 4A3 (PTP4A3) Promotes Human Uveal Melanoma Aggressiveness through Membrane Accumulation of Matrix Metalloproteinase 14 (MMP14). *Investig. Opthalmology Vis. Sci.* **2016**, *57*, 1982. [CrossRef]

124. Wang, H.; Quah, S.Y.; Dong, J.M.; Manser, E.; Tang, J.P.; Zeng, Q. PRL-3 Down-regulates PTEN Expression and Signals through PI3K to Promote Epithelial-Mesenchymal Transition. *Cancer Res.* **2007**, *67*, 2922–2926. [CrossRef]

125. Li, Q.; Bai, Y.; Lyle, L.T.; Yu, G.; Amarasinghe, O.; Nguele Meke, F.; Carlock, C.; Zhang, Z.-Y. Mechanism of PRL2 phosphatase-mediated PTEN degradation and tumorigenesis. *Proc. Natl. Acad. Sci. USA* **2020**, *117*, 20538–20548. [CrossRef] [PubMed]

126. Wu, X.; Zeng, H.; Zhang, X.; Zhao, Y.; Sha, H.; Ge, X.; Zhang, M.; Gao, X.; Xu, Q. Phosphatase of regenerating liver-3 promotes motility and metastasis of mouse melanoma cells. *Am. J. Pathol.* **2004**, *164*, 2039–2054. [CrossRef]

127. Qian, F.; Li, Y.-P.; Sheng, X.; Zhang, Z.-C.; Song, R.; Dong, W.; Cao, S.-X.; Hua, Z.-C.; Xu, Q. PRL-3 siRNA inhibits the metastasis of B16-BL6 mouse melanoma cells in vitro and in vivo. *Mol. Med.* **2007**, *13*, 151–159. [CrossRef]

128. Laurent, C.; Valet, F.; Planque, N.; Silveri, L.; Maacha, S.; Anezo, O.; Hupe, P.; Plancher, C.; Reyes, C.; Albaud, B.; et al. High PTP4A3 phosphatase expression correlates with metastatic risk in uveal melanoma patients. *Cancer Res.* **2011**, *71*, 666–674. [CrossRef]

129. Duciel, L.; Anezo, O.; Mandal, K.; Laurent, C.; Planque, N.; Coquelle, F.M.; Gentien, D.; Manneville, J.-B.; Saule, S. Protein tyrosine phosphatase 4A3 (PTP4A3/PRL-3) promotes the aggressiveness of human uveal melanoma through dephosphorylation of CRMP2. *Sci. Rep.* **2019**, *9*, 2990. [CrossRef] [PubMed]

130. Pathak, M.K.; Dhawan, D.; Lindner, D.J.; Borden, E.C.; Farver, C.; Yi, T. Pentamidine is an inhibitor of PRL phosphatases with anticancer activity. *Mol. Cancer Ther.* **2002**, *1*, 1255–1264. [PubMed]

131. Sun, J.-P.; Luo, Y.; Yu, X.; Wang, W.-Q.; Zhou, B.; Liang, F.; Zhang, Z.-Y. Phosphatase activity, trimerization, and the C-terminal polybasic region are all required for PRL1-mediated cell growth and migration. *J. Biol. Chem.* **2007**, *282*, 29043–29051. [CrossRef] [PubMed]

132. Bai, Y.; Yu, Z.H.; Liu, S.; Zhang, L.; Zhang, R.Y.; Zeng, L.F.; Zhang, S.; Zhang, Z.Y. Novel anticancer agents based on targeting the trimer interface of the PRL phosphatase. *Cancer Res.* **2016**, *76*, 4805–4815. [CrossRef] [PubMed]

133. Giménez-Mascarell, P.; González-Recio, I.; Fernández-Rodríguez, C.; Oyenarte, I.; Müller, D.; Martínez-Chantar, M.; Martínez-Cruz, L. Current Structural Knowledge on the CNNM Family of Magnesium Transport Mediators. *Int. J. Mol. Sci.* **2019**, *20*, 1135. [CrossRef] [PubMed]

134. Hardy, S.; Kostantin, E.; Hatzihristidis, T.; Zolotarov, Y.; Uetani, N.; Tremblay, M.L. Physiological and oncogenic roles of the PRL phosphatases. *FEBS J.* **2018**, *285*, 3886–3908. [CrossRef]

135. Hardy, S.; Uetani, N.; Wong, N.; Kostantin, E.; Labbé, D.P.; Bégin, L.R.; Mes-Masson, A.; Miranda-Saavedra, D.; Tremblay, M.L. The protein tyrosine phosphatase PRL-2 interacts with the magnesium transporter CNNM3 to promote oncogenesis. *Oncogene* **2015**, *34*, 986–995. [CrossRef]

136. Thura, M.; Al-Aidaroos, A.Q.; Gupta, A.; Chee, C.E.; Lee, S.C.; Hui, K.M.; Li, J.; Guan, Y.K.; Yong, W.P.; So, J.; et al. PRL3-zumab as an immunotherapy to inhibit tumors expressing PRL3 oncoprotein. *Nat. Commun.* **2019**, *10*, 2484. [CrossRef]

137. Freeman, R.M.; Plutzky, J.; Neel, B.G. Identification of a human src homology 2-containing protein-tyrosine-phosphatase: A putative homolog of Drosophila corkscrew. *Proc. Natl. Acad. Sci. USA* **1992**, *89*, 11239–11243. [CrossRef]

138. Hof, P.; Pluskey, S.; Dhe-Paganon, S.; Eck, M.J.; Shoelson, S.E. Crystal Structure of the Tyrosine Phosphatase SHP-2. *Cell* **1998**, *92*, 441–450. [CrossRef]

139. QU, C.K. The SHP-2 tyrosine phosphatase: Signaling mechanisms and biological functions. *Cell Res.* **2000**, *10*, 279–288. [CrossRef] [PubMed]

140. Tartaglia, M.; Niemeyer, C.M.; Fragale, A.; Song, X.; Buechner, J.; Jung, A.; Hählen, K.; Hasle, H.; Licht, J.D.; Gelb, B.D. Somatic mutations in PTPN11 in juvenile myelomonocytic leukemia, myelodysplastic syndromes and acute myeloid leukemia. *Nat. Genet.* **2003**, *34*, 148–150. [CrossRef] [PubMed]

141. Tartaglia, M.; Mehler, E.L.; Goldberg, R.; Zampino, G.; Brunner, H.G.; Kremer, H.; van der Burgt, I.; Crosby, A.H.; Ion, A.; Jeffery, S.; et al. Mutations in PTPN11, encoding the protein tyrosine phosphatase SHP-2, cause Noonan syndrome. *Nat. Genet.* **2001**, *29*, 465–468. [CrossRef]

142. Hu, Z.-Q.; Ma, R.; Zhang, C.; Li, J.; Li, L.; Hu, Z.-T.; Gao, Q.; Li, W.-M. Expression and clinical significance of tyrosine phosphatase SHP2 in thyroid carcinoma. *Oncol. Lett.* **2015**, *10*, 1507–1512. [CrossRef] [PubMed]

143. Dong, S.; Li, F.-Q.; Zhang, Q.; Lv, K.-Z.; Yang, H.-L.; Gao, Y.; Yu, J.-R. Expression and Clinical Significance of SHP2 in Gastric Cancer. *J. Int. Med. Res.* **2012**, *40*, 2083–2089. [CrossRef]

144. Leibowitz, M.S.; Srivastava, R.M.; Andrade Filho, P.A.; Egloff, A.M.; Wang, L.; Seethala, R.R.; Ferrone, S.; Ferris, R.L. SHP2 Is Overexpressed and Inhibits pSTAT1-Mediated APM Component Expression, T-cell Attracting Chemokine Secretion, and CTL Recognition in Head and Neck Cancer Cells. *Clin. Cancer Res.* **2013**, *19*, 798–808. [CrossRef]

145. Hu, Z.; Fang, H.; Wang, X.; Chen, D.; Chen, Z.; Wang, S. Overexpression of SHP2 tyrosine phosphatase promotes the tumorigenesis of breast carcinoma. *Oncol. Rep.* **2014**, *32*, 205–212. [CrossRef]

146. Xie, H.; Huang, S.; Li, W.; Zhao, H.; Zhang, T.; Zhang, D. Upregulation of Src homology phosphotyrosyl phosphatase 2 (Shp2) expression in oral cancer and knockdown of Shp2 expression inhibit tumor cell viability and invasion in vitro. *Oral Surg. Oral Med. Oral Pathol. Oral Radiol.* **2014**, *117*, 234–242. [CrossRef]

147. Han, T.; Xiang, D.-M.; Sun, W.; Liu, N.; Sun, H.-L.; Wen, W.; Shen, W.-F.; Wang, R.-Y.; Chen, C.; Wang, X.; et al. PTPN11/Shp2 overexpression enhances liver cancer progression and predicts poor prognosis of patients. *J. Hepatol.* **2015**, *63*, 651–660. [CrossRef]

148. Zheng, J.; Huang, S.; Huang, Y.; Song, L.; Yin, Y.; Kong, W.; Chen, X.; Ouyang, X. Expression and prognosis value of SHP2 in patients with pancreatic ductal adenocarcinoma. *Tumor Biol.* **2016**, *37*, 7853–7859. [CrossRef]

149. Zhang, K.; Zhao, H.; Ji, Z.; Zhang, C.; Zhou, P.; Wang, L.; Chen, Q.; Wang, J.; Zhang, P.; Chen, Z.; et al. Shp2 promotes metastasis of prostate cancer by attenuating the PAR3/PAR6/aPKC polarity protein complex and enhancing epithelial-to-mesenchymal transition. *Oncogene* **2016**, *35*, 1271–1282. [CrossRef] [PubMed]

150. The Human Protein Atlas. Available online: https://www.proteinatlas.org/ENSG00000179295-PTPN11/pathology/melanoma (accessed on 27 August 2020).

151. Yuan, X.; Bu, H.; Zhou, J.; Yang, C.-Y.; Zhang, H. Recent Advances of SHP2 Inhibitors in Cancer Therapy: Current Development and Clinical Application. *J. Med. Chem.* **2020**. [CrossRef] [PubMed]

152. Berman, J.D. Human leishmaniasis: Clinical, diagnostic, and chemotherapeutic developments in the last 10 years. *Clin. Infect. Dis.* **1997**, *24*, 684–703. [CrossRef]

153. Pathak, M.K.; Yi, T. Sodium stibogluconate is a potent inhibitor of protein tyrosine phosphatases and augments cytokine responses in hemopoietic cell lines. *J. Immunol.* **2001**, *167*, 3391–3397. [CrossRef] [PubMed]

154. Yi, T.; Pathak, M.K.; Lindner, D.J.; Ketterer, M.E.; Farver, C.; Borden, E.C. Anticancer Activity of Sodium Stibogluconate in Synergy with IFNs. *J. Immunol.* **2002**, *169*, 5978–5985. [CrossRef]

155. Win-Piazza, H.; Schneeberger, V.E.; Chen, L.; Pernazza, D.; Lawrence, H.R.; Sebti, S.M.; Lawrence, N.J.; Wu, J. Enhanced anti-melanoma efficacy of interferon alfa-2b via inhibition of Shp2. *Cancer Lett.* **2012**, *320*, 81–85. [CrossRef] [PubMed]

156. Soong, J.; Scott, G. Plexin B1 inhibits MET through direct association and regulates Shp2 expression in melanocytes. *J. Cell Sci.* **2013**, *126*, 688–695. [CrossRef]

157. Li, J.; Reed, S.A.; Johnson, S.E. Hepatocyte growth factor (HGF) signals through SHP2 to regulate primary mouse myoblast proliferation. *Exp. Cell Res.* **2009**, *315*, 2284–2292. [CrossRef]

158. Schaeper, U.; Gehring, N.H.; Fuchs, K.P.; Sachs, M.; Kempkes, B.; Birchmeier, W. Coupling of Gab1 to C-Met, Grb2, and Shp2 Mediates Biological Responses. *J. Cell Biol.* **2000**, *149*, 1419–1432. [CrossRef]

159. Wilson, T.R.; Fridlyand, J.; Yan, Y.; Penuel, E.; Burton, L.; Chan, E.; Peng, J.; Lin, E.; Wang, Y.; Sosman, J.; et al. Widespread potential for growth-factor-driven resistance to anticancer kinase inhibitors. *Nature* **2012**, *487*, 505–509. [CrossRef]

160. Ahmed, T.A.; Adamopoulos, C.; Karoulia, Z.; Wu, X.; Sachidanandam, R.; Aaronson, S.A.; Poulikakos, P.I. SHP2 Drives Adaptive Resistance to ERK Signaling Inhibition in Molecularly Defined Subsets of ERK-Dependent Tumors. *Cell Rep.* **2019**, *26*, 65–78. [CrossRef]

161. Dance, M.; Montagner, A.; Salles, J.-P.; Yart, A.; Raynal, P. The molecular functions of Shp2 in the Ras/Mitogen-activated protein kinase (ERK1/2) pathway. *Cell. Signal.* **2008**, *20*, 453–459. [CrossRef] [PubMed]

162. Grossmann, K.S.; Rosário, M.; Birchmeier, C.; Birchmeier, W. The Tyrosine Phosphatase Shp2 in Development and Cancer. *Adv. Cancer Res.* **2010**, *106*, 53–89. [PubMed]

163. Zeng, L.-F.; Zhang, R.-Y.; Yu, Z.-H.; Li, S.; Wu, L.; Gunawan, A.M.; Lane, B.S.; Mali, R.S.; Li, X.; Chan, R.J.; et al. Therapeutic Potential of Targeting the Oncogenic SHP2 Phosphatase. *J. Med. Chem.* **2014**, *57*, 6594–6609. [CrossRef] [PubMed]

164. LaMarche, M.J.; Acker, M.G.; Argintaru, A.; Bauer, D.; Boisclair, J.; Chan, H.; Chen, C.; Chen, Y.-N.P.; Chen, Z.; Deng, Z.; et al. Identification of TNO155, an Allosteric SHP2 Inhibitor for the Treatment of Cancer. *J. Med. Chem.* **2020**. [CrossRef] [PubMed]

165. Tucci, M.; Passarelli, A.; Mannavola, F.; Felici, C.; Stucci, L.S.; Cives, M.; Silvestris, F. Immune System Evasion as Hallmark of Melanoma Progression: The Role of Dendritic Cells. *Front. Oncol.* **2019**, *9*, 1148. [CrossRef] [PubMed]

166. Mahmoud, F.; Shields, B.; Makhoul, I.; Avaritt, N.; Wong, H.K.; Hutchins, L.F.; Shalin, S.; Tackett, A.J. Immune surveillance in melanoma: From immune attack to melanoma escape and even counterattack. *Cancer Biol. Ther.* **2017**, *18*, 451–469. [CrossRef]

167. Shresta, S.; Pham, C.T.; Thomas, D.A.; Graubert, T.A.; Ley, T.J. How do cytotoxic lymphocytes kill their targets? *Curr. Opin. Immunol.* **1998**, *10*, 581–587. [CrossRef]

168. Erdag, G.; Schaefer, J.T.; Smolkin, M.E.; Deacon, D.H.; Shea, S.M.; Dengel, L.T.; Patterson, J.W.; Slingluff, C.L. Immunotype and immunohistologic characteristics of tumor-infiltrating immune cells are associated with clinical outcome in metastatic melanoma. *Cancer Res.* **2012**, *72*, 1070–1080. [CrossRef]

169. Haanen, J.B.A.G. Immunotherapy of melanoma. *EJC Suppl.* **2013**, *11*, 97–105. [CrossRef]

170. Deeks, E.D. Pembrolizumab: A Review in Advanced Melanoma. *Drugs* **2016**, *76*, 375–386. [CrossRef] [PubMed]

171. Atrash, S.; Makhoul, I.; Mizell, J.S.; Hutchins, L.; Mahmoud, F. Response of metastatic mucosal melanoma to immunotherapy: It can get worse before it gets better. *J. Oncol. Pharm. Pract.* **2017**, *23*, 215–219. [CrossRef] [PubMed]

172. Hodi, F.S.; O'Day, S.J.; McDermott, D.F.; Weber, R.W.; Sosman, J.A.; Haanen, J.B.; Gonzalez, R.; Robert, C.; Schadendorf, D.; Hassel, J.C.; et al. Improved survival with ipilimumab in patients with metastatic melanoma. *N. Engl. J. Med.* **2010**, *363*, 711–723. [CrossRef]

173. Johnson, D.B.; Peng, C.; Sosman, J.A. Nivolumab in melanoma: Latest evidence and clinical potential. *Ther. Adv. Med. Oncol.* **2015**, *7*, 97–106. [CrossRef] [PubMed]

174. Restifo, N.P.; Smyth, M.J.; Snyder, A. Acquired resistance to immunotherapy and future challenges. *Nat. Rev. Cancer* **2016**, *16*, 121–126. [CrossRef]

175. O'Donnell, J.S.; Long, G.V.; Scolyer, R.A.; Teng, M.W.L.; Smyth, M.J. Resistance to PD1/PDL1 checkpoint inhibition. *Cancer Treat. Rev.* **2017**, *52*, 71–81. [CrossRef]

176. Lorenz, U. SHP-1 and SHP-2 in T cells: Two phosphatases functioning at many levels. *Immunol. Rev.* **2009**, *228*, 342–359. [CrossRef]

177. Dempke, W.C.M.; Uciechowski, P.; Fenchel, K.; Chevassut, T. Targeting SHP-1, 2 and SHIP Pathways: A novel strategy for cancer treatment? *Oncology* **2018**, *95*, 257–269. [CrossRef]

178. Yang, J.; Liu, L.; He, D.; Song, X.; Liang, X.; Zhao, Z.J.; Zhou, G.W. Crystal structure of human protein-tyrosine phosphatase SHP-1. *J. Biol. Chem.* **2003**, *278*, 6516–6520. [CrossRef]

179. Varone, A.; Spano, D.; Corda, D. Shp1 in Solid Cancers and Their Therapy. *Front. Oncol.* **2020**, *10*, 935. [CrossRef]

180. Takeuchi, S.; Matsushita, M.; Zimmermann, M.; Ikezoe, T.; Komatsu, N.; Seriu, T.; Schrappe, M.; Bartram, C.R.; Koeffler, H.P. Clinical significance of aberrant DNA methylation in childhood acute lymphoblastic leukemia. *Leuk. Res.* **2011**, *35*, 1345–1349. [CrossRef] [PubMed]

181. Li, Y.; Yang, L.; Pan, Y.; Yang, J.; Shang, Y.; Luo, J. Methylation and decreased expression of SHP-1 are related to disease progression in chronic myelogenous leukemia. *Oncol. Rep.* **2014**, *31*, 2438–2446. [CrossRef] [PubMed]

182. Küçük, C.; Hu, X.; Jiang, B.; Klinkebiel, D.; Geng, H.; Gong, Q.; Bouska, A.; Iqbal, J.; Gaulard, P.; McKeithan, T.W.; et al. Global promoter methylation analysis reveals novel candidate tumor suppressor genes in natural killer cell lymphoma. *Clin. Cancer Res.* **2015**, *21*, 1699–1711. [CrossRef]

183. Ding, K.; Chen, X.; Wang, Y.; Liu, H.; Song, W.; Li, L.; Wang, G.; Song, J.; Shao, Z.; Fu, R. Plasma DNA methylation of p16 and shp1 in patients with B cell non-Hodgkin lymphoma. *Int. J. Clin. Oncol.* **2017**, *22*, 585–592. [CrossRef] [PubMed]

184. Liu, J.; Yaming, W.; Sun, X.; Ji, N.; Sun, S.; Yajie, W.; Liu, F.; Cui, Q.; Chen, W.; Liu, Y. Promoter methylation attenuates SHP1 expression and function in patients with primary central nervous system lymphoma. *Oncol. Rep.* **2017**, *37*, 887–894. [CrossRef] [PubMed]

185. Joo, M.K.; Park, J.-J.; Yoo, H.S.; Lee, B.J.; Chun, H.J.; Lee, S.W.; Bak, Y.-T. Epigenetic regulation and anti-tumorigenic effects of SH2-containing protein tyrosine phosphatase 1 (SHP1) in human gastric cancer cells. *Tumor Biol.* **2016**, *37*, 4603–4612. [CrossRef]

186. Sheng, Y.; Wang, H.; Liu, D.; Zhang, C.; Deng, Y.; Yang, F.; Zhang, T.; Zhang, C. Methylation of tumor suppressor gene CDH13 and SHP1 promoters and their epigenetic regulation by the UHRF1/PRMT5 complex in endometrial carcinoma. *Gynecol. Oncol.* **2016**, *140*, 145–151. [CrossRef]

187. Tassidis, H.; Brokken, L.J.; Jirström, K.; Ehrnström, R.; Pontén, F.; Ulmert, D.; Bjartell, A.; Härkönen, P.; Wingren, A.G. Immunohistochemical detection of tyrosine phosphatase SHP-1 predicts outcome after radical prostatectomy for localized prostate cancer. *Int. J. Cancer* **2010**, *126*, 2296–2307. [CrossRef]

188. Zhang, Y.; Zhao, D.; Zhao, H.; Wu, X.; Zhao, W.; Wang, Y.; Xia, B.; Da, W. Hypermethylation of SHP-1 promoter in patient with high-risk myelodysplastic syndrome and it predicts poor prognosis. *Med. Oncol.* **2012**, *29*, 2359–2363. [CrossRef]

189. Chanida, V.; Poonchavist, C.; Virote, S.; Apiwat, M. The role of SHP-1 promoter 2 hypermethylation detection of lymph node micrometastasis in resectable stage I non-small cell lung cancer as a prognostic marker of disease recurrence. *Int. J. Clin. Oncol.* **2014**, *19*, 586–592. [CrossRef]

190. Tibaldi, E.; Zonta, F.; Bordin, L.; Magrin, E.; Gringeri, E.; Cillo, U.; Idotta, G.; Pagano, M.A.; Brunati, A.M. The tyrosine phosphatase SHP-1 inhibits proliferation of activated hepatic stellate cells by impairing PDGF receptor signaling. *Biochim. Biophys. Acta Mol. Cell Res.* **2014**, *1843*, 288–298. [CrossRef] [PubMed]

191. Bousquet, C.; Delesque, N.; Lopez, F.; Saint-Laurent, N.; Estève, J.-P.; Bedecs, K.; Buscail, L.; Vaysse, N.; Susini, C. sst2 Somatostatin Receptor Mediates Negative Regulation of Insulin Receptor Signaling through the Tyrosine Phosphatase SHP-1. *J. Biol. Chem.* **1998**, *273*, 7099–7106. [CrossRef] [PubMed]

192. Nakagami, H.; Cui, T.-X.; Iwai, M.; Shiuchi, T.; Takeda-Matsubara, Y.; Wu, L.; Horiuchi, M. Tumor Necrosis Factor-α Inhibits Growth Factor–Mediated Cell Proliferation Through SHP-1 Activation in Endothelial Cells. *Arterioscler. Thromb. Vasc. Biol.* **2002**, *22*, 238–242. [CrossRef] [PubMed]

193. Insabato, L.; Amelio, I.; Quarto, M.; Zannetti, A.; Tolino, F.; de Mauro, G.; Cerchia, L.; Riccio, P.; Baumhoer, D.; Condorelli, G.; et al. Elevated Expression of the Tyrosine Phosphatase SHP-1 Defines a Subset of High-Grade Breast Tumors. *Oncology* **2009**, *77*, 378–384. [CrossRef] [PubMed]

194. Mok, S.C.; Kwok, T.T.; Berkowitz, R.S.; Barrett, A.J.; Tsui, F.W.L. Overexpression of the Protein Tyrosine Phosphatase, Nonreceptor Type 6 (PTPN6), in Human Epithelial Ovarian Cancer. *Gynecol. Oncol.* **1995**, *57*, 299–303. [CrossRef] [PubMed]

195. Croker, B.A.; Lawson, B.R.; Rutschmann, S.; Berger, M.; Eidenschenk, C.; Blasius, A.L.; Moresco, E.M.Y.; Sovath, S.; Cengia, L.; Shultz, L.D.; et al. Inflammation and autoimmunity caused by a SHP1 mutation depend on IL-1, MyD88, and a microbial trigger. *Proc. Natl. Acad. Sci. USA* **2008**, *105*, 15028–15033. [CrossRef]

196. Fowler, C.C.; Pao, L.I.; Blattman, J.N.; Greenberg, P.D. SHP-1 in T cells limits the production of CD8 effector cells without impacting the formation of long-lived central memory cells. *J. Immunol.* **2010**, *185*, 3256–3267. [CrossRef]

197. Kilgore, N.E.; Carter, J.D.; Lorenz, U.; Evavold, B.D. Cutting edge: Dependence of TCR antagonism on Src homology 2 domain-containing protein tyrosine phosphatase activity. *J. Immunol.* **2003**, *170*, 4891–4895. [CrossRef]

198. Watson, H.A.; Dolton, G.; Ohme, J.; Ladell, K.; Vigar, M.; Wehenkel, S.; Hindley, J.; Mohammed, R.N.; Miners, K.; Luckwell, R.A.; et al. Purity of transferred CD8(+) T cells is crucial for safety and efficacy of combinatorial tumor immunotherapy in the absence of SHP-1. *Immunol. Cell Biol.* **2016**, *94*, 802–808. [CrossRef]

199. Sathish, J.G.; Dolton, G.; Leroy, F.G.; Matthews, R.J. Loss of Src homology region 2 domain-containing protein tyrosine phosphatase-1 increases CD8+ T cell-APC conjugate formation and is associated with enhanced in vivo CTL function. *J. Immunol.* **2007**, *178*, 330–337. [CrossRef]

200. Stromnes, I.M.; Fowler, C.; Casamina, C.C.; Georgopolos, C.M.; McAfee, M.S.; Schmitt, T.M.; Tan, X.; Kim, T.-D.; Choi, I.; Blattman, J.N.; et al. Abrogation of SRC homology region 2 domain-containing phosphatase 1 in tumor-specific T cells improves efficacy of adoptive immunotherapy by enhancing the effector function and accumulation of short-lived effector T cells in vivo. *J. Immunol.* **2012**, *189*, 1812–1825. [CrossRef] [PubMed]

201. Bollu, L.R.; Mazumdar, A.; Savage, M.I.; Brown, P.H. Molecular Pathways: Targeting Protein Tyrosine Phosphatases in Cancer. *Clin. Cancer Res.* **2017**, *23*, 2136–2142. [CrossRef] [PubMed]

202. Yu, Z.-H.; Zhang, Z.-Y. Regulatory Mechanisms and Novel Therapeutic Targeting Strategies for Protein Tyrosine Phosphatases. *Chem. Rev.* **2018**, *118*, 1069–1091. [CrossRef] [PubMed]

203. Naing, A.; Reuben, J.M.; Camacho, L.H.; Gao, H.; Lee, B.-N.; Cohen, E.N.; Verschraegen, C.; Stephen, S.; Aaron, J.; Hong, D.; et al. Phase I Dose Escalation Study of Sodium Stibogluconate (SSG), a Protein Tyrosine Phosphatase Inhibitor, Combined with Interferon Alpha for Patients with Solid Tumors. *J. Cancer* **2011**, *2*, 81–89. [CrossRef] [PubMed]

204. Wang, S.-F.; Fouquet, S.; Chapon, M.; Salmon, H.; Regnier, F.; Labroquère, K.; Badoual, C.; Damotte, D.; Validire, P.; Maubec, E.; et al. Early T cell signalling is reversibly altered in PD-1+ T lymphocytes infiltrating human tumors. *PLoS ONE* **2011**, *6*, e17621. [CrossRef]

205. Kundu, S.; Fan, K.; Cao, M.; Lindner, D.J.; Zhao, Z.J.; Borden, E.; Yi, T. Novel SHP-1 inhibitors tyrosine phosphatase inhibitor-1 and analogs with preclinical anti-tumor activities as tolerated oral agents. *J. Immunol.* **2010**, *184*, 6529–6536. [CrossRef]

206. Ramachandran, I.R.; Song, W.; Lapteva, N.; Seethammagari, M.; Slawin, K.M.; Spencer, D.M.; Levitt, J.M. The Phosphatase Src Homology Region 2 Domain-Containing Phosphatase-1 Is an Intrinsic Central Regulator of Dendritic Cell Function. *J. Immunol.* **2011**, *186*, 3934–3945. [CrossRef]

207. Snook, J.P.; Soedel, A.J.; Ekiz, H.A.; O'Connell, R.M.; Williams, M.A. Inhibition of SHP-1 Expands the Repertoire of Antitumor T Cells Available to Respond to Immune Checkpoint Blockade. *Cancer Immunol. Res.* **2020**, *8*, 506–517. [CrossRef]

208. Weiss, S.A.; Wolchok, J.D.; Sznol, M. Immunotherapy of Melanoma: Facts and Hopes. *Clin. Cancer Res.* **2019**, *25*, 5191–5201. [CrossRef]

209. Ai, L.; Xu, A.; Xu, J. Roles of PD-1/PD-L1 pathway: Signaling, cancer, and beyond. In *Regulation of Cancer Immune Checkpoints*; Springer: Singapore, 2020; pp. 33–59.

210. Wu, Y.; Deng, W.; McGinley, E.C.; Klinke, D.J. Melanoma exosomes deliver a complex biological payload that upregulates PTPN11 to suppress T lymphocyte function. *Pigment. Cell Melanoma Res.* **2017**, *30*, 203–218. [CrossRef]

211. Xiao, P.; Guo, Y.; Zhang, H.; Zhang, X.; Cheng, H.; Cao, Q.; Ke, Y. Myeloid-restricted ablation of Shp2 restrains melanoma growth by amplifying the reciprocal promotion of CXCL9 and IFN-γ production in tumor microenvironment. *Oncogene* **2018**, *37*, 5088–5100. [CrossRef] [PubMed]

212. Ramesh, A.; Kumar, S.; Nandi, D.; Kulkarni, A. CSF1R- and SHP2-Inhibitor-Loaded Nanoparticles Enhance Cytotoxic Activity and Phagocytosis in Tumor-Associated Macrophages. *Adv. Mater.* **2019**, *31*, e1904364. [CrossRef] [PubMed]

213. Zhang, T.; Guo, W.; Yang, Y.; Liu, W.; Guo, L.; Gu, Y.; Shu, Y.; Wang, L.; Wu, X.; Hua, Z.; et al. Loss of SHP-2 activity in CD4+ T cells promotes melanoma progression and metastasis. *Sci. Rep.* **2013**, *3*, 2845. [CrossRef] [PubMed]

214. Dubé, N.; Tremblay, M.L. Involvement of the small protein tyrosine phosphatases TC-PTP and PTP1B in signal transduction and diseases: From diabetes, obesity to cell cycle, and cancer. *Biochim. Biophys. Acta Proteins Proteom.* **2005**, *1754*, 108–117. [CrossRef]

215. Doody, K.M.; Bourdeau, A.; Tremblay, M.L. T-cell protein tyrosine phosphatase is a key regulator in immune cell signaling: Lessons from the knockout mouse model and implications in human disease. *Immunol. Rev.* **2009**, *228*, 325–341. [CrossRef]

216. Heinonen, K.M.; Nestel, F.P.; Newell, E.W.; Charette, G.; Seemayer, T.A.; Tremblay, M.L.; Lapp, W.S. T-cell protein tyrosine phosphatase deletion results in progressive systemic inflammatory disease. *Blood* **2004**, *103*, 3457–3464. [CrossRef]

217. Bourdeau, A.; Trop, S.; Doody, K.M.; Dumont, D.J.; Tremblayef, M.L. Inhibition of T Cell Protein Tyrosine Phosphatase Enhances Interleukin-18-Dependent Hematopoietic Stem Cell Expansion. *Stem Cells* **2013**, *31*, 293–304. [CrossRef]

218. Galic, S.; Hauser, C.; Kahn, B.B.; Haj, F.G.; Neel, B.G.; Tonks, N.K.; Tiganis, T. Coordinated Regulation of Insulin Signaling by the Protein Tyrosine Phosphatases PTP1B and TCPTP. *Mol. Cell. Biol.* **2005**, *25*, 819–829. [CrossRef]

219. Loh, K.; Fukushima, A.; Zhang, X.; Galic, S.; Briggs, D.; Enriori, P.J.; Simonds, S.; Wiede, F.; Reichenbach, A.; Hauser, C.; et al. Elevated Hypothalamic TCPTP in Obesity Contributes to Cellular Leptin Resistance. *Cell Metab.* **2011**, *14*, 684–699. [CrossRef]

220. Morales, L.D.; Archbold, A.K.; Olivarez, S.; Slaga, T.J.; DiGiovanni, J.; Kim, D.J. The role of T-cell protein tyrosine phosphatase in epithelial carcinogenesis. *Mol. Carcinog.* **2019**, *58*, 1640–1647. [CrossRef]

221. Nishiyama-Fujita, Y.; Shimizu, T.; Sagawa, M.; Uchida, H.; Kizaki, M. The role of TC-PTP (PTPN2) in modulating sensitivity to imatinib and interferon-α in CML cell line, KT-1 cells. *Leuk. Res.* **2013**, *37*, 1150–1155. [CrossRef] [PubMed]

222. Karlsson, E.; Veenstra, C.; Emin, S.; Dutta, C.; Pérez-Tenorio, G.; Nordenskjöld, B.; Fornander, T.; Stål, O. Loss of protein tyrosine phosphatase, non-receptor type 2 is associated with activation of AKT and tamoxifen resistance in breast cancer. *Breast Cancer Res. Treat.* **2015**, *153*, 31–40. [CrossRef]

223. Karlsson, E.; Veenstra, C.; Gårsjö, J.; Nordenskjöld, B.; Fornander, T.; Stål, O. PTPN2 deficiency along with activation of nuclear Akt predict endocrine resistance in breast cancer. *J. Cancer Res. Clin. Oncol.* **2019**, *145*, 599–607. [CrossRef] [PubMed]

224. Young, R.M.; Polsky, A.; Refaeli, Y. TC-PTP is required for the maintenance of MYC-driven B-cell lymphomas. *Blood* **2009**, *114*, 5016–5023. [CrossRef] [PubMed]

225. Blanquart, C.; Karouri, S.-E.; Issad, T. Protein tyrosine phosphatase-1B and T-cell protein tyrosine phosphatase regulate IGF-2-induced MCF-7 cell migration. *Biochem. Biophys. Res. Commun.* **2010**, *392*, 83–88. [CrossRef]

226. Manguso, R.T.; Pope, H.W.; Zimmer, M.D.; Brown, F.D.; Yates, K.B.; Miller, B.C.; Collins, N.B.; Bi, K.; LaFleur, M.W.; Juneja, V.R.; et al. In vivo CRISPR screening identifies Ptpn2 as a cancer immunotherapy target. *Nature* **2017**, *547*, 413–418. [CrossRef]

227. LaFleur, M.W.; Nguyen, T.H.; Coxe, M.A.; Miller, B.C.; Yates, K.B.; Gillis, J.E.; Sen, D.R.; Gaudiano, E.F.; Al Abosy, R.; Freeman, G.J.; et al. PTPN2 regulates the generation of exhausted CD8+ T cell subpopulations and restrains tumor immunity. *Nat. Immunol.* **2019**, *20*, 1335–1347. [CrossRef]

228. Davies, H.; Bignell, G.R.; Cox, C.; Stephens, P.; Edkins, S.; Clegg, S.; Teague, J.; Woffendin, H.; Garnett, M.J.; Bottomley, W.; et al. Mutations of the BRAF gene in human cancer. *Nature* **2002**, *417*, 949–954. [CrossRef]

229. Amann, V.C.; Ramelyte, E.; Thurneysen, S.; Pitocco, R.; Bentele-Jaberg, N.; Goldinger, S.M.; Dummer, R.; Mangana, J. Developments in targeted therapy in melanoma. *Eur. J. Surg. Oncol.* **2017**, *43*, 581–593. [CrossRef]

230. Arozarena, I.; Wellbrock, C. Phenotype plasticity as enabler of melanoma progression and therapy resistance. *Nat. Rev. Cancer* **2019**, *19*, 377–391. [CrossRef]

231. Mullard, A. Phosphatases start shedding their stigma of undruggability. *Nat. Rev. Drug Discov.* **2018**, *17*, 847–849. [CrossRef]

232. Hangan-Steinman, D.; Ho, W.C.; Shenoy, P.; Chan, B.M.; Morris, V.L. Differences in phosphatase modulation of alpha4beta1 and alpha5beta1 integrin-mediated adhesion and migration of B16F1 cells. *Biochem. Cell Biol.* **1999**, *77*, 409–420. [CrossRef] [PubMed]

233. Bottini, N.; Stefanini, L.; Williams, S.; Alonso, A.; Jascur, T.; Abraham, R.T.; Couture, C.; Mustelin, T. Activation of ZAP-70 through specific dephosphorylation at the inhibitory Tyr-292 by the low molecular weight phosphotyrosine phosphatase (LMPTP). *J. Biol. Chem.* **2002**, *277*, 24220–24224. [CrossRef] [PubMed]

234. Mustelin, T.; Vang, T.; Bottini, N. Protein tyrosine phosphatases and the immune response. *Nat. Rev. Immunol.* **2005**, *5*, 43–57. [CrossRef] [PubMed]

235. Yokosuka, T.; Takamatsu, M.; Kobayashi-Imanishi, W.; Hashimoto-Tane, A.; Azuma, M.; Saito, T. Programmed cell death 1 forms negative costimulatory microclusters that directly inhibit T cell receptor signaling by recruiting phosphatase SHP2. *J. Exp. Med.* **2012**, *209*, 1201–1217. [CrossRef]

236. Zhao, M.; Guo, W.; Wu, Y.; Yang, C.; Zhong, L.; Deng, G.; Zhu, Y.; Liu, W.; Gu, Y.; Lu, Y.; et al. SHP2 inhibition triggers anti-tumor immunity and synergizes with PD-1 blockade. *Acta Pharm. Sin. B* **2019**, *9*, 304–315. [CrossRef]

Review

Mechanisms of Acquired BRAF Inhibitor Resistance in Melanoma: A Systematic Review

Ilaria Proietti [1,*], Nevena Skroza [1], Nicoletta Bernardini [1], Ersilia Tolino [1], Veronica Balduzzi [1], Anna Marchesiello [1], Simone Michelini [1], Salvatore Volpe [1], Alessandra Mambrin [1], Giorgio Mangino [2], Giovanna Romeo [2,3,4], Patrizia Maddalena [1], Catherine Rees [5] and Concetta Potenza [1]

[1] Dermatology Unit "Daniele Innocenzi", Department of Medical-Surgical Sciences and Bio-Technologies, Sapienza University of Rome, Fiorini Hospital, Polo Pontino, 04019 Terracina, Italy; nevena.skroza@uniroma1.it (N.S.); nicoletta.bernardini@hotmail.it (N.B.); ersiliatolino@gmail.com (E.T.); veronica.balduzzi@gmail.com (V.B.); anna.marchesiello90@gmail.com (A.M.); simo.mik@hotmail.it (S.M.); salvatore.volpe@uniroma1.it (S.V.); mambrinalessandra@gmail.com (A.M.); patrizia.maddalena@gmail.com (P.M.); concetta.potenza@uniroma1.it (C.P.)

[2] Department of Medico-Surgical Sciences and Biotechnologies, Sapienza University of Rome, 00185 Rome, Italy; giorgio.mangino@uniroma1.it (G.M.); Giovanna.romeo@uniroma1.it (G.R.)

[3] Department of Infectious, Parasitic and Immune-Mediated Diseases, Istituto Superiore di Sanità, 00185 Rome, Italy

[4] Institute of Molecular Biology and Pathology, Consiglio Nazionale delle Ricerche, 00185 Rome, Italy

[5] Springer Healthcare, Auckland 0627, New Zealand; Catherine.rees@springer.com

* Correspondence: proiettilaria@gmail.com; Tel.: +(39)333-468-4342

Received: 27 August 2020; Accepted: 25 September 2020; Published: 29 September 2020

Simple Summary: Patients with advanced melanoma are often treated with v-raf murine sarcoma viral oncogene homolog B1 (BRAF) inhibitors. Although these agents prolong life, patients inevitably develop resistance and their cancer progresses. This review examines all of the potential ways that melanoma cells develop resistance to BRAF inhibitors. These mechanisms involve genetic and epigenetic changes that activate different signaling pathways, thereby bypassing the effect of BRAF inhibition, but they also involve a change in cell phenotype and the suppression of anticancer immune responses. Currently, BRAF inhibitor resistance can be partially overcome by combining a BRAF inhibitor with a mitogen-activated protein kinase kinase (MEK) inhibitor, but many other combinations are being tested. Eventually, it may be possible to choose the best combination of drugs based on the genetic profile of an individual's cancer.

Abstract: This systematic review investigated the literature on acquired v-raf murine sarcoma viral oncogene homolog B1 (BRAF) inhibitor resistance in patients with melanoma. We searched MEDLINE for articles on BRAF inhibitor resistance in patients with melanoma published since January 2010 in the following areas: (1) genetic basis of resistance; (2) epigenetic and transcriptomic mechanisms; (3) influence of the immune system on resistance development; and (4) combination therapy to overcome resistance. Common resistance mutations in melanoma are BRAF splice variants, *BRAF* amplification, neuroblastoma RAS viral oncogene homolog (NRAS) mutations and mitogen-activated protein kinase kinase 1/2 (MEK1/2) mutations. Genetic and epigenetic changes reactivate previously blocked mitogen-activated protein kinase (MAPK) pathways, activate alternative signaling pathways, and cause epithelial-to-mesenchymal transition. Once BRAF inhibitor resistance develops, the tumor microenvironment reverts to a low immunogenic state secondary to the induction of programmed cell death ligand-1. Combining a BRAF inhibitor with a MEK inhibitor delays resistance development and increases duration of response. Multiple other combinations based on known mechanisms of resistance are being investigated. BRAF inhibitor-resistant cells develop a range of 'escape routes', so multiple different treatment targets will probably be required to overcome resistance. In the future, it may be possible to personalize combination therapy towards the specific resistance pathway in individual patients.

Keywords: BRAF inhibitors; melanoma; metastasis; microenvironment; resistance; therapy

---

1. Introduction

Although melanoma is the least common type of skin cancer, it is the most deadly [1], causing approximately 61,000 deaths per year around the world [2]. An estimated 42–45% of melanomas harbor mutations of the gene for v-Raf murine sarcoma viral oncogene homolog B (BRAF) [3,4], an activating serine/threonine protein kinase in the mitogen-activated protein kinase (MAPK) signaling pathway. In patients with cutaneous melanoma, almost all of these mutations affect codon 600 of exon 15 [3]. The most common mutations are V600E (accounting for ~80% of BRAF mutations), in which a single nucleotide substitution (GTG to GAG) results in valine being substituted for glutamate, and V600K (accounting for ~16% of BRAF mutations), where two nucleotides are affected (GTG to AAG), resulting in valine being substituted for lysine [4]. Other BRAF mutations include V600D and V600R (together accounting for ~3%).

These mutations are oncogenic drivers that cause tumor progression and metastasis, and their discovery led to the development of small molecule inhibitors of BRAF, including vemurafenib, dabrafenib and encorafenib, for the treatment of melanoma [5,6]. Testing for BRAF mutations is now globally recommended in order to choose the most appropriate therapy for patients with stage III or IV melanoma [7,8].

BRAF inhibitors dramatically improved response rate and survival compared with standard chemotherapy in patients with BRAF-mutated melanoma, but these benefits were not durable, and most patients developed progressive disease as a result of resistance development [5]. Consequently, the recommended treatment approach for patients with advanced or metastatic BRAF-mutated melanoma is now a combination of a BRAF inhibitor and a MAPK kinase (MEK) inhibitor [7,8].

Only through thorough understanding of the mechanisms of BRAF inhibitor resistance can we hope to develop strategies for achieving the full therapeutic potential of contemporary treatments in patients with melanoma. Therefore, the aim of the current systematic review was to thoroughly investigate the literature on acquired BRAF inhibitor resistance, in order to identify future potential treatment strategies. Based on what is known about mechanisms of BRAF inhibitor resistance [9], four predefined topics were examined: (1) the genetic basis of resistance; (2) epigenetic and transcriptomic mechanisms of resistance; (3) influence of the immune system on BRAF inhibitor resistance; and (4) the potential of combination therapy to overcome resistance.

2. Methods

We undertook a search of the MEDLINE database on May 18, 2020 for any articles on BRAF inhibitor resistance, when used alone or in combination with a MEK inhibitor, in patients with cutaneous melanoma (a full description of the searches is shown in Supplementary Materials) published since the beginning of 2010. Four separate searches were undertaken on each predefined topic. We excluded conference reports/abstracts, news items and case reports, and articles published in languages other than English.

One author (IP) reviewed the search results and chose potential articles based on the title and abstracts. Articles were rejected from the review if they were not specific to acquired BRAF resistance (i.e., the focus was on primary resistance) in cutaneous melanoma (i.e., excluding ocular or mucosal melanoma).

From this group of articles, those most likely to contain pertinent information about BRAF inhibitor resistance mechanisms were extracted for discussion.

3. Results

The search identified the following number of articles: (1) 406 on genetic mechanisms, (2) 46 on epigenetic mechanisms; (3) 82 on immune mechanisms; and (4) 499 on overcoming resistance with combination therapy.

A more in-depth review of the articles meant that additional articles were excluded as being either not relevant or not an included article type, and some articles were reclassified into another topic section. The final number of articles included in each section were: (1) 106 articles on genetic mechanisms; (2) 61 articles on epigenetic or transcriptomic mechanisms; (3) 23 articles on immune mechanisms; and (4) 189 articles on overcoming resistance. There was some overlap between sections as some articles were relevant to more than one topic.

3.1. Genetic Mechanisms of Resistance

Advances in genetic analytical techniques, such as next generation sequencing (NGS) and clustered regularly interspaced short palindromic repeats (CRISPR) have considerably expanded our knowledge of the genetic changes involved in BRAF inhibitor resistance, and raise the possibility of incorporating mutational information into predictive or prognostic models [10–12].

A range of genetic mutations have been identified as causing acquired BRAF resistance (recently reviewed by Tian and Guo and summarized in Table 1) [11,13–70]. The most common mutations are BRAF splice variants, BRAF amplification, neuroblastoma RAS viral oncogene homolog (NRAS) mutations and MEK1/2 mutations [51,61,71,72], and these appear to be associated with different disease phenotypes. For example, brain metastases appear to be associated with NRAS mutations whereas hepatic progression is associated with MEK1/2 mutations [72]. Longitudinal assessments indicate that patients tend to accumulate resistance-related mutational changes over time [72]. However, the rate of resistance development does not appear to be related to the antitumor activity of the BRAF inhibitor (i.e., the speed at which the BRAF inhibitor kills treatment-sensitive cells has no bearing on the speed at which resistant clones develop) [73].

Splice variants of BRAF mediate resistance by affecting BRAF dimerization. In cells with wild-type BRAF, activation by RAS leads to the formation of homodimers (BRAF-BRAF) or heterodimers with CRAF (BRAF-CRAF), whereas cells with V600E mutations do not form dimers and activate MEK via monomeric BRAF. BRAF inhibitors are ineffective in melanoma with wild-type BRAF because the homo- and heterodimers retain their signaling capacity, whereas these inhibitors block the action of monomeric BRAF. Splice variants of BRAF V600E are also able to form dimers and therefore to activate MEK in the presence of BRAF inhibitors [50,74].

Genetic changes to key molecules in the NRAS/BRAF/MEK pathway lead to reactivation of the previously blocked MAPK pathways or activation of alternative signaling pathways, such as the phosphatidylinositol 3-kinase (PI3K)/protein kinase B (AKT) pathway (Figure 1) [11,36,75–78]. In some patients with BRAF inhibitor resistance, the activation of PI3K/AKT is driven by loss of phosphatase and tensin homolog (PTEN) expression [79].

The genetic changes also contribute to the increased cytoprotective autophagy seen in BRAF inhibitor-resistant melanoma cells [80–83], allowing tumor cell proliferation to continue unchecked. The enhanced autophagic-lysosomal activity of BRAF-resistant melanoma cells exacerbates adenosine triphosphate (ATP) secretion, which in turn increases melanoma cell invasion [82].

However, some mutations are independent of downstream pathways. Overexpressed genes in BRAF inhibitor-resistant cells are often associated with growth factors and their receptors, cell adhesion molecules and extracellular matrix binding [84]. Common mutations involve effects on receptor tyrosine kinases (RTKs), such as epidermal growth factor receptor (EGFR), platelet-derived growth factor receptor (PDGFR), hepatocyte growth factor (HGF), or insulin-like growth factor (IGF) receptor, which in turn activate parallel pathways [11,19,20,25,29,58,59,84–86]. Research has shown extensive redundancy in RTK-mediated signaling pathways, whereby a broad range of widely expressed RTKs are upregulated in cells with BRAF inhibitor resistance [86]. These changes are mediated post-translationally via the inhibition of proteolytic 'shedding' of cell surface receptors [87]. This shedding is a normal part of the negative feedback loop that limits intracellular signaling, but is blocked by BRAF inhibitors. As a result, there is an increase in cell surface receptor levels in the tumor during treatment, causing activation or enhancement of alternative signaling pathways.

Table 1. Genetic mutations contributing to acquired BRAF inhibitor resistance (adapted from Tian and Guo, 2020) [13].

Mutation	Mechanism
NRAS mutations [15,31,44,47,49,51,60,61]	Constitutively active RAS mutants enhance BRAF V600E dimerization, reactivate the ERK pathway, and confer resistance to BRAF inhibitor which only block monomeric BRAF V600E
CRAF overexpression, RAF paradox and dimerization of RAF proteins [17,24,36]	BRAF inhibitors can paradoxically activate wild-type BRAF kinase through the induction of dimerization or MAP3K8/COT and CRAF activation, resulting in MEK/ERK phosphorylation and eventually promoting cell proliferation
Secondary BRAF mutations [63,69]	Secondary mutations in V600E (single-nucleotide substitution) or L505H have been detected in patients with BRAF inhibitor resistance. The mutations in V600E increases BRAF kinase activity and causes cross-resistance with MEK inhibitors
BRAF gene amplification and splicing [11,47,50,51,61,62,64,70]	The amplification of the BRAF gene led to significant upregulation of BRAF protein expression, contributing to the reactivation of ERK in the presence of BRAF inhibitors. Alternative splicing can lead to the expression of truncated BRAF proteins that lack the N-terminal RAS-binding domain but retain the kinase domain, which can form homodimers that are resistant to BRAF inhibitor
MEK1/2 mutations [21,44,51,61,62,64]	MEK1/2 mutations could reactivate downstream ERK signaling without the need for BRAF stimulation
Upregulation of membrane receptors, RTKs, or receptor interaction proteins [11,19,20,25,28,29,35, 38–41,46,48,49,52,53,57–59,61,66,67]	Overexpression or hyperactivation of membrane receptors/RTKs could promote acquired resistance through the activation of parallel pathways or by direct induction of the RAS pathway; partly mediated by MITF copy gain
Aberrations in the PI3K -AKT pathway [11,14,16,18,23,32–34,55,56,68]	PI3K and AKT-activating mutations enhance AKT signaling, which promotes anti-apoptotic signals and upregulates expression of essential proliferative genes, allowing survival signals independently of BRAF
Down-regulation of STAG2 or STAG3 expression [54,68]	Down-regulation of STAG2 or STAG3 expression suppressed CTCF-mediated expression of DUSP6, resulting in the reactivation of ERK
Activation of the YAP/TAZ pathway [14,22,42,45]	The activation of YAP/TAZ pathway after actin remodeling renders resistance to BRAF targeted therapy
Down-regulation of expression of DUSPs [30]	DUSPs are the largest group of phosphatases for dephosphorylating ERK1/2 kinase, DUSPs are considered to be the negative feedback loop of MAPK signaling in response to BRAF-targeted therapy
RAC1 mutation [43,65]	Single-nucleotide variant in RAC1 maintains activation of MAPK pathway via PAK1-mediated co-activation
Somatic mutations in NF1 [27]	Usually a negative regulator of the RAS pathway, inactivation of NF1 expression leads to increased activity in downstream pathways such as PI3K/AKT
Downregulation of expression of RNF125 [37]	Deficiency of RNF125 suppresses ubiquitination and degradation of JAK1, thereby promoting the expression of EGFR that activates downstream ERK signaling and conferring resistance to BRAF-targeted therapy
DBL guanosine exchange factors [26]	Gain-of-function mutations in genes regulating the DBL/RAC1/PAK signaling axis drive resistance to BRAF inhibitors

AKT, protein kinase B; BRAF, v-Raf murine sarcoma viral oncogene homolog B; COT, cancer Osaka thyroid oncogene); CRAF, RAF proto-oncogene serine/threonine-protein kinase; DUSP, dual-specificity phosphatase; EGRF, epidermal growth factor receptor; ERK, extracellular signal-regulated kinase; JAK, Janus kinase; MAPK, mitogen-activated protein kinase; NF1, neurofibromin 1; PAK1, human p21-activated kinase; PI3K, phosphatidylinositol 3-kinase; RAC1, Ras-related C3 botulinum toxin substrate 1; RAF, rapidly accelerated fibrosarcoma; RNF, ring finger protein; RTK, receptor tyrosine kinase; STAG, small T-antigen; TAZ, transcriptional coactivator with PDZ-binding motif; YAP, yes-associated protein.

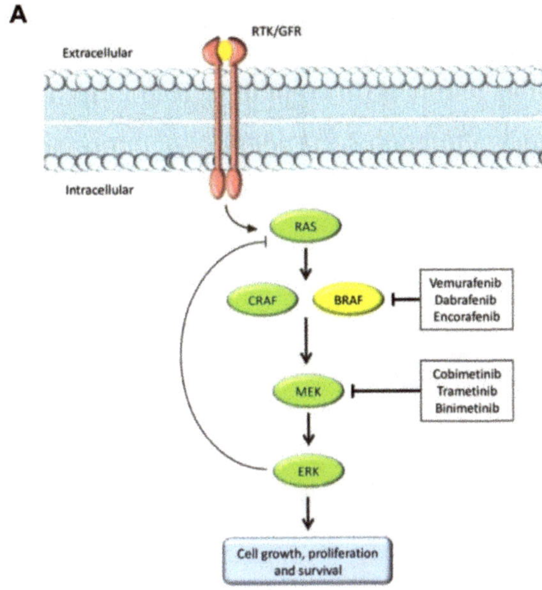
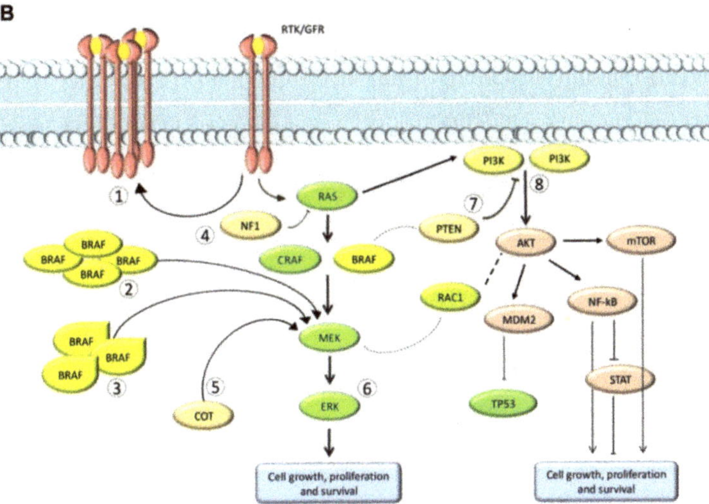

Figure 1. The MAPK signaling pathway (**A**) in a BRAF inhibitor-sensitive cell showing sites of action of BRAF inhibitors and MEK inhibitors, and (**B**) after BRAF inhibitor resistance development [78]. Mechanisms of resistance are numbered: (1) upregulation of RTK; (2) BRAF amplification; (3) BRAF alternative splicing; (4) loss of NF1; (6) ERK activation; (7) loss of PTEN; and (8) activation of alternative signaling pathways. BRAF, v-Raf murine sarcoma viral oncogene homolog B; ERK, extracellular signal-regulated kinase; GFR, growth factor receptor; mTOR, mammalian target of rapamycin; NF1, neurofibromin 1; PTEN, phosphatase and tensin homolog; RTK, receptor tyrosine kinase. From Tanda et al. Frontiers in Molecular Biosciences 2020, 7, 154 [78]. © 2020 Tanda, Vanni, Boutros, Andreotti, Bruno, Ghiorzo and Spagnolo.

Transcriptomic analysis showed that, compared with BRAF inhibitor-sensitive cells, those with acquired resistance had differences in 887 upregulated genes and 1014 downregulated genes [88]. Upregulated genes were mainly Group IV genes involved in inflammatory response, cell migration, exocrine system development, regulation of peptidase activity and tissue development, or Group V genes involved in cellular response to lipopolysaccharide, regulation of epithelial cell apoptosis, processes involved in the ovulation cycle, and regulation of interleukin (IL)-1β production (Figure 2) [88].

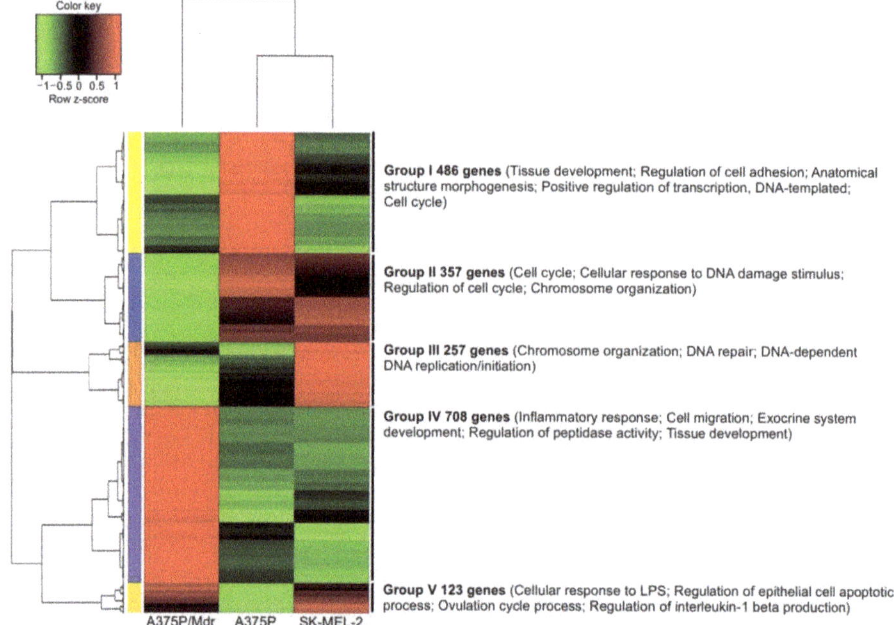

Figure 2. Heat map analysis of genes differentially expressed in cells with BRAF inhibitor sensitivity (A375P), acquired BRAF inhibitor resistance (A375P/Mdr) and innate BRAF inhibitor resistance (SK-MEL-2) [88]. SK-MEL-2 cells have wild-type BRAF and are resistant to BRAF inhibitors because these agents lack activity against wild-type BRAF. From Ahn et al. Biomol 2019, 27, 302–310 [88]. Copyright ©2019, The Korean Society of Applied Pharmacology.

The contribution to resistance of genetic alterations in cell cycle regulators is an interesting finding. One such change is upregulation of the cell cycle genes CDK6 and CCND1 [89]. This appears to open up the possibility of using CDK inhibitors to overcome resistance; preclinical data indicate that adding palbociclib to treatment with BRAF inhibitors and/or MEK inhibitors prevented resistance development in treatment-naïve melanoma cells and animal models, but did not overcome resistance in cells and animals with acquired BRAF inhibitor resistance [90].

Other genetic changes implicated in the development of vemurafenib resistance are loss of genes that encode NF1 (a negative regulator of RAS) and CUL3 (a key protein in the ubiquitin ligase complex) [27,91,92]. Loss of CUL3 is associated with increased RAC1 activity [91]. RAC1 is a member of Rho family of GTPases that regulate actin dynamics, cytoskeleton organization, and cell motility [93]. Rho GTPases regulate gene transcription via the downstream co-activator YAP1 [45,94]. Inhibition of actin remodeling, possibly by inhibiting YAP1, has been suggested as a potential target for overcoming BRAF inhibitor resistance [94,95]. Actin is not the only cytoskeleton protein implicated in BRAF inhibitor resistance: differentially regulated genes coding for microtubules and the intermediate filament nestin are also involved [96,97].

The net effect of these genetic alterations is a shift in the phenotype of BRAF inhibitor-resistant melanoma cells, resulting in an epithelial-to-mesenchymal transition (EMT) [97–99], characterized by changes in cell–cell adhesion, cell-matrix adhesion, cellular polarity, and the cytoskeleton [100]. As well as being mediated by the genetic changes within the cell, this EMT shift is also stimulated by nearby fibroblasts in the tumoral stroma [101], which secrete growth factors (such as HGF) that strongly activate the MAPK cascade [102].

3.2. Epigenetic and Transcriptomic Mechanisms

Plasticity of the melanoma cell phenotype is often driven by changes in the tumor microenvironment, such as hypoxia, pH, and nutrient supply [103,104]. BRAF inhibitor treatment is also associated with a change in the metabolic profile of cells, with a shift towards more mitochondrial respiration and the formation of reactive oxygen species [103,105–110].

These changes in the cellular metabolic profile and in the tumor microenvironment affect the activity of histone-modifying enzymes, including histone methyltransferase, histone demethylase, and histone deacetylase (HDAC) [111]. These changes modulate transcription by modifying chromatin structure. For example, the increased oxidative metabolism in BRAF inhibitor-treated cells causes a shift from glucose to glutamine metabolism [103], with increased glutamine catabolism. Low glutamine levels in the core of a melanoma induce histone hypermethylation and BRAF inhibitor resistance [112]. BRAF inhibitor-resistant cells show increased expression of KDM5B, a histone demethylase enzyme [113].

Downregulation of a range of HDAC genes have been reported to be associated with BRAF inhibitor resistance, including SIRT2 (encoding for sirtuin 2) [114], SIRT6 (encoding for sirtuin 6) [115], and HAT1 (encoding for histone acetyltransferase 1) [116]. On the other hand, HDAC8 is upregulated in BRAF inhibitor-resistant cells [117].

Noncoding portions of DNA are also implicated in the development of resistance, particularly loci that are involved in transcription factor recruitment and occupancy [118]. Long non-coding RNA (lncRNA) loci are also implicated; transcriptional activation of one such lncRNA, EMICERI, activates a neighboring gene that confers resistance to BRAF inhibitors [119].

BRAF inhibitor resistance involves a range of transcription factors (Table 2) [14,16,22,28,45,46,48,52,58, 66,81,89,97,117,120–140]. A transcriptomic analysis of BRAF inhibitor-resistant melanoma cells showed that some transcription factors were upregulated whereas others were downregulated [141]. These researchers highlighted a set of transcriptional 'master regulators' including STAT, FOXO, ZEB1 (upregulated) and MITF, HIF1A and MYB (downregulated) that control a range of effector pathways (Figure 3). One of the key processes regulated by these transcription factors is ErbB3 phosphorylation, which leads to PI3K/AKT activation [52]. Downstream effects of these transcription-related events include inhibition of apoptosis [142] and the EMT phenotypic change seen in BRAF inhibitor-resistant cells [99,135].

Resistance is also mediated at a post-transcriptional level through effects on translation mediated by RNA binding proteins (e.g., human antigen R and 4E-BP) [143,144], translation initiation complexes (e.g., eIF4F or eIF4E) [144,145], micro-RNAs (Table 2) [59,146–154], and by modifications in wobble tRNA [155]. Some of these micro-RNAs (e.g., miR-199b-5p) are regulators of cell-cell signaling via VEGF and HIF-1α, while others (e.g., miR-4488 and miR-4443) are involved in autophagy regulation [149].

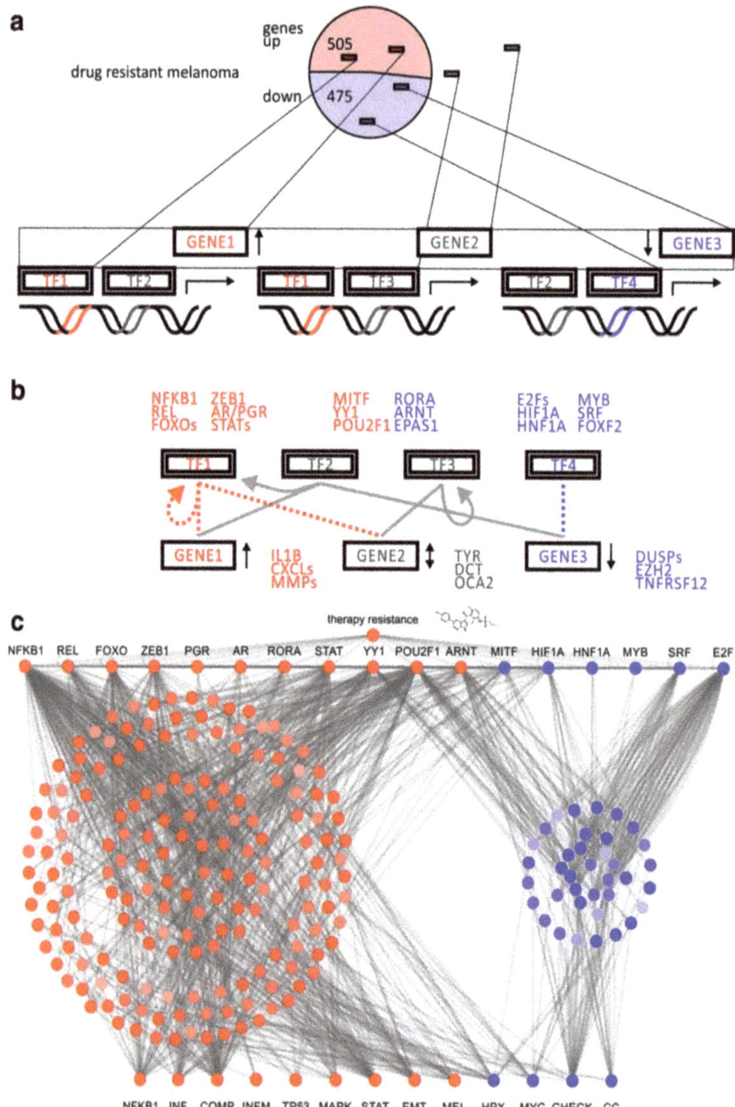

Figure 3. Transcription factor motif analysis of BRAF inhibitor resistance in cellular models of malignant melanoma [141]. Red indicates transcription factors relevant to upregulated genes and blue indicates transcription factors relevant to downregulated genes. (**a**) Schematic representation of differentially expressed genes in a drug resistance model and transcription factor motifs associated with regulated target genes. Upregulated factors are depicted in red and downregulated factors in blue. (**b**) Hierarchical transcription factor network with master regulators on top and downstream targets at bottom. Sets of transcription factor target genes are identified in enrichment analysis based on sequence motifs. (**c**) Hierarchical network model illustrates how therapy resistance in cancer selects for specific transcriptional master regulators to rewire target genes in effector pathways in a concerted fashion. From Zecena et al. BMC Syst Biol 2018, 12, 33 [141]. Copyright © 2018, Zecena, Tveit, Wang, Farhat, Panchal, Liu, Singh, Sanghera, Bainiwal, Teo, Meyskens, Liu-Smith, and Filipp.

Table 2. Transcription factors and microRNA implicated in BRAF inhibitor resistance.

Transcription Factors	MicroRNA
STAT3 [28,122,131,139]	miR-7 [59]
FLI1 [121]	miR-92a-15p [150,152]
RUNX [123,129]	miR-204-5p [147,149]
YAP [14,22,45]	miR-211-5p [147]
c-MYC [138]	miR-126-3p [146]
Aryl hyodrocarbon receptor [125]	miR-514a [154]
SOX proteins (SOX2, SOX10) [58,126,130,131]	miR-579-3p [148]
β-catenin [120,124,139]	miR-4443 [149]
MITF [46,48,127,130,132]	miR-4488 [149]
MRTF [45]	miR-1246 [151]
JUN [89,97,117,128,136]	miR-200c [153]
ZEB-1 or -2 [66,137]	miR-708-5p [152]
WNT5 [16,120,124,134]	miR-199-5p [149]
NFATc2 [135]	
NRF-1 [133]	
FOXD3 [52]	
E2F1 [140]	
TFEB [81]	

3.3. Immune Mechanisms

The genetic and epigenetic changes associated with BRAF inhibitor treatment affect the interaction between the melanoma tumor and the immune system in various ways. Untreated melanoma is not particularly immunogenic, but treatment with BRAF inhibitors causes a transient increase in antitumor immunogenicity, with recruitment of T-cells and natural killer (NK) cells and a reduction in regulatory T-cells (Tregs) [156].

Treated patients may develop immune-related adverse events (e.g., arthralgia, immune-related skin reactions), which can be a marker of response to BRAF inhibitors [157–159]. However, once resistance to BRAF inhibitors develops, the tumor microenvironment reverts to its low immunogenic state, with fewer infiltrating T-cells and NK cells [9,160], more double-negative T-cells [9,161], and restoration of myeloid-derived suppressor cells [162]. Moreover, the T-cells and NK cells that are present are functionally impaired compared with BRAF inhibitor-sensitive tumors [160], such that cytotoxic T-lymphocytes and NK cells are less effective at recognizing and killing BRAF inhibitor-resistant cells [163,164].

In the tumor cell, this reversion to a low immunogenic state is caused by induced expression of programmed cell death ligand-1 (PD-L1) [165–169], increased expression of the immunoregulatory protein Galactin-1 [170], and increased expression of CD47, an immunoregulatory cell marker, on tumor cells, leading to reduced killing by cytotoxic T-cells and macrophages [133]. BRAF inhibitor-resistant tumor cells also show reduced expression of target antigens [164], and increased production of IL-10 [33]. They also show upregulation of the IGF receptor, which sensitizes BRAF inhibitor-resistant cells to the effects of cytotoxic T-lymphocytes by increasing the cellular uptake of granzyme B [171].

Another change which may have therapeutic implications is the restoration of carcinoembryonic antigen-related cell adhesion molecule 1 (CEACAM1) expression. CEACAM1 is an intercellular adhesion molecule that regulates cell proliferation, cellular energetics, and inflammation in cancer cells, including melanoma, and has a role in regulating the immune cells in the tumor microenvironment [172,173]. CEACAM1 is downregulated prior to resistance development [174], suggesting that it may be a future target for the treatment of BRAF inhibitor-resistant melanoma.

In addition to the changes within the tumor cell, secreted soluble factors originating from macrophages in the tumor microenvironment contribute to resistance, including tumor necrosis factor-α (TNF-α) and vascular endothelial growth factor (VEGF), which contribute to tumor cell growth and invasion, and further infiltration of macrophages and T-lymphocytes into the tumor microenvironment [175,176].

3.4. Overcoming Resistance via Treatment Combinations

The standard of care for BRAF-mutated melanoma is now the combination of BRAF inhibitors and MEK inhibitors [7,8], as there is considerable high-quality evidence from randomized comparative studies that this approach prolongs progression-free survival (PFS) and overall survival (OS) compared with BRAF inhibitor monotherapy [177–187]. Some patients are able to achieve a durable response lasting $\geq$ 5 years with such combinations [183,186].

MEK inhibitors block the MEK/ERK signaling pathway that is activated in cells during treatment with BRAF inhibitors, thereby delaying the development of resistance and increasing the duration of response [188,189]. MEK inhibitors also suppress the production of PD-L1 [167], and so help to suppress immune activation that occurs during BRAF inhibition. The most extensively researched BRAF inhibitor + MEK inhibitor combinations are vemurafenib + cobimetinib [178], dabrafenib + trametinib [184,185,187,190–192], and encorafenib + binimetinib [179,180].

While there are no head-to-head comparisons of these combinations, some differences do exist. Encorafenib has the highest paradox index of the three BRAF inhibitors [193]. This index is a ratio of the EC_{80} for ERK activation relative to the IC_{80} of BRAF inhibitor-resistant cell growth, and provides a measure of the therapeutic window for antitumor activity before paradoxical ERK activation [193]. The paradox index of encorafenib is 50, compared with 10 for dabrafenib and 5.5 for vemurafenib. This difference may help to explain the higher efficacy of encorafenib monotherapy compared with vemurafenib monotherapy in the COLUMBUS study [180].

Although these combinations have not been directly compared, the combination of encorafenib + binimetinib was associated with numerically longer OS and PFS in the COLUMBUS study than was seen with vemurafenib + cobimetinib in the coBRIM study or dabrafenib + trametinib in the COMBI-v study [181]. While the overall rate of adverse events with each combination was similar, encorafenib + binimetinib was associated with a lower incidence of pyrexia compared with dabrafenib + trametinib and of photosensitivity compared with vemurafenib + cobimetinib [181].

Once patients progress after BRAF inhibitor + MEK inhibitor treatment, the recommended second-line approach is immunotherapy with an immune checkpoint inhibitor [7]. Some patients who progress on BRAF inhibitor treatment may benefit from a rechallenge with BRAF inhibitor + MEK inhibitor therapy [192].

Preliminary (phase Ib and II) clinical trials have been conducted using triple therapy with a BRAF inhibitor + MEK inhibitor + PD-L1 inhibitor in mostly BRAF inhibitor-naïve patients, with high rates of response (63–73%) [194–196]. In the comparative phase II study, triplet therapy prolonged PFS compared with the BRAF inhibitor + MEK inhibitor doublet combination (median 16.0 vs. 10.3 months; p = 0.043), but was associated with more than two times the rate of grade 3 or 4 adverse events (58.3% vs. 26.7%) [194].

The diversity of genetic changes driving resistance, and the inter- and intra-individual heterogeneity of resistance-associated mutations complicates treatment of BRAF inhibitor-resistant melanoma [130,197,198]. One potential approach is to silence the expression of specific genes using small interfering RNA (siRNA) [199]. However, the clinical application of this approach is hampered by difficulties in delivering siRNA into the cytoplasm of tumor cells without the use of viral vectors, which can cause mutation, immune activation, or inflammation [200]. To the best of our knowledge, none of the non-viral vectors for siRNA have reached clinical investigation in melanoma.

In addition to heterogeneous genetic changes, there are a range of phenotypic adaptations to BRAF inhibition, indicating that several alternative signaling routes are involved in overcoming BRAF resistance [201]. Because BRAF inhibitor-resistant cells have developed a range of 'escape routes', it is likely that multiple different treatment targets will be required to overcome resistance. Therefore, researchers have investigated targeting node points in the activated pathways. Based on the activation/overexpression of transmembrane receptors or RTKs and activation of alternative signaling pathways, some of the novel combinations being tested in vitro include BRAF inhibitors with: TGF-β receptor inhibitors [154], inhibitors of growth factor receptors or RTKs (e.g., gefitinib, sorafenib, dovinitib) [28,53,202–204], PI3K/mTOR inhibitors [18,101,205–216], bifunctional MAPK/PI3K antagonists [217], anaplastic lymphoma kinase (ALK) inhibitors [35], novel MEK and/or aurora kinase inhibitors [218–222], MAPK activators [223], glucocorticoid receptor antagonists [224], CDK4/6 inhibitors [225], ERK1/2 inhibitors [226,227], CRAF inhibitors [24], pan-RAF inhibitors [228–230], heat shock protein (HSP) inhibitors [231–237], or MCL1 inhibitors [238]. However, few of these combinations have been investigated clinically.

One of the few combinations to be investigated in clinical trials is the combination of a BRAF inhibitor (vemurafenib) with an HSP90 inhibitor (XL888) [239]. This combination was tested in an open-label phase I study in 21 patients with BRAF V600E-mutated metastatic melanoma, but none of the patients had received BRAF inhibitors before, so none had acquired resistance. The objective response rate in the 20 evaluable patients was 75%, with three complete responses and 12 partial responses, and the 1-year OS rate was 60% [239]. The most common grade 3 adverse events were skin toxicities (rash, squamous cell carcinoma, and new primary melanoma), diarrhea, headache, and fatigue [239]. The authors reported that they will be undertaking another clinical trial with the triplet combination of vemurafenib, cobimetinib, and XL888, but note that reduced doses of the BRAF and MEK inhibitors may be needed to limit toxicity with this combination [239].

Epigenetic mechanisms may be targeted by inhibiting key transcription factors, such as WNT5 or STAT3 [134,135,222,240], or by the use of microRNA mimetics [59] or HDAC inhibitors [111,163,241–245]. HDAC inhibitors have also been investigated in combination with a CDK inhibitor with promising in vitro and in vivo activity [233,244,246]. A single-arm, open-label, proof-of-concept study is underway in the Netherlands to investigate the effect of the HDAC inhibitor vorinostat in patients with BRAF V600E-mutated resistant melanoma (NCT02836548) [247]. Patients who progress on treatment with a BRAF inhibitor and/or a combination of a BRAF inhibitor + MEK inhibitor will receive 14 days of treatment with vorinostat, before reinitiating their earlier BRAF/MEK inhibitor treatment. The aim of the study is to see whether vorinostat can purge the resistant clones and re-establish responsiveness to BRAF/MEK inhibitor treatment [247].

Agents targeting the immune system are also being investigated for the treatment of BRAF inhibitor resistance. As described earlier, the use of immune checkpoint inhibitors in combination with a BRAF inhibitor + MEK inhibitor appears to be effective but is associated with a high rate of toxicity [194]. Other potential immune-targeted therapies undergoing preclinical investigation include adoptive T-cell therapy [171,248,249], dendritic cell vaccination [250], and combining BRAF inhibitor treatment with a toll-like receptor 7 agonist (e.g., imiquimod) [160].

A considerable number of preclinical studies are investigating other novel targets for overcoming BRAF inhibitor resistance. These include combining BRAF and/or MEK inhibitors with inhibitors of pre-mRNA splicing (to counteract resistance caused by BRAF splicing) [251], BH3-mimetics [252,253], BCL2 inhibitors [254], mitochondrial-targeted agents [255,256], inhibitors of p90 ribosomal S6 kinases [257,258], pro-caspase activating compounds [259], Rho kinase 1 (ROCK1) inhibitors [260], protein kinase Cδ inhibitors [261], tubulin inhibitors [262], ErbB2 or ErbB3 inhibitors [222,263, 264], activators of the liver-X nuclear hormone receptor [265], an antibody conjugate targeting the endothelin B receptor [266], monoclonal antibodies against chondroitin sulfate proteoglycan 4 [267], inhibitors of sterol regulator element binding protein I (SREBP-1) [268], copper chelators [269], polo-like 3 kinase inhibitors (including in models of BRAF + MEK inhibitor resistance) [270,271], anti-nodal antibodies [272], PAK1 inhibitors [273], GLI1/2 inhibitors [274], inhibitors of IQ motif-containing GTPase activating protein 1 (IQGAP1) [275], serotonin agonists [276], CK2 inhibitors [277], p53 activators [278], metformin [279], statins [280], non-steroidal anti-inflammatory drugs [281], mibefradil [282], hydroxychloroquine (an autophagy inhibitor) [83], and A100 (a reactive oxygen species-activated prodrug) [283].

4. Future Directions

As described above, considerable research is being undertaken to identify potential new combinations of treatments that may limit or prevent the development of BRAF inhibitor resistance, or overcome resistance once developed. In addition to this, investigations are underway to improve existing therapies, such as employing nanovehicle technology to enhance the safety or targeted delivery of drugs, which may enable the use of higher doses or more potent combinations of existing agents [284,285]. Preclinical studies have investigated the encapsulation of MEK inhibitors in pegylated nanoliposomes for oral administration [284], or the topical administration of BRAF inhibitor-loaded nanovehicles into melanoma lesions using a microneedling technique [285].

The multiplicity of potential targets raises the possibility of using genomic and proteomic data to personalize combination therapy towards the specific pathway that is activated during BRAF inhibitor resistance in individual patients with melanoma [211]. Investigators are developing an MAPK pathway activity score from aggregated gene expression data that could help to determine the best drug combination to use [64], but further validation is needed.

5. Conclusions

Given the complexity and heterogeneity of pathways involved in BRAF inhibitor resistance, a 'one size fits all' approach to overcoming acquired resistance is unlikely to succeed. However, the plethora of research in this field means that multiple promising leads are being identified and investigated, which bodes well for the development of new treatment approaches for patients with acquired BRAF inhibitor resistance.

Supplementary Materials: The following are available online at http://www.mdpi.com/2072-6694/12/10/2801/ s1, Supplementary material: Literature search protocols, (1) Genetics; (2) Epigenetics; (3) Immune system; (4) Overcoming resistance.

Author Contributions: Content planning and search strategy development, I.P., N.S., N.B., E.T., V.B., A.M. (Anna Marchesiello), S.M., S.V., A.M. (Alessandra Mambrin), G.M., G.R., P.M., and C.P.; Search results review and article selection: I.P. and C.R.; writing—outline and first draft preparation: C.R.; writing—review: I.P., N.S., N.B., E.T., V.B., A.M. (Anna Marchesiello), S.M., S.V., A.M. (Alessandra Mambrin), G.M., G.R., P.M., and C.P. All authors have read and agreed to the published version of the manuscript.

Funding: Medical writing assistance was funded by Pierre Fabre.

Conflicts of Interest: The authors declare no conflict of interest. Catherine Rees is a professional medical writer employed by Springer Healthcare Communications. Her medical writing services were funded by Pierre Fabre. The funder had no role in the design of the review; in the collection, analyses, or interpretation of data; or in the decision to publish the results.

References

1.   Leiter, U.; Eigentler, T.; Garbe, C. Epidemiology of skin cancer. In *Sunlight, Vitamin D and Skin Cancer*, 2nd ed.; Reichrath, J., Ed.; Landes Bioscience: Austin, TX, USA, 2014; pp. 120–140.

2.   Bray, F.; Ferlay, J.; Soerjomataram, I.; Siegel, R.L.; Torre, L.A.; Jemal, A. Global cancer statistics 2018: GLOBOCAN estimates of incidence and mortality worldwide for 36 cancers in 185 countries. *CA Cancer J. Clin.* **2018**, *68*, 394–424. [CrossRef] [PubMed]

3.   Greaves, W.O.; Verma, S.; Patel, K.P.; Davies, M.A.; Barkoh, B.A.; Galbincea, J.M.; Yao, H.; Lazar, A.J.; Aldape, K.D.; Medeiros, L.J.; et al. Frequency and spectrum of BRAF mutations in a retrospective, single-institution study of 1112 cases of melanoma. *J. Mol. Diagn.* **2013**, *15*, 220–226. [CrossRef] [PubMed]

4.   Rubinstein, J.C.; Sznol, M.; Pavlick, A.C.; Ariyan, S.; Cheng, E.; Bacchiocchi, A.; Kluger, H.M.; Narayan, D.; Halaban, R. Incidence of the V600K mutation among melanoma patients with BRAF mutations, and potential therapeutic response to the specific BRAF inhibitor PLX4032. *J. Transl. Med.* **2010**, *8*, 67. [CrossRef]

5.   Holderfield, M.; Deuker, M.M.; McCormick, F.; McMahon, M. Targeting RAF kinases for cancer therapy: BRAF-mutated melanoma and beyond. *Nat. Rev. Cancer* **2014**, *14*, 455–467. [CrossRef] [PubMed]

6.   Proietti, I.; Skroza, N.; Michelini, S.; Mambrin, A.; Balduzzi, V.; Bernardini, N.; Marchesiello, A.; Tolino, E.; Volpe, S.; Maddalena, P.; et al. BRAF Inhibitors: Molecular Targeting and Immunomodulatory Actions. *Cancers* **2020**, *12*, 1823. [CrossRef]

7.   Michielin, O.; van Akkooi, A.C.J.; Ascierto, P.A.; Dummer, R.; Keilholz, U.; ESMO Guidelines Committee. Cutaneous melanoma: ESMO Clinical Practice Guidelines for diagnosis, treatment and follow-updagger. *Ann. Oncol.* **2019**, *30*, 1884–1901. [CrossRef]

8.   National Comprehensive Cancer Network. *NCCN Clinical Practice Guidelines in Oncology: Cutaneous Melanoma. Version 3. 2020*; National Comprehensive Cancer Network: Plymouth Meeting, PA, USA, 2020; 18 May 2020.

9. Song, C.; Piva, M.; Sun, L.; Hong, A.; Moriceau, G.; Kong, X.; Zhang, H.; Lomeli, S.; Qian, J.; Yu, C.C.; et al. Recurrent Tumor Cell-Intrinsic and -Extrinsic Alterations during MAPKi-Induced Melanoma Regression and Early Adaptation. *Cancer Discov.* **2017**, *7*, 1248–1265. [CrossRef]

10. Johnson, D.B.; Menzies, A.M.; Zimmer, L.; Eroglu, Z.; Ye, F.; Zhao, S.; Rizos, H.; Sucker, A.; Scolyer, R.A.; Gutzmer, R.; et al. Acquired BRAF inhibitor resistance: A multicenter meta-analysis of the spectrum and frequencies, clinical behaviour, and phenotypic associations of resistance mechanisms. *Eur. J. Cancer* **2015**, *51*, 2792–2799. [CrossRef]

11. Louveau, B.; Delyon, J.; De Moura, C.R.; Battistella, M.; Jouenne, F.; Golmard, L.; Sadoux, A.; Podgorniak, M.P.; Chami, I.; Marco, O.; et al. A targeted genomic alteration analysis predicts survival of melanoma patients under BRAF inhibitors. *Oncotarget* **2019**, *10*, 1669–1687. [CrossRef]

12. Olbryt, M.; Piglowski, W.; Rajczykowski, M.; Pfeifer, A.; Student, S.; Fiszer-Kierzkowska, A. Genetic Profiling of Advanced Melanoma: Candidate Mutations for Predicting Sensitivity and Resistance to Targeted Therapy. *Target. Oncol.* **2020**, *15*, 101–113. [CrossRef]

13. Tian, Y.; Guo, W. A Review of the Molecular Pathways Involved in Resistance to BRAF Inhibitors in Patients with Advanced-Stage Melanoma. *Med. Sci. Monit.* **2020**, *26*, e920957. [CrossRef]

14. Elmageed, Z.Y.A.; Moore, R.F.; Tsumagari, K.; Lee, M.M.; Sholl, A.B.; Friedlander, P.; Al-Qurayshi, Z.; Hassan, M.; Wang, A.R.; Boulares, H.A.; et al. Prognostic Role of BRAF^{V600E} Cellular Localization in Melanoma. *J. Am. Coll. Surg.* **2018**, *226*, 526–537. [CrossRef] [PubMed]

15. Ahn, J.H.; Lee, M. The siRNA-mediated downregulation of N-Ras sensitizes human melanoma cells to apoptosis induced by selective BRAF inhibitors. *Mol. Cell. Biochem.* **2014**, *392*, 239–247. [CrossRef] [PubMed]

16. Anastas, J.N.; Kulikauskas, R.M.; Tamir, T.; Rizos, H.; Long, G.V.; von Euw, E.M.; Yang, P.T.; Chen, H.W.; Haydu, L.; Toroni, R.A.; et al. WNT5A enhances resistance of melanoma cells to targeted BRAF inhibitors. *J. Clin. Investig.* **2014**, *124*, 2877–2890. [CrossRef] [PubMed]

17. Antony, R.; Emery, C.M.; Sawyer, A.M.; Garraway, L.A. C-RAF mutations confer resistance to RAF inhibitors. *Cancer Res.* **2013**, *73*, 4840–4851. [CrossRef] [PubMed]

18. Atefi, M.; von Euw, E.; Attar, N.; Ng, C.; Chu, C.; Guo, D.; Nazarian, R.; Chmielowski, B.; Glaspy, J.A.; Comin-Anduix, B.; et al. Reversing melanoma cross-resistance to BRAF and MEK inhibitors by co-targeting the AKT/mTOR pathway. *PLoS ONE* **2011**, *6*, e28973. [CrossRef] [PubMed]

19. Atzori, M.G.; Ceci, C.; Ruffini, F.; Trapani, M.; Barbaccia, M.L.; Tentori, L.; D'Atri, S.; Lacal, P.M.; Graziani, G. Role of VEGFR-1 in melanoma acquired resistance to the BRAF inhibitor vemurafenib. *J. Cell. Mol. Med.* **2020**, *24*, 465–475. [CrossRef]

20. Caenepeel, S.; Cooke, K.; Wadsworth, S.; Huang, G.; Robert, L.; Moreno, B.H.; Parisi, G.; Cajulis, E.; Kendall, R.; Beltran, P.; et al. MAPK pathway inhibition induces MET and GAB1 levels, priming BRAF mutant melanoma for rescue by hepatocyte growth factor. *Oncotarget* **2017**, *8*, 17795–17809. [CrossRef]

21. Carlino, M.S.; Fung, C.; Shahheydari, H.; Todd, J.R.; Boyd, S.C.; Irvine, M.; Nagrial, A.M.; Scolyer, R.A.; Kefford, R.F.; Long, G.V.; et al. Preexisting MEK1P124 mutations diminish response to BRAF inhibitors in metastatic melanoma patients. *Clin. Cancer Res.* **2015**, *21*, 98–105. [CrossRef]

22. Choe, M.H.; Yoon, Y.; Kim, J.; Hwang, S.G.; Han, Y.H.; Kim, J.S. miR-550a-3-5p acts as a tumor suppressor and reverses BRAF inhibitor resistance through the direct targeting of YAP. *Cell Death Dis.* **2018**, *9*, 640. [CrossRef]

23. Coppe, J.P.; Mori, M.; Pan, B.; Yau, C.; Wolf, D.M.; Ruiz-Saenz, A.; Brunen, D.; Prahallad, A.; Cornelissen-Steijger, P.; Kemper, K.; et al. Mapping phospho-catalytic dependencies of therapy-resistant tumours reveals actionable vulnerabilities. *Nat. Cell Biol.* **2019**, *21*, 778–790. [CrossRef] [PubMed]

24. Doudican, N.A.; Orlow, S.J. Inhibition of the CRAF/prohibitin interaction reverses CRAF-dependent resistance to vemurafenib. *Oncogene* **2017**, *36*, 423–428. [CrossRef] [PubMed]

25. Dugo, M.; Nicolini, G.; Tragni, G.; Bersani, I.; Tomassetti, A.; Colonna, V.; Del Vecchio, M.; De Braud, F.; Canevari, S.; Anichini, A.; et al. A melanoma subtype with intrinsic resistance to BRAF inhibition identified by receptor tyrosine kinases gene-driven classification. *Oncotarget* **2015**, *6*, 5118–5133. [CrossRef] [PubMed]

26. Feddersen, C.R.; Schillo, J.L.; Varzavand, A.; Vaughn, H.R.; Wadsworth, L.S.; Voigt, A.P.; Zhu, E.Y.; Jennings, B.M.; Mullen, S.A.; Bobera, J.; et al. Src-Dependent DBL Family Members Drive Resistance to Vemurafenib in Human Melanoma. *Cancer Res.* **2019**, *79*, 5074–5087. [CrossRef]

27. Gibney, G.T.; Smalley, K.S. An unholy alliance: Cooperation between BRAF and NF1 in melanoma development and BRAF inhibitor resistance. *Cancer Discov.* **2013**, *3*, 260–263. [CrossRef]

28. Girotti, M.R.; Pedersen, M.; Sanchez-Laorden, B.; Viros, A.; Turajlic, S.; Niculescu-Duvaz, D.; Zambon, A.; Sinclair, J.; Hayes, A.; Gore, M.; et al. Inhibiting EGF receptor or SRC family kinase signaling overcomes BRAF inhibitor resistance in melanoma. *Cancer Discov.* **2013**, *3*, 158–167. [CrossRef]

29. Gross, A.; Niemetz-Rahn, A.; Nonnenmacher, A.; Tucholski, J.; Keilholz, U.; Fusi, A. Expression and activity of EGFR in human cutaneous melanoma cell lines and influence of vemurafenib on the EGFR pathway. *Target. Oncol.* **2015**, *10*, 77–84. [CrossRef]

30. Gupta, R.; Bugide, S.; Wang, B.; Green, M.R.; Johnson, D.B.; Wajapeyee, N. Loss of BOP1 confers resistance to BRAF kinase inhibitors in melanoma by activating MAP kinase pathway. *Proc. Natl. Acad. Sci. USA* **2019**, *116*, 4583–4591. [CrossRef]

31. Hintzsche, J.; Kim, J.; Yadav, V.; Amato, C.; Robinson, S.E.; Seelenfreund, E.; Shellman, Y.; Wisell, J.; Applegate, A.; McCarter, M.; et al. IMPACT: A whole-exome sequencing analysis pipeline for integrating molecular profiles with actionable therapeutics in clinical samples. *J. Am. Med. Inform. Assoc.* **2016**, *23*, 721–730. [CrossRef]

32. Hu, W.; Jin, L.; Jiang, C.C.; Long, G.V.; Scolyer, R.A.; Wu, Q.; Zhang, X.D.; Mei, Y.; Wu, M. AEBP1 upregulation confers acquired resistance to BRAF (V600E) inhibition in melanoma. *Cell Death Dis.* **2013**, *4*, e914. [CrossRef]

33. Inozume, T.; Tsunoda, T.; Morisaki, T.; Harada, K.; Shirasawa, S.; Kawamura, T. Acquisition of resistance to vemurafenib leads to interleukin-10 production through an aberrant activation of Akt in a melanoma cell line. *J. Dermatol.* **2018**, *45*, 1434–1439. [CrossRef]

34. Jain, A.; Tripathi, R.; Turpin, C.P.; Wang, C.; Plattner, R. Abl kinase regulation by BRAF/ERK and cooperation with Akt in melanoma. *Oncogene* **2017**, *36*, 4585–4596. [CrossRef]

35. Janostiak, R.; Malvi, P.; Wajapeyee, N. Anaplastic Lymphoma Kinase Confers Resistance to BRAF Kinase Inhibitors in Melanoma. *iScience* **2019**, *16*, 453–467. [CrossRef]

36. Johannessen, C.M.; Boehm, J.S.; Kim, S.Y.; Thomas, S.R.; Wardwell, L.; Johnson, L.A.; Emery, C.M.; Stransky, N.; Cogdill, A.P.; Barretina, J.; et al. COT drives resistance to RAF inhibition through MAP kinase pathway reactivation. *Nature* **2010**, *468*, 968–972. [CrossRef]

37. Kim, H.; Frederick, D.T.; Levesque, M.P.; Cooper, Z.A.; Feng, Y.; Krepler, C.; Brill, L.; Samuels, Y.; Hayward, N.K.; Perlina, A.; et al. Downregulation of the Ubiquitin Ligase RNF125 Underlies Resistance of Melanoma Cells to BRAF Inhibitors via JAK1 Deregulation. *Cell Rep.* **2015**, *11*, 1458–1473. [CrossRef]

38. Konermann, S.; Brigham, M.D.; Trevino, A.E.; Joung, J.; Abudayyeh, O.O.; Barcena, C.; Hsu, P.D.; Habib, N.; Gootenberg, J.S.; Nishimasu, H.; et al. Genome-scale transcriptional activation by an engineered CRISPR-Cas9 complex. *Nature* **2015**, *517*, 583–588. [CrossRef]

39. Krayem, M.; Aftimos, P.; Najem, A.; van den Hooven, T.; van den Berg, A.; Hovestad-Bijl, L.; de Wijn, R.; Hilhorst, R.; Ruijtenbeek, R.; Sabbah, M.; et al. Kinome Profiling to Predict Sensitivity to MAPK Inhibition in Melanoma and to Provide New Insights into Intrinsic and Acquired Mechanism of Resistance Short Title: Sensitivity Prediction to MAPK Inhibitors in Melanoma. *Cancers* **2020**, *12*, 512. [CrossRef]

40. Lehraiki, A.; Cerezo, M.; Rouaud, F.; Abbe, P.; Allegra, M.; Kluza, J.; Marchetti, P.; Imbert, V.; Cheli, Y.; Bertolotto, C.; et al. Increased CD271 expression by the NF-kB pathway promotes melanoma cell survival and drives acquired resistance to BRAF inhibitor vemurafenib. *Cell Discov.* **2015**, *1*, 15030. [CrossRef]

41. Lei, F.X.; Jin, L.; Liu, X.Y.; Lai, F.; Yan, X.G.; Farrelly, M.; Guo, S.T.; Zhao, X.H.; Zhang, X.D. RIP1 protects melanoma cells from apoptosis induced by BRAF/MEK inhibitors. *Cell Death Dis.* **2018**, *9*, 679. [CrossRef]

42. Lin, L.; Bivona, T.G. The Hippo effector YAP regulates the response of cancer cells to MAPK pathway inhibitors. *Mol. Cell. Oncol.* **2016**, *3*, e1021441. [CrossRef]

43. Lionarons, D.A.; Hancock, D.C.; Rana, S.; East, P.; Moore, C.; Murillo, M.M.; Carvalho, J.; Spencer-Dene, B.; Herbert, E.; Stamp, G.; et al. RAC1^{P29S} Induces a Mesenchymal Phenotypic Switch via Serum Response Factor to Promote Melanoma Development and Therapy Resistance. *Cancer Cell* **2019**, *36*, 68–83.e9. [CrossRef]

44. Long, G.V.; Fung, C.; Menzies, A.M.; Pupo, G.M.; Carlino, M.S.; Hyman, J.; Shahheydari, H.; Tembe, V.; Thompson, J.F.; Saw, R.P.; et al. Increased MAPK reactivation in early resistance to dabrafenib/trametinib combination therapy of BRAF-mutant metastatic melanoma. *Nat. Commun.* **2014**, *5*, 5694. [CrossRef]

45. Misek, S.A.; Appleton, K.M.; Dexheimer, T.S.; Lisabeth, E.M.; Lo, R.S.; Larsen, S.D.; Gallo, K.A.; Neubig, R.R. Rho-mediated signaling promotes BRAF inhibitor resistance in de-differentiated melanoma cells. *Oncogene* **2020**, *39*, 1466–1483. [CrossRef]

46. Molnar, E.; Garay, T.; Donia, M.; Baranyi, M.; Rittler, D.; Berger, W.; Timar, J.; Grusch, M.; Hegedus, B. Long-Term Vemurafenib Exposure Induced Alterations of Cell Phenotypes in Melanoma: Increased Cell Migration and Its Association with EGFR Expression. *Int. J. Mol. Sci.* **2019**, *20*, 4484. [CrossRef]

47. Monsma, D.J.; Cherba, D.M.; Eugster, E.E.; Dylewski, D.L.; Davidson, P.T.; Peterson, C.A.; Borgman, A.S.; Winn, M.E.; Dykema, K.J.; Webb, C.P.; et al. Melanoma patient derived xenografts acquire distinct Vemurafenib resistance mechanisms. *Am. J. Cancer Res.* **2015**, *5*, 1507–1518.

48. Muller, J.; Krijgsman, O.; Tsoi, J.; Robert, L.; Hugo, W.; Song, C.; Kong, X.; Possik, P.A.; Cornelissen-Steijger, P.D.; Geukes Foppen, M.H.; et al. Low MITF/AXL ratio predicts early resistance to multiple targeted drugs in melanoma. *Nat. Commun.* **2014**, *5*, 5712. [CrossRef]

49. Nazarian, R.; Shi, H.; Wang, Q.; Kong, X.; Koya, R.C.; Lee, H.; Chen, Z.; Lee, M.K.; Attar, N.; Sazegar, H.; et al. Melanomas acquire resistance to B-RAF(V600E) inhibition by RTK or N-RAS upregulation. *Nature* **2010**, *468*, 973–977. [CrossRef]

50. Poulikakos, P.I.; Persaud, Y.; Janakiraman, M.; Kong, X.; Ng, C.; Moriceau, G.; Shi, H.; Atefi, M.; Titz, B.; Gabay, M.T.; et al. RAF inhibitor resistance is mediated by dimerization of aberrantly spliced BRAF(V600E). *Nature* **2011**, *480*, 387–390. [CrossRef]

51. Rizos, H.; Menzies, A.M.; Pupo, G.M.; Carlino, M.S.; Fung, C.; Hyman, J.; Haydu, L.E.; Mijatov, B.; Becker, T.M.; Boyd, S.C.; et al. BRAF inhibitor resistance mechanisms in metastatic melanoma: Spectrum and clinical impact. *Clin. Cancer Res.* **2014**, *20*, 1965–1977. [CrossRef]

52. Ruggiero, C.F.; Malpicci, D.; Fattore, L.; Madonna, G.; Vanella, V.; Mallardo, D.; Liguoro, D.; Salvati, V.; Capone, M.; Bedogni, B.; et al. ErbB3 Phosphorylation as Central Event in Adaptive Resistance to Targeted Therapy in Metastatic Melanoma: Early Detection in CTCs during Therapy and Insights into Regulation by Autocrine Neuregulin. *Cancers* **2019**, *11*, 1425. [CrossRef]

53. Sabbatino, F.; Wang, Y.; Wang, X.; Flaherty, K.T.; Yu, L.; Pepin, D.; Scognamiglio, G.; Pepe, S.; Kirkwood, J.M.; Cooper, Z.A.; et al. PDGFRα up-regulation mediated by sonic hedgehog pathway activation leads to BRAF inhibitor resistance in melanoma cells with BRAF mutation. *Oncotarget* **2014**, *5*, 1926–1941. [CrossRef] [PubMed]

54. Shen, C.H.; Kim, S.H.; Trousil, S.; Frederick, D.T.; Piris, A.; Yuan, P.; Cai, L.; Gu, L.; Li, M.; Lee, J.H.; et al. Loss of cohesin complex components STAG2 or STAG3 confers resistance to BRAF inhibition in melanoma. *Nat. Med.* **2016**, *22*, 1056–1061. [CrossRef]

55. Shi, H.; Hong, A.; Kong, X.; Koya, R.C.; Song, C.; Moriceau, G.; Hugo, W.; Yu, C.C.; Ng, C.; Chodon, T.; et al. A novel AKT1 mutant amplifies an adaptive melanoma response to BRAF inhibition. *Cancer Discov.* **2014**, *4*, 69–79. [CrossRef]

56. Shull, A.Y.; Latham-Schwark, A.; Ramasamy, P.; Leskoske, K.; Oroian, D.; Birtwistle, M.R.; Buckhaults, P.J. Novel somatic mutations to PI3K pathway genes in metastatic melanoma. *PLoS ONE* **2012**, *7*, e43369. [CrossRef]

57. Straussman, R.; Morikawa, T.; Shee, K.; Barzily-Rokni, M.; Qian, Z.R.; Du, J.; Davis, A.; Mongare, M.M.; Gould, J.; Frederick, D.T.; et al. Tumour micro-environment elicits innate resistance to RAF inhibitors through HGF secretion. *Nature* **2012**, *487*, 500–504. [CrossRef]

58. Sun, C.; Wang, L.; Huang, S.; Heynen, G.J.; Prahallad, A.; Robert, C.; Haanen, J.; Blank, C.; Wesseling, J.; Willems, S.M.; et al. Reversible and adaptive resistance to BRAF(V600E) inhibition in melanoma. *Nature* **2014**, *508*, 118–122. [CrossRef]

59. Sun, X.; Li, J.; Sun, Y.; Zhang, Y.; Dong, L.; Shen, C.; Yang, L.; Yang, M.; Li, Y.; Shen, G.; et al. miR-7 reverses the resistance to BRAFi in melanoma by targeting EGFR/IGF-1R/CRAF and inhibiting the MAPK and PI3K/AKT signaling pathways. *Oncotarget* **2016**, *7*, 53558–53570. [CrossRef]

60. Teh, J.L.F.; Cheng, P.F.; Purwin, T.J.; Nikbakht, N.; Patel, P.; Chervoneva, I.; Ertel, A.; Fortina, P.M.; Kleiber, I.; HooKim, K.; et al. In Vivo E2F Reporting Reveals Efficacious Schedules of MEK1/2-CDK4/6 Targeting and mTOR-S6 Resistance Mechanisms. *Cancer Discov.* **2018**, *8*, 568–581. [CrossRef]

61. Van Allen, E.M.; Wagle, N.; Sucker, A.; Treacy, D.J.; Johannessen, C.M.; Goetz, E.M.; Place, C.S.; Taylor-Weiner, A.; Whittaker, S.; Kryukov, G.V.; et al. The genetic landscape of clinical resistance to RAF inhibition in metastatic melanoma. *Cancer Discov.* **2014**, *4*, 94–109. [CrossRef]

62. Villanueva, J.; Infante, J.R.; Krepler, C.; Reyes-Uribe, P.; Samanta, M.; Chen, H.Y.; Li, B.; Swoboda, R.K.; Wilson, M.; Vultur, A.; et al. Concurrent MEK2 mutation and BRAF amplification confer resistance to BRAF and MEK inhibitors in melanoma. *Cell Rep.* **2013**, *4*, 1090–1099. [CrossRef]

63. Wagenaar, T.R.; Ma, L.; Roscoe, B.; Park, S.M.; Bolon, D.N.; Green, M.R. Resistance to vemurafenib resulting from a novel mutation in the BRAFV600E kinase domain. *Pigment Cell Melanoma Res.* **2014**, *27*, 124–133. [CrossRef] [PubMed]

64. Wagle, M.C.; Kirouac, D.; Klijn, C.; Liu, B.; Mahajan, S.; Junttila, M.; Moffat, J.; Merchant, M.; Huw, L.; Wongchenko, M.; et al. A transcriptional MAPK Pathway Activity Score (MPAS) is a clinically relevant biomarker in multiple cancer types. *NPJ Precis. Oncol.* **2018**, *2*, 7. [CrossRef] [PubMed]

65. Watson, I.R.; Li, L.; Cabeceiras, P.K.; Mahdavi, M.; Gutschner, T.; Genovese, G.; Wang, G.; Fang, Z.; Tepper, J.M.; Stemke-Hale, K.; et al. The RAC1 P29S hotspot mutation in melanoma confers resistance to pharmacological inhibition of RAF. *Cancer Res.* **2014**, *74*, 4845–4852. [CrossRef] [PubMed]

66. Xue, G.; Kohler, R.; Tang, F.; Hynx, D.; Wang, Y.; Orso, F.; Pretre, V.; Ritschard, R.; Hirschmann, P.; Cron, P.; et al. mTORC1/autophagy-regulated MerTK in mutant BRAFV600 melanoma with acquired resistance to BRAF inhibition. *Oncotarget* **2017**, *8*, 69204–69218. [CrossRef] [PubMed]

67. Yadav, V.; Zhang, X.; Liu, J.; Estrem, S.; Li, S.; Gong, X.Q.; Buchanan, S.; Henry, J.R.; Starling, J.J.; Peng, S.B. Reactivation of mitogen-activated protein kinase (MAPK) pathway by FGF receptor 3 (FGFR3)/Ras mediates resistance to vemurafenib in human B-RAF V600E mutant melanoma. *J. Biol. Chem.* **2012**, *287*, 28087–28098. [CrossRef] [PubMed]

68. Perna, D.; Karreth, F.A.; Rust, A.G.; Perez-Mancera, P.A.; Rashid, M.; Iorio, F.; Alifrangis, C.; Arends, M.J.; Bosenberg, M.W.; Bollag, G.; et al. BRAF inhibitor resistance mediated by the AKT pathway in an oncogenic BRAF mouse melanoma model. *Proc. Natl. Acad. Sci. USA* **2015**, *112*, E536–E545. [CrossRef]

69. Hoogstraat, M.; Gadellaa-van Hooijdonk, C.G.; Ubink, I.; Besselink, N.J.; Pieterse, M.; Veldhuis, W.; van Stralen, M.; Meijer, E.F.; Willems, S.M.; Hadders, M.A.; et al. Detailed imaging and genetic analysis reveal a secondary BRAF(L505H) resistance mutation and extensive intrapatient heterogeneity in metastatic BRAF mutant melanoma patients treated with vemurafenib. *Pigment Cell Melanoma Res.* **2015**, *28*, 318–323. [CrossRef]

70. Pupo, G.M.; Boyd, S.C.; Fung, C.; Carlino, M.S.; Menzies, A.M.; Pedersen, B.; Johansson, P.; Hayward, N.K.; Kefford, R.F.; Scolyer, R.A.; et al. Clinical significance of intronic variants in BRAF inhibitor resistant melanomas with altered BRAF transcript splicing. *Biomark. Res.* **2017**, *5*, 17. [CrossRef]

71. Shi, H.; Moriceau, G.; Kong, X.; Lee, M.K.; Lee, H.; Koya, R.C.; Ng, C.; Chodon, T.; Scolyer, R.A.; Dahlman, K.B.; et al. Melanoma whole-exome sequencing identifies (V600E)B-RAF amplification-mediated acquired B-RAF inhibitor resistance. *Nat. Commun.* **2012**, *3*, 724. [CrossRef]

72. Johnson, G.L.; Stuhlmiller, T.J.; Angus, S.P.; Zawistowski, J.S.; Graves, L.M. Molecular pathways: Adaptive kinome reprogramming in response to targeted inhibition of the BRAF-MEK-ERK pathway in cancer. *Clin. Cancer Res.* **2014**, *20*, 2516–2522. [CrossRef]

73. Mistry, H.B.; Orrell, D.; Eftimie, R. Model based analysis of the heterogeneity in the tumour size dynamics differentiates vemurafenib, dabrafenib and trametinib in metastatic melanoma. *Cancer Chemother. Pharmacol.* **2018**, *81*, 325–332. [CrossRef] [PubMed]

74. Molina-Arcas, M.; Downward, J. How to fool a wonder drug: Truncate and dimerize. *Cancer Cell* **2012**, *21*, 7–9. [CrossRef] [PubMed]

75. Benito-Jardon, L.; Diaz-Martinez, M.; Arellano-Sanchez, N.; Vaquero-Morales, P.; Esparis-Ogando, A.; Teixido, J. Resistance to MAPK Inhibitors in Melanoma Involves Activation of the IGF1R-MEK5-Erk5 Pathway. *Cancer Res.* **2019**, *79*, 2244–2256. [CrossRef] [PubMed]

76. Goetz, E.M.; Garraway, L.A. Mechanisms of Resistance to Mitogen-Activated Protein Kinase Pathway Inhibition in BRAF-Mutant Melanoma. *Am. Soc. Clin. Oncol. Educ. Book* **2012**, *32*, 680–684. [CrossRef]

77. Lidsky, M.; Antoun, G.; Speicher, P.; Adams, B.; Turley, R.; Augustine, C.; Tyler, D.; Ali-Osman, F. Mitogen-activated protein kinase (MAPK) hyperactivation and enhanced NRAS expression drive acquired vemurafenib resistance in V600E BRAF melanoma cells. *J. Biol. Chem.* **2014**, *289*, 27714–27726. [CrossRef]

78. Tanda, E.T.; Vanni, I.; Boutros, A.; Andreotti, V.; Bruno, W.; Ghiorzo, P.; Spagnolo, F. Current state of target treatment in BRAF mutated melanoma. *Front. Mol. Biosci.* **2020**, *7*, 154. [CrossRef]

79. Paraiso, K.H.; Xiang, Y.; Rebecca, V.W.; Abel, E.V.; Chen, Y.A.; Munko, A.C.; Wood, E.; Fedorenko, I.V.; Sondak, V.K.; Anderson, A.R.; et al. PTEN loss confers BRAF inhibitor resistance to melanoma cells through the suppression of BIM expression. *Cancer Res.* **2011**, *71*, 2750–2760. [CrossRef]

80. Ahn, J.H.; Lee, M. Autophagy-Dependent Survival of Mutant B-Raf Melanoma Cells Selected for Resistance to Apoptosis Induced by Inhibitors against Oncogenic B-Raf. *Biomol. Ther.* **2013**, *21*, 114–120. [CrossRef]

81. Li, S.; Song, Y.; Quach, C.; Guo, H.; Jang, G.B.; Maazi, H.; Zhao, S.; Sands, N.A.; Liu, Q.; In, G.K.; et al. Transcriptional regulation of autophagy-lysosomal function in BRAF-driven melanoma progression and chemoresistance. *Nat. Commun.* **2019**, *10*, 1693. [CrossRef]

82. Martin, S.; Dudek-Peric, A.M.; Garg, A.D.; Roose, H.; Demirsoy, S.; Van Eygen, S.; Mertens, F.; Vangheluwe, P.; Vankelecom, H.; Agostinis, P. An autophagy-driven pathway of ATP secretion supports the aggressive phenotype of BRAF^{V600E} inhibitor-resistant metastatic melanoma cells. *Autophagy* **2017**, *13*, 1512–1527. [CrossRef]

83. Ma, X.H.; Piao, S.F.; Dey, S.; McAfee, Q.; Karakousis, G.; Villanueva, J.; Hart, L.S.; Levi, S.; Hu, J.; Zhang, G.; et al. Targeting ER stress-induced autophagy overcomes BRAF inhibitor resistance in melanoma. *J. Clin. Investig.* **2014**, *124*, 1406–1417. [CrossRef] [PubMed]

84. Szasz, I.; Koroknai, V.; Kiss, T.; Vizkeleti, L.; Adany, R.; Balazs, M. Molecular alterations associated with acquired resistance to BRAFV600E targeted therapy in melanoma cells. *Melanoma Res.* **2019**, *29*, 390–400. [CrossRef] [PubMed]

85. Wang, J.; Sinnberg, T.; Niessner, H.; Dolker, R.; Sauer, B.; Kempf, W.E.; Meier, F.; Leslie, N.; Schittek, B. PTEN regulates IGF-1R-mediated therapy resistance in melanoma. *Pigment Cell Melanoma Res.* **2015**, *28*, 572–589. [CrossRef] [PubMed]

86. Wilson, T.R.; Fridlyand, J.; Yan, Y.; Penuel, E.; Burton, L.; Chan, E.; Peng, J.; Lin, E.; Wang, Y.; Sosman, J.; et al. Widespread potential for growth-factor-driven resistance to anticancer kinase inhibitors. *Nature* **2012**, *487*, 505–509. [CrossRef]

87. Miller, M.A.; Oudin, M.J.; Sullivan, R.J.; Wang, S.J.; Meyer, A.S.; Im, H.; Frederick, D.T.; Tadros, J.; Griffith, L.G.; Lee, H.; et al. Reduced Proteolytic Shedding of Receptor Tyrosine Kinases Is a Post-Translational Mechanism of Kinase Inhibitor Resistance. *Cancer Discov.* **2016**, *6*, 382–399. [CrossRef]

88. Ahn, J.H.; Hwang, S.H.; Cho, H.S.; Lee, M. Differential Gene Expression Common to Acquired and Intrinsic Resistance to BRAF Inhibitor Revealed by RNA-Seq Analysis. *Biomol. Ther.* **2019**, *27*, 302–310. [CrossRef]

89. Li, K.; Zhao, S.; Long, J.; Su, J.; Wu, L.; Tao, J.; Zhou, J.; Zhang, J.; Chen, X.; Peng, C. A novel chalcone derivative has antitumor activity in melanoma by inducing DNA damage through the upregulation of ROS products. *Cancer Cell. Int.* **2020**, *20*, 36. [CrossRef]

90. Martin, C.A.; Cullinane, C.; Kirby, L.; Abuhammad, S.; Lelliott, E.J.; Waldeck, K.; Young, R.J.; Brajanovski, N.; Cameron, D.P.; Walker, R.; et al. Palbociclib synergizes with BRAF and MEK inhibitors in treatment naive melanoma but not after the development of BRAF inhibitor resistance. *Int. J. Cancer* **2018**, *142*, 2139–2152. [CrossRef]

91. Vanneste, M.; Feddersen, C.R.; Varzavand, A.; Zhu, E.Y.; Foley, T.; Zhao, L.; Holt, K.H.; Milhem, M.; Piper, R.; Stipp, C.S.; et al. Functional Genomic Screening Independently Identifies CUL3 as a Mediator of Vemurafenib Resistance via Src-Rac1 Signaling Axis. *Front. Oncol.* **2020**, *10*, 442. [CrossRef]

92. Whittaker, S.R.; Theurillat, J.P.; Van Allen, E.; Wagle, N.; Hsiao, J.; Cowley, G.S.; Schadendorf, D.; Root, D.E.; Garraway, L.A. A genome-scale RNA interference screen implicates NF1 loss in resistance to RAF inhibition. *Cancer Discov.* **2013**, *3*, 350–362. [CrossRef]

93. Parri, M.; Chiarugi, P. Rac and Rho GTPases in cancer cell motility control. *Cell Commun. Signal.* **2010**, *8*, 23. [CrossRef] [PubMed]

94. Kim, M.H.; Kim, J.; Hong, H.; Lee, S.H.; Lee, J.K.; Jung, E.; Kim, J. Actin remodeling confers BRAF inhibitor resistance to melanoma cells through YAP/TAZ activation. *EMBO J.* **2016**, *35*, 462–478. [CrossRef] [PubMed]

95. Fisher, M.L.; Grun, D.; Adhikary, G.; Xu, W.; Eckert, R.L. Inhibition of YAP function overcomes BRAF inhibitor resistance in melanoma cancer stem cells. *Oncotarget* **2017**, *8*, 110257–110272. [CrossRef]

96. Schmitt, M.; Sinnberg, T.; Nalpas, N.C.; Maass, A.; Schittek, B.; Macek, B. Quantitative Proteomics Links the Intermediate Filament Nestin to Resistance to Targeted BRAF Inhibition in Melanoma Cells. *Mol. Cell. Proteom.* **2019**, *18*, 1096–1109. [CrossRef] [PubMed]

97. Titz, B.; Lomova, A.; Le, A.; Hugo, W.; Kong, X.; Hoeve, J.T.; Friedman, M.; Shi, H.; Moriceau, G.; Song, C.; et al. JUN dependency in distinct early and late BRAF inhibition adaptation states of melanoma. *Cell Discov.* **2016**, *2*, 16028. [CrossRef]

98. Cordaro, F.G.; De Presbiteris, A.L.; Camerlingo, R.; Mozzillo, N.; Pirozzi, G.; Cavalcanti, E.; Manca, A.; Palmieri, G.; Cossu, A.; Ciliberto, G.; et al. Phenotype characterization of human melanoma cells resistant to dabrafenib. *Oncol. Rep.* **2017**, *38*, 2741–2751. [CrossRef]

99. Wang, J.; Huang, S.K.; Marzese, D.M.; Hsu, S.C.; Kawas, N.P.; Chong, K.K.; Long, G.V.; Menzies, A.M.; Scolyer, R.A.; Izraely, S.; et al. Epigenetic changes of EGFR have an important role in BRAF inhibitor-resistant cutaneous melanomas. *J. Investig. Dermatol.* **2015**, *135*, 532–541. [CrossRef]

100. Roche, J. The Epithelial-to-Mesenchymal Transition in Cancer. *Cancers* **2018**, *10*, 52. [CrossRef]

101. Seip, K.; Fleten, K.G.; Barkovskaya, A.; Nygaard, V.; Haugen, M.H.; Engesaeter, B.O.; Maelandsmo, G.M.; Prasmickaite, L. Fibroblast-induced switching to the mesenchymal-like phenotype and PI3K/mTOR signaling protects melanoma cells from BRAF inhibitors. *Oncotarget* **2016**, *7*, 19997–20015. [CrossRef]

102. Smith, M.P.; Wellbrock, C. Molecular Pathways: Maintaining MAPK Inhibitor Sensitivity by Targeting Nonmutational Tolerance. *Clin. Cancer Res.* **2016**, *22*, 5966–5970. [CrossRef]

103. Baenke, F.; Chaneton, B.; Smith, M.; Van Den Broek, N.; Hogan, K.; Tang, H.; Viros, A.; Martin, M.; Galbraith, L.; Girotti, M.R.; et al. Resistance to BRAF inhibitors induces glutamine dependency in melanoma cells. *Mol. Oncol.* **2016**, *10*, 73–84. [CrossRef] [PubMed]

104. Osrodek, M.; Hartman, M.L.; Czyz, M. Physiologically Relevant Oxygen Concentration (6% O_2) as an Important Component of the Microenvironment Impacting Melanoma Phenotype and Melanoma Response to Targeted Therapeutics In Vitro. *Int. J. Mol. Sci.* **2019**, *20*, 4203. [CrossRef]

105. Aloia, A.; Mullhaupt, D.; Chabbert, C.D.; Eberhart, T.; Fluckiger-Mangual, S.; Vukolic, A.; Eichhoff, O.; Irmisch, A.; Alexander, L.T.; Scibona, E.; et al. A Fatty Acid Oxidation-dependent Metabolic Shift Regulates the Adaptation of BRAF-mutated Melanoma to MAPK Inhibitors. *Clin. Cancer Res.* **2019**, *25*, 6852–6867. [CrossRef] [PubMed]

106. Audrito, V.; Manago, A.; La Vecchia, S.; Zamporlini, F.; Vitale, N.; Baroni, G.; Cignetto, S.; Serra, S.; Bologna, C.; Stingi, A.; et al. Nicotinamide Phosphoribosyltransferase (NAMPT) as a Therapeutic Target in BRAF-Mutated Metastatic Melanoma. *J. Natl. Cancer Inst.* **2018**, *110*, 290–303. [CrossRef] [PubMed]

107. Corazao-Rozas, P.; Guerreschi, P.; Andre, F.; Gabert, P.E.; Lancel, S.; Dekiouk, S.; Fontaine, D.; Tardivel, M.; Savina, A.; Quesnel, B.; et al. Mitochondrial oxidative phosphorylation controls cancer cell's life and death decisions upon exposure to MAPK inhibitors. *Oncotarget* **2016**, *7*, 39473–39485. [CrossRef]

108. Corazao-Rozas, P.; Guerreschi, P.; Jendoubi, M.; Andre, F.; Jonneaux, A.; Scalbert, C.; Garcon, G.; Malet-Martino, M.; Balayssac, S.; Rocchi, S.; et al. Mitochondrial oxidative stress is the Achille's heel of melanoma cells resistant to Braf-mutant inhibitor. *Oncotarget* **2013**, *4*, 1986–1998. [CrossRef]

109. Figarola, J.L.; Singhal, J.; Singhal, S.; Kusari, J.; Riggs, A. Bioenergetic modulation with the mitochondria uncouplers SR4 and niclosamide prevents proliferation and growth of treatment-naive and vemurafenib-resistant melanomas. *Oncotarget* **2018**, *9*, 36945–36965. [CrossRef]

110. Su, Y.; Bintz, M.; Yang, Y.; Robert, L.; Ng, A.H.C.; Liu, V.; Ribas, A.; Heath, J.R.; Wei, W. Phenotypic heterogeneity and evolution of melanoma cells associated with targeted therapy resistance. *PLoS Comput. Biol.* **2019**, *15*, e1007034. [CrossRef]

111. Wang, L.; de Oliveira, R.L.; Huijberts, S.; Bosdriesz, E.; Pencheva, N.; Brunen, D.; Bosma, A.; Song, J.Y.; Zevenhoven, J.; Los-de Vries, G.T.; et al. An Acquired Vulnerability of Drug-Resistant Melanoma with Therapeutic Potential. *Cell* **2018**, *173*, 1413–1425.e1414. [CrossRef]

112. Pan, M.; Reid, M.A.; Lowman, X.H.; Kulkarni, R.P.; Tran, T.Q.; Liu, X.; Yang, Y.; Hernandez-Davies, J.E.; Rosales, K.K.; Li, H.; et al. Regional glutamine deficiency in tumours promotes dedifferentiation through inhibition of histone demethylation. *Nat. Cell Biol.* **2016**, *18*, 1090–1101. [CrossRef]

113. Liu, X.; Zhang, S.M.; McGeary, M.K.; Krykbaeva, I.; Lai, L.; Jansen, D.J.; Kales, S.C.; Simeonov, A.; Hall, M.D.; Kelly, D.P.; et al. KDM5B Promotes Drug Resistance by Regulating Melanoma-Propagating Cell Subpopulations. *Mol. Cancer Ther.* **2019**, *18*, 706–717. [CrossRef] [PubMed]

114. Bajpe, P.K.; Prahallad, A.; Horlings, H.; Nagtegaal, I.; Beijersbergen, R.; Bernards, R. A chromatin modifier genetic screen identifies SIRT2 as a modulator of response to targeted therapies through the regulation of MEK kinase activity. *Oncogene* **2015**, *34*, 531–536. [CrossRef] [PubMed]

115. Strub, T.; Ghiraldini, F.G.; Carcamo, S.; Li, M.; Wroblewska, A.; Singh, R.; Goldberg, M.S.; Hasson, D.; Wang, Z.; Gallagher, S.J.; et al. SIRT6 haploinsufficiency induces BRAF^{V600E} melanoma cell resistance to MAPK inhibitors via IGF signalling. *Nat. Commun.* **2018**, *9*, 3440. [CrossRef] [PubMed]

116. Bugide, S.; Parajuli, K.R.; Chava, S.; Pattanayak, R.; Manna, D.L.D.; Shrestha, D.; Yang, E.S.; Cai, G.; Johnson, D.B.; Gupta, R. Loss of HAT1 expression confers BRAFV600E inhibitor resistance to melanoma cells by activating MAPK signaling via IGF1R. *Oncogenesis* **2020**, *9*, 44. [CrossRef]

117. Emmons, M.F.; Faiao-Flores, F.; Sharma, R.; Thapa, R.; Messina, J.L.; Becker, J.C.; Schadendorf, D.; Seto, E.; Sondak, V.K.; Koomen, J.M.; et al. HDAC8 Regulates a Stress Response Pathway in Melanoma to Mediate Escape from BRAF Inhibitor Therapy. *Cancer Res.* **2019**, *79*, 2947–2961. [CrossRef]

118. Sanjana, N.E.; Wright, J.; Zheng, K.; Shalem, O.; Fontanillas, P.; Joung, J.; Cheng, C.; Regev, A.; Zhang, F. High-resolution interrogation of functional elements in the noncoding genome. *Science* **2016**, *353*, 1545–1549. [CrossRef]

119. Joung, J.; Engreitz, J.M.; Konermann, S.; Abudayyeh, O.O.; Verdine, V.K.; Aguet, F.; Gootenberg, J.S.; Sanjana, N.E.; Wright, J.B.; Fulco, C.P.; et al. Genome-scale activation screen identifies a lncRNA locus regulating a gene neighbourhood. *Nature* **2017**, *548*, 343–346. [CrossRef]

120. Atkinson, J.M.; Rank, K.B.; Zeng, Y.; Capen, A.; Yadav, V.; Manro, J.R.; Engler, T.A.; Chedid, M. Activating the Wnt/beta-Catenin Pathway for the Treatment of Melanoma–Application of LY2090314, a Novel Selective Inhibitor of Glycogen Synthase Kinase-3. *PLoS ONE* **2015**, *10*, e0125028. [CrossRef]

121. Azimi, A.; Tuominen, R.; Svedman, F.C.; Caramuta, S.; Pernemalm, M.; Frostvik Stolt, M.; Kanter, L.; Kharaziha, P.; Lehtio, J.; Hertzman Johansson, C.; et al. Silencing FLI or targeting CD13/ANPEP lead to dephosphorylation of EPHA2, a mediator of BRAF inhibitor resistance, and induce growth arrest or apoptosis in melanoma cells. *Cell Death Dis.* **2017**, *8*, e3029. [CrossRef]

122. Becker, T.M.; Boyd, S.C.; Mijatov, B.; Gowrishankar, K.; Snoyman, S.; Pupo, G.M.; Scolyer, R.A.; Mann, G.J.; Kefford, R.F.; Zhang, X.D.; et al. Mutant B-RAF-Mcl-1 survival signaling depends on the STAT3 transcription factor. *Oncogene* **2014**, *33*, 1158–1166. [CrossRef]

123. Boregowda, R.K.; Medina, D.J.; Markert, E.; Bryan, M.A.; Chen, W.; Chen, S.; Rabkin, A.; Vido, M.J.; Gunderson, S.I.; Chekmareva, M.; et al. The transcription factor RUNX2 regulates receptor tyrosine kinase expression in melanoma. *Oncotarget* **2016**, *7*, 29689–29707. [CrossRef] [PubMed]

124. Chien, A.J.; Haydu, L.E.; Biechele, T.L.; Kulikauskas, R.M.; Rizos, H.; Kefford, R.F.; Scolyer, R.A.; Moon, R.T.; Long, G.V. Targeted BRAF inhibition impacts survival in melanoma patients with high levels of Wnt/beta-catenin signaling. *PLoS ONE* **2014**, *9*, e94748. [CrossRef] [PubMed]

125. Corre, S.; Tardif, N.; Mouchet, N.; Leclair, H.M.; Boussemart, L.; Gautron, A.; Bachelot, L.; Perrot, A.; Soshilov, A.; Rogiers, A.; et al. Sustained activation of the Aryl hydrocarbon Receptor transcription factor promotes resistance to BRAF-inhibitors in melanoma. *Nat. Commun.* **2018**, *9*, 4775. [CrossRef] [PubMed]

126. Cronin, J.C.; Loftus, S.K.; Baxter, L.L.; Swatkoski, S.; Gucek, M.; Pavan, W.J. Identification and functional analysis of SOX10 phosphorylation sites in melanoma. *PLoS ONE* **2018**, *13*, e0190834. [CrossRef] [PubMed]

127. Czyz, M.; Sztiller-Sikorska, M.; Gajos-Michniewicz, A.; Osrodek, M.; Hartman, M.L. Plasticity of Drug-Naive and Vemurafenib- or Trametinib-Resistant Melanoma Cells in Execution of Differentiation/Pigmentation Program. *J. Oncol. Print* **2019**, *2019*, 1697913. [CrossRef]

128. Delmas, A.; Cherier, J.; Pohorecka, M.; Medale-Giamarchi, C.; Meyer, N.; Casanova, A.; Sordet, O.; Lamant, L.; Savina, A.; Pradines, A.; et al. The c-Jun/RHOB/AKT pathway confers resistance of BRAF-mutant melanoma cells to MAPK inhibitors. *Oncotarget* **2015**, *6*, 15250–15264. [CrossRef]

129. Giricz, O.; Mo, Y.; Dahlman, K.B.; Cotto-Rios, X.M.; Vardabasso, C.; Nguyen, H.; Matusow, B.; Bartenstein, M.; Polishchuk, V.; Johnson, D.B.; et al. The RUNX1/IL-34/CSF-1R axis is an autocrinally regulated modulator of resistance to BRAF-V600E inhibition in melanoma. *JCI Insight* **2018**, *3*, 26. [CrossRef]

130. Hartman, M.L.; Sztiller-Sikorska, M.; Gajos-Michniewicz, A.; Czyz, M. Dissecting Mechanisms of Melanoma Resistance to BRAF and MEK Inhibitors Revealed Genetic and Non-Genetic Patient- and Drug-Specific Alterations and Remarkable Phenotypic Plasticity. *Cells* **2020**, *9*, 142. [CrossRef]

131. Huser, L.; Sachindra, S.; Granados, K.; Federico, A.; Larribere, L.; Novak, D.; Umansky, V.; Altevogt, P.; Utikal, J. SOX2-mediated upregulation of CD24 promotes adaptive resistance toward targeted therapy in melanoma. *Int. J. Cancer* **2018**, *143*, 3131–3142. [CrossRef]

132. Ji, Z.; Chen, Y.E.; Kumar, R.; Taylor, M.; Njauw, C.N.J.; Miao, B.; Frederick, D.T.; Wargo, J.A.; Flaherty, K.T.; Tsao, H.; et al. MITF Modulates Therapeutic Resistance through EGFR Signaling. *J. Investig. Dermatol.* **2015**, *135*, 1863–1872. [CrossRef]

133. Liu, F.; Jiang, C.C.; Yan, X.G.; Tseng, H.Y.; Wang, C.Y.; Zhang, Y.Y.; Yari, H.; La, T.; Farrelly, M.; Guo, S.T.; et al. BRAF/MEK inhibitors promote CD47 expression that is reversible by ERK inhibition in melanoma. *Oncotarget* **2017**, *8*, 69477–69492. [CrossRef] [PubMed]

134. Mohapatra, P.; Prasad, C.P.; Andersson, T. Combination therapy targeting the elevated interleukin-6 level reduces invasive migration of BRAF inhibitor-resistant melanoma cells. *Mol. Oncol.* **2019**, *13*, 480–494. [CrossRef] [PubMed]

135. Perotti, V.; Baldassari, P.; Molla, A.; Nicolini, G.; Bersani, I.; Grazia, G.; Benigni, F.; Maurichi, A.; Santinami, M.; Anichini, A.; et al. An actionable axis linking NFATc2 to EZH2 controls the EMT-like program of melanoma cells. *Oncogene* **2019**, *38*, 4384–4396. [CrossRef] [PubMed]

136. Ramsdale, R.; Jorissen, R.N.; Li, F.Z.; Al-Obaidi, S.; Ward, T.; Sheppard, K.E.; Bukczynska, P.E.; Young, R.J.; Boyle, S.E.; Shackleton, M.; et al. The transcription cofactor c-JUN mediates phenotype switching and BRAF inhibitor resistance in melanoma. *Sci. Signal.* **2015**, *8*, ra82. [CrossRef]

137. Richard, G.; Dalle, S.; Monet, M.A.; Ligier, M.; Boespflug, A.; Pommier, R.M.; de la Fouchardiere, A.; Perier-Muzet, M.; Depaepe, L.; Barnault, R.; et al. ZEB1-mediated melanoma cell plasticity enhances resistance to MAPK inhibitors. *EMBO Mol. Med.* **2016**, *8*, 1143–1161. [CrossRef]

138. Singleton, K.R.; Crawford, L.; Tsui, E.; Manchester, H.E.; Maertens, O.; Liu, X.; Liberti, M.V.; Magpusao, A.N.; Stein, E.M.; Tingley, J.P.; et al. Melanoma Therapeutic Strategies that Select against Resistance by Exploiting MYC-Driven Evolutionary Convergence. *Cell Rep.* **2017**, *21*, 2796–2812. [CrossRef]

139. Sinnberg, T.; Makino, E.; Krueger, M.A.; Velic, A.; Macek, B.; Rothbauer, U.; Groll, N.; Potz, O.; Czemmel, S.; Niessner, H.; et al. A Nexus Consisting of Beta-Catenin and Stat3 Attenuates BRAF Inhibitor Efficacy and Mediates Acquired Resistance to Vemurafenib. *EBioMedicine* **2016**, *8*, 132–149. [CrossRef]

140. Liu, X.; Mi, J.; Qin, H.; Li, Z.; Chai, J.; Li, M.; Wu, J.; Xu, J. E2F1/IGF-1R Loop Contributes to BRAF Inhibitor Resistance in Melanoma. *J. Investig. Dermatol.* **2020**, *140*, 1295–1299.e1291. [CrossRef]

141. Zecena, H.; Tveit, D.; Wang, Z.; Farhat, A.; Panchal, P.; Liu, J.; Singh, S.J.; Sanghera, A.; Bainiwal, A.; Teo, S.Y.; et al. Systems biology analysis of mitogen activated protein kinase inhibitor resistance in malignant melanoma. *BMC Syst. Biol.* **2018**, *12*, 33. [CrossRef]

142. Fofaria, N.M.; Frederick, D.T.; Sullivan, R.J.; Flaherty, K.T.; Srivastava, S.K. Overexpression of Mcl-1 confers resistance to BRAFV600E inhibitors alone and in combination with MEK1/2 inhibitors in melanoma. *Oncotarget* **2015**, *6*, 40535–40556. [CrossRef]

143. Fernandez, M.; Sutterluty-Fall, H.; Schwarzler, C.; Lemeille, S.; Boehncke, W.H.; Merat, R. Overexpression of the human antigen R suppresses the immediate paradoxical proliferation of melanoma cell subpopulations in response to suboptimal BRAF inhibition. *Cancer Med.* **2017**, *6*, 1652–1664. [CrossRef] [PubMed]

144. Zhan, Y.; Dahabieh, M.S.; Rajakumar, A.; Dobocan, M.C.; M'Boutchou, M.N.; Goncalves, C.; Lucy, S.L.; Pettersson, F.; Topisirovic, I.; van Kempen, L.; et al. The role of eIF4E in response and acquired resistance to vemurafenib in melanoma. *J. Investig. Dermatol.* **2015**, *135*, 1368–1376. [CrossRef] [PubMed]

145. Boussemart, L.; Malka-Mahieu, H.; Girault, I.; Allard, D.; Hemmingsson, O.; Tomasic, G.; Thomas, M.; Basmadjian, C.; Ribeiro, N.; Thuaud, F.; et al. eIF4F is a nexus of resistance to anti-BRAF and anti-MEK cancer therapies. *Nature* **2014**, *513*, 105–109. [CrossRef]

146. Caporali, S.; Amaro, A.; Levati, L.; Alvino, E.; Lacal, P.M.; Mastroeni, S.; Ruffini, F.; Bonmassar, L.; Cappellini, G.C.A.; Felli, N.; et al. miR-126-3p down-regulation contributes to dabrafenib acquired resistance in melanoma by up-regulating ADAM9 and VEGF-A. *J. Exp. Clin. Cancer Res.* **2019**, *38*, 272. [CrossRef] [PubMed]

147. Diaz-Martinez, M.; Benito-Jardon, L.; Alonso, L.; Koetz-Ploch, L.; Hernando, E.; Teixido, J. miR-204-5p and miR-211-5p Contribute to BRAF Inhibitor Resistance in Melanoma. *Cancer Res.* **2018**, *78*, 1017–1030. [CrossRef]

148. Fattore, L.; Mancini, R.; Acunzo, M.; Romano, G.; Lagana, A.; Pisanu, M.E.; Malpicci, D.; Madonna, G.; Mallardo, D.; Capone, M.; et al. miR-579-3p controls melanoma progression and resistance to target therapy. *Proc. Natl. Acad. Sci. USA* **2016**, *113*, E5005–E5013. [CrossRef]

149. Fattore, L.; Ruggiero, C.F.; Pisanu, M.E.; Liguoro, D.; Cerri, A.; Costantini, S.; Capone, F.; Acunzo, M.; Romano, G.; Nigita, G.; et al. Reprogramming miRNAs global expression orchestrates development of drug resistance in BRAF mutated melanoma. *Cell Death Differ.* **2019**, *26*, 1267–1282. [CrossRef]

150. Hwang, S.H.; Ahn, J.H.; Lee, M. Upregulation of S100A9 contributes to the acquired resistance to BRAF inhibitors. *Genes Genom.* **2019**, *41*, 1273–1280. [CrossRef]

151. Kim, J.H.; Ahn, J.H.; Lee, M. Upregulation of MicroRNA-1246 Is Associated with BRAF Inhibitor Resistance in Melanoma Cells with Mutant BRAF. *Cancer Res. Treat.* **2017**, *49*, 947–959. [CrossRef]

152. Kozar, I.; Cesi, G.; Margue, C.; Philippidou, D.; Kreis, S. Impact of BRAF kinase inhibitors on the miRNomes and transcriptomes of melanoma cells. *Biochim. Biophys. Acta Gen. Subj.* **2017**, *1861*, 2980–2992. [CrossRef]

153. Liu, S.; Tetzlaff, M.T.; Wang, T.; Yang, R.; Xie, L.; Zhang, G.; Krepler, C.; Xiao, M.; Beqiri, M.; Xu, W.; et al. miR-200c/Bmi1 axis and epithelial-mesenchymal transition contribute to acquired resistance to BRAF inhibitor treatment. *Pigment Cell Melanoma Res.* **2015**, *28*, 431–441. [CrossRef] [PubMed]

154. Stark, M.S.; Bonazzi, V.F.; Boyle, G.M.; Palmer, J.M.; Symmons, J.; Lanagan, C.M.; Schmidt, C.W.; Herington, A.C.; Ballotti, R.; Pollock, P.M.; et al. miR-514a regulates the tumour suppressor NF1 and modulates BRAFi sensitivity in melanoma. *Oncotarget* **2015**, *6*, 17753–17763. [CrossRef] [PubMed]

155. Rapino, F.; Delaunay, S.; Rambow, F.; Zhou, Z.; Tharun, L.; De Tullio, P.; Sin, O.; Shostak, K.; Schmitz, S.; Piepers, J.; et al. Codon-specific translation reprogramming promotes resistance to targeted therapy. *Nature* **2018**, *558*, 605–609. [CrossRef] [PubMed]

156. Frederick, D.T.; Piris, A.; Cogdill, A.P.; Cooper, Z.A.; Lezcano, C.; Ferrone, C.R.; Mitra, D.; Boni, A.; Newton, L.P.; Liu, C.; et al. BRAF inhibition is associated with enhanced melanoma antigen expression and a more favorable tumor microenvironment in patients with metastatic melanoma. *Clin. Cancer Res.* **2013**, *19*, 1225–1231. [CrossRef]

157. Hopkins, A.M.; Van Dyk, M.; Rowland, A.; Sorich, M.J. Effect of early adverse events on response and survival outcomes of advanced melanoma patients treated with vemurafenib or vemurafenib plus cobimetinib: A pooled analysis of clinical trial data. *Pigment Cell Melanoma Res.* **2019**, *32*, 576–583. [CrossRef]

158. Ben-Betzalel, G.; Baruch, E.N.; Boursi, B.; Steinberg-Silman, Y.; Asher, N.; Shapira-Frommer, R.; Schachter, J.; Markel, G. Possible immune adverse events as predictors of durable response to BRAF inhibitors in patients with BRAF V600-mutant metastatic melanoma. *Eur. J. Cancer* **2018**, *101*, 229–235. [CrossRef]

159. Consoli, F.; Manganoni, A.M.; Grisanti, S.; Petrelli, F.; Venturini, M.; Rangoni, G.; Guarneri, F.; Incardona, P.; Vermi, W.; Pinton, P.G.C.; et al. Panniculitis and vitiligo occurring during BRAF and MEK inhibitors combination in advanced melanoma patients: Potential predictive role of treatment efficacy. *PLoS ONE* **2019**, *14*, e0214884. [CrossRef]

160. Bellmann, L.; Cappellano, G.; Schachtl-Riess, J.F.; Prokopi, A.; Seretis, A.; Ortner, D.; Tripp, C.H.; Brinckerhoff, C.E.; Mullins, D.W.; Stoitzner, P. A TLR7 agonist strengthens T and NK cell function during BRAF-targeted therapy in a preclinical melanoma model. *Int. J. Cancer* **2020**, *146*, 1409–1420. [CrossRef]

161. Greenplate, A.R.; McClanahan, D.D.; Oberholtzer, B.K.; Doxie, D.B.; Roe, C.E.; Diggins, K.E.; Leelatian, N.; Rasmussen, M.L.; Kelley, M.C.; Gama, V.; et al. Computational Immune Monitoring Reveals Abnormal Double-Negative T Cells Present across Human Tumor Types. *Cancer Immunol. Res.* **2019**, *7*, 86–99. [CrossRef]

162. Steinberg, S.M.; Shabaneh, T.B.; Zhang, P.; Martyanov, V.; Li, Z.; Malik, B.T.; Wood, T.A.; Boni, A.; Molodtsov, A.; Angeles, C.V.; et al. Myeloid Cells That Impair Immunotherapy Are Restored in Melanomas with Acquired Resistance to BRAF Inhibitors. *Cancer Res.* **2017**, *77*, 1599–1610. [CrossRef]

163. Jazirehi, A.R.; Nazarian, R.; Torres-Collado, A.X.; Economou, J.S. Aberrant apoptotic machinery confers melanoma dual resistance to BRAF(V600E) inhibitor and immune effector cells: Immunosensitization by a histone deacetylase inhibitor. *Am. J. Clin. Exp. Immunol.* **2014**, *3*, 43–56. [PubMed]

164. Pieper, N.; Zaremba, A.; Leonardelli, S.; Harbers, F.N.; Schwamborn, M.; Lubcke, S.; Schrors, B.; Baingo, J.; Schramm, A.; Haferkamp, S.; et al. Evolution of melanoma cross-resistance to CD8$^+$ T cells and MAPK inhibition in the course of BRAFi treatment. *Oncoimmunology* **2018**, *7*, e1450127. [CrossRef] [PubMed]

165. Atefi, M.; Avramis, E.; Lassen, A.; Wong, D.J.; Robert, L.; Foulad, D.; Cerniglia, M.; Titz, B.; Chodon, T.; Graeber, T.G.; et al. Effects of MAPK and PI3K pathways on PD-L1 expression in melanoma. *Clin. Cancer Res.* **2014**, *20*, 3446–3457. [CrossRef] [PubMed]

166. Gowrishankar, K.; Gunatilake, D.; Gallagher, S.J.; Tiffen, J.; Rizos, H.; Hersey, P. Inducible but not constitutive expression of PD-L1 in human melanoma cells is dependent on activation of NF-kappaB. *PLoS ONE* **2015**, *10*, e0123410. [CrossRef]

167. Jiang, X.; Zhou, J.; Giobbie-Hurder, A.; Wargo, J.; Hodi, F.S. The activation of MAPK in melanoma cells resistant to BRAF inhibition promotes PD-L1 expression that is reversible by MEK and PI3K inhibition. *Clin. Cancer Res.* **2013**, *19*, 598–609. [CrossRef]

168. Kim, M.H.; Kim, C.G.; Kim, S.K.; Shin, S.J.; Choe, E.A.; Park, S.H.; Shin, E.C.; Kim, J. YAP-Induced PD-L1 Expression Drives Immune Evasion in BRAFi-Resistant Melanoma. *Cancer Immunol. Res.* **2018**, *6*, 255–266. [CrossRef]

169. Sottile, R.; Pangigadde, P.N.; Tan, T.; Anichini, A.; Sabbatino, F.; Trecroci, F.; Favoino, E.; Orgiano, L.; Roberts, J.; Ferrone, S.; et al. HLA class I downregulation is associated with enhanced NK-cell killing of melanoma cells with acquired drug resistance to BRAF inhibitors. *Eur. J. Immunol.* **2016**, *46*, 409–419. [CrossRef]

170. Gorniak, P.; Wasylecka-Juszczynska, M.; Lugowska, I.; Rutkowski, P.; Polak, A.; Szydlowski, M.; Juszczynski, P. BRAF inhibition curtails IFN-gamma-inducible PD-L1 expression and upregulates the immunoregulatory protein galectin-1 in melanoma cells. *Mol. Oncol.* **2020**, *24*, 1817–1832. [CrossRef]

171. Atay, C.; Kwak, T.; Lavilla-Alonso, S.; Donthireddy, L.; Richards, A.; Moberg, V.; Pilon-Thomas, S.; Schell, M.; Messina, J.L.; Rebecca, V.W.; et al. BRAF Targeting Sensitizes Resistant Melanoma to Cytotoxic T Cells. *Clin. Cancer Res.* **2019**, *25*, 2783–2794. [CrossRef]

172. Beauchemin, N.; Arabzadeh, A. Carcinoembryonic antigen-related cell adhesion molecules (CEACAMs) in cancer progression and metastasis. *Cancer Metastasis Rev.* **2013**, *32*, 643–671. [CrossRef]

173. Turcu, G.; Nedelcu, R.I.; Ion, D.A.; Brinzea, A.; Cioplea, M.D.; Jilaveanu, L.B.; Zurac, S.A. CEACAM1: Expression and Role in Melanocyte Transformation. *Dis. Markers* **2016**, *2016*, 9406319. [CrossRef] [PubMed]

174. Kfir-Elirachman, K.; Ortenberg, R.; Vizel, B.; Besser, M.J.; Barshack, I.; Schachter, J.; Nemlich, Y.; Markel, G. Regulation of CEACAM1 Protein Expression by the Transcription Factor ETS-1 in BRAF-Mutant Human Metastatic Melanoma Cells. *Neoplasia* **2018**, *20*, 401–409. [CrossRef] [PubMed]

175. Mandala, M.; Massi, D. Immunotolerance as a Mechanism of Resistance to Targeted Therapies in Melanoma. *Handb. Exp. Pharmacol.* **2018**, *249*, 129–143. [CrossRef]

176. Wang, T.; Xiao, M.; Ge, Y.; Krepler, C.; Belser, E.; Lopez-Coral, A.; Xu, X.; Zhang, G.; Azuma, R.; Liu, Q.; et al. BRAF Inhibition Stimulates Melanoma-Associated Macrophages to Drive Tumor Growth. *Clin. Cancer Res.* **2015**, *21*, 1652–1664. [CrossRef] [PubMed]

177. Abdel-Rahman, O.; ElHalawani, H.; Ahmed, H. Doublet BRAF/MEK inhibition versus single-agent BRAF inhibition in the management of BRAF-mutant advanced melanoma, biological rationale and meta-analysis of published data. *Clin. Transl. Oncol.* **2016**, *18*, 848–858. [CrossRef]

178. Ascierto, P.A.; McArthur, G.A.; Dreno, B.; Atkinson, V.; Liszkay, G.; Di Giacomo, A.M.; Mandala, M.; Demidov, L.; Stroyakovskiy, D.; Thomas, L.; et al. Cobimetinib combined with vemurafenib in advanced BRAF(V600)-mutant melanoma (coBRIM): Updated efficacy results from a randomised, double-blind, phase 3 trial. *Lancet Oncol.* **2016**, *17*, 1248–1260. [CrossRef]

179. Dummer, R.; Ascierto, P.A.; Gogas, H.J.; Arance, A.; Mandala, M.; Liszkay, G.; Garbe, C.; Schadendorf, D.; Krajsova, I.; Gutzmer, R.; et al. Overall survival in patients with BRAF-mutant melanoma receiving encorafenib plus binimetinib versus vemurafenib or encorafenib (COLUMBUS): A multicentre, open-label, randomised, phase 3 trial. *Lancet Oncol.* **2018**, *19*, 1315–1327. [CrossRef]

180. Dummer, R.; Ascierto, P.A.; Gogas, H.J.; Arance, A.; Mandala, M.; Liszkay, G.; Garbe, C.; Schadendorf, D.; Krajsova, I.; Gutzmer, R.; et al. Encorafenib plus binimetinib versus vemurafenib or encorafenib in patients with BRAF-mutant melanoma (COLUMBUS): A multicentre, open-label, randomised phase 3 trial. *Lancet Oncol.* **2018**, *19*, 603–615. [CrossRef]

181. Hamid, O.; Cowey, C.L.; Offner, M.; Faries, M.; Carvajal, R.D. Efficacy, Safety, and Tolerability of Approved Combination BRAF and MEK Inhibitor Regimens for BRAF-Mutant Melanoma. *Cancers* **2019**, *11*, 1642. [CrossRef]

182. Hauschild, A.; Larkin, J.; Ribas, A.; Dreno, B.; Flaherty, K.T.; Ascierto, P.A.; Lewis, K.D.; McKenna, E.; Zhu, Q.; Mun, Y.; et al. Modeled Prognostic Subgroups for Survival and Treatment Outcomes in BRAF V600-Mutated Metastatic Melanoma: Pooled Analysis of 4 Randomized Clinical Trials. *JAMA Oncol.* **2018**, *4*, 1382–1388. [CrossRef]

183. Long, G.V.; Eroglu, Z.; Infante, J.; Patel, S.; Daud, A.; Johnson, D.B.; Gonzalez, R.; Kefford, R.; Hamid, O.; Schuchter, L.; et al. Long-Term Outcomes in Patients With BRAF V600-Mutant Metastatic Melanoma Who Received Dabrafenib Combined With Trametinib. *J. Clin. Oncol.* **2018**, *36*, 667–673. [CrossRef] [PubMed]

184. Long, G.V.; Stroyakovskiy, D.; Gogas, H.; Levchenko, E.; de Braud, F.; Larkin, J.; Garbe, C.; Jouary, T.; Hauschild, A.; Grob, J.J.; et al. Combined BRAF and MEK inhibition versus BRAF inhibition alone in melanoma. *N. Engl. J. Med.* **2014**, *371*, 1877–1888. [CrossRef] [PubMed]

185. Long, G.V.; Stroyakovskiy, D.; Gogas, H.; Levchenko, E.; de Braud, F.; Larkin, J.; Garbe, C.; Jouary, T.; Hauschild, A.; Grob, J.J.; et al. Dabrafenib and trametinib versus dabrafenib and placebo for Val600

BRAF-mutant melanoma: A multicentre, double-blind, phase 3 randomised controlled trial. *Lancet* **2015**, *386*, 444–451. [CrossRef]

186. Robert, C.; Grob, J.J.; Stroyakovskiy, D.; Karaszewska, B.; Hauschild, A.; Levchenko, E.; Chiarion Sileni, V.; Schachter, J.; Garbe, C.; Bondarenko, I.; et al. Five-Year Outcomes with Dabrafenib plus Trametinib in Metastatic Melanoma. *N. Engl. J. Med.* **2019**, *381*, 626–636. [CrossRef]

187. Robert, C.; Karaszewska, B.; Schachter, J.; Rutkowski, P.; Mackiewicz, A.; Stroiakovski, D.; Lichinitser, M.; Dummer, R.; Grange, F.; Mortier, L.; et al. Improved overall survival in melanoma with combined dabrafenib and trametinib. *N. Engl. J. Med.* **2015**, *372*, 30–39. [CrossRef] [PubMed]

188. Paraiso, K.H.; Fedorenko, I.V.; Cantini, L.P.; Munko, A.C.; Hall, M.; Sondak, V.K.; Messina, J.L.; Flaherty, K.T.; Smalley, K.S. Recovery of phospho-ERK activity allows melanoma cells to escape from BRAF inhibitor therapy. *Br. J. Cancer* **2010**, *102*, 1724–1730. [CrossRef] [PubMed]

189. Wongchenko, M.J.; McArthur, G.A.; Dreno, B.; Larkin, J.; Ascierto, P.A.; Sosman, J.; Andries, L.; Kockx, M.; Hurst, S.D.; Caro, I.; et al. Gene Expression Profiling in BRAF-Mutated Melanoma Reveals Patient Subgroups with Poor Outcomes to Vemurafenib That May Be Overcome by Cobimetinib Plus Vemurafenib. *Clin. Cancer Res.* **2017**, *23*, 5238–5245. [CrossRef]

190. Long, G.V.; Grob, J.J.; Nathan, P.; Ribas, A.; Robert, C.; Schadendorf, D.; Lane, S.R.; Mak, C.; Legenne, P.; Flaherty, K.T.; et al. Factors predictive of response, disease progression, and overall survival after dabrafenib and trametinib combination treatment: A pooled analysis of individual patient data from randomised trials. *Lancet Oncol.* **2016**, *17*, 1743–1754. [CrossRef]

191. Long, G.V.; Hauschild, A.; Santinami, M.; Atkinson, V.; Mandala, M.; Chiarion-Sileni, V.; Larkin, J.; Nyakas, M.; Dutriaux, C.; Haydon, A.; et al. Adjuvant Dabrafenib plus Trametinib in Stage III BRAF-Mutated Melanoma. *N. Engl. J. Med.* **2017**, *377*, 1813–1823. [CrossRef]

192. Schreuer, M.; Jansen, Y.; Planken, S.; Chevolet, I.; Seremet, T.; Kruse, V.; Neyns, B. Combination of dabrafenib plus trametinib for BRAF and MEK inhibitor pretreated patients with advanced BRAF^{V600}-mutant melanoma: An open-label, single arm, dual-centre, phase 2 clinical trial. *Lancet Oncol.* **2017**, *18*, 464–472. [CrossRef]

193. Adelmann, C.H.; Ching, G.; Du, L.; Saporito, R.C.; Bansal, V.; Pence, L.J.; Liang, R.; Lee, W.; Tsai, K.Y. Comparative profiles of BRAF inhibitors: The paradox index as a predictor of clinical toxicity. *Oncotarget* **2016**, *7*, 30453–30460. [CrossRef]

194. Ascierto, P.A.; Ferrucci, P.F.; Fisher, R.; Del Vecchio, M.; Atkinson, V.; Schmidt, H.; Schachter, J.; Queirolo, P.; Long, G.V.; Di Giacomo, A.M.; et al. Dabrafenib, trametinib and pembrolizumab or placebo in BRAF-mutant melanoma. *Nat. Med.* **2019**, *25*, 941–946. [CrossRef]

195. Ribas, A.; Lawrence, D.; Atkinson, V.; Agarwal, S.; Miller, W.H., Jr.; Carlino, M.S.; Fisher, R.; Long, G.V.; Hodi, F.S.; Tsoi, J.; et al. Combined BRAF and MEK inhibition with PD-1 blockade immunotherapy in BRAF-mutant melanoma. *Nat. Med.* **2019**, *25*, 936–940. [CrossRef]

196. Sullivan, R.J.; Hamid, O.; Gonzalez, R.; Infante, J.R.; Patel, M.R.; Hodi, F.S.; Lewis, K.D.; Tawbi, H.A.; Hernandez, G.; Wongchenko, M.J.; et al. Atezolizumab plus cobimetinib and vemurafenib in BRAF-mutated melanoma patients. *Nat. Med.* **2019**, *25*, 929–935. [CrossRef]

197. Gowrishankar, K.; Snoyman, S.; Pupo, G.M.; Becker, T.M.; Kefford, R.F.; Rizos, H. Acquired resistance to BRAF inhibition can confer cross-resistance to combined BRAF/MEK inhibition. *J. Investig. Dermatol.* **2012**, *132*, 1850–1859. [CrossRef]

198. Roller, D.G.; Capaldo, B.; Bekiranov, S.; Mackey, A.J.; Conaway, M.R.; Petricoin, E.F.; Gioeli, D.; Weber, M.J. Combinatorial drug screening and molecular profiling reveal diverse mechanisms of intrinsic and adaptive resistance to BRAF inhibition in V600E BRAF mutant melanomas. *Oncotarget* **2016**, *7*, 2734–2753. [CrossRef]

199. Matheis, F.; Heppt, M.V.; Graf, S.A.; Duwell, P.; Kammerbauer, C.; Aigner, A.; Besch, R.; Berking, C. A Bifunctional Approach of Immunostimulation and uPAR Inhibition Shows Potent Antitumor Activity in Melanoma. *J. Investig. Dermatol.* **2016**, *136*, 2475–2484. [CrossRef]

200. Chalbatani, G.M.; Dana, H.; Gharagouzloo, E.; Grijalvo, S.; Eritja, R.; Logsdon, C.D.; Memari, F.; Miri, S.R.; Rad, M.R.; Marmari, V. Small interfering RNAs (siRNAs) in cancer therapy: A nano-based approach. *Int. J. Nanomed.* **2019**, *14*, 3111–3128. [CrossRef]

201. Sharma, R.; Fedorenko, I.; Spence, P.T.; Sondak, V.K.; Smalley, K.S.; Koomen, J.M. Activity-Based Protein Profiling Shows Heterogeneous Signaling Adaptations to BRAF Inhibition. *J. Proteome Res.* **2016**, *15*, 4476–4489. [CrossRef]

202. Cheng, H.; Chang, Y.; Zhang, L.; Luo, J.; Tu, Z.; Lu, X.; Zhang, Q.; Lu, J.; Ren, X.; Ding, K. Identification and optimization of new dual inhibitors of B-Raf and epidermal growth factor receptor kinases for overcoming resistance against vemurafenib. *J. Med. Chem.* **2014**, *57*, 2692–2703. [CrossRef]

203. Langdon, C.G.; Held, M.A.; Platt, J.T.; Meeth, K.; Iyidogan, P.; Mamillapalli, R.; Koo, A.B.; Klein, M.; Liu, Z.; Bosenberg, M.W.; et al. The broad-spectrum receptor tyrosine kinase inhibitor dovitinib suppresses growth of BRAF-mutant melanoma cells in combination with other signaling pathway inhibitors. *Pigment Cell Melanoma Res.* **2015**, *28*, 417–430. [CrossRef]

204. Srivastava, A.; Moorthy, A. Sorafenib induces synergistic effect on inhibition of vemurafenib resistant melanoma growth. *Front. Biosci. Schol. Ed.* **2019**, *11*, 193–202.

205. Bonnevaux, H.; Lemaitre, O.; Vincent, L.; Levit, M.N.; Windenberger, F.; Halley, F.; Delorme, C.; Lengauer, C.; Garcia-Echeverria, C.; Virone-Oddos, A. Concomitant Inhibition of PI3Kbeta and BRAF or MEK in PTEN-Deficient/BRAF-Mutant Melanoma Treatment: Preclinical Assessment of SAR260301 Oral PI3Kbeta-Selective Inhibitor. *Mol. Cancer Ther.* **2016**, *15*, 1460–1471. [CrossRef]

206. Byron, S.A.; Loch, D.C.; Wellens, C.L.; Wortmann, A.; Wu, J.; Wang, J.; Nomoto, K.; Pollock, P.M. Sensitivity to the MEK inhibitor E6201 in melanoma cells is associated with mutant BRAF and wildtype PTEN status. *Mol. Cancer* **2012**, *11*, 75. [CrossRef]

207. Calero, R.; Morchon, E.; Martinez-Argudo, I.; Serrano, R. Synergistic anti-tumor effect of 17AAG with the PI3K/mTOR inhibitor NVP-BEZ235 on human melanoma. *Cancer Lett.* **2017**, *406*, 1–11. [CrossRef]

208. Deuker, M.M.; Durban, V.M.; Phillips, W.A.; McMahon, M. PI3'-kinase inhibition forestalls the onset of MEK1/2 inhibitor resistance in BRAF-mutated melanoma. *Cancer Discov.* **2015**, *5*, 143–153. [CrossRef]

209. Gao, M.Z.; Wang, H.B.; Chen, X.L.; Cao, W.T.; Fu, L.; Li, Y.; Quan, H.T.; Xie, C.Y.; Lou, L.G. Aberrant modulation of ribosomal protein S6 phosphorylation confers acquired resistance to MAPK pathway inhibitors in BRAF-mutant melanoma. *Acta Pharmacol. Sin.* **2019**, *40*, 268–278. [CrossRef]

210. Greger, J.G.; Eastman, S.D.; Zhang, V.; Bleam, M.R.; Hughes, A.M.; Smitheman, K.N.; Dickerson, S.H.; Laquerre, S.G.; Liu, L.; Gilmer, T.M. Combinations of BRAF, MEK, and PI3K/mTOR inhibitors overcome acquired resistance to the BRAF inhibitor GSK2118436 dabrafenib, mediated by NRAS or MEK mutations. *Mol. Cancer Ther.* **2012**, *11*, 909–920. [CrossRef]

211. Krepler, C.; Xiao, M.; Sproesser, K.; Brafford, P.A.; Shannan, B.; Beqiri, M.; Liu, Q.; Xu, W.; Garman, B.; Nathanson, K.L.; et al. Personalized Preclinical Trials in BRAF Inhibitor-Resistant Patient-Derived Xenograft Models Identify Second-Line Combination Therapies. *Clin. Cancer Res.* **2016**, *22*, 1592–1602. [CrossRef]

212. Lassen, A.; Atefi, M.; Robert, L.; Wong, D.J.; Cerniglia, M.; Comin-Anduix, B.; Ribas, A. Effects of AKT inhibitor therapy in response and resistance to BRAF inhibition in melanoma. *Mol. Cancer* **2014**, *13*, 83. [CrossRef]

213. Ruzzolini, J.; Peppicelli, S.; Andreucci, E.; Bianchini, F.; Margheri, F.; Laurenzana, A.; Fibbi, G.; Pimpinelli, N.; Calorini, L. Everolimus selectively targets vemurafenib resistant BRAF^{V600E} melanoma cells adapted to low pH. *Cancer Lett.* **2017**, *408*, 43–54. [CrossRef]

214. Sweetlove, M.; Wrightson, E.; Kolekar, S.; Rewcastle, G.W.; Baguley, B.C.; Shepherd, P.R.; Jamieson, S.M. Inhibitors of pan-PI3K Signaling Synergize with BRAF or MEK Inhibitors to Prevent BRAF-Mutant Melanoma Cell Growth. *Front. Oncol.* **2015**, *5*, 135. [CrossRef]

215. Tsukamoto, S.; Huang, Y.; Umeda, D.; Yamada, S.; Yamashita, S.; Kumazoe, M.; Kim, Y.; Murata, M.; Yamada, K.; Tachibana, H. 67-kDa laminin receptor-dependent protein phosphatase 2A (PP2A) activation elicits melanoma-specific antitumor activity overcoming drug resistance. *J. Biol. Chem.* **2014**, *289*, 32671–32681. [CrossRef]

216. Wallin, J.J.; Edgar, K.A.; Guan, J.; Berry, M.; Prior, W.W.; Lee, L.; Lesnick, J.D.; Lewis, C.; Nonomiya, J.; Pang, J.; et al. GDC-0980 is a novel class I PI3K/mTOR kinase inhibitor with robust activity in cancer models driven by the PI3K pathway. *Mol. Cancer Ther.* **2011**, *10*, 2426–2436. [CrossRef]

217. Galban, S.; Apfelbaum, A.A.; Espinoza, C.; Heist, K.; Haley, H.; Bedi, K.; Ljungman, M.; Galban, C.J.; Luker, G.D.; Dort, M.V.; et al. A Bifunctional MAPK/PI3K Antagonist for Inhibition of Tumor Growth and Metastasis. *Mol. Cancer Ther.* **2017**, *16*, 2340–2350. [CrossRef]

218. Narita, Y.; Okamoto, K.; Kawada, M.I.; Takase, K.; Minoshima, Y.; Kodama, K.; Iwata, M.; Miyamoto, N.; Sawada, K. Novel ATP-competitive MEK inhibitor E6201 is effective against vemurafenib-resistant melanoma harboring the MEK1-C121S mutation in a preclinical model. *Mol. Cancer Ther.* **2014**, *13*, 823–832. [CrossRef]

219. Park, S.J.; Hong, S.W.; Moon, J.H.; Jin, D.H.; Kim, J.S.; Lee, C.K.; Kim, K.P.; Hong, Y.S.; Choi, E.K.; Lee, J.S.; et al. The MEK1/2 inhibitor AS703026 circumvents resistance to the BRAF inhibitor PLX4032 in human malignant melanoma cells. *Am. J. Med. Sci.* **2013**, *346*, 494–498. [CrossRef]
220. Pathria, G.; Garg, B.; Borgdorff, V.; Garg, K.; Wagner, C.; Superti-Furga, G.; Wagner, S.N. Overcoming MITF-conferred drug resistance through dual AURKA/MAPK targeting in human melanoma cells. *Cell Death Dis.* **2016**, *7*, e2135. [CrossRef]
221. Phadke, M.S.; Sini, P.; Smalley, K.S. The Novel ATP-Competitive MEK/Aurora Kinase Inhibitor BI-847325 Overcomes Acquired BRAF Inhibitor Resistance through Suppression of Mcl-1 and MEK Expression. *Mol. Cancer Ther.* **2015**, *14*, 1354–1364. [CrossRef]
222. Uitdehaag, J.C.; de Roos, J.A.; van Doornmalen, A.M.; Prinsen, M.B.; Spijkers-Hagelstein, J.A.; de Vetter, J.R.; de Man, J.; Buijsman, R.C.; Zaman, G.J. Selective Targeting of CTNBB1-, KRAS- or MYC-Driven Cell Growth by Combinations of Existing Drugs. *PLoS ONE* **2015**, *10*, e0125021. [CrossRef]
223. Graziani, G.; Artuso, S.; De Luca, A.; Muzi, A.; Rotili, D.; Scimeca, M.; Atzori, M.G.; Ceci, C.; Mai, A.; Leonetti, C.; et al. A new water soluble MAPK activator exerts antitumor activity in melanoma cells resistant to the BRAF inhibitor vemurafenib. *Biochem. Pharmacol.* **2015**, *95*, 16–27. [CrossRef]
224. Estrela, J.M.; Salvador, R.; Marchio, P.; Valles, S.L.; Lopez-Blanch, R.; Rivera, P.; Benlloch, M.; Alcacer, J.; Perez, C.L.; Pellicer, J.A.; et al. Glucocorticoid receptor antagonism overcomes resistance to BRAF inhibition in BRAF^{V600E}-mutated metastatic melanoma. *Am. J. Cancer Res.* **2019**, *9*, 2580–2598.
225. Yadav, V.; Burke, T.F.; Huber, L.; Van Horn, R.D.; Zhang, Y.; Buchanan, S.G.; Chan, E.M.; Starling, J.J.; Beckmann, R.P.; Peng, S.B. The CDK4/6 inhibitor LY2835219 overcomes vemurafenib resistance resulting from MAPK reactivation and cyclin D1 upregulation. *Mol. Cancer Ther.* **2014**, *13*, 2253–2263. [CrossRef]
226. Basken, J.; Stuart, S.A.; Kavran, A.J.; Lee, T.; Ebmeier, C.C.; Old, W.M.; Ahn, N.G. Specificity of Phosphorylation Responses to Mitogen Activated Protein (MAP) Kinase Pathway Inhibitors in Melanoma Cells. *Mol. Cell. Proteom.* **2018**, *17*, 550–564. [CrossRef]
227. Morris, E.J.; Jha, S.; Restaino, C.R.; Dayananth, P.; Zhu, H.; Cooper, A.; Carr, D.; Deng, Y.; Jin, W.; Black, S.; et al. Discovery of a novel ERK inhibitor with activity in models of acquired resistance to BRAF and MEK inhibitors. *Cancer Discov.* **2013**, *3*, 742–750. [CrossRef]
228. Nakamura, A.; Arita, T.; Tsuchiya, S.; Donelan, J.; Chouitar, J.; Carideo, E.; Galvin, K.; Okaniwa, M.; Ishikawa, T.; Yoshida, S. Antitumor activity of the selective pan-RAF inhibitor TAK-632 in BRAF inhibitor-resistant melanoma. *Cancer Res.* **2013**, *73*, 7043–7055. [CrossRef]
229. Whittaker, S.R.; Cowley, G.S.; Wagner, S.; Luo, F.; Root, D.E.; Garraway, L.A. Combined Pan-RAF and MEK Inhibition Overcomes Multiple Resistance Mechanisms to Selective RAF Inhibitors. *Mol. Cancer Ther.* **2015**, *14*, 2700–2711. [CrossRef]
230. Wang, L.; Zhu, G.; Zhang, Q.; Duan, C.; Zhang, Y.; Zhang, Z.; Zhou, Y.; Lu, T.; Tang, W. Rational design, synthesis, and biological evaluation of Pan-Raf inhibitors to overcome resistance. *Org. Biomol. Chem.* **2017**, *15*, 3455–3465. [CrossRef]
231. Acquaviva, J.; Smith, D.L.; Jimenez, J.P.; Zhang, C.; Sequeira, M.; He, S.; Sang, J.; Bates, R.C.; Proia, D.A. Overcoming acquired BRAF inhibitor resistance in melanoma via targeted inhibition of Hsp90 with ganetespib. *Mol. Cancer Ther.* **2014**, *13*, 353–363. [CrossRef]
232. Budina-Kolomets, A.; Webster, M.R.; Leu, J.I.; Jennis, M.; Krepler, C.; Guerrini, A.; Kossenkov, A.V.; Xu, W.; Karakousis, G.; Schuchter, L.; et al. HSP70 Inhibition Limits FAK-Dependent Invasion and Enhances the Response to Melanoma Treatment with BRAF Inhibitors. *Cancer Res.* **2016**, *76*, 2720–2730. [CrossRef]
233. Haarberg, H.E.; Paraiso, K.H.; Wood, E.; Rebecca, V.W.; Sondak, V.K.; Koomen, J.M.; Smalley, K.S. Inhibition of Wee1, AKT, and CDK4 underlies the efficacy of the HSP90 inhibitor XL888 in an in vivo model of NRAS-mutant melanoma. *Mol. Cancer Ther.* **2013**, *12*, 901–912. [CrossRef]
234. Mielczarek-Lewandowska, A.; Sztiller-Sikorska, M.; Osrodek, M.; Czyz, M.; Hartman, M.L. 17-Aminogeldanamycin selectively diminishes IRE1alpha-XBP1s pathway activity and cooperatively induces apoptosis with MEK1/2 and BRAF^{V600E} inhibitors in melanoma cells of different genetic subtypes. *Apoptosis* **2019**, *24*, 596–611. [CrossRef]
235. Rebecca, V.W.; Wood, E.; Fedorenko, I.V.; Paraiso, K.H.; Haarberg, H.E.; Chen, Y.; Xiang, Y.; Sarnaik, A.; Gibney, G.T.; Sondak, V.K.; et al. Evaluating melanoma drug response and therapeutic escape with quantitative proteomics. *Mol. Cell. Proteom.* **2014**, *13*, 1844–1854. [CrossRef]

236. Smyth, T.; Paraiso, K.H.T.; Hearn, K.; Rodriguez-Lopez, A.M.; Munck, J.M.; Haarberg, H.E.; Sondak, V.K.; Thompson, N.T.; Azab, M.; Lyons, J.F.; et al. Inhibition of HSP90 by AT13387 delays the emergence of resistance to BRAF inhibitors and overcomes resistance to dual BRAF and MEK inhibition in melanoma models. *Mol. Cancer Ther.* **2014**, *13*, 2793–2804. [CrossRef]

237. Wu, X.; Marmarelis, M.E.; Hodi, F.S. Activity of the heat shock protein 90 inhibitor ganetespib in melanoma. *PLoS ONE* **2013**, *8*, e56134. [CrossRef]

238. Sale, M.J.; Minihane, E.; Monks, N.R.; Gilley, R.; Richards, F.M.; Schifferli, K.P.; Andersen, C.L.; Davies, E.J.; Vicente, M.A.; Ozono, E.; et al. Targeting melanoma's MCL1 bias unleashes the apoptotic potential of BRAF and ERK1/2 pathway inhibitors. *Nat. Commun.* **2019**, *10*, 5167. [CrossRef]

239. Eroglu, Z.; Chen, Y.A.; Gibney, G.T.; Weber, J.S.; Kudchadkar, R.R.; Khushalani, N.I.; Markowitz, J.; Brohl, A.S.; Tetteh, L.F.; Ramadan, H.; et al. Combined BRAF and HSP90 Inhibition in Patients with Unresectable BRAF^{V600E}-Mutant Melanoma. *Clin. Cancer Res.* **2018**, *24*, 5516–5524. [CrossRef]

240. Kaoud, T.S.; Mohassab, A.M.; Hassan, H.A.; Yan, C.; Van Ravenstein, S.X.; Abdelhamid, D.; Dalby, K.N.; Abdel-Aziz, M. NO-releasing STAT3 inhibitors suppress BRAF-mutant melanoma growth. *Eur. J. Med. Chem.* **2020**, *186*, 111885. [CrossRef]

241. Booth, L.; Roberts, J.L.; Sander, C.; Lee, J.; Kirkwood, J.M.; Poklepovic, A.; Dent, P. The HDAC inhibitor AR42 interacts with pazopanib to kill trametinib/dabrafenib-resistant melanoma cells in vitro and in vivo. *Oncotarget* **2017**, *8*, 16367–16386. [CrossRef]

242. Borst, A.; Haferkamp, S.; Grimm, J.; Rosch, M.; Zhu, G.; Guo, S.; Li, C.; Gao, T.; Meierjohann, S.; Schrama, D.; et al. BIK is involved in BRAF/MEK inhibitor induced apoptosis in melanoma cell lines. *Cancer Lett.* **2017**, *404*, 70–78. [CrossRef]

243. Gallagher, S.J.; Gunatilake, D.; Beaumont, K.A.; Sharp, D.M.; Tiffen, J.C.; Heinemann, A.; Weninger, W.; Haass, N.K.; Wilmott, J.S.; Madore, J.; et al. HDAC inhibitors restore BRAF-inhibitor sensitivity by altering PI3K and survival signalling in a subset of melanoma. *Int. J. Cancer* **2018**, *142*, 1926–1937. [CrossRef]

244. Rowdo, F.P.M.; Baron, A.; Gallagher, S.J.; Hersey, P.; Emran, A.A.; Von Euw, E.M.; Barrio, M.M.; Mordoh, J. Epigenetic inhibitors eliminate senescent melanoma BRAFV600E cells that survive long-term BRAF inhibition. *Int. J. Oncol.* **2020**, *56*, 1429–1441. [CrossRef]

245. Peng, U.; Wang, Z.; Pei, S.; Ou, Y.; Hu, P.; Liu, W.; Song, J. ACY-1215 accelerates vemurafenib induced cell death of BRAF-mutant melanoma cells via induction of ER stress and inhibition of ERK activation. *Oncol. Rep.* **2017**, *37*, 1270–1276. [CrossRef]

246. Heijkants, R.; Willekens, K.; Schoonderwoerd, M.; Teunisse, A.; Nieveen, M.; Radaelli, E.; Hawinkels, L.; Marine, J.C.; Jochemsen, A. Combined inhibition of CDK and HDAC as a promising therapeutic strategy for both cutaneous and uveal metastatic melanoma. *Oncotarget* **2018**, *9*, 6174–6187. [CrossRef] [PubMed]

247. Huijberts, S.; Wang, L.; de Oliveira, R.L.; Rosing, H.; Nuijen, B.; Beijnen, J.; Bernards, R.; Schellens, J.; Wilgenhof, S. Vorinostat in patients with resistant BRAF^{V600E} mutated advanced melanoma: A proof of concept study. *Fut. Oncol.* **2020**, *16*, 619–629. [CrossRef]

248. Dorrie, J.; Babalija, L.; Hoyer, S.; Gerer, K.F.; Schuler, G.; Heinzerling, L.; Schaft, N. BRAF and MEK Inhibitors Influence the Function of Reprogrammed T Cells: Consequences for Adoptive T-Cell Therapy. *Int. J. Mol. Sci.* **2018**, *19*, 289. [CrossRef]

249. Gargett, T.; Fraser, C.K.; Dotti, G.; Yvon, E.S.; Brown, M.P. BRAF and MEK inhibition variably affect GD2-specific chimeric antigen receptor (CAR) T-cell function in vitro. *J. Immunother.* **2015**, *38*, 12–23. [CrossRef]

250. Tel, J.; Koornstra, R.; de Haas, N.; van Deutekom, V.; Westdorp, H.; Boudewijns, S.; van Erp, N.; Di Blasio, S.; Gerritsen, W.; Figdor, C.G.; et al. Preclinical exploration of combining plasmacytoid and myeloid dendritic cell vaccination with BRAF inhibition. *J. Transl. Med.* **2016**, *14*, 88. [CrossRef]

251. Salton, M.; Kasprzak, W.K.; Voss, T.; Shapiro, B.A.; Poulikakos, P.I.; Misteli, T. Inhibition of vemurafenib-resistant melanoma by interference with pre-mRNA splicing. *Nat. Commun.* **2015**, *6*, 7103. [CrossRef]

252. Frederick, D.T.; Fragomeni, R.A.S.; Schalck, A.; Ferreiro-Neira, I.; Hoff, T.; Cooper, Z.A.; Haq, R.; Panka, D.J.; Kwong, L.N.; Davies, M.A.; et al. Clinical profiling of BCL-2 family members in the setting of BRAF inhibition offers a rationale for targeting de novo resistance using BH3 mimetics. *PLoS ONE* **2014**, *9*, e101286. [CrossRef]

253. Garandeau, D.; Noujarede, J.; Leclerc, J.; Imbert, C.; Garcia, V.; Bats, M.L.; Rambow, F.; Gilhodes, J.; Filleron, T.; Meyer, N.; et al. Targeting the Sphingosine 1-Phosphate Axis Exerts Potent Antitumor Activity in BRAFi-Resistant Melanomas. *Mol. Cancer Ther.* **2019**, *18*, 289–300. [CrossRef] [PubMed]

254. Haq, R.; Yokoyama, S.; Hawryluk, E.B.; Jonsson, G.B.; Frederick, D.T.; McHenry, K.; Porter, D.; Tran, T.N.; Love, K.T.; Langer, R.; et al. BCL2A1 is a lineage-specific antiapoptotic melanoma oncogene that confers resistance to BRAF inhibition. *Proc. Natl. Acad. Sci. USA* **2013**, *110*, 4321–4326. [CrossRef] [PubMed]

255. Hong, S.K.; Starenki, D.; Wu, P.K.; Park, J.I. Suppression of B-Raf^{V600E} melanoma cell survival by targeting mitochondria using triphenyl-phosphonium-conjugated nitroxide or ubiquinone. *Cancer Biol. Ther.* **2017**, *18*, 106–114. [CrossRef] [PubMed]

256. Serasinghe, M.N.; Gelles, J.D.; Li, K.; Zhao, L.; Abbate, F.; Syku, M.; Mohammed, J.N.; Badal, B.; Rangel, C.A.; Hoehn, K.L.; et al. Dual suppression of inner and outer mitochondrial membrane functions augments apoptotic responses to oncogenic MAPK inhibition. *Cell Death Dis.* **2018**, *9*, 29. [CrossRef]

257. Kosnopfel, C.; Sinnberg, T.; Sauer, B.; Niessner, H.; Schmitt, A.; Makino, E.; Forschner, A.; Hailfinger, S.; Garbe, C.; Schittek, B. Human melanoma cells resistant to MAPK inhibitors can be effectively targeted by inhibition of the p90 ribosomal S6 kinase. *Oncotarget* **2017**, *8*, 35761–35775. [CrossRef]

258. Theodosakis, N.; Micevic, G.; Langdon, C.G.; Ventura, A.; Means, R.; Stern, D.F.; Bosenberg, M.W. p90RSK Blockade Inhibits Dual BRAF and MEK Inhibitor-Resistant Melanoma by Targeting Protein Synthesis. *J. Investig. Dermatol.* **2017**, *137*, 2187–2196. [CrossRef]

259. Peh, J.; Fan, T.M.; Wycislo, K.L.; Roth, H.S.; Hergenrother, P.J. The Combination of Vemurafenib and Procaspase-3 Activation Is Synergistic in Mutant BRAF Melanomas. *Mol. Cancer Ther.* **2016**, *15*, 1859–1869. [CrossRef]

260. Smit, M.A.; Maddalo, G.; Greig, K.; Raaijmakers, L.M.; Possik, P.A.; van Breukelen, B.; Cappadona, S.; Heck, A.J.; Altelaar, A.F.; Peeper, D.S. ROCK1 is a potential combinatorial drug target for BRAF mutant melanoma. *Mol. Syst. Biol.* **2014**, *10*, 772. [CrossRef]

261. Takashima, A.; English, B.; Chen, Z.; Cao, J.; Cui, R.; Williams, R.M.; Faller, D.V. Protein kinase Cdelta is a therapeutic target in malignant melanoma with NRAS mutation. *ACS Chem. Biol.* **2014**, *9*, 1003–1014. [CrossRef]

262. Wang, J.; Chen, J.; Miller, D.D.; Li, W. Synergistic combination of novel tubulin inhibitor ABI-274 and vemurafenib overcome vemurafenib acquired resistance in BRAFV600E melanoma. *Mol. Cancer Ther.* **2014**, *13*, 16–26. [CrossRef]

263. Fattore, L.; Malpicci, D.; Marra, E.; Belleudi, F.; Noto, A.; De Vitis, C.; Pisanu, M.E.; Coluccia, P.; Camerlingo, R.; Roscilli, G.; et al. Combination of antibodies directed against different ErbB3 surface epitopes prevents the establishment of resistance to BRAF/MEK inhibitors in melanoma. *Oncotarget* **2015**, *6*, 24823–24841. [CrossRef] [PubMed]

264. Kugel, C.H., 3rd; Hartsough, E.J.; Davies, M.A.; Setiady, Y.Y.; Aplin, A.E. Function-blocking ERBB3 antibody inhibits the adaptive response to RAF inhibitor. *Cancer Res.* **2014**, *74*, 4122–4132. [CrossRef] [PubMed]

265. Pencheva, N.; Buss, C.G.; Posada, J.; Merghoub, T.; Tavazoie, S.F. Broad-spectrum therapeutic suppression of metastatic melanoma through nuclear hormone receptor activation. *Cell* **2014**, *156*, 986–1001. [CrossRef] [PubMed]

266. Asundi, J.; Lacap, J.A.; Clark, S.; Nannini, M.; Roth, L.; Polakis, P. MAPK pathway inhibition enhances the efficacy of an anti-endothelin B receptor drug conjugate by inducing target expression in melanoma. *Mol. Cancer Ther.* **2014**, *13*, 1599–1610. [CrossRef] [PubMed]

267. Yu, L.; Favoino, E.; Wang, Y.; Ma, Y.; Deng, X.; Wang, X. The CSPG4-specific monoclonal antibody enhances and prolongs the effects of the BRAF inhibitor in melanoma cells. *Immunol. Res.* **2011**, *50*, 294–302. [CrossRef]

268. Talebi, A.; Dehairs, J.; Rambow, F.; Rogiers, A.; Nittner, D.; Derua, R.; Vanderhoydonc, F.; Duarte, J.A.G.; Bosisio, F.; Van den Eynde, K.; et al. Sustained SREBP-1-dependent lipogenesis as a key mediator of resistance to BRAF-targeted therapy. *Nat. Commun.* **2018**, *9*, 2500. [CrossRef]

269. Brady, D.C.; Crowe, M.S.; Greenberg, D.N.; Counter, C.M. Copper Chelation Inhibits BRAF^{V600E}-Driven Melanomagenesis and Counters Resistance to BRAF^{V600E} and MEK1/2 Inhibitors. *Cancer Res.* **2017**, *77*, 6240–6252. [CrossRef]

270. Babagana, M.; Kichina, J.V.; Slabodkin, H.; Johnson, S.; Maslov, A.; Brown, L.; Attwood, K.; Nikiforov, M.A.; Kandel, E.S. The role of polo-like kinase 3 in the response of BRAF-mutant cells to targeted anticancer therapies. *Mol. Carcinog.* **2020**, *59*, 5–14. [CrossRef]

271. Sanchez, I.M.; Purwin, T.J.; Chervoneva, I.; Erkes, D.A.; Nguyen, M.Q.; Davies, M.A.; Nathanson, K.L.; Kemper, K.; Peeper, D.S.; Aplin, A.E. In Vivo ERK1/2 Reporter Predictively Models Response and Resistance to Combined BRAF and MEK Inhibitors in Melanoma. *Mol. Cancer Ther.* **2019**, *18*, 1637–1648. [CrossRef]

272. Hendrix, M.J.; Kandela, I.; Mazar, A.P.; Seftor, E.A.; Seftor, R.E.; Margaryan, N.V.; Strizzi, L.; Murphy, G.F.; Long, G.V.; Scolyer, R.A. Targeting melanoma with front-line therapy does not abrogate Nodal-expressing tumor cells. *Lab. Investig.* **2017**, *97*, 176–186. [CrossRef]

273. Babagana, M.; Johnson, S.; Slabodkin, H.; Bshara, W.; Morrison, C.; Kandel, E.S. P21-activated kinase 1 regulates resistance to BRAF inhibition in human cancer cells. *Mol. Carcinog.* **2017**, *56*, 1515–1525. [CrossRef] [PubMed]

274. Vlckova, K.; Reda, J.; Ondrusova, L.; Krayem, M.; Ghanem, G.; Vachtenheim, J. GLI inhibitor GANT61 kills melanoma cells and acts in synergy with obatoclax. *Int. J. Oncol.* **2016**, *49*, 953–960. [CrossRef] [PubMed]

275. Jameson, K.L.; Mazur, P.K.; Zehnder, A.M.; Zhang, J.; Zarnegar, B.; Sage, J.; Khavari, P.A. IQGAP1 scaffold-kinase interaction blockade selectively targets RAS-MAP kinase-driven tumors. *Nat. Med.* **2013**, *19*, 626–630. [CrossRef] [PubMed]

276. Liu, W.; Stachura, P.; Xu, H.C.; Ganesh, N.U.; Cox, F.; Wang, R.; Lang, K.S.; Gopalakrishnan, J.; Haussinger, D.; Homey, B.; et al. Repurposing the serotonin agonist Tegaserod as an anticancer agent in melanoma: Molecular mechanisms and clinical implications. *J. Exp. Clin. Cancer Res.* **2020**, *39*, 38. [CrossRef]

277. Parker, R.; Vella, L.J.; Xavier, D.; Amirkhani, A.; Parker, J.; Cebon, J.; Molloy, M.P. Phosphoproteomic Analysis of Cell-Based Resistance to BRAF Inhibitor Therapy in Melanoma. *Front. Oncol.* **2015**, *5*, 95. [CrossRef]

278. Krayem, M.; Journe, F.; Wiedig, M.; Morandini, R.; Najem, A.; Sales, F.; van Kempen, L.C.; Sibille, C.; Awada, A.; Marine, J.C.; et al. p53 Reactivation by PRIMA-1(Met) (APR-246) sensitises (V600E/K)BRAF melanoma to vemurafenib. *Eur. J. Cancer* **2016**, *55*, 98–110. [CrossRef]

279. Ryabaya, O.; Prokofieva, A.; Akasov, R.; Khochenkov, D.; Emelyanova, M.; Burov, S.; Markvicheva, E.; Inshakov, A.; Stepanova, E. Metformin increases antitumor activity of MEK inhibitor binimetinib in 2D and 3D models of human metastatic melanoma cells. *Biomed. Pharmacother.* **2019**, *109*, 2548–2560. [CrossRef]

280. Theodosakis, N.; Langdon, C.G.; Micevic, G.; Krykbaeva, I.; Means, R.E.; Stern, D.F.; Bosenberg, M.W. Inhibition of isoprenylation synergizes with MAPK blockade to prevent growth in treatment-resistant melanoma, colorectal, and lung cancer. *Pigment Cell Melanoma Res.* **2019**, *32*, 292–302. [CrossRef]

281. Brummer, C.; Faerber, S.; Bruss, C.; Blank, C.; Lacroix, R.; Haferkamp, S.; Herr, W.; Kreutz, M.; Renner, K. Metabolic targeting synergizes with MAPK inhibition and delays drug resistance in melanoma. *Cancer Lett.* **2019**, *442*, 453–463. [CrossRef]

282. Barcelo, C.; Siso, P.; Maiques, O.; Garcia-Mulero, S.; Sanz-Pamplona, R.; Navaridas, R.; Megino, C.; Felip, I.; Urdanibia, I.; Eritja, N.; et al. T-Type Calcium Channels as Potential Therapeutic Targets in Vemurafenib-Resistant BRAF^{V600E} Melanoma. *J. Investig. Dermatol.* **2020**, *140*, 1253–1265. [CrossRef]

283. Yuan, L.; Mishra, R.; Patel, H.; Abdulsalam, S.; Greis, K.D.; Kadekaro, A.L.; Merino, E.J.; Garrett, J.T. Utilization of Reactive Oxygen Species Targeted Therapy to Prolong the Efficacy of BRAF Inhibitors in Melanoma. *J. Cancer* **2018**, *9*, 4665–4676. [CrossRef] [PubMed]

284. Fu, Y.; Rathod, D.; Abo-Ali, E.M.; Dukhande, V.V.; Patel, K. EphA2-Receptor Targeted PEGylated Nanoliposomes for the Treatment of BRAF^{V600E} Mutated Parent- and Vemurafenib-Resistant Melanoma. *Pharmaceutics* **2019**, *11*, 504. [CrossRef] [PubMed]

285. Tham, H.P.; Xu, K.; Lim, W.Q.; Chen, H.; Zheng, M.; Thng, T.G.S.; Venkatraman, S.S.; Xu, C.; Zhao, Y. Microneedle-Assisted Topical Delivery of Photodynamically Active Mesoporous Formulation for Combination Therapy of Deep-Seated Melanoma. *ACS Nano* **2018**, *12*, 11936–11948. [CrossRef] [PubMed]

Review

Molecular Insights and Emerging Strategies for Treatment of Metastatic Uveal Melanoma

Fabiana Mallone , Marta Sacchetti , Alessandro Lambiase * and Antonietta Moramarco

Department of Sense Organs, Sapienza University of Rome, 00161 Rome, Italy;
fabiana.mallone@uniroma1.it (F.M.); marta.sacchetti@uniroma1.it (M.S.);
antonietta.moramarco@uniroma1.it (A.M.)
* Correspondence: alessandro.lambiase@uniroma1.it; Tel.: +39-335-7046-521

Received: 23 August 2020; Accepted: 23 September 2020; Published: 25 September 2020

Simple Summary: Around 50% of patients with uveal melanoma (UM) still develop metastatic disease. Despite recent advances in the diagnosis and prognosis of UM, improvements in overall survival have not been achieved. At present, there is no available standard of care for adjuvant and metastatic settings. The aim of our review article was to discuss the latest advances in understanding the molecular mechanisms underlying uveal melanoma and novel treatment options for metastatic disease. We provided a detailed analysis of the most recently published works in the Literature along with a number of ongoing clinical trials for adjuvant and metastatic treatment of uveal melanoma. New insights into the pathogenesis of UM and promising results from the study of innovative tailored therapies could offer viable opportunities for translating in clinical practice.

Abstract: Uveal melanoma (UM) is the most common intraocular cancer. In recent decades, major advances have been achieved in the diagnosis and prognosis of UM allowing for tailored treatments. However, nearly 50% of patients still develop metastatic disease with survival rates of less than 1 year. There is currently no standard of adjuvant and metastatic treatment in UM, and available therapies are ineffective resulting from cutaneous melanoma protocols. Advances and novel treatment options including liver-directed therapies, immunotherapy, and targeted-therapy have been investigated in UM-dedicated clinical trials on single compounds or combinational therapies, with promising results. Therapies aimed at prolonging or targeting metastatic tumor dormancy provided encouraging results in other cancers, and need to be explored in UM. In this review, the latest progress in the diagnosis, prognosis, and treatment of UM in adjuvant and metastatic settings are discussed. In addition, novel insights into tumor genetics, biology and immunology, and the mechanisms underlying metastatic dormancy are discussed. As evident from the numerous studies discussed in this review, the increasing knowledge of this disease and the promising results from testing of novel individualized therapies could offer future perspectives for translating in clinical use.

Keywords: uveal melanoma (UM); metastatic uveal melanoma (mUM); prognostication; adjuvant therapy; metastatic therapy; metastatic dormancy; liver-directed-therapies; immunotherapy; targeted-therapy; combined therapy

---

1. Introduction

Uveal melanoma (UM) is the most common intraocular malignancy. UM originates from melanocytes of the uveal tract of the eye, including the iris, ciliary body, and retinal choroid. Despite successful control of the primary tumor and the significant improvements in early identification of patients at risk of metastatic progression, metastatic disease still occurs frequently and is invariably lethal. In addition, to date, there is no available standard of care for the treatment of metastatic UM (mUM). Current treatment protocols are mainly adapted from cutaneous melanoma, although they differ in terms of clinical and genetic profile, and ocular melanoma is often excluded from most clinical trials.

In recent decades, the increasing understanding of tumor genetics, biology, and immunology has allowed for a better insight into the pathogenesis of UM; as a consequence, several clinical trials have been performed to investigate novel therapeutic targets for the treatment of UM in an attempt to change the disease course.

In this paper, we reviewed recent advances in diagnosis, prognosis, and innovative treatment options for UM in adjuvant and metastatic settings, and perspectives for translating in clinical practice. Special emphasis was directed towards the mechanisms underlying metastatic dormancy and related therapeutic applications, as promising preventative strategies for metastatic growth.

2. Epidemiology and Uveal Melanoma Characteristics

UM is a relatively rare malignancy accounting for 5.3% of all melanoma cases recorded in the USA [1]; however, it represents the most common primary intraocular cancer in adults. Among ocular melanomas, 85% originates from the uvea, 4.8% from the conjunctiva, and the remaining 10.2% from other sites [1]. UM mostly arises from melanocytes located in the choroid (90%) and to a lesser extent in the iris (4%) and ciliary body (6%) [2]. The annual overall incidence of UM remained stable in recent decades at approximately 5.1 cases per million individuals in the USA and between 1.3 and 8.6 cases per million in Europe [3–10]. The disease is more frequent in Caucasian ethnicity, with a median age of presentation of approximately 60 years and 30% greater incidence in males [3–5,7,11–13]. Most relevant predisposing factors for the development of UM are the presence of dysplastic nevus syndrome, choroidal nevi, ocular or oculodermal melanocytosis, familial syndromes including germline BAP1 (BRCA1-associated protein 1) mutations, and neurofibromatosis [14–17]. Of note, cutaneous melanoma is not a risk factor for UM [18,19]. One in 8000 choroidal nevi transform into melanoma. The elements suggestive of a malign lesion are listed in Table 1 [20–22]. The chance of transformation is 4% if any characteristic is present, and it is more than 50% if three or more features are combined [20,21]. Choroidal nevi that do not exhibit any malignant feature require initial monitoring twice a year and then once a year if stability persists. Those showing 1 or 2 features need strict monitoring at least every 4–6 months, while those with 3 or more features should be referred to a specialized center for possible primary treatment and prognostic stratification [20,21]. According to the Collaborative Ocular Melanoma Study (COMS), the diagnosis of UM can be exclusively clinical, with a clinical misdiagnosis rate of only 0.48% [23]. However, other studies highlight the importance of fine-needle aspiration biopsy (FNAB) for diagnostic accuracy in selected cases [24]. In recent decades, treatment of primary UM has been evolving from enucleation towards effective eye-conserving modalities inclusive of radiation, surgical, and laser therapy [25]. However, the five-year survival rate has not registered substantial improvements during the past four decades, and it is still estimated at 70–80% irrespective of the type of treatment [4,9,26–30]. Radiotherapy and surgery achieve local disease control exceeding 90%, but approximately 50% of patients ultimately develop metastases, with UM showing a considerably worse prognosis than its cutaneous counterpart [4,26,31–33]. The estimates of metastatic progression are reported from 32% at 5 years, to 50% at 15 years, and 62% at 35 years [26]. The uveal tract is rich in vascular structure, and UM is peculiar for its almost exclusive dissemination via the hematogenous route with a propensity for the liver as the first site of metastasis in over 90% of cases. However, conjunctival lymphatic infiltration following direct invasion of the sclera has also been described [34]. The main predictors of metastatic progression of UM include clinical (tumor thickness and basal diameter, ciliary body involvement, degree of extraocular extension), histopathologic (epithelioid cytomorphology, infiltrating lymphocytes and macrophages, fibrovascular loops and networks, high mitotic activity), and genetic factors [35–40].

Table 1. Features predicting malignant transformation of choroidal nevus into melanoma are listed in the mnemonic "TFSOM UHHD 'to find small ocular melanoma using helpful hints daily'". The percentages of choroidal nevus growth into melanoma based on number of involved features are reported along with recommended clinical monitoring [20,21].

Mnemonic	Feature	N of Features	Choroidal Nevus Growth into Melanoma (%)	Monitoring
To Find Small Ocular Melanoma Using helpful Hints Daily	Thickness > 2 mm	None	4%	Every 6 months → Once a year (if stability persists)
	Fluid (subretinal)			
	Symptoms - decreased vision - flashes/floaters	1 Feature	36%	Every 4–6 months
	Orange pigment			
	Margin ≤ 3 mm to disc	2 Features	45%	Every 4–6 months
	Ultrasonographic hollowness			
	Halo absence Drusen absence	3 or more Features	50%	Referral to Experienced Center Primary treatment Prognosis

3. Prognosis of UM

Recent studies showed that UM characterization based on cytogenetics and gene expression profiling (GEP) significantly improved the prognosis in UM [38,41–45]. As consequence, great efforts have been made to identify karyotype or gene alterations which are associated with higher tendency to metastatic spread.

Based on the increased understanding of UM genetics, some authors proposed integrating the American Joint Committee on Cancer (AJCC) Tumor-Node-Metastasis (TNM) clinical staging system with genetic parameters to improve the prediction of metastatic progression [39,46–50]. These studies showed that the prognostic significance of tumor basal diameter was considerably enhanced when considered together with chromosome mutational profile and histological grade, leading to UM clustering into low/high-risk models from large cohort studies [39,46,47,50]. Likewise, the group of Vaquero-Garcia developed a model for personalized Prediction of Risk of Metastasis in UM (PRiMeUM) based on clinical and chromosomal information, showing 85% prediction accuracy [48]. Concomitantly, efforts were dedicated to the study of classifications solely based on genetic data [43–45,51]. Recently, The Cancer Genome Atlas (TCGA) classification has been proposed to improve the identification of high-risk patients for metastatic disease with respect to an AJCC-TNM clinical staging system [52]. Specifically, the TCGA project, starting in 2005, was designed to conduct an expression analysis of mRNA, micro RNA, and long noncoding RNA and catalog genetic mutations of 33 different cancer types, including UM [53]. Based on TCGA results of chromosome 3 disomy or monosomy and degree of chromosome 8q gain, UM was categorized into 4 classes (A, B, C, and D) of progressive worsening prognosis [51]. Tumor clinical features and outcomes of metastatic risk and death were evaluated in a large sample of UM categorized on the basis of TCGA system, and the more advanced classes corresponded to older age, greater tumor size, and worst prognosis [54].

Cytogenetic Alterations in UM

The most studied karyotype alterations in UM include chromosome 3 and 8. Chromosome 3 complete monosomy is the most frequent karyotype abnormality observed in almost half of all UM [38,55–57]. The presence of this monosomy is associated with a five-year survival rate of 39%, whereas a 90% five-year survival rate is observed for UM without monosomy 3 [58]. Interestingly, BRCA1-associated protein-1 (BAP1) is a tumor-suppressor gene placed on chromosome 3, and it is mutated in 47% of primary UM and up to 91% of metastatic UM [59–64]. The splicing factor 3B subunit 1 (SF3B1) is consistent with disomy 3 and shows more favorable prognosis, although an association with delayed metastases has been reported [65–67]. Eukaryotic translation initiation factor 1A, X-linked (EIF1AX) gene mutations are also described along with SF3B1 in UM with disomy 3, but metastatic tendency is less frequent [68]. Among other chromosome alterations, rearrangements in chromosome 8q have been described in approximately 40% of UM. Specifically, UM with normal 8q profile have 93% five-year survival, while those with one additional copy have 67% and those with 8q amplification have 29% [58,69]. A recent study on 1059 UM patients over 8 years follow-up, showed that the concurrent presence of 3 complete monosomy and 8q gain resulted in an increased risk of metastasis and death [43]. Interestingly, the most severe mutational events consisting of chromosome 3 loss and chromosome 8q gain correlated positively with ciliary body involvement, tumor thickness and basal diameter, proximity to the optic disc, extraocular spread, epithelioid cells, and age [70]. This evidence suggests that early intervention, when tumor growth is limited and the genetic profile more favorable, could prevent tumor dedifferentiation into a more aggressive type [54,70]. Further cytogenetic aberrations associated with an increased risk of distant recurrences include 1p loss, 6q loss, and 8p loss [71,72]. Due to advances in knowledge on cancer biology, over the last two decades, genetic tests are routinely performed in clinical practice, ranging from single chromosome 3 evaluation to arrays of analyses for chromosomes 1, 3, 6, and 8 and GEP. However, both cytogenetic testing and GEP for prognosis of UM require invasive procedures to harvest the specimens from either enucleation or intraoperative FNAB [44,73–76]. Specifically, among cytogenetic tests, analysis of a karyotype using fluorescence in situ hybridization (FISH) and the array comparative genomic hybridization (aCGH) allows for the detection of translocations and partial deletions of the chromosome 3 and requires a large amount of tissue with a technical failure rate with FNAB of approximately 50% [77,78]. Among other genetic analyses, multiplex ligand-dependent probe amplification (MLPA) consists of 31 probes to analyze loci on chromosomes 1p, 3, 6, and 8 and, thus, to identify high- and low-risk patients. Similar to FISH and aCGH, MLPA requires large tissue samples with increased risk of biopsy complications [47]. An alternative cytogenetic method requiring a lower number of samples is represented by microsatellite analysis (MSA). This technique combines fluorescent probes with PCR, and has proven accurate for the identification of aberrations on chromosome 3, but not for those on chromosomes 8 and 6 [57]. Besides these limitations, cytogenetic tests are prone to sampling errors in UM due to its dense cellularity and elongated nuclei that weave in and out of the plane of section, and to significant intratumoral heterogeneity [47,79,80]. GEP, using an RNA-based assay, currently represents the gold standard in molecular prognosis. This test has a technical failure rate of only 3%, and can be performed on fine-needle biopsies even when the quantity of RNA is below detectable limits. It allows for the categorization of UMs in Class 1 and Class 2 based on low and high risk of metastatic potential, respectively [44,45]. The prognostic accuracy of GEP classification has proven superior over cytogenetic methods, and this would be related to the heterogeneous distribution of chromosomal markers throughout the tumor. GEP analysis is indeed very sensitive for detecting the proper class signature in heterogeneous tumors as it performs simultaneous evaluation of several genes representative of the tumor microenvironment [45]. There is evidence reporting that GEP is more capable of capturing the overall tumor functional complexity than a chromosomal marker [47,80]. The group of Onken et al. developed a clinically feasible platform for analyzing GEP by a 15-gene PCR-based assay [81]. This assay is now commercially available in the United States as DecisionDx-UM® (Castle Biosciences, Friendswood, TX, USA), and it is highly accurate, easy to interpret, and independent from additional analyses. In detail, GEP Class 1 is subdivided into class 1A and class 1B, with 2 and 21% five-year metastatic risk, respectively, whereas GEP Class 2 is associated with a five-year metastatic risk of 72% [66,82]. Recently, GEP classification has been revised based on PRAME (preferentially expressed antigen in melanoma) status. PRAME has been reported as an independent prognostic biomarker for UM that identifies increased metastatic risk in patients with Class 1 or disomy 3 tumors. When combined with a 12-gene expression panel, PRAME messenger-RNA expression predicted a five-year metastatic rate of 0 in class 1/PRAME−, 38% in class 1/PRAME+, and 71% in class 2 tumors. Interestingly, PRAME+ status was positively correlated with larger tumor diameter after analysis of the TCGA Research Network dataset. PRAME expression positively correlated with larger tumor diameter and SF3B1 mutations as well as gain of 1q, 6p, 8q, and 9q and loss of 6q and 11q [83].

4. Adjuvant Therapies and Surveillance of UM

Despite significant advances in prognosis and identification of high-risk patients, adjuvant systemic treatments effective in preventing metastases or improving outcomes in UM are not yet available in clinical practice [84,85]. Therefore, intensified surveillance appears crucial for the early detection of oligometastatic disease manageable with liver-directed therapies, as well as to enroll patients eligible for clinical trials [86]. Specifically, patients identified at a high-risk of metastatic progression based on cytogenetics or GEP should have six-monthly life-long surveillance including a clinical review, nurse specialist support, and liver-specific imaging by a nonionizing modality [87]. It is reported that magnetic resonance imaging (MRI) obtained every six months is capable of detecting the metastases before the onset of symptoms in 92% of cases [86]. A few adjuvant therapy trials have been tested in UM based on favorable results in cutaneous melanoma. A randomized controlled clinical trial investigated the effects of dacarbazine (DTIC)—an intravenous alkylating agent—demonstrating no survival advantage over observation in UM [88]. Likewise, a randomized study on methanol-extraction residue of bacille Calmette-Guérin (BCG) reported no benefit in improving the survival rate [89].

Two nonrandomized studies failed to demonstrate beneficial effects on survival rate with systemic adjuvant low-dose interferon-alpha (IFN-α) compared with matched historical controls [90,91]. Fotemustine, an alkylating agent with high hepatic uptake, has been studied as an adjuvant therapy for UM by intra-arterial hepatic delivery, with a trend towards improved survival but not statistical significance compared with matched historical controls [92]. However, these adjuvant studies were conducted before the introduction of molecular methods of prognosis. In this respect, a multicenter randomized phase III clinical trial (FOTEADJ) based on genomic analysis in high risk UM patients treated with adjuvant fotemustine versus observation was performed, but it was stopped earlier for futility [93]. Similarly, a nonrandomized prospective phase II clinical trial designed to evaluate sequential low-dose DTIC and IFN-α-2b in cytogenetic high-risk patients was completed, but it failed to reach the primary outcome of progression free survival (PFS) or overall survival (OS) increase at five-year follow-up [94]. Novel classes of molecules have been investigated in the adjuvant setting for UM with promising results. The tyrosine-kinase receptors c-Met and c-Kit are highly expressed in UM and activate the Ras/Erk, and PI3-kinase pathways following binding to the hepatocyte growth factor (HGF) and stem cell factor (SCF), respectively. These pathways have been definitely demonstrated to be involved in cancer occurrence and progression [95]. Crizotinib—a tyrosine-kinase inhibitor that inhibits the phosphorylation of c-Met—was shown to significantly reduce the development of distant metastases in a murine model of metastatic UM when compared with an untreated control group [96]. Of note, crizotinib is currently under evaluation in patients with UM [97]. A retrospective cohort study based on adjuvant sunitinib—a tyrosine-kinase inhibitor that inhibits c-Kit—and conducted in high-risk patients stratified according to cytogenetics and GEP, resulted in three- and five-year improvement of OS estimates [98]. To confirm such results, a phase II pilot clinical trial evaluating sunitinib, tamoxifen, and cisplatin in patients with high-risk ocular melanoma is ongoing [99]. Similarly, a randomized, noncomparative phase II clinical trial investigating sunitinib and the histone deacetylase (HDAC) inhibitor, valproic acid, for high-risk tumors in an adjuvant setting is currently ongoing [100]. The assumption for using HDAC inhibitors in the adjuvant setting for UM is based on their ability to reverse the phenotypic effects of BAP1 loss in cultured UM cells [101,102]. Another interesting approach for the treatment of UM is the use of adjuvant dendritic cell (DC) vaccination. An open-label phase II clinical trial was performed to investigate immunologic responses after adjuvant DC vaccination in patients defined at high-risk based on cytogenetics. This study showed an increase in OS in patients with a detectable tumor antigen-specific immune response [103]. In addition, a multicenter, randomized, two-armed, open-label phase III study to evaluate the adjuvant vaccination with tumor RNA-loaded autologous DCs in patients with resected monosomy 3 UM is currently ongoing [104]. A similar phase I/II study on mRNA transfected autologous DCs in high-risk uveal melanoma patients has recently been closed due to slow accrual [105]. Immunotherapy with monoclonal antibodies, such as antiprogrammed cell death protein 1 (PD-1) nivolumab, anticytotoxic T-lymphocyte-associated Protein 4 (CTLA-4) ipilimumab, and anti lymphocyte activation gene 3 (LAG-3) relatlimab, modulated the immune responses in the tumor microenvironment and interfered with tumor growth and spread [106,107]. A small sample size phase I/II pilot trial of adjuvant ipilimumab in high-risk primary uveal melanoma demonstrated that 80% of patients were disease-free at 36 months [108]. A randomized phase II trial, evaluating nivolumab with or without ipilimumab or relatlimab in neoadjuvant and adjuvant settings, is currently ongoing in patients with AJCC stage IIIB–IV posterior UM [109]. Similarly, a phase II single-arm multicenter study to evaluate the effects of adjuvant ipilimumab treatment in combination with nivolumab in subjects with high-risk UM is currently recruiting [110]. An innovative approach is represented by prophylactic radiation therapy to the liver; however, a phase II trial on external-beam hepatic irradiation in high-risk patients has recently been closed for lack of accrual [111]. Moreover, there is recent evidence of potent antitumor effects in UM cells following nonselective beta-blocker administration, and concurrent expression of β1 and β2 adrenoceptors in UM specimens. These findings suggest that further investigation is needed in the context of clinical trials for adjuvant scope [112]. Current clinical trials in the adjuvant setting are summarized in Table 2.

Table 2. Current adjuvant trials in uveal melanoma.

Clinical Trials N	Tested Agent and Mechanism of Action	Phase	Status
NCT02223819	Crizotinib (c-Met inhibitor)	II	Recruiting
NCT02068586	Sunitinib (c-Kit inhibitor) vs. Valproic acid (HDAC inhibitor)	II	Recruiting
NCT00489944	Suntinib (c-Kit inhibitor) + Tamoxifen (estrogen receptor modulator) + Cisplatin (alkylating agent)	II	Unknown
NCT01983748	Dendritic cell vaccination (immunotherapy)	III	Recruiting
NCT00929019	Dendritic cell vaccination (immunotherapy)	I/II	Terminated, slow accrual
NCT02519322	Nivolumab (anti-PD1) with or without Ipilimumab (anti-CTLA4) or Relatlimab (anti-LAG3)	II	Recruiting
NCT03528408	Ipilimumab (anti-CTLA4) + Nivolumab(anti-PD1)	II	Recruiting
NCT02336763	Prophylactic External-Beam Radiation Therapy to the liver	II	Terminated, lack of accrual

5. Metastatic Dormancy and Therapeutic Opportunities

A proportion of only 1–2% of patients with UM presents metastatic disease at the time of diagnosis [113,114]. However, mathematical models of cell doubling times and direct histopathologic evaluation suggest that UM hepatic micrometastases would be present since the time of initial diagnosis [115–118]. The identification of circulating tumor cells in the bloodstream of patients clinically free of metastasis supports this consideration [117,119]. It is worth noting that early detection and primary treatment showed to have an impact on disease-related morbidity but not on patients' OS [7,118,120,121]. Moreover, there is evidence that UM liver metastases may remain stable for years until an exponential proliferation occurs [122]. Based on these observations, a growing interest has been focused in the study of the mechanisms underlying UM cell dormancy in the liver and potential ways to prolong or specifically target them [115–118]. The identification of the key factors involved in the transition to rapidly growing tumors is crucial for the development of novel therapeutic strategies in order to prolong metastatic dormancy or eliminate dormant cancer cells in a controlled fashion. In fact, the state of dormancy confers refractoriness to conventional therapies aimed at targeting rapidly proliferating cells. Several studies have been performed to identify novel therapeutic targets able to control or eliminate metastatic dormant cells in different cancers; however, metastatic dormancy has not yielded appropriate clinical investigation in melanoma, and specifically in UM. Dormant tumor cells are supposed to be in a quiescent state, prevented from proliferating exponentially due to blockage of the cell cycle. The dormancy of disseminated tumor cells is supposed to be the result of a balance between anti- and protumorigenic immune and inflammatory responses, failure in activating the angiogenic switch, genetic modulation by metastasis suppressor genes (MSGs), and associated signaling pathways [123–125]. The arrest of circulating cancer cells adhering to the sinusoidal endothelium of the hepatic lobules leads to avascular micrometastasis development in periportal areas [126]. Simultaneously, tumor lytic M1 phenotype macrophages are recruited along with activated hepatic stellate cells (HSCs) secreting extracellular matrix and proinflammatory mediators [126–128]. The paracrine signaling on UM cells would then lead to a reduction in D-type cyclin and a deregulation in the interaction of the cyclin-dependent kinases (CDKs) with the CDK inhibitors (CKIs), with consequent arrest of UM cells proliferation [129,130]. The potential to evade the immune system in cancer is associated with increasing tumor genetic instability and related reduction in immunogenicity following progression [119,122]. In accordance, CD4+ tumor infiltrating lymphocytes (TILs) and CD8+ TILs have been described in advanced metastatic disease with perivascular and peritumoral distribution, respectively. Thus, suggesting their inability to infiltrate the tumor mass as disease progresses [127,131,132]. Furthermore, a clear prevalence of proangiogenic and protumorigenic M2 phenotype macrophages has been demonstrated within hepatic mUM in late stages of the disease [127,131]. Interestingly, BAP1 loss has recently been correlated with upregulation of several genes associated with suppressive immune responses and, at the protein level, with entrapment of infiltrating immune cells within peritumoural fibrotic areas surrounding mUM [62]. There is evidence suggesting that dormant cells express mutant neoantigens, which can occur naturally or rather derive from treatment [133,134]. The low frequency of naturally occurring neoantigen-specific

T cells clones, anyway, has favored the advent of specific adoptive T-cells transfer therapies [135,136]. Adoptive T-cells transfer therapies may be of interest for UM since the primary tumor originates in the immune-privileged environment of the eye, and it may present tumor neoantigens for which the host's immune system is not prepared. T-cells targeting glycoprotein gp100 were tested for this scope in vitro and in vivo in human UM, demonstrating homing to the eye and effective tumoricidal function [137]. Recently, cell-based vaccines modified to express MHC II alleles syngeneic to the recipient and the costimulatory molecule CD80 have been studied in UM. These vaccines were capable of activating CD4+ T cells specific to uveal melanoma neoantigens, that in turn reacted with primary UM cells and cross-reacted with mUM cells. Moreover, CD4+ T cells activated CD8+ T cell-mediated immunity against primary and mUM cells [138]. In addition, IFN-γ production by CD4+ T cells activated by UM vaccines, promoted an antitumor response by inhibiting neovascularization and tumor cell proliferation, as well as upregulating tumor-expressed MHC molecules [138]. In detail, IFN-γ showed the ability to mediate long-term cell growth arrest in vitro and in vivo nude mice models via STAT1 and p27-dependent mechanisms, and it also raised the hypothesis of inducing specific T-cells capable of IFN-γ production upon recognition of tumor cells [139–141]. For this purpose, a phase I/II clinical trial to evaluate the efficacy of immunotherapy with TILs in combination with intratumoral injections of IFN-γ-adenovirus in cutaneous metastatic melanoma is ongoing [142]. Moreover, a study investigating tumor microenvironment, demonstrated that the use of low doses of the anti-VEGF receptor 2 (VEGFR2) antibody was able to polarize the immune inhibitory M2-like phenotype towards the immune stimulatory M1-like phenotype and to recruit CD4+ and CD8+ T-cells. These mechanisms suggested that low-dose antiangiogenic treatment in adjunct to vaccine therapy could enhance anticancer efficacy [143]. Among other factors regulating the awakening of dormant cells, the angiogenetic sprout allowing the shift from prevascular to highly vascularized lesions is known to play a critical role [123]. The group of Grossniklaus et al. identified pseudo-sinusoidal spaces between the sinusoidal endothelium and hepatocytes (space of Disse), developing from stellate cells to nourish large infiltrative pattern metastases [144,145]. The hepatic fibrosis/stellate cell activation and the mTORC1/S6K signaling axis have been fully characterized while profiling secretome from high-risk metastatic UM compared to normal choroidal melanocytes [146]. In a study on GEP analysis in experimental animal models of different human types of cancer, a downregulation of angiogenesis inhibitor thrombospondin and decreased sensitivity to angiostatin in switched fast-growing versus dormant tumors were described [147]. Other genes associated with the angiogenic process were observed to contribute to tumor dormancy, including tropomyosin, transforming growth factor beta 2 (TGF-b2), Eph receptor A5 (EphA5), histone H2BK, proline 4-hydroxylase alpha polypeptide I, and insulin-like growth factor binding protein 5 (IGFBP-5) [123]. Several studies not including UM investigated the role of MSGs in preventing the formation of metastases and favoring dormancy [124,125,141,148–159]. Specifically, the MSGs KISS1, RhoG-DI2, and Nm23-H1 showed to be able to suppress the development of distant metastases without significantly affecting tumor growth at the primary site [124,125]. Interestingly, MSGs rarely mutate, and their downregulation in highly metastatic tumors would rather be associated with epigenetic modifications. In this regard, possible therapeutic targets could be represented by DNA methyl-transferases and histone deacetylases [102,160,161].

6. Treatment of Metastatic Disease

UM patients with metastatic disease are hardly candidates for curative treatments, with a reported 15% one-year survival and an average life expectancy varying in literature from 6 to 12 months [1,2,6,26,30]. There is currently no standard of care for the treatment of mUM, and available treatments are mostly adapted from cutaneous melanoma protocols in spite of their different clinical and genetic profiles [162]. Furthermore, patients presenting with ocular or mucosal melanoma are frequently excluded from clinical trials. Thus far, systemic chemotherapy has provided poor response rates (0–15%) in clinical trials for UM metastatic disease [163–165]. Liver-directed therapies have shown limited improvements in response rates, but no benefit in OS [166–169]. Unlike the positive results achieved in the treatment of metastatic cutaneous melanoma, immunotherapy and targeted therapies have failed to improve OS in metastatic UM.

6.1. Chemotherapy

Systemic chemotherapy, adopted from cutaneous melanoma, has been evaluated in the context of single-arm phase II studies in mUM as monotherapy, including dacarbazine, temozolomide, fotemustine, and docosahexaenoic acid (DHA)-paclitaxel, or combined treatments (BOLD regimen+interferon α-2b, dacarbazine and treosulfan or cisplatin, gemcitabine and treosulfan, cisplatin, dacarbazine and vinblastine). Similar response rates of less than 15%, PFS limited to 4 months, and OS of no more than 12 months were reported [163–165,170–173]. Moreover, significant hematological, pulmonary and neurological toxicities were observed. Therefore, research is rather addressed to the development and testing of targeted and immune therapies, as well as liver locoregional approaches.

6.2. Liver-Directed Therapies

The liver is the first and the sole site of metastatic spread in more than 50% of mUM patients, and liver-directed therapies including surgical resection, regional perfusion, and embolization have been investigated in patients with mUM confined to the liver [26]. Available knowledge on the efficacy of liver-directed therapies is mostly based on down-sized, retrospective, single-institution studies. A recent meta-analysis determining benchmarks of PFS and OS for mUM suggested more favorable outcomes with liver-directed therapies compared with chemotherapy, immunotherapy, and targeted therapy, even after adjusting for prognostic factors [174].

6.2.1. Surgery

It is reported from retrospective cohort studies that surgical resection of oligometastases is effective for curative intent in mUM. Curative (R0) resection is the most important positive prognostic factor following liver resection [87]. However, only 5–10% of patients are candidates for surgery based on liver metastase distribution and size [175–177]. Comparative studies in patients with mUM showed that median OS ranges between 10 and 35 months in patients treated with surgery (differing between R0 resection or debulk of metastases), between 9 and 15 months with any systemic treatment, and between 2 and 6 months with the best supportive care, from comparative studies on mUM [86,121,175–183]. In the largest series currently available, longer survival was associated with metastasis-free intervals longer than 24 months; R0 resection, number of liver metastases ≤ 4, and absence of miliary disease were associated with prolonged survival [175]. The group of Akyuz et al. demonstrated five-year survival exceeding 20% after complete tumor destruction under laparoscopic resection or laparoscopic radiofrequency ablation (RFA) including a nonsurgical comparator group [177]. Confirming rates were reported in several noncomparative studies [176,182–186]. Other studies evaluated RFA or hepatic intra-arterial chemotherapy (HIA) as an adjunct to liver surgery to increase the number of patients with bilobar metastases achieving R0 resection [181,187,188]. Importantly, RFA associated with liver surgery and liver surgery alone demonstrated similar survival outcomes [187]. In addition, a recent study evaluated combined surgery and RFA in a small group of patients relapsing after complete first liver resection, showing prolonged survival outcomes [189]. However, postresection local and distant recurrences are frequent exhibiting rates of 75%, and there are no data from randomized clinical trials which demonstrate survival benefit over systemic therapy [183,186].

6.2.2. Regional Perfusion Therapies

Direct targeting of hepatic arterial circulation represents an attractive strategy for unresectable isolated liver disease. Metastases to the liver are indeed preferentially supplied by hepatic artery branches unlike normal hepatic circulation, receiving blood mainly from the portal vein. Regional approaches allow the direct delivery of high doses of chemotherapy with minimal systemic exposure and include hepatic intra-arterial chemotherapy (HIA), isolated hepatic perfusion (IHP), percutaneous hepatic perfusion (PHP), and hepatic transarterial chemoembolization (TACE). A phase III randomized clinical trial from the European Organization for the Research and Treatment of Cancer (EORTC) assigned 171 patients with UM and liver metastases to receive fotemustine via HIA or intravenously (IV). Significant improvements were registered in PFS (4.5 vs. 3.5 months) and response rate (10.5 vs. 2.4%) with HIA compared with IV administration, but no difference was demonstrated between the two arms in terms of OS (median 14.6 months for HIA vs. 13.8 months for IV fotemustine) [166]. IHP, as open or percutaneous procedure (PHP), is a form of intra-arterial chemotherapy requiring a temporary extracorporeal filtration system to surgically isolate the liver from systemic circulation. Results from a phase II clinical trial suggested a survival advantage of 14 months for patients treated with IHP using melphalan compared with the longest survival rate of patients with UM liver metastases not treated with IHP, associated with tolerable morbidity [190]. The SCANDIUM study—a randomized multicenter phase III clinical trial—is currently ongoing in patients with UM and isolated liver metastases to evaluate the efficacy of IHP melphalan compared with the best alternative care in OS [191]. Results from a randomized phase III trial including 93 patients with melanoma metastatic to the liver (88% ocular, 12% cutaneous) treated with either PHP with melphalan or best available care—showed that PHP was effective in significantly improving median PFS (245 days vs. 49 days, $P < 0.001$) and overall response rate (34.1 vs. 2% $P < 0.001$). This study failed to demonstrate survival overall benefit; however, the crossover design of the study may confound the survival data [167]. The phase III multicenter FOCUS clinical trial is currently ongoing in patients with metastatic disease and hepatic-dominant UM treated with either PHP with melphalan or distinct options under the best alternative care (transarterial chemoembolization, dacarbazine, ipilimumab, or pembrolizumab)—randomly assigned. However, due to accrual issues, this study was later modified to remove randomization [192]. Another strategy among liver-locoregional treatments for mUM is TACE with infusion of chemotherapeutics including cisplatin, carboplatine, mytomicin, fotoemustine, and 1,3-bis (2-cholorethyl)-1-nitrosourea (BCNU), followed by embolization agents such as iodized oil or polyvenylalcohol particles. In a large, retrospective cohort study, chemoembolization was effective when compared with systemic therapies in inducing 33% response rate versus 1%; however, no survival benefit was demonstrated [193]. Similar findings were reported in noncomparative studies, with overall response rates varying from 20.4% to 46% [193–198]. From a phase II study, improved response rates and survival were demonstrated in patients with less than 20% liver involvement, suggesting that small and well demarcated tumors receiving their supply solely from the hepatic artery represent the best targets for embolization [196]. In accordance, other studies reported that an extent of liver involvement > 50% predicted poor outcomes with arterial chemoembolization [197]. Importantly, two-thirds of patients with stabilization of hepatic metastases following TACE developed dissemination in extrahepatic sites within a short time, thus raising the issue of combining systemic immuno-chemotherapy with local treatments [196]. In a pilot clinical trial, platinum-based TACE with polyvenylalcohol (PVA)-particle embolization in combination with systemic immuno-chemotherapy achieved a 57% partial response rate and survival benefit [199]. A phase-I/II randomized trial evaluated HIA with cisplatin and TACE with cisplatin and polyvinyl sponge (PVS) in 19 patients with UM metastatic to the liver, reporting a modest overall response rate (16%) and dose-limiting toxicities [200].

6.2.3. Radioembolization

Among other techniques, radioembolization (RE) using yttrium-90 (^{90}Y)-labeled microspheres was evaluated as salvage therapy in the context of small, retrospective cohorts, reporting median OS rates ranging from 9 to 24 months. Partial response or a stabilization of the disease was reported for 57 and 77% of patients, respectively [168,169]. Recently, a prospective phase II clinical trial evaluated the efficacy of RE in treatment-naïve patients with mUM (group A) and in participants who progressed after immunoembolization (IE) (group B). This study demonstrated similar median OS (18.5 months and 19.2 months) and 1-year survival rate (60.9 and 69.6%) between the two groups, respectively. Interestingly, the stabilization of hepatic disease was achieved in 87.0% of participants in groups A versus 58.3% in group B [201]. In a recent single arm, open labeled, nonrandomized study, the combination of yttrium-90 microspheres and intravenous cisplatin was well tolerated in mUM, but it failed at demonstrating sustained disease control with a median PFS of 3 months and median OS of 10 months [202]. A nonrandomized phase I clinical trial investigating ^{90}Y -labelled microspheres in combination with sorafenib—a multikinase inhibitor of cell proliferation and angiogenesis—was concluded, but the results have not yet been published [203].

6.2.4. Immunoembolization (IE)

The increased release of tumor antigens after tumor destruction via embolization leads to the development of immunoembolization (IE) using granulocyte-macrophage colony-stimulating factor (GM-CSF). A randomized phase II study, investigating IE versus bland embolization (BE) in patients with mUM, demonstrated similar OS rates (21.5 months in IE group versus 17.2 months in BE group), with a significant survival advantage in patients with at least 20% of liver involvement within the IE cohort. Moreover, the intense inflammatory reaction in response correlated positively with delayed progression of extrahepatic metastases [204].

6.3. Immunotherapy

Immunological checkpoint inhibitors targeting the cytotoxic T-lymphocyte associated antigen (CTLA)-4 (ipilimumab), the programmed cell death 1 (PD)-1 protein (pembrolizumab, nivolumab), or the programmed cell death 1 ligand (PDL)-1 (durvalumab, atezolizumab) aim at stimulating endogenous antitumor cytotoxic T cell response. The efficacy achieved in the management of metastatic cutaneous melanoma and other cancers with reported durable response rates ranging from 20 to over 60%, has not been observed in mUM [205–207]. Studies reported a response rate below 10%, and a median survival less of than 1 year with a single-agent checkpoint block have been widely described [208–213]. This is likely related to the immune privilege of the eye, which establishes mechanisms to evade the immune system, and with the low mutational load with limited potential neoepitopes of UM if compared with cutaneous melanoma. However, clinical benefit from PD-1 block has been reported in selected UM patients with biallelic MBD4 loss showing a high mutational burden [214]. Two phase II clinical trials are investigating combinatorial checkpoint blockade with nivolumab and ipilimumab in treatment-naïve or pretreated patients with mUM [215,216]. Specifically, preliminary results from the clinical trial NCT01585194 showed a partial remission (PR) in 17% of patients, and a stable disease in 53% of patients. The median OS was estimated at 1.6 years, and the 1-year OS was 62%. However, 40% of patients experienced treatment-related adverse events (TRAEs), with 29% of treatment discontinuation [217]. A multicenter phase II open label study evaluating the concomitant use of pembrolizumab and entinostat (HDAC inhibitor) in adult patients with metastatic mUM (PEMDAC study) is currently ongoing [218,219]. A phase Ib/II clinical trial demonstrated that combination treatment with RFA and ipilimumab in uveal melanoma (SECIRA-UM) was well tolerated, but with very limited clinical activity [220]. Importantly, a randomized phase I/II study is currently ongoing in mUM patients to evaluate the safety and efficacy of combining melphalan PHP with ipilimumab and nivolumab [221]. In addition, the efficacy of the combination of ipilimumab and nivolumab has been investigated in association with IE, and following ^{90}Y radioembolization, respectively, in two ongoing phase II trials [222,223]. Novel immune-based therapies include different modalities of adoptive T cell therapy such as TILs, engineered T cell receptors (TCRs) and chimeric antigen receptors (CAR) on T cells, and T cell redirection [224–227]. Adoptive transfer of autologous TILs has shown promise in mediating tumor regression in a single-center phase II study on refractory mUM patients who showed progression after both anti-CTLA-4 and anti-PD-1 checkpoint blockade [224]. A phase II study on immunotherapy using autologous TILs in mUM is currently ongoing [228]. In this direction, a phase Ib study combining adoptive T cell therapy using autologous CD8+ antigen-specific T cells and anti-CTLA-4 for patients with mUM is ongoing [229]. IMCgp100 is a bispecific ImmTAC

(Immune-mobilizing monoclonal TCRs against cancer) molecule, targeting gp100 peptide on UM cells and the CD3 protein complex on the surface of T cells. It redirects the recruitment of CD8+ cytotoxic T lymphocytes against melanoma cells. This molecule has shown a favorable safety profile and durable responses in mUM, and it is currently under investigation in advanced UM in a single-arm, phase I/II dose-escalating clinical trial and in a randomized, controlled phase II trial versus the investigator's choice of therapy [230–233]. As an additional promising strategy, CAR-T cells directed against human epidermal growth factor receptor 2 (HER2) were demonstrated to be able to kill uveal and cutaneous melanoma cells in vitro and in vivo settings [226]. A phase I clinical trial is evaluating the effect of autologous T-lymphocytes expressing GD2-specific chimeric antigen on different GD2-expressing cancers including UM [234]. Among immunotherapeutic options for mUM, cell-based and peptide vaccines are currently being investigated in several ongoing clinical trials as single therapy or in combination with immunomodulatory agents, based on favorable preclinical and clinical studies [138,235–238]. As innovative approach, liver intralesional PV-10 chemoablation (10% rose bengal disodium) allowed for a rapid lysis of tumor cells followed by a secondary tumor-specific T cell-mediated antitumor immune response [239]. Based on high rates of complete response and durable local control achieved in metastatic cutaneous melanoma, PV-10 chemoablation has been evaluated in a phase I safety and tolerability study in mUM [240].

6.4. Targeted Therapy

Recent advances in the molecular profiling of UM provide a rationale for treatments that selectively target the effectors of the molecular pathways which regulate tumor growth. Specifically, mutations of the GNAQ and GNA11 genes encoding for Gα subunits of G-proteins drive oncogenesis in most of primary and mUM, whereas mutations in the phospholipase C4 (PLCB4) or in the Cysteinyl Leukotriene Receptor 2 (CYSLTR2) genes occur less frequently [241,242]. The development of therapies aimed at directly targeting Gα proteins is still in an initial phase in mUM, whereas BRAF (v-raf murine sarcoma viral oncogene homolog B1)-targeted therapy has achieved substantial results in cutaneous melanoma [243,244]. The cyclic depsipeptide FR900359 (FR) was observed to allosterically inhibit the GDP/GTP exchange to obtain inactive Gαβγ heterotrimers from constitutively active GαQ and 11, thus promoting cell cycle arrest in UM cells in culture and inhibiting tumor growth in UM mouse xenografts [245,246]. The design of simplified analogues of FR900359 capable of effective GαQ/11 inhibition, including the small molecule YM-19, opens new perspectives for pharmaceutical development [247]. Among gene regulatory approaches, a combination therapy of oncolytic adenovirus H101 and siRNA mediating GNAQ downregulation was shown to induce UM cells apoptosis in in vivo activating UM cell apoptosis [248]. Moreover, a system for conjugating siRNAs to functionalized gold nanoparticles (AuNPs) able to recognize transcripts of mutant GNAQ mRNA was developed. This approach resulted in greater intracellular release of siRNA and decreased cancer cell viability [249]. GNAQ/GNA11 mutations drive the constitutive activation of the mitogen-activated protein kinase (MAPK) pathway, and therapies targeting downstream effectors of Gα at the level of MEK, PKC, and AKT have been investigated. In the phase III clinical trial SUMIT, naïve mUM patients were randomized to receive either selumetinib—a selective MEK inhibitor—or placebo, in combination with dacarbazine. Results of this study did not show a difference in the primary endpoint of PFS between the two groups of treatment [250]. The combination of selumetinib with the AKT inhibitor MK2206 resulted in synergistic suppression of GNAQ mutant cell viability in vitro and in xenograft mouse models of UM [251]. Based on these encouraging preclinical results, a randomized phase II clinical trial was performed to investigate the efficacy of trametinib—a selective MEK inhibitor—with or without AKT inhibition. However, this study did not demonstrate any substantial improvement in the primary endpoint of response (PFS) for combinational treatment [252]. Sorafenib—a kinase inhibitor targeting RAF/MEK/ERK pathway and VEGFR/PDGFR—was investigated in a phase II study by the Southwest Oncology Group (SWOG) cooperative group in combination with carboplatin and paclitaxel in mUM, but the limited overall efficacy did not warrant further clinical tests [253]. Among other strategies, a phase I study is ongoing evaluating the preliminary antitumor activity of LXS196, a PKC inhibitor, as monotherapy and in combination with HDM201 (MDM2 inhibitor) in patients with mUM [254]. A number of phase I and II trials on other targeted therapies in mUM have been completed, demonstrating no impact on survival indicators, including lenalidomide (TNF-α secretion inhibitor), gefitinib (epidermal growth factor inhibitor), bevacizumab and aflibercept (vascular endothelial growth factor inhibitors), imatinib (KIT inhibitor), sunitinib (tyrosine kinase inhibitor), vorinostat (histone deacetylase inhibitor), carbozantinib (tyrosine kinases c-Met and VEGFR2 inhibitor), cixutumumab (IGF1R inhibitor), everoliumus (mTOR inhibitor) plus pasireotide (a somatostatin analog), and ganetespib (heat-shock protein 90 inhibitor) ([196,255–269]. Currently active phase I/II trials in mUM, targeting molecules other than MEK, AKT, and PKC, are based on BVD-523 (ERK1/ERK2 inhibitor), BPX-701 (a genetically modified autologous T cell product incorporating an HLA-A2-restricted PRAME-directed TCR), and cabozantinib (multikinase inhibitor) versus temozolomide or dacarbazine [270–272]. The main trials currently evaluating liver-directed therapies, immunotherapies, and targeted therapies for mUM are listed in Table 3.

Table 3. Main ongoing trials for metastatic uveal melanoma.

Clinical Trials N	Tested Agent and Mechanism of Action	Phase	Status
NCT01785316	IHP with melphalan or best alternative care	III	Recruiting
NCT02678572	PHP with melphalan or best alternative care	III	Recruiting
NCT01893099	^{90}Y-labelled microspheres and sorafenib (inhibitor of RAF/MEK/ERK and VEGFR/PDGFR)	I	Complete, no results
NCT02626962	Nivolumab (anti-PD1) + ipilimumab (anti-CTLA4)	II	Active, not recruiting
NCT01585194	Nivolumab (anti-PD1) + ipilimumab (anti-CTLA4)	II	Active, not recruiting
NCT02697630	Pembrolizumab (anti-PD1) + Entinostat (HDAC inhibitor)	II	Active, not recruiting
NCT04283890	PHP with melphalan + ipilimumab (anti-CTLA4) and nivolumab (anti-PD1)	I/II	Recruiting
NCT02913417	^{90}Y-labelled microspheres + Ipilimumab (anti-CTLA4) and nivolumab (anti-PD1)	I/II	Recruiting
NCT03472586	Immunoembolization + Ipilimumab (anti-CTLA4) and nivolumab (anti-PD1)	II	Recruiting
NCT03467516	TILs	II	Recruiting
NCT03068624	TILs + cyclophosphamide (alkylating agent), aldesleukin (human recombinant IL-2), and ipilimumab (anti-CTLA4)	Ib	Active, not recruiting
NCT02570308	ImmTAC molecule (IMCgp100) targeting gp100	I/II	Active, not recruiting
NCT03070392	ImmTAC molecule (IMCgp100) targeting gp100Vs. investigator's choice	II	Active, not recruiting.
NCT03635632	C7R-GD2.CAR T cells	I	Recruiting
NCT00219843	Intralesional (IL) PV-10 chemoablation(rose bengal disodium, 10%)	I	Complete, no results
NCT01979523	Trametinib (MEK inhibitor) ± AKT inhibition	II	Complete, has results
NCT02601378	LXS196 (PKC inhibitor) ± HDM201 (MDM2 inhibitor)	I	Active, not recruiting
NCT01413191	Cixutumumab (IGF1R inhibitor)	II	Complete, has results
NCT01252251	Everoliumus (mTOR inhibitor) and pasireotide (somatostatin analog)	II	Complete, has results
NCT03417739	BVD-523(ERK1/ERK2 inhibitor)	II	Active, not recruiting
NCT02743611	BPX-701 (PRAME-targeting T-cell receptor)	I/II	Active, not recruiting
NCT01835145	Carbozantinib (c-MET, c-KIT, VEGFR2 inhibitor) vs. temozolomide (alkylating agent) or dacarbazine (alkylating agent)	II	Active, not recruiting, has results

Abbreviations: Isolated hepatic perfusion (IHP), Percutaneous hepatic perfusion (PHP), immunoembolization (IE), tumor-infiltrating lymphocytes (TILs), chimeric antigen receptors (CAR), Immune-mobilizing monoclonal TCRs against cancer (ImmTAC).

7. Conclusions

UM represents a challenge for oncologists and ophthalmologists in terms of early diagnosis, clinical and genetic characterization, and treatments. In recent decades, considerable advances have been made in the diagnosis and classification of patients at low/high-risk of metastatic progression in UM, thereby facilitating early and tailored intervention. However, 50% of patients still develop metastatic disease, and survival rates do not show substantial improvements. Thus far, there is no accepted standard of care for the treatment of UM in adjuvant and metastatic settings, and most of current treatments for UM are adapted from results observed in cutaneous melanoma, although UM shows different clinical and molecular features from its cutaneous counterpart. The increasing knowledge of tumor biology, genetics, and immunology has recently led to UM-specific clinical trials for adjuvant and metastatic scope. Specifically, insights into the primary and metastatic UM microenvironment and tumor immune surveillance, as well as the mechanisms regulating metastatic tumor dormancy, paved the way for the identification of targets for future therapies. Therefore, research should be focused on testing novel promising therapies, and continued participation in clinical trials should be encouraged. This will hopefully increase the survival benefit of UM patients similarly to what has recently been observed for cutaneous melanoma.

Author Contributions: Conceptualization, F.M. and A.M.; methodology, F.M.; software, F.M.; validation, M.S., A.L. and A.M.; formal analysis, M.S.; investigation, A.M.; resources, F.M.; data curation, F.M.; writing—original draft preparation, F.M.; writing—review and editing, M.S.; visualization, A.M.; supervision, A.L.; project administration, A.L. All authors have read and agreed to the published version of the manuscript.

Funding: This research received no external funding.

Conflicts of Interest: The authors declare no conflict of interest.

References

1. Chang, A.E.; Karnell, L.H.; Menck, H.R. The national cancer data base report on cutaneous and noncutaneous melanoma: A summary of 84,836 cases from the past decade. *Cancer* **1998**, *83*, 1664–1678. [CrossRef]

2. Shields, C.L.; Kaliki, S.; Furuta, M.; Mashayekhi, A.; Shields, J.A. Clinical spectrum and prognosis of uveal melanoma based on age at presentation in 8033 cases. *Retina* **2012**, *32*, 1363–1372. [CrossRef] [PubMed]

3. McLaughlin, C.C.; Wu, X.C.; Jemal, A.; Martin, H.J.; Roche, L.M.; Chen, V.W. Incidence of noncutaneous melanomas in the U.S. *Cancer* **2005**, *103*, 1000–1007. [CrossRef] [PubMed]

4. Singh, A.D.; Turell, M.E.; Topham, A.K. Uveal melanoma: Trends in incidence, treatment, and survival. *Ophthalmology* **2011**, *118*, 1881–1885. [CrossRef]

5. Virgili, G.; Gatta, G.; Ciccolallo, L.; Capocaccia, R.; Biggeri, A.; Crocetti, E.; Lutz, J.M.; Paci, E. Incidence of Uveal Melanoma in Europe. *Ophthalmology* **2007**, *114*, 2309–2315. [CrossRef]

6. Mallone, S.; De Vries, E.; Guzzo, M.; Midena, E.; Verne, J.; Coebergh, J.W.; Marcos-Gragera, R.; Ardanaz, E.; Martinez, R.; Chirlaque, M.D.; et al. Descriptive epidemiology of malignant mucosal and uveal melanomas and adnexal skin carcinomas in Europe. *Eur. J. Cancer* **2012**, *48*, 1167–1175. [CrossRef]

7. Damato, E.M.; Damato, B.E. Detection and time to treatment of uveal melanoma in the United Kingdom: An evaluation of 2384 patients. *Ophthalmology* **2012**, *119*, 1582–1589. [CrossRef]

8. Mahendraraj, K.; Lau, C.S.M.; Lee, I.; Chamberlain, R.S. Trends in incidence, survival, and management of uveal melanoma: A population-based study of 7,516 patients from the surveillance, epidemiology, and end results database (1973–2012). *Clin. Ophthalmol.* **2016**, *10*, 2113–2119. [CrossRef]

9. Aronow, M.E.; Topham, A.K.; Singh, A.D. Uveal Melanoma: 5-Year Update on Incidence, Treatment, and Survival (SEER 1973-2013). *Ocul. Oncol. Pathol.* **2018**, *4*, 145–151. [CrossRef]

10. Xu, Y.; Lou, L.; Wang, Y.; Miao, Q.; Jin, K.; Chen, M.; Ye, J. Epidemiological Study of Uveal Melanoma from US Surveillance, Epidemiology, and End Results Program (2010–2015). *J. Ophthalmol.* **2020**. [CrossRef]

11. Hu, D.N.; Yu, G.P.; McCormick, S.A.; Schneider, S.; Finger, P.T. Population-based incidence of uveal melanoma in various races and ethnic groups. *Am. J. Ophthalmol.* **2005**, *140*, 612.e1–612.e8. [CrossRef] [PubMed]

12. Yu, G.-P.; Hu, D.-N.; McCormick, S.A. Latitude and Incidence of Ocular Melanoma. *Photochem. Photobiol.* **2006**, *82*, 1621. [CrossRef] [PubMed]

13. Shields, C.L.; Kaliki, S.; Cohen, M.N.; Shields, P.W.; Furuta, M.; Shields, J.A. Prognosis of uveal melanoma based on race in 8100 patients: The 2015 Doyne Lecture. *Eye* **2015**, *29*, 1027–1035. [CrossRef] [PubMed]

14. Shields, C.L.; Kaliki, S.; Livesey, M.; Walker, B.; Garoon, R.; Bucci, M.; Feinstein, E.; Pesch, A.; Gonzalez, C.; Lally, S.E.; et al. Association of ocular and oculodermal melanocytosis with the rate of uveal melanoma metastasis analysis of 7872 consecutive eyes. *JAMA Ophthalmol.* **2013**, *131*, 993–1003. [CrossRef] [PubMed]

15. Singh, A.D.; De Potter, P.; Fijal, B.A.; Shields, C.L.; Shields, J.A.; Elston, R.C. Lifetime prevalence of uveal melanoma in white patients with oculo(dermal) melanocytosis. *Ophthalmology* **1998**, *105*, 195–198. [CrossRef]

16. Hammer, H.; Oláh, J.; Tóth-Molnár, E. Dysplastic nevi are a risk factor for uveal melanoma. *Eur. J. Ophthalmol.* **1996**, *6*, 472–474. [CrossRef]

17. McDonald, K.A.; Krema, H.; Chan, A.-W. Cutaneous Signs and Risk Factors for Ocular Melanoma. *J. Am. Acad. Dermatol.* **2020**. [CrossRef]

18. Barker, C.A.; Salama, A.K. New NCCN guidelines for uveal melanoma and treatment of recurrent or progressive distant metastatic melanoma. *J. Natl. Compr. Cancer Netw.* **2018**, *16*, 646–650. [CrossRef]

19. Rodrigues, M.; de Koning, L.; Coupland, S.E.; Jochemsen, A.G.; Marais, R.; Stern, M.H.; Valente, A.; Barnhill, R.; Cassoux, N.; Evans, A.; et al. So close, yet so far: Discrepancies between uveal and other melanomas. a position paper from UM cure 2020. *Cancers* **2019**, *11*, 1032. [CrossRef]

20. Shields, C.L.; Cater, J.; Shields, J.A.; Singh, A.D.; Santos, M.C.M.; Carvalho, C. Combination of clinical factors predictive of growth of small choroidal melanocytic tumors. *Arch. Ophthalmol.* **2000**, *118*, 360–364. [CrossRef]

21. Shields, C.L.; Furuta, M.; Berman, E.L.; Zahler, J.D.; Hoberman, D.M.; Dinh, D.H.; Mashayekhi, A.; Shields, J.A. Choroidal nevus transformation into melanoma: Analysis of 2514 consecutive cases. *Arch. Ophthalmol.* **2009**, *127*, 981–987. [CrossRef] [PubMed]

22. Melia, B.M.; Diener-West, M.; Bennett, S.R.; Folk, J.C.; Montague, P.R.; Weingeist, T.A.; Hawkins, B.S. Factors predictive of growth and treatment of small choroidal melanoma: COMS report no. 5. *Arch. Ophthalmol.* **1997**, *115*, 1537–1544. [CrossRef]

23. Accuracy of Diagnosis of Choroidal Melanomas in the Collaborative Ocular Melanoma Study: COMS Report No. 1. *Arch. Ophthalmol.* **1990**, *108*, 1268–1273. [CrossRef] [PubMed]

24. Shields, J.A.; Shields, C.L.; Ehya, H.; Eagle, R.C.; De Potter, P. Fine-needle aspiration biopsy of suspected intraocular tumors. *Int. Ophthalmol. Clin.* **1993**, *33*, 77–82. [CrossRef]

25. Hawkins, B.S. The COMS randomized trial of iodine 125 brachytherapy for choroidal melanoma: V. Twelve-year mortality rates and prognostic factors: COMS report no. 28. *Arch. Ophthalmol.* **2006**, *124*, 1684–1693. [CrossRef]

26. Kujala, E.; Mäkitie, T.; Kivelä, T. Very Long-Term Prognosis of Patients with Malignant Uveal Melanoma. *Investig. Ophthalmol. Vis. Sci.* **2003**, *44*, 4651–4659. [CrossRef]

27. Hawkins, B.S. The Collaborative Ocular Melanoma Study (COMS) randomized trial of pre-enucleation radiation of large choroidal melanoma: IV. Ten-year mortality findings and prognostic factors. COMS report number 24. *Am. J. Ophthalmol.* **2004**, *138*, 936–951. [CrossRef]

28. Bishop, K.D.; Olszewski, A.J. Epidemiology and survival outcomes of ocular and mucosal melanomas: A population-based analysis. *Int. J. Cancer* **2014**, *134*, 2961–2971. [CrossRef]

29. Burr, J.M.; Mitry, E.; Rachet, B.; Coleman, M.P. Survival from uveal melanoma in England and Wales 1986 to 2001. *Ophthalmic Epidemiol.* **2007**, *14*, 3–8. [CrossRef]

30. Kuk, D.; Shoushtari, A.N.; Barker, C.A.; Panageas, K.S.; Munhoz, R.R.; Momtaz, P.; Ariyan, C.E.; Brady, M.S.; Coit, D.G.; Bogatch, K.; et al. Prognosis of Mucosal, Uveal, Acral, Nonacral Cutaneous, and Unknown Primary Melanoma From the Time of First Metastasis. *Oncologist* **2016**, *21*, 848–854. [CrossRef]

31. Seibel, I.; Cordini, D.; Rehak, M.; Hager, A.; Riechardt, A.I.; Böker, A.; Heufelder, J.; Weber, A.; Gollrad, J.; Besserer, A.; et al. Local Recurrence after Primary Proton Beam Therapy in Uveal Melanoma: Risk Factors, Retreatment Approaches, and Outcome. *Am. J. Ophthalmol.* **2015**, *160*, 628–636. [CrossRef] [PubMed]

32. Coupland, S.E.; Sidiki, S.; Clark, B.J.; McClaren, K.; Kyle, P.; Lee, W.R. Metastatic choroidal melanoma to the contralateral orbit 40 years after enucleation. *Arch. Ophthalmol.* **1996**, *114*, 751–756. [CrossRef] [PubMed]

33. Shields, J.A.; Augsburger, J.J.; Donoso, L.A.; Bernardino, V.B.; Portenar, M. Hepatic metastasis and orbital recurrence of uveal melanoma after 42 years. *Am. J. Ophthalmol.* **1985**, *100*, 666–668. [CrossRef]

34. Dithmar, S.; Diaz, C.E.; Grossniklaus, H.E. Intraocular melanoma spread to regional lymph nodes. *Retina* **2000**, *20*, 76–79. [CrossRef]

35. Badve, S.S.; Fisher, C. AJCC 8th edition—A step forward. *Breast J.* **2020**, *26*, 1263–1264. [CrossRef]

36. Shields, C.L.; Furuta, M.; Thangappan, A.; Nagori, S.; Mashayekhi, A.; Lally, D.R.; Kelly, C.C.; Rudich, D.S.; Nagori, A.V.; Wakade, O.A.; et al. Metastasis of uveal melanoma millimeter-by-millimeter in 8033 consecutive eyes. *Arch. Ophthalmol.* **2009**, *127*, 989–998. [CrossRef]

37. Shields, C.L.; Kaliki, S.; Furuta, M.; Fulco, E.; Alarcon, C.; Shields, J.A. American Joint Committee on Cancer classification of posterior uveal melanoma (tumor size category) predicts prognosis in 7731 patients. *Ophthalmology* **2013**, *120*, 2066–2071. [CrossRef]

38. Gill, H.S.; Char, D.H. Uveal melanoma prognostication: From lesion size and cell type to molecular class. *Can. J. Ophthalmol.* **2012**, *47*, 246–253. [CrossRef]

39. Bagger, M.; Andersen, M.T.; Andersen, K.K.; Heegaard, S.; Kiilgaard, J.F. The prognostic effect of American joint committee on cancer staging and genetic status in patients with choroidal and ciliary body melanoma. *Investig. Ophthalmol. Vis. Sci.* **2015**, *56*, 438–444. [CrossRef]

40. Amin, M.B.; Greene, F.L.; Edge, S.B.; Compton, C.C.; Gershenwald, J.E.; Brookland, R.K.; Meyer, L.; Gress, D.M.; Byrd, D.R.; Winchester, D.P. The Eighth Edition AJCC Cancer Staging Manual: Continuing to build a bridge from a population-based to a more "personalized" approach to cancer staging. *CA Cancer J. Clin.* **2017**, *67*, 93–99. [CrossRef]

41. Cassoux, N.; Rodrigues, M.J.; Plancher, C.; Asselain, B.; Levy-Gabriel, C.; Lumbroso-Le Rouic, L.; Piperno-Neumann, S.; Dendale, R.; Sastre, X.; Desjardins, L.; et al. Genome-wide profiling is a clinically relevant and affordable prognostic test in posterior uveal melanoma. *Br. J. Ophthalmol.* **2014**, *98*, 769–774. [CrossRef] [PubMed]

42. Damato, B.; Duke, C.; Coupland, S.E.; Hiscott, P.; Smith, P.A.; Campbell, I.; Douglas, A.; Howard, P. Cytogenetics of Uveal Melanoma. A 7-Year Clinical Experience. *Ophthalmology* **2007**, *114*, 1925–1931. [CrossRef] [PubMed]

43. Shields, C.L.; Say, E.A.T.; Hasanreisoglu, M.; Saktanasate, J.; Lawson, B.M.; Landy, J.E.; Badami, A.U.; Sivalingam, M.D.; Hauschild, A.J.; House, R.J.; et al. Personalized Prognosis of Uveal Melanoma Based on Cytogenetic Profile in 1059 Patients over an 8-Year Period: The 2017 Harry S. Gradle Lecture. *Ophthalmology* **2017**, *124*, 1523–1531. [CrossRef] [PubMed]

44. Onken, M.D.; Worley, L.A.; Ehlers, J.P.; Harbour, J.W. Gene expression profiling in uveal melanoma reveals two molecular classes and predicts metastatic death. *Cancer Res.* **2004**, *64*, 7205–7209. [CrossRef]

45. Onken, M.D.; Worley, L.A.; Char, D.H.; Augsburger, J.J.; Correa, Z.M.; Nudleman, E.; Aaberg, T.M.; Altaweel, M.M.; Bardenstein, D.S.; Finger, P.T.; et al. Collaborative ocular oncology group report number 1: Prospective validation of a multi-gene prognostic assay in uveal melanoma. *Ophthalmology* **2012**, *119*, 1596–1603. [CrossRef]

46. Damato, B.; Coupland, S.E. A reappraisal of the significance of largest basal diameter of posterior uveal melanoma. *Eye* **2009**, *23*, 2152–2162. [CrossRef] [PubMed]

47. Damato, B.; Dopierala, J.; Klaasen, A.; van Dijk, M.; Sibbring, J.; Coupland, S.E. Multiplex ligation-dependent probe amplification of uveal melanoma: Correlation with metastatic death. *Investig. Ophthalmol. Vis. Sci.* **2009**, *50*, 3048–3055. [CrossRef]

48. Vaquero-Garcia, J.; Lalonde, E.; Ewens, K.G.; Ebrahimzadeh, J.; Richard-Yutz, J.; Shields, C.L.; Barrera, A.; Green, C.J.; Barash, Y.; Ganguly, A. PRiMeUM: A model for predicting risk of metastasis in uveal melanoma. *Investig. Ophthalmol. Vis. Sci.* **2017**, *58*, 4096–4105. [CrossRef]

49. Walter, S.D.; Chao, D.L.; Feuer, W.; Schiffman, J.; Char, D.H.; Harbour, J.W. Prognostic implications of tumor diameter in association with gene expression profile for uveal melanoma. *JAMA Ophthalmol.* **2016**, *134*, 734–740. [CrossRef]

50. Eleuteri, A.; Damato, B.; Coupland, S.E.; Taktak, A.F.G. Enhancing survival prognostication in patients with choroidal melanoma by integrating pathologic clinical and genetic predictors of metastasis. *Int. J. Biomed. Eng. Technol.* **2012**, *8*, 18–35. [CrossRef]

51. Jager, M.J.; Brouwer, N.J.; Esmaeli, B. The Cancer Genome Atlas Project: An Integrated Molecular View of Uveal Melanoma. *Ophthalmology* **2018**, *125*, 1139–1142. [CrossRef] [PubMed]

52. Mazloumi, M.; Vichitvejpaisal, P.; Dalvin, L.A.; Yaghy, A.; Ewens, K.G.; Ganguly, A.; Shields, C.L. Accuracy of the Cancer Genome Atlas Classification vs American Joint Committee on Cancer Classification for Prediction of Metastasis in Patients with Uveal Melanoma. *JAMA Ophthalmol.* **2020**, *138*, 260–267. [CrossRef] [PubMed]

53. Robertson, A.G.; Shih, J.; Yau, C.; Gibb, E.A.; Oba, J.; Mungall, K.L.; Hess, J.M.; Uzunangelov, V.; Walter, V.; Danilova, L.; et al. Integrative Analysis Identifies Four Molecular and Clinical Subsets in Uveal Melanoma. *Cancer Cell* **2017**, *32*, 204–220.e15. [CrossRef] [PubMed]

54. Vichitvejpaisal, P.; Dalvin, L.A.; Mazloumi, M.; Ewens, K.G.; Ganguly, A.; Shields, C.L. Genetic Analysis of Uveal Melanoma in 658 Patients Using the Cancer Genome Atlas Classification of Uveal Melanoma as A, B, C, and D. *Ophthalmology* **2019**, *126*, 1445–1453. [CrossRef] [PubMed]

55. Harbour, J.W. The genetics of uveal melanoma: An emerging framework for targeted therapy. *Pigment Cell Melanoma Res.* **2012**, *25*, 171–181. [CrossRef]

56. Prescher, G.; Bornfeld, N.; Hirche, H.; Horsthemke, B.; Jöckel, K.H.; Becher, R. Prognostic implications of monosomy 3 in uveal melanoma. *Lancet* **1996**, *347*, 1222–1225. [CrossRef]

57. Tschentscher, F.; Prescher, G.; Zeschnigk, M.; Horsthemke, B.; Lohmann, D.R. Identification of chromosomes 3, 6, and 8 aberrations in uveal melanoma by microsatellite analysis in comparison to comparative genomic hybridization. *Cancer Genet. Cytogenet.* **2000**, *122*, 13–17. [CrossRef]

58. Versluis, M.; De Lange, M.J.; Van Pelt, S.I.; Ruivenkamp, C.A.L.; Kroes, W.G.M.; Cao, J.; Jager, M.J.; Luyten, G.P.M.; Van Der Velden, P.A. Digital PCR validates 8q dosage as prognostic tool in uveal melanoma. *PLoS ONE* **2015**, *10*, e0116371. [CrossRef]

59. Harbour, J.W.; Onken, M.D.; Roberson, E.D.O.; Duan, S.; Cao, L.; Worley, L.A.; Council, M.L.; Matatall, K.A.; Helms, C.; Bowcock, A.M. Frequent mutation of BAP1 in metastasizing uveal melanomas. *Science* **2010**, *330*, 1410–1413. [CrossRef]

60. Carbone, M.; Ferris, L.K.; Baumann, F.; Napolitano, A.; Lum, C.A.; Flores, E.G.; Gaudino, G.; Powers, A.; Bryant-Greenwood, P.; Krausz, T.; et al. BAP1 cancer syndrome: Malignant mesothelioma, uveal and cutaneous melanoma, and MBAITs. *J. Transl. Med.* **2012**, *10*, 179. [CrossRef]

61. Laíns, I.; Bartosch, C.; Mondim, V.; Healy, B.; Kim, I.K.; Husain, D.; Miller, J.W. Second Primary Neoplasms in Patients With Uveal Melanoma: A SEER Database Analysis. *Am. J. Ophthalmol.* **2016**, *165*, 54–64. [CrossRef] [PubMed]

62. Figueiredo, C.R.; Kalirai, H.; Sacco, J.J.; Azevedo, R.A.; Duckworth, A.; Slupsky, J.R.; Coulson, J.M.; Coupland, S.E. Loss of BAP1 expression is associated with an immunosuppressive microenvironment in uveal melanoma, with implications for immunotherapy development. *J. Pathol.* **2020**, *250*, 420–439. [CrossRef] [PubMed]

63. Karlsson, J.; Nilsson, L.M.; Mitra, S.; Alsén, S.; Shelke, G.V.; Sah, V.R.; Forsberg, E.M.V.; Stierner, U.; All-Eriksson, C.; Einarsdottir, B.; et al. Molecular profiling of driver events in metastatic uveal melanoma. *Nat. Commun.* **2020**, *11*, 1–13. [CrossRef] [PubMed]

64. Louie, B.H.; Kurzrock, R. BAP1: Not Just a BRCA1-Associated Protein. *Cancer Treat. Rev.* **2020**, *90*, 102091. [CrossRef]

65. Yavuzyigitoglu, S.; Koopmans, A.E.; Verdijk, R.M.; Vaarwater, J.; Eussen, B.; Van Bodegom, A.; Paridaens, D.; Kiliç, E.; De Klein, A. Uveal Melanomas with SF3B1 Mutations: A Distinct Subclass Associated with Late-Onset Metastases. *Ophthalmology* **2016**, *123*, 1118–1128. [CrossRef]

66. Harbour, J.W.; Chen, R. The DecisionDx-UM Gene Expression Profile Test Provides Risk Stratification and Individualized Patient Care in Uveal Melanoma. *PLoS Curr.* **2013**, *5*. [CrossRef]

67. Luscan, A.; Just, P.A.; Briand, A.; Burin Des Roziers, C.; Goussard, P.; Nitschké, P.; Vidaud, M.; Avril, M.F.; Terris, B.; Pasmant, E. Uveal melanoma hepatic metastases mutation spectrum analysis using targeted next-generation sequencing of 400 cancer genes. *Br. J. Ophthalmol.* **2015**, *99*, 437–439. [CrossRef]

68. Martin, M.; Maßhöfer, L.; Temming, P.; Rahmann, S.; Metz, C.; Bornfeld, N.; Van De Nes, J.; Hitpass, L.K.; Hinnebusch, A.G.; Horsthemke, B.; et al. Exome sequencing identifies recurrent somatic mutations in EIF1AX and SF3B1 in uveal melanoma with disomy 3. *Nat. Genet.* **2013**, *45*, 933–936. [CrossRef]

69. Höglund, M.; Gisselsson, D.; Hansen, G.B.; White, V.A.; Säll, T.; Mitelman, F.; Horsman, D. Dissecting karyotypic patterns in malignant melanomas: Temporal clustering of losses and gains in melanoma karyotypic evolution. *Int. J. Cancer* **2004**, *108*, 57–65. [CrossRef]

70. Damato, B.E.; Heimann, H.; Kalirai, H.; Coupland, S.E. Age, survival predictors, and metastatic death in patients with choroidal melanoma tentative evidence of a therapeutic effect on survival. *JAMA Ophthalmol.* **2014**, *132*, 605–613. [CrossRef]

71. Aalto, Y.; Eriksson, L.; Seregard, S.; Larsson, O.; Knuutila, S. Concomitant loss of chromosome 3 and whole arm losses and gains of chromosome 1, 6, or 8 in metastasizing primary uveal melanoma. *Investig. Ophthalmol. Vis. Sci.* **2001**, *42*, 313–317.

72. Ewens, K.G.; Kanetsky, P.A.; Richards-Yutz, J.; Al-Dahmash, S.; de Luca, M.C.; Bianciotto, C.G.; Shields, C.L.; Ganguly, A. Genomic profile of 320 uveal melanoma cases: Chromosome 8p-loss and metastatic outcome. *Investig. Ophthalmol. Vis. Sci.* **2013**, *54*, 5721–5729. [CrossRef] [PubMed]

73. Naus, N.C.; Verhoeven, A.C.A.; van Drunen, E.; Slater, R.; Mooy, C.M.; Paridaens, D.A.; Luyten, G.P.M.; de Klein, A. Detection of Genetic Prognostic Markers in Uveal Melanoma Biopsies Using Fluorescence in Situ Hybridization. *Clin. Cancer Res.* **2002**, *8*, 534–539.

74. Sisley, K.; Nichols, C.; Parsons, M.A.; Farr, R.; Rees, R.C.; Rennie, I.G. Clinical applications of chromosome analysis, from fine needle aspiration biopsies, of posterior uveal melanomas. *Eye* **1998**, *12*, 203–207. [CrossRef] [PubMed]

75. Shields, C.L.; Ganguly, A.; Materin, M.A.; Teixeira, L.; Mashayekhi, A.; Swanson, L.A.; Marr, B.P.; Shields, J.A. Chromosome 3 analysis of uveal melanoma using fine-needle aspiration biopsy at the time of plaque radiotherapy in 140 consecutive cases: The Deborah Iverson, MD, Lectureship. *Arch. Ophthalmol.* **2007**, *125*, 1017–1024. [CrossRef] [PubMed]

76. Klofas, L.K.; Bogan, C.M.; Coogan, A.; Schultenover, S.J.; Weiss, V.L.; Daniels, A.B. Instrument gauge and type in uveal melanoma fine needle biopsy: Implications for diagnostic yield and molecular prognostication. *Am. J. Ophthalmol.* **2020**. [CrossRef]

77. Young, T.A.; Rao, N.P.; Glasgow, B.J.; Moral, J.N.; Straatsma, B.R. Fluorescent In Situ Hybridization for Monosomy 3 via 30-Gauge Fine-Needle Aspiration Biopsy of Choroidal Melanoma In Vivo. *Ophthalmology* **2007**, *114*, 142–146. [CrossRef]

78. Cross, N.A.; Ganesh, A.; Parpia, M.; Murray, A.K.; Rennie, I.G.; Sisley, K. Multiple locations on chromosome 3 are the targets of specific deletions in uveal melanoma. *Eye* **2006**, *20*, 476–481. [CrossRef]

79. Lake, S.L.; Kalirai, H.; Dopierała, J.; Damato, B.E.; Coupland, S.E. Comparison of Formalin-Fixed and Snap-Frozen Samples Analyzed by Multiplex Ligation-Dependent Probe Amplification for Prognostic Testing in Uveal Melanoma. *Investig. Ophthalmol. Vis. Sci.* **2012**, *53*, 2647–2652. [CrossRef]

80. Worley, L.A.; Onken, M.D.; Person, E.; Robirds, D.; Branson, J.; Char, D.H.; Perry, A.; Harbour, J.W. Transcriptomic versus chromosomal prognostic markers and clinical outcome in uveal melanoma. *Clin. Cancer Res.* **2007**, *13*, 1466–1471. [CrossRef]

81. Onken, M.D.; Worley, L.A.; Tuscan, M.D.; Harbour, J.W. An accurate, clinically feasible multi-gene expression assay for predicting metastasis in uveal melanoma. *J. Mol. Diagn.* **2010**, *12*, 461–468. [CrossRef] [PubMed]

82. Field, M.G.; Harbour, J.W. Recent developments in prognostic and predictive testing in uveal melanoma. *Curr. Opin. Ophthalmol.* **2014**, *25*, 234–239. [CrossRef]

83. Field, M.G.; Decatur, C.L.; Kurtenbach, S.; Gezgin, G.; Van Der Velden, P.A.; Jager, M.J.; Kozak, K.N.; Harbour, J.W. PRAME as an independent biomarker for metastasis in uveal melanoma. *Clin. Cancer Res.* **2016**, *22*, 1234–1242. [CrossRef] [PubMed]

84. Triozzi, P.L.; Singh, A.D. Adjuvant Therapy of Uveal Melanoma: Current Status. *Ocul. Oncol. Pathol.* **2014**, *1*, 54–62. [CrossRef] [PubMed]

85. Seth, R.; Messersmith, H.; Kaur, V.; Kirkwood, J.M.; Kudchadkar, R.; McQuade, J.L.; Provenzano, A.; Swami, U.; Weber, J.; Alluri, K.C.; et al. Systemic Therapy for Melanoma: ASCO Guideline. *J. Clin. Oncol.* **2020**, JCO.20.00198. [CrossRef] [PubMed]

86. Marshall, E.; Romaniuk, C.; Ghaneh, P.; Wong, H.; McKay, M.; Chopra, M.; Coupland, S.E.; Damato, B.E. MRI in the detection of hepatic metastases from high-risk uveal melanoma: A prospective study in 188 patients. *Br. J. Ophthalmol.* **2013**, *97*, 159–163. [CrossRef]

87. Nathan, P.; Cohen, V.; Coupland, S.; Curtis, K.; Damato, B.; Evans, J.; Fenwick, S.; Kirkpatrick, L.; Li, O.; Marshall, E.; et al. Uveal Melanoma UK National Guidelines. *Eur. J. Cancer* **2015**, *51*, 2404–2412. [CrossRef]

88. Desjardins, L.; Dorval, T.; Levy, C.; Cojean, I.; Schlienger, P.; Salmon, R.J.; Validire, P.; Asselain, B. Etude randomisée de chimiothérapie adjuvante par le Déticène dans le mélanome choroïdien. *Ophtalmologie* **1998**, *12*, 168–173.

89. McLean, I.W.; Berd, D.; Mastrangelo, M.J.; Shields, J.A.; Davidorf, F.H.; Grever, M.; Makley, T.A.; Gamel, J.W. A randomized study of methanol-extraction residue of bacille Calmette-Guerin as postsurgical adjuvant therapy of uveal melanoma. *Am. J. Ophthalmol.* **1990**, *110*, 522–526. [CrossRef]

90. Richtig, E.; Langmann, G.; Schlemmer, G.; Müllner, K.; Papaefthymiou, G.; Bergthaler, P.; Smolle, J. Verträglichkeit und wirksamkeit einer adjuvanten interferon-alfa-2b-behandlung beim aderhautmelanom. *Ophthalmologe* **2006**, *103*, 506–511. [CrossRef]

91. Lane, A.M.; Egan, K.M.; Harmon, D.; Holbrook, A.; Munzenrider, J.E.; Gragoudas, E.S. Adjuvant Interferon Therapy for Patients with Uveal Melanoma at High Risk of Metastasis. *Ophthalmology* **2009**, *116*, 2206–2212. [CrossRef]

92. Voelter, V.; Schalenbourg, A.; Pampallona, S.; Peters, S.; Halkic, N.; Denys, A.; Goitein, G.; Zografos, L.; Leyvraz, S. Adjuvant intra-arterial hepatic fotemustine for high-risk uveal melanoma patients. *Melanoma Res.* **2008**, *18*, 220–224. [CrossRef] [PubMed]

93. Piperno-Neumann, S.; Rodrigues, M.J.; Servois, V.; Pierron, G.; Gastaud, L.; Negrier, S.; Levy-Gabriel, C.; Lumbroso, L.; Cassoux, N.; Bidard, F.-C.; et al. A randomized multicenter phase 3 trial of adjuvant fotemustine versus surveillance in high risk uveal melanoma (UM) patients (FOTEADJ). *J. Clin. Oncol.* **2017**, *35*, 9502. [CrossRef]

94. Binkley, E.; Triozzi, P.L.; Rybicki, L.; Achberger, S.; Aldrich, W.; Singh, A. A prospective trial of adjuvant therapy for high-risk uveal melanoma: Assessing 5-year survival outcomes. *Br. J. Ophthalmol.* **2020**, *104*, 524–528. [CrossRef] [PubMed]

95. Sato, T.; Han, F.; Yamamoto, A. The biology and management of uveal melanoma. *Curr. Oncol. Rep.* **2008**, *10*, 431–438. [CrossRef]

96. Surriga, O.; Rajasekhar, V.K.; Ambrosini, G.; Dogan, Y.; Huang, R.; Schwartz, G.K. Crizotinib, a c-Met Inhibitor, Prevents Metastasis in a Metastatic Uveal Melanoma Model. *Mol. Cancer Ther.* **2013**, *12*, 2817–2826. [CrossRef]

97. Search of: NCT02223819—List Results—ClinicalTrials.gov. Available online: https://clinicaltrials.gov/ct2/results?cond=&term=NCT02223819&cntry=&state=&city=&dist= (accessed on 22 August 2020).

98. Valsecchi, M.E.; Orloff, M.; Sato, R.; Chervoneva, I.; Shields, C.L.; Shields, J.A.; Mastrangelo, M.J.; Sato, T. Adjuvant Sunitinib in High-Risk Patients with Uveal Melanoma: Comparison with Institutional Controls. *Ophthalmology* **2018**, *125*, 210–217. [CrossRef]

99. Search of: NCT00489944—List Results—ClinicalTrials.gov. Available online: https://clinicaltrials.gov/ct2/results?cond=&term=NCT00489944&cntry=&state=&city=&dist= (accessed on 22 August 2020).

100. Search of: NCT02068586—List Results—ClinicalTrials.gov. Available online: https://clinicaltrials.gov/ct2/results?cond=&term=NCT02068586&cntry=&state=&city=&dist= (accessed on 22 August 2020).

101. Landreville, S.; Agapova, O.A.; Matatall, K.A.; Kneass, Z.T.; Onken, M.D.; Lee, R.S.; Bowcock, A.M.; Harbour, J.W. Histone Deacetylase Inhibitors Induce Growth Arrest and Differentiation in Uveal Melanoma. *AACR* **2011**, *18*, 408–416. [CrossRef]

102. Fagone, P.; Caltabiano, R.; Russo, A.; Lupo, G.; Anfuso, C.D.; Basile, M.S.; Longo, A.; Nicoletti, F.; De Pasquale, R.; Libra, M.; et al. Identification of novel chemotherapeutic strategies for metastatic uveal melanoma. *Sci. Rep.* **2017**, *7*, 44564. [CrossRef]

103. Bol, K.; van den Bosch, T.; Schreibelt, G.; Punt, C.; Figdor, C.; Paridaens, D.; de Vries, J. Adjuvant dendritic cell vaccination in high-risk uveal melanoma patients. *J. Immunother. Cancer* **2015**, *3*, 127. [CrossRef]

104. Search of: NCT01983748—List Results—ClinicalTrials.gov. Available online: https://clinicaltrials.gov/ct2/results?cond=&term=NCT01983748&cntry=&state=&city=&dist= (accessed on 22 August 2020).

105. Search of: NCT00929019—List Results—ClinicalTrials.gov. Available online: https://clinicaltrials.gov/ct2/results?cond=&term=NCT00929019&cntry=&state=&city=&dist= (accessed on 22 August 2020).

106. Schank, T.E.; Hassel, J.C. cancers Immunotherapies for the Treatment of Uveal Melanoma-History and Future. *Cancers* **2019**, *11*, 1048. [CrossRef] [PubMed]

107. Durante, M.A.; Rodriguez, D.A.; Kurtenbach, S.; Kuznetsov, J.N.; Sanchez, M.I.; Decatur, C.L.; Snyder, H.; Feun, L.G.; Livingstone, A.S.; Harbour, J.W. Single-cell analysis reveals new evolutionary complexity in uveal melanoma. *Nat. Commun.* **2020**, *11*, 1–10. [CrossRef] [PubMed]

108. Fountain, E.; Bassett, R.L.; Cain, S.; Posada, L.; Gombos, D.S.; Hwu, P.; Bedikian, A.; Patel, S.P. Adjuvant ipilimumab in high-risk Uveal melanoma. *Cancers* **2019**, *11*, 152. [CrossRef] [PubMed]

109. Search of: NCT02519322—List Results—ClinicalTrials.gov. Available online: https://clinicaltrials.gov/ct2/results?cond=&term=NCT02519322&cntry=&state=&city=&dist= (accessed on 22 August 2020).

110. Search of: NCT03528408—List Results—ClinicalTrials.gov. Available online: https://clinicaltrials.gov/ct2/results?cond=&term=NCT03528408&cntry=&state=&city=&dist= (accessed on 11 September 2020).

111. Search of: NCT02336763—List Results—ClinicalTrials.gov. Available online: https://clinicaltrials.gov/ct2/results?cond=&term=NCT02336763&cntry=&state=&city=&dist= (accessed on 22 August 2020).

112. Bustamante, P.; Miyamoto, D.; Goyeneche, A.; de Alba Graue, P.G.; Jin, E.; Tsering, T.; Dias, A.B.; Burnier, M.N.; Burnier, J.V. Beta-blockers exert potent anti-tumor effects in cutaneous and uveal melanoma. *Cancer Med.* **2019**, *8*, 7265–7277. [CrossRef]

113. Croock, D.L. Metastatic uveal melanoma: Diagnosis and treatment. A literature review. *Bull. Société Belg. Ophtalmol.* **2002**, *286*, 59.

114. Freton, A.; Chin, K.J.; Raut, R.; Tena, L.B.; Kivelä, T.; Finger, P.T. Initial PET/CT staging for choroidal melanoma: AJCC correlation and second nonocular primaries in 333 patients. *Eur. J. Ophthalmol.* **2011**, *22*, 236–243. [CrossRef]

115. Grossniklaus, H.E. Understanding Uveal Melanoma Metastasis to the Liver: The Zimmerman Effect and the Zimmerman Hypothesis. *Ophthalmology* **2019**, *126*, 483–487. [CrossRef]

116. Singh, A.D. Uveal melanoma: Implications of tumor doubling time. *Ophthalmology* **2001**, *108*, 829–830. [CrossRef]

117. Torres, V.; Triozzi, P.; Eng, C.; Tubbs, R.; Schoenfiled, L.; Crabb, J.W.; Saunthararajah, Y.; Singh, A.D. Circulating tumor cells in uveal melanoma. *Futur. Oncol.* **2011**, *7*, 101–109. [CrossRef]

118. Ossowski, L.; Aguirre-Ghiso, J.A. Dormancy of metastatic melanoma. *Pigment Cell Melanoma Res.* **2010**, *23*, 41–56. [CrossRef]

119. Blanco, P.L.; Lim, L.A.; Miyamoto, C.; Burnier, M.N. Uveal melanoma dormancy: An acceptable clinical endpoint? *Melanoma Res.* **2012**, *22*, 334–340. [CrossRef] [PubMed]

120. Ah-Fat, F.G.; Damato, B.E. Delays in the diagnosis of uveal melanoma and effect on treatment. *Eye* **1998**, *12*, 781–782. [CrossRef] [PubMed]

121. Rietschel, P.; Panageas, K.S.; Hanlon, C.; Patel, A.; Abramson, D.H.; Chapman, P.B. Variates of survival in metastatic uveal melanoma. *J. Clin. Oncol.* **2005**, *23*, 8076–8080. [CrossRef] [PubMed]

122. Mouriaux, F.; Zaniolo, K.; Bergeron, M.A.; Weidmann, C.; De La Fouchardière, A.; Fournier, F.; Droit, A.; Morcos, M.W.; Landreville, S.; Guérin, S.L. Effects of long-term serial passaging on the characteristics and properties of cell lines derived from uveal melanoma primary tumors. *Investig. Ophthalmol. Vis. Sci.* **2016**, *57*, 5288–5301. [CrossRef]

123. Almog, N. Molecular mechanisms underlying tumor dormancy. *Cancer Lett.* **2010**, *294*, 139–146. [CrossRef]

124. Hedley, B.D.; Allan, A.L.; Chambers, A.F. Tumor dormancy and the role of metastasis suppressor genes in regulating ectopic growth. *Futur. Oncol.* **2006**, *2*, 627–641. [CrossRef]

125. Horak, C.E.; Lee, J.H.; Marshall, J.C.; Shreeve, S.M.; Steeg, P.S. The role of metastasis suppressor genes in metastatic dormancy. *Apmis* **2008**, *116*, 586–601. [CrossRef]

126. Vidal-Vanaclocha, F. The Prometastatic Microenvironment of the Liver. *Cancer Microenviron.* **2008**, *1*, 113–129. [CrossRef]

127. Krishna, Y.; Mccarthy, C.; Kalirai, H.; Coupland, S.E.; Yamini, K.; Conni, M.; Helen, K. Inflammatory cell infiltrates in advanced metastatic uveal melanoma. *Hum. Pathol.* **2017**, *66*, 159–166. [CrossRef]

128. Eyles, J.; Puaux, A.; Wang, X. Tumor cells disseminate early, but immunosurveillance limits metastatic outgrowth, in a mouse model of melanoma. *J. Clin. Investig.* **2010**, *120*, 2030–2039. [CrossRef] [PubMed]

129. Mouriaux, F.; Casagrande, F.; Pillaire, M.-J.; Manenti, S.; Malecaze, F.; Darbon, J.-M. Differential Expression of Gl Cyclins and Cyclin-Dependent Kinase Inhibitors in Normal and Transformed Melanocytes. *Investig. Ophthalmol. Vis. Sci.* **1998**, *39*, 876–884.

130. Mouriaux, F.; Maurage, C.; science, P.L. Cyclin-dependent kinase inhibitory protein expression in human choroidal melanoma tumors. *Investig. Ophthalmol. Vis. Sci.* **2000**, *41*, 2837–2843.

131. Bronkhorst, I.H.G.; Jager, M.J. Uveal melanoma: The inflammatory microenvironment. *J. Innate Immun.* **2012**, *4*, 454–462. [CrossRef]

132. Ly, L.; Baghat, A.; Versluis, M.; Jordanova, E.; Immunol, G.L.-J.; Luyten, G.P.M.; van Rooijen, N.; van Hall, T.; van der Velden, P.A.; Jager, M.J. In aged mice, outgrowth of intraocular melanoma depends on proangiogenic M2-type macrophages. *J. Immunol.* **2010**, *185*, 3481–3488. [CrossRef]

133. Ward, J.; Gubin, M.; Schreiber, R.D. The role of neoantigens in naturally occurring and therapeutically induced immune responses to cancer. In *Advances in Immunology*; Elsevier: Amsterdam, The Netherlands, 2016.

134. Tran, E.; Turcotte, S.; Gros, A.; Robbins, P.F.; Lu, Y.C.; Dudley, M.E.; Wunderlich, J.R.; Somerville, R.P.; Hogan, K.; Hinrichs, C.S.; et al. Cancer immunotherapy based on mutation-specific CD4+ T cells in a patient with epithelial cancer. *Science* **2014**, *344*, 641–645. [CrossRef]

135. Turcotte, S.; Gros, A.; Hogan, K.; Tran, E.; Hinrichs, C.S.; Wunderlich, J.R.; Dudley, M.E.; Rosenberg, S.A. Phenotype and Function of T Cells Infiltrating Visceral Metastases from Gastrointestinal Cancers and Melanoma: Implications for Adoptive Cell Transfer Therapy. *J. Immunol.* **2013**, *191*, 2217–2225. [CrossRef]

136. Alexandrov, L.B.; Nik-Zainal, S.; Wedge, D.C.; Aparicio, S.A.J.R.; Behjati, S.; Biankin, A.V.; Bignell, G.R.; Bolli, N.; Borg, A.; Børresen-Dale, A.L.; et al. Signatures of mutational processes in human cancer. *Nature* **2013**, *500*, 415–421. [CrossRef]

137. Sutmuller, R.P.M.; Schurmans, L.R.H.M.; van Duivenvoorde, L.M.; Tine, J.A.; van der Voort, E.I.H.; Toes, R.E.M.; Melief, C.J.M.; Jager, M.J.; Offringa, R. Adoptive T Cell Immunotherapy of Human Uveal Melanoma Targeting gp100. *J. Immunol.* **2000**, *165*, 7308–7315. [CrossRef] [PubMed]

138. Kittler, J.M.; Sommer, J.; Fischer, A.; Britting, S.; Karg, M.M.; Bock, B.; Atreya, I.; Heindl, L.M.; Mackensen, A.; Bosch, J.J. Characterization of CD4+ T cells primed and boosted by MHCII primary uveal melanoma cell-based vaccines. *Oncotarget* **2019**, *10*, 1812–1828. [CrossRef] [PubMed]

139. Chen, W.; Wang, W.; Chen, L.; Chen, J.; Lu, X.; Li, Z.; Wu, B.; Yin, L.; Guan, Y.-Q. Long-term G 1 cell cycle arrest in cervical cancer cells induced by co-immobilized TNF-α plus IFN-γ polymeric drugs. *J. Mater. Chem. B* **2018**, *6*, 327–336. [CrossRef]

140. Kortylewski, M.; Komyod, W.; Kauffmann, M.E.; Bosserhoff, A.; Heinrich, P.C.; Behrmann, I. Interferon-γ-Mediated Growth Regulation of Melanoma Cells: Involvement of STAT1-Dependent and STAT1-Independent Signals. *J. Investig. Dermatol.* **2004**, *122*, 414–422. [CrossRef] [PubMed]

141. Schmitt, M.J.; Philippidou, D.; Reinsbach, S.E.; Margue, C.; Wienecke-Baldacchino, A.; Nashan, D.; Behrmann, I.; Kreis, S. Interferon-γ-induced activation of Signal Transducer and Activator of Transcription 1 (STAT1) up-regulates the tumor suppressing microRNA-29 family in melanoma cells. *Cell Commun. Signal.* **2012**, *10*, 1–14. [CrossRef] [PubMed]

142. Search of: NCT01082887—List Results—ClinicalTrials.gov. Available online: https://clinicaltrials.gov/ct2/results?cond=&term=NCT01082887&cntry=&state=&city=&dist= (accessed on 22 August 2020).

143. Huang, Y.; Yuan, J.; Righi, E.; Kamoun, W.S.; Ancukiewicz, M.; Nezivar, J.; Santosuosso, M.; Martin, J.D.; Martin, M.R.; Vianello, F.; et al. Vascular normalizing doses of antiangiogenic treatment reprogram the immunosuppressive tumor microenvironment and enhance immunotherapy. *Proc. Natl. Acad. Sci. USA* **2012**, *109*, 17561–17566. [CrossRef]

144. Grossniklaus, H.E. Progression of ocular melanoma metastasis to the liver: The 2012 Zimmerman lecture. *JAMA Ophthalmol.* **2013**, *131*, 462–469. [CrossRef]

145. Grossniklaus, H.E.; Zhang, Q.; You, S.; McCarthy, C.; Heegaard, S.; Coupland, S.E. Metastatic ocular melanoma to the liver exhibits infiltrative and nodular growth patterns. *Hum. Pathol.* **2016**, *57*, 165–175. [CrossRef] [PubMed]

146. Angi, M.; Kalirai, H.; Prendergast, S.; Simpson, D.; Hammond, D.E.; Madigan, M.C.; Beynon, R.J.; Coupland, S.E. In-depth proteomic profiling of the uveal melanoma secretome. *Oncotarget* **2016**, *7*, 49623–49635. [CrossRef]

147. Almog, N.; Ma, L.; Raychowdhury, R.; Schwager, C.; Erber, R.; Short, S.; Hlatky, L.; Vajkoczy, P.; Huber, P.E.; Folkman, J.; et al. Transcriptional Switch of Dormant Tumors to Fast-Growing Angiogenic Phenotype. *Cancer Res.* **2009**, *69*, 836–880. [CrossRef]

148. Lee, J.H.; Miele, M.E.; Hicks, D.J.; Phillips, K.K.; Trent, J.M.; Weissman, B.E.; Welch, D.R. KiSS-1, a novel human malignant melanoma metastasis-suppressor gene. *J. Natl. Cancer Inst.* **1996**, *88*, 1731–1737. [CrossRef]

149. Li, J.; Zhou, J.; Chen, G.; Wang, H.; Wang, S.; Xing, H.; Gao, Q.; Lu, Y.; He, Y.; Ma, D. Inhibition of ovarian cancer metastasis by adeno-associated virus-mediated gene transfer of nm23H1 in an orthotopic implantation model. *Cancer Gene Ther.* **2006**, *13*, 266–272. [CrossRef]

150. Liu, F.; Qi, H.-L.; Chen, H.-L. Effects of all-trans retinoic acid and epidermal growth factor on the expression of nm23-H1 in human hepatocarcinoma cells. *J. Cancer Res. Clin. Oncol.* **2000**, *126*, 85–90.

151. Search of: NCT03572387—List Results—ClinicalTrials.gov. Available online: https://clinicaltrials.gov/ct2/results?cond=&term=NCT03572387&cntry=&state=&city=&dist= (accessed on 22 August 2020).

152. Jiang, Y.; Berk, M.; Singh, L.S.; Tan, H.; Yin, L.; Powell, C.T.; Xu, Y. KiSS1 suppresses metastasis in human ovarian cancer via inhibition of protein kinase C alpha. *Clin. Exp. Metastasis* **2005**, *22*, 369–376. [CrossRef] [PubMed]

153. Takino, T.; Koshikawa, N.; Miyamori, H.; Tanaka, M.; Sasaki, T.; Okada, Y.; Seiki, M.; Sato, H. Cleavage of metastasis suppressor gene product KiSS-1 protein/metastin by matrix metalloproteinases. *Oncogene* **2003**, *22*, 4617–4626. [CrossRef] [PubMed]

154. Theodorescu, D.; Sapinoso, L.M.; Conaway, M.R.; Oxford, G.; Hampton, G.M.; Frierson, H.F. Reduced Expression of Metastasis Suppressor RhoGDI2 Is Associated with Decreased Survival for Patients with Bladder Cancer. *Clin. Cancer Res.* **2004**, *10*, 3800–3806. [CrossRef] [PubMed]

155. Drake, J.M.; Danke, J.R.; Henry, M.D. Bone-specific growth inhibition of prostate cancer metastasis by atrasentan. *Cancer Biol. Ther.* **2010**, *9*, 607–614. [CrossRef]

156. Armstrong, A.J.; Creel, P.; Turnbull, J.; Moore, C.; Jaffe, T.A.; Haley, S.; Petros, W.; Yenser, S.; Gockerman, J.P.; Sleep, D.; et al. A Phase I-II Study of Docetaxel and Atrasentan in Men with Castration-Resistant Metastatic Prostate Cancer. *Clin. Cancer Res.* **2008**, *14*, 6270–6276. [CrossRef]

157. Quinn, D.I.; Tangen, C.M.; Hussain, M.; Lara, P.N.; Goldkorn, A.; Moinpour, C.M.; Garzotto, M.G.; Mack, P.C.; Carducci, M.A.; Monk, J.P.; et al. Docetaxel and atrasentan versus docetaxel and placebo for men with advanced castration-resistant prostate cancer (SWOG S0421): A randomised phase 3 trial. *Lancet Oncol.* **2013**, *14*, 893–900. [CrossRef]

158. Witteveen, P.; van der Mijn, K.; Neoplasia, M.L.; Los, M.; Kronemeijer, R.H.; Groenewegen, G.; Voest, E.E. Phase 1/2 study of atrasentan combined with pegylated liposomal doxorubicin in platinum-resistant recurrent ovarian cancer. *Neoplasia* **2010**, *12*, 941-IN20. [CrossRef]

159. Carducci, M.A.; Manola, J.; Nair, S.; Liu, G.; Rousey, S.; Dutcher, J.P.; Wilding, G. Atrasentan in Patients With Advanced Renal Cell Carcinoma: A Phase 2 Trial of the ECOG-ACRIN Cancer Research Group (E6800). *Clin. Genitourin. Cancer* **2015**, *13*, 531–539.e1. [CrossRef]

160. Santini, V.; Gozzini, A.; Ferrari, G. Histone deacetylase inhibitors: Molecular and biological activity as a premise to clinical application. *Current Drug Metab.* **2007**, *8*, 383–394. [CrossRef]

161. Baradaran, P.C.; Kozovska, Z.; Furdova, A.; Smolková, B. Targeting Epigenetic Modifications in Uveal Melanoma. *Int. J. Mol. Sci.* **2020**, *21*, 5314. [CrossRef]

162. Van Der Kooij, M.K.; Speetjens, F.M.; Van Der Burg, S.H.; Kapiteijn, E. Uveal Versus Cutaneous Melanoma; Same Origin, Very Distinct Tumor Types. *Cancers* **2019**, *11*, 845. [CrossRef] [PubMed]

163. Homsi, J.; Bedikian, A.; Papadopoulos, N.E.; Kim, K.B.; Hwu, W.-J.; Mahoney, S.L.; Hwu, P. Phase 2 open-label study of weekly docosahexaenoic acid–paclitaxel in patients with metastatic uveal melanoma. *Melanoma Res.* **2010**, *20*, 507–510. [CrossRef] [PubMed]

164. Kivelä, T.T.; Suciu, S.; Hansson, J.; Kruit, W.H.J.; Vuoristo, M.-S.; Kloke, O.; Gore, M.; Hahka-Kemppinen, M.; Parvinen, L.-M.; Kumpulainen, E.; et al. Bleomycin, vincristine, lomustine and dacarbazine (BOLD) in combination with recombinant interferon alpha-2b for metastatic uveal melanoma. *Eur. J. Cancer* **2003**, *39*, 1115–1120. [CrossRef]

165. Schmittel, A.; Scheulen, M.E.; Bechrakis, N.E.; Strumberg, D.; Baumgart, J.; Bornfeld, N.; Foerster, M.H.; Thiel, E.; Keilholz, U. Phase II trial of cisplatin, gemcitabine and treosulfan in patients with metastatic uveal melanoma. *Melanoma Res.* **2005**, *15*, 205–207. [CrossRef] [PubMed]

166. Leyvraz, S.; Piperno-Neumann, S. Hepatic intra-arterial versus intravenous fotemustine in patients with liver metastases from uveal melanoma (EORTC 18021): A multicentric randomized. *Ann. Oncol.* **2014**, *25*, 742–746. [CrossRef]

167. Pingpank, J.F.; Hughes, M.S.; Alexander, H.R.; Faries, M.B.; Zager, J.S.; Royal, R.; Whitman, E.D.; Nutting, C.W.; Siskin, G.P.; Agarwala, S.S. A phase III random assignment trial comparing percutaneous hepatic perfusion with melphalan (PHP-mel) to standard of care for patients with hepatic metastases from metastatic ocular or cutaneous melanoma. *J. Clin. Oncol.* **2010**, *28*, LBA8512. [CrossRef]

168. Gonsalves, C.F.; Eschelman, D.J.; Sullivan, K.L.; Anne, P.R.; Doyle, L.; Sato, T. Radioembolization as Salvage Therapy for Hepatic Metastasis of Uveal Melanoma: A Single-Institution Experience. *Am. J. Roentgenol.* **2011**, *196*, 468–473. [CrossRef]

169. Klingenstein, A.; Haug, A.; Zech, C.J.; Schaller, U.C. Radioembolization as Locoregional Therapy of Hepatic Metastases in Uveal Melanoma Patients. *Cardiovasc. Interv. Radiol.* **2012**, *36*, 158–165. [CrossRef]

170. Bedikian, A.; Papadopoulos, N.; Plager, C.; Eton, O.; Ring, S. Phase II evaluation of temozolomide in metastatic choroidal melanoma. *Melanoma Res.* **2003**, *13*, 303–306. [CrossRef]

171. Spagnolo, F.; Grosso, M.; Picasso, V.; Tornari, E.; Pesce, M.; Queirolo, P. Treatment of metastatic uveal melanoma with intravenous fotemustine. *Melanoma Res.* **2013**, *23*, 196–198. [CrossRef]

172. O'Neill, P.; Butt, M.; Eswar, C.; Gillis, P.; Marshall, E. A prospective single arm phase II study of dacarbazine and treosulfan as first-line therapy in metastatic uveal melanoma. *Melanoma Res.* **2006**, *16*, 245–248. [CrossRef] [PubMed]

173. Schinzari, G.; Rossi, E.; Cassano, A.; Dadduzio, V.; Quirino, M.; Pagliara, M.; Blasi, M.A.; Barone, C. Cisplatin, dacarbazine and vinblastine as first line chemotherapy for liver metastatic uveal melanoma in the era of immunotherapy: A single institution phase II study. *Melanoma Res.* **2017**, *27*, 591–595. [CrossRef] [PubMed]

174. Khoja, L.; Atenafu, E.G.; Suciu, S.; Leyvraz, S.; Sato, T.; Marshall, E.; Keilholz, U.; Zimmer, L.; Patel, S.; Piperno-Neumann, S.; et al. Meta-analysis in metastatic uveal melanoma to determine progression free and overall survival benchmarks: An international rare cancers initiative (IRCI) ocular melanoma study. *Ann. Oncol.* **2019**, *30*, 1370–1380. [CrossRef] [PubMed]

175. Mariani, P.; Piperno-Neumann, S.; Servois, V.; Berry, M.; Dorval, T.; Plancher, C.; Couturier, J.; Levy-Gabriel, C.; Rouic, L.L.-L.; Desjardins, L.; et al. Surgical management of liver metastases from uveal melanoma: 16 years' experience at the Institut Curie. *Eur. J. Surg. Oncol. (EJSO)* **2009**, *35*, 1192–1197. [CrossRef] [PubMed]

176. Frenkel, S.; Nir, I.; Hendler, K.; Lotem, M.; Eid, A.; Jurim, O.; Pe'Er, J. Long-term survival of uveal melanoma patients after surgery for liver metastases. *Br. J. Ophthalmol.* **2009**, *93*, 1042–1046. [CrossRef]

177. Akyuz, M.; Yazici, P.; Dural, A.C.; Yigitbas, H.; Okoh, A.; Bucak, E.; McNamara, M.; Singh, A.; Berber, E. Laparoscopic management of liver metastases from uveal melanoma. *Surg. Endosc.* **2015**, *30*, 2567–2571. [CrossRef]

178. Rivoire, M.; Kodjikian, L.; Baldo, S.; Kaemmerlen, P.; Négrier, S.; Grange, J.-D. Treatment of Liver Metastases From Uveal Melanoma. *Ann. Surg. Oncol.* **2005**, *12*, 422–428. [CrossRef]

179. Augsburger, J.J.; Correa, Z.M.; Shaikh, A.H. Effectiveness of Treatments for Metastatic Uveal Melanoma. *Am. J. Ophthalmol.* **2009**, *148*, 119–127. [CrossRef]

180. Ripley, R.T.; Davis, J.L.; Klapper, J.A.; Mathur, A.; Kammula, U.; Royal, R.E.; Yang, J.C.; Sherry, R.M.; Hughes, M.S.; Libutti, S.K.; et al. Liver Resection for Metastatic Melanoma with Postoperative Tumor-Infiltrating Lymphocyte Therapy. *Ann. Surg. Oncol.* **2009**, *17*, 163–170. [CrossRef]

181. Salmon, R.; Levy, C.; Plancher, C.; Dorval, T.; Desjardins, L.; Leyvrazi, S.; Pouillart, P.; Schlienger, P.; Servois, V.; Asselain, B. Treatment of liver metastases from uveal melanoma by combined surgery—chemotherapy. *Eur. J. Surg. Oncol. (EJSO)* **1998**, *24*, 127–130. [CrossRef]

182. Aoyama, T.; Mastrangelo, M.J.; Berd, D.; Nathan, F.E.; Shields, C.L.; Shields, J.A.; Rosato, E.L.; Rosato, F.E.; Sato, T. Protracted survival after resection of metastatic uveal melanoma. *Cancer* **2000**, *89*, 1561–1568. [CrossRef]

183. Pawlik, T.M.; Zorzi, D.; Abdalla, E.K.; Clary, B. Hepatic Resection for Metastatic Melanoma: Distinct Patterns of Recurrence and Prognosis for Ocular Versus Cutaneous Disease Endomicroscopy View project geografical disparities in liver transplant View project. *Ann. Surg. Oncol.* **2006**, *13*, 712–720. [CrossRef] [PubMed]

184. Adam, R.; Chiche, L.; Aloia, T.; Elias, D.; Salmon, R.; Rivoire, M.; Jaeck, D.; Saric, J.; Le Treut, Y.P.; Belghiti, J.; et al. Hepatic resection for noncolorectal nonendocrine liver metastases: Analysis of 1452 patients and development of a prognostic model. *Ann. Surg.* **2006**, *244*, 524. [CrossRef]

185. de Ridder, J.A.M. Liver Metastases Incidence, Treatment & Prognostic Factors. Ph.D. Thesis, Radboud University of Nijmegen, Nijmegen, The Netherlands, 19 May 2017.

186. Groeschl, R.T.; Nachmany, I.; Steel, J.L.; Reddy, S.K.; Glazer, E.S.; De Jong, M.C.; Pawlik, T.M.; Geller, D.A.; Tsung, A.; Marsh, J.W.; et al. Hepatectomy for Noncolorectal Non-Neuroendocrine Metastatic Cancer: A Multi-Institutional Analysis. *J. Am. Coll. Surg.* **2012**, *214*, 769–777. [CrossRef]

187. Mariani, P.; Almubarak, M.M.; Kollen, M.; Wagner, M.; Plancher, C.; Audollent, R.; Piperno-Neumann, S.; Cassoux, N.; Servois, V. Radiofrequency ablation and surgical resection of liver metastases from uveal melanoma. *Eur. J. Surg. Oncol. (EJSO)* **2016**, *42*, 706–712. [CrossRef] [PubMed]

188. Derek, E.; Matsuoka, L.; Alexopoulos, S.; Fedenko, A.A.; Genyk, Y.; Selby, R. Combined surgical resection and radiofrequency ablation as treatment for metastatic ocular melanoma. *Surg. Today* **2012**, *43*, 367–371. [CrossRef] [PubMed]

189. Servois, V.; Bouhadiba, T.; Dureau, S.; Da Costa, D.; Almubarak, M.M.; Foucher, R.; Savignoni, A.; Cassoux, N.; Pierron, G.; Mariani, P. Iterative treatment with surgery and radiofrequency ablation of uveal melanoma liver metastasis: Retrospective analysis of a series of very long-term survivors. *Eur. J. Surg. Oncol.* **2019**, *45*, 1717–1722. [CrossRef]

190. Olofsson Bagge, R.; Cahlin, C.; All-Ericsson, C.; Hashimi, F. Isolated Hepatic Perfusion for Ocular Melanoma Metastasis: Registry Data Suggests a Survival Benefit Reducing radiation-induced side effects in organs at risk View project PDX v2.0 View project. *Artic. Ann. Surg. Oncol.* **2013**, *21*, 466–472. [CrossRef]

191. Search of: NCT01785316—List Results—ClinicalTrials.gov. Available online: https://clinicaltrials.gov/ct2/results?cond=&term=NCT01785316&cntry=&state=&city=&dist= (accessed on 22 August 2020).

192. Search of: NCT02678572—List Results—ClinicalTrials.gov. Available online: https://clinicaltrials.gov/ct2/results?cond=&term=NCT02678572&cntry=&state=&city=&dist= (accessed on 22 August 2020).

193. Bedikian, A.Y.; Legha, S.S.; Mavligit, G.; Carrasco, C.H.; Khorana, S.; Plager, C.; Papadopoulos, N.; Benjamin, R.S. Treatment of uveal melanoma metastatic to the liver. A review of the M. D. Anderson cancer center experience and prognostic factors. *Cancer* **1995**, *76*, 1665–1670. [CrossRef]

194. Feun, L.G.; Reddy, K.R.; Yrizarry, J.M.; Savaraj, N.; Guerra, J.J.; Purser, R.K.; Waldman, S.; Levi, J.U.; Moffatt, F.; Morrell, L.; et al. A Phase I Study of Chemoembolization with Cisplatin and Lipiodol for Primary and Metastatic Liver Cancer. *Am. J. Clin. Oncol.* **1994**, *17*, 405–410. [CrossRef]

195. Mavligit, G.; Charnsangavej, C.; Carrasco, C.; Jama, Y.P.; Benjamin, R.S.; Wallace, S. Regression of ocular melanoma metastatic to the liver after hepatic arterial chemoembolization with cisplatin and polyvinyl sponge. *JAMA* **1988**, *260*, 974–976. [CrossRef] [PubMed]

196. Patel, S.P.; Kim, K.B.; Papadopoulos, N.E.; Hwu, W.-J.; Hwu, P.; Prieto, V.G.; Bar-Eli, M.; Zigler, M.; Dobroff, A.; Bronstein, Y.; et al. A phase II study of gefitinib in patients with metastatic melanoma. *Melanoma Res.* **2011**, *21*, 357–363. [CrossRef] [PubMed]

197. Gupta, S.; Bedikian, A.; Ahrar, J.; Ensor, J.; Ahrar, K.; Madoff, D.C.; Wallace, M.J.; Murthy, R.; Tam, A.; Hwu, P. Hepatic artery chemoembolization in patients with ocular melanoma metastatic to the liver: Response, survival, and prognostic factors. *Am. J. Clin. Oncol.* **2010**, *33*, 474–480. [CrossRef] [PubMed]

198. Vogl, T.J.; Eichler, K.; Zangos, S.; Herzog, C.; Hammerstingl, R.; Balzer, J.; Gholami, A. Preliminary experience with transarterial chemoembolization (TACE) in liver metastases of uveal malignant melanoma: Local tumor control and survival. *J. Cancer Res. Clin. Oncol.* **2006**, *133*, 177–184. [CrossRef] [PubMed]

199. Huppert, P.; Fierlbeck, G.; Pereira, P.L.; Schanz, S.; Duda, S.; Wietholtz, H.; Rozeik, C.; Claussen, C.D. Transarterial chemoembolization of liver metastases in patients with uveal melanoma. *Eur. J. Radiol.* **2010**, *74*, e38–e44. [CrossRef] [PubMed]

200. Agarwala, S.; Panikkar, R.; Kirkwood, J.M. Phase I/II randomized trial of intrahepatic arterial infusion chemotherapy with cisplatin and chemoembolization with cisplatin and polyvinyl sponge in patients with. *Melanoma Res.* **2004**, *14*, 217–222. [CrossRef] [PubMed]

201. Gonsalves, C.F.; Eschelman, D.J.; Adamo, R.D.; Anne, P.R.; Orloff, M.M.; Terai, M.; Hage, A.N.; Yi, M.; Chervoneva, I.; Sato, T. A Prospective Phase II Trial of Radioembolization for Treatment of Uveal Melanoma Hepatic Metastasis. *Radiology* **2019**, *293*, 223–231. [CrossRef]

202. Arulananda, S.; Parakh, S.; Palmer, J.; Goodwin, M.; Andrews, M.C.; Cebon, J. A pilot study of intrahepatic yttrium-90 microsphere radioembolization in combination with intravenous cisplatin for uveal melanoma liver-only metastases. *Cancer Rep.* **2019**, *2*, e1183. [CrossRef]

203. Search of: NCT01893099—List Results—ClinicalTrials.gov. Available online: https://clinicaltrials.gov/ct2/results?cond=&term=NCT01893099+&cntry=&state=&city=&dist= (accessed on 22 August 2020).

204. Valsecchi, M.; Terai, M.; Eschelman, D.J.; Gonsalves, C.F.; Chervoneva, I.; Shields, J.A.P.; Shields, C.L.; Yamamoto, A.; Sullivan, K.L.; Laudadio, M.; et al. Double-Blinded, Randomized Phase II Study Using Embolization with or without Granulocyte–Macrophage Colony-Stimulating Factor in Uveal Melanoma with. *J. Vasc. Interv. Radiol.* **2015**, *26*, 523–532. [CrossRef]

205. Hodi, F.S.; O'Day, S.J.; McDermott, D.F.; Weber, R.W.; Sosman, J.A.; Haanen, J.B.; Gonzalez, R.; Robert, C.; Schadendorf, D.; Hassel, J.C.; et al. Improved Survival with Ipilimumab in Patients with Metastatic Melanoma. *N. Engl. J. Med.* **2010**, *363*, 711–723. [CrossRef]

206. Garon, E.B.; Rizvi, N.A.; Hui, R.; Leighl, N.; Balmanoukian, A.S.; Eder, J.P.; Patnaik, A.; Aggarwal, C.; Gubens, M.; Horn, L.; et al. Pembrolizumab for the Treatment of Non–Small-Cell Lung Cancer. *N. Engl. J. Med.* **2015**, *372*, 2018–2028. [CrossRef] [PubMed]

207. Brahmer, J.; Reckamp, K.L.; Baas, P.; Crino, L.; Eberhardt, W.E.; Poddubskaya, E.; Antonia, S.; Pluzanski, A.; Vokes, E.E.; Holgado, E.; et al. Nivolumab versus Docetaxel in Advanced Squamous-Cell Non–Small-Cell Lung Cancer. *N. Engl. J. Med.* **2015**, *373*, 123–135. [CrossRef] [PubMed]

208. Zimmer, L.; Vaubel, J.; Mohr, P.; Hauschild, A.; Utikal, J.; Šimon, J.; Garbe, C.; Herbst, R.; Enk, A.; Kämpgen, E.; et al. Phase II DeCOG-Study of Ipilimumab in Pretreated and Treatment-Naïve Patients with Metastatic Uveal Melanoma. *PLoS ONE* **2015**, *10*, e0118564. [CrossRef] [PubMed]

209. Rodriguez, J.M.P.; De Olza, M.O.; Codes, M.; Lopez-Martin, J.A.; Berrocal, A.; García, M.; Gurpide, A.; Homet, B.; Martin-Algarra, S. Phase II study evaluating ipilimumab as a single agent in the first-line treatment of adult patients (Pts) with metastatic uveal melanoma (MUM): The GEM-1 trial. *J. Clin. Oncol.* **2014**, *32*, 9033. [CrossRef]

210. Joshua, A.M.; Monzon, J.G.; Mihalcioiu, C.; Hogg, D.; Smylie, M.; Cheng, T. A phase 2 study of tremelimumab in patients with advanced uveal melanoma. *Melanoma Res.* **2015**, *25*, 342–347. [CrossRef]

211. Algazi, A.P.; Tsai, K.K.; Shoushtari, A.N.; Munhoz, R.R.; Eroglu, Z.; Piulats, J.M.; Ott, P.A.; Johnson, D.B.; Hwang, J.; Daud, A.I.; et al. Clinical outcomes in metastatic uveal melanoma treated with PD-1 and PD-L1 antibodies. *Cancer* **2016**, *122*, 3344–3353. [CrossRef]

212. Mignard, C.; Huvier, A.D.; Gillibert, A.; Modeste, A.B.D.; Dutriaux, C.; Khammari, A.; Avril, M.-F.; Kramkimel, N.; Mortier, L.; Marcant, P.; et al. Efficacy of Immunotherapy in Patients with Metastatic Mucosal or Uveal Melanoma. *J. Oncol.* **2018**, *2018*, 1–9. [CrossRef]

213. Johnson, D.B.; Bao, R.; Ancell, K.K.; Daniels, A.B.; Wallace, D.; Sosman, J.A.; Luke, J.J. Response to Anti–PD-1 in Uveal Melanoma Without High-Volume Liver Metastasis. *J. Natl. Compr. Cancer Netw.* **2019**, *17*, 114–117. [CrossRef]

214. Rodrigues, M.; Mobuchon, L.; Houy, A.; Fiévet, A.; Gardrat, S.; Barnhill, R.L.; Popova, T.; Servois, V.; Rampanou, A.; Mouton, A.; et al. Outlier response to anti-PD1 in uveal melanoma reveals germline MBD4 mutations in hypermutated tumors. *Nat. Commun.* **2018**, *9*, 1–6. [CrossRef]

215. Search of: NCT02626962—List Results—ClinicalTrials.gov. Available online: https://clinicaltrials.gov/ct2/results?cond=&term=NCT02626962+&cntry=&state=&city=&dist= (accessed on 22 August 2020).

216. Search of: NCT01585194—List Results—ClinicalTrials.gov. Available online: https://clinicaltrials.gov/ct2/results?cond=&term=NCT01585194&cntry=&state=&city=&dist= (accessed on 22 August 2020).

217. Pelster, M.; Gruschkus, S.K.; Bassett, R.; Gombos, D.S.; Shephard, M.; Posada, L.; Glover, M.; Diab, A.; Hwu, P.; Patel, S.P. Phase II study of ipilimumab and nivolumab (ipi/nivo) in metastatic uveal melanoma (UM). *J. Clin. Oncol.* **2019**, *37*, 9522. [CrossRef]

218. Search of: NCT02697630—List Results—ClinicalTrials.gov. Available online: https://clinicaltrials.gov/ct2/results?cond=&term=NCT02697630&cntry=&state=&city=&dist= (accessed on 11 September 2020).

219. Jespersen, H.; Bagge, R.O.; Ullenhag, G.; Carneiro, A.; Helgadottir, H.; Ljuslinder, I.; Levin, M.; All-Eriksson, C.; Andersson, B.; Stierner, U.; et al. Concomitant use of pembrolizumab and entinostat in adult patients with metastatic uveal melanoma (PEMDAC study): Protocol for a multicenter phase II open label study. *BMC Cancer* **2019**, *19*, 415. [CrossRef] [PubMed]

220. Rozeman, E.A.; Prevoo, W.; Meier, M.A.; Sikorska, K.; Van, T.M.; Van De Wiel, B.A.; Van Der Wal, J.E.; Mallo, H.A.; Grijpink-Ongering, L.G.; Broeks, A.; et al. Phase Ib/II trial testing combined radiofrequency ablation and ipilimumab in uveal melanoma (SECIRA-UM). *Melanoma Res.* **2020**, *30*, 252–260. [CrossRef] [PubMed]

221. Search of: NCT04283890—List Results—ClinicalTrials.gov. Available online: https://clinicaltrials.gov/ct2/results?cond=&term=NCT04283890+&cntry=&state=&city=&dist= (accessed on 22 August 2020).

222. Search of: NCT03472586—List Results—ClinicalTrials.gov. Available online: https://clinicaltrials.gov/ct2/results?cond=&term=NCT03472586&cntry=&state=&city=&dist= (accessed on 22 August 2020).

223. Search of: NCT02913417—List Results—ClinicalTrials.gov. Available online: https://clinicaltrials.gov/ct2/results?cond=&term=NCT02913417&cntry=&state=&city=&dist= (accessed on 22 August 2020).

224. Chandran, S.; Somerville, R.; Yang, J.; Sherry, R.M.; Klebanoff, C.A.; Goff, S.L.; Wunderlich, J.R.; Danforth, D.N.; Zlott, D.; Paria, B.C. Treatment of metastatic uveal melanoma with adoptive transfer of tumour-infiltrating lymphocytes: A single-centre, two-stage, single-arm, phase 2 study. *Lancet Oncol.* **2017**, *18*, 792–802. [CrossRef]

225. van Loenen, M.M.; de Boer, R.; Hagedoorn, R.S.; Jankipersadsing, V.; Amir, A.L.; Falkenburg, J.H.F.; Heemskerk, M.H.M. Multi-cistronic vector encoding optimized safety switch for adoptive therapy with T-cell receptor-modified T cells. *Gene Ther.* **2013**, *20*, 861–867. [CrossRef] [PubMed]

226. Forsberg, E.M.V.; Lindberg, M.F.; Jespersen, H.; Alsén, S.; Bagge, R.O.; Donia, M.; Svane, I.M.; Nilsson, O.; Ny, L.; Nilsson, L.M.; et al. HER2 CAR-T Cells Eradicate Uveal Melanoma and T-cell Therapy–Resistant Human Melanoma in IL2 Transgenic NOD/SCID IL2 Receptor Knockout Mice. *Cancer Res.* **2019**, *79*, 899–904. [CrossRef]

227. Tavera, R.J.; Forget, M.A.; Kim, Y.U.; Sakellariou-Thompson, D.; Creasy, C.A.; Bhatta, A.; Fulbright, O.J.; Ramachandran, R.; Thorsen, S.T.; Flores, E.; et al. Utilizing T-cell Activation Signals 1, 2, and 3 for Tumor-infiltrating Lymphocytes (TIL) Expansion: The Advantage over the Sole Use of Interleukin-2 in Cutaneous and Uveal Melanoma. *J. Immunother.* **2018**, *41*, 399–405. [CrossRef]

228. Search of: NCT03467516—List Results—ClinicalTrials.gov. Available online: https://clinicaltrials.gov/ct2/results?cond=&term=NCT03467516&cntry=&state=&city=&dist= (accessed on 22 August 2020).

229. Search of: NCT03068624—List Results—ClinicalTrials.gov. Available online: https://clinicaltrials.gov/ct2/results?cond=&term=NCT03068624&cntry=&state=&city=&dist= (accessed on 22 August 2020).

230. Damato, B.; Dukes, J.; Goodall, H.; Carvajal, R.D. Tebentafusp: T Cell Redirection for the Treatment of Metastatic Uveal Melanoma. *Cancers* **2019**, *11*, 971. [CrossRef]

231. Middleton, M.; Steven, N.M.; Evans, T.J.; Infante, J.R.; Sznol, M.; Mulatero, C.; Hamid, O.; Shoushtari, A.N.; Shingler, W.; Johnson, A.; et al. Safety, pharmacokinetics and efficacy of IMCgp100, a first-in-class soluble TCR-antiCD3 bispecific t cell redirector with solid tumour activity: Results from the FIH study in melanoma. *J. Clin. Oncol.* **2016**, *34*, 3016. [CrossRef]

232. Search of: NCT02570308—List Results—ClinicalTrials.gov. Available online: https://clinicaltrials.gov/ct2/results?cond=&term=NCT02570308&cntry=&state=&city=&dist= (accessed on 22 August 2020).

233. Search of: NCT03070392—List Results—ClinicalTrials.gov. Available online: https://clinicaltrials.gov/ct2/results?cond=&term=NCT03070392&cntry=&state=&city=&dist= (accessed on 22 August 2020).

234. Search of: NCT03635632—List Results—ClinicalTrials.gov. Available online: https://clinicaltrials.gov/ct2/results?cond=&term=NCT03635632&cntry=&state=&city=&dist= (accessed on 22 August 2020).

235. Bol, K.F.; Mensink, H.W.; Aarntzen, E.H.; Schreibelt, G.; Keunen, J.E.; Coulie, P.G.; De Klein, A.; Punt, C.J.; Paridaens, D.; Figdor, C.G.; et al. Long Overall Survival After Dendritic Cell Vaccination in Metastatic Uveal Melanoma Patients. *Am. J. Ophthalmol.* **2014**, *158*, 939–947. [CrossRef]

236. Sarnaik, A.A.; Yu, B.; Yu, D.; Morelli, D.; Hall, M.; Bogle, D.; Yan, L.; Targan, S.; Solomon, J.; Nichol, G.; et al. Extended Dose Ipilimumab with a Peptide Vaccine: Immune Correlates Associated with Clinical Benefit in Patients with Resected High-Risk Stage IIIc/IV Melanoma. *Clin. Cancer Res.* **2010**, *17*, 896–906. [CrossRef]

237. Lesterhuis, W.J.; Schreibelt, G.; Scharenborg, N.M.; Brouwer, H.M.H.; Gerritsen, M.P.; Croockewit, S.; Coulie, P.G.; Torensma, R.; Adema, G.J.; Figdor, C.G.; et al. Wild-type and modified gp100 peptide-pulsed dendritic cell vaccination of advanced melanoma patients can lead to long-term clinical responses independent of the peptide used. *Cancer Immunol. Immunother.* **2011**, *60*, 249–260. [CrossRef] [PubMed]

238. Bosch, J.J.; Iheagwara, U.K.; Reid, S.; Srivastava, M.K.; Wolf, J.; Lotem, M.; Ksander, B.R.; Ostrand-Rosenberg, S. Uveal melanoma cell-based vaccines express MHC II molecules that traffic via the endocytic and secretory pathways and activate CD8+ cytotoxic, tumor-specific T cells. *Cancer Immunol. Immunother.* **2009**, *59*, 103–112. [CrossRef] [PubMed]

239. Thompson, J.F.; Agarwala, S.S.; Smithers, B.; Ross, M.I.; Scoggins, C.R.; Coventry, B.J.; Neuhaus, S.J.; Minor, D.R.; Singer, J.M.; Wachter, E.A. Phase 2 Study of Intralesional PV-10 in Refractory Metastatic Melanoma. *Ann. Surg. Oncol.* **2014**, *22*, 2135–2142. [CrossRef] [PubMed]

240. Search of: NCT00219843—List Results—ClinicalTrials.gov. Available online: https://clinicaltrials.gov/ct2/results?cond=&term=NCT00219843&cntry=&state=&city=&dist= (accessed on 22 August 2020).

241. Van Raamsdonk, C.D.; Bezrookove, V.; Green, G.; Bauer, J.; Gaugler, L.; O'Brien, J.M.; Simpson, E.M.; Barsh, G.S.; Bastian, B.C. Frequent somatic mutations of GNAQ in uveal melanoma and blue naevi. *Nature* **2008**, *457*, 599–602. [CrossRef]

242. Van Raamsdonk, C.D.; Griewank, K.; Crosby, M.B.; Garrido, M.C.; Vemula, S.; Wiesner, T.; Obenauf, A.C.; Wackernagel, W.; Green, G.; Bouvier, N.; et al. Mutations inGNA11in Uveal Melanoma. *N. Engl. J. Med.* **2010**, *363*, 2191–2199. [CrossRef]

243. Robert, C.; Grob, J.J.; Stroyakovskiy, D.; Karaszewska, B.; Hauschild, A.; Levchenko, E.; Chiarion-Sileni, V.; Schachter, J.; Garbe, C.; Bondarenko, I.; et al. Five-Year Outcomes with Dabrafenib plus Trametinib in Metastatic Melanoma. *N. Engl. J. Med.* **2019**, *381*, 626–636. [CrossRef]

244. Ascierto, P.A.; Dummer, R.; Gogas, H.J.; Flaherty, K.T.; Arance, A.; Mandala, M.; Liszkay, G.; Garbe, C.; Schadendorf, D.; Krajsova, I.; et al. Update on tolerability and overall survival in COLUMBUS: Landmark analysis of a randomised phase 3 trial of encorafenib plus binimetinib vs vemurafenib or encorafenib in patients with BRAF V600-mutant melanoma. *Eur. J. Cancer* **2020**, *126*, 33–44. [CrossRef]

245. Onken, M.; Makepeace, C. Targeting nucleotide exchange to inhibit constitutively active G protein α subunits in cancer cells. *Sci. Signal.* **2018**, *11*. [CrossRef]

246. Lapadula, D.; Farias, E.; Randolph, C.E.; Purwin, T.J.; McGrath, D.; Charpentier, T.H.; Zhang, L.; Wu, S.; Terai, M.; Sato, T.; et al. Effects of Oncogenic Gαq and Gα11 Inhibition by FR900359 in Uveal Melanoma. *Mol. Cancer Res.* **2018**, *17*, 963–973. [CrossRef]

247. Xiong, X.-F.; Zhang, H.; Boesgaard, M.W.; Underwood, C.R.; Bräuner-Osborne, H.; Strømgaard, K. Structure–Activity Relationship Studies of the Natural Product G q/11 Protein Inhibitor YM-254890. *ChemMedChem* **2019**, *14*, 865–870. [CrossRef]

248. Li, Y.; He, J.; Qiu, C.; Shang, Q.; Qian, G.; Fan, X.; Ge, S.; Jia, R. The oncolytic virus H101 combined with GNAQ siRNA-mediated knockdown reduces uveal melanoma cell viability. *J. Cell. Biochem.* **2018**, *120*, 5766–5776. [CrossRef] [PubMed]

249. Posch, C.; Latorre, A.; Crosby, M.B.; Celli, A.; Latorre, A.; Vujic, I.; Sanlorenzo, M.; Green, G.A.; Weier, J.; Zekhtser, M.; et al. Detection of GNAQ mutations and reduction of cell viability in uveal melanoma cells with functionalized gold nanoparticles. *Biomed. Microdevices* **2015**, *17*, 15. [CrossRef] [PubMed]

250. Carvajal, R.D.; Piperno-Neumann, S.; Kapiteijn, E.; Chapman, P.B.; Frank, S.; Joshua, A.M.; Piulats, J.M.; Wolter, P.; Cocquyt, V.; Chmielowski, B. Selumetinib in combination with dacarbazine in patients with metastatic uveal melanoma: A phase III, multicentre, randomised trial (SUMIT). *J. Clin. Oncol.* **2018**, *36*, 1232–1239. [CrossRef] [PubMed]

251. Ambrosini, G.; Musi, E.; Ho, A.L.; De Stanchina, E.; Schwartz, G.K. Inhibition of Mutant GNAQ Signaling in Uveal Melanoma Induces AMPK-Dependent Autophagic Cell Death. *Mol. Cancer Ther.* **2013**, *12*, 768–776. [CrossRef]

252. Search of: NCT01979523—List Results—ClinicalTrials.gov. Available online: https://clinicaltrials.gov/ct2/results?cond=&term=NCT01979523&cntry=&state=&city=&dist= (accessed on 22 August 2020).

253. Bhatia, S.; Moon, J.; Margolin, K.A.; Weber, J.S.; Lao, C.D.; Othus, M.; Aparicio, A.M.; Ribas, A.; Sondak, V.K. Phase II Trial of Sorafenib in Combination with Carboplatin and Paclitaxel in Patients with Metastatic Uveal Melanoma: SWOG S0512. *PLoS ONE* **2012**, *7*, e48787. [CrossRef]

254. Search of: NCT02601378—List Results—ClinicalTrials.gov. Available online: https://clinicaltrials.gov/ct2/results?cond=&term=+NCT02601378&cntry=&state=&city=&dist= (accessed on 22 August 2020).

255. Zeldis, J.; Heller, C.; Seidel, G.; Yuldasheva, N.; Stirling, D.; ShutackS, Y.; Libutti, K. A randomized phase II trial comparing two doses of lenalidomide for the treatment of stage IV ocular melanoma. *J. Clin. Oncol.* **2009**, *27*, e20012. [CrossRef]

256. Luke, J.J.; Olson, D.J.; Allred, J.B.; Strand, C.A.; Bao, R.; Zha, Y.; Carll, T.; Labadie, B.W.; Bastos, B.R.; Butler, M.O.; et al. Randomized Phase II Trial and Tumor Mutational Spectrum Analysis from Cabozantinib versus Chemotherapy in Metastatic Uveal Melanoma (Alliance A091201). *Clin. Cancer Res.* **2019**, *26*, 804–811. [CrossRef]

257. Search of: NCT01413191—List Results—ClinicalTrials.gov. Available online: https://clinicaltrials.gov/ct2/results?cond=&term=NCT01413191&cntry=&state=&city=&dist= (accessed on 22 August 2020).

258. Search of: NCT01252251—List Results—ClinicalTrials.gov. Available online: https://clinicaltrials.gov/ct2/results?cond=&term=NCT01252251&cntry=&state=&city=&dist= (accessed on 22 August 2020).

259. Shoushtari, A.N.; Ong, L.T.; Schoder, H.; Singh-Kandah, S.; Abbate, K.T.; Postow, M.A.; Callahan, M.K.; Wolchok, J.; Chapman, P.B.; Panageas, K.S.; et al. A phase 2 trial of everolimus and pasireotide long-acting release in patients with metastatic uveal melanoma. *Melanoma Res.* **2016**, *26*, 272–277. [CrossRef]

260. Daud, A.I.; Kluger, H.M.; Kurzrock, R.; Schimmoller, F.; Weitzman, A.L.; Samuel, T.A.; Moussa, A.H.; Gordon, M.S.; Shapiro, G. I Phase II randomised discontinuation trial of the MET/VEGF receptor inhibitor cabozantinib in metastatic melanoma. *Br. J. Cancer* **2017**, *116*, 432–440. [CrossRef]

261. Shah, S.; Luke, J.J.; Jacene, H.A.; Chen, T.; Giobbie-Hurder, A.; Ibrahim, N.; Buchbinder, E.I.; McDermott, D.F.; Flaherty, K.T.; Sullivan, R.J.; et al. Results from phase II trial of HSP90 inhibitor, STA-9090 (ganetespib), in metastatic uveal melanoma. *Melanoma Res.* **2018**, *28*, 605–610. [CrossRef]

262. Guenterberg, K.; Grignol, V.; ROlenckielekar, K.V.; Varker, K.A.; Chen, H.X.; Kendra, K.L.; Olencki, T.L.; Carson, W.E. A pilot study of bevacizumab and interferon-α2b in ocular melanoma. *Am. J. Clin. Oncol.* **2011**, *34*, 87. [CrossRef]

263. Piperno-Neumann, S.; Servois, V.; Bidard, F.-C.; Mariani, P.; Plancher, C.; Diallo, A.; Vago-Ady, N.; Desjardins, L. BEVATEM: Phase II study of bevacizumab (B) in combination with temozolomide (T) in patients (pts) with first-line metastatic uveal melanoma (MUM): Final results. *J. Clin. Oncol.* **2013**, *31*, 9057. [CrossRef]

264. Spitler, L.E.; Boasberg, P.; Day, S.O.; Hamid, O.; Cruickshank, S.; Mesko, S.; Weber, R.W. Phase II Study of Nab-Paclitaxel and Bevacizumab as First-line Therapy for Patients with Unresectable Stage III and IV Melanoma. *Am. J. Clin. Oncol.* **2015**, *38*, 61–67. [CrossRef] [PubMed]

265. Tarhini, A.A.; Frankel, P.; Margolin, K.A.; Christensen, S.; Ruel, C.; Shipe-Spotloe, J.; Gandara, D.R.; Chen, A.; Kirkwood, J.M. Aflibercept (VEGF Trap) in Inoperable Stage III or Stage IV Melanoma of Cutaneous or Uveal Origin. *Clin. Cancer Res.* **2011**, *17*, 6574–6581. [CrossRef] [PubMed]

266. Hofmann, U.B.; Kauczok-Vetter, C.S.; Houben, R.; Becker, J.C. Overexpression of the KIT/SCF in Uveal Melanoma Does Not Translate into Clinical Efficacy of Imatinib Mesylate. *Clin. Cancer Res.* **2009**, *15*, 324–329. [CrossRef] [PubMed]

267. Penel, N.; Delcambre, C.; Durando, X.; Clisant, S.; Hebbar, M.; Négrier, S.; Fournier, C.; Isambert, N.; Mascarelli, F.; Mouriaux, F. O-Mel-Inib: A Cancéro-pôle Nord-Ouest multicenter phase II trial of high-dose Imatinib mesylate in metastatic uveal melanoma. *Investig. New Drugs* **2008**, *26*, 561–565. [CrossRef]

268. Mahipal, A.; Tijani, L.; Chan, K.; Laudadio, M.; Mastrangelo, M.J.; Sato, T. A pilot study of sunitinib malate in patients with metastatic uveal melanoma. *Melanoma Res.* **2012**, *22*, 440–446. [CrossRef]

269. Haas, N.B.; Quirt, I.; Hotte, S.; McWhirter, E.; Polintan, R.; Litwin, S.; Adams, P.D.; McBryan, T.; Wang, L.; Martin, L.P.; et al. Phase II trial of vorinostat in advanced melanoma. *Investig. New Drugs* **2014**, *32*, 526–534. [CrossRef]

270. Search of: NCT03417739—List Results—ClinicalTrials.gov. Available online: https://clinicaltrials.gov/ct2/results?cond=&term=NCT03417739&cntry=&state=&city=&dist= (accessed on 22 August 2020).

271. Search of: NCT02743611—List Results—ClinicalTrials.gov. Available online: https://clinicaltrials.gov/ct2/results?cond=&term=NCT02743611+&cntry=&state=&city=&dist= (accessed on 22 August 2020).

272. Search of: NCT01835145—List Results—ClinicalTrials.gov. Available online: https://clinicaltrials.gov/ct2/results?cond=&term=NCT01835145&cntry=&state=&city=&dist= (accessed on 22 August 2020).

 **cancers**

Review

Essential Oils and Their Main Chemical Components: The Past 20 Years of Preclinical Studies in Melanoma

Marta Di Martile [1,*], **Stefania Garzoli [2]**, **Rino Ragno [2,3]** and **Donatella Del Bufalo [1,*]**

[1] Preclinical Models and New Therapeutic Agents Unit, IRCCS Regina Elena National Cancer Institute, Via Elio Chianesi 53, 00144 Rome, Italy

[2] Department of Chemistry and Technologies of Drugs, Sapienza University, Piazzale Aldo Moro 5, 00185 Rome, Italy; stefania.garzoli@uniroma1.it (S.G.); rino.ragno@uniroma1.it (R.R.)

[3] Rome Center for Molecular Design, Department of Drug Chemistry and Technology, Sapienza University, Piazzale Aldo Moro 5, 00185 Rome, Italy

* Correspondence: marta.dimartile@ifo.gov.it (M.D.M.); donatella.delbufalo@ifo.gov.it (D.D.B.); Tel.: +39-0652666891 (M.D.M.); +39-0652662575 (D.D.B.)

Received: 31 July 2020; Accepted: 14 September 2020; Published: 16 September 2020

Simple Summary: In the last years, targeted therapy and immunotherapy modified the landscape for metastatic melanoma treatment. These therapeutic approaches led to an impressive improvement in patients overall survival. Unfortunately, the emergence of drug resistance and side effects occurring during therapy strongly limit the long-term efficacy of such treatments. Several preclinical studies demonstrate the efficacy of essential oils as antitumoral agents, and clinical trials support their use to reduce side effects emerging during therapy. In this review we have summarized studies describing the molecular mechanism through which essential oils induce in vitro and in vivo cell death in melanoma models. We also pointed to clinical trials investigating the use of essential oils in reducing the side effects experienced by cancer patients or those undergoing anticancer therapy. From this review emerged that further studies are necessary to validate the effectiveness of essential oils for the management of melanoma.

Abstract: The last two decades have seen the development of effective therapies, which have saved the lives of a large number of melanoma patients. However, therapeutic options are still limited for patients without BRAF mutations or in relapse from current treatments, and severe side effects often occur during therapy. Thus, additional insights to improve treatment efficacy with the aim to decrease the likelihood of chemoresistance, as well as reducing side effects of current therapies, are required. Natural products offer great opportunities for the discovery of antineoplastic drugs, and still represent a useful source of novel molecules. Among them, essential oils, representing the volatile fraction of aromatic plants, are always being actively investigated by several research groups and show promising biological activities for their use as complementary or alternative medicine for several diseases, including cancer. In this review, we focused on studies reporting the mechanism through which essential oils exert antitumor action in preclinical wild type or mutant BRAF melanoma models. We also discussed the latest use of essential oils in improving cancer patients' quality of life. As evidenced by the many studies listed in this review, through their effect on apoptosis and tumor progression-associated properties, essential oils can therefore be considered as potential natural pharmaceutical resources for cancer management.

Keywords: melanoma; essential oils; angiogenesis; apoptosis; metastasis

1. Introduction

Melanoma is the third most common cutaneous malignancy and one of the most dangerous forms of skin cancer. It is increasing worldwide and is caused by several factors, including environmental and genetic ones. In primary tumors, the most frequent driver oncogenic mutations (BRAF, NRAS, KIT), as well as mutations responsible for intrinsic resistance, have been identified with genomic analyses. Genetic alterations developed during the acquired resistance and genetic lesions leading to metastasis spreading were also identified [1].

The use of surgery in case of primary tumors is considered curative. With the advent of targeted therapy (MAPK pathway inhibitors) and immunotherapy (immune checkpoint inhibitors), treatment of metastatic melanoma has changed dramatically in recent years. These therapeutic strategies generated unprecedented improvement in patients' survival [2,3]. Targeted therapy, mostly represented by BRAF inhibitors, shows efficacy in the treatment of BRAF mutated metastatic melanoma when administered as single agents or in combination with MEK inhibitors [4]. Immune checkpoint inhibitors, such as anti-CTLA4 and anti-PD1 monoclonal antibodies, are able to activate the immune response in the host through their ability to unmask the inhibition of the host immune cells [5]. Targeted therapy in combination with immune checkpoint inhibitors has also showed encouraging results in the last few years [6]. Patients who are not eligible candidates for targeted therapy or immunotherapy are treated with systemic chemotherapy, such as dacarbazine, taxanes or temozolomide, while radiation therapy is used for brain metastases or for the treatment of oligometastatic disease [7]. Despite this clinical success, treatment of metastatic melanoma is still ineffective in about half of the treated patients and is often limited by numerous side effects and by the emergence of resistance to both targeted therapy and immunotherapy [8–10].

The diagnosis of melanoma and the therapeutic approaches used to contrast the disease, strongly affect the patient's quality of life. In fact, melanoma patients often experience the risk of disease progression, or even of new primary disease, for several years after diagnosis. Often, anxiety, depression and fear, leading to impaired social functioning, are evidenced in melanoma survivors. For these reasons, there is an urgent need to identify new therapeutic strategies for melanoma patients who do not respond or relapse after therapy, as well as to define new options to improve their quality of life.

In the last two decades, many new anticancer agents have been discovered from natural products as an alternative option to current cancer therapy. This is due to high cost, emergence of drug resistance and the side effects of standard therapies. Several plant-derived drugs, such as camptothecin, vinblastine, vincristine, etoposide, taxol, and paclitaxel have found wide applications in cancer therapy [11,12]. Current cancer research has brought about many promising preclinical results regarding the antiproliferative, anti-inflammatory, antioxidant, antiangiogenic and antimetastatic effect of essential oils (EOs), which are relevant components of aromatic plants. In addition, chemoprevention by EOs could represent a potentially effective option in the fight against cancer and, in particular, against melanoma. At present, EOs are used as a non-invasive therapy with minimal risk intervention that could potentially improve the quality of life of cancer patients and could alleviate the severity of treatment-related liver injuries or, more generally, symptoms in cancer patients undergoing chemotherapy. Alleviation of collateral effects would enhance prognosis status and patients' survival.

From a survey in scopus [13] with the "essential oil" and "melanoma" keywords, a list of almost 170 references were retrieved and analyzed. Focusing our attention on papers published in the last 20 years, we will summarize and discuss the chemical composition of EOs used in melanoma models, and the molecular mechanism through which EOs and their main components exert an antitumor effect in preclinical melanoma models carrying wild type or mutated BRAF. One section is dedicated to the use of EOs in clinical trials for managing cancer symptoms. This comprehensive summary could be a useful source for a better understanding of EOs' mechanism of action. It could also help researchers to appreciate and consider the importance of EOs being a potential adjuvant to enhance the efficacy of current available therapy, as well as, for improving patients' quality of life.

Taking into consideration the studies presented in this review, EOs showing antitumor activity could represent a good opportunity for combination therapy or, particularly, for reducing those side effects caused by current treatments.

2. Essential Oils

A wide number of studies have reported that plants contain valuable compounds, including bioactive EOs [14]. EOs are a complex combination of chemical aromatic-smelling low-molecular weight compounds. They are derived from the plants' secondary metabolism [15], leading predominately to monoterpenes [16], sesquiterpenes [17] and their oxygenated derivatives [18]. They are biosynthesized in different plant organs and parts such as flowers, leaves, fruits and roots [15] and are industrially produced mainly by hydro- or steam-distillation [19,20]. Chemical qualitative and quantitative composition of EOs, which are composed of volatile compounds, is determined using a combination of Gas Chromatography/Flame Ionization Detection, Gas Chromatography/Mass Spectrometry and determination of their Kovats index [21] and/or Linear Retention Indices [22]. The chemical composition of aromatic plants furnishing EOs active against melanoma models is reported in Table 1. Common and family names of the plants, as well as the percentage of the main components identified in each EO, are also included.

Table 1. Chemical composition of essential oils (EOs) active against melanoma models.

Plant Name from Which EOs Were Extracted	Plant Common Name	Plant Family Name	Main EO Chemical Components	Reference
Achillea millefolium	Yarrow, common yarrow, thousand-leaf	Asteraceae	Artemisia ketone (14.92%), camphor (11.64%), linalyl acetate (11.51%), 1,8-cineole (10.15%)	[23]
Alpinia zerumbet	Light Galangal, shell ginger	Zingiberaceae	γ-Terpinene (14.5%), cineole (13.8%), p-cymene (13.5%), sabinene (12.5%), terpinen-4-ol (11.9%), caryophyllene oxide (4.96%), methyl cinnamate (4.24%), caryophyllene (2.4%), γ-terpineol (1.28%)	[24]
Annona vepretorum	Araticum, pinha da caatinga, araticum-da-Bahia	Annonaceae	Bicyclogermacrene (35.7%), spathulenol (18.89%), α-phellandrene (8.08%), α-pinene (2.18%), o-cymene (6.24%)	[25]
Anthemis wiedemanniana	-	Asteraceae	9,12-Octadecadienoic acid (12.2%), hexadecanoic acid (10.5%), hexahydrofarnesyl acetone (8.3%), 1,8-cineol (6.2%), carvacrol (5.8%)	[26]
Artemisia anomala	-	Asteraceae or Compositae	Camphor (18.3%), 1,8-cineole (17.3%), β-caryophyllene oxide (12.7%), borneol (9.5%)	[27]
Artemisia argyi	-	Asteraceae or Compositae	Caryophyllene (10.19%), eucalyptol (23.66%)	[28]
Atriplex undulata	-	Chenopodiaceae	p-Acetanisole (28.1%), β-damascenone (9.3%), β-ionone (5.1%), viridiflorene (4.7%), 3-oxo-α-ionol (2.2%)	[29]
Casearia lasiophylla	-	Salicaceae	Germacrene D (18.6%), E-caryophyllene (14.7%), δ-cadinene (6.2%), α-cadinol (5.4%)	[30]
Chrysanthemum boreale Makino	-	Asteraceae	Germacrene D (10.6–34.9%), β-caryophyllene (10.8%), (–)-camphor (10.8–18.0%), β-thujone (11.7%), α-thujone (9.8%)	[31]

Table 1. *Cont.*

Plant Name from Which EOs Were Extracted	Plant Common Name	Plant Family Name	Main EO Chemical Components	Reference
Cinnamomum cassia	-	Lauraceae	Cis-2-methoxycinnamic acid (43.06%), cinnamaldehyde (42.37%)	[32]
Cinnamomum zeylanicum	-	Lauraceae	Eugenol (70%), β-caryophyllene (2.4%)	[33]
Citrus bergamia	Acid lemon	Rutaceae	Limonene (38.1%), linalyl acetate (28.9%), γ-terpinene (7.3%), linalool (6.4%), β-pinene (5.4%), bergapten (1.7%)	[34]
Citrus medica	Citron	Rutaceae	Limonene (35.4%), γ-terpinene (24.5%), geranial (5.5%), neral (4.4%), β-pinene (2.6%), α-pinene (2.5%), β-myrcene (2.1%), terpinen-4-ol (1.5%)	[34]
Cuminum cyminum	Cumin-jeera	Apiaceae or Umbelliferae	Cuminaldehyde (39.48%), γ-terpinene (15.21%), O-cymene (11.82%), β-pinene (11.13%), 2-caren-10-al (7.93%), trans-carveol (4.49%), and myrtenal (3.5%)	[35]
Curcuma aromatica	Wild turmeric	Zingiberaceae	8,9-Dehydro-9- formyl-cycloisolongifolene (2.66–36.83%), germacrone (4.31–16.53%), ar-turmerone (2.52–17.69%), turmerone (2.62–18.38%), ermanthin (0.75–13.26%), β-sesquiphyllandrene (0.33–11.32%), ar-curcumene (0.29–10.52%)	[36]
Curcuma kwangsiensis	Mango-ginger	Zingiberaceae	8,9-Dehydro-9-formyl-cycloisolongifolene (2.37–42.59%), germacrone (6.53–22.20%), L-camphor (0.19–6.12%)	[37]
Curcuma zedoaria	Kua-zedoary	Zingiberaceae	8,9-Dehydro-9-formyl-cycloisolongifolene (60%), 6-ethenyl-4,5,6,7-tetra-hydro-3,6-dimethyl-5-isopropenyl-trans-benzofuran (12%)	[38]
Dalbergia pinnata	Laleng-chali	Fabaceae	Elemicin (91.06%), methyl eugenol (3.69%), 4-allyl-2,6-dimethoxyphenol (1.16%), whiskey lactone (0.55%)	[39]
Eryngium amethystinum	-	Apiaceae	Germacrene D (56.7%), β-elemene (4.7%), bicyclogermacrene (3.3%), α-copaene (2.2%), (E)-caryophyllene (1.9%), germacrene B (1.8%), germacra-4(15),5,10(14)-trien-1-α-ol (1.7%), cadin-4-en-10-ol (1.6%)	[40]
Eryngium campestre	Eryngo, field eryngo, sea-holly	Apiaceae	Germacrene D (13.8%), allo-aromadendrene (7.7%), spathulenol (7.0%), ledol (5.7%), cadin-4-en-10-ol (3.9%), γ-cadinene (3.6%), epi-α-muurolol (2.1%), germacra-4(15),5,10(14)-trien-1-α-ol (2.0%), δ-cadinene (1.9%), caryophyllene oxide (1.5%)	[40]
Eucalyptus camaldulensis	Murray red gum, red gum, red river gum	Myrtaceae	1,8-Cineole (23.9%), α-eudesmol (11.6%), γ-eudesmol (8.0%), and elemol (5.0%)	[41]

Table 1. *Cont.*

Plant Name from Which EOs Were Extracted	Plant Common Name	Plant Family Name	Main EO Chemical Components	Reference
Eugenia cuspidifolia	-	Myrtaceae	Caryophyllene oxide (57.46%), α-copaene (3.75%)	[42]
Eugenia tapacumensis	-	Myrtaceae	Caryophyllene oxide (55.95%), α-copaene (13.67%)	[42]
Eugenia uniflora	Brazil cherry	Myrtaceae	Curzerene (13.4–50.6%), selina-1,3,7(11)-trien-2-one (18.1–43.1%), selina-1,3,7(11)-trien-2-one epoxidem(16.0–30.4%), germacrene B (5.0–18.4%), caryophyllene oxide(1.2–18.1%), (E)-caryophyllene (0.3–9.1%), β-elemene (3.5–8.9%), γ-elemene (2.0–7.8%)	[43]
Glechoma hederacea	Ground ivy, field balm, gill over the ground, runaway robin	Lamiaceae or Labiatae	Trans-3-pinanone (41.4%), 4,5,6,7-tetrahydro-5-isopropenyl-3,6-β-dimethyl-6-α-vinylbenzofuran (10.8%), β-caryophyllene (10.2%), and spathulenol (4.3%)	[44]
Helichrysum microphyllum	-	Asteraceae or Compositae	Neryl acetate (18.2%), rosifoliol (11.3%), δ-cadinene (8.4%), γ-cadinene (6.7%)	[45]
Heracleum sphondylium	Cow parsnip, eltrot	Apiaceae or Umbelliferae	Octyl acetate (54.9–60.2%), octyl butyrate (10.1–13.4%)	[46]
Hypericum hircinum	-	Hypericaceae	Cis-β-guaiene (29.3%), δ-selinene (11.3%), isolongifolan-7-α-ol (9.8%), (E)-caryophyllene (7.2%)	[47]
Laurus nobilis	Bay Tree, sweet bay, Grecian Laurel, true laurel	Lauraceae	1,8-cineole (35.15%)	[48]
Lavandula augustifolia	English lavender, true lavender	Lamiaceae or Labiatae	α-Pipene, β-pipene, camphene, eucalyptol, D-limonene	[49]
Lippia gracilis	-	Verbenaceae	Thymol (55.50%), p-cymene (10.80%), γ-terpinene (5.53%), myrcene (4.03%)	[50]
Liriodendron tulipifera	Tulip tree, tulip poplar, yellow poplar, canary whitewood	Magnoliaceae	(Z)-β-Ocimene (12.5–25.2%), (E)-β-ocimene (3.7–6.8%), β-elemene (16.4–17.1%), germacrene D (18.9–27.2%)	[51]
Melaleuca alternifolia	Tea Tree	Myrtaceae	Terpinen-4-ol (42.35%), γ-terpinene (20.65%), α-terpinene (9.76%)	[52]
Melaleuca quinquenervia	-	Myrtaceae	1,8-Cineole (21.06%), α-pinene (15.93%), viridiflorol (14.55%), α-terpineol (13.73%)	[53]
Mentha aquatica	-	Lamiaceae or Labiatae	β-Ocimene (22.18%), β-pinene (15.41%), 1,8-cineole (12.87%), α-pinene (10.49%)	[54]
Myrcia laruotteana	-	Myrtaceae	α-Bisabolol (23.6%), α-bisabolol oxide B (11.5%)	[55]
Myristica fragrans	Mace, nutmeg	Myristicaceae	Myristicin, limonene, eugenol and terpinen-4-ol	[56]
Nectandra leucantha	-	Lauraceae	Bicyclogermacrene (28.44%), germacrene A (7.34%)	[57]

Table 1. *Cont.*

Plant Name from Which EOs Were Extracted	Plant Common Name	Plant Family Name	Main EO Chemical Components	Reference
Ocimum basilicum	Sweet basil, common basil, thai basil, tropical basil	Lamiaceae or Labiatae	1,8 Cineole (11.0%), linalool (42.5%), estragole (33.1%)	[58]
Ocimum gratissimum	African basil, east Indian basil, russian basil, shrubby basil	Lamiaceae or Labiatae	Eugenol (54.0%), 1,8 cineole (21.6%), β-selinene (5.5%), β-caryophyllene (5.3%), (Z)-ocimene (4.0%)	[58]
Ocimum micranthum	-	Lamiaceae or Labiatae	Eugenol (64.8%), β-caryophyllene (14,3%), bicyclogermacrene (8.1%)	[58]
Ocimum tenuiflorum	Sacred basil	Lamiaceae or Labiatae	Eugenol (59.4%), β-caryophyllene (29.4%), germacrene A (8.1%)	[58]
Origanum ehrenbergii	-	Lamiaceae or Labiatae	Carvacrol, thymoquinone	[59]
Origanum syriacum	Bible hyssop	Lamiaceae or Labiatae	Carvacrol, thymoquinone	[59]
Perilla frutescens	Shiso, beefsteakplant, spreading beefsteak plant	Lamiaceae or Labiatae	isoegomaketone	[60]
Piper aleyreanum	-	Piperaceae	β-Elemene (16.3%), bicyclogermacrene (9.2%), δ-elemene (8.2%), germacrene D (6.9%), β-caryophyllene (6.2%), spathulenol (5.2%)	[61]
Piper cernuum	-	Piperaceae	α-Pinene, camphene, limonene, carvacrol, tymol, myrcene, p-cymene, aterpineol, linalol	[62]
Piper klotzschianum	-	Piperaceae	Germacrene D (7.3–22.8%), bicyclogermacrene (13.4–21.6%), E)-caryophyllene (11.9–16.8%), β-pinene (2.3–27.2%), α-pinene (1.4–7.2%)	[63]
Pistacia lentiscus	Chios mastictree, aroeira, lentiscus, lentisk, mastic, mastictree	Anacardiaceae	Perillyl alcohol	[64]
Pituranthos tortuosus	-	Apiaceae	Sabinene (24.24%), α-pinene (17.98%), limonene (16.12%), and terpinen-4-ol (7.21%)	[65,66]
Plectranthus amboinicus	Country borage, Indian borage	Lamiaceae or Labiatae	Carvacrol thymol, cis-caryophyllene, trans-caryophyllene, and p-cymene	[67]
Pomelo peel	-	Rutaceae	Limonene (55.92%), β-myrcene (31.17%), β-pinene (3.16%), ocimene (1.42%), β-copaene (1.24%)	[68]
Porcelia macrocarpa	-	Annonaceae	Germacrene D (47%), bicyclogermacrene (37%), verbanyl acetate (0.5%), phytol (1.2%)	[69]
Pterodon emarginatus	Faveiro, sucupira, sucupira-branca	Fabaceae	β-Elemene (15.3%), trans-caryophyllene (35.9%), α-humulene (6.8%), germacrene-D (9.8%), bicyclogermacrene (5.5%), spathulenol (5.9%)	[70]

Table 1. *Cont.*

Plant Name from Which EOs Were Extracted	Plant Common Name	Plant Family Name	Main EO Chemical Components	Reference
Salvia aurea	-	Lamiaceae or Labiatae	Caryophyllene oxide (12.5%), α-amorphene (12.0%), aristolone (11.4%), aro-madendrene (10.7%), elemenone (6.0%)	[71]
Salvia breacteata	-	Lamiaceae or Labiatae	Caryophyllene oxide (16.6%)	[72]
Salvia judaica	-	Lamiaceae or Labiatae	Caryophyllene oxide (12.8%)	[71]
Salvia libanotica	-	Lamiaceae or Labiatae	Cineole (57.4%), camphor (8.4%), β-pinene (5.1%), α-pinene (3.9%), camphene (3.0%)	[73]
Salvia officinalis	Sage, kitchen sage, small leaf sage, garden sage	Lamiaceae or Labiatae	Caryophyllene (25.634%), camphene (14.139%), eucalyptol (13.902%)	[74]
Salvia rubifolia	-	Lamiaceae or Labiatae	γ-Muurolene (11.8%)	[72]
Salvia verbenaca	Wild clary	Lamiaceae or Labiatae	Hexadecanoic acid (11–23.1%), Z)-9-octadecenoic acid (5.6–11.1%), benzaldehyde (1.1–7.3%)	[75]
Salvia viscosa	-	Lamiaceae or Labiatae	Caryophyllene oxide (12.7%)	[71]
Santalum album	White sandal tree, sandalwood, sandal tree, sandal	Santalaceae	α-Santalol (61%), β-santalol (28%)	[76]
Satureja hortensis	Summer savory	Lamiaceae or Labiatae	γ-Terpinene (37.862%), o-cymene (15.113%), thymol (13.491%), carvacrol (13.225%)	[77]
Schinus terebinthifolius Raddi	Brazilian pepper tree	Anacardiaceae	β-Longipinene (8.1%), germacrene D (23.8%), biclyclogermacrene (15.0%), α-pinene (5.7%), β-pinene (9.1%)	[78]
Stachys germanica	Downy woundwort, German hedgenettle	Lamiaceae or Labiatae	(Z,Z,Z)-9,12,15-octa-decatrienoic acid methyl ester (33.3%), exadecanoic acid (22.1%)	[79]
Stachys parviflora	-	Lamiaceae or Labiatae	α-Terpenyl acetate (23.6%), β-caryophyllene (16.8%), bicyclogermacrene (9.3%), spathulenol (4.9%), α-pinene (4.2%)	[80]
Syzygium aromaticum	Clove, Zanzibar redhead	Myrtaceae	Eugenol (61%), (β-carophillene 5.7%)	[33,81]
Tagetes erecta	African marigold, Aztec marigold, big marigold, American marigold	Asteraceae or Compositae	Limonene (10.4%), α-terpinolene (18.1%), (E)-ocimenone (13.0%), dihydrotagetone (11.8%)	[82]
Tanacetum macrophyllum	Tansy, rayed tansy, tansy chrysanthemum	Asteraceae or Compositae	Germacrene D (6.9–30.9%), 10-epi-γ-eudesmol (3.9–13.5%), camphor (11.1%), linalool (0.6–10.8%), 1,8-cineole (5.5–8.8%)	[83]
Thymus alternans	-	Lamiaceae or Labiatae	(E)-Nerolidol (15.8–31.4%), germacrene D (6.7–7.4%), geranial (6.8–7.7%), (E)-β-ocimene (2.6–7.0%), linalool (1.7–6.4%), geraniol (3.3–6.2%), neral (4.9–5.4%)	[84]

Table 1. *Cont.*

Plant Name from Which EOs Were Extracted	Plant Common Name	Plant Family Name	Main EO Chemical Components	Reference
Thymus munbyanus	-	Lamiaceae or Labiatae	Borneol (31.2–44.8%), camphor (5.7–13.6%), camphene (3.6–7.5%), 1,8-cineole (4.2–6.0%), germacrene D (3.1–5.0%)	[85]
Thymus vulgaris	English thyme, French thyme, garden thyme, thyme	Lamiaceae or Labiatae	γ-Terpinene (68.415%), thymol (24.721%), caryophyllene (5.5%), α-pinene (4.816%)	[74]
Tridax procumbens	Coat buttons, coat-button, Mexican daisy	Asteraceae or Compositae	α-Pipene, β-pinene, phellandrene, sabinene	[86]
Vetiveria zizanioides	Cuscus grass, khus-khus, khas-khas, vetiver	Poaceae	Cedr-8-en-13-ol (12.4%), α-amorphene (7.80%), β-vatirenene (5.94%), α-gurjunene (5.91%), dehydroaromadendrene (5.45%)	[87]
Vitex Negundo	Common chaste tree, negundo, five keaved chaste tree, negundo chastetree, chaste tree	Lamiaceae or Labiatae	Sabinene (19.04%), caryophyllene (18.27%)	[88]
Vitex Trifolia	Indian privet, Arabian lilac, Indian three-leaf vitex, hand of mary	Lamiaceae or Labiatae	α-Pinene (11.38%), β-pinene (2.84%), sabinene (10.25%), eucaluptol (8.60%), camphene (12.69%), manoyl oxide (16.11%), abietatriene (9.03%)	[89]
Wedelia chinensis	Chinese wedelia	Asteraceae or Compositae	Carvocrol, trans-caryophyllene	[90]
Zornia brasiliensis	-	Fabaceae	Trans-nerolidol (48.0%), germacrene D (13.9%), α-humulene (9.3%), trans-caryophyllene (8.4%), and (Z,E)-α-farnesene (7.3%)	[91]

where not indicated, some plant common names and EOs composition were not available from the cited paper. EOs from *Boswellia carterii, Citrus grandis, Citrus hystix, Citrus reticulate, Psiadia terebinthina* were not included in the table as their chemical compositions were not reported in the corresponding cited articles.

Anticancer activities of EO mixtures as a whole and only a hypothetical correlation between the chemical profile and the anticancer activity have been described for some EOs. In some cases, EOs' bioactive properties were related to the anticancer activity of specific components. Due to EOs' complex chemical composition, the additive, synergistic or anti-synergistic roles of individual EO constituents are currently being investigated to establish the possible pharmacological activity thereof [92,93]. This categorization is even more complicated due to the 'chemotype' concept, in which the same plant could produce different EOs characterized by different chemical composition profiles and, hence, different biological properties [94]. *Ocimum tenuiflorum* (holy basil), *Thymus vulgaris* (thyme), *Lavandula angustifolia* (lavender) and *Mentha piperita* (peppermint) are examples of plants with several chemotypes [95]. Despite this mix up, an effort to characterize EOs is currently taking place in medical and pharmaceutical fields. This characterization could help to obtain a clearer indication of EOs' uses in traditional medicine, chemical or pharmaceutical, as witnessed by almost 5000 articles published in PubMed (http://pubmed.ncbi.nlm.gov/). In the last decade, an average positive increment of more than 7% per year was observed in this field [96].

When referring to EOs, their chemical composition and biological activities strictly depend on habitat, climate condition, season, agronomic practices, soil type, extraction procedures, as well as the harvesting stages and storage conditions of plants [31,36,37,51,75,97–103]. All these elements should be taken into account. By analyzing gene expression patterns and metabolic fingerprints, recently Spring's group identified environmental factors as regulatory factors of biosynthetic pathways [104].

Substantial variability was also reported according to the part of plant used for extraction of EOs [83]. Examples are EOs from *Helichrysum microphyllum* [45] and from *Liriodendron tulipifera*, and their main components β-elemene and (E)-nerolidol, showing antiproliferative activity in human melanoma cells strictly depending on harvesting period [51]. Moreover, EOs from *Chrysanthemum boreale* Makino, showed different levels of their component contents and bioactivities among the harvesting stages [31], while phytoconstituents and bioactivities of EOs from *Curcuma kwangsiensis*, strictly depended on the natural habitat [37]. Another example is provided by chemotaxonomical analysis of *Artemisia absinthium, Salvia officinalis, Tanacetum vulgare* and *Thuja occidentalis*, the amount of thujones (α-thujone and β-thujone) present in the EOs of the four species being strictly related to the plant organ and to its developmental phase [105]. Moreover, exposure to ultraviolet (UV) light was reported to induce deterioration of EOs' biochemical profiles [49] or activation of some EOs [34]. In this regard, the antiproliferative effect of both *Citrus medica* and *Citrus bergamia* EOs and their constituent bergapten, was observed in human melanoma cells after exposure to UV irradiation, thus indicating UV irradiation's ability to activate EOs [34]. In some cases, derivatives of EOs are designed to increase EOs' half-life. This is the case of farnesyl-O-acetylhydroquinone, geranyl-O-acetylhydroquinone, geranyl ester and farnesyl ester derived from geraniol and farnesol. Structure–activity relationship studies reported the ability of these derivatives to reduce proliferation of mouse melanoma cells more efficiently than the parental compounds [106,107]. Furthermore, 6-(menthoxybutyryl)thymoquinone, the terpene conjugate derivative of thymoquinone, was shown to be more active in human melanoma cells than its parental compound [108]. A cumulative impact of some EOs' components was also evidenced: farnesol and nerolidol in combination showed an enhanced antiproliferative effect in mouse melanoma cells when compared to exposure to single treatments [109].

For all their biological effects, EOs can be considered as an interesting source for therapeutic, food preservation and/or nutraceutical uses [110–114].

3. Mechanism of Action of EOs in Melanoma

EOs and some of their components exert antitumor activity in melanoma models by affecting multiple pathways, including inhibition of in vitro cell proliferation, alteration of cell distribution in the different cell cycle phases, induction of apoptosis, inhibition of in vitro cell invasion and migration, in vivo tumor growth and metastasization and in vitro/in vivo angiogenesis. Several EOs also act as chemopreventive agents in melanoma, reduce melanogenesis and show antioxidant properties.

The most frequently used human melanoma models are represented by M14, A2058, A375 cells and SK-MEL variants. Murine melanoma B16 cells, that originate in the syngeneic C57BL/6 (H-2b) mouse strain, and its derivatives B16-F1, B16-F10, B16-F10-Nex2, B16-Bl6, B164A5 [115–118] represent the most used in vivo models. They are employed to evaluate the effect of EOs on tumor growth and metastasization, as well as tumor angiogenesis, after subcutaneous or intravenous (lateral tail vein) cell injection [119]. In some cases subretinal, intradermal, intracerebral injection of melanoma cells is employed [38,86,120,121]. Routes of EO administration include oral, intraperitoneal, intravitreal, peritumoral, topical as well as inhalation (fragrant environmental).

3.1. Inhibition of Cell Proliferation

Cell viability and proliferation can be detected by a wide range of assays based on several cell functions, including mitochondrial enzyme and cellular uptake activity, cell membrane permeability and ATP production. The most used assay is a colorimetric test that evaluates the reduction of yellow 3-(4,5-dimethythiazol-2-yl)-2,5-diphenyl tetrazolium bromide by mitochondrial succinate dehydrogenase [122]. Analysis of cell proliferation can also be performed by alamarBlue and PrestoBlue assays, both using reduction of resazurin as an indicator of cell viability [122]. Detection of 5-bromo-2′-deoxyuridine (BrdU)- or 5-ethynyl-2′-deoxyuridine (EdU)- labeled DNA also represents a valid method to detect viable cells: BrdU or EdU are efficiently incorporated into DNA of replicating cells by substituting thymidine and their binding to DNA can be detected with specific antibodies [123].

Moreover, the thymidine incorporation assay, based on measuring the incorporation of methyl-[3H]-thymidine into the DNA of dividing cancer cells, is frequently used to estimate cell proliferation [124].

A high number of EOs and their components have been found to reduce in vitro proliferation/viability of melanoma cells, and in some cases the cytotoxic potential has been predicted by in silico studies [125].

EOs and their main components demonstrated to reduce in vitro proliferation of melanoma cells are reported in Table 2. The used melanoma models are also indicated in the Table.

Table 2. EOs and their components that have been demonstrated to affect in vitro or in vivo melanoma growth and metastasization.

Pathway Affected	Plant Name from Which EOs Were Extracted	EO Active Components	In Vitro and In Vivo Models	Reference
In vitro cell proliferation	*Annona Vepretorum*	Spathulenol, o-cymene, α-pinene	B16-F10	[25]
	Anthemis wiedemanniana	-	C32	[26]
	Artemisia anomala	-	BRO	[27]
	Casearia lasiophylla	-	UACC-62	[30]
	Citrus bergamia	Bergapten	A375	[34]
	Citrus medica	Limonene	A375	[34,126]
	Coleus aromaticus	Carvacrol	A375	[127]
	Curcuma aromatica	-	B16	[36]
	Curcuma kwangsiensis	-	B16	[37]
	Curcuma zedoaria	-	B16-Bl6	[38]
	Eryngium amethystinum	-	A375	[40]
	Eryngium campestre	-	A375	[40]
	Eugenia cuspidifolia	-	SK-MEL-19	[42]
	Eugenia tapacumensis	-	SK-MEL-19	[42]
	Eugenia uniflora	Curzerene	SK-MEL-19	[43]
	Helichrysum microphyllum	-	A375	[45]
	Heracleum sphondylium	Octyl butyrate	A375	[46]
	Hypericum hircinum	-	B16-F1	[47]
	Laurus nobilis	-	C32	[48]
	Lippia gracilis	-	B16-F10	[50,128]
	Liriodendron tulipifera	β-Elemene	A375	[51]
	Melaleuca alternifolia	Terpinen-4-ol	A375, M14, B16-F10	[52,129,130]
	Melaleuca quinquenervia	1,8-Cineole, α-Pipene, α-Terpineol	B16	[53]
	Myrcia laruotteana	-	UACC-62	[55]
	Nectandra leucantha	Bicyclogermacrene	B16-F10-Nex2	[57]
	Perilla frutescens	Isoegomaketone	B16	[60]
	Piper aleyreanum	-	SK-MEL-19	[61]
	Piper cernuum	Camphene	B16-F10-Nex2	[62]
	Piper klotzschianum	-	B16-F10	[63]
	Porcelia macrocarpa	-	B16-F10-Nex2	[69]

Table 2. *Cont.*

Pathway Affected	Plant Name from Which EOs Were Extracted	EO Active Components	In Vitro and In Vivo Models	Reference
	Pterodon emarginatus	-	MeWo	[70]
	Salvia aurea	-	M14, A375, A2058	[71]
	Salvia bracteata	-	M14	[72]
	Salvia judaica	-	M14, A375, A2058	[71]
	Salvia officinalis	-	A375, M14, A2058, B164A5	[74,101]
	Salvia rubifolia	-	M14	[72]
	Salvia verbenaca	-	M14	[75]
	Salvia viscosa	-	M14, A375, A2058	[71]
	Satureja hortensis	-	B164A5, A375	[77]
	Schinus terebinthifolius Raddi	α-Pipene, β-pipene, pipane	B16-F10-Nex2, A2058	[78]
	Stachys germanica	-	C32	[79]
	Stachys parviflora	-	B16-F10	[80]
	Syzygium aromaticum	Eugenol	B16	[81]
	Tagetes erecta	-	B16-F10	[82]
	Tanacetum macrophyllum	-	A375	[83]
	Thuja occidentalis	Thujone	A375	[131]
	Thymus alternans	-	A375	[84]
	Thymus munbyanus	-	A375	[85]
	Thymus vulgaris	-	B164A5, A375	[74]
	Vitex Trifolia	Abietatriene	B16-F10	[89]
		Carvacrol	SK-MEL-2	[132]
		Citral	B16-F10, SK-MEL-147, UACC-257	[133]
		Eugenol	SK-MEL-2, A2058, SK-MEL-28, Sbcl2, WM3211, WM98-1, WM1205Lu, LCM-MEL GR-MEL, 13443	[132,134–136]
		Farnesol	B16, B16-F10	[106,109]
		Farnesyl anthranilate	B16	[106,107]
		Farnesyl-O-acetylhydroquinone	B16	[106]
		Menthol	A375	[137]
		Neridol	B16	[109]
		Thymol	SK-MEL-2	[132]
		Zerumbone	CHL-1, A375	[138,139]
		β-Caryophyllene	B16-F10	[140]

Table 2. *Cont.*

Pathway Affected	Plant Name from Which EOs Were Extracted	EO Active Components	In Vitro and In Vivo Models	Reference
In vitro tumor progression-associated functions	*Alpinia zerumbet*	-	HUVEC	[141]
	Eugenia uniflora	Curzerene	SK-MEL-19	[43]
	Melaleuca alternifolia	Terpinen-4-Ol	M14	[142]
	Pituranthos tortuosus	-	B16-F10	[66]
	Satureja hortensis	-	B164A5, A375	[77]
		Myrtenal	B16-F0, B16-F10, SK-MEL-5	[143]
		Thujone	B16-F10	[144]
		Thymoquinone	B16-F10, A375	[145]
		Zerumbone	CHL-1	[138]
In vivo tumor growth and metastasization	*Annona Vepretorum*	Spathulenol, o-cymene, α-pinene	B16-F10 (C57BL/6J)	[25]
	Boswellia carterii	-	B16-F10 (C57BL/6)	[146]
	Curcuma zedoaria	-	B16-Bl6 (C57BL/6)	[38]
	Melaleuca alternifolia	-	B16-F10 (C57BL/6J)	[147]
	Perilla frutescens	Isoegomaketone	B16 (C57BL/6N)	[60]
	Piper cernuum	Camphene	B16-F10-Nex2 (C57BL/6)	[62]
	Pituranthos tortuosus	-	B16-F10 (BALB/c)	[65,66]
	Plectranthus amboinicus	-	B16-F10 (C57BL/6)	[67]
	Salvia officinalis	β-Ursolic acid	B16 (C57BL/6)	[148]
	Schinus terebinthifolius Raddi	α-Pipene	B16-F10-Nex2 (C57BL/6)	[149]
	Tridax procumbens	-	B16-F10 (C57BL/6)	[86]
	Zornia brasiliensis	-	B16-F10 (C57BL/6)	[91]
		Eugenol	B16 (B6D2F1)	[135]
		Limonene	B16-F10 (C57BL/6)	[150]
		Myrtenal	B16-F10 (C57BL/6)	[143]
		Perillic Acid	B16-F10 (C57BL/6)	[150]
		Thujone	B16-F10 (C57BL/6)	[144]
		Thymoquinone	B16-F10 (C57BL/6)	[121,151]
		α-Pinene	B16-F10-Nex2 (C57BL/6)	[149]
		β-Caryophyllene	B16-F10 (C57BL/6N)	[140]
		β-Elemene	B16-F10 (C57BL/6)	[120,152]

where not indicated, EO active components were not available from the cited articles. When referred to in vivo studies, the murine strain used is indicated in brackets. Human Umbilical Vein Endothelial Cells (HUVEC).

3.2. Alteration of Cell Cycle Distribution

Cell cycle transition is a process involving multiple checkpoints, which control growth signals, cell size and DNA integrity. It is regulated by active forms of cyclin-dependent kinases, which control the passage of cells from one phase of the cell cycle to another. Cyclin-dependent kinases act as cell cycle regulators to elicit cell cycle arrest in response to DNA damage [153]. Cytofluorimetric analysis of cells stained with the DNA intercalator, propidium iodide, represents the most used methods to analyze cell cycle transition.

Several authors demonstrated the ability of EOs or their constituents to induce DNA damage in melanoma cells, and consequently inducing delay/arrest in the different cell cycle phases. Recently, Ramadam et al. reported the *Melaleuca alternifolia* EO (Tea tree oil, TTO) ability to induce cell cycle arrest at the G2/M phase of A375 cells [129], while in a previously published study, Beilharz's group described the TTO ability to elicit G1 cell cycle arrest in B16 cells [130]. In addition, α-santalol, the main component of *Santalum album* (Sandalwood oil) [154], in UACC-62 cells induced G2/M phase arrest through down-regulation of proteins critical for G2/M transition, such as cyclin A/Cdk2 and cyclin B/Cdc2 complexes, as well as microtubule depolymerisation. It also increased expression of wild-type p53 [155]. Eugenol, a component present in many EOs including *Syzygium aromaticum* (Clove oil) and *Cinnamomum zeylanicum* [156], abrogated the G2/M phase in A2058 but not in SK-MEL-28 cells. It was also reported to arrest WM1205Lu cells in the S phase of the cell cycle through the inhibition of E2F1 transcription factor activity. E2F1 is a key cell cycle regulator, targeting genes that encode proteins involved in G1/S transition [134,135]. Eugenol also reduced the expression of proliferation cell nuclear antigen in A2058 cells [134]. Given the role of E2F1 in melanoma progression and resistance to therapy [157,158], the authors also suggested that eugenol could be developed as an E2F-targeted agent for melanoma treatment. Farnesol (a mixture of trans, trans and cis, trans isomers) and nerolidol, present in different EOs including that from *Psidium guajava* [159], induced an increase in cells in the G0/G1 phase, concomitant with a reduction in the S phase, and a cumulative impact of the two compounds was evidenced in B16 cells [109].

The different response between cell lines sometimes observed after exposure to the same EO indicate a non-generalizable and cell-type specific effect or is due to differences in the composition of the EOs' used in the distinct studies. The genetic background of the cell lines tested should be also considered: some melanoma cells harbor activating BRAF^{V600E} mutations (A375, M14, A2058, SK-MEL-5, SK-MEL-19, SK-MEL-28, UACC-62, UACC-257, 518A2, G-361, WM266) not present in other ones (B16, Sbcl2, CHL1, WM3211, SK-MEL-147, MeWo, RPMI-7932) (https://web.expasy.org/cellosaurus/), or harbor different mutations (i.e., NRAS, CDKN2A, TP53, TERT). A better understanding of these cell-to-cell differences is crucial for a deeper comprehension of the EO mechanism of action.

EOs and their main components demonstrated to induce cell cycle perturbation of melanoma cells are reported in Table 3. The used melanoma models are also listed in the Table.

Table 3. EOs and their components that were demonstrated to affect cell cycle distribution, apoptosis, necrosis and autophagy of melanoma cells.

Pathway Affected	Plant Name from Which EOs Were Extracted	EO Active Components	In Vitro and In Vivo Melanoma Models	Reference
Cell cycle	*Melaleuca alternifolia*	Terpinen-4-ol	A375, B16	[129,130]
	Santalum album	α-Santol	UACC-62	[155]
		Eugenol	A2058, WM1205Lu, Sbcl2, WM3211	[134,135]
		Farnesol	B16	[109]
		Neridol	B16	[109]
	Annona Vepretorum	Spathulenol, o-cymene, α-pinene	B16-F10	[25]
	Boswellia carterii		B16-F10, FM94	[146]
	Coleus aromaticus	Carvacrol	A375	[127]
	Eugenia uniflora	Curzerene	SK-MEL-19	[43]
	Melaleuca alternifolia	Terpinen-4-Ol	A375, M14	[52,129]
	Perilla frutescens	Isoegomaketone	B16	[60]

Table 3. *Cont.*

Pathway Affected	Plant Name from Which EOs Were Extracted	EO Active Components	In Vitro and In Vivo Melanoma Models	Reference
Apoptosis	*Piper cernuum*	Camphene	B16-F10-Nex2	[62]
	Pituranthos tortuosus	-	B16-F10	[66]
	Salvia aurea	-	M14	[71]
	Salvia bracteata	-	M14	[72]
	Salvia judaica	-	M14	[71]
	Salvia officinalis	-	A375, M14, A2058	[101]
	Salvia verbenaca	-	M14	[75]
	Salvia viscosa	-	M14	[71]
	Salvia rubifolia	-	M14	[72]
	Schinus terebinthifolius Raddi	α-Pipene	B16-F10-Nex2, A2058	[78,149]
	Thuja occidentalis	Thujone	A375	[131]
	Tridax procumbens	-	B16-F10 (C57BL/6)	[86]
		Citral	B16-F10, SK-MEL-147, UACC-257	[133]
		Eugenol	A2058, SK-MEL-28	[134–136]
		Linalool	RPMI-7932	[160]
		Menthol	A375, G-361	[161,162]
		Thymoquinone	B16-F10	[121]
		Zerumbone	CHL-1, A375	[138,139]
		β-Carophyllene	B16-F10 (C57BL/6N)	[140]
Necrosis and autophagy	*Melaleuca alternifolia*	Terpinen-4-Ol	B16	[130]
	Salvia aurea	-	M14	[71]
	Salvia bracteata	-	M14	[72]
	Salvia judaica	-	M14	[71]
	Salvia viscosa	-	M14	[71]
	Salvia rubifolia	-	M14	[72]
		Citral	B16-F10	[133]

3.3. Induction of Apoptosis

Apoptosis is a programmed cell death characterized by the activation of a group of intracellular caspases leading to a cascade of events connected with various substrates, including poly(ADP-ribose) polymerase-I (PARP), and internalization by phagocytes [163,164]. Hallmarks of apoptosis include cell shrinkage, formation of apoptotic bodies, DNA fragmentation, heterochromatin aggregation and activation of caspases and substrates such as PARP. Exposure of phosphatidyl serine on the cell surface, change in the mitochondrial membrane potential, as well as change in the ratio of mRNA expression of pro- and anti-apoptotic proteins, also represent features of apoptotic cells. The presence of a sub-G0/G1 population in the cell cycle is considered indicative of DNA damage and apoptosis [165]. Comet assay, a genotoxic test, can be used as an indicator of early apoptosis, since cells entering apoptosis undergo DNA fragmentation resulting in the characteristic images. In these images the tail and the head indicate, respectively, fragmented and intact DNA [166].

Most EOs have been reported to cause cell death of melanoma cells, primarily inducing apoptosis, and hallmarks of apoptosis were recognized in melanoma cells treated with a huge number of EOs or their components. TTO induced apoptosis in A375 cells through the activation of caspases 3, 7 and 9, upregulation of p53 and Bax proapoptotic proteins and downregulation of bcl-2 antiapoptotic protein [129]. TTO and its main active component, terpinen-4-ol, also induced apoptosis in both adriamycin-sensitive and -resistant M14 cells: interaction with plasma membrane and subsequent reorganization of membrane lipid architecture has been identified as a possible mechanism through which TTO induced caspases dependent apoptosis [52,167]. Moreover, EO from *Salvia verbenaca* showed proapoptotic effects in M14 cells, and the EOs from cultivated plants exhibited major effects when compared to those growing in natural sites [75], thus further confirming the relevance of cultivar conditions on EOs activity. The same group also pointed a proapoptotic activity of *Salvia officinalis* EO in A375, M14 and A2058 cells, the percentages of main components depending on environmental factors [101]. The authors also suggested activation of apoptosis by *Salvia rubifolia* and *Salvia bracteata* EOs in M14 cells using a Comet assay [72,166]. *Boswellia carterii* EO (frankincense oil) was found to induce apoptosis through downregulation of Bcl-2 and up regulation of Bax proteins in B16-F10 cells, while inducing down regulation of Mcl-1 and cleavage of caspase 3 and 9 and PARP in human melanoma FM94 cells. On the contrary, proliferation of normal human epithelial melanocytes was not affected [146]. EOs extracted from *Pituranthos tortuosus* and *Annona Vepretorum* and their major constituents spathulenol, o-cymene and α-pinene, induced apoptosis of B16-F10 cells [25,66].

Hallmarks of apoptosis were also recognized in cells treated with several EO components, including camphene, α-pinene, eugenol, linalool, zerumbone, carvacrol, thujone, curzerene, citral, thymoquinone, isoegomaketone and menthol. Camphene isolated from *Piper cernuum* EO, induced apoptosis in B16-F10-Nex2 cells through loss of mitochondrial membrane potential, activation of caspases 3, endoplasmic reticulum stress, release of calcium, increased expression of high mobility group box 1 (HMGB1) and cell surface calreticulin [62]. The induction of HMGB1 and calreticulin after treatment with camphene could elicit immunogenic cell death, a relevant pathway for the activation of the immune system [168]. In B16-F10-Nex2 cells, α-pinene, a component present in many EOs including those from *Schinus terebinthifolius*, *Tridax procumbens*, *Pituranthos tortuosus*, *Annona Vepretorum* and *Boswellia carterii* [169], induced disruption of the mitochondrial potential, production of reactive oxygen species (ROS), activation of caspase 3, aggregation of heterochromatin, fragmentation of DNA and exposure of phosphatidyl serine on the cell surface [149,169]. Experiments performed to identify the structure/activity relationship, indicated the presence of a double bond in the α-pinene structure as crucial for its cytotoxic potential against both B16-F10-Nex2 and A2058 cells [78].

Proapoptotic properties, in terms of DNA fragmentation, phosphatidylserine exposure, and mitochondrial damage were reported by eugenol and the 6,6′-dibromo-dehydrodieugenol (S) enantiomeric form, in established and primary melanoma cells from patient tissue samples, with no effect on fibroblasts. Clastogenesis analysis and clonogenic assay also pointed out the ability of eugenol, respectively, to induce DNA breaks and to reduce colony forming potential, through a direct cytotoxicity or a lingering antiproliferative effect, in A2058 and SK-MEL-28 cells [134,136,170]. Scanning electron microscopy and transmission electron microscopy demonstrated that linalool, present in several EOs including those from *Citrus bergamia*, induced morphological changes and apoptosis in RPMI-7932 human melanoma cells while not affecting proliferation of normal keratinocytes [160].

The proapoptotic effect of zerumbone, one of the main constituents of EOs from *Zingiber zerumbet* and *Cheilocostus speciosus* [171] was proved in human melanoma CHL-1 cells through induction of ROS, reduction of mitochondrial matrix potential and mitochondrial biogenesis mediated by reduced mitochondrial ATP synthesis, mitochondrial DNA levels, and mRNA expression of mitochondrial transcription factor A level, a mitochondrial biogenesis factor [138]. Activation of apoptosis through downregulation of Bcl-2 and upregulation of Bax and cytochrome c gene and protein levels, as well as activation of caspases 3, was also observed in A375 cells after treatment with zerumbone [139]. Morphological and biochemical features of apoptosis in A375 cells were also induced by carvacrol, the main constituent isolated from *Coleus aromaticus* and present in other EOs, such as those from *Origanum ehrenbergii*, *Origanum syriacum* and *Satureja hortensis* [127]. EO fractions rich in thujone, isolated from *Thuja occidentalis* [172], induced apoptosis in A375 cells, with minimal growth inhibitory responses when exposed to normal cells [131]. Curzerene from *Eugenia uniflora* activated apoptosis in SK-MEL-19 cells [43].

Induction of apoptosis, necrosis, and autophagy was observed in B16-F10 cells after treatment with citral, a key component of EOs from *Cymbopogon citrates, Melissa officinalis* and *Verbena officinalis*. Apoptosis was associated with reduction of extracellular signal-regulated kinase-1 (ERK-1) and -2 (ERK-2), AKT and nuclear factor kappa B (NF-kB), as well as, induction of oxidative stress, DNA lesions, ROS and lipid peroxidation. SK-MEL-147 and UACC-257 cells showed a lower sensitivity to citral, when compared to B16-F10 cells [133].

Data pointing toward a proapoptotic activity of thymoquinone, a constituent of the EOs from *Nigella sativa* and *Thymus species,* have been also reported in B16-F10 cells through JAK2/STAT signal transduction [121]. Isoegomaketone from *Perilla frutescens* trigged ROS-mediated caspase-dependent and -independent apoptosis in B16 cells. In vitro studies were supported by in vivo experiments demonstrating that oral gavage of isoegomaketone in mice subcutaneously carrying B16 melanoma inhibited tumor growth, induced apoptosis, as well as increased Bax/Bcl-2 ratio [60].

Menthol, a compound present in EOs such as peppermint and mint has been reported to exert in vitro cytotoxic effect in A375 cells and to induce morphological changes, such as cell shrinkage and ruptured membranes, indicative of apoptosis. Decrease in transient receptor potential melastatin 8 (TRPM8), at the transcript level, was also evidenced following treatment with menthol. TRPM8 is a membrane receptor involved in the regulation of calcium ion influx and melanocytic behavior, and upregulated in melanoma [173]. The authors also hypothesized that the effect of menthol on TRPM8 expression could be linked to both decrease in cell proliferation and increase in cell death [137,161]. Menthol induced cytotoxicity was also pointed out in G-361 melanoma cells through a TRPM8-dependent mechanism only when using high doses of menthol [162], thus indicating a significant difference between A375 and G-361 cells in the sensitivity to menthol.

EOs and their main components demonstrated to induce apoptosis in melanoma cells are reported in Table 3. The in vitro and in vivo melanoma models used are also listed in the Table.

3.4. Induction of Necrosis or Modulation of Autophagy

Necrosis is a non-programmed cell death that, contrary to apoptosis, does not use a highly regulated intracellular program. Necrotic cells have usually lost cell membrane integrity and release products and enzymes in the extracellular space, with consequent activation of an inflammatory response. They are taken up and internalized by macropinocytotic mechanisms [163].

Autophagy is a recycling process playing a relevant role in cell survival and maintenance. Through the analysis of LC3II protein expression, presence of dot-like formations of endogenous LC3 protein and its colocalization with the lysosome marker LAMP-1, degradation of the specific autophagy substrate p62, use of early and late autophagy inhibitors, it is possible to analyze whether a particular compound affects autophagy, inducing autophagy rather than decreasing autophagosomal turnover [174,175].

Only a few EOs have been reported to induce necrosis in melanoma cells. Among them, TTO and its major active component, terpinen-4-ol, induced necrotic cell death coupled with low level apoptotic cell death in B16 cells. Necrosis was evidenced by ultrastructural features, including cell and organelle swelling, identified by video time lapse microscopy and transmission electron microscopy [130]. Measurement of lactate dehydrogenase (LDH) release from the cytosol into the supernatant demonstrated cellular membrane damage associated with necrosis in B16-F10 cells exposed to citral. Apoptosis induction and signs of autophagy were also evident after treatment with citral [133]. EOs distilled from *Salvia bracteata, Salvia rubifolia, Salvia aurea, Salvia judaica* and *Salvia viscosa* induced apoptosis and necrosis in M14 cells [71,72].

Regarding autophagy, some EOs or their components, have been found to affect basal autophagy or to trigger autophagy induced by serum starvation and/or rapamycin in cancer cells from different origin, indicating an mTOR independent mechanism [176]. α-Thujone, D-limonene, terpinel-4-ol and β-elemene are all components of EOs reported to affect in vitro and/or in vivo melanoma growth or melanogenesis [120,131,144,152,177] and to induce autophagy, respectively, in glioblastoma [178], neuroblastoma [179], leukemic [180] and breast carcinoma [181] cells. To the best of our knowledge, no studies exist regarding the ability of these or other EO components to induce autophagy or to affect autophagic flux in melanoma cells. Generation of autophagic vacuoles in B16-F10 cells was observed after treatment with citral. The authors suggested that B16-F10 cells shifted cellular metabolism, trying to recycle damaged structures by oxidative stress under treatment with citral, but no further characterization of autophagy was provided [133].

EOs and their main components demonstrated to induce necrosis or to modulate autophagy in melanoma cells are reported in Table 3. The in vitro melanoma models used are also listed in the Table.

3.5. Inhibition of Angiogenesis and Lymphangiogenesis

Angiogenesis, the formation of new blood vessels from existing microvessels, plays a relevant role in the growth and metastasization of many tumors, including melanoma [182]. Lesions at the beginning stage do not grow in the absence of angiogenesis or inflammation [183]. Furthermore, vasculogenic mimicry (VM) and lymphangiogenesis contribute to the metastatic spread of melanoma [184]. VM and lymphangiogenesis represent, respectively, the ability of cells to form networks of vessel-like channels and the formation of lymphatic vessels from pre-existing ones. Thus, inhibition of angiogenesis, VM or lymphangiogenesis, could represent a valid strategy for melanoma prevention and treatment. Vascular endothelial growth factor (VEGF) is one of the most important angiogenic factors that, through its binding to the receptor tyrosine kinase VEGFR, and the formation of a VEGF–VEGFR complex, induces angiogenesis (VEGF-A) or lymphangiogenesis (VEGF-C, VEGF-D) [185,186].

The ability to reduce angiogenesis has been highlighted in several EOs and their components. In particular, in vitro endothelial cells and in vivo/ex vivo assays such as, rat aortic ring, matrigel plug and chick embryo chorioallantoic membrane (CAM), have been used to study the effect of EOs on the formation of new blood vessels from pre-existing ones. Figure 1A summarizes pro- and antiangiogenic factors regulated by EOs or their components.

EO from *Pistacia lentiscus* (Mastic oil) has been extensively studied due to its antiangiogenic effect attributed to both its activities on in vitro endothelial cell proliferation and in vivo microvessel formation. It also inhibited VEGF release by B16 cells. Investigation of underlying mechanism by *Pistacia lentiscus* EOs in endothelial cells demonstrated activation of RhoA, a regulator of neovessel organization [64]. *Tridax procumbens* EO, when administered intraperitoneally, inhibited capillary formation in B16-F10 injected intradermally on the shaven ventral skin of C57BL/6 mice [86]. Using the same experimental model, the *Plectranthus amboinicus* EO ability to reduce tumor-directed blood vessel formation was also evidenced [67]. In addition, EO from *Curcuma zedoaria* was reported to suppress in vitro proliferation of human umbilical vein endothelial cells, sprouting vessels of aortic ring and formation of microvessels in CAM. The antiangiogenic effect was also confirmed by immunohistochemical analysis of tumors showing a reduced expression of the endothelial marker CD34 after oral administration of *Curcuma zedoaria* EO in C57BL/6 mice carrying B16-Bl6 melanoma [38].

Regarding EO components, several groups reported that β-elemene, one of the most active constituents of *Curcuma zedoaria* and *Curcuma wenyujin* EOs, inhibited the VEGF-induced sprouting vessel of rat aortic ring, microvessel formation of CAM, as well as CD34 and VEGF expression in C57BL/6 mice carrying B16-F10 melanoma after subretinal injection. VEGF expression in serum and lung of mice was also inhibited following treatment with β-elemene [120,152]. In vitro and in vivo studies carried out by Jung's group highlighted the ability of dietary β-caryophyllene, a component found in many EOs (basil, black pepper, cinnamon, cannabis, lavender, rosemary, cloves, oregano), to suppress high fat diet (HDF)-induced in vitro and in vivo angiogenesis and lymphangiogenesis. In particular, dietary β-caryophyllene reduced the expression at transcriptional and protein level of hypoxia inducible factors 1α, VEGF-A, CD31 and VE-cadherin observed in the tumors of HFD-fed mice. In addition, the HFD-stimulated expression of VEGF-C, VEGF-D, VEGF-R3 and lymphatic vessel endothelial receptor was prevented by β-caryophyllene supplementation in the diet. The authors indicate the effect of β-caryophyllene in angiogenesis as one of the most important mechanisms for reduced tumor growth in β-caryophyllene-fed mice [140].

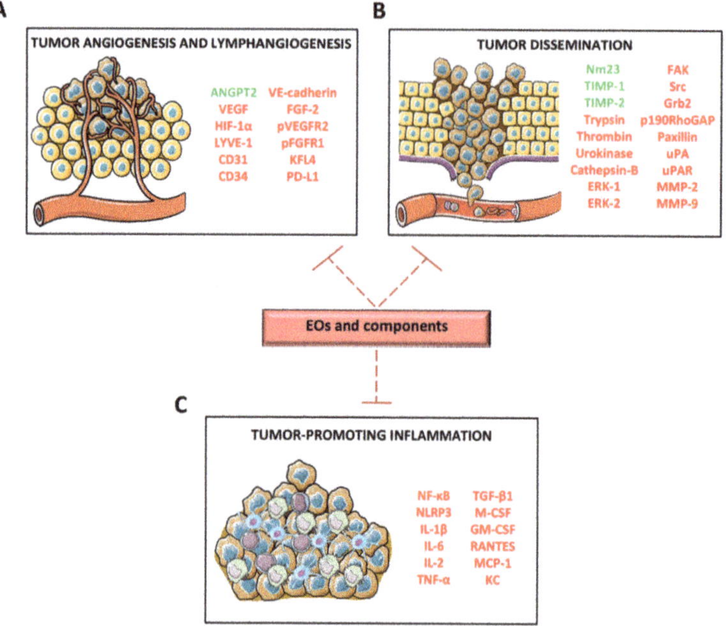

Figure 1. EOs and their components reduce tumor angiogenesis, lymphangiogenesis (**A**) and tumor metastasization, by targeting proteins responsible for tumor dissemination (**B**) and tumor-promoting inflammation (**C**). Angiopoietin 2 (ANGPT2), vascular endothelial growth factor (VEGF), hypoxia responsive factor 1α (HIF-1α), lymphatic vessel endothelial receptor (LYVE-1), cluster of differentiation 31 (CD31), cluster of differentiation 34 (CD34), vascular endothelial cadherin (VE-cadherin), fibroblast growth factor 2 (FGF-2), phospho vascular endothelial growth factor receptor 2 (pVEGFR2), phospho fibroblast growth factor receptor-1 (pFGFR1), krüppel-like factor 4 (KFL4), programmed death-ligand 1 (PD-L1), tissue inhibitor of metalloproteinase-1 (TIMP-1) and -2 (TIMP-2), extracellular signal-regulated kinase-1 (ERK-1) and -2 (ERK-2), focal adhesion kinase (FAK), growth factor receptor-bound protein 2 (Grb2), urokinase-type plasminogen activator (uPA), uPA receptor (uPAR), metalloproteinases-2 (MMP-2) and -9 (MMP-9), nuclear factor kappa B (NF-kB), NLR family pyrin domain containing 3 (NLRP3), interleukin 1β (IL-1 β), interleukin 6 (IL-6), interleukin 2 (IL-2), tumor necrosis factor-α (TNF-α), transforming growth factor β1 (TGF-β1), macrophage colony stimulating factor (M-CSF), granulocyte-macrophage colony stimulating factor (GM-CSF), monocyte chemoattractant protein-1 (MCP-1), keratinocyte chemoattractant (KC). Proteins that are upregulated by EOs or their components are reported in green, proteins that are downregulated by EOs or their components are reported in red. Parts of the figure are drawn using pictures from Servier Medical Art (https://smart.servier.com).

Even if a direct effect on melanoma angiogenesis has not yet been reported, several EOs or their components have been found to affect in vitro endothelial functions and/or in vivo neovascularisation. Among them, EOs from *Hypericum perforatum* [187] and *Myristica fragrans* [56] should be mentioned, together with coated-dacarbazine eugenol liposomes, for their ability to reduce proliferation and migration of endothelial-like cells [188]. *Citrus lemon* EO nanoemulsions have been found to decrease angiogenesis in CAM [189]. Interestingly, when analyzing the effect of the ointment prepared from *Salvia officinalis* EO on the healing process of an infected wound mouse model, Farahpour et al. demonstrated that, in addition to antioxidant and anti-inflammatory properties, *Salvia officinalis* EO accelerates the wound healing process. This process requires coordination of overlapping distinct cellular activities including angiogenesis. Promotion of the healing process by *Salvia officinalis* EO was attributed to enhanced angiogenesis through upregulation of VEGF and fibroblast growth factor-2

(FGF-2) expression. An increase in the number of blood vessels and fibroblasts, through cyclin-D1 pathway activation and enhanced expression of Bcl-2, was also observed [190]. A positive effect on wound healing process has also been described for other EOs such as lavender EO, TTO, *Alpinia zerumbet* and *Chrysantemum boreale* Makino EO and their use has been suggested for the treatment of wounds, burns, abscesses or diseases such as diabetes [191–194]. Thus, the antiangiogenic property or the positive effect on wound healing showed by some EOs should be considered when proposing them for treatment of cancer or other diseases.

The ability to interfere with the process of angiogenesis has also been attributed to zerumbone, perillyl alcohol and curcumol, three components of many EOs. Zerumbone, has been found to inhibit proliferation, migration, and morphogenesis, but not viability of endothelial cells, as well as the outgrowth of new blood vessels in rat aortic rings, vessel formation in the matrigel plug and CAM assays. All these effects were mediated by downregulation of phosphorylation of VEGFR2 and fibroblast growth factor receptor-1, two essential signaling pathways for angiogenesis [195,196]. Perillyl alcohol, a component of *Anethum graveolens*, *Conyza newii* and *Citrus limon* EOs [197], blocked the growth of new blood vessel in the in vivo CAM assay and inhibited the in vitro morphogenic differentiation of endothelial cells. Perillyl alcohol also reduced proliferation and induced both apoptosis and the expression of the antiangiogenic molecule, angiopoietin 2, in endothelial cells, indicating that it exerted its effect through both vessel regression and neovascularization suppression [198]. Curcumol, a representative component for the quality control of the EO of *Curcuma wenyujin*, inhibited angiogenesis by reducing PD-L1 expression in endothelial cells. In particular, addition of curcumol reduced the expression of VEGF and metalloproteinase-9 (MMP-9) and tube formation induced by PD-L1. It also cooperated with the ability of PD-L1 silencing to downregulate VEGF and MMP-9 expression and morphogenesis of endothelial cells [199].

EOs and their main components demonstrated to affect angiogenesis by using both endothelial and melanoma models are reported in Table 4. The used in vitro and in vivo models are also listed.

Table 4. EOs and their components that have been demonstrated to affect in vitro and in vivo angiogenesis and lymphangiogenesis.

Plant Name from Which EOs Were Extracted	EO Active Components	In Vitro and In Vivo Models	Reference
Citrus lemon	-	CAM	[189]
Curcuma zedoaria	-	HUVEC, CAM, rat aortic ring assay, B16-Bl6 (C57BL/6)	[38]
Hypericum perforatum	-	EAhy.926	[187]
Myristica fragrans	-	EAhy.926	[56]
Pistacia lentiscus	-	B16	[64]
Plectranthus amboinicus	-	B16-F10 (C57BL/6)	[67]
Salvia officinalis	-	Infected wound model (BALB/c)	[190]
Tridax procumbens	-	B16-F10 (C57BL/6)	[86]
	Curcumol	HUVEC	[199]
	Eugenol	EAhy.926	[188]
	Perillyl Alcohol	BLMVEC, HUVEC, B16-F10	[198]
	Zerumbone	HUVEC, CAM, rat aortic ring assay	[195,196]
	β-Caryophyllene	B16-F10 (C57BL/6N)	[140]
	β-Elemene	CAM, rat aortic ring assay, B16-F10 (C57BL/6)	[120,152]

Human Umbilical Vein Endothelial Cells (HUVEC), transformed human umbilical vein endothelial cells produced by fusion of A549/8 lung adenocarcinoma with human umbilical endothelial cells (EAhy.926), Bovine Lung Microvascular Endothelial Cells (BLMVEC), chick embryo chorioallantoic membrane (CAM). Where not reported, EO active components were not available in the cited paper. When referrring to in vivo studies, the murine strain used is indicated in brackets.

3.6. Alteration of In Vitro Tumor Progression-Associated Functions and Inhibition of In Vivo Tumor Growth and Metastasization

Tumor metastasization, the spread of tumor cells from the primary site to distant organs, represents the main cause of death of cancer patients, including those affected by melanoma. Thus, new therapeutic approaches, which are able to block functions associated to tumor progression, or even to prevent metastasization, represent a big turning point for cancer therapy. Several studies reported that the antimetastatic potential of EOs goes across the regulation of inflammatory cytokines and chemokines. Figure 1B,C summarizes factors regulated by EOs and responsible for tumor dissemination and tumor-promoting inflammation.

Several studies demonstrated EOs' ability to affect in vitro tumor progression-associated functions and in vivo tumor growth and metastasization. EO from *Alpinia zerumbet* was shown to inhibit transforming growth factor (TGF)-β1-induced endothelial-to-mesenchymal transition in endothelial cells through regulation of Krüppel-like factor 4. Activation of endothelial-to-mesenchymal transition, a process in which endothelial cells switch from the endothelial to a mesenchymal-like phenotype, cell marker and functions, contribute to cancer progression [141].

Treatment of A375 cells with EOs from *Satureja hortensis* inhibited the in vitro cell migration process, while not affecting migration of normal keratinocytes and fibroblasts [77]. TTO and its main active component, terpinen-4-ol, interfered with in vitro cell migration and invasion of adriamycin-sensitive and -resistant M14 cells by inhibiting the intracellular pathway induced by the multidrug transporter p170 glycoprotein [142]. In vivo results also reported that topical TTO formulation was able to slow the growth of B16-F10 melanoma subcutaneously injected in C57BL/6J mice. The treatment was accompanied by a quick and complete disappearance of skin irritation together with recruitment of neutrophils and other immune effector cells in the treated area, while it did not induce systemic toxicity [147]. Both studies highlighted the potential of TTO in topical formulations as a promising chemopreventive candidate or as an alternative topical antitumor treatment against melanoma.

Through its component β-ursolic acid, *Salvia officinalis* EO, inhibited proteases implicated in the mechanisms by which tumor cells metastasize, such as serine proteases (trypsin, thrombin and urokinase) and the cysteine protease cathepsin B. In vivo inhibition of lung colonization of B16 mouse melanoma cells by intraperitoneal administration of β-ursolic acid was also highlighted [148]. After oral administration, *Curcuma zedoaria* EO was reported to suppress in vivo growth of B16-Bl6 tumors after subcutaneous cell injection into the left oxter of C57BL/6 mice, and their metastasization to the lung. A reduced expression of MMP-2 and MMP-9 in serum of treated mice was also evidenced after treatment with *Curcuma zedoaria* EO [38].

Intraperitoneal treatment with *Boswellia carterii* EO reduced the tumor burden in C57BL/6 mice carrying the B16-F10 model, while it did not elicit a detrimental effect on body weight. The authors also reported hepatoprotection by the EO [146]. Considering that liver injury is a frequent consequence of melanoma drug treatments [200,201], this represents an important remark. In the same experimental model, EO from *Pituranthos tortuosus* inhibited in vitro cell migration and invasion, focal adhesion and invadopodia formation. It also induced downregulation of kinases, and molecules involved in cell movement and migration, such as FAK, Src, ERK, p130Cas and paxillin. A decreased expression of p190RhoGAP and Grb2, which impaired cell migration and actin assembly, was also induced by the *Pituranthos tortuosus* EO. In vivo treatment of B16-F10 carrying mice with intraperitoneal administered *Pituranthos tortuosus* EO led to impaired tumor growth with no sign of abnormal behavior or adverse toxicity [66].

Zornia brasiliensis [91] and *Annona vepretorum* [25] EOs intraperitoneally administered in C57BL/6 mice subcutaneously carrying B16-F10 melanoma elicited antitumor activity. Importantly, microencapsulation of the *Annona vepretorum* EO with β-cyclodextrin, used to form inclusion complexes with EO and to improve their characteristics, further increased in vivo tumor growth inhibition with respect to free-EO to the level induced by dacarbazine [202]. While not showing any lethal effect/abnormality on mice when injected intraperitoneally, EO from *Tridax procumbens* elicited a reduction of tumor lung nodules of B16-F10 cells injected through the tail vein. Increased apoptosis, associated with enhanced p53 expression, was also observed after treatment. Decrease in body weight, increase in white blood cells and decrease in haemoglobin observed in untreated group, were almost normalized in the EO treated group [86]. Using the same experimental model, the same group also evidenced the ability of EO from *Plectranthus amboinicus* to decrease experimental metastase formation [67].

Moreover, some EO components were reported to inhibit tumor progression-associated properties and in vivo metastasization. In this regard, zerumbone [138] and curzerene from *Eugenia uniflora* EO [43] inhibited in vitro migration of CHL-1 and SK-MEL-19 cells, respectively.

In the Section 3.5, we reported that β-caryophyllene reduced angiogenesis and lymphangiogenesis. In addition to these effects, β-caryophyllene in tumor tissues also reduced M2 macrophages and macrophage mannose receptors. Reduction of cytokines promoting macrophage recruitment and differentiation toward M2 type, such as keratinocyte chemoattractant, monocyte chemoattractant protein-1 (MCP-1) and macrophage colony stimulating factor, were also observed. β-Caryophyllene also increased the number of apoptotic cells and the expression of apoptosis related proteins Bax and activated caspases 3. In the adipose tissues surrounding the lymphnode, β-caryophyllene reduced M2 macrophages and blocked the CCL19-CCL21/CCR7 axis, a signaling pathway important for recruitment of CCR7-expressing cancer cells or leukocytes to lymphnodes. The authors suggested the use of β-caryophyllene for people with high risk of melanoma and/or consuming a high-fat diet regimen [140].

The antimetastatic potential of thujone was demonstrated after injection of B16-F10 cells in the lateral tail vein of C57BL/6 mice. In addition to the reduction of lung nodules, thujone administration also reduced expression of MMP-2, MMP-9, ERK-1, ERK-2, and VEGF and upregulated the expression of nm-23, tissue inhibitor of metalloproteinase-1 (TIMP-1), and TIMP-2 in the lung tissues and the production of pro-inflammatory cytokines such as tumor necrosis factor-α (TNF-α), interleukin (IL)-1β, IL-6, IL-2 and granulocyte-monocyte colony-stimulating factor. In the same model, thujone also inhibited in vitro secretion of MMP-2 and MMP-9 and the adhesion of tumor cells to the collagen-coated plate, as well as cell invasion and migration [144].

Intraperitoneally administered β-elemene was reported to inhibit in vivo growth and metastasization of C57BL/6 mice carrying B16-F10 melanoma through downregulation of tumor promoting factors such as MMP-2, MMP-9, VEGF, urokinase-type plasminogen activator (uPA) and uPA receptor. Reduction of melanin content in lung confirmed the antimetastatic effect of β-elemene [120,152]. The ability of intravitreal administered β-elemene to block the growth of subretinal injected B16-F10 cells in C57BL/6J mice was also reported [120].

Intraperitoneal administration of limonene and perillic acid remarkably reduced the experimental metastatic tumor nodule formation of C57BL/6 mice intravenously injected with B16-F10 cells and increased the life span of animals. Limonene and perillic acid treatment also induced an increased expression in lung tissues and an enhanced serum content of sialic acid and uronic acid, two biochemical markers playing important roles in tumor growth and metastasis, including cell–cell communication and tumor cell escape from immune surveillance [150,203,204]. An antimetastatic effect by α-pinene from *Schinus terebinthifolius* Raddi was reported when B16-F10-Nex2 cells were injected intravenously and C57BL/6 mice treated intraperitoneally [149]. Recent findings provided important insights into the mechanism through which α-pinene induced tumor regression in melanoma models. Some authors supposed the relevance of environment in minimizing cancer growth and reported that α-pinene has no inhibitory effect on melanoma cell proliferation in vitro, but indicated activation in the hypothalamus/sympathetic nerve/leptin axis tumor growth (decreased plasma leptin concentration) and in the immune system (increased the numbers of B cells, CD4+ T cells, CD8+ T cells, and NK cells) as possible mechanisms through which exposure to a fragrant environment containing α-pinene suppressed B16 tumor growth in C57BL/6 mice [205,206].

Camphene derived from *Piper cernuum* EO, when injected peritumorally, exerted antitumor activity in vivo by inhibiting subcutaneous growth of B16-F10-Nex2 in C57BL/6 mice, while it did not induce toxic effects, weight loss, or behavior alterations in mice [62]. Oral administration of thymoquinone reduced experimental metastases of B16-F10 model through destabilization of the oncogene MUC4 mRNA by tristetraprolin, a RNA binding protein regulating the MUC4 transcript [151], while intraperitoneal administration inhibited the growth of the B16-F10 intracerebral model and increased the median overall survival of C57BL/6J mice. Reduction by thymoquinone of inflammatory cytokines, such as MCP-1, TGF-β1, and RANTES, as well as induction of apoptosis, were identified as possible causes of tumor inhibition [121]. Another study evidenced thymoquinone's ability to inhibit the in vitro migration of both human (A375) and mouse (B16-F10) melanoma cells and to suppress B16-F10 metastasis in C57BL/6 mice by inhibition of NLRP3 inflammasome and NF-κB activity, thus indicating thymoquinone ability to act as a potential immunotherapeutic agent [145].

Intraperitoneal administration of eugenol in B6D2F1 mice bearing B16 melanoma, reduced tumor sizes, extended the mice median survival and reduced metastasis [135]. Myrtenal, one of the most abundant components in the *Teucrium polium* EO, reduced in vitro invasion and migration of both murine (B16-F0 and B16-F10) and human (SK-MEL-5) melanoma cells and metastasis induced in C57BL/6 mice bearing B16-F10 melanoma, through inhibition of the proton pump V-ATPases [143].

EOs and their main components demonstrated to affect in vitro tumor progression-associated functions and in vivo tumor growth and metastasization are reported in Table 2. The in vitro and in vivo models used are also listed in the Table.

3.7. Sensitization of Antitumor Agents

In addition to their ability to affect in vitro and in vivo tumor growth, several EO constituents have been reported to act synergistically with conventional chemotherapy and radiotherapy [207]. While some particular EO constituents, such as eugenol, geraniol, β-elemene, limonene, β-caryophyllene, have been shown to synergize with chemotherapy or radiotherapy in leukemia [208] or solid tumors [207,208], their efficacy in combination therapy for melanoma models are rare.

β-Elemene, one of the most active constituents of EOs from *Curcuma zedoaria* and *Curcuma wenyujin*, remarkably decreased A375 cell proliferation and enhanced apoptosis induced by radiation [209]. The effect of thymoquinone on the sensitivity of human 518A2 melanoma cells to doxorubicin was also analyzed. The authors demonstrated a cell line-dependent effect of thymoquinone on doxorubicin sensitivity inducing a synergistic proapoptotic effect on leukemia and multi-drug-resistant breast cancer cells, but not in 518A2 cells where the combination did not affect caspase kinetics or mitochondrial membrane potential but induced an antagonistic effect [208]. The authors indicate alteration in the apoptotic machinery of 518A2 cells [210] as a possible explanation of their response to thymoquinone/doxorubicin combination, thus indicating that further studies are needed to explore the effect of thymoquinone on doxorubicin sensitivity of melanoma cells. Surprisingly, the authors also reported a significant growth inhibitory effect of thymoquinone/doxorubicin combination in normal foreskin fibroblasts [208]. The ability of thymoquinone to further enhance the in vitro apoptotic effect of Gamma Knife irradiation has been reported in B16-F10, while it did not add any survival benefit to Gamma Knife treatment in C57BL/6J mice with intracerebral B16-F10 melanoma [121].

Liposomes loaded with dacarbazine and eugenol, and coated with hyaluronic acid in order to enable the active targeting of the CD44 receptor that is overexpressed by tumor cells [211], have been reported to inhibit proliferation of both human (SK-MEL-28) and mouse (B16-F10) melanoma cells and to induce both apoptosis and necrosis. The authors suggested the use of this liposome formulation to reduce dacarbazine dose during chemotherapy and consequently toxicity on normal cells [188].

EOs and their main components demonstrated to sensitize antitumor agents are reported in Table 5. The used melanoma models are also listed in the Table.

Table 5. Effect of EOs and their components in the sensitization of antitumor agents, chemoprevention and melanogenesis.

Pathway Affected	Plant Name from Which EOs Were Extracted	EO Active Components	In Vitro and In Vivo Models	Reference
Sensitization of antitumor agents		Eugenol	SK-MEL-28, B16-F10	[188]
		Thymoquinone	B16-F10	[121]
		β-Elemene	A375	[209]
Chemoprevention	*Mentha aquatica*	-	DMBA/TPA (FVB/NJ)	[54]
	Salvia libanotica	-	DMBA/TPA (BALB/c)	[73]
	Santalum album	α-Santol	DMBA/TPA (CD1, SENCAR)	[76]
		Eugenol	DMBA/TPA, DMBA/croton oil (Swiss albino)	[212,213]
		Farnesol	DMBA/TPA (Swiss albino)	[214]
		Geraniol	DMBA/TPA (Swiss albino)	[215]
		Limonene	DMBA/TPA (Swiss albino)	[177]
		Menthol	DMBA/TPA (ICR)	[216]
		Perillyl Alcohol	DMBA/TPras mut, DMBA/TPA (Swiss albino)	[217,218]
Melanogenesis	*Achillea millefolium*	Linalyl Acetate	B16	[23]
	Alpinia zerumbet	-	B16-F10	[24]
	Artemisia argyi	-	B16-F10	[28]
	Cinnamomum cassia	Cinnamaldehyde	B16	[32]
	Cinnamomum zeylanicum	-	B16	[33]
	Citrus grandis	-	B16-F10	[219]
	Citrus hystrix	-	B16-F10	[219]
	Citrus reticulata	-	B16-F10	[219]
	Cryptomeria japonica	-	B16	[220]
	Chrysanthemum boreale Makino	Cuminaldehyde	B16-Bl6	[31]
	Dalbergia pinnata	-	Zebrafish embryos	[39]
	Eucalyptus camaldulensis	-	B16-F10	[41]
	Glechoma hederacea	-	B16	[44]
	Melaleuca quinquenervia	1,8-Cineole, α-pipene, α-terpineol	B16	[53]
	Mentha aquatica	β-Caryophyllene	B16-F10	[221]
	Origanum syriacum	Carvacrol	B16-F1	[59,222]
	Origanum ehrenbergii	Carvacrol	B16-F1	[59]
	Pomelo peel	-	B16	[68]
	Psiadia terebinthina	-	B16-F10	[219]
	Syzygium aromaticum	Eugenol	B16	[81]
	Vetiveria zizanioides	Cedr-8-En-13-Ol	B16	[87]
	Vitex Negundo	-	B16-F10	[88]
	Vitex Trifolia	Abietatriene	B16-F10	[89]
		Phytol	B16-F10	[223]
		Thymoquinone	B16-F10	[224]
		Valencene	B16-F10	[225]
		Zerumbone	B16-F10	[226]

where not indicated, EO active components were not available from the cited articles. When referred to in vivo studies, the murine strain used is indicated in brackets. 7,12-dimethylbenz[a]anthracene (DMBA), 12-O-tetradecanoylphobol-13-acetate (TPA), HaRas gene driven by the tyrosinase promoter (TPras).

3.8. Chemopreventive Activity

Exposure to artificial or natural UV rays are among the major risks for the development of both non-melanoma and melanoma skin cancer. Other risk factors for melanoma development include the number of naevi and dysplastic naevi, phenotype characteristics and family history [227–229]. Chemopreventive agents prevent formation of cancer by multiple mechanisms, interfering with the initiation, promotion, or progression steps. Although mouse models that spontaneously develop melanoma are extremely rare, different chemically- or genetically-induced melanoma mouse models have been developed to evaluate the chemopreventive potential of compounds. Among them, the two-stage skin carcinogenesis 7,12-dimethylbenz[a]anthracene (DMBA)/12-O-tetradecanoylphobol-13-acetate (TPA) model, that fully recapitulates the multistage tumorigenesis of the skin, is the most commonly used to evaluate the chemopreventive potential of several compounds, including EOs. In particular, tumor initiation in BALB/c, CD1, ICR, SENCAR and Swiss albino models can be obtained with a single topical application of DMBA, whereas tumor promotion can be triggered by repeated applications of TPA [230]. To decrease the latency of melanoma appearance, these compounds are often administered in combination with other agents such as UV rays, or used in genetically engineered mice that harbor activating mutations in BRAF and NRAS, two oncogenes frequently mutated in human melanoma [230]. However, a drawback of DMBA/TPA models exists in the evidence that they induce the development of papillomas and small naevi more frequently than that of melanoma, making their use more accurate in the study of skin cancer on the whole, rather than of cutaneous melanoma. Thus, the chemopreventive effect of EOs or their constituents in skin cancer is discussed in this paragraph.

The chemopreventive potential of *Santalum album* EO and its major constituent α-santalol has been demonstrated in DMBA/TPA-induced skin tumors in CD1 mice: topical application of *Santalum album* EO reduced tumor incidence and multiplicity in animals. Furthermore, the chemopreventive potential of α-santalol was similar to that of *Santalum album* in DMBA/TPA-induced skin tumors in CD1 and SENCAR mice [76]. Topical application of *Salvia libanotica* EO (sage oil) delayed tumor appearance and inhibited tumor incidence and yield in DMBA/TPA in BALB/c mice, whereas decreasing EO concentration reduced only tumor yield [73]. By using the same DMBA/TPA model, *Mentha aquatica* var. *Kenting Water Mint* EO has been identified as a chemopreventive agent against cutaneous side effects induced by vemurafenib, a BRAF inhibitor used for treatment of melanoma patients carrying the BRAF^{V600} mutation. The results of this study evidenced that *Mentha aquatica* EO induces G2/M cell-cycle arrest and apoptosis, reduces cell viability, colony formation and the invasive and migratory functions of the mouse keratinocyte bearing HRAS^{Q61L} mutation. This mutation is found in melanoma patients with a higher probability of developing keratoacanthomas and squamous cell carcinoma after treatment with vemurafenib. In vivo treatment with *Mentha aquatica* EO decreased the formation of cutaneous papilloma and the expression of keratin14 and COX-2 observed in FVB/NJ mice exposed to DMBA/TPA and treated with vemurafenib [54].

Several studies investigated the chemopreventive potential of some components of EOs. Among them, Pal and colleagues demonstrated that oral administration of eugenol produced a reduction in the incidence and size of skin cancer in Swiss albino mice treated with DMBA and croton oil, along with an increase in the overall survival of mice. The carcinogenic process prevention by eugenol was due to reduction in cell proliferation and induction of apoptosis through the downregulation of bcl-2, c-Myc and H-ras expression along with the upregulation of active caspase 3, Bax and p53 in the skin lesions [213]. In the same year, Kaur and colleagues confirmed the inhibitory effect of eugenol in DMBA/TPA-induced skin cancer in Swiss albino mice. In particular, eugenol treatment delayed tumor onset, incidence and multiplicity when applied during the initiation, as well as during the promotion phase. The chemopreventive effect of eugenol was due to the induction of apoptosis, prevention of oxidative stress, decrease in ornithine decarboxylase activity, attenuation of tumor inflammation caused by reduction in NF-kB pathway, COX-2 and iNOS in tumor samples and in pro-inflammatory cytokines in mice serum (e.g., IL-6, TNF-α and PGE$_2$) [212]. However, both groups demonstrated that mice treated with DMBA and TPA developed squamous cell carcinoma of the skin in their experimental system [212,213].

Perillyl alcohol topical treatment showed the ability to delay development and incidence of DMBA-induced melanoma in transgenic mice harboring a mutated HaRas gene driven by the tyrosinase promoter (TPras). Moreover, perillyl alcohol treatment reduced the UV-induced ROS, the levels of Ras protein and inhibited the activation of MAPK and AKT pathways in a cell line derived from DMBA-induced melanoma of TPras mice [217]. A few years later, another two studies by Chaudhary and colleagues confirmed the chemopreventive effect of perillyl alcohol and its precursor D-limonene in DMBA/TPA-induced skin cancer in Swiss albino mice. In particular, topical application of perillyl alcohol or D-limonene elicited a significant reduction in tumor incidence and tumor burden with extension of the latency period of tumor development. In agreement with Prevatt et al., perillyl alcohol or D-limonene treatment effectively reduced the skin tumorigenesis by inducing apoptosis, reducing ROS production, inflammation, Ras/Raf/ERK1/2 pathway and Bcl-2 expression along with an increase in Bax levels [177,218]. Moreover, a phase 2a clinical trial showed a modest reduction in nuclear chromatin abnormality caused by twice-daily topical application of perillyl alcohol in participants with sun-damaged skin [231]. A phase 1 clinical trial demonstrated that daily topical application of perillyl alchol cream for 30 days was well tolerated in participants with normal appearing skin [232].

The chemopreventive potential of farnesol, geraniol and menthol was analyzed by several investigators using DMBA/TPA-promoted skin tumorigenesis in Swiss albino or in ICR, a strain of albino mice. These monoterpenes reduced tumor incidence and tumor volume with an extension of latency period during the promotion phase. The mechanism of action of these three components is the same as is described for perillyl alchol, D-limonene and geraniol. In particular, their chemopreventive effects occurring through alteration of phase II detoxification agents [215], reduction in inflammation and ROS production, suppression of the Ras/Raf/ERK1/2, p38 and NF-kB signaling pathways, reduction in Bcl-2 and induction of Bax expression [177,214].

EOs and their main components demonstrated to show chemopreventive potential are reported in Table 5. The used melanoma models are also listed in the Table.

3.9. Antioxidant Effect

Oxidative stress producing ROS, such as superoxide, hydrogen peroxide, and hydroxyl radical, are associated with many cancer types, including non-melanoma and melanoma skin cancer, where ROS are reported to cause free radical damage to the skin [233]. Oxidative stress is involved in all stages of melanoma development, and modulates intracellular pathways related to cellular proliferation and death. Several factors, including inadequate lifestyle and/or diet or UV-irradiation lead to the formation of ROS, which are often associated with alterations in the DNA, proteins and lipids, and consequent induction of cellular aging, mutagenicity and carcinogenicity. By chelating oxidation-catalytic metals or by scavenging free radicals and ROS, natural enzymatic and non-enzymatic antioxidant defense counteracts the dangerous effects of free radicals and other oxidants. Hence the relevance of antioxidant compounds to decrease oxidative stress or damage. In this context, many plant-derived components, including EOs, are reported as a useful antioxidant source able to remove excessive free radicals and to prevent free radical-induced damage.

To evaluate the antioxidant activity of a compound of interest, several test procedures should be carried out including cell-free methods, such as 2,2′-azino-bis-3-ethylbenzthiazoline-6-sulphonic acid (ABTS) and 1,1-diphenyl-2-picrylhydrazyl (DPPH) that are the most popular and commonly employed. Other cell-free assays include hydrogen peroxide, nitric oxide, peroxynitrite radical, superoxide radical, hydroxyl radical scavenging activity, metal-ion chelating assay, lipid peroxidation and the xanthine oxidase method. Analysis of oxidation status performed in in vitro cell systems or in vivo animal models (blood or tissues) include glutathione, glutathione peroxidase, glutathione-S-transferase, superoxide dismutase and catalase activity [234].

When performing a search on PubMed, about 2800 results in the last 20 years are shown for "essential oil" and "antioxidant". Analysis of the PubMed results identified only few research articles reporting studies on the evaluation of EO antioxidant activity in melanoma models. Among these studies, the antioxidant activity of *Melaleuca quinquenervia* EO and its constituents 1,8 cineole, α-pinene and α-terpineol [53,235], *Vetiveria zizanioides* EO and its most abundant compound, cedr-8-en-13-ol [87], *Cinnamomum cassia* EO and its major component cinnamaldehyde [32] and *Achillea Millefolium* EO with its main component linalyl acetate [23], was reflected in the recovered activities of glutathione peroxidase, superoxide dismutase, and catalase in α-MSH stimulated B16 cells. Protection from cell oxidative damage by *Lavandula Angustifolia* EO [49] and by high concentrations of *Eucalyptus camaldulensis* EO [41] was also observed in B16-F10 cells. Worthy of mention, is the in vivo study demonstrating that *Wedelia chinensis* EO (Osbeck oil) showed in vivo antioxidant activity by scavenging free radicals (lipid peroxidation and nitric oxide) and enhancing the level of endogenous antioxidants (catalase, superoxide dismutase, glutathione peroxidase, glutathione) in lung, liver and blood tissues of B16-F10-carrying C57BL/6 mice [90]. Protective effects against oxidative stress and apoptosis also in bovine aortic endothelial cells was induced by *Crocus Sativus* EO (Saffron oil) through SAPK/JNK and ERK1/2 signaling pathways, supporting its use in endothelial dysfunctions [236].

Table 6 shows studies in which the antioxidant effects of EOs and their components have been analyzed using cell-free assays or melanoma models.

Table 6. EOs and their components showing antioxidant effect in cell-free assay or in melanoma models.

Plant Name from Which EOs Were Extracted	EO Active Components	Cell Free Assay and Melanoma Models	Reference
Achillea millefolium	Linalyl Acetate	B16	[23]
Alpinia zerumbet	-	DPPH, ABTS, nitric oxide, hydroxyl radical scavenging activity, xanthine oxidase	[24]
Artemisia argyi	-	DPPH, ABTS, metal-ion chelation	[28]
Atriplex undulata	-	Crocin bleaching inhibition, DPPH	[29]
Cinnamomum cassia	Cinnamaldehyde	B16	[32]
Chrysanthemum boreale Makino	-	DPPH, ABTS	[31]
Cryptomeria japonica	-	B16	[220]
Cuminum Cyminum	-	DPPH, superoxide anion radical-scavenging activity, β-carotene/linoleic acid	[35]
Curcuma aromatica	-	DPPH	[36]
Curcuma kwangsiensis	-	DPPH	[37]
Dalbergia pinnata	-	DPPH, ABTS	[39]
Eucalyptus camaldulensis	-	DPPH, ABTS, B16-F10	[41]
Eugenia uniflora	-	DPPH, β-carotene/linoleic acid	[43]
Glechoma hederacea	-	β-carotene/linoleic acid, B16	[44]
Helichrysum microphyllum	-	DPPH, ABTS	[45]
Hypericum hircinum	-	DPPH, ABTS	[47]
Lavandula augustifolia	-	B16-F10	[49]
Melaleuca quinquenervia	1,8-Cineole, α-pipene, α-terpineol	B16	[53]
Ocimum basilicum	-	DPPH, hypoxanthine/xanthine oxidase	[58]
Ocimum gratissimum	-	DPPH, hypoxanthine/xanthine oxidase	[58,237]
Ocimum micranthum	-	DPPH, hypoxanthine/xanthine oxidase	[58]
Ocimum tenuiflorum	-	DPPH, hypoxanthine/xanthine oxidase	[58]

Table 6. *Cont.*

Plant Name from Which EOs Were Extracted	EO Active Components	Cell Free Assay and Melanoma Models	Reference
Piper aleyreanum	-	DPPH	[61]
Psidium guineense	-	DPPH, ABTS	[238]
Pomelo peel	-	DPPH, ABTS	[68]
Satureja hortensis	-	DPPH	[77]
Stachys cretica	-	DPPH	[79]
Stachys hydrophila	-	DPPH	[79]
Stachys palustri	-	DPPH	[79]
Stachys parviflora	-	DPPH, β-carotene/linoleic acid	[80]
Tanacetum macrophyllum	-	DPPH, ABTS, FRAP	[83]
Thymus munbyanus	-	DPPH, ABTS, FRAP	[85]
Thymus vulgaris	-	DPPH	[125]
Vetiveria zizanioides	-	β-carotene/linoleic acid, B16	[87]
Vitex Negundo	-	DPPH, ABTS, metal-ion chelation	[88]
Wedelia chinensis	-	B16-F10 (C57BL/6)	[90,239]
	Eugenol	A2058, SK-MEL 28	[134]

where not indicated, EO active components were not available from the cited articles. When referred to in vivo studies, the murine strain used is indicated in brackets. 1,1-diphenyl-2-picrylhydrazyl (DPPH), 2,2′-azino-bis-3-ethylbenzthiazoline-6-sulphonic acid (ABTS), ferric reducing/antioxidant power (FRAP).

3.10. Antimelanogenic Activity

Several EOs and their components have been reported to suppress melanogenesis, the secretion of melanin by epidermal melanocytes, and their use in skin-whitening materials has been suggested. Melanin is synthesized as a normal defense to diverse stimuli. An excessive production of melanin can be associated with hyperpigmentation and melanoma. Melanogenesis is stimulated by the melanogenic factors α-Melanocyte Stimulating Hormone (α-MSH) and Stem Cell Factor. Through the α-MSH/MC1R binding and the cyclic adenosine monophosphate-protein kinase A-cAMP response element binding protein (cAMP-PKA-CREB) signaling pathway, α-MSH stimulation activates microphthalmia-associated transcription factor (MITF), that in turn increases the expression of its target melanogenic genes, such as tyrosinase, tyrosinase-related protein-1 (TRP-1) and -2 (TRP-2). Furthermore, ERK1/2 are involved in the regulation of MITF expression through their ability to promote MITF phosphorylation and degradation [240,241]. α-MSH-induced melanogenesis is associated with ROS generation [242]. This is the reason why oxidation and melanogenesis are strictly interconnected, and EOs-induced decrease in melanin production is often attributed to EOs antioxidant property [243,244]. Figure 2 shows the effect of EOs and their components in melanogenesis and oxidation.

Several EOs, including those from *Pomelo Peel*, *Glechoma heredacea*, *Eucalyptus camaldulensis*, *Melaleuca quinquenervia*, *Chrysanthemum boreale Makino*, *Alpinia zerumbert*, *Achillea millefolium*, *Cinnamomum cassia*, *Artemisia argyi*, *Cryptomeria japonica* and *Vetiveria zizanioides* showed antimelanogenic activity, alongside their antioxidant properties. For these properties, they can be widely used in food, pharmaceutical, as well as in cosmetic industries as natural compounds.

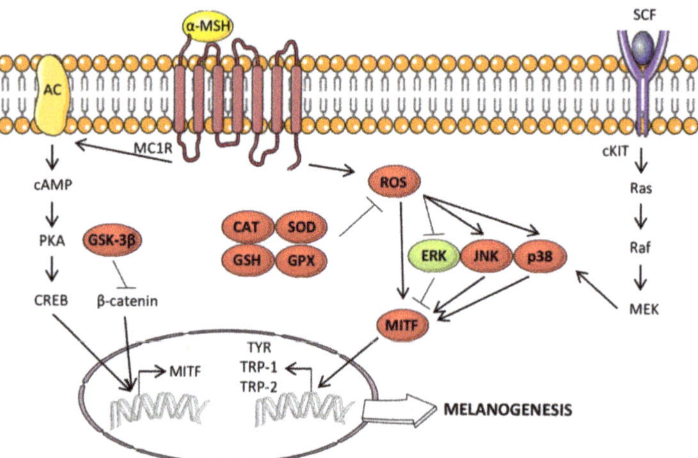

Figure 2. EOs and their components reduce melanogenesis and oxidation through interconnected mechanisms. Superoxide dismutase (SOD), glutathione peroxidase (GPX), catalase (CAT), glutathione (GSH), reactive oxygen species (ROS), α-melanocyte stimulating hormone (α-MSH), melanocortin 1 receptor (MC1R), adenylyl cyclase (AC), stem cell factor (SCF), microphthalmia-associated transcription factor (MITF), tyrosinase (TYR), tyrosinase-related protein -1 (TRP-1) and -2 (TRP-2), glycogen synthase kinase 3β (GSK3β), c-Jun N-terminal kinase (JNK), extracellular signal-regulated kinase (ERK). Proteins that are upregulated by EOs or their components are reported in green, proteins that are downregulated by EOs or their components are reported in red. Parts of the figure are drawn using pictures from Servier Medical Art (https://smart.servier.com).

The antimelanogenic properties of EO from *Pomelo Peel* have demonstrated in B16 cells where it blocked the synthesis pathway of melanin through the decrease of expression and activity of intracellular tyrosinase, without affecting cell viability and morphology [68]. Moreover, EOs from *Cryptomeria japonica* (Yakushima native cedar), *Syzygium aromaticum* (Clove oil) and *Cinnamomum zeylanicum* demonstrated antimelanogenesis activity in the same experimental model [33,81,245]. The last two EOs contained a high level of eugenol, which was also found to inhibit tyrosinase and melanogenesis in the same model [81]. An antimelanogenic characteristic of *Eucalyptus camaldulensis* EO was evidenced by its ability to inhibit the activity of mushroom tyrosinase, often used as the target enzyme in screening potential inhibitors of melanogenesis, and to decrease tyrosinase activity and intracellular melanin content of B16-F10 cells. It also reduced the expression level of MC1R, tyrosinase, TRP-1 and TRP-2, and of MAPK, JNK, PKA and ERK signaling pathways, thus suggesting the involvement of these pathways in *Eucalyptus camaldulensis* EO-mediated inhibition of melanogenesis [41]. In addition, EOs from *Psiadia terebinthina*, *Citrus grandis*, *Citrus hystrix*, and *Citrus reticulata* inhibited both intracellular and extracellular melanin production when tested against the B16-F10 model [219]. EOs from *Glechoma hederacea* [44], *Vetiveria zizanioides* [87], *Cinnamomum cassia* and its main component cinnamaldehyde [32], and *Melaleuca quinquenervia* and its main constituents, 1,8-cineole, α-pinene, and α-terpineol [53], showed potent anti-tyrosinase and antimelanogenic activities in α-MSH-stimulated B16 cells.

Other EOs showing antimelanogenic properties, in terms of reduction of melanin synthesis, intracellular tyrosinase activity and MITF expression, when tested in B16-F10 cells, include those extracted from *Vitex negundo* [88], *Origanum syriacum* and *Origanum ehrenbergii* and their main component, carvacrol [59], *Artemisia argyi* [28], *Chrysanthemum boreale* Makino [31], two varieties of *Alpinia Zerumbet* EOs, shima and tairin oils [24], *Achillea millefolium* and its component lynalyl acetate blocking melanin production through the regulation of the JNK and ERK signaling pathways [23],

Mentha aquatica (lime mint oil) and one of its main compounds, β-caryophyllene, [221], *Vitex Trifolia* and its main component abietatriene [89].

Regarding the antimelanogenic effect of EO components, the study of Lin JH's group reported the ability of zerumbrone to decrease melanin accumulation in B16-F10 cells, and to suppress the expression of MITF and its target genes, TYRP1 and TYRP2, after MSH stimulation with a mechanism involving ERK1/2, but not the PKA-CREB signaling pathway [226]. Antimelanogenic activity in the same melanoma model through ubiquitination and proteasomal degradation of MITF, in a ROS/ERK-dependent way, was also reported for phytol [223]. Valencene one of the major constituents of *Ocotea dispersa* EO decreased melanin content after UVB irradiation in B16-F10 cells [225]. A zebrafish embryo experimental model was also employed to demonstrate antimelanogenic activity of some EOs or their components, such as that from *Dalbergia pinnata*, which reduced tyrosinase activity and body pigmentation in zebrafish embryos [39], and thymoquinone which blocked melanogenesis also in B16-F10 mouse melanoma cells through the inhibition of the glycogen synthase kinase 3β (GSK3β)/β-catenin pathway [224]. Contrasting results were reported by the Zaidi KU group using the same experimental model and the same assays (tyrosinase activity and melanin production). The authors demonstrated that thymoquinone plays a protective role for melanogenesis [246].

Table 5 shows studies demonstrating the antimelanogenic activity of EOs and their active components in melanoma models. The used melanoma models are also listed.

4. Clinical Use of EOs for Cancer Patients

The side effects experienced by patients diagnosed with cancer or undergoing radio- or chemotherapy can be debilitating and can be challenging in the management of the disease. In the last few years, EOs have gained popularity as supportive therapies for cancer patients [247]. Some EOs have been reported to improve the quality of life of patients affected by cancer and showed efficacy in several side effects, such as chemotherapy-induced nausea and vomiting, mucositis, ulcer of skin, distress, depression and anxiety.

A descriptive systematic review, carried out by Boehm K et al. in 2012, highlighted short-term effects of aromatherapy, the use of EOs, on depression, anxiety, and overall wellbeing. Minimal adverse effects were reported for EO use and potential risks, including ingesting large amounts, local skin irritation, allergic contact dermatitis and phototoxicity [248]. A small feasibility study performed to evaluate the effects of mouthwash with EOs from *Leptospermum scoparium* and *Kunzea ericoides*, on mucositis of the oropharyngeal area induced by radiation during treatment for head and neck carcinoma, provided a positive effect on the development of radiation induced mucositis [249]. Topical application of the *Boswellia carterii* EO demonstrated its efficacy as supportive therapy for cancer-related fatigue in a case study [250]. Inhalation of *Citrus aurantium* EO exhibited an anxiolytic effect and reduced the symptoms associated with anxiety in patients with chronic myeloid leukaemia [251]. A cool damp washcloth with *Mentha x piperita* EO was reported to be effective in decreasing the intensity of nausea experienced by patients receiving chemotherapy [252]. Antiemetic activity of volatile oil from *Mentha spicata* and *Mentha piperita* has been also reported in chemotherapy-treated patients [253]. A recent randomized controlled trial with 120 patients evidenced the efficacy of aromatherapy with lavender and peppermint EOs in improving the sleep quality of cancer patients [254]. Linalool, linalyl acetate and menthol present in lavender or peppermint EOs [255,256] can be responsible of EOs effect on sleep due to their sedative effects. A previously performed study by M. Lisa Blackburn showed the positive effect of aromatherapy with lavender, peppermint and chamomile on insomnia and other common symptoms among 50 patients with acute leukaemia [257].

Some studies also reported the lack of effect of aromatherapy in improving sleep quality in cancer patients or in reducing chemotherapy side effects. In this regard, while non-toxic, non-invasive and well received and tolerated, the inhaled *Zingiber officinale* EO was not an effective complementary therapy for chemotherapy-induced nausea and vomiting and health-related quality of life neither in children with cancer [258] nor in women with breast cancer [259]. While not showing any harm or adverse events, the study by Sasano's group evidenced no effects of aromatherapy on quality of life, sleep quality, and vital sign during perioperative periods of breast cancer patients [260]. Meta-analysis of three randomized controlled trials including a total of 278 participants did not show any clinical effect of aromatherapy massage on reducing pain in cancer patients [261]. Inhalation of *Lavandula angustifolia*, *Citrus bergamia*, *Cedrus atlantica* EOs administered during radiotherapy did not reduce anxiety [262]. Sample size, duration of intervention, tools used for measuring the different symptoms, use of different chemotherapies and cancer stage could account for the differences observed in the different studies.

At present, several EOs are used in clinical trials to evaluate their efficacy for symptom management in patients during cancer therapy. In order to evaluate the effect of inhaled EOs on common quality of life issues during chemotherapy, targeted therapy, and/or immunotherapy administered intravenously, a single blind, randomized controlled trial study was completed few months ago. In particular, the effect of inhalated *Zingiber officinale*, *German chamomile*, or *Citrus bergamia* EOs on nausea, anxiety, loss of appetite, and fatigue has been evaluated with no available information yet about the results (NCT03858855). Two clinical trials are currently open and actively recruiting patients undergoing chemotherapy to assess the perceived effectiveness to relieve symptoms of nausea or vomiting and/or anxiety, by using peppermint and lavender EOs (NCT02163369, NCT03449511). A clinical trial is actively recruiting breast cancer patients in order to test the hypothesis that an EOs blend composed of *Curcuma longa*, *Piper nigrum*, *Pelargonium asperum*, *Zingiber officinale*, *Mentha x piperita*, and *Rosmarinus officinalis*, could reduce symptoms of chemotherapy induced peripheral neuropathy, a painful, debilitating consequence of cancer treatment considered the most adverse of non-hematologic events (NCT03449303). A randomized, controlled, single blind and longitudinal study is recruiting women with breast cancer undergoing chemotherapy in order to evaluate the impact of a hedonic aroma (inhalation) on the clinical, emotional and neurocognitive variables that contribute to reduce the side effects of chemotherapy and promote quality of life (NCT03585218). A not yet recruiting clinical trial aimed at evaluating (1) the ability of peppermint and lavender aromatherapy (sniff) to promote, respectively, relief of nausea or anxiety, in an outpatient oncology setting (NCT04449315); (2) the effects of inhaled peppermint and ginger EOs, or pure vanilla extract on chemotherapy induced nausea and vomiting in men and women with breast cancer (NCT04478630) is going to start in August/September of the current year 2020.

5. Conclusions

Natural products have always played a pivotal role in drug discovery and in the development of many potent anticancer agents. It is, thus, desirable to continue the efforts aiming at identifying new antitumor compounds from natural sources. Based on the studies mentioned in this review, different EOs and some of their constituents appear to be suitable as a part of effective melanoma prevention, prevention of metastatic melanoma, or as complementary therapies to supplement patient care. Preclinical data also indicate the possibility of using some EO components as adjuvant agents to reduce the toxicity of drugs used in cancer prevention or therapy, such as statin [263]. Pharmacokinetic studies are needed to validate the safety and efficacy of EOs and their bioactive compounds. In fact, even if several papers indicated EOs and their components as safe and not toxic [258–260], hepatotoxicity described for monoterpenes and sesquiterpenes, major components of many EOs, should be also considered and should deserve more attention [264]. Further studies on terpene metabolism and toxicity need to be performed to avoid the risk of eventual liver injury. In addition, the relevance of the BRAF status of melanoma cells in response to EOs is worthy of further investigation to shed light on this issue. In fact, to the best of our knowledge, no studies have evaluated whether the antitumor effect of EOs is related with BRAF status.

Since EOs composition is affected by several factors, including geographical position and agricultural practices, an important issue to be considered is the standardization of their composition. In this context, multidisciplinary applications, including machine learning, can constitute a possible tool able to predict the bioactivity of complex mixtures and to design EOs characterized by high antineoplastic efficacy and low toxicity [93,96,265].

The studies presented in this review hold promise for further analysis of EOs as new anticancer drugs and as a source of potential anticancer supplement against melanoma. Further investigations in this area are certainly necessary, desirable, and warranted to validate the results, to ascertain the therapeutic spectrum of biological studies and to determine the clinical efficacy and safety of EOs on patients affected by melanoma.

Author Contributions: Conceptualization, M.D.M., R.R. and D.D.B.; Figures and Tables preparation: M.D.M.; Writing—original draft preparation, M.D.M. and D.D.B.; Analysis of essential oils composition, S.G. and R.R. All authors have read and agreed to the published version of the manuscript.

Funding: M.D.M. is recipient of a fellowship from Italian Foundation for Cancer Research. The manuscript was supported by grants from Italian Association for Cancer Research (DDB, IG 18560); IRCCS Regina Elena National Cancer Institute (MDM, Ricerca Corrente 2018–2019); Progetti di Rilevante Interesse Nazionale 2017 (RR, prot. 2017JL8SRX); Ateneo Sapienza 2019 (RR, prot. RM11916B8876093E) and Ateneo Sapienza 2018 (RR, prot. RM118164361B425B).

Acknowledgments: This review is dedicated to the memory of our marvelous and joyful colleague and friend Marianna Desideri, who enthusiastically worked on our project on essential oils before she passed away. We thank Adele Petricca for preparation of the manuscript and Virginia Filacchione for English revision.

Conflicts of Interest: The authors declare no conflict of interest.

References

1. Tímár, J.; Vizkeleti, L.; Doma, V.; Barbai, T.; Rásó, E. Genetic progression of malignant melanoma. *Cancer Metastasis Rev.* **2016**, *35*, 93–107. [CrossRef] [PubMed]

2. Nograwdy, B. Game-changing class of immunotherapy drugs lengthens melanoma survival rates. *Nature* **2020**, *580*, S14–S16. [CrossRef] [PubMed]

3. Ambrosi, L.; Khan, S.; Carvajal, R.D.; Yang, J. Novel Targets for the Treatment of Melanoma. *Curr. Oncol. Rep.* **2019**, *21*, 97. [CrossRef] [PubMed]

4. Subbiah, V.; Baik, C.; Kirkwood, J.M. Clinical Development of BRAF plus MEK Inhibitor Combinations. *Trends Cancer* **2020**. [CrossRef] [PubMed]

5. Leonardi, G.C.; Candido, S.; Falzone, L.; Spandidos, D.A.; Libra, M. Cutaneous melanoma and the immunotherapy revolution (Review). *Int. J. Oncol.* **2020**. [CrossRef]

6. Shin, M.H.; Kim, J.; Lim, S.A.; Kim, J.; Lee, K.M. Current Insights into Combination Therapies with MAPK Inhibitors and Immune Checkpoint Blockade. *Int. J. Mol. Sci.* **2020**, *21*, 2531. [CrossRef]

7. Becco, P.; Gallo, S.; Poletto, S.; Frascione, M.P.M.; Crotto, L.; Zaccagna, A.; Paruzzo, L.; Caravelli, D.; Carnevale-Schianca, F.; Aglietta, M. Melanoma Brain Metastases in the Era of Target Therapies: An Overview. *Cancers* **2020**, *12*, 1640. [CrossRef]

8. Ishizuka, J.J.; Manguso, R.T.; Cheruiyot, C.K.; Bi, K.; Panda, A.; Iracheta-Vellve, A.; Miller, B.C.; Du, P.P.; Yates, K.B.; Dubrot, J.; et al. Loss of ADAR1 in tumours overcomes resistance to immune checkpoint blockade. *Nature* **2019**, *565*, 43–48. [CrossRef]

9. Mangan, B.L.; McAlister, R.K.; Balko, J.M.; Johnson, D.B.; Moslehi, J.J.; Gibson, A.; Phillips, E.J. Evolving Insights into the Mechanisms of Toxicity Associated with Immune Checkpoint Inhibitor Therapy. *Br. J. Clin. Pharmacol.* **2020**. [CrossRef]

10. Lu, H.; Liu, S.; Zhang, G.; Bin, W.; Zhu, Y.; Frederick, D.T.; Hu, Y.; Zhong, W.; Randell, S.; Sadek, N.; et al. PAK signalling drives acquired drug resistance to MAPK inhibitors in BRAF-mutant melanomas. *Nature* **2017**, *550*, 133–136. [CrossRef]

11. Cheung, M.K.; Yue, G.G.L.; Chiu, P.W.Y.; Lau, C.B.S. A Review of the Effects of Natural Compounds, Medicinal Plants, and Mushrooms on the Gut Microbiota in Colitis and Cancer. *Front. Pharmacol.* **2020**, *11*, 744. [CrossRef] [PubMed]

12. Pradhan, D.; Biswasroy, P.; Sahu, A.; Sahu, D.K.; Ghosh, G.; Rath, G. Recent advances in herbal nanomedicines for cancer treatment. *Curr. Mol. Pharmacol.* **2020**. [CrossRef] [PubMed]

13. Burnham, J.F. Scopus database: A review. *Biomed. Digit. Libr.* **2006**, *3*, 1–8. [CrossRef] [PubMed]

14. Zuzarte, M.; Salgueiro, L. Essential Oils Chemistry. In *Bioactive Essential Oils and Cancer*; de Sousa, D.o.P., Ed.; Springer International Publishing: Cham, Switzerland, 2015; pp. 19–61. [CrossRef]

15. Rehman, R.; Hanif, M.A.; Mushtaq, Z.; Al-Sadi, A.M. Biosynthesis of essential oils in aromatic plants: A review. *Food Rev. Int.* **2016**, *32*, 117–160. [CrossRef]

16. Dehsheikh, A.B.; Sourestani, M.M.; Dehsheikh, P.B.; Mottaghipisheh, J.; Vitalini, S.; Iriti, M. Monoterpenes: Essential Oil Components with Valuable Features. *Mini Rev. Med. Chem.* **2020**. [CrossRef]

17. De Cássia Da Silveira e Sá, R.; Andrade, L.N.; De Sousa, D.P. Sesquiterpenes from Essential Oils and Anti-Inflammatory Activity. *Nat. Prod. Commun.* **2015**, *10*. [CrossRef]

18. Dhifi, W.; Bellili, S.; Jazi, S.; Bahloul, N.; Mnif, W. Essential Oils' Chemical Characterization and Investigation of Some Biological Activities: A Critical Review. *Medicines* **2016**, *3*, 25. [CrossRef]

19. Tongnuanchan, P.; Benjakul, S. Essential Oils: Extraction, Bioactivities, and Their Uses for Food Preservation. *J. Food Sci.* **2014**, *79*, R1231–R1249. [CrossRef]

20. Masango, P. Cleaner production of essential oils by steam distillation. *J. Clean. Prod.* **2005**, *13*, 833–839. [CrossRef]

21. Kováts, E. Gas-chromatographische Charakterisierung organischer Verbindungen. Teil 1: Retentionsindices aliphatischer Halogenide, Alkohole, Aldehyde und Ketone. *Helv. Chim. Acta* **1958**, *41*, 1915–1932. [CrossRef]

22. Zellner, B.D.; Bicchi, C.; Dugo, P.; Rubiolo, P.; Dugo, G.; Mondello, L. Linear retention indices in gas chromatographic analysis: A review. *Flavour Frag. J.* **2008**, *23*, 297–314. [CrossRef]

23. Peng, H.Y.; Lin, C.C.; Wang, H.Y.; Shih, Y.; Chou, S.T. The melanogenesis alteration effects of *Achillea millefolium* L. Essential oil and linalyl acetate: Involvement of oxidative stress and the JNK and ERK signaling pathways in melanoma cells. *PLoS ONE* **2014**, *9*, e95186. [CrossRef] [PubMed]

24. Tu, P.T.; Tawata, S. Anti-Oxidant, Anti-Aging, and Anti-Melanogenic Properties of the Essential Oils from Two Varieties of Alpinia zerumbet. *Molecules* **2015**, *20*, 16723–16740. [CrossRef] [PubMed]

25. Bomfim, L.M.; Menezes, L.R.A.; Rodrigues, A.C.B.C.; Dias, R.B.; Gurgel Rocha, C.A.; Soares, M.B.P.; Neto, A.F.S.; Nascimento, M.P.; Campos, A.F.; Silva, L.C.R.C.E.; et al. Antitumour Activity of the Microencapsulation of *Annona vepretorum* Essential Oil. *Basic Clin. Pharmacol. Toxicol.* **2016**, *118*, 208–213. [CrossRef]

26. Conforti, F.; Menichini, F.; Formisano, C.; Rigano, D.; Senatore, F.; Bruno, M.; Rosselli, S.; Çelik, S. Anthemis wiedemanniana essential oil prevents LPS-induced production of NO in RAW 264.7 macrophages and exerts antiproliferative and antibacterial activities invitro. *Nat. Prod. Res.* **2012**, *26*, 1594–1601. [CrossRef]

27. Zhao, J.; Zheng, X.; Newman, R.A.; Zhong, Y.; Liu, Z.; Nan, P. Chemical composition and bioactivity of the essential oil of *Artemisia anomala* from China. *J. Essent. Oil Res.* **2013**, *25*, 520–525. [CrossRef]

28. Huang, H.C.; Wang, H.F.; Yih, K.H.; Chang, L.Z.; Chang, T.M. Dual bioactivities of essential oil extracted from the leaves of *Artemisia argyi* as an antimelanogenic versus antioxidant agent and chemical composition analysis by GC/MS. *Int. J. Mol. Sci.* **2012**, *13*, 14679–14697. [CrossRef]

29. Rodriguez, S.A.; Murray, A.P. Antioxidant activity and chemical composition of essential oil from *Atriplex undulata*. *Nat. Prod. Commun.* **2010**, *5*, 1841–1844. [CrossRef]

30. Salvador, M.J.; de Carvalho, J.E.; Wisniewski, A., Jr.; Kassuya, C.A.L.; Santos, E.P.; Riva, D.; Stefanello, M.E.A. Chemical composition and cytotoxic activity of the essential oil from the leaves of *Casearia lasiophylla*. *Rev. Bras. Farmacogn.* **2011**, *21*, 864–868. [CrossRef]

31. Kim, D.Y.; Won, K.J.; Hwang, D.I.; Park, S.M.; Kim, B.; Lee, H.M. Chemical Composition, Antioxidant and Anti-melanogenic Activities of Essential Oils from *Chrysanthemum boreale* Makino at Different Harvesting Stages. *Chem. Biodivers.* **2018**, *15*, e1700506. [CrossRef]

32. Chou, S.T.; Chang, W.L.; Chang, C.T.; Hsu, S.L.; Lin, Y.C.; Shih, Y. Cinnamomum cassia essential oil inhibits α-MSH-induced melanin production and oxidative stress in murine B16 melanoma cells. *Int. J. Mol. Sci.* **2013**, *14*, 19186–19201. [CrossRef] [PubMed]

33. Fiocco, D.; Arciuli, M.; Arena, M.P.; Benvenuti, S.; Gallone, A. Chemical composition and the anti-melanogenic potential of different essential oils. *Flavour Frag. J.* **2016**, *31*, 255–261. [CrossRef]

34. Menichini, F.; Tundis, R.; Loizzo, M.R.; Bonesi, M.; Provenzano, E.; Cindio, B.D.; Menichini, F. In vitro photo-induced cytotoxic activity of *Citrus bergamia* and *C. medica* L. cv. Diamante peel essential oils and identified active coumarins. *Pharm. Biol.* **2010**, *48*, 1059–1065. [CrossRef] [PubMed]

35. Hajlaoui, H.; Mighri, H.; Noumi, E.; Snoussi, M.; Trabelsi, N.; Ksouri, R.; Bakhrouf, A. Chemical composition and biological activities of *Tunisian Cuminum cyminum* L. essential oil: A high effectiveness against *Vibrio* spp. strains. *Food Chem. Toxicol.* **2010**, *48*, 2186–2192. [CrossRef] [PubMed]

36. Xiang, H.; Zhang, L.; Yang, Z.; Chen, F.; Zheng, X.; Liu, X. Chemical compositions, antioxidative, antimicrobial, anti-inflammatory and antitumor activities of *Curcuma aromatica* Salisb. essential oils. *Ind. Crop. Prod.* **2017**, *108*, 6–16. [CrossRef]

37. Zhang, L.; Yang, Z.; Huang, Z.; Zhao, M.; Li, P.; Zhou, W.; Zhang, K.; Zheng, X.; Lin, L.; Tang, J.; et al. Variation in Essential Oil and Bioactive Compounds of *Curcuma kwangsiensis* Collected from Natural Habitats. *Chem. Biodivers.* **2017**, *14*, e1700020. [CrossRef]

38. Chen, W.; Lu, Y.; Gao, M.; Wu, J.; Wang, A.; Shi, R. Anti-angiogenesis effect of essential oil from *Curcuma zedoaria* in vitro and in vivo. *J. Ethnopharmacol.* **2011**, *133*, 220–226. [CrossRef]

39. Zhou, W.; He, Y.; Lei, X.; Liao, L.; Fu, T.; Yuan, Y.; Huang, X.; Zou, L.; Liu, Y.; Ruan, R.; et al. Chemical composition and evaluation of antioxidant activities, antimicrobial, and anti-melanogenesis effect of the essential oils extracted from *Dalbergia pinnata* (Lour.) Prain. *J. Ethnopharmacol.* **2020**, *254*, 112731. [CrossRef]

40. Cianfaglione, K.; Blomme, E.E.; Quassinti, L.; Bramucci, M.; Lupidi, G.; Dall'Acqua, S.; Maggi, F. Cytotoxic Essential Oils from *Eryngium campestre* and *Eryngium amethystinum* (Apiaceae) Growing in Central Italy. *Chem. Biodivers.* **2017**, *14*, e1700096. [CrossRef]

41. Huang, H.C.; Ho, Y.C.; Lim, J.M.; Chang, T.Y.; Ho, C.L.; Chang, T.M. Investigation of the anti-melanogenic and antioxidant characteristics of Eucalyptus camaldulensis flower essential oil and determination of its chemical composition. *Int. J. Mol. Sci.* **2015**, *16*, 10470–10490. [CrossRef]

42. Aranha, E.S.P.; de Azevedo, S.G.; dos Reis, G.G.; Silva Lima, E.; Machado, M.B.; de Vasconcellos, M.C. Essential oils from *Eugenia* spp.: In vitro antiproliferative potential with inhibitory action of metalloproteinases. *Ind. Crop. Prod.* **2019**, *141*, 111736. [CrossRef]

43. Figueiredo, P.L.B.; Pinto, L.C.; da Costa, J.S.; da Silva, A.R.C.; Mourão, R.H.V.; Montenegro, R.C.; da Silva, J.K.R.; Maia, J.G.S. Composition, antioxidant capacity and cytotoxic activity of *Eugenia uniflora* L. chemotype-oils from the Amazon. *J. Ethnopharmacol.* **2019**, *232*, 30–38. [CrossRef] [PubMed]

44. Chou, S.T.; Lai, C.C.; Lai, C.P.; Chao, W.W. Chemical composition, antioxidant, anti-melanogenic and anti-inflammatory activities of *Glechoma hederacea* (Lamiaceae) essential oil. *Ind. Crop. Prod.* **2018**, *122*, 675–685. [CrossRef]

45. Ornano, L.; Venditti, A.; Sanna, C.; Ballero, M.; Maggi, F.; Lupidi, G.; Bramucci, M.; Quassinti, L.; Bianco, A. Chemical composition and biological activity of the essential oil from *Helichrysum microphyllum* cambess. ssp. tyrrhenicum bacch., brullo e giusso growing in la maddalena archipelago, Sardinia. *J. Oleo Sci.* **2015**, *64*, 19–26. [CrossRef]

46. Maggi, F.; Quassinti, L.; Bramucci, M.; Lupidi, G.; Petrelli, D.; Vitali, L.A.; Papa, F.; Vittori, S. Composition and biological activities of hogweed [*Heracleum sphondylium* L. subsp. ternatum (Velen.) Brummitt] essential oil and its main components octyl acetate and octyl butyrate. *Nat. Prod. Res.* **2014**, *28*, 1354–1363. [CrossRef]

47. Quassinti, L.; Lupidi, G.; Maggi, F.; Sagratini, G.; Papa, F.; Vittori, S.; Bianco, A.; Bramucci, M. Antioxidant and antiproliferative activity of *Hypericum hircinum* L. subsp. majus (Aiton) N. Robson essential oil. *Nat. Prod. Res.* **2013**, *27*, 862–868. [CrossRef]

48. Loizzo, M.R.; Tundis, R.; Menichini, F.; Saab, A.M.; Statti, G.A.; Menichini, F. Cytotoxic activity of essential oils from Labiatae and Lauraceae families against in vitro human tumor models. *Anticancer Res.* **2007**, *27*, 3293–3299.

49. Gismondi, A.; Canuti, L.; Grispo, M.; Canini, A. Biochemical composition and antioxidant properties of *Lavandula angustifolia* Miller essential oil are shielded by propolis against UV radiations. *Photochem. Photobiol.* **2014**, *90*, 702–708. [CrossRef]

50. Ferraz, R.P.; Bomfim, D.S.; Carvalho, N.C.; Soares, M.B.; da Silva, T.B.; Machado, W.J.; Prata, A.P.; Costa, E.V.; Moraes, V.R.; Nogueira, P.C.; et al. Cytotoxic effect of leaf essential oil of *Lippia gracilis* Schauer (Verbenaceae). *Phytomedicine* **2013**, *20*, 615–621. [CrossRef]

51. Quassinti, L.; Maggi, F.; Ortolani, F.; Lupidi, G.; Petrelli, D.; Vitali, L.A.; Miano, A.; Bramucci, M. Exploring new applications of tulip tree (*Liriodendron tulipifera* L.): Leaf essential oil as apoptotic agent for human glioblastoma. *Environ. Sci. Pollut. Res.* **2019**, *26*, 30485–30497. [CrossRef]

52. Calcabrini, A.; Stringaro, A.; Toccacieli, L.; Meschini, S.; Marra, M.; Colone, M.; Salvatore, G.; Mondello, F.; Arancia, G.; Molinari, A. Terpinen-4-ol, the main component of *Melaleuca alternifolia* (tea tree) oil inhibits the in vitro growth of human melanoma cells. *J. Investig. Dermatol.* **2004**, *122*, 349–360. [CrossRef] [PubMed]

53. Chao, W.W.; Su, C.C.; Peng, H.Y.; Chou, S.T. Melaleuca quinquenervia essential oil inhibits α-melanocyte-stimulating hormone-induced melanin production and oxidative stress in B16 melanoma cells. *Phytomedicine* **2017**, *34*, 191–201. [CrossRef] [PubMed]

54. Chang, C.T.; Soo, W.N.; Chen, Y.H.; Shyur, L.F. Essential Oil of Mentha aquatica var. Kenting water mint suppresses two-stage skin carcinogenesis accelerated by BRAF inhibitor vemurafenib. *Molecules* **2019**, *24*, 2344. [CrossRef] [PubMed]

55. Stefanello, M.É.A.; Riva, D.; De Carvalho, J.E.; Ruiz, A.L.T.G.; Salvador, M.J. Chemical Composition and Cytotoxic Activity of Essential Oil from *Myrcia laruotteana* Fruits. *J. Essent. Oil Res.* **2011**, *23*, 7–10. [CrossRef]

56. Piaru, S.P.; Mahmud, R.; Abdul Majid, A.M.; Mahmoud Nassar, Z.D. Antioxidant and antiangiogenic activities of the essential oils of *Myristica fragrans* and *Morinda citrifolia*. *Asian Pac. J. Trop. Med.* **2012**, *5*, 294–298. [CrossRef]

57. Grecco Sdos, S.; Martins, E.G.; Girola, N.; de Figueiredo, C.R.; Matsuo, A.L.; Soares, M.G.; Bertoldo Bde, C.; Sartorelli, P.; Lago, J.H. Chemical composition and in vitro cytotoxic effects of the essential oil from *Nectandra leucantha* leaves. *Pharm. Biol.* **2015**, *53*, 133–137. [CrossRef]

58. Trevisan, M.T.; Vasconcelos Silva, M.G.; Pfundstein, B.; Spiegelhalder, B.; Owen, R.W. Characterization of the volatile pattern and antioxidant capacity of essential oils from different species of the genus Ocimum. *J. Agric. Food Chem.* **2006**, *54*, 4378–4382. [CrossRef]

59. El Khoury, R.; Michael-Jubeli, R.; Bakar, J.; Dakroub, H.; Rizk, T.; Baillet-Guffroy, A.; Lteif, R.; Tfayli, A. Origanum essential oils reduce the level of melanin in B16-F1 melanocytes. *Eur. J. Dermatol.* **2019**, *29*, 596–602. [CrossRef]

60. Kwon, S.J.; Lee, J.H.; Moon, K.D.; Jeong, I.Y.; Ahn, D.U.K.; Lee, M.K.; Seo, K.I. Induction of apoptosis by isoegomaketone from *Perilla frutescens* L. in B16 melanoma cells is mediated through ROS generation and mitochondrial-dependent, -independent pathway. *Food Chem. Toxicol.* **2014**, *65*, 97–104. [CrossRef]

61. Da Silva, J.K.R.; Pinto, L.C.; Burbano, R.M.R.; Montenegro, R.C.; Guimarães, E.F.; Andrade, E.H.A.; Maia, J.G.S. Essential oils of Amazon Piper species and their cytotoxic, antifungal, antioxidant and anti-cholinesterase activities. *Ind. Crop. Prod.* **2014**, *58*, 55–60. [CrossRef]

62. Girola, N.; Figueiredo, C.R.; Farias, C.F.; Azevedo, R.A.; Ferreira, A.K.; Teixeira, S.F.; Capello, T.M.; Martins, E.G.A.; Matsuo, A.L.; Travassos, L.R.; et al. Camphene isolated from essential oil of Piper cernuum (Piperaceae) induces intrinsic apoptosis in melanoma cells and displays antitumor activity in vivo. *Biochem. Biophys. Res. Commun.* **2015**, *467*, 928–934. [CrossRef] [PubMed]

63. Lima, R.N.; Ribeiro, A.S.; Santiago, G.M.P.; D'S. Costa, C.O.; Soares, M.B.; Bezerra, D.P.; Shanmugam, S.; Freitas, L.D.S.; Alves, P.B. Antitumor and Aedes aegypti Larvicidal Activities of Essential Oils from Piper klotzschianum, *P. hispidum*, and *P. arboreum*. *Nat. Prod. Commun.* **2019**, *14*. [CrossRef]

64. Loutrari, H.; Magkouta, S.; Pyriochou, A.; Koika, V.; Kolisis, F.N.; Papapetropoulos, A.; Roussos, C. Mastic oil from *Pistacia lentiscus* var. chia inhibits growth and survival of human K562 leukemia cells and attenuates angiogenesis. *Nutr. Cancer* **2006**, *55*, 86–93. [CrossRef] [PubMed]

65. Krifa, M.; El Mekdad, H.; Bentouati, N.; Pizzi, A.; Ghedira, K.; Hammami, M.; El Meshri, S.E.; Chekir-Ghedira, L. Immunomodulatory and anticancer effects of *Pituranthos tortuosus* essential oil. *Tumor Biol.* **2015**, *36*, 5165–5170. [CrossRef] [PubMed]

66. Krifa, M.; El Meshri, S.E.; Bentouati, N.; Pizzi, A.; Sick, E.; Chekir-Ghedira, L.; Rondé, P. In Vitro and in Vivo Anti-Melanoma Effects of *Pituranthos tortuosus* Essential Oil Via Inhibition of FAK and Src Activities. *J. Cell. Biochem.* **2016**, *117*, 1167–1175. [CrossRef]

67. Manjamalai, A.; Grace, V.M.B. The chemotherapeutic effect of essential oil of *Plectranthus amboinicus* (Lour) on lung metastasis developed by B16F-10 cell line in C57BL/6 mice. *Cancer Investig.* **2013**, *31*, 74–82. [CrossRef]

68. He, W.; Li, X.; Peng, Y.; He, X.; Pan, S. Anti-oxidant and anti-melanogenic properties of essential oil from peel of Pomelo cv. Guan XI. *Molecules* **2019**, *24*, 242. [CrossRef]

69. Da Silva, E.B.; Matsuo, A.L.; Figueiredo, C.R.; Chaves, M.H.; Sartorelli, P.; Lago, J.H. Chemical constituents and cytotoxic evaluation of essential oils from leaves of *Porcelia macrocarpa* (Annonaceae). *Nat. Prod. Commun.* **2013**, *8*, 277–279. [CrossRef]

70. Dutra, R.C.; Pittella, F.; Dittz, D.; Marcon, R.; Pimenta, D.S.; Lopes, M.T.P.; Raposo, N.R.B. Chemical composition and cytotoxicity activity of the essential oil of *Pterodon emarginatus*. *Rev. Bras. Farmacogn.* **2012**, *22*, 971–978. [CrossRef]

71. Russo, A.; Formisano, C.; Rigano, D.; Cardile, V.; Arnold, N.A.; Senatore, F. Comparative phytochemical profile and antiproliferative activity on human melanoma cells of essential oils of three lebanese Salvia species. *Ind. Crop. Prod.* **2016**, *83*, 492–499. [CrossRef]

72. Cardile, V.; Russo, A.; Formisano, C.; Rigano, D.; Senatore, F.; Arnold, N.A.; Piozzi, F. Essential oils of *Salvia bracteata* and *Salvia rubifolia* from Lebanon: Chemical composition, antimicrobial activity and inhibitory effect on human melanoma cells. *J. Ethnopharmacol.* **2009**, *126*, 265–272. [CrossRef] [PubMed]

73. Gali-Muhtasib, H.U.; Affara, N.I. Chemopreventive effects of sage oil on skin papillomas in mice. *Phytomedicine* **2000**, *7*, 129–136. [CrossRef]

74. Alexa, E.; Sumalan, R.M.; Danciu, C.; Obistioiu, D.; Negrea, M.; Poiana, M.A.; Rus, C.; Radulov, I.; Pop, G.; Dehelean, C. Synergistic antifungal, allelopatic and anti-proliferative potential of *Salvia officinalis* L., and *Thymus vulgaris* L. Essential oils. *Molecules* **2018**, *23*, 185. [CrossRef] [PubMed]

75. Russo, A.; Cardile, V.; Graziano, A.C.E.; Formisano, C.; Rigano, D.; Canzoneri, M.; Bruno, M.; Senatore, F. Comparison of essential oil components and in vitro anticancer activity in wild and cultivated *Salvia verbenaca*. *Nat. Prod. Res.* **2015**, *29*, 1630–1640. [CrossRef] [PubMed]

76. Santha, S.; Dwivedi, C. Anticancer Effects of Sandalwood (*Santalum album*). *Anticancer Res.* **2015**, *35*, 3137–3145.

77. Popovici, R.A.; Vaduva, D.; Pinzaru, I.; Dehelean, C.A.; Farcas, C.G.; Coricovac, D.; Danciu, C.; Popescu, I.; Alexa, E.; Lazureanu, V.; et al. A comparative study on the biological activity of essential oil and total hydro-alcoholic extract of *Satureja hortensis* L. *Exp. Ther. Med.* **2019**, *18*, 932–942. [CrossRef] [PubMed]

78. Santana, J.S.; Sartorelli, P.; Guadagnin, R.C.; Matsuo, A.L.; Figueiredo, C.R.; Soares, M.G.; Da Silva, A.M.; Lago, J.H.G. Essential oils from *Schinus terebinthifolius* leaves chemical composition and in vitro cytotoxicity evaluation. *Pharm. Biol.* **2012**, *50*, 1248–1253. [CrossRef]

79. Conforti, F.; Menichini, F.; Formisano, C.; Rigano, D.; Senatore, F.; Arnold, N.A.; Piozzi, F. Comparative chemical composition, free radical-scavenging and cytotoxic properties of essential oils of six Stachys species from different regions of the Mediterranean Area. *Food Chem.* **2009**, *116*, 898–905. [CrossRef]

80. Shakeri, A.; D'Urso, G.; Taghizadeh, S.F.; Piacente, S.; Norouzi, S.; Soheili, V.; Asili, J.; Salarbashi, D. LC-ESI/LTQOrbitrap/MS/MS and GC–MS profiling of *Stachys parviflora* L. and evaluation of its biological activities. *J. Pharm. Biomed. Anal.* **2019**, *168*, 209–216. [CrossRef]

81. Arung, E.T.; Matsubara, E.; Kusuma, I.W.; Sukaton, E.; Shimizu, K.; Kondo, R. Inhibitory components from the buds of clove (*Syzygium aromaticum*) on melanin formation in B16 melanoma cells. *Fitoterapia* **2011**, *82*, 198–202. [CrossRef]

82. De Oliveira, P.F.; Alves, J.M.; Damasceno, J.L.; Oliveira, R.A.M.; Júnior Dias, H.; Crotti, A.E.M.; Tavares, D.C. Cytotoxicity screening of essential oils in cancer cell lines. *Rev. Bras. Farmacogn.* **2015**, *25*, 183–188. [CrossRef]

83. Venditti, A.; Frezza, C.; Sciubba, F.; Serafini, M.; Bianco, A.; Cianfaglione, K.; Lupidi, G.; Quassinti, L.; Bramucci, M.; Maggi, F. Volatile components, polar constituents and biological activity of tansy daisy (*Tanacetum macrophyllum* (Waldst. et Kit.) Schultz Bip.). *Ind. Crop. Prod.* **2018**, *118*, 225–235. [CrossRef]

84. Dall'Acqua, S.; Peron, G.; Ferrari, S.; Gandin, V.; Bramucci, M.; Quassinti, L.; Mártonfi, P.; Maggi, F. Phytochemical investigations and antiproliferative secondary metabolites from *Thymus alternans* growing in Slovakia. *Pharm. Biol.* **2017**, *55*, 1162–1170. [CrossRef]

85. Bendif, H.; Boudjeniba, M.; Miara, M.D.; Biqiku, L.; Bramucci, M.; Lupidi, G.; Quassinti, L.; Vitali, L.A.; Maggi, F. Essential Oil of *Thymus munbyanus* subsp. coloratus from Algeria: Chemotypification and in vitro Biological Activities. *Chem. Biodivers.* **2017**, *14*, e1600299. [CrossRef]

86. Manjamalai, A.; Mahesh Kumar, M.J.; Berlin Grace, V.M. Essential oil of tridax procumbens L induces apoptosis and suppresses angiogenesis and lung metastasis of the B16F-10 cell line in C57BL/6 mice. *Asian Pac. J. Cancer Prev.* **2012**, *13*, 5887–5895. [CrossRef] [PubMed]

87. Peng, H.Y.; Lai, C.C.; Lin, C.C.; Chou, S.T. Effect of Vetiveria zizanioides essential oil on melanogenesis in melanoma cells: Downregulation of tyrosinase expression and suppression of oxidative stress. *Sci. World J.* **2014**, *2014*. [CrossRef]

88. Huang, H.C.; Chang, T.Y.; Chang, L.Z.; Wang, H.F.; Yih, K.H.; Hsieh, W.Y.; Chang, T.M. Inhibition of melanogenesis Versus antioxidant properties of essential oil extracted from leaves of vitex negundo linn and chemical composition analysis by GC-MS. *Molecules* **2012**, *17*, 3902–3916. [CrossRef] [PubMed]

89. Lee, H.G.; Kim, T.Y.; Jeon, J.H.; Lee, S.H.; Hong, Y.K.; Jin, M.H. Inhibition of melanogenesis by abietatriene from Vitex trifolia leaf oil. *Nat. Prod. Sci.* **2016**, *22*, 252–258. [CrossRef]

90. Manjamalai, A.; Berlin Grace, V.M. Antioxidant activity of essential oils from *Wedelia chinensis* (Osbeck) in vitro and in vivo lung cancer bearing C57BL/6 mice. *Asian Pac. J. Cancer Prev.* **2012**, *13*, 3065–3071. [CrossRef]

91. Costa, E.V.; Menezes, L.R.A.; Rocha, S.L.A.; Baliza, I.R.S.; Dias, R.B.; Rocha, C.A.G.; Soares, M.B.P.; Bezerra, D.P. Antitumor properties of the leaf essential oil of *Zornia brasiliensis*. *Planta Med.* **2015**, *81*, 563–567. [CrossRef]

92. Harris, R. Synergism in the essential oil world. *Int. J. Aromather.* **2002**, *12*, 179–186. [CrossRef]

93. Patsilinakos, A.; Artini, M.; Papa, R.; Sabatino, M.; Bozovic, M.; Garzoli, S.; Vrenna, G.; Buzzi, R.; Manfredini, S.; Selan, L.; et al. Machine Learning Analyses on Data including Essential Oil Chemical Composition and In Vitro Experimental Antibiofilm Activities against Staphylococcus Species. *Molecules* **2019**, *24*, 890. [CrossRef] [PubMed]

94. Freitas, J.V.B.; Alves Filho, E.G.; Silva, L.M.A.; Zocolo, G.J.; de Brito, E.S.; Gramosa, N.V. Chemometric analysis of NMR and GC datasets for chemotype characterization of essential oils from different species of Ocimum. *Talanta* **2018**, *180*, 329–336. [CrossRef] [PubMed]

95. Clarke, S. Chapter 8-Handling, safety and practical applications for use of essential oils. In *Essential Chemistry for Aromatherapy*, 2nd ed.; Clarke, S., Ed.; Churchill Livingstone: Edinburgh, UK, 2008; pp. 231–264. [CrossRef]

96. Sabatino, M.; Fabiani, M.; Bozovic, M.; Garzoli, S.; Antonini, L.; Marcocci, M.E.; Palamara, A.T.; De Chiara, G.; Ragno, R. Experimental Data Based Machine Learning Classification Models with Predictive Ability to Select in Vitro Active Antiviral and Non-Toxic Essential Oils. *Molecules* **2020**, *25*, 2452. [CrossRef]

97. Delfine, S.; Marrelli, M.; Conforti, F.; Formisano, C.; Rigano, D.; Menichini, F.; Senatore, F. Variation of Malva sylvestris essential oil yield, chemical composition and biological activity in response to different environments across Southern Italy. *Ind. Crop. Prod.* **2017**, *98*, 29–37. [CrossRef]

98. Djarri, L.; Medjroubi, K.; Akkal, S.; Elomri, A.; Seguin, E.; Groult, M.L.; Verite, P. Variability of two essential oils of *Kundmannia sicula* (L.) DC., a traditional Algerian medicinal plant. *Molecules* **2008**, *13*, 812–817. [CrossRef]

99. Sefidkon, F.; Abbasi, K.; Khaniki, G.B. Influence of drying and extraction methods on yield and chemical composition of the essential oil of Satureja hortensis. *Food Chem.* **2006**, *99*, 19–23. [CrossRef]

100. Mohtashami, S.; Rowshan, V.; Tabrizi, L.; Babalar, M.; Ghani, A. Summer savory (*Satureja hortensis* L.) essential oil constituent oscillation at different storage conditions. *Ind. Crop. Prod.* **2018**, *111*, 226–231. [CrossRef]

101. Russo, A.; Formisano, C.; Rigano, D.; Senatore, F.; Delfine, S.; Cardile, V.; Rosselli, S.; Bruno, M. Chemical composition and anticancer activity of essential oils of Mediterranean sage (*Salvia officinalis* L.) grown in different environmental conditions. *Food Chem. Toxicol.* **2013**, *55*, 42–47. [CrossRef]

102. Garzoli, S.; Pirolli, A.; Vavala, E.; Di Sotto, A.; Sartorelli, G.; Bozovic, M.; Angiolella, L.; Mazzanti, G.; Pepi, F.; Ragno, R. Multidisciplinary Approach to Determine the Optimal Time and Period for Extracting the Essential Oil from Mentha suaveolens Ehrh. *Molecules* **2015**, *20*, 9640–9655. [CrossRef]

103. Bozovic, M.; Navarra, A.; Garzoli, S.; Pepi, F.; Ragno, R. Esential oils extraction: A 24-hour steam distillation systematic methodology. *Nat. Prod. Res.* **2017**, *31*, 2387–2396. [CrossRef] [PubMed]

104. Padilla-González, G.F.; Frey, M.; Gómez-Zeledón, J.; Da Costa, F.B.; Spring, O. Metabolomic and gene expression approaches reveal the developmental and environmental regulation of the secondary metabolism of yacón (*Smallanthus sonchifolius*, Asteraceae). *Sci. Rep.* **2019**, *9*, 1–15. [CrossRef] [PubMed]

105. Zámboriné Németh, É.; Thi Nguyen, H. Thujone, a widely debated volatile compound: What do we know about it? *Phytochem. Rev.* **2020**, *19*, 405–423. [CrossRef]

106. McAnally, J.A.; Jung, M.; Mo, H. Farnesyl-O-acetylhydroquinone and geranyl-O-acetylhydroquinone suppress the proliferation of murine B16 melanoma cells, human prostate and colon adenocarcinoma cells, human lung carcinoma cells, and human leukemia cells. *Cancer Lett.* **2003**, *202*, 181–192. [CrossRef] [PubMed]

107. Mo, H.; Tatman, D.; Jung, M.; Elson, C.E. Farnesyl anthranilate suppresses the growth, in vitro and in vivo, of murine B16 melanomas. *Cancer Lett.* **2000**, *157*, 145–153. [CrossRef]

108. Effenberger, K.; Breyer, S.; Schobert, R. Terpene conjugates of the *Nigella sativa* seed-oil constituent thymoquinone with enhanced efficacy in cancer cells. *Chem. Biodivers.* **2010**, *7*, 129–139. [CrossRef]

109. Tatman, D.; Mo, H. Volatile isoprenoid constituents of fruits, vegetables and herbs cumulatively suppress the proliferation of murine B16 melanoma and human HL-60 leukemia cells. *Cancer Lett.* **2002**, *175*, 129–139. [CrossRef]

110. Fabbri, J.; Maggiore, M.A.; Pensel, P.E.; Denegri, G.M.; Elissondo, M.C. In vitro efficacy study of *Cinnamomum zeylanicum* essential oil and cinnamaldehyde against the larval stage of *Echinococcus granulosus*. *Exp. Parasitol.* **2020**, *214*, 107904. [CrossRef]

111. Mediouni, S.; Jablonski, J.A.; Tsuda, S.; Barsamian, A.; Kessing, C.; Richard, A.; Biswas, A.; Toledo, F.; Andrade, V.M.; Even, Y.; et al. Oregano Oil and Its Principal Component, Carvacrol, Inhibit HIV-1 Fusion into Target Cells. *J. Virol.* **2020**, *94*. [CrossRef]

112. Liu, Q.; Meng, X.; Li, Y.; Zhao, C.N.; Tang, G.Y.; Li, H.B. Antibacterial and Antifungal Activities of Spices. *Int. J. Mol. Sci.* **2017**, *18*, 1283. [CrossRef]

113. Kokoska, L.; Kloucek, P.; Leuner, O.; Novy, P. Plant-Derived Products as Antibacterial and Antifungal Agents in Human Health Care. *Curr. Med. Chem.* **2019**, *26*, 5501–5541. [CrossRef] [PubMed]

114. Deyno, S.; Mtewa, A.G.; Abebe, A.; Hymete, A.; Makonnen, E.; Bazira, J.; Alele, P.E. Essential oils as topical anti-infective agents: A systematic review and meta-analysis. *Complement. Ther. Med.* **2019**, *47*, 102224. [CrossRef] [PubMed]

115. Fidler, I.J. Selection of successive tumour lines for metastasis. *Nat. New Biol.* **1973**, *242*, 148–149. [CrossRef] [PubMed]

116. Danciu, C.; Oprean, C.; Coricovac, D.E.; Andreea, C.; Cimpean, A.; Radeke, H.; Soica, C.; Dehelean, C. Behaviour of four different B16 murine melanoma cell sublines: C57BL/6J skin. *Int. J. Exp. Pathol.* **2015**, *96*, 73–80. [CrossRef]

117. Teicher, B.A. *Tumor Models in Cancer Research*; Springer Science & Business Media: Berlin/Heidelberg, Germany, 2010.

118. Paschoalin, T.; Carmona, A.K.; Rodrigues, E.G.; Oliveira, V.; Monteiro, H.P.; Juliano, M.A.; Juliano, L.; Travassos, L.R. Characterization of thimet oligopeptidase and neurolysin activities in B16F10-Nex2 tumor cells and their involvement in angiogenesis and tumor growth. *Mol. Cancer* **2007**, *6*, 44. [CrossRef]

119. Nakamura, K.; Yoshikawa, N.; Yamaguchi, Y.; Kagota, S.; Shinozuka, K.; Kunitomo, M. Characterization of mouse melanoma cell lines by their mortal malignancy using an experimental metastatic model. *Life Sci.* **2002**, *70*, 791–798. [CrossRef]

120. Shi, H.; Liu, L.; Liu, L.M.; Geng, J.; Chen, L. Inhibition of tumor growth by beta-elemene through downregulation of the expression of uPA, uPAR, MMP-2, and MMP-9 in a murine intraocular melanoma model. *Melanoma Res.* **2015**, *25*, 15–21. [CrossRef]

121. Hatiboglu, M.A.; Kocyigit, A.; Guler, E.M.; Akdur, K.; Nalli, A.; Karatas, E.; Tuzgen, S. Thymoquinone Induces Apoptosis in B16-F10 Melanoma Cell Through Inhibition of p-STAT3 and Inhibits Tumor Growth in a Murine Intracerebral Melanoma Model. *World Neurosurg.* **2018**, *114*, e182–e190. [CrossRef]

122. Xu, M.; McCanna, D.J.; Sivak, J.G. Use of the viability reagent PrestoBlue in comparison with alamarBlue and MTT to assess the viability of human corneal epithelial cells. *J. Pharmacol. Toxicol.* **2015**, *71*, 1–7. [CrossRef]

123. Mead, T.J.; Lefebvre, V. Proliferation assays (BrdU and EdU) on skeletal tissue sections. *Methods Mol. Biol.* **2014**, *1130*, 233–243.

124. Griffiths, M.; Sundaram, H. Drug design and testing: Profiling of antiproliferative agents for cancer therapy using a cell-based methyl-[3H]-thymidine incorporation assay. *Methods Mol. Biol.* **2011**, *731*, 451–465. [PubMed]

125. Alsaraf, S.; Hadi, Z.; Al-Lawati, W.M.; Al Lawati, A.A.; Khan, S.A. Chemical composition, in vitro antibacterial and antioxidant potential of Omani Thyme essential oil along with in silico studies of its major constituent. *J. King Saud Univ. Sci.* **2020**, *32*, 1021–1028. [CrossRef]

126. Mitropoulou, G.; Fitsiou, E.; Spyridopoulou, K.; Tiptiri-Kourpeti, A.; Bardouki, H.; Vamvakias, M.; Panas, P.; Chlichlia, K.; Pappa, A.; Kourkoutas, Y. Citrus medica essential oil exhibits significant antimicrobial and antiproliferative activity. *LWT Food Sci. Technol.* **2017**, *84*, 344–352. [CrossRef]

127. Govindaraju, S.; Arulselvi, P.I. Characterization of Coleus aromaticus essential oil and its major constituent carvacrol for in vitro antidiabetic and antiproliferative activities. *J. Herbs Spices Med. Plants* **2018**, *24*, 37–51. [CrossRef]

128. Melo, J.O.; Fachin, A.L.; Rizo, W.F.; Jesus, H.C.R.; Arrigoni-Blank, M.F.; Alves, P.B.; Marins, M.A.; França, S.C.; Blank, A.F. Cytotoxic effects of essential oils from three lippia gracilis schauer genotypes on HeLa, B16, and MCF-7 cells and normal human fibroblasts. *Genet. Mol. Res.* **2014**, *13*, 2691–2697. [CrossRef]

129. Ramadan, M.A.; Shawkey, A.E.; Rabeh, M.A.; Abdellatif, A.O. Expression of P53, BAX, and BCL-2 in human malignant melanoma and squamous cell carcinoma cells after tea tree oil treatment in vitro. *Cytotechnology* **2019**, *71*, 461–473. [CrossRef]

130. Greay, S.J.; Ireland, D.J.; Kissick, H.T.; Levy, A.; Beilharz, M.W.; Riley, T.V.; Carson, C.F. Induction of necrosis and cell cycle arrest in murine cancer cell lines by *Melaleuca alternifolia* (tea tree) oil and terpinen-4-ol. *Cancer Chemother. Pharmacol.* **2010**, *65*, 877–888. [CrossRef]

131. Biswas, R.; Mandal, S.K.; Dutta, S.; Bhattacharyya, S.S.; Boujedaini, N.; Khuda-Bukhsh, A.R. Thujone-Rich Fraction of Thuja occidentalis Demonstrates Major Anti-Cancer Potentials: Evidences from In Vitro Studies on A375 Cells. *Evid. Based Complement. Altern. Med.* **2011**, *2011*, 568148. [CrossRef]

132. Rajput, J.; Bagul, S.; Tadavi, S.; Bendre, R. Comparative Anti-Proliferative Studies of Natural Phenolic Monoterpenoids on Human Malignant Tumour Cells. *Med. Aromat. Plants* **2016**, *5*, 1–4. [CrossRef]

133. Sanches, L.J.; Marinello, P.C.; Panis, C.; Fagundes, T.R.; Morgado-Diaz, J.A.; de-Freitas-Junior, J.C.; Cecchini, R.; Cecchini, A.L.; Luiz, R.C. Cytotoxicity of citral against melanoma cells: The involvement of oxidative stress generation and cell growth protein reduction. *Tumour Biol.* **2017**, *39*. [CrossRef]

134. Junior, P.L.; Camara, D.A.; Costa, A.S.; Ruiz, J.L.; Levy, D.; Azevedo, R.A.; Pasqualoto, K.F.; de Oliveira, C.F.; de Melo, T.C.; Pessoa, N.D.; et al. Apoptotic effect of eugenol envolves G2/M phase abrogation accompanied by mitochondrial damage and clastogenic effect on cancer cell in vitro. *Phytomedicine* **2016**, *23*, 725–735. [CrossRef] [PubMed]

135. Ghosh, R.; Nadiminty, N.; Fitzpatrick, J.E.; Alworth, W.L.; Slaga, T.J.; Kumar, A.P. Eugenol causes melanoma growth suppression through inhibition of E2F1 transcriptional activity. *J. Biol. Chem.* **2005**, *280*, 5812–5819. [CrossRef]

136. Pisano, M.; Pagnan, G.; Loi, M.; Mura, M.E.; Tilocca, M.G.; Palmieri, G.; Fabbri, D.; Dettori, M.A.; Delogu, G.; Ponzoni, M.; et al. Antiproliferative and pro-apoptotic activity of eugenol-related biphenyls on malignant melanoma cells. *Mol. Cancer* **2007**, *6*, 8. [CrossRef] [PubMed]

137. Kijpornyongpan, T.; Sereemaspun, A.; Chanchao, C. Dose-dependent cytotoxic effects of menthol on human malignant melanoma A-375 cells: Correlation with TRPM8 transcript expression. *Asian Pac. J. Cancer Prev.* **2014**, *15*, 1551–1556. [CrossRef]

138. Yan, H.; Ren, M.Y.; Wang, Z.X.; Feng, S.J.; Li, S.; Cheng, Y.; Hu, C.X.; Gao, S.Q.; Zhang, G.Q. Zerumbone inhibits melanoma cell proliferation and migration by altering mitochondrial functions. *Oncol. Lett.* **2017**, *13*, 2397–2402. [CrossRef]

139. Wang, S.D.; Wang, Z.H.; Yan, H.Q.; Ren, M.Y.; Gao, S.Q.; Zhang, G.Q. Chemotherapeutic effect of Zerumbone on melanoma cells through mitochondria-mediated pathways. *Clin. Exp. Dermatol.* **2016**, *41*, 858–863. [CrossRef]

140. Jung, J.I.; Kim, E.J.; Kwon, G.T.; Jung, Y.J.; Park, T.; Kim, Y.; Yu, R.; Choi, M.S.; Chun, H.S.; Kwon, S.H.; et al. β-Caryophyllene potently inhibits solid tumor growth and lymph node metastasis of B16F10 melanoma cells in high-fat diet-induced obese C57BL/6N mice. *Carcinogenesis* **2015**, *36*, 1028–1039. [CrossRef]

141. Zhang, Y.; Li, C.; Huang, Y.; Zhao, S.; Xu, Y.; Chen, Y.; Jiang, F.; Tao, L.; Shen, X. EOFAZ inhibits endothelialtomesenchymal transition through downregulation of KLF4. *Int. J. Mol. Med.* **2020**, *46*, 300–310.

142. Bozzuto, G.; Colone, M.; Toccacieli, L.; Stringaro, A.; Molinari, A. Tea tree oil might combat melanoma. *Planta Med.* **2011**, *77*, 54–56. [CrossRef]

143. Martins, B.X.; Arruda, R.F.; Costa, G.A.; Jerdy, H.; de Souza, S.B.; Santos, J.M.; de Freitas, W.R.; Kanashiro, M.M.; de Carvalho, E.C.Q.; Sant'Anna, N.F.; et al. Myrtenal-induced V-ATPase inhibition-A toxicity mechanism behind tumor cell death and suppressed migration and invasion in melanoma. *Biochim. Biophys. Acta Gen. Subj.* **2019**, *1863*, 1–12. [CrossRef]

144. Siveen, K.S.; Kuttan, G. Thujone inhibits lung metastasis induced by B16F-10 melanoma cells in C57BL/6 mice. *Can. J. Physiol. Pharmacol.* **2011**, *89*, 691–703. [CrossRef] [PubMed]

145. Ahmad, I.; Muneer, K.M.; Tamimi, I.A.; Chang, M.E.; Ata, M.O.; Yusuf, N. Thymoquinone suppresses metastasis of melanoma cells by inhibition of NLRP3 inflammasome. *Toxicol. Appl. Pharmacol.* **2013**, *270*, 70–76. [CrossRef] [PubMed]

146. Hakkim, F.L.; Bakshi, H.A.; Khan, S.; Nasef, M.; Farzand, R.; Sam, S.; Rashan, L.; Al-Baloshi, M.S.; Abdo Hasson, S.S.A.; Jabri, A.A.; et al. Frankincense essential oil suppresses melanoma cancer through down regulation of Bcl-2/Bax cascade signaling and ameliorates heptotoxicity via phase I and II drug metabolizing enzymes. *Oncotarget* **2019**, *10*, 3472–3490. [CrossRef]

147. Greay, S.J.; Ireland, D.J.; Kissick, H.T.; Heenan, P.J.; Carson, C.F.; Riley, T.V.; Beilharz, M.W. Inhibition of established subcutaneous murine tumour growth with topical *Melaleuca alternifolia* (tea tree) oil. *Cancer Chemother. Pharmacol.* **2010**, *66*, 1095–1102. [CrossRef]

148. Jedinak, A.; Muckova, M.; Kost'alova, D.; Maliar, T.; Masterova, I. Antiprotease and antimetastatic activity of ursolic acid isolated from Salvia officinalis. *Z. Naturforsch. C J. Biosci.* **2006**, *61*, 777–782. [CrossRef]

149. Matsuo, A.L.; Figueiredo, C.R.; Arruda, D.C.; Pereira, F.V.; Scutti, J.A.; Massaoka, M.H.; Travassos, L.R.; Sartorelli, P.; Lago, J.H. Alpha-Pinene isolated from *Schinus terebinthifolius* Raddi (Anacardiaceae) induces apoptosis and confers antimetastatic protection in a melanoma model. *Biochem. Biophys. Res. Commun.* **2011**, *411*, 449–454. [CrossRef]

150. Raphael, T.J.; Kuttan, G. Effect of naturally occurring monoterpenes carvone, limonene and perillic acid in the inhibition of experimental lung metastasis induced by B16F-10 melanoma cells. *J. Exp. Clin. Cancer Res.* **2003**, *22*, 419–424.

151. Lee, S.R.; Mun, J.Y.; Jeong, M.S.; Lee, H.H.; Roh, Y.G.; Kim, W.T.; Kim, M.H.; Heo, J.; Choi, Y.H.; Kim, S.J.; et al. Thymoquinone-Induced Tristetraprolin Inhibits Tumor Growth and Metastasis through Destabilization of MUC4 mRNA. *Int. J. Mol. Sci.* **2019**, *20*, 2614. [CrossRef] [PubMed]

152. Chen, W.; Lu, Y.; Wu, J.; Gao, M.; Wang, A.; Xu, B. Beta-elemene inhibits melanoma growth and metastasis via suppressing vascular endothelial growth factor-mediated angiogenesis. *Cancer Chemother. Pharmacol.* **2011**, *67*, 799–808. [CrossRef]

153. Lim, S.; Kaldis, P. Cdks, cyclins and CKIs: Roles beyond cell cycle regulation. *Development* **2013**, *140*, 3079. [CrossRef]

154. Bommareddy, A.; Brozena, S.; Steigerwalt, J.; Landis, T.; Hughes, S.; Mabry, E.; Knopp, A.; VanWert, A.L.; Dwivedi, C. Medicinal properties of alpha-santalol, a naturally occurring constituent of sandalwood oil. *Nat. Prod. Res.* **2019**, *33*, 527–543. [CrossRef]

155. Zhang, X.; Dwivedi, C. Skin cancer chemoprevention by α-santalol. *Front. Biosci. Schol. Ed.* **2011**, *3*, S777–S787.

156. Barboza, J.N.; da Silva Maia Bezerra Filho, C.; Silva, R.O.; Medeiros, J.V.R.; de Sousa, D.P. An overview on the anti-inflammatory potential and antioxidant profile of eugenol. *Oxid. Med. Cell. Longev.* **2018**, *2018*. [CrossRef] [PubMed]

157. Liu, X.; Mi, J.; Qin, H.; Li, Z.; Chai, J.; Li, M.; Wu, J.; Xu, J. E2F1/IGF-1R Loop Contributes to BRAF Inhibitor Resistance in Melanoma. *J. Investig. Dermatol.* **2020**, *140*, 1295–1299. [CrossRef]

158. Meng, P.; Bedolla, R.G.; Yun, H.; Fitzpatrick, J.E.; Kumar, A.P.; Ghosh, R. Contextual role of E2F1 in suppression of melanoma cell motility and invasiveness. *Mol. Carcinog.* **2019**, *58*, 1701–1710. [CrossRef] [PubMed]

159. Delmondes, G.A.; Santiago Lemos, I.C.; Dias, D.Q.; Cunha, G.L.D.; Araújo, I.M.; Barbosa, R.; Coutinho, H.D.M. Pharmacological applications of farnesol (C(15)H(26)O): A patent review. *Expert Opin. Ther. Pat.* **2020**, *30*, 227–234. [CrossRef]

160. Cerchiara, T.; Straface, S.V.; Brunelli, E.; Tripepi, S.; Gallucci, M.C.; Chidichimo, G. Antiproliferative effect of linalool on RPMI 7932 human melanoma cell line: Ultrastructural studies. *Nat. Prod. Commun.* **2015**, *10*, 547–549. [CrossRef]

161. Slominski, A. Cooling skin cancer: Menthol inhibits melanoma growth. Focus on "TRPM8 activation suppresses cellular viability in human melanoma". *Am. J. Physiol. Cell Physiol.* **2008**, *295*, C293–C295. [CrossRef]

162. Yamamura, H.; Ugawa, S.; Ueda, T.; Morita, A.; Shimada, S. TRPM8 activation suppresses cellular viability in human melanoma. *Am. J. Physiol. Cell Physiol.* **2008**, *295*, C296–C301. [CrossRef] [PubMed]

163. D'Arcy, M.S. Cell death: A review of the major forms of apoptosis, necrosis and autophagy. *Cell Biol. Int.* **2019**, *43*, 582–592. [CrossRef] [PubMed]

164. Krysko, D.V.; Berghe, T.V.; Parthoens, E.; D'Herde, K.; Vandenabeele, P. Methods for distinguishing apoptotic from necrotic cells and measuring their clearance. *Method. Enzymol.* **2008**, *442*, 307–341.

165. Galluzzi, L.; Vitale, I.; Aaronson, S.A.; Abrams, J.M.; Adam, D.; Agostinis, P.; Alnemri, E.S.; Altucci, L.; Amelio, I.; Andrews, D.W.; et al. Molecular mechanisms of cell death: Recommendations of the Nomenclature Committee on Cell Death 2018. *Cell Death Differ.* **2018**, *25*, 486–541. [CrossRef]

166. Choucroun, P.; Gillet, D.; Dorange, G.; Sawicki, B.; Dewitte, J.D. Comet assay and early apoptosis. *Mutat. Res.* **2001**, *478*, 89–96. [CrossRef]

167. Giordani, C.; Molinari, A.; Toccacieli, L.; Calcabrini, A.; Stringaro, A.; Chistolini, P.; Arancia, G.; Diociaiuti, M. Interaction of tea tree oil with model and cellular membranes. *J. Med. Chem.* **2006**, *49*, 4581–4588. [CrossRef] [PubMed]

168. Galluzzi, L.; Vitale, I.; Warren, S.; Adjemian, S.; Agostinis, P.; Martinez, A.B.; Chan, T.A.; Coukos, G.; Demaria, S.; Deutsch, E.; et al. Consensus guidelines for the definition, detection and interpretation of immunogenic cell death. *J. Immunother. Cancer* **2020**, *8*. [CrossRef]

169. Salehi, B.; Upadhyay, S.; Orhan, I.E.; Jugran, A.K.; Jayaweera, S.L.D.; Dias, D.A.; Sharopov, F.; Taheri, Y.; Martins, N.; Baghalpour, N.; et al. Therapeutic potential of α-and β-pinene: A miracle gift of nature. *Biomolecules* **2019**, *9*, 738. [CrossRef]

170. Jaganathan, S.K.; Supriyanto, E. Antiproliferative and molecular mechanism of eugenol-induced apoptosis in cancer cells. *Molecules* **2012**, *17*, 6290–6304. [CrossRef]

171. Girisa, S.; Shabnam, B.; Monisha, J.; Fan, L.; Halim, C.E.; Arfuso, F.; Ahn, K.S.; Sethi, G.; Kunnumakkara, A.B. Potential of zerumbone as an anti-cancer agent. *Molecules* **2019**, *24*, 734. [CrossRef]

172. Pelkonen, O.; Abass, K.; Wiesner, J. Thujone and thujone-containing herbal medicinal and botanical products: Toxicological assessment. *Regul. Toxicol. Pharm.* **2013**, *65*, 100–107. [CrossRef]

173. Guo, H.; Carlson, J.A.; Slominski, A. Role of TRPM in melanocytes and melanoma. *Exp. Dermatol.* **2012**, *21*, 650–654. [CrossRef] [PubMed]

174. Parzych, K.R.; Klionsky, D.J. An overview of autophagy: Morphology, mechanism, and regulation. *Antioxid. Redox. Signal.* **2014**, *20*, 460–473. [CrossRef]

175. Kuma, A.; Komatsu, M.; Mizushima, N. Autophagy-monitoring and autophagy-deficient mice. *Autophagy* **2017**, *13*, 1619–1628. [CrossRef] [PubMed]

176. Athamneh, K.; Alneyadi, A.; Alsamri, H.; Alrashedi, A.; Palakott, A.; El-Tarabily, K.A.; Eid, A.H.; Al Dhaheri, Y.; Iratni, R. Origanum majorana Essential Oil Triggers p38 MAPK-Mediated Protective Autophagy, Apoptosis, and Caspase-Dependent Cleavage of P70S6K in Colorectal Cancer Cells. *Biomolecules* **2020**, *10*, 412. [CrossRef] [PubMed]

177. Chaudhary, S.C.; Siddiqui, M.S.; Athar, M.; Alam, M.S. D-Limonene modulates inflammation, oxidative stress and Ras-ERK pathway to inhibit murine skin tumorigenesis. *Hum. Exp. Toxicol.* **2012**, *31*, 798–811. [CrossRef] [PubMed]

178. Pudełek, M.; Catapano, J.; Kochanowski, P.; Mrowiec, K.; Janik-Olchawa, N.; Czyż, J.; Ryszawy, D. Therapeutic potential of monoterpene α-thujone, the main compound of Thuja occidentalis L. essential oil, against malignant glioblastoma multiforme cells in vitro. *Fitoterapia* **2019**, *134*, 172–181. [CrossRef]

179. Russo, R.; Cassiano, M.G.; Ciociaro, A.; Adornetto, A.; Varano, G.P.; Chiappini, C.; Berliocchi, L.; Tassorelli, C.; Bagetta, G.; Corasaniti, M.T. Role of D-Limonene in autophagy induced by bergamot essential oil in SH-SY5Y neuroblastoma cells. *PLoS ONE* **2014**, *9*, e113682. [CrossRef] [PubMed]

180. Banjerdpongchai, R.; Khaw-On, P. Terpinen-4-ol induces autophagic and apoptotic cell death in human leukemic HL-60 cells. *Asian Pac. J. Cancer Prev.* **2013**, *14*, 7537–7542. [CrossRef]

181. Ding, X.F.; Shen, M.; Xu, L.Y.; Dong, J.H.; Chen, G. 13,14-bis(cis-3,5-dimethyl-1-piperazinyl)-beta-elemene, a novel beta-elemene derivative, shows potent antitumor activities via inhibition of mTOR in human breast cancer cells. *Oncol. Lett.* **2013**, *5*, 1554–1558. [CrossRef]

182. Streit, M.; Detmar, M. Angiogenesis, lymphangiogenesis, and melanoma metastasis. *Oncogene* **2003**, *22*, 3172–3179. [CrossRef]

183. Albini, A.; Tosetti, F.; Li, V.W.; Noonan, D.M.; Li, W.W. Cancer prevention by targeting angiogenesis. *Nat. Rev. Clin. Oncol.* **2012**, *9*, 498–509. [CrossRef]

184. Mabeta, P. Paradigms of vascularization in melanoma: Clinical significance and potential for therapeutic targeting. *Biomed. Pharmacother.* **2020**, *127*, 110135. [CrossRef]

185. Stacker, S.A.; Caesar, C.; Baldwin, M.E.; Thornton, G.E.; Williams, R.A.; Prevo, R.; Jackson, D.G.; Nishikawa, S.; Kubo, H.; Achen, M.G. VEGF-D promotes the metastatic spread of tumor cells via the lymphatics. *Nat. Med.* **2001**, *7*, 186–191. [CrossRef] [PubMed]

186. Jour, G.; Ivan, D.; Aung, P.P. Angiogenesis in melanoma: An update with a focus on current targeted therapies. *J. Clin. Pathol.* **2016**, *69*, 472–483. [CrossRef] [PubMed]

187. Kiyan, H.T.; Demirci, B.; Baser, K.H.; Demirci, F. The in vivo evaluation of anti-angiogenic effects of Hypericum essential oils using the chorioallantoic membrane assay. *Pharm. Biol.* **2014**, *52*, 44–50. [CrossRef] [PubMed]

188. Mishra, H.; Mishra, P.K.; Iqbal, Z.; Jaggi, M.; Madaan, A.; Bhuyan, K.; Gupta, N.; Gupta, N.; Vats, K.; Verma, R.; et al. Co-Delivery of Eugenol and Dacarbazine by Hyaluronic Acid-Coated Liposomes for Targeted Inhibition of Survivin in Treatment of Resistant Metastatic Melanoma. *Pharmaceutics* **2019**, *11*, 163. [CrossRef]

189. Yousefian Rad, E.; Homayouni Tabrizi, M.; Ardalan, P.; Seyedi, S.M.R.; Yadamani, S.; Zamani-Esmati, P.; Haghani Sereshkeh, N. Citrus lemon essential oil nanoemulsion (CLEO-NE), a safe cell-depended apoptosis inducer in human A549 lung cancer cells with anti-angiogenic activity. *J. Microencapsul.* **2020**, *37*, 394–402. [CrossRef]

190. Farahpour, M.R.; Pirkhezr, E.; Ashrafian, A.; Sonboli, A. Accelerated healing by topical administration of Salvia officinalis essential oil on *Pseudomonas aeruginosa* and *Staphylococcus aureus* infected wound model. *Biomed. Pharmacother.* **2020**, *128*, 110120. [CrossRef]

191. Santos-Júnior, L.; Oliveira, T.V.C.; Cândido, J.F.; Santana, D.S.; Pereira, R.N.F.; Pereyra, B.B.S.; Gomes, M.Z.; Lima, S.O.; Albuquerque-Júnior, R.L.C.; Cândido, E.A.F. Effects of the essential oil of *Alpinia zerumbet* (Pers.) B.L. Burtt & R.M. Sm. on healing and tissue repair after partial Achilles tenotomy in rats. *Acta Cir. Bras.* **2017**, *32*, 449–458. [PubMed]

192. Kim, D.Y.; Won, K.J.; Yoon, M.S.; Hwang, D.I.; Yoon, S.W.; Park, J.H.; Kim, B.; Lee, H.M. Chrysanthemum boreale Makino essential oil induces keratinocyte proliferation and skin regeneration. *Nat. Prod. Res.* **2015**, *29*, 562–564. [CrossRef]

193. Avola, R.; Granata, G.; Geraci, C.; Napoli, E.; Eleonora Graziano, A.C.; Cardile, V. Oregano (*Origanum vulgare* L.) essential oil provides anti-inflammatory activity and facilitates wound healing in a human keratinocytes cell model. *Food Chem. Toxicol.* **2020**. [CrossRef] [PubMed]

194. Labib, R.M.; Ayoub, I.M. Appraisal on the wound healing potential of Melaleuca alternifolia and *Rosmarinus officinalis* L. essential oil-loaded chitosan topical preparations. *PLoS ONE* **2019**, *14*, e0219561. [CrossRef] [PubMed]

195. Park, J.H.; Park, G.M.; Kim, J.K. Zerumbone, Sesquiterpene Photochemical from Ginger, Inhibits Angiogenesis. *Korean J. Physiol. Pharmacol.* **2015**, *19*, 335–340. [CrossRef] [PubMed]

196. Samad, N.A.; Abdul, A.B.; Rahman, H.S.; Rasedee, A.; Tengku Ibrahim, T.A.; Keon, Y.S. Zerumbone Suppresses Angiogenesis in HepG2 Cells through Inhibition of Matrix Metalloproteinase-9, Vascular Endothelial Growth Factor, and Vascular Endothelial Growth Factor Receptor Expressions. *Pharmacogn. Mag.* **2018**, *13*, S731–S736. [PubMed]

197. Shojaei, S.; Kiumarsi, A.; Moghadam, A.R.; Alizadeh, J.; Marzban, H.; Ghavami, S. Perillyl Alcohol (Monoterpene Alcohol), Limonene. *Enzymes* **2014**, *36*, 7–32. [PubMed]

198. Loutrari, H.; Hatziapostolou, M.; Skouridou, V.; Papadimitriou, E.; Roussos, C.; Kolisis, F.N.; Papapetropoulos, A. Perillyl alcohol is an angiogenesis inhibitor. *J. Pharmacol. Exp. Ther.* **2004**, *311*, 568–575. [CrossRef]

199. Zuo, H.X.; Jin, Y.; Wang, Z.; Li, M.Y.; Zhang, Z.H.; Wang, J.Y.; Xing, Y.; Ri, M.H.; Jin, C.H.; Xu, G.H.; et al. Curcumol inhibits the expression of programmed cell death-ligand 1 through crosstalk between hypoxia-inducible factor-1alpha and STAT3 (T705) signaling pathways in hepatic cancer. *J. Ethnopharmacol.* **2020**, *257*, 112835. [CrossRef]

200. Tanaka, R.; Fujisawa, Y.; Sae, I.; Maruyama, H.; Ito, S.; Hasegawa, N.; Sekine, I.; Fujimoto, M. Severe hepatitis arising from ipilimumab administration, following melanoma treatment with nivolumab. *Jpn. J. Clin. Oncol.* **2017**, *47*, 175–178. [CrossRef]

201. Scarpati, G.D.; Fusciello, C.; Perri, F.; Sabbatino, F.; Ferrone, S.; Carlomagno, C.; Pepe, S. Ipilimumab in the treatment of metastatic melanoma: Management of adverse events. *Oncotargets Ther.* **2014**, *7*, 203–209. [CrossRef]

202. Marques, H.M.C. A review on cyclodextrin encapsulation of essential oils and volatiles. *Flavour Frag. J.* **2010**, *25*, 313–326. [CrossRef]

203. Zhou, X.; Yang, G.; Guan, F. Biological Functions and Analytical Strategies of Sialic Acids in Tumor. *Cells* **2020**, *9*, 273. [CrossRef]

204. Wang, X.; Liu, R.; Zhu, W.; Chu, H.; Yu, H.; Wei, P.; Wu, X.; Zhu, H.; Gao, H.; Liang, J.; et al. UDP-glucose accelerates SNAI1 mRNA decay and impairs lung cancer metastasis. *Nature* **2019**, *571*, 127–131. [CrossRef] [PubMed]

205. Kusuhara, M.; Urakami, K.; Masuda, Y.; Zangiacomi, V.; Ishii, H.; Tai, S.; Maruyama, K.; Yamaguchi, K. Fragrant environment with alpha-pinene decreases tumor growth in mice. *Biomed. Res.* **2012**, *33*, 57–61. [CrossRef]

206. Kusuhara, M.; Maruyama, K.; Ishii, H.; Masuda, Y.; Sakurai, K.; Tamai, E.; Urakami, K. A Fragrant Environment Containing α-Pinene Suppresses Tumor Growth in Mice by Modulating the Hypothalamus/Sympathetic Nerve/Leptin Axis and Immune System. *Integr. Cancer Ther.* **2019**, *18*, 1534735419845139. [CrossRef] [PubMed]

207. Lesgards, J.F.; Baldovini, N.; Vidal, N.; Pietri, S. Anticancer activities of essential oils constituents and synergy with conventional therapies: A review. *Phytother. Res.* **2014**, *28*, 1423–1446. [CrossRef]

208. Effenberger-Neidnicht, K.; Schobert, R. Combinatorial effects of thymoquinone on the anti-cancer activity of doxorubicin. *Cancer Chemother. Pharmacol.* **2011**, *67*, 867–874. [CrossRef]

209. Balavandi, Z.; Neshasteh-Riz, A.; Koosha, F.; Eynali, S.; Hoormand, M.; Shahidi, M. The Use of ss-Elemene to Enhance Radio Sensitization of A375 Human Melanoma Cells. *Cell J.* **2020**, *21*, 419–425. [PubMed]

210. Benimetskaya, L.; Lai, J.C.; Khvorova, A.; Wu, S.; Hua, E.; Miller, P.; Zhang, L.M.; Stein, C.A. Relative Bcl-2 independence of drug-induced cytotoxicity and resistance in 518A2 melanoma cells. *Clin. Cancer Res.* **2004**, *10*, 8371–8379. [CrossRef]

211. Senbanjo, L.T.; Chellaiah, M.A. CD44: A Multifunctional Cell Surface Adhesion Receptor Is a Regulator of Progression and Metastasis of Cancer Cells. *Front. Cell Dev. Biol.* **2017**, *5*, 18. [CrossRef]

212. Kaur, G.; Athar, M.; Alam, M.S. Eugenol precludes cutaneous chemical carcinogenesis in mouse by preventing oxidative stress and inflammation and by inducing apoptosis. *Mol. Carcinog.* **2010**, *49*, 290–301. [CrossRef]

213. Pal, D.; Banerjee, S.; Mukherjee, S.; Roy, A.; Panda, C.K.; Das, S. Eugenol restricts DMBA croton oil induced skin carcinogenesis in mice: Downregulation of c-Myc and H-ras, and activation of p53 dependent apoptotic pathway. *J. Dermatol. Sci.* **2010**, *59*, 31–39. [CrossRef]

214. Chaudhary, S.C.; Alam, M.S.; Siddiqui, M.S.; Athar, M. Chemopreventive effect of farnesol on DMBA/TPA-induced skin tumorigenesis: Involvement of inflammation, Ras-ERK pathway and apoptosis. *Life Sci.* **2009**, *85*, 196–205. [CrossRef] [PubMed]

215. Manoharan, S.; Selvan, M.V. Chemopreventive potential of geraniol in 7,12-dimethylbenz(a) anthracene (DMBA) induced skin carcinogenesis in Swiss albino mice. *J. Environ. Biol.* **2012**, *33*, 255–260. [PubMed]

216. Liu, Z.; Shen, C.; Tao, Y.; Wang, S.; Wei, Z.; Cao, Y.; Wu, H.; Fan, F.; Lin, C.; Shan, Y.; et al. Chemopreventive efficacy of menthol on carcinogen-induced cutaneous carcinoma through inhibition of inflammation and oxidative stress in mice. *Food Chem. Toxicol.* **2015**, *82*, 12–18. [CrossRef] [PubMed]

217. Lluria-Prevatt, M.; Morreale, J.; Gregus, J.; Alberts, D.S.; Kaper, F.; Giaccia, A.; Powell, M.B. Effects of perillyl alcohol on melanoma in the TPras mouse model. *Cancer Epidemiol. Biomark. Prev.* **2002**, *11*, 573–579.

218. Chaudhary, S.C.; Alam, M.S.; Siddiqui, M.S.; Athar, M. Perillyl alcohol attenuates Ras-ERK signaling to inhibit murine skin inflammation and tumorigenesis. *Chem. Biol. Interact.* **2009**, *179*, 145–153. [CrossRef]

219. Aumeeruddy-Elalfi, Z.; Lall, N.; Fibrich, B.; Blom van Staden, A.; Hosenally, M.; Mahomoodally, M.F. Selected essential oils inhibit key physiological enzymes and possess intracellular and extracellular antimelanogenic properties in vitro. *J. Food Drug Anal.* **2018**, *26*, 232–243. [CrossRef]

220. Horiba, H.; Nakagawa, T.; Zhu, Q.; Ashour, A.; Watanabe, A.; Shimizu, K. Biological Activities of Extracts from Different Parts of Cryptomeria japonica. *Nat. Prod. Commun.* **2016**, *11*, 1337–1342. [CrossRef]

221. Yang, C.H.; Huang, Y.C.; Tsai, M.L.; Cheng, C.Y.; Liu, L.L.; Yen, Y.W.; Chen, W.L. Inhibition of melanogenesis by β-caryophyllene from lime mint essential oil in mouse B16 melanoma cells. *Int. J. Cosmet. Sci.* **2015**, *37*, 550–554. [CrossRef]

222. El Khoury, R.; Michael Jubeli, R.; El Beyrouthy, M.; Baillet Guffroy, A.; Rizk, T.; Tfayli, A.; Lteif, R. Phytochemical screening and antityrosinase activity of carvacrol, thymoquinone, and four essential oils of Lebanese plants. *J. Cosmet. Dermatol.* **2019**, *18*, 944–952. [CrossRef]

223. Ko, G.A.; Cho, S.K. Phytol suppresses melanogenesis through proteasomal degradation of MITF via the ROS-ERK signaling pathway. *Chem. Biol. Interact.* **2018**, *286*, 132–140. [CrossRef]

224. Jeong, H.; Yu, S.M.; Kim, S.J. Inhibitory effects on melanogenesis by thymoquinone are mediated through the betacatenin pathway in B16F10 mouse melanoma cells. *Int. J. Oncol.* **2020**, *56*, 379–389. [PubMed]

225. Nam, J.H.; Nam, D.Y.; Lee, D.U. Valencene from the Rhizomes of Cyperus rotundus Inhibits Skin Photoaging-Related Ion Channels and UV-Induced Melanogenesis in B16F10 Melanoma Cells. *J. Nat. Prod.* **2016**, *79*, 1091–1096. [CrossRef] [PubMed]

226. Oh, T.I.; Jung, H.J.; Lee, Y.M.; Lee, S.; Kim, G.H.; Kan, S.Y.; Kang, H.; Oh, T.; Ko, H.M.; Kwak, K.C.; et al. Zerumbone, a Tropical Ginger Sesquiterpene of Zingiber officinale Roscoe, Attenuates alpha-MSH-Induced Melanogenesis in B16F10 Cells. *Int. J. Mol. Sci.* **2018**, *19*, 3149. [CrossRef] [PubMed]

227. Gandini, S.; Sera, F.; Cattaruzza, M.S.; Pasquini, P.; Abeni, D.; Boyle, P.; Melchi, C.F. Meta-analysis of risk factors for cutaneous melanoma: I. Common and atypical naevi. *Eur. J. Cancer* **2005**, *41*, 28–44. [CrossRef] [PubMed]

228. Gandini, S.; Sera, F.; Cattaruzza, M.S.; Pasquini, P.; Picconi, O.; Boyle, P.; Melchi, C.F. Meta-analysis of risk factors for cutaneous melanoma: II. Sun exposure. *Eur. J. Cancer* **2005**, *41*, 45–60. [CrossRef] [PubMed]

229. Gandini, S.; Sera, F.; Cattaruzza, M.S.; Pasquini, P.; Zanetti, R.; Masini, C.; Boyle, P.; Melchi, C.F. Meta-analysis of risk factors for cutaneous melanoma: III. Family history, actinic damage and phenotypic factors. *Eur. J. Cancer* **2005**, *41*, 2040–2059. [CrossRef]

230. McKinney, A.J.; Holmen, S.L. Animal models of melanoma: A somatic cell gene delivery mouse model allows rapid evaluation of genes implicated in human melanoma. *Chin. J. Cancer* **2011**, *30*, 153–162. [CrossRef]

231. Stratton, S.P.; Alberts, D.S.; Einspahr, J.G.; Sagerman, P.M.; Warneke, J.A.; Curiel-Lewandrowski, C.; Myrdal, P.B.; Karlage, K.L.; Nickoloff, B.J.; Brooks, C.; et al. A phase 2a study of topical perillyl alcohol cream for chemoprevention of skin cancer. *Cancer Prev. Res. (Phila)* **2010**, *3*, 160–169. [CrossRef] [PubMed]

232. Stratton, S.P.; Saboda, K.L.; Myrdal, P.B.; Gupta, A.; McKenzie, N.E.; Brooks, C.; Salasche, S.J.; Warneke, J.A.; Ranger-Moore, J.; Bozzo, P.D.; et al. Phase 1 study of topical perillyl alcohol cream for chemoprevention of skin cancer. *Nutr. Cancer* **2008**, *60*, 325–330. [CrossRef]

233. Nishigori, C.; Hattori, Y.; Toyokuni, S. Role of reactive oxygen species in skin carcinogenesis. *Antioxid. Redox. Signal.* **2004**, *6*, 561–570. [CrossRef]

234. Alam, M.N.; Bristi, N.J.; Rafiquzzaman, M. Review on in vivo and in vitro methods evaluation of antioxidant activity. *Saudi Pharm. J.* **2013**, *21*, 143–152. [CrossRef] [PubMed]

235. Cui, X.; Gong, J.; Han, H.; He, L.; Teng, Y.; Tetley, T.; Sinharay, R.; Chung, K.F.; Islam, T.; Gilliland, F.; et al. Relationship between free and total malondialdehyde, a well-established marker of oxidative stress, in various types of human biospecimens. *J. Thorac. Dis.* **2018**, *10*, 3088–3097. [CrossRef] [PubMed]

236. Rahiman, N.; Akaberi, M.; Sahebkar, A.; Emami, S.A.; Tayarani-Najaran, Z. Protective effects of saffron and its active components against oxidative stress and apoptosis in endothelial cells. *Microvasc. Res.* **2018**, *118*, 82–89. [CrossRef] [PubMed]

237. Onyebuchi, C.; Kavaz, D. Chitosan And N, N, N-Trimethyl Chitosan Nanoparticle Encapsulation Of Ocimum Gratissimum Essential Oil: Optimised Synthesis, In Vitro Release And Bioactivity. *Int. J. Nanomed.* **2019**, *14*, 7707–7727. [CrossRef]

238. Do Nascimento, K.F.; Moreira, F.M.F.; Alencar Santos, J.; Kassuya, C.A.L.; Croda, J.H.R.; Cardoso, C.A.L.; Vieira, M.D.C.; Gois Ruiz, A.L.T.; Ann Foglio, M.; de Carvalho, J.E.; et al. Antioxidant, anti-inflammatory, antiproliferative and antimycobacterial activities of the essential oil of *Psidium guineense* Sw. and spathulenol. *J. Ethnopharmacol.* **2018**, *210*, 351–358. [CrossRef]

239. Manjamalai, A.; Grace, B. Chemotherapeutic effect of essential oil of Wedelia chinensis (Osbeck) on inducing apoptosis, suppressing angiogenesis and lung metastasis in C57BL/6 mice model. *J. Cancer Sci. Ther.* **2013**, *5*, 271–281. [CrossRef]

240. Lambert, M.W.; Maddukuri, S.; Karanfilian, K.M.; Elias, M.L.; Lambert, W.C. The physiology of melanin deposition in health and disease. *Clin. Dermatol.* **2019**, *37*, 402–417. [CrossRef]

241. Jackett, L.A.; Scolyer, R.A. A Review of Key Biological and Molecular Events Underpinning Transformation of Melanocytes to Primary and Metastatic Melanoma. *Cancers* **2019**, *11*, 2041. [CrossRef]

242. Liu, G.S.; Peshavariya, H.; Higuchi, M.; Brewer, A.C.; Chang, C.W.; Chan, E.C.; Dusting, G.J. Microphthalmia-associated transcription factor modulates expression of NADPH oxidase type 4: A negative regulator of melanogenesis. *Free Radic. Biol. Med.* **2012**, *52*, 1835–1843. [CrossRef]

243. Gillbro, J.M.; Olsson, M.J. The melanogenesis and mechanisms of skin-lightening agents–existing and new approaches. *Int. J. Cosmet. Sci.* **2011**, *33*, 210–221. [CrossRef]

244. Pillaiyar, T.; Manickam, M.; Jung, S.H. Downregulation of melanogenesis: Drug discovery and therapeutic options. *Drug Discov. Today* **2017**, *22*, 282–298. [CrossRef] [PubMed]

245. Nakagawa, T.; Zhu, Q.; Ishikawa, H.; Ohnuki, K.; Kakino, K.; Horiuchi, N.; Shinotsuka, H.; Naito, T.; Matsumoto, T.; Minamisawa, N. Multiple uses of essential oil and by-products from various parts of the Yakushima native Cedar (Cryptomeria Japonica). *J. Wood Chem. Technol.* **2016**, *36*, 42–55. [CrossRef]

246. Zaidi, K.U.; Khan, F.N.; Ali, S.A.; Khan, K.P. Insight into Mechanistic Action of Thymoquinone Induced Melanogenesis in Cultured Melanocytes. *Protein Pept. Lett.* **2019**, *26*, 910–918. [CrossRef] [PubMed]

247. Dyer, J.; Cleary, L.; Ragsdale-Lowe, M.; McNeill, S.; Osland, C. The use of aromasticks at a cancer centre: A retrospective audit. *Complement. Ther. Clin. Pract.* **2014**, *20*, 203–206. [CrossRef]

248. Boehm, K.; Büssing, A.; Ostermann, T. Aromatherapy as an adjuvant treatment in cancer care—A descriptive systematic review. *Afr. J. Tradit. Complement. Altern. Med.* **2012**, *9*, 503–518. [CrossRef] [PubMed]

249. Maddocks-Jennings, W.; Wilkinson, J.M.; Cavanagh, H.M.; Shillington, D. Evaluating the effects of the essential oils *Leptospermum scoparium* (manuka) and Kunzea ericoides (kanuka) on radiotherapy induced mucositis: A randomized, placebo controlled feasibility study. *Eur. J. Oncol. Nurs.* **2009**, *13*, 87–93. [CrossRef]

250. Reis, D.; Jones, T.T. Frankincense Essential Oil as a Supportive Therapy for Cancer-Related Fatigue: A Case Study. *Holist. Nurs. Pract.* **2018**, *32*, 140–142. [CrossRef]

251. Pimenta, F.C.; Alves, M.F.; Pimenta, M.B.; Melo, S.A.; de Almeida, A.A.; Leite, J.R.; Pordeus, L.C.; Diniz Mde, F.; de Almeida, R.N. Anxiolytic Effect of *Citrus aurantium* L. on Patients with Chronic Myeloid Leukemia. *Phytother. Res.* **2016**, *30*, 613–617. [CrossRef]

252. Mapp, C.P.; Hostetler, D.; Sable, J.F.; Parker, C.; Gouge, E.; Masterson, M.; Willis-Styles, M.; Fortner, C.; Higgins, M. Peppermint Oil: Evaluating Efficacy on Nausea in Patients Receiving Chemotherapy in the Ambulatory Setting. *Clin. J. Oncol. Nurs.* **2020**, *24*, 160–164. [CrossRef] [PubMed]

253. Tayarani-Najaran, Z.; Talasaz-Firoozi, E.; Nasiri, R.; Jalali, N.; Hassanzadeh, M. Antiemetic activity of volatile oil from Mentha spicata and Mentha x piperita in chemotherapy-induced nausea and vomiting. *Ecancermedicalscience* **2013**, *7*, 290.

254. Hamzeh, S.; Safari-Faramani, R.; Khatony, A. Effects of Aromatherapy with Lavender and Peppermint Essential Oils on the Sleep Quality of Cancer Patients: A Randomized Controlled Trial. *Evid. Based Complement. Altern. Med.* **2020**, *2020*, 7480204. [CrossRef] [PubMed]

255. Re, L.; Barocci, S.; Sonnino, S.; Mencarelli, A.; Vivani, C.; Paolucci, G.; Scarpantonio, A.; Rinaldi, L.; Mosca, E. Linalool modifies the nicotinic receptor–ion channel kinetics at the mouse neuromuscular junction. *Pharmacol. Res.* **2000**, *42*, 177–181. [CrossRef] [PubMed]

256. Gedney, J.J.; Glover, T.L.; Fillingim, R.B. Sensory and affective pain discrimination after inhalation of essential oils. *Psychosom. Med.* **2004**, *66*, 599–606. [CrossRef] [PubMed]

257. Blackburn, L.; Achor, S.; Allen, B.; Bauchmire, N.; Dunnington, D.; Klisovic, R.B.; Naber, S.J.; Roblee, K.; Samczak, A.; Tomlinson-Pinkham, K.; et al. The Effect of Aromatherapy on Insomnia and Other Common Symptoms Among Patients With Acute Leukemia. *Oncol. Nurs. Forum* **2017**, *44*, E185–E193. [CrossRef]

258. Evans, A.; Malvar, J.; Garretson, C.; Pedroja Kolovos, E.; Baron Nelson, M. The Use of Aromatherapy to Reduce Chemotherapy-Induced Nausea in Children With Cancer: A Randomized, Double-Blind, Placebo-Controlled Trial. *J. Pediatr. Oncol. Nurs.* **2018**, *35*, 392–398. [CrossRef]

259. Lua, P.L.; Salihah, N.; Mazlan, N. Effects of inhaled ginger aromatherapy on chemotherapy-induced nausea and vomiting and health-related quality of life in women with breast cancer. *Complement. Ther. Med.* **2015**, *23*, 396–404. [CrossRef]

260. Tamaki, K.; Fukuyama, A.K.; Terukina, S.; Kamada, Y.; Uehara, K.; Arakaki, M.; Yamashiro, K.; Miyashita, M.; Ishida, T.; McNamara, K.M. Randomized trial of aromatherapy versus conventional care for breast cancer patients during perioperative periods. *Breast Cancer Res. Treat.* **2017**, *162*, 523–531. [CrossRef]

261. Chen, T.H.; Tung, T.H.; Chen, P.S.; Wang, S.H.; Chao, C.M.; Hsiung, N.H.; Chi, C.C. The Clinical Effects of Aromatherapy Massage on Reducing Pain for the Cancer Patients: Meta-Analysis of Randomized Controlled Trials. *Evid. Based Complement. Altern. Med.* **2016**, *2016*, 9147974. [CrossRef]

262. Graham, P.H.; Browne, L.; Cox, H.; Graham, J. Inhalation aromatherapy during radiotherapy: Results of a placebo-controlled double-blind randomized trial. *J. Clin. Oncol.* **2003**, *21*, 2372–2376. [CrossRef]

263. Mo, H.; Jeter, R.; Bachmann, A.; Yount, S.T.; Shen, C.L.; Yeganehjoo, H. The Potential of Isoprenoids in Adjuvant Cancer Therapy to Reduce Adverse Effects of Statins. *Front. Pharmacol.* **2018**, *9*, 1515. [CrossRef]

264. Zárybnický, T.; Boušová, I.; Ambrož, M.; Skálová, L. Hepatotoxicity of monoterpenes and sesquiterpenes. *Arch. Toxicol.* **2018**, *92*, 1–13. [CrossRef] [PubMed]
265. Artini, M.; Patsilinakos, A.; Papa, R.; Bozovic, M.; Sabatino, M.; Garzoli, S.; Vrenna, G.; Tilotta, M.; Pepi, F.; Ragno, R.; et al. Antimicrobial and Antibiofilm Activity and Machine Learning Classification Analysis of Essential Oils from Different Mediterranean Plants against Pseudomonas aeruginosa. *Molecules* **2018**, *23*, 482. [CrossRef] [PubMed]

Article

Loss of Two-Pore Channel 2 (TPC2) Expression Increases the Metastatic Traits of Melanoma Cells by a Mechanism Involving the Hippo Signalling Pathway and Store-Operated Calcium Entry

Antonella D'Amore [1,†], Ali Ahmed Hanbashi [2,3,†], Silvia Di Agostino [4], Fioretta Palombi [1], Andrea Sacconi [5], Aniruddha Voruganti [2], Marilena Taggi [1], Rita Canipari [1], Giovanni Blandino [4], John Parrington [2,*,‡] and Antonio Filippini [1,‡]

[1] Department of Anatomy, Histology, Forensic Medicine and Orthopaedics, Unit of Histology and Medical Embryology, SAPIENZA University of Rome, 16 Via A. Scarpa, 00161 Roma, Italy; antonella.damore@uniroma1.it (A.D.); fioretta.palombi@uniroma1.it (F.P.); marilena.taggi@uniroma1.it (M.T.); rita.canipari@uniroma1.it (R.C.); antonio.filippini@uniroma1.it (A.F.)

[2] Department of Pharmacology, University of Oxford, Mansfield Road, Oxford OX1 3QT, UK; ali.hanbashi@worc.ox.ac.uk (A.A.H.); aniruddha.voruganti@exeter.ox.ac.uk (A.V.)

[3] Department of Pharmacology, College of Pharmacy, Jazan University, Jazan 45142, Saudi Arabia

[4] Oncogenomic and Epigenetic Unit, Istituto di Ricovero e Cura a Carattere Scientifico (IRCCS), Regina Elena National Cancer Institute, 00144 Rome, Italy; silvia.diagostino@ifo.gov.it (S.D.A.); giovanni.blandino@ifo.gov.it (G.B.)

[5] Biostatistica e Bioinformatica, University of Southern Denmark (UOSD) Clinical Trial Center, Istituto di Ricovero e Cura a Carattere Scientifico (IRCCS), Regina Elena National Cancer Institute-IFO, 00144 Rome, Italy; sacconiandrea@hotmail.com

[*] Correspondence: john.parrington@worc.ox.ac.uk; Tel.: +44-01865-271591; Fax: +44-01865-271853

[†] The authors contributed equally to this article.

[‡] The authors share senior authorship.

Received: 3 August 2020; Accepted: 18 August 2020; Published: 24 August 2020

Abstract: Melanoma is one of the most aggressive and treatment-resistant human cancers. The two-pore channel 2 (TPC2) is located on late endosomes, lysosomes and melanosomes. Here, we characterized how TPC2 knockout (KO) affected human melanoma cells derived from a metastatic site. TPC2 KO increased these cells' ability to invade the extracelullar matrix and was associated with the increased expression of mesenchymal markers ZEB-1, Vimentin and N-Cadherin, and the enhanced secretion of MMP9. TPC2 KO also activated genes regulated by YAP/TAZ, which are key regulators of tumourigenesis and metastasis. Expression levels of ORAI1, a component of store-operated Ca^{2+} entry (SOCE), and PKC-βII, part of the HIPPO pathway that negatively regulates YAP/TAZ activity, were reduced by TPC2 KO and RNA interference knockdown. We propose a cellular mechanism mediated by ORAI1/Ca^{2+}/PKC-βII to explain these findings. Highlighting their potential clinical significance, patients with metastatic tumours showed a reduction in TPC2 expression. Our research indicates a novel role of TPC2 in melanoma. While TPC2 loss may not activate YAP/TAZ target genes in primary melanoma, in metastatic melanoma it could activate such genes and increase cancer aggressiveness. These findings aid the understanding of tumourigenesis mechanisms and could provide new diagnostic and treatment strategies for skin cancer and other metastatic cancers.

Keywords: TPC2; HIPPO; melanoma; SOCE; metastasis

---

1. Introduction

Melanoma originates from the tumoral transformation of melanocytes in the skin [1]. The depth of the invasion and the metastasis of the lymph nodes are very important prognostic features for the classification of melanoma; typically, this type of cancer is associated with a very heterogeneous tumour characterized by high mutational burden (e.g., *BRAF, NRAS, PTEN, TP53, CDKN2A* are the most frequently muted genes) and because of this, it is considered a particularly aggressive type of cancer [2]. For this reason, a greater understanding of the molecular and cellular mechanisms underlying melanoma tumourigenesis and metastasis and the identification of new therapeutic targets are both important goals for cancer research. Two-pore channels (TPCs) are members of the voltage-gated cation channel superfamily [3]. Only two isoforms are present in humans and other primates: TPC1 and TPC2. TPC2 is present in the membranes of lysosomes, late endosomes and also melanosomes. There is currently a debate about the precise mechanism of action and mode of regulation of TPC2, both in terms of the protein's biophysical characteristics, and whether it is primarily activated by nicotinic acid adenine dinucleotide phosphate (NAADP), or by phosphatidylinositol 3,5-bisphosphate (PI(3,5)P$_2$) [4,5].

Thus contrasting studies have indicated that TPCs are highly selective for Na$^+$ and are regulated by [PI(3,5)P$_2$] but not NAADP [4,5], or alternatively that TPCs mediate Ca^{2+} release evoked by NAADP [6–9]. Possible explanations for such contrasting data are that NAADP may not directly interact with TPCs, but through an accessory binding protein [10,11], and that Mg^{2+} can affect NAADP-evoked TPC2 Na$^+$ currents [12]; thus, different experimental conditions may influence the mode of TPC activation. In line with such a possibility, a recent study demonstrated the different effect of two novel TPC2 agonists; one of which induced Ca^{2+} release and the other Na$^+$ release. These agonists appear to mimic the actions of NAADP and PI(3,5)P$_2$, suggesting that the two physiological agonists bind different sites on the TPC2 ion channel and regulate different processes mediated by this protein [13].

We have previously demonstrated the involvement of TPC2 in one potential aspect of tumourigenesis, namely neo-angiogenesis. By using TPC1 and TPC2 knock-out in vivo models, we reported that VEGF-induced neo-angiogenesis occurred through the regulation of the VEGFR2/NAADP/ TPC2/Ca^{2+} signalling pathway [14]. More generally, Ca^{2+} signals have been shown to play a role in cancer, including melanoma, in various ways [15]. Store-operated calcium entry (SOCE) is an important Ca^{2+} signalling mechanism. The depletion of Ca^{2+} from the endoplasmic reticulum (ER) induces plasma membrane Ca^{2+} influx [16]. SOCE is mediated by the endoplasmic Ca^{2+} depletion sensor (STIM1), which translocates to the plasma membrane and activates the Ca^{2+} influx channel (ORAI1). Previous studies have shown that the role of SOCE may be different between metastatic melanoma cells derived from patients, and those without metastatic activities [17]. Intriguingly, a known link exists between endolysosomal Ca^{2+} and SOCE, since it has also been shown that Ca^{2+} release from the endolysosomal compartment can trigger SOCE in primary cultured neurons [18]. Moreover, TFEB has been identified as a novel regulator of intracellular Ca^{2+} homeostasis, with the lysosomal control of SOCE being proposed as an important aspect of lysosomal function and cellular health [19]. TFEB-dependent reduction of SOCE is dependent on lysosomal dynamics. TFEB knockdown reduced the number of lysosomes and their localization to the plasma membrane.

In recent years, the HIPPO pathway has been demonstrated to modulate cell proliferation and differentiation, and to contribute to the progression of a number of diseases, including cancer [20]. Mammalian Ste20-like kinases 1/2 (MST1/2) and the large tumour suppressor 1/2 (LATS1/2) are components of the kinase cascade of the HIPPO pathway [21]. Following the activation of the HIPPO pathway, MST1/2 is phosphorylated and activates LATS1/2, which can then phosphorylate the yes association protein (YAP) and its paralogue, the transcriptional coactivator with PDZ-binding motif (TAZ), resulting in the inhibition of the transcriptional activity of YAP/TAZ [20]. This phosphorylation promotes YAP/TAZ cytoplasmic localisation; when not inhibited in this way, the translocation of YAP/TAZ to the nucleus promotes the activation of their target genes which play key roles in cell growth, proliferation and organ development [22]. Interestingly, the hyperactivation of YAP has been shown to be associated with poor cancer prognosis [23].

Here, we studied whether TPC2 may play different roles in melanoma in terms of primary and metastatic tumours, and also demonstrate for the first time a connection between TPC2, Ca^{2+} signalling, and the activation of the HIPPO signalling pathway.

2. Results

2.1. Association between TPC2 Expression and the Prognosis of Melanoma Patients

In recent years, with the development of high-throughput RNA sequencing (RNA-Seq), large amounts of RNA-Seq data have emerged. To evaluate the prognostic role of TPC2 (gene name *TPCN2*) in skin cutaneous melanoma (SKCM), we analysed *TPCN2* mRNA expression levels in the Cancer Genome Atlas (TCGA). We identified a significant difference in *TPCN2* mRNA expression between primary and metastatic patients. The reduction of TPC2 expression in the metastatic patients could indicate a different prognostic role for TPC2 in these two stages of cancer (Figure 1A).

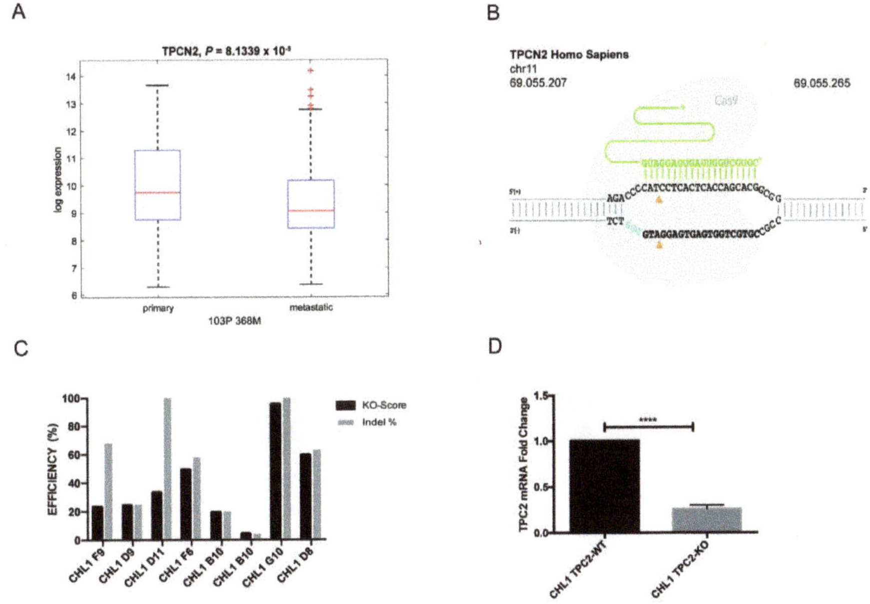

Figure 1. TPC2 expression in the melanoma patient dataset and the generation of a human melanoma cell model TPC2 knockout. (**A**) Analysis of expression and prognostic value for TPC2 between primary and metastatic patients in human skin cutaneous melanoma (SKCM). Dashed line indicates 95% confidence interval (*p*-value: 8.1339 × 10^{-5}). (**B**) Schematic of CRISPR-Cas9 strategy used to target CHL1 human melanoma cell line. (**C**) Interference of CRISPR Edits (ICE) Analysis of possible knockout (KO) clones, from which the CHL1 clone G10 was selected. KO score indicates the percentage of sequences that are putative knockouts. (**D**) qPCR analysis of TPC2 transcript levels in CHL1 (wild type (WT) and KO). Data in bar charts represent the mean ± s.e.m. of three independent experiments. **** *p* < 0.0001.

2.2. TPC2 Depletion Impairs Human Melanoma Cell Adhesion Mediated by α2β1 Integrin Receptor

Amelanotic melanoma is a rare form of human melanoma cancer and generally difficult to diagnose in the first stage due to its lack of pigmentation. Therefore, finding new markers to recognize this form of melanoma at an early stage of tumourigenesis is an important goal that could have major therapeutic benefits. CHL1 cells are a model of a human amelanotic melanoma and are derived from a metastatic site.

We first used CRISPR/Cas9 genome editing to knockout (KO) the *TPCN2* gene in CHL1 and B16-F0 (murine primary melanoma) cell lines. Three different single guide RNAs (sgRNAs) were used to target *TPCN2* exon 4 with the aim of generating a frame shift in the TPC2 open reading frame (ORF) via non-homologous end-joining (NHEJ), in each cell line (Figure 1B and Figure S1A); this targeting was followed by CRISPR editing analysis using an Interference of CRISPR Edits (ICE) tool to select clones with the highest KO score (Figure 1C and Figure S1B). Furthermore, *TPCN2* mRNA expression in each putative TPC2 KO cell line was assessed (Figure 1D and Figure S1C).

Then, we investigated the proliferation capacity of CHL1/TPC2 KO cells. TPC2 KO cells showed a slower G2 phase compared to wild type (WT) cells, after 24 h from seeding (Figure 2A). After 48 h from seeding, the TPC2 KO cell line recovered the proliferation rate of the control cell line (Figure 2A). These data are further corroborated by the analysis of the cell cycle markers cyclin D1 (CCND1) for the G1/S phase, cyclin B1 (CCNB1) for the G2/M phase, and of the cyclin-dependent kinase inhibitor 1 (p21) as a marker of cell cycle arrest. These transcripts showed no significant difference in expression (Figure 2B).

Subsequently, we performed an adhesion assay by using cell plates coated with collagen type I matrix to assess metastatic traits after TPC2 KO. Interestingly, TPC2 KO cells showed a drastically reduced ability to bind this matrix (Figure 2C). Looking further, we investigated the expression of α2 and β1 integrin chains that together form the main receptor to collagen type I. Plasma membrane expression of both these adhesion proteins in live cells showed a significant reduction in the TPC2 KO cells (Figure 2D). In contrast, the whole cell expression of β1-integrin did not change (Figure 2E). These findings show similarity to those of Nguyen et al. [24] who showed that TPC2 down regulation in bladder carcinoma cells is related to a halted trafficking of β1-integrins. These combined findings strongly suggest that TPC2 is involved in tumour metastasis rather than in cancer cell proliferation.

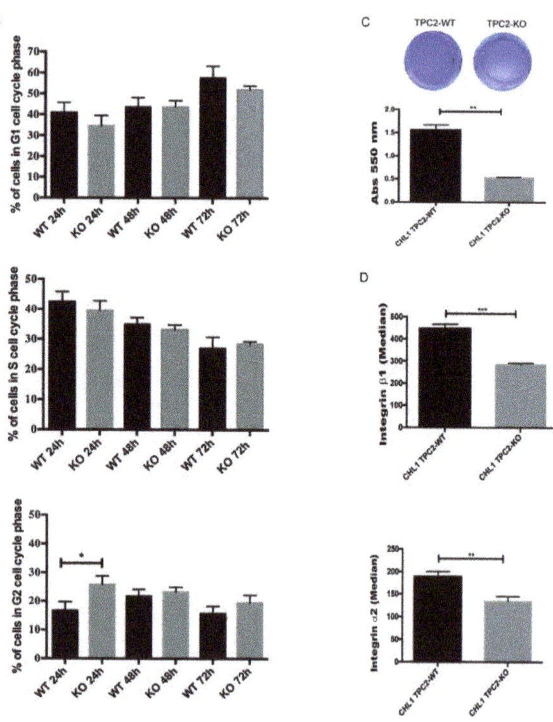

Figure 2. *Cont.*

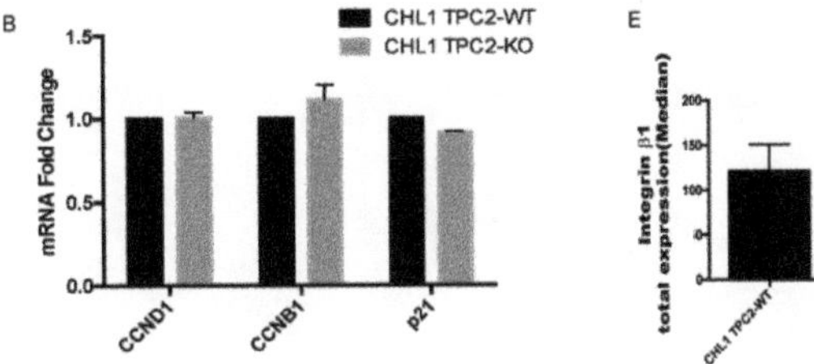

Figure 2. TPC2 KO impairs cell adhesion to collagen type I matrix. (**A**) Cell cycle analysis at 24, 48, and 72 h of CHL1 WT and TPC2 KO cells. (**B**) qPCR analysis of cell cycle markers CCND1, CCNB1 and p21; H3 was used as an internal control. (**C**) Adhesion assay on collagen type I matrix, cells stained with crystal violet. (**D**) Flow cytometric detection of $\alpha2\beta1$ integrin chain in CHL1 WT and TPC2 KO cells. (**E**) Flow cytometric detection of Integrin-$\beta1$ on CHL1 WT and TPC2 KO fixed and permeabilized cells. Data in bar charts represent the mean ± s.e.m. of three independent experiments (* $p < 0.05$; ** $p < 0.01$; *** $p < 0.001$).

2.3. TPC2 Inhibition in Human Melanoma Cells Increases Their Invasion Capability

To further explore the role of TPC2 in metastasis, we compared WT and TPC2 KO CHL1 cells' ability to invade a matrigel substrate. After 24 h, TPC2 KO cells were more invasive (Figure 3A). In line with this, the TPC2 KO cells showed an increased secretion of matrix metalloproteinase 9 (MMP9), as highlighted by zymography assays (Figure 3B).

The epithelial–mesenchymal transition (EMT) is an evolutionarily conserved developmental process that confers metastatic properties upon cancer cells by enhancing their mobility, invasion, and resistance to apoptotic stimuli. A common feature of EMT is the loss of epithelial cadherin (E-cadherin) expression and the concomitant up regulation or de novo expression of neural cadherin (N-cadherin). Analysing E- and N-cadherin expression, we found that neither WT nor TPC2 KO cells expressed E-cadherin but in TPC2 KO cells the expression of N-cadherin was increased (Figure 3C). We also analysed the expression of the transcription factor zinc-finger E-box-binding homeobox1 (ZEB-1), a known inducer of EMT [25], and vimentin, the major component of the intermediate filaments, which is normally expressed in mesenchymal cells [26], and considered a mesenchymal marker. ZEB-1 and vimentin were up regulated in TPC2 KO cells (Figure 3D). All these data suggest that TPC2 KO cells are more mesenchymal and so more invasive. Melanocyte Inducing Transcription Factor (MITF) expression promotes melanoma cell survival and migration [27,28]. TPC2 KO cells showed a higher level of expression of the MITF protein (Figure 3E).

Moreover, we analysed the clinical correlation of TPC2 expression with that of some of the genes which correlated with TPC2 expression in our in vitro studies. There was a significant negative correlation between TPC2 and ZEB1 (R = −0.535, $p < 0.05$); N-cadherin (CDH2) (R = − 0.317, $p < 0.05$) and MMP9 (R = −0.213, $p < 0.05$) (Figure 3F) in the human SKCM dataset. We also analysed this correlation among primary and metastatic patients (Figures S3 and S4; Table S1).

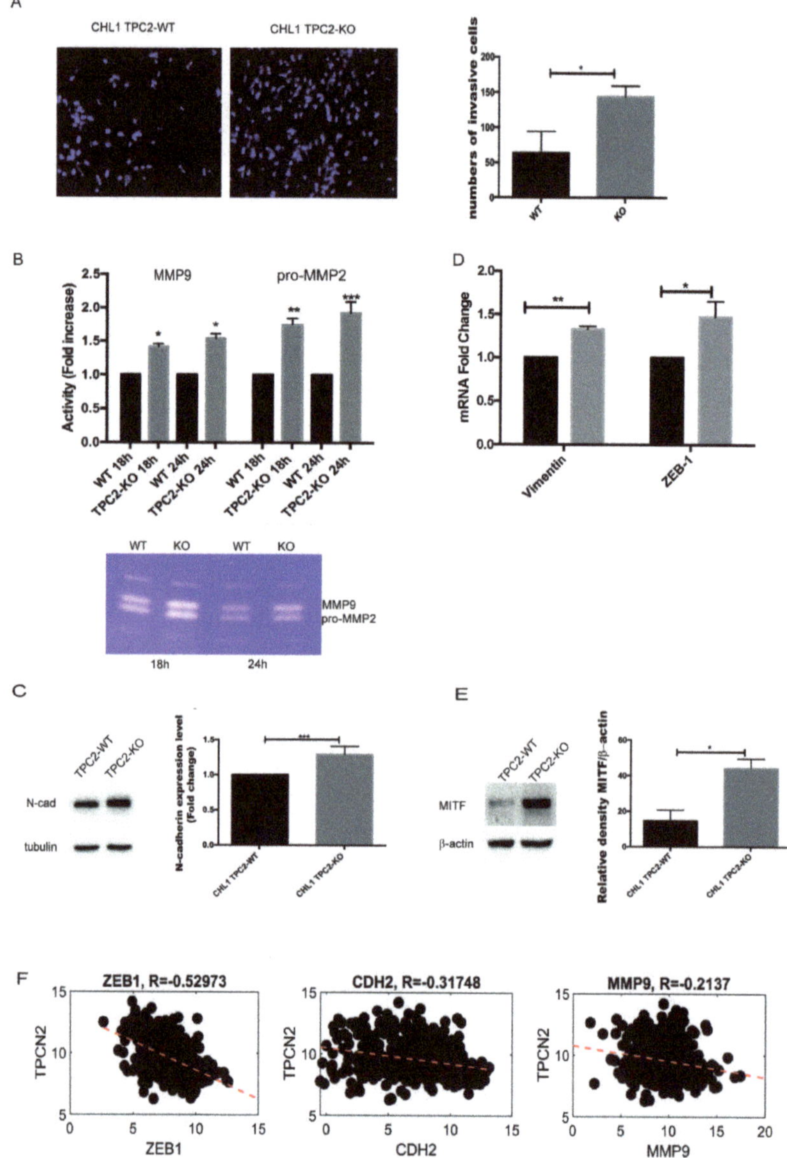

Figure 3. TPC2 depletion increases metastatic melanoma cell invasion. (**A**) Transwell assay after 24 h incubation on matrigel matrix, cell nuclei stained with DAPI. (**B**) Proteolytic Matrix Metalloproteinases (MMPs) activity in CHL1 WT and TPC2 KO cells detected by zymography. Active MMP-9 (82 kDa) and proMMP-2 (72 kDa). (**C**) N-cadherin protein expression in CHL1 WT and TPC2 KO cells, normalised to tubulin. Relative density of N-cadherin/tubulin. (**D**) qPCR analysis of vimentin and ZEB1 in CHL1 WT and TPC2 KO cell lines. H3 was used as an internal control. (**E**) MITF protein expression in CHL1 WT and TPC2 KO normalised to β-actin. Relative density of MITF/β-actin. (**F**) Correlation between TPCN2 and ZEB1, N-Cadherin (CDH2), and MMP-9 in human SKCM. Data in bar charts represent the mean ± s.e.m. of three independent experiments (* $p < 0.05$; ** $p < 0.001$; *** $p < 0.001$).

2.4. TPC2 and HIPPO Pathway Connection in CHL1 and MeWo Metastatic Cell Lines

To gain insight into the molecular mechanisms underlying the apparent increased aggressiveness of CHL1 TPC2 KO cells, we investigated the potential links with an important emerging pathway associated with tumour initiation and progression. The emerging role of YAP/TAZ and the transcriptional effectors of HIPPO signalling, as the key drivers of tumour initiation and growth, are major recent discoveries in cancer research. Their activity can control proliferation, invasion and metastasis and is usually associated with poor prognosis [29]. High YAP levels correlate with decreased survival in melanoma patients [23] and TAZ activation has been linked to lung cancer brain metastasis [30]. To assess the activation status of YAP/TAZ, we analysed the expression of ankyrin repeat domain-containing protein (ANKRD1), cysteine-rich 61 (CYR61), and connective tissue growth factor (CTGF), which are considered to be the bona fide YAP/TAZ target genes [31]. Notably, all these target genes were strongly up regulated in TPC2 KO cells, indicating YAP/TAZ transcriptional activation (Figure 4A).

In order to independently study whether there is a correlation between a reduction of TPC2 expression and the enhanced expression of YAP/TAZ target genes, we used RNA interference (RNAi) to transiently silence TPC2 expression for 24 h in the CHL1 and MeWo metastatic melanoma cell lines (Figure 4B). Importantly, using this completely different approach to supress TPC2 expression, we observed increased expression levels of ANKRD1 (but only for CHL1, not for MeWo), CTGF, and CYR61 mRNAs, in TPC2-silenced cells (Figure 4C,D). We also studied whether the activation of YAP/TAZ target genes that occurs in TPC2 KO cells is associated with the increased movement of YAP/TAZ from the cytoplasm to the nucleus. By differential nucleus-cytoplasm protein extraction, we confirmed an increase in YAP and TAZ in the nucleus in the TPC2 KO cells (Figure 4E).

Collectively, these data confirm a link between the loss of, or reduction of, TPC2 expression and the transcriptional activation of YAP/TAZ target genes linked to the movement of these transcriptional activators from the cytoplasm to the nucleus. These results shed a light on a novel role for TPC2 at a late stage of cancer.

To further validate these findings in melanoma patients, we studied the clinical correlation of TPC2 expression with that of YAP/TAZ target genes, in the skin melanoma TCGA database. There was a significant negative correlation between TPC2 and CTGF ($R = -0.311$, $p < 0.05$); CYR61 ($R = -0.134$, $p < 0.05$), but not with ANKDR1 ($R = -0.0932$, $p = 0.07$) (Figure 4F). We also analysed this correlation among primary and metastatic patients (Figures S3 and S4; Table S1).

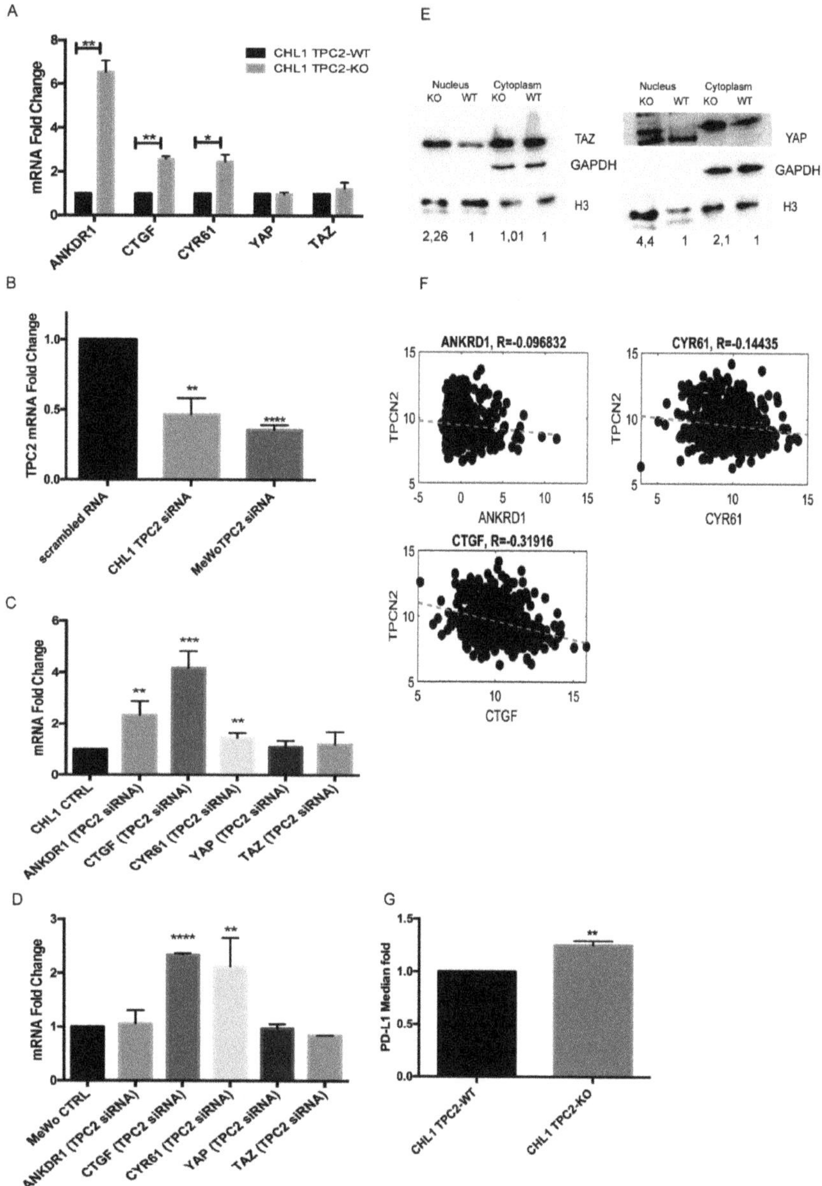

Figure 4. TPC2 KO and silencing induces YAP/TAZ target genes. (**A**) qPCR analysis of YAP/TAZ target genes in CHL1 WT and TPC2 KO cells. H3 was used as an internal control. (**B**) TPC2 transient inhibition after 24 h from CHL1 transfection. (**C**) qPCR analysis of YAP/TAZ target genes after the transient silencing of TPC2 in CHL1 cells. Glyceraldehyde 3-phosphate dehydrogenase (GAPDH) was used as an internal control. (**D**) qPCR analysis of YAP/TAZ target genes after the transient silencing of TPC2 in the MeWo cell line. GAPDH was used as an internal control. (**E**) Nucleus–cytoplasm extraction to analyse YAP and TAZ nuclear localization. (**F**) Correlation between TPCN2 and CTGF, CYR61, and ANKRD1 in human SKCM (for the *p*-value see Table S1). (**G**) Flow cytometric detection of PD-L1 in CHL1 WT and TPC2 KO cells after IFN-γ induction. Data in bar charts represent the mean ± s.e.m. of three independent experiments (* $p < 0.05$; ** $p < 0.01$; *** $p < 0.001$; **** $p < 0.0001$).

2.5. PD-L1 Induced by INTERFERON-γ Is Significantly Higher in CHL1 TPC2 KO Cells

It has previously been demonstrated that TAZ can regulate the transcription of PD-L1 [32]. PD-L1 is one of the ligands of PD-1, which represents a target for advanced melanoma immunotherapy [33]. PD-L1 shows abnormally high expression in tumour cells and is considered the main factor responsible for promoting the ability of tumour immune escape [34]. After treatment with 100 ng/mL Interferon-γ (IFN-γ), CHL1 TPC2 KO cells displayed an increased level of surface expression of PD-L1 (Figure 4G), providing further indirect confirmation of the activation of the YAP/TAZ pathway after TPC2 KO.

2.6. TPC2 Regulates YAP/TAZ Activity Modulating ORAI 1 Channel Expression

An important question is how loss of expression of an endolysosomal ion channel like TPC2 affects YAP/TAZ activity. Previously, we and others have shown that TPC2 is a mediator of Ca^{2+} signalling responses in cells [7,14]. To date, the only known link between YAP/TAZ activity and Ca^{2+} signalling is via store-operated Ca^{2+} entry (SOCE), a process that replenishes the endoplasmic reticulum (ER) Ca^{2+} store through an interaction between the ER protein STIM1 and the plasma membrane ion channel ORAI1 [35]. A study in glioblastoma cells showed that the stimulation of SOCE inhibited YAP/TAZ activity via the activation of PKC-βII [36]. This kinase regulates MST1/2 and LATS 1/2 phosphorylation, both signalling components in the HIPPO pathway. When they are activated by phosphorylation, YAP/TAZ are phosphorylated and sequestered in the cytosol. The glioblastoma study also showed that the elevated level of Ca^{2+} induced by SOCE led to actin cytoskeleton remodelling mediated by INF2 [36]. In the current study we compared the expression levels of ORAI1 and PKC-βII. Interestingly, TPC2 KO cells showed a substantial down regulation of both ORAI1 and PKC-βII expression (Figure 5A–C); in contrast, STIM1 and INF2 expression were unchanged. We also observed that ORAI1 expression is reduced after TPC2 transient silencing (Figure 5D). To confirm the connection between ORAI1, TPC2 and YAP/TAZ, we performed a rescue of ORAI1 protein in our TPC2 KO model, and we saw a reduction of YAP/TAZ target gene expression compared to the mock-transfected cells when ORAI1 is overexpressed (Figure 5E,F). We also analysed vimentin and ZEB1 levels and showed that they are decreased when ORAI1 is overexpressed (Figure 5G). To study whether mechanotransduction may play a role in YAP/TAZ activation following TPC2 KO, we investigated the activation of YAP/TAZ target genes and N-Cadherin in ultra-low cell attachment conditions and found no such activation in these circumstances (Figure S2). These findings suggest that the loss of TPC2 may lead to the activation of YAP/TAZ target genes, and therefore increased metastasis, because of changes in the capacity of both SOCE and the cytoskeleton to mediate the inhibition of the HIPPO pathway.

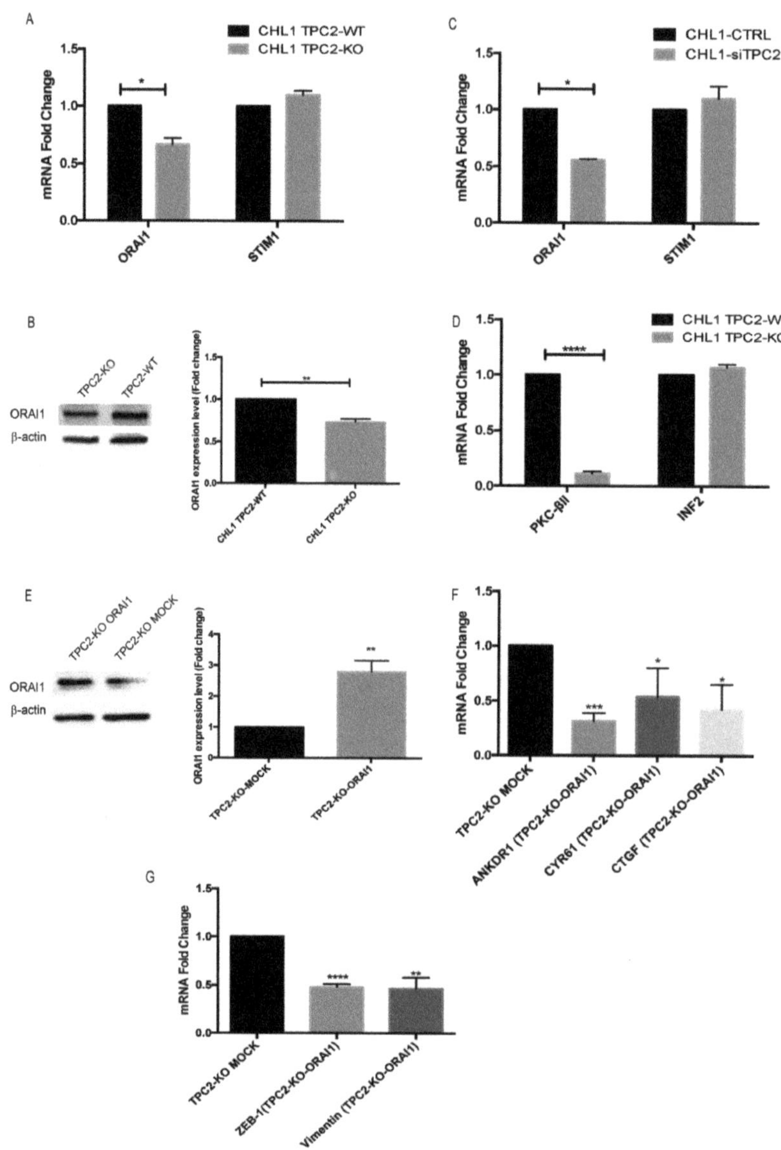

Figure 5. Possible mechanism involved in YAP/TAZ regulation mediated by TPC2. (**A**) qPCR analysis of ORAI1 and STIM1 in CHL1 WT and TPC2 KO cells. (**B**) ORAI1 protein expression in CHL1 WT and TPC2 KO cells normalised to β-actin. Relative density of ORAI1/β-actin. (**C**) qPCR analysis of ORAI1 and STIM1 expression after the transient silencing of TPC2. (**D**) qPCR analysis of PKC-βII and INF2 expression in CHL1 WT and TPC2 KO cells. (**E**) ORAI1 protein expression after 24 h from transfection in CHL1 TPC2 KO-MOCK and ORAI1 expressing cells, normalised to β-actin. Relative density of ORAI1/β-actin. (**F**) qPCR analysis of YAP/TAZ target gene expression after ORAI1 overexpression in CHL1 TPC2 KO cells. H3 was used as an internal control. (**G**) qPCR analysis of vimentin and ZEB1 expression after ORAI1 overexpression in CHL1 TPC2 KO cells. H3 was used as an internal control. Data in bar charts represent the mean ± s.e.m. of three independent experiments (* $p < 0.05$; ** $p < 0.01$; *** $p < 0.001$; **** $p < 0.0001$).

2.7. *TPC2 Is Overexpressed while YAP/TAZ Target Genes Are Down Regulated after Vemurafenib Treatment in A375 Melanoma Cells*

To further support the link between the reduction of TPC2 expression and YAP/TAZ target gene activation, we also analysed a dataset from another study [37] that investigated the effect of the BRAF inhibitor Vemurafenib on the BRAFV600E A375 melanoma cell line, the mutant form of BRAF in this cell line being differentially sensitive to inhibition by Vemurafenib (Figure 6). Studying this dataset, we noticed that after Vemurafenib treatment, TAZ and its target genes CTGF, CYR61, and ANKRD1 were down regulated, while TPC2 was overexpressed. Moreover, vimentin and ZEB1 were down regulated after Vemurafenib treatment, indicating a reduction of mesenchymal phenotype [37]. By indicating that YAP/TAZ activity is decreased when TPC2 expression is increased, such findings from a completely independent study are in line with our findings that indicate an inverse relationship between TPC2 expression and YAP/TAZ activity.

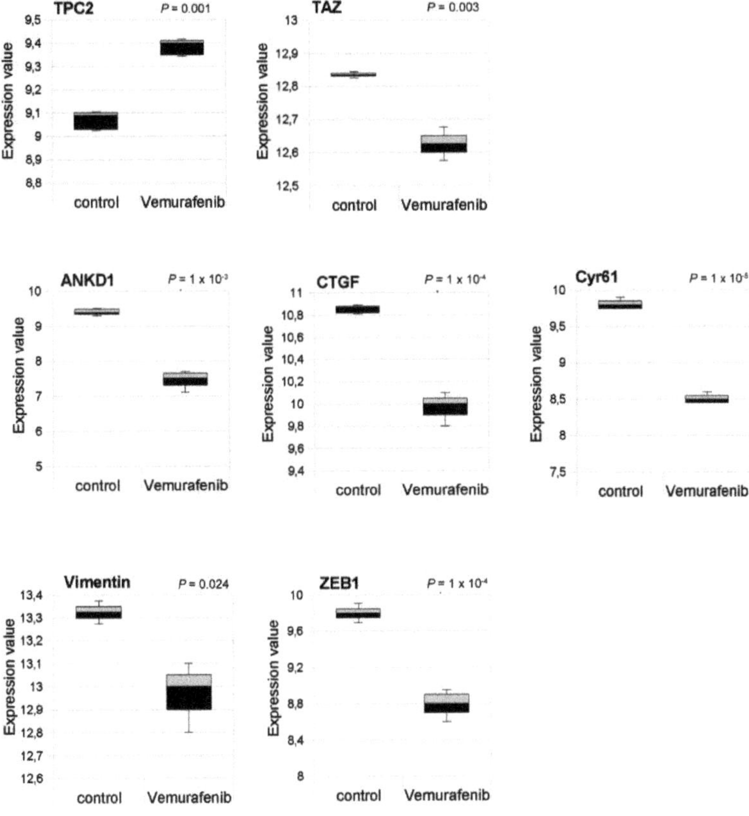

Figure 6. Analysis of A375 melanoma cells harbouring the BRAF V600E oncogenic mutation following treatment with the BRAF inhibitor Vemurafenib.

3. Discussion

TPC2 has been previously demonstrated to play a role in cancer progression. In particular, Nguyen et al. [24] have shown that silencing and pharmacologically inhibiting this channel impaired the migration of urinary bladder and hepatic human carcinoma cells in vitro, and of murine breast cancer cells both in vitro and in an in vivo mouse model. In contrast, our current study identifies for the first time a role for TPC2 in the aggressiveness of metastatic melanoma, and strongly indicates that in human metastatic melanoma cells, this role appears to differ from what is currently known from studying human cancer cells derived from primary tumours.

At the outset of this study, we performed a bioinformatic analysis to evaluate the potential prognostic role of TPC2 in human SKCM. Interestingly, we identified a significant difference in the expression of TPC2 between primary and metastatic patients, with TPC2 expression being significantly reduced in the latter. This suggests that the role played by TPC2 in primary and metastatic tumours might be qualitatively different.

To study the role of TPC2 in melanoma tumourigenesis and metastasis, we generated a CHL1 TPC2 KO cell line, CHL1 cells being a model of human amelanotic melanoma and derived from a metastatic site. They expressed a wild-type BRAF kinase and a mutant TP53 protein. Given these premises, CHL1 is considered a metastatic melanoma cell line. In this model, we observed that the loss of TPC2 expression reduced the ability of these cells to bind the collagen type I matrix. This binding is mediated by the $\alpha2\beta1$ integrin receptor, which was also reduced in the TPC2 KO cells. Moreover, the ability of CHL1 cells to invade a matrigel matrix was also enhanced by TPC2 KO, and this was associated with an increased activity of MMP9, a protein involved in cell invasion. Combined, these findings strongly indicate that TPC2 KO cells are more invasive than their WT counterparts, a highly unexpected finding given the existing data on TPC2 in different cancer cell models [24].

The analysis of markers of the EMT, a key process in the onset of metastasis, also supports this hypothesis. Thus, the TPC2 KO cells showed increased expression levels of ZEB-1, N-cadherin and vimentin, indicating that TPC2 KO cells are more mesenchymal. Furthermore, MITF, which is amplified in up to 20% of melanomas, with higher incidence among metastatic melanoma samples [38], was more highly expressed in TPC2 KO cells. Interestingly, a significant negative correlation between TPC2 and ZEB1, N-cadherin, and MMP9 was also found in the human SKCM dataset.

Significantly, some of the YAP/TAZ target genes, such as ANKRD1, CYR61, and CTGF, were also strongly up regulated in the TPC2 KO cells. YAP/TAZ are drivers of tumour initiation and growth. Their activity can control proliferation, invasion and metastasis, and is usually associated with poor prognosis [29]. It has also been shown that high levels of YAP are correlated with a decrease in survival for melanoma patients [23].

Studies of the effect of gene knockout on the properties of cells in culture can benefit from using parallel, independent approaches to achieving a reduction of a gene's expression. Importantly, we also confirmed that the transient inhibition of TPC2 in the CHL1 and MeWo metastatic human cell lines via RNAi using anti-TPC2 siRNAs, has similar effects on the expression of YAP/TAZ target genes. We also showed the translocation of YAP and TAZ from the cytoplasm to the nucleus in TPC2 KO cells, indicating the activation of this pathway. Moreover, this activation did not occur in ultra-low cell attachment conditions, suggesting a link with mechanotransduction in the activation of these factors that is associated with the reduction in TPC2 expression. Thus, these data suggest an aggressive phenotype for TPC2 KO cells. Furthermore, looking to the potential clinical significance of these findings by the use of the skin melanoma TCGA database, we also found a significant negative correlation between TPC2 expression and that of the YAP/TAZ target genes CTGF and CYR61, although intriguingly, there was no significant correlation with another target gene, ANKRD1. Interestingly, we also found a lack of ANKRD1 expression induction in MeWo TPC2-silenced cells, maybe indicating a differential effect of reduced TPC2 expression on this gene in distinct types of metastatic cancer cells.

Further studies will be required in the future to explore why there may be such a differential effect of TPC2 KO on the expression of CTGF, CYR61, and ANKRD1 in some circumstances, which could be revealing in terms of its implications for our understanding of the mechanisms of activation of these genes by YAP/TAZ.

Further indications of the increased aggressiveness of TPC2 KO cells comes from our findings relating to the induction of the expression of PD-L1. This suggests the increased ability of these cells to escape the immune system [34].

The transcription of PD-L1 has been shown to be regulated by TAZ [32], but this relationship between TAZ and PD-L1 is not conserved in a number of mouse cell lines, likely due to differences between the human and mouse PD-L1 gene promoters [32]. In the murine model of melanoma B16-F0 cells, the amount of PD-L1 exposed on the membrane of TPC2 KO cells was reduced (Figure S1D). Furthermore, in this model we did not see any YAP/TAZ pathway activation (Figure S1E). These data indicate a difference between the human and murine melanoma models studied but could also support a different role for TPC2 in melanoma according to the stage of the pathology as B16-F0 cells are derived from a primary tumour. One could speculate that the breast cancer murine line 4T1 in Nguyen et al. [24] would also not be associated with the activation of the YAP/TAZ pathway, due to the different regulation of this pathway between murine and human species. Further studies of animal and human WT and TPC2 KO melanoma cell lines derived from tumours at different stages of metastatic progression will be required to address this issue more definitively. However, our analysis of an independent study by Parmenter et al. [37] showed that after Vemurafenib treatment, TPC2 is overexpressed, while YAP/TAZ target genes are down regulated as well as ZEB1 and Vimentin, confirming our data in a different model of melanoma.

A key outstanding question is how a reduction in the expression of an endolysosomal ion channel would manifest itself in an increase in the metastatic traits of melanoma cells mediated by YAP/TAZ. Possibly, this occurs through a change in endolysosomal pH. Changing TPC2 expression could affect lysosomal or melanosomal pH either by increasing melanosomal and lysosomal pH when TPC2 is genetically inhibited [39–41] or decreasing lysosomal pH if TPC2 is overexpressed or pharmacologically activated [13,42]. However, it remains to be shown how such a change in endolysosomal pH would affect the HIPPO pathway, which is the primary mediator of YAP/TAZ activity.

Given previous evidence that TPC2 is a mediator of Ca^{2+} signalling responses, one obvious possibility is that changes in such responses underlie how a loss of, or a reduction in, TPC2 expression levels are associated with an increase in YAP/TAZ activity, and as a consequence, an enhancement of metastatic traits. However, the only published link between Ca^{2+} signalling responses and TAP/TAZ activity has implicated SOCE, a process that replenishes the endoplasmic reticulum (ER) Ca^{2+} store through an interaction between the ER protein STIM1 and the plasma membrane ion channel ORAI1 [35]. Thus, a study in glioblastoma cells has shown that the stimulation of SOCE inhibited YAP/TAZ activity via the activation of PKC-βII [36].

To see whether changes in SOCE might underlie the link between changes in TPC2 expression and YAP/TAZ activity, we analysed the expression of ORAI1 and PKC-βII; when PKC-βII is active, this kinase phosphorylates MST1/2 and LATS 1/2, which subsequently phosphorylate YAP/TAZ to remain in the cytosol. We found that in TPC2 KO or anti-TPC2 siRNA-treated cells, both ORAI1 and PKC-βII levels decreased. This suggests that an inhibition of SOCE associated with a loss of, or reduction of, TPC2 expression, may allow YAP/TAZ to translocate into the nucleus and activate their target genes. Indeed, when ORAI1 was overexpressed in TPC2 KO cells, YAP/TAZ target genes were down regulated. This raises the question of how changes in the expression of TPC2, an endolysosomal ion channel, could cause such a change in the levels of expression of ORAI1 and PKC-βII. One obvious possibility is that such changes are a consequence of a functional inhibition of SOCE, which would be in line with a previously described connection between SOCE and endolysosomal Ca^{2+} release [18,19].

4. Materials and Methods

4.1. Cell Lines

Human CHL1 and MeWo (ATCC, Manassas, VA, USA), and mouse B16-F0 (ATCC, Manassas, VA, USA), melanoma cell lines were maintained in Dulbecco's modified Eagle's medium (Sigma, St. Louis, MO, USA) supplemented with 2 mmol/L glutamine (Sigma, St. Louis, MO, USA), 10% foetal bovine serum (Gibco by Life Technologies, Carlsbad, CA, USA), and antibiotics (P/S, Sigma, St. Louis, MO, USA).

4.2. Western Blot Analysis

Following cell lysis, the proteins were separated on SDS-PAGE gels. Subsequently, the gels were blotted overnight at 4 °C onto nitrocellulose membranes (Amersham, Buckinghamshire, UK). The membranes were incubated with the following primary antibodies overnight at 4 °C: anti-MITF (Santacruz Biotch, Dallas, TX, USA), anti-YAP (Cell Signalling, Danvers, MA, USA); anti-TAZ (Sigma, St. Louis, MO, USA); anti-N-Cadherin (Abcam, Cambridge, UK); and anti-ORAI1 (ProSci Inc^{TM}, San Diego, CA, USA). The intensity of Western blot bands was quantified by Image Lab (Biorad, Hercules, CA, USA) and normalised using β-actin and tubulin (Sigma, St. Louis, MO, USA), GAPDH (Millipore, Burlington, MA, USA), and H3 (Abcam, Cambridge, UK).

4.3. CRISPR Design

Single guide RNAs (sgRNAs) were designed using the Synthego (Menlo Park, CA, USA) Design Tool to minimize off-target effects. The cells were transfected and analysed using Synthego (Menlo Park, CA, USA) protocols. Genotyping was performed using the following TPC2 primer sequences:

Human:
Forward
GATAGGGCGGTTACCATCATC
Reverse
CTCACCGGGTCAAAGTACAA
Mouse:
Forward
GGTATTACTTCGAACGTCTGCCAACGGT
Reverse
TTCAAAGCGCCAAAAGCTCACTAGCAA

4.4. Cell Transfection

1×10^5 cells were seeded in a 12-well plate. hTPC2 siRNA (Qiagen, Hilden, Germany) was used at 10 nM. Empty-plasmid and RFP-ORAI1 plasmid ([43] gifted by Prof. Vincenzo Sorrentino) were used at 500 ng/mL. Cells were transfected using Lipofectamine 2000 (Life Technologies, Carlsbad, CA, USA) following the manufacturer's protocol.

4.5. Adhesion Assay

Twelve-well plates were coated with collagen type I (Sigma, St. Louis, MO, USA) and incubated overnight at 4 °C. The following day, the excess matrix was removed and blocked with 0.1% bovine serum albumin (BSA) in Calcium Magnesium free Dulbecco's PBS (CMF-DPBS) for 1 h at room temperature. Subsequently, 2×10^5 cells were seeded and incubated for 90 min at 37 °C. Each well was stained by crystal violet which was dissolved using 10% acetic acid. The absorbance was read at 550 nm.

4.6. Invasion Assay

We seeded 1×10^5 cells on Transwell (Corning, Corning, NY, USA) that was coated with reduced growth factor Matrigel (BD, Franklin Lakes, NJ, USA) and a chemo-attractant gradient of FBS (1% to 20%) was used to favour cell movement. After 24 h incubation, the cells were fixed and stained with DAPI. For each Transwell, 5 fields were analysed by counting the cell numbers using Image J 2.0.0-rc-67/1.52c (US National Institutes of Health, Bethesda, MD, USA).

4.7. Flow Cytometry and Cell Cycle Analysis

1×10^5 cells were incubated for 30 min with anti-Integrinβ1 (Santacruz Biotech, Dallas, TX, USA) or anti-Integrinα2 (BD Horizon, Franklin Lakes, NJ, USA) antibodies, and with the secondary antibody anti-mouse Alexa-fluor 488, for another 30 min. For PDL1 (Biolegend, San Diego, CA, USA), after 24 h IFN-γ treatment, 1×10^5 cells were incubated for 30 min in the same buffer with anti-PDL1 antibody. Dead cells were excluded by Sytox Blue Stain (Life Technologies, Carlsbad, CA, USA) or propidium iodide (Sigma, St. Louis, MO, USA). For cell cycle analysis, the cells were fixed with 70% ethanol, washed three times with PBS and stained for 3 h at room temperature with PBS–propidium iodide, then analysed using a CyAn ADP flow cytometer (Beckman Coulter, Brea, CA, USA) and FCS express 5 (De Novo software, Glendale, CA, USA).

4.8. Bioinformatic Analysis

Normalized gene expression of skin cutaneous melanoma patients was obtained from the Broad Institute TCGA Genome Data Analysis Center (2016): TCGA data from Broad GDAC Firehose 2016_01_28 run; Broad Institute of MIT and Harvard Dataset. https://doi.org/10.7908/C11G0KM9. The statistical significance of the differential modulation of *TPCN2* gene between subgroups of patients was inferred by Student's *t*-test (*p*-value = 8.13×10^{-5}) and with a nonparametric Wilcoxon Rank-Sum test (*p*-value = 8.58×10^{-5}). Positive and negative association between the genes was evaluated by Spearman's rank correlation. High and low expression values of a gene for each subgroup of patients were assessed by positive and negative z-scores, respectively. Analyses were performed using MATLAB R2019b software. We analysed the GSE42872 dataset [37] from the Gene Expression Omnibus database (GEO: https://www.ncbi.nlm.nih.gov/gds). GEO2R, an online analysing tool of GEO DataSets, was utilized to analyse differentially expressed genes between A375 melanoma cells harbouring the BRAF V600E oncogenic mutation, and which had been treated, or not, with the BRAF inhibitor Vemurafenib. A *p*-value of <0.05 was used as the cut off criterion to identify significant differential expression between the two groups.

4.9. Quantitative Real-Time PCR

Total RNA was isolated using an RNeasy mini kit (Qiagen Hilden, Germany). First-strand cDNA synthesis was performed using a SuperScript III reagent kit (Invitrogen, Carlsbad, CA, USA). Real-time PCR was then carried out with primers specific for ANKDR1, CTGF, CYR61, YAP, TAZ, TPC2, ORAI1, STIM1, PKC-βII, INF2, ZEB1, VIMENTIN, CCND1, CCNB1, and p21 (Table 1), using a Powerup SYBR green master mix (Applied Biosystems, Waltham, Massachusetts, USA). GAPDH and H3 were used as internal controls. Relative mRNA expression levels were calculated using the ΔΔCT method.

Table 1. Primer sequences.

Gene	Forward	Reverse
CTGF	AGGAGTGGGTGTGTGACGA	CCAGGCAGTTGGCTCTAATC
CYR61	CCTTGTGGACAGCCAGTGTA	ACTTGGGCCGGTATTTCTTC
ANKRD1	AGTAGAGGAACTGGTCACTGG	TGGGCTAGAAGTGTCTTCAGAT
YAP	GCACCTCTGTGTTTTAAGGGTCT	CAACTTTTGCCCTCCTCCAA
TAZ(WWTR1)	GGCTGGGAGATGACCTTCAC	CTGAGTGGGGTGGTTCTGCT
GAPDH	GGAGCGAGATCCCTCCAAAAT	GGCTGTTGTCATACTTCTCATGG
STIM	ATCTCACAGCTCATGGTATGCTCC	GGAAGGTGCCAAAGAGTGTGTTTC
ORAI	TACTTGAGCCGCGCCAAGCTTAAA	ACCGAGTTGAGATTGTGCACGTTG
PKCβII	GACCAAACACCCAGGCAAAC	GATGGCGGGTGAAAAATCGG
INF2	CACATCCAACGTGATGGTGAAG	GGAGAGCTCGTTCATGACAATG

4.10. Gelatin Zymography for Matrix Metalloproteinases (MMPs) Detection

The supernatant was collected after 18 and 24 h after cell starvation with 1% FBS. Gelatinolytic activity of conditioned media was assayed as previously described [44], followed by gelatine zymography as previously described [14]. The values were normalized to protein content.

4.11. Statistical Analysis

Data are presented as the mean ± s.e.m. of results from at least three independent experiments. Student's *t*-test was used for statistical comparison between the means where applicable (two groups) or ordinary one-way ANOVA (for groups of three or more) * $p < 0.05$; ** $p < 0.01$; *** $p < 0.001$; **** $p < 0.0001$.

5. Conclusions

In summary, our data indicate that a loss of, or reduction of, TPC2 expression in a metastatic model of human melanoma increased the aggressiveness of melanoma cells. In particular, we identified a previously undisclosed connection between TPC2 and the YAP/TAZ pathway, which could be an emerging focus for cancer research. One must keep in mind that the interaction of cells with the extracellular surroundings can influence the activation of YAP/TAZ, since the latter are also considered to be mechanosensors and mechanotransducers. Indeed, in line with this, we found no activation of YAP/TAZ target genes in TPC2 KO cells in ultra-low cell attachment conditions. It will therefore be interesting in future studies to investigate the possible role of TPC2 in the modulation of cancer cell mechanotransduction and its different role in primary or metastatic tumours. Together with the findings discussed here, such studies will be important in further characterising the mechanistic link between the endolysomal ion channel TPC2 and metastatic cancer, as well as identifying new ways to diagnose and treat this type of cancer.

Supplementary Materials: The following are available online at http://www.mdpi.com/2072-6694/12/9/2391/s1, Figure S1: YAP/TAZ target genes' induction in B16-F0 murine cell line, Figure S2: Ultra low attachment conditions, Figure S3: Correlation analysis between TPCN2 and Vimentin; MITF; ZEB1; CTGF; ANKRD1; CYR61; CHD-2(N-Cadherin); MMP9 in primary patients in human skin melanoma TCGA database, Figure S4: Correlation analysis between TPCN2 and Vimentin; MITF; ZEB1; CTGF; ANKRD1; CYR61; CHD-2(N-Cadherin); MMP9 in metastatic patients in human skin melanoma TCGA database, Table S1: Statistical analysis of TPC2 and other genes' correlation in skin melanoma TGCA dataset.

Author Contributions: Conceptualization A.D., A.A.H., S.D.A., F.P., J.P., and A.F.; validation A.D., A.A.H., and S.D.A.; investigation A.D., A.A.H., S.D.A., A.V., M.T., and R.C.; resources A.A.H. and A.F.; data curation A.D., A.A.H., and A.S.; writing—the original draft preparation A.D. and J.P.; writing—review and editing A.D., A.A.H., F.P., J.P., and A.F.; visualization F.P., G.B., and J.P.; supervision J.P. and A.F.; project administration J.P. and A.F.; software A.D. and A.S.; funding acquisition J.P. and A.F. All authors have read and agreed to the published version of the manuscript.

Funding: This research received no external funding.

Acknowledgments: This work was supported by Progetti di Ricerca di Ateneo, "La Sapienza" University of Rome (Italy) to A.F. and by the Saudi Cultural Bureau on behalf of Jazan University (Ph.D. scholarship for A.A.H.).

Conflicts of Interest: The authors declare no conflict of interest.

References

1. Uong, A.; Zon, L.I. Melanocytes in development and cancer. *J. Cell Physiol.* **2010**, *222*, 38–41. [CrossRef] [PubMed]

2. Shain, A.H.; Bastian, B.C. From melanocytes to melanomas. *Nat. Rev. Cancer* **2016**, *16*, 345–358. [CrossRef] [PubMed]

3. Zhu, M.X.; Ma, J.; Parrington, J.; Calcraft, P.J.; Galione, A.; Evans, A.M. Calcium signaling via two-pore channels: Local or global, that is the question. *Am. J. Physiol. Cell Physiol.* **2010**, *298*, C430–C441. [CrossRef] [PubMed]

4. Wang, X.; Zhang, X.; Dong, X.P.; Samie, M.; Li, X.; Cheng, X.; Goschka, A.; Shen, D.; Zhou, Y.; Harlow, J.; et al. TPC proteins are phosphoinositide- activated sodium-selective ion channels in endosomes and lysosomes. *Cell* **2012**, *151*, 372–383. [CrossRef] [PubMed]

5. Cang, C.; Zhou, Y.; Navarro, B.; Seo, Y.J.; Aranda, K.; Shi, L.; Battaglia-Hsu, S.; Nissim, I.; Clapham, D.E.; Ren, D. mTOR regulates lysosomal ATP-sensitive two-pore Na(+) channels to adapt to metabolic state. *Cell* **2013**, *152*, 778–790. [CrossRef]

6. Pitt, S.J.; Reilly-O'Donnell, B.; Sitsapesan, R. Exploring the biophysical evidence that mammalian two-pore channels are NAADP-activated calcium-permeable channels. *J. Physiol.* **2016**, *594*, 4171–4179. [CrossRef]

7. Calcraft, P.J.; Ruas, M.; Pan, Z.; Cheng, X.; Arredouani, A.; Hao, X.; Tang, J.; Rietdorf, K.; Teboul, L.; Chuang, K.T.; et al. NAADP mobilizes calcium from acidic organelles through two-pore channels. *Nature* **2009**, *459*, 596–600. [CrossRef]

8. Galione, A. NAADP receptors. *Cold Spring Harb. Perspect. Biol.* **2011**, *3*, a004036. [CrossRef]

9. Ruas, M.; Davis, L.C.; Chen, C.C.; Morgan, A.J.; Chuang, K.T.; Walseth, T.F.; Grimm, C.; Garnham, C.; Powell, T.; Platt, N.; et al. Expression of Ca(2)(+)-permeable two-pore channels rescues NAADP signalling in TPC-deficient cells. *EMBO J.* **2015**, *34*, 1743–1758. [CrossRef]

10. Lin-Moshier, Y.; Walseth, T.F.; Churamani, D.; Davidson, S.M.; Slama, J.T.; Hooper, R.; Brailoiu, E.; Patel, S.; Marchant, J.S. Photoaffinity labeling of nicotinic acid adenine dinucleotide phosphate (NAADP) targets in mammalian cells. *J. Biol. Chem.* **2012**, *287*, 2296–2307. [CrossRef]

11. Walseth, T.F.; Lin-Moshier, Y.; Jain, P.; Ruas, M.; Parrington, J.; Galione, A.; Marchant, J.S.; Slama, J.T. Photoaffinity labeling of high affinity nicotinic acid adenine dinucleotide phosphate (NAADP)-binding proteins in sea urchin egg. *J. Biol. Chem.* **2012**, *287*, 2308–2315. [CrossRef] [PubMed]

12. Jha, A.; Ahuja, M.; Patel, S.; Brailoiu, E.; Muallem, S. Convergent regulation of the lysosomal two-pore channel-2 by Mg(2)(+), NAADP, PI(3,5)P(2) and multiple protein kinases. *EMBO J.* **2014**, *33*, 501–511. [CrossRef]

13. Gerndt, S.; Chen, C.C.; Chao, Y.K.; Yuan, Y.; Burgstaller, S.; Scotto Rosato, A.; Krogsaeter, E.; Urban, N.; Jacob, K.; Nguyen, O.N.P.; et al. Agonist-mediated switching of ion selectivity in TPC2 differentially promotes lysosomal function. *Elife* **2020**, *9*, e54712. [CrossRef] [PubMed]

14. Favia, A.; Desideri, M.; Gambara, G.; D'Alessio, A.; Ruas, M.; Esposito, B.; Del Bufalo, D.; Parrington, J.; Ziparo, E.; Palombi, F.; et al. VEGF-induced neoangiogenesis is mediated by NAADP and two-pore channel-2-dependent Ca2+ signaling. *Proc. Natl. Acad. Sci. USA* **2014**, *111*, E4706–E4715. [CrossRef]

15. Islam, M.S. Molecular Regulations and Functions of the Transient Receptor Potential Channels of the Islets of Langerhans and Insulinoma Cells. *Cells* **2020**, *9*, 685. [CrossRef] [PubMed]

16. Lewis, R.S. The molecular choreography of a store-operated calcium channel. *Nature* **2007**, *446*, 284–287. [CrossRef]

17. Hooper, R.; Zaidi, M.R.; Soboloff, J. The heterogeneity of store-operated calcium entry in melanoma. *Sci. China Life Sci.* **2016**, *59*, 764–769. [CrossRef]

18. Hui, L.; Geiger, N.H.; Bloor-Young, D.; Churchill, G.C.; Geiger, J.D.; Chen, X. Release of calcium from endolysosomes increases calcium influx through N-type calcium channels: Evidence for acidic store-operated calcium entry in neurons. *Cell Calcium* **2015**, *58*, 617–627. [CrossRef]

19. Sbano, L.; Bonora, M.; Marchi, S.; Baldassari, F.; Medina, D.L.; Ballabio, A.; Giorgi, C.; Pinton, P. TFEB-mediated increase in peripheral lysosomes regulates store-operated calcium entry. *Sci. Rep.* **2017**, *7*, 40797. [CrossRef]

20. Han, Y. Analysis of the role of the Hippo pathway in cancer. *J. Transl. Med.* **2019**, *17*, 116. [CrossRef]

21. Yu, F.X.; Zhao, B.; Guan, K.L. Hippo Pathway in Organ Size Control, Tissue Homeostasis, and Cancer. *Cell* **2015**, *163*, 811–828. [CrossRef] [PubMed]

22. Zanconato, F.; Battilana, G.; Forcato, M.; Filippi, L.; Azzolin, L.; Manfrin, A.; Quaranta, E.; Di Biagio, D.; Sigismondo, G.; Guzzardo, V.; et al. Transcriptional addiction in cancer cells is mediated by YAP/TAZ through BRD4. *Nat. Med.* **2018**, *24*, 1599–1610. [CrossRef] [PubMed]

23. Menzel, M.; Meckbach, D.; Weide, B.; Toussaint, N.C.; Schilbach, K.; Noor, S.; Eigentler, T.; Ikenberg, K.; Busch, C.; Quintanilla-Martinez, L.; et al. In melanoma, Hippo signaling is affected by copy number alterations and YAP1 overexpression impairs patient survival. *Pigment Cell Melanoma Res.* **2014**, *27*, 671–673. [CrossRef] [PubMed]

24. Nguyen, O.N.; Grimm, C.; Schneider, L.S.; Chao, Y.K.; Atzberger, C.; Bartel, K.; Watermann, A.; Ulrich, M.; Mayr, D.; Wahl-Schott, C.; et al. Two-Pore Channel Function Is Crucial for the Migration of Invasive Cancer Cells. *Cancer Res.* **2017**, *77*, 1427–1438. [CrossRef] [PubMed]

25. Ran, J.; Lin, D.L.; Wu, R.F.; Chen, Q.H.; Huang, H.P.; Qiu, N.X.; Quan, S. ZEB1 promotes epithelial-mesenchymal transition in cervical cancer metastasis. *Fertil. Steril.* **2015**, *103*, 1606–1614.e2. [CrossRef] [PubMed]

26. Satelli, A.; Li, S. Vimentin in cancer and its potential as a molecular target for cancer therapy. *Cell Mol. Life Sci.* **2011**, *68*, 3033–3046. [CrossRef]

27. Beuret, L.; Flori, E.; Denoyelle, C.; Bille, K.; Busca, R.; Picardo, M.; Bertolotto, C.; Ballotti, R. Up-regulation of MET expression by alpha-melanocyte-stimulating hormone and MITF allows hepatocyte growth factor to protect melanocytes and melanoma cells from apoptosis. *J. Biol. Chem.* **2007**, *282*, 14140–14147. [CrossRef]

28. Gentile, A.; Trusolino, L.; Comoglio, P.M. The Met tyrosine kinase receptor in development and cancer. *Cancer Metastasis Rev.* **2008**, *27*, 85–94. [CrossRef]

29. Dobrokhotov, O.; Samsonov, M.; Sokabe, M.; Hirata, H. Mechanoregulation and pathology of YAP/TAZ via Hippo and non-Hippo mechanisms. *Clin. Transl. Med.* **2018**, *7*, 23. [CrossRef]

30. Hoj, J.P.; Mayro, B.; Pendergast, A.M. A TAZ-AXL-ABL2 Feed-Forward Signaling Axis Promotes Lung Adenocarcinoma Brain Metastasis. *Cell Rep.* **2019**, *29*, 3421–3434. [CrossRef]

31. Dupont, S.; Morsut, L.; Aragona, M.; Enzo, E.; Giulitti, S.; Cordenonsi, M.; Zanconato, F.; Le Digabel, J.; Forcato, M.; Bicciato, S.; et al. Role of YAP/TAZ in mechanotransduction. *Nature* **2011**, *474*, 179–183. [CrossRef] [PubMed]

32. Janse van Rensburg, H.J.; Azad, T.; Ling, M.; Hao, Y.; Snetsinger, B.; Khanal, P.; Minassian, L.M.; Graham, C.H.; Rauh, M.J.; Yang, X. The Hippo Pathway Component TAZ Promotes Immune Evasion in Human Cancer through PD-L1. *Cancer Res.* **2018**, *78*, 1457–1470. [CrossRef] [PubMed]

33. Luo, N.; Formisano, L.; Gonzalez-Ericsson, P.I.; Sanchez, V.; Dean, P.T.; Opalenik, S.R.; Sanders, M.E.; Cook, R.S.; Arteaga, C.L.; Johnson, D.B.; et al. Melanoma response to anti-PD-L1 immunotherapy requires JAK1 signaling, but not JAK2. *Oncoimmunology* **2018**, *7*, e1438106. [CrossRef] [PubMed]

34. Jiang, Y.; Sun, A.; Zhao, Y.; Ying, W.; Sun, H.; Yang, X.; Xing, B.; Sun, W.; Ren, L.; Hu, B.; et al. Proteomics identifies new therapeutic targets of early-stage hepatocellular carcinoma. *Nature* **2019**, *567*, 257–261. [CrossRef]

35. Mercer, J.C.; Dehaven, W.I.; Smyth, J.T.; Wedel, B.; Boyles, R.R.; Bird, G.S.; Putney, J.W., Jr. Large store-operated calcium selective currents due to co-expression of Orai1 or Orai2 with the intracellular calcium sensor, Stim1. *J. Biol. Chem.* **2006**, *281*, 24979–24990. [CrossRef]

36. Liu, Z.; Wei, Y.; Zhang, L.; Yee, P.P.; Johnson, M.; Zhang, X.; Gulley, M.; Atkinson, J.M.; Trebak, M.; Wang, H.G.; et al. Induction of store-operated calcium entry (SOCE) suppresses glioblastoma growth by inhibiting the Hippo pathway transcriptional coactivators YAP/TAZ. *Oncogene* **2019**, *38*, 120–139. [CrossRef]

37. Parmenter, T.J.; Kleinschmidt, M.; Kinross, K.M.; Bond, S.T.; Li, J.; Kaadige, M.R.; Rao, A.; Sheppard, K.E.; Hugo, W.; Pupo, G.M.; et al. Response of BRAF-mutant melanoma to BRAF inhibition is mediated by a network of transcriptional regulators of glycolysis. *Cancer Discov.* **2014**, *4*, 423–433. [CrossRef]

38. Garraway, L.A.; Widlund, H.R.; Rubin, M.A.; Getz, G.; Berger, A.J.; Ramaswamy, S.; Beroukhim, R.; Milner, D.A.; Granter, S.R.; Du, J.; et al. Integrative genomic analyses identify MITF as a lineage survival oncogene amplified in malignant melanoma. *Nature* **2005**, *436*, 117–122. [CrossRef]

39. Ambrosio, A.L.; Boyle, J.A.; Aradi, A.E.; Christian, K.A.; Di Pietro, S.M. TPC2 controls pigmentation by regulating melanosome pH and size. *Proc. Natl. Acad. Sci. USA* **2016**, *113*, 5622–5627. [CrossRef]

40. Bellono, N.W.; Escobar, I.E.; Oancea, E. A melanosomal two-pore sodium channel regulates pigmentation. *Sci. Rep.* **2016**, *6*, 26570. [CrossRef]

41. Lin, P.H.; Duann, P.; Komazaki, S.; Park, K.H.; Li, H.; Sun, M.; Sermersheim, M.; Gumpper, K.; Parrington, J.; Galione, A.; et al. Lysosomal two-pore channel subtype 2 (TPC2) regulates skeletal muscle autophagic signaling. *J. Biol. Chem.* **2015**, *290*, 3377–3389. [CrossRef] [PubMed]

42. Lu, Y.Y.; Hao, B.X.; Graeff, R.; Wong, C.W.M.; Wu, W.T.; Yue, J. Two pore channel 2 (TPC2) inhibits autophagosomal-lysosomal fusion by alkalinizing lysosomal pH. *J. Biol. Chem.* **2017**, *292*, 12088. [CrossRef] [PubMed]

43. Giurisato, E.; Gamberucci, A.; Ulivieri, C.; Marruganti, S.; Rossi, E.; Giacomello, E.; Randazzo, D.; Sorrentino, V. The KSR2-calcineurin complex regulates STIM1-ORAI1 dynamics and store-operated calcium entry (SOCE). *Mol. Biol. Cell* **2014**, *25*, 1769–1781. [CrossRef] [PubMed]

44. Piovesana, R.; Faroni, A.; Taggi, M.; Matera, A.; Soligo, M.; Canipari, R.; Manni, L.; Reid, A.J.; Tata, A.M. Muscarinic receptors modulate Nerve Growth Factor production in rat Schwann-like adipose-derived stem cells and in Schwann cells. *Sci. Rep.* **2020**, *10*, 7159. [CrossRef]

Article

Plasmacytoid Dendritic Cell Impairment in Metastatic Melanoma by Lactic Acidosis

Matilde Monti [1,†], **Raffaella Vescovi** [1,†], **Francesca Consoli** [2], **Davide Farina** [3], **Daniele Moratto** [4], **Alfredo Berruti** [2], **Claudia Specchia** [1] and **William Vermi** [1,5,*]

[1] Department of Molecular and Translational Medicine, University of Brescia, 25123 Brescia, Italy; m.monti002@unibs.it (M.M.); raffaella.vescovi@gmail.com (R.V.); claudia.specchia@unibs.it (C.S.)

[2] Oncology Unit, ASST Spedali Civili di Brescia, 25123 Brescia, Italy; francesca.consoli@icloud.com (F.C.); alfredo.berruti@unibs.it (A.B.)

[3] Radiology Unit, Department of Medical and Surgical Specialties, Radiological Sciences and Public Health, University of Brescia, 25123 Brescia, Italy; davide.farina@unibs.it

[4] Laboratory of Genetic Disorders of Childhood, Angelo Nocivelli Institute for Molecular Medicine, ASST Spedali Civili di Brescia, 25123 Brescia, Italy; daniele.moratto@gmail.com

[5] Department of Pathology and Immunology, Washington University School of Medicine, Saint Louis, MO 63101, USA

* Correspondence: william.vermi@unibs.it; Tel.: +39-030-399-8425

† These authors equally contributed to the work.

Received: 30 June 2020; Accepted: 24 July 2020; Published: 28 July 2020

Abstract: The introduction of targeted therapies and immunotherapies has significantly improved the outcome of metastatic melanoma (MM) patients. These approaches rely on immune functions for their anti-melanoma response. Plasmacytoid dendritic cells (pDCs) exhibit anti-tumor function by production of effector molecules, type I interferons (I-IFNs), and cytokines. Tissue and blood pDCs result compromised in MM, although these findings are still partially conflicting. This study reports that blood pDCs were dramatically depleted in MM, particularly in patients with high lactate dehydrogenase (LDH) and high tumor burden; the reduced pDC frequency was associated with poor overall survival. Circulating pDCs resulted also in significant impairment in interferon alpha (IFN-α) and C-X-C motif chemokine 10 (CXCL10) production in response to toll-like receptor (TLR)-7/8 agonists; on the contrary, the response to TLR-9 agonist remained intact. In the BRAF^{V600+} subgroup, no recovery of pDC frequency could be obtained by BRAF and MEK inhibitors (BRAFi; MEKi), whereas their function was partially rescued. Mechanistically, in vitro exposure to lactic acidosis impaired both pDC viability and function. In conclusion, pDCs from MM patients were found to be severely impaired, with a potential role for lactic acidosis. Short-term responses to treatments were not associated with pDC recovery, suggesting long-lasting effects on their compartment.

Keywords: plasmacytoid dendritic cells; melanoma; lactate dehydrogenase; TLR; interferon; CXCL10

---

1. Introduction

The prognosis of metastatic melanoma (MM) patients has been dramatically improved by novel therapeutic strategies including targeted therapies and immune checkpoint blockades (ICB) [1,2]. In fact, more than half of melanoma patients (about 50–60%) harbor BRAF mutation, together with the corresponding downstream signal transduction in the MAPK (mitogen-activated protein kinase) pathway [3]. Historically, the development of high selective targeted agents such as vemurafenib or dabrafenib has dramatically improved overall survival (OS), progression-free survival (PFS), and response rate in BRAF^{V600+} advanced melanoma patients, in comparison to standard chemotherapy [4,5]. Unfortunately, the great majority of patients treated with BRAF inhibitor (BRAFi) monotherapy developed secondary resistance to treatment within 6–8 months [4,5]. Blockade of CTLA-4 and PD-1 receptors expressed by lymphocytes leads to their activation against tumor cells. The anti-CTLA-4 antibody ipilimumab was the first discovered ICB, showing a plateau in the survival curve in 21% of patients [6,7]. Anti-PD-1 agents such as nivolumab and pembrolizumab improve PFS and OS in comparison to ipilimumab, with an objective response rate of about 40% [6,8]. Anti-PD-1 treatments are considered an effective option in advanced melanoma patients, regardless of BRAF mutation [9].

The MAPK pathway hyper-activation is associated with increased metastatic behavior, reduced apoptosis, and modulation of interaction between melanoma cells and the immune system [10,11]. Moreover, BRAF^{V600+} melanoma cell lines secrete immunosuppressive cytokines such as interleukin 10 (IL-10), vascular endothelial growth factor (VEGF), and interleukin 6 (IL-6), which promote the recruitment of regulatory T cells and myeloid-derived suppressor cells [12]. Mutant BRAF also downregulates the expression of melanoma differentiation antigens (MDA) and class I major histocompatibility complex (MHC-I) molecules on tumor cells, preventing their recognition by CD8$^+$ T cells [13,14]. Consequently, MAPK-targeted therapy affects melanoma cell immunogenicity and immune contexture involving different effector and regulatory mechanisms [10,15]. The MAPK pathway inhibition suppresses the secretion of immunosuppressive cytokines and leads to upregulation of MDA in melanoma cell lines [12,14], improving T cell recognition and increasing intra-tumoral CD4$^+$ and CD8$^+$ T lymphocytes [11]. The inhibition of MAPK pathway in melanoma cells also restores cytokine secretion and co-stimulatory molecule expression by dendritic cells (DCs), rescuing their compromised function [12]. On the contrary, MEK inhibition negatively affects in vitro DC and T cell viability and function [11,16]. However, the systemic administration of BRAFi does not alter leukocyte subset frequencies [17], and the combination of BRAFi and MEKi enhances immunological activation [10]. A pre-existing immunological signature predicts the response to BRAFi/MEKi [18]. Furthermore, the MAPK activation downregulates interferon alpha receptor 1 (IFNAR1) signaling; accordingly, BRAFi reverses IFNAR1 inhibition in melanoma biopsies, providing a rationale for the combination of interferon alpha (IFN-α) with BRAFi [19].

All these findings suggest a significant role for immune cell components in mediating responses, not only to ICB, but also to targeted therapy. Among cells involved in melanoma immunity, plasmacytoid dendritic cells (pDCs) exert an important role in shaping the anti-tumor immune response. A distinctive feature of pDCs is the production of a large amount of type I interferon (I-IFN) after nucleic acid sensing through toll-like receptor (TLR) 7- and 9-dependent signaling pathways [20,21]. IFN-α production not only affects tumor cell proliferation, angiogenesis, and metastasis [22,23], but also acts on different immune cell populations involved in anticancer immunity such as NK, T, and B lymphocytes [24–26]. Autocrine IFN-α/β signaling also regulates the induction of interferon signature genes, among which pro-inflammatory chemokines (i.e., CXCL9, CXCL10, and CXCL11) [27], with potential antitumor activity driving TH1 polarization of immune cells [28].

By analyzing a large cohort of primary and metastatic cutaneous melanomas, we recently monitored pDC dynamic during melanoma evolution. In MM patients, pDCs become almost absent in the tumor tissues and severely reduced in their circulation, particularly in the advanced M1c group [29]. The subversion of the mechanisms leading to the systemic pDC collapse might potentiate spontaneous and drug-induced anti-melanoma immunity, providing additional benefit to the current systemic therapies. Mechanistically, exposure to melanoma cell supernatants resulted in significant death of terminally differentiated pDCs and in defective generation of pDCs from CD34$^+$ progenitors [29]. This effect is dependent on soluble components released by melanoma cells, and a role by lactic acidosis microenvironment has been proposed [30,31]. The neoplastic cell metabolism shifts toward high glucose uptake and enhanced lactate production, regardless of oxygen availability, known as the Warburg effect [32]. Lactate dehydrogenase (LDH-A) is a key enzyme that converts pyruvate to lactate in the final step of the glycolytic pathway. High levels of lactate within tumor cells are exported by monocarboxylate transporters (MCT) coupled with protons exported across the plasma membrane, leading to lactic acidosis in the tumor microenvironment [33]. This leads to enhancement of tumor-associated immune-suppressive functions and inhibition of effector cells in the tumor milieu [30,31,34]. This study analyzes circulating pDCs in a prospective cohort of chemo-naïve MM patients. pDCs from MM patients are severely impaired in their frequency and function, with a potential role for lactic acidosis. Moreover, short-term responses to BRAFi and MEKi treatments are not associated with a full pDC recovery, suggesting long lasting effects on their compartment.

2. Results

2.1. Clinical Features and Outcome of the MM Cohort

Patients' clinical characteristics at baseline (T0) are reported in Table 1 and Table S1. Patients had a median age of 60 years (range: 23–79 years) at the time of clinical diagnosis, while healthy donors (HD; $N = 25$) had a median age of 44 years (range: 25–56 years). Sixteen patients had BRAF-mutated melanomas, including BRAF p.V600E ($n = 12$) or p.V600K ($n = 4$). Seven MM were NRAS-mutated, and included p.Q61K ($n = 2$) or p.Q61R ($n = 5$). The remaining six patients were BRAF/NRAS wild-type. The median LDH level at baseline was 220 IU/L (range: 134–236 IU/L). Patients were sub-grouped into M1a ($n = 5$) + M1b ($n = 7$) and M1c ($n = 17$) categories, accordingly to the "American Joint Committee on Cancer (AJCC) Melanoma Staging and Classification 7th edition". Sixteen patients (55.17%) were treated with targeted therapies (vemurafenib; dabrafenib; vemurafenib + cobimetinib; dabrafenib + trametinib), and 13 patients received ICB (ipilimumab or pembrolizumab) (44.83%). Complete response (CR), partial response (PR), and stable disease (SD) were achieved in 6.89%, 31.03%, and 6.89% of patients, respectively. Progression disease (PD) was observed in 55.17% of patients (43.75% in the targeted therapy group and 56.25% in the ICB group).

Table 1. Clinical and molecular features of the study cohort at baseline (T0; $N = 29$).

Clinical/ Molecular Features	BRAF^{V600+} MM ($N = 16$)			NRAS^{Q61+} MM ($N = 7$)			BRAF^{wt}/NRAS^{wt} ($N = 6$)			
	N	n	%	N	n	%	N	n	%	p
Gender (males)	16	12	75	7	5	71.4	6	5	83.3	1
Stage	16			7			6			0.5
M1a		3	18.7		1	14.3		1	16.7	
M1b		5	31.3		0	0		2	33.3	
M1c		8	50		6	85.7		3	50	
Brain metastases	16	3	18.7	7	2	28.6	6	2	33.3	0.6
Tumor sites (≥3)	16	6	37.5	7	3	42.9	6	4	66.7	0.5
LDH (high *)	16	4	25	6	3	50	5	2	40	0.5
Therapy	16			7			6			
BRAFi		4	25							
BRAFi + MEKi		12	75							
Anti-CTLA-4					7	100		5	83.3	
Anti-PD-1								1	16.7	
	N	Median (min–max)	SD	N	Median (min–max)	SD	N	Median (min–max)	SD	p
Age	16	58.5 (23–76)	14.0	7	58.0 (48–76)	10.0	6	62.0 (53–79)	8.8	0.2
Tumor burden (mm)	15	116.5 (0–408.2)	127.6	7	77.9 (8–260.3)	89.0	6	136.4 (52.2–309.5)	98.0	0.8

* above the normal range; p: p-value; SD: standard deviation.

2.2. Peripheral Blood Immune Populations and pDC Function Were Impaired in Chemo-Naïve MM Patients

As previously reported, circulating pDCs are reduced in various human cancer patients compared to healthy subjects, particularly in the advanced stage of disease [35]. By flow cytometry on fresh whole blood, we previously documented the collapse of pDC and myeloid DC (mDC) subsets in advanced MM [29]. We extended this finding, showing that CD3$^+$ and CD4$^+$ T lymphocytes remained unchanged in MM (Table 2). We could not detect significant differences on the basis of the molecular profile of the tumor both for absolute number of total leukocytes and for the frequencies of all analyzed immune cell populations (Table S2). Altogether, these data suggest that MM patients are characterized by an impairment of blood DCs, but not T lymphocytes.

Table 2. Peripheral blood immune populations in healthy donors (HD; $N = 25$) and metastatic melanoma (MM) at baseline (T0; $N = 29$).

Immune Cell Population		MM ($N = 29$)			HD ($N = 25$)			p *
		N	Median	IQR	N	Median	IQR	
% pDCs on PBMCs		29	0.3	0.2–0.3	24	0.4	0.4–0.6	**0.0006**
% mDCs on PBMCs		29	0.3	0.2–0.5	15	0.5	0.4–0.7	**0.03**
% CD3$^+$ on PBMCs		29	58.9	52.6–62.3	15	58	50.1–66.8	0.9
% CD4$^+$ on PBMCs		29	28.7	20.1–39.0	15	34.6	27.4–40.8	0.1
% IFN-α^+ pDCs	R848	27	48.5	30.2–65.0	25	71.5	64.3–74.7	**0.004**
	IMQ	27	21.1	8.7–41.7	24	38.3	29.4–45.1	**0.02**
	CpG	26	9.3	4.3–13.9	22	11.6	5–15.2	0.4
% CXCL10$^+$ pDCs	R848	24	78.4	64.8–85.3	19	80.9	79.1–86.8	0.08
	IMQ	24	47.9	22.4–61.5	19	61.6	48.9–71.4	**0.03**
	CpG	25	7.1	3.7–17.1	20	10.2	2.9–14.8	0.7

IQR (interquartile range): Q1–Q3; * p: p-value. The significant p-values are highlighted in bold.

We subsequently tested the response of pDCs to TLR-7/9 agonist stimulation. I-IFN exerts cancer cell intrinsic effects and modulates the immunoediting process [36–38], while CXCL10, also known as interferon gamma inducible protein 10 (IP-10), is a pro-inflammatory chemokine that is relevant for the recruitment of antigen-specific T cells into the tumor tissues [39]. Through using intracellular flow cytometry, we analyzed the pDC proficiency to produce IFN-α and CXCL10, after in vitro stimulation of peripheral blood mononuclear cells (PBMCs) with TLR-7/9 agonists (Figure 1A). On a background of patients' heterogeneity, stimulation with two different TLR-7 agonists resulted in a significant reduction of the frequency of IFN-α-producing pDCs in MM (R848 $p = 0.004$; imiquimod (IMQ) $p = 0.02$) (Table 2 and Figure 1B). Moreover, the percentage of CXCL10 producing pDCs was also significantly reduced after IMQ exposure ($p = 0.03$) (Table 2 and Figure 1C). On the contrary, under CpG-A oligodeoxynucleotides (ODN) stimulation, IFN-α and CXCL10 production were not significantly different between MM and HD (Table 2 and Figure 1B,C). The defect in TLR-7 signaling response was not correlated with the melanoma molecular profile in the cohort analyzed herein (Table S2).

According to the melanoma staging system (seventh edition, AJCC), our cohort of MM patients was classified into M1a+b and M1c categories (Table S1) [40]. Compared to HD and M1a+b, the M1c subgroup showed a significant decrease in IFN-α-producing pDCs after TLR-7 stimulation (R848: HD vs. M1c, $p < 0.0001$; IMQ: HD vs. M1c, $p = 0.0014$; M1a+b vs. M1c, $p = 0.0190$) (Figure 1D). Similarly, the percentage of CXCL10-producing pDCs resulted in a significant decrease in advanced disease stages (R848: HD vs. M1c, $p = 0.0133$; IMQ: HD vs. M1c, $p = 0.0485$) after TLR-7 agonist administration (Figure 1E). Altogether, these data highlight an impairment of the TLR-7 signaling pathway in fully differentiated pDCs that are associated with disease progression.

2.3. LDH Level and Tumor Burden Were Associated with Decreased Frequency of Peripheral Blood Immune Cells in Chemo-Naïve MM Patients

Elevated level of serum LDH is a relevant independent negative prognostic biomarker in melanoma that indicates an active metastatic disease [41]. The analysis of immune cells in the peripheral blood of chemo-naïve patients revealed that levels of LDH above the normal range (details are found in the Materials and Methods section) are associated with a significant decreased frequency of circulating lymphocytes ($p = 0.02$), particularly of CD4$^+$ T cells ($p = 0.03$) (Table 3 and Figure 2A,B). Similarly, a decline in the frequencies of pDCs and mDCs resulted in association with elevated serum levels of the enzyme ($p = 0.003$ and $p = 0.02$, respectively) (Table 3, Figure 2C,D). These data suggest that the systemic release of LDH in advanced melanoma might affect the generation or viability of blood T lymphocytes, pDCs, and mDCs. On the contrary, serum LDH levels are not associated with a reduced frequency of IFN-α^+ and CXCL10$^+$ pDCs in response to TLR-7/9 agonists in chemo-naïve MM patients (Table 3).

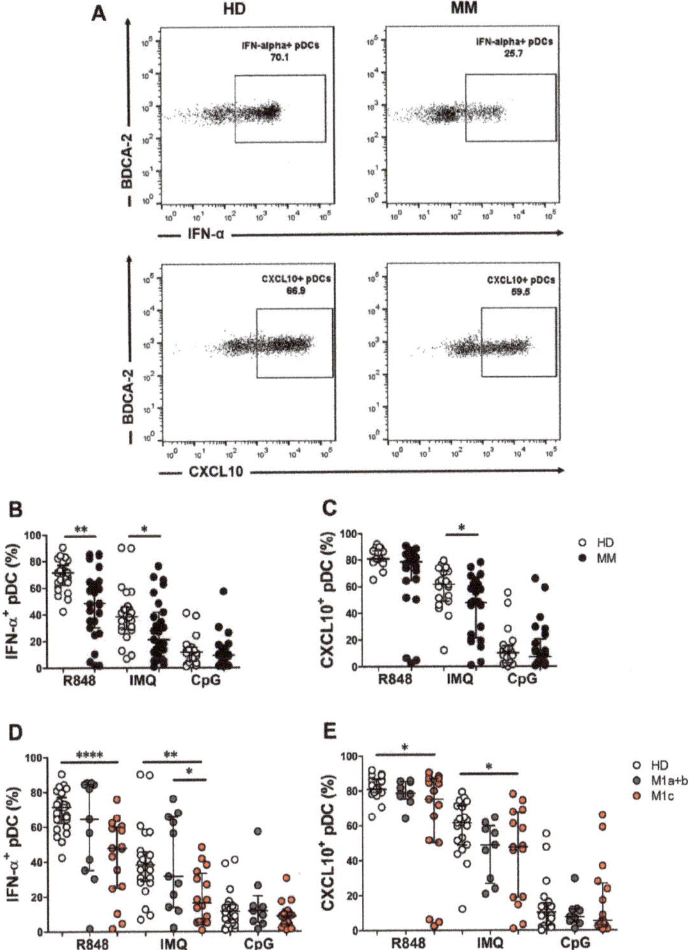

Figure 1. Frequency of interferon alpha (IFN-α) and CXCL10-producing plasmacytoid dendritic cells (pDCs) in chemo-naïve MM patients and HD. Representative dot plots of R848-stimulated IFN-α^+ and CXCL10$^+$ pDC subsets obtained from HD and MM patients are shown (**A**). PBMCs were isolated from peripheral blood of HD ($n = 25$) and MM patients ($n = 29$). Total PBMCs were cultured in RPMI 1640 medium supplemented with 10% FBS and IL-3 and stimulated with R848 or IMQ for 2 h (**B,D**) and 6 h (**C,E**), and with CpG-ODN 2216 for 6 h (**B–E**). IFN-α (**B,D**) and CXCL10 (**C,E**) were analyzed by intracellular flow cytometry staining. Scatter dot plot graphs illustrate the percentages of positive pDCs evaluated on BDCA-2$^+$/CD123$^+$ cells. Subgroup analysis of the MM cohort illustrating the frequency of IFN-α^+ and CXCL10$^+$ pDCs in M1a-c categories (**D,E**). Median and IQR are shown in (**B,C**). Mean and SD are shown in (**D,E**). The statistical significance was calculated by Wilcoxon–Mann–Whitney test (**B,C**) and by a Student's *t*-test (**D,E**). * $p < 0.05$; ** $p < 0.01$; **** $p < 0.0001$.

Table 3. Peripheral blood immune populations and lactate dehydrogenase (LDH) level in MM patients cohort at baseline (T0; $N = 29$).

Immune Cell Population		Normal LDH ($N = 18$)			High LDH * ($N = 9$)			
		N	Median	IQR	N	Median	IQR	p
$n°$ leukocytes/µl		14	6805	5830–9270	9	9110	6860–12,050	0.08
% neutrophils on LK		14	54.7	47.2–68.5	8	66.2	54.9–74.3	0.1
% lymphocytes on LK		14	32.3	20.8–39	8	19.9	12.1–26.1	**0.02**
% monocytes on LK		14	8.3	7.5–11	8	7.5	6.1–10.3	0.2
% eosinophils on LK		14	2.1	1.6–2.5	8	0.8	0.1–3	0.1
% basophils on LK		14	0.6	0.4–0.8	8	0.5	0.2–0.6	0.3
% pDCs on PBMCs		18	0.3	0.2–0.4	9	0.1	0.1–0.2	**0.003**
% mDCs on PBMCs		18	0.4	0.3–0.5	9	0.2	0.1–0.2	**0.02**
% CD3$^+$ on PBMCs		18	60.2	54.6–68.9	9	46.7	40.7–59.8	0.06
% CD4$^+$ on PBMCs		18	30.9	25.5–42.6	9	23.8	17.8–27	**0.03**
% IFN-α^+ pDCs	R848	17	56.7	35.1–82	9	43.0	25.9–58	0.2
	IMQ	17	27.7	14–56.4	9	13.6	5.4–27	0.08
	CpG	16	7.8	4.2–14.2	9	9.9	8.3–10.9	1
% CXCL10$^+$ pDCs	R848	14	78.4	71.9–85.5	9	75.2	51.6–84.1	0.7
	IMQ	14	47.3	23.9–57.9	9	47.5	19–69	0.4
	CpG	15	5.2	2.9–11.7	9	17.1	5.7–36.8	0.1

IQR (interquartile range): Q1–Q3; *p*: *p*-value; LK: leukocytes; * referred to values above the normal range. The significant *p*-values are highlighted in bold.

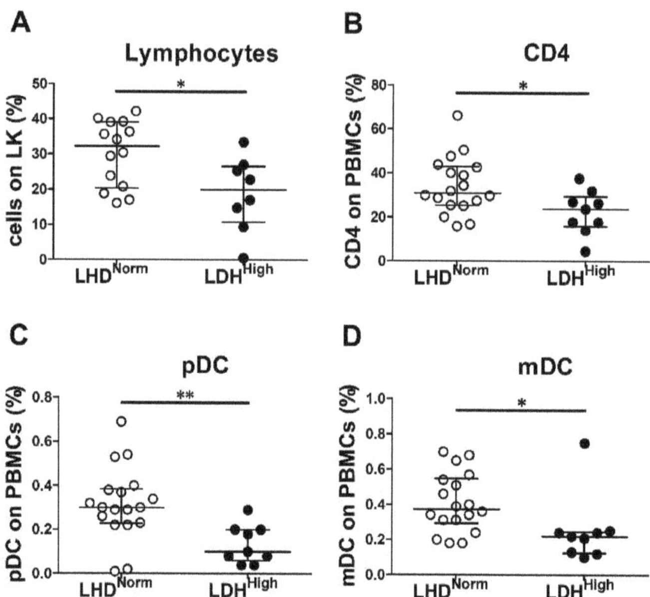

Figure 2. High LDH serum level associated with decreased frequencies of peripheral blood immune cells in MM patients. Immune cell populations were identified on whole blood samples of chemo-naïve MM patients by cell counts or flow cytometry. Scatter dot plot graphs represent the percentages of lymphocytes on total leukocytes (LK) (**A**), CD4$^+$ T lymphocytes (**B**), pDCs (**C**), and mDCs (**D**) on total PBMCs. LDH levels are indicated as LDH^{Norm} (for values within the normal range) and LDH^{High} (for values greater than the normal range). Error bars represent the median with IQR ($n = 27$). The statistical significance was calculated by Wilcoxon–Mann–Whitney test. * $p < 0.05$; ** $p < 0.01$.

Tumor burden at baseline has been identified as a relevant predictor for MM outcome [42]. Soluble products or metabolic competition by massive amounts of melanoma cells could explain the defective immune cell frequencies and function. Accordingly, tumor burden resulted in inverse correlated with the frequencies of circulating lymphocytes ($p = 0.03$), including $CD3^+$ and $CD4^+$ T cells ($p = 0.04$ and $p = 0.02$, respectively), as well as pDC and mDC frequencies ($p = 0.006$ and $p = 0.001$, respectively) (Table 4). On the contrary, no correlation was detected between the tumor burden and the pDC function (Table 4).

Table 4. Correlation between tumor burden (mm) and peripheral blood immune cells of the MM patient cohort at baseline (T0; $N = 29$).

Immune Cell Population		N	Rho	p
$n°$ leukocytes/µl		22	0.41	0.06
% neutrophils on LK		21	0.24	0.30
% lymphocytes on LK		21	−0.48	**0.03**
% monocytes on LK		21	−0.08	0.72
% eosinophils on LK		21	−0.22	0.34
% basophils on LK		21	−0.44	**0.05**
% pDCs on PBMCs		28	−0.51	**0.006**
% mDCs on PBMCs		28	−0.59	**0.001**
% $CD3^+$ on PBMCs		28	−0.39	**0.04**
% $CD4^+$ on PBMCs		28	−0.45	**0.02**
% IFN-α^+ pDCs	R848	26	−0.36	0.07
	IMQ	26	−0.34	0.09
	CpG	25	−0.10	0.62
% CXCL10$^+$ pDCs	R848	23	−0.28	0.20
	IMQ	23	−0.11	0.63
	CpG	24	0.11	0.61

Rho: Pearson correlation coefficient; LK: leukocytes; The significant p-values are highlighted in bold.

In conclusion, systemic LDH and bulky disease are associated with impaired immune cell generation and viability, particularly in DCs.

2.4. In Vitro Exposure to Lactic Acidosis Impaired the Viability and Function of Fully Differentiated pDCs

The oncometabolite lactate and acidosis have immunosuppressive effects on various immune cells, including pDCs [43–45]. Lactate, glucose concentrations, and the pH level were measured in the supernatants of human melanoma cell lines (SN-mel; collected as previously described by Vescovi et al. [29]). Compared to RPMI control, SN-mel contained higher levels of lactate, inversely associated with glucose levels (Figure S1A), and slight acidosis (pH range: 6.7–7.2 versus 8.0) (Figure S1B), indicating metabolic lactic acidosis. We tested the viability of pDCs and T lymphocytes in three different culture conditions: (i) lactic acidosis, (ii) lactosis, and (iii) acidosis. To this end, we exposed pDCs and T lymphocytes purified from HD to increasing concentrations of lactic acid (LA), sodium lactate (NaL), and hydrochloric acid (HCl). The pH values of the medium containing LA 10 mM, 15 mM, and 20 mM corresponded to 6.5, 6.0, and 5.5, respectively. pDC death, in the form of late apoptosis or necrosis, was significantly increased by high concentrations of LA (20 mM; $p = 0.01$) (Figure 3A–C). Similarly, the viability of T cells was significantly reduced by the highest lactic acidosis condition ($p = 0.03$), in terms of both apoptotic ($p = 0.03$) and necrotic cell death ($p = 0.03$) (Figure 3D–F). The highest concentration of HCl (pH = 5.5) was also associated with a significant increase of the pDC late apoptosis or necrosis ($p = 0.03$), while affecting both late ($p = 0.01$) and early apoptosis ($p = 0.01$) of

T cells (Figure 3G–L). On the contrary, the lactosis condition did not affect pDC and T cell viability (Figure S2). Together these results indicate that acidosis, but not lactosis, induces pDC and T cell death.

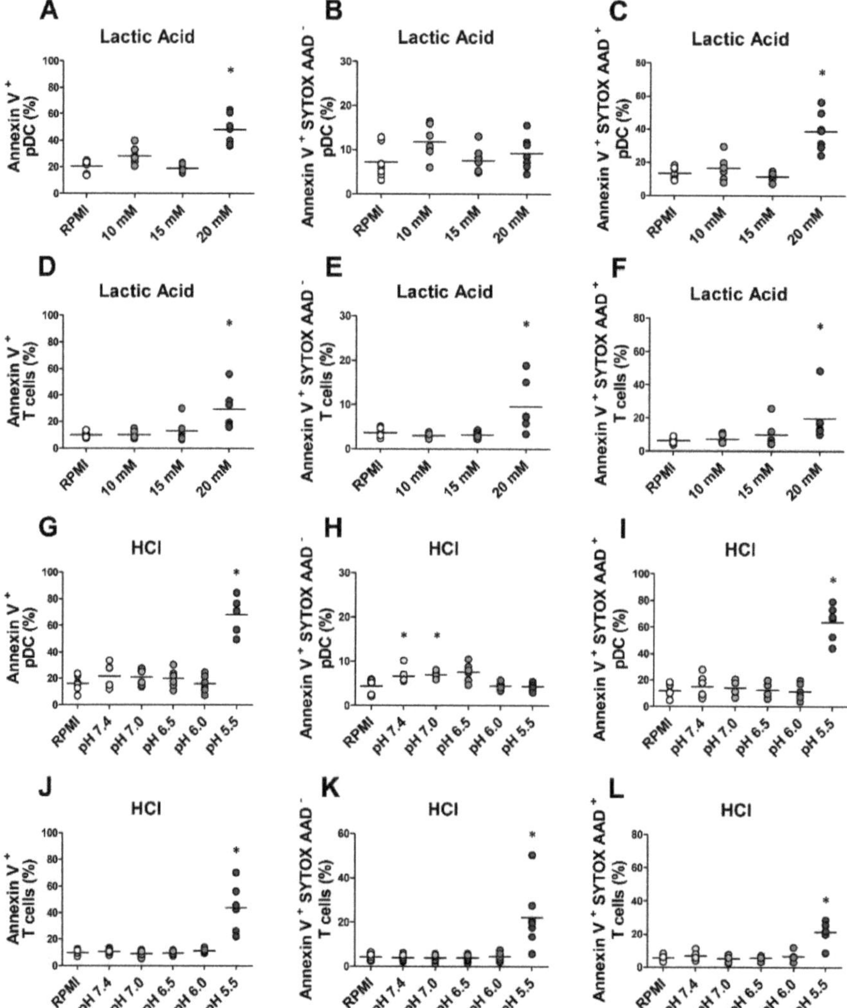

Figure 3. Lactic acidosis affects the viability of pDCs and T cells. pDCs and T cells purified from buffy coats of HD were cultured in RPMI 1640 medium supplemented with 10% FBS plus lactic Acid (10 mM; 15 mM; 20 mM) (n = 7 (**A–C**); n = 6, (**D–F**)) or hydrochloric acid (pH = 7.4; 7.0; 6.5; 6.0; 5.5 (n = 6 (**G–I**); n = 7, (**J–L**)) for 24 h. IL-3 was added to pDCs' culture. The cellular viability was analyzed by annexin V/SYTOX AADvanced staining in flow cytometry. Aligned dot plot graphs show the percentages of dead (**A,D,G,J**), early apoptotic (**B,E,H,K**), and late apoptotic or necrotic cells (**C,F,I,L**). Bars represent the mean of biological replicates. The statistical significance was calculated by two-sample paired sign test. * $p < 0.05$.

Lactate has been recently proposed as an inhibitor of the pDC function in the tumor microenvironment [45,46]. Hence, we assessed the IFN-α production on pDCs isolated from HD and exposed to increasing concentrations of LA. The percentage of IFN-α-producing pDCs under R848 stimulation was progressively reduced by increasing concentrations of LA (10 mM, p = 0.01; 15 mM,

$p = 0.01$; 20 mM, $p = 0.01$) and was completely abolished at a concentration of 20 mM (Figure 4A). pDCs were subsequently exposed to increasing concentration of HCl and stimulated with R848. The percentage of IFN-α^+ pDCs was progressively reduced by decreasing pH levels as well (pH 6.0, $p = 0.01$; pH 5.5, $p = 0.01$) (Figure 4B). In conclusion, in vitro acidosis dramatically impaired the pDC viability and their proficiency to produce IFN-α.

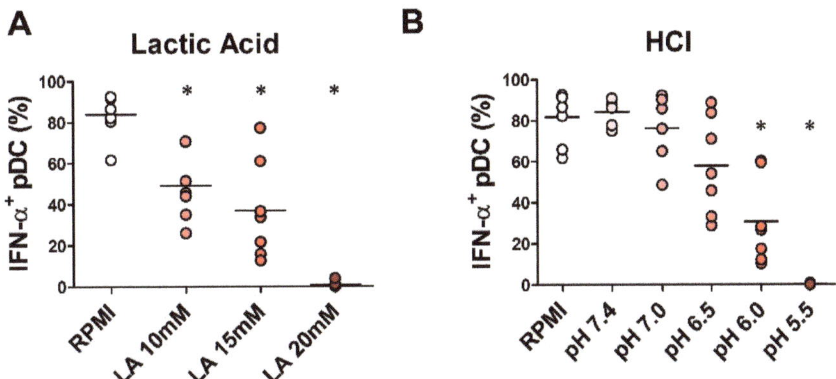

Figure 4. In vitro lactic acidosis affects IFN-α production by pDCs. pDCs purified from buffy coats of HD were cultured in RPMI 1640 medium supplemented with 10% FBS and IL-3 plus lactic acid (10 mM; 15 mM; 20 mM) ($n = 7$; (**A**)) or hydrochloric acid (pH = 7.4; 7.0; 6.5; 6.0; 5.5) ($n = 7$; (**B**)) for 24 h. pDCs were stimulated with R848 for 2 h. Intracellular IFN-α was analyzed by flow cytometry. Aligned dot plot graphs show the percentages of IFN-α^+ pDCs evaluated on BDCA-2$^+$/CD123$^+$ cells. Bars represent the mean of biological replicates. The statistical significance was calculated by two-sample paired sign test. * $p < 0.05$.

2.5. Baseline Lymphocyte and pDC Frequencies Predicted MM Outcome

On the basis of the clinical relevance of recent treatment options for advanced stage of disease, additional prognosticators might help in patient's selection. The analysis of survival revealed that among 29 MM patients, the median OS time was 14 months (range: 1–35 months), with a total of 62% of patients dying, while the median PFS time was 4 months (range: 1–35 months) (Table S3). Among the significant prognosticators, NRAS mutations ($p = 0.03$), M1c stage ($p = 0.0016$), presence of brain metastases ($p = 0.0001$), high LDH levels ($p = 0.0002$), and elevated tumor burden ($p = 0.01$) resulted in a significant correlation with poor outcome in univariate analysis (Figure 5A–E). In terms of immune profile, a reduced frequency of lymphocytes ($p = 0.01$), pDCs ($p = 0.01$), CD3$^+$ ($p = 0.03$), and CD4$^+$ T cells ($p = 0.01$) predict poor OS in an univariate Cox regression model (Table 5). By subgrouping the MM cohort on the basis of the frequencies of immune cell subsets, only the low frequency of pDCs resulted in an association with worse prognosis ($p = 0.03$; Figure 6). Finally, a decrease of IFN-α and CXCL10-positive pDCs after stimulation with R848 ($p = 0.04$ and $p = 0.03$, respectively) also predicted a worse outcome. In our cohort, the PFS probability calculated in an univariate Cox regression model was significantly correlated with patient's molecular profile ($p = 0.0006$), stage ($p = 0.004$), brain metastasis ($p = 0.001$), and LDH serum level ($p = 0.003$) (Figure 5F–J); on the contrary, the immune profile failed to predict PFS (Table S4).

These data suggest that profiling circulating immune cells, particularly pDCs, might offer additional biomarkers of outcome.

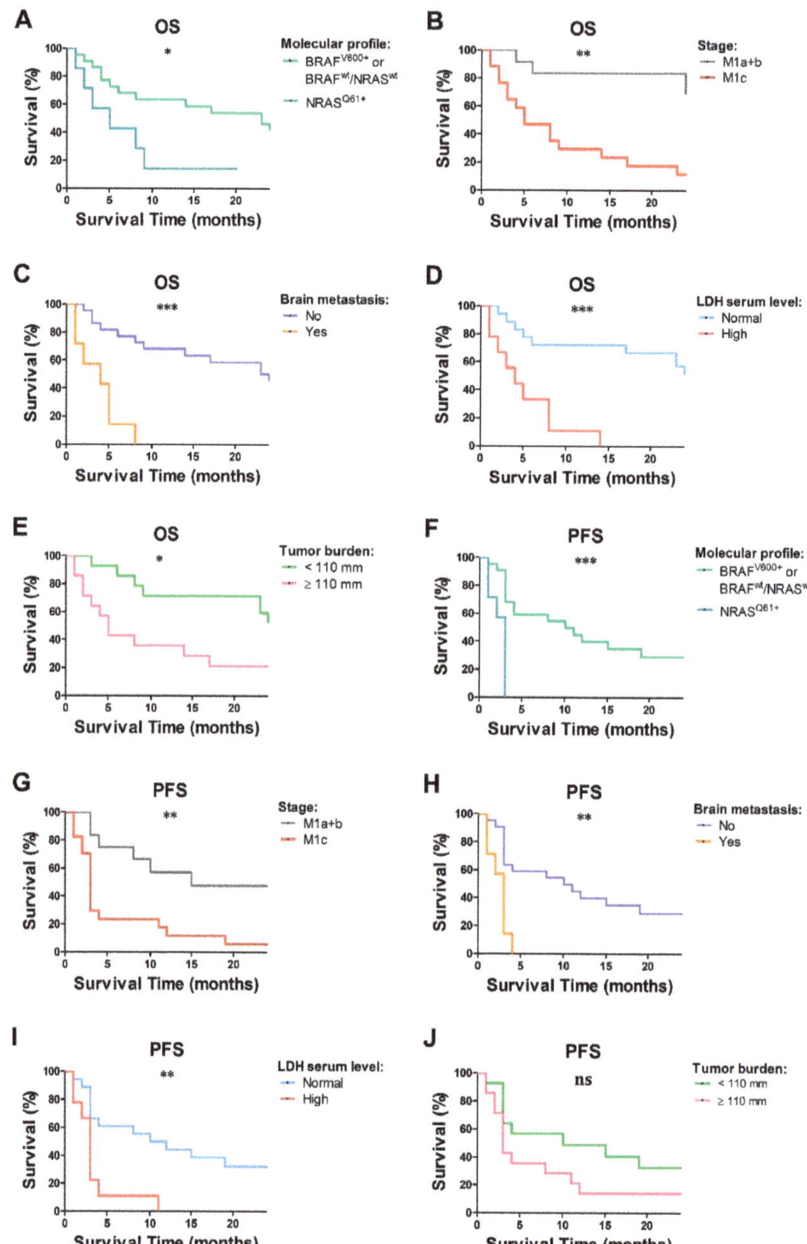

Figure 5. Survival analysis of MM patients. Overall survival (OS) (**A**–**E**) and progression-free survival (PFS) (**F**–**J**) according to molecular profile (*N* = 29; (**A,F**)), stage of melanoma (*N* = 29; (**B,G**)), presence of brain metastases (*N* = 29; (**C,H**)), LDH serum level (*N* = 27; (**D,I**)), and tumor burden (*N* = 28; (**E,J**)) are reported. Survival analysis was performed using the Kaplan–Meier method and the statistical significance was calculated by the log-rank test. * $p < 0.05$; ** $p < 0.01$; *** $p < 0.001$; ns = not statistically significant *p*-value.

Table 5. Univariate and multivariate Cox regression models for OS in MM patients at baseline (T0; $N = 29$).

Immune Cell Population		Univariate			Multivariate [°]		
		HR	p	95% CI	HR	p	95% CI
$n°$ leukocytes/µl		1.06 **	0.27	0.96–1.18			
% neutrophils on LK		1.01	0.61	0.96–1.07			
% lymphocytes on LK		**0.92**	**0.01**	0.87–0.98	0.95	0.06	0.90–1.00
% monocytes on LK		1.04	0.73	0.82–1.33			
% eosinophils on LK		0.79	0.33	0.49–1.27			
% basophils on LK		0.85 *	0.14	0.02–1.75			
% pDCs on PBMCs		**0.61 ***	**0.01**	0.68–1.06	0.72	0.06	0.52–1.01
% mDCs on PBMCs		0.84 *	0.24	0.63–1.12			
% CD3[+] on PBMCs		**0.95**	**0.03**	0.91–1.00	0.98	0.54	0.94–1.04
% CD4[+] on PBMCs		**0.93**	**0.01**	0.88–0.98	0.97	0.30	0.90–1.03
% IFN-α^+ pDCs	R848	**0.98**	**0.04**	0.96–1.00	0.98	0.19	0.96–1.01
	IMQ	0.98	0.07	0.95–1.00			
	CpG	0.98	0.44	0.93–1.03			
% CXCL10[+] pDCs	R848	**0.98**	**0.03**	0.96–1.00	0.99	0.25	0.97–1.01
	IMQ	1.00	0.72	0.97–1.02			
	CpG	**1.04**	**0.01**	1.01–1.07	**1.05**	**0.008**	1.01–1.08

LK: leukocytes; * HR associated with a 0.1 unit increase; ** HR associated with a 1000 unit increase; ° adjusted for molecular profile (NRAS[Q61+] vs. BRAF[V600+] or BRAF[wt]/NRAS[wt]) and stage of the disease (M1c vs. M1a or M1b). p: p-value. The significant p-values and relative hazard ratio are highlighted in bold.

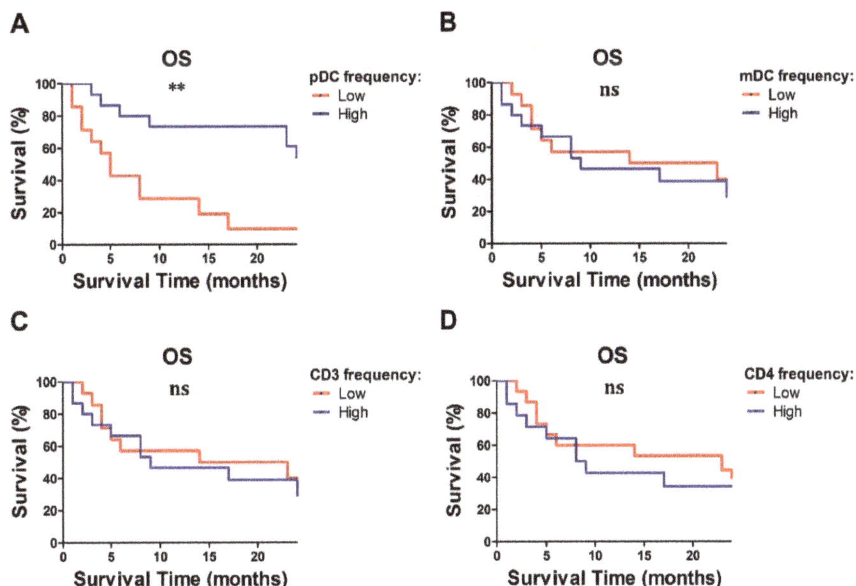

Figure 6. Survival analysis of MM patients according to immune cell subsets. Overall survival analysis (OS) according to the frequency of pDCs (**A**), mDCs (**B**), CD3[+] (**C**), and CD4[+] T lymphocytes (**D**) ($N = 29$) are reported. Survival analysis was performed using the Kaplan–Meier method and the statistical significance was calculated by the log-rank test. ** $p < 0.01$; ns = not statistically significant p-value.

2.6. Partial pDC Recovery after BRAFi and MEKi Administration

A combination of BRAFi and MEKi represents the standard of care for BRAF^{V600+} MM patients. These agents inhibit the constitutively activated RAF–RAS–MEK–ERK pathway in melanoma cells [47] and are associated with potent clinical responses [48,49], partially mediated by the immune system [10,15], including DCs [16]. We monitored peripheral blood leukocyte frequencies and pDC function in BRAF^{V600+} MM patients treated with a combination of MEKi and BRAFi or BRAFi alone. For this purpose, blood samples were obtained at different time points over treatment, on the basis of the expected time to the clinical response (details in the Materials and Methods Section 4.1). Compared to the values found in HD, no recovery in the frequencies of peripheral blood immune populations was observed. However, within the MM cohort, some differences emerged by comparing various time points. A significant increase of circulating lymphocytes was observed over 120 days of treatment ($p = 0.05$) (Table 6 and Figure 7A). Instead, pDC frequency was significantly reduced after 30 days of therapy ($p = 0.01$) but returned to the baseline level (T0) after 120 days of treatment (Table 6 and Figure 7B). No significant variation in cell frequency was detected for the other immune populations (Table 6 and Figure 7C).

Table 6. Variation of the peripheral blood immune populations at 1 month (T1; $N = 12$) and 4 months (T2; $N = 9$) from therapy's initiation compared to the baseline (T0; $N = 16$) in the group of BRAF^{V600+} MM patients.

Immune Cell Population		Variation T1–T0				Variation T2–T0			
		N	Mean	SD	p	N	Mean	SD	p
$n°$ leukocytes/µl		13	−810	1891	0.2	6	−1161	1260	0.1
% neutrophils on LK		13	−3.2	10.8	0.3	6	−10.3	12.7	0.1
% lymphocytes on LK		13	3.6	8.9	0.1	6	11.0	12.2	**0.05**
% monocytes on LK		13	1.0	3.6	0.2	6	0.1	1.9	0.8
% eosinophils on LK		13	0.7	1.8	0.2	6	−0.7	1.6	0.4
% basophils on LK		13	0.4	0.5	0.07	6	0.0	0.2	0.6
% pDCs on PBMCs		16	−0.1	0.1	**0.01**	9	−0.1	0.1	0.3
% mDCs on PBMCs		16	0.0	0.2	1.0	9	0.0	0.2	1.0
% CD3$^+$ on PBMCs		16	−1.4	10.2	0.6	9	−4.5	9.7	0.5
% CD4$^+$ on PBMCs		16	−0.1	6.6	1.0	9	−0.5	20.6	0.4
% IFN-α$^+$ pDCs	R848	14	0.3	37.1	1.0	7	19.1	17.7	**0.03**
	IMQ	14	−2.0	34.0	1.0	7	8.6	14.6	0.1
	CpG	12	3.6	7.6	0.2	5	−0.1	9.1	0.9
% CXCL10$^+$ pDCs	R848	12	−9.2	31.9	0.2	5	−4.1	9.1	0.4
	IMQ	12	−4.9	32.4	0.5	5	8.9	19.5	0.2
	CpG	12	−1.6	15.3	0.8	5	1.9	1.5	0.08

LK: leukocytes; p: p-value. The significant p-values are highlighted in bold.

Recent studies suggest that the MAPK pathway inhibition by targeted therapies in BRAF^{V600+} melanoma patients can improve the melanoma-specific immune responses [15]. In monocyte-derived DCs (moDC), interleukin 12 (IL-12) and tumor necrosis factor alpha (TNF-α) production and co-stimulatory molecule expression is impaired after co-culture with melanoma cells, but it can be easily restored by pre-treatment with BRAFi [16]. Moreover, inhibitors of MEK1/2 (i.e., PD0325901 and U0126) significantly increase the TLR-9-mediated production of I-IFN in pDCs, restoring the I-IFN production previously blocked via B cell receptor (BCR)-like signaling [50]. We examined the function of pDCs from MM patients treated with BRAFi alone or in combination with MEKi. A significant increase in the percentage of IFN-α$^+$ pDCs was registered, comparing T2 versus T0 patients ($p = 0.03$) after R848 stimulation (Table 6 and Figure 7D). In contrast, no differences were detected in the pool of IFN-α-producing pDCs after IMQ or CpG stimulation. No recovery of the percentage of CXCL10-producing pDCs following TLR-7 and TLR-9 stimulation was obtained after

1 month and 4 months of therapy administration (Table 6). Taken together, these data suggest a partial rescue of the pDC function, in terms of the percentage of IFN-α producing pDCs after R848 stimulation. The inhibitory effects of MEKi on T cells and moDC cytokine production, co-stimulatory molecule expression, and viability have been previously demonstrated [10,51]; on the contrary, no effects have been reported by direct exposure of immune cells to BRAFi. No data are available on the pDC compartment. In light of these reports, we exposed fully differentiated pDCs from HD to the BRAFi PLX4032 (vemurafenib) treatment alone or in combination with the MEKi U0126. The percentage of dead pDCs, as measured by annexin V/SYTOX AADvanced staining, did not significantly increase after 24 h of treatment with PLX4032 with or without U0126 compared to the vehicle control (Figure S3A–C). Furthermore, after 24 h of BRAFi and MEKi treatment, purified pDCs were stimulated with R848, IMQ, and CpG and the IFN-α and CXCL10 intracellular production was evaluated by flow cytometry analysis. Although variable under R848 and IMQ stimuli, the percentage of IFN-α$^+$ pDCs was not significantly affected by direct exposure to PLX4032 and U0126 compared to vehicle control (Figure S3D). Similarly, the percentage of CXCL10$^+$ pDCs was unchanged after PLX4032 in vitro administration both alone or combined with U0126 (Figure S3E).

Altogether, these results rule out inhibitory effects of direct exposure of BRAFi and MEKi on the pDC recovery in patients undergoing systemic cancer treatment.

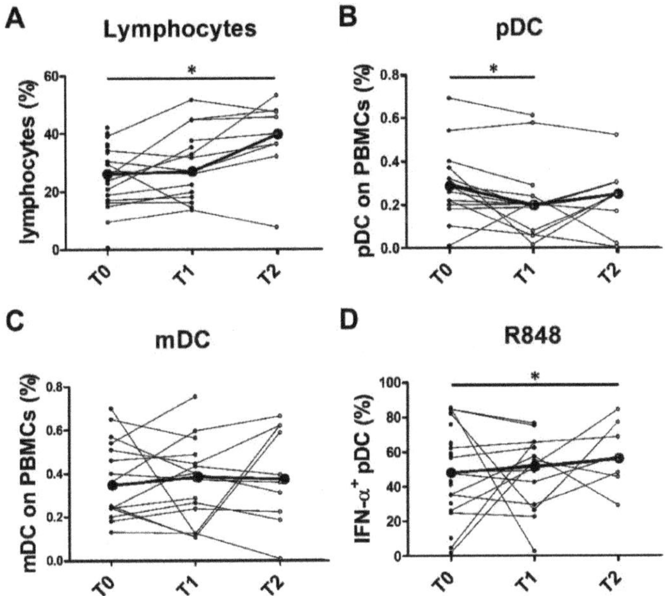

Figure 7. Frequency of peripheral blood immune cells and IFN-α$^+$ pDCs in BRAF^{V600+} MM patients over therapy administration. Cell counts (**A**) and flow cytometry (**B,C**) analysis were performed on whole blood from BRAF^{V600+} MM patients before therapy initiation (T0; n = 16), after 30 days (T1; n = 12), and after 120 days (T2; n = 9) from therapy administration (**A–C**). Total PBMCs isolated from peripheral blood of MM patients were cultured in RPMI 1640 medium and stimulated with R848 for 2 h (**D**). IFN-α was analyzed by intracellular flow cytometry staining (**D**). Before–after graphs illustrate the frequency of lymphocytes (**A**), pDCs (**B**), and mDCs (**C**) on total PBMCs, and the frequency of IFN-α$^+$ pDCs on BDCA-2$^+$/CD123$^+$ cells (**D**) for each subject. Bold black lines represent the median values. The statistical significance was calculated by Wilcoxon signed-rank test. * p < 0.05.

3. Discussion

The frequency of circulating pDCs is dramatically reduced in melanoma patients with systemic spread [29,52,53], but their clinical significance as well as their functional state have been poorly characterized [35]. Findings from retrospective analysis are conflicting [35]. This study reports the analysis of the circulating pDCs in a prospective chemo-naïve MM patient cohort. Results indicate that pDCs are dramatically depleted in MM patients with elevated serum LDH and high tumor burden. By comparison with HD, we found that the residual pDCs result in severe impairment in IFN-α and CXCL10 production in response to TLR-7/8 agonists, particularly in more advanced disease stage. On the contrary, the ability to respond to TLR-9 agonists remained intact. By multiple time point monitoring, we also found that in the BRAF^{V600+} subgroup, the pDC frequency was not recovered by BRAFi/MEKi treatment, whereas pDC function was partially restored. Finally, in vitro exposure to lactic acidosis negatively affected both the viability and function of terminally differentiated pDCs. One limitation of this study was the small sample size of the study cohort; however, the obtained findings provide the rationale for a prospective large-scale study.

The role of pDCs in cancer immunity is relevant due to their capability to produce large amounts of I- and III-IFNs, when properly activated through TLR-7 and -9 agonists, linking the innate and adaptive immune responses [54]. However, tumor-conditioned pDCs contribute to the establishment of an immunosuppressive milieu in several types of cancer and are associated with poor outcome [55–57]. IFNs participate in the host anti-tumor immune responses by exerting various regulatory functions on tumor cells as well as on cells of the microenvironment, especially on immune cells [36,58]. Here, we found that circulating pDCs are severely impaired in IFN-α and CXCL10 production in MM patients, suggesting that tumor-mediated I-IFN inhibitory mechanisms might take place in the blood. Even though the pDC response is largely heterogeneous in human subjects, we found a reduced fraction of IFN-α-producing pDCs after stimulation with TLR-7 agonists (i.e., resiquimod and imiquimod), in chemo-naïve MM patients. On the contrary, pDCs remained proficient after stimulation with a TLR-9 agonist (i.e., CpG-A ODN 2216). From a mechanistic point of view and as extension to the current study, we envisage the analysis of TLR-7 and TLR-9 protein expression on peripheral blood pDCs from MM patients. Despite the structural and functional similarities between TLR-7 and TLR-9, recent findings suggest that they are distinctly regulated in intracellular localization and trafficking by the molecular chaperone UNC93B1 [59,60]. Moreover, a contaminating pre-DC subpopulation, unable to produce high amounts of I-IFN in response to TLR7/8 and TLR9 stimulation, has been recently documented within the pDC fraction [61]. Although CpG-ODN is considered the most efficient stimulus for the TLR-9 signaling pathway activation [62], only a limited fraction of CpG-activated pDCs in HD showed an intracellular positivity for IFN-α. Human blood pDCs diversify into functionally distinct subsets after activation by CpG-ODN; the production of IFN-α by individually stimulated pDCs is controlled by stochastic gene regulation and paracrine I-IFN signaling in the microenvironment and might play a protective role reducing I-IFN levels and tissue damage [63,64].

It is worth noting that the M1c subgroup of MM was characterized by a more severe impairment of TLR-7-activated pDCs in the IFN-α and CXCL10 production, indicating a progressive reduction of the pDC function associated with advanced disease stages. Moreover, the functional impairment of the R848-activated pDCs predicted a worse prognosis in term of OS. On the other hand, neither the high LDH serum level nor the elevated tumor burden directly affected the pDC capability to produce IFN-α and CXCL10, recognizing their poor values as reliable biomarkers for pDC dysfunction in melanoma patients. On the basis of these results, assays measuring pDC function should be incorporated in future analysis to confirm their clinical relevance and utility.

Lactate and acidosis, generated by high glycolytic tumor metabolism, lead to a local suppressive effect on various immune cells, including macrophages, myeloid derived suppressor cells, and DCs [30,34,65]. The oncometabolite lactate has been recently proposed as a suppressor of IFN-α production, capable of reprogramming intra-tumoral pDCs to tolerogenic function [45,46]. High serum LDH predicts poor survival [66] and poor clinical response to anti-PD-1 treatment in melanoma patients [67], suggesting effects on cancer cell elimination by adaptive immunity [31]. A collapse of the circulating DC compartments has been previously demonstrated in advanced melanomas, more dramatically in the systemic disease (M1c) [29,52,53]. In the MM cohort analyzed in this study, the low frequencies of pDC, mDC, and T lymphocyte subsets were also associated with high serum LDH and elevated tumor burden as predictors of shorter OS. Accordingly, in vitro studies presented here hint that the exposure of pDCs and of T cells to severe acidosis promote their cell death. At the same time, increasing concentration of lactic acid as well as HCl have progressively reduced IFN-α production by fully differentiated pDCs. We found that increased lactate production was associated with a reduction of the glucose concentration in the melanoma supernatants, suggesting an active glycolytic activity by melanoma cells and nutrient deprivation in cell culture medium. On the contrary, only a slight acidosis was measured on SN-mel. Altogether, these findings support the hypothesis that an increased lactic acidosis induced by high glycolytic tumor metabolism is involved in the melanoma-mediated pDC collapse. On the other hand, as previously reported, glucose deprivation does not interfere with viability of human pDCs, whereas glycolysis is an essential metabolic pathway for pDCs to execute innate immune functions (e.g., IFN-α secretion) [68,69], suggesting that nutrient deprivation or glycolysis inhibition could impair pDC function in MM. Like many other cancer types, rapidly proliferating melanoma cells utilize aerobic glycolysis at high rates. Aerobic glycolysis, the resulting lactate and proton secretion in the tumor microenvironment as well as the levels of MCT1 transporter in melanoma cells cooperate to promote metastatization in several ways [70,71], as well as to suppress the immune surveillance. The I-IFN deficiency of pDCs might likely result from defective TLR-7/9 signaling pathways induced by soluble factors released in high amounts in melanoma patients, such as cytokines [72,73]. In addition, the I-IFN response is fine-tuned by a set of surface interferon inhibitory receptors, such as the immunoglobulin-like transcript 7 (ILT-7) and the type II C-type lectin BDCA-2 [74,75], and their physiological ligands have been identified on tumor cells [76,77]. Tumor-associated pDCs can be reprogrammed to their anti-tumor function upon appropriate re-activation with TLR-7 and TLR-9 agonists [78–81], and numerous clinical trials (phase I-III) are ongoing (Clinicaltrials.gov study identifiers: NCT02644967, NCT03445533, NCT03052205, NCT03084640, NCT03618641, NCT02680184, NCT03831295, and NCT02521870). Our study could help to identify a proper window for clinical intervention of these compounds in combination with the current therapies in MM patients. In particular, our data suggest a defective signaling through TLR-7, but not TLR-9. On the basis of these findings, administration of synthetic ODNs might represent a valid alternative to topical administration of IMQ.

The molecular profile of cancer cells can modify cancer cell immunogenicity [82] and the surrounding immune contexture [11,83]. The occurrence of BRAF^{V600+} mutation is associated with an increased pDC density in melanoma metastasis compared with BRAF wild-type tumors [84], whereas in locally advanced PCM, a collapse of the pDC compartment particularly occurs in NRAS-mutated tumors [29]. In this study, the frequencies of peripheral blood DCs or T lymphocytes were not associated with the tumor molecular profile, as was the impaired production of IFN-α by pDCs. The analysis was extended to BRAF^{V600+} MM patients treated with dabrafenib plus trametinib, vemurafenib plus cobimetinib, or vemurafenib/dabrafenib alone, which act on the constitutively activated RAF–RAS–MEK–ERK pathway [47]. Targeted therapies represent the treatment of choice in patients with bulky and symptomatic disease who deserve a rapid clinical response [9]. The therapeutic activity of BRAFi and MEKi partially depends on host immune cell activation. Indeed, BRAFi post-treatment biopsies are characterized by an increased T cell infiltration [15], and immune checkpoint molecules are increasingly expressed by tumor cells and T cells within 2 weeks of therapy [10]. Moreover, in vitro study has demonstrated that combination of vemurafenib and MEKi U0126 promotes the recovery of DC functions impaired by melanoma cells [16]. Although an objective clinical response by RECIST was observed in many BRAF^{V600+} MM patients undergoing targeted therapy, no circulating pDC and mDC recovery was noticed after 120 days of treatment by time point analysis, while a slight lymphocyte restoration was achieved. It would be relevant to test long-term responder for a better understanding of the mechanisms sustaining DC recovery. On the contrary, a moderate increase of the frequency of IFN-α-producing pDCs was obtained in patients after 4 months from therapy initiation, although the pDC function was not fully recovered. The lack of DC recovery in BRAF^{V600+} patients might be explained by some direct effects of MEKi on DC compartment, as previously demonstrated [16]. However, in vitro experiments from this study showed that pDC viability and function were not significantly affected by direct exposure to BRAFi and MEKi. Microscopic residual disease might interfere with a full pDC recovery; this suggests that combined circulating tumor DNA (ctDNA) and DC monitoring in the setting of a prospective analysis might provide support to this notion.

4. Materials and Methods

4.1. Experimental Design

This study included a prospective cohort (BRAF-mutated ($n = 16$) and BRAF/NRAS wild-type patients ($n = 13$)) for a total of 29 histologically confirmed metastatic melanoma (MM) patients (AJCC, Stage IV, Chicago, IL, USA) and 25 healthy donors (HD), enrolled between December 2014 and November 2017. The local ethics committee provided formal approval to this project (WV-Immunocancer 2014 to W.V., institutional review board code NP906). Exclusion criteria included immune deficiency (steroid administration); bone marrow transplant; and known history of human immunodeficiency virus 1 and 2 (HIV1; HIV2), hepatitis B virus (HBV), and hepatitis C virus (HCV) infection.

Patients with BRAF p.V600E/K mutant MM were monitored at different time points (T0 = 0 day, $n = 16$; T1 = 30 days, $n = 12$; T2 = 120 days, $n = 9$). HD were enrolled at the same time as chemo-naïve MM patients. A primary stratification according to the "AJCC Melanoma Staging and Classification 7th edition" was applied to our MM cohort (Table S1). Furthermore, patients were stratified on the basis of baseline LDH: values below or equal to 1 × the upper limit of normal (ULN) were identified as "normal" ($n = 18$), and values above 1 × ULN were identified as "high" ($n = 9$) (Table S1). The "tumor burden" of 28 patients who had at least 10 measurable lesions was analyzed on computed tomography scans of the chest and abdomen (Table S1). Lung nodules ≤ 5 mm in axial diameter were excluded. Baseline disease burden was determined by the sum of the product of axial diameter for the biggest metastatic lesions. The objective response was defined as complete response (CR), partial response (PR), stable disease (SD), or progressive disease (PD), according to RECIST 1.1 criteria (Table S1).

4.2. Human Subjects and Blood Specimen Processing

A total of 10 mL of whole blood was collected from 29 MM patients and 25 HDs. Blood was drawn directly into S-Monovette 2.7 mL K3E (1.6 mg EDTA/mL; Sarstedt, Nümbrecht, Germany), gently rocked at room temperature until processing. Additionally, the blood counts were performed on MM samples as part of their clinical routine hematology. Clinical features of the MM patients previously reported by Vescovi R. et al. [29] are included as Table S1.

4.3. Peripheral Blood Mononuclear Cell Stimulation

Peripheral blood mononuclear cells (PBMCs) were obtained from HD and MM patients by Ficoll gradient. PBMCs (1×10^6 cells/mL) were cultured in RPMI 1640 medium (Biochrom GmbH, Berlin, Germany) with 10% fetal bovine serum (FBS) (Biochrom GmbH, Holliston, MA, USA) and 20 ng/mL human IL-3 (Miltenyi Biotec, Bergisch Gladbach, Germany). Total PBMCs (1×10^6 cells/mL) were stimulated with resiquimod (R848) or imiquimod (IMQ) 5 μg/mL (Invivogen, San Diego, CA, USA) and CpG-ODN 2216 6 μg/mL (Miltenyi Biotec, Bergisch Gladbach, Germany) for 2 h and 6 h, respectively. Brefeldin A (1 μg/mL; Sigma-Aldrich, St. Louis, MO, USA) was added after 1 h or 4 h (in samples stimulated for 2 h or 6 h, respectively).

4.4. Purification, Culture, and Stimulation of Peripheral Blood pDCs and T Lymphocytes

PBMCs were obtained from buffy coats of healthy volunteer blood donors (courtesy of the Centro Trasfusionale, ASST Spedali Civili, Brescia, Italy) by Ficoll gradient. Peripheral blood pDCs and T cells were magnetically sorted with the Plasmacytoid Dendritic Cell Isolation Kit II and the human Pan T Cell Isolation Kit (Miltenyi Biotec, Bergisch Gladbach, Germany), respectively. Isolated pDCs and T lymphocytes (5×10^5 cells/mL) were cultured in RPMI 1640 medium (Biochrom GmbH, Holliston, MA, USA) with 10% FBS (Biochrom GmbH, Holliston, MA, USA), and 20 ng/mL human IL-3 (Miltenyi Biotec, Bergisch Gladbach, Germany) was added to pDCs' culture medium.

pDCs and T lymphocytes were cultured for 24 h in RPMI 1640 medium supplemented with 10% FBS (Biochrom GmbH, Holliston, MA, USA), containing lactic acid (LA) (Sigma-Aldrich, St. Louis, MO, USA) or sodium lactate (NaL) (Alfa Aesar, Haverhill, MA, USA) at the concentrations of 10 mM, 15 mM, and 20 mM. The pH levels of the media were measured. In addition, pDCs and T lymphocytes were cultured in RPMI 1640 medium supplemented with 10% FBS (Biochrom GmbH, Holliston, MA, USA), titrated to pH $\approx$ 7.4, 7.0, 6.5, 6.0, and 5.5 using HCl. A total of 20 ng/mL human IL-3 (Miltenyi Biotec, Bergisch Gladbach, Germany) was added to the pDC culture medium.

pDCs were treated with 1 μM of the BRAF Inhibitor PLX4032 (Selleck Biochem, Houston, TX), with or without 12.5 μM of the MEK inhibitor U0126 (Merck Millipore, Darmstadt, Germany), and the 0.2% of DMSO in RPMI 1640 medium was used as vehicle control.

pDCs were stimulated with 5 μg/mL of R848 or IMQ (Invivogen) and 6 μg/mL of CpG-ODN 2216 (Miltenyi-Biotec) for 2 h and 6 h. Brefeldin A (1 μg/mL, Sigma-Aldrich, St. Louis, MO, USA) was added after 1 h and 4 h, respectively.

4.5. Flow-Cytometric Analysis

Fluorescence minus one (FMO) was used to set the marker for positive cells. A baseline fluorescence control was used as a reference to set the fluorescence thresholds for positivity. The results were expressed as the percentage of positive cells. For measurement of the spectral overlaps, the fluorescence detected on all measurement channels was evaluated for single-labelled "compensation control" samples prior to the performed flow cytometryanalysis.

For whole blood staining, 200 μL of whole blood was incubated with a panel of fluorochrome-conjugated antibodies (panel #1 reported in Table S5) for 15 min in the dark at 4 °C. Red blood cells were lysed and leukocytes were fixed by adding FACS Lysing Solution (BD Bioscience, San Jose, CA, USA) following the manufacturer's instructions. A minimum of 2×10^5 PBMCs were acquired according to the forward light scatter versus side light scatter profile, and doublet discrimination was performed. The gating strategy is reported in Figure S4. Samples were processed on FACS Canto II system (Becton Dickinson, San Jose, CA, USA).

The purified cell viability assay was performed using Pacific Blue or APC Annexin V/SYTOX AADvanced Apoptosis Kit, for flow cytometry (Thermo Fisher Scientific, Waltham, MA, USA), following the manufacturer's instructions. Briefly, this assay identifies the early apoptotic cells as annexin V^+/SYTOX AADvanced$^-$ and late apoptotic/necrotic cells as annexin V^+/SYTOX AADvanced$^+$. Samples were processed on MACS Quant Citofluorimeter (Miltenyi-Biotec). Results were analyzed by FlowJo X software (Tree Star Inc, Ashland, Wilmington, NC, USA).

For the evaluation of IFN-α and CXCL10 intracellular production, PBMCs and purified pDCs, stimulated as described above, were surface labelled with anti-BDCA-2 and anti-CD123 fluorochrome-conjugated antibodies. Subsequently, cells were fixed and permeabilized using the Inside Stain Kit (Miltenyi Biotec, Bergisch Gladbach, Germany) and the intracellular cytokine labelling was performed using anti IFN-α and anti-CXCL10/IP-10 fluorochrome-conjugated antibodies (panel #2 reported in Table S5). Samples were processed on MACS Quant Citofluorimeter (Miltenyi Biotec, Bergisch Gladbach, Germany). Results were analysed by FlowJo X software (Tree Star Inc, Wilmington, NC, USA).

4.6. Statistical Analysis

Patient characteristics were described at therapy initiation (T0). Categorical variables were reported as absolute frequencies and percentages and were compared across groups using the Fisher's exact test. Continuous variables were expressed as median and interquartile range (IQR). The Wilcoxon–Mann–Whitney test and the Kruskal–Wallis test were used to compare variable distributions across two and more than two subgroups of patients, respectively. The association between the peripheral blood immune cell frequencies and tumor burden at T0 among MM patients was evaluated using the Spearman's rank correlation coefficient. Peripheral blood leukocyte populations were measured at 1 month (T1) and at 4 months (T2) after therapy initiation (T0) in the group of MM patients that were BRAF-mutated. Changes in the peripheral blood leukocyte frequencies at 1 month (T1–T0) and at 4 months (T2–T0) were tested using the Wilcoxon signed-rank test. All patients were followed up after T0. Two endpoints (disease progression and death) were used to calculate the PFS and OS probability, respectively. PFS was defined as the time interval between T0 and the date of identification of progressive disease; OS was defined as the time interval between T0 and the date of death. Survival curves were calculated using Kaplan–Meier method, and differences in survival between subgroups of patients were tested using the log-rank test. For continuous variables, subgroups were defined using the median value as cut-off. Univariate Cox proportional hazard model were fitted to evaluate the role of the peripheral blood leukocyte populations and other established prognostic factors on the considered outcomes. Multivariable regression models were used to adjust the estimates for molecular profile and stage of the disease. The hazard ratios (HR), 95% confidence intervals (CI), and p-values from a Wald test were reported. The two-sample paired sign test and two-sample paired or unpaired Student's t-test were used to compare groups from in vitro experiments. A two-tailed p-value < 0.05 was considered statistically significant.

The statistical analysis was performed using STATA 15 (StataCorp. 2017. Stata Statistical Software: Release 15. College Station, TX: StataCorp LLC.) or GraphPad Prism Software version 5 (GraphPad Software, San Diego, CA, USA).

5. Conclusions

In conclusion, pDCs from MM patients are severely impaired in their frequency and function. Findings emerged here suggest a relevant role for melanoma metabolism promoting lactic acidosis. In BRAF^{V600+} MM, short-term treatment is not associated to a full pDC recovery; however, TLR-9 agonists as adjuvant remain a valid therapeutic strategy for a proficient pDC activation. Monitoring the pDC compartment and functions might represent a clinically relevant tool for the selection of MM cases, which likely benefit from TLR-9 agonists as a completion of their treatment plan.

Supplementary Materials: The following are available online at http://www.mdpi.com/2072-6694/12/8/2085/s1, Figure S1: Lactate, glucose concentrations, and pH levels measured on melanoma cell line supernatants (SN-mel), Figure S2: Lactosis did not affect the viability of pDCs and T cells, Figure S3: BRAF and MEK inhibitors (BRAFi; MEKi) did not affect pDC viability and function, Figure S4: Gating strategy for the identification of peripheral blood immune populations, Table S1: Clinical data of the metastatic melanoma (MM) cohort. Table S2: Peripheral blood immune populations in the MM molecular groups at baseline (T0; $N = 29$), Table S3: Analysis of the overall survival (OS) and the progression-free survival (PFS) among MM patients ($N = 29$) during the follow-up, Table S4: Univariate and multivariate Cox regression models for PFS in MM patients at baseline (T0; $N = 29$), Table S5: Antibodies used for flow cytometry.

Author Contributions: Conceptualization, W.V., M.M., and R.V.; methodology, M.M., R.V., and D.M.; formal analysis, M.M., R.V., F.C., D.F., D.M., and C.S.; investigation, M.M. and R.V.; resources, F.C., A.B., and D.F.; data curation, F.C. and C.S.; writing—original draft preparation, M.M., R.V., and W.V.; writing—review and editing, F.C., D.F., D.M., and C.S.; supervision, W.V.; funding acquisition, W.V. All authors have read and agreed to the published version of the manuscript.

Funding: This research was funded by "Associazione Italiana per la Ricerca sul Cancro" (AIRC), Italy, to W. Vermi, grant number IG-15378 and IG-23179.

Acknowledgments: We are grateful to Silvia Lonardi for reading the manuscript. We would like to thank personnel of the Servizio Immuno-Trasfusionale ASST Spedali Civili di Brescia for their help in providing buffy coats. We thank Nurses of the Oncology Unit ASST Spedali Civili di Brescia that provided support for patient care, especially Eleonora Lombardi.

Conflicts of Interest: The authors declare no conflict of interest.

References

1. Corrie, P.; Hategan, M.; Fife, K.; Parkinson, C. Management of melanoma. *Br. Med. Bull.* **2014**, *111*, 149–162. [CrossRef] [PubMed]
2. Ugurel, S.; Rohmel, J.; Ascierto, P.A.; Flaherty, K.T.; Grob, J.J.; Hauschild, A.; Larkin, J.; Long, G.V.; Lorigan, P.; McArthur, G.A.; et al. Survival of patients with advanced metastatic melanoma: The impact of novel therapies-update 2017. *Eur. J. Cancer* **2017**, *83*, 247–257. [CrossRef] [PubMed]
3. Davies, H.; Bignell, G.R.; Cox, C.; Stephens, P.; Edkins, S.; Clegg, S.; Teague, J.; Woffendin, H.; Garnett, M.J.; Bottomley, W.; et al. Mutations of the BRAF gene in human cancer. *Nature* **2002**, *417*, 949–954. [CrossRef] [PubMed]
4. Chapman, P.B.; Hauschild, A.; Robert, C.; Haanen, J.B.; Ascierto, P.; Larkin, J.; Dummer, R.; Garbe, C.; Testori, A.; Maio, M.; et al. Improved survival with vemurafenib in melanoma with BRAF V600E mutation. *N. Engl. J. Med.* **2011**, *364*, 2507–2516. [CrossRef]
5. Hauschild, A.; Grob, J.J.; Demidov, L.V.; Jouary, T.; Gutzmer, R.; Millward, M.; Rutkowski, P.; Blank, C.U.; Miller, W.H., Jr.; Kaempgen, E.; et al. Dabrafenib in BRAF-mutated metastatic melanoma: A multicentre, open-label, phase 3 randomised controlled trial. *Lancet* **2012**, *380*, 358–365. [CrossRef]
6. Wolchok, J.D.; Weber, J.S.; Maio, M.; Neyns, B.; Harmankaya, K.; Chin, K.; Cykowski, L.; de Pril, V.; Humphrey, R.; Lebbe, C. Four-year survival rates for patients with metastatic melanoma who received ipilimumab in phase II clinical trials. *Ann. Oncol.* **2013**, *24*, 2174–2180. [CrossRef]
7. Schadendorf, D.; Hodi, F.S.; Robert, C.; Weber, J.S.; Margolin, K.; Hamid, O.; Patt, D.; Chen, T.T.; Berman, D.M.; Wolchok, J.D. Pooled Analysis of Long-Term Survival Data From Phase II and Phase III Trials of Ipilimumab in Unresectable or Metastatic Melanoma. *J. Clin. Oncol.* **2015**, *33*, 1889–1894. [CrossRef]
8. Topalian, S.L.; Sznol, M.; McDermott, D.F.; Kluger, H.M.; Carvajal, R.D.; Sharfman, W.H.; Brahmer, J.R.; Lawrence, D.P.; Atkins, M.B.; Powderly, J.D.; et al. Survival, durable tumor remission, and long-term safety in patients with advanced melanoma receiving nivolumab. *J. Clin. Oncol.* **2014**, *32*, 1020–1030. [CrossRef]
9. Garbe, C.; Amaral, T.; Peris, K.; Hauschild, A.; Arenberger, P.; Bastholt, L.; Bataille, V.; Del Marmol, V.; Dreno, B.; Fargnoli, M.C.; et al. European consensus-based interdisciplinary guideline for melanoma. Part 2: Treatment—Update 2019. *Eur. J. Cancer* **2019**. [CrossRef]
10. Frederick, D.T.; Piris, A.; Cogdill, A.P.; Cooper, Z.A.; Lezcano, C.; Ferrone, C.R.; Mitra, D.; Boni, A.; Newton, L.P.; Liu, C.; et al. BRAF inhibition is associated with enhanced melanoma antigen expression and a more favorable tumor microenvironment in patients with metastatic melanoma. *Clin. Cancer Res.* **2013**, *19*, 1225–1231. [CrossRef]
11. Boni, A.; Cogdill, A.P.; Dang, P.; Udayakumar, D.; Njauw, C.N.; Sloss, C.M.; Ferrone, C.R.; Flaherty, K.T.; Lawrence, D.P.; Fisher, D.E.; et al. Selective BRAFV600E inhibition enhances T-cell recognition of melanoma without affecting lymphocyte function. *Cancer Res.* **2010**, *70*, 5213–5219. [CrossRef]
12. Sumimoto, H.; Imabayashi, F.; Iwata, T.; Kawakami, Y. The BRAF-MAPK signaling pathway is essential for cancer-immune evasion in human melanoma cells. *J. Exp. Med.* **2006**, *203*, 1651–1656. [CrossRef]
13. Bradley, S.D.; Chen, Z.; Melendez, B.; Talukder, A.; Khalili, J.S.; Rodriguez-Cruz, T.; Liu, S.; Whittington, M.; Deng, W.; Li, F.; et al. BRAFV600E Co-opts a Conserved MHC Class I Internalization Pathway to Diminish Antigen Presentation and CD8 + T-cell Recognition of Melanoma. *Cancer Immunol. Res.* **2015**, *3*, 602–609. [CrossRef] [PubMed]
14. Kono, M.; Dunn, I.S.; Durda, P.J.; Butera, D.; Rose, L.B.; Haggerty, T.J.; Benson, E.M.; Kurnick, J.T. Role of the mitogen-activated protein kinase signaling pathway in the regulation of human melanocytic antigen expression. *Mol. Cancer Res.* **2006**, *4*, 779–792. [CrossRef] [PubMed]
15. Wilmott, J.S.; Long, G.V.; Howle, J.R.; Haydu, L.E.; Sharma, R.N.; Thompson, J.F.; Kefford, R.F.; Hersey, P.; Scolyer, R.A. Selective BRAF inhibitors induce marked T-cell infiltration into human metastatic melanoma. *Clin. Cancer Res.* **2012**, *18*, 1386–1394. [CrossRef] [PubMed]

16. Ott, P.A.; Henry, T.; Baranda, S.J.; Frleta, D.; Manches, O.; Bogunovic, D.; Bhardwaj, N. Inhibition of both BRAF and MEK in BRAF(V600E) mutant melanoma restores compromised dendritic cell (DC) function while having differential direct effects on DC properties. *Cancer Immunol. Immunother.* **2013**, *62*, 811–822. [CrossRef]

17. Hong, D.S.; Vence, L.; Falchook, G.; Radvanyi, L.G.; Liu, C.; Goodman, V.; Legos, J.J.; Blackman, S.; Scarmadio, A.; Kurzrock, R.; et al. BRAF(V600) inhibitor GSK2118436 targeted inhibition of mutant BRAF in cancer patients does not impair overall immune competency. *Clin. Cancer Res.* **2012**, *18*, 2326–2335. [CrossRef]

18. Hauschild, A.; Larkin, J.; Ribas, A.; Dreno, B.; Flaherty, K.T.; Ascierto, P.A.; Lewis, K.D.; McKenna, E.; Zhu, Q.; Mun, Y.; et al. Modeled Prognostic Subgroups for Survival and Treatment Outcomes in BRAF V600-Mutated Metastatic Melanoma: Pooled Analysis of 4 Randomized Clinical Trials. *JAMA Oncol.* **2018**, *4*, 1382–1388. [CrossRef]

19. Sabbatino, F.; Wang, Y.; Scognamiglio, G.; Favoino, E.; Feldman, S.A.; Villani, V.; Flaherty, K.T.; Nota, S.; Giannarelli, D.; Simeone, E.; et al. Antitumor Activity of BRAF Inhibitor and IFNalpha Combination in BRAF-Mutant Melanoma. *J. Natl. Cancer Inst.* **2016**, *108*. [CrossRef]

20. Iwasaki, A.; Medzhitov, R. Toll-like receptor control of the adaptive immune responses. *Nat. Immunol.* **2004**, *5*, 987–995. [CrossRef]

21. Takeda, K.; Kaisho, T.; Akira, S. Toll-like receptors. *Annu. Rev. Immunol.* **2003**, *21*, 335–376. [CrossRef] [PubMed]

22. Von Marschall, Z.; Scholz, A.; Cramer, T.; Schäfer, G.; Schirner, M.; Oberg, K.; Wiedenmann, B.; Höcker, M.; Rosewicz, S. Effects of interferon alpha on vascular endothelial growth factor gene transcription and tumor angiogenesis. *J. Natl. Cancer Inst.* **2003**, *95*, 437–448. [CrossRef] [PubMed]

23. Singh, R.K.; Gutman, M.; Bucana, C.D.; Sanchez, R.; Llansa, N.; Fidler, I.J. Interferons alpha and beta down-regulate the expression of basic fibroblast growth factor in human carcinomas. *Proc. Natl. Acad. Sci. USA* **1995**, *92*, 4562–4566. [CrossRef] [PubMed]

24. Liang, S.; Wei, H.; Sun, R.; Tian, Z. IFNalpha regulates NK cell cytotoxicity through STAT1 pathway. *Cytokine* **2003**, *23*, 190–199. [CrossRef]

25. Marrack, P.; Kappler, J.; Mitchell, T. Type I interferons keep activated T cells alive. *J. Exp. Med.* **1999**, *189*, 521–530. [CrossRef]

26. Curtsinger, J.M.; Valenzuela, J.O.; Agarwal, P.; Lins, D.; Mescher, M.F. Type I IFNs provide a third signal to CD8 T cells to stimulate clonal expansion and differentiation. *J. Immunol.* **2005**, *174*, 4465–4469. [CrossRef]

27. Blackwell, S.E.; Krieg, A.M. CpG-A-induced monocyte IFN-gamma-inducible protein-10 production is regulated by plasmacytoid dendritic cell-derived IFN-alpha. *J. Immunol.* **2003**, *170*, 4061–4068. [CrossRef]

28. Wildbaum, G.; Netzer, N.; Karin, N. Plasmid DNA encoding IFN-gamma-inducible protein 10 redirects antigen-specific T cell polarization and suppresses experimental autoimmune encephalomyelitis. *J. Immunol.* **2002**, *168*, 5885–5892. [CrossRef]

29. Vescovi, R.; Monti, M.; Moratto, D.; Paolini, L.; Consoli, F.; Benerini, L.; Melocchi, L.; Calza, S.; Chiudinelli, M.; Rossi, G.; et al. Collapse of the Plasmacytoid Dendritic Cell Compartment in Advanced Cutaneous Melanomas by Components of the Tumor Cell Secretome. *Cancer Immunol. Res.* **2019**, *7*, 12–28. [CrossRef]

30. Gottfried, E.; Kunz-Schughart, L.A.; Ebner, S.; Mueller-Klieser, W.; Hoves, S.; Andreesen, R.; Mackensen, A.; Kreutz, M. Tumor-derived lactic acid modulates dendritic cell activation and antigen expression. *Blood* **2006**, *107*, 2013–2021. [CrossRef]

31. Ding, J.; Karp, J.E.; Emadi, A. Elevated lactate dehydrogenase (LDH) can be a marker of immune suppression in cancer: Interplay between hematologic and solid neoplastic clones and their microenvironments. *Cancer Biomark.* **2017**, *19*, 353–363. [CrossRef] [PubMed]

32. Warburg, O.; Wind, F.; Negelein, E. THE METABOLISM OF TUMORS IN THE BODY. *J. Gen. Physiol.* **1927**, *8*. [CrossRef] [PubMed]

33. Pavlova, N.N.; Thompson, C.B. The Emerging Hallmarks of Cancer Metabolism. *Cell Metab.* **2016**, *23*. [CrossRef]

34. Bronte, V. Tumor cells hijack macrophages via lactic acid. *Immunol. Cell Biol.* **2014**, *92*, 647–649. [CrossRef] [PubMed]

35. Monti, M.; Consoli, F.; Vescovi, R.; Bugatti, M.; Vermi, W. Human Plasmacytoid Dendritic Cells and Cutaneous Melanoma. *Cells* **2020**, *9*, 417. [CrossRef] [PubMed]

36. Dunn, G.P.; Bruce, A.T.; Sheehan, K.C.; Shankaran, V.; Uppaluri, R.; Bui, J.D.; Diamond, M.S.; Koebel, C.M.; Arthur, C.; White, J.M.; et al. A critical function for type I interferons in cancer immunoediting. *Nat. Immunol.* **2005**, *6*, 722–729. [CrossRef]

37. Borden, E.C. Interferons alpha and beta in cancer: Therapeutic opportunities from new insights. *Nat. Rev. Drug Discov.* **2019**, *18*, 219–234. [CrossRef]

38. Le Mercier, I.; Poujol, D.; Sanlaville, A.; Sisirak, V.; Gobert, M.; Durand, I.; Dubois, B.; Treilleux, I.; Marvel, J.; Vlach, J.; et al. Tumor promotion by intratumoral plasmacytoid dendritic cells is reversed by TLR7 ligand treatment. *Cancer Res.* **2013**, *73*, 4629–4640. [CrossRef]

39. Megjugorac, N.J.; Young, H.A.; Amrute, S.B.; Olshalsky, S.L.; Fitzgerald-Bocarsly, P. Virally stimulated plasmacytoid dendritic cells produce chemokines and induce migration of T and NK cells. *J. Leukoc. Biol.* **2004**, *75*, 504–514. [CrossRef]

40. Zbytek, B.; Carlson, J.A.; Granese, J.; Ross, J.; Mihm, M.C., Jr.; Slominski, A. Current concepts of metastasis in melanoma. *Expert Rev. Dermatol.* **2008**, *3*, 569–585. [CrossRef]

41. Karagiannis, P.; Fittall, M.; Karagiannis, S.N. Evaluating biomarkers in melanoma. *Front. Oncol.* **2014**, *4*, 383. [CrossRef] [PubMed]

42. Long, G.V.; Grob, J.J.; Nathan, P.; Ribas, A.; Robert, C.; Schadendorf, D.; Lane, S.R.; Mak, C.; Legenne, P.; Flaherty, K.T.; et al. Factors predictive of response, disease progression, and overall survival after dabrafenib and trametinib combination treatment: A pooled analysis of individual patient data from randomised trials. *Lancet Oncol.* **2016**, *17*, 1743–1754. [CrossRef]

43. Caronni, N.; Simoncello, F.; Stafetta, F.; Guarnaccia, C.; Ruiz-Moreno, J.S.; Opitz, B.; Galli, T.; Proux-Gillardeaux, V.; Benvenuti, F. Downregulation of Membrane Trafficking Proteins and Lactate Conditioning Determine Loss of Dendritic Cell Function in Lung Cancer. *Cancer Res.* **2018**, *78*, 1685–1699. [CrossRef]

44. Comito, G.; Iscaro, A.; Bacci, M.; Morandi, A.; Ippolito, L.; Parri, M.; Montagnani, I.; Raspollini, M.R.; Serni, S.; Simeoni, L.; et al. Lactate modulates CD4(+) T-cell polarization and induces an immunosuppressive environment, which sustains prostate carcinoma progression via TLR8/miR21 axis. *Oncogene* **2019**, *38*, 3681–3695. [CrossRef] [PubMed]

45. Raychaudhuri, D.; Bhattacharya, R.; Sinha, B.P.; Liu, C.S.C.; Ghosh, A.R.; Rahaman, O.; Bandopadhyay, P.; Sarif, J.; D'Rozario, R.; Paul, S.; et al. Lactate Induces Pro-tumor Reprogramming in Intratumoral Plasmacytoid Dendritic Cells. *Front. Immunol.* **2019**, *10*, 1878. [CrossRef] [PubMed]

46. Zhang, W.; Wang, G.; Xu, Z.G.; Tu, H.; Hu, F.; Dai, J.; Chang, Y.; Chen, Y.; Lu, Y.; Zeng, H.; et al. Lactate Is a Natural Suppressor of RLR Signaling by Targeting MAVS. *Cell* **2019**, *178*, 176–189. [CrossRef] [PubMed]

47. Wan, P.T.; Garnett, M.J.; Roe, S.M.; Lee, S.; Niculescu-Duvaz, D.; Good, V.M.; Jones, C.M.; Marshall, C.J.; Springer, C.J.; Barford, D.; et al. Mechanism of activation of the RAF-ERK signaling pathway by oncogenic mutations of B-RAF. *Cell* **2004**, *116*, 855–867. [CrossRef]

48. Johannessen, C.M.; Boehm, J.S.; Kim, S.Y.; Thomas, S.R.; Wardwell, L.; Johnson, L.A.; Emery, C.M.; Stransky, N.; Cogdill, A.P.; Barretina, J.; et al. COT drives resistance to RAF inhibition through MAP kinase pathway reactivation. *Nature* **2010**, *468*, 968–972. [CrossRef]

49. Wagle, N.; Emery, C.; Berger, M.F.; Davis, M.J.; Sawyer, A.; Pochanard, P.; Kehoe, S.M.; Johannessen, C.M.; Macconaill, L.E.; Hahn, W.C.; et al. Dissecting therapeutic resistance to RAF inhibition in melanoma by tumor genomic profiling. *J. Clin. Oncol.* **2011**, *29*, 3085–3096. [CrossRef]

50. Janovec, V.; Aouar, B.; Font-Haro, A.; Hofman, T.; Trejbalova, K.; Weber, J.; Chaperot, L.; Plumas, J.; Olive, D.; Dubreuil, P.; et al. The MEK1/2-ERK Pathway Inhibits Type I IFN Production in Plasmacytoid Dendritic Cells. *Front. Immunol.* **2018**, *9*, 364. [CrossRef]

51. Ott, P.A.; Bhardwaj, N. Impact of MAPK Pathway Activation in BRAF(V600) Melanoma on T Cell and Dendritic Cell Function. *Front. Immunol.* **2013**, *4*, 346. [CrossRef] [PubMed]

52. Chevolet, I.; Speeckaert, R.; Schreuer, M.; Neyns, B.; Krysko, O.; Bachert, C.; Van Gele, M.; van Geel, N.; Brochez, L. Clinical significance of plasmacytoid dendritic cells and myeloid-derived suppressor cells in melanoma. *J. Transl. Med.* **2015**, *13*, 9. [CrossRef] [PubMed]

53. Failli, A.; Legitimo, A.; Orsini, G.; Romanini, A.; Consolini, R. Numerical defect of circulating dendritic cell subsets and defective dendritic cell generation from monocytes of patients with advanced melanoma. *Cancer Lett.* **2013**, *337*, 184–192. [CrossRef]

54. Kadowaki, N.; Antonenko, S.; Lau, J.Y.; Liu, Y.J. Natural interferon alpha/beta-producing cells link innate and adaptive immunity. *J. Exp. Med.* **2000**, *192*, 219–226. [CrossRef] [PubMed]

55. Aspord, C.; Leccia, M.T.; Charles, J.; Plumas, J. Plasmacytoid dendritic cells support melanoma progression by promoting Th2 and regulatory immunity through OX40L and ICOSL. *Cancer Immunol. Res.* **2013**, *1*, 402–415. [CrossRef] [PubMed]

56. Faget, J.; Bendriss-Vermare, N.; Gobert, M.; Durand, I.; Olive, D.; Biota, C.; Bachelot, T.; Treilleux, I.; Goddard-Leon, S.; Lavergne, E.; et al. ICOS-ligand expression on plasmacytoid dendritic cells supports breast cancer progression by promoting the accumulation of immunosuppressive CD4+ T cells. *Cancer Res.* **2012**, *72*, 6130–6141. [CrossRef]

57. Hartmann, E.; Wollenberg, B.; Rothenfusser, S.; Wagner, M.; Wellisch, D.; Mack, B.; Giese, T.; Gires, O.; Endres, S.; Hartmann, G. Identification and functional analysis of tumor-infiltrating plasmacytoid dendritic cells in head and neck cancer. *Cancer Res.* **2003**, *63*, 6478–6487.

58. Dunn, G.P.; Koebel, C.M.; Schreiber, R.D. Interferons, immunity and cancer immunoediting. *Nat. Rev. Immunol.* **2006**, *6*, 836–848. [CrossRef]

59. Majer, O.; Liu, B.; Woo, B.J.; Kreuk, L.S.M.; Van Dis, E.; Barton, G.M. Release from UNC93B1 reinforces the compartmentalized activation of select TLRs. *Nature* **2019**, *575*, 371–374. [CrossRef]

60. Majer, O.; Liu, B.; Kreuk, L.S.M.; Krogan, N.; Barton, G.M. UNC93B1 recruits syntenin-1 to dampen TLR7 signalling and prevent autoimmunity. *Nature* **2019**, *575*, 366–370. [CrossRef]

61. See, P.; Dutertre, C.A.; Chen, J.; Günther, P.; McGovern, N.; Irac, S.E.; Gunawan, M.; Beyer, M.; Händler, K.; Duan, K.; et al. Mapping the Human DC Lineage Through the Integration of High-Dimensional Techniques. *Science* **2017**, *356*. [CrossRef] [PubMed]

62. Siegal, F.P.; Kadowaki, N.; Shodell, M.; Fitzgerald-Bocarsly, P.A.; Shah, K.; Ho, S.; Antonenko, S.; Liu, Y.J. The nature of the principal type 1 interferon-producing cells in human blood. *Science* **1999**, *284*, 1835–1837. [CrossRef] [PubMed]

63. Alculumbre, S.G.; Saint-Andre, V.; Di Domizio, J.; Vargas, P.; Sirven, P.; Bost, P.; Maurin, M.; Maiuri, P.; Wery, M.; Roman, M.S.; et al. Diversification of human plasmacytoid predendritic cells in response to a single stimulus. *Nat. Immunol.* **2018**, *19*, 63–75. [CrossRef] [PubMed]

64. Wimmers, F.; Subedi, N.; van Buuringen, N.; Heister, D.; Vivie, J.; Beeren-Reinieren, I.; Woestenenk, R.; Dolstra, H.; Piruska, A.; Jacobs, J.F.M.; et al. Single-cell analysis reveals that stochasticity and paracrine signaling control interferon-alpha production by plasmacytoid dendritic cells. *Nat. Commun.* **2018**, *9*, 3317. [CrossRef] [PubMed]

65. Corzo, C.A.; Condamine, T.; Lu, L.; Cotter, M.J.; Youn, J.I.; Cheng, P.; Cho, H.I.; Celis, E.; Quiceno, D.G.; Padhya, T.; et al. HIF-1α regulates function and differentiation of myeloid-derived suppressor cells in the tumor microenvironment. *J. Exp. Med.* **2010**, *207*, 2439–2453. [CrossRef] [PubMed]

66. Balch, C.M.; Soong, S.J.; Atkins, M.B.; Buzaid, A.C.; Cascinelli, N.; Coit, D.G.; Fleming, I.D.; Gershenwald, J.E.; Houghton, A., Jr.; Kirkwood, J.M.; et al. An evidence-based staging system for cutaneous melanoma. *CA Cancer J. Clin.* **2004**, *54*, 131–149; quiz 134–182. [CrossRef]

67. Diem, S.; Kasenda, B.; Spain, L.; Martin-Liberal, J.; Marconcini, R.; Gore, M.; Larkin, J. Serum lactate dehydrogenase as an early marker for outcome in patients treated with anti-PD-1 therapy in metastatic melanoma. *Br. J. Cancer* **2016**, *114*, 256–261. [CrossRef]

68. Bajwa, G.; DeBerardinis, R.J.; Shao, B.; Hall, B.; Farrar, J.D.; Gill, M.A. Cutting Edge: Critical Role of Glycolysis in Human Plasmacytoid Dendritic Cell Antiviral Responses. *J. Immunol.* **2016**, *196*, 2004–2009. [CrossRef]

69. Saas, P.; Varin, A.; Perruche, S.; Ceroi, A. Recent insights into the implications of metabolism in plasmacytoid dendritic cell innate functions: Potential ways to control these functions. *F1000Research* **2017**, *6*, 456. [CrossRef]

70. Fischer, G.M.; Vashisht Gopal, Y.N.; McQuade, J.L.; Peng, W.; DeBerardinis, R.J.; Davies, M.A. Metabolic strategies of melanoma cells: Mechanisms, interactions with the tumor microenvironment, and therapeutic implications. *Pigment Cell Melanoma Res.* **2018**, *31*, 11–30. [CrossRef]

71. Tasdogan, A.; Faubert, B.; Ramesh, V.; Ubellacker, J.M.; Shen, B.; Solmonson, A.; Murphy, M.M.; Gu, Z.; Gu, W.; Martin, M.; et al. Metabolic heterogeneity confers differences in melanoma metastatic potential. *Nature* **2020**, *577*, 115–120. [CrossRef] [PubMed]

72. Han, N.; Zhang, Z.; Jv, H.; Hu, J.; Ruan, M.; Zhang, C. Culture supernatants of oral cancer cells induce impaired IFN-alpha production of pDCs partly through the down-regulation of TLR-9 expression. *Arch. Oral Biol.* **2018**, *93*, 141–148. [CrossRef] [PubMed]

73. Sisirak, V.; Vey, N.; Goutagny, N.; Renaudineau, S.; Malfroy, M.; Thys, S.; Treilleux, I.; Labidi-Galy, S.I.; Bachelot, T.; Dezutter-Dambuyant, C.; et al. Breast cancer-derived transforming growth factor-beta and tumor necrosis factor-alpha compromise interferon-alpha production by tumor-associated plasmacytoid dendritic cells. *Int. J. Cancer* **2013**, *133*, 771–778. [CrossRef]

74. Dzionek, A.; Sohma, Y.; Nagafune, J.; Cella, M.; Colonna, M.; Facchetti, F.; Gunther, G.; Johnston, I.; Lanzavecchia, A.; Nagasaka, T.; et al. BDCA-2, a novel plasmacytoid dendritic cell-specific type II C-type lectin, mediates antigen capture and is a potent inhibitor of interferon alpha/beta induction. *J. Exp. Med.* **2001**, *194*, 1823–1834. [CrossRef] [PubMed]

75. Cao, W.; Rosen, D.B.; Ito, T.; Bover, L.; Bao, M.; Watanabe, G.; Yao, Z.; Zhang, L.; Lanier, L.L.; Liu, Y.J. Plasmacytoid dendritic cell-specific receptor ILT7-Fc epsilonRI gamma inhibits Toll-like receptor-induced interferon production. *J. Exp. Med.* **2006**, *203*, 1399–1405. [CrossRef] [PubMed]

76. Cao, W.; Bover, L.; Cho, M.; Wen, X.; Hanabuchi, S.; Bao, M.; Rosen, D.B.; Wang, Y.H.; Shaw, J.L.; Du, Q.; et al. Regulation of TLR7/9 responses in plasmacytoid dendritic cells by BST2 and ILT7 receptor interaction. *J. Exp. Med.* **2009**, *206*, 1603–1614. [CrossRef]

77. Riboldi, E.; Daniele, R.; Parola, C.; Inforzato, A.; Arnold, P.L.; Bosisio, D.; Fremont, D.H.; Bastone, A.; Colonna, M.; Sozzani, S. Human C-type lectin domain family 4, member C (CLEC4C/BDCA-2/CD303) is a receptor for asialo-galactosyl-oligosaccharides. *J. Biol. Chem.* **2011**, *286*, 35329–35333. [CrossRef]

78. Krieg, A.M. Therapeutic potential of Toll-like receptor 9 activation. *Nat. Rev. Drug Discov.* **2006**, *5*, 471–484. [CrossRef]

79. Pashenkov, M.; Goess, G.; Wagner, C.; Hormann, M.; Jandl, T.; Moser, A.; Britten, C.M.; Smolle, J.; Koller, S.; Mauch, C.; et al. Phase II trial of a toll-like receptor 9-activating oligonucleotide in patients with metastatic melanoma. *J. Clin. Oncol.* **2006**, *24*, 5716–5724. [CrossRef]

80. Aspord, C.; Tramcourt, L.; Leloup, C.; Molens, J.P.; Leccia, M.T.; Charles, J.; Plumas, J. Imiquimod inhibits melanoma development by promoting pDC cytotoxic functions and impeding tumor vascularization. *J. Investig. Dermatol.* **2014**, *134*, 2551–2561. [CrossRef]

81. Teulings, H.E.; Tjin, E.P.M.; Willemsen, K.J.; van der Kleij, S.; Ter Meulen, S.; Kemp, E.H.; Krebbers, G.; van Noesel, C.J.M.; Franken, C.; Drijfhout, J.W.; et al. Anti-Melanoma immunity and local regression of cutaneous metastases in melanoma patients treated with monobenzone and imiquimod; a phase 2 a trial. *Oncoimmunology* **2018**, *7*, e1419113. [CrossRef] [PubMed]

82. Van den Hout, M.; Koster, B.D.; Sluijter, B.J.R.; Molenkamp, B.G.; van de Ven, R.; van den Eertwegh, A.J.M.; Scheper, R.J.; van Leeuwen, P.A.M.; van den Tol, M.P.; de Gruijl, T.D. Melanoma Sequentially Suppresses Different DC Subsets in the Sentinel Lymph Node, Affecting Disease Spread and Recurrence. *Cancer Immunol. Res.* **2017**, *5*, 969–977. [CrossRef] [PubMed]

83. Sunaga, N.; Imai, H.; Shimizu, K.; Shames, D.S.; Kakegawa, S.; Girard, L.; Sato, M.; Kaira, K.; Ishizuka, T.; Gazdar, A.F.; et al. Oncogenic KRAS-induced interleukin-8 overexpression promotes cell growth and migration and contributes to aggressive phenotypes of non-small cell lung cancer. *Int. J. Cancer* **2012**, *130*, 1733–1744. [CrossRef]

84. Dabrosin, N.; Sloth Juul, K.; Baehr Georgsen, J.; Andrup, S.; Schmidt, H.; Steiniche, T.; Heide Ollegaard, T.; Bonnelykke Behrndtz, L. Innate immune cell infiltration in melanoma metastases affects survival and is associated with BRAFV600E mutation status. *Melanoma Res.* **2019**, *29*, 30–37. [CrossRef] [PubMed]

*cancers* 

Review
Sex and Gender Disparities in Melanoma

Maria Bellenghi [1,†], **Rossella Puglisi** [1,†], **Giada Pontecorvi** [1], **Alessandra De Feo** [2], **Alessandra Carè** [1,*]  **and Gianfranco Mattia** [1]

[1] Center for Gender-specific Medicine, Istituto Superiore di Sanità, 00161 Rome, Italy;
maria.bellenghi@iss.it (M.B.); rossella.puglisi@iss.it (R.P.); giada.pontecorvi@iss.it (G.P.);
gianfranco.mattia@iss.it (G.M.)

[2] Laboratory of Experimental Oncology, IRCCS Istituto Ortopedico Rizzoli, 40136 Bologna, Italy;
alessandra.defeo@ior.it

[*] Correspondence: alessandra.care@iss.it; Tel.: +39-0649902411

[†] These authors contributed equally.

Received: 4 June 2020; Accepted: 3 July 2020; Published: 7 July 2020

Abstract: Worldwide, the total incidence of cutaneous melanoma is higher in men than in women, with some differences related to ethnicity and age and, above all, sex and gender. Differences exist in respect to the anatomic localization of melanoma, in that it is more frequent on the trunk in men and on the lower limbs in women. A debated issue is if—and to what extent—melanoma development can be attributed to gender-specific behaviors or to biologically intrinsic differences. In the search for factors responsible for the divergences, a pivotal role of sex hormones has been observed, although conflicting results indicate the involvement of other mechanisms. The presence on the X chromosome of numerous miRNAs and coding genes playing immunological roles represents another important factor, whose relevance can be even increased by the incomplete X chromosome random inactivation. Considering the known advantages of the female immune system, a different cancer immune surveillance efficacy was suggested to explain some sex disparities. Indeed, the complexity of this picture emerged when the recently developed immunotherapies unexpectedly showed better improvements in men than in women. Altogether, these data support the necessity of further studies, which consider enrolling a balanced number of men and women in clinical trials to better understand the differences and obtain actual gender-equitable healthcare.

Keywords: melanoma; sex/gender; sex-hormones; immunity; microRNAs; immunotherapy

---

1. Introduction

Melanoma is the most aggressive type of skin cancer, at present accounting for 1% of total cancer deaths in Italy. For a long time, only the surgical resection of early lesions was associated with long-term survival in more than 90% of patients, whereas advanced melanomas were mostly incurable. Although in the last decades a steadily increasing incidence of cutaneous melanoma was observed worldwide, an important 18% decrease in mortality was recently associated with improved knowledge of biological data and the introduction of novel therapeutic approaches, melanoma reduction being the highest among the other major cancers [1].

The incidence and mortality rate of the disease differ widely across the globe depending on the country of residence, ethnicity, and socioeconomic conditions and, chiefly, access to early detection and primary care [2]. It is also of note that incidence gradually decreases going from Northern to Southern Italy [1].

An additional key variable in melanoma is gender, in that a female advantage has been generally reported. Among the younger Italian population (under 50 years old), melanoma represents the 2nd most frequent tumor in men and the 3rd in women, the risk of developing this type of cancer during the life course being 1:66 and 1:85, respectively. In both sexes, the incidence is rising, with a 4.4% increase in men and a 3.1% increase in women per year. In 2019, 12,300 new cases were expected, with little prevalence in males [1].The mechanisms underlying gender disparity in melanoma development are not clear enough. Lifestyles play a role, with ultraviolet exposure representing an important risk factor, as women are more interested in sun exposure and tanning [3]. Conversely, males are generally less likely to engage in preventive behaviors [4] or to self-detect their melanomas [5]. Indeed, a different readiness of detection might be associated with the gender body-site distribution being primary melanomas more truncal in males and localized on the lower extremities in females. Thus, also an earlier diagnosis can partly explain the better survival rate of women.

As for the histological features, although thicker and ulcerated tumors were more frequently observed in men, these elements do not seem responsible for the unfavorable prognosis compared to women [6]. A large part of the female survival advantage could be explained with lower dissemination, resulting in a reduction in both lymph nodes and distant metastases when compared with males [7], and even after spreading to a visceral organ, a better prognosis seems to persist for women [8].

Looking for genetic differences, it is important to note that in women, the random—and sometimes incomplete—inactivation of one X chromosome in each single cell leads to mosaicism, and in turn, to the advantages associated with female genetic heterogeneity [9].

A significantly higher number of missense mutations was found among men with a mutational load ratio Men-to-Women of 1.85. Although the number of mutations is lower in melanoma female patients, their presence appears more relevant for increasing the overall survival, suggesting the functional pressure of the more efficient female immune system [10]. Furthermore, a study conducted in a Hispanic population identified several Single Nucleotide Polymorphisms (SNPs) differently associated with pigmentation, sun tolerance and melanoma risk in a sex-related manner [11].

Sex hormones play a fundamental role, as several studies demonstrated the association of estrogen and estrogen receptor expressions with melanoma survival in women. Notably, the female immune system is more efficient than the male one, and women mount both innate and adaptive immune responses stronger than men do. This higher effectiveness on the one hand is an advantage against infectious diseases and cancers, while on the other, it makes women more prone to autoimmune diseases [12].

Here we report the main disparities between men and women in an attempt to understand the sex- or gender-related effects on melanoma development, progression and response to therapy. We also speculate on the regulatory function possible played by miRNA that affects sex differences in melanoma pathogenesis at hormonal and immune levels.

2. Sex Steroid Hormone Receptors in Melanoma

Sex hormones belong to the steroid hormone family, mainly synthesized by the adrenal cortex and gonads, and in minor part by several peripheral tissues, such as the skin. In fact, starting from blood precursors or mobilizing cholesterol from cellular stores, the skin is able to produce several biologically active steroids, such as estrogens, testosterone (T) and dihydrotestosterone (DHT) [13]. Of course, the endocrine function of the skin becomes of particular relevance in men and in postmenopausal women, when almost all estrogens are made through the extraglandular conversion of androgens into estrogens. Estrogens exert their biological effects by binding to and activating two members of the nuclear steroid receptor superfamily, the estrogen receptor α (ERα) and β (ERβ), as well as the more recently discovered G protein-coupled estrogen receptor (GPER). While both genomic and non-genomic pathways have been described for the signaling activities of ERs, GPER is believed to mediate a rapid and non-genomic response upon hormone binding. Estrogens regulate the growth and differentiation of normal and several neoplastic tissues (such as breast, ovarian and endometrial tumors). Indeed, in cancer growth, ERs exert the opposite effects, ERα being pro- and ERβ anti-proliferative [14]. It is believed that cutaneous ER levels are generally higher in women than in men [15,16]. Some immunohistochemical analyses indicated that in melanocytic nevi and malignant melanoma cells, ERβ was present but ERα was not [17,18], even if both ERα and ERβ mRNAs were found in several melanocytic lesions [19]. Indeed, the presence or not of ERα either in primary or in metastatic melanoma is an unresolved issue. Downmodulation of this receptor has been shown to be under epigenetic control and appears to be directly proportional to disease progression [20]. In this perspective, the detection of hypermethylated ERα in melanoma patient sera was proposed as a predictive marker of bio-chemo-therapy response, thus becoming a negative prognostic factor [20]. Noteworthy, ERα is also able to synergize with the insulin-like growth factor 1 receptor (IGF1R) in response to 17β-estradiol (E2) and IGF1 stimuli, as showed in the MCF7 breast cancer cell line [21]. A recent work on melanoma shows the presence of one SNP in IGF1 and another in IGF1R, likely associated with increased risk or protective effect, respectively, especially in men. In line with this, the possible role of ESR1 SNPs in melanoma requires further investigation [22]. Conversely ERβ expression is inversely correlated with Breslow thickness, the most important and independent predictive marker of melanoma [16]. According to the survival advantage of female melanoma patients, men show significantly lower levels of ERβ in both melanoma and healthy tissues [16]. A more recent study on melanoma survival ratios from the Human Protein Atlas and The Cancer Genome Atlas Genomic Data Commons (GDC) showed that while low ERβ expression was associated with shortened relapse-free survival (RFS), ERα and GPER were not [23]. Concerning in vitro studies, several melanoma cell lines express ERβ, irrespective of genetic background [24]. GPER is also expressed in melanoma [25] its co-expression with ERβ being associated with better outcomes, especially in pregnancy-associated melanoma [26]. The nuclear receptor superfamily also includes the androgen receptor (AR), consisting of α and β isoforms encoded by a gene located on the X chromosome [27]. AR respond to androgenic hormones by using the same genomic and non-genomic pathway of ERs and evoking similar effects [28]. Although less studied than ERs, the expression of AR was assessed in several melanoma cell lines and in human metastatic specimens, where high receptor levels were detected [29]. Concerning progesterone (P), to date, different receptors (PRs) that are able to activate both genomic and non-genomic pathways have been identified [27]. Since melanocytic lesions seem to change during pregnancy, several attempts have been made to associate PR expression to melanoma course in pregnant and in non-pregnant women, without reaching definitive results [30].

3. Female Hormone Activity

A large body of evidence supports the beneficial role played by estrogens against melanoma progression. Epidemiological studies have pointed out how menarche and menopause with consequent changes in the endogenous estrogen exposure can influence melanoma risk [31]. However, the persistence of female benefit in older postmenopausal women is an unresolved issue due to conflicting results published over time [32–34]. Controversial data were reported for exogenous estrogens, such as oral contraceptive [35] and hormone-replacement therapy (HRT). Concerning the latter point, some European cohort studies recently showed an increased risk in melanoma associated with use of HRT [36–40], whereas others did not [31,41]. Likewise, the attempt to associate pregnancy with melanoma outcomes did not reach conclusive results [42,43].

In melanoma cell lines, in vitro response to 17β-estradiol (E2) treatment produced different effects on either proliferation or invasion ability [44–46]. In other reports, only in vivo mouse models (Figure 1, left) were able to show the estrogenic capability to contrast disease progression. As demonstrated in athymic nude mice, the same melanoma cell line unresponsive to estradiol treatment in vitro displayed significantly reduced growth in castrated mice treated with 17-β estradiol pellets, suggesting other possible indirect hormone actions [47]. A growing interest in the use of 2-methoxyestradiol (2ME2), a non-toxic endogenous metabolite of E2, highlighted its ability to block the human melanoma cell-cycle, inducing apoptosis both in vitro [48] and in male severe combined immunodeficient (SCID) mouse models [49]. Sex-related differences in metastasis formation were principally observed in the liver, the organ mainly active in estrogen conversion into 2ME2 [50]. This was in perfect agreement with results previously observed in SCID mice where, as a consequence of intrasplenic injection of ER positive human melanoma cells, a greater number of metastases was observed in the male liver than in that of the female [51]. Indeed, the estrous cycle of female mice could also affect the capability of B16 melanoma cells to metastasize in different organs [52]. More recently, the antiproliferative and cytotoxic effects of 2ME2 were shown in melanoma cells with different genetic backgrounds, as well as in the counterparts resistant to either BRAFi (v-raf murine sarcoma viral oncogene homolog B1) or the BRAFi+MEKi (Mitogen-Activated Protein Kinase Kinase) combination [53].

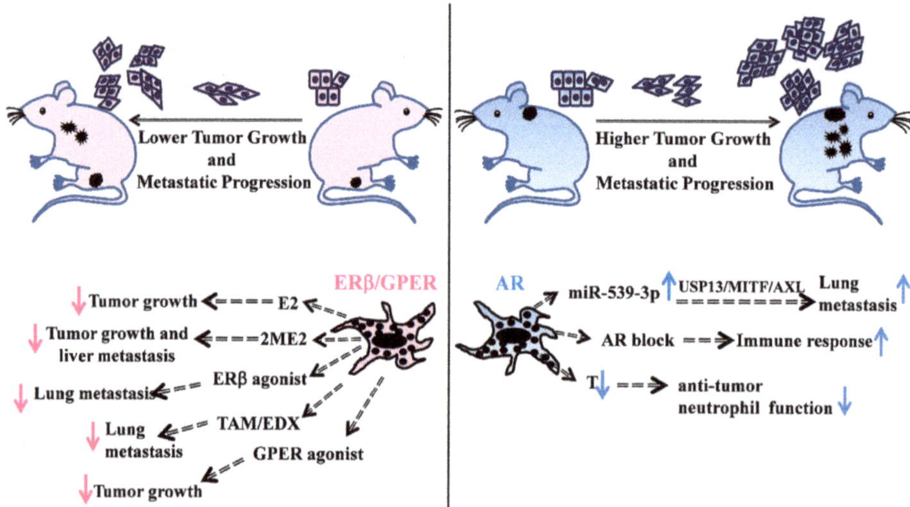

Figure 1. Schematic overview of some representative in vivo studies supporting the hormonal involvement in melanoma disease. (**Left**) Evidences of female hormonal treatments in in vivo mice models. E2 treatment of castrated mice s.c. injected with ER positive melanoma cells reduces tumor growth [47]. 2ME2 treatment of SCID mice, previously intrasplenically injected with melanoma cells, decreases primary tumor growth and liver metastasis number [49]. Treatment of melanoma cells with GPER agonist affects tumor growth in host mice improving response to immunotherapy [54]. Administration of the ERβ agonist LY500307 in female mice, previously i.v. injected with melanoma cells, decreases lung nodules [55]. Oral administration of TAM [56] and EDX [57] exert inhibitory effects on tumor metastatization into the mouse lungs. (**Right**) AR action is responsible for the increased number of lung metastases through the miR-539-3p/USP13/MITF/AXL axis [58], whereas AR blockade mediates an increase in the immune response [59]. T loss in castrated male mice causes a decrease in the anti-tumor neutrophil function [60]. E2: 17β-estradiol; 2-ME2: 2-methoxyestradiol; TAM: Tamoxifen; EDX: endoxifen; USP13: ubiquitin specific peptidase 13; MITF: microphthalmia-associated transcription factor; AXL: receptor tyrosine kinase AXL; T: testosterone. s.c. subcutaneous, i.v. intravenous.

Besides hormones and their metabolites, several natural and synthetic compounds with different affinities to ERs and divergent activities, i.e., agonist/antagonist, were tested in vitro and in vivo. It is worth remembering the ERβ agonist diarylpropionitrile (DPN) efficacy in inhibiting NRAS (Neuroblastoma RAS Viral Oncogene Homolog)-mutated melanoma cell proliferation [24], as well as many specific ERβ agonists of natural origin able to exert an antitumor function in different human melanoma cell lines in vitro and in vivo [61]. Recently, the synthetic non-steroidal estrogen and selective ERβ agonist LY500307 showed the capability to suppress melanoma lung metastasis in the B16 murine melanoma in vivo model by up-regulating innate immunity in a tumor microenvironment [55]. Tamoxifen (TAM), widely used for the treatment of both early and advanced breast cancer, belongs to the selective estrogen receptor modulator (SERM) class of drugs acting on the ERs [62]. TAM's effectiveness in melanoma treatment is due to the discordant data derived from both in vitro and in vivo studies. Numerous results have described SERM-dependent inhibition of melanoma cell proliferation, suggesting the possible involvement of IGF1R inactivation [63] as well as the reduction of invasion and metastasis through the inhibition of protein kinase C (PKC) downstream pathways [56]. Unfortunately, the activity of TAM against melanoma progression is extremely poor in vivo [64], unless combined treatments with chemotherapy are considered. In the latter condition, a better response to chemotherapy treatment was observed in advanced melanoma, with female patients being more likely to respond, albeit with increased toxicity and a doubtful survival advantage [65]. TAM's clinical efficacy, at least in part, depends on its metabolization in the liver, resulting in variable concentrations of active metabolites in the patient plasma. Therefore, to induce melanoma cell death in vitro, endoxifen (EDX), a metabolite of TAM that is safer and has more cytostatic activity than the parent drug has been used [66]. Furthermore, EDX orally administered in mice for four weeks reduced lung metastatic nodules without any side effects [57]. Another reason for different epidemiological evidence in TAM's effectiveness could lie in its possible different and tissue-related effects on the two estrogen receptors, similarly to E2 [67,68]. The recent GPER inclusion among TAM-responsive receptors in melanoma further complicated the scenario [25]. Several studies described the ability of GPER to react with E2 and its specific agonist G-1 to evoke a suppressive action against melanoma progression [25,69,70]. In addition, GPER's ability to make melanoma more vulnerable to immune-mediated eradication upon hormone treatment was reported [54]. Several in vitro studies showed the concentration-dependent progesterone activity on several human melanoma cell line growth, adhesion and migration abilities [71–75]. According to these in vitro results, repeated pregnancies inhibited the development of genetically defined BRAF-driven human melanocytic xenografts when compared with non-pregnant females [54].

4. Male Hormone Activity

A large body of evidence supported AR involvement in melanoma incidence and progression, although a clear correlation between AR expression in melanoma and bad prognosis is lacking. Many reports demonstrated AR involvement in melanoma growth and invasion in vitro [44,76] and in vivo (Figure 1, right), eventually involving immune response blockage [59]. A recent work conducted on a small number of patients correlated AR-positive melanoma patients with worse survival when compared to those AR negative [58]. In this study, AR was shown to induce miR-539-3p expression that, targeting ubiquitin-specific peptidase 13 (USP13), abolished its de-ubiquitination activity on the microphthalmia-associated transcription factor (MITF) and determined the induction of the receptor tyrosine kinase AXL, with a consequent increase in metastases [58]. AR has been recently involved in other specific molecular pathways, for example its recruitment by the SRA-like long non-coding RNA (SLNCR) to the early growth response 1 (EGR1)-bound chromatin loci to repress p21 expression [77]. Furthermore, the activation of the non-genomic pathway, via the combination of the epidermal growth factor receptor (EGFR) and AR, enhanced AR activity itself and modified the melanoma-associated antigen protein-A11 (MAGE-A11), improving melanoma proliferation [28]. Another study described a decreased AR level in several tumors, including melanoma, when compared to the normal tissue counterparts, and with intratumoral receptor levels higher in males than in females. In contrast to most literature data on melanoma, high AR protein expression levels were associated with increased overall survival (OS) and progression-free survival (PFS) [78]. Recently, testosterone levels also gained some relevance as a possible cause of increased melanoma incidence in aging males. Multiple syngeneic metastatic mouse models demonstrated the importance of testosterone signaling on neutrophil maturation and function, since castration or androgen inhibition significantly increased melanoma burden [60].

5. Sex and Immunity

In the early stages of development, a high immunogenic phenotype characterizes melanomas, thus providing an ideal cancer model to understand the complex cross talk between tumors and immune cells and to give a possible explanation to sex differences in the host immune response. A scientific breakthrough in oncology research has been the recognition of immune system participation in the initiation, progression and, in some cases, the resolution of melanoma. Moreover, the recent introduction of the immune checkpoint inhibitors (ICI) in the therapeutic plan for melanoma patients has certified the crucial role of immunity in anticancer therapy, highlighting sex disparities [12,79–81]. Immunological sex differences concern both innate and adaptive immune responses and are influenced by either sex hormones or different specific genetic backgrounds between males and females [12]. In general, estrogens exert an immune-enhancing effect, contrasting to the immune-suppressive one of testosterone. Many different cell types express estrogen receptors (α and β ERs) (i.e., epithelial cells, lymphoid tissues and immune cells) that allow estrogen binding and signal activation [82–84]. Direct implication of estrogen-dependent effects on innate immunity have been recently evidenced in a syngeneic mouse model of melanoma where in vivo studies showed that the estrogen agonist erteberel, activating the ERβ signaling, was capable of augmenting innate immunity and suppressing lung metastatic colonization by recruitment of antitumor neutrophils to the metastatic niche [55]. On the contrary, castration in male mice led to an increased autoimmune response by the induction of the major histocompatibility complex II (MHCII) [85]. Moreover, estrogens enhanced the expression of MHCII on dendritic cells (DCs), while testosterone decreased it [86].

The E2/ERα axis plays a pivotal role in controlling functional responses of DC subgroups, and it is responsible for the epigenetic regulation in DC precursors, driving their differentiation also by modulating Interferon I (IFN-I) secretion [87]. In melanoma, DCs of tumor microenvironment act by modulating T cell activity and take part in the immune infiltration, which is considered an indicator of immune-therapy response [88]. In addition, a specific subgroup of DCs, named Langerhans' cells and belonging to the skin immune system (SIS), is regulated by ERβ signaling [89,90]. Among the DC cells, the plasmacytoid DCs (pDCs) display major differences between women and men, their activity being under control of the E2/ERα axis and the X- linked Toll -like receptor 7 (TLR7) [91,92]. TLR7 belongs to the Toll-like receptor signaling and participates in the innate response against microbial infectious, favoring a better response in female association with a higher IFNα production [93]. The TLR signaling is an important autoregulatory mechanism that maintains tissue homeostasis, whose members are expressed on various skin cells, such as keratinocytes and melanocytes [94]. In particular, melanocytes express TLR 2-5 and TLR7, 9 and 10 [95]. Recent literature data evidenced their implication both in melanoma and non-melanoma skin cancer in supporting the immune escape [96–98]. Indeed, TLR agonists, targeting TLR7, 8, and 9, have been described as successful treatment options for melanoma and basal cell carcinoma (BCC), enhancing DC recruitment and T cell responses [99–101]. It is also important to note that several X chromosome genes take part in the innate immune function [102–106].

During the acute inflammatory response, the phagocytic activities of neutrophils and macrophages as well as the microbial killing by reactive oxygen species (ROS) are more efficient in females than in males. This female advantage has been also shown among melanoma patients. Melanoma cells exhibited significantly higher oxidative stress and produced larger amounts of ROS when compared to melanocytes and surrounding normal tissue [107]. ROS stimulated melanoma progression and metastatization through a number of changes, including (i) DNA modification, (ii) cell proliferation, (iii) tissue remodeling, (iv) immune surveillance escape, (v) pro-metastatic processes activation [108,109]. Malorni and colleagues demonstrated that males express lower levels of anti-oxidants, such as glutathione (GSH), catalase and superoxide dismutase (SOD) when compared to females, thus exhibiting a higher rate of oxidative stress [110,111]. Looking at the systemic influence of ROS on the metastatic phase of melanoma, a "ROS-sex issue" could be another reason for the differences eventually resulting in lower male survival rates [112–114].

In humans, natural killer cells (NKs) express both ERs and PRs, but not AR. Female hormones promote the induction of IFN γ, secretion of granzyme B and favor caspase-dependent apoptosis. In spite of these apparently contradictory data, male subjects exhibit a higher number of NK cells [12]. NKs participate in the first line of response against melanoma. One of the mechanisms used by melanoma for avoiding the CD8+ T cell antitumor action is the downregulation of MHC I. This reduction should support the removal of melanoma cells due to the capability of NKs to recognize and specifically eliminate cells expressing low levels of MHC I. Therefore, NKs appear a good target population for melanoma immune therapy [115–117].

The activation of adaptive immune responses has been demonstrated to counteract melanoma progression, metastatic spreading and therapy-related resistance. Two main different groups of melanoma-associated antigens have been characterized: (i) antigens expressed by normal and malignant melanocytes (as Gp100, tyrosinase, Melan-A and the isoform of tyrosinase-related protein, TRP-2 (INT2), and (ii) cancer testis antigens mainly expressed by transformed cells, such as melanoma antigen-1 (MAGE-A1), the highly immunogenic tumor antigen NY-ESO-1, and the preferentially expressed antigen of melanoma (PRAME) [118–122]. Female antigen-presenting cells (APCs) are more efficient than the male ones in both the presentation and initiation of a secondary response in primed lymphocytes. Indeed, androgen treatment of female mice reduced the efficiency of APC function, while estrogens exalted this function [123,124].

E2 concentration is considered the central rheostat for different adaptive immune response regulations: low E2 concentration induces Th1 response and cell mediated immunity, while high E2 concentration favors Th2 response and humoral immunity.

Beyond the hormonal role, an additional explanation supporting sex differences in the immune system is the different genetic background, as females carry two X chromosomes and males just one. A mechanism to re-equilibrate gene expression is the random inactivation of one X copy in each female cell, thus making every woman a mosaic for X-linked expression [125]. An advantage of female mosaicism is the possibility to tolerate gene mutations responsible for X-linked diseases that severely affect males, such as those named X-linked primary immune deficiencies [126]. Furthermore, a percentage proximal to 15% of X-linked regions fails inactivation in women and, consequently, some genes in these regions might display a level of expression double than men [9]. It is important to highlight the presence of a high number of genes with immunological function on the X chromosome, possibly underlying not only the higher female immune response to infections, but also a positive effect on the anticancer immune responses. Among these genes, we can ascribe the Interleukin 2 Receptor Subunit Gamma (IL2Rγ) chain, the Interleukin 3 Receptor Subunit Alpha (IL3Rα) chain, and the Interleukin 13 Alpha (IL13α) chain, GATA-binding protein 1 (GATA1), Forkhead Box P3 (Foxp3) and CD40 Ligand (CD40L) [127]. On the contrary, this more reactive immune system, increasing female susceptibility to develop autoimmune disorders, represents a disadvantage [128,129].

Sex differences in lymphocyte subsets, including B cells, CD4+ and CD8+ T cells, have been demonstrated among adults. Females have a higher CD4+ T cell count and a higher CD4/CD8 ratio than age-matched males; whereas males have a higher CD8+ T cell frequency [130,131]. Studies in humans suggested a higher number of T regulatory cells (Treg) in healthy adult males compared to females, although some conflictual results regarding Treg frequency were reported [132].

Wesa and colleagues demonstrated that female melanoma patients have a high frequency of CD4+ TAA (tumor-associated antigen)-specific T cells compared to male patients, and that these cells are more prone to express an apoptotic phenotype in the presence of active disease [133]. The increase in and/or improved functions of tumor-specific T-helper (Th) cells could be a biological response to these differences due to it being involved in anti-tumoral responses [134].

In recent years, in the immunological context of tumor biology, the immune checkpoint inhibitors emerged as promising therapeutic targets in different cancers, including melanoma [135]. The relationship between PD-1 and sex hormones recently emerged, despite the fact that data literature offered limited in vivo studies and conflicting results. Different data demonstrated that PD-1 is able to respond to sex steroids and that the hormone-mediated effect on PD-1 signaling might influence the regulation of autoimmune diseases [136]. In addition, estrogenic hormones were capable to modulate the programmed cell death ligand 1 (PD-L1) and B7-costimulatory molecules. Due to its contribution to immune evasion and induction of T regulatory cells, B7-H1 was associated with cells with pro-tumoral activity [137–139]. Lin and colleagues demonstrated that sexual hormones, in particular estrogens, modulated the Treg-linked B7-H1 immune suppression function. In a B7-H1 KO mouse model, female mice showed a better response to a B16 murine melanoma cell injection because a reduced Treg activation allowed a strong antitumor response when compared to males. Finally, in the same model, E2-mediated inhibition of the Treg function reduced primary tumor growth in female mice when compared to their male counterparts [140]. These results should induce a more careful estimation of sex differences in immune response and a sex-based interpretation of the therapeutic plans for anticancer immunotherapy (Figure 2).

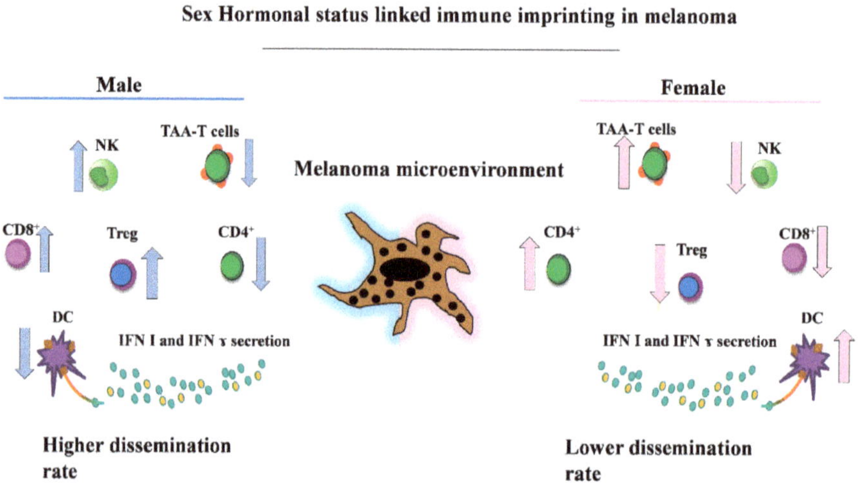

Sex Hormonal status linked immune imprinting in melanoma

Figure 2. Key gender-related differences of the immune cell populations involved in melanoma. This representative picture shows the main immune cell populations present with higher frequencies in males (i.e., CD8+, Treg and NK cells [12,123–125]) (**left**) and in female patients (i.e., the immune phenotype enriched in TAA-T cells, CD4+ and DCs cells [80,123,124,126] with higher levels of circulating Interferon I (IFN I) and IFN γ [80,84,85]) (**right**). NK: natural killer; TAA: tumor-associated antigen; DCs: dendritic cells.

6. Sex Differences and MiRNAs

The sex-linked differences affecting regulatory pathways in melanoma pathogenesis and the associated immune responses are further controlled by the emerging interaction between these functional signals and miRNA-specific epigenetic regulation. In these two decades of studies, microRNAs resulted as major post-transcriptional modulators of gene expression. In their mature form, miRNAs are 19–25 nucleotide long single strand RNAs, generally expressed in all cell types. A great number of works have described miRNA biogenesis and transcriptional regulation, and we refer to them for detailed explanation [141–143]. The high number of targets for a single miRNA and the capability of more than one miRNA to effectively repress the same gene result in complex and pleiotropic regulatory effects on cell physiology. MiRNAs regulate numerous cellular pathways and their alteration in pathological conditions determines a strong dysregulation of these pathways. In fact, in cancer pathogenesis, miRNAs can act as either oncosuppressors or oncogenes [144,145]. Of note, different tumors as well as diverse stages in the same cancer type display specific miRNA signatures [146]. In melanoma, specific expression profiles characterizing differences in disease progression, mutational state, as well as miRNA "facilitators" of drug resistance have been described [147].

Several studies, mainly in breast cancer, have demonstrated the direct correlation between miRNA expression and sexual hormones [148]. The miRNA-dependent regulation by direct targeting of ER mRNA at its 3′UTR and, vice versa, the capability of estrogen-specific pathways to modulate miRNA expression have been described and strictly depend on the specific activated receptors [149–151].

In melanoma, the effect of E2 treatment on miRNA expression and, more in general, their real functional link remain essentially unknown. Nonetheless, the elucidation of miRNA and ER-functional networks might help to understand some controversial aspects of female survival advantage compared to male survival in initial phases of melanoma development, or the reduction in these advantages in postmenopausal women when female estrogenic concentration vertically declines [32,34]. It is here important to consider that melanoma could be classified among the hormone-sensitive tumors according to complex overlapping actions played by estrogens and androgens, particularly by the opposite effects of α and β estrogen receptors (ER) [28]. Some functional parallelisms with miRNAs controlled by the estrogenic action in breast cancer might help to understand if some miRNA alterations in melanoma might be under hormonal control. An important example is the feedback loop involving the miR-221 and -222 cluster and ERα in breast cancer. A regulatory circuitry exists on one side based on the direct interaction of ERα with the estrogen receptor-binding site in the promoter region of miR-221&222 to induce their expression [152], on the other side on miR-221&222 direct targeting of ERα. Indeed, these two miRNAs and ERα are negatively related [153].

In breast cancer, miR-221 and -222 overexpression, and in turn ERα reduction, were shown to trigger cancer cell proliferation and invasion [154]. Furthermore, the miR-221 and -222-dependent decreased expression of ERα reduced cell sensitivity to the tamoxifen endocrine therapy, favoring resistance and eventually exacerbating malignant progression of disease. Accordingly, xenograft tumors treated with a specific antagomir, the down-modulating of miR-221 and -222 expression removed this effect [155]. Although the possible estrogenic regulation of miR-221 and- 222 expression in melanoma progression remains to be defined, the roles and regulation of these two miRNAs have been deeply studied. Similarly, to breast cancer, melanoma progression and spread require miR-221 and -222 expression, while ERα expression is lost. At least six target genes were revealed, including p27^{Kip1}, tyrosine-protein kinase Kit (c-KIT), ETS proto-oncogene 1 (ETS-1), AP-1 transcription factor subunit (c-FOS), Activating enhancer binding Protein 2 α (AP2α) and Stearoyl-CoA Desaturase 5 (SCD5), all with direct or indirect tumor suppressor roles according to the central roles of miR-221 and -222 in melanoma proliferation and dissemination [156–160].

Thanks to the high number of miRNAs present on the X chromosome (approximately 120 miRNAs) compared to both autosomes and mainly to the Y chromosome (at present only 4 miRNAs), X-linked miRNAs might have a role in sex differences evocated in melanoma immune response. Potentially, miRNAs might escape dosage compensation in association with the genomic co-localized genes evading X inactivation [161,162]. Indeed, a sex different susceptibility to cancer development concerns different solid tumors, with males in some cases at higher risk compared to females [163–165]. Several miRNAs involved in hematopoietic lineage differentiation and in pathological conditions play a role in chronic inflammation as a predisposing factor in the onset and progression of cancer [166,167]. Thus, it might be relevant to consider the possible functional interconnections between X-linked miRNAs and immune responses underlying sex differences in melanoma (see Figure 3).

Once more, miR-221 and -222 are the most studied X-located oncomiRs that result in strongly deregulated different forms of cancer, including breast cancer, prostate cancer, liver cancer, bladder cancer, thyroid cancer, glioblastoma and melanoma [168]. Their important role and sex-related modulation was assessed in cardiovascular diseases where miR-222 indirectly reduced the endothelial nitric oxide synthase formation through ETS-1 targeting in cardiac vascular cells [169]. The ETS-1-miR-222 circuitry was also associated with melanoma progression, miR-222 induction being strictly dependent on the constitutive ETS-1 phosphorylation and compartmentalization in metastatic tumors and ETS1 directly targeted by miR-222 [157]. In hematopoiesis, miR-221 and -222 were shown to reduce differentiation of the embryonic hematopoietic progenitor cells because of their direct targeting of c-KIT mRNA [170]. We imagine that increased miR-221 and -222 levels, representing either melanoma cell-intrinsic or immune system-extrinsic factors, can contribute via exosome release to disease progression [171–173].

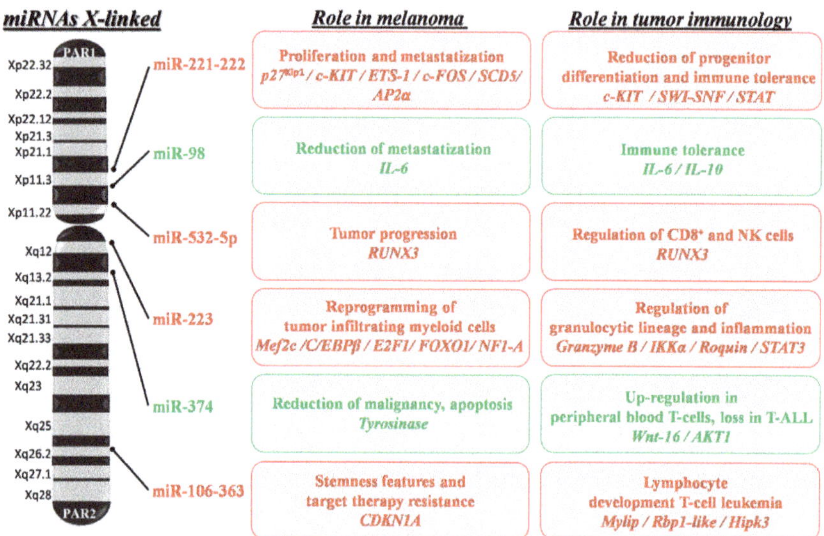

Figure 3. X-linked miRNA mapping and key roles in melanoma development and progression. Schematic depiction of miRNA localization on the human X chromosome: miRNAs with a confirmed role in melanoma pathogenesis are shown. Red writing inside boxes indicate miRNAs, target genes (italics) and their downstream oncogenic effects; green writing indicates those with oncosuppressive functions. PAR1 and 2: pseudoautosomal region 1 and 2; $p27^{Kip1}$: cyclin-dependent kinase inhibitor 1B; ETS-1: ETS proto-oncogene 1; SCD5: Stearoyl-CoA Desaturase 5; SWI-SNF: Switch/Sucrose Non-Fermentable; STAT: Signal Transducer and Activator of Transcription; IL-6 and 10: Interleukin 6 and 10; RUNX3: Runt-related transcription factor 3; IKKα: Inhibitory-ΚB Kinase α; STAT3: Signal Transducer and Activator of Transcription 3; MEF2C: Myocyte Enhancer Factor 2C; C/EBPβ: CCAAT/enhancer-Binding Protein β; E2F1: E2F transcription factor 1; FOXO1: Forkhead Box protein O1; NF1-A: Nuclear Factor 1 A; Wnt-16: Wnt family member 16; AKT1: AKT serine/threonine kinase 1; CDKN1A: Cyclin-Dependent Kinase Inhibitor 1A; Rbp1-like: Retinol Binding Protein 1-like; Hipk3: Homeodomain-Interacting Protein Kinase 3.

The miR-17-92 cluster includes 15 miRNAs organized in three paralogue groups on different chromosomes. Among them, the miR-106-363 paralogues are localized on the X chromosome and include six miRNAs (miR-106a,-18b,-20b,-19b2,-92-2 and -363). Due to sequence similarity, genomic organization and functional connections, these miRNAs are further organized in a different family, named miR-17,-18,-19 and -92. All were designated with oncogenic potential in different forms of cancer [174]. Mainly, the miR-17 family was linked to melanoma progression and individuated in non-responders or patients expressing high levels of PD-L1 and resistance to BRAF/ MEK inhibitors [147,175]. In addition, miR-106a and miR-363 play a role in the immune regulation by controlling lymphocyte development. When overexpressed, these miRNAs act as oncogenes and are strongly implicated in T cell leukemia development by targeting myosin regulatory light chain-interacting protein (Mylip), retinoblastoma-binding protein 1-like (Rbp1-like), and homeodomain-interacting protein kinase 3 (Hipk3) [176]. In melanoma, miR-363-3p was indicated in the targeting of the inhibitor of cell cycle progression of cyclin-dependent kinase inhibitor 1A (CDKN1A). This targeting was associated with the parallel expression of Hif-1α and the acquisition of stemness features on melanoma cells as specified by CD133, CD271, Jarid1B, and Nanog expression [177].

Tumor-infiltrating myeloid cells are immature tumor-infiltrating cells of hematopoietic origin that favor cancer progression by a mechanism of immune suppression described in the breast and in melanoma. The myeloid-derived suppressive cells were also crucial for the metastatic niche formation, inducing expression of factors with an invasive-promoting action. Different X-linked miRNAs have been described to have a role in this suppressive mechanism. The more studied is miR-223, which when overexpressed, reduced maturation of the granulocytic cell lineage that regulates the Mef2c transcription factor necessary for myeloid progenitor proliferation, and it acted as negative modulator of inflammatory responses in animal models [178,179]. MiR-223 is highly expressed in melanoma. In a mouse model of melanoma metastasis, miR-223 in association with miR-21, miR-29a and miR-142-3p, was demonstrated to support the reprogramming of tumor-infiltrating myeloid cells and the acquisition of pro-metastatic and angiogenic properties [180].Validated targets for miR-223, during inflammation, include granzyme B, Inhibitory-KB Kinase α (IKKα), Roquin and Signal Transducer and Activator of Transcription 3 (STAT3), while cancer-associated targets include CCAAT/enhancer-Binding Protein β (C/EBPβ), E2F transcription factor 1 (E2F1), Forkhead Box I A (FOXO1) and Nuclear Factor I A (NFI-A) [181].

MiR-532-5p is upregulated in primary and metastatic melanoma cell lines and in tumor samples. This upregulation was associated with down-modulation of its target Runt-related transcription factor 3 (RUNX3), a member of the runt-related family of genes that are known as developmental transcriptional regulators. These genes are important in the progression of a variety of human cancers and are involved in the differentiation program of normal tissues [182]. It is interesting to note that RUNX proteins regulated the transcription of the MDR1 gene in cytotoxic CD8+ lymphocytes and NK cells. The MDR1 protein was required for the accumulation of cytotoxic T lymphocytes (CTL) cells after acute viral infections or in the protective function of memory-T cells following microbial challenge. In view of this new research, the counterproductive effect of the MDR1 inhibition as an anticancer therapeutic strategy was suggested. In this new context, miR-532-5p-dependent epigenetic regulation might be co-responsible for the reduction of tumor-infiltrating CD8+ cells in the tumor microenvironment [183].

Among other X-linked miRNAs, miR-374, part of the human miR-374 family, down-regulates tyrosinase expression and reduces melanoma malignancy by attenuation of WNT signaling, thus promoting melanoma cell apoptosis in a mouse model of melanoma [184]. In a combined expression profile, the miR-374 family was found upregulated in monocytes and granulocytes vs. progenitor cells, while in the immune context, miR-374b was up-regulated in peripheral blood where T cells targeted Wnt-16 and AKT1 [185,186]. Finally, several X-linked miRNAs contributing to the melanoma-associated immune regulation act by modulating the Toll-like receptor-mediated responses. Indeed, the development of agonists for the TLR signal has been evocated to trigger new therapeutic strategies for the induction of anticancer immune response [187]. An interesting example is miR-98. This miRNA is down-modulated in melanoma progression where it plays a negative feedback loop with the pro-inflammatory IL6 [188], but it also acts as a regulator of IL10 expression in response to microbial infection or after lipopolysaccharide (LPS) stimulation [189]. Furthermore, miR-221 and -222 oncomiRs, besides a number of other functions, participate in the regulation of TLR response-inducing tolerance to microbial infection essentially through TNFα (Tumor Necrosis Factor α) degradation by regulation of Switch/Sucrose Non-Fermentable (SWI/SNF) and Signal Transducer and Activator of Transcription (STAT) [190]. It is important to underline the parallelism between TLR and miR-221 and -222 functions whose expressions might contribute to the increased immune impairment evidenced in melanoma patients eventually favoring tumor metastatization.

7. Sex Differences and Response to Therapies

Starting from Clark's observation in 1969, melanoma seemed less malignant in women than in men [191]. Since then, many other large studies confirmed this result, reporting a 20–30% female advantage worldwide. A recent study based on a patient cohort of approximately 16.000 stage II–IV cutaneous melanoma patients confirmed the female advantage in melanoma survival by using the latest prognostic staging system for risk stratification [192].

Moving forward, in regard to the relevance of sex and gender in clinical research, we should start from considering the underrepresentation of women in clinical trials. At present, despite many invitations for inclusion of a balanced number of men and women, no significant improvements were obtained and women, particularly in the initial phases of the studies, still represent 20–25% of the total [193].

In addition, even when a significant number of women are enrolled, often data are not properly stratified for the analyses [194,195]. In accordance with the sex and gender equity in research (SAGER) guidelines, differences between men and women should be always considered in the evaluation of response to therapies, toxicities and outcomes [196]. More important, in Italy, the Law 3/2018, particularly article 1, specifically includes the concept of gender medicine in clinical trials for human medicinal products, highlighting the importance of a gender-specific approach [197]. On this basis, we should expect to see some gradual progress in the near future, starting from the point that that men and women are not the same [198].

In the last few years, melanoma therapeutic approaches showed a rapid evolution with the main goal of increasing clinical efficacy and the duration of responses. The first promising result was obtained by targeted therapy against the activating mutations of BRAF, present in 40–60% of cutaneous melanomas with a similar distribution among men and women [199,200]. Major benefits were achieved with the combination of BRAF and MEK inhibitors in order to avoid the so-called "paradoxical activation" of the MAPK pathway downstream to BRAF [201,202]. More recently, important results were obtained through the inhibition of some molecules involved in T cell suppression. In particular, the impairment of CTLA-4, PD-1 or PDL-1 functions through the immune checkpoint inhibitors (ICIs) gave impressive responses in several types of cancers, including metastatic melanoma [203]. Although some sex differences were observed in the efficacy of more traditional therapies [204], according to the known disparities in the immune system with women sustaining stronger responses [12], it was reasonable to expect diverse sex-associated anti-tumor effects of ICIs [205]. In fact, in the initial studies in proper in vivo models, immune checkpoint inhibitors were more effective in female mice than in their male counterparts [140]. Conversely, in human clinical trials, unexpected better outcomes were shown in men. Several meta-analyses evidenced this result in cancer-randomized clinical trials treated with ICIs, initially suggesting a more evident benefit associated with the anti-CTLA4 treatment for males vs. females [206]. Looking for a higher statistical power, an elevated number of patients (>11,000) was analyzed as part of 20 randomized controlled trials based either on CTLA-4 or PD-1 inhibitors. Results confirmed and extended the increased efficacy of immunotherapy in men vs. women [80]. Many hypotheses can be proposed to explain these disparities, including the sex-related differences of the immune system at baseline conditions, as indicated by the higher number of CD8+ T lymphocytes and the lower CD4+/CD8+ ratio in males than in females [12]. In addition, the higher amount of Treg cells in men, the subpopulation preferentially depleted by anti-CTLA-4 antibodies, could play a role in the different responses, subsequently producing a more significant male benefit [207].

It is important to note that the results associated with ICI plus chemotherapy appear different, giving more benefits to women than to men. This might possibly depend on the chemotherapy-derived increase in the mutational load of tumors, their antigenicity and the stronger immune response of women [10]. The complexity is further augmented by disparities between the ICI agents, for example, the anti-CTLA4 antibodies being more sex-related than the anti-PD-1 ones. Indeed, sex-related differences seem more relevant to overall survival in melanoma patients than in other cancers [208,209].

Another intriguing result reported the association between obesity and outcomes in male metastatic melanoma patients treated with targeted or immunotherapies. In men, a high BMI was associated with a doubled survival time, whereas this result was not detected in women. One explanation might be that obesity in men results in higher levels of estrogens due to aromatase activity converting androgens into estrogens [210]. A role might be also played by the sex hormone-binding globulin (SHBG), a glycoprotein acting as a transporter of testosterone and estradiol in plasma. SHBG is inversely associated with obesity, displaying gender-associated differences. Indeed, women have higher levels of SHBG when compared to men, and SHBG amounts are further reduced in obese men [211].

To conclude, it is important to note that even in adverse events, some sex-related differences can be observed. Generally, women have more adverse reactions to chemotherapy than men, which might derive from differences in the pharmacokinetics and pharmacodynamics of the drugs [212]. Furthermore, worst might be the expected risks of severe symptomatic adverse effects experienced by women after immunotherapy due to their stronger immune responses [213,214]. Nonetheless, because of the short follow up of immunotherapy, some discrepancy still exist in regard to this point [215].

Altogether, these data strongly support the relevance of balanced enrolment of men and women in clinical trials in order to consistently confirm differences, evaluating disparities between populations, ages, and sites [216]. Sex and/or gender represent a variable to consider in order to obtain impartial approaches, selecting the best therapeutic option with the lowest adverse effect for each person.

8. Conclusions

The increment of data on the pathogenic mechanisms made melanoma finally attackable also in its metastatic phase by target therapy and, more recently, by immunotherapy. However, many shadows persist that need to be dissipated. Among them, the mechanisms underlying sex and/or gender differences represent a key point requiring further studies. Epidemiological data confirm female advantages in both incidence and mortality, and these differences are present in all the analyzed racial and ethnic groups. Although disparities can be partly ascribed to different gender-related behaviors, recent results indicate biological variations, including genetic and epigenetic aspects, as key players.

An important action might depend on X chromosome regulation. The random inactivation of one of the two X chromosomes in women, together with the possibility of some X-regions escaping this regulation, surely contribute to female benefits. In addition, the X chromosome is enriched with both coding and regulatory non coding genes, - such as miRNAs, playing immunological functions. The complex crosstalk among different cells of the tumor microenvironment and the influence of the hormone regulatory signals on these cells might affect immunotherapy response, with a profound impact on results. Therefore, evaluating gender variance and connecting these aspects will improve patient stratification and assist in tailoring the provided healthcare to the individual patient.

Funding: This work was in part supported by The Italian Association for Cancer Research (IG18815 to AC) and by The Italian Ministry of Health.

Conflicts of Interest: The authors declare no conflict of interest.

References

1. Italian Association of Medical Oncology (AIOM); Italian Association of Tumor Registries (Airtum). The Numbers of Cancer in Italy 2019. Available online: http://www.salute.gov.it/portale/news/p3_2_1_1_1.jsp?lingua=italiano&menu=notizie&p=dalministero&id=3897 (accessed on 4 July 2020).

2. Ferlay, J.; Colombet, M.; Soerjomataram, I.; Mathers, C.; Parkin, D.M.; Piñeros, M.; Znaor, A.; Bray, F. Estimating the global cancer incidence and mortality in 2018: GLOBOCAN sources and methods. *Int. J. Cancer* **2019**, *144*, 1941–1953. [CrossRef] [PubMed]

3. Holman, D.M.; Ding, H.; Guy, G.P.; Watson, M.; Hartman, A.M.; Perna, F.M. Prevalence of sun protection use and sunburn and association of demographic and behaviorial characteristics with sunburn among US adults. *JAMA Dermatol.* **2018**, *154*, 561–568. [CrossRef] [PubMed]

4. Courtenay, W.H. Behavioral Factors Associated with Disease, Injury, and Death among Men: Evidence and Implications for Prevention. *J. Men Stud.* **2000**, *9*, 81–142. [CrossRef]

5. Paddock, L.E.; Lu, S.E.; Bandera, E.V.; Rhoads, G.G.; Fine, J.; Paine, S.; Barnhill, R.; Berwick, M. Skin self-examination and long-term melanoma survival. *Melanoma Res.* **2016**, *26*, 401–408. [CrossRef]

6. Joosse, A.; Collette, S.; Suciu, S.; Nijsten, T.; Lejeune, F.; Kleeberg, U.R.; Coebergh, J.W.; Eggermont, A.M.; de Vries, E. Superior outcome of women with stage I/II cutaneous melanoma: Pooled analysis of four European organization for research and treatment of cancer phase III trials. *J. Clin. Oncol.* **2012**, *30*, 2240–2247. [CrossRef] [PubMed]

7. Joosse, A.; de Vries, E.; Eckel, R.; Nijsten, T.; Eggermont, A.M.; Hölze, L.D.; Coebergh, J.W.; Engel, J. Munich Melanoma Group. Gender Differences in Melanoma Survival: Female Patients Have a Decreased Risk of Metastasis. *J. Investig. Dermatol.* **2011**, *131*, 719–726. [CrossRef] [PubMed]

8. Joosse, A.; Collette, S.; Suciu, S.; Nijsten, T.; Patel, P.M.; Keilholz, U.; Eggermont, A.M.; Coebergh, J.W.; de Vries, E. Sex is an independent prognostic indicator for survival and relapse/progression free-survival in metastasized stage III to stage IV melanoma: A pooled analysis of five European organization for research and treatment on cancer randomized controlled trials. *J. Clin. Oncol.* **2013**, *31*, 2337–2346. [CrossRef]

9. Balaton, B.P.; Dixon-McDougal, T.; Peeters, S.B.; Brown, C.J. The eXceptional nature of the X chromosome. *Hum. Mol. Genet.* **2018**, *27*, R242–R249. [CrossRef]

10. Gupta, S.; Artomov, M.; Goggins, W.; Daly, M.; Tsao, H. Gender Disparity and Mutation Burden in Metastatic Melanoma. *J. Natl. Cancer Inst.* **2015**, *107*, djv221. [CrossRef]

11. Hernando, B.; Ibarrola-Villava, M.; Fernandez, L.P.; Peña-Chilet, M.; Llorca-Cardeñosa, M.; Oltra, S.S.; Alonso, S.; Boyano, M.D.; Martinez-Cadenas, C.; Ribas, G. Sex-specific genetic effects associated with pigmentation, sensitivity to sunlight, and melanoma in a population of Spanish origin. *Biol. Sex Differ.* **2016**, *7*, 17. [CrossRef]

12. Klein, S.L.; Flanagan, K.L. Sex differences in immune responses. *Nat. Rev. Immunol.* **2016**, *16*, 626–638. [CrossRef]

13. Slominski, A.; Zbytek, B.; Nikolakis, G.; Manna, P.R.; Skobowiat, C.; Zmijewski, M.; Li, W.; Janjetovic, Z.; Postlethwaite, A.; Zouboulis, C.C.; et al. Steroidogenesis in the skin: Implications for local immune functions. *J. Steroid Biochem. Mol. Biol.* **2013**, *137*, 107–123. [CrossRef] [PubMed]

14. Jia, M.; Dahlman-Wright, K.; Gustafsson, J.Å. Estrogen receptor alpha and beta in health and disease. *Best Pract. Res. Clin. Endocrinol. Metab.* **2015**, *29*, 557–568. [CrossRef]

15. Zhou, J.H.; Kim, K.B.; Myers, J.N.; Fox, P.S.; Ning, J.; Bassett, R.L.; Hasanein, H.; Prieto, V.G. Immunohistochemical expression of hormone receptors in melanoma of pregnant women, nonpregnant women, and men. *Am. J. Derm.* **2014**, *36*, 74–79. [CrossRef]

16. De Giorgi, V.; Gori, A.; Gandini, S.; Papi, F.; Grazzini, M.; Rossari, S.; Simoni, A.; Maio, V.; Massi, D. Oestrogen receptor beta and melanoma: A comparative study. *Br. J. Dermatol.* **2013**, *168*, 513–519. [CrossRef]

17. Ohata, C.; Tadokoro, T.; Itami, S. Expression of estrogen receptor beta in normal skin, melanocytic nevi and malignant melanomas. *J. Dermatol.* **2008**, *35*, 215–221. [CrossRef]

18. Schmidt, A.N.; Nanney, L.B.; Boyd, A.S.; King, L.E., Jr.; Ellis, D.L. Oestrogen receptor beta expression in melanocytic lesions. *Exp. Dermatol.* **2006**, *15*, 971–980. [CrossRef]

19. De Giorgi, V.; Mavilia, C.; Massi, D.; Gozzini, A.; Aragona, P.; Tanini, A.; Sestini, S.; Paglierani, M.; Boddi, V.; Brandi, M.L.; et al. Estrogen receptor expression in cutaneous melanoma: A real-time reverse transcriptase-polymerase chain reaction and immunohistochemical study. *Arch. Dermatol.* **2009**, *145*, 30–36. [CrossRef]

20. Mori, T.; Martinez, S.R.; O'Day, S.J.; Morton, D.L.; Umetani, N.; Kitago, M.; Tanemura, A.; Nguyen, S.L.; Tran, A.N.; Wang, H.J.; et al. Estrogen receptor-alpha methylation predicts melanoma progression. *Cancer Res.* **2006**, *66*, 6692–6698. [CrossRef]

21. Yu, Z.; Gao, W.; Jiang, E.; Lu, F.; Zhang, L.; Shi, Z.; Wang, X.; Chen, L.; Lv, T. Interaction between IGF-IR and ER induced by E2 and IGF-I. *PLoS ONE* **2013**, *8*, e62642. [CrossRef]

22. Yuan, T.-A.; Yourk, V.; Farhat, A.; Guo, K.L.; Garcia, A.; Meyskens, F.L.; Liu-Smith, F. A Possible Link of Genetic Variations in ER/IGF1R Pathway and Risk of Melanoma. *Int. J. Mol. Sci.* **2020**, *21*, 1776. [CrossRef]

23. Liu, M.; Du, Y.; Li, H.; Wang, L.; Ponikwicka-Tyszko, D.; Lebiedzinska, W.; Pilaszewicz-Puza, A.; Liu, H.; Zhou, L.; Fan, H.; et al. Cyanidin-3-o-Glucoside Pharmacologically Inhibits Tumorigenesis via Estrogen Receptor β in Melanoma Mice. *Front. Oncol.* **2019**, *9*, 1110. [CrossRef]

24. Marzagalli, M.; Casati, L.; Moretti, R.M.; Montagnani Marelli, M.; Limonta, P. Estrogen Receptor β Agonists Differentially Affect the Growth of Human Melanoma Cell Lines. *PLoS ONE* **2015**, *10*, e0134396. [CrossRef] [PubMed]

25. Ribeiro, M.P.C.; Santos, A.E.; Custódio, J.B.A. The activation of the G protein-coupled estrogen receptor (GPER) inhibits the proliferation of mouse melanoma K1735-M2 cells. *Chem. Biol. Interact.* **2017**, *277*, 176–184. [CrossRef] [PubMed]

26. Fábián, M.; Rencz, F.; Krenács, T.; Brodszky, V.; Hársing, J.; Németh, K.; Balogh, P.; Kárpáti, S. Expression of G protein-coupled oestrogen receptor in melanoma and in pregnancy-associated melanoma. *J. Eur. Acad. Derm. Venereol.* **2017**, *31*, 1453–1461. [CrossRef]

27. Contrò, V.; Basile, J.R.; Proia, P. Sex steroid hormone receptors, their ligands, and nuclear and non-nuclear pathways. *AIMS Mol.Sci.* **2015**, *2*, 294–310. [CrossRef]

28. Mitkov, M.; Joseph, R.; Copland, J., III. Steroid hormone influence on melanomagenesis. *Mol. Cell. Endocrinol.* **2015**, *417*, 94–102. [CrossRef]

29. Morvillo, V.; Lüthy, I.A.; Bravo, A.I.; Capurro, M.I.; Portel, P.; Calandra, R.S.; Mordoh, J. Androgen receptors in human melanoma cell lines IIB-MEL-LES and IIB-MEL-IAN and in human melanoma metastases. *Melanoma Res.* **2002**, *12*, 529–538. [CrossRef]

30. Still, R.; Brennecke, S. Melanoma in pregnancy. *Obstet. Med.* **2017**, *10*, 107–112. [CrossRef]

31. Donley, G.M.; Liu, W.T.; Pfeiffer, R.M.; McDonald, E.C.; Peters, K.O.; Tucker, M.A.; Cahoon, E.K. Reproductive factors, exogenous hormone use and incidence of melanoma among women in the United States. *Br. J. Cancer* **2019**, *120*, 754–760. [CrossRef] [PubMed]

32. Kemeny, M.M.; Busch, E.; Stewart, A.K.; Menck, H.R. Superior survival of young women with malignant melanoma. *Am. J. Surg.* **1998**, *175*, 437–444. [CrossRef]

33. Lasithiotakis, K.; Leiter, U.; Meier, F.; Eigentler, T.; Metzler, G.; Moehrle, M.; Breuninger, H.; Garbe, C. Age and gender are significant independent predictors of survival in primary cutaneous melanoma. *Cancer* **2008**, *112*, 1795–1804. [CrossRef] [PubMed]

34. Mervic, L.; Leiter, U.; Meier, F.; Eigentler, T.; Forschner, A.; Metzler, G.; Bartenje, I.; Büttner, P.; Garbe, C. Sex differences in survival of cutaneous melanoma are age dependent: An analysis of 7338 patients. *Melanoma Res.* **2011**, *21*, 244–252. [CrossRef]

35. Cervenka, I.; Mahamat-Saleh, Y.; Savoye, I.; Dartois, L.; Boutron-Ruault, M.C.; Fournier, A.; Kvaskoff, M. Oral contraceptive use and cutaneous melanoma risk: A French prospective cohort study. *Int. J. Cancer* **2018**, *143*, 2390–2399. [CrossRef] [PubMed]

36. Botteri, E.; Støer, N.C.; Sakshaug, S.; Graff-Iversen, S.; Vangen, S.; Hofvind, S.; Ursin, G.; Weiderpass, E. Menopausal hormone therapy and risk of melanoma: Do estrogens and progestins have a different role? *Int. J. Cancer* **2017**, *141*, 1763–1770. [CrossRef] [PubMed]

37. Simin, J.; Tamimi, R.; Lagergren, J.; Adami, H.O.; Brusselaers, N. Menopausal hormone therapy and cancer risk: An overestimated risk? *Eur. J. Cancer* **2017**, *84*, 60–68. [CrossRef]

38. Hicks, B.M.; Kristensen, K.B.; Pedersen, S.A.; Hölmich, L.R.; Pottegård, A. Hormone replacement therapy and the risk of melanoma in post-menopausal women. *Hum. Reprod.* **2019**, *34*, 2418–2429. [CrossRef]

39. Botteri, E.; Støer, N.C.; Weiderpass, E.; Pukkala, E.; Ylikorkala, O.; Lyytinen, H. Menopausal Hormone Therapy and Risk of Melanoma: A Nationwide Register-Based Study in Finland. *Cancer Epidemiol. Biomark. Prev.* **2019**, *28*, 1857–1860. [CrossRef]

40. Cervenka, I.; Rahmoun, M.A.; Mahamat-Saleh, Y.; Boutron-Ruault, M.C.; Fournier, A.; Kvaskoff, M. Premenopausal Use of Progestogens and Cutaneous Melanoma Risk: A French Prospective Cohort Study. *Am. J. Epidemiol.* **2019**. [CrossRef]

41. Roh, M.R.; Eliades, P.; Gupta, S.; Grant-Kels, J.M.; Tsao, H. Cutaneous melanoma in women. *Int. J. Womens Dermatol.* **2017**, *3*, S11–S15. [CrossRef]

42. Byrom, L.; Olsen, C.M.; Knight, L.; Khosrotehrani, K.; Green, A.C. Does pregnancy after a diagnosis of melanoma affect prognosis? Systematic review and meta-analysis. *Dermatol. Surg.* **2015**, *41*, 875–882. [CrossRef] [PubMed]

43. Kyrgidis, A.; Lallas, A.; Moscarella, E.; Longo, C.; Alfano, R.; Argenziano, G. Does pregnancy influence melanoma prognosis? A meta-analysis. *Melanoma Res.* **2017**, *27*, 289–299. [CrossRef] [PubMed]

44. Richardson, B.; Price, A.; Wagner, M.; Williams, V.; Lorigan, P.; Browne, S.; Miller, J.G.; Mac Neil, S. Investigation of female survival benefit in metastatic melanoma. *Br. J. Cancer* **1999**, *80*, 2025–2033. [CrossRef] [PubMed]

45. Kanda, N.; Watanabe, S. 17beta-estradiol, progesterone, and dihydrotestosterone suppress the growth of human melanoma by inhibiting interleukin-8 production. *J. Investig. Dermatol.* **2001**, *117*, 274–283. [CrossRef]

46. Sarti, M.S.; Visconti, M.A.; Castrucci, A.M. Biological activity and binding of estradiol to SK-Mel 23 human melanoma cells. *Braz. J. Med. Biol. Res.* **2004**, *37*, 901–905. [CrossRef]

47. Feucht, K.A.; Walker, M.J.; Das Gupta, T.K.; Beattie, C.W. Effect of 17 beta-estradiol on the growth of estrogen receptor positive human melanoma in vitro and in athymic mice. *Cancer Res.* **1988**, *48*, 7093–7101.

48. Ghosh, R.; Ott, A.M.; Seetharam, D.; Slaga, T.J.; Kumar, A.P. Cell cycle block and apoptosis induction in a human melanoma cell line following treatment with 2-methoxyostradiol: Therapeutic implications? *Melanoma Res.* **2003**, *13*, 119–127. [CrossRef]

49. Dobos, J.; Timar, J.; Bocsi, J.; Burian, Z.; Nagy, K.; Barna, G.; Petak, I.; Ladanyi, A. In vitro and in vivo antitumor effect of 2-Methoxyestradiol on human melanoma. *Int. J. Cancer* **2004**, *112*, 771–776. [CrossRef]

50. Dobos, J. Endocrine factors influencing melanoma progression. *MagyOnkol* **2009**, *53*, 47–50.

51. Ladányi, A.; Tímár, J.; Bocsi, J.; Tóvári, J.; Lapis, K. Sex-dependent liver metastasis of human melanoma lines in SCID mice. *Melanoma Res.* **1995**, *5*, 83–86. [CrossRef]

52. Vantyghem, S.A.; Postenka, C.O.; Chambers, A.F. Estrous cycle influences organ-specific metastasis of B16F10 melanoma cells. *Cancer Res.* **2003**, *63*, 4763–4765. [PubMed]

53. Massaro, R.R.; Faião-Flores, F.; Rebecca, V.W.; Sandri, S.; Alves-Fernandes, D.K.; Pennacchi, P.C.; Smalley, K.S.M.; Maria-Engler, S.S. Inhibition of proliferation and invasion in 2D and 3D models by 2-methoxyestradiol in human melanoma cells. *Pharmacol. Res.* **2017**, *119*, 242–250. [CrossRef] [PubMed]

54. Natale, C.A.; Li, J.; Zhang, J.; Dahal, A.; Dentchev, T.; Stanger, B.Z.; Ridky, T.W. Activation of G protein-coupled estrogen receptor signaling inhibits melanoma and improves response to immune checkpoint blockade. *eLife* **2018**, *7*, e31770. [CrossRef] [PubMed]

55. Zhao, L.; Huang, S.; Mei, S.; Yang, Z.; Xu, L.; Zhou, N.; Yang, Q.; Shen, Q.; Wang, W.; Le, X.; et al. Pharmacological activation of estrogen receptor beta augments innate immunity to suppress cancer metastasis. *Proc. Natl. Acad. Sci. USA* **2018**, *115*, E3673–E3681. [CrossRef] [PubMed]

56. Matsuoka, H.; Tsubaki, M.; Yamazoe, Y.; Ogaki, M.; Satou, T.; Itoh, T.; Kusunoki, T.; Nishida, S. Tamoxifen inhibits tumor cell invasion and metastasis in mouse melanoma through suppression of PKC/MEK/ERK and PKC/PI3K/Akt pathways. *Exp. Cell. Res.* **2009**, *315*, 2022–2032. [CrossRef]

57. Chen, P.; Sheikh, S.; Ahmad, A.; Ali, S.M.; Ahmad, M.U.; Ahmad, I. Orally administered endoxifen inhibits tumor growth in melanoma-bearing mice. *Cell. Mol. Biol. Lett.* **2018**, *23*, 3. [CrossRef]

58. Wang, Y.; Ou, Z.; Sun, Y.; Yeh, S.; Wang, X.; Long, J.; Chang, C. Androgen receptor promotes melanoma metastasis via altering the miRNA-539-3p/USP13/MITF/AXL signals. *Oncogene* **2017**, *36*, 1644–1654. [CrossRef]

59. Hsueh, E.C.; Gupta, R.K.; Lefor, A.; Reyzin, G.; Ye, W.; Morton, D.L. Androgen blockade enhances response to melanoma vaccine. *J. Surg. Res.* **2003**, *110*, 393–398. [CrossRef]

60. Markman, J.L.; Porritt, R.A.; Wakita, D.; Lane, M.E.; Martinon, D.; Noval Rivas, M.; Luu, M.; Posadas, E.M.; Crother, T.R.; Arditi, M. Loss of testosterone impairs anti-tumor neutrophil function. *Nat. Commun.* **2020**, *11*, 1613. [CrossRef]

61. Marzagalli, M.; Montagnani Marelli, M.; Casati, L.; Fontana, F.; Moretti, R.M.; Limonta, P. Estrogenreceptor β in melanoma: From molecularinsights to potentialclinical utility. *Front. Endocrinol. (Lausanne)* **2016**, *7*, 140. [CrossRef]

62. Shagufta; Ahmad, I. Tamoxifen a pioneering drug: An update on the therapeutic potential of tamoxifen derivatives. *Eur. J. Med. Chem.* **2018**, *143*, 515–531. [CrossRef] [PubMed]

63. Kanter-Lewensohn, L.; Girnita, L.; Girnita, A.; Dricu, A.; Olsson, G.; Leech, L.; Nilsson, G.; Hilding, A.; Wejde, J.; Brismar, K.; et al. Tamoxifen-induced cell death in malignant melanoma cells: Possible involvement of the insulin-like growth factor-1 (IGF-1) pathway. *Mol. Cell. Endocrinol.* **2000**, *165*, 131–137. [CrossRef]

64. Lens, M.B.; Reiman, T.; Husain, A.F. Use of tamoxifen in the treatment of malignant melanoma. *Cancer* **2003**, *98*, 1355–1361. [CrossRef] [PubMed]

65. Beguerie, J.R.; Xingzhong, J.; Valdez, R.P. Tamoxifen vs. non-tamoxifen treatment for advanced melanoma: A meta-analysis. *Int. J. Dermatol.* **2010**, *49*, 1194–1202. [CrossRef] [PubMed]

66. Ribeiro, M.P.; Santos, A.E.; Custódio, J.B. Rethinking tamoxifen in the management of melanoma: New answers for an old question. *Eur. J. Pharmacol.* **2015**, *764*, 372–378. [CrossRef]

67. Madeira, M.; Mattar, A.; Logullo, A.F.; Soares, F.A.; Gebrim, L.H. Estrogen receptor alpha/beta ratio and estrogen receptor beta as predictors of endocrine therapy responsiveness-a randomized neoadjuvant trial comparison between anastrozole and tamoxifen for the treatment of postmenopausal breast cancer. *BMC Cancer* **2013**, *13*, 425. [CrossRef]

68. Horner-Glister, E.; Maleki-Dizaji, M.; Guerin, C.; Johnson, S.; Styles, J.; White, I. Influence of oestradiol and tamoxifen on oestrogen receptors-α and -β protein degradation and non-genomic signalling pathways in uterine and breast carcinoma cells. *J. Mol. Endocrinol.* **2005**, *35*, 421–432. [CrossRef]

69. Sun, M.; Xie, H.F.; Tang, Y.; Lin, S.Q.; Li, J.M.; Sun, S.N.; Hu, X.L.; Huang, Y.X.; Shi, W.; Jian, D. G protein-coupled estrogen receptor enhances melanogenesis via cAMP-protein kinase (PKA) by upregulating microphthalmia-related transcription factor-tyrosinase in melanoma. *J. Steroid Biochem. Mol. Biol.* **2017**, *165*, 236–246. [CrossRef]

70. Natale, C.A.; Duperret, E.K.; Zhang, J.; Sadeghi, R.; Dahal, A.; O'Brien, K.T.; Cookson, R.; Winkler, J.D.; Ridky, T.W. Sex steroids regulate skin pigmentation through nonclassical membrane-bound receptors. *eLife* **2016**, *5*, e15104. [CrossRef]

71. Fang, X.; Zhang, X.; Zhou, M.; Li, J. Effects of Progesterone on the Growth Regulation in Classical Progesterone Receptor-negative Malignant Melanoma Cells. *J. Huazhong Univ. Sci. Technol. Med. Sci.* **2010**, *30*, 231–234. [CrossRef]

72. Ramaraj, P.; Cox, J.L. In vitro effect of progesterone on human melanoma (BLM) cell growth. *Int. J. Clin. Exp. Med.* **2014**, *7*, 3941–3953.

73. Leder, D.C.; Brown, J.R.; Ramaraj, P. In-vitro rescue and recovery studies of human melanoma (BLM) cell growth, adhesion and migration functions after treatment with progesterone. *Int. J. Clin. Exp. Med.* **2015**, *8*, 12275–12285.

74. Ramaraj, P.; Cox, J. In-Vitro Effect of Sex Steroids on Mouse Melanoma (B16F10) Cell Growth. *CellBio* **2014**, *3*, 60–71. [CrossRef]

75. Moroni, G.; Gaziano, R.; Buè, C.; Agostini, M.; Perno, C.F.; Sinibaldi-Vallebona, P.; Pica, F. Progesterone and Melanoma Cells: An Old Story Suspended between Life and Death. *J. Steroids Horm. Sci.* **2015**, *S13*, 158.

76. Morvillo, V.; Luthy, I.A.; Bravo, A.I.; Capurro, M.I.; Donaldson, M.; Quintans, C.; Calandra, R.S.; Mordoh, J. Atypical androgen receptor in the human melanoma cell line IIB-MEL-J. *Pigment Cell Res.* **1995**, *8*, 135–141. [CrossRef]

77. Schmidt, K.; Carroll, J.S.; Yee, E.; Thomas, D.D.; Wert-Lamas, L.; Neier, S.C.; Sheynkman, G.; Ritz, J.; Novina, C.D. The lncRNA SLNCR Recruits the Androgen Receptor to EGR1-Bound Genes in Melanoma and Inhibits Expression of Tumor Suppressor p21. *Cell Rep.* **2019**, *27*, 2493–2507. [CrossRef] [PubMed]

78. Hu, C.; Fang, D.; Xu, H.; Wang, Q.; Xia, H. The androgen receptor expression and association with patient's survival in different cancers. *Genomics* **2020**, *112*, 1926–1940. [CrossRef]

79. Clocchiatti, A.; Cora, E.; Zhang, Y.; Dotto, G.P. Sexual dimorphism in cancer. *Nat. Rev. Cancer* **2016**, *16*, 330–339. [CrossRef]

80. Conforti, F.; Pala, L.; Bagnardi, V.; De Pas, T.; Martinetti, M.; Viale, G.; Gelber, R.D.; Goldhirsch, A. Cancer immunotherapy efficacy and patients' sex: A systematic review and meta-analysis. *Lancet Oncol.* **2018**, *19*, 737–746. [CrossRef]

81. Wallis, C.J.D.; Butaney, M.; Satkunasivam, R.; Freedland, S.J.; Patel, S.P.; Hamid, O.; Pal, S.K.; Klaassen, Z. Association of PatientSex With Efficacy of Immune Checkpoint Inhibitors and Overall Survival in Advanced Cancers: A Systematic Review and Meta-analysis. *JAMA Oncol.* **2019**, *5*, 529–536. [CrossRef]

82. Khan, D.; Ansar Ahmed, S. The ImmuneSystem Is a Natural Target for Estrogen Action: Opposing Effects of Estrogen in Two Prototypical Autoimmune Diseases. *Front. Immunol.* **2016**, *6*, 635. [CrossRef] [PubMed]

83. Kovats, S. Estrogen receptors regulate innate immune cells and signaling pathways. *Cell Immunol.* **2015**, *294*, 63–69. [CrossRef]

84. Pierdominici, M.; Maselli, A.; Colasanti, T.; Giammarioli, A.M.; Delunardo, F.; Vacirca, D.; Sanchez, M.; Giovannetti, A.; Malorni, W.; Ortona, E. Estrogen receptor profiles in human peripheral blood lymphocytes. *Immunol. Lett.* **2010**, *132*, 79–85. [CrossRef] [PubMed]

85. Behrens, M.; Trejo, T.; Luthra, H.; Griffiths, M.; David, C.S.; Taneja, V. Mechanism by which HLA-DR4 regulates sex-bias of arthritis in humanized mice. *J. Autoimmun.* **2010**, *35*, 1–9. [CrossRef]

86. Kovats, S.; Carreras, E. Regulation of dendritic cell differentiation and function by estrogen receptor ligands. *Cell Immunol.* **2008**, *252*, 81–90. [CrossRef] [PubMed]

87. Capone, I.; Marchetti, P.; Ascierto, P.A.; Malorni, W.; Gabriele, L. Sexual Dimorphism of Immune Responses: A New Perspective in Cancer Immunotherapy. *Front. Immunol.* **2018**, *9*, 552. [CrossRef] [PubMed]

88. Barry, K.C.; Hsu, J.; Broz, M.L.; Cueto, F.J.; Binnewies, M.; Combes, A.J.; Nelson, A.E.; Loo, K.; Kumar, R.; Rosenblum, M.D.; et al. A naturalkiller-dendriticcell axis defines checkpoint therapy-responsive tumor microenvironments. *Nat. Med.* **2018**, *24*, 1178–1191. [CrossRef]

89. De Giorgi, V.; Gori, A.; Grazzini, M.; Rossari, S.; Scarfi, F.; Corciova, S.; Verdelli, A.; Lotti, T.; Massi, D. Estrogens, estrogen receptors and melanoma. *Expert Rev. Anticancer Ther.* **2011**, *11*, 739–747. [CrossRef]

90. Krahn-Bertil, E.; Dos Santos, M.; Damour, O.; Andre, V.; Bolzinger, M.A. Expression of estrogen-related receptor beta (ERRβ) in human skin. *Eur. J. Dermatol.* **2010**, *20*, 719–723.

91. Seillet, C.; Laffont, S.; Trémollières, F.; Rouquié, N.; Ribot, C.; Arnal, J.F.; Douin-Echinard, V.; Gourdy, P.; Guéry, J.C. The TLR-mediated response of plasmacytoid dendritic cells is positively regulated by estradiol in vivo through cell-intrinsic estrogen receptor α signaling. *Blood* **2012**, *119*, 454–464. [CrossRef]

92. Laffont, S.; Rouquié, N.; Azar, P.; Seillet, C.; Plumas, J.; Aspord, C.; Guéry, J.C. X-Chromosome complement and estrogen receptor signaling independently contribute to the enhanced TLR7-mediated IFN-α production of plasmacytoid dendritic cells from women. *J. Immunol.* **2014**, *193*, 5444–5452. [CrossRef] [PubMed]

93. Berghöfer, B.; Frommer, T.; Haley, G.; Fink, L.; Bein, G.; Hackstein, H. TLR7 Ligands Induce Higher IFN-alpha Production in Females. *J. Immunol.* **2006**, *177*, 2088–2096. [CrossRef] [PubMed]

94. Song, P.I.; Park, Y.M.; Abraham, T.; Harten, B.; Zivony, A.; Neparidze, N.; Armstrong, C.A.; Ansel, J.C. Human keratinocytes express functional CD14 and toll-like receptor 4. *J. Investig. Dermatol.* **2002**, *119*, 424–432. [CrossRef]

95. Jin, S.H.; Kang, H.Y. Activation of Toll-like Receptors 1, 2, 4, 5, and 7 on Human Melanocytes Modulate Pigmentation. *Ann. Dermatol.* **2010**, *22*, 486–489. [CrossRef] [PubMed]

96. Goto, Y.; Arigami, T.; Kitago, M.; Nguyen, S.L.; Narita, N.; Ferrone, S.; Morton, D.L.; Irie, R.F.; Hoon, D.S. Activation of Toll-like receptors 2, 3, and 4 on human melanoma cells induces inflammatory factors. *Mol. Cancer Ther.* **2008**, *7*, 3642–3653. [CrossRef]

97. Saint-Jean, M.; Knol, A.C.; Nguyen, J.M.; Khammari, A.; Dréno, B. TLR expression in human melanoma cells. *Eur. J. Dermatol.* **2011**, *21*, 899–905. [CrossRef]

98. Mittal, D.; Saccheri, F.; Vénéreau, E.; Pusterla, T.; Bianchi, M.E.; Rescigno, M. TLR4-mediated skin carcinogenesis is dependent on immune and radioresistant cells. *EMBO J.* **2010**, *29*, 2242–2252. [CrossRef]

99. Hemmi, H.; Kaisho, T.; Takeuchi, O.; Sato, S.; Sanjo, H.; Hoshino, K.; Horiuchi, T.; Tomizawa, H.; Takeda, K.; Akira, S. Small anti-viral compounds activate immune cells via the TLR7 MyD88-dependent signaling pathway. *Nat. Immunol.* **2002**, *3*, 196–200. [CrossRef]

100. Stockfleth, E.; Trefzer, U.; Garcia-Bartels, C.; Wegner, T.; Schmook, T.; Sterry, W. The use of Toll-like receptor-7 agonist in the treatment of basal cell carcinoma: An overview. *Br. J. Dermatol.* **2003**, *149*, 53–56. [CrossRef]

101. Schön, M.P.; Schön, M. Immune modulation and apoptosis induction: Two sides of the antitumoral activity of imiquimod. *Apoptosis* **2004**, *9*, 291–298. [CrossRef]

102. Stiff, A.; Trikha, P.; Wesolowski, R.; Kendra, K.; Hsu, V.; Uppati, S.; McMichael, E.; Duggan, M.; Campbell, A.; Keller, K.; et al. Myeloid-Derived Suppressor Cells Express Bruton's Tyrosine Kinase and Can Be Depleted in Tumor-Bearing Hosts by IbrutinibTreatment. *Cancer Res.* **2016**, *76*, 2125–2136. [CrossRef] [PubMed]

103. Srivastava, R.; Geng, D.; Liu, Y.; Zheng, L.; Li, Z.; Joseph, M.A.; McKenna, C.; Bansal, N.; Ochoa, A.; Davila, E. Augmentation of therapeutic responses in melanoma by inhibition of IRAK-1,-4. *Cancer Res.* **2012**, *72*, 6209–6216. [CrossRef]

104. Romano, S.; Xiao, Y.; Nakaya, M.; D'Angelillo, A.; Chang, M.; Jin, J.; Hausch, F.; Masullo, M.; Feng, X.; Romano, M.F.; et al. FKBP51 employs both scaffold and isomerase functions to promote NF-κB activation in melanoma. *Nucleic Acids Res.* **2015**, *43*, 6983–6993. [CrossRef]

105. Touil, Y.; Segard, P.; Ostyn, P.; Begard, S.; Aspord, C.; El Machhour, R.; Masselot, B.; Vandomme, J.; Flamenco, P.; Idziorek, T.; et al. Melanoma dormancy in a mouse model is linked to GILZ/FOXO3A-dependent quiescence of disseminated stem-like cells. *Sci. Rep.* **2016**, *6*, 30405. [CrossRef] [PubMed]

106. Aydin, E.; Johansson, J.; Nazir, F.H.; Hellstrand, K.; Martner, A. Role of NOX2-Derived Reactive Oxygen Species in NK Cell-Mediated Control of Murine Melanoma Metastasis. *Cancer Immunol. Res.* **2017**, *5*, 804–811. [CrossRef] [PubMed]

107. Sander, C.S.; Hamm, F.; Elsner, P.; Thiele, J.J. Oxidative stress in malignant melanoma and non-melanoma skin cancer. *Br. J. Dermatol.* **2003**, *148*, 913–922. [CrossRef]

108. Joosse, A.; De Vries, E.; van Eijck, C.H.; Eggermont, A.M.; Nijsten, T.; Coebergh, J.W. Reactive oxygen species and melanoma: An explanation for gender differences in survival? *Pigment Cell Melanoma Res.* **2010**, *23*, 352–364. [CrossRef]

109. Obrador, E.; Liu-Smith, F.; Dellinger, R.W.; Salvador, R.; Meyskens, F.L.; Estrela, J.M. Oxidative stress and antioxidants in the pathophysiology of malignant melanoma. *Biol. Chem.* **2019**, *400*, 589–612. [CrossRef]

110. Malorni, W.; Straface, E.; Matarrese, P.; Ascione, B.; Coinu, R.; Canu, S.; Galluzzo, P.; Marino, M.; Franconi, F. Redox state and gender differences in vascular smooth muscle cells. *FEBS Lett.* **2008**, *582*, 635–642. [CrossRef]

111. Miller, A.A.; De Silva, T.M.; Jackman, K.A.; Sobey, C.G. Effect of gender and sex hormones on vascular oxidativestress. *Clin. Exp. Pharmacol. Physiol.* **2007**, *34*, 1037–1043. [CrossRef]

112. Cheng, G.C.; Schulze, P.C.; Lee, R.T.; Sylvan, J.; Zetter, B.R.; Huang, H. Oxidative stress and thioredoxin-interacting protein promote intravasation of melanoma cells. *Exp. Cell Res.* **2004**, *300*, 297–307. [CrossRef] [PubMed]

113. Nishikawa, M. Reactive oxygen species in tumor metastasis. *Cancer Lett.* **2008**, *266*, 53–59. [CrossRef] [PubMed]

114. Offner, F.A.; Wirtz, H.C.; Schiefer, J.; Bigalke, I.; Klosterhalfen, G.; Bittinger, F.; Mittermayer, C.; Kirkpatrick, C.J. Interaction of Human Malignant Melanoma (ST-ML-12) Tumor Spheroids with Endothelial Cell Monolayers. Damage to Endothelium by Oxygen-Derived Free Radicals. *Am. J. Pathol.* **1992**, *141*, 601–610. [PubMed]

115. Mendez, R.; Aptsiauri, N.; Del Campo, A.; Maleno, I.; Cabrera, T.; Ruiz-Cabello, F.; Garrido, F.; Garcia-Lora, A. HLA and melanoma: Multiple alterations in HLA class I and II expression in human melanoma cell lines from ESTDAB cell bank. *Cancer Immunol. Immunother.* **2009**, *58*, 1507–1515. [CrossRef]

116. Garrido, F.; Algarra, I.; García-Lora, A.M. The escape of cancer from T lymphocytes: Immunoselection of MHC class I loss variants harboring structural-irreversible "hard" lesions. *Cancer Immunol. Immunother.* **2010**, *59*, 1601–1606. [CrossRef]

117. Hölsken, O.; Miller, M.; Cerwenka, A. Exploiting natural killer cells for therapy of melanoma. *J. Dtsch. Dermatol. Ges.* **2015**, *13*, 23–29. [CrossRef]

118. Castelli, C.; Rivoltini, L.; Andreola, G.; Carrabba, M.; Renkvist, N.; Parmiani, G. T-cell recognition of melanoma-associated antigens. *J. Cell. Physiol.* **2000**, *182*, 323–331. [CrossRef]

119. Epping, M.T.; Bernards, R. A causal role for the human tumor antigen preferentially expressed antigen of melanoma in cancer. *Cancer Res.* **2006**, *66*, 10639–10642. [CrossRef]

120. Lupetti, R.; Pisarra, P.; Verrecchia, A.; Farina, C.; Nicolini, G.; Anichini, A.; Bordignon, C.; Sensi, M.; Parmiani, G.; Traversari, C. Translation of a retained intron in tyrosinase-related protein (TRP) 2 mRNA generates a new cytotoxic T lymphocyte (CTL)-defined and shared human melanoma antigen not expressed in normal cells of the melanocytic lineage. *J. Exp. Med.* **1998**, *188*, 1005–1016. [CrossRef]

121. Scanlan, M.J.; Gure, A.O.; Jungbluth, A.A.; Old, L.J.; Chen, Y.T. Cancer/testis antigens: An expanding family of targets for cancer immunotherapy. *Immunol. Rev.* **2002**, *188*, 22–32. [CrossRef]

122. Passarelli, A.; Mannavola, F.; Stucci, L.S.; Tucci, M.; Silvestris, F. Immune system and melanoma biology: A balance between immunosurveillance and immune escape. *Oncotarget* **2017**, *8*, 106132–106142. [CrossRef]

123. Spitzer, J.A. Gender differences in some host defense mechanisms. *Lupus* **1999**, *8*, 380–383. [CrossRef]

124. Weinstein, Y.; Ran, S.; Segal, S. Sex-associated Differences in the Regulation of Immune Responses Controlled by the MHC of the Mouse. *J. Immunol.* **1984**, *132*, 656–661.

125. Orstavik, K.H. X chromosome inactivation in clinical practice. *Hum. Genet.* **2009**, *126*, 363–373. [CrossRef]

126. Migeon, B.R. The role of X inactivation and cellular mosaicism in women's health and sex-specific diseases. *JAMA* **2006**, *295*, 1428–1433. [CrossRef]

127. Libert, C.; Dejager, L.; Pinheiro, I. The X chromosome in immune functions: When a chromosome makes the difference. *Nat. Rev. Immunol.* **2010**, *10*, 594–604. [CrossRef] [PubMed]

128. Tedeschi, S.K.; Bermas, B.; Costenbader, K.H. Sexual disparities in the incidence and course of SLE and RA. *Clin. Immunol.* **2013**, *149*, 211–218. [CrossRef] [PubMed]

129. Ortona, E.; Pierdominici, M.; Maselli, A.; Veroni, C.; Aloisi, F.; Shoenfeld, Y. Sex-based differences in autoimmune diseases. *Ann. Ist. Super. Sanita* **2016**, *52*, 205–212. [PubMed]

130. Abdullah, M.; Chai, P.S.; Chong, M.Y.; Tohit, E.R.; Ramasamy, R.; Pei, C.P.; Vidyadaran, S. Gender effect on in vitro lymphocyte subset levels of healthy individuals. *Cell Immunol.* **2012**, *272*, 214–219. [CrossRef] [PubMed]

131. Uppal, S.S.; Verma, S.; Dhot, P.S. Normal values of CD4 and CD8 lymphocyte subsets in healthy indian adults and the effects of sex, age, ethnicity, and smoking. *Cytom. B Clin. Cytom.* **2003**, *52*, 32–36. [CrossRef]

132. Afshan, G.; Afzal, N.; Qureshi, S. CD4+CD25(hi) Regulatory T Cells in Healthy Males and Females Mediate Gender Difference in the Prevalence of Autoimmune Diseases. *Clin. Lab.* **2012**, *58*, 567–571.

133. Wesa, A.K.; Mandic, M.; Taylor, J.L.; Moschos, S.; Kirkwood, J.M.; Kwok, W.W.; Finke, J.H.; Storkus, W.J. Circulating Type-1 Anti-Tumor CD4(+) T Cells are Preferentially Pro-Apoptotic in Cancer Patients. *Front. Oncol.* **2014**, *4*, 266. [CrossRef]

134. Knutson, K.L.; Disis, M.L. Tumor antigen-specific Thelper cells in cancer immunity and immunotherapy. *Cancer Immunol. Immunother.* **2005**, *54*, 721–728. [CrossRef]

135. Dong, H.; Strome, S.E.; Salomao, D.R.; Tamura, H.; Hirano, F.; Flies, D.B.; Roche, P.C.; Lu, J.; Zhu, G.; Tamada, K.; et al. Tumor-associated B7-H1 promotes T-cell apoptosis: A potential mechanism of immune evasion. *Nat. Med.* **2002**, *8*, 793–800. [CrossRef]

136. Wang, C.; Dehghani, B.; Li, Y.; Kaler, L.J.; Vandenbark, A.A.; Offner, H. Oestrogen modulates experimental autoimmune encephalomyelitis and interleukin-17 production via programmed death 1. *Immunology* **2009**, *126*, 329–335. [CrossRef]

137. Curiel, T.J.; Wei, S.; Dong, H.; Alvarez, X.; Cheng, P.; Mottram, P.; Krzysiek, R.; Knutson, K.L.; Daniel, B.; Zimmermann, M.C.; et al. Blockade of B7-H1 improves myeloid dendritic cell-mediated antitumor immunity. *Nat. Med.* **2003**, *9*, 562–567. [CrossRef]

138. Dong, H.; Chen, L. B7-H1 pathway and its role in the evasion of tumor immunity. *J. Mol. Med. (Berl.)* **2003**, *81*, 281–287. [CrossRef]

139. Wang, Y.; Zhuang, Q.; Zhou, S.; Hu, Z.; Lan, R. Costimulatory molecule B7-H1 on the immune escape of bladder cancer and its clinical significance. *J. Huazhong Univ. Sci. Technol. Med. Sci.* **2009**, *29*, 77–79. [CrossRef]

140. Lin, P.Y.; Sun, L.; Thibodeaux, S.R.; Ludwig, S.M.; Vadlamudi, R.K.; Hurez, V.J.; Bahar, R.; Kious, M.J.; Livi, C.B.; Wall, S.R.; et al. B7-H1-dependent sex-related differences in tumor immunity and immunotherapy responses. *J. Immunol.* **2010**, *185*, 2747–2753. [CrossRef]

141. Zeng, Y. Principles of micro-RNA production and maturation. *Oncogene* **2006**, *25*, 6156–6162. [CrossRef]

142. Bartel, D.P. MicroRNAs: Target recognition and regulatory functions. *Cell* **2009**, *136*, 215–233. [CrossRef] [PubMed]

143. Kim, V.N.; Han, J.; Siomi, M.C. Biogenesis of small RNAs in animals. *Nat. Rev. Mol. Cell Biol.* **2009**, *10*, 126–139. [CrossRef]

144. Jansson, M.D.; Lund, A.H. MicroRNA and cancer. *Mol. Oncol.* **2012**, *6*, 590–610. [CrossRef]

145. Calin, G.A.; Sevignani, C.; Dumitru, C.D.; Hyslop, T.; Noch, E.; Yendamuri, S.; Shimizu, M.; Rattan, S.; Bullrich, F.; Negrini, M.; et al. Human microRNA genes are frequently located at fragile sites and genomic regions involved in cancers. *Proc. Natl. Acad. Sci. USA* **2004**, *101*, 2999–3004. [CrossRef] [PubMed]

146. Calin, G.A.; Croce, C.M. MicroRNA signatures in human cancers. *Nat. Rev. Cancer* **2006**, *6*, 857–866. [CrossRef] [PubMed]

147. Fattore, L.; Costantini, S.; Malpicci, D.; Ruggiero, C.F.; Ascierto, P.A.; Croce, C.M.; Mancini, R.; Ciliberto, G. MicroRNAs in melanoma development and resistance to target therapy. *Oncotarget* **2017**, *8*, 22262–22278. [CrossRef]

148. Collins, L.C.; Marotti, J.D.; Gelber, S.; Cole, K.; Ruddy, K.; Kereakoglow, S.; Brachtel, E.F.; Schapira, L.; Come, S.E.; Winer, E.P.; et al. Pathologic features and molecular phenotype by patient age in a large cohort of young women with breast cancer. *Breast Cancer Res. Treat.* **2012**, *131*, 1061–1066. [CrossRef]

149. Maillot, G.; Lacroix-Triki, M.; Pierredon, S.; Gratadou, L.; Schmidt, S.; Benes, V.; Roche, H.; Dalenc, F.; Auboeuf, D.; Millevoi, S.; et al. Widespread estrogen-dependent repression of microRNAs involved in breast tumor cell growth. *Cancer Res.* **2009**, *69*, 8332–8340. [CrossRef]

150. Iorio, M.V.; Ferracin, M.; Liu, C.G.; Veronese, A.; Spizzo, R.; Sabbioni, S.; Magri, E.; Pedriali, M.; Fabbri, M.; Campiglio, M.; et al. MicroRNA gene expression deregulation in human breast cancer. *Cancer Res.* **2005**, *65*, 7065–7070. [CrossRef]

151. Bottner, M.; Thelen, P.; Jarry, H. Estrogen receptor beta: Tissue distribution and the still largely enigmatic physiological function. *J. Steroid Biochem. Mol. Biol.* **2014**, *139*, 245–251. [CrossRef]

152. Di Leva, G.; Gasparini, P.; Piovan, C.; Ngankeu, A.; Garofalo, M.; Taccioli, C.; Iorio, M.V.; Li, M.; Volinia, S.; Alder, H. MicroRNA Cluster 221-222 and Estrogen Receptor α Interactions in Breast Cancer. *J. Natl. Cancer Inst.* **2010**, *102*, 706–721. [CrossRef] [PubMed]

153. Liu, B.; Che, Q.; Qiu, H.; Bao, W.; Chen, X.; Lu, W.; Li, B.; Wan, X. Elevated MiR-222-3p Promotes Proliferation and Invasion of Endometrial Carcinoma via Targeting ERα. *PLoS ONE* **2014**, *9*, e87563. [CrossRef] [PubMed]

154. Di Leva, G.; Cheung, D.G.; Croce, C.M. miRNA clusters as therapeutic targets for hormone-resistant breast cancer. *Expert Rev. Endocrinol. Metab.* **2015**, *10*, 607–617. [CrossRef]

155. Zhao, J.J.; Lin, J.; Yang, H.; Kong, W.; He, L.; Ma, X.; Coppola, D.; Cheng, J.Q. MicroRNA-221/222 negatively regulates estrogen receptor alpha and is associated with tamoxifen resistance in breast cancer. *J. Biol. Chem.* **2008**, *283*, 31079–31086. [CrossRef]

156. Felicetti, F.; Errico, M.C.; Bottero, L.; Segnalini, P.; Stoppacciaro, A.; Biffoni, M.; Felli, N.; Mattia, G.; Petrini, M.; Colombo, M.P.; et al. The promyelocyticleukemiazinc finger-microRNA-221/-222 pathwaycontrols melanoma progressionthrough multiple oncogenicmechanisms. *Cancer Res.* **2008**, *68*, 2745–2754. [CrossRef] [PubMed]

157. Mattia, G.; Errico, M.C.; Felicetti, F.; Petrini, M.; Bottero, L.; Tomasello, L.; Romania, P.; Boe, A.; Segnalini, P.; Di Virgilio, A.; et al. Constitutive activation of the ETS-1-miR-222 circuitry in metastatic melanoma. *Pigment Cell Melanoma Res.* **2011**, *24*, 953–965. [CrossRef] [PubMed]

158. Errico, M.C.; Felicetti, F.; Bottero, L.; Mattia, G.; Boe, A.; Felli, N.; Petrini, M.; Bellenghi, M.; Pandha, H.S.; Calvaruso, M.; et al. The abrogation of the HOXB7/PBX2 complex induces apoptosis in melanoma through the miR-221&222-c-FOS pathway. *Int. J. Cancer* **2013**, *133*, 879–892.

159. Puglisi, R.; Bellenghi, M.; Pontecorvi, G.; Gulino, A.; Petrini, M.; Felicetti, F.; Bottero, L.; Mattia, G.; Carè, A. SCD5 restored expression favors differentiation and epithelial-mesenchymal reversion in advanced melanoma. *Oncotarget* **2018**, *9*, 7567–7581. [CrossRef]

160. Felli, N.; Errico, M.C.; Pedini, F.; Petrini, M.; Puglisi, R.; Bellenghi, M.; Boe, A.; Felicetti, F.; Mattia, G.; De Feo, A.; et al. AP2α controls the dynamic balance between miR-126&126* and miR-221&222 during melanoma progression. *Oncogene* **2016**, *35*, 3016–3026.

161. Ohno, S.; Labhart, A.; Mann, T.; Samuels, L.T.; Zander, J. (Eds.) Sex Chromosomes and Sex-Linked Genes. In *Monographs in Endocrinology*; Springer: Berlin, Germany, 1967.

162. Matarrese, P.; Tieri, P.; Anticoli, S.; Ascione, B.; Conte, M.; Franceschi, C.; Malorni, W.; Salvioli, S.; Ruggieri, A. X-chromosome-linked miR548am-5p is a key regulator of sex disparity in the susceptibility to mitochondria-mediated apoptosis. *Cell Death Dis.* **2019**, *10*, 673–685. [CrossRef]

163. Pearce, M.S.; Parker, L. Childhood cancer registrations in the developing world: Still more boys than girls. *Int. J. Cancer* **2001**, *91*, 402–406. [CrossRef]

164. Cartwright, R.A.; Gurney, K.A.; Moorman, A.V. Sex ratios and the risks of haematological malignancies. *Brit. J. Haematol.* **2002**, *118*, 1071–1077. [CrossRef] [PubMed]

165. Cook, M.B.; Dawsey, S.M.; Freedman, N.D.; Inskip, P.D.; Wichner, S.M.; Quraishi, S.M.; Devesa, S.S.; McGlynn, K.A. Sex disparities in cancer incidence by period and age. *Cancer Epidem. Biomar.* **2009**, *18*, 1174–1182. [CrossRef] [PubMed]

166. Lindsay, M.A. microRNAs and the immune response. *Trends Immunol.* **2008**, *29*, 343–351. [CrossRef] [PubMed]

167. Zhao, J.L.; Rao, D.S.; Boldin, M.P.; Taganov, K.D.; O'Connell, R.M.; Baltimore, D. NF-kappa B dysregulation in microRNA-146a-deficient mice drives the development of myeloid malignancies. *Proc. Natl. Acad. Sci. USA* **2011**, *108*, 9184–9189. [CrossRef]

168. Song, Q.; An, Q.; Niu, B.; Lu, X.; Zhang, N.; Cao, X. Role of miR-221/222 in Tumor Development and the Underlying Mechanism. *J. Oncol.* **2019**, *2019*, e7252013. [CrossRef]

169. Urbich, C.; Kuehbacher, A.; Dimmeler, S. Role of microRNAs in vascular diseases, inflammation, and angiogenesis. *Cardiovasc. Res.* **2008**, *79*, 581–588. [CrossRef] [PubMed]

170. Lee, J.Y.; Kim, M.; Heo, H.R.; Ha, K.S.; Han, E.T.; Park, W.S.; Yang, S.R.; Hong, S.H. Inhibition of MicroRNA-221 and 222 Enhances Hematopoietic Differentiation from Human Pluripotent Stem Cells via c-KIT Upregulation. *Mol. Cells* **2018**, *41*, 971–978.

171. Pontecorvi, G.; Bellenghi, M.; Puglisi, R.; Carè, A.; Mattia, G. Tumor-derived extracellular vesicles and microRNAs: Functional roles, diagnostic, prognostic and therapeutic options. *Cytokine Growth Factor Rev.* **2020**, *51*, 75–83. [CrossRef]

172. Felicetti, F.; De Feo, A.; Coscia, C.; Puglisi, R.; Pedini, F.; Pasquini, L.; Bellenghi, M.; Errico, M.C.; Pagani, E.; Carè, A. Exosome-mediated transfer of miR-222 is sufficient to increase tumor malignancy in melanoma. *J. Transl. Med.* **2016**, *24*, 56. [CrossRef]

173. Korabecna, M.; Koutova, L.; Tesarova, P. The potential roles of vesicle-enclosed miRNAs in communication between macrophages and cancer cells in tumor microenvironment. *Neoplasma* **2017**, *64*, 406–411. [CrossRef] [PubMed]

174. Gruszka, R.; Zakrzewska, M. The Oncogenic Relevance of miR-17-92 Cluster and Its Paralogous miR-106b-25 and miR-106a-363 Clusters in Brain Tumors. *Int. J. Mol. Sci.* **2018**, *19*, 879. [CrossRef] [PubMed]

175. Audrito, V.; Serra, S.; Stingi, A.; Orso, F.; Gaudino, F.; Bologna, C.; Neri, F.; Garaffo, G.; Nassini, R.; Baroni, G.; et al. PD-L1 up-regulation in melanoma increases disease aggressiveness and is mediated through miR-17-5p. *Oncotarget* **2017**, *8*, 15894–15911. [CrossRef]

176. Landais, S.; Landry, S.; Legault, P.; Rassart, E. Oncogenic potential of the miR-106-363 cluster and its implication in human T-cell leukemia. *Cancer Res.* **2007**, *67*, 5699–5707. [CrossRef]

177. Hao, T.; Li, C.X.; Ding, X.Y.; Xing, X.J. MicroRNA-363-3p/p21 (Cip1/Waf1) Axis Is Regulated by HIF-2α in Mediating Stemness of Melanoma Cells. *Neoplasma* **2019**, *23*, 66–427. [CrossRef] [PubMed]

178. Johnnidis, J.B.; Harris, M.H.; Wheeler, R.T.; Stehling, S.S.; Lam, M.H.; Kirak, O.; Brummelkamp, T.R.; Fleming, M.D.; Camargo, F.D. Regulation of progenitor cell proliferation and granulocyte function by microRNA-223. *Nature* **2008**, *451*, 1125–1129. [CrossRef] [PubMed]

179. Moschos, S.A.; Williams, A.E.; Perry, M.M.; Birrell, M.A.; Belvisi, M.G.; Lindsay, M.A. Expression profiling in vivo demonstrates rapid changes in lung microRNA levels following lipopolysaccharide induced inflammation but not in the anti-inflammatory action of glucocorticoids. *BMC Genom.* **2007**, *8*, 240. [CrossRef] [PubMed]

180. Mathsyaraja, H.; Thies, K.; Taffany, D.A.; Deighan, C.; Liu, T.; Yu, L.; Fernandez, S.A.; Shapiro, C.; Otero, J.; Timmers, C. CSF1-ETS2 Induced microRNA in Myeloid Cells Promote Metastatic Tumor Growth. *Oncogene* **2015**, *34*, 3651–3661. [CrossRef]

181. Haneklaus, M.; Gerlic, M.; O'Neill, L.A.; Masters, S.L. miR-223: Infection, inflammation and cancer. *J. Intern. Med.* **2013**, *274*, 215–226. [CrossRef]

182. Kitago, M.; Martinez, S.R.; Nakamura, T.; Sim, M.S.; Hoon, D.S. Regulation of RUNX3 Tumor Suppressor Gene Expression in Cutaneous Melanoma. *Clin. Cancer Res.* **2009**, *15*, 2988–2994. [CrossRef]

183. Chen, M.L.; Sun, A.; Cao, W.; Eliason, A.; Mendez, K.M.; Getzler, A.J.; Tsuda, S.; Diao, H.; Mukori, C.; Bruno, N.E. Physiological expression and function of the MDR1 transporter in cytotoxic T lymphocytes. *J. Exp. Med.* **2020**, *217*, e201991388. [CrossRef] [PubMed]

184. Li, X.J.; Li, Z.F.; Xu, Y.Y.; Han, Z.; Liu, Z.J. microRNA-374 inhibits proliferation and promotes apoptosis of mouse melanoma cells by inactivating the Wnt signalling pathway through its effect on tyrosinase. *J. Cell Mol. Med.* **2019**, *238*, 4991–5005. [CrossRef] [PubMed]

185. Rajasekhar, M.; Schmitz, U.; Flamant, S.; Wong, J.J.-L.; Bailey, C.G.; Ritchie, W.; Hols, J.; Rasko, J.E.J. Identifying microRNA determinants of human myelopoiesis. *Sci. Rep.* **2018**, *8*, 7264. [CrossRef] [PubMed]

186. Qian, D.; Chen, K.; Deng, H.; Rao, H.; Huang, H.; Liao, Y.; Sun, X.; Lu, S.; Yuan, Z.; Xie, D. MicroRNA-374b suppresses proliferation and promotes apoptosis in T-cell lymphoblastic lymphoma by repressing AKT1 and Wnt-16. *Clin. Cancer Res.* **2015**, *21*, 4881–4891. [CrossRef] [PubMed]

187. Yang, Y.; Feng, R.; Wang, Y.Z.; Sun, H.W.; Zou, Q.M.; Li, H.B. Toll-like receptors: Triggers of regulated cell death and promising targets for cancer therapy. *Immunol. Lett.* **2020**, *17*. [CrossRef]

188. Li, F.; Li, X.J.; Qiao, L.; Shi, F. miR-98 suppresses melanoma metastasis through a negative feedback loop with its target gene IL-6. *Exp. Mol. Med.* **2014**, *46*, e116. [CrossRef]

189. Li, L.; Sun, P.; Zhang, C.; Li, Z.; Zhou, W. MiR-98 suppresses the effects of tumor-associated macrophages on promoting migration and invasion of hepatocellular carcinoma cells by regulating IL-10. *Biochimie* **2018**, *150*, 23–30. [CrossRef]

190. Seeley, J.J.; Baker, R.G.; Mohamed, G.; Bruns, T.; Hayden, M.S.; Deshmukh, S.D.; Freedberg, D.E.; Ghosh, S. Induction of innate immune memory via microRNA targeting of chromatin remodelling factors. *Nature* **2018**, *559*, 114–119. [CrossRef]

191. Clark, W.H.; From, L.; Bernardino, E.A.; Mihm, M.C. The histogenesis and biological behavior of primary human malignant melanomas of the skin. *Cancer Res.* **1969**, *29*, 705–727.

192. Hieken, T.J.; Glasgow, A.E.; Enninga, E.A.L.; Kottschade, L.A.; Dronca, R.S.; Markovic, S.N.; Block, M.S.; Habermann, E.B. Sex-Based Differences in Melanoma Survival in a Contemporary Patient Cohort. *J. Womens Health (Larchmt)* **2020**. [CrossRef]

193. Yakerson, A. Women in clinical trials: A review of policy development and health equity in the Canadian context. *Int. J. Equity Health* **2019**, *18*, 56. [CrossRef] [PubMed]

194. Olsen, C.M.; Pandeya, N.; Thompson, B.S.; Dusingize, J.C.; Webb, P.M.; Green, A.C.; Neale, R.E.; Whiteman, D.C. QSkin Study. Risk Stratification for Melanoma: Models Derived and Validated in a Purpose-Designed Prospective Cohort. *J. Natl. Cancer Inst.* **2018**, *110*, 1075–1083. [CrossRef] [PubMed]

195. Avery, E.; Clark, J. Sex-related reporting in randomised controlled trials in medical journals. *Lance* **2016**, *388*, 2839–2840. [CrossRef]

196. Heidari, S.; Babor, T.F.; De Castro, P.; Tort, S.; Curno, M. Sex and Gender Equity in Research: Rationale for the SAGER guidelines and recommended use. *Res. Integr. Peer. Rev.* **2016**, *1*, 2. [CrossRef]

197. Italian Official Gazette GU Series General n.25. Delegation to the Government Concerning Clinical Trials of Medicinal Products and Provisions for the Reorganization of the Health Professions and for the Healthcare Officials of the Ministry of Health. 2018. Available online: https://www.gazzettaufficiale.it/eli/gu/2018/01/31/25/sg/pdf (accessed on 4 July 2020).

198. Unger, J.M.; Vaidya, R.; Albain, K.S.; LeBlanc, M.L.; Minasian, L.M.; Gotay, C.; Henry, N.L.; Fisch, M.J.; Ramsey, S.D.; Blanke, C.D.; et al. Sex differences in adverse event reporting in SWOG chemotherapy, biologic/immunotherapy, and targeted agent cancer clinical trials. *J. Clin. Oncol.* **2019**, *37*, 11588. [CrossRef]

199. Kim, S.Y.; Kim, S.N.; Hahn, H.J.; Lee, Y.W.; Choe, Y.B.; Ahn, K.J. Meta-analysis of BRAF mutations and clinicopathologic characteristics in primary melanoma. *J. Am. Acad. Dermatol.* **2015**, *72*, 1036–1046. [CrossRef]

200. Schadendorf, D.; van Akkooi, A.C.J.; Berking, C.; Griewank, K.G.; Gutzmer, R.; Hauschild, A.; Stang, A.; Roesch, A.; Ugurel, S. Melanoma. *Lancet* **2018**, *392*, 971–984. [CrossRef]

201. Ascierto, P.A.; Ferrucci, P.F.; Fisher, R.; Del Vecchio, M.; Atkinson, V.; Schmidt, H.; Schachter, J.; Queirolo, P.; Long, G.V.; Di Giacomo, A.M.; et al. Dabrafenib, trametinib and pembrolizumab or placebo in BRAF-mutant melanoma. *Nat. Med.* **2019**, *25*, 941–946. [CrossRef]

202. Broman, K.K.; Dossett, L.A.; Sun, J.; Eroglu, Z.; Zager, J.S. Update on BRAF and MEK inhibition for treatment of melanoma in metastatic, unresectable, and adjuvant settings. *Expert Opin. Drug. Saf.* **2019**, *18*, 381–392. [CrossRef]

203. Khalil, D.N.; Smith, E.L.; Brentjens, R.J.; Wolchok, J.D. The future of cancer treatment: Immunomodulation, CARs and combination immunotherapy. *Nat. Rev. Clin. Oncol.* **2016**, *13*, 273–290. [CrossRef]

204. Robert, C.; Grob, J.J.; Stroyakovskiy, D.; Karaszewska, B.; Hauschild, A.; Levchenko, E.; ChiarionSileni, V.; Schachter, J.; Garbe, C.; Bondarenko, I.; et al. Five-Year Outcomes with Dabrafenib plus Trametinib in Metastatic Melanoma. *N. Engl. J. Med.* **2019**, *381*, 626–636. [CrossRef] [PubMed]

205. Oertelt-Prigione, S. The influence of women sex and gender on the immune response. *Autoimmun. Rev.* **2012**, *11*, A479–A485. [CrossRef] [PubMed]

206. Botticelli, A.; Onesti, C.E.; Zizzari, I.; Cerbelli, B.; Sciattella, P.; Occhipinti, M.; Roberto, M.; Di Pietro, F.; Bonifacino, A.; Ghidini, M.; et al. The sexist behaviour of immune checkpoint inhibitors in cancer therapy? *Oncotarget* **2017**, *8*, 99336–99346. [CrossRef] [PubMed]

207. Walker, L.S. Treg and CTLA-4: Two intertwining pathways to immune tolerance. *J. Autoimmun.* **2013**, *45*, 49–57. [CrossRef] [PubMed]

208. Wu, Y.; Ju, Q.; Jia, K.; Yu, J.; Shi, H.; Wu, H.; Jiang, M. Correlation between sex and efficacy of immune checkpoint inhibitors (PD-1 and CTLA-4 inhibitors). *Int. J. Cancer* **2018**, *143*, 45–51. [CrossRef]

209. Grassadonia, A.; Sperduti, I.; Vici, P.; Iezzi, L.; Brocco, D.; Gamucci, T.; Pizzuti, L.; Maugeri-Saccà, M.; Marchetti, P.; Cognetti, G.; et al. Effect of Gender on the Outcome of Patients Receiving Immune Checkpoint Inhibitors for Advanced Cancer: A Systematic Review and Meta-Analysis of Phase III Randomized Clinical Trials. *J. Clin. Med.* **2018**, *7*, 542. [CrossRef]

210. McQuade, J.L.; Daniel, C.R.; Hess, K.R.; Mak, C.; Wang, D.Y.; Rai, R.R.; Park, J.J.; Haydu, L.E.; Spencer, C.; Wongchenko, M.; et al. Association of body-mass index and outcomes in patients with metastatic melanoma treated with targeted therapy, immunotherapy, or chemotherapy: A retrospective, multicohort analysis. *Lancet Oncol.* **2018**, *19*, 310–322. [CrossRef]

211. Grasa, M.D.M.; Gulfo, J.; Camps, N.; Alcalà, R.; Monserrat, L.; Moreno-Navarrete, J.M.; Ortega, F.J.; Esteve, M.; Remesar, X.; Fernàndez-Lòpez, J.A.; et al. Modulation of SHBG Binding to Testosterone and Estradiol by Sex and Morbid Obesity. *Eur. J. Endocrinol.* **2017**, *176*, 393–404. [CrossRef]

212. Wang, J.; Huang, Y. Pharmacogenomics of sex difference in chemotherapeutic toxicity. *Curr. Drug. Discov. Technol.* **2007**, *4*, 59–68. [CrossRef]

213. Ozdemir, B.C.; Coukos, G.; Wagner, A.D. Immune-related adverse events of immune checkpoint inhibitors and the impact of sex-what we know and what we need to learn. *Ann. Oncol.* **2018**, *29*, 1067. [CrossRef]

214. Wang, S.; Cowley, L.A.; Liu, X.S. Sex Differences in Cancer Immunotherapy Efficacy, Biomarkers, and Therapeutic Strategy. *Molecules* **2019**, *24*, 3214. [CrossRef] [PubMed]

215. Klein, S.L.; Morgan, R. The impact of sex and gender on immunotherapy outcomes. *Biol. Sex Differ.* **2020**, *11*, 24. [CrossRef] [PubMed]

216. Olsen, C.M.; Thompson, F.C.; Pandeya, N.; Whiteman, D.C. Evaluation of Sex-Specific Incidence of Melanoma. *JAMA Dermatol.* **2020**, *156*, 553–560. [CrossRef] [PubMed]

Review

BRAF Inhibitors: Molecular Targeting and Immunomodulatory Actions

Ilaria Proietti [1,*], Nevena Skroza [1], Simone Michelini [1], Alessandra Mambrin [1], Veronica Balduzzi [1], Nicoletta Bernardini [1], Anna Marchesiello [1], Ersilia Tolino [1], Salvatore Volpe [1], Patrizia Maddalena [1], Marco Di Fraia [1], Giorgio Mangino [2,3], Giovanna Romeo [2,3] and Concetta Potenza [1]

[1] Department of Medical-Surgical Sciences and Biotechnologies, Dermatology Unit "Daniele Innocenzi", Sapienza University of Rome, Polo Pontino, 04100 Latina, Italy; nevena.skroza@uniroma1.it (N.S.); simo.mik@hotmail.it (S.M.); mambrinalessandra@gmail.com (A.M.); veronica.balduzzi@gmail.com (V.B.); nicoletta.bernardini@libero.it (N.B.); anna.marchesiello90@gmail.com (A.M.); ersiliatolino@gmail.com (E.T.); salvatore.volpe@uniroma1.it (S.V.); patrizia.maddalena@uniroma1.it (P.M.); 16191@gmail.com (M.D.F.); concetta.potenza@uniroma1.it (C.P.)

[2] Department of Medico-Surgical Sciences and Biotechnologies, Sapienza University of Rome, 04100 Latina, Italy; giorgio.mangino@uniroma1.it (G.M.); giovanna.romeo@uniroma1.it (G.R.)

[3] Institute of Molecular Biology and Pathology, Consiglio Nazionale delle Ricerche, 00100 Rome, Italy

* Correspondence: proiettilaria@gmail.com; Tel.: +39-3334684342 or +39-0773708811

Received: 27 May 2020; Accepted: 26 June 2020; Published: 7 July 2020

Abstract: The BRAF inhibitors vemurafenib, dabrafenib and encorafenib are used in the treatment of patients with BRAF-mutant melanoma. They selectively target BRAF kinase and thus interfere with the mitogen-activated protein kinase (MAPK) signalling pathway that regulates the proliferation and survival of melanoma cells. In addition to their molecularly targeted activity, BRAF inhibitors have immunomodulatory effects. The MAPK pathway is involved in T-cell receptor signalling, and interference in the pathway by BRAF inhibitors has beneficial effects on the tumour microenvironment and anti-tumour immune response in BRAF-mutant melanoma, including increased immune-stimulatory cytokine levels, decreased immunosuppressive cytokine levels, enhanced melanoma differentiation antigen expression and presentation of tumour antigens by HLA 1, and increased intra-tumoral T-cell infiltration and activity. These effects promote recognition of the tumour by the immune system and enhance anti-tumour T-cell responses. Combining BRAF inhibitors with MEK inhibitors provides more complete blockade of the MAPK pathway. The immunomodulatory effects of BRAF inhibition alone or in combination with MEK inhibition provide a rationale for combining these targeted therapies with immune checkpoint inhibitors. Available data support the synergy between these treatment approaches, indicating such combinations provide an additional beneficial effect on the tumour microenvironment and immune response in BRAF-mutant melanoma.

Keywords: BRAF-mutant melanoma; BRAF inhibitor; mechanism of action; melanoma; targeted therapy; tumour microenvironment

---

1. Introduction

Melanoma is the deadliest form of skin cancer and results from the uncontrolled division of melanocytes. This is highlighted by the fact that melanoma accounts for only 4 percent of all dermatological cancers, but it is responsible for 80 percent of deaths [1]. Globally, over the last 50 years there has been a steady increase in the prevalence of melanoma with nearly 200,000 new cases diagnosed annually, with the highest rates recorded in Australia and New Zealand [2]. This has been paralleled by an improvement in survival rates likely as a result of earlier and better diagnosis, and the introduction of modern therapies including immunotherapy and targeted therapy [1,2]. In the US melanoma is the fifth most common cancer in men and the seventh most common cancer in women, and over the last 20 years 5-year survival rates have increased from approximately 80% to >90% [1,2]. In Europe in 2018, melanoma of the skin was ranked as the seventh most prevalent cancer with 144,209 new cases and 27,147 deaths [3]. In a previous report it was noted that there was a wide variation in incidence between countries, ranging from 2.2 in Greece to 19.2 in Switzerland (per 100,000 person years). The highest incidences tended to be recorded in Northern European countries (Estonia, Latvia, Lithuania, Scandinavia, the UK, and Ireland) and Western European (Austria, Belgium, France (metropolitan), Germany, Luxenberg, Switzerland and the Netherlands) [4].

Melanoma occurs as consequence of genetic mutations and environmental factors. Gene mutation lead to cell proliferation and growth, and the development of an invasive cell phenotype. Key risk factors predisposing to cutaneous melanoma include family history, male gender, older age, racial skin phenotype (Caucasian race increases risk), lighter coloured eyes (blue and hazel eyes have a greater risk than dark brown eyes), the presence of multiple dysplastic nevi, immunosuppression, sun sensitivity, and exposure to ultraviolet (UV) radiation in sunlight [1,2]. Two high risk groups are fair skinned individuals with red or blond hair with many freckles, and persons with darker hair and skin with high melanocytic nevi numbers [2]. Prevention (avoiding exposure to harmful UV radiation) and early detection are key initiatives to reduce the risk of developing a melanoma.

In recent years, the treatment of patients with melanoma has been advanced by the development of molecular targeted therapies and immunotherapies. This is based upon the finding that carcinogenesis is dependent on genetic mutations that activate mitogenic signalling, and that established tumours usually remain dependent on these signalling pathways. Thus, in-depth knowledge of the melanocyte signalling cascade offers therapeutic potential in patients with melanoma [5].

Recent advances in molecular genetics have provided a better understanding of the genetic mutations underpinning the pathogenesis and proliferation of melanocytes [6]. The majority of these mutations affecting the mitogen-activated protein kinase (MAPK) pathway, which is involved in the regulation of cell growth and proliferation [7,8]. Most notably, mutations affecting BRAF are found in more than half of patients diagnosed with cutaneous melanoma [9]. In BRAF-mutated melanoma, BRAF kinase becomes hyperactivated, resulting in increased cell proliferation and survival [7,10]. A better knowledge of the genetic and epigenetic changes of BRAF mutations through gene sequencing, as well as a fuller understanding of the underlying pathogenicity of melanoma, should help direct future targeted treatments and the development of a personalised approach to patient management. At the present time, three selective BRAF kinase inhibitors have been approved for use in the treatment of BRAF-mutant melanoma, including vemurafenib, dabrafenib and encorafenib [8]. Other genetic drivers in the pathogenesis of melanoma include *NRAS* and *neurofibromatosis 1 (NF1)* mutations [11], but these will not be discussed in detail in this article.

This review summarises the mechanisms underlying the efficacy of BRAF inhibitors in the treatment of BRAF-mutant melanoma, including their inhibitory effect on constitutively activated BRAF kinase and consequent interference with the MAPK pathway, and their immunomodulatory role in the tumour microenvironment, leading to enhanced tumour recognition by the immune system and anti-tumour T-cell responses.

2. MAPK Pathway in Melanoma

The MAPK/extracellular signal-related kinase (ERK) pathway, under normal physiological conditions, plays a key role in the regulation of fundamental cellular processes, including cell growth, development, division, transformation, proliferation, migration and death (apoptosis). This is achieved through a broad-spectrum of interactions involving mitogens, growth factors and cytokines [7,8]. MAPK signaling is initiated via cell surface tyrosine kinase receptors, and subsequent activation of RAS, a membrane-bound GTPase. Transduction of the extracellular signals to the intracellular environment occurs through a hierarchical cascade of phosphorylation reactions which lead to the activation of specific kinases [7,8]. In particular, RAF (rapidly accelerated fibrosarcoma) protein kinases [of which there are three isoforms (A,B,C); namely, ARAF, BRAF and CRAF] are involved in the phosphorylation and activation of the MAP/ERK 1 and 2 kinases (MEK 1 and 2), which in turn phosphorylate the substrates ERK 1 and 2 [8]. Activated ERK is responsible for the phosphorylation of a range of substrates that are involved in the regulation of the gene expression which is essential for tumour growth and cytoskeletal functioning. These include effects on metabolism differentiation, proliferation, senescence and, ultimately, cell death.

RAF kinases, which are encoded by the *Raf* gene, exhibit serine/threonine protein kinase activity, with BRAF having the strongest activity and ARAF the weakest [7]. Of these three kinases, BRAF has the highest mutation rate and this has been reported to be up to 90% in melanoma tumours [7]. Activation of RAF is regulated by RAS. Upon stimulation by upstream factors, inactive RAS-GDP is converted to active RAS-GTP in the plasma membrane, which results in translocation of RAF to the membrane and the formation of RAF–RAS-GTP complexes [12]. Bound RAF kinases are activated through priming, including phosphorylation of key residues and dimerization through the kinase domain. RAF kinase domains have inactive and active conformations; dimerization helps produce the active conformation of RAFs, including BRAF, in the normal physiological setting [12].

Over-activating mutations affecting the MAPK/ERK pathway can lead to uncontrolled cell growth and this has been associated with a number of different neoplasms, including involvement in the pathogenesis/progression of melanomas [8]. Mutations of BRAF are the most common mutations leading to overactivation of the MAPK pathway, and BRAF-activating mutations are found in more than half of cutaneous melanomas [8,13]. The most common is the V600E mutation which accounts for 80–90% of BRAF mutations in melanomas and is present in almost 60% of cutaneous cases but is present in only 5% of mucosal melanomas [8]. The second most common mutation involves V600K which has been reported in approximately 8% of melanomas [8,13]. The p.V600E mutation, which leads to substitution of valine by glutamic acid at amino acid position 600 in the BRAF protein has been reported to result from the transversion c.1799T > A in exon 15 [8,13]. This mutation increases BRAF kinase activity ~700-fold compared with wild type BRAF [14]. While the base changes involved in this V600E mutation are not typical of those associated with UV radiation, there is clear evidence that they might result from error-prone replication of UV-damaged DNA, possibly as a consequence of multiple acute episodes of sun exposure [15]. The mutated gene leads to the production of a constitutively activated BRAF protein that dysregulates downstream MAPK signalling, promoting cellular proliferation and inhibiting apoptosis [7,10]. Oncogenic BRAF mutations promote spontaneous BRAF activation by various means, such as enhancing RAS-GTP-associated dimerization, or causing BRAF dimerization in the absence of RAS-GTP [12]. The structural mechanism underlying BRAF activation caused by the common V600E mutation is not fully understood, but there is evidence that the V600E substitution may promote conformational changes in BRAF leading to dimerization-induced activation [12].

3. Tumour Microenvironment in Melanoma

A number of studies have demonstrated that alongside genetic mutations, alterations in the tumour microenvironment (characterized by increased levels of proteins able to favour tumour invasion and infiltration) are responsible for melanoma proliferation [11]. In this regard, matrix metalloproteinases (MMPs), particularly MMP-9 and MMP-2, play a key role. These MMPs increase the degradation of components of the extracellular matrix, thus favouring tumour cell infiltration. Melanomas are associated with numerous mutations involving genes controlling cellular processes such us, proliferation (BRAF, NRAS and NF1), growth and metabolism [phosphatase and tensin homolog (PTEN) and KIT proto-oncogene receptor tyrosine kinase (KIT)], resistance to apoptosis [tumour protein p53 (TP53)], cell cycle control [cyclin-dependent kinase inhibitor 2A (CDKN2A)] and replicative lifespan [telomerase reverse transcriptase (TERT)] [11]. These genetic alterations typically lead to the aberrant activation of two main signalling pathways in melanoma: the RAS/RAF/MEK/ERK signalling cascade [also known as the mitogen-activated protein kinase (MAPK) pathway] and the phosphoinositol-3-kinase (PI3K)/AKT pathway [11]. The MAPK pathway is involved in T-cell receptor signalling. BRAF-mutant tumours with constitutive upregulation of the MAPK pathway, such as melanoma, can induce immune-escape mechanisms that make them immunologically "cold" and able to evade T-cell immune responses [16,17]. The tumours employ various mechanisms affecting different stages of the cancer–immunity cycle.

BRAF-mutant tumours create an immunosuppressive microenvironment that prevents the presentation of tumour antigens by antigen-presenting cells (such as dendritic cells and macrophages), and the T-cell priming that follows [16,18,19]. For example, BRAF^{V600E} cells adversely affect the maturation of dendritic cells and reduce their ability to produce the cytokines necessary for T-cell activation and expansion [16,19,20]. Tumour T-cell infiltrates are reduced in BRAF-mutated melanomas [21].

BRAF-mutant melanoma can escape recognition by effector T cells; for example, through low expression of melanoma differentiation antigens and by down-regulating the expression of human leucocyte antigen (HLA) class I molecules on melanoma cell surfaces [16,22]. HLA 1 is necessary for the presentation of antigens for recognition by T cells and, therefore, by reducing its expression, BRAF-mutant melanoma cells can escape recognition.

The microenvironment of BRAF-mutant melanomas can also inhibit effector T-cell functions; for example, by promoting the accumulation of regulatory T cells and myeloid-derived suppressor cells (MDSCs) [16,17,19,20]. The regulatory T cells, via cell to cell contact-dependent mechanisms and immunosuppressive cytokines (interleukin (IL)-6 and IL-10) limit responses to effector T cells [16,21]. MDSCs include a variety of cells of myeloid origin which are able to potently suppress T cells

In summary, BRAF-mutant melanomas have a tumour microenvironment involving upregulation of the MAPK signalling pathway that creates a pro-tumorigenic environment and an ineffective anti-tumour immune response [16].

4. BRAF Inhibitors: Mechanism of Action

4.1. Kinase Repression

The BRAF inhibitors vemurafenib, dabrafenib and encorafenib are approved for use in the treatment of patients with BRAF^{V600}-mutant advanced melanoma (Table 1) [23–30]. These drugs are orally available, small molecule, selective inhibitors of BRAF kinase. By inhibiting BRAF they interfere with the MAPK signalling pathway that regulates the proliferation and survival of melanoma cells.

They exhibit high specificity for BRAF^{V600}-mutant cell lines, with this specificity thought to be due to preferential inhibition of the active conformation of BRAF, achieved by competitive occupation of the ATP binding pocket which stabilizes the kinase in its active conformation [8].

Preclinical studies demonstrated that vemurafenib and dabrafenib provide potent selective inhibition of kinase activity in BRAF^{V600}-mutant melanoma cell lines, blocking ERK phosphorylation and cellular proliferation, and inducing G1 cell-cycle arrest and apoptosis [23–26,29,30]. Vemurafenib has confirmed activity against V600E, V600D and V600R mutant cell lines [23] and dabrafenib against V600E, V600D, V600R and V600K [24,25]. Inhibition is not seen with either drug in cells expressing wild type BRAF or non-V600 mutations [23,25,29]. In contrast, encorafenib, which targets V600E and V600K mutants, also displays some inhibitory effect in wild type BRAF [8,27,28]. In xenograft models of BRAF^{V600E}-expressing melanoma, administration of these agents inhibits tumour growth, and at higher doses induces tumour regression [23,24,26,28,29].

Table 1. BRAF Inhibitors [8,23–30].

INN	Chemical Name	Activity	FDA/EMA Approved Indications/Date of First Approval
Vemurafenib	*N*-[3-[5-(4-chlorophenyl)-1*H*-pyrrolo[2,3-b]pyridine-3-carbonyl]-2,4-difluorophenyl]propane-1-sulfonamide entry 1	Selectively binds to the ATP-binding site of BRAF kinase and inhibits its activity. Activity against BRAF$^{V600E, V600D, V600R}$ mutant cell lines [23].	Monotherapy in adults with BRAF^{V600}-mutation-positive unresectable or metastatic melanoma. FDA 2011 / EMA 2012
Dabrafenib	*N*-[3-[5-(2-aminopyrimidin-4-yl)-2-*tert*-butyl-1,3-thiazol-4-yl]-2-fluorophenyl]-2,6-difluorobenzenesulfonamide	Selectively binds to and inhibits the activity of BRAF. Activity against BRAF$^{V600E, V600D, V600R, V600K}$ mutant cell lines [24–26].	Monotherapy in adults with unresectable or metastatic melanoma with BRAF^{V600E} mutation (FDA) or BRAF^{V600} mutations (EMA). Combination therapy with trametinib in patients with unresectable or metastatic melanoma with BRAF^{V600E} or BRAF^{V600K} mutations (FDA) or BRAF^{V600} mutations (EMA), and also adjuvant treatment of adults with stage III melanoma with a BRAF^{V600} mutation, following complete resection (EMA). FDA 2013 / EMA 2013
Encorafenib	Methyl *N*-[(2*S*)-1-[[4-[3-[5-chloro-2-fluoro-3-(methanesulfonamido)phenyl]-1-propan-2-ylpyrazol-4-yl]pyrimidin-2-yl]amino]propan-2-yl]carbamate	Selective ATP-competitive RAF kinase inhibitor. Activity against BRAF$^{V600E, V600D, V600K}$ mutant cell lines [8,27,28]	Combination therapy with binimetinib in adults with unresectable or metastatic melanoma with a BRAF^{V600} mutation (EMA) or BRAF^{V600E} or BRAF^{V600K} mutation detected by FDA-approved test (FDA). FDA 2018 / EMA 2018

EMA = European Medicines Agency; FDA = US Food and Drug Administration.

In BRAF wild type cells, vemurafenib, dabrafenib and encorafenib can cause RAS-dependent paradoxical activation of the MAPK pathway, especially in cells that have pre-existing RAS mutations [10,12,24,31]. It may also select for the survival of non-BRAF^{V600} cells, causing drug resistance [10]. Combining a BRAF inhibitor with a MEK inhibitor (which acts by inhibiting kinases further downstream of BRAF in the MAPK pathway) prevents some of this increased MAPK signalling and provides more potent and durable inhibition of ERK signalling [24,31,32]. Dual MAPK pathway inhibition is a standard treatment option for BRAF-mutated melanoma.

4.2. Immunomodulatory Actions (Tumour Microenvironment)

BRAF inhibitors can reverse some of the immunosuppressive effects associated with the BRAF-mutant tumour microenvironment (discussed in Section 3), augmenting the immune response and turning them back into immunologically "hot" tumours [16,17,21,33]. This is achieved via several different mechanisms.

Firstly, BRAF inhibitors help restore an immune-stimulatory microenvironment in BRAF-mutant melanomas [16,17,21,33]. This is done by enhancing the expression of immune-stimulatory molecules/cytokines [19,34], reducing the expression of immunosuppressant molecules/cytokines [20,35–37], and reducing the accumulation of regulatory immune cells (e.g., regulatory T cells and MDSCs) [11,13,19,20,34–41] (Table 2).

Table 2. Mechanisms by which BRAF Inhibitors Help Restore an Immune-stimulatory Tumour Microenvironment in BRAF-mutant Melanoma [11,13,18,20,34–41].

Effect	Findings
Increased expression of immune-stimulatory molecules/cytokines	Increased CD40L and IFNγ expression on intratumoural CD4+ TILs in a murine model of melanoma [19] Increased IL-12 and TNFα production and surface marker expression (CD80, CD83, CD86) in DCs co-cultured with BRAF-mutant melanoma cells [20,34] Increased levels of IFNγ, TNFα and the chemokine CCL4 in melanoma patient serum samples [38]
Reduced expression of immunosuppressive molecules/cytokines	Decreased expression of IL-1, IL-6, IL-8, IL-10, VEGF in BRAF-mutant melanoma cells [20,35,36] Decreased expression of IL-6 and IL-8 in melanoma patient biopsies [37] Decreased levels of IL8 (CXCL8) in melanoma patient serum samples [38]
Reduced accumulation of regulatory immune cells and regulatory chemokines	Reduced accumulation of Tregs and MDSCs in murine models of melanoma [19,40] Decreased MDSC level in melanoma patient blood samples [39] Decreased expression of CCL2, increased ratio of CD8+ T cells to Tregs in a murine BRAF-mutant melanoma model [41]

CCL2 = CC-chemokine ligand 2; DC = dendritic cell; IFN = interferon; IL = interleukin; MDSC = myeloid-derived suppressor cell; NK = natural killer; TIL = tumour-infiltrating cell; Treg = regulatory T cell; VEGF = vascular endothelial growth factor.

Secondly, BRAF inhibitors increase T-cell infiltration into the tumour microenvironment of BRAF-mutant melanomas [16,17,21]. Increased CD8+ T-cell infiltration after BRAF-inhibitor treatment has been demonstrated in animal models, as well as from biopsies from patients [37,38,42–44]. Concurrent administration of a BRAF inhibitor improves the activity and infiltration of melanoma-specific adoptive T cells [45]. There is evidence from in vitro and in vivo studies that the enhanced tumour infiltration by T cells might be due to a reduction in vascular endothelial growth factor (VEGF) expression during BRAF inhibitor administration [20,22,35,36,45]. High VEGF levels cause blood vessel abnormalities that can restrict the entry of drugs and immune cells into tumours. In addition to T cells, an increased number of natural killer cells have been demonstrated in tumour infiltrates in a mouse model of BRAF-mutant melanoma after treatment with a BRAF inhibitor [41].

Thirdly, BRAF inhibitors enhance the recognition of melanoma cells by the immune system and so reduce the likelihood of the tumour escaping recognition by T cells [16,17,21,33]. Evidence from in vitro and in vivo studies indicate that BRAF inhibitors do this by increasing the expression of melanoma differentiation antigens, such as melanoma antigen recognised by T cells (MART-1), glycoprotein 100 (gp100), tyrosinase-related protein-1 (TYRP-1), TYRP-2 and dopachrome tautomerase (DCT) on BRAF-mutant cells [37,46], and also by increasing the expression of HLA 1 on the surface of BRAF-mutant melanoma cells [22,35].

Finally, BRAF inhibitors can improve the activity of effector T cells [16,21,33]. Studies using BRAF-mutant cell lines have shown that BRAF inhibitors increase T-cell activity, as indicated by an increase in IFNγ release [19,46]. Analysis of biopsies from patients with BRAF-mutant melanoma has shown that BRAF inhibitor treatment increases expression of markers of T-cell cytotoxicity, such as perforin and granzyme B [37].

MEK inhibitors also display some immunomodulatory activity, including in BRAF wild type melanoma [16]. In line with the effects of BRAF inhibitor monotherapy discussed above, the combination of a BRAF inhibitor and a MEK inhibitor has been shown to increase immune-stimulatory molecules/cytokines [35], reduce immunosuppressive cytokines [35–37], reduce VEGF expression [36], increase T-cell infiltrates [37], increase HLA 1 [35] and melanoma antigen [37] expression, and increase markers of T-cell cytotoxicity [37] in BRAF-mutant melanoma. Combination BRAF inhibitor plus MEK inhibitor treatment has also been shown to promote the cleavage of gasdermin E and release of HMGB1, suggesting that it may in part regulate the tumour immune microenvironment through pyroptosis (an inflammatory type of programmed cell death) [47].

The ability of BRAF inhibitors (and MEK inhibitors) to modify the tumour microenvironment and enhance anti-tumour immune responses in BRAF-mutated melanoma supports the idea that a combination of targeted therapy and immunotherapy could provide greater anti-tumour activity. Examples of immunotherapies used in melanoma include adoptive immunotherapy (passive transfer of activated T cells) and immune checkpoint inhibitors (which prevent the activation of T cells) [48]. Checkpoint inhibitors include monoclonal antibodies directed against programmed death receptor-1 (PD1), PD ligand-1 (PDL1), or cytotoxic T-lymphocyte-associated protein 4 (CTLA-4). These molecules are negative regulators of T-cell activity, and blocking their actions strengthens effector T cell functioning and anti-tumour responses [48]. Analysis of sequential biopsies from a patient with melanoma found a transient increase in T-cell infiltrate (followed by a decrease) during BRAF inhibitor monotherapy, which increased again and persisted after a dose of anti-CTLA4 antibody [42,43]. The combination of a BRAF inhibitor with either an anti-PD1 or anti-PDL1 checkpoint inhibitor has been shown to increase T-cell infiltration, the ratio of CD8+ T cells to regulatory T cells, and T-cell activity, compared with either agent alone in a murine model of BRAF-mutant melanoma [42,43]. Triple combination therapy with a BRAF inhibitor, a MEK inhibitor and pmel-1 adoptive cell transfer immunotherapy has been shown to increase T-cell infiltration into tumours and improve cytotoxicity in a murine model [49]. Studies in mice have also found that triple combination therapy with a BRAF inhibitor, MEK inhibitor and an anti-PD1 antibody produces greater anti-tumour activity than anti-PD1 monotherapy or any double combination amongst these therapies [49–51]. This triple combination therapy is associated with increased CD8+ T-cell infiltration, CD4+ cells and tumour-associated macrophages compared with anti-PD1 monotherapy [51]. Overall, these data suggest that the combination of targeted therapy and immunotherapy has a beneficial effect on the tumour microenvironment and T-cell response.

5. Clinical Implications

Most patients with newly diagnosed melanoma have early stage disease and surgical excision is the treatment of choice and is usually curative. This highlights the importance of regular skin checks and early detection. Some of these patients will relapse during the course of the disease, and others (about 10%) will present with advanced unresectable disease which may have already metastasized [11]. Medical treatment in these cases has been revolutionised in the last decade with the availability of several new therapies such as BRAF and MEK inhibitors and anti-PD1 and anti-CTLA4 immunotherapies. As a consequence of these developments, these newer therapies have become the mainstay of advanced melanoma therapy, and chemotherapy is now considered second-line at best [11].

BRAF inhibitors have been shown to rapidly suppress melanoma growth and control the malignancy in a large proportion of patients [11,52,53]. Although BRAF inhibitors can induce good responses in many patients with BRAF-mutant melanoma, in some cases, a reduction in effectiveness can be observed after 7–8 months of therapy [32,54,55]. Several mechanisms account for the reduction in tumour response to BRAF inhibitor therapy [52,53]. These include both primary (intrinsic) and secondary (acquired) mechanisms: primary applies to patients who do not respond to BRAF inhibitor therapy from the outset (approximately 15% of patients) [52]; whereas secondary mechanisms involve individuals who initially responded to BRAF inhibitor therapy, but subsequently relapsed [52]. Various pathways/mechanisms have been shown to potentially be involved in the development of acquired resistance [52]:

Activation of the PI3K-Akt pathway by upregulation of specific receptor tyrosine kinases (including insulin growth factor receptor 1 and platelet derived growth factor receptor β) in a non-ERK dependent manner, as well as induction of the pathway by epigenetically changed epidermal growth factor (EGFR). This dual activation of the P13K-Akt pathway promotes resistance, cell survival and proliferation

Activation of the MAPK pathway as a result of NRAS activating mutations, changes leading to maintenance of RAF dimerization, the 'BRAF inhibitor paradox' (in which the BRAF inhibitor blocks MAPK signaling in mutant cells, but activates the MAPK pathway in non-mutant cells), and resistance to RAF inhibition through activation of HGF and its receptor MET (leading to the reactivation of the MAPK and P13K-Akt pathways)

Secondary mutations in MEK1 and 2 have both been linked to acquired resistance

The AXL receptor tyrosine kinase has been shown to be overexpressed in patients who have relapsed after treatment with BRAF and MEK inhibitors.

Combining a BRAF inhibitor with a MEK inhibitor (which acts further downstream on the MAPK pathway) can delay the development of resistance compared with BRAF inhibitor monotherapy and provides improved response and survival outcomes [32,56–61]. Consequently, dual MAPK inhibition with the combination of a BRAF inhibitor plus an MEK inhibitor is a standard-of-care approach for patients with unresectable or metastatic BRAF-mutant melanoma [62]. Approved combinations for patients with unresectable or metastatic melanoma include darafenib/trametinib, vemurafenib/cobimetinib and encorafenib/binimetinib [63]. Combination targeted therapy can also be beneficial in an adjuvant setting [64], and dabrefenib/trametinib has been approved for use as adjuvant treatment after complete resection of stage III melanoma [65]. However, in many patients, melanoma progression eventually occurs and resistance to treatment develops.

This led to an interest in immunotherapeutic agents to extend the therapeutic effect and induce long-acting anti-melanoma effects, and immune checkpoint inhibitors are another important treatment option for patients with advanced melanoma [11,16,52]. It has been reported that single-agent checkpoint inhibitors produced a clear clinical benefit over chemotherapy in metastatic melanoma, and these benefits appeared to be consistent across patient subgroups [66]. Moreover, combination of nivolumab plus ipilimumab, an anti- CTLA-4, was significantly more effective in term of objective response rate, progression-free survival and overall survival relative to ipilimumab used alone [66]. However, because checkpoint receptors play important roles in regulating autoimmunity, the major toxicities associated with the use of these drugs include autoimmune symptoms. The incidence of immune-related adverse events is relatively high, varying from 70% in patients treated with anti-PD-1/anti-PD-L1 antibodies to 90% in patients treated with anti-CTLA-4 [67]. Therefore, it is important to assess which patients we are dealing with and how they will respond to the therapy. Despite the noteworthy successes of immune check point blockade, to date, only a subset of patients achieve durable clinical responses [68]. Indeed, more than half of patients treated with anti-PD1 therapy fail to respond or eventually develop progression [69]. Some of this resistance may be due to immune mechanisms and, given that BRAF inhibitors (and MEK inhibitors) have favourable effects on antitumor immunity, switching to combination targeted therapy could help overcome some anti-PD1 resistance mechanisms. Clinical trials to date have demonstrated variable results with regard to efficacy, but BRAF/MEK-targeted therapy can be considered in patients with BRAF-mutant melanoma who do not respond to anti-PD1 therapy [69].

The safety and tolerability of treatment is an important clinical consideration. Paradoxical activation of the MAPK pathway by BRAF inhibitors can increase the risk of the other cutaneous malignancies developing [31]. Combination therapy with a BRAF inhibitor plus a MEK inhibitor is associated with reduced dermatological toxicity, although a slightly worse gastrointestinal adverse event profile, compared with monotherapy [70]. Combination therapy with BRAF inhibitors and some immunotherapeutic agents can also be associated with increased toxicity; in particular, the combination of vemurafenib and ipilimumab (an anti-CTLA4 monoclonal antibody) was associated with marked hepatotoxicity [21,48]. In contrast, the combination of a BRAF inhibitor with an anti-PDL1 antibody appears to be well tolerated [21,48].

Looking ahead, clinical trials evaluating the combination of a BRAF inhibitor (with or without a MEK inhibitor) and immunotherapy are ongoing [21,48]. Given that some studies found promising clinical responses but a problem with toxicity with certain combinations [71], other trials are evaluating sequential treatment with targeted therapy and immunotherapy in patients with BRAF-mutant melanoma, in order to identify the optimal sequencing and timing of treatment [21,48].

6. Conclusions

Advances in the treatment of metastatic melanoma have expanded over the last decade to include approaches such as targeted molecular therapy and immunotherapy, including T-cell checkpoint inhibition. These have been driven by the development of resistance to one or more of these therapeutic methods. The BRAF inhibitors vemurafenib, dabrafenib and encorafenib are used in the treatment of patients with $BRAF^{V600}$-mutant advanced melanoma. They selectively target BRAF kinase and thus interfere with the MAPK signalling pathway that regulates the proliferation and survival of melanoma cells.

In addition to their molecularly targeted activity, BRAF inhibitors also exhibit immunomodulatory effects. The MAPK pathway is involved in T-cell receptor signalling, and interference with the pathway by BRAF inhibitors has beneficial effects on the tumour microenvironment and anti-tumour immune response in BRAF-mutant melanoma. This is achieved through several different mechanisms including: increasing immune-stimulatory cytokine levels; decreasing immunosuppressive cytokine levels; enhancing melanoma differentiation antigen expression and presentation of tumour antigens by HLA 1; and increasing intra-tumoral T-cell infiltration and activity. Overall, these effects promote recognition of the tumour by the immune system and enhance anti-tumour T-cell responses in BRAF-mutant melanoma.

Combining BRAF inhibitors with MEK inhibitors provides more complete blockade of the MAPK pathway. Moreover, the immunomodulatory effects of BRAF inhibition alone or in combination with MEK inhibition provide a rationale for combining these targeted therapies with immune checkpoint inhibitors. Available preclinical data support the synergy between these treatment approaches, indicating such combinations provide an additional beneficial effect on the tumour microenvironment and immune response in BRAF-mutant melanoma. Clinical trials are evaluating whether this translates into an improved clinical response and duration of response in patients. Areas for future research to address specific issues with the goal of improving treatment outcomes include:

Gaining a better understanding of resistance mechanisms for all potential therapies, through clinical trials and parallel translational/preclinical studies

Increasing the focus on personalised approaches to patient management, so as to maximise the benefits of available therapies whilst avoiding unnecessary toxicity

Studies to help define which treatment to start with, optimal dosage regimens/schedules, and when to consider adding another therapy

Identification of appropriate biomarkers to help physicians make more reliable predictions of likely response and possible toxicity.

Author Contributions: I.P., N.S., S.M., A.M. (Alessandra Mambrin), V.B., N.B., A.M. (Anna Marchesiello), E.T., S.V., P.M., M.D.F., G.M., G.R. and C.P. have drafted the manuscript and has revised it critically. All authors have read and agreed to the published version of the manuscript.

Funding: This research received no external funding.

Acknowledgments: Writing and editing support was provided by Content Ed Net. Authors thank Novartis Farma Italy for funding editorial support.

Conflicts of Interest: The authors declare no conflict of interest. Novartis Farma Italy had no role in the writing of the manuscript or in the decision to publish the results.

References

1. Miller, A.J.; Mishmi, M.C., Jr. Melanoma. *N. Engl. J. Med.* **2006**, *355*, 51–65. [CrossRef]
2. Azoury, S.C.; Lange, J.R. Epidemiology, risk factors, prevention, and early detection of melanoma. *Surg. Clin. N. Am.* **2014**, *94*, 945–962. [CrossRef] [PubMed]
3. Globocan 2018. Melanoma of the skin fact sheet. Available online: https://gco.iarc.fr/today/data/factsheets/cancers/16-Melanoma-of-skin-fact-sheet.pdf (accessed on 30 May 2020).
4. Forsea, A.; del Marmol, V.; de Vries, E.; Bailey, E.; Geller, A. Melanoma incidence and mortality in Europe: New estimates, persistent disparities. *Br. J. Dermatol.* **2012**, *167*, 1124–1130. [CrossRef] [PubMed]
5. Poulikakos, P.I.; Rosen, N. Mutant BRAF melanomas–dependence and resistance. *Cancer Cell.* **2011**, *19*, 11–15. [CrossRef]
6. Chudnovsky, Y.; Khavari, P.A.; Adams, A.E. Melanoma genetics and the development of rational therapeutics. *J. Clin. Investig.* **2005**, *115*, 813–824. [CrossRef] [PubMed]
7. Guo, Y.J.; Pan, W.W.; Liu, S.B.; Shen, Z.F.; Xu, Y.; Hu, L.L. ERK/MAPK signalling pathway and tumorigenesis. *Exp. Ther. Med.* **2020**, *19*, 1997–2007. [CrossRef]
8. Savoia, P.; Fava, P.; Casoni, F.; Cremona, O. Targeting the ERK signaling pathway in melanoma. *Int. J. Mol. Sci.* **2019**, *20*, 1483. [CrossRef] [PubMed]
9. Sullivan, R.; LoRusso, P.; Boerner, S.; Dummer, R. Achievements and challenges of molecular targeted therapy in melanoma. *Am. Soc. Clin. Oncol. Educ. Book* **2015**, *35*, 177–186. [CrossRef] [PubMed]
10. Kim, A.; Cohen, M.S. The discovery of vemurafenib for the treatment of BRAF-mutated metastatic melanoma. *Expert Opin. Drug. Discov.* **2016**, *11*, 907–916. [CrossRef]
11. Leonardi, G.C.; Falzone, L.; Salemi, R.; Zanghì, A.; Spandidos, D.A.; Mccubrey, J.A.; Candido, S.; Libra, M. Cutaneous melanoma: From pathogenesis to therapy (Review). *Int. J. Oncol.* **2018**, *52*, 1071–1080. [CrossRef]
12. Karoulia, Z.; Gavathiotis, E.; Poulikakos, P.I. New perspectives for targeting RAF kinase in human cancer. *Nat. Rev. Cancer.* **2017**, *17*, 676–691. [CrossRef] [PubMed]
13. Davies, H.; Bignell, G.R.; Cox, C.; Stephens, P.; Edkins, S.; Clegg, S.; Teague, J.; Woffendin, H.; Garnett, M.J.; Bottomley, W.; et al. Mutations of the BRAF gene in human cancer. *Nature* **2002**, *417*, 949–954. [CrossRef] [PubMed]
14. Wan, P.T.; Garnett, M.J.; Roe, S.M.; Lee, S.; Niculescu-Duvaz, D.; Good, V.M.; Jones, C.M.; Marshall, C.J.; Springer, C.J.; Barford, D.; et al. Cancer Genome Project. Mechanism of activation of the RAF-ERK signaling pathway by oncogenic mutations of B-RAF. *Cell* **2004**, *116*, 855–867. [CrossRef]
15. Thomas, N.E.; Berwick, M.; Cordeiro-Stone, M. Could BRAF mutations in melanocytic lesions arise from DNA damage induced by ultraviolet radiation? *J. Invest Dermatol.* **2006**, *126*, 1693–1696. [CrossRef]
16. Kuske, M.; Westphal, D.; Wehner, R.; Schmitz, M.; Beissert, S.; Praetorius, C.; Meier, F. Immunomodulatory effects of BRAF and MEK inhibitors: Implications for Melanoma therapy. *Pharmacol. Res.* **2018**, *136*, 151–159. [CrossRef] [PubMed]
17. Croce, L.; Coperchini, F.; Magri, F.; Chiovato, L.; Rotondi, M. The multifaceted anti-cancer effects of BRAF-inhibitors. *Oncotarget* **2019**, *10*, 6623–6640. [CrossRef] [PubMed]
18. Chen, D.S.; Mellman, I. Oncology meets immunology: The cancer-immunity cycle. *Immunity* **2013**, *39*, 1–10. [CrossRef]
19. Ho, P.C.; Meeth, K.M.; Tsui, Y.C.; Srivastava, B.; Bosenberg, M.W.; Kaech, S.M. Immune-based antitumor effects of BRAF inhibitors rely on signaling by CD40L and IFNγ. *Cancer Res.* **2014**, *74*, 3205–3217. [CrossRef]
20. Sumimoto, H.; Imabayashi, F.; Iwata, T.; Kawakami, Y. The BRAF-MAPK signaling pathway is essential for cancer-immune evasion in human melanoma cells. *J. Exp. Med.* **2006**, *203*, 1651–1656. [CrossRef]
21. Pelster, M.S.; Amaria, R.N. Combined targeted therapy and immunotherapy in melanoma: A review of the impact on the tumor microenvironment and outcomes of early clinical trials. *Ther. Adv. Med. Oncol.* **2019**, *11*, 1758835919830826. [CrossRef]
22. Bradley, S.D.; Chen, Z.; Melendez, B.; Talukder, A.; Khalili, J.S.; Rodriguez-Cruz, T.; Liu, S.; Whittington, M.; Deng, W.; Li, F.; et al. BRAFV600E Co-opts a Conserved MHC Class I Internalization Pathway to Diminish Antigen Presentation and CD8+ T-cell Recognition of Melanoma. *Cancer Immunol. Res.* **2015**, *3*, 602–609. [CrossRef]

23. Yang, H.; Higgins, B.; Kolinsky, K.; Packman, K.; Go, Z.; Iyer, R.; Kolis, S.; Zhao, S.; Lee, R.; Grippo, J.F.; et al. RG7204 (PLX4032), a selective BRAFV600E inhibitor, displays potent antitumor activity in preclinical melanoma models. *Cancer Res.* **2010**, *70*, 5518–5527. [CrossRef] [PubMed]

24. King, A.J.; Arnone, M.R.; Bleam, M.R.; Moss, K.G.; Yang, J.; Fedorowicz, K.E.; Smitheman, K.N.; Erhardt, J.A.; Hughes-Earle, A.; Kane-Carson, L.S.; et al. Dabrafenib; preclinical characterization, increased efficacy when combined with trametinib, while BRAF/MEK tool combination reduced skin lesions. *PLoS ONE* **2013**, *8*, e67583. [CrossRef]

25. Gentilcore, G.; Madonna, G.; Mozzillo, N.; Ribas, A.; Cossu, A.; Palmieri, G.; Ascierto, P.A. Effect of daBRAFenib on melanoma cell lines harbouring the BRAF(V600D/R) mutations. *BMC Cancer* **2013**, *13*, 17. [CrossRef]

26. Rheault, T.R.; Stellwagen, J.C.; Adjabeng, G.M.; Hornberger, K.R.; Petrov, K.G.; Waterson, A.G.; Dickerson, S.H.; Mook, R.A., Jr.; Laquerre, S.G.; King, A.J.; et al. Discovery of dabrafenib: A selective inhibitor of Raf kinases with antitumor activity against B-Raf-driven tumors. *ACS Med. Chem. Lett.* **2013**, *4*, 358–362. [CrossRef]

27. Delord, J.P.; Robert, C.; Nyakas, M.; McArthur, G.A.; Kudchakar, R.; Mahipal, A.; Yamada, Y.; Sullivan, R.; Arance, A.; Kefford, R.F.; et al. Phase I dose-escalation and -expansion study of the BRAF inhibitor encorafenib (LGX818) in metastatic BRAF-mutant melanoma. *Clin Cancer Res.* **2017**, *23*, 5339–5348. [CrossRef]

28. Koelblinger, P.; Thuerigen, O.; Dummer, R. Development of encorafenib for BRAF-mutated advanced melanoma. *Curr. Opin. Oncol.* **2018**, *30*, 125–133. [CrossRef]

29. Tsai, J.; Lee, J.T.; Wang, W.; Zhang, J.; Cho, H.; Mamo, S.; Bremer, R.; Gillette, S.; Kong, J.; Haass, N.K. Discovery of a selective inhibitor of oncogenic B-Raf kinase with potent antimelanoma activity. *Proc. Natl. Acad. Sci. USA* **2008**, *105*, 3041–3046. [CrossRef] [PubMed]

30. Joseph, E.W.; Pratilas, C.A.; Poulikakos, P.I.; Tadi, M.; Wang, W.; Taylor, B.S.; Halilovic, E.; Persaud, Y.; Xing, F.; Viale, A.; et al. The RAF inhibitor PLX4032 inhibits ERK signaling and tumor cell proliferation in a V600E BRAF-selective manner. *Proc. Natl. Acad. Sci. USA* **2010**, *107*, 14903–14908. [CrossRef] [PubMed]

31. Adelmann, C.H.; Ching, G.; Du, L.; Saporito, R.C.; Bansal, V.; Pence, L.J.; Liang, R.; Lee, W.; Tsai, K.Y. Comparative profiles of BRAF inhibitors: The paradox index as a predictor of clinical toxicity. *Oncotarget* **2016**, *7*, 30453–30460. [CrossRef]

32. Larkin, J.; Ascierto, P.A.; Dréno, B.; Atkinson, V.; Liszkay, G.; Maio, M.; Mandalà, M.; Demidov, L.; Stroyakovskiy, D.; Thomas, L.; et al. Combined vemurafenib and cobimetinib in BRAF-mutated melanoma. *N. Engl. J. Med.* **2014**, *371*, 1867–1876. [CrossRef] [PubMed]

33. Kelley, M.C. Immune responses to braf-targeted therapy in melanoma: Is targeted therapy immunotherapy? *Crit. Rev. Oncog.* **2016**, *21*, 83–91. [CrossRef] [PubMed]

34. Ott, P.A.; Henry, T.; Baranda, S.J.; Frleta, D.; Manches, O.; Bogunovic, D.; Bhardwaj, N. Inhibition of both BRAF and MEK in BRAF(V600E) mutant melanoma restores compromised dendritic cell (DC) function while having differential direct effects on DC properties. *Cancer Immunol. Immunother.* **2013**, *62*, 811–822. [CrossRef]

35. Liu, L.; Mayes, P.A.; Eastman, S.; Shi, H.; Yadavilli, S.; Zhang, T.; Yang, J.; Seestaller-Wehr, L.; Zhang, S.Y.; Hopson, C.; et al. The BRAF and MEK inhibitors daBRAFenib and trametinib: Effects on immune function and in combination with immunomodulatory antibodies targeting PD-1, PD-L1, and CTLA-4. *Clin. Cancer Res.* **2015**, *21*, 1639–1651. [CrossRef]

36. Whipple, C.A.; Boni, A.; Fisher, J.L.; Hampton, T.H.; Tsongalis, G.J.; Mellinger, D.L.; Yan, S.; Tafe, L.J.; Brinckerhoff, C.E.; Turk, M.J.; et al. The mitogen-activated protein kinase pathway plays a critical role in regulating immunological properties of BRAF mutant cutaneous melanoma cells. *Melanoma Res.* **2016**, *26*, 223–235. [CrossRef] [PubMed]

37. Frederick, D.T.; Piris, A.; Cogdill, A.P.; Cooper, Z.A.; Lezcano, C.; Ferrone, C.R.; Mitra, D.; Boni, A.; Newton, L.P.; Liu, C.; et al. BRAF inhibition is associated with enhanced melanoma antigen expression and a more favorable tumor microenvironment in patients with metastatic melanoma. *Clin. Cancer Res.* **2013**, *19*, 1225–1231. [CrossRef] [PubMed]

38. Wilmott, J.S.; Haydu, L.E.; Menzies, A.M.; Lum, T.; Hyman, J.; Thompson, J.F.; Hersey, P.; Kefford, R.F.; Scolyer, R.A.; Long, G.V. Dynamics of chemokine, cytokine, and growth factor serum levels in BRAF-mutant melanoma patients during BRAF inhibitor treatment. *J. Immunol.* **2014**, *192*, 2505–2513. [CrossRef]

39. Schilling, B.; Paschen, A. Immunological consequences of selective BRAF inhibitors in malignant melanoma: Neutralization of myeloid-derived suppressor cells. *Oncoimmunology* **2013**, *2*, e25218. [CrossRef]

40. Steinberg, S.M.; Zhang, P.; Malik, B.T.; Boni, A.; Shabaneh, T.B.; Byrne, K.T.; Mullins, D.W.; Brinckerhoff, C.E.; Ernstoff, M.S.; Bosenberg, M.W.; et al. BRAF inhibition alleviates immune suppression in murine autochthonous melanoma. *Cancer Immunol. Res.* **2014**, *2*, 1044–1050. [CrossRef]

41. Knight, D.A.; Ngiow, S.F.; Li, M.; Parmenter, T.; Mok, S.; Cass, A.; Haynes, N.M.; Kinross, K.; Yagita, H.; Koya, R.C.; et al. Host immunity contributes to the anti-melanoma activity of BRAF inhibitors. *J. Clin. Investig.* **2013**, *123*, 1371–1381. [CrossRef]

42. Cooper, Z.A.; Juneja, V.R.; Sage, P.T.; Frederick, D.T.; Piris, A.; Mitra, D.; Lo, J.A.; Hodi, F.S.; Freeman, G.J.; Bosenberg, M.W.; et al. Response to BRAF inhibition in melanoma is enhanced when combined with immune checkpoint blockade. *Cancer Immunol. Res.* **2014**, *2*, 643–654. [CrossRef] [PubMed]

43. Cooper, Z.A.; Reuben, A.; Amaria, R.N.; Wargo, J.A. Evidence of synergy with combined BRAF-targeted therapy and immune checkpoint blockade for metastatic melanoma. *Oncoimmunology* **2014**, *3*, e954956. [CrossRef]

44. Wilmott, J.S.; Long, G.V.; Howle, J.R.; Haydu, L.E.; Sharma, R.N.; Thompson, J.F.; Kefford, R.F.; Hersey, P.; Scolyer, R.A. Selective BRAF inhibitors induce marked T-cell infiltration into human metastatic melanoma. *Clin. Cancer Res.* **2012**, *18*, 1386–1394. [CrossRef] [PubMed]

45. Liu, C.; Peng, W.; Xu, C.; Lou, Y.; Zhang, M.; Wargo, J.A.; Chen, J.Q.; Li, H.S.; Watowich, S.S.; Yang, Y.; et al. BRAF inhibition increases tumor infiltration by T cells and enhances the antitumor activity of adoptive immunotherapy in mice. *Clin. Cancer Res.* **2013**, *19*, 393–403. [CrossRef] [PubMed]

46. Boni, A.; Cogdill, A.P.; Dang, P.; Udayakumar, D.; Njauw, C.N.; Sloss, C.M.; Ferrone, C.R.; Flaherty, K.T.; Lawrence, D.P.; Fisher, D.E.; et al. Selective BRAFV600E inhibition enhances T-cell recognition of melanoma without affecting lymphocyte function. *Cancer Res.* **2010**, *70*, 5213–5219. [CrossRef]

47. Erkes, D.A.; Cai, W.; Sanchez, I.M.; Purwin, T.J.; Rogers, C.; Field, C.O.; Berger, A.C.; Hartsough, E.J.; Rodeck, U.; Alnemri, E.S.; et al. Mutant BRAF and MEK inhibitors regulate the tumor immune microenvironment via pyroptosis. *Cancer Discov.* **2020**, *10*, 254–269. [CrossRef]

48. Yu, C.; Liu, X.; Yang, J.; Zhang, M.; Jin, H.; Ma, X.; Shi, H. Combination of immunotherapy with targeted therapy: Theory and practice in metastatic melanoma. *Front Immunol.* **2019**, *10*, 990. [CrossRef]

49. Hu-Lieskovan, S.; Mok, S.; Homet Moreno, B.; Tsoi, J.; Robert, L.; Goedert, L.; Pinheiro, E.M.; Koya, R.C.; Graeber, T.G.; Comin-Anduix, B.; et al. Improved antitumor activity of immunotherapy with BRAF and MEK inhibitors in BRAF(V600E) melanoma. *Sci. Transl. Med.* **2015**, *7*, 279ra41. [CrossRef]

50. Deken, M.A.; Gadiot, J.; Jordanova, E.S.; Lacroix, R.; van Gool, M.; Kroon, P.; Pineda, C.; Geukes Foppen, M.H.; Scolyer, R.; Song, J.Y.; et al. Targeting the MAPK and PI3K pathways in combination with PD1 blockade in melanoma. *Oncoimmunology* **2016**, *5*, e1238557. [CrossRef]

51. Homet Moreno, B.; Mok, S.; Comin-Anduix, B.; Hu-Lieskovan, S.; Ribas, A. Combined treatment with daBRAFenib and trametinib with immune-stimulating antibodies for BRAF mutant melanoma. *Oncoimmunology* **2015**, *5*, e1052212. [CrossRef]

52. Patel, H.; Yacoub, N.; Mishra, R.; White, A.; Yuan, L.; Alanazi, S.; Garrett, J.T. Current Advances in the Treatment of BRAF-Mutant Melanoma. *Cancers (Basel)* **2020**, *12*, 482. [CrossRef]

53. Fujimura, T.; Fujisawa, Y.; Kambayashi, Y.; Aiba, S. Significance of BRAF Kinase Inhibitors for Melanoma Treatment: From Bench to Bedside. *Cancers (Basel)* **2019**, *11*, 1342. [CrossRef] [PubMed]

54. Griffin, M.; Scotto, D.; Josephs, D.H.; Mele, S.; Crescioli, S.; Bax, H.J.; Pellizzari, G.; Wynne, M.D.; Nakamura, M.; Hoffmann, R.M.; et al. BRAF inhibitors: Resistance and the promise of combination treatments for melanoma. *Oncotarget* **2017**, *8*, 78174–78192. [CrossRef] [PubMed]

55. Sosman, J.A.; Kim, K.B.; Schuchter, L.; Gonzalez, R.; Pavlick, A.C.; Weber, J.S.; McArthur, G.A.; Hutson, T.E.; Moschos, S.J.; Flaherty, K.T.; et al. Survival in BRAF V600-mutant advanced melanoma treated with vemurafenib. *N. Engl. J. Med.* **2012**, *366*, 707–714. [CrossRef]

56. Robert, C.; Karaszewska, B.; Schachter, J.; Rutkowski, P.; Mackiewicz, A.; Stroiakovski, D.; Lichinitser, M.; Dummer, R.; Grange, F.; Mortier, L.; et al. Improved overall survival in melanoma with combined daBRAFenib and trametinib. *N. Engl. J. Med.* **2015**, *372*, 30–39. [CrossRef] [PubMed]

57. Robert, C.; Grob, J.J.; Stroyakovskiy, D.; Karaszewska, B.; Hauschild, A.; Levchenko, E.; Chiarion Sileni, V.; Schachter, J.; Garbe, C.; Bondarenko, I.; et al. Five-year outcomes with daBRAFenib plus trametinib in metastatic melanoma. *N. Engl. J. Med.* **2019**, *381*, 626–636. [CrossRef] [PubMed]

58. Long, G.V.; Stroyakovskiy, D.; Gogas, H.; Levchenko, E.; de Braud, F.; Larkin, J.; Garbe, C.; Jouary, T.; Hauschild, A.; Grob, J.J.; et al. DaBRAFenib and trametinib versus daBRAFenib and placebo for Val600 BRAF-mutant melanoma: A multicentre, double-blind, phase 3 randomised controlled trial. *Lancet* **2015**, *386*, 444–451. [CrossRef]

59. Ascierto, P.A.; McArthur, G.A.; Dréno, B.; Atkinson, V.; Liszkay, G.; Di Giacomo, A.M.; Mandalà, M.; Demidov, L.; Stroyakovskiy, D.; Thomas, L.; et al. Cobimetinib combined with vemurafenib in advanced BRAF(V600)-mutant melanoma (coBRIM): Updated efficacy results from a randomised, double-blind, phase 3 trial. *Lancet Oncol.* **2016**, *17*, 1248–1260. [CrossRef]

60. Dummer, R.; Ascierto, P.A.; Gogas, H.J.; Arance, A.; Mandala, M.; Liszkay, G.; Garbe, C.; Schadendorf, D.; Krajsova, I.; Gutzmer, R.; et al. Encorafenib plus binimetinib versus vemurafenib or encorafenib in patients with BRAF-mutant melanoma (COLUMBUS): A multicentre, open-label, randomised phase 3 trial. *Lancet Oncol.* **2018**, *19*, 603–615. [CrossRef]

61. Dummer, R.; Ascierto, P.A.; Gogas, H.J.; Arance, A.; Mandala, M.; Liszkay, G.; Garbe, C.; Schadendorf, D.; Krajsova, I.; Gutzmer, R.; et al. Overall survival in patients with BRAF-mutant melanoma receiving encorafenib plus binimetinib versus vemurafenib or encorafenib (COLUMBUS): A multicentre, open-label, randomised, phase 3 trial. *Lancet Oncol.* **2018**, *19*, 1315–1327. [CrossRef]

62. Sun, J.; Carr, M.J.; Khushalani, N.I. Principles of targeted therapy for melanoma. *Surg. Clin. N. Am.* **2020**, *100*, 175–188. [CrossRef] [PubMed]

63. Coit, D.G.; Thompson, J.A.; Albertini, M.R.; Barker, C.; Carson, W.E.; Contreras, C.; Daniels, G.A.; DiMaio, D.; Fields, R.C.; Fleming, M.D.; et al. Cutaneous Melanoma, Version 2.2019, NCCN Clinical Practice Guidelines in Oncology. *J. Natl. Compr. Cancer Netw.* **2019**, *17*, 367–402. [CrossRef] [PubMed]

64. Poklepovic, A.S.; Luke, J.J. Considering adjuvant therapy for stage II melanoma. *Cancer* **2020**, *126*, 1166–1174. [CrossRef] [PubMed]

65. Schummer, P.; Schilling, B.; Gesierich, A. Long-term outcomes in BRAF-mutated melanoma treated with combined targeted therapy or immune checkpoint blockade: Are we approaching a true cure? *Am. J. Clin. Dermatol.* **2020**, 1–12. [CrossRef]

66. Luke, J.J. Comprehensive clinical trial data summation for BRAF-MEK inhibition and checkpoint immunotherapy in metastatic melanoma. *Oncologist* **2019**, *24*, e1197–e1211. [CrossRef]

67. Michot, J.M.; Bigenwald, C.; Champiat, S.; Collins, M.; Carbonnel, F.; Postel-Vinay, S.; Berdelou, A.; Varga, A.; Bahleda, R.; Hollebecque, A.; et al. Immune-related adverse events with immune checkpoint blockade: A comprehensive review. *Eur. J. Cancer* **2016**, *54*, 139–148. [CrossRef]

68. Sadozai, H.; Gruber, T.; Hunger, R.E.; Schenk, M. Recent successes and future directions in immunotherapy of cutaneous melanoma. *Front. Immunol.* **2017**, *8*, 1617. [CrossRef]

69. Babacan, N.A.; Eroglu, Z. Treatment options for advanced melanoma after anti-pd-1 therapy. *Curr. Oncol. Rep.* **2020**, *22*, 38. [CrossRef]

70. Greco, A.; Safi, D.; Swami, U.; Ginader, T.; Milhem, M.; Zakharia, Y. Efficacy and adverse events in metastatic melanoma patients treated with combination BRAF plus MEK inhibitors versus BRAF inhibitors: A systematic review. *Cancers (Basel)* **2019**, *11*, 1950. [CrossRef]

71. Haugh, A.M.; Johnson, D.B. Management of V600E and V600K BRAF-Mutant Melanoma. *Curr. Treat. Options Oncol.* **2019**, *20*, 81. [CrossRef]

Article

WNT5A-Induced Activation of the Protein Kinase C Substrate MARCKS Is Required for Melanoma Cell Invasion

Purusottam Mohapatra *, Vikas Yadav, Maren Toftdahl and Tommy Andersson *

Cell and Experimental Pathology, Department of Translational Medicine, Clinical Research Centre, Skåne University Hospital, Lund University, SE-202 13 Malmö, Sweden; vikas.yadav@med.lu.se (V.Y.); maren.toftdahl@gmail.com (M.T.)

* Correspondence: purusottam.mohapatra@med.lu.se (P.M.); tommy.andersson@med.lu.se (T.A.); Tel.: +46-40-391167 (P.M. & T.A.)

Received: 4 January 2020; Accepted: 27 January 2020; Published: 4 February 2020

Abstract: WNT5A is a well-known mediator of melanoma cell invasion and metastasis via its ability to activate protein kinase C (PKC), which is monitored by phosphorylation of the endogenous PKC substrate myristoylated alanine-rich c-kinase substrate (MARCKS). However, a possible direct contribution of MARCKS in WNT5A-mediated melanoma cell invasion has not been investigated. Analyses of melanoma patient databases suggested that similar to *WNT5A* expression, *MARCKS* expression appears to be associated with increased metastasis. A relationship between the two is suggested by the findings that recombinant WNT5A (rWNT5A) induces both increased expression and phosphorylation of MARCKS, whereas WNT5A silencing does the opposite. Moreover, WNT5A-induced invasion of melanoma cells was blocked by siRNA targeting MARCKS, indicating a crucial role of MARCKS expression and/or its phosphorylation. Next, we employed a peptide inhibitor of MARCKS phosphorylation that did not affect MARCKS expression and found that it abolished WNT5A-induced melanoma cell invasion. Similarly, rWNT5A induced the accumulation of phosphorylated MARCKS in membrane protrusions at the leading edge of melanoma cells. Our results demonstrate that WNT5A-induced phosphorylation of MARCKS is not only an indicator of PKC activity but also a crucial regulator of the metastatic behavior of melanoma and therefore an attractive future antimetastatic target in melanoma patients.

Keywords: melanoma; invasion; WNT5A; MARCKS; phosphorylation; MANS peptide

---

1. Introduction

Melanoma is an aggressive skin cancer in which rapid metastasis leads to a short overall median survival in patients. Initiation of melanoma has been shown to be related to environmental factors but also to genetic factors [1,2]. Most reasonably, it is in most cases a combination of both environmental and genetic factors that causes melanoma initiation and its progression to a metastatic state [3]. Research suggests that multiple pathways are involved not only in the initiation but also in the fatal metastatic spread of melanoma [1]. Therefore, a detailed understanding of different metastatic-related signaling pathways associated with invasive events is crucial for the development of novel therapeutics that would prevent the dissemination of melanoma cells. In this context, WNT5A signaling has been identified as an important cascade in melanoma progression and metastasis [4–8]. However, characterizing the mechanism of WNT5A signaling in melanoma metastasis has proven to be challenging and, consequently, knowledge of the downstream molecular partners of WNT5A signaling is still incomplete. The identification of downstream targets of WNT5A signaling would allow the development of a combined and more effective treatment strategy in which not only WNT5A signaling is directly targeted, but also essential downstream effector(s) of this signaling pathway.

WNT5A modulates melanoma cell behavior via multiple cell surface receptors/co-receptors, including ROR2 [9] and Frizzleds [10], and downstream molecular targets, such as APT1 [11], IL-6 [12], and PKC-STAT3 [13], which in turn affect various melanoma cell functions, including migration and invasion. One clear example of the essential role of WNT5A signaling in melanoma cell dissemination is the observation that simultaneous inhibition of WNT5A expression (with an anti-IL-6 antibody) and downstream WNT5A signaling (with the WNT5A antagonistic peptide Box5) effectively impairs melanoma cell migration and invasion [12]. An important pathway that has been shown to be crucial for WNT5A-mediated melanoma invasion and metastasis is the protein kinase C (PKC) signaling pathway, as demonstrated in several studies by Weeraratna and coworkers [8,9,14]. One plausible explanation for why PKC signaling is important for WNT5A-induced invasion of melanoma cells comes from the finding that WNT5A/PKC signaling causes epithelial-mesenchymal-transition (EMT)-like changes in melanoma cells [14], a transformation well known to increase tumor cell invasiveness and metastasis [15]. Another alternative downstream target of WNT5A-induced PKC signaling in melanoma cells that so far has only been used to monitor PKC activity in these cells is the phosphorylation of the endogenous PKC substrate myristoylated alanine-rich C-kinase substrate (MARCKS) [16].

MARCKS is a membrane-bound protein that functions in various important cellular processes, such as cytoskeletal remodeling, motility, secretion and exocytosis [17]. For instance, MARCKS has been shown to regulate the actin cytoskeleton and, consequently, the number and length of filopodia [18]. Interestingly, the phosphorylated form of MARCKS has been shown to promote cell motility and membrane protrusions and significantly affect the invasiveness of several cancer cells [17]. However, a possible functional role of the MARCKS protein in WNT5A-induced melanoma cell migration and invasion has not yet been studied.

The aim of the present study was to investigate a potential downstream regulatory role of the MARCKS protein in WNT5A-mediated invasion of melanoma cells. Our results demonstrate that WNT5A-induced activation of MARCKS is a necessary molecular event that is crucial for the metastatic behavior of melanoma cells.

2. Results

2.1. MARCKS Expression and Prognostic Value in Melanoma Tissue and Cells

To evaluate a possible role of MARCKS in melanoma progression, we first investigated its expression in normal versus melanoma tissue from patients included in the Oncomine database [19]. The Oncomine database contains cancer microarray results, in which transcriptome data can be compared with respective normal tissue data for most cancers and their subtypes. Our Oncomine-based analyses revealed that MARCKS mRNA expression is higher in patient-derived melanoma tissue compared to normal skin tissue samples (Figure S1A,B). To answer the question of whether and how the melanoma expression of MARCKS is related to prognosis, we analyzed the TCGA melanoma (TCGA-SKCM) cases using the "R2: genomics analysis and visualization platform" (http://r2.amc.nl). We have used the "scan modus" cutoff mode of the Kaplan Meier (KM) plot option of the "R2: genomics analysis and visualization platform" which according to the developers of the R2: platform, conducts the sample grouping in a more significant manner compared to other cutoff modes included in the R2: platform. Unfortunately, with this cutoff mode one only obtains a small number of samples with high MARCKS expression (Figure S1C). Therefore, it is very difficult to draw a solid conclusion with regard to how MARCKS expression relates to overall survival of melanoma patients. However, to gain further insight into the cellular mechanisms behind the observation that MARCKS expression could predict the prognosis of melanoma patients, we used the HOPP melanoma cell line database and analyzed the association of MARCKS expression with melanoma cell proliferation and invasion [20]. As shown in Figure S1D, MARCKS expression was significantly related to melanoma cell invasion, as evidenced by the increased expression levels of MARCKS mRNA in the invasive melanoma cell lines in comparison with those in proliferative melanoma cell lines (Figure S1D). Taken together, our in silico analyses suggest a possible role for MARCKS in promoting melanoma metastasis and thereby reducing the survival of melanoma patients.

2.2. Correlation between MARCKS and WNT5A Expression

The finding that MARCKS expression relates to the progression of melanoma is comparable to that found for WNT5A expression in relation to melanoma progression. Therefore, we next explored a possible association between MARCKS and WNT5A expression. We first analyzed the correlation between MARCKS and WNT5A in the TCGA database melanoma samples using the www.cbioportal.org data visualization and analysis website. Interestingly, our results revealed a weak positive correlation between MARCKS and WNT5A mRNA expression in melanoma patient tissue (Figure S2A). The above observation led us to investigate how the expression of MARCKS and WNT5A proteins correlated in melanoma cell lines with different WNT5A levels. The results revealed that all four melanoma cell lines (WM852, HTB63, A375 and A2058) used in our study expressed significant levels of MARCKS irrespective of their WNT5A levels (Figure S2B–D). Interestingly, WM852 melanoma cells which had higher WNT5A levels (Figure S2B,C) expressed significantly elevated amounts of MARCKS when compared to the other melanoma cell lines (HTB63, A375 and A2058 cells) (Figure S2B,D). Although we did not find a strong correlation between MARCKS and WNT5A expression, there are other possibilities whereby these two molecules can be related, for example, via a WNT5A-induced phosphorylation of the MARCKS protein [16].

2.3. WNT5A Signaling Increases MARCKS Phosphorylation in Melanoma Cells

First, we evaluated the levels of MARCKS phosphorylation at Ser-159/163 and Ser-167/170 sites in different melanoma cell lines via Western blotting approach (Figure S3). Interestingly, we observed that the Ser-159 and Ser-163 phosphorylation levels of MARCKS were significantly higher when compared to the Ser-167 and Ser-170 phosphorylation levels in all tested melanoma cell lines (Figure S3). Our results support previous reports on MARCKS Ser-159 and -163 phosphorylation [21]. Based on these results, we decided to focus on the Ser-159/163 phosphorylation of the MARCKS protein and how it relates to WNT5A signaling and melanoma cell invasion.

Although it has been previously reported that WNT5A/PKC signaling is directly involved in melanoma cell metastasis via an epithelial-mesenchymal-like transition [8,9,14], the possibility that WNT5A signaling promotes the invasiveness of human melanoma cells via altered expression or phosphorylation of MARCKS has not yet been studied. Here, A2058 (Figure 1A–C) and A375 (Figure S4A–C) melanoma cells were exposed to 0.2 µg/mL recombinant WNT5A (rWNT5A) protein for different time periods (starting at 15 minutes to a maximum of 24 hr) to observe any changes in total expression and Ser-159/163 phosphorylation levels of MARCKS. Phorbol myristate acetate (PMA), a PKC activator, was used as a positive control for MARCKS phosphorylation. Interestingly, the expression of both total MARCKS and phosphorylated MARCKS (Ser-159/163) increased with increasing time periods of rWNT5A treatment in both A2058 (Figure 1A–C) and A375 (Figure S4A–C) cells. The WNT5A-mediated increase in total MARCKS expression was statistically significant in A2058 cells after 1 h (Figure 1B) and in A375 melanoma cells after 3 h (Figure S4B) of treatment with rWNT5A. However, the increase in MARCKS phosphorylation at Ser-159/163 was statistically significant after 15 minutes of rWNT5A treatment in both A2058 (Figure 1C) and A375 melanoma cell lines (Figure S4C). These results clearly suggest that MARCKS phosphorylation at the Ser-159/163 residues is regulated by WNT5A signaling in metastatic melanoma cells.

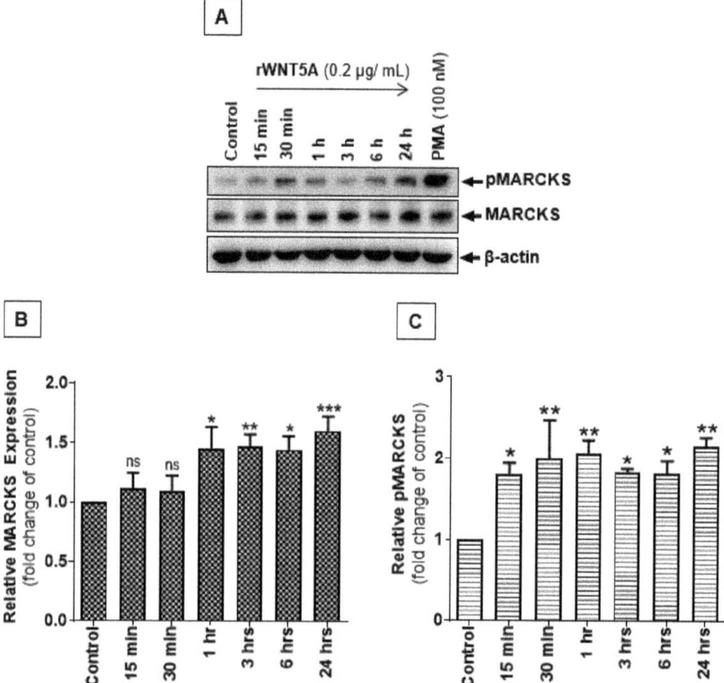

Figure 1. WNT5A signaling increases the expression and phosphorylation of MARCKS in melanoma cells. Western blotting was performed as detailed in the materials and methods section to demonstrate the importance of WNT5A signaling in MARCKS regulation. (**A**) Western blots showing the effect of 0.2 µg/mL rWNT5A treatment in A2058 melanoma cells at the indicated time points on both the expression and phosphorylation of MARCKS. β-Actin was used as a loading control. (**B,C**) The graphs represent the densitometry analysis of (**B**) total MARCKS and (**C**) pMARCKS Ser-159/163 levels in A2058 melanoma cells. The results (n = 4) are presented as the means ± S.E.M.; *, $p < 0.05$, **, $p < 0.001$, and ***, $p < 0.001$.

2.4. The MARCKS Protein Is Important for WNT5A-Mediated Invasion of Melanoma Cells

Based on the above results, we speculated that WNT5A-mediated melanoma cell invasion could be directly dependent on MARCKS expression and/or its phosphorylation. A2058 melanoma cells expressing very low amounts of WNT5A but with significant expression of the MARCKS protein (Figure S2B–D) were used to test whether the WNT5A-induced melanoma cell invasion was dependent on the presence of the MARCKS protein. MARCKS expression was reduced in A2058 melanoma cells by two different MARCKS siRNAs treatments (Figure 2A–C). Interestingly, stimulation with rWNT5A caused an increase in the numbers of invasive cells, whereas MARCKS silencing led to a 30–40% reduction in A2058 melanoma cell invasion compared to the control siRNA-transfected cells (Figure 2D). Induction of WNT5A signaling via treatment with rWNT5A significantly increased the number of invasive A2058 cells. Interestingly, however, we observed that rWNT5A exposure could not rescue the anti-invasive effect of MARCKS siRNA silencing in A2058 melanoma cells (Figure 2D). Importantly, these results did not discriminate as to whether it was the expression or the phosphorylation status of MARCKS that is crucial for WNT5A-induced melanoma cell invasion.

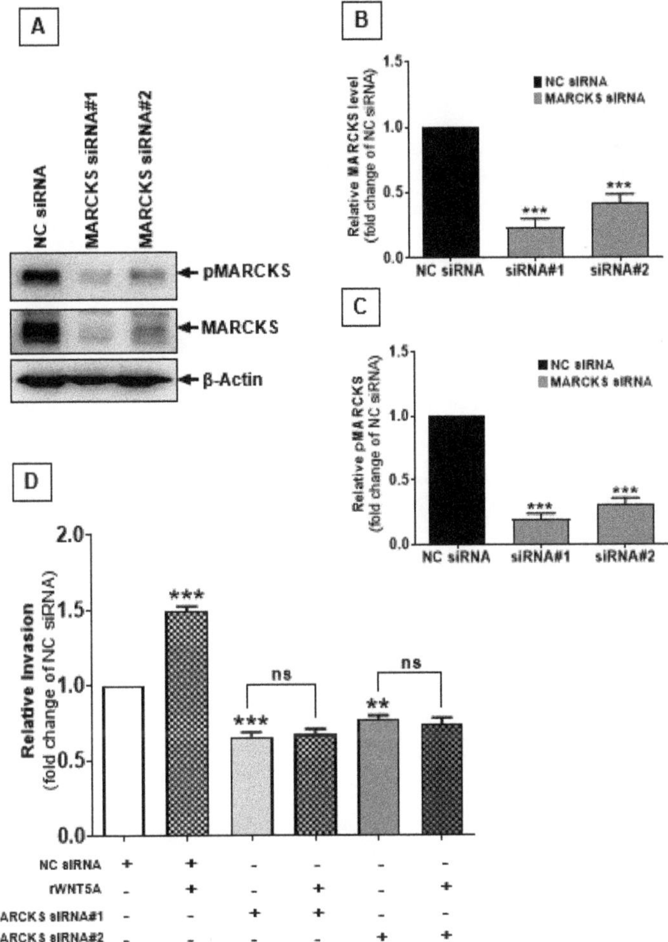

Figure 2. MARCKS is important for WNT5A-mediated melanoma cell invasion. (**A**) Western blot analysis of MARCKS and pMARCKS Ser-159/163 in A2058 melanoma cells transfected with two different MARCKS siRNAs as described in the materials and methods section. β-Actin was used as a loading control. (**B,C**) The graphs represent densitometry analyses of (**B**) MARCKS and (**C**) pMARCKS S159/163 levels. The results (n = 4) are presented as the means ± S.E.M.; ***, $p < 0.001$. (**D**) Transwell invasion assays were performed to determine the effect of rWNT5A (0.2 μg/mL) on the invasive capacity of MARCKS-silenced A2058 melanoma cells. The numbers of invaded cells were quantified using the NIH ImageJ software, and the results are presented as relative invasion. The results (n = 3) are presented as the means ± S.E.M.; **, $p < 0.001$, and ***, $p < 0.001$.

To test the above results, we decided to take an opposite approach—that is, we reduced WNT5A signaling and studied its effect on MARCKS expression and phosphorylation. At the same time, we checked the effect of WNT5A silencing on melanoma cell invasion. We silenced WNT5A in HTB63 melanoma cells with two different WNT5A siRNAs (Figure 3) and observed that there was only a minor effect on the total MARCKS level (Figure 3A,C). Interestingly, the Ser-159/163 phosphorylation of MARCKS (Figure 3A,D) was significantly decreased after WNT5A knockdown in HTB63 melanoma cells. As expected, our invasion assay revealed that WNT5A silencing decreased the invasive capacity of HTB63 melanoma cells (Figure 3E).

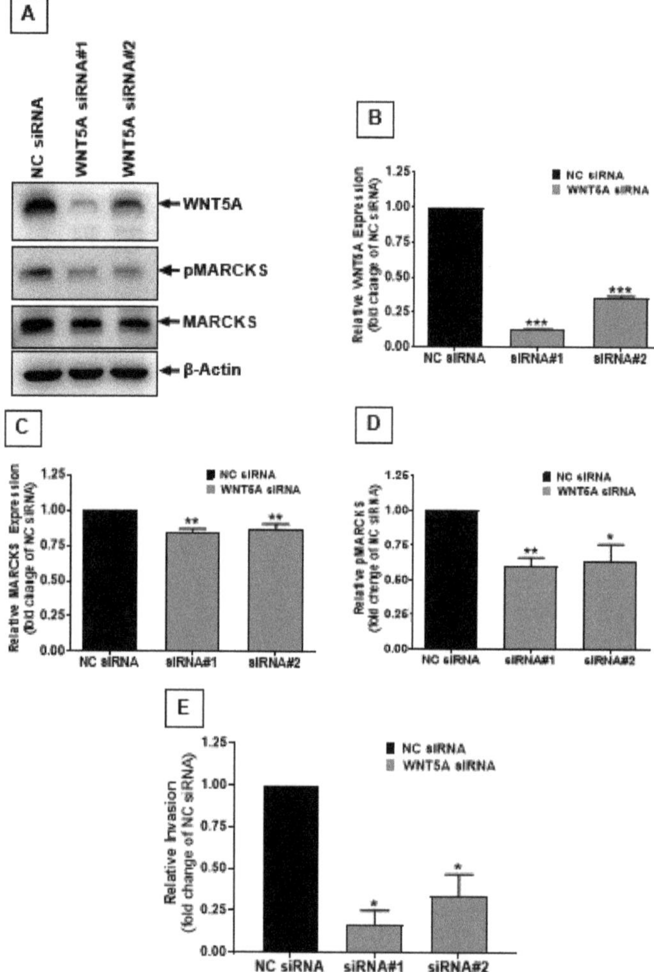

Figure 3. Inhibition of WNT5A signaling simultaneously reduced cell invasion and the expression and phosphorylation of MARCKS in melanoma cells. (**A**) Western blot analyses of MARCKS and pMARCKS Ser-159/163 in HTB63 melanoma cells transfected with two different WNT5A siRNAs as described in the materials and methods section. β-Actin was used as a loading control. (**B–D**) The graphs represent the densitometry analysis of (**B**) WNT5A expression, (**C**) MARCKS expression and (**D**) pMARCKS Ser-159/163 levels in WNT5A siRNA-transfected HTB63 melanoma cells. The results (n = 4) are presented as the means ± S.E.M.; *, $p < 0.05$, **, $p < 0.001$, and ***, $p < 0.001$. (**E**) Transwell invasion assays were performed to study the effect of siRNA-mediated inhibition of WNT5A signaling on the invasive capacity of HTB63 melanoma cells. The numbers of invaded cells were counted using the NIH ImageJ software, and the results are presented as the relative invasion compared to control siRNA. The results (n = 5) are presented as the means ± S.E.M.; *, $p < 0.05$.

2.5. Direct Inhibition of MARCKS Phosphorylation Blocks WNT5A-Mediated Melanoma Cell Invasion

To evaluate whether it is the ability of WNT5A to increase the expression of MARCKS or whether it is its ability to elevate the phosphorylation level of MARCKS that is crucial for melanoma cell invasion, we took a direct approach to inhibit MARCKS phosphorylation with a cell-permeable peptide identical to the MARCKS N-terminus sequence (the MANS peptide), a peptide that does not affect MARCKS expression. We tested the effect of MANS and a control peptide on WNT5A-induced A2058 melanoma cell invasion. Previously, the MANS peptide has been shown to inhibit basal lung cancer cell migration and invasion by specifically inhibiting MARCKS Ser-159/163 phosphorylation [22]. We first verified the expression of total MARCKS and phosphorylated MARCKS (Ser-159/163) after treatment with 100 µM RNS (control peptide) or 100 µM MANS (MARCKS phosphorylation inhibitory peptide) (Figure 4A–C). The phosphorylated MARCKS Ser-159/163 levels were reduced by approximately 35% in MANS peptide-treated A2058 melanoma cells compared to cells treated with the RNS control peptide (Figure 4A,C). However, treatment with either of these peptides did not change the expression levels of MARCKS in A2058 cells (Figure 4A,B). Our observations suggest that the MANS peptide can effectively and specifically reduce the phosphorylation of MARCKS in melanoma cells. Interestingly, we observed that the MANS peptide reduced the invasion of A2058 melanoma cells by 50% compared to that of vehicle-treated cells (Figure 4D). Most importantly, the presence of the MARCKS phosphorylation inhibitory peptide MANS abolished WNT5A-induced A2058 melanoma cell invasion (Figure 4D). The RNS control peptide had a small effect on A2058 cell invasion, which was reversed by rWNT5A treatment (Figure 4D). These observations strongly suggest that WNT5A specifically phosphorylates MARCKS to promote melanoma cell invasion.

2.6. WNT5A Increases Phosphorylated MARCKS Levels at the Cell edge and the Leading Front, Including Cell Protrusions of Melanoma Cells

It is well known that lamellipodia-like structures at the cell leading edge are essential for the ability of cells to migrate and invade. Based on a previous finding that the MARCKS protein has been implicated in the formation of such structures [23] and our present results, we speculated that WNT5A signaling could promote melanoma cell migration and invasion via translocation of phosphorylated MARCKS to the leading edge of melanoma cells. We investigated A2058 melanoma cells exposed to rWNT5A for any changes in the subcellular localization of MARCKS by confocal microscopy (Figure 5). Interestingly, our quantitative results further revealed that the fluorescence intensity of phosphorylated MARCKS was significantly higher at the cell periphery (Figure 5A,B) and at the leading edge (Figure 5A,C) of A2058 melanoma cells exposed to rWNT5A compared to the vehicle (control)-treated cells. Since MARCKS phosphorylation hinders its actin cross-linking ability, we wanted to study the consequence(s) of WNT5A-mediated MARCKS phosphorylation for the F-actin network of melanoma cells. We quantified the F-actin content at the cell leading edge with high phospho-MARCKS and noticed a significantly decreased fluorescence intensity of F-actin (Figure 5A,D). Overall, these findings add further support for an essential role of phosphorylated MARCKS protein in WNT5A-mediated melanoma cell migration and invasion and indicate that WNT5A-induced translocation of MARCKS to the leading edge of a melanoma cell is an essential step for how WNT5A participates in melanoma cell migration and invasion.

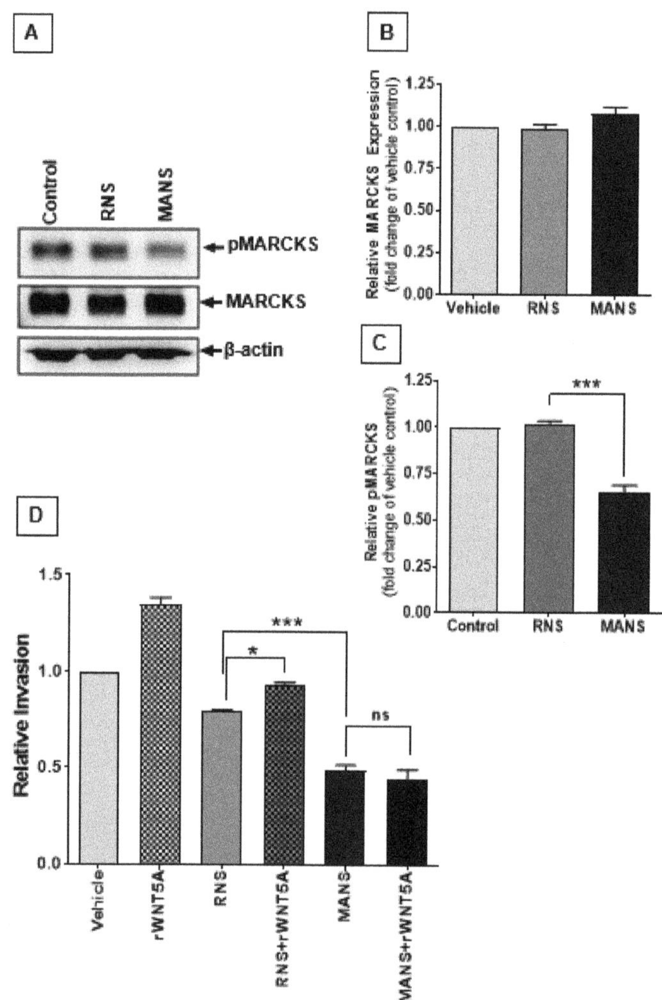

Figure 4. MARCKS phosphorylation-specific peptide MANS blocks WNT5A-induced melanoma cell invasion. Western blotting and transwell invasion assays were performed to evaluate the specific role of MARCKS phosphorylation in WNT5A-promoted melanoma cell invasion. (**A**) Western blots showing the expression of total MARCKS and levels of pMARCKS Ser-159/163 in 100 μM RNS- or MANS peptide-treated A2058 melanoma cells. Representative blots of five separate experiments are shown here. (**B,C**) The graph represents the densitometry analysis of (**B**) MARCKS and (**C**) pMARCKS Ser-159/163 levels normalized against total MARCKS levels. The results (n = 5) are presented as the means ± S.E.M.; ***, $p < 0.001$. (**D**) Effect of the direct inhibition of pMARCKS on WNT5A-induced melanoma cell invasion was evaluated by transwell cell invasion assay as mentioned in the materials and methods section. (**D**) Graph showing the relative melanoma cell invasion of rWNT5A unstimulated/stimulated and RNS peptide- or MANS peptide-treated A2058 melanoma cells. The results (n = 4) are presented as the means ± S.E.M.; *, $p < 0.05$, and ***, $p < 0.001$.

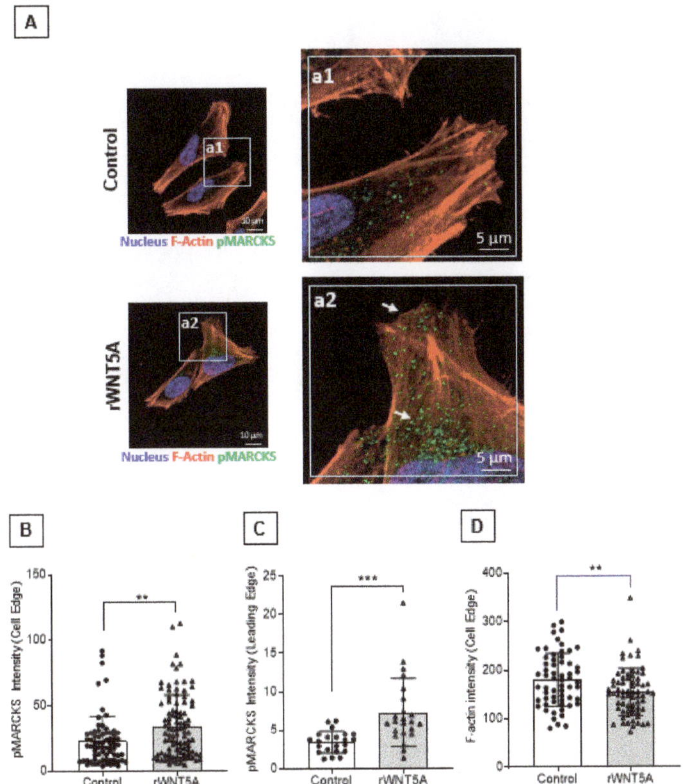

Figure 5. WNT5A signaling promotes the localization of phosphorylated MARCKS to the edge of melanoma cells. (**A**) A2058 melanoma cells were treated with 0.2 µg/mL rWNT5A, and immunofluorescence staining experiments were performed to study the localization of pMARCKS Ser-159/163 as described in the materials and methods section. (**A**) The images on the left show the pMARCKS Ser-159/163 localization in untreated and rWNT5A-treated A2058 melanoma cells. The images to the right identified as a1 and a2 are the magnified region of the marked areas/regions of the left (main) images. Representative images of three independent experiments are shown here. The graphs show the (**B**) integrated fluorescence intensity of pMARCKS Ser-159/163 at the cell edge (including the protrusions) in control *vs* WNT5A-treated A2058 melanoma cells quantified using CellProfiler software. (**C**) Integrated fluorescence intensity of pMARCKS Ser-159/163 at the leading edge of control *vs* WNT5A-treated A2058 melanoma cells was quantified using the Zeiss LSM-700 microscope image analysis software (**D**) F-actin intensity at the cell edge of control *vs* WNT5A-treated A2058 melanoma cells was quantified using CellProfiler software. Two-tailed unpaired Student's t test was performed, and the above results are presented as the means ± S.E.M., **, $p < 0.05$, ***, $p < 0.001$.

2.7. RhoA-ROCK Signaling Also Contributes to WNT5A-Induced Phosphorylation of MARCKS to Increase Invasiveness of Melanoma Cells

Although MARCKS phosphorylation has previously been used as an indication of WNT5A-mediated PKC activation in melanoma [16], a recent report by Tanabe et al. indicated that RhoA-ROCK signaling can also contribute to MARCKS phosphorylation in human neuroblastoma cells [24]. This prompted us to directly test whether RhoA-ROCK signaling is also involved in WNT5A-mediated phosphorylation of MARCKS in melanoma cells. Our results showed that both a PKC inhibitor and a ROCK inhibitor inhibited approximately 50% of WNT5A-induced MARCKS phosphorylation in A2058 (Figure 6A,B) and A375 (Figure S5A,B) melanoma cells compared to the vehicle-stimulated controls. Interestingly, the combined inhibition of PKC and ROCK signaling caused more than 90% inhibition of MARCKS phosphorylation in both A2058 (Figure 6A,B) and A375 (Figure S5A,B) melanoma cell lines, and this inhibition could not be rescued by cotreatment with rWNT5A. These results indicate that both PKC and RhoA-ROCK are important signals for WNT5A-mediated phosphorylation of MARCKS. Next, the invasion assay results revealed that a combination of a PKC and a ROCK inhibitor blocked the basal invasion of A2058 melanoma cells by at least 50%. More importantly, we observed that rWNT5A treatment had no significant effect on the invasion of A2058 melanoma cells in the simultaneous presence of a PKC and a ROCK inhibitor (Figure 6C). These observations demonstrate a dual signaling phenomenon whereby both the WNT5A-PKC and WNT5A/RhoA-ROCK signaling pathway contribute to MARCKS phosphorylation, thereby dictating its essential role in melanoma cell invasion.

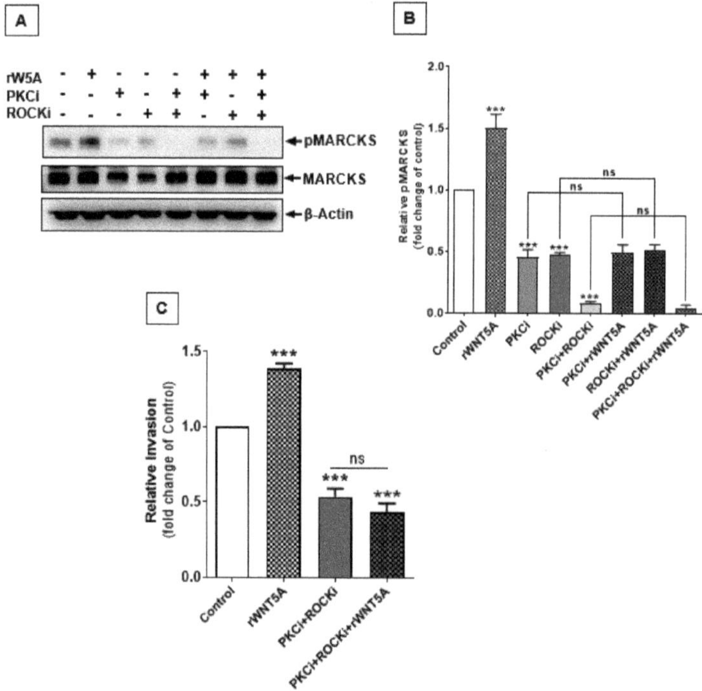

Figure 6. WNT5A-induced RhoA-ROCK signaling also contributes to MARCKS phosphorylation and melanoma cell invasion. Western blotting and transwell invasion assays were performed as described in the materials and methods section to investigate the intermediate signaling regulator of WNT5A-mediated phosphorylation of MARCKS in melanoma cell lines. (**A**) Western blot analysis of A2058 melanoma cells showing the rWNT5A-induced phosphorylation of MARCKS Ser-159/163 in untreated cells and A2058 melanoma cells treated with individual or combination of a PKC inhibitor (Gö6983) and/or ROCK inhibitor (Y-27632). β-Actin was used as a loading control. (**B**) The graph represents the densitometry analysis of phosphorylated-MARCKS normalized against total MARCKS using Bio-Rad ImagePro 6.0 software. The results (n = 5) are presented as the means ± S.E.M.; ***, $p < 0.001$. (**C**) Transwell invasion assays were performed to evaluate the effect of rWNT5A on the invasion capacity of PKC inhibitor- and ROCK inhibitor-treated A2058 melanoma cells. The number of invaded cells was counted using the NIH ImageJ software, and the results are presented as relative invasion. The results (n = 4) are presented as the means ± S.E.M.; ***, $p < 0.001$.

3. Discussion

Melanoma is well known for its pronounced metastatic behavior, making this cancer type one of the most aggressive. Several different regulators, including WNT5A and their complex downstream signaling pathways, have been identified to control various aspects of melanoma metastasis. In this study, we identified MARCKS phosphorylation as a novel WNT5A downstream signal crucial for melanoma cell invasion, a key event in the metastatic process.

Although WNT5A/PKC signaling has been previously reported to promote melanoma cell migration and invasion resulting in increased metastasis [14,25], a functional role of the PKC substrate MARCKS has not been studied. Rather, the WNT5A/PKC signal has been shown in melanoma cells to result in an EMT process that was suggested to explain how this signaling pathway caused elevated invasion leading to increased metastasis [14]. Previously, the phosphorylation status of the MARCKS protein has only been used as an indicator of WNT5A-mediated PKC activation in melanoma cells [16]. In the present study, we used online melanoma database analyses to demonstrate that MARCKS expression was increased in melanoma tissue and that it was directly associated with melanoma patient survival and melanoma cell invasiveness. Despite the fact that these databases are based on mRNA analyses and do not necessarily reflect the actual protein levels, these observations suggest that MARCKS expression could be associated with melanoma progression. Interestingly, the database findings on MARCKS are very similar to those of the WNT5A ligand, which is well known for its ability to promote melanoma invasion and metastasis [8] and has been shown to activate PKC [13,14].

By investigating a possible role of the MARCKS protein in melanoma progression and its relation to WNT5A, we found that siRNA silencing of MARCKS resulted in markedly reduced expression of not only total MARCKS but also of phosphorylated MARCKS, in parallel with a reduced invasive capacity of melanoma cells. Most interestingly, the ability of WNT5A to stimulate the invasiveness of MARCKS siRNA-silenced melanoma cells was completely abolished, suggesting an essential WNT5A-downstream signaling role of MARCKS in its ability to induce melanoma cell invasion. This notion was further supported by our findings that siRNA silencing of WNT5A reduced not only the total expression of MARCKS but also the level of phosphorylated MARCKS. Similarly, reconstitution of WNT5A signaling by rWNT5A in cells with very low expression of WNT5A increased not only the total expression of MARCKS but also its phosphorylation level. Although the effector domain of MARCKS, also called the phosphorylation site domain, has, once it is phosphorylated, been reported to mediate the association of the MARCKS protein with other cellular components leading to altered cellular functions [24,26,27], siRNA silencing of MARCKS does not determine if it is the total expression of MARCKS or its phosphorylation that is essential for WNT5A-induced melanoma cell invasion.

Using the MANS peptide, a cell permeable inhibitor of MARCKS phosphorylation [22,28], we were able to specifically abolish MARCKS phosphorylation without altering its expression. This approach allowed us to completely abolish WNT5A-induced melanoma cell invasion, thus demonstrating for the first time a crucial role of MARCKS phosphorylation in WNT5A-driven melanoma cell invasion. Previously, it has been documented that the MANS peptide could effectively block endogenous fibroblast migration by blocking MARCKS phosphorylation [28]. Interestingly, in another report, Chen and coworkers demonstrated that the MANS peptide could reduce the basal level of lung cancer cell invasion [22]. However, these studies focused on the importance of MARCKS for basal invasiveness of cells and not on the possibility that an external ligand could overcome the inhibition of basal cancer cell migration or invasion.

Since MARCKS is an actin-binding protein and phosphorylation of MARCKS could affect actin dynamics and thereby modulate melanoma cell motility and/or invasion, we further studied the molecular and subcellular events during WNT5A-mediated MARCKS phosphorylation in melanoma cells. Our observation of a crucial role of WNT5A-induced MARCKS phosphorylation in regulation of melanoma cell migration and invasion was further substantiated by the findings that MARCKS localized at the cell leading edge and lamellipodia and filopodia-like protrusions, indicating that WNT5A, by inducing MARCKS phosphorylation at the cellular edge, could modulate the cytoskeleton to influence melanoma cell invasion. It has been recently reported that PKC-induced MARCKS phosphorylation impairs cell polarization by destabilizing and redistributing the F-actin network [29]. Although there are no reports showing that WNT5A-induced MARCKS phosphorylation could affect F-actin rearrangement in cancer cells, an interesting study by Iioka et al. demonstrated that noncanonical WNT11 signaling via MARCKS regulates cortical actin dynamics and the formation of lamellipodia- and filopodia-like protrusions [30]. Our observation of a decreased F-actin content at the cell edge suggests that WNT5A signaling by phosphorylating MARCKS interferes with the MARCKS F-actin binding ability and destabilizes the F-actin network at the cell edge to favor melanoma cell invasion.

Interestingly, we demonstrated that inhibition of PKC and ROCK signaling blocked WNT5A-mediated MARCKS phosphorylation and melanoma cell invasion, which indicated that both PKC and RhoA-ROCK, as intermediate signaling pathways, are crucial during WNT5A-regulated MARCKS function. Interestingly, both PKC and RhoA-ROCK signaling are known to lead to MARCKS phosphorylation [17,22,24,29]. Furthermore, PKC and ROCK signaling has been documented as an important event in actin cytoskeletal remodeling, which is essential for cellular rearrangements during convergence, extensions movements [31] and human adipose stem cell regeneration [32]. The present finding that a signaling ligand mediates MARCKS phosphorylation via a simultaneous activation of PKC and RhoA-ROCK pathways has not been previously described in any cancer model. Since MARCKS phosphorylation is important for the control of actin dynamics, our novel observations of PKC and RhoA-ROCK signaling being involved in melanoma cell invasion mediated by WNT5A-induced MARCKS phosphorylation, suggest the importance of these signaling pathways in melanoma. Our study shed light on the previously unknown molecular events associated with WNT5A signaling that regulate MARCKS phosphorylation, which is crucial for melanoma metastasis.

4. Materials and Methods

4.1. Cell Culture and Reagents

The WM852 melanoma cell line was procured from the Coriell Institute (Camden NJ, USA) and maintained in RPMI-1640 medium. The HTB63, A375 and A2058 melanoma cell lines were purchased from ATCC (Old Town Manassas, VA, USA). HTB63 melanoma cells were cultured in McCoy's 5A medium, and both A375 and A2058 melanoma cell lines were cultured in DMEM. The culture medium for all the cell lines was supplemented with 10% FBS, antibiotics and L-glutamine. The supplier confirmed the genetic authentications of all the cell lines, and no cell line was used for more than four years. All cell lines were routinely screened for the absence of mycoplasma infection. The PKC-inhibitor (Gö6983) and ROCK-inhibitor (Y-27632) were procured from Selleckchem (Munich, Germany). The predesigned siRNAs used in this study were obtained from Thermo Fischer Scientific (Waltham, MA, USA).

4.2. Western Blotting

Briefly, treated and untreated cells from separate experiments were lysed, and the protein concentration was quantified using the Bradford's method (Sigma-Aldrich, Stockholm, Sweden). SDS gel electrophoresis was performed, and the protein bands were transferred to a PVDF membrane for Western blotting. Western blotting was conducted using primary mouse anti-MARCKS (Santa Cruz Biotechnology, Dallas, TX, USA; dilution 1:1000), rabbit anti-phospho-Ser-159/163 MARCKS (Cell Signaling Technology, Danvers, MA, USA; dilution 1:1000), rabbit anti-phospho-Ser-167/170 MARCKS (Cell Signaling Technology, dilution 1:500), goat anti-WNT5A (R&D Systems, Minneapolis, MN, USA; dilution 1:100), and anti-β-actin (Sigma-Aldrich, dilution 1:30,000) antibodies and secondary HRP-conjugated rabbit anti-goat, goat anti-mouse or goat anti-rabbit antibodies (Dako, Santa Clara, CA, USA; dilution 1:10,000). Western blot images were obtained using a ChemiDoc™ imaging system (Bio-Rad, Hercules, CA, USA). The densitometry quantifications of relative protein expression were conducted using Image Lab software (version 6.0, Bio-Rad).

4.3. MANS and RNS Peptide Synthesis and Treatment

The MANS and RNS peptides were custom synthesized by CASLO ApS (Lyngby, Denmark). The peptides were obtained as lyophilized powder. The MANS peptide is a specific MARCKS phosphorylation inhibitory peptide that consists of a sequence matching the first 24 amino acids of the MARCKS myristoylated N-terminal region [22]. The sequence of the MANS peptide was MA-GAQFSKTAAKGEAAAERPGEAAVA (where MA is the N-terminal myristate chain), and the RNS control peptide sequence was MA-GTAPAAEGAGAEVKRASAEAKQAF, meaning that it had the same amino acid content as the MANS peptide but randomly rearranged. The peptides were dissolved in 1× sterile PBS. A2058 melanoma cells were pretreated for 24 h with 100 µM RNS or MANS peptide before the different experiments were performed as described in the results section and in the figure legends.

4.4. Transwell Cell Invasion Assay

The transwell cell invasion method was performed to quantify the number of invaded cells after different treatment conditions. Briefly, 50,000 melanoma cells/inserts were used to compare cell invasive capacities after different treatment/transfection conditions. After treatment, the invaded melanoma cells were fixed in 70% ethanol and stained with crystal violet. The noninvaded cells were removed from the insert membrane using wet cotton swabs followed by washing. Images of the invaded cells were captured using an inverted microscope (Nikon, Tokyo, Japan), and the cell numbers were counted using cell counter plugins of the NIH ImageJ software.

4.5. Transient Gene Silencing Using siRNA

The different melanoma cells were transfected with either MARCKS or WNT5A siRNAs using Lipofectamine 2000 transfection reagent (Invitrogen, Carlsbad, CA, USA) in accordance with the manufacturer's instructions. All predesigned siRNA oligonucleotides were purchased from Thermo Fischer Scientific (Waltham, MA, USA). The siRNAs were provided as lyophilized powder and resuspended in nuclease-free water (supplied by the manufacturer) prior to transfection. Approximately 400,000 cells/well of either A2058 or HTB63 melanoma cells were transfected with 100 nM negative control (NC) siRNA (#4390843), anti-MARCKS siRNA#1 (s8636), anti-MARCKS siRNA#2 (243176), anti-WNT5A siRNA#1 (s14871), or anti-WNT5A siRNA#2 (s14872) siRNA oligonucleotides.

4.6. Immunofluorescence Staining, Confocal Imaging and Image Analysis

Approximately 5000 cells/coverslip of adherent A2058 melanoma cells were treated with 0.2 µg/mL rWNT5A for 1 hr. For phosphorylated MARCKS staining, the cells on the coverslips were fixed with 4% paraformaldehyde for 10 minutes at room temperature, washed with cold 1xPBS, and then blocked in 2% BSA and 0.1% Triton X-100 in 1× PBS for 1 h. After blocking, the cells were incubated in the presence of an anti-phospho-MARCKS (Ser-159/163) antibody (dilution 1:100) overnight at 4 °C. After the primary antibody incubation, the cells were washed three times with cold 1× PBS and incubated with a secondary anti-rabbit antibody conjugated to Alexa Fluor® 488 (Invitrogen, dilution 1:250) and phalloidin-TRITC (Sigma-Aldrich, dilution 1:400) for 1 hr at room temperature in the dark. Cells were washed three times with cold 1× PBS, and their nuclei were counterstained with DAPI. The coverslips containing stained cells were mounted with a fluorescent mounting solution (Dako) on glass slides and kept overnight at 4 °C for curing. The fluorescence images were captured under a 63× oil objective using a confocal microscope (LSM 700, Carl Zeiss, Oberkochen, Germany). Image analyses were performed using the microscope software and CellProfiler image analysis software [33]. The cells clearly showing a leading cell edge were selected from independent experiments to quantify the phospho-MARCKS (Ser-159/163) fluorescence intensity at the leading edge by LSM700 microscope image acquisition and analysis software (Carl Zeiss). Regions of interest (ROIs) at the cell leading edge were marked manually, and the fluorescence intensity values obtained in those regions were used for quantification. The CellProfiler pipelines for fluorescence intensity measurement at the cell edge were set to quantify the fluorescence intensity at the edge of the cell membrane and cell protrusions (integrated fluorescence intensity at the edge) for the localization of phosphorylated MARCKS in rWNT5A untreated and treated cells.

4.7. Statistical Analyses

Statistical analyses were performed using GraphPad Prism 8.0 software. All multiple group analyses were verified for significance using ANOVA complemented with Dunnett's test (when comparing with control) or Bonferroni's multiple comparison test (when comparing all the groups). Significant differences between two groups were assessed using unpaired two-tailed Students' *t*-tests. The results are presented as the mean ± S.E.M. Differences were considered significant when $p < 0.05$.

5. Conclusions

Our study demonstrates multiple novel findings outlining an essential role of MARCKS phosphorylation in WNT5A-induced melanoma cell invasion. The basis for this conclusion comes from our findings that MARCKS expression might be associated with melanoma progression and patient survival, WNT5A signaling triggers increased expression and phosphorylation of MARCKS, WNT5A increased the localization of phosphorylated MARCKS at the cell leading edge and in its protrusions and, most importantly, blocking MARCKS phosphorylation with the MANS peptide without affecting its expression abolished WNT5A-mediated melanoma cell invasion. These findings provide the basis for a new antimetastatic treatment strategy that involves the simultaneous targeting of WNT5A and MARCKS signaling in malignant melanoma patients.

Supplementary Materials: The following are available online at http://www.mdpi.com/2072-6694/12/2/346/s1, Figure S1: MARCKS is positively associated with melanoma progression and poor overall survival of melanoma patients, Figure S2: Metastatic melanoma cells express significant levels of MARCKS and WNT5A, Figure S3: Phosphorylation levels of MARCKS at Ser-159 and Ser-163 are predominant in metastatic melanoma cells, Figure S4: WNT5A signaling increases the expression and phosphorylation of MARCKS in A375 melanoma cells, Figure S5: PKC and RhoA-ROCK signaling are also important for WNT5A-mediated phosphorylation of MARCKS in A375 melanoma cells, Additional Materials Related to Western Blotting Experiments: Whole blots for Western blotting experiments in the main figures are added to this section.

Author Contributions: Conceptualization, P.M. and T.A.; methodology, P.M. and T.A.; validation, P.M. and T.A.; formal analysis, P.M., V.Y. and T.A.; investigation, P.M., V.Y. and M.T.; resources, T.A.; data curation, P.M. and T.A.; writing—original draft preparation, P.M.; writing—review and editing, P.M., V.Y., M.T. and T.A.; visualization, P.M. and V.Y.; supervision, T.A.; project administration, T.A.; funding acquisition, T.A. All authors have read and agree to the published version of the manuscript.

Funding: This work was supported by the Swedish Cancer Foundation (No. 17 0537), the Swedish Research Council (No. 2015-02540), the Skåne University Hospital Research Foundation, the Malmö Cancer Foundation, Sweden and the Governmental Funding of Clinical Research within the National Health Services (all to T.A.).

Acknowledgments: The authors would like to thank Kenneth Adler (North Carolina State University, USA) for providing useful information on MANS and RNS peptide synthesis.

Conflicts of Interest: T.A. is a shareholder of WntResearch and is a part-time Chief Scientific Officer of WntResearch. This does not alter the author's adherence to all guidelines for publication in this journal.

References

1. Prasad, C.P.; Mohapatra, P.; Andersson, T. Therapy for BRAFi-Resistant Melanomas: Is WNT5A the Answer? *Cancers (Basel)* **2015**, *7*, 1900–1924. [CrossRef]

2. Laikova, K.V.; Oberemok, V.V.; Krasnodubets, A.M.; Gal'chinsky, N.V.; Useinov, R.Z.; Novikov, I.A.; Temirova, Z.Z.; Gorlov, M.V.; Shved, N.A.; Kumeiko, V.V.; et al. Advances in the Understanding of Skin Cancer: Ultraviolet Radiation, Mutations, and Antisense Oligonucleotides as Anticancer Drugs. *Molecules* **2019**, *24*, 1516. [CrossRef]

3. Leclerc, J.; Ballotti, R.; Bertolotto, C. Pathways from senescence to melanoma: Focus on MITF sumoylation. *Oncogene* **2017**, *36*, 6659–6667. [CrossRef]

4. Bittner, M.; Meltzer, P.; Chen, Y.; Jiang, Y.; Seftor, E.; Hendrix, M.; Radmacher, M.; Simon, R.; Yakhini, Z.; Ben-Dor, A.; et al. Molecular classification of cutaneous malignant melanoma by gene expression profiling. *Nature* **2000**, *406*, 536–540. [CrossRef]

5. Carr, K.M.; Bittner, M.; Trent, J.M. Gene-expression profiling in human cutaneous melanoma. *Oncogene* **2003**, *22*, 3076–3080. [CrossRef]

6. Da Forno, P.D.; Pringle, J.H.; Hutchinson, P.; Osborn, J.; Huang, Q.; Potter, L.; Hancox, R.A.; Fletcher, A.; Saldanha, G.S. WNT5A expression increases during melanoma progression and correlates with outcome. *Clin. Cancer Res.* **2008**, *14*, 5825–5832. [CrossRef]

7. Paluncic, J.; Kovacevic, Z.; Jansson, P.J.; Kalinowski, D.; Merlot, A.M.; Huang, M.L.; Lok, H.C.; Sahni, S.; Lane, D.J.; Richardson, D.R. Roads to melanoma: Key pathways and emerging players in melanoma progression and oncogenic signaling. *Biochim. Biophys. Acta* **2016**, *1863*, 770–784. [CrossRef]

8. Weeraratna, A.T.; Jiang, Y.; Hostetter, G.; Rosenblatt, K.; Duray, P.; Bittner, M.; Trent, J.M. Wnt5a signaling directly affects cell motility and invasion of metastatic melanoma. *Cancer Cell* **2002**, *1*, 279–288. [CrossRef]

9. O'Connell, M.P.; Fiori, J.L.; Xu, M.; Carter, A.D.; Frank, B.P.; Camilli, T.C.; French, A.D.; Dissanayake, S.K.; Indig, F.E.; Bernier, M.; et al. The orphan tyrosine kinase receptor, ROR2, mediates Wnt5A signaling in metastatic melanoma. *Oncogene* **2010**, *29*, 34–44. [CrossRef]

10. Sato, A.; Yamamoto, H.; Sakane, H.; Koyama, H.; Kikuchi, A. Wnt5a regulates distinct signalling pathways by binding to Frizzled2. *EMBO J.* **2010**, *29*, 41–54. [CrossRef]

11. Sadeghi, R.S.; Kulej, K.; Kathayat, R.S.; Garcia, B.A.; Dickinson, B.C.; Brady, D.C.; Witze, E.S. Wnt5a signaling induced phosphorylation increases APT1 activity and promotes melanoma metastatic behavior. *Elife* **2018**, *7*. [CrossRef] [PubMed]

12. Linnskog, R.; Mohapatra, P.; Moradi, F.; Prasad, C.P.; Andersson, T. Demonstration of a WNT5A-IL-6 positive feedback loop in melanoma cells: Dual interference of this loop more effectively impairs melanoma cell invasion. *Oncotarget* **2016**, *7*, 37790–37802. [CrossRef] [PubMed]

13. Dissanayake, S.K.; Olkhanud, P.B.; O'Connell, M.P.; Carter, A.; French, A.D.; Camilli, T.C.; Emeche, C.D.; Hewitt, K.J.; Rosenthal, D.T.; Leotlela, P.D.; et al. Wnt5A regulates expression of tumor-associated antigens in melanoma via changes in signal transducers and activators of transcription 3 phosphorylation. *Cancer Res* **2008**, *68*, 10205–10214. [CrossRef]

14. Dissanayake, S.K.; Wade, M.; Johnson, C.E.; O'Connell, M.P.; Leotlela, P.D.; French, A.D.; Shah, K.V.; Hewitt, K.J.; Rosenthal, D.T.; Indig, F.E.; et al. The Wnt5A/protein kinase C pathway mediates motility in melanoma cells via the inhibition of metastasis suppressors and initiation of an epithelial to mesenchymal transition. *J. Biol. Chem.* **2007**, *282*, 17259–17271. [CrossRef]

15. Pearson, G.W. Control of Invasion by Epithelial-to-Mesenchymal Transition Programs during Metastasis. *J. Clin. Med.* **2019**, *8*, 646. [CrossRef]

16. Jenei, V.; Sherwood, V.; Howlin, J.; Linnskog, R.; Safholm, A.; Axelsson, L.; Andersson, T. A t-butyloxycarbonyl-modified Wnt5a-derived hexapeptide functions as a potent antagonist of Wnt5a-dependent melanoma cell invasion. *Proc. Natl. Acad. Sci. USA* **2009**, *106*, 19473–19478. [CrossRef]

17. Fong, L.W.R.; Yang, D.C.; Chen, C.H. Myristoylated alanine-rich C kinase substrate (MARCKS): A multirole signaling protein in cancers. *Cancer Metastasis Rev.* **2017**, *36*, 737–747. [CrossRef]

18. Li, H.; Chen, G.; Zhou, B.; Duan, S. Actin filament assembly by myristoylated alanine-rich C kinase substrate-phosphatidylinositol-4,5-diphosphate signaling is critical for dendrite branching. *Mol. Biol. Cell* **2008**, *19*, 4804–4813. [CrossRef]

19. Rhodes, D.R.; Yu, J.; Shanker, K.; Deshpande, N.; Varambally, R.; Ghosh, D.; Barrette, T.; Pandey, A.; Chinnaiyan, A.M. ONCOMINE: A cancer microarray database and integrated data-mining platform. *Neoplasia* **2004**, *6*, 1–6. [CrossRef]

20. Widmer, D.S.; Cheng, P.F.; Eichhoff, O.M.; Belloni, B.C.; Zipser, M.C.; Schlegel, N.C.; Javelaud, D.; Mauviel, A.; Dummer, R.; Hoek, K.S. Systematic classification of melanoma cells by phenotype-specific gene expression mapping. *Pigment Cell Melanoma Res.* **2012**, *25*, 343–353. [CrossRef]

21. Herget, T.; Oehrlein, S.A.; Pappin, D.J.; Rozengurt, E.; Parker, P.J. The myristoylated alanine-rich C-kinase substrate (MARCKS) is sequentially phosphorylated by conventional, novel and atypical isotypes of protein kinase C. *Eur. J. Biochem.* **1995**, *233*, 448–457. [CrossRef]

22. Chen, C.H.; Thai, P.; Yoneda, K.; Adler, K.B.; Yang, P.C.; Wu, R. A peptide that inhibits function of Myristoylated Alanine-Rich C Kinase Substrate (MARCKS) reduces lung cancer metastasis. *Oncogene* **2014**, *33*, 3696–3706. [CrossRef]

23. Yamaguchi, H.; Shiraishi, M.; Fukami, K.; Tanabe, A.; Ikeda-Matsuo, Y.; Naito, Y.; Sasaki, Y. MARCKS regulates lamellipodia formation induced by IGF-I via association with PIP2 and beta-actin at membrane microdomains. *J. Cell Physiol.* **2009**, *220*, 748–755. [CrossRef]

24. Tanabe, A.; Kamisuki, Y.; Hidaka, H.; Suzuki, M.; Negishi, M.; Takuwa, Y. PKC phosphorylates MARCKS Ser159 not only directly but also through RhoA/ROCK. *Biochem. Biophys. Res. Commun.* **2006**, *345*, 156–161. [CrossRef]

25. Medrano, E.E. Wnt5a and PKC, a deadly partnership involved in melanoma invasion. *Pigment Cell Res.* **2007**, *20*, 258–259. [CrossRef]

26. Rodriguez Pena, M.J.; Castillo Bennett, J.V.; Soler, O.M.; Mayorga, L.S.; Michaut, M.A. MARCKS protein is phosphorylated and regulates calcium mobilization during human acrosomal exocytosis. *PLoS ONE* **2013**, *8*, e64551. [CrossRef]

27. Gatlin, J.C.; Estrada-Bernal, A.; Sanford, S.D.; Pfenninger, K.H. Myristoylated, alanine-rich C-kinase substrate phosphorylation regulates growth cone adhesion and pathfinding. *Mol. Biol. Cell* **2006**, *17*, 5115–5130. [CrossRef]

28. Ott, L.E.; Sung, E.J.; Melvin, A.T.; Sheats, M.K.; Haugh, J.M.; Adler, K.B.; Jones, S.L. Fibroblast Migration Is Regulated by Myristoylated Alanine-Rich C-Kinase Substrate (MARCKS) Protein. *PLoS ONE* **2013**, *8*, e66512. [CrossRef]

29. Aparicio, G.; Arruti, C.; Zolessi, F.R. MARCKS phosphorylation by PKC strongly impairs cell polarity in the chick neural plate. *Genesis* **2018**, *56*, e23104. [CrossRef]

30. Iioka, H.; Ueno, N.; Kinoshita, N. Essential role of MARCKS in cortical actin dynamics during gastrulation movements. *J. Cell Biol.* **2004**, *164*, 169–174. [CrossRef]

31. Zhu, S.; Liu, L.; Korzh, V.; Gong, Z.; Low, B.C. RhoA acts downstream of Wnt5 and Wnt11 to regulate convergence and extension movements by involving effectors Rho kinase and Diaphanous: Use of zebrafish as an in vivo model for GTPase signaling. *Cell Signal.* **2006**, *18*, 359–372. [CrossRef]

32. Santos, A.; Bakker, A.D.; de Blieck-Hogervorst, J.M.; Klein-Nulend, J. WNT5A induces osteogenic differentiation of human adipose stem cells via rho-associated kinase ROCK. *Cytotherapy* **2010**, *12*, 924–932. [CrossRef]

33. Kamentsky, L.; Jones, T.R.; Fraser, A.; Bray, M.A.; Logan, D.J.; Madden, K.L.; Ljosa, V.; Rueden, C.; Eliceiri, K.W.; Carpenter, A.E. Improved structure, function and compatibility for CellProfiler: Modular high-throughput image analysis software. *Bioinformatics* **2011**, *27*, 1179–1180. [CrossRef]

MDPI
St. Alban-Anlage 66
4052 Basel
Switzerland
Tel. +41 61 683 77 34
Fax +41 61 302 89 18
www.mdpi.com

Cancers Editorial Office
E-mail: cancers@mdpi.com
www.mdpi.com/journal/cancers

www.ingramcontent.com/pod-product-compliance
Lightning Source LLC
LaVergne TN
LVHW071627170726
843515LV00009B/2501